Handbook of RNA
Biochemistry

Edited by
Roland Karl Hartmann,
Albrecht Bindereif,
Astrid Schön, and
Eric Westhof

Further Reading

Gjerde, D. T., Hoang, L., Hornby, D.

RNA Purification and Analysis
Sample Preparation, Extraction, Chromatography

2009
ISBN: 978-3-527-32116-2

Müller, S. (ed.)

Nucleic Acids from A to Z
A Concise Encyclopedia

2008
ISBN: 978-3-527-31211-5

Courey, A.

Mechanisms in Transcriptional Regulation

2008
ISBN: 978-1-4051-0370-1

Smith, H. C. (ed.)

RNA and DNA Editing
Molecular Mechanisms and Their Integration into Biological Systems

2008
ISBN: 978-0-470-10991-5

Schepers, U.

RNA Interference in Practice
Principles, Basics, and Methods for Gene Silencing in C.elegans, Drosophila, and Mammals

2008
ISBN: 978-3-527-31020-3

Handbook of RNA Biochemistry

Student Edition

Edited by
Roland Karl Hartmann, Albrecht Bindereif,
Astrid Schön, and Eric Westhof

WILEY-VCH

WILEY-VCH Verlag GmbH & Co. KGaA

The Editors

Prof. Dr. Roland Karl Hartmann
Philipps-Universität Marburg
Institut für Pharmazeutische Chemie
Marbacher Weg 6
35037 Marburg
Germany
roland.hartmann@staff.uni-marburg.de

Prof. Dr. Albrecht Bindereif
Justus-Liebig-Universität Giessen
Institut für Biochemie
Heinrich-Buff-Ring 58
35392 Giessen
Germany
albrecht.bindereif@chemie.bio.uni-giessen.de

Dr. Astrid Schön
Universität Leipzig
Institut für Biochemie
Brüderstr. 34
04103 Leipzig
Germany
schoena@uni-leipzig.de

Prof. Dr. Eric Westhof
CNRS – UPR 9002
Institut de Biologie Moléculaire et Cellulaire
15 rue René Descartes
67084 Strasbourg
France
e.westhof@ibmc.u-strasbg.fr

1st Edition 2005
1st Student Edition 2009

■ All books published by Wiley-VCH are carefully produced. Nevertheless, authors, editors, and publisher do not warrant the information contained in these books, including this book, to be free of errors. Readers are advised to keep in mind that statements, data, illustrations, procedural details or other items may inadvertently be inaccurate.

Library of Congress Card No.: applied for

British Library Cataloguing-in-Publication Data
A catalogue record for this book is available from the British Library.

Bibliographic information published by the Deutsche Nationalbibliothek
The Deutsche Nationalbibliothek lists this publication in the Deutsche Nationalbibliografie; detailed bibliographic data are available on the Internet at http://dnb.d-nb.de.

© 2009 WILEY-VCH Verlag GmbH & Co. KGaA, Weinheim, Germany

All rights reserved (including those of translation into other languages). No part of this book may be reproduced in any form – by photoprinting, microfilm, or any other means – nor transmitted or translated into a machine language without written permission from the publishers. Registered names, trademarks, etc. used in this book, even when not specifically marked as such, are not to be considered unprotected by law.

Typesetting Asco Typesetters, Hong Kong
Printing Strauss GmbH, Moerlenbach
Binding Litges & Dopf Buchbinderei GmbH, Heppenheim

Printed in the Federal Republic of Germany
Printed on acid-free paper

ISBN 978-3-527-32534-4

Short Contents

Part I	**RNA Synthesis** *1*	
I.1	Enzymatic RNA Synthesis, Ligation and Modification	*3*
I.2	Chemical RNA Synthesis *95*	
Part II	**Structure Determination** *131*	
II.1	Molecular Biology Methods *133*	
II.2	Biophysical Methods *385*	
II.3	Fluorescence and Single Molecule Studies *453*	
Part III	**RNA Genomics and Bioinformatics** *489*	
Part IV	**Analysis of RNA Function** *665*	
IV.1	RNA–Protein Interactions *in vitro* *667*	
IV.2	RNA–Protein Interactions *in vivo* *729*	
IV.3	SELEX *783*	
Part V	**RNAi** *895*	
Appendix: UV Spectroscopy for the Quantitation of RNA *910*		

Handbook of RNA Biochemistry: Student Edition. Edited by R. K. Hartmann, A. Bindereif, A. Schön, E. Westhof
Copyright © 2009 WILEY-VCH Verlag GmbH & Co. KGaA, Weinheim
ISBN: 978-3-527-32534-4

Contents

Preface *XXXI*

List of Contributors *XXXV*

Part I RNA Synthesis *1*

I.1 Enzymatic RNA Synthesis, Ligation and Modification *3*

1 Enzymatic RNA Synthesis using Bacteriophage T7 RNA Polymerase *3*
   Heike Gruegelsiepe, Astrid Schön, Leif A. Kirsebom and Roland K. Hartmann
1.1 Introduction *3*
1.2 Description of Method – T7 Transcription *in vitro* *4*
1.2.1 Templates *5*
1.2.2 Special Demands on the RNA Product *6*
1.2.2.1 Homogeneous 5′ and 3′ Ends, Small RNAs, Functional Groups at the 5′ End *6*
1.2.2.2 Modified Substrates *7*
1.3 Transcription Protocols *8*
1.3.1 Transcription with Unmodified Nucleotides *8*
1.3.2 Transcription with 2′-Fluoro-modified Pyrimidine Nucleotides *14*
1.3.3 Purification *15*
1.4 Troubleshooting *17*
1.4.1 Low or No Product Yield *17*
1.4.2 Side-products and RNA Quality *17*
1.5 Rapid Preparation of T7 RNA Polymerase *17*
1.5.1 Required Material *18*
1.5.2 Procedure *18*
1.5.2.1 Cell Growth, Induction and Test for Expression of T7 RNAP *18*
1.5.2.2 Purification of T7 RNAP *19*
1.5.3 Notes and Troubleshooting *20*
Acknowledgement *21*
References *21*

Handbook of RNA Biochemistry: Student Edition. Edited by R. K. Hartmann, A. Bindereif, A. Schön, E. Westhof
Copyright © 2009 WILEY-VCH Verlag GmbH & Co. KGaA, Weinheim
ISBN: 978-3-527-32534-4

2	**Production of RNAs with Homogeneous 5′ and 3′ Ends** *22*	

Mario Mörl, Esther Lizano, Dagmar K. Willkomm and Roland K. Hartmann

2.1	Introduction *22*	
2.2	Description of Approach *23*	
2.2.1	*Cis*-cleaving Autocatalytic Ribozyme Cassettes *23*	
2.2.1.1	The 5′ Cassette *23*	
2.2.1.2	The 3′ Cassette *23*	
2.2.1.3	Purification of Released RNA Product and Conversion of End Groups *26*	
2.2.2	*Trans*-cleaving Ribozymes for the Generation of Homogeneous 3′ Ends *26*	
2.2.3	Further Strategies Toward Homogeneous Ends *29*	
2.3	Critical Experimental Steps, Changeable Parameters, Troubleshooting *29*	
2.3.1	Construction of *Cis*-cleaving 5′ and 3′ Cassettes *29*	
2.3.2	Dephosphorylation Protocols *33*	
2.3.3	Protocols for RNase P Cleavage *34*	
2.3.4	Potential Problems *34*	
	References *35*	

3	**RNA Ligation using T4 DNA Ligase** *36*	

Mikko J. Frilander and Janne J. Turunen

3.1	Introduction *36*	
3.2	Overview of the RNA Ligation Method using the T4 DNA Ligase *37*	
3.3	Large-scale Transcription and Purification of RNAs *38*	
3.4	Generating Homogeneous Acceptor 3′ Ends for Ligation *40*	
3.5	Site-directed Cleavage with RNase H *42*	
3.6	Dephosphorylation and Phosphorylation of RNAs *43*	
3.7	RNA Ligation *44*	
3.8	Troubleshooting *45*	
3.9	Protocols *46*	
	Acknowledgments *51*	
	References *51*	

4	**T4 RNA Ligase** *53*	

Tina Persson, Dagmar K. Willkomm and Roland K. Hartmann

4.1	Introduction *53*	
4.2	Mechanism and Substrate Specificity *54*	
4.2.1	Reaction Mechanism *54*	
4.2.2	Early Studies *56*	
4.2.3	Substrate Specificity and Reaction Conditions *57*	
4.3	Applications of T4 RNA Ligase *58*	
4.3.1	End-labeling *58*	
4.3.2	Circularization *59*	
4.3.3	Intermolecular Ligation of Polynucleotides *59*	

VIII Contents

4.4 T4 RNA Ligation of Large RNA Molecules 61
4.5 Application Examples and Protocols 64
4.5.1 Production of Full-length tRNAs 64
4.5.2 Specific Protocols 65
4.5.3 General Methods (GM) 69
4.5.4 Chemicals and Enzymes 70
4.5.4.1 Chemical Synthesis and Purification of Oligoribonucleotides 70
4.5.4.2 Chemicals 71
4.5.4.3 Enzymes 72
4.6 Troubleshooting 72
Acknowledgments 72
References 72

5 Co- and Post-Transcriptional Incorporation of Specific Modifications Including Photoreactive Groups into RNA Molecules 75
Nathan H. Zahler and Michael E. Harris
5.1 Introduction 75
5.1.1 Applications of RNA Modifications 75
5.1.2 Techniques for Incorporation of Modified Nucleotides 77
5.2 Description 79
5.2.1 5'-End Modification by Transcription Priming 79
5.2.2 Chemical Phosphorylation of Nucleosides to Generate 5'-Monophosphate or 5'-Monophosphorothioate Derivatives 80
5.2.3 Attachment of an Arylazide Photo-crosslinking Agent to a 5'-Terminal Phosphorothioate 82
5.2.4 3'-Addition of an Arylazide Photo-crosslinking Agent 83
5.3 Troubleshooting 84
References 84

6 3'-Terminal Attachment of Fluorescent Dyes and Biotin 86
Dagmar K. Willkomm and Roland K. Hartmann
6.1 Introduction 86
6.2 Description of Method 87
6.3 Protocols 88
6.3.1 3' Labeling 88
6.3.1.1 Biotin Attachment [12] 88
6.3.1.2 Fluorescence Labeling [5] 89
6.3.2 Preparatory Procedures: Dephosphorylation of RNA Produced with 3' Hammerheads 89
6.3.3 RNA Downstream Purifications 90
6.3.3.1 Gel Chromatography 90
6.3.3.2 Purification on Denaturing Polyacrylamide Gels 90
6.3.4 Quality Control 91
6.4 Troubleshooting 91

6.4.1	Problems Caused Prior to the Labeling Reaction	*91*
6.4.2	Problems with the Labeling Reaction Itself	*92*
6.4.3	Post-labeling Problems	*93*
	References	*93*

I.2	**Chemical RNA Synthesis**	*95*
7	**Chemical RNA Synthesis, Purification and Analysis**	*95*
	Brian S. Sproat	
7.1	Introduction	*95*
7.2	Description	*97*
7.2.1	The Solid-phase Synthesis of RNA	*97*
7.2.1.1	Manual RNA Synthesis	*99*
7.2.1.2	Automated RNA Synthesis	*100*
7.2.2	Deprotection	*101*
7.2.2.1	Deprotection of Base Labile Protecting Groups	*101*
7.2.2.2	Desilylation of Trityl-off RNA	*102*
7.2.2.3	Desilylation of Trityl-on RNA	*102*
7.2.3	Purification	*103*
7.2.3.1	Anion-exchange HPLC Purification	*103*
7.2.3.2	Reversed-phase HPLC Purification of Trityl-on RNA	*104*
7.2.3.3	Detritylation of Trityl-on RNA	*105*
7.2.3.4	Desalting by HPLC	*106*
7.2.4	Analysis of the Purified RNA	*107*
7.3	Troubleshooting	*107*
	References	*110*

8	**Modified RNAs as Tools in RNA Biochemistry**	*112*
	Thomas E. Edwards and Snorri Th. Sigurdsson	
8.1	Introduction	*112*
8.1.1	Modification Strategy: The Phosphoramidite Method	*113*
8.1.2	Modification Strategy: Post-synthetic Labeling	*115*
8.2	Description of Methods	*116*
8.2.1	Post-synthetic Modification: The 2′-Amino Approach	*116*
8.2.1.1	Reaction of 2′-Amino Groups with Succinimidyl Esters	*119*
8.2.1.2	Reaction of 2′-Amino Groups with Aromatic Isothiocyanates	*119*
8.2.1.3	Reaction of 2′-Amino Groups with Aliphatic Isocyanates	*120*
8.3	Experimental Protocols	*120*
8.3.1	Synthesis of Aromatic Isothiocyanates and Aliphatic Isocyanates	*120*
8.3.2	Post-synthetic Labeling of 2′-Amino-modified RNA	*122*
8.3.3	Post-synthetic Labeling of 4-Thiouridine-modified RNA	*125*
8.3.4	Verification of Label Incorporation	*125*
8.3.5	Potential Problems and Troubleshooting	*126*
	References	*127*

Part II	Structure Determination 131
II.1	Molecular Biology Methods 133
9	**Direct Determination of RNA Sequence and Modification by Radiolabeling Methods** 133
	Olaf Gimple and Astrid Schön
9.1	Introduction 133
9.2	Methods 133
9.2.1	Isolation of Pure RNA Species from Biological Material 134
9.2.1.1	Preparation of Size-fractionated RNA 134
9.2.1.2	Isolation of Single Unknown RNA Species Following a Functional Assay 134
9.2.1.3	Isolation of Single RNA Species with Partially Known Sequence 135
9.2.2	Radioactive Labeling of RNA Termini 137
9.2.2.1	5' Labeling of RNAs 137
9.2.2.2	3' Labeling of RNAs 138
9.2.3	Sequencing of End-labeled RNA 140
9.2.3.1	Sequencing by Base-specific Enzymatic Hydrolysis of End-labeled RNA 141
9.2.3.2	Sequencing by Base-specific Chemical Modification and Cleavage 144
9.2.4	Determination of Modified Nucleotides by Post-labeling Methods 146
9.2.4.1	Analysis of Total Nucleotide Content 146
9.2.4.2	Determination of Position and Identity of Modified Nucleotides 148
9.3	Conclusions and Outlook 149
	Acknowledgments 149
	References 149
10	**Probing RNA Structures with Enzymes and Chemicals *In Vitro* and *In Vivo*** 151
	Eric Huntzinger, Maria Possedko, Flore Winter, Hervé Moine, Chantal Ehresmann and Pascale Romby
10.1	Introduction 151
10.2	The Probes 153
10.2.1	Enzymes 153
10.2.2	Chemical Probes 153
10.2.3	Lead(II) 155
10.3	Methods 155
10.3.1	Equipment and Reagents 155
10.3.2	RNA Preparation and Renaturation Step 156
10.3.3	Enzymatic and Lead(II)-induced Cleavage Using End-labeled RNA 157
10.3.4	Chemical Modifications 160
10.3.5	Primer Extension Analysis 161
10.3.6	*In Vivo* RNA Structure Mapping 163

10.3.6.1	*In Vivo* DMS Modification	*163*
10.3.6.2	*In Vivo* Lead(II)-induced RNA Cleavages	*165*
10.4	Commentary	*166*
10.4.1	Critical Parameters	*166*
10.4.2	*In Vivo* Mapping	*168*
10.5	Troubleshooting	*168*
10.5.1	*In Vitro* Mapping	*168*
10.5.2	*In Vivo* Probing	*169*
	Acknowledgments	*169*
	References	*170*
11	**Study of RNA–Protein Interactions and RNA Structure in Ribonucleoprotein Particles** *172*	
	Virginie Marchand, Annie Mougin, Agnès Méreau and Christiane Branlant	
11.1	Introduction *172*	
11.2	Methods *175*	
11.2.1	RNP Purification *175*	
11.2.2	RNP Reconstitution *176*	
11.2.2.1	Equipment, Materials and Reagents *176*	
11.2.2.2	RNA Preparation and Renaturation Step *177*	
11.2.3	EMSA *178*	
11.2.3.1	EMSA Method *179*	
11.2.3.2	Supershift Method *180*	
11.2.3.3	Identification of Proteins Contained in RNP by EMSA Experiments Coupled to a Second Gel Electrophoresis and Western Blot Analysis *184*	
11.2.4	Probing of RNA Structure *185*	
11.2.4.1	Properties of the Probes Used *185*	
11.2.4.2	Equipment, Material and Reagents *186*	
11.2.4.3	Probing Method *187*	
11.2.5	UV Crosslinking and Immunoselection *195*	
11.2.5.1	Equipment, Materials and Reagents *195*	
11.2.5.2	UV Crosslinking Method *196*	
11.3	Commentaries and Pitfalls *196*	
11.3.1	RNP Purification and Reconstitution *198*	
11.3.1.1	RNA Purification and Renaturation *198*	
11.3.1.2	EMSA *199*	
11.3.2	Probing Conditions *199*	
11.3.2.1	Choice of the Probes Used *199*	
11.3.2.2	Ratio of RNA/Probes *200*	
11.3.3	UV Crosslinking *200*	
11.3.3.1	Photoreactivity of Individual Amino Acids and Nucleotide Bases *200*	
11.3.3.2	Labeled Nucleotide in RNA *201*	
11.3.4	Immunoprecipitations *201*	
11.3.4.1	Efficiency of Immunoadsorbents for Antibody Binding *201*	

11.4	Troubleshooting 201
11.4.1	RNP Reconstitution 201
11.4.2	RNA Probing 201
11.4.3	UV Crosslinking 202
11.4.4	Immunoprecipitations 202
	Acknowledgments 202
	References 202

12	**Terbium(III) Footprinting as a Probe of RNA Structure and Metal-binding Sites** 205
	Dinari A. Harris and Nils G. Walter
12.1	Introduction 205
12.2	Protocol Description 206
12.2.1	Materials 206
12.3	Application Example 210
12.4	Troubleshooting 212
	References 213

13	**Pb^{2+}-induced Cleavage of RNA** 214
	Leif A. Kirsebom and Jerzy Ciesiolka
13.1	Introduction 214
13.2	Pb^{2+}-induced Cleavage to Probe Metal Ion Binding Sites, RNA Structure and RNA–Ligand Interactions 216
13.2.1	Probing High-affinity Metal Ion Binding Sites 216
13.2.2	Pb^{2+}-induced Cleavage and RNA Structure 220
13.2.3	Pb^{2+}-induced Cleavage to Study RNA–Ligand Interactions 221
13.3	Protocols for Metal Ion-induced Cleavage of RNA 222
13.4	Troubleshooting 225
13.4.1	No Pb^{2+}-induced Cleavage Detected 225
13.4.2	Complete Degradation of the RNA 225
	Acknowledgments 226
	References 226

14	***In Vivo* Determination of RNA Structure by Dimethylsulfate** 229
	Christina Waldsich and Renée Schroeder
14.1	Introduction 229
14.2	Description of Method 230
14.2.1	Cell Growth and *In Vivo* DMS Modification 230
14.2.2	RNA Preparation 231
14.2.3	Reverse Transcription 232
14.3	Evidence for Protein-induced Conformational Changes within RNA *In Vivo* 233
14.4	Troubleshooting 235
	References 237

15	**Probing Structure and Binding Sites on RNA by Fenton Cleavage** *238*
	Gesine Bauer and Christian Berens
15.1	Introduction *238*
15.2	Description of Methods *240*
15.2.1	Fe^{2+}-mediated Cleavage of Native Group I Intron RNA *240*
15.2.2	Fe^{2+}-mediated Tetracycline-directed Hydroxyl Radical Cleavage Reactions *241*
15.3	Comments and Troubleshooting *245*
	References *247*

16	**Measuring the Stoichiometry of Magnesium Ions Bound to RNA** *250*
	A. J. Andrews and Carol Fierke
16.1	Introduction *250*
16.2	Separation of Free Magnesium from RNA-bound Magnesium *251*
16.3	Forced Dialysis is the Preferred Method for Separating Bound and Free Magnesium Ions *252*
16.4	Alternative Methods for Separating Free and Bound Magnesium Ions *254*
16.5	Determining the Concentration of Free Magnesium in the Flow-through *255*
16.6	How to Determine the Concentration of Magnesium Bound to the RNA and the Number of Binding Sites on the RNA *256*
16.7	Conclusion *257*
16.8	Troubleshooting *258*
	References *258*

17	**Nucleotide Analog Interference Mapping and Suppression: Specific Applications in Studies of RNA Tertiary Structure, Dynamic Helicase Mechanism and RNA–Protein Interactions** *259*
	Olga Fedorova, Marc Boudvillain, Jane Kawaoka and Anna Marie Pyle
17.1	Background *259*
17.1.1	The Role of Biochemical Methods in Structural Studies *259*
17.1.2	NAIM: A Combinatorial Approach for RNA Structure–Function Analysis *262*
17.1.2.1	Description of the Method *262*
17.1.2.2	Applications *265*
17.1.3	NAIS: A Chemogenetic Tool for Identifying RNA Tertiary Contacts and Interaction Interfaces *268*
17.1.3.1	General Concepts *268*
17.1.3.2	Applications: Elucidating Tertiary Contacts in Group I and Group II Ribozymes *269*
17.2	Experimental Protocols for NAIM *271*
17.2.1	Nucleoside Analog Thiotriphosphates *271*
17.2.2	Preparation of Transcripts Containing Phosphorothioate Analogs *271*
17.2.3	Radioactive Labeling of the RNA Pool *273*

17.2.4	The Selection Step of NAIM: Three Applications for Studies of RNA Function *273*	
17.2.4.1	Group II Intron Ribozyme Activity: Selection through Transesterification *273*	
17.2.4.2	Reactivity of RNA Helicases: Selection by RNA Unwinding *277*	
17.2.4.3	RNA–Protein Interactions: A One-pot Reaction for Studying Transcription Termination *279*	
17.2.5	Iodine Cleavage of RNA Pools *283*	
17.2.6	Analysis and Interpretation of NAIM Results *284*	
17.2.6.1	Quantification of Interference Effects *284*	
17.3	Experimental Protocols for NAIS *287*	
17.3.1	Design and Creation of Mutant Constructs *287*	
17.3.2	Functional Analysis of Mutants for NAIS Experiments *289*	
17.3.3	The Selection Step for NAIS *289*	
17.3.4	Data Analysis and Presentation *290*	
	Acknowledgments *291*	
	References *291*	
18	**Nucleotide Analog Interference Mapping: Application to the RNase P System** *294*	
	Simona Cuzic and Roland K. Hartmann	
18.1	Introduction *294*	
18.1.1	Nucleotide Analog Interference Mapping (NAIM) – The Approach *294*	
18.1.2	Critical Aspects of the Method *296*	
18.1.2.1	Analog Incorporation *296*	
18.1.2.2	Functional Assays *297*	
18.1.2.3	Factors Influencing the Outcome of NAIM Studies *297*	
18.1.3	Interpretation of Results *298*	
18.1.4	Nucleotide Analog Interference Suppression (NAIS) *300*	
18.2	NAIM Analysis of *Cis*-cleaving RNase P RNA–tRNA Conjugates *300*	
18.2.1	Characterization of a *Cis*-cleaving *E. coli* RNase P RNA–tRNA Conjugate *300*	
18.2.2	Application Example *301*	
18.2.3	Materials *305*	
18.2.4	Protocols *306*	
18.2.5	Data Evaluation *311*	
18.3	Troubleshooting *313*	
	References *317*	
19	**Identification and Characterization of Metal Ion Binding by Thiophilic Metal Ion Rescue** *319*	
	Eric L. Christian	
19.1	Introduction *319*	
19.2	General Considerations of Experimental Conditions *323*	
19.2.1	Metal Ion Stocks and Conditions *323*	

19.2.2	Consideration of Buffers and Monovalent Salt *324*
19.2.3	Incorporation of Phosphorothioate Analogs *325*
19.2.4	Enzyme–Substrate Concentration *327*
19.2.5	General Kinetic Methods *328*
19.2.6	Measurement of Apparent Metal Ion Affinity *329*
19.2.7	Characterization of Metal Ion Binding *333*
19.2.8	Further Tests of Metal Ion Cooperativity *336*
19.3	Additional Considerations *337*
19.3.1	Verification of k_{rel} *337*
19.3.2	Contributions to Complexity of Reaction Kinetics *338*
19.3.3	Size and Significance of Observed Effects *339*
19.4	Conclusion *340*
	Acknowledgments *341*
	References *341*

20 Identification of Divalent Metal Ion Binding Sites in RNA/DNA-metabolizing Enzymes by Fe(II)-mediated Hydroxyl Radical Cleavage *345*

Yan-Guo Ren, Niklas Henriksson and Anders Virtanen

20.1	Introduction *345*
20.2	Probing Divalent Metal Ion Binding Sites *346*
20.2.1	Fe(II)-mediated Hydroxyl Radical Cleavage *346*
20.2.2	How to Map Divalent Metal Ion Binding Sites *347*
20.2.3	How to Use Aminoglycosides as Functional and Structural Probes *349*
20.3	Protocols *350*
20.4	Notes and Troubleshooting *351*
	References *352*

21 Protein–RNA Crosslinking in Native Ribonucleoprotein Particles *354*

Henning Urlaub, Klaus Hartmuth and Reinhard Lührmann

21.1	Introduction *354*
21.2	Overall Strategy *354*
21.3	UV Crosslinking *355*
21.4	Identification of UV-induced Protein–RNA Crosslinking Sites by Primer Extension Analysis *357*
21.5	Identification of Crosslinked Proteins *361*
21.6	Troubleshooting *364*
21.7	Protocols *367*
	Acknowledgments *372*
	References *372*

22 Probing RNA Structure by Photoaffinity Crosslinking with 4-Thiouridine and 6-Thioguanosine *374*

Michael E. Harris and Eric L. Christian

22.1	Introduction *374*
22.2	Description *377*

22.2.1	General Considerations: Reaction Conditions and Concentrations of Interacting Species *377*
22.2.2	Generation and Isolation of Crosslinked RNAs *380*
22.2.3	Primer Extension Mapping of Crosslinked Nucleotides *381*
22.3	Troubleshooting *382*
	References *384*

II.2 Biophysical Methods *385*

23 Structural Analysis of RNA and RNA–Protein Complexes by Small-angle X-ray Scattering *385*
Tao Pan and Tobin R. Sosnick

23.1	Introduction *385*
23.2	Description of the Method *387*
23.2.1	General Requirements *387*
23.2.2	An Example for the Application of SAXS *389*
23.3	General Information *389*
23.4	Question 1: The Oligomerization State of P RNA and the RNase P Holoenzyme *390*
23.5	Question 2: The Overall Shape *392*
23.6	Question 3: The Holoenzyme–Substrate Complexes *392*
23.7	Troubleshooting *395*
23.7.1	Problem 1: Radiation Damage and Aggregation *395*
23.7.2	Problem 2: High Scattering Background *395*
23.7.3	Problem 3: Scattering Results cannot be Fit to Simple Models *396*
23.8	Conclusions/Outlook *396*
	Acknowledgments *396*
	References *397*

24 Temperature-Gradient Gel Electrophoresis of RNA *398*
Detlev Riesner and Gerhard Steger

24.1	Introduction *398*
24.2	Method *399*
24.2.1	Principle *399*
24.2.2	Instruments *400*
24.2.3	Handling *400*
24.3	Optimization of Experimental Conditions *401*
24.3.1	Attribution of Secondary Structures to Transition Curves in TGGE *401*
24.3.2	Pore Size of the Gel Matrix *402*
24.3.3	Electric Field *402*
24.3.4	Ionic Strength and Urea *402*
24.4	Examples *402*
24.4.1	Analysis of Different RNA Molecules in a Single TGGE *403*
24.4.2	Analysis of Structure Distributions of a Single RNA – Detection of Specific Structures by Oligonucleotide Labeling *405*

24.4.3	Analysis of Mutants *409*	
24.4.4	Retardation Gel Electrophoresis in a Temperature Gradient for Detection of Protein–RNA Complexes *409*	
24.4.5	Outlook *413*	
	References *414*	

25 UV Melting, Native Gels and RNA Conformation *415*
Andreas Werner

25.1	Monitoring RNA Folding in Solution *415*
25.2	Methods *417*
25.3	Data Analysis *420*
25.4	Energy Calculations and Limitations *422*
25.5	RNA Concentration *424*
25.6	Salt and pH Dependence *424*
25.7	Native Gels *426*
	References *427*

26 Sedimentation Analysis of Ribonucleoprotein Complexes *428*
Jan Medenbach, Andrey Damianov, Silke Schreiner and Albrecht Bindereif

26.1	Introduction *428*
26.2	Glycerol Gradient Centrifugation *429*
26.2.1	Equipment *429*
26.2.2	Reagents *429*
26.2.3	Method *430*
26.2.3.1	Preparation of the Glycerol Gradient *430*
26.2.3.2	Sample Preparation and Centrifugation *430*
26.2.3.3	Preparation of RNA from Gradient Fractions *431*
26.2.3.4	Simultaneous Preparation of RNA and Proteins *431*
26.2.3.5	Control Gradient with Sedimentation Markers *432*
26.2.3.6	Notes and Troubleshooting *433*
26.3	Fractionation of RNPs by Cesium Chloride Density Gradient Centrifugation *434*
26.3.1	Equipment *434*
26.3.2	Reagents *434*
26.3.3	Method *435*
26.3.3.1	Preparation of the Gradient and Ultracentrifugation *435*
26.3.3.2	Preparation of RNA from the Gradient Fractions *435*
26.3.3.3	Control Gradient for Density Calculation *435*
26.3.3.4	Notes and Troubleshooting *435*
	Acknowledgments *437*
	References *437*

27 Preparation and Handling of RNA Crystals *438*
Boris François, Aurélie Lescoute-Phillips, Andreas Werner and Benoît Masquida

27.1	Introduction *438*

27.2	Design of Short RNA Constructs 439
27.3	RNA Purification 439
27.3.1	HPLC Purification 439
27.3.2	Gel Electrophoresis 440
27.3.3	RNA Recovery 441
27.3.3.1	Elution of the RNA from the Gel 441
27.3.3.2	Concentration and Desalting 441
27.4	Setting Crystal Screens for RNA 442
27.4.1	Renaturing the RNA 447
27.4.2	Setting-up Crystal Screens 447
27.4.3	Forming Complexes with Organic Ligands: The Example of Aminoglycosides 447
27.4.4	Evaluate Screening Results 449
27.4.5	The Optimization Process 449
27.5	Conclusions 451
	References 452

II.3 Fluorescence and Single Molecule Studies 453

28 Fluorescence Labeling of RNA for Single Molecule Studies 453
Filipp Oesterhelt, Enno Schweinberger and Claus Seidel

28.1	Introduction 453
28.2	Fluorescence Resonance Energy Transfer (FRET) 456
28.2.1	Measurement of Distances via FRET 456
28.3	Questions that can be Addressed by Single Molecule Fluorescence 458
28.3.1	RNA Structure and Dynamics 459
28.3.2	Single Molecule Fluorescence in Cells 460
28.3.2.1	Techniques used for Fluorescent Labeling RNA in Cells 460
28.3.2.2	Intracellular Mobility 462
28.3.3	Single Molecule Detection in Nucleic Acid Analysis 462
28.3.3.1	Fragment Sizing 462
28.3.3.2	Single Molecule Sequencing 462
28.4	Equipment for Single Molecule FRET Measurements 463
28.4.1.1	Excitation of the Fluorophores 463
28.4.1.2	Fluorescence Detection 464
28.4.1.3	Data Analysis 465
28.5	Sample Preparation 466
28.5.1	Fluorophore–Nucleic Acid Interaction 466
28.5.2	RNA Labeling 466
28.5.2.1	Fluorophores for Single Molecule Fluorescence Detection 466
28.5.2.2	Fluorophores used for FRET Experiments 467
28.5.2.3	Attaching Fluorophores to RNA 467
28.5.2.4	Linkers 468
28.5.3	Fluorescence Background 468

28.5.3.1	Raman Scattered Light	468
28.5.3.2	Cleaning Buffers	468
28.5.3.3	Clean Surfaces	469
28.5.4	Surface Modification	469
28.5.4.1	Coupling Single Molecules to Surfaces	469
28.5.4.2	Surface Passivation	470
28.5.5	Preventing Photodestruction	470
28.6	Troubleshooting	470
28.6.1.1	Orientation Effects	470
28.6.1.2	Dissociation of Molecular Complexes	471
28.6.1.3	Adsorption to the Surface	471
28.6.1.4	Diffusion Limited Observation Times	471
28.6.1.5	Intensity Fluctuations	472
	References	472

29 Scanning Force Microscopy and Scanning Force Spectroscopy of RNA 475
Wolfgang Nellen

29.1	Introduction	475
29.2	Questions that could be Addressed by SFM	477
29.3	Statistics	481
29.4	Scanning Force Spectroscopy (SFS)	481
29.5	Questions that may be Addressed by SFS	483
29.6	Protocols	483
29.7	Troubleshooting	485
29.8	Conclusions	486
	Acknowledgments	487
	References	487

Part III   RNA Genomics and Bioinformatics 489

30 Comparative Analysis of RNA Secondary Structure: 6S RNA 491
James W. Brown and J. Christopher Ellis

30.1	Introduction	491
30.1.1	RNA Secondary Structure	492
30.1.2	Comparative Sequence Analysis	492
30.1.3	Strengths and Weakness of Comparative Analysis	493
30.1.4	Comparison with Other Methods	494
30.2	Description	495
30.2.1	Collecting Sequence Data	495
30.2.2	Thermodynamic Predictions	498
30.2.3	Initial Alignment	500
30.2.4	Terminal Helix (P1a)	502
30.2.5	Subterminal Helix (P1b)	506
30.2.6	Apical Helix (P2a)	506

30.2.7	Subapical Helices (P2b and P2c)	507
30.2.8	Potential Interior Stem–loop (P3)	508
30.2.9	Is There Anything Else?	508
30.2.10	Where To Go From Here	509
30.3	Troubleshooting	510
	Acknowledgments	511
	References	511

31 Secondary Structure Prediction 513
Gerhard Steger

31.1	Introduction	513
31.2	Thermodynamics	513
31.3	Formal Background	516
31.4	mfold	518
31.4.1	Input to the mfold Server	518
31.4.1.1	Sequence Name	518
31.4.1.2	Sequence	518
31.4.1.3	Constraints	519
31.4.1.4	Further Parameters	521
31.4.1.5	Immediate versus Batch Jobs	522
31.4.2	Output from the mfold Server	524
31.4.2.1	Energy Dot Plot	524
31.4.2.2	RNAML (RNA Markup Language) Syntax	524
31.4.2.3	Extra Files	524
31.4.2.4	Download All Foldings	525
31.4.2.5	View ss-count Information	525
31.4.2.6	View Individual Structures	525
31.4.2.7	Dot Plot Folding Comparisons	527
31.5	RNAfold	527
31.5.1	Input to the RNAfold Server	528
31.5.1.1	Sequence and Constraints	528
31.5.1.2	Further Parameters	529
31.5.1.3	Immediate versus Batch Jobs	530
31.5.2	Output from the RNAfold Server	531
31.5.2.1	Probability Dot Plot	531
31.5.2.2	Text Output of Secondary Structure	531
31.5.2.3	Graphical Output of Secondary Structure	531
31.5.2.4	Mountain Plot	533
31.6	Troubleshooting	533
	References	534

32 Modeling the Architecture of Structured RNAs within a Modular and Hierarchical Framework 536
Benoît Masquida and Eric Westhof

32.1	Introduction	536

32.2	Modeling Large RNA Assemblies 537
32.2.1	The Modeling Process 538
32.2.1.1	Getting the Right Secondary Structure 539
32.2.1.2	Extrusion of the Secondary Structure in 3-D 540
32.2.1.3	Interactive Molecular Modeling 540
32.2.1.4	Refinement of the Model 542
32.3	Conclusions 543
	References 544
33	**Modeling Large RNA Assemblies using a Reduced Representation** 546
	Jason A. Mears, Scott M. Stagg and Stephen C. Harvey
33.1	Introduction 546
33.2	Basic Modeling Principles 547
33.2.1	Pseudo-atoms and Reduced Representation 549
33.2.2	Implementing RNA Secondary Structure 550
33.2.3	Protein Components 551
33.2.4	Implementing Tertiary Structural Information 551
33.2.5	Modeling Protocol 552
33.3	Application of Modeling Large RNA Assemblies 554
33.3.1	Modeling the Ribosome Structure at Low Resolution 554
33.3.2	Modeling Dynamic Assembly of the Ribosome with Reduced Representation 556
33.4	Conclusion 557
33.5	Troubleshooting 557
	References 559
34	**Molecular Dynamics Simulations of RNA Systems** 560
	Pascal Auffinger and Andrea C. Vaiana
34.1	Introduction 560
34.2	MD Methods 560
34.3	Simulation Setups 562
34.3.1	Choosing the Starting Structure 562
34.3.1.1	Model Built Structures 563
34.3.1.2	X-ray Structures 563
34.3.1.3	NMR Structures 563
34.3.2	Checking the Starting Structure 563
34.3.2.1	Conformational Checks 563
34.3.2.2	Protonation Issues 564
34.3.2.3	Solvent 564
34.3.3	Adding Hydrogen Atoms 564
34.3.4	Choosing the Environment (Crystal, Liquid) and Ions 564
34.3.5	Setting the Box Size and Placing the Ions 565
34.3.5.1	Box Size 565
34.3.5.2	Monovalent Ions 565
34.3.5.3	Divalent Ions 565

34.3.6	Choosing the Program and Force Field	565
34.3.6.1	Programs	565
34.3.6.2	Force Fields	566
34.3.6.3	Parameterization of Modified Nucleotides and Ligands	566
34.3.6.4	Water Models	567
34.3.7	Treatment of Electrostatic Interactions	567
34.3.8	Other Simulation Parameters	568
34.3.8.1	Thermodynamic Ensemble	568
34.3.8.2	Temperature and Pressure	568
34.3.8.3	Shake, Time Steps and Update of the Non-bonded Pair List	568
34.3.8.4	The Flying Ice Cube Problem	568
34.3.9	Equilibration	569
34.3.10	Sampling	569
34.3.10.1	How Long Should a Simulation Be?	569
34.3.10.2	When to Stop a Simulation	570
34.3.10.3	Multiple MD (MMD) Simulations	570
34.4	Analysis	570
34.4.1	Evaluating the Quality of the Trajectories	570
34.4.1.1	Consistency Checks	571
34.4.1.2	Comparison with Experimental Data	571
34.4.1.3	Visualization	571
34.4.2	Convergence Issues	571
34.4.3	Conformational Parameters	572
34.4.4	Solvent Analysis	572
34.5	Perspectives	572
	Acknowledgments	573
	References	573
35	**Seeking RNA Motifs in Genomic Sequences**	**577**
	Matthieu Legendre and Daniel Gautheret	
35.1	Introduction	577
35.2	Choosing the Right Search Software: Limitations and Caveats	578
35.3	Retrieving Programs and Sequence Databases	581
35.4	Organizing RNA Motif Information	581
35.5	Evaluating Search Results	583
35.6	Using the RNAMOTIF Program	585
35.7	Using the ERPIN Program	589
35.8	Troubleshooting	592
35.8.1	RNAMOTIF	592
35.8.1.1	Too Many Solutions	592
35.8.1.2	Program Too Slow	592
35.8.2	ERPIN	592
35.8.2.1	Too Many Solutions	592
35.8.2.2	Program Too Slow	593
	Acknowledgments	593
	References	593

36	**Approaches to Identify Novel Non-messenger RNAs in Bacteria and to Investigate their Biological Functions: RNA Mining** *595*
	Jörg Vogel and E. Gerhart H. Wagner
36.1	Introduction *595*
36.2	Searching for Small, Untranslated RNAs *597*
36.2.1	Introduction *597*
36.2.2	Direct Labeling and Direct Cloning *598*
36.2.3	Functional Screens *599*
36.2.4	Biocomputational Screens *602*
36.2.5	Microarray Detection *605*
36.2.6	Shotgun Cloning (RNomics) *606*
36.2.7	Co-purification with Proteins or Target RNAs *609*
36.2.8	Screens for *Cis*-encoded Antisense RNAs *610*
36.3	Conclusions *610*
	Acknowledgments *611*
	References *611*
37	**Approaches to Identify Novel Non-messenger RNAs in Bacteria and to Investigate their Biological Functions: Functional Analysis of Identified Non-mRNAs** *614*
	E. Gerhart H. Wagner and Jörg Vogel
37.1	Introduction *614*
37.2	Approaches for Elucidation of Bacterial sRNA Function *615*
37.2.1	Large-scale Screening for Function *615*
37.2.2	Preparing for Subsequent Experiments: Strains and Plasmids *615*
37.2.3	Experimental Approaches *618*
37.2.4	Physiological Phenotypes (Lethality, Growth Defects, etc.) *618*
37.2.5	Analyzing sRNA Effects on Specific mRNA Levels by Microarrays *619*
37.2.6	Analyzing sRNA Effects by Proteomics *620*
37.2.7	Analyzing sRNA Effects by Metabolomics *621*
37.2.8	Finding Targets by Reporter Gene Approaches *621*
37.2.9	Bioinformatics-aided Approaches *623*
37.2.10	Prediction of Regulatory Sequences in the Vicinity of sRNA Gene Promoters *623*
37.2.11	Finding Interacting Sites (Complementarity/Antisense) *624*
37.3	Additional Methods Towards Functional and Mechanistic Characterizations *625*
37.3.1	Finding sRNA-associated Proteins *625*
37.3.2	Regulation of the Target RNA – Use of Reporter Gene Fusions *626*
37.3.3	Northern Analyses *627*
37.3.4	Analysis of sRNAs – RACE and Primer Extensions *627*
37.3.5	Structures of sRNAs and Target RNAs *628*
37.4	Conclusions *629*
37.5	Protocols *629*
	Acknowledgments *639*
	References *640*

38		Experimental RNomics: A Global Approach to Identify Non-coding RNAs in Model Organisms 643
		Alexander Hüttenhofer
38.1		Introduction 643
38.2		Materials 644
38.2.1		Oligonucleotide Primers 644
38.2.2		Enzymes 644
38.2.3		Buffers 644
38.2.4		Reagents, Kits, Vectors and Bacterial Cells 645
38.3		Protocols for Library Construction and Analysis 645
38.4		Computational Analysis of ncRNA Sequences 652
38.5		Troubleshooting 653
		Acknowledgments 653
		References 653
39		Large-scale Analysis of mRNA Splice Variants by Microarray 655
		Young-Soo Kwon, Hai-Ri Li and Xiang-Dong Fu
39.1		Introduction 655
39.2		Overview of RASL Technology 655
39.3		Description of Methods 657
39.3.1		Preparation of Index Arrays 657
39.3.2		Annotation of Alternative Splicing 658
39.3.3		Target Design 658
39.3.4		Preparation of Target Pool 658
39.3.5		The RASL Assay Protocol 659
39.3.6		PCR Amplification 659
39.3.7		Hybridization on Index Array 660
39.3.8		Data Analysis 661
39.4		Troubleshooting 661
39.4.1		System Limitation and Pitfalls 661
39.4.2		Potential Experimental Problems 662
		References 663
IV		**Analysis of RNA Function** 665
IV.1		**RNA–Protein Interactions in vitro** 667
40		Use of RNA Affinity Matrices for the Isolation of RNA-binding Proteins 667
		Steffen Schiffer, Sylvia Rösch, Bettina Späth, Markus Englert, Hildburg Beier and Anita Marchfelder
40.1		Introduction 667
40.2		Materials 668
40.2.1		CNBr-activated Sepharose 4B Affinity Column 668
40.2.2		NHS-activated HiTrap Columns 669
40.3		Methods 669

40.3.1	Coupling of tRNAs to CNBr-activated Sepharose 4B 669
40.3.2	Coupling of tRNAs to a 5-ml NHS-activated HiTrap Column 671
40.4	Application 672
40.4.1	Purification of the Nuclear RNase Z from Wheat Germ 672
40.5	Notes 674
	References 674
41	**Biotin-based Affinity Purification of RNA–Protein Complexes** 676
	Zsofia Palfi, Jingyi Hui and Albrecht Bindereif
41.1	Introduction 676
41.2	Materials 677
41.2.1	Oligonucleotides 677
41.2.2	Affinity Matrices 678
41.2.3	Cell Extracts 678
41.2.4	Buffers and Solutions 679
41.2.5	Additional Materials 679
41.3	Methods 680
41.3.1	Affinity Purification of RNPs 680
41.3.1.1	Depletion of Total Cell Lysate from SAg-binding Material (Pre-clearing) 680
41.3.1.2	Pre-blocking SAg Beads 681
41.3.1.3	Affinity Selection of RNPs for Structural Studies 681
41.3.1.4	Affinity Selection of RNPs for Functional Studies by Displacement Strategy 685
41.3.2	Affinity Purification of Specific RNA-binding Proteins by Biotinylated RNAs 689
41.4	Troubleshooting 691
41.4.1	Biotinylated 2′OMe RNA Oligonucleotides 691
41.4.2	Extracts and Buffers 691
41.4.3	Optimization of the Experimental Conditions: When Yields are Low 691
41.4.4	Optimization of the Experimental Conditions: When Non-specific Background is Too High 692
	References 692
42	**Immunoaffinity Purification of Spliceosomal and Small Nuclear Ribonucleoprotein Complexes** 694
	Cindy L. Will, Evgeny M. Makarov, Olga V. Makarova and Reinhard Lührmann
42.1	Introduction 694
42.2	Generation of Anti-peptide Antibodies: Peptide Selection Criteria 694
42.3	Immunoaffinity Selection of U4/U6.U5 Tri-snRNPs 697
42.4	Immunoaffinity Purification of 17S U2 snRNPs 699
42.5	Approaches for the Isolation of Native, Human Spliceosomal Complexes 702
42.6	Isolation of Activated Spliceosomes by Immunoaffinity Selection with Anti-peptide Antibodies against the SKIP Protein 703

Acknowledgments 709
References 709

43 Northwestern Techniques for the Identification of RNA-binding Proteins from cDNA Expression Libraries and the Analysis of RNA–Protein Interactions 710

Ángel Emilio Martínez de Alba, Michela Alessandra Denti and Martin Tabler

43.1	Introduction 710	
43.2	Methods 712	
43.2.1	Preparation of Probes and Buffers 712	
43.2.1.1	Preparation of ^{32}P-labeled RNA Probes 712	
43.2.1.2	Preparation of Blocking RNA 712	
43.2.1.3	Preparation of the Northwestern Buffer 713	
43.2.2	Protocol 1: Northwestern Screening for Identification of RNA-binding Proteins from cDNA Expression Libraries 713	
43.2.2.1	Preparation of the Host Plating Culture 714	
43.2.2.2	Plating of the cDNA Phage Expression Library 714	
43.2.2.3	Adsorbing Recombinant Proteins to Nitrocellulose Membranes 715	
43.2.2.4	Incubation with an RNA Ligand 716	
43.2.2.5	Washing of Membranes 717	
43.2.2.6	Identification of True Positives 717	
43.2.3	Protocol 2: Northwestern Techniques to Detect and Analyze RNA–Protein Interactions 719	
43.2.3.1	Protein Sample Preparation 719	
43.2.3.2	Protein Electrophoresis and Transfer 721	
43.2.3.3	Incubation of the Membranes with an RNA Probe 722	
43.2.3.4	Washing of Membranes and Autoradiography 722	
43.3	Troubleshooting 723	
43.3.1	Probe Quality 723	
43.3.2	Background Signals 724	
43.3.3	Signal-to-Background Ratio 724	
43.3.4	Protein Conformation 726	
43.3.5	Weak Binding Signals 726	
43.3.6	False Positives 726	
43.3.7	Quality of the cDNA Library 727	
43.3.8	Fading Signals 727	
43.3.9	Supplementary 727	
	References 727	

IV.2 RNA–Protein Interactions *in vivo* 729

44 Fluorescent Detection of Nascent Transcripts and RNA-binding Proteins in Cell Nuclei 729

Jennifer A. Geiger and Karla M. Neugebauer

44.1 Introduction 729
44.2 Description of the Methods 730

44.2.1	Overview	*730*
44.2.2	Preparation of Fluorescent DNA Probes for *In Situ* Hybridization	*731*
44.2.2.1	Method 1: Nick Translation of Plasmid DNA	*731*
44.2.2.2	Method 2: PCR Amplification and DNase I Digestion	*732*
44.2.3	Performing Combined Immunocytochemistry and FISH	*733*
44.2.4	Troubleshooting	*735*
	Acknowledgments	*735*
	References	*735*
45	**Identification and Characterization of RNA-binding Proteins through Three-hybrid Analysis**	*737*
	Felicia Scott and David R. Engelke	
45.1	Introduction	*737*
45.2	Basic Strategy of the Method	*738*
45.3	Detailed Components	*739*
45.3.1	Yeast Reporter Strain	*740*
45.3.2	Plasmids	*740*
45.3.3	Hybrid RNA	*740*
45.3.3.1	Technical Considerations for the Hybrid RNA	*742*
45.3.4	Activation Domain FP2	*743*
45.3.4.1	Technical Considerations for the Activation Domain FP2	*743*
45.3.5	Positive Controls	*744*
45.4	Protocols	*745*
45.4.1	Transformation of Yeast	*745*
45.4.2	Assaying for *HIS3* Expression	*748*
45.4.3	Assaying for β-Galactosidase Activity	*748*
45.5	Troubleshooting	*749*
45.6	Additional Applications	*751*
45.7	Summary	*752*
	Acknowledgments	*752*
	References	*752*
46	**Analysis of Alternative Splicing *In Vivo* using Minigenes**	*755*
	Yesheng Tang, Tatyana Novoyatleva, Natalya Benderska, Shivendra Kishore, Alphonse Thanaraj and Stefan Stamm	
46.1	Introduction	*755*
46.2	Overview of the Method	*755*
46.3	Methods	*757*
46.3.1	Construction of Minigenes	*757*
46.3.2	Transfection of Cells	*757*
46.3.3	Analysis	*771*
46.3.3.1	RT-PCR	*771*
46.3.3.2	Other Analysis Methods	*771*
46.3.4	Necessary Controls	*773*
46.3.5	Advantages and Disadvantages of the Method	*773*
46.3.6	Related Methods	*774*

46.4	Troubleshooting	*774*
46.5	Bioinformatic Resources	*775*
46.6	Protocols	*775*
	References	*780*

IV.3	**SELEX**	*783*

47 Artificial Selection: Finding Function amongst Randomized Sequences *783*
Ico de Zwart, Catherine Lozupone, Rob Knight, Amanda Birmingham, Mali Illangasekare, Vasant Jadhav, Michal Legiewicz, Irene Majerfeld, Jeremy Widmann and Michael Yarus

47.1	The SELEX Method	*783*
47.2	Understanding a Selection	*784*
47.2.1	Sequence Motif Representation and Abundance	*786*
47.2.2	The Recovery Efficiency of Different RNAs	*787*
47.2.3	Stringency	*789*
47.2.4	Amplification and Transcription Biases	*789*
47.3	Isolation of RNAs that Bind Small Molecules	*790*
47.3.1	Stringency and K_D	*792*
47.3.2	Selection for Multiple Targets in One Column	*793*
47.3.3	Characterizing Motif Activity	*793*
47.4	Techniques for Selecting Ribozymes	*794*
47.4.1	Making the RNA a Substrate for the Reaction	*795*
47.4.2	The Inherent Reactivity of RNA	*795*
47.4.3	Selecting Active RNAs	*797*
47.4.4	Negative Selections	*798*
47.4.5	Stringency	*799*
47.4.6	Analysis of the Product	*799*
47.4.7	Determining the Scope of the Reaction	*799*
47.5	Sequence Analysis	*800*
47.5.1	Identifying Related Sequences	*800*
47.5.2	Predicting Structure	*802*
47.5.3	Chemical and Enzymatic Mapping	*803*
47.5.4	Finding Minimal Requirements	*804*
47.5.5	Three-dimensional Structural Modeling	*804*
	Acknowledgments	*805*
	References	*805*

48 Aptamer Selection against Biological Macromolecules: Proteins and Carbohydrates *807*
C. Stefan Vörtler and Maria Milovnikova

48.1	Introduction	*807*
48.2	General Strategy	*808*
48.2.1	Choosing a Suitable Target	*810*
48.2.1.1	Protein Targets	*810*
48.2.1.2	Carbohydrate Targets	*811*

48.2.2	Immobilization of the Target	*812*
48.2.3	Selection Assays	*812*
48.2.4	Design and Preparation of the Library	*813*
48.3	Running the *In Vitro* Selection Cycle	*814*
48.4	Analysis of the Selection Outcome	*815*
48.5	Troubleshooting	*816*
48.6	Protocols	*817*
	Acknowledgments	*836*
	References	*836*
49	***In Vivo* SELEX Strategies**	*840*
	Thomas A. Cooper	
49.1	Introduction	*840*
49.2	Procedure Overview	*841*
49.2.1	Design of the Randomized Exon Cassette	*843*
49.2.2	Design of the Minigene	*844*
49.2.3	RT-PCR Amplification	*846*
49.2.4	Monitoring for Enrichment of Exon Sequences that Function as Splicing Enhancers	*846*
49.2.5	Troubleshooting	*847*
49.3	Protocols	*848*
	Acknowledgments	*852*
	References	*852*
50	***In Vitro* Selection against Small Targets**	*853*
	Dirk Eulberg, Christian Maasch, Werner G. Purschke and Sven Klussmann	
50.1	Introduction	*853*
50.2	Target Immobilization	*856*
50.2.1	Covalent Immobilization	*857*
50.2.1.1	Epoxy-activated Matrices	*857*
50.2.1.2	NHS-activated Matrices	*859*
50.2.1.3	Pyridyl Disulfide-activated Matrices	*860*
50.2.2	Non-covalent Immobilization	*861*
50.3	Nucleic Acid Libraries	*862*
50.3.1	Library Design	*862*
50.3.2	Starting Pool Preparation	*863*
50.4	Enzymatics	*865*
50.4.1	Reverse Transcription	*865*
50.4.2	PCR	*866*
50.4.3	*In Vitro* Transcription	*867*
50.5	Partitioning	*868*
50.6	Binding Assays	*873*
50.6.1	Equilibrium Dialysis	*873*
50.6.2	Equilibrium Filtration Analysis	*874*
50.6.3	Isocratic Competitive Affinity Chromatography	*875*
	References	*877*

51		SELEX Strategies to Identify Antisense and Protein Target Sites in RNA or Heterogeneous Nuclear Ribonucleoprotein Complexes 878
		Martin Lützelberger, Martin R. Jakobsen and Jørgen Kjems
51.1		Introduction 878
51.1.1		Applications for Antisense 879
51.1.2		Selecting Protein-binding Sites 879
51.2		Construction of the Library 879
51.2.1		Generation of Random DNA Fragments from Genomic or Plasmid DNA 881
51.2.2		Preparing RNA Libraries from Plasmid, cDNA or Genomic DNA 881
51.3		Identification of Optimal Antisense Annealing Sites in RNAs 882
51.4		Identification of Natural RNA Substrates for Proteins and Other Ligands 884
51.5		Cloning, Sequencing and Validating the Selected Inserts 884
51.6		Troubleshooting 885
51.6.1		Sonication of Plasmid DNA does not Yield Shorter Fragments 885
51.6.2		Inefficient Ligation 885
51.6.3		Inefficient *Mme*I Digestion 885
51.6.4		The Amplification of the Unselected Library is Inefficient 886
51.6.5		The Library Appears to be Non-random in the Unselected Pool 886
51.6.6		The Selected RNAs do not Bind Native Protein 886
51.7		Protocols 886
		References 894

V   RNAi 895

52		Gene Silencing Methods for Mammalian Cells: Application of Synthetic Short Interfering RNAs 897
		Matthias John, Anke Geick, Philipp Hadwiger, Hans-Peter Vornlocher and Olaf Heidenreich
52.1		Introduction 897
52.2		Background Information 898
52.3		Ways to Induce RNAi in Mammalian Cells 900
52.3.1		Important Parameters 901
52.3.1.1		siRNA Design 901
52.3.1.2		Target Site Selection 901
52.3.1.3		Preparation of siRNA Samples 902
52.3.2		Transfection of Mammalian Cells with siRNA 902
52.3.3		Electroporation of Mammalian Cells with siRNA 904
52.3.4		Induction of RNAi by Intracellular siRNA Expression 905
52.4		Troubleshooting 908
		References 908

Appendix: UV Spectroscopy for the Quantitation of RNA 910

Index 915

Preface

The field of RNA research has experienced an incredible boom in recent years, with no calming down in sight. Stimulated by a common fascination for RNA, we became convinced about two years ago that something is missing in the RNA field: a comprehensive handbook of RNA biochemistry that combines protocols of methods and techniques in RNA chemistry, biochemistry and bioinformatics – a handbook that is not a guide for specialists only, but also for graduate and PhD students as well as experienced scientists who wish to embark on RNA research projects. We had in mind to merge several core features into our book concept, making it unique among related publications. These include a thorough introduction to the individual approach or method, providing the necessary background for non-specialists and addressing the scientific potential as well as the limitations in terms of applicability, resolution and interpretation. A second feature is the detailed description of experimental protocols, such that the reader should be able to apply the technique(s) directly, without extensive further reading of the original literature. Related to this is the incorporation of troubleshooting sections, describing pitfalls and discussing critical experimental steps as well as ideas for adequate problem solving in cases of failure.

A substantial fraction of the handbook (Part I) describes basic approaches and methods of RNA synthesis and modification (e. g. T7 transcription, co- and post-transcriptional modification, enzymatic RNA ligation, chemical RNA synthesis, co- and post-synthetic modifications), many of which may be considered timeless as they have been continuously applied and developed during the last four decades. Several of these basic techniques are routinely used in many RNA laboratories, but in most cases PhD students just apply a protocol inherited from former lab members. This fulfils the purpose as long as the method works smoothly. In cases of failure, though, the devil is usually in the details, and a deeper insight into the procedure or enzymatic reaction is then required. Likewise, techniques have often been tailored to a specific task, but variation of application for other purposes is not always straightforward. An example is T4 RNA ligase primarily used for [^{32}P]pCp end-labeling. When [^{32}P]pCp is replaced with longer donor substrates, the reaction is often inefficient and more liable to disturbance, and unwanted ligation products tend to be the rule rather than the exception. The handbook will be a valuable guide in such cases.

Handbook of RNA Biochemistry: Student Edition. Edited by R. K. Hartmann, A. Bindereif, A. Schön, E. Westhof
Copyright © 2009 WILEY-VCH Verlag GmbH & Co. KGaA, Weinheim
ISBN: 978-3-527-32534-4

The biological insight gained from any RNA or RNA-protein structure as determined, for example, by X-ray crystallography is thoroughly enhanced when flanked by biochemical experiments that investigate RNA structure in solution, or even in its cellular environment. Thus, in Part II, a string of chapters is dedicated to the investigation of RNA structure in solution, naked as well as in the context of RNA-protein complexes. Both *in vitro* and *in vivo* approaches are presented that use enzymatic, chemical or metal ion probes. Elaborate crosslinking methods to investigate higher-order RNA structure or RNA-protein interactions are covered as well, and two contributions outline approaches that are based on complex nucleic acid libraries and allow to screen RNA molecules for important functional groups (NAIM, NAIS). In general, the techniques described in this handbook have been developed to a prodigious level of detail and sophistication by expert laboratories in recent years, and timely RNA research depends on the accuracy and reproducibility of experiments conducted on the basis of such protocols.

The integral role of metal ions in RNA architecture and catalysis was first put in a nutshell by Mike Yarus in his *Cheshire Cat* metaphor (*FASEB J.*, 1993). Any RNA researcher will consequently devote at least some experiments to the role of metal ions. Thus, a cornucopia of approaches and protocols in Part II of the handbook is dedicated to the study of metal ion binding to RNA molecules and, as an extension, to RNA/DNA-metabolizing enzymes. This includes the characterization of high affinity metal ion binding sites and strategies to unravel the localization and functional role of catalytic metal ions, as well as a method to quantify Mg^{2+} binding to RNA.

Understanding RNA function is unimaginable without biophysical and computer-based approaches. Thus, Part II further contains several chapters that deal with physico-chemical techniques: various types of spectroscopy (e. g. CD, fluorescence, UV, small angle X-ray scattering) as applied to RNA are described; a number of special techniques address the physical properties of naked or protein-complexed RNAs, such as their melting and sedimentation behaviour. Fluorescently labeled RNAs are nowadays indispensable tools for the study of RNA structure and dynamics, and single molecule studies in living cells give detailed insight into molecular distribution and dynamic processes. Single molecule research has been further addressed by a contribution on scanning force microscopy (SFM) and a derived technique, scanning force spectroscopy (SFS), used to determine inter- and intramolecular binding forces. Although X-ray crystallography *per se* is not covered in this book, an introduction to the preparation and handling of RNA crystals is included.

Bioinformatics, traditionally relegated to the application of BLAST-type programs for the search and annotation of unknown genes, more and more extends to the understanding of the relationships between sequences and three-dimensional structures in the context of molecular Darwinian evolution. Part III begins with the powerful applications of phylogenetic and sequence comparisons in the RNA field, especially the determination of secondary structures. Theoretical prediction of secondary structures, the foundations of which are explained extensively in one chapter, can be successful, although on the basis of *in silico* approaches only it is often difficult to decide between several possible solutions. The search for RNA motifs in genomic sequences is exemplified for motifs recognized by protein cofactors in the maturation or translation of messenger RNAs. The identification of such motifs contributes to the proper annotation of genes and the sorting out of proteomics data. In the end, RNA molecules exert their functions because they adopt a specific tertiary fold. Three chapters tackle this question, focusing first on how to assemble complex tertiary structures *in silico* starting from fragments which have been structurally characterized, then giving reasons for the advantages of increasing the coarseness of the representations, and finally describing the calculations of the molecular dynamics simulations in order to apprehend movements, and binding of water molecules, ions and ligands. The very recent field of RNomics has led to the discovery of a plethora of potential regulatory RNAs, whose functional and structural analysis will keep us busy for the next decades. Several pioneers in the field address the strategic as well as the bioinformatic and experimental aspects of RNomics. Part III is completed by a contribution introducing a novel genome-wide approach, the analysis of alternative splicing variants by microarrays. Although still at an early stage of development, this will certainly become an important tool in the future.

Since nakedness of RNA in most cases reflects an artificial *in vitro* state, Part IV of the handbook focuses on RNA-protein interaction. A queue of chapters present different techniques directly related to the analysis of RNA function in conjunction with proteins – both *in vitro* and *in vivo*. Addressed are various affinity purification methods, Northwestern techniques, three-hybrid screening, fluorescent detection of RNA and RNA-binding proteins, and the *in vivo* analysis of alternative splicing. Another series of functional approaches is dedicated to *in vitro* and *in vivo* SELEX strategies, encompassing the different applications of this technique and detailing the experimental steps and associated pitfalls. This provides a wealth of in-depth information in this still developing area of research.

The vast field of RNA interference is represented by a contribution (Part V) written by authors who have pioneered the field of gene silencing in mammalian cells. We have refrained from extending the RNAi part because its specific methodology is currently under rapid evolution and varies greatly between different biological systems. Furthermore, specialized monographs have become available recently.

In the end, we hope that this comprehensive collection of protocols spanning RNA research will be a helpful and useful toolbox for all researchers already working with RNA as well as for those planning to foray in the RNA world.

Last but not least, we are indebted to all authors for their engagement, patience and the high quality of their contributions. Special thanks go to Dagmar K. Willkomm for her continuous assistance in the editing process. The editors would also like to thank Frank Weinreich and the staff of Wiley-VCH for making the production of this handbook possible.

Roland K. Hartmann
Albrecht Bindereif
Astrid Schön
Eric Westhof

October 2004

List of Contributors

A. J. Andrews
University of Michigan
Department of Chemistry
930 North University
Ann Arbor, MI 48189-1055
USA

Pascal Auffinger
CNRS – UPR 9002
Institut de Biologie Moléculaire et Cellulaire
15 rue René Descartes
67084 Strasbourg
France

Gesine Bauer
Friedrich-Alexander-Universität Erlangen-Nürnberg
Institut für Mikrobiologie, Biochemie und Genetik
Staudtstr. 5
91058 Erlangen
Germany

Hildburg Beier
Universität Würzburg
Institut für Biochemie
Biozentrum, Am Hubland
97074 Würzburg
Germany

Natalya Benderska
University of Erlangen
Institute of Biochemistry
Fahrstrasse 17
91054 Erlangen
Germany

Christian Berens
Friedrich-Alexander-Universität Erlangen-Nürnberg
Institut für Mikrobiologie, Biochemie und Genetik
Staudtstr. 5
91058 Erlangen
Germany

Albrecht Bindereif
Justus-Liebig-Universität Giessen
Institut für Biochemie
Heinrich-Buff-Ring 58
35392 Giessen
Germany

Amanda Birmingham
University of Colorado
Department of Molecular, Cellular and Developmental Biology
Boulder, CO 80309-0347
USA

Marc Boudvillain
Centre de Biophysique Moleculaire, CNRS
rue Charles Sadron
45071 Orleans cedex 2
France

Christiane Branlant
UMR 7567 CNRS-UHP
Université Herni Poincaré Nancy 1
Boulevard des Aiguillettes, BP 239
54506 Vandoeuvre-Lès-Nancy
France

James W. Brown
North Carolina State University
Department of Microbiology
Raleigh, NC 27695-7615
USA

List of Contributors

Eric L. Christian
Case Western Reserve University
Center for RNA Molecular Biology
Cleveland, OH 44106-4943
USA

Jerzy Ciesiolka
Polish Academy of Sciences
Institute of Bioorganic Chemistry
Noskowskiego 12/14
61-704 Poznan
Poland

Thomas A. Cooper
Baylor College of Medicine
Departments of Pathology and Molecular
and Cellular Biology
One Baylor Plaza
Houston, TX 77030
USA

Simona Cuzic
Philipps-Universität Marburg
Institut für Pharmazeutische Chemie
Marbacher Weg 6
35037 Marburg
Germany

Andrey Damianov
Justus-Liebig-Universität Giessen
Institut für Biochemie
Heinrich-Buff-Ring 58
35392 Giessen
Germany

Michaela Alessandra Denti
University of Rome "La Sapienza"
Department of Genetics and Molecular
Biology
P.le Aldo Moro 5
00185 Rome
Italy

Ico de Zwart
University of Colorado
Department of Molecular, Cellular and
Developmental Biology
Boulder, CO 80309-0347
USA

Thomas E. Edwards
University of Iceland
Science Institute
Dunhaga 3
107 Reykjavik
Iceland

Chantal Ehresmann
CNRS – UPR 9002
Institut de Biologie Moléculaire et Cellulaire
15 rue René Descartes
67084 Strasbourg
France

J. Christopher Ellis
North Carolina State University
Department of Microbiology
Raleigh, NC 27695-7615
USA

David R. Engelke
University of Michigan
Department of Biological Chemistry
Ann Arbor, MI 48109-0606
USA

Markus Englert
Universität Würzburg
Institut für Biochemie
Biozentrum, Am Hubland
97074 Würzburg
Germany

Dirk Eulberg
NOXXON Pharma AG
Max-Dohrn-Strasse 8–10
10589 Berlin
Germany

Olga Fedorova
Yale University
Department of Molecular Biophysics and
Biochemistry
266 Whitney Ave
New Haven, CT 06520
USA

Carol Fierke
University of Michigan
Department of Chemistry
930 North University
Ann Arbor, MI 48189-1055
USA

Boris François
CNRS – UPR 9002
Institut de Biologie Moléculaire et Cellulaire
15 rue René Descartes

67084 Strasbourg
France

Mikko J. Frilander
University of Helsinki
Institute of Biotechnology
Viikinkaari 9
000014 University of Helsinki
Finland

Xiang-Dong Fu
University of California, San Diego
Department of Cellular and Molecular
Medicine
San Diego, CA 92093-0651
USA

Daniel Gautheret
INSERM ERM-206
Université de la Méditerranée
Luminy Case 906
13288 Marseille Cedex 09
France

Anke Geick
Alnylam Europe AG
Fritz-Hornschuch-Str. 9
95326 Kulmbach
Germany

Jennifer A. Geiger
Max-Planck-Institute of Molecular Cell Biology
and Genetics
Pfotenhauerstrasse 108
01307 Dresden
Germany

Olaf Gimple
Universität Würzburg
Institut für Biochemie
Am Hubland
97074 Würzburg
Germany

Heike Gruegelsiepe
Philipps-Universität Marburg
Institut für Pharmazeutische Chemie
Marbacher Weg 6
35037 Marburg
Germany

Philipp Hadwiger
Alnylam Europe AG
Fritz-Hornschuch-Str. 9

95326 Kulmbach
Germany

Dinari A. Harris
University of Michigan
Department of Chemistry
Ann Arbor, MI 48109
USA

Michael E. Harris
Case Western Reserve University
Center for RNA Molecular Biology and
Department of Molecular Biology and
Microbiology
Cleveland OH 44106-4973
USA

Roland K. Hartmann
Philipps-Universität Marburg
Institut für Pharmazeutische Chemie
Marbacher Weg 6
35037 Marburg
Germany

Klaus Hartmuth
Max-Planck-Institute for Biophysical
Chemistry
Department of Cellular Biochemistry
Am Fassberg 11
37077 Göttingen
Germany

Stephen C. Harvey
Georgia Institute of Technology
Department of Biology
310 Ferst Dr.
Atlanta, GA 30332
USA

Olaf Heidenreich
University of Tübingen
Department of Molecular Biology
Auf der Morgenstelle 15
72076 Tübingen
Germany

Niklas Henriksson
Uppsala University
Department of Cell and Molecular Biology
BMC Box 596
751 24 Uppsala
Sweden

List of Contributors

Jingyi Hui
Justus-Liebig-Universität Giessen
Institut für Biochemie
Heinrich-Buff-Ring 58
35392 Giessen
Germany

Eric Huntzinger
CNRS – UPR 9002
Institut de Biologie Moléculaire et Cellulaire
15 rue René Descartes
67084 Strasbourg
France

Alexander Hüttenhofer
Innsbruck Medical University
Division of Genomics and RNomics
Peter-Mayr-Str. 4b
6020 Innsbruck
Austria

Mali Illangasekare
University of Colorado
Department of Molecular, Cellular and Developmental Biology
Boulder, CO 80309-0347
USA

Vasant Jadhav
University of Colorado
Department of Molecular, Cellular and Developmental Biology
Boulder, CO 80309-0347
USA

Martin R. Jakobsen
University of Aarhus
Department of Molecular Biology
Møllers Allé
8000 Åarhus C
Denmark

Matthias John
Alnylam Europe AG
Fritz-Hornschuch-Str. 9
95326 Kulmbach
Germany

Jane Kawaoka
Yale University
Department of Molecular Biophysics and Biochemistry
266 Whitney Ave
New Haven, CT 06520
USA

Leif A. Kirsebom
Uppsala University
Department of Cell and Molecular Biology
Biomedical Center
751 24 Uppsala
Sweden

Shivendra Kishore
University of Erlangen
Institute of Biochemistry
Fahrstrasse 17
91054 Erlangen
Germany

Jørgen Kjems
University of Aarhus
Department of Molecular Biology
Møllers Allé
8000 Åarhus C
Denmark

Sven Klussmann
NOXXON Pharma AG
Max-Dohrn-Strasse 8–10
10589 Berlin
Germany

Rob Knight
University of Colorado
Department of Molecular, Cellular and Developmental Biology
Boulder, CO 80309-0347
USA

Young-Soo Kwon
University of California, San Diego
Department of Cellular and Molecular Medicine
San Diego, CA 92093-0651
USA

Matthieu Legendre
INSERM ERM-206
Université de la Méditerranée
Luminy Case 906
13288 Marseille Cedex 09
France

Mihail Legiewicz
University of Colorado
Department of Molecular, Cellular and Developmental Biology
Boulder, CO 80309-0347
USA

Aurélie Lescoute-Philipps
CNRS – UPR 9002
Institut de Biologie Moléculaire et Cellulaire
15 rue René Descartes
67084 Strasbourg
France

Hai-Ri Li
University of California, San Diego
Department of Cellular and Molecular Medicine
San Diego, CA 92093-0651
USA

Ester Lizano
Max-Planck-Institute for Evolutionary Anthropology
Deutscher Platz 6
04103 Leipzig
Germany

Catherine Lozupone
University of Colorado
Department of Molecular, Cellular and Developmental Biology
Boulder, CO 80309-0347
USA

Reinhard Lührmann
Max-Planck-Institute for Biophysical Chemistry
Department of Cellular Biochemistry
Am Fassberg 11
37077 Göttingen
Germany

Martin Lützelberger
University of Aarhus
Department of Molecular Biology
Møllers Allé
8000 Aarhus C
Denmark

Christian Maasch
NOXXON Pharma AG
Max-Dohrn-Strasse 8–10
10589 Berlin
Germany

Irene Majerfeld
University of Colorado
Department of Molecular, Cellular and Developmental Biology
Boulder, CO 80309-0347
USA

Evgeny M. Makarov
Max-Planck-Institute for Biophysical Chemistry
Department of Cellular Biochemistry
Am Fassberg 11
37077 Göttingen
Germany

Olga V. Makarova
Max-Planck-Institute for Biophysical Chemistry
Department of Cellular Biochemistry
Am Fassberg 11
37077 Göttingen
Germany

Virginie Marchand
UMR 7567 CNRS-UHP
Université Herni Poincaré Nancy 1
Boulevard des Aiguillettes, BP 239
54506 Vandoeuvre-Lès-Nancy
France

Anita Marchfelder
Universität Ulm
Molekulare Botanik
Albert-Einstein-Allee 11
89069 Ulm
Germany

Ángel Emilio Martínez de Alba
Universidad Politécnica de Valencia
Instituto de Biología Molecular y Celular de Plantas
Avenida de los Naranjos s/n
46022 Valencia
Spain

Benoît Masquida
CNRS – UPR 9002
Institut de Biologie Moléculaire et Cellulaire
15 rue René Descartes
67084 Strasbourg
France

Jason A. Mears
Georgia Institute of Technology
Department of Biology
310 Ferst Dr.
Atlanta, GA 30332
USA

Jan Medenbach
Justus-Liebig-Universität Giessen

Institut für Biochemie
Heinrich-Buff-Ring 58
35392 Giessen
Germany

Agnès Méreau
UMR 7567 CNRS-UHP
Université Herni Poincaré Nancy 1
Boulevard des Aiguillettes, BP 239
54506 Vandoeuvre-Lès-Nancy
France

Mario Mörl
Max-Planck-Institute for Evolutionary
Anthropology
Deutscher Platz 6
04103 Leipzig
Germany

Hervé Moine
CNRS – UPR 9002
Institut de Biologie Moléculaire et Cellulaire
15 rue René Descartes
67084 Strasbourg
France

Annie Mougin
UMR 7567 CNRS-UHP
Université Herni Poincaré Nancy 1
Boulevard des Aiguillettes, BP 239
54506 Vandoeuvre-Lès-Nancy
France

Wolfgang Nellen
Universität Kassel
Abteilung Genetik
Heinrich-Plett-Straße 40
34132 Kassel
Germany

Karla M. Neugebauer
Max-Planck-Institute of Molecular Cell Biology
and Genetics
Pfotenhauerstrasse 108
01307 Dresden
Germany

Tatyana Novoyatleva
University of Erlangen
Institute of Biochemistry
Fahrstrasse 17
91054 Erlangen
Germany

Filipp Oesterhelt
Heinrich-Heine-Universität Düsseldorf
Institute for Molecular Physical Chemistry
Universitätsstr. 1
40225 Düsseldorf
Germany

Zsofia Palfi
Justus-Liebig-Universität Giessen
Institut für Biochemie
Heinrich-Buff-Ring 58
35392 Giessen
Germany

Tao Pan
University of Chicago
Department of Biochemistry and Molecular
Biology
920 E 58th St
Chicago, IL 60637
USA

Maria Parisova
Biochemie
Universität Bayreuth
Universitätsstraße 30
95447 Bayreuth
Germany

Tina Persson
Lund University
Department of Chemistry
P.O. Box 124
221 00 Lund
Sweden

Maria Possedko
CNRS – UPR 9002
Institut de Biologie Moléculaire et Cellulaire
15 rue René Descartes
67084 Strasbourg
France

Werner G. Purschke
NOXXON Pharma AG
Max-Dohrn-Strasse 8–10
10589 Berlin
Germany

Anna Marie Pyle
Yale University
Department of Molecular Biophysics and
Biochemistry
266 Whitney Ave

New Haven, CT 06520
USA

Yan-Guo Ren
Genomics Institute of the Novartis Research
Foundation
10675 John Jay Hopkins Dr.
San Diego, CA 92121
USA

Detlev Riesner
Heinrich-Heine-Universität Düsseldorf
Institut für Physikalische Biologie
Universitätsstr. 1
40225 Düsseldorf
Germany

Pascale Romby
CNRS – UPR 9002
Institut de Biologie Moléculaire et Cellulaire
15 rue René Descartes
67084 Strasbourg
France

Sylvia Rösch
Universität Ulm
Molekulare Botanik
Albert-Einstein-Allee 11
89069 Ulm
Germany

Steffen Schiffer
Universität Ulm
Molekulare Botanik
Albert-Einstein-Allee 11
89069 Ulm
Germany

Astrid Schön
Universität Leipzig
Institut für Biochemie
Brüderstr. 34
04103 Leipzig
Germany

Silke Schreiner
Justus-Liebig-Universität Giessen
Institut für Biochemie
Heinrich-Buff-Ring 58
35392 Giessen
Germany

Renée Schroeder
Department of Microbiology and Genetics

University of Vienna
Dr. Bohrgasse 9/4
1030 Vienna
Austria

Enno Schweinberger
Heinrich-Heine-Universität Düsseldorf
Institute for Molecular Physical Chemistry
Universitätsstr. 1
40225 Düsseldorf
Germany

Felicia Scott
University of Michigan
Department of Biological Chemistry
Ann Arbor, MI 48109-0606
USA

Claus Seidel
Heinrich-Heine-Universität Düsseldorf
Institute for Molecular Physical Chemistry
Universitätsstr. 1
40225 Düsseldorf
Germany

Snorri Th. Sigurdsson
University of Iceland
Science Institute
Dunhaga 3
107 Reykjavik
Iceland

Tobin R. Sosnick
University of Chicago
Department of Biochemistry and Molecular
Biology
920 E 58th St
Chicago, IL 60637
USA

Bettina Späth
Universität Ulm
Molekulare Botanik
Albert-Einstein-Allee 11
89069 Ulm
Germany

Brian S. Sproat
RNA-TEC NV
Minderbroedersstraat 17–19
3000 Leuven
Belgium

List of Contributors

Scott M. Stagg
Georgia Institute of Technology
Department of Biology
310 Ferst Dr.
Atlanta, GA 30332
USA

Stefan Stamm
University of Erlangen
Institute of Biochemistry
Fahrstrasse 17
91054 Erlangen
Germany

Gerhard Steger
Heinrich-Heine-Universität Düsseldorf
Institut für Physikalische Biologie
Universitätsstr. 1
40225 Düsseldorf
Germany

Martin Tabler
Foundation for Research and Technology
Hellas
Institute of Molecular Biology and
Biotechnology
P.O.Box 1527
71110 Heraclion (Crete)
Greece

Yesheng Tang
University of Erlangen
Institute of Biochemistry
Fahrstrasse 17
91054 Erlangen
Germany

Alphonse Thanaraj
European Bioinformatics Institute
Wellcome Trust Genome Campus
Hinxton
Cambridge, CB10 1SD
United Kingdom

Janne J. Turunen
University of Helsinki
Institute of Biotechnology
Viikinkaari 9
000014 University of Helsinki
Finland

Henning Urlaub
Max-Planck-Institute for Biophysical
Chemistry
Department of Cellular Biochemistry
Am Fassberg 11
37077 Göttingen
Germany

Andrea C. Vaiana
CNRS – UPR 9002
Institut de Biologie Moléculaire et Cellulaire
15 rue René Descartes
67084 Strasbourg
France

Anders Virtanen
Uppsala University
Department of Cell and Molecular Biology
BMC Box 596
751 24 Uppsala
Sweden

Jörg Vogel
Max Planck Institute for Infection Biology
Schumannstr. 21/22
10117 Berlin
Germany

Hans-Peter Vornlocher
Alnylam Europe AG
Fritz-Hornschuch-Str. 9
95326 Kulmbach
Germany

C. Stefan Vörtler
Biochemie
Universität Bayreuth
Universitätsstraße 30
95447 Bayreuth
Germany

E. Gerhart H. Wagner
Uppsala University Biomedical Center
Department of Cell and Molecular Biology
Box 596
751 24 Uppsala
Sweden

Christina Waldsich
Yale University
Department of Molecular Biophysics
and Biochemistry
266 Whitney Ave
New Haven, CT 06520
USA

Nils G. Walter
University of Michigan
Department of Chemistry
Ann Arbor, MI 48109
USA

Andreas Werner
CNRS – UPR 9002
Institut de Biologie Moléculaire et Cellulaire
15 rue René Descartes
67084 Strasbourg
France

Eric Westhof
CNRS – UPR 9002
Institut de Biologie Moléculaire et Cellulaire
15 rue René Descartes
67084 Strasbourg
France

Jeremy Widmann
University of Colorado
Department of Molecular, Cellular and
Developmental Biology
Boulder, CO 80309-0347
USA

Cindy L. Will
Max-Planck-Institute for Biophysical
Chemistry
Department of Cellular Biochemistry
Am Fassberg 11
37077 Göttingen
Germany

Dagmar K. Willkomm
Philipps-Universität Marburg
Institut für Pharmazeutische Chemie
Marbacher Weg 6
35037 Marburg
Germany

Flore Winter
CNRS – UPR 9002
Institut de Biologie Moléculaire et Cellulaire
15 rue René Descartes
67084 Strasbourg
France

Michael Yarus
University of Colorado
Department of Molecular, Cellular and
Developmental Biology
Boulder, CO 80309-0347
USA

Nathan H. Zahler
University of Michigan
Department of Chemistry
930 N. University
Ann Arbor, MI 48109-1055
USA

Part I
RNA Synthesis

I.1
Enzymatic RNA Synthesis, Ligation and Modification

1
Enzymatic RNA Synthesis using Bacteriophage T7 RNA Polymerase

Heike Gruegelsiepe, Astrid Schön, Leif A. Kirsebom and Roland K. Hartmann

1.1
Introduction

Bacteriophage T7 RNA polymerase (T7 RNAP) was first cloned and overexpressed from bacteriophage T7-infected *Escherichia coli* cells in 1984 [1]. In contrast to multi-subunit DNA-dependent RNA polymerases from eukaryotes and prokaryotes, T7 RNAP consists of a single subunit of about 100 kDa [2]. The subdomains adopt a hand-like shape with the palm, thumb and fingers around a central cleft where the active site containing the functionally essential amino acid residues is located, creating a binding cavity for magnesium ions and ribonucleotide substrates. For RNA synthesis, the unwound template strand is positioned such that the template base −1 is anchored in a hydrophobic pocket in direct vicinity to the active site [3].

T7 RNAP is highly specific for its own promoters and exhibits no affinity even to closely related phage T3 promoters, although the 23-bp consensus sequences are very similar (Fig. 1.1A). During the initiation process, the polymerase goes through several elongation attempts, generating short abortive oligoribonucleotides. Only when the nascent RNA transcript exceeds 9–12 nt do initiation complexes convert to stable elongation complexes. Transcription proceeds with an average rate of 200–260 nt/s until the elongation complex encounters a termination signal or falls off the template end during *in vitro* run-off transcription [4, 5]. The error frequency in transcripts of wild-type (wt) T7 RNAP is about 6×10^{-5} [6].

In the following sections, we will describe protocols that have been used routinely for T7 transcription. Further, a robust and simple protocol for the partial purification of T7 RNAP is included, which yields an enzyme preparation that fully satisfies all *in vitro* transcription demands. The transcription protocols given suffice for most purposes. However, in special cases, such as the synthesis of milligram quantities, modified RNAs or very A,U-rich RNAs, it may be worthwhile to further optimize transcription conditions. We also draw the reader's attention to the paper by Milligan and Uhlenbeck [7], which briefly discusses many fundamental aspects of T7 transcription and is still handed out to every new member of our groups.

Handbook of RNA Biochemistry: Student Edition. Edited by R. K. Hartmann, A. Bindereif, A. Schön, E. Westhof
Copyright © 2009 WILEY-VCH Verlag GmbH & Co. KGaA, Weinheim
ISBN: 978-3-527-32534-4

A

```
                              -17                        +1
                               |                          |
T7 (class III)          5'- TAA TAC GAC TCA CTA TAG GGA GA -3'
T7 (Φ2.5, class II)     5'- TAA TAC GAC TCA CTA TTA GGG AA -3'
T3 (class III)          5'- AAT TAA CCC TCA CTA AAG GGA GA -3'
Sp6 (class III)         5'- ATT TAG GTG ACA CTA TAG AAG AG -3'
                              ‾‾‾‾‾‾‾‾‾‾‾‾‾‾‾‾‾‾   ‾‾‾‾‾‾‾‾‾‾‾‾‾
                                binding domain     initiating domain
```

B

+1   +6
| |
pppGGGAGA
 ‾‾‾‾‾
 smaller
 effects

Nucleotide +1	Relative yield	Nucleotide +2	Relative yield
C	0.1	C	0.5
A	0.2	A	0.5
U	n.d.	U	n.d.
G	1.0	G	1.0

C

```
      -17                        +1
       |                          |
5'- TAA TAC GAC TCA CTA TAG GNN (N) ⟶
3'- ATT ATG CTG AGT GAT ATC CNN (N)ₓ
```

Fig. 1.1. (A) Consensus sequences of class III promoters of bacteriophages T7, T3 and SP6, and sequence of the T7 ϕ2.5 class II promoter [5, 34–36]. Phage polymerase initiating domains also include the first 5–6 nt of the transcribed template strand. The transcription start (position +1) is indicated by the arrow. The phage T7 genome encodes a total of 17 promoters, including five class III promoters and one replication promoter (ϕOR), which are all completely conserved in the region from nt −17 to +6. In addition, there are 10 T7 class II promoters plus one more replication promoter (ϕOL); among these 11 promoters, which display some sequence variation within the −17 to +6 region, only the ϕ2.5 and ϕOL promoter initiate transcription with an A instead of G residue [35] (B) Effect of sequence variations in the +1 to +6 region of the T7 class III promoter on transcription efficiency (adapted from Milligan and Uhlenbeck [7]); n.d.: not determined. (C) T7 class III promoter region with the recommended G identities at positions +1 and +2 of the RNA transcript shown in grey.

1.2
Description of Method – T7 Transcription *in vitro*

T7 RNAP can be used *in vitro* to produce milligram amounts of RNA polymers ranging from less than 100 to 30 000 nt [7, 8]. Since the commonly used T7 class III promoter, usually referred to as the T7 promoter, is also strictly conserved in the transcribed region of nt +1 to +6, sequence variations especially at nt +1 and +2 influence transcription yields significantly (Fig. 1B and C [7]).

1.2.1
Templates

Templates can be generated in three different ways: by insertion into a plasmid [double-stranded (ds) DNA], by polymerase chain reaction (PCR) (dsDNA) or by annealing a T7 promoter DNA oligonucleotide to a single-stranded template DNA oligonucleotide.

Strategy (a) Insertion into a Plasmid
We prefer to work with plasmid dsDNA templates, because once the correct sequence of a plasmid clone has been confirmed, the DNA can be conveniently amplified by *in vivo* plasmid replication exploiting the high fidelity of bacterial DNA polymerases. The RNA expression cassette (either with or without the T7 promoter sequence) is usually obtained by PCR and cloned into a bacterial plasmid. Since PCR amplification is error-prone, plasmid inserts ought to be sequenced. When the T7 RNAP promoter region from −17 to −1 is not encoded in the PCR fragment, one can use commercially available T7 transcription vectors (e.g. pGEM®3Z and derivatives from Promega or the pPCR-Script series from Stratagene) containing the T7 promoter and a multiple cloning site for insertion of the RNA expression cassette. If there are no sequence constraints at the transcript 5′ end, we routinely design templates encoding 5′-GGA at positions +1 to +3 of the RNA transcript, which usually results in high transcription yields. Whenever possible, at least the nucleotide preferences at positions +1 and +2 should be taken into account (Fig. 1.1B and C). The plasmid-encoded RNA expression cassette ought to be followed by a single restriction site (avoid restriction enzymes yielding 3′ overhangs [7]) for producing run-off transcripts. Templates with 5′ overhangs have successfully been used in several laboratories. Among those have been templates linearized with restriction enzymes that cleave several residues away from their binding/recognition site, such as *Fok*I. The advantage of using this type of restriction enzymes is independence from the sequence at the cleavage site. This permits the design of RNA transcript 3′ ends of complete identity to natural counterparts. Individual steps of template preparation are (1) ligation of (PCR) insert into plasmid, (2) cloning in *E. coli*, purification and sequencing of plasmid, (3) linearization of plasmid DNA for run-off transcription, (4) phenol/chloroform extraction and ethanol precipitation of template DNA before (5) use in T7 transcription assays.

Strategy (b) Direct use of Templates Generated by PCR
Direct use of PCR fragments as templates is faster than insertion into a plasmid and preferred if only minor amounts of RNA are required.

Strategy (c) Annealing of a T7 Promoter DNA Oligonucleotide to a Single-stranded Template
This strategy is the fastest and we have used it to synthesize small amounts of an RNA 31mer for 5′-end-labeling purposes (see Protocol 6).

1.2.2
Special Demands on the RNA Product

1.2.2.1 Homogeneous 5′ and 3′ Ends, Small RNAs, Functional Groups at the 5′ End

While T7 RNAP usually initiates transcription at a defined position, it tends to append one or occasionally even a few more non-templated nucleotides to the product 3′ terminus [7, 9]. Also, 5′ end heterogeneity may become a problem when the template encodes a transcript with more than three consecutive guanosines at the 5′ end [10], as well as in the case of unusual 5′-terminal sequences, such as 5′-CACUGU, 5′-CAGAGA or 5′-GAAAAA [11]. Yet, 5′ end heterogeneity seems to be a problem associated with T7 class III promoters (Fig. 1.1A). Almost complete 5′ end homogeneity of T7 transcripts has been achieved with templates directing transcription from the more rarely used T7 ϕ2.5 class II promoter (Fig. 1.1A), at which T7 RNAP initiates synthesis with an A instead of a G residue. Transcription yields from this promoter were reported to equal those of the commonly used T7 class III promoter [12].

However, for the production of RNAs with 100% 5′ and 3′ end homogeneity, several methods are available (see Chapters 2 and 3). For example, hammerhead or Hepatitis delta virus (HDV) ribozymes can be tethered to the RNA of interest on one or both sides (see Chapter 2). The ribozyme(s) will release the RNA product by self-cleavage during transcription. Such a *cis*-acting ribozyme placed upstream releases the RNA of interest with a 5′-OH terminus directly accessible to 5′-end-labeling (see Chapter 9) and simultaneously eliminates the problem of 5′ end heterogeneity as well as constraints on the identity of the 5′-terminal nucleotide of the RNA of interest (Chapter 2). The same strategy may also be considered for synthesis of large amounts of smaller RNAs. Chemical synthesis and purification of 10 mg of, for example, an RNA 15mer by a commercial supplier can be quite expensive. In such a case, a cheaper alternative would be to transcribe the 15mer sandwiched between two *cis*-cleaving ribozymes, resulting in post-transcriptional release of the 15mer with uniform 5′ and 3′ ends. Purification of the 15mer (and separation from the released ribozyme fragments) can then be achieved either by denaturing polyacrylamide gel electrophoresis (PAGE), UV shadowing or staining and gel elution (see Chapters 3 and 9), or by preparative HPLC if available (see Chapters 7 and 27). If T7 RNAP is self-prepared according to the protocol described in this article, synthesis of 10 mg of a 15mer will become quite affordable.

Normally, transcription by T7 RNAP is initiated with GTP, resulting in 5′-triphosphate ends. If, however, 5′-OH ends or 5′-monophosphate termini are preferred, T7 RNAP can be prompted to initiate transcripts with guanosine or 5′-ApG (to generate 5′-OH ends for direct end-labeling with ^{32}P) or 5′-GMP (to generate 5′-monophosphates), when these components are added to reaction mixtures in excess over GTP [13]. RNA transcripts with 5′-GMP ends are preferred when the RNA is used for ligation with other RNA molecules.

Tab. 1.1. Modified substrates for T7 transcription.

NTP	wt T7 RNAP	Reference
NTPαS (Sp)	+	14
NTPαS (Rp)	−	14
5-Br-UTP	+	7
5-F-UTP	+	7
5-Hexamethyleneamino-UTP	+	7
6-Aza-UTP	+	7
4-Thio-UTP	+	7
Pseudo-UTP	+	7
8-Br-ATP	+	7
7-Me-GTP	−	7
ITP (with initiator)[1]	+	15
2′-dNTP	+/−	7
2′-dNTPαS	+/−	16, 17
2′-O-Me-NTP or -NTPαS	+/−	16
GTPγS	+	18

+/−: low incorporation efficiency.
[1] Inosine triphosphate (ITP) cannot be used to start transcription, but can substitute for GTP during elongation if a primer, such as 5′-ApG or 5′-GMP, is present as initiator of transcription.

1.2.2.2 Modified Substrates

There are a number of modified nucleoside-5′-triphosphates known to be substrates for T7 RNAP. Table 1.1 has been adopted from Milligan and Uhlenbeck [7] and expanded by addition of more recent information.

Due to discrimination of rNTPs and dNTPs by wt T7 RNAP, the polymerase incorporates rNTPs 70- to 80-fold more efficiently than dNTPs in the presence of Mg^{2+} as the metal ion cofactor. However, a T7 RNAP mutant (Y639F) carrying a tyrosine to phenylalanine exchange at position 639 [19] was shown to have only about 4-fold higher preference for rNTPs than dNTPs [19, 20] and thus permits more efficient incorporation of substrates lacking the ribose 2′-hydroxyl, such as 2′-deoxy-2′-fluoro or 2′-deoxy-2′-amino nucleotides [20]. Incorporation of substrate analogs with 2′-ribose modifications can also be stimulated to some extent in reactions catalyzed by wt T7 RNAP upon addition of Mn^{2+} [16]. Likewise, dNTPαS analogs were partially incorporated into RNase P RNA in a sequence-specific manner under mixed metal ion conditions (Mg^{2+}/Mn^{2+} [21]). Despite these achievements, the Y639F mutant T7 RNAP is nowadays the enzyme of choice for the incorporation of all nucleotides with 2′-ribose modifications (available from Epicentre, WI, USA). For detailed protocols, the reader is referred to [20, 22].

Further modifications can be introduced into transcripts by tailored initiator (oligo)nucleotides. Di- to hexanucleotides with a 3′-terminal guanine base, including di- to tetranucleotides with internal or terminal 2′-deoxy- or 2′-O-methylated residues, were tested as initiators of transcription by wt T7 RNAP [23]. 5′-Terminal incorporation varied between 20% (hexamer) and 80–95% in the case of 5′-ApG or a 5′-biotinylated ApG (e.g. custom-synthesized by IBA, Göttingen, Germany). Also, transcription by T7 RNAP, in this case from the T7 ϕ2.5 class II promoter, was initiated with coenzymes containing an adenosine moiety, such as CoA (3′-dephospho-coenzyme A), NAD or FAD. Reduced NADH and oxidized FAD are highly fluorescent, which opens up the perspective to employ coenzyme-linked RNAs for the study of RNA–RNA or RNA–protein interactions by fluorescence techniques [24].

1.3
Transcription Protocols

1.3.1
Transcription with Unmodified Nucleotides

The protocols given below have been applied to template DNAs directing transcription from the T7 class III promoter. Transcription yields can differ substantially, depending on the individual DNA template and the origin of T7 RNAP. In Protocols 1–6 (Hartmann lab), T7 RNAP from MBI Fermentas has been used. These protocols may be suboptimal with T7 RNAP from other sources. This has been accounted for by including protocols from the Kirsebom (Protocols 7 and 8) and Schön (Protocol 9) labs. Nevertheless, it is advisable to put some effort into the optimization of the transcription protocol if large amounts of RNA are to be produced or if the transcript represents a "standard RNA" in the laboratory, used over longer periods, which may require repeated synthesis. In addition to commercially available T7 RNAP, protocols for partial purification of T7 RNAP from bacterial overexpression strains are available ([1, 25, 26] and Section 1.5).

When a new DNA template is used for the first time, one approach is to perform small test transcriptions on a 50-µl scale according to different basic protocols (e.g. Protocols 1–3). Reaction mixtures should be prepared at room temperature, since DNA may precipitate in the presence of spermidine at low temperatures. In the case of plasmid DNA, template amounts of 40–80 µg/ml (final assay concentration) are used as a rule of thumb, whereas a PCR template of 400 bp is adjusted to about 5 µg/ml final assay concentration. In the Hartmann lab, we usually incubate transcription mixes for 4–6 h at 37 °C, although overnight incubations have been used as well. A variation is to add another aliquot (e.g. 2 U/µl) of T7 RNAP after 2 h at 37 °C, followed by a further 2-h incubation period at 37 °C. In transcriptions according to Protocol 3, a white precipitate will appear because of pyrophosphate accumulation. This is avoided in Protocols 1 and 2 where pyrophosphate is hydrolyzed due to the presence of pyrophosphatase. Extension of incubation periods beyond 4–6 h did not prove advantageous in our hands, and may be associated with some product degradation since T7 RNAP has a DNase and RNase function, which is normally inhibited by NTP substrates added in excess to *in vitro* transcription assays. However, after extended incubation periods, the NTP concentration may drop under a critical limit, thereby favoring RNA degradation [27]. Protocol 4 represents an inexpensive strategy to incorporate a 5′-terminal guanosine. Since guanosine has a low solubility, a 30 mM solution is prepared and kept at 75 °C; the reaction mixture – except for guanosine and T7 RNAP – is prepared at room temperature and prewarmed to 37 °C before addition of guanosine.

We like to note here that transcription from the T7 ϕ2.5 class II promoter is initiated with A instead of G (Fig. 1.1A), opening up the perspective to incorporate an adenosine at the 5′ end. Adenosine would fulfill the same purpose as the aforementioned guanosine used in the case of the class III promoter, but is advantageous because of its better water solubility [12, 24].

As an alternative to starting class III-promoter-directed transcripts with guanosine, the dinucleotide 5′-ApG (see above) may be employed. The 5′-terminal incorporation of the dinucleotide leads to a −1 adenosine extension of the transcript (Protocol 5). ApG, which is more convenient to use than guanosine, is also available from Sigma, but about 10 000-fold more expensive than guanosine. Transcripts can further be initiated with 5′-GMP if present in excess over 5′-GTP (15 versus 3.75 mM). Although the majority of RNA products should possess a 5′-terminal monophosphate, a dephosphorylation/phosphorylation strategy (see Chapter 3) may be preferred to obtain RNA products with 100% 5′-monophosphates.

Protocol 6 is a quick protocol for the synthesis of small amounts of shorter RNAs for 5′-end-labeling purposes. The promoter and template DNA oligonucleotides are simply added to the reaction mixture which is then preincubated for 1 h at 37 °C before starting transcription by addition of T7 RNAP.

Protocol 7 (from the Kirsebom lab) has the characteristics of high T7 RNAP concentrations and the presence of RNase inhibitor, suitable for large-scale transcriptions. Protocol 8 is used in the Kirsebom lab for the production of internally labeled RNA.

Protocol 9 (from the Schön lab) has been used for standard transcriptions as well as for the synthesis of extremely A,U-rich RNAs, with the ratio of individual NTPs adapted to their proportion in the final transcript.

Protocol 1: Hartmann lab

	Final concentration	1000 μl
HEPES pH 7.5, 1 M	80 mM	80 μl
DTT 100 mM	5 mM	50 μl
MgCl$_2$ 3 M	22 mM	7.3 μl
Spermidine 100 mM	1 mM	10 μl
BSA[1] 20 mg/ml	0.12 mg/ml	6 μl
rNTP mix (25 mM each)	3.75 mM (each)	150 μl
Template (linearized plasmid 3.2 kb) 1 μg/μl	40 μg/ml	40 μl
Pyrophosphatase[2] 200 U/ml	1 U/ml	5 μl
T7 RNAP 200 U/μl	1000–2000 U/ml	5–10 μl
RNase-free water		to 1000 μl

For small-scale transcriptions (50 µl final volume), reaction mixes are incubated for 4–6 h at 37 °C. For preparative transcription according to this protocol and Protocols 2–5, usually 1-ml reaction mixtures are prepared and then incubated in 200-µl aliquots (for better thermal equilibration) for 2 h at 37 °C; then a second aliquot of T7 RNAP is added (400 U/200 µl reaction mix), followed by another 2 h of incubation at 37 °C. Efficient transcription reactions in the 1-ml scale result in a product yield of about 3 nmol.

[1] BSA (Sigma, minimum purity 98% based on electrophoretic analysis, pH 7).
[2] Pyrophosphatase from yeast (Roche, EC 3.6.1.1, 200 U/mg, <0.01% ATPase and phosphatases each).

Protocol 2: Hartmann lab

	Final concentration	1000 µl
HEPES pH 7.5, 1 M	80 mM	80 µl
DTT 100 mM	15 mM	150 µl
MgCl$_2$ 3 M	33 mM	11 µl
Spermidine 100 mM	1 mM	10 µl
rNTP mix (25 mM each)	3.75 mM (each)	150 µl
Template (linearized plasmid 3.2 kb) 1 µg/µl	80 µg/ml	80 µl
Pyrophosphatase 200 U/ml	2 U/ml	10 µl
T7 RNAP 200 U/µl	2000–3000 U/ml	10–15 µl
RNase-free water		to 1000 µl

For incubation, see Protocol 1.

Protocol 3: Hartmann lab

	Final concentration	1000 µl
5 × transcription buffer (MBI)[1]	1 × buffer	200 µl
MgCl$_2$ 3 M	40 mM	13.3 µl
rNTP mix (25 mM each)	3 mM (each)	120 µl
Template (linearized plasmid 3.2 kb) 1 µg/µl	80 µg/ml	80 µl
T7 RNAP 200 U/µl	2000–3000 U/ml	10–15 µl
RNase-free water		to 1000 µl

[1] 5 × transcription buffer (MBI Fermentas): 200 mM Tris–HCl (pH 7.9 at 25 °C), 30 mM MgCl$_2$, 50 mM DTT, 50 mM NaCl and 10 mM spermidine. For incubation, see Protocol 1.

Protocol 4: Hartmann lab

	Final concentration	1000 µl
HEPES pH 7.5, 1 M	80 mM	80 µl
DTT 100 mM	5 mM	50 µl
MgCl$_2$ 3 M	22 mM	7.3 µl

1.3 Transcription Protocols

Spermidine 100 mM	1 mM	10 µl
BSA 20 mg/ml	0.12 mg/ml	6 µl
rNTP mix (25 mM each)	3.75 mM (each)	150 µl
Template (linearized plasmid 3.2 kb) 1 µg/µl	40 µg/ml	40 µl
Pyrophosphatase 200 U/ml	5 U/ml	25 µl
RNase-free water		321.7 µl
• Prewarm mixture to 37 °C, then add:		
guanosine (30 mM, kept at 75 °C)	9 mM	300 µl
T7 RNAP 200 U/µl	2000 U/ml	10 µl

For incubation, see Protocol 1.

Protocol 5: Initiation with 5'-GMP or 5'-ApG (Hartmann lab)

	Final concentration	1000 µl
HEPES pH 7.5, 1 M	80 mM	80 µl
DTT 100 mM	5 mM	50 µl
MgCl$_2$ 3 M	22 mM	7.3 µl
Spermidine 100 mM	1 mM	10 µl
BSA 20 mg/ml	0.12 mg/ml	6 µl
rNTP mix (25 mM each)	3.75 mM (each)[1]	150 µl
5'-GMP 100 mM (initiator)[1]	15 mM	150 µl
Template (linearized plasmid 3.2 kb) 1 µg/µl	40 µg/ml	40 µl
Pyrophosphatase 200 U/ml	5 U/ml	25 µl
T7 RNAP 200 U/µl	2000 U/ml	10 µl
RNase-free water		to 1000 µl

[1] When 5'-GMP is replaced with the dinucleotide 5'-ApG for transcription initiation, adjust 5'-ApG to 7.5 mM and rNTPs to 2.5 mM each (final concentrations). For incubation, see Protocol 1.

Protocol 6: Hartmann lab

	Final concentration	500 µl
HEPES pH 8.0, 1 M	160 mM	80 µl
DTT 100 mM	15 mM	75 µl
MgCl$_2$ 3 M	33 mM	5.5 µl
Spermidine 100 mM	1 mM	5 µl
rNTP mix (25 mM each)	3.75 mM (each)[3]	75 µl
Promoter DNA oligonucleotide 3.3 µg/µl[1]	132 µg/ml	20 µl
Template DNA oligonucleotide 2.1 µg/µl[2]	84 µg/ml	20 µl
BSA 20 mg/ml	0.12 mg/ml	3 µl
Pyrophosphatase 200 U/ml	2 U/ml	5 µl

1 Enzymatic RNA Synthesis using Bacteriophage T7 RNA Polymerase

RNase-free water		51.5 µl
• Prewarm mixture to 37 °C		
Then add guanosine (30 mM, kept at 75 °C)	9 mM	150 µl
• Mix and preincubate for 1 h at 37 °C		
Then add T7 RNAP 200 U/µl	4000 U/ml	10 µl

Incubate at 37 °C for 4 h.

[1] Promoter DNA oligonucleotide: 5'-TAA TAC GAC TCA CTA TAG.

[2] In this example, the template DNA oligonucleotide had the sequence 5'-GGT CAT AGG TAT TCC CCC TCT CTC CAT TCC TAT AGT GAG TCG TAT TAA, resulting in an RNA product with the sequence 5'-GGA AUG GAG AGA GGG GGA AUA CCU AUG ACC.

[3] To increase the percentage of transcripts initiated with guanosine, the ratio of guanosine to rNTPs may be increased, e.g. by reducing the rNTP concentration to 1.5 mM each.

10 × transcription buffer (TRX), Kirsebom lab, for transcription Protocols 7 and 8

	Final concentration	1000 µl
Tris–HCl pH 7.5, 1 M	200 mM	200 µl
Tris–HCl pH 7.9, 1 M	200 mM	200 µl
MgCl$_2$ 3 M	240 mM	80 µl
Spermidine 100 mM	20 mM	200 µl
RNase-free water		320 µl

Protocol 7: Kirsebom lab, non-radioactive transcription, volume sufficient for 4 reactions; however, note that the mix is prepared for 4.5 reactions

	Final concentration[1]	432 µl
10 × TRX	1×	45 µl
DTT 0.5 M	10 mM	9 µl
0.2% Triton X-100	0.01%	22.5 µl
ATP (100 mM)	2 mM	9 µl
GTP (100 mM)	2 mM	9 µl
CTP (100 mM)	2 mM	9 µl
UTP (100 mM)	2 mM	9 µl
RNase inhibitor 24 U/µl	32 U/ml	0.6 µl
T7 RNAP 200 U/µl	10 000 U/ml	22.5 µl
RNase-free water		296.4 µl

• To 96 µl of this mix add:

template (linearized plasmid ≈ 3.2 kb) 1 µg/µl	40 µg/ml	4 µl

Incubate at 37 °C for ≤10 h.

[1,2,3] Final concentrations after addition of template.

Protocol 8: Kirsebom lab; internal radioactive labeling mix, volume sufficient for 9 reactions; however, note that the mix is prepared for 10 reactions

	Final concentration[1]	230 μl
10 × TRX	1×	25 μl
DTT 0.5 M	10 mM	5 μl
0.2% Triton X-100	0.01%	12.5 μl
ATP (100 mM)	2 mM	5 μl
GTP (100 mM)	2 mM	5 μl
CTP (100 mM)	2 mM	5 μl
UTP 1 mM	0.2 mM	50 μl
[α-^{32}P]UTP 800 Ci/mmol (20 mCi/ml)	10 Ci/mmol	25 μl
RNase inhibitor 24 U/μl	31.7 U/ml	0.33 μl
T7 RNAP 200 U/μl	10 000 U/ml	12.5 μl
RNase-free water		84.7 μl

- To 23 μl of this mix add:

template (linearized plasmid ≈ 3.2 kb) 1 μg/μl	80 μg/ml	2 μl

Incubate at 37 °C for ≤10 h.
[1] Final concentrations after addition of template.

To account for a severely biased nucleotide composition of the template, such as in RNase P RNAs from the *Cyanophora paradoxa* cyanelle [28] or from a plant-pathogenic phytoplasma [29], the relative concentrations of rNTPs are adjusted accordingly. For phytoplasma RNase P RNA (around 73% A + U), the composition of the nucleotide mix was calculated as follows:

	Calculated mol% of each nucleotide in transcript	Concentration of each rNTP in nucleotide mix (mM)	Final concentration of each rNTP in reaction mix (mM)
rATP	41.08	33	3.3
rCTP	11.06	9	0.9
rGTP	16.03	12.5	1.25
rUTP	31.83	25.5	2.55
Total	100	80	8.0

The following sample protocol is routinely used for preparation of large amounts of RNA and can be easily adjusted to the transcription of templates with a biased nucleotide composition. In such cases, the "standard" rNTP mix (20 mM each rNTP) is replaced by the template-specific rNTP mix with adjusted nucleotide concentrations.

Protocol 9: Preparative transcription, nucleotide composition from the example above (Schön lab)

	Final concentration	250 µl
10 × transcription buffer[1]	1 × buffer	25 µl
rNTP mix (here: 33 mM ATP/9 mM CTP/12.5 mM GTP/25.5 mM UTP)	0.1 × rNTP mix	25 µl
Template (linearized plasmid 3.2 kb) 0.1 µg/µl	50 µg/ml	125 µl
T7 RNAP (own preparation, 5–10 µg/µl total protein; see Section 1.5)	40–200 µg/ml[2]	2–5 µl[2]
RNase-free water		to 250 µl

[1] 10 × transcription buffer: 400 mM Tris–HCl (pH 7.9 at 25 °C), 120 mM $MgCl_2$, 50 mM DTT, 50 mM NaCl and 10 mM spermidine.

[2] Depending on the specific activity of the individual T7 RNAP preparation; see Section 1.5.2.2.

For phytoplasma RNase P RNA, a 250-µl reaction performed under these conditions results in up to 150 µg RNA after gel purification. Note that due to the high concentrations of rNTPs and Mg^{2+} in the reaction, insoluble precipitates may form if the complete mix is kept on ice. It is thus advisable to start with water before adding the other components, and to prewarm the mix to 37 °C before the addition of template and polymerase. We let the reactions proceed for at least 2 h (preferably overnight) at 37 °C and quench the excess Mg^{2+} by addition of Na_2EDTA to a final concentration of 25 mM before phenol extraction and EtOH precipitation (see Section 1.3.3, Step 1).

After transcription, check product yield and quality (5–10-µl aliquot plus equal volume gel loading buffer) by PAGE in the presence of 8 M urea and stain the gel with ethidium bromide. Load at least one reference RNA on the gel to identify the genuine product, since sometimes a complex mixture of bands is observed. A low number of bands in addition to the product band points to good transcription performance, and high yields of transcription correlate with the observation that the RNA product appears as a prominent band, while the DNA template is faintly visible. However, with some templates one has to be satisfied with product amounts exceeding that of the template by a factor of only 5. Aberrant transcripts of similar size and abundance as the desired product, sometimes even appearing as a smear, can make identification and gel purification of the RNA product of interest impossible. In view of such potential problems, the best transcription protocol will be the one generating the highest amount of specific product at the lowest cost of incorrect products. If RNA yields are not satisfactory, vary concentrations of template, $MgCl_2$, T7 RNAP or DTT for further optimization.

1.3.2
Transcription with 2′-Fluoro-modified Pyrimidine Nucleotides

To produce nuclease-resistant 2′-fluoro-modified RNAs with the Y639F mutant T7 RNAP, we replaced rUTP and rCTP with the corresponding 2′-fluoro-analogs (IBA,

Göttingen, Germany; or Epicentre, WI, USA). Protocol 10 has been employed in an *in vitro* selection study using a 117-bp double-stranded PCR template including an internal segment of 60 randomized positions. Transcription assays were incubated for 3–4 h at 37 °C.

Protocol 10: Transcription of 2′-fluoro-modified RNA (Hartmann lab)

		150 µl
10 × transcription buffer[1]	1 × buffer	15 µl
DTT 100 mM	5 mM	7.5 µl
2′-F-CTP, 2′-F-UTP (10 mM each)	1.25 mM (each)	18.75 µl
rATP, rGTP (10 mM each)	1.25 mM (each)	18.75 µl
[α-^{32}P]ATP 800 Ci/mmol (10 mCi/ml)	0.16 Ci/mmol	3 µl
PCR template	1–3.33 nmol/ml	0.15–0.5 nmol
Y639F mutant T7 RNAP (3.7 µg/µl)	93.73 µg/ml	3.8 µl
RNase-free water		to 150 µl

[1] 10 × transcription buffer: 400 mM Tris–HCl (pH 8.0), 200 mM $MgCl_2$, 10 mM spermidine, 0.1% Triton X-100.

1.3.3
Purification

Some purification steps are optional, and depend on transcription quality and the demands on product purity. Often, a purification procedure only including Steps 5–8 is sufficient. Another protocol for gel purification of RNA is described in Chapter 3.

(1) Insoluble pyrophosphate complexes. In transcription reactions without pyrophosphatase or when pyrophosphatase activity is low, a white pyrophosphate precipitate may form. In such cases, remove the precipitate before Step 2 by centrifugation at 14 000 *g* for about 5 min directly after transcription. Carefully remove the clear supernatant and transfer to a new Eppendorf tube for further sample processing. Also, Na_2EDTA (500 mM, pH 7.5) may be added immediately after transcription to give a final concentration of 50–100 mM. By chelating Mg^{2+}, the formation of insoluble precipitates is substantially reduced.
(2) DNase I digestion to remove template DNA. Add 10 U DNase I (RNase-free, Roche) per 200 µl and incubate for 20 min at 37 °C.
(3) Phenol and chloroform extractions to remove the enzyme(s). For extraction with phenol (Tris-saturated, stabilized with hydroxy-quinoline, pH 7.7; Biomol), mix the sample with 0.5–1.0 volumes phenol, vortex for 30 s, centrifuge 1–5 min (until phases have cleared) at 12 000 *g*, withdraw the aqueous upper phase and mix it with 0.5–1.0 volumes chloroform, vortex for 30 s, centrifuge 3 min at 12 000 *g* and transfer the aqueous upper phase to a new tube, avoiding to withdraw any chloroform.

(4) Removal of salt for better gel resolution. For example, use NAP 10 columns (Amersham Biosciences, now part of GE Healthcare), column material: Sephadex G-25.
(5) Ethanol precipitation. Ethanol precipitation is performed to remove residual chloroform and salts, and to concentrate the RNA. Mix sample with 2.5 volumes ethanol, 0.1 volumes 3 M NaOAc (pH 4.7) and 1 µl glycogen (20 µg/µl); leave for 10–20 min at −70 °C or at least 2 h at −20 °C. Centrifuge for 30–45 min at 4 °C and 16 000 g. Wash the pellet with 70% ethanol and centrifuge again for 10 min. After ethanol precipitation and air-drying of the pellet, redissolve it in a small volume of RNase-free water and add an equal volume of gel loading buffer [0.33 × TBE (see Step 6), 2.7 M urea, 67% formamide, 0.01% (w/v) each bromophenol blue (BPB) and xylene cyanol blue (XCB)].
(6) Preparative denaturing PAGE. A gel well, 6- to 7-cm wide and 1-mm thick, is appropriate for loading the product RNA from an efficient 1-ml transcription reaction. The pocket size is of some importance, as an overloaded gel may cause separation problems; on the other hand, if the pocket is too large, RNA bands may be barely visible and elution efficiency may decrease. After electrophoresis, the desired RNA band is visualized at 254 nm by UV-shadowing and marked for gel excision (for details, see Chapter 3); gel running buffer: 1 × TBE (89 mM Tris base, 89 mM boric acid, 2 mM EDTA).
(7) Elution of RNA product.
   (a) *Diffusion elution.* Cover the excised gel pieces with elution buffer and shake overnight at 4 °C; in the case of efficient transcriptions, a second gel elution step in fresh elution buffer may substantially increase the yield. Different elution buffers can be used: buffer A: 1 mM EDTA, 200 mM Tris–HCl (pH 7)/buffer B: 1 M NaOAc (pH 4.7)/buffer C (successfully used for the elution of phosphorothioate-modified RNAs): 1 M NH$_4$OAc (pH 7). Usually, buffer A is used; buffer B was found to be advantageous in cases where elution efficiency with buffer A was low. After elution, RNA is concentrated by ethanol precipitation.
   (b) *Alternatively: electro-elution.* Excised gel pieces containing the RNA are placed in a BIOTRAP chamber (Schleicher & Schuell, Dassel, Germany; in USA and Canada: ELUTRAP) following the manufacturer's protocol. The RNA is eluted in 0.5 × TBE buffer (see Step 6). The final volume of RNA solution after elution is approximately 600 µl, depending on the extent of evaporation during the elution process. The elution is permitted to proceed overnight at 150 V/20 mA or for 4–6 h at 200–300 V/30 mA. During the elution process there is evaporation resulting in condensation on the lid of the BIOTRAP chamber. This lid has to be closed when the BIOTRAP is running. To minimize evaporation/condensation, the BIOTRAP should not be run at higher voltage than 150 V overnight. After elution, the RNA is extracted once with phenol and twice with chloroform/isoamylalcohol, followed by ethanol precipitation.
(8) Quantification (UV spectroscopy, see Chapter 4 and Appendix) and quality check (denaturing PAGE).

1.4
Troubleshooting

1.4.1
Low or No Product Yield

- If product yields are low with a protocol that had already been successfully used for the same template, repeat transcription assay once without any alteration on 50-µl scale; if unsuccessful as well, test different enzyme batches or enzymes from alternative suppliers. Differences between enzyme preparations can be considerable.
- Be sure that all components (except enzymes) have been warmed up to ambient temperature before preparation of reaction mixtures at ambient temperature.
- Check that thawed stock solutions, particularly concentrated transcription buffers, do not contain precipitated ingredients. For nucleotide solutions, limit freeze–thawing cycles, store in aliquots at −20 °C, and adjust stock solutions (in H_2O) to pH 7.0 or consider to buffer with 10–40 mM Tris–HCl adjusted to the pH used in transcription reactions; use lithium salts if available and be aware that diluted working solutions may degrade rapidly.
- Prepare new template DNA; take particular care to effectively remove salts as well as traces of phenol and chloroform.
- For templates with a highly biased nucleotide composition (e.g. coding for RNAs with extremely high A + U content [28, 29]), adjust the composition of the rNTP solution according to the nucleotide ratio of the RNA. However, do not alter the total nucleotide concentration of the reaction mix.

1.4.2
Side-products and RNA Quality

- Usually we get the least artifact products, in addition to the correct RNA product, with Protocol 3, but redissolving the RNA after ethanol precipitation may become a severe problem due to pyrophosphate precipitates. To alleviate such solubility problems, see Section 1.3.3, Step 1.
- Gel entry problems or smear on gel: perform Steps 1–5 of Section 1.3.3 before proceeding to gel purification (Section 1.3.3, Step 6).

1.5
Rapid Preparation of T7 RNA Polymerase

This protocol is based on the publications of Grodberg and Dunn [30] and Zawadzki and Gross [25], and provides a fast and efficient procedure for the preparation of a highly stable T7 RNAP which is sufficiently pure for most purposes. The chromatography is described for FPLC, but any standard low-pressure equipment will give satisfactory results if the procedure is adapted accordingly.

1.5.1
Required Material

E. coli BL21 pAR1219 (obtained from F. W. Studier, Biology Department, Brookhaven National Laboratory, Upton, NY 11973, USA).

Medium
LB (Luria-Bertani) medium [31] supplemented with 50 µg/ml Ampicillin.

Buffers and solutions
100 mM IPTG
TEN buffer (50 mM Tris–HCl, pH 8.1; 2 mM Na_2EDTA; 20 mM NaCl)
Phenylmethylsulfonyl fluoride (PMSF), 20 mg/ml in isopropanol
Leupeptin, 5 mg/ml
Egg white lysozyme, 1.5 mg/ml in TEN (freshly prepared)
0.8% sodium deoxycholate solution
2 M ammonium sulfate (enzyme grade)
Polymin P: 10% solution, adjusted to pH 8 with HCl
Saturated ammonium sulfate solution (4.1 M; adjust pH to 7 with some drops of concentrated Tris base, keep at 4 °C where a precipitate will form)
Buffer C [20 mM sodium phosphate pH 7.7; 1 mM Na_2EDTA; 1 mM DTT; 5% glycerol (w/v)]
4 × Laemmli gel loading buffer: 100 mM Tris–HCl, pH 6.8, 8% (w/v) SDS, 30% glycerol (w/v), 8% (v/v) β-mercaptoethanol, 0.04% (w/v) bromophenol blue; adjust pH before addition of bromophenol blue
Buffer C-10, C-100: buffer C supplemented with 10 or 100 mM NaCl, respectively.

Electrophoresis and chromatography
Laemmli-type SDS gel for protein separation under denaturing conditions (10% PAA)
Merck EMD Fractogel SO_3^-, equilibrated in buffer C-100 in a 2 × 10 cm column

1.5.2
Procedure

1.5.2.1 Cell Growth, Induction and Test for Expression of T7 RNAP

(1) Inoculate 25 ml LB supplemented with 50 µg/ml Ampicillin with a colony from a fresh plate culture and grow overnight at 37 °C.
(2) Four 2-l flasks with 500 ml of the same medium are inoculated 1:100 from this culture; grow at 37 °C under vigorous shaking.
(3) When the cultures have reached an OD_{600} of about 0.6 (which should take not longer than 3 h), transfer 1 ml to an Eppendorf tube, centrifuge for 5 min at 5000 r.p.m. in a desktop centrifuge and keep the sediment as a control.
(4) Then induce the remaining culture for expression by addition of IPTG to a

1.5 Rapid Preparation of T7 RNA Polymerase

final concentration of 0.5 mM. At 1.5, 2 and 3 h after induction, take 1-ml samples as in Step 3.

(5) Harvest the flask cultures by centrifugation (10 min, 5000 g), wash once with TEN buffer, shock-freeze in liquid N_2 or dry ice and keep at −80 °C until needed.

(6) Analyze the 1-ml samples from Steps 3 and 4 for expression of T7 RNAP as follows: resuspend the cell sediment in 100 µl of 1 × concentrated Laemmli gel loading buffer (note that this buffer is usually prepared as a more concentrated stock solution; see Section 1.5.1) and denature for 2 min at 95 °C. Then load 10–25 µl of each sample onto an SDS–10% polyacrylamide gel with appropriate size markers. If expression is sufficient (a strong band of about 100 kDa should appear 2–3 h after induction), proceed with enzyme purification.

1.5.2.2 Purification of T7 RNAP

Generally, all steps are performed on ice or at 4 °C and all buffers are supplemented with the protease inhibitor PMSF (20 µg/ml, if not stated otherwise). From each purification step, a small sample should be retained for the determination of protein concentration and purity. Protein concentrations are most conveniently determined by dye binding [32] or by direct measurement of extinction at 280 nm.

(1) Resuspend cells in 24 ml of TEN buffer supplemented with 50 µl of PMSF and 20 µl of leupeptin stock solutions (see Section 1.5.1).
(2) Add 6 ml lysozyme solution; after 20 min of incubation, add 2.5 ml of 0.8% sodium deoxycholate solution and incubate for another 20 min.
(3) Shear the DNA in this viscous lysate by sonication (4 × 15 s with an immersible probe; 2–5 min for sonication in a water bath).
(4) Add 5 ml 2 M ammonium sulfate and adjust the total volume to 50 ml with TEN buffer.
(5) Remove DNA by slow addition of 5 ml Polymin P and stirring for 20 min.
(6) After centrifugation (15 min, 39 000 g), keep the supernatant and determine its volume.
(7) Precipitate the enzyme from the supernatant by slow addition of 0.82 volumes of saturated ammonium sulfate and stirring for another 15 min.
(8) After centrifugation for 15 min at 12 000 g, resuspend the sediment in 15 ml buffer C-100, and dialyze for at least 8 h against 2 × 1 l of the same buffer. Dialysis should be extensive in order to completely remove the ammonium sulfate contained in the sediment; otherwise, T7 RNAP will not bind to the cation exchange column.
(9) Remove insoluble material by centrifugation as in Step 8, and apply the supernatant to the EMD-SO_3^- column at a flow rate of about 20 ml/h. Wash the column with 10 volumes of buffer C-100 or until protein is no longer detectable in the flow-through. Then apply a 250-ml gradient from 100 to 500 mM NaCl in buffer C and collect 5-ml fractions. T7 RNAP elutes between 300 and 400 mM NaCl from the EMD-SO_3^- column, as visible by the high

protein content in these fractions. Finally, wash the column with 1 M NaCl in buffer C; the resin can be reused after equilibration in buffer C-100.
(10) 10–20 µl of each fraction and the flow-through, and 2–5 µl of the applied sample, are then analyzed by SDS–PAGE as described in Section 1.5.2.1, Step 6.
(11) The fractions containing T7 RNAP are pooled and dialyzed for at least 8 h against 2 × 2 l buffer C-10.
(12) The resulting precipitate, enriched in T7 RNAP, is collected by centrifugation as in Step 8, resuspended in 1–2 ml of buffer C-100 and adjusted to 50% glycerol (w/v) for storage at −20 °C.

Specific activity of T7 RNAP may be determined by incorporation of ^{3}H- or ^{32}P-labeled nucleotides into acid-precipitable material and can reach 400 000 U/mg (1 U is defined as the incorporation of 1 nmol AMP into acid precipitable material in 1 h at 37 °C [9]). For most purposes, it is sufficient to titrate the amount of enzyme preparation needed to give good transcription yields without too many side products. To test a typical preparation of T7 RNAP (5–10 µg/µl total protein content), we vary the amount of polymerase (usually between 0.1 and 1 µl) in a series of analytical transcriptions, for example, in a variation of Protocol 9 scaled down to 25 µl, containing 2 mM of each rNTP and a standard template DNA (around 50% G + C). Transcription products are then separated by gel electrophoresis and the efficiency of transcription is evaluated after toluidine blue staining of the gel (see Chapter 9). Alternatively, for easier detection of abortive transcripts and degradation products, a radioactive tracer (e.g. 10 µCi of [α-^{32}P]GTP) can be added to the transcription reaction and the products visualized on a PhosphoImager or by autoradiography (see Protocol 8).

1.5.3
Notes and Troubleshooting

(1) If protein gel electrophoresis of crude cell samples (Section 1.5.2.1, Step 6) yields badly smeared bands, try to shear the DNA by sonication as described for the enzyme purification. Alternatively, samples can be squeezed with a syringe through a thin (0.7 mm, or 22 gauge) needle.
(2) If expression of T7 RNAP is not sufficient, vary ampicillin or IPTG concentration, change growth times before and after induction, or switch to other growth media, such as TB or 2 × YT [31], or M9 Minimal Medium supplemented with trace elements [33].
(3) The Fractogel EMD-SO$_3^-$ column can be substituted by any strong cation exchanger of the SO$_3^-$ type. However, with most other column matrices, T7 RNAP elutes much earlier (around 200 mM NaCl).
(4) If T7 RNAP does not precipitate after the dialysis step with buffer C-10 (Section 1.5.2.2, Step 11), repeat dialysis with fresh buffer C (without NaCl).
(5) If a substantial portion of the precipitate from Step 12 above (Section 1.5.2.2) cannot be re-dissolved in buffer C-100, one may gradually increase the NaCl concentration up to 500 mM.

Acknowledgement

We thank Dagmar K. Willkomm for critical reading of the manuscript.

References

1 P. Davanloo, A. H. Rosenberg, J. J. Dunn, F. W. Studier, Proc. Natl. Acad. Sci. USA **1984**, *81*, 2035–2039.
2 R. H. Ebright, J. Mol. Biol. **2000**, *304*, 687–698.
3 G. M. T. Cheetham, T. A. Steitz, Curr. Opin. Struct. Biol. **2000**, *10*, 117–123.
4 G. A. Diaz, M. Rong, W. T. McAllister, R. K. Durbin, Biochemistry **1996**, *35*, 10837–10843.
5 S. N. Kochetkov, E. E. Rusakova, V. L. Tunitskaya, FEBS Lett. **1998**, *440*, 264–267.
6 S. Brakmann, S. Grzeszik, ChemBioChem **2001**, *2*, 212–219.
7 J. F. Milligan, O. C. Uhlenbeck, Methods Enzymol. **1989**, *180*, 51–63.
8 I. D. Pokrovskaya, V. V. Gurevich, Anal. Biochem. **1994**, *220*, 420–423.
9 J. F. Milligan, D. R. Groebe, G. W. Witherell, O. C. Uhlenbeck, Nucleic Acids Res. **1987**, *15*, 8783–8798.
10 J. A. Pleiss, M. L. Derrick, O. C. Uhlenbeck, RNA **1998**, *4*, 1313–1317.
11 M. Helm, H. Brulé, R. Giegé, C. Florentz, RNA **1999**, *5*, 618–621.
12 T. M. Coleman, G. Wang, F. Huang, Nucleic Acids Res. **2004**, *32*, e14.
13 J. R. Sampson, O. C. Uhlenbeck, Proc. Natl. Acad. Sci. USA **1988**, *85*, 1033–1037.
14 A. D. Griffiths, B. V. Potter, I. C. Eperon, Nucleic Acids Res. **1987**, *15*, 4145–4162.
15 V. D. Axelrod, F. D. Kramer, Biochemistry **1985**, *24*, 5716–5723.
16 F. Conrad, A. Hanne, R. K. Gaur, G. Krupp, Nucleic Acids Res. **1995**, *23*, 1845–1853.
17 R. K. Gaur, G. Krupp, FEBS Lett. **1993**, *315*, 56–60.
18 N. Logsdon, C. G. L. Lee, J. W. Harper, Anal. Biochem. **1992**, *205*, 36–41.
19 R. Sousa, R. Padilla, EMBO J. **1995**, *14*, 4609–4621.
20 Y. Huang, F. Eckstein, R. Padilla, R. Sousa, Biochemistry **1997**, *36*, 8231–8242.
21 W. D. Hardt, V. A. Erdmann, R. K. Hartmann, RNA **1996**, *2*, 1189–1198.
22 S. P. Ryder, S. A. Strobel, Methods **1999**, *18*, 38–50.
23 C. Pitulle, R. G. Kleineidam, B. Sproat, G. Krupp, Gene **1992**, *112*, 101–105.
24 F. Huang, Nucleic Acids Res. **2003**, *31*, e8.
25 V. Zawadzki, H. J. Gross, Nucleic Acids Res. **1991**, *19*, 1948.
26 T. Ellinger, R. Ehricht, BioTechniques **1998**, *24*, 718–720.
27 S. S. Sastry, B. M. Ross, J. Biol. Chem. **1997**, *272*, 8644–8652.
28 M. Baum, A. Cordier, A. Schön, J. Mol. Biol. **1996**, *257*, 43–52.
29 M. Wagner, C. Fingerhut, H. J. Gross, A. Schön, Nucleic Acids Res. **2001**, *29*, 2661–2665.
30 J. Grodberg, J. J. Dunn, J. Bacteriol. **1988**, *170*, 1245–1253.
31 J. Sambrook, E. F. Fritsch, T. Maniatis, Molecular Cloning: A Laboratory Manual, Cold Spring Harbor Laboratory Press, Cold Spring Harbor, NY, **1989**.
32 M. M. Bradford, Anal. Biochem. **1976**, *72*, 248–254.
33 D. E. Mossakawska, R. A. G. Smith, in: Protein NMR Techniques, D. G. Reid (ed.), Humana Press, Totowa, NJ, **1997**.
34 F. Huang, C. W. Bugg, M. Yarus, Biochemistry **2000**, *39*, 15548–15555.
35 J. J. Dunn, F. W. Studier, J. Mol. Biol. **1983**, *166*, 477–535.
36 S. S. Lee, C. Kang, J. Biol. Chem. **1993**, *268*, 19299–19304.

2
Production of RNAs with Homogeneous 5′ and 3′ Ends

Mario Mörl, Esther Lizano, Dagmar K. Willkomm and Roland K. Hartmann

2.1
Introduction

Synthesis of RNA molecules by *in vitro* transcription, primarily utilizing T7 RNA polymerase (T7 RNAP) for reasons of price and efficiency, is a widely used method in biochemistry and molecular biology. For most purposes, it is sufficient to purify transcripts by electrophoretic separation on denaturing PAA (polyacrylamide) gels. Thereby, template DNA, enzyme, nucleotides as well as incomplete transcripts (resulting from premature transcription termination) can be removed efficiently. However, the seemingly uniform population of full-length transcripts is in most cases heterogeneous because T7 RNAP tends to add one or occasionally even a few additional, non-encoded nucleotides to the 3′ end [1, 2]. Also, non-templated nucleotide incorporation at the 5′ end can affect up to 30% of the transcribed RNA molecules when the template encodes a T7 transcript with more than three consecutive guanosines at the 5′ end [3, 4]. Such micro-heterogeneous full-length transcripts can rarely be separated efficiently by denaturing PAGE. Thus, if applications require complete homogeneity at the transcript 5′ and 3′ ends, e.g. in the context of *in vitro* aminoacylation of tRNAs, ligation of RNA fragments (see Chapters 3 and 4), RNA crystallization or NMR studies, refined approaches are needed.

Apart from product end heterogeneity, another problem associated with *in vitro* transcription stems from the fact that the sequence requirements of the core promoter regions recognized by phage RNA polymerases also put some constraints on the 5′-terminal nucleotide identities of the transcript. For example, transcription by T7 RNAP from the predominantly used (class III) T7 promoter only proceeds with reasonable efficiency if the transcript is initiated with a G residue (for more details, see Chapter 1). As a consequence, synthesis yields will be low for transcripts with a different nucleotide identity at the +1 position.

An approach toward eliminating 5′ and 3′ end heterogeneity as well as sequence constraints at the 5′ end is the use of catalytic RNA entities acting in *cis* or in *trans*, which precisely release the RNA of interest from a primary transcript with extra

Handbook of RNA Biochemistry: Student Edition. Edited by R. K. Hartmann, A. Bindereif, A. Schön, E. Westhof
Copyright © 2009 WILEY-VCH Verlag GmbH & Co. KGaA, Weinheim
ISBN: 978-3-527-32534-4

5′- and 3′-flanking sequences. These strategies permit to produce RNA without any sequence restrictions and with essentially complete 5′ and 3′ end homogeneity.

2.2
Description of Approach

2.2.1
Cis-cleaving Autocatalytic Ribozyme Cassettes

2.2.1.1 The 5′ Cassette

An elegant and efficient way to produce transcripts with uniform 5′ ends is to use a construct consisting of a self-cleaving hammerhead sequence immediately upstream of the RNA sequence of interest. The hammerhead ribozyme is a small structural element, originally identified in pathogenic plant viroids and virusoids, which catalyzes phosphodiester bond hydrolysis at a single defined position. With a length of about 50 nt, the hammerhead sequence can easily be inserted as a DNA cassette at the 5′ end of the DNA sequence to be transcribed (Fig. 2.1). In the primary transcript (Fig. 2.1, middle), the 5′ part of the RNA of interest is integrated into the secondary structure of the hammerhead ribozyme. Since this ribozyme sequence is able to fold into the catalytically active form immediately after synthesis, cleavage occurs already during transcription – provided that Mg^{2+} ions are present at sufficiently high concentrations (5 mM or above), as is the case under standard transcription conditions. In addition to 5′ end uniformity, another advantage of a 5′ hammerhead is that optimal transcription start sequences can be used upstream of the ribozyme cassette in favor of high transcription yields.

Within a 5′-flanking hammerhead cassette, the sequence of helix P1 is dictated by the 5′-terminal nucleotides of the RNA of interest (Fig. 2.1, middle). Since the only requirement for P1 is helicity, there are basically no sequence constraints for this structural element. Thus, any efficient minimal hammerhead variant matching the consensus sequence shown in Fig. 2.1 can be utilized [5–8]. Two 5′-hammerhead constructs which in our hands have proved efficient in such strategies are shown in Fig. 2.2(A and B).

2.2.1.2 The 3′ Cassette

For the production of RNAs with homogeneous 3′ ends, again hammerhead cassettes (Fig. 2.2C) can be tethered to the RNA of interest. A disadvantage of these structures are some sequence constraints imposed on the product RNA 3′ end, such as the requirement for a 3′-terminal NUH (N = any nucleotide; H = C, U or A), preferentially GUC, sequence motif in hammerhead constructs (Fig. 2.2C [5]).

The autocatalytic domain of the Hepatitis delta virus (HDV ribozyme), however, has no such sequence requirements upstream of its cleavage site and can therefore be employed for any transcript sequence of interest (Fig. 2.1 and 2.2A). This ribo-

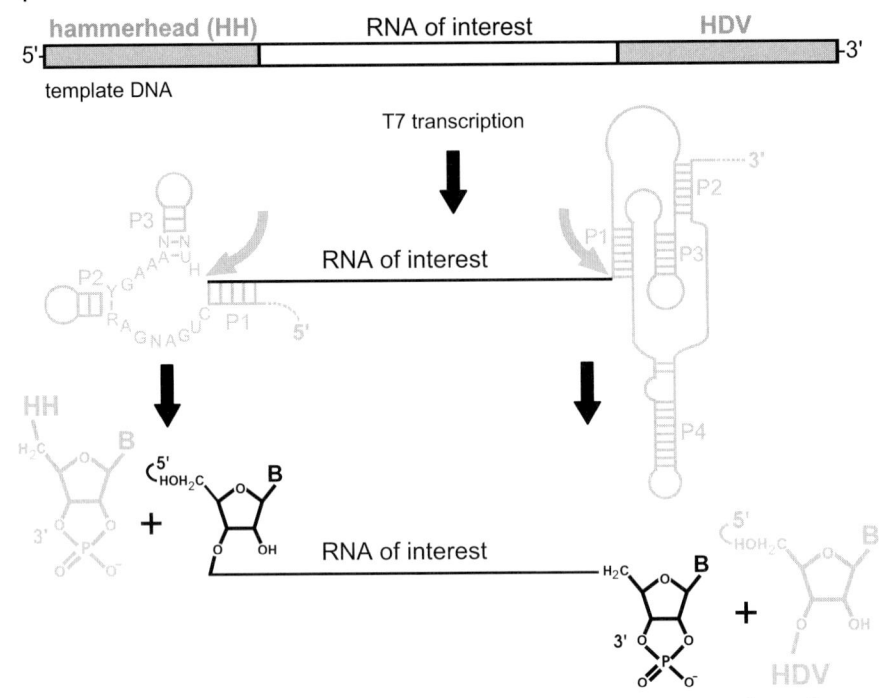

Fig. 2.1. The 5′ and 3′ cassettes for the creation of homogeneous RNA ends. The DNA molecule to be transcribed (linearized plasmid or PCR product) includes extra sequence cassettes upstream and downstream of the sequence encoding the RNA of interest. These cassettes (indicated in grey at the top) are transcribed into self-cleaving ribozyme structures (hammerhead at the 5′ end, HDV ribozyme at the 3′ end) flanking the RNA of interest. Conserved nucleotides in the hammerhead core structure are indicated; Y = pyrimidine; R = purine; NUH (N = any nucleotide; H = C, U or A), preferentially GUC. Both cassettes are designed to permit insertion of virtually any sequence between the ribozymes, rendering this strategy applicable to essentially any RNA of interest. The resulting cleavage product carries a 5′-hydroxyl group, which can be converted to a (radioactively labeled) 5′-phosphate. At the 3′ terminus, a 2′,3′-cyclic phosphate group is generated which interferes with some RNA functions. Protocols for the removal of the 3′-terminal phosphate group are described in Section 2.3.2.

zyme folds into four helical domains that form a pseudoknotted structure, with the cleavage site located immediately upstream of helix P1 at the 5′ end of the ribozyme [9, 10]. Using an optimized version of the HDV ribozyme, homogeneous RNA 3′ ends can easily be generated [11]. The HDV ribozyme is already sufficiently active at Mg^{2+} concentrations of about 1 mM, usually resulting in efficient self-cleavage during transcription. A frequently observed problem is that sequences of the RNA of interest interfere with proper folding of the HDV ribozyme, leading to reduced cleavage efficiency. In our hands, this problem has been solved by incubating the primary transcript repeatedly (10 cycles) for 3 min at 60 °C and 3 min at 25 °C. This procedure allows the ribozyme cassette of the transcript to adopt,

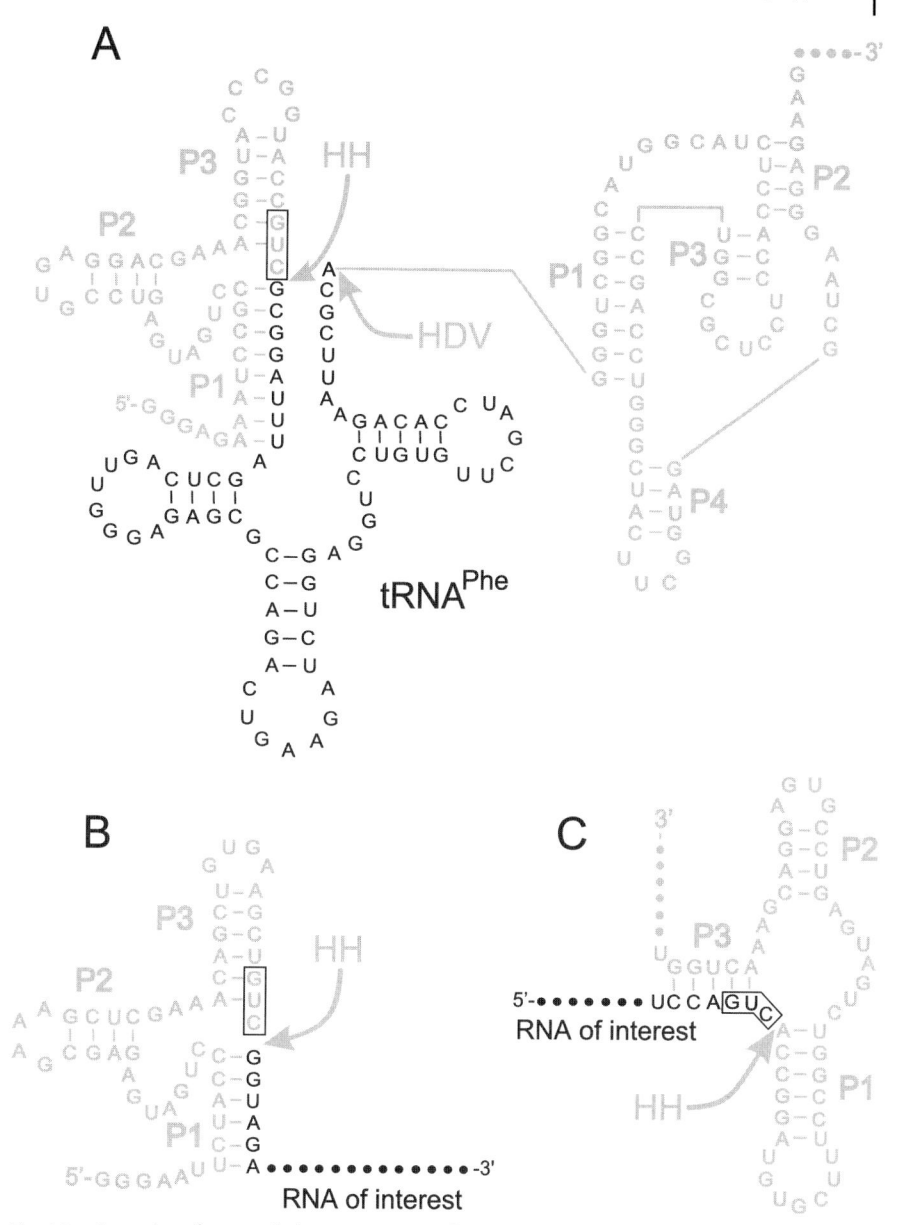

Fig. 2.2. Examples of terminal ribozyme cassettes in primary transcripts. The RNA of interest is shown in black, the ribozyme modules are depicted in grey. Ribozyme self-cleavage sites are indicated by arrows. The preferred GUC triad of hammerhead ribozymes is boxed. The constructs shown have displayed efficient and precise autocatalytic cleavage in the authors' laboratories. (A) Sequence of a primary transcript consisting of a 5'-hammerhead, a central tRNA^{Phe} as the RNA of interest and a 3'-terminal HDV ribozyme. (B and C) Examples of a 5'- and 3'-terminal hammerhead, respectively.

at least transiently, its active structure and has in our hands led to quantitative cleavage.

2.2.1.3 Purification of Released RNA Product and Conversion of End Groups

After co-transcriptional self-cleavage of the terminal ribozyme cassette(s), it is advisable to purify the released RNA of interest by denaturing PAGE and subsequent gel elution in, for example, 500 mM ammonium acetate, pH 5.7, 0.1 mM EDTA and 0.1% SDS at 4 °C overnight, followed by ethanol precipitation ([12]; for detailed protocols, see Chapters 1 and 3). If, for example, a tRNA product is released from the primary transcript, the ribozyme cassettes will have similar lengths of about 50–70 nt, complicating purification of the tRNA. A practical solution to the problem is to extend the 5′ and/or 3′ termini of the primary transcripts, thereby increasing the size of the released terminal fragments containing the ribozyme core structures.

Another important point to be considered is the nature of the 5′ and 3′ ends generated by these ribozymes. Both hammerhead and HDV ribozyme cassettes catalyze transesterification reactions induced by a nucleophilic attack of the 2′-OH group of the neighboring ribose, resulting in 5′-hydroxyl and 2′,3′-cyclic phosphate termini instead of 5′-phosphates and 3′-hydroxyls. A terminal 5′-OH group is for most purposes neutral to RNA function and may even be advantageous, since it is directly accessible to 5′-end-labeling with $[\gamma\text{-}^{32/33}P]ATP$ by T4 polynucleotide kinase (T4 PNK). The 3′-terminal cyclophosphate, however, interferes with a variety of downstream experiments, such as aminoacylation of tRNAs, ligation of RNA molecules or 3′-end-labeling using $[5'\text{-}^{32}P]pCp$. To remove the 2′,3′-cyclic phosphate, the phosphatase activity of T4 PNK can be employed with excellent efficiency (see protocols below and [13, 14]).

2.2.2
Trans-cleaving Ribozymes for the Generation of Homogeneous 3′ Ends

As an alternative to the *cis*-cleaving ribozyme cassettes, homogeneous 3′ ends can be obtained by *trans*-cleavage, e.g. using a *trans*-acting hammerhead ribozyme [15], the *Neurospora* Varkud satellite (VS) ribozyme [15] or the catalytic RNA subunit of a bacterial RNase P [16] as detailed below. A *trans*-acting VS ribozyme has the advantage that the RNA of interest requires only a short stem–loop 3′-extension of 24 nt which serves as the ribozyme substrate recognized by tertiary interactions [15]. Like the HDV ribozyme, the VS ribozyme has no specific sequence requirements upstream of the cleavage site and generates 2′,3′-cyclic phosphate ends [17]. Similar to the VS ribozyme strategy, it is possible to utilize the HDV ribozyme in a *trans*-cleavage reaction. Here, the transcripts are extended by as few as seven nucleotides complementary to the 3′ part of helix P1 of the ribozyme. A correspondingly 5′-truncated HDV RNA can base-pair with this extension, thereby restoring the P1 helix and a functional HDV ribozyme structure. Cleavage of the target RNA then occurs seven nucleotides upstream from its 3′ end [18].

In vitro, the roughly 380-nt long bacterial RNase P RNA, with and without the

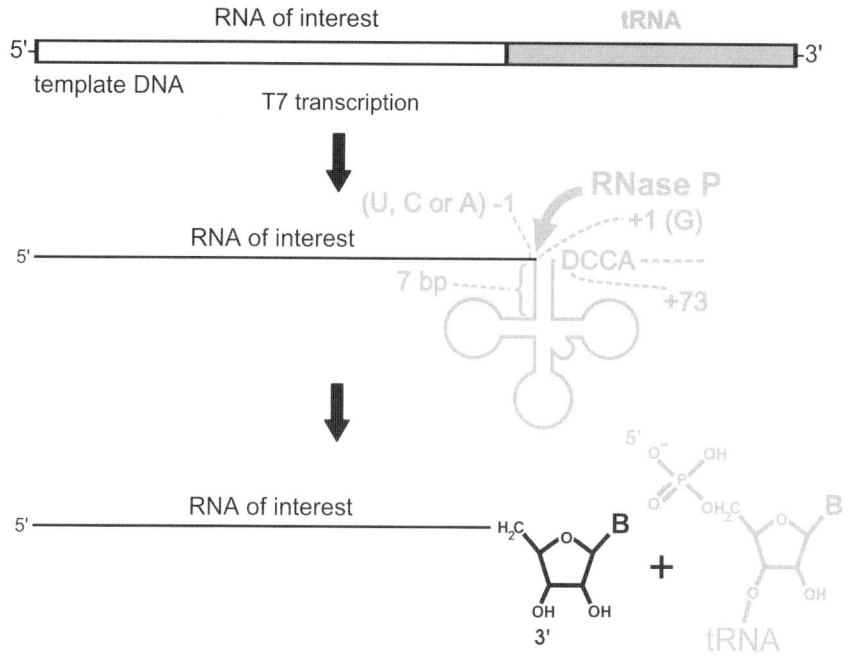

Fig. 2.3. RNase P cleavage to generate homogeneous 3′ ends. An alternative strategy to generate homogeneous 3′ ends is the use of bacterial RNase P (RNA). The DNA template for transcription is designed in a way that a tRNA gene (in grey) is positioned immediately downstream of the sequence encoding the RNA molecule of interest. By incubating the primary transcript, derived from run-off transcription, with RNase P (RNA), the tRNA molecule is cleaved off at its 5′ end and the upstream portion of the primary transcript is released. Aspects of importance: avoid a G residue at position −1; avoid base-pairing between nt −1 and discriminator (+73); the discriminator should be D (= G, A or U) and not C when using *E. coli* RNase P RNA [21]; the first acceptor stem base pair should be G_{+1}–C_{+72}; choose a class I tRNA with 7-bp acceptor stem and short variable arm; encode CCA at the tRNA 3′ terminus; a 3′ extension of up to 6 nt beyond CCA does not interfere with RNase P cleavage, but a second CCA trinucleotide should be avoided within this extension.

RNase P protein, catalyzes accurate and efficient removal of the 5′-flanking sequences of tRNA precursor transcripts. To exploit this system for the production of RNAs with homogeneous 3′ ends, the RNA of interest is fused to a downstream tRNA sequence. As a result, the construct mimics a tRNA precursor molecule, whose 5′-leader sequence represents the RNA of interest (Fig. 2.3). Precise endonucleolytic cleavage by the RNase P ribozyme then releases the RNA product with uniform 3′-OH termini (Fig. 2.3) which, in contrast to the 2′,3′-cyclic phosphates generated by hammerhead and HDV ribozymes, are suitable for many downstream applications.

Since the substrate recognition elements for RNase P RNA are located predominantly in the tRNA structure, there are little upstream sequence requirements to be fulfilled. One restriction is to avoid a G residue at position −1 relative to the tRNA, which otherwise may cause some aberrant cleavage between positions −1 and −2. Also, for reasons of cleavage fidelity, use of a bacterial class I tRNA moiety with a 7-bp acceptor stem and short variable arm is recommended (such as *Bacillus subtilis* tRNA^{Asp} [16, 19] or *Thermus thermophilus* tRNA^{Gly} [20]). Further, any base-pairing potential between nt −1 and the discriminator base (+73) should be avoided, the discriminator should be D (= G, A or U) and not C when using *Escherichia coli* RNase P RNA [21], and the first acceptor stem base pair should be G$_{+1}$–C$_{+72}$. However, note that some class I tRNA transcripts, despite meeting all the aforementioned requirements, may elicit some miscleavage, depending on their sequence context [22]. We have successfully used RNase P RNAs from *E. coli* and *T. thermophilus*, which are available from the authors as T7 expression plasmids. Yet, since achieving 100% 3′ end homogeneity – and not 99% – by use of this approach may require some optimization, we suggest considering the use of an HDV-based strategy as a quicker alternative if such an extent of homogeneity is essential for downstream procedures.

In applications where a short 3′ appendage to the RNA of interest is preferable (e.g. when incorporating costly isotopically labeled nucleotides for NMR studies), the tethered RNase P substrate can be reduced to a hairpin structure of less than half the tRNA size. One such substrate is pATSerU$_{−1}$G$_{+73}$ and its variants pATSerU$_{−1}$U$_{+73}$, pATSerC$_{−1}$A$_{+73}$ and pATSerA$_{−1}$U$_{+73}$, which release an RNA of interest with a 3′-terminal U, C or A residue, respectively [23]. These substrates showed >99% cleavage at the canonical RNase P cleavage site under conditions of 50 mM Tris–HCl, pH 7.2, 5% (w/v) PEG 6000, 100 mM NH$_4$Cl and 40 mM MgCl$_2$ ([23]; Leif A. Kirsebom, personal communication).

The RNase P RNA-based approach could have advantages over a *cis*-cleaving HDV cassette in cases where interactions between the RNA of interest and the HDV cassette prevent the ribozyme from adopting an active conformation. Such folding interference is less likely for a tRNA cassette because tRNAs are among the most stable autonomous RNA folding units. If folding of the downstream tRNA is impeded by the RNA of interest at 37 °C, thermostable RNase P RNA from *T. thermophilus* may be used for cleavage at elevated temperatures, assuming that folding interference is abolished under such conditions. This ribozyme, when acting on a transcript with, for example, a GC-rich tRNA^{Gly} cassette from the same organism [24], will cleave off the tRNA moiety with high precision at temperatures of up to 75 °C [25].

The *trans*-cleaving ribozyme approach may be somewhat laborious if the ribozyme is prepared independently of the transcript of interest and if *in vitro* transcription is followed by a second, independent incubation step for the cleavage reaction. However, the procedure can be simplified by transcribing both RNAs simultaneously from two different templates added to the same reaction mix [15]. Irrespective of whether the ribozyme is co-transcribed or added post-transcriptionally, the transcription mixture containing the RNase P substrate transcript has to be adjusted to RNase P cleavage assay conditions, since RNase P RNA requires elevated mono- and divalent metal ion concentrations for efficient processing (see Section 2.3.3). Alternatively, substrate transcripts may be concentrated by ethanol precipitation before RNase P RNA processing. As for the *cis*-ribozyme cassette systems, purification of the desired cleavage product by denaturing PAGE is the method of choice.

2.2.3
Further Strategies Toward Homogeneous Ends

In addition to those described above, other approaches for generating homogeneous ends are available as well. Two methods, an RNase H-based strategy and another making use of T7 transcription templates with two consecutive 2′-*O*-methyl nucleotides at the 5′ end of the template strand, are detailed in Chapter 3. The RNase H-based strategy can be employed to generate homogeneous 5′ and 3′ ends, while the 2′-*O*-methyl approach is suited for the production of homogeneous 3′ ends only.

Recently, almost complete 5′ end homogeneity of T7 transcripts was demonstrated with templates directing transcription from the less frequently used T7 class II promoter, at which T7 RNAP initiates synthesis with an A instead of a G residue. Transcription yields from this promoter were reported to equal those of the commonly used T7 class III promoter (see Chapter 1 and [26]).

An elegant variation of *trans*-cleavage concepts, usually relying on ribozymes, involves a 10–23 DNA enzyme [26–28]. This type of DNAzyme has a 15-nt core DNA sequence flanked by two arms that form 8- to 10-bp long hybrid helices with the substrate RNA. Cleavage occurs within a 5′-RY motif (R = A, G; Y = C, U; for details, see [29]) at the junction of the two hybrid helices, resulting in 5′-OH and 2′,3′-cyclic phosphate termini. Sequence-tailored versions of the 10–23 DNAzyme, representing simple DNA oligonucleotides (around 30 nt), can be easily obtained from commercial suppliers and provide an inexpensive and effortless, although less explored alternative to ribozymes in the production of RNAs with homogeneous ends.

2.3
Critical Experimental Steps, Changeable Parameters, Troubleshooting

2.3.1
Construction of *Cis*-cleaving 5′ and 3′ Cassettes

A critical step in the construction of hammerhead and/or HDV ribozyme cassettes is to establish an efficient overlap extension PCR (Fig. 2.4). Usually, the 5′ hammerhead cassette is created by two overlapping oligonucleotides that cover the complete hammerhead domain (Fig. 2.4A). In addition, the upstream primer can carry the sequence for the T7 RNAP promoter at the 5′ end (and/or a terminal restriction enzyme cleavage site if plasmid cloning of the PCR fragment is intended). A minor disadvantage is that such an oligonucleotide will be extended by at least 17 nt, which is associated with lower yields of chemical synthesis. Therefore, a

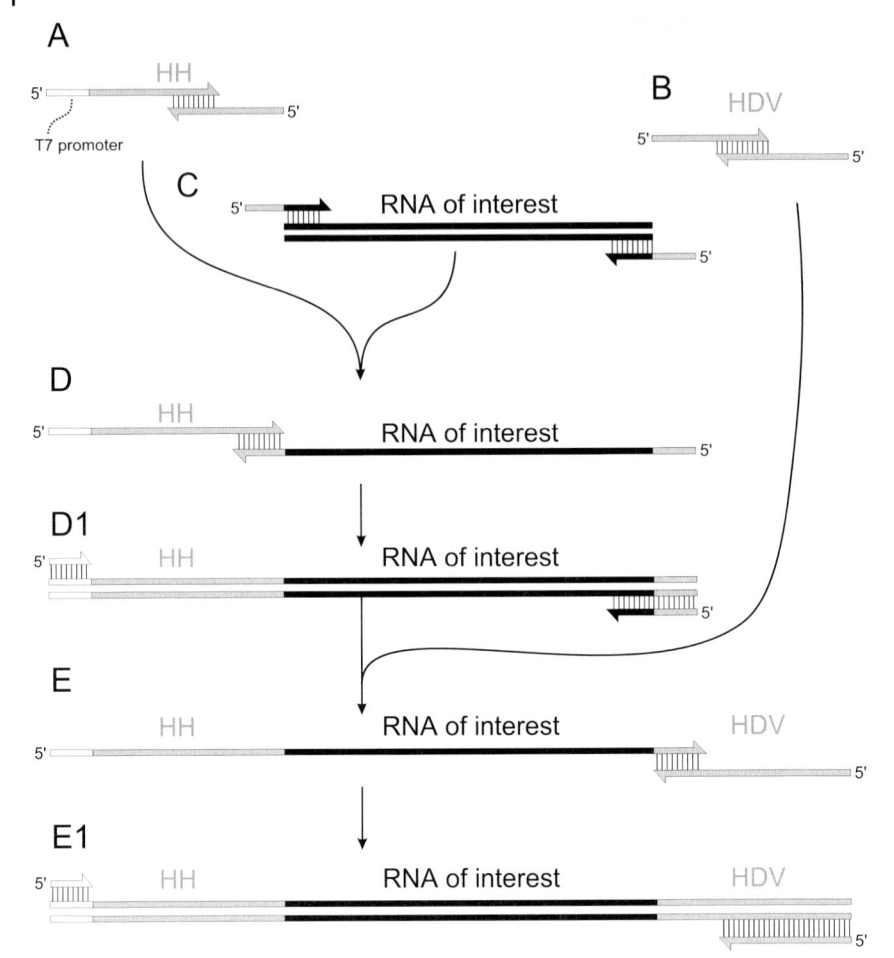

Fig. 2.4. Construction of transcription templates carrying autocatalytic ribozyme cassettes at the 5′ and 3′ termini. (A and B) By PCR extension of overlapping primer pairs, the initial cassettes (HH, hammerhead; HDV, HDV ribozyme) are created. (C) Using primers with 5′ extensions overlapping the ribozyme cassettes, the sequence of interest is amplified in a third PCR reaction. (D) Subsequently, PCR products from reactions A and C are combined (without addition of primers), leading to the fusion of the hammerhead sequence to the sequence encoding the RNA of interest. (D1) The resulting overlap extension product is further amplified using the indicated primers. (E) Eventually, the same strategy is applied to append the HDV sequence. (E1) The final product carrying both cassettes at the corresponding ends is further amplified with terminal primers.

reasonable alternative is to use an upstream primer without the T7 promoter sequence and to insert the PCR product into a cloning vector that encodes the T7 promoter immediately upstream of the cloning site. Otherwise, the T7 promoter can be introduced at a later PCR step.

In a second PCR reaction, the 3′-cassette representing the HDV ribozyme domain is synthesized, again by the use of overlapping primers (Fig. 2.4B). In a third step, the sequence encoding the RNA of interest is amplified (Fig. 2.4C). Here, the upstream primer includes a region overlapping the hammerhead sequence, such that the resulting product can be used for an overlap extension in combination with the hammerhead PCR product (Fig. 2.4D). Likewise, the downstream primer introduces an extension corresponding to the 5′ part of the HDV ribozyme cassette. In a final overlap extension (Fig. 2.4E), the amplified product from Fig. 2.4(D1) is combined with the HDV PCR product in order to generate the full-length construct. Before cloning this product into an appropriate plasmid, it is further amplified using terminal primers (Fig. 2.4E1). After cloning, inserted sequences need to be verified, since the numerous PCR steps involved may lead to aberrant products or sequence deviations.

A complete protocol representing an example of the PCR strategy depicted in Fig. 2.4 is detailed below. The resulting final primary transcript with upstream hammerhead and downstream HDV ribozyme cassettes is illustrated in Fig. 2.2(A). PCR reactions outlined below have been successfully performed with *Taq* DNA polymerase. However, one may consider to use a thermostable DNA polymerase with 3′–5′ proofreading activity, such as *Pfu* polymerase, for the generation of all PCR products that are subsequently used in overlap extension reactions. The reason is that *Taq* polymerase (which has no proofreading activity) tends to add a single non-templated A residue to the 3′ end of PCR products. While this activity is exploited in some cloning strategies (TA-cloning kits), it potentially interferes with overlap extensions: the additional 3′-terminal A does not base pair with the complementary strand. As a consequence, the fraction of strands carrying this extra A residue may not be extended, thus decreasing the overall yield of extension product [30].

PCR protocols
PCR reactions were performed with 2.5 U *Taq* DNA polymerase per 50 µl standard reaction volume in 10 mM Tris–HCl, pH 8.3, 1.5 mM MgCl$_2$, 50 mM KCl. For overlap extensions, the complementary stretches of primer pairs are underlined. The T7 promoter is given in italics, a terminal *Bgl*II site in lowercase letters. Reactions A–E1 correspond to the steps shown in Fig. 2.4. Note that when including terminal restriction enzyme recognition sites in the PCR product for cloning purposes, a few flanking nucleotides beyond the restriction sites have to be added for efficient restriction enzyme cleavage after PCR amplification (for details, see New England Biolabs catalogue "Reference Appendix").

(A) Overlap extension of regions P2, P3 and the 5′ part of P1 of the hammerhead cassette (see Fig. 2.1), including an upstream T7 promoter and a *Bgl*II site
Primer 1 *sense*: 5′-GGa gat ctA *ATA CGA CTC ACT ATA GGG AGA A*AT CCG
  CCT GAT GAG-3′
Primer 2 *antisense*: 5′-GAC GGT ACC GGG TAC CGT TTC GTC CTC ACG GAC
  TCA TCA GGC GGA-3′
PCR profile: 20 cycles: 1 min 94 °C/1 min 40 °C/30 s 72 °C

Resulting sequence: 5'-GGa gat ctA ATA CGA CTC ACT ATA GGG AGA AAT CCG CCT GAT GAG TCC GTG AGG ACG AAA CGG TAC CCG GTA CCG TC-3'; 77 bp

(B) Overlap extension of the HDV ribozyme cassette
Primer 3 *sense*: 5'-GGG TCG GCA TGG CAT CTC CAC CTC CTC GCG GTC CGA CCT GGG CTA-3'
Primer 4 *antisense*: 5'-CTT CTC CCT TAG CCT ACC GAA GTA GCC CAG GTC GGA CCG CGA GGA-3'
PCR profile: 20 cycles: 1 min 94 °C/1 min 60 °C/30 s 72 °C
Resulting sequence: 5'-GGG TCG GCA TGG CAT CTC CAC CTC CTC GCG GTC CGA CCT GGG CTA CTT CGG TAG GCT AAG GGA GAA G-3'; 67 bp

(C) Amplification of the template encoding the RNA of interest (here: yeast tRNA^{Phe}, Fig. 2.2A) using primers overlapping with hammerhead and HDV sequence
Primer 5 *sense* (5' extension into the hammerhead sequence underlined): 5'-GTA CCC GGT ACC GTC GCG GAT TTA GCT CAG-3'
Primer 6 *antisense* (5' extension into the HDV ribozyme sequence underlined): 5'-TGG AGA TGC CAT GCC GAC CCT GCG AAT TCT GTG G-3'
PCR profile: 2 min 94 °C
30 cycles: 1 min 94 °C/1 min 42 °C/30 s 72 °C
Resulting sequence (regions overlapping with hammerhead and HDV ribozyme sequences are underlined): 5'-GTA CCC GGT ACC GTC GCG GAT TTA GCT CAG TTG GGA GAG CGC CAG ACT GAA GAT CTG GAG GTC CTG TGT TCG ATC CAC AGA ATT CGC AGG GTC GGC ATG GCA TCT CCA-3'; 108 bp

(D) Overlap extension of products from A and C
PCR profile: 4 min 94 °C
10 cycles: 1 min 94 °C/2 min 40 °C/45 s 72 °C

(D1) Addition of primers, product amplification
Primer 7 *sense*: 5'-GGa gat ctA ATA CGA CTC ACT ATA GGG-3'
Primer 6 *antisense*: 5'-TGG AGA TGC CAT GCC GAC CCT GCG AAT TCT GTG G-3'
PCR profile: 30 cycles; 1 min 94 °C/2 min 55 °C/45 s 72 °C
Resulting sequence (hammerhead region and overlap with HDV ribozyme cassette underlined): 5'-GGa gat ct A ATA CGA CTC ACT ATA GGG AGA AAT CCG CCT GAT GAG TCC GTG AGG ACG AAA CGG TAC CCG GTA CCG TCG CGG ATT TAG CTC AGT TGG GAG AGC GCC AGA CTG AAG ATC TGG AGG TCC TGT GTT CGA TCC ACA GAA TTC GCA GGG TCG GCA TGG CAT CTC CA-3'; 170 bp

(E) Overlap extension using the product obtained in D1 (carrying a 3' extension into the HDV-coding sequence) and the PCR product for the HDV ribozyme cassette (B)
PCR profile: 4 min 94 °C
10 cycles: 1 min 94 °C/2 min 60 °C/45 s 72 °C

2.3 Critical Experimental Steps, Changeable Parameters, Troubleshooting

(E1) Addition of primers, product amplification

Primer 7 *sense*: 5'-GGa gat ctA ATA CGA CTC ACT ATA GGG-3'
Primer 4 *antisense*: 5'-CTT CTC CCT TAG CCT ACC GAA GTA GCC CAG GTC GGA CCG CGA GGA-3'
PCR profile: 30 cycles; 1 min 94 °C/2 min 60 °C/45 s 72 °C
Resulting sequence (hammerhead and HDV ribozyme regions underlined): 5'-GGa gat ctA ATA CGA CTC ACT ATA <u>GGG AGA AAT CCG CCT GAT GAG TCC GTG AGG ACG AAA CGG TAC CCG GTA CCG TCG CGG ATT</u> TAG CTC AGT TGG GAG AGC GCC AGA CTG AAG ATC TGG AGG TCC TGT GTT CGA TCC ACA GAA TTC GCA <u>GGG TCG GCA TGG CAT CTC CAC CTC CTC GCG GTC GAC CTG GGC TA</u> CTT CGG TAG GCT AAG GAA GAA G-3'; 217 bp

2.3.2
Dephosphorylation Protocols

As discussed above, the activities of both hammerhead and HDV ribozymes lead to the release of RNA molecules that carry 5'-OH and 2',3'-cyclic phosphate groups at their termini. While the 5' ends can be phosphorylated by standard T4 PNK procedures, several efficient and robust protocols can be used in order to remove the 2',3'-cyclic phosphate group [11]. However, the efficiency of these protocols may vary with the RNA substrate to be dephosphorylated. Hence, it is recommended to test different dephosphorylation procedures if one method does not give satisfying results. An easy test for the removal of the terminal 2',3'-cyclic phosphate group is to analyze aliquots of the RNA before and after treatment with T4 PNK by denaturing PAGE: The removal of the phosphate group leads to a reduced net charge of the transcript, which, for small RNAs (less than 100 nt), can be monitored by a lower electrophoretic mobility of the RNA in comparison to the untreated RNA (see Chapter 6).

Dephosphorylation protocol 1
Up to 50 pmol RNA are incubated with 6 U T4 PNK (New England Biolabs) in a final volume of 50 µl for 6 h at 37 °C in the following buffer:
100 mM Tris–HCl, pH 6.5
100 mM magnesium acetate
5 mM β-mercaptoethanol

Dephosphorylation protocol 2
100 pmol RNA are incubated with 1 U T4 PNK (New England Biolabs) in a final volume of 50 µl for 6 h at 37 °C in a buffer containing:
100 mM imidazole–HCl, pH 6.0
10 mM $MgCl_2$
0.1 mM ATP
10 mM β-mercaptoethanol
20 µg/ml BSA (RNase–free)

Dephosphorylation protocol 3
Up to 300 pmol RNA are incubated with 10 U T4 PNK (New England Biolabs) in a final volume of 20 µl for 6 h at 37 °C in the following buffer:
100 mM morpholinoethanesulfonate/NaOH, pH 5.5
300 mM NaCl
10 mM $MgCl_2$
10 mM β-mercaptoethanol

2.3.3
Protocols for RNase P Cleavage

Preincubation of *E. coli* RNase P RNA
The preincubation step described below is not essential when using *E. coli* RNase P RNA, but it can increase the proportion of ribozyme molecules competent for substrate binding [31]. The procedure is the following:

20 pmol of *E. coli* RNase P RNA (produced by *in vitro* transcription) incubated in a volume of 15 µl for 1 h at 37 °C in RNase P cleavage buffer:
50 mM Tris–HCl, pH 7.5
0.1 mM EDTA
100 mM ammonium acetate
100 mM magnesium acetate
5% PEG 6000

Cleavage reaction
10 pmol of an RNA–tRNA primary transcript of the type shown in Fig. 2.3 (in 15 µl RNase P cleavage buffer) is added to the preincubation mixture and incubated for 1.5 h at 37 °C. It is recommended to adjust the ratio of ribozyme:substrate concentration to 2:1 as a compromise between cleavage efficiency and saving of ribozyme material. When using RNase P RNA from *T. thermophilus*, cleavage is usually performed in the same buffer as used for *E. coli* RNase P RNA (see above). However, a typical incubation temperature is 55 °C [31]. Preincubation of *T. thermophilus* RNase P RNA (20 min at 55 °C) in RNase P cleavage buffer is essential for ribozyme activation if the cleavage reaction is to be performed at 37 °C, but can be omitted for cleavage assays at 55–75 °C.

2.3.4
Potential Problems

Although the described systems do not have any known restrictions concerning the sequence of the RNA of interest, one should keep in mind that some primary structures might interfere with the correct folding of the ribozyme cassettes or the linked tRNA molecule. Such misfolding may result in a low reaction efficiency or even in no cleavage at all. As a first approach, a temperature cycling procedure, such as the one described in Section 2.2.1.2, should be attempted. If unsuccessful, it is advisable to change the linked tRNA sequence (in the case of an RNase P substrate) or to switch from *cis*-cleaving ribozyme cassettes to the *trans*-cleaving RNase P strategy or *vice versa*. To avoid such failures from the beginning, we recommend readers scrutinize the structure of the primary transcript by Mfold [32] before experimental work.

References

1 D. E. Draper, S. A. White, J. M. Kean, *Methods Enzymol.* **1988**, *164*, 221–237.
2 J. F. Milligan, O. C. Uhlenbeck, *Methods Enzymol.* **1989**, *180*, 51–62.
3 J. A. Pleiss, M. L. Derrick, O. C. Uhlenbeck, *RNA* **1998**, *4*, 1313–1317.
4 M. Helm, H. Brulé, R. Giegé, C. Florentz, *RNA* **1999**, *5*, 618–621.
5 K. R. Birikh, P. A. Heaton, F. Eckstein, *Eur. J. Biochem.* **1997**, *245*, 1–16.
6 B. Clouet-d'Orval, O. C. Uhlenbeck, *Biochemistry* **1997**, *36*, 9087–9092.
7 T. K. Stage-Zimmermann, O. C. Uhlenbeck, *RNA* **1998**, *4*, 875–889.
8 T. Persson, R. K. Hartmann, F. Eckstein, *ChemBioChem* **2002**, *3*, 1066–1071.
9 M. D. Been, A. T. Perrotta, S. P. Rosenstein, *Biochemistry* **1992**, *31*, 11843–11852.
10 B. M. Chowrira, P. A. Pavco, J. A. McSwiggen, *J. Biol. Chem.* **1994**, *269*, 25856–25864.
11 H. Schürer, K. Lang, J. Schuster, M. Mörl, *Nucleic Acids Res.* **2002**, *30*, e56.
12 D. A. Peattie, *Proc. Natl. Acad. Sci. USA* **1979**, *76*, 1760–1764.
13 V. Cameron, O. C. Uhlenbeck, *Biochemistry* **1977**, *16*, 5120–5126.
14 L. F. Povirk, R. J. Steighner, *Biotechniques* **1990**, *9*, 562.
15 A. R. Ferré-D'Amaré, J. A. Doudna, *Nucleic Acids Res.* **1996**, *24*, 977–978.
16 W. A. Ziehler, D. R. Engelke, *Biotechniques* **1996**, *20*, 622–624.
17 R. A. Collins, *Biochem. Soc. Trans.* **2002**, *30*, 1122–1126.
18 A. Wichlacz, M. Legiewicz, J. Ciesiolka, *Nucleic Acids Res.* **2004**, *32*, E39.
19 B. K. Oh, D. N. Frank, N. R. Pace, *Biochemistry* **1998**, *37*, 7277–7283.
20 S. Busch, L. A. Kirsebom, H. Notbohm, R. K. Hartmann, *J. Mol. Biol.* **2000**, *299*, 941–951.
21 L. A. Kirsebom, *Biochem. Soc. Trans.* **2002**, *30*, 1153–1158.
22 A. Loria, T. Pan, *Biochemistry* **1998**, *37*, 10126–10133.
23 M. Brännvall, B. M. Pettersson, L. A. Kirsebom, *J. Mol. Biol.* **2003**, *325*, 697–709.
24 W. D. Hardt, J. Schlegl, V. A. Erdmann, R. K. Hartmann, *J. Mol. Biol.* **1995**, *247*, 161–172.
25 R. K. Hartmann, V. Erdmann, *Nucleic Acids Res.* **1991**, *19*, 5957–5964.
26 T. M. Coleman, G. Wang, F. Huang, *Nucleic Acids Res.* **2004**, *32*, e14.
27 S. W. Santoro, G. F. Joyce, *Proc. Natl. Acad. Sci. USA* **1997**, *94*, 4262–4266.
28 S. W. Santoro, G. F. Joyce, *Biochemistry* **1998**, *37*, 13330–13342.
29 M. J. Cairns, A. King, L. Q. Sun, *Nucleic Acids Res.* **2003**, *31*, 2883–2889.
30 N. A. Shevchuk, A. V. Bryksin, Y. A. Nusinovich, F. C. Cabello, M. Sutherland, S. Ladisch, *Nucleic Acids Res.* **2004**, *32*, E19.
31 W. D. Hardt, J. Schlegl, V. A. Erdmann, R. K. Hartmann, *Nucleic Acids Res.* **1993**, *21*, 3521–3527.
32 M. Zuker, *Nucleic Acids Res.* **2003**, *31*, 3406–3415.

3
RNA Ligation using T4 DNA Ligase

Mikko J. Frilander and Janne J. Turunen

3.1
Introduction

Efficient RNA ligation methods for generating site-specifically modified long RNA molecules using the T4 DNA ligase were initially described some 10 years ago [1]. More recently, this method has been widely used to provide chimeric RNAs to study various RNA–RNA and RNA–protein interactions in diverse biochemical reconstitution systems as well as in live cells (e.g. in *Xenopus* oocytes). The modifications introduced include simple insertion of a single radioactive group at a specific location of an RNA molecule and more complicated alterations in which different nucleotide analogs, crosslinking groups or RNA backbone modifiers have been inserted into long RNA molecules [2–10].

Currently the efficiency of chemical RNA synthesis is such that high-quality RNA molecules up to 80-nt long can be obtained from commercial sources. Since the yield of chemical synthesis is significantly higher than that of RNA ligation, the latter should be carried out only when chemical synthesis cannot be used to obtain the desired molecules. Such cases include (1) the introduction of radioactive groups or modified nucleotides in the middle of a long RNA molecule, (2) the need for capped RNAs, or (3) requirement for modified RNA molecules that are longer than what can be synthesized chemically. As the repertoire of various phosphoramidite monomers used in chemical RNA synthesis is constantly growing, RNA oligonucleotides containing modified nucleotides are frequently utilized in the construction of a long chimeric RNA molecule. In such cases, the RNA oligonucleotides are ligated to other RNA molecules, which are often produced by *in vitro* transcription reactions. This combination provides a way to modify functional groups in virtually any position in a given RNA molecule.

Despite the seemingly simple overall reaction, i.e. the joining of two or three RNA molecules together with the aid of T4 DNA ligase, the execution of a ligation experiment can be a laborious task if large amounts of the ligated products are needed. Here we will review the current methods for RNA ligation. As several excellent reviews on this subject have already been published [11–13], we will here, in addition to concentrating on the RNA ligation itself, describe special methods to

Handbook of RNA Biochemistry: Student Edition. Edited by R. K. Hartmann, A. Bindereif, A. Schön, E. Westhof
Copyright © 2009 WILEY-VCH Verlag GmbH & Co. KGaA, Weinheim
ISBN: 978-3-527-32534-4

generate high-quality *in vitro* transcripts with homogeneous 5′ or 3′ termini, which are required for large-scale production of chimeric RNAs. Other widely used strategies for the generation of homogeneous 5′ and 3′ ends are detailed in Chapter 2.

3.2
Overview of the RNA Ligation Method using the T4 DNA Ligase

Although T4 DNA ligase is ordinarily used to ligate DNA molecules, it can also catalyze the formation of a phosphodiester bond between two RNA molecules or between RNA and DNA molecules, as its original name "T4 polynucleotide ligase" indicates [14]. The principle of the RNA ligation by T4 DNA ligase is depicted in Fig. 3.1. Typical applications are the so-called two-way (Fig. 3.1A) and three-way

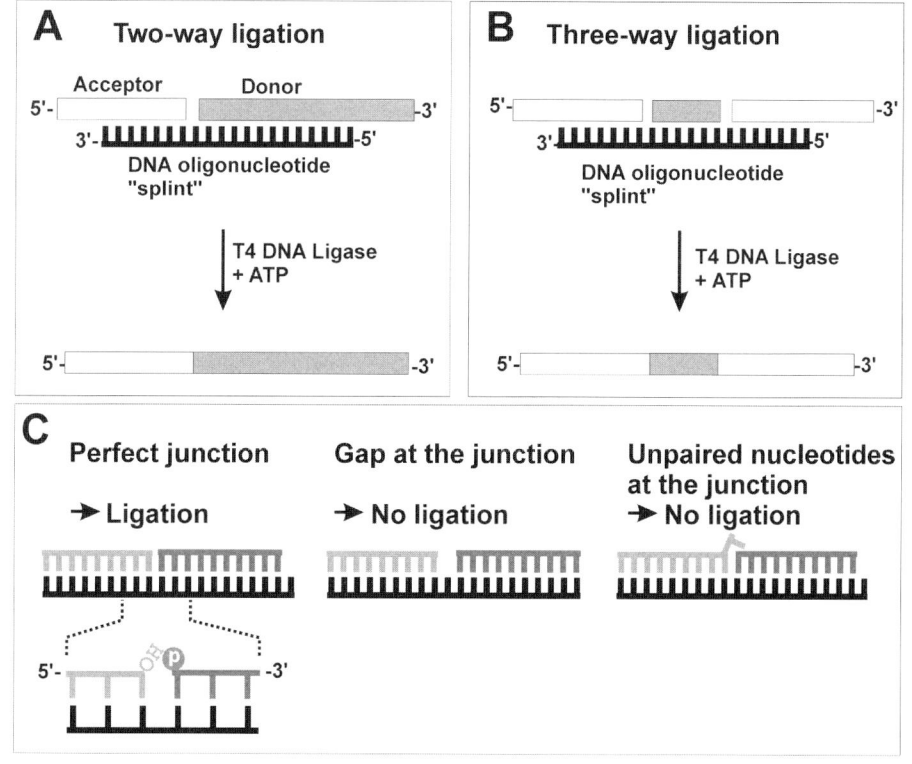

Fig. 3.1. The principle of the RNA ligation with T4 DNA ligase. DNA splint oligonucleotide (black) hybridizes with two (A: two-way ligation) or three (B: three-way ligation) RNA molecules (grey and white) and forms a double-helical structure. (C) Ligation requires 3′-OH and 5′-monophosphate on acceptor and donor molecules, respectively. A gap in the double-stranded helix structure or unpaired nucleotides at the junction (such as $n+1$ products resulting from T7 transcription) will inhibit the ligation.

(Fig. 3.1B) ligations, in which either two or three RNA pieces, respectively, are joined together with the aid of T4 DNA ligase. Normally one of the pieces contains the desired modification(s), while the other(s) are used to reconstitute the full-length RNA molecule under study. The RNA pieces to be ligated are aligned and held together with a complementary bridging DNA oligonucleotide, also known as "DNA splint" or "cDNA template". The ligase catalyses phosphodiester bond formation between the 5'-phosphate of the donor (3' substrate RNA) and the 3'-hydroxyl of the acceptor (5' substrate RNA). Therefore, RNAs containing, for example, a 3'-phosphate or a 5'-triphosphate are not ligated. Furthermore, the RNA/DNA double helix formed by the two RNA pieces and the DNA splint has to be consecutive without any bulges or gaps, especially at the point of the junction of the two RNA molecules.

Although the RNA molecules can also be joined with T4 RNA ligase (see Chapter 4), the use of T4 DNA ligase has several advantages in the construction of long RNA molecules. The main advantage is that, as a consequence of the strict requirement for uninterrupted/unbulged double-helical structure by the T4 DNA ligase, the ligation reaction takes place only for such RNA molecules that have correct termini at the point of the junction (Fig. 3.1C). Furthermore, the use of T4 DNA ligase does not lead to an unwanted formation of circular RNA structures that can be a problem when using T4 RNA ligase with RNA molecules containing unprotected 3' or 5' ends. Finally, as compared to the T4 RNA ligase, the T4 DNA ligase has a lower K_m for polynucleotides, resulting in higher ligation efficiency at lower RNA concentrations, and it does not display any sequence specificity on either donor or acceptor molecules.

A simplified flowchart for a three-way ligation experiment is presented in Fig. 3.2 to illustrate the typical steps needed for the preparation of individual RNA pieces for the RNA ligation. In this example, the final product has been capped with GpppG and the modified nucleotides are located in the middle of the molecule. The modified nucleotides introduced into the RNA molecule are incorporated in the central piece, which is chemically synthesized. The 5' and 3' pieces are produced by *in vitro* transcription using T7 RNA polymerase. In this scheme the preparation of the 5' piece is relatively simple, while the 3' piece requires further processing to generate a 5' monophosphate and/or correct 5'-terminal sequence (dephosphorylation followed by phosphorylation to generate 5'-monophosphate termini or, alternatively, site-specific cleavage with RNase H). As the quality of the chemically synthesized pieces is often relatively high and does not necessarily require extensive processing, we will concentrate in the following sections on the production of high-quality transcripts by T7 RNA polymerase.

3.3
Large-scale Transcription and Purification of RNAs

When preparing RNA for the ligation it is necessary to start with a relatively large initial amount of RNA to account for the unavoidable losses at the various stages of

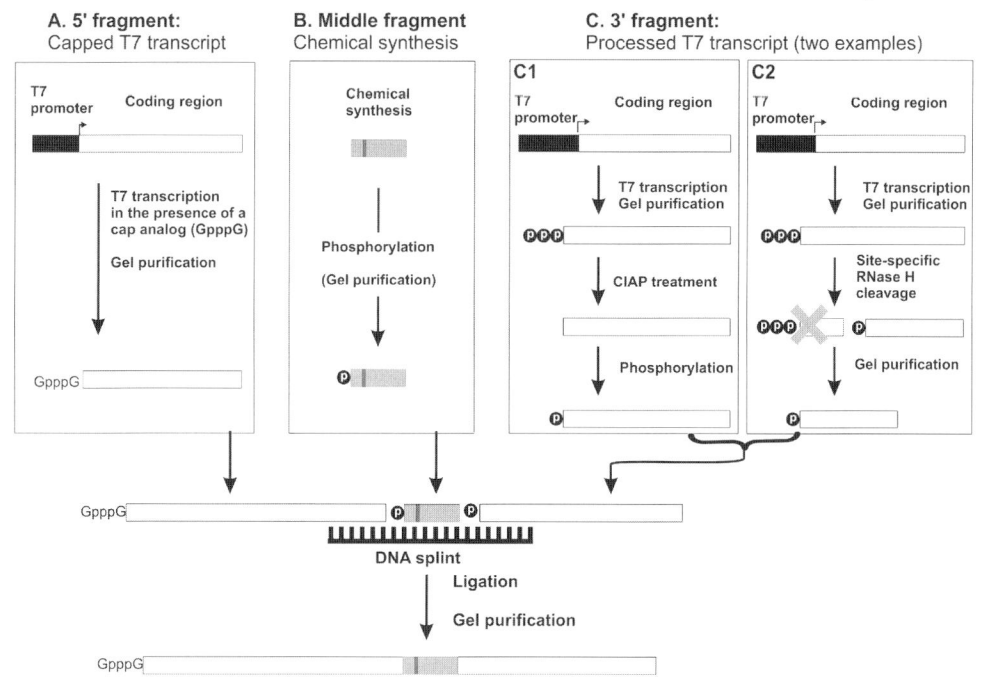

Fig. 3.2. An example of steps needed for the production of RNA pieces for a three-way ligation with T4 DNA ligase. (A) 5' Fragment: capped RNA is produced by transcription with T7 RNA polymerase. (B) Middle fragment: RNA oligonucleotide containing a modified nucleotide (black stripe) is produced by chemical synthesis. (C) 3' Fragment: two alternative examples are presented. (C1) The 3' fragments is produced by T7 transcription, followed by dephosphorylation (to remove the 5' triphosphate) and phosphorylation (to add a single phosphate to the 5' terminus). (C2) The 3' fragment is initially produced as a longer precursor, which is subsequently cleaved and gel purified.

the procedure (substrate manipulations, RNA ligation itself and the subsequent purification steps). A practical rule of thumb used in our laboratory is to start with an at least 10-fold excess of each individual RNA piece compared to the amount of the ligated product needed in the final experiments. As described in Chapter 1, transcription by T7 phage polymerase can be used to generate large quantities of RNAs from defined DNA templates containing phage-specific promoters. The templates can be linearized plasmids or PCR products or even annealed oligonucleotides [15]. In ligation experiments, the use of PCR products is preferred because they provide an easy way to specify nucleotides at the point of junction of the two RNAs to be ligated. Here we describe conditions for generating large amounts of RNA (several nanomoles) from a single transcription reaction (Protocol 1). This reaction is suitable for RNA molecules that fit the promoter consensus of the phage polymerase, where at least the first transcribed nucleotide or, if possible, both the first and the second nucleotides should be G residues. If other than a G residue is required as the initial nucleotide, the transcript can be initially produced as a longer precursor which is subsequently cleaved at a specific site with a ribozyme (Chapter 2) or RNase H (see below). Alternatively, the transcription can be primed with a suitable dinucleotide ([1, 16, 17] and Chapter 1).

We have successfully used the buffer and reaction conditions indicated in Protocol 1 with T7 RNA polymerase, with yields up to 4–5 nmol of gel-purified RNA from a 200-µl reaction. The reaction conditions described in Protocol 1 are not recommended for the production of capped RNAs as the high concentrations of divalent cations and spermidine tend to precipitate the cap analogs. Instead, modified conditions described in Protocol 2 should be used for the transcription of capped RNAs.

Following the transcription reaction, the DNA template is degraded with DNase to ensure that the contaminating template DNA (which will have almost the same mobility in the gel as the transcript itself) will not interfere with the further steps of the ligation procedure. Subsequently, the full-length transcript RNA will be purified from prematurely terminated products by denaturing gel electrophoresis. Following electrophoresis, the bands are visualized using the "UV shadow" technique, excised, and eluted by passive diffusion (see Protocol 3).

3.4
Generating Homogeneous Acceptor 3′ Ends for Ligation

PCR-based template generation followed by transcription with phage polymerases is a simple and efficient method for generating large amounts of RNA fragments for ligation purposes. However, a problem with the phage polymerases is that they can add non-templated nucleotides to the 3′ end of the synthesized RNA molecule. In the worst cases more than 50% of the synthesized RNA molecules can contain these so-called $n+1$ and $n+2$ nucleotides [18, 19]. As the T4 DNA ligase requires an absolute match between the DNA splint and the two RNA molecules to be ligated, the non-template addition of extra nucleotides can significantly reduce the efficiency of the RNA ligation. This is not necessarily a severe problem in small-scale or initial screening experiments, in which limited amounts of the ligated products are often sufficient. However, if large amounts of the ligated products are needed, the inefficient ligation may become a major limitation. In such cases methods producing RNAs with specific 3′ terminus can often lead to a several-fold increase in the quantity of the ligated products.

A simple method for reducing non-templated nucleotide addition has been described recently [20]. In this method the very 5′ end of the downstream PCR primer used in the synthesis of the template DNA for T7 transcription is modified: instead of standard deoxyribonucleotides, it contains two 2′-O-methyl RNA residues (Fig. 3.3A). During the PCR reaction the primers are incorporated into the synthesized DNA fragments which subsequently serve as templates in transcription by T7 RNA polymerase. During transcription the modified nucleotides at the 5′ end of the template strand will lead to a significant reduction of the

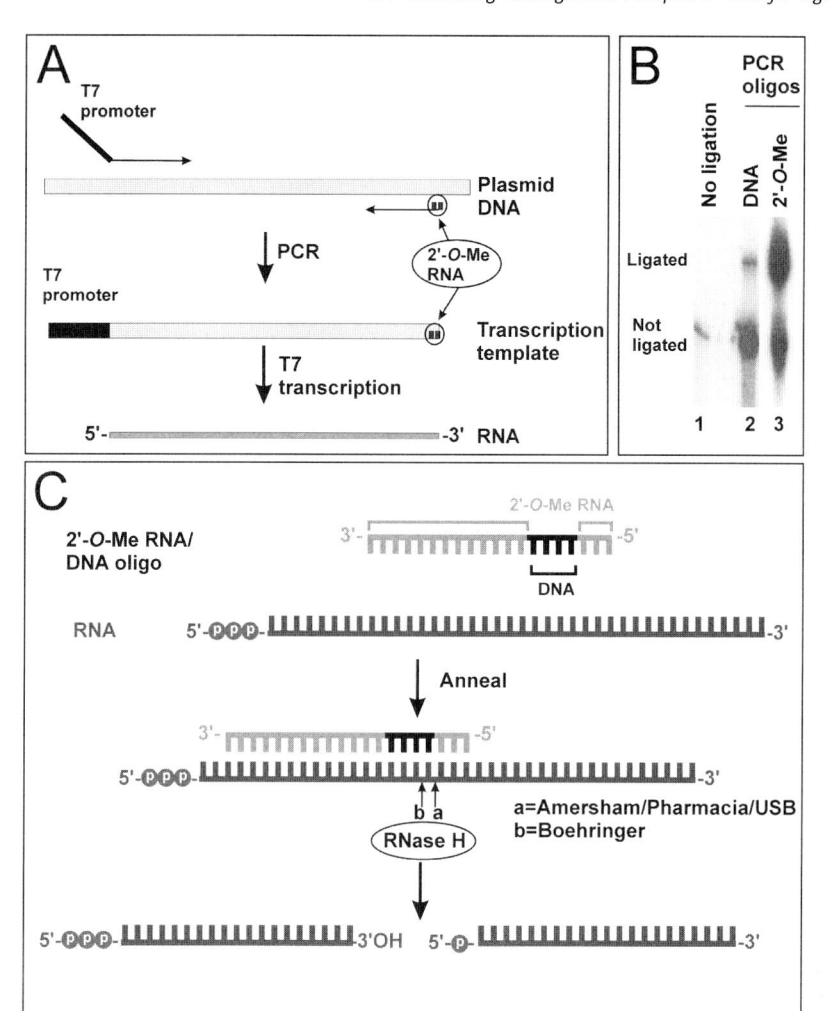

Fig. 3.3. (A) Production of T7 transcripts with homogeneous 3′ ends. The template for T7 transcription is produced by PCR. The upstream primer contains a promoter for T7 RNA polymerase, while the downstream primer contains two 2′-O-methyl RNA residues at the 5′ end of the oligonucletotide. Both primers are incorporated into the PCR product and during transcription with T7 RNA polymerase, the 2′-O-methyl RNA residues prevent the addition of non-template nucleotides to the 3′ end of the RNA molecules. (B) Comparison of the ligation efficiencies when a standard DNA oligonucleotide and a hybrid DNA/2′-O-Methyl RNA oligonucleotide (depicted in panel A) were used as the downstream primer in the PCR reaction to produce a template for T7 transcription. Lane 1: control lane containing a ^{32}P-labeled donor molecule; lane 2: ligation with an RNA fragment derived from transcription using an all-DNA template; lane 3: ligation with an RNA fragment derived from transcription using a template with two 5′ terminal 2′-O-methyl modifications. In the ligation reactions (lanes 2 and 3) the donor RNA fragments were radioactively labeled, while the acceptor RNAs were unlabeled. (C) The principle of site-specific RNase H cleavage. The cleavage sites of Amersham/Pharmacia/USB (a) or Boehringer (b) RNase H [24] are indicated.

non-templated addition of extra nucleotides to the 3' end of transcripts [20]. We have used this method successfully, and have been able to raise the ligation efficiency from 20 to 70% with no other alterations in the procedure (Fig. 3.3B). The drawback of the hybrid DNA/2'-O-methyl RNA oligonucleotides is that they are more expensive and are readily available only from few commercial sources. We have purchased our oligonucleotides either from Dharmacon Research (www.dharmacon.com) or from Keck oligonucleotide synthesis facility at Yale University (info.med.yale.edu/wmkeck/oligos.htm).

Another way of creating homogenous 3' ends is to synthesize the RNA as a longer precursor and cut it into the desired length using site-directed cleavage with RNase H or with a specific ribozyme. The ribozyme cleavage has been described in Chapter 2, while the site-specific RNase H cleavage, which can be used to trim both the 3' and 5' end of the RNA molecule, will be described in detail in the next paragraph.

3.5
Site-directed Cleavage with RNase H

At times it may not be possible to transcribe the desired RNA directly, e.g. when the first nucleotide is not a guanosine. One possible solution for this is the site-specific cleavage of a longer precursor RNA molecule using RNase H and chimeric 2'-O-methyl RNA/DNA oligonucleotides. This method may also be used to solve problems with 5' or 3' end heterogeneity.

RNase H recognizes and binds nucleic acids that are duplexes of DNA and RNA and cleaves the backbone of the RNA strand leaving a 5'-phosphate and a 3'-hydroxyl [21]. The site of cleavage may be specified when using oligonucleotides containing a short DNA stretch (3 or 4 nt) which is flanked by 2'-O-methyl-RNA sequences [22, 23]. We have successfully used hybrid 20mer oligonucleotides which contain three 2'-O-methyl RNA residues at the 5' end, followed by four DNA residues and thirteen 2'-O-methyl residues (see Fig. 3.3C). An important observation to be noted [24] is that the exact position of cleavage is, for unknown reasons, dependent on the commercial source of the RNase H. The enzymes supplied by Pharmacia, Sigma and Takarashuzo were reported to cleave the phosphodiester bond which is located 3' to the ribonucleotide base-paired with the 5'-most deoxyribonucleotide of the oligonucleotide, whereas the enzyme from Boehringer Mannheim cleaved the bond located one nucleotide upstream (5' direction) in the RNA molecule (Fig. 3.3C). After the report was made, however, there have been mergers of the aforementioned companies, and it is unclear which enzyme sources and purification protocols are now used by the merged companies. In our studies we have used RNase H supplied by Amersham and found that it functions as the one supplied earlier by Pharmacia. If other enzyme sources are used it is advisable to map the exact cleavage site by primer extension analysis.

Protocol 4 describes a general cleavage strategy using RNase H and hybrid DNA/

2′-O-methyl RNA oligonucleotides. In the first step the RNA and the hybrid oligonucleotides are allowed to anneal, after which the appropriate buffers and enzyme are added. For efficient annealing, the RNA and oligonucleotides should be initially denatured completely by heating them to 95 °C and then, by lowering the temperature slowly, allowed to anneal. Addition of a monovalent salt, such as KCl, further enhances the annealing, but the concentration in the final cleavage reaction should not exceed 50 mM. Small amounts of EDTA are included to chelate any traces of divalent cations, e.g. Mg^{2+}, to reduce chemical degradation of the RNAs at high temperatures. The amounts of RNA and oligonucleotides can be adjusted to suit the particular experiment, but the reaction should always contain close to equimolar amounts of the oligonucleotide relative to the RNA to be cleaved. Large excess of the hybrid oligonucleotide may lead to aberrant cleavage at additional sites. We typically use 10–15% excess of the oligonucleotide to ensure efficient annealing. If the 5′ end of the fragment is to be dephosphorylated for a subsequent labeling with a radioactive phosphate, the final yield of the cleaved RNA product may be increased by carrying out the dephosphorylation prior to the gel purification.

3.6
Dephosphorylation and Phosphorylation of RNAs

The 5′-triphosphate resulting from the transcription reaction (see Fig. 3.2) must be converted to 5′-monophosphate if the RNA is to be used as a donor RNA in the ligation reactions. This is achieved by first dephosphorylating the RNA and then phosphorylating the resulting 5′-hydroxyl with unlabeled or radioactively labeled phosphate. The dephosphorylation catalyzed by the calf intestinal alkaline phosphatase (CIAP) is carried out at 50 °C (see Protocol 5). The elevated temperature is used to reduce the effect of RNA secondary structure on dephosphorylation. Following the dephosphorylation it is necessary to completely remove the CIAP, as it could seriously inhibit the further downstream steps. As very large amounts of CIAP are used, we typically remove it by proteinase K digestion. It is also possible to use other phosphatases [such as shrimp alkaline phosphatase (SAP)] which are easier o inactivate compared to CIAP. However, our experience with SAP is that at least with some substrates it does not work as efficiently as CIAP.

As mentioned previously, the donor RNA must have a 5′-phosphate for successful ligation. Dephosphorylated transcripts or chemically synthesized oligonucleotides can be phosphorylated by T4 polynucleotide kinase (PNK) in the presence of ATP (note that RNA oligonucleotides can also be phosphorylated during the chemical synthesis). T4 PNK can also be used in site-specific labeling to insert a single radioactive phosphorus at the junction between the acceptor and donor. Protocol 6 describes phosphorylation with [γ-^{32}P]ATP resulting in a specific activity of approximately 0.5×10^6 c.p.m./pmol. The protocol can also be used for non-radioactive phosphorylation to create the 5′ phosphate required for ligation, in which case 500 μM unlabeled ATP should be used.

3.7
RNA Ligation

Following successful production of the individual RNA pieces they will finally be joined by RNA ligation. The ligation protocol, much like the RNase H cleavage discussed previously, is divided into two parts. First, the RNA fragments are aligned together with a bridging oligonucleotide, also known as the "DNA splint" or "cDNA template". Second, the reaction mix is added and the RNA ends at the junction are joined by the T4 DNA ligase in the presence of ATP. The major consideration when planning both steps is to know the molar concentrations of each individual RNA piece and the DNA splint oligonucleotide. Furthermore, the ligation volume should be kept as small as possible, as the ligation proceeds more efficiently when the reactants are concentrated. The final volume of the reaction should therefore be kept at approximately 10–20 µl. Finally, to help the detection of the ligated products in purification gels and their subsequent quantification one should always include trace amounts of radioactively labeled RNAs in the ligation reaction. This should be done even when the aim is to produce unlabeled chimeric RNAs.

Let us consider first the annealing of the RNA fragments and the splint. Splints spanning the sequence for 20 nt on both sides of the junction align the substrates efficiently, although splints down to about 20 nt (10 on each side) can be used [11]. For efficient ligation, the three polynucleotides should be present in equal molar amounts. If one of the RNA fragments is scarce, the splint and the other fragment may be added in excess to drive the reaction. Adding donor RNA in excess can also be used to decrease the negative effect of the 3′ end heterogeneity of the acceptor RNA. However, it is important that the concentration of at least one of the RNA fragments is greater than that of the DNA splint. If large amounts of the splint are used, the individual RNA fragments may hybridize to different splint molecules and thus be sequestered from their ligation partners.

Some monovalent salt and EDTA may be added to the annealing reaction, for the same reasons as previously described for annealing prior to RNase H cleavage (see Section 3.5). Performing the annealing in a thermal cycler with a heated lid has also the added advantage that little condensation of water occurs on the lid of the tube, which is an important consideration when small volumes, which are susceptible to drying, are used.

The buffer used for the ligation reaction may be made by the researcher, as the one described in Protocol 7, or a commercial one supplied with the enzyme can be used. Macromolecular crowding agents, such as polyvinyl alcohol (PVA), polyethylene glycol (PEG) or polyvinyl pyrrolidine (PVP) should be added to increase the effective concentration of the reactants. Some thought should be given to the amount of ligase used, as it has been reported that the T4 DNA ligase does not turn over efficiently on RNA-containing duplexes [11]. Therefore, a stoichiometric amount of T4 DNA ligase should be used, with 1 Weiss unit corresponding approximately to 1 pmol.

Finally, the incubation time and temperature should be considered. Incubating the reaction at 30 °C for 4 h is a widely used approach. In our research we have also performed the incubation overnight at room temperature (around 20–25 °C) and this seems to result in somewhat higher yields, probably due to both increased incubation time and the reduced temperature which can stabilize the double-stranded structures at the junction.

3.8 Troubleshooting

A typical problem with ligations is a low yield of the final ligated product. With two-way ligation the efficiency should be at least 20%, but even nearly stoichiometric ligations are possible with high-quality RNA. With three-way ligations the efficiencies can sometimes be less than 10%, but one should be able to increase the efficiency to approximately 30% relatively easily. If the ligation yield is very low, the first thing to do is to determine which one of the RNA or DNA fragments is responsible for the low efficiency. This can be done easily by setting up small-scale test ligations which contain only about 1 pmol of each fragment. Short DNA oligonucleotides can also be used as acceptors or donors during troubleshooting instead of the actual RNA fragments.

If the problem with the ligation can be pinpointed to the acceptor RNA the most obvious question would be the quality of the 3′ end of the RNA molecule: is it homogeneous or does it contain non-templated nucleotides? Non-templated nucleotide additions can often be avoided using the techniques described in Sections 3.4 and 3.5. Another possible problem with the acceptor is the presence of a stable RNA secondary structure that could prevent the hybridization of the DNA splint oligonucleotide. The best way to resolve this is to destabilize the structure with site-specific mutations. The mutations near the junction are often easy to incorporate by means of the PCR oligonucleotides that are used in the generation of templates for T7 transcription. If the strategy does not allow mutations, an alternative is to use an additional "disrupter" oligonucleotide which binds adjacent to the splint oligonucleotide and prevents the formation of stable RNA secondary structures.

A typical problem with the donor RNA, in addition to the stable RNA secondary structure described above, is inefficient dephosphorylation (or phosphorylation) at the 5′ end of the RNA. This can be caused for example by a stable secondary structure at the 5′ end of the molecule. A larger amount of phosphatase (or kinase) may be used to overcome this problem or, alternatively, an analogous disrupter oligonucleotide strategy could be designed.

Apart from the problems with ligations, one can also experience difficulties at stages that are further downstream of the actual RNA ligation reaction, but which are the results of the ligation procedure. One, at least theoretical, possibility is the cleavage of the ligated RNA during incubation with cellular extract due to endogenous RNase H activity and residual amounts of the DNA splint oligonucleotide in the ligated RNA sample. We have not observed any RNase H activity resulting from a contaminating DNA splint in any of our studies. However, if this is of concern, it can be avoided by treating the ligation reactions with DNase before purification.

3.9
Protocols

Protocol 1: Transcription
5 × T7 transcription buffer: 600 mM HEPES–KOH, pH 7.5; 120 mM $MgCl_2$; 100 mM DTT; 5 mM spermidine.

T7 transcription 200 µl		Final concentration
40 µl	5 × T7 transcription buffer	1×
40 µl	25 mM each rNTP	5 mM
5 µl	40 U/µl RNase inhibitor (Promega)	1 U/µl
2–10 µl	100 U/µl T7 RNA polymerase	1–5 U/µl
40 µl	50–250 ng/µl PCR-product	10–50 ng/µl
65–73 µl	RNase-free water	

(1) Combine reaction components, add the DNA last to avoid precipitation by high concentrations of spermidine present in the transcription buffer.
(2) Incubate at 37 °C for 1.5 h, afterwards add more enzyme (0.5–1 × the original amount) and incubate for an additional 1.5 h. Typically a white pyrophosphate precipitate will start to accumulate at the later stages of the transcription reaction.
(3) Add 1 U of RNase-free DNase, such as RQ1 DNase (Promega) for each microgram of template DNA and continue the incubation for an additional 15–30 min.
(4) Extract RNA once by phenol:chloroform:isoamyl alcohol (25:24:1).
(5) Precipitate RNA by adding 20 µl 3 M NaOAc (pH 5.2) or NaCl and 2.5 volumes of ethanol.
(6) Dissolve the pellet in 5 µl water and gel-purify the RNA by denaturing polyacrylamide gel electrophoresis as described in Protocol 3.

Protocol 2: Transcription of capped RNAs
5 × transcription buffer for capped RNA transcription (Promega): 200 mM Tris–HCl, pH 7.9; 30 mM $MgCl_2$; 10 mM spermidine; 50 mM NaCl.

T7 transcription 200 µl		Final concentration
40 µl	5 × transcription buffer	1×
2 µl	100 mM ATP	1 mM
2 µl	100 mM CTP	1 mM
2 µl	100 mM UTP	1 mM
1 µl	100 mM GTP	0.5 mM
40 µl	10 mM G(5′)ppp(5′)G	2 mM[1]

40 µl	50–250 ng/µl PCR product	10–50 ng/µl
5 µl	40 U/µl RNase inhibitor (Promega)	1 U/µl
2–10 µl	100 U/µl T7 RNA polymerase	1–5 U/µl
58–66 µl	RNase-free water	

[1] Cap analogs should be at least in 4-fold molar excess relative to GTP to ensure efficient initiation with the cap analog. Similarly, 5′-hydroxyl- or 5′-monophosphate-containing RNAs can be produced by priming the transcription reaction with guanosine or GMP, respectively, under reaction conditions comparable to those used with the cap analog (for details, see Chapter 1).

Carry out the reaction as described in Protocol 1 and gel-purify according to Protocol 3.

Protocol 3: Purification of RNA by denaturing gel electrophoresis

Gel loading buffer: 0.01% bromophenol blue; 0.006% xylene cyanol in 7.5 M urea/ 1 × TBE.

RNA elution buffer: 50 mM Tris–HCl, pH 7.5; 10 mM EDTA; 0.1% SDS; 0.3 M NaCl.

Before electrophoresis, the reaction mixture should be extracted once with an equal volume of phenol:chloroform:isoamyl alcohol (25:24:1) followed by ethanol precipitation, as proteins in the sample may cause smearing of the bands and retain some of the RNA in the wells. With large amounts (several nanomoles) of RNA, the pellet should be dissolved in a small volume of water (5–10 µl) before applying the gel loading buffer in order to minimize loss.

Prepare the polyacrylamide gel (19:1 acrylamide:N,N'-methylene bisacrylamide) in 7.5 M urea/1 × TBE. The percentage of acrylamide should be adjusted to the size of RNA fragment to be purified. The thickness of the gel should be adjusted according to the amount of RNA to be loaded. As the transcription reactions usually contain several nanomoles of RNA, a relatively thick gel (approximately 1 mm) should be used to avoid smearing of the bands. Additionally, RNA samples should be distributed between several wells, even though this may reduce the RNA yields after the elution. In later steps, when dealing with smaller amounts of RNA, a thinner gel (0.3–0.5 mm) may be used to increase the elution yields.

(1) Pre-run the denaturing gel at least 20–30 min at 60 W.
(2) Dissolve the RNA pellet in a small amount of water (5–10 µl) and 1 volume of gel loading buffer. The final urea concentration should be at least 3.5 M but loading buffers containing up to 7.5 M urea can be used.
(3) Heat the RNA samples at 95 °C for 3–5 min. Immediately put the samples on ice. Centrifuge the samples if there is water condensation on the lid. This is especially important if a high concentration urea loading buffer is used, as otherwise the loading buffer (and the sample) can crystallize in the pipette tip during loading.
(4) Load the samples and run the gel at approximately 60 W for at least 30–60 min, or longer if necessary.

(5) Separate the gel plates. Place the gel between two Saran wrap (or alike) sheets.
(6) Visualize the RNA bands using UV shadowing. The gel (between two sheets of Saran wrap) is placed on a fluorescent TLC plate or intensifying screen (a sheet of white paper will do as well) and illuminated briefly with UV light (254 nm). RNA bands are visualized as dark bands on a fluorescent background. Use a pen or marker to indicate the location of each band. The exposure time should be minimized to avoid damage to RNA by the UV light.
(7) Cut out the bands and add approximately 5 volumes of elution buffer to the excised bands. Carry out the elution overnight at room temperature using a tube rotator, "rocking table" or a similar device to provide for gentle shaking during elution.
(8) Collect the supernatant. For increased yield, replace the buffer and continue elution for a further 4–6 h. Extract the eluates once with phenol:chloroform: isoamyl alcohol and once with chloroform:isoamyl alcohol. This helps to remove any impurities present in the gel, as well as any remaining gel fragments. The RNA is concentrated by ethanol precipitation. Note that there is no need to add salt as the elution buffer already contains a sufficient amount of salt for the precipitation. Wash the pellet at least 3 times with 70% ethanol to remove any traces of SDS and dissolve it in RNase-free water. After purification the concentration of the transcript should be measured by UV spectrometry (see Appendix).
(9) If purifying radiolabeled RNAs, the bands can be visualized by autoradiography instead of UV shadowing. In this case the gel should be left on one of the glass plates and covered with Saran wrap. Pieces of fluorescent tape (such as Rad-Tape from Diversified Biotech) serving as alignment marks are attached to the wrapped gel and illuminated briefly with light. In the darkroom, place an X-ray film on top of the gel. Expose the film (depending on the activity of the RNA to be purified, this can be anything from 10 s to several hours). Use the markings from the fluorescent tape to align the gel and the film, mark the bands, cut them out and elute as described in Step 7.

Protocol 4: Site-directed cleavage with RNase H
5 × RNase H buffer (Amersham): 100 mM Tris–HCl, pH 7.5; 100 mM KCl; 50 mM $MgCl_2$; 0.5 mM EDTA; 0.5 mM DTT.

Annealing reaction 13 µl		Final concentration
5 µl	400 µM RNA (2 nmol)	154 µM
5.5 µl	400 µM 2′-O-methyl RNA/DNA oligo (2.2 nmol)	169 µM
1.3 µl	1 M KCl	100 mM
1.2 µl	1 mM EDTA	92 µM

In a thermal cycler, run the following program: 95 °C 5 min, 85 °C 10 s, decrease the temperature with a slope of −0.1 °C/s until the temperature of 35 °C is reached. Alternatively, the annealing may be carried out in a heating block. In this case place the samples first in a hot heating block (95 °C) for 3–5 min, then remove the block from the heating unit and allow to cool slowly to room temperature. In each case, the tubes should be checked for any condensed water on the lid after the annealing and centrifuged if necessary.

Cleavage reaction 50 μl		Final concentration
13 μl	annealing mix	
10 μl	5 × RNase H buffer	1×
1.5 μl	40 U/μl RNase inhibitor (Promega)	~1 U/μl
0.5 μl	100 mM DTT	1 mM
15 μl	5 U/μl RNase H (Amersham)	1.5 U/μl
10 μl	RNase-free water	

(1) Incubate at 37 °C for 3–4 h, extract once with phenol:choloroform:isoamyl alcohol and ethanol-precipitate. If dephosphorylation is to be performed before electrophoresis, carry out chloroform:isoamyl alcohol extraction after the phenol extraction to remove any traces of the phenol.
(2) Separate the cleavage products in a denaturing polyacrylamide gel and visualize the bands by UV shadowing. Use approximately 200 pmol of uncut RNA as a control to distinguish the full-length RNA. Elute and purify as described in Protocol 3.

Protocol 5: Dephosphorylation
10 × CIAP buffer (Finnzymes): 100 mM Tris–HCl, pH 7.9; 100 mM $MgCl_2$; 10 mM DTT; 500 mM NaCl.

Dephosphorylation reaction 100 μl		Final concentration
10 μl	10 × CIAP buffer	1×
10 μl	200 μM RNA (2 nmol)	20 μM
2.5 μl	40 U/μl RNase inhibitor (Promega)	1 U/μl
10 μl	10 U/μl CIAP (Finnzymes)	1 U/μl
68 μl	RNase-free water	

(1) Incubate at 50 °C for 1 h.
(2) Add at least 1 volume of an appropriate 1 × buffer for proteinase K (the RNA elution buffer described earlier also works very well) and 100 μg of proteinase K to the reaction and continue incubation at 50 °C for an additional hour.
(3) Extract RNA with phenol:chloroform:isoamyl alcohol followed by extraction with chloroform: isoamyl alcohol, and ethanol precipitation. If the RNA fragment has been cleaved with RNase H prior to the dephosphorylation and has not been purified by gel electrophoresis, this should be performed at this stage.

Protocol 6: Site-specific labeling with radioactive phosphorus at the donor 5′ end
10 × T4 polynucleotide kinase buffer: 700 mM Tris–HCl, pH 7.5; 100 mM $MgCl_2$; 50 mM DTT.

Phosphorylation reaction 40 μl		Final concentration
4 μl	10 × T4 polynucleotide kinase buffer	1×
5 μl	20 μM RNA (100 pmol)	2.5 μM
6 μl	200 μM ATP[1]	30 μM
1 μl	40 U/μl RNase inhibitor (Promega)	1 U/μl
20 μl	[γ-^{32}P]ATP (10 μCi/μl, 6000 Ci/mmol)[1]	1.7 μM
1 μl	10 U/μl T4 polynucleotide kinase	0.25 U/μl
3 μl	RNase-free water	

[1] The ratio of radioactively labeled versus cold ATP can be adjusted depending on the particular experiments. However, the total concentration of ATP in the reaction should be kept at least 2× higher than the concentration of the RNA. If large amounts of highly radioactive RNA are needed, more crude, but concentrated [γ-^{32}P]ATP can be used [such as NEG 035C (New England Nuclear) or PB15068 (Amersham) – both having activities of about 150 μCi/μl]. If RNA is not to be radioactively labeled, the ^{32}P-labeled ATP can be replaced with 500 μM unlabeled ATP.

(1) Incubate at 37 °C for 30–60 min. Subsequently raise the concentration of cold ATP to 45 μM and continue incubation for 15 min.
(2) Extract once with phenol:chloroform:isoamyl alcohol and once with chloroform, and ethanol-precipitate.
(3) Dissolve the pellet in a few microliters of water. Alternatively, the dry pellet can be directly used in the ligation if dividing the RNA donor into smaller aliquots is not needed.

Protocol 7: Ligation
10 × ligation buffer: 500 mM Tris–HCl, pH 7.5; 100 mM MgCl$_2$; 200 mM DTT; 10 mM ATP.

Annealing mix for a two-way ligation, 5 μl[1]		Final concentration
1.2 μl	50 μM donor RNA (50 pmol)	10 μM[2]
1.4 μl	50 μM acceptor RNA (70 pmol)	14 μM[2]
1.2 μl	50 μM splint DNA oligo (60 pmol)	12 μM
0.5 μl	1 mM EDTA	0.1 mM
0.5 μl	500 mM KCl	50 mM

[1] Even if the aim is to produce unlabeled product it is a good idea to include trace amounts of a radioactively labeled piece to the ligation reaction corresponding to one of the RNA fragments. This provides an easy way to quantify the total yield of the ligation reaction. Simply compare the amount of radioactivity included in the reaction with the amount of radioactivity after the gel purification (by measuring small samples with liquid scintillation counting) and multiply this percentage with the total amount of this particular RNA fragment in the reaction.
[2] We have successfully used RNA concentrations ranging from 100 nM to 20 μM for each individual RNA segment in the ligation reactions.

(1) In a thermal cycler, run the following program: 95 °C 5 min, 85 °C 10 s, a slope of −0.1 °C/s until the temperature of 35 °C is reached.
(2) Check for condensation at the lid and centrifuge the tube(s) if necessary. Combine the ligation reaction as follows:

Ligation reaction 10 µl		Final concentration
5 µl	annealing reaction	
1 µl	10 × ligation buffer	1×
1.5 µl	13% polyvinyl alcohol (PVA)	∼2%
0.5 µl	40 U/µl RNase inhibitor	2 U/µl
2 µl	30 Weiss U/µl T4 DNA ligase (Fermentas)	6 U/µl

(3) Incubate at room temperature overnight or at 30 °C for at least 4 h.
(4) Extract once with phenol:chloroform:isoamyl alcohol and ethanol-precipitate. Separate the ligated products from the unligated donor and acceptor in a denaturing polyacrylamide gel. Use either radioactively labeled donor, acceptor, or a full-length transcript as a control. Due to the small amount of RNA present, the bands must be visualized using autoradiography. Elute and purify the excised samples as described in Protocol 3. As the amount of the RNA can be relatively small after the ligation, care must be taken to maximize the yields of the gel-purified product. At this stage we have used siliconized microcentrifuge tubes in the elution step and carried out the RNA precipitation in the presence of carrier (10 µg of glycogen).

Acknowledgments

We thank Heli Pessa, Minni Laurila and Xiaojuan Meng for critical reading of the manuscript. This work was supported by Academy of Finland (grant 50527 to M. J. F.).

References

1 M. J. MOORE, P. A. SHARP, *Science* 1992, 256, 992–997.
2 A. M. MACMILLAN, C. C. QUERY, C. R. ALLERSON, S. CHEN, G. L. VERDINE, P. A. SHARP, *Genes Dev.* 1994, 8, 3008–3020.
3 C. C. QUERY, M. J. MOORE, P. A. SHARP, *Genes Dev.* 1994, 8, 587–597.
4 O. GOZANI, R. FELD, R. REED, *Genes Dev.* 1996, 10, 233–243.
5 Z. PASMAN, M. A. GARCIA-BLANCO, *Nucleic Acids Res.* 1996, 24, 1638–1645.
6 H. LE HIR, E. IZAURRALDE, L. E. MAQUAT, M. J. MOORE, *EMBO J.* 2000, 19, 6860–6869.
7 H. LE HIR, M. J. MOORE, L. E. MAQUAT, *Genes Dev.* 2000, 14, 1098–1108.
8 P. A. MARONEY, C. M. ROMFO, T. W. NILSEN, *Mol. Cell* 2000, 6, 317–328.
9 M. J. FRILANDER, J. A. STEITZ, *Mol. Cell* 2001, 7, 217–226.
10 T. S. MCCONNELL, J. A. STEITZ, *EMBO J.* 2001, 20, 3577–3586.
11 M. J. MOORE, C. C. QUERY, in: *RNA: Protein Interactions. A Practical*

Approach, C. W. J. Smith (ed.), Oxford University Press, Oxford, 1998, pp. 75–108.
12 R. Reed, M. D. Chiara, *Methods Enzymol.* 1999, *18*, 3–12.
13 Y.-T. Yu, *Methods Enzymol.* 1999, *18*, 13–21.
14 K. Kleppe, J. H. Van de Sande, H. G. Khorana, *Proc. Natl. Acad. Sci. USA* 1970, *67*, 68–73.
15 J. F. Milligan, O. C. Uhlenbeck, *Methods Enzymol.* 1989, *180*, 51–62.
16 J. R. Wyatt, E. J. Sontheimer, J. A. Steitz, *Genes Dev.* 1992, *6*, 2542–2553.
17 E. J. Sontheimer, J. A. Steitz, *Science* 1993, *262*, 1989–1996.
18 J. F. Milligan, D. R. Groebe, G. W. Witherell, O. C. Uhlenbeck, *Nucleic Acids Res.* 1987, *15*, 8783–8798.
19 G. Krupp, *Gene* 1988, *72*, 75–89.
20 C. Kao, M. Zheng, S. Rüdisser, *RNA* 1999, *5*, 1268–1272.
21 I. Berkower, J. Leis, J. Hurwitz, *J. Biol. Chem.* 1973, *248*, 5914–5921.
22 H. Inoue, Y. Hayase, S. Iwai, E. Ohtsuka, *FEBS Lett.* 1987, *215*, 327–330.
23 J. Lapham, D. M. Crothers, *RNA* 1996, *2*, 289–296.
24 J. Lapham, Y.-T. Yu, M.-D. Shu, J. A. Steitz, D. M. Crothers, *RNA* 1997, *3*, 950–951.

4
T4 RNA Ligase

Tina Persson, Dagmar K. Willkomm and Roland K. Hartmann

4.1
Introduction

The growing interest in RNA–RNA and RNA–protein interactions has led to an increased demand for the production of RNA molecules with chain lengths usually in the range of 75–500 nt. In particular, chemogenetics (exchange of single functional groups) has become more and more popular in studies of RNA function. The most efficient RNA synthesis method used today is *in vitro* run-off transcription by T7 RNA polymerase (see Chapter 1). Although this technique produces large amounts of RNA at relatively low cost with reasonable effort, synthesis of RNA by *in vitro* transcription suffers from essentially two drawbacks: (1) potential 5′ and/or 3′ end heterogeneity of the product, and (2) the inability to introduce internal modifications at specific sites, apart from the limited scope of modifications that can be introduced by the polymerase. End heterogeneity can be overcome by several approaches described in detail in Chapters 2 and 3, but for the site-specific incorporation of nucleotide modifications, chemical RNA synthesis is in most instances inevitable [1].

Development of new commercially available phosphoramidites of natural and unnatural nucleosides, improvement of 2′-OH protecting groups, and the use of more efficient activators are innovations that have paved the way for chemical synthesis of RNA molecules in better and more reproducible yields [2–5]; see also Chapters 7 and 8). Despite these advances, substantial amounts of longer RNA molecules are still very difficult and expensive to synthesize. Current techniques permit efficient routine chemical synthesis of RNA molecules of up to 50 nt. As a consequence, longer RNA molecules with site-specific modifications are usually prepared by chemical synthesis of an RNA oligonucleotide (less than 50 nt) carrying the modification(s), which is then ligated to one or two other RNA molecules preferably produced by *in vitro* transcription. For such applications, the "DNA splint" ligation technique employing T4 DNA ligase (described in Chapter 3) is widely used. We will focus here on an alternative method based on T4 RNA ligase.

T4 RNA ligase (EC 6.5.1.3; also named Rnl1 or RnlA RNA ligase 1), a 347-aa

Handbook of RNA Biochemistry: Student Edition. Edited by R. K. Hartmann, A. Bindereif, A. Schön, E. Westhof
Copyright © 2009 WILEY-VCH Verlag GmbH & Co. KGaA, Weinheim
ISBN: 978-3-527-32534-4

polypeptide encoded by gene 63 of bacteriophage T4, belongs to a family of oligonucleotide end-joining enzymes involved in RNA repair, splicing and editing pathways [6–8]. The enzyme, introduced into molecular biology laboratories about 30 years ago [9–11], catalyzes formation of phosphodiester bonds between 5′-phosphate and 3′-hydroxyl ends of preferentially single-stranded RNA (ssRNA) and, less efficiently, ssDNA. The oligo(ribo)nucleotide carrying the terminal 3′-hydroxyl group is termed acceptor substrate, and the one providing the terminal 5′-monophosphate is described as the donor substrate (Fig. 4.1A). In intramolecular circularization reactions, both end groups (3′-OH, 5′-phosphate) are located on the same oligo(ribo)nucleotide molecule. The biological role of T4 RNA ligase seems to have its seeds in the intricate antagonisms of T4 phage and *Escherichia coli* host strains: T4 infection was shown to induce activation of a nuclease that cleaves bacterial tRNA^{Lys} in the anticodon loop. T4 RNA ligase in concert with T4 polynucleotide kinase catalyze the repair of the damaged tRNA^{Lys} [12, 13].

4.2
Mechanism and Substrate Specificity

4.2.1
Reaction Mechanism

The reaction catalyzed by T4 RNA ligase consists of three distinct and reversible steps (Fig. 4.1A [14]). In the first step, the ligase reacts with ATP to form an adenylated enzyme intermediate. In the second step, a donor substrate with a 5′ terminal

Fig. 4.1. (A) Mechanism of the ligation reaction catalyzed by T4 RNA ligase. In the ligation reaction, an acceptor substrate with a 3′-hydroxyl group reacts with a donor substrate carrying a 5′-phosphate group, resulting in a 3′–5′-phosphodiester bond. A reaction cycle consists of three distinct and reversible steps [14]: initially, the ligase reacts with ATP to form an adenylated enzyme intermediate (at lysine 99; [44, 45]), with concomitant release of pyrophosphate (1.). Then a donor substrate with a 5′-terminal monophosphate is bound by the enzyme and converted to the adenylated donor A(5′)pp(5′)Np(Np)$_n$, an intermediate in which the terminal adenosine moiety is attached via a 5′,5′ diphosphate bridge to the donor RNA (2.). In a final transesterification step, the phosphodiester bond connecting the two 5′,5′-linked phosphates is cleaved and a phosphodiester bond is formed between the donor and acceptor substrate, with concomitant release of AMP (3.). (B) End groups of donor and acceptor substrates that prevent formation of alternative products (intramolecular donor or acceptor cyclization, or formation of donor or acceptor tandems) in intermolecular ligation reactions of two oligo(ribo)nucleotides. In the case of the acceptor substrate, a 5′-OH terminus excludes that this substrate can act as a donor. A 3′-OH terminus is mandatory for acceptor function, whereas blockage of this end group in the case of donor substrates precludes that they can function as an acceptor; 3′ end blockage is achieved by introducing a terminal 2′,3′-cyclic phosphate (a), 3′-phosphate (b), a dideoxy residue (c; [25]), a 3′-inverted deoxythymidine (d; modification available from Dharmacon) or a periodate-oxidized 3′-terminal ribose (e; for details, see Chapter 6).

4.2 Mechanism and Substrate Specificity | 55

monophosphate is bound by the enzyme and the enzyme-linked 5'-AMP moiety is now transferred to the 5'-phosphate of the donor substrate, yielding the adenylated donor product in which the terminal adenosine moiety is attached via a 5',5'-diphosphate bridge [A(5')pp(5')Np(Np)$_n$]. In a final transesterification step, the phosphodiester bond connecting the two 5',5'-linked phosphates is cleaved and a phosphodiester bond is formed between the donor and acceptor substrate, with concomitant release of AMP.

4.2.2
Early Studies

Toward understanding the substrate specificity of T4 RNA ligase, it is instructive to briefly review results of early studies. The enzyme was first described by its ability to circularize 5'-^{32}P-labeled tRNA and polyhomoribonucleotides, such as poly(A), in a reaction requiring ATP and Mg^{2+} [11]. Relative to poly(A), the reaction occurred about 4-fold less efficient with poly(I), around 100-fold less efficient with poly(C) and poly(U), and at least 800-fold less efficient with poly(dA). Circularization of poly(A) molecules with an average chain length between 34 and 40 nt was about twice as efficient as for those 70–100 nt in length and no intermolecular ligation products were observed in this early study [11]. Subsequently, the shortest circularizing polyadenylate was shown to be (pA)$_8$, the optimal chain length for this reaction being 10–30 [15]. Cyclization was generally found to be the preferred reaction by orders of magnitude over intermolecular joining [16, 17], with four exceptions representing conditions that favor intermolecular ligation: (1) when the donor is too short to cyclize, (2) when a DNA donor is combined with a 5'-dephosphorylated RNA acceptor, (3) when the acceptor carries 5'- and 3'-hydroxyls and the donor a 5'-phosphate and a blocked 3'-terminus (Fig. 4.1B [18]), and (4) when the donor 5'-phosphate of one nucleic acid molecule is juxtaposed to the 3'-hydroxyl of a second acceptor nucleic acid molecule by base-pairing interactions, resulting in a *quasi*-intramolecular reaction (see below).

For ligation of DNA, studies with 5'-^{32}P-labeled oligodeoxythymidylates of various length ([5'-^{32}P]dT$_n$) revealed that ligase-catalyzed cyclization requires a minimal chain length of 6 dT residues and the best efficiency was obtained with chains of 20 [17]. Since DNA is a less efficient acceptor than RNA, adenylated A(5')pp(5')dT$_n$ intermediate accumulated to some extent – an observation not (or less) observed with RNA substrates. Interestingly, addition of the ribotrinucleotide ApApA, beyond serving as acceptor substrate, also stimulated cyclization of [5'-^{32}P]dT$_n$, suggesting that acceptors not only function as substrates for ligation, but also as cofactors for adenylation of the donor. ApA instead of ApApA neither stimulated donor adenylation nor was it joined [16, 17]. Likewise, ApA, IpI or UpU were found to be inactive as acceptors in the overall reaction with pAp as donor [18].

Analysis of minimal donor substrates of the pNp type (nucleoside-3',5'-bisphosphate) in the adenylation partial reaction revealed the highest efficiency of A(5')pp(5')Np formation for pCp, whereas the reaction with pUp and pAp was 3-fold and with pGp 10-fold less efficient [14]. A similar hierarchy was seen in the overall reaction [18, 19]. On the acceptor side, ApApA was a much better substrate than UpUpU. Ligation of the "poor" pGp donor to the "good" ApApA acceptor could be stimulated to proceed to almost completion by increasing the pGp concentration from 1 to 10 mM [14]. Furthermore, this study led to the conclusion that the enzyme exhibits more specificity in the donor adenylation reaction than in the subsequent joining of adenylated donor to acceptor. Formation of the UpUpUpGp product was much more efficient in the reaction of UpUpU with the adenylated donor A(5′)pp(5′)Gp than in the overall reaction starting from UpUpU, pGp and the ATP cofactor. Thus, pre-adenylation of the donor substrate may serve as a strategy to improve product yields when dealing with "poor" substrates [14]. However, this principle may not be generalized since results of two other studies investigating the synthesis of UpUpUpAp from UpUpU and pAp suggested rate limitation at the level of transfer of adenylated donor A(5′)pp(5′)Ap to acceptor [18, 19].

4.2.3
Substrate Specificity and Reaction Conditions

Substrate specificity of T4 RNA ligase can be summarized as follows. The minimal donor substrate is a nucleoside-3′,5′-bisphosphate, with efficiency decreasing in the order pCp > pUp ≈ pAp > pGp. Isocytidine-3′,5′-bisphosphate was reactive as well, indicating that also modified bases are tolerated [20]. Except for p(dCp), the deoxyribonucleoside-3′,5′-bisphosphates were found to be poorer donor substrates than the corresponding pNp ribonucleosides in the overall reaction [18]. 5′-AMP (pA) is not a donor and also nucleoside-2′,5′-bisphosphates are neither donor substrates nor effective inhibitors. Thus, in the donor substrate the enzyme specifically recognizes the 5′-terminal phosphate and ribonucleoside plus the next 3′-linked phosphate; the chain length of the donor exerts only marginal effects on reaction extent [18].

The smallest reactive acceptors are trinucleoside diphosphates (NpNpN) with a 3′-terminal hydroxyl [16]. Recently, also dinucleoside polyphosphates with at least four bridging phosphates, such as Gp_4G, were identified as acceptor substrates [21]. The 3′-terminal ribose moiety is important for acceptor recognition, and a 3′-terminal adenosine is preferred over cytidine and guanosine, showing intermediate reactivity, while a uridine residue is a relatively poor substrate (A > C ≥ G > U [18, 19]). However, intermolecular ligation yields are not simply dependent on the identity of the 3′-terminal base. For example, lower yields were observed with trimeric NpNpN acceptors containing a U residue at any of the three positions [18]. In another study, comparing trimeric acceptors equal in base composition, two consecutive purines enhanced ligation yields with pCp as donor (e.g. GpApU > UpApG > ApUpG [19]).

Regarding DNA oligonucleotides as substrates for T4 RNA ligase, combined appreciation of several studies [17, 22, 23] revealed the following rules of thumb: DNAs are less efficient substrates than RNAs, but discrimination against DNA occurs mainly at the acceptor substrate level.

Despite these substrate preferences, several strategies have been successfully used to optimize reaction conditions for poorer substrates. Reductions in Mg^{2+} concentration or addition of dimethylsulfoxide (DMSO) improved product (UpUpUpAp) yields for a UpUpU acceptor and a pAp donor [18]. Stimulation by DMSO, however, was not observed in ligation of UpUpU with pCp [19], suggesting that the effect of such additives may not be generalized. Also, joining of UpUpU to pUpUpUpCp at high enzyme concentration (350 U/ml) was stimulated from 6 to 60% product yield when ligation mixtures were incubated for 18 h at 15 °C instead of 1 h at 37 °C [19]. Ligation efficiencies with DNA were increased to some extent by elevated enzyme concentration, (partial) replacement of Mg^{2+} with Mn^{2+}, reduction of incubation temperature to around 17 °C, variation of donor:substrate:ATP ratio as well as their individual concentrations or low ATP concentration plus an ATP regeneration system [22, 23].

In summary, the substrate specificity of T4 RNA ligase is rather broad, permitting to ligate essentially any RNA or DNA sequence, mostly with satisfactory efficiency. Best reaction yields are commonly obtained at pH 7.2–7.8, 10–20 mM Mg^{2+}, 10–20% DMSO and often at temperatures as low as 5 °C, with substantial activity even exerted at 0 °C [24]. Several of the abovementioned parameters may be varied to optimize ligation yields. Furthermore, additives beyond DMSO, such as PEG 8000 or hexamine cobalt chloride, have been shown to improve product formation [25]. Another aspect is to prevent formation of unwanted byproducts. Thus, for intermolecular joining of two oligo(ribo)nucleotides, one should generally bear the following aspects in mind, illustrated in Fig. 4.1(B): the acceptor substrate ought to carry a 5′-hydroxyl terminus to prevent acceptor cyclization or joining of two acceptor substrates; likewise, the donor substrate should be blocked at its 3′-terminus by a 2′,3′-cyclic phosphate, a 3′-phosphate, a dideoxy residue [25], a 3′-inverted deoxythymidine or a periodate-oxidized terminal ribose (see Chapter 6) to avoid donor cyclization or formation of donor tandems.

4.3
Applications of T4 RNA Ligase

4.3.1
End-labeling

A common application of T4 RNA ligase is 3′-end-labeling with [5′-^{32}P]pCp. A ribocytidine dinucleotide bearing a 5′-phosphate and a 3′-terminal non-radioactive label, such as a fluorescein group (pCpC$_{3′\text{-fluorescein}}$), was also shown to be efficiently attached to the 3′ end of RNA substrates by the enzyme [26]. In a related application, a 5′-phosphorylated pCpC dinucleotide with a 3′-terminal polyethylene glycol linker including an internal photocleavage site and a terminal primary aliphatic amino group for coupling purposes was attached to RNA acceptor 3′ ends [27]. Kinoshita et al. [28] have made use of the fact that the ligase attaches an

AMP residue via a 5′,5′-pyrophosphate linkage to the 5′-phosphate end of the donor RNA or DNA, representing the intermediate donor activation step on the reaction pathway (Fig. 4.1A). In the absence of an acceptor nucleic acid, this intermediate was shown to be stable, and the authors demonstrated that fluorescent 2-aminopurine riboside triphosphate or 3′-amino ATP (for subsequent biotinylation) could replace the normal ATP in this 5′-end-labeling reaction catalyzed by T4 RNA ligase.

4.3.2
Circularization

The enzyme has further been exploited for intramolecular circularization of linear RNA to produce an authentic infectious circular RNA (371 nt) of the citrus exocortis viroid strain A (CEV-A [29]). One site of ligation was located within a dinucleotide internal loop of the viroid's rod-like structure ([29, 30] and Mfold secondary structure prediction of CEV-A), juxtaposing the reacting end groups (Fig. 4.2A). T4 RNA ligase was further employed to produce circular versions of hammerhead ribozyme strands as small as 15 nt, which exhibited increased activity, a reduced requirement for divalent metal ions, as well as increased resistance against nucleolytic degradation [31]. Interestingly, efficient T4 RNA ligase-catalyzed circularization of such hammerhead ribozyme oligoribonucleotides was achieved after their internal 7–8 nt had been annealed to a complementary DNA oligonucleotide, either linear or presented within a DNA hairpin loop (Fig. 4.2B and C). This setup favored circularization over formation of linear dimers, and the short central RNA–DNA duplex constrained the overall flexibility of the RNA oligonucleotide, while simultaneously juxtaposing the single-stranded 5′ and 3′ ends to be ligated.

4.3.3
Intermolecular Ligation of Polynucleotides

Ligation of RNA oligonucleotides to the 5′-end of mRNAs or other RNAs is used in so-called 5′-RACE (rapid amplification of cDNA ends) strategies to map RNA 5′ ends [32, 33]. Likewise, ligation of DNA or RNA oligonucleotides to RNA 3′ ends is the initial step before reverse transcription and PCR in approaches to map RNA 3′ ends (3′-RACE) or to determine the length of poly(A) tails and to identify polyadenylation sites [32, 33].

Nishigaki et al. [34] utilized T4 RNA ligase to tie two DNA single strands together (the 3′-terminal nucleotide of the acceptor oligonucleotide was a riboC in these experiments to increase ligation efficiency). To bring the reacting ends in proximity to each other, they equipped the two oligonucleotides with 12-nt long complementary sequences at one terminus, such that they formed "Y"-like hybrid structures, with the blunt-ended helix representing the stem and the two single-stranded arms presenting the donor and acceptor groups at their tips (Fig. 4.2D). This setup converted the intermolecular into a *quasi*-intramolecular reaction, and

4 T4 RNA Ligase

Fig. 4.2. Substrates that have been successfully used in the reaction catalyzed by T4 RNA ligase. (A) Intramolecular circularization of linear RNA to produce an authentic infectious circular RNA (371 nt) of CEV-A [29]. (B and C) Setups to produce circular versions of hammerhead ribozyme strands, using complementary DNA oligonucleotides that are either linear (B) or that present the complementary sequence within a hairpin loop (C) [31]. (D) "Y"-shape design for intermolecular ligation [34]. (E)

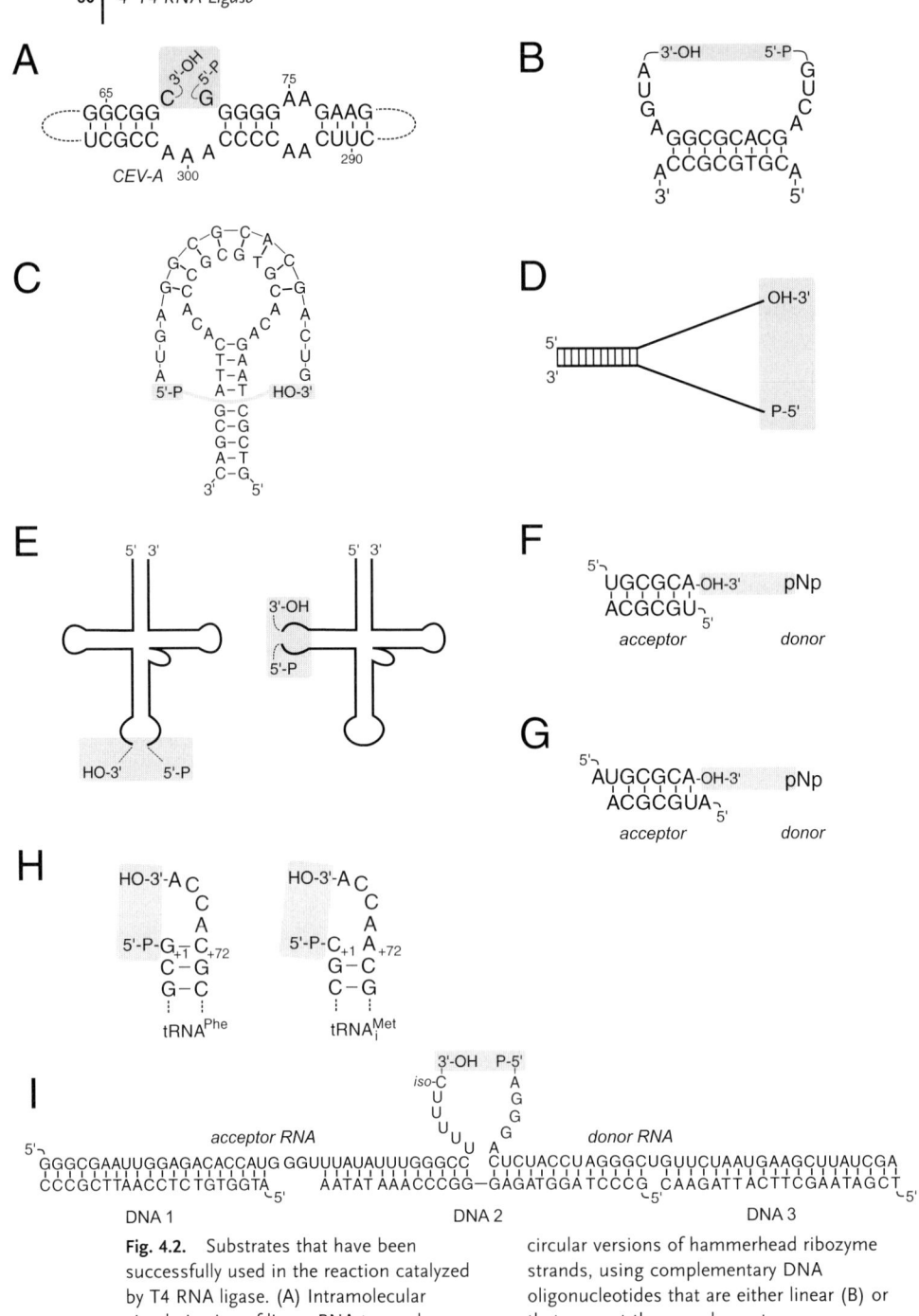

the blunt-ended stem of the "Y" ensured that only the unpaired end of each oligonucleotide reacted. Even dangling "Y" arms of up to ca. 50 nt each still gave ligation yields of around 20%.

Tessier et al. [25] have optimized the T4 RNA ligase reaction for the joining of pure DNA oligonucleotides, in this case a 25mer and a 23mer. Ligation yields of more than 50% were achieved by including PEG 8000 and hexamine cobalt chloride in the reaction. Oligonucleotide joining was favored over accumulation of the adenylated donor intermediate by restricting the ATP cofactor concentration to 20 µM. In this setup, the acceptor carried 5'- and 3'-OH end groups, whereas the donor oligonucleotide carried the 5'-phosphoryl group, but its 3' end was blocked via a single dideoxy analog (Fig. 4.1B), added by terminal transferase, to avoid ligation of two donor oligonucleotides.

4.4
T4 RNA Ligation of Large RNA Molecules

For the ligation of larger RNAs, several aspects should be kept in mind.

1. Proximity of Ends: To increase the probability for an enzyme molecule to simultaneously bind both reacting end groups, single-stranded RNA acceptor and donor ought to be brought in close proximity to each other. Generally, this is accomplished by letting the ends protrude from a helical region, resulting in a hairpin loop as the reaction product. Such a design converts intermolecular into *quasi-intramolecular* reactions. The importance of the structural context of the ligation site is easily illustrated for tRNA molecules. The secondary structure of tRNA is built from three stem–loop structures and one stem, which together form what is known as a cloverleaf structure (Fig. 4.3). Based on the enzyme's preference for single-stranded RNA termini, the three loops are expected to be favorable ligation areas in a tRNA molecule. Indeed, corresponding ligation strategies have been established for the anticodon loop (Fig. 4.2E and 4.3A [35–37]) and the D loop (Fig. 4.2E and 4.3B [38]), while so far T loop ligation in the context of a full-length tRNA structure has not been described. The size of the product hairpin loop will

---

Documented strategies used for the ligation of broken tRNA structures. (F and G) Double-stranded acceptor substrates [39] with blunt ends (F) or with a single nucleotide 5' overhang (G). (H) Donor termini in the context of tRNA structures. The 5'-terminal phosphate of tRNA^{Phe} is an inefficient donor; in contrast, the 5'-terminal phosphate of tRNA$_i^{Met}$ has excellent donor substrate quality due to the mispairing between C_{+1} and A_{+72} [24]. (I) Adaptation of the "DNA splint oligonucleotide" principle to the T4 RNA ligase reaction [20]. DNA oligonucleotides DNA 1 and DNA 3, used to prevent formation of unwanted ligation byproducts, are likely to be dispensable when the acceptor RNA carries a 5'-OH terminus and the donor RNA is blocked at its 3' end, as specified in Fig. 4.1(B). IsoC: *iso*-cytidine used instead of a natural nucleoside as the 3'-terminal nucleotide of the acceptor RNA in this particular experiment.

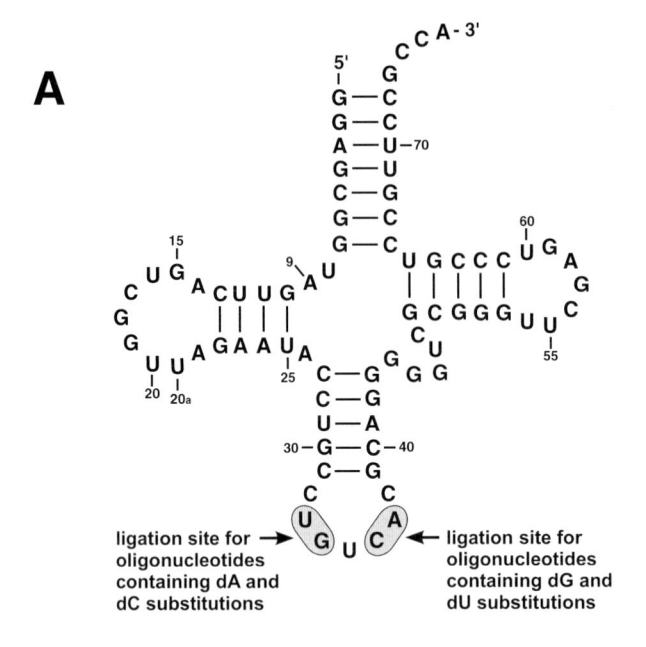

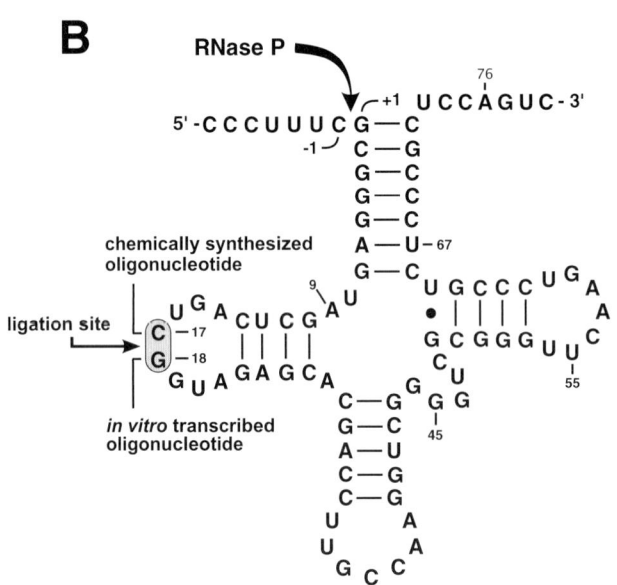

Fig. 4.3. (A) Schematic illustration of the T4 RNA ligation strategy for *E. coli* tRNA^Asp. The two alternative ligation sites (highlighted in grey) were placed in the anticodon loop, between U33 and G34 for oligonucleotides containing 2′-deoxyA or 2′-deoxyC modifications, and between C36 and A37 for those containing 2′-deoxyG or 2′-deoxyU modifications. (B) Schematic representation of the T4 RNA ligation strategy for *Thermus thermophilus* ptRNA^Gly. The arrow between positions −1 and +1 marks the canonical RNase P cleavage site; the ligation site between C17 and G18 is highlighted in grey.

influence ligation efficiency: increased loop size means more flexibility of the dangling single strands and increased average distance of the reactive groups before ligation. This is expected to reduce the efficiency of product formation. On the other hand, ligation within a tetraloop may occur less efficiently because the acceptor and donor ends may already be conformationally restricted before end joining.

2. Acceptor Substrates: Earlier studies with short acceptor oligoribonucleotide duplexes (6 bp) and pNp donors have indicated that acceptor substrates can be double stranded or even blunt ended (Fig. 4.2F and G) [39]. Moreover, duplexes with single nucleotide 5′ overhangs reacted most efficiently with the pNp donor that can base-pair with the duplex overhang (Fig. 4.2G). For example, with single cytidine 5′ overhangs on each end of the acceptor duplex, best yields were obtained for the pGp donor [39]. Regarding blunt-ended acceptor duplexes, it should, however, be mentioned that a dA_8 DNA acceptor oligonucleotide was joined less efficiently to a donor substrate in the presence of a complementary dT_8 DNA oligonucleotide [23], suggesting that a blunt-ended duplex, which is expected to form from the two oligonucleotides, is a less efficient type of acceptor substrate.

3. Donor Substrates: Donor termini that are part of a helix are inefficient substrates [24]. For example, the 5′-terminal phosphate of G_{+1} of yeast tRNA^{Phe} was a relatively inefficient donor because G_{+1} is base-paired to C_{+72}, forming the terminal acceptor stem base pair (Fig. 4.2H). In comparison, the 5′-terminal monophosphate at C_{+1} of *E. coli* initiator tRNA^{Met}, carrying a single mismatch at the acceptor stem terminus (C_{+1} and A_{+72}, Fig. 4.2H), was a much better donor in the T4 RNA ligase reaction. Here, intramolecular cyclization between the 5′-terminal phosphate and the 3′-hydroxyl at the 3′ end (A_{+76}) was the favored reaction, already occurring with high efficiency at very low enzyme concentrations [24].

4. Accessibility of Ligation Sites: In the case of large RNAs, the helix structure that clamps the preferably single-stranded acceptor and donor substrates should be positioned at the surface of the RNA complex to ensure enzyme access. In the case of RNAs for which the ligation site is embedded in higher order structures, experimenters usually switch to the "DNA splint" ligation technique using T4 DNA ligase, since it involves disruption of RNA structure (see Chapter 3 and [40]). However, it may sometimes have escaped attention that the "splint principle" has also been successfully adapted to the T4 RNA ligase reaction by annealing donor and acceptor RNAs to a bridging DNA oligonucleotide for juxtaposition of reacting end groups [20]. Yet, in this setup the bridging oligonucleotide design excluded 5–6 nt of each, acceptor 3′ end and donor 5′ end, from the RNA–DNA hybrid, creating a broken bulge loop structure in the RNA strand, while the DNA strand was entirely engaged in base pairing (Fig. 4.2I).

5. RNA End Homogeneity: A problem connected with *in vitro* transcription is 3′ and/or 5′ heterogeneity of RNA products, which may reduce ligation efficiency

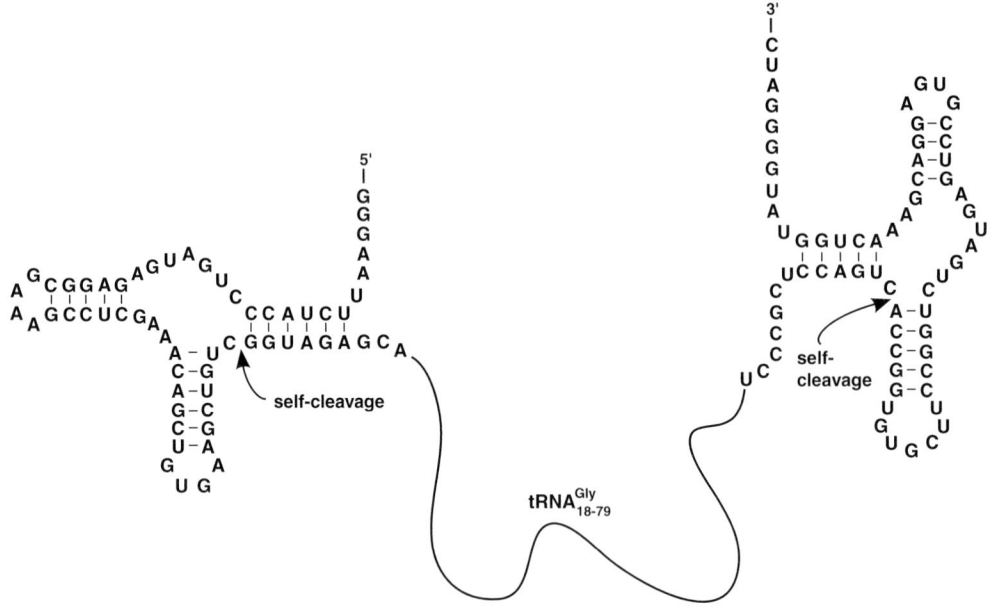

Fig. 4.4. A primary transcript with self-cleaving hammerhead ribozyme structures at its 5′ and 3′ termini. During transcription, the flanking regions are removed by ribozyme self-cleavage, releasing the internal RNA molecule of interest, in this case the donor substrate for the ligation illustrated in Fig. 4.3(B) (nt 18–79 of tRNA^{Gly}).

and compromise product homogeneity (for details, see Chapter 1). This problem can be solved either by sandwiching the RNA of interest between terminal *cis*-cleaving ribozymes (Fig. 4.4 and Chapter 2), or by use of alternative approaches described in Chapter 3. When involving *cis*-cleaving ribozymes, however, it is important to keep in mind that the cleavage reaction produces a 2′,3′-cyclic phosphate at the 3′ end and a hydroxyl group at the 5′-terminus. A 5′-hydroxyl is optimal for acceptor substrates, but requires phosphorylation when present on the donor substrate. Conversely, a 2′,3′-cyclic phosphate nicely blocks the donor 3′ end, but has to be removed when present on an acceptor substrate. Protocols to remove 2′,3′-cyclic phosphates are described in Chapters 2 and 6.

4.5
Application Examples and Protocols

4.5.1
Production of Full-length tRNAs

In our hands, the T4 RNA ligation procedure was successfully used for the production of ca. 80-nt long tRNA derivatives with site-specific modifications. In the first case, a 77-nt long *E. coli* tRNA^{Asp} was prepared by enzymatic ligation of two chemically synthesized oligonucleotides, each between 34 and 43 nt in length. The ligation sites were placed in the anticodon loop, the most explored region of tRNA for ligation by T4 RNA ligase. In this study [37] analyzing the effect of 2′-deoxy modifications on aminoacylation, tRNA variants were prepared that either contained single-site 2′-deoxy modifications or had all, for example, A residues in the 5′ or 3′ half or even in the entire tRNA replaced with the 2′-deoxyA analog. The ligation position for tRNA halves containing 2′-deoxyA- and 2′-deoxyC modifications at every A and C position, respectively, was placed between anticodon nucleotides U33 and G34, while nucleotides C36 and A37 were selected as ligation site when the substrate halves contained 2′-deoxyG- and 2′-deoxyU modifications (Fig. 4.3A). Ligation yields were between 30 and 50%. These findings illustrate that satisfactory product yields can be obtained despite seemingly unfavorable identities (see Section 4.2) of the acceptor 3′ terminus (U33) and the donor 5′ terminus (G34).

In the second application, a bacterial precursor tRNA^{Gly} (ptRNA^{Gly} from *Thermus thermophilus*) was prepared from a 24-nt acceptor substrate obtained by chemical synthesis (representing the 5′ portion of the ptRNA) and a 62-nt donor substrate representing the 3′-proximal portion of the ptRNA and generated by T7 RNA transcription (Fig. 4.3B). The 24-nt acceptor oligonucleotide carried 5′- and 3′-terminal hydroxyl groups, as routinely present in chemically synthesized RNAs. The 62-nt donor RNA was released from a primary transcript with terminal *cis*-hammerheads (Fig. 4.4), generating the aforementioned 5′-hydroxyl and 2′,3′-cyclic phosphate end groups. Before ligation, the 5′ end was phosphorylated (Protocol 2; see also Chapter 3) using T4 polynucleotide kinase. Here, the purpose was to study the effect of ribose modifications at nt −1 of ptRNA^{Gly} on catalysis by *E. coli* RNase P RNA [38]. The ligation site was placed in the D loop, between positions C17 and G18 (Fig. 4.3B), to minimize the length of the chemically synthesized RNA oligonucleotides carrying the single-site modification and thus to reduce the costs of chemical synthesis. The ligation yield was about 50% (see Protocol 3), again despite unfavorable identity of the donor 5′-terminus (G18).

4.5.2
Specific Protocols

In addition to the specific protocols given below, some routine buffers and procedures used in these protocols are detailed in Section 4.5.3, *General methods*, referred to by the abbreviation "GM" in the following.

Protocol 1: *In vitro* transcription
(1) *In vitro* transcription reactions (1 ml) using T7 RNA polymerase (T7 RNAP) include the following components:

		Final concentration
80 μl	1 M HEPES–KOH, pH 7.5	80 mM
11 μl	2 M MgCl$_2$	22 mM

6 µl	20 mg/ml BSA[1]	120 µg/ml
10 µl	100 mM spermidine	1 mM
50 µl	100 mM DTT	5 mM
37.5 µl	100 mM each NTP (pH 7)	3.75 mM of each NTP
5 µl	1 U/µl pyrophosphatase[2]	5 U/ml
30 µl	template 1 µg/µl (linearized plasmid 3.2 kb)	30 µg/ml
760.5 µl	RNase-free water	

[1] BSA (Sigma, minimum purity 98% based on electrophoretic analysis, pH 7).
[2] Pyrophosphatase from yeast (Roche, EC 3.6.1.1, 200 U/mg, <0.01% each ATPase and phosphatases).

The total volume of this reaction mix without enzyme is 990 µl. Before start of transcription, the reaction mix is divided into 5 portions of 198 µl, to each of which 2 µl T7 RNAP (MBI Fermentas, stock solution 100 U/µl; final assay concentration 1 U/µl) is added, followed by incubation at 37 °C overnight. Shorter incubation periods (2–5 hours) have been used as well and may even be preferable (Chapters 1 and 3).

In transcription reactions lacking pyrophosphatase or when pyrophosphate activity is insufficient, a white pyrophosphate precipitate may be observed. In such cases, it is advisable to remove the precipitate before Step 2 by centrifugation at 14 000 g for about 5 min directly after incubation of transcription mixtures has been stopped. The clear supernatant is then carefully removed and transferred to a new Eppendorf tube for further sample processing. Another way to reduce the amount of precipitate is to add Na_2EDTA immediately after transcription to give a final concentration of 50–100 mM (use a 500 mM stock solution at pH 7.5). By chelating Mg^{2+}, the formation of insoluble precipitates is substantially reduced.

(2) Before isolation of transcription product, the DNA template is degraded by addition of 1 µl DNase I (RNase-free, Roche; stock solution 10 U/µl; final concentration 50 U/ml) to 200 µl reaction mix and incubation at 37 °C for 20 min.
(3) Extract the RNA solution (200 µl) once with a phenol/chloroform mixture (GM). It may further be advisable to remove excess NTPs and/or salt components before EtOH precipitation (Step 4) by use of NAP 10 columns (Pharmacia Biotech) or equivalent matrices. Removal of excess NTPs is in general necessary only when using ^{32}P-radiolabeled nucleotides to avoid background radioactivity in purification gels. Salt components may impair band separation by denaturing polyacrylamide gel electrophoresis (PAGE), particularly when optimal gel resolution is essential, such as in separations of full-length RNA product from $n+1$ and $n-1$ species.
(4) Precipitate the RNA with ethanol (GM).
(5) Dissolve the RNA pellet in 20–40 µl RNase-free water and purify the RNA of interest by denaturing PAGE according to Protocol 4.
(6) Finally, determine the RNA concentration (GM).

Protocol 2: 5′-Phosphorylation of donor oligonucleotide
Many manufacturers provide convenient protocols for 5′-phosphorylation. We prefer to use a T4 polynucleotide kinase (T4 PNK) protocol from New England Biolabs. T4 PNK requires a free hydroxyl group at the 5′ terminus.
The protocol for the phosphorylation reaction is based on a total volume of 50 μl.

Phosphorylation buffer (10×)	Final concentration
500 mM Tris–HCl, pH 7.6	50 mM
100 mM MgCl$_2$	10 mM
50 mM DTT	5 mM
5 mM ATP	0.5 mM

(1) Combine 3 nmol of donor RNA with 5 μl 10 × phosphorylation buffer, 1 μl RNase Inhibitor (MBI Fermentas; stock solution 25 U/μl; final concentration 0.5 U/μl), 2.5 μl ATP (100 mM) and RNase-free water to a final volume of 47.5 μl; mix gently.
(2) Add 2.5 μl T4 PNK (MBI Fermentas; stock solution 10 U/μl; final concentration 0.5 U/μl) and incubate the mixture at 37 °C for 30 min.
(3) To increase 5′-phosphorylation efficiency, add another 2.5 μl of ATP (100 mM) and also another 2.5 μl T4 PNK (stock solution 10 U/μl), and incubate the mixture for another 20 min at 37 °C. Either directly add 55 μl 2 × gel loading buffer (GM) for PAGE purification (Protocol 4) or proceed to Step 4.
(4) Precipitate the RNA with ethanol (GM).
(5) Dissolve the RNA pellet in about 20 μl RNase-free water and purify the RNA by denaturing PAGE according to Protocol 4.
(6) Determine the RNA concentration (GM).

Protocol 3: Enzymatic ligation
The protocol described below was used to generate full-length *E. coli* tRNA^{Asp} and *T. thermophilus* ptRNA^{Gly}. The tRNA^{Asp} was generated by ligation of pairs of chemically synthesized oligoribonucleotides, 34–43 nt in length (Fig. 4.3A); ptRNA^{Gly} was generated by ligation of its 3′ portion (nt 18–79), transcribed *in vitro*, to a chemically synthesized 24meric oligoribonucleotide contributing the ptRNA^{Gly} 5′ portion (Fig. 4.3B).

(1) If the donor 3′-oligonucleotide does not contain a 5′-phosphate, 5′-phosphorylate according to Protocol 2 prior to ligation. It is recommended to check the purity of the oligoribonucleotides to be used in the ligation reaction on an analytical denaturing polyacrylamide (PAA) gel. If necessary, also gel-purify the freshly phosphorylated donor oligonucleotide as described in Protocol 4.
(2) Ligation reactions are performed in a total volume of 200 μl (reached in step 5). As the first step, combine 3 nmol of 5′-phosphorylated donor oligoribonucleotide and 4.5 nmol of the corresponding acceptor oligoribonucleotide with:

		Final concentration in 200 μl
20 μl	1 M HEPES–KOH, pH 7.5	100 mM
20 μl	100 mM DTT	10 mM
x μl	RNase-free water to a volume of 156 μl	

(3) Denature for 3 min at 90 °C, followed by incubation for 10 min at 65 °C and slow cooling (45 min) to ambient temperature in a metal block removed from the heating apparatus to anneal the oligoribonucleotides.
(4) After the cooling step, add:

		Final concentration in 200 μl
12 μl	250 mM MgCl$_2$	15 mM
1 μl	100 mM ATP	0.5 mM
20 μl	100% DMSO	10% (v/v)
3.75 μl	25–40 U/μl RNase inhibitor (MBI Fermentas or Promega)	0.5–0.75 U/μl

Mix gently.
(5) Add 7 μl of T4 RNA ligase (stock solution 20 000 U/ml; final concentration 0.7 U/μl, New England Biolabs), resulting in the final volume of 200 μl. Incubate at 16 °C for about 12–15 h.
(6) To analyze ligation efficiency, withdraw 1.5 μl from the ligation reaction and mix with 10 μl loading buffer; load onto a denaturing 8–12% PAA gel, stain the gel with ethidium bromide and visualize RNA bands by exposure to UV light.
(7) Ethanol precipitate the bulk of the ligation reaction (GM).
(8) Dissolve the resulting RNA pellet in 15–30 μl RNase-free water and purify the RNA by denaturing PAGE according to Protocol 4.
(9) Calculate the RNA concentration (GM).

Protocol 4: Preparative purification of RNA by denaturing PAGE
The appropriate gel concentration depends, as usually, on the size of the RNA; a 12% PAA/8 M urea sequencing gel (1-mm thick) was used for purification of ligated tRNA^{Asp} (77 nt), a corresponding 10% gel for purification of ligated ptRNA^{Gly} (86 nt). A somewhat lower urea concentration, such as 7.5 M (Chapter 3), may be used, but we recommend concentrations of at least 7 M. To facilitate localizing the RNA product on the gel, it is advisable to increase the concentration of the product band by pooling three separate ligation reactions (3 × 15–30 μl), each based on 3 nmol of input donor RNA (Step 8 above).

(1) Mix the 45–90 μl of pooled ligation product with 45–90 μl 2 × gel loading buffer and pipette into a 2-cm broad gel pocket of the PAA/8 M urea gel. As an example, the above-mentioned 86-nt ligation product was run on a roughly 40-cm long, 1-mm thick 10% PAA/8 M urea gel until the xylene cyanol marker had migrated 22–25 cm from the top.
(2) Detect the ligated product by UV shadowing, excise from the gel and elute from crushed gel slices in elution buffer overnight at 4 °C. A detailed description of this procedure is given in Chapter 3, Protocol 3.
(3) Concentrate the eluted RNA by ethanol precipitation and redissolve in 20–40 μl RNase-free water (GM).
(4) Determine the concentration by UV spectroscopy (GM).

4.5.3
General Methods (GM)

Several of the methods described here are detailed in the handbook *Molecular Cloning: A Laboratory Manual* [41]. Also, for some of the methods given below, slightly variant protocols exist and might be used just as well.

Preparation of RNase-free water

A major cause of RNA degradation is due to ribonuclease contamination in the water used for the preparation of buffers and solutions. It is therefore recommended to use double-distilled water, which may be further treated with DEPC (add 1/1000 volume DEPC and stir vigorously for 2 h followed by autoclaving for DEPC decomposition). For longer storage, the water may be additionally filtered through a 0.6-μm filter and kept in 1 ml aliquots in Eppendorf tubes, preferably at −20 °C. It should be noted that DEPC remnants may interfere with enzymatic reactions. Thus, fresh double-distilled water, frozen at −20 °C for storage, may be preferred; store water at ambient temperature only when freshly autoclaved, but freeze in aliquots once the bottles have been opened.

Gel running buffer (5 × TBE)

54 g of Tris base (446 mM), 27.5 g of boric acid (445 mM) and 20 ml of 0.5 M EDTA (pH 8.0)/l; no pH adjustment required; store at room temperature.

2 × gel loading buffer for denaturing PAGE

Weigh 48 g urea, 50 mg bromophenol blue, 50 mg xylene cyanol and 3.72 g EDTA; add 20 ml 5 × TBE (see above), adjust the volume to 80 ml with double-distilled water and finally to 100 ml after complete dissolving [41]. Store the solution at room temperature.

Staining buffer (ethidium bromide solution)

PAA/8 M urea gels are stained in 1 × TBE containing 0.5–1.0 μg/ml ethidium bromide (EtBr). An EtBr stock solution at a concentration of 10 mg/ml in water is preferred. The solution should be stored in dark bottles (e.g. bottles covered with aluminum foil) at room temperature or 4 °C.

Phenol/chloroform extraction

Add 200 μl of phenol/chloroform (ratio 5:1, pH 4.7, AMRESCO), cooled to 4 °C, to 200 μl of aqueous RNA solution. Vortex the resulting mixture vigorously before centrifugation at 14 000 g for about 2 min. Transfer the aqueous phase carefully to a new Eppendorf tube and extract twice with 100 μl chloroform in the same manner. Remove the RNA-containing aqueous phase (upper layer) and recover the RNA by ethanol precipitation.

Ethanol precipitation of RNA

- *Method A*: RNA precipitation by method A applies to RNA in buffer solution or in RNase-free double-distilled water. Add 100 μl NaOAc (3 M, pH 4.7), 1 μl glycogen (20 mg/ml) and 900 μl EtOH to 100 μl of RNA solution and mix vigorously. Store the sample at −20 °C for 2–3 h or at −80 °C for 30 min. Centrifuge the cooled sample at 14 000 g for about 10–30 min at around 10 °C. Carefully remove the supernatant and wash the RNA pellet with ice-cold 70 or 80% EtOH, followed by a short centrifugation step (e.g. 5 min, as above). Prior to dissolving the pellet in RNase-free water, air-dry the pellet for 10–15 min.
- *Method B*: This method of RNA precipitation is used for RNA recovered by gel elution in buffer B (see below). Add 1 μl glycogen (20 mg/ml) and 1 ml EtOH to 450 μl of the RNA solution and proceed as in Method A.

Elution of RNA from PAA/8 M urea gel slices

We routinely use two different elution buffers (A and B, see below). There is no general rule for the choice between these buffers and no systematic differences in yield have been noticed. However, if the RNA shows some degradation on the analytical gel after elution, it is recommended to switch to buffer B that has a lower pH. Both buffers should be stored at 4 or −20 °C.

Elution buffer A: 200 mM Tris–HCl (pH 7).
Elution buffer B: 1 M NaOAc (pH 4.7).

Calculation of RNA concentration

Two alternative formulas can be used for the calculation of RNA concentrations based on UV spectrometry measurement (for more details, see Appendix):

(a) $c = A_{260} \times$ dilution factor of cuvette solution$/(\alpha_{m260} \times$ number of nucleotides $\times l)$, where A_{260} is the absorbance at 260 nm, α_{m260} is the average molar absorption coefficient of the 4 nucleotides at 260 nm (for single-stranded DNA and RNA, an average α_{m260} of 8 500 M^{-1} cm^{-1} is appropriate) and l is the path length of cuvette (normally 1 cm).

(b) 1 A_{260} unit (absorbance of 1 measured at 260 nm in a 1-cm cuvette) corresponds to approximately 40 μg/ml single-stranded RNA. Total amount of RNA (μg) = 40 × A_{260} units × dilution factor of cuvette solution × total volume of RNA stock solution in milliliters.

4.5.4
Chemicals and Enzymes

4.5.4.1 Chemical Synthesis and Purification of Oligoribonucleotides

Oligo(ribo)nucleotides were purchased from IBA (Göttingen, Germany) or self-synthesized using standard phosphoramidite chemistry on an Applied Biosystems 394A DNA synthesizer. The coupling time was 16.6 min for ribonucleoside and 30 s for 2′-deoxyribonucleoside building blocks. Oligoribonucleotides were removed from the solid support and purified according to the protocol outlined below [42]. An alternative work-up procedure for RNAs longer than 50 nt has been described [2].

Purification of oligoribonucleotides

(1) Transfer the polymer-bound oligoribonucleotide from the column to a 4-ml vial and suspend in 3:1 (v/v) NH_3:EtOH at 55 °C for 16 h. Before opening the vial, cool the mixture on ice for about 15 min; then reduce the volume to around 1 ml by use of a Speed Vac system. Remove the clear aqueous supernatant carefully and transfer to a new 2-ml Eppendorf tube; wash the solid support with 100 µl water. Combine the aqueous phases and concentrate to dryness in a Speed Vac; add 500 µl EtOH and again concentrate the oligoribonucleotide to dryness to remove any water left from the preceding steps. Resuspend the base-deprotected oligoribonucleotide in 500 µl 1 M tetrabutylammonium fluoride (TBAF) in tetrahydrofurane (THF) and let the mixture react under gentle shaking at room temperature for at least 20 h. Then add 500 µl 2 M NaOAc (pH 6), and concentrate to a volume of about 0.5–0.6 ml (30–60 min in a Speed Vac). Extract the resulting mixture with 2–3 × 800 µl ethyl acetate (EtOAc) and centrifuge again in the Speed Vac for about 5–15 min until a clear solution is obtained. Do not concentrate the mixture to dryness in this step. Add 1.6 ml EtOH and store the mixture at −20 °C for 2 h or overnight. Centrifuge at 14 000 g for 15 min and remove the supernatant carefully. Air-dry the oligoribonucleotide pellet for 15 min and dissolve in 600 µl RNase-free water.
(2) Purify the oligoribonucleotide on a PAA/8 M urea sequencing gel (2-mm thick); a 12–15% PAA/8 M urea gel should be used for purification of 30- to 40-nt long oligoribonucleotides, a 20% gel for shorter ones. Mix the 600 µl product sample with 600 µl loading buffer and apply to the gel.
(3) Detect the oligoribonucleotide by UV shadowing, excise from the gel and elute from gel pieces by using a Biotrap system (Schleicher & Schuell Biotrap elution chambers and membranes BT1 and BT2 [42]; see also Chapter 1). Alternatively, diffusion elution from crushed gel slices may be employed (see Section 4.5.3). However, in our hands the Biotrap technique was more reliable and efficient when eluting oligoribonucleotides from a preparative gel (gel thickness of 2 mm).
(4) After gel elution, further purify the oligoribonucleotide by using Sep-Pak cartridges according to the manufacturer's instructions (Waters Sep-Pak cartridges, Millipore). In this step, any salt components left from the previous steps are removed. Then concentrate the oligoribonucleotide to dryness using a Speed Vac system.
(5) Dissolve the resulting pellet in 400–600 µl RNase-free water and determine the concentration as described in Section 4.5.3.

4.5.4.2 Chemicals
Ribonucleoside phosphoramidites for chemical nucleic acid synthesis were purchased from PerSeptive Biosystems (Hamburg, Germany); NTPs and glyco-

gen were obtained from Roche; 48% polyacrylamide/bisacrylamide and the phenol/chloroform mix (5:1, pH 4.7) were purchased from AMRESCO; diethylpyrocarbonate (DEPC) was obtained from FLUKA; 1 M TBAF in THF was purchased from Aldrich.

4.5.4.3 Enzymes

T4 polynucleotide kinase and T4 RNA ligase were purchased from New England Biolabs. RNase Inhibitor and T7 RNA polymerase were purchased from MBI Fermentas. DNase I and pyrophosphatase were obtained from Roche.

4.6
Troubleshooting

Low yields of the ligation reaction may have the following reasons:

(1) Check for unfavorable secondary structure formation of the RNA fragments, particularly at the ligation joint. For this purpose, software such as OLIGO version 4.0 (National Bioscience) or Mfold [43] can be employed.
(2) Heterogeneous 3′ ends of RNA transcripts: for RNA fragments of up to about 40 nt it is usually sufficient to purify the RNA by preparative PAGE in the presence of 8 M urea prior to the ligation reaction. However, for RNA transcripts longer than around 50 nt, RNAs slightly differing in length from the main product are hard to get rid of by preparative gel purification. Methods to eliminate the problem of 5′- and 3′ end heterogeneities are described in Chapters 2 and 3.
(3) RNA degradation – this is usually due to RNase contamination in water or solutions. Prepare and store RNase-free water as described in Section 4.5.3. Check individual solutions for RNase activity; prepare fresh buffers and solutions.

Acknowledgments

The authors thank Erik Tullberg, Ulf Berg and Anders Barfod for critical reading of the manuscript. This work was supported by Crafoordska stiftelsen, FLÄK (Forskarskolan i Läkemedelsvetenskap), Carl Tryggers Stiftelse, Kungliga Fysiografiska Sällskapet i Lund, Schybers stiftelse and the Deutsche Forschungsgemeinschaft.

References

1 S. VERMA, F. ECKSTEIN, Annu. Rev. Biochem. **1998**, 67, 99–134.
2 T. PERSSON, U. KUTZKE, S. BUSCH, R. HELD, R. K. HARTMANN, Bioorg. Med. Chem. **2001**, 9, 51–56.
3 C. VARGEESE, J. CARTER, J. YEGGE, S. KRIVJANSKY, A. SETTLE, E. KROPP, K. PETERSON, W. PIEKEN, Nucleic Acids Res. **1998**, 26, 1046–1050.
4 R. I. HOGREFE, A. P. MCCAFFREY, L.

U. Borozdina, E. S. McCampbell, M. M. Vaghefi, *Nucleic Acids Res.* **1993**, *21*, 4739–4741.

5 F. Wincott, A. DiRenzo, C. Shaffer, S. Grimm, D. Tracz, C. Workman, D. Sweedler, C. Gonzales, S. Scaringe, N. Usman, *Nucleic Acids Res.* **1995**, *23*, 2677–2684.

6 E. A. Arn, J. Abelson, RNA ligases: function, mechanism, and sequence conservation, in: *RNA Structure and Function*, R. W. Simons, M. Grunberg-Manago (eds), Cold Spring Harbor Laboratory Press, Cold Spring Harbor, NY, **1998**, pp. 695–726.

7 A. Schnaufer, A. K. Panigrahi, B. Panicucci, R. P. Igo, Jr, E. Wirtz, R. Salavati, K. Stuart, *Science* **2001**, *291*, 2159–2162.

8 J. Abelson, C. R. Trotta, H. Li, *J. Biol. Chem.* **1998**, *273*, 12685–12688.

9 G. Kaufmann, U. Z. Littauer, *Proc. Natl. Acad. Sci. USA* **1974**, *71*, 3741–3745.

10 G. C. Walker, O. C. Uhlenbeck, E. Bedows, R. I. Gumport, *Proc. Natl. Acad. Sci. USA* **1975**, *72*, 122–126.

11 R. Silber, V. G. Malathi, J. Hurwitz, *Proc. Natl. Acad. Sci. USA* **1972**, *69*, 3009–3013.

12 M. Amitsur, R. Levitz, G. Kaufmann, *EMBO J.* **1987**, *6*, 2499–2503.

13 C. Tyndall, J. Meister, T. A. Bickle, *J. Mol. Biol.* **1994**, *237*, 266–274.

14 L. W. McLaughlin, N. Piel, E. Graeser, *Biochemistry* **1985**, *24*, 267–273.

15 G. Kaufmann, T. Klein, U. Z. Littauer, *FEBS Lett.* **1974**, *46*, 271–275.

16 G. Kaufmann, N. R. Kallenbach, *Nature* **1975**, *254*, 452–454.

17 A. Sugino, T. J. Snopek, N. R. Cozzarelli, *J. Biol. Chem.* **1977**, *252*, 1732–1738.

18 T. E. England, O. C. Uhlenbeck, *Biochemistry* **1978**, *17*, 2069–2076.

19 E. Romaniuk, L. W. McLaughlin, T. Neilson, P. J. Romaniuk, *Eur. J. Biochem.* **1982**, *125*, 639–643.

20 J. D. Bain, C. Switzer, *Nucleic Acids Res.* **1992**, *20*, 4372.

21 E. A. Atencia, M. Montes, M. A. G. Sillero, A. Sillero, *Eur. J. Biochem.* **2000**, *267*, 1707–1714.

22 D. M. Hinton, J. A. Baez, R. I. Gumport, *Biochemistry* **1978**, *17*, 5091–5097.

23 M. I. Moseman McCoy, R. I. Gumport, *Biochemistry* **1980**, *19*, 635–642.

24 A. G. Bruce, O. C. Uhlenbeck, *Nucleic Acids Res.* **1978**, *5*, 3665–3677.

25 D. C. Tessier, R. Brousseau, T. Vernet, *Anal. Biochem.* **1986**, *158*, 171–178.

26 G. L. Igloi, *Anal. Biochem.* **1996**, *233*, 124–129.

27 F. Hausch, A. Jäschke, *Bioconjugate Chem.* **1997**, *8*, 885–890.

28 Y. Kinoshita, K. Nishigaki, Y. Husimi, *Nucleic Acids Res.* **1997**, *25*, 3747–3748.

29 J. E. Rigden, M. A. Rezaian, *Virology* **1992**, *186*, 201–206.

30 D. Skoric, M. Conerly, J. A. Szychowski, J. S. Semancik, *Virology* **2001**, *280*, 115–123.

31 L. Wang, D. E. Ruffner, *Nucleic Acids Res.* **1998**, *26*, 2502–2504.

32 L. Argaman, R. Hershberg, J. Vogel, G. Bejerano, E. G. H. Wagner, H. Margalit, S. Altuvia, *Curr. Biol.* **2001**, *11*, 941–950.

33 X. Liu, M. A. Gorovsky, *Nucleic Acids Res.* **1993**, *21*, 4954–4960.

34 K. Nishigaki, K. Taguchi, Y. Kinoshita, T. Aita, Y. Husimi, *Mol. Diversity* **1998**, *4*, 187–190.

35 T. Ohtsuki, G. Kawai, K. Watanabe, *J. Biochem.* **1998**, *124*, 28–34.

36 L. D. Sherlin, T. L. Bullock, T. A. Nissan, J. J. Perona, F. J. Lariviere, O. C. Uhlenbeck, S. A. Scaringe, *RNA* **2001**, *7*, 1671–1678.

37 C. S. Vörtler, O. Fedorova, T. Persson, U. Kutzke, F. Eckstein, *RNA* **1998**, *4*, 1444–1454.

38 T. Persson, S. Cuzic, R. K. Hartmann, *J. Biol. Chem.* **2003**, *278*, 43394–43401.

39 N. Sugimoto, A. Matsumura, K. Hasegawa, M. Sasaki, *Bull. Chem. Soc. Jpn.* **1991**, *64*, 2978–2982.

40 M. J. Moore, P. A. Sharp, *Science* **1992**, *256*, 992–997.

41 J. Sambrook, D. W. Russel, *Molecular Cloning: A Laboratory Manual*, Cold Spring Habor Laboratory Press, Cold Spring Harbor, NY, **2001**.

42 T. Tuschl, F. Eckstein, *Proc. Natl. Acad. Sci. USA* **1993**, *90*, 6991–6994.

43 M. Zuker, *Nucleic Acids Res.* **2003**, *31*, 3406–3415.

44 S. Heaphy, M. Singh, M. J. Gait, *Biochemistry* **1987**, *26*, 1688–1696.

45 L. K. Wang, C. K. Ho, Y. Pei, S. Shuman, *J. Biol. Chem.* **2003**, *278*, 29454–29462.

5
Co- and Post-Transcriptional Incorporation of Specific Modifications Including Photoreactive Groups into RNA Molecules

Nathan H. Zahler and Michael E. Harris

5.1
Introduction

A great deal of modern RNA biochemistry and molecular biology involves the incorporation of modified nucleotides into RNA molecules. Nucleotides bearing base and backbone modifications can be introduced randomly or at specific locations, and the variety of modifications available can facilitate both mechanistic and structural studies of RNA. The goal of this chapter is to outline some common techniques by which modified nucleotides and photo-crosslinking agents can be introduced into an RNA molecule and to provide examples of experiments in which these techniques have proven useful. Techniques discussed in detail include 5′-end modification by transcription priming, generation of nucleotide monophosphates and monophosphorothioates for use in 5′-end modification, derivatization of a 5′-phosphorothioate modification with a photo-crosslinking agent, and post-transcriptional 3′ end modification.

5.1.1
Applications of RNA Modifications

The availability of a wide range of modified nucleotides provides several key advantages for the study of RNA. Modified nucleotides can be used to alter individual functional groups, or in some cases individual atoms, and as such allow for more straightforward interpretation of experimental results than is possible with base mutation alone. Available analogs can be used to alter RNA base and backbone functional groups, and even remove nucleotide bases altogether [1]. Additionally, modifications with useful chemistry, such as photo-crosslinking agents, can be introduced. Combined with techniques to either randomly modify a population of RNAs or position modifications in a site-specific manner, the use of chemical modifications is fundamental to the study of RNA.

Site-specific incorporation of modified nucleotides is useful when a single nucleotide or functional group is the focus of study. Site-specific placement of modified nucleotides is therefore an essential first step in a variety of further tech-

Handbook of RNA Biochemistry: Student Edition. Edited by R. K. Hartmann, A. Bindereif, A. Schön, E. Westhof
Copyright © 2009 WILEY-VCH Verlag GmbH & Co. KGaA, Weinheim
ISBN: 978-3-527-32534-4

niques. For example, site-specific fluorescent markers can be introduced into RNA molecules for kinetic and folding studies (see Chapter 28 [2–4]). Site-specific placement of phosphorothioate modifications is also necessary for many thiophillic metal ion rescue experiments (see Chapter 19). 5′-End modification with m^7Gp^3G or other cap analog is an essential step in the use of *in vitro* transcribed RNAs in translation experiments [5–8]. In addition, affinity tags can also be placed at specific positions in an RNA structure to facilitate isolation of ribonucleoprotein complexes (see Part IV.1).

Site-specific modification of RNA can also be used to position photo-crosslinking agents for studies of RNA structure. A number of modifications are available for RNA which can form crosslinks to both RNA and proteins [9–12]. Several commonly used photo-crosslinking agents such as 4-thiouridine and 6-thioguanosine can be internally incorporated into an RNA molecule (see Chapter 22, e.g. [13, 14]). In addition, the 5′ and 3′ ends of RNA molecules can be post-transcriptionally modified with arylazide containing photo-crosslinking agents (see below and [11, 15, 16]). Together with the use of circular permutation to move the 5′ end of complex RNA molecules to different positions in the RNA structure, these modifications have been used to provide distance constraints for the structures of complex RNAs such as bacterial ribonuclease P RNA and 16S ribosomal RNA [17–19].

Another powerful aspect of site-specific modification is the ability to systematically vary nucleotide functional groups. For example, this approach has been used to examine the effects of phosphorothioate modification on MS2 coat protein binding to RNA and to investigate the contributions of 2′-hydroxyl groups to group I intron substrate binding [20–22]. Similarly, if a nucleotide base is known to contribute to structure or function, an array of modified nucleotides can be used to examine the contributions of individual base functional groups. Such studies have been undertaken to investigate the role of the conserved GU base pair immediately adjacent to the group I intron cleavage site and to assess the similarities in branch point adenosine recognition by the spliceosome and the group II intron [23–25]. Furthermore, this approach can be extended to the level of individual functional groups for 2′-hydroxyls and other functional groups which can be replaced with a number of chemically distinct modifications (e.g. [26, 27]).

Unlike site-specific modification, which is useful when a location or functional group is known to be of interest, random incorporation of modified nucleotides can be used to survey for important nucleotides and functional groups. The primary example of this approach is nucleotide analog interference mapping (NAIM), which has been used successfully in a number of systems to identify functionally important base and backbone functional groups (see Chapters 17 and 18 [28–31]). Photo-crosslinking agents can also be positioned randomly throughout an RNA. 4-Thiouridine and 6-thioguanosine can be randomly incorporated through transcription and have been used in the study of RNA structure and RNA–protein interactions in such diverse systems as the HIV-I Rev protein, the ribosome, RNA polymerase and the group II self-splicing intron [10, 32]. Finally, an RNA can be completely substituted with modifications. This approach can be used to introduce modifications that protect therapeutic RNAs from ribonucleases or increase their cellular uptake (for review, see [33]).

5.1.2
Techniques for Incorporation of Modified Nucleotides

The choice between techniques for incorporation of modifications is dependent primarily on the location of the desired modification and thus on the experiment in which the modified RNA will be used. Site-specific incorporation of modified nucleotides can be accomplished by a number of methods. Internal modifications are often introduced by chemical synthesis of all or part of the desired RNA. If the RNA in question is short enough, the simplest solution for internal site-specific incorporation is often chemical synthesis of the entire RNA. If not, chemical synthesis can be combined with RNA ligation (see Chapters 3 and 4). Ligation, however, has limitations which can often lead to low yields and circular permutation combined with either 5′- or 3′-end modification can be a viable alternative, especially for highly structured RNAs [11].

RNAs with site-specific 5′-end modifications can be prepared in a straightforward manner by transcription priming, an example protocol for which is given below. During transcription with T7 RNA polymerase, nucleotides lacking a 5′-triphosphate cannot be incorporated into an elongating RNA chain, but can be used to initiate a transcript ([34]; see also Chapter 1). Therefore modified nucleosides or nucleotide monophosphates included in an *in vitro* transcription reaction will be incorporated only at the 5′ ends of transcripts. However, transcription priming can produce a mixed population of modified and unmodified RNAs. To maximize analog incorporation, it is essential to include a large excess of modified nucleotide over the corresponding unmodified nucleotide triphosphate. While this has the effect of increasing the fraction of the population with the desired 5′ modification, it also tends to lower transcription efficiency. It is therefore generally necessary to empirically determine the appropriate analog concentration to balance these factors.

Random incorporation of modified nucleotides into RNA molecules can also be performed co-transcriptionally (e.g. [28–32]). In this case, modified nucleotide triphosphates are included in the reaction mixture and are added randomly to the elongating RNA chain. As with transcription priming, the modified nucleotide competes with the corresponding unmodified nucleotide for incorporation. Unlike transcription priming, however, the goal of random incorporation is generally not to produce a completely modified RNA. For experiments such as NAIM, which call for interference or selection, it is preferable to have at most one modification per transcript.

It is also important to note that the range of modified nucleotides that can be incorporated during transcription, both internally and at the 5′ terminus, is limited by the specificity of T7 RNA polymerase. Wild-type T7 RNA polymerase is unable to efficiently incorporate some modified nucleotides, including those with 2′ modifications that alter hydrogen-bonding ability or introduce large substituents. However, some of these restrictions can be relaxed through the use of mutant polymerases which allow for the use of 2′ modifications and possibly minor groove modifications [35, 36].

Site-specific 3′-terminal modifications are generally introduced post-transcriptionally. T4 RNA ligase can be used to add a modified nucleotide to the 3′ end of a RNA (see Chapter 4). In addition, post-transcriptional chemical modification can be used to attach affinity selection or photo-crosslinking agents to the 3′ terminus using the method of Oh and Pace [16], an example protocol for which is given below. In this method, the unique 2′,3′-cis-diol of the 3′ terminus is oxidized to form a dialdehyde, which is then reacted with an alkyldiamine under reducing conditions to yield a unique primary aliphatic amine (Fig. 5.1; see also Chapter 6). This primary amine can then be further derivatized with N-hydroxysuccinimidyl esters to introduce photo-crosslinking agents or other useful modifications. The modifica-

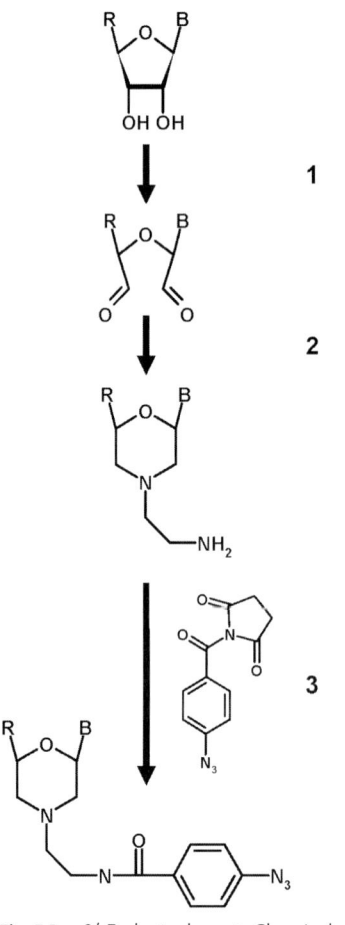

Fig. 5.1. 3′-End attachment. Chemical scheme for 3′-end modification by the method of Oh and Pace showing the three reactions steps. (1) Oxidation of the 2′,3′-cis-diol of the 3′-terminal ribose to a dialdehyde. (2) Reduction of the dialdehyde in the presence of ethylenediamine to yield a primary amine with a two-carbon linker. (3) Reaction with the N-hydroxysuccinimidyl ester derivative of an azidophenacyl crosslinking agent to yield the final, 3′-crosslinking construct.

tions that can be introduced to the 3′ end in this manner are only limited by the availability of reagents which will react with primary amines.

5.2
Description

5.2.1
5′-End Modification by Transcription Priming

As described above, 5′-end modification can be accomplished co-transcriptionally by transcription priming [34]. The following protocol describes the procedure for incorporation of a 5′-phosphorothioate modification with guanosine 5′-monophosphorothioate (GMPS). GMPS modified RNA can be further derivatized and an example of modification with an arylazide photo-crosslinking agent is described below. Other nucleotide monophosphates or nucleosides can be substituted in this protocol if a different 5′ modification is required. GMPS to GTP ratios of 10:1 to 40:1 result in a 70–90% yield of 5′-modified transcripts [11]. The following protocol was designed to maximize transcription efficiency of a 76-nt bacterial tRNA and utilizes a ratio of 4.8:1. If modified nucleotides other than GMPS are used, the optimal ratio for efficient incorporation and maximal transcription efficiency will have to be determined empirically.

This protocol predominantly requires widely available reagents. Nucleotide triphosphates, as well as T7 RNA polymerase and its associated buffer are available from a number of sources including Ambion. Yeast pyrophosphatase is available from Sigma-Aldrich. GMPS is not currently commercially available, but can be generated by chemical phosphorylation of guanosine (see below).

Begin by mixing the following:
10 × transcription buffer	10 µl
0.2 M DTT	3 µl
1 M MgCl$_2$	2 µl
100 mM ATP	4 µl
100 mM CTP	4 µl
100 mM UTP	4 µl
100 mM GTP	1.25 µl
30 mM GMPS	20 µl
Linearized DNA template	5 µg
40 U/µl RNase inhibitor	1 µl
5 U/µl yeast pyrophosphatase	5 µl
40 U/µl T7 RNA polymerase	5 µl
Water	to 100 µl total reaction volume

Incubate the reaction mixture overnight at 37 °C, recover products by ethanol precipitation and gel purify on an appropriate percentage denaturing polyacrylamide gel.

5.2.2
Chemical Phosphorylation of Nucleosides to Generate 5′-Monophosphate or 5′-Monophosphorothioate Derivatives

Generation of 5′-phosphorylated nucleosides can be achieved in a straightforward manner and in high yield by chemical phosphorylation of nucleosides with phosphoryl chloride and its derivatives. The example protocol given below describes the generation of GMPS by reacting unphosphorylated guanosine with thiophosphoryl chloride. The reaction of phosphoryl chloride with nucleosides in triethylamine occurs almost exclusively with the 5′-hydroxyl, making protection of the 2′- and 3′-hydroxyl unnecessary [37]. In addition, workup to generate nucleotide triphosphates from the resulting monophosphates has been well described [38]. Because many useful analogs are only available as nucleosides, this general procedure can provide opportunities for probing RNA structure and function that might not otherwise be available.

This protocol begins with the generation of a saturated solution and slurry of the nucleoside in triethylamine. Phosphoryl chloride or one of its derivatives is then added to the slurry. A nucleophilic reaction results in the 5′ attachment of phosphoryl chloride to the nucleoside. As the nucleoside becomes phosphorylated its solubility increases and the pH of the solution is lowered, resulting in more of the nucleoside becoming soluble until a clear solution is achieved. When this reaction is complete, the nucleotide 5′-phosphoryl chloride is hydrolyzed to the nucleotide monophosphate by simple addition of an excess of water. The nucleotide is then purified by ion exchange chromatography [39, 40].

A wide variety of nucleoside analogs and phosphoryl chloride derivatives are commercially available and can be substituted for thiophosphoryl chloride in this protocol. Methyl-, phenyl- and ethyl-dichlorophosphite, thiophosphoryl chloride, and 4-nitrophenyl phosphodichloridate are available from Sigma-Aldrich, and can be used to yield useful analogs. For example, in addition to GMPS, we have used this procedure to generate other 5′-phosphorylated guanosine derivatives including 6-thioguanosine monophosphate and guanosine 5′-*p*-nitrophenylphosphate [39, 40]. All other reagents used in the following procedure are commonly available from a variety of sources.

To synthesize GMPS, begin by mixing 2 mmol of guanosine and 5 ml triethylamine in a small round-bottom flask. Use a heating mantle to gently warm the flask to 50 °C and stir at that temperature for 10 min. Next, cool the guanosine solution on wet ice for at least 10 min. With the resulting guanosine slurry on ice, add 0.6 ml (5.8 mmol) of thiophosphoryl chloride. Continue stirring on ice in a cold room at 4 °C overnight (or at least 7 h) until the guanosine slurry has become a clear solution. The reaction mixture is next mixed with 500 ml of water to hydrolyze the resultant to guanosine 5′-thiophosphoryl chloride to GMPS. Because the hydrolysis reaction is exothermic, the reaction mixture should be added to the water in small aliquots and the solution should be stirred on ice between additions. Prior to chromatographic purification, adjust the pH of the solution to 7.5.

For purification of GMPS and other analogs, we have used a 2.5 × 18-cm col-

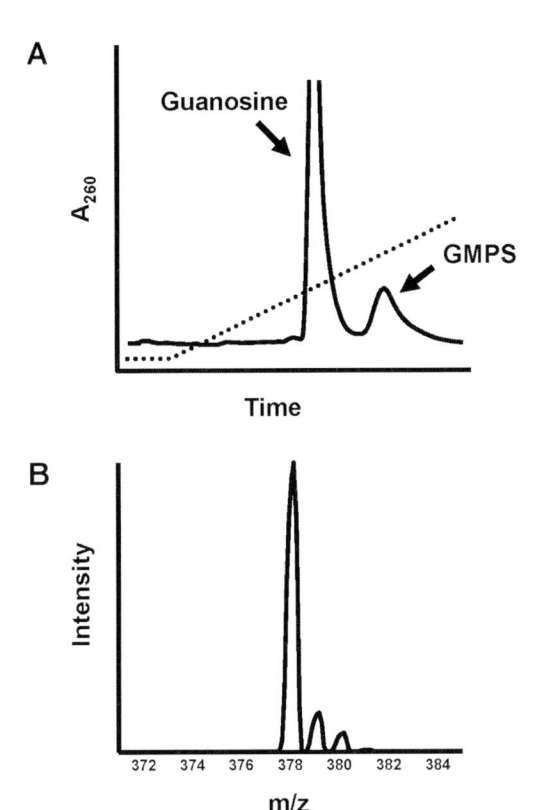

Fig. 5.2. GMPS synthesis. (A) Absorbance trace showing the purification of GMPS from guanosine by liquid chromatography. The dotted line indicates the fraction of buffer B (0.5 M ammonium bicarbonate, pH 7.0) in water. (B) Mass spectrographic analysis of GMPS from the indicated peak in (A). Data were acquired with a Thermoquest TSQ quadrapole mass spectrometer equipped with an electrospray ion source.

umn containing Supelco TSK-gel Toyopearl DEAE-650M resin prepared according to the manufacturers instructions. Load the sample in 50-ml aliquots at a flow rate of 1 ml/min and wash the column with three column volumes of water. Elute the nucleoside and phosphorylated compound with a linear gradient of 0–100% B in 60 ml (A = water; B = 0.5 M ammonium bicarbonate, pH 7.0). A sample chromatogram with good separation between the nucleoside, which elutes first, and the thiophosphorylated nucleotide is shown in Fig. 5.2(A). The column can be regenerated by washing with several column volumes of 100% B followed by water. Pool peak fractions and recover products by rotary evaporation. Next, resuspend the residue in 200 ml of 10% ethanol and dry again. Repeat this step at least 4 times. Finally, resuspend the GMPS product in 5 ml of water.

The identity of the individual peaks can be confirmed by running the appropriate fractions on polyethyleneimine TLC plates developed in 1 M lithium chloride relative to standards. In addition, samples can be analyzed using mass spectrome-

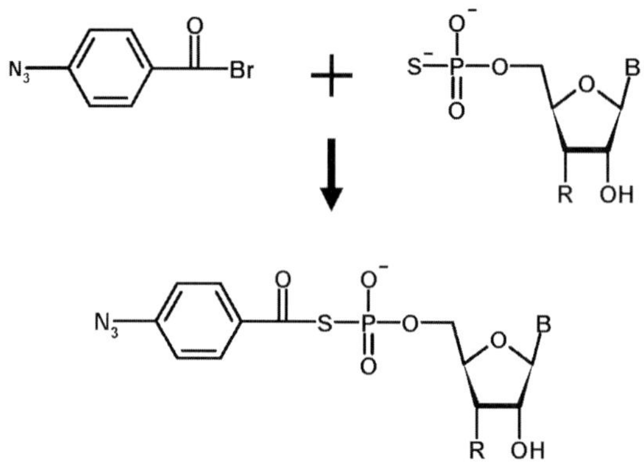

Fig. 5.3. Azidophenacyl bromide attachment to a 5′-phosphorothioate. Reaction scheme for modification of a 5′-monophosphorothioate modification with azidophenacyl bromide. Reaction with an acid bromide generates a linkage through the sulfur atom of the phosphorothioate.

try. A sample of mass chromatogram of final pooled GMPS fractions is shown in Fig. 5.2(B). A clear peak at 378 m.u. indicates that GMPS was the sole component of the second peak shown in Fig. 5.2(A). Peaks at 379 and 380 m.u. are due to the natural abundance of carbon and oxygen isotopes. The final product concentration can be determined by UV absorbance ($\varepsilon_{260} = 11.7 \times 10^3$ M^{-1} cm^{-1} at pH 7 [41]).

5.2.3
Attachment of an Arylazide Photo-crosslinking Agent to a 5′-Terminal Phosphorothioate

RNAs which are modified with a 5′-terminal phosphorothioate as described above can be further derivatized with reagents containing an acid bromide [11, 15]. The following example describes the attachment of an azidophenacyl photo-crosslinking agent to a 5′-phosphorothioate modified RNA (see Fig. 5.3). This crosslinking agent has been used to investigate the structures of various RNAs and has an effective crosslinking radius of approximately 10 Å.

Sodium bicarbonate, Tris(hydroxymethyl)aminomethane (Tris), ethylenediaminetetraacetic acid (EDTA), sodium dodecyl sulfate (SDS) and methanol are commonly available from a number of suppliers. Azidophenacyl bromide (APBr) is available from Sigma-Aldrich. The methanol solution of APBr used for this reaction should be freshly prepared. In addition, reducing agents such as DTT may reduce the azide moiety of this crosslinker to an amine, and should be avoided. Finally, as this protocol involves the use of photosensitive reagents, care should be taken to avoid exposure to light.

Begin by resuspending 10–40 μg of GMPS primed RNA in 10 μl of water. Add the following:

Water	24 µl
100 mM sodium bicarbonate, pH 9.0	20 µl
0.05% SDS	22 µl
3 mg/ml APBr in methanol	40 µl

Incubate the reaction mixture in the dark at room temperature 1 h. Add 100 µl of 10 mM Tris, 1 mM EDTA, pH 8.0. Next, extract the reaction mixture with an equal volume of 1:1 phenol:chloroform to remove any unreacted APBr. Finally, extract the reaction with an equal volume of chloroform and recover the RNA products by ethanol precipitation.

5.2.4
3′-Addition of an Arylazide Photo-crosslinking Agent

Photo-crosslinking agents and other modification can be post-transcriptionally added to the 3′ end of RNA molecules by chemical modification [11, 42]. As described above, this procedure takes advantage of the unique *cis*-diol of the 3′-terminal ribose. The procedure described below is a three-step process. In the first step, the 3′-*cis*-diol is oxidized to a dialdehyde (Fig. 5.1) [43]. In the second step, a primary amine is introduced at the 3′ end. The example below uses ethylenediame in this step of the reaction [44, 45]. If a longer carbon chain is desired, other compounds such as 1,6-diaminohexane can be used [42]. In the third and final step, an N-hydroxysuccinimidyl containing reagent is reacted with the modified RNA. The example given here utilizes N-hydroxysuccinimidyl 4-azidobenzoate for the addition of the arylazide crosslinking agent. Other N-hydroxysuccinimidyl containing reagents, such as a number of available fluorescent labels, should also be usable in this protocol.

As with the other protocols presented in this section, most of the reagents called for are commonly available. N-hydroxysuccinimidyl 4-azidobenzoate is available from Sigma (http://www.sigmaaldrich.com). Like azidophenacyl bromide used in the protocols above, care should be taken to avoid exposure of the photoagent to light and to reducing agents such as DTT. In addition, appropriate safety precautions should be taken when handling ethylenediamine.

Begin by mixing the following:

1 M sodium acetate, pH 5.4	10 µl
30 mM NaIO$_4$	10 µl
RNA	5–10 µg
Water	to 100 µl

Incubate the reaction mixture for 1 h at room temperature in the dark. Next, recover the RNA by ethanol precipitation. Resuspend the precipitate in 72 µl and add the following:

200 mM imidazole, pH 8.0	10 µl
50 mM NaCNBH$_3$	10 µl
100 mM EDTA	1 µl
15 mM ethylenediamine	6.7 µl

Incubate this solution at 37 °C for 1 h. Add 10 µl of 50 mM NaBH$_4$ and continue incubating at 37 °C for an additional 10 min. Once again, recover reaction products by ethanol precipitation. Resuspend the precipitate in 50 µl of 100 mM HEPES, pH 9.0 and add 50 µl of 20 mM N-hydroxysuccinimidyl 4-azidobenzoate. Allow this mixture to react at room temperature for 1 h in the dark. Finally, ethanol precipitate to recover the 3′-modified RNA product.

5.3
Troubleshooting

The 5′-end modification protocol described in this section is dependent on T7 RNA polymerase to incorporate modified nucleotides. As such, a major concern is likely to be the efficiency of incorporation. As described above, the level of incorporation can be adjusted to suit the experiment at hand by altering the ratio of modified nucleotide to its unmodified counterpart in the transcription reaction. However, increasing concentrations of modified nucleotide, especially if the modification does not lend itself to efficient incorporation is likely to also reduce overall yield. Mutant versions of T7 RNA polymerase may also be useful in increasing the incorporation of modifications, especially those with 2′ or minor groove constituents [35, 36].

References

1 M. Takeshita, C. N. Chang, F. Johnson, S. Will, A. P. Grollman, J. Biol. Chem. 1987, 262, 10171–10179.
2 N. G. Walter, Methods 2001, 25, 19–30.
3 X. Zhuang, L. E. Bartley, H. P. Babcock, R. Russell, T. Ha, D. Herschlag, S. Chu, Science 2000, 288, 2048–2051.
4 K. K. Singh, T. Rucker, A. Hanne, R. Parwaresch, G. Krupp, Biotechniques 2000, 29, 344–348, 350–341.
5 R. Contreras, H. Cheroutre, W. Degrave, W. Fiers, Nucleic Acids Res. 1982, 10, 6353–6362.
6 M. M. Konarska, R. A. Padgett, P. A. Sharp, Cell 1984, 38, 731–736.
7 J. K. Yisraeli, D. A. Melton, Methods Enzymol. 1989, 180, 42–50.
8 J. Jemielity, T. Fowler, J. Zuberek, J. Stepinski, M. Lewdorowicz, A. Niedzwiecka, R. Stolarski, E. Darzynkiewicz, R. E. Rhoads, RNA 2003, 9, 1108–1122.
9 S. H. Hixson, S. S. Hixson, Biochemistry 1975, 14, 4251–4254.
10 A. Favre, C. Saintome, J. L. Fourrey, P. Clivio, P. Laugaa, J. Photochem. Photobiol. B 1998, 42, 109–124.
11 M. E. Harris, E. L. Christian, Methods 1999, 18, 51–59.
12 M. M. Hanna, S. Dissinger, B. D. Williams, J. E. Colston, Biochemistry 1989, 28, 5814–5820.
13 J. R. Wyatt, E. J. Sontheimer, J. A. Steitz, Genes Dev. 1992, 6, 2542–2553.
14 E. L. Christian, M. E. Harris, Biochemistry 1999, 38, 12629–12638.
15 A. B. Burgin, N. R. Pace, EMBO J. 1990, 9, 4111–4118.
16 B. K. Oh, N. R. Pace, Nucleic Acids Res. 1994, 22, 4087–4094.
17 M. E. Harris, A. V. Kazantsev, J. L. Chen, N. R. Pace, RNA 1997, 3, 561–576.
18 M. E. Harris, N. R. Pace, Mol. Biol. Rep. 1995, 22, 115–123.
19 A. Montpetit, C. Payant, J. M.

Nolan, L. Brakier-Gingras, *RNA* **1998**, *4*, 1455–1466.
20 D. Dertinger, L. S. Behlen, O. C. Uhlenbeck, *Biochemistry* **2000**, *39*, 55–63.
21 A. M. Pyle, T. R. Cech, *Nature* **1991**, *350*, 628–631.
22 S. A. Strobel, T. R. Cech, *Biochemistry* **1993**, *32*, 13593–13604.
23 S. A. Strobel, T. R. Cech, *Science* **1995**, *267*, 675–679.
24 S. A. Strobel, T. R. Cech, *Biochemistry* **1996**, *35*, 1201–1211.
25 R. K. Gaur, L. W. McLaughlin, M. R. Green, *RNA* **1997**, *3*, 861–869.
26 D. J. Earnshaw, M. L. Hamm, J. A. Piccirilli, A. Karpeisky, L. Beigelman, B. S. Ross, M. Manoharan, M. J. Gait, *Biochemistry* **2000**, *39*, 6410–6421.
27 D. Herschlag, F. Eckstein, T. R. Cech, *Biochemistry* **1993**, *32*, 8312–8321.
28 R. K. Gaur, G. Krupp, *Nucleic Acids Res.* **1993**, *21*, 21–26.
29 S. A. Strobel, *Curr. Opin. Struct. Biol.* **1999**, *9*, 346–352.
30 E. L. Christian, N. H. Zahler, N. M. Kaye, M. E. Harris, *Methods* **2002**, *28*, 307–322.
31 Y. L. Chiu, T. M. Rana, *RNA* **2003**, *9*, 1034–1048.
32 S. Dokudovskaya, O. Dontsova, O. Shpanchenko, A. Bogdanov, R. Brimacombe, *RNA* **1996**, *2*, 146–152.
33 J. Kurreck, *Eur. J. Biochem.* **2003**, *270*, 1628–1644.
34 J. R. Sampson, O. C. Uhlenbeck, *Proc. Natl. Acad. Sci. USA* **1988**, *85*, 1033–1037.
35 R. Padilla, R. Sousa, *Nucleic Acids Res.* **2002**, *30*, e138.
36 R. Padilla, R. Sousa, *Nucleic Acids Res.* **1999**, *27*, 1561–1563.
37 M. Yoshikawa, T. Kato, T. Takenishi, *Tetrahedron Lett.* **1967**, *50*, 5065–5068.
38 A. Arabshahi, P. A. Frey, *Biochem. Biophys. Res. Commun.* **1994**, *204*, 150–155.
39 E. L. Christian, D. S. McPheeters, M. E. Harris, *Biochemistry* **1998**, *37*, 17618–17628.
40 A. G. Cassano, V. E. Anderson, M. E. Harris, *J. Am. Chem. Soc.* **2002**, *124*, 10964–10965.
41 R. M. C. Dawson, D. C. Elliott, W. G. Elliott, K. M. Jones, *Data for Biochemical Research*, Clarendon Press, Oxford, **1986**.
42 B. K. Oh, D. N. Frank, N. R. Pace, *Biochemistry* **1998**, *37*, 7277–7283.
43 S. B. Easterbrook-Smith, J. C. Wallace, D. B. Keech, *Eur. J. Biochem.* **1976**, *62*, 125–130.
44 U. C. Krieg, P. Walter, A. E. Johnson, *Proc. Natl. Acad. Sci. USA* **1986**, *83*, 8604–8608.
45 R. Rayford, D. D. Anthony, Jr, R. E. O'Neill, Jr, W. C. Merrick, *J. Biol. Chem.* **1985**, *260*, 15708–15713.

6
3′-Terminal Attachment of Fluorescent Dyes and Biotin

Dagmar K. Willkomm and Roland K. Hartmann

6.1
Introduction

A large number of experimental approaches in RNA biochemistry are based on some sort of label or tag within RNA molecules. Several of the methods at hand for RNA labeling are widely used, such as incorporation of modified nucleotides during solid-phase synthesis (Chapters 7 and 8) or transcription (Chapter 1), as well as post-transcriptional random chemical labeling by commercially available kits (e.g. Biotin Chem-Link from Roche or ULYSIS Nucleic Acid Labeling Kits from Molecular Probes). In contrast, the selective chemical attachment of a label to periodate-oxidized RNA 3′ ends described here is less common, despite some advantages over alternative labeling techniques. Primarily, the procedure is efficient, inexpensive and suitable for large-scale preparations. Further, it is not restricted to newly synthesized RNA and can therefore also be applied to commercially available RNAs or molecules isolated from cells and tissue. Moreover, the tag is incorporated at the RNA end, where it is less likely to impede proper folding than elsewhere in the molecule. Finally, as a single labeling group is added to each RNA molecule's 3′ end, this technique yields a rather uniform RNA population in terms of structure and RNA:label ratio.

Both fluorescent and biotin RNA tags are versatile tools in a broad array of techniques. One of the traditional domains of 3′ fluorescence of labeling RNA is fluorescence resonance energy transfer (see also Chapter 28) and related methods which have initially been used to probe RNA positioning within the ribosome [1, 2] and to analyze RNA–protein interactions within the signal recognition particle [3, 4]; further applications range from RNA–RNA binding measurements [5] to the use of fluorescent RNAs as hybridization probes [6, 7] and for microinjection experiments [8, 9].

Biotin as a tag is generally used because of its exceptionally tight binding to streptavidin, a 60-kDa bacterial protein, and avidin, the related protein from egg white. With an estimated dissociation constant of about 10^{-14} M, the complex formation between biotin and streptavidin is essentially irreversible under a wide variety of conditions [10 and references therein, 11]. More recently, modified avidins which allow reversible binding have also become available [6]. Accordingly, biotin tags have their uses in immobilization of RNA to solid supports as a prerequisite for capture assays and affinity chromatography [12, 13] or surface plasmon resonance measurements [14, 15]. Finally, among many further applications (reviewed in [16]), biotin-labeled RNAs are also employed as probes, with a subsequent detection mostly based on streptavidin-alkaline phosphatase conjugates [17].

6.2 Description of Method

The underlying reaction mechanism of 3′-label chemical attachment is based on the selective periodate-mediated oxidation of the RNA 3′-terminal ribose *cis*-diol (Fig. 6.1A; [18] and references therein). Oxidation results in a dialdehyde which is highly susceptible to nucleophilic attack and will therefore readily react with nucleophilic amino components such as hydrazine derivatives. As a final step, the reaction product can be stabilized by borohydride reduction.

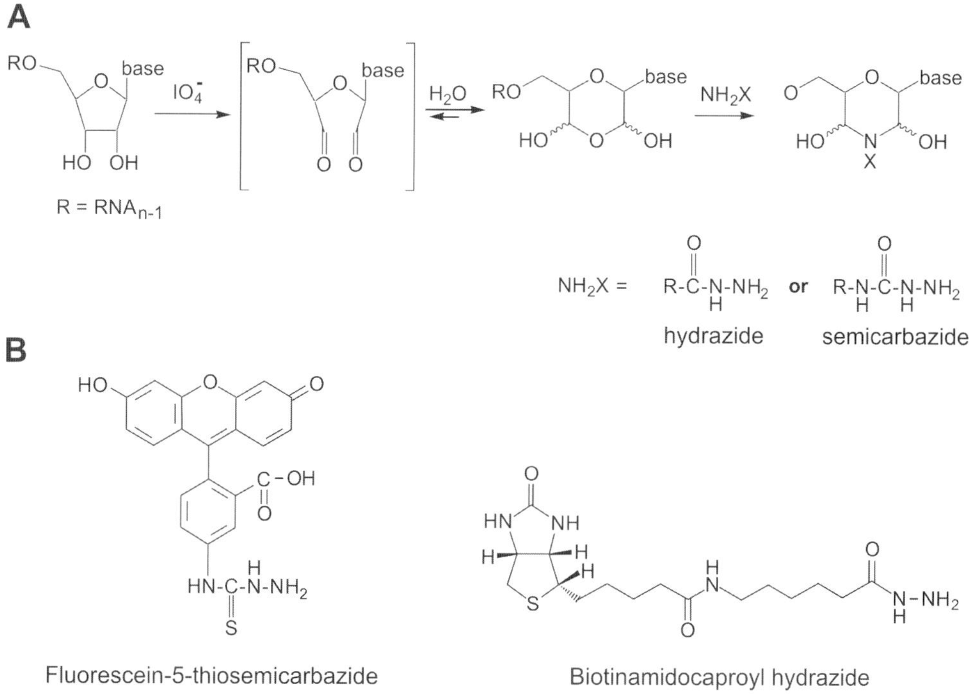

Fig. 6.1. (A) Reaction mechanism for modifications at the 3′-terminal ribose of RNAs. Oxidation of the ribose *cis*-diol with periodate results in a reactive dialdehyde which is then attacked by the hydrazide amino group (adapted from [18]). (B) Examples of coupling reagents: fluorescein-5-thiosemicarbazide, the hydrazide most frequently used as an RNA fluorescence tag, and biotinamidocaproyl hydrazide, to be used for biotin labeling.

With regard to the requirements of the labeling chemistry, a variety of hydrazine derivatives have been developed and are commercially available. Biotin can be purchased as biotin hydrazide, a direct hydrazide conjugate, from a number of sources. In two other biotin derivatives on the market, biotinamidocaproyl hydrazide (= biotin-X-hydrazide = biotin-6-aminohexanoic hydrazide, Fig. 6.1(B); from Sigma-Aldrich or Calbiochem) and biotin-XX-hydrazide (Calbiochem, Molecular Probes), the carboxylic acid hydrazide moiety serving as nucleophilic reactant is separated from the biotin moiety by a 7- or 14-atom spacer, respectively, to minimize steric interference between RNA and tag. Biotin-X hydrazide is the more commonly used reagent due to better water solubility.

Regarding fluorescent dyes, the fluorophore is in most cases directly conjugated to the semicarbazide or hydrazide moiety, as for example in fluorescein-5-thiosemicarbazide (Fig. 6.1B) and Alexa Fluor^{TM} hydrazides (both from Molecular Probes), as well as eosin-5-thiosemicarbazide (Sigma-Aldrich). However, reagents with a short extra spacer between the fluorophore and the reactive group, advantageous to some applications [19], are also available. For a compilation of the diverse fluorophore structures and fluorescence properties, see [6] and Molecular Probes' extensive online list at http://www.interchim.fr/bio/molprobes/cd/docs/tables/0301.htm.

6.3
Protocols

6.3.1
3′ Labeling

6.3.1.1 Biotin Attachment [12]

(1) Incubate up to about 3–4 nmol RNA in a total volume of 100 µl 40 mM KIO$_4$ for 1 h at room temperature in the dark.
(2) Stop the reaction with 100 µl of 50% ethylene glycol, then add 1/10th volume of 3 M NaOAc (pH 5.2) and 2.5 volumes of 96% ethanol to precipitate the RNA. After centrifugation, wash the pellet with 70% ethanol and dry.
(3) Dissolve the dried pellet in 100 µl of 10 mM biotinamidocaproyl hydrazide and incubate for 2 h at 37 °C.
(4) Add 100 µl of 0.2 M NaBH$_4$ and 200 µl of 1 M Tris–HCl, pH 8.2. Incubate for 30 min on ice in the dark.
(5) Purify the RNA (see Section 6.3.3) to remove salt and coupling reagent.

Special care has to be taken with some of the solutions:

- KIO$_4$ needs to be prepared as a 50 mM aqueous solution adjusted to pH 7.0 with 10 N NaOH. The KIO$_4$ will dissolve only upon NaOH addition; at pH 7.0,

50 mM is at the limit of solubility. A more alkaline pH, although helpful for dissolving the KIO$_4$, reduces the yield of the overall reaction.
- Biotinamidocaproyl hydrazide has low solubility in water; to prepare the required 10 mM solution, incubate for 2 min at 95 °C.
- 0.2 M NaBH$_4$: the solution sets free hydrogen. Therefore, prepare small portions (Eppendorf tube scale is fine).

6.3.1.2 Fluorescence Labeling [5]

While similar in chemistry, a slightly different protocol has been used in our lab for fluorescence labeling. In order to minimize photobleaching of the fluorescent dye, all reactions need to be carried out in the dark.

(1) Incubate 20 nmol of RNA in a volume of 400 µl 2.5 mM NaIO$_4$, 100 mM NaOAc (pH 5.0) for 50 min on ice.
(2) Precipitate with ethanol, wash the pellet with 70% ethanol and dry.
(3) For the coupling reaction, dissolve the RNA in 400 µl of 100 mM NaOAc (pH 5.0) and 1 mM fluorescein-5-thiosemicarbazide (200 mM stock solution in dimethyl formamide) and incubate on ice overnight.
(4) Ethanol-precipitate and redissolve in 50 µl double-distilled water.
(5) Purify (see Section 6.3.3).

6.3.2 Preparatory Procedures: Dephosphorylation of RNA Produced with 3′ Hammerheads

Transcription of RNA with 3′ *cis*-hammerheads is particularly attractive because it gives defined homogeneous 3′ ends as opposed to the heterogeneous 3′ ends which result from run-off transcription (see Chapter 1). 3′ Labeling of these RNAs released by hammerhead self-cleavage, however, poses a problem because their 3′ end is masked by a 2′,3′-cyclic phosphate. Prior to the labeling reaction, these RNAs therefore require 3′ dephosphorylation, which can be done in a kinase reaction at low concentration of ATP, thus making use of the T4 polynucleotide kinase phosphatase activity [20]. The method can also be applied to other RNA substrates with a 3′-phosphate.

(1) Incubate 300 pmol RNA with 2′,3′-cyclic phosphate ends for 6 h at 37 °C in a total volume of 100 µl containing:
   0.1 mM ATP
   100 mM imidazole–HCl (pH 6.0)
   10 mM MgCl$_2$
   10 mM β-mercaptoethanol
   2 µg BSA
   20 U T4 polynucleotide kinase
(2) Extract first with an equal volume of phenol/chloroform (1:1) and then with chloroform.
(3) Desalt by gel chromatography (see below) and precipitate with ethanol.

6.3.3
RNA Downstream Purifications

6.3.3.1 Gel Chromatography
Among the diverse options, Sephadex G-50 gel exclusion chromatography with self-made spin columns [21] has in our hands proved a cheap and efficient means to remove unincorporated label, particularly biotin, with almost no loss of RNA material.

(1) Prepare Sephadex G-50 slurry: add an equal volume of double-distilled water to the Sephadex, let swell for several hours with occasional gentle shaking. Wash twice with double-distilled water: let sediment (or spin briefly at low speed), exchange the water, resuspend and repeat once more. Perform a third wash with $0.1 \times$ TE (1 mM Tris, 0.1 mM EDTA) pH 7.4, let sediment and adjust the final volume of the supernatant liquid phase to 30% of the total volume (Sephadex plus supernatant). Resuspend before use.
(2) Remove the barrel from a 2-ml syringe, plug the syringe with a tiny amount of siliconized glass wool and fill with 1.5 ml Sephadex G-50 slurry.
(3) Place in a 15-ml disposable tube and centrifuge in a swinging-bucket rotor for 2 min at 550 g. Discard flowthrough.
(4) Place a decapped Eppendorf tube at the bottom of the 15-ml tube. Apply 200 µl of sample to the column and spin exactly as before. The flowthrough collected in the Eppendorf contains the purified RNA.

Depending on the centrifuge, the centrifugation step might need optimization. The aim is that the volume of flowthrough finally collected will be identical to the sample volume applied to the column. Alternatively, ready-made columns are available from a number of suppliers (e.g. Amersham Biosciences).

6.3.3.2 Purification on Denaturing Polyacrylamide Gels
With chromatographic purifications, there is the risk of low amounts of coupling reagents not being removed. For many applications it might therefore prove useful to check the purification by running a sample on a denaturing polyacrylamide gel, where unincorporated fluorescent dye can be seen to migrate below the bromophenol blue band, or to do a gel purification in the first place.

The method of gel purification, described extensively elsewhere in this handbook (e.g. Chapter 3), is convenient and efficient for RNAs of 400 nt or less. In addition to complete removal of coupling reagent, for shorter RNAs, such as tRNAs, it also allows elimination of unmodified RNA from the RNA pool because the attached dye slows down RNA migration (Fig. 6.2). For larger RNAs, the difference in electrophoretic mobility caused by the 3' label might well be too small to allow discrimination of modified and unmodified molecules.

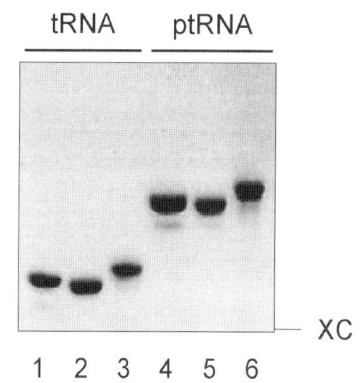

Fig. 6.2. Denaturing 9% polyacrylamide gel showing a tRNA^{Gly} (79 nt, lanes 1–3) and its 93-nt precursor (lanes 4–6) – both with homogeneous 3' ends due to release from a primary transcript carrying a 3'-terminal *cis*-hammerhead – after hammerhead self-removal and gel purification (lanes 2 and 5), after subsequent 3' dephosphorylation (lanes 1 and 4) and finally after 3' labeling with fluorescein-5-thiosemicarbazide (lanes 3 and 6). Both the dephosphorylated molecules (lacking the negative charge of one phosphate group) and the tagged molecules (of increased molecular weight) run slightly slower than the unmodified RNAs. XC: position of xylene cyanol blue, at 20 cm from the slot.

6.3.4
Quality Control

Particularly when setting up the procedure in one's lab, it is advisable to run controls for labeling efficiency. For fluorescence labeling, the expected efficiency of the reaction is over 90%, and for homogeneous populations of small RNAs this can be monitored by denaturing polyacrylamide gel electrophoresis (Fig. 6.2). Biotin attachment should also proceed almost quantitatively. Biotinylation efficiency can be analyzed by gel shift assays with saturating amounts of streptavidin (commercially available for example from Sigma-Aldrich) on native agarose gels (see Fig. 6.3). As opposed to fluorescence labeling analysis, the shift caused by streptavidin is fairly large, so that the labeling of slightly heterogeneous RNA populations and larger RNAs can also be monitored.

6.4
Troubleshooting

When labeling and/or downstream application efficiencies are low, check the following aspects.

6.4.1
Problems Caused Prior to the Labeling Reaction

Quality of the RNA 3' Ends
While the RNA 2'- and 3'-hydroxyls are a prerequisite for the reaction chemistry, a number of cleavage activities generate RNA 3'-phosphate or 2',3'-cyclic phosphate

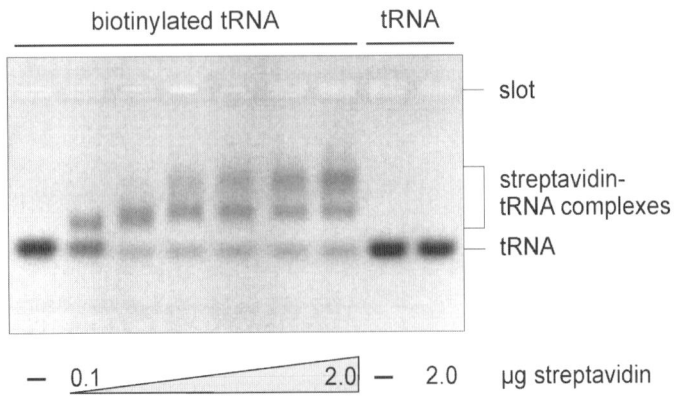

Fig. 6.3. Streptavidin retardation experiment to monitor the efficiency of biotinylation of yeast total tRNA (Roche). 140 ng of biotinylated tRNA were incubated with up to 2 µg of streptavidin for 15 min in 10 mM Tris, pH 7.4, 2.5 mM MgCl$_2$ and 100 mM NaCl. Samples were run on a 0.8% ethidium bromide-stained agarose gel. Different streptavidin–tRNA complexes were resolved on the gel, which may be related to the fact that streptavidin forms tetramers. At saturating amounts of streptavidin, over 80% of the tRNA showed reduced gel mobility. Non-biotinylated tRNA incubated with 2 µg streptavidin is shown as a control on the right.

ends, including several RNases (e.g. RNase A and RNase T1 [22]), metal ions [23] and small ribozymes (reviewed in [24]). The respective RNAs will need to be dephosphorylated prior to the labeling reaction (see Section 6.3.2). If the labeling efficiency is low even after a dephosphorylation step, the extent of dephosphorylation can be checked by denaturing PAGE (Fig. 6.2). However, PAGE resolution limits this kind of analysis to RNAs smaller than 100 nt.

Purity of the RNA to be Labeled
Nucleotides also react with the labeling reagents. When preparing the RNA to be labeled by transcription, thorough purification by either Sephadex columns or, preferably, polyacrylamide gel electrophoresis is therefore recommended. Impurities such as proteins or salts might impede the labeling reaction as well.

When the labeling efficiency for the RNA of interest is low, even though 2′,3′-hydroxyl ends should be present and the RNA has been thoroughly purified and desalted prior to labeling, a most likely cause is poor RNA quality due to degradation. In this case, an entirely new preparation of RNA should be used.

6.4.2
Problems with the Labeling Reaction Itself

pH of Reagents
A crucial and limiting aspect to the overall yield of the labeling reaction is the lability of the dialdehyde reaction intermediate. Since it is destabilized at basic pH, the

reaction conditions need to be kept non-alkaline. In particular, for the periodate oxidation according to the biotin-labeling protocol, neutral pH of the 50 mM KIO_4 solution is essential. At higher pH the reaction will be markedly less efficient.

Stability of Reagents

In general, it is advisable to store all labeling reagents at −20 °C. Nevertheless, especially the sodium borohydride solution will suffer from repeated freeze–thaw cycles and should be stored in aliquots. We have further prepared the periodate solutions freshly after three freeze–thaw cycles, and have kept the borohydride and periodate solutions for no longer than a few months.

6.4.3
Post-labeling Problems

Removal of Labeling Reagents

For most downstream applications, special care has to be taken to efficiently remove excess labeling reagents. Ethanol precipitation, even repeatedly, will generally not suffice. Substantial quantities of (anionic) contaminant fluorescent dyes will be visible on polyacrylamide gels, utilizing an appropriate excitation light source. Such gels need to be run very shortly in order to prevent the low molecular weight dye from exiting the gel and to minimize lateral diffusion. Unincorporated biotin derivatives can be detected by competition with the biotinylated RNA for binding to streptavidin, resulting in a considerable increase in the amount of streptavidin needed for saturation in a control shift assay.

Loss of RNA Material during Downstream Purification

The risk of losing RNA during purification is not significantly altered compared to unmodified RNA. A very important exception is phenol extraction of biotinylated material: since biotin increases hydrophobicity, the biotinylated RNA may be retained to some extent at the water/phenol interphase. Phenol extraction of biotinylated RNA should therefore be avoided.

Stability of Labeled RNA

Thiosemicarbazide adducts tend to degrade above pH 8 and at elevated temperatures [25], and thus should strictly be kept cold and dark, as well as below pH 8.

References

1. O. W. ODOM, D. J. ROBBINS, J. LYNCH, D. DOTTAVIO-MARTIN, G. KRAMER, B. HARDESTY, *Biochemistry* **1980**, *19*, 5947–5954.
2. M. STOFFLER-MEILICKE, G. STOFFLER, O. W. ODOM, A. ZINN, G. KRAMER, B. HARDESTY, *Proc. Natl. Acad. Sci. USA* **1981**, *78*, 5538–5542.
3. F. JANIAK, P. WALTER, A. E. JOHNSON, *Biochemistry* **1992**, *31*, 5830–5840.
4. G. LENTZEN, B. DOBBERSTEIN, W.

Wintermeyer, *FEBS Lett.* **1994**, *348*, 233–238.
5 S. Busch, L. A. Kirsebom, H. Notbohm, R. K. Hartmann, *J. Mol. Biol.* **2000**, *299*, 941–951.
6 R. P. Haugland, *Handbook of Fluorescent Probes and Research Products*, 9th edn, Molecular Probes, Eugene, OR, **2002**. This handbook, available from Molecular Probes or online (www.probes.com/handbook) provides comprehensive up-to-date information on labeling and detection reagents and techniques currently used.
7 D. Egger, R. Bolten, C. Rahner, K. Bienz, *Histochem. Cell Biol.* **1999**, *111*, 319–324.
8 C. Kruse, D. K. Willkomm, A. Grünweller, T. Vollbrandt, S. Sommer, S. Busch, T. Pfeiffer, J. Brinkmann, R. K. Hartmann, P. K. Müller, *Biochem. J.* **2000**, *346*, 107–115.
9 T. Pederson, *Nucleic Acids Res.* **2001**, *29*, 1013–1016.
10 M. L. Jones, G. P. Kurzban, *Biochemistry* **1995**, *34*, 11750–11756.
11 M. Wilchek, E. A. Bayer (eds), *Avidin–Biotin Technology (Methods Enzymol 184)*, Academic Press, San Diego, CA, **1990**.
12 U. Von Ahsen, H. F. Noller, *Science* **1995**, *267*, 234–237.
13 J.-L. Jestin, E. Dème, A. Jacquier, *EMBO J.* **1997**, *16*, 2945–2954.
14 M. Buckle, R. M. Williams, M. Negroni, H. Buc, *Proc. Natl. Acad. Sci. USA* **1996**, *93*, 889–894.
15 M. Hendrix, E. S. Priestley, G. F. Joyce, C.-H. Wong, *J. Am. Chem. Soc.* **1997**, *119*, 3641–3648.
16 M. Wilchek, E. A. Bayer, *Biomol. Eng.* **1999**, *16*, 1–4.
17 F. M. Ausubel, R. Brent, R. E. Kingston, D. D. Moore, J. G. Seidmann, J. A. Smith, K. Struhl, *Current Protocols in Molecular Biology*, Wiley, New York, **1994**.
18 F. Hansske, F. Cramer, *Methods Enzymol.* **1979**, *59*, 172–181.
19 D. Klostermeier, D. P. Millar, *Methods* **2001**, *23*, 240–254.
20 V. Cameron, O. C. Uhlenbeck, *Biochemistry* **1977**, *16*, 5120–5126.
21 J. Sambrook, E. F. Fritsch, T. Maniatis, *Molecular Cloning: A Laboratory Manual*. Cold Spring Harbor Laboratory, Cold Spring Harbor Press, NY, **1989**.
22 J. N. Davidson, *The Biochemistry of the Nucleic Acids*, 7th edn. Academic Press, New York, **1972**.
23 T. Pan, D. M. Long, O. C. Uhlenbeck, Divalent metal ions in RNA folding and catalysis, in: *The RNA World*, R. F. Gesteland, J. F. Atkins (eds), Cold Spring Harbor Laboratory Press, Cold Spring Harbor Press, NY, **1983**.
24 Y. Takagi, M. Warashina, W. J. Stec, K. Yoshinari, K. Taira, *Nucleic Acids Res.* **2001**, *29*, 1815–1834.
25 R. Dulbecco, J. D. Smith, *Biochim. Biophys. Acta* **1960**, *39*, 358–361.

I.2
Chemical RNA Synthesis

7
Chemical RNA Synthesis, Purification and Analysis

Brian S. Sproat

7.1
Introduction

The interest in chemically synthesized RNA took a dramatic leap forward with the discovery and application of the small interfering RNA (siRNA) technology [1, 2], a technique which has revolutionized functional genomics and target validation during the past 2–3 years and equals antisense technology in its applicability. However, the chemical synthesis of RNA until recently lagged a long way behind the well-established DNA synthesis technology. A few pioneers in the field have contributed to the three solid-phase RNA synthesis chemistry variants that are now used by commercial suppliers of RNA, i.e. the 2′-O-TBDMS method [3], the TOM method [4], which is a variant of the TBDMS method, and the 2′-ACE method [5, 6]. The abbreviations refer to the ether protecting groups used for the ribose 2′-hydroxyl group: TBDMS is *tert*-butyldimethylsilyl, TOM is triisopropylsilyloxymethyl and ACE is *bis*(acetoxyethoxy)methyl. The TOM and ACE variants are quite recent. In the past, problems with RNA synthesis were largely caused by poor-quality RNA phosphoramidites (the building blocks for solid phase synthesis), inappropriate protecting groups taken from DNA synthesis, poor activating agents and suboptimal deprotection protocols. This combination of largely unavoidable obstacles combined with the intrinsic chemical and biological instability of RNA led in most cases to failed syntheses. However, the boom in usage of synthetic RNA both for siRNA and other applications such as ribozymes and aptamers has had a very positive effect in that the speciality reagent suppliers have improved the quality of the building blocks leading to healthy competition and affordable prices. Moreover, the use of optimized protecting groups, coupling agents and deprotection protocols has revolutionized chemical RNA synthesis.

Since most commercially available synthesizers are not compatible with the highly specialized 2′-ACE chemistry, the methods described here have been restricted to the standard TBDMS chemistry, but also apply to the closely related TOM chemistry. The synthesis method described here is of course one of many variant methods, but all methods are in the end a variation of the basic methods described here. Since synthesis will be performed in the solid-phase using well-

Handbook of RNA Biochemistry: Student Edition. Edited by R. K. Hartmann, A. Bindereif, A. Schön, E. Westhof
Copyright © 2009 WILEY-VCH Verlag GmbH & Co. KGaA, Weinheim
ISBN: 978-3-527-32534-4

established phosphoramidite chemistry [7, 8] it can be performed manually or on any of the commercially available instruments. Synthesis starts from the 3′ terminus starting with the 3′-terminal nucleoside anchored most commonly via a succinyl linkage to an insoluble matrix, generally aminopropyl or long-chain alkylamine functionalized controlled pore glass, or polystyrene, contained in an appropriate reaction vessel. The nucleobases of the phosphoramidites and functionalized supports are preferably protected with N-phenoxyacetyl (pac) [9] or N-tert-butylphenoxyacetyl (tac) [10] groups enabling very mild deprotection of the RNA at the end of the synthesis, however the use of N^6-benzoyl A, N^4-acetyl C and N^2-isobutyryl G phosphoramidites leads to similar results regarding yield and purity of the RNA. The structure of the tac-protected cytidine building block is illustrated in Fig. 7.1. The advantage of the solid-phase method is that reagents are introduced into the vessel for removing protecting groups and enabling chain extension of the RNA 1 nt at a time and excess reagents are simply flushed away with a suitable solvent, in this case acetonitrile. The cyclical process is repeated until the desired length of RNA is obtained. Since no intermediate purification steps are possible, purification is done at the end of the assembly when most of the protecting groups have been removed. In practice, all reactions proceed close

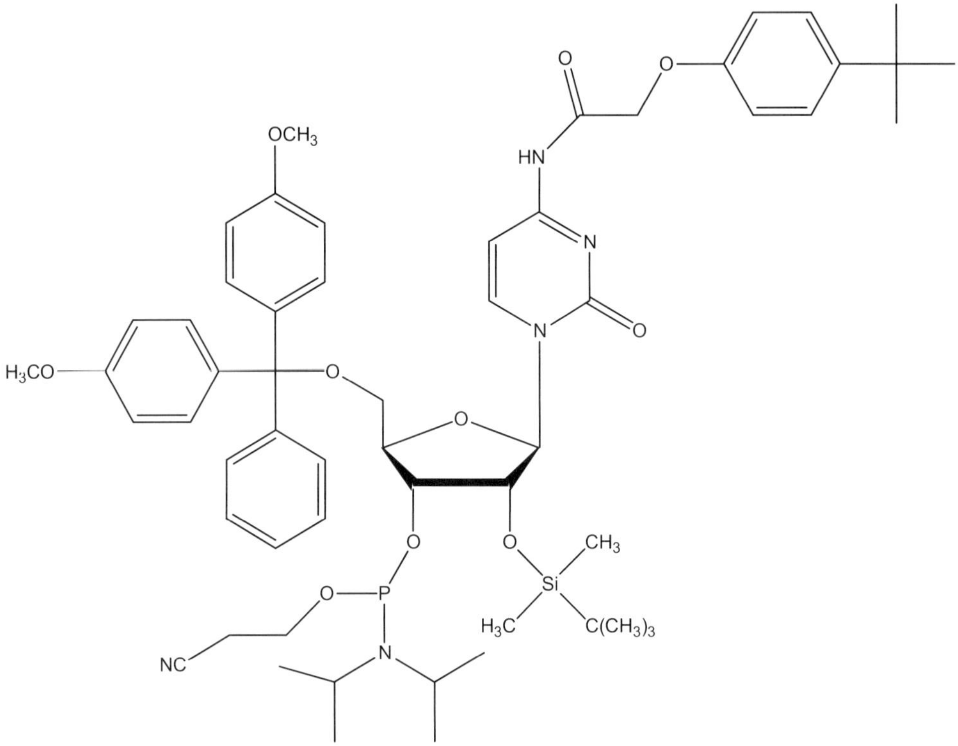

Fig. 7.1. Structure of a standard cytidine phosphoramidite building block, carrying 4-t-butylphenoxyacetyl protection of the exocyclic amino group.

to 100% yield and the chain extension reaction has a yield in the range of 98.5–99%, thus enabling in most cases a straightforward purification of the crude product.

Upon completion of the synthesis the fully protected support-bound RNA is deprotected in a stepwise fashion. In the first step the linkage to the solid-phase and the nucleobase and phosphate protecting groups are cleaved. In the second step the TBDMS groups are cleaved using triethylamine *tris*(hydrofluoride) [11, 12]. When RNAs longer than about 25 nt are synthesized it is best to leave the 5'-terminal dimethoxytrityl group attached as it is lipophilic and can be used as a purification aid. The crude RNA is then purified by anion-exchange and/or reversed-phase HPLC according to the length of the RNA and the purity required. For applications such as NMR spectroscopy and X-ray crystallography, purities of greater than 98% are desirable.

7.2 Description

7.2.1 The Solid-phase Synthesis of RNA

This section is devoted to the synthesis of the fully protected RNA in the solid-phase. The various steps involved in each cycle of the synthesis are illustrated in Fig. 7.2. Each cycle comprises a detritylation step that unmasks the 5'-hydroxyl group for chain extension, a coupling step in which the desired nucleotide as a phosphoramidite building block activated with 5-(benzylthio)-1*H*-tetrazole [13] is added, a capping step that acylates any unreacted 5'-hydroxyl group, an oxidation step that converts the phosphite triester to a phosphate triester, a second capping step that removes any occluded iodine and of course in between washing steps with acetonitrile to remove excess reagents. 5-(Benzylthio)-1*H*-tetrazole (BTT) for activation of the sterically hindered 2'-O-TBDMS-protected phosphoramidites is strongly preferred over conventional 1*H*-tetrazole with regard to both speed and coupling efficiency [13]; however, 4,5-dicyanoimidazole (DCI) and 5-(ethylthio)-1*H*-tetrazole can also be used with similar efficiency. Moreover, syntheses can be performed manually or machine-assisted using the following reagents and equipment:

(1) RNA monomers: 5'-O-Dimethoxytrityl-N(pac or tac)-2'-O-TBDMS-3'-O-(β-cyanoethylphosphoramidites) of adenosine (A), uridine (U), cytidine (C) and guanosine (G). These compounds are available for instance from Pierce (Milwaukee, USA) or Proligo (Hamburg, Germany) and should be stored dry under argon at $-20\,°C$. Alternative suppliers of fast deprotecting RNA phosphoramidites are Transgenomic, Promega JBL, Glen Research and ChemGenes.
(2) Solid-phase supports, either CPG (Proligo, Pierce, ChemGenes, Glen Research and Transgenomic) or polystyrene (available from Amersham Biosciences, now part of GE Healthcare) functionalized with A, U, C and G.

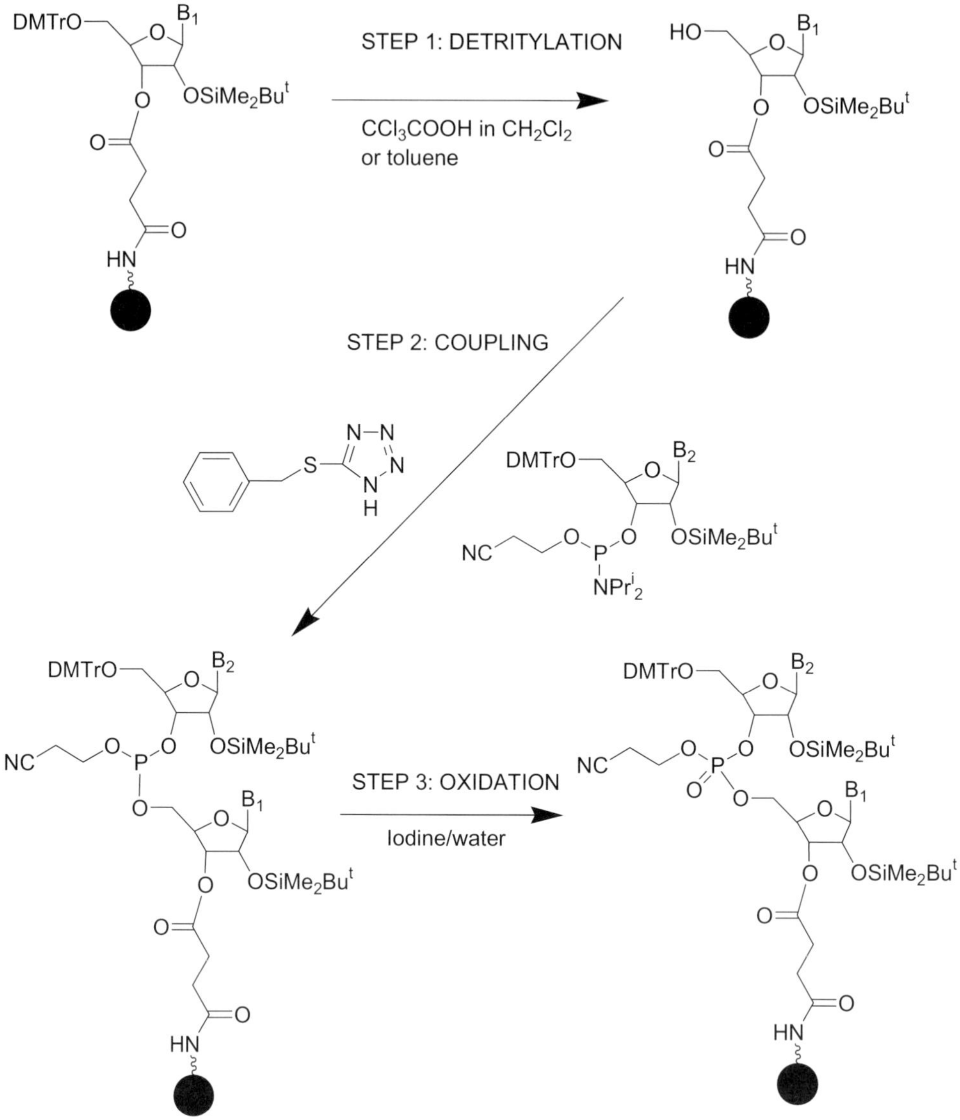

Fig. 7.2. Scheme illustrating a single cycle of solid-phase RNA synthesis via the phosphoramidite method. The black circle represents the controlled pore glass support. B_1 and B_2 represent protected nucleobases, e.g. uracil-1-yl, N^4-(4-t-butylphenoxyacetyl) cytosine-1-yl, N^2-(4-t-butylphenoxyacetyl) guanin-9-yl or N^6-(4-t-butylphenoxyacetyl) adenin-9-yl.

(3) 5-Benzylthio-1H-tetrazole (BTT), the activating agent which is available with a very low residual water content from emp Biotech (Berlin, Germany) or CMS (Oxford, UK).
(4) Capping solutions A (fast deprotection since it is based on 4-*tert*-butylphenoxy-

acetic anhydride) and B from Proligo. Capping solutions are also available from Merck, Riedel-de-Haen, Biosolve and Malinckrodt-Baker, for example.

(5) Oxidation solution containing iodine from Proligo, Merck, Riedel-de-Haen, Biosolve or Malinckrodt-Baker, for example. Since the iodine concentration is often only 17 mM, the concentration should be adjusted to 50 mM when performing large-scale syntheses by adding the correct amount of high-purity iodine.

(6) Deblock solution comprising 3% trichloroacetic acid in dichloromethane from Proligo or another supplier. For large-scale syntheses an alternative deblock solution containing up to 6% dichloroacetic acid in toluene is usually used, in particular on the Äkta OligoPilot 10.

(7) DNA synthesis grade acetonitrile containing less than 30 p.p.m. water (Malinckrodt-Baker, Merck, Riedel-de-Haen and Biosolve, for example).

(8) Assorted 1000 series gas-tight syringes with volumes of 0.5, 1 and 2.5 ml, which can be purchased from the Hamilton Company (Reno, NV, USA).

(9) DNA/RNA synthesizer (Applied Biosystems, Amersham Biosciences or other manufacturer) or, for manual synthesis, a glass reaction vessel fitted with a B14 ground glass joint at the top and a fine porosity glass frit and a tap at the bottom [14]. A set of suitable vessels can be made by any laboratory glass blower.

7.2.1.1 Manual RNA Synthesis

(1) Weigh out the requisite amounts of the tac- or pac-protected monomers required in small vials that can be closed with a septum and dry them overnight *in vacuo* over separate containers of phosphorus pentaoxide and potassium hydroxide pellets to remove traces of water. Suitable vials for this purpose are those amber glass bottles, which are used by suppliers of DNA and RNA phosphoramidites. To perform syntheses in the range of 1–3 µmol it is recommended to use 8–10 equivalents of monomer per coupling relative to the amount of support used. For synthesis scales above 5 µmol the monomer excess can be reduced to 5-fold.

(2) Carefully release the vacuum with dry argon and seal the bottles with tight fitting rubber septa.

(3) Using a gas tight syringe dissolve each of the monomers in the requisite volume of dry acetonitrile to give a 0.1 M solution and seal with Parafilm. It is not recommended to store the monomer solutions for more than 2–3 days at room temperature.

(4) Prepare an adequate volume of a 0.3 M solution of BTT in very dry acetonitrile in a tightly closed bottle under argon.

(5) Prepare 100 ml of capping mixture comprising 1 volume of fast deprotection capping solution A and 1.1 volumes of capping solution B in a tightly stoppered flask. Fresh capping mixture should be made each day.

(6) Weigh out the requisite amount of CPG carrying the desired 3'-terminal ribonucleoside into the glass reaction vessel. For a 1-µmol scale synthesis the ves-

sel should have a volume of about 5 ml, whereas for 10-µmol scale a volume of 20 ml is more appropriate to allow good washing.

(7) Using a Pasteur pipette add 3% TCA in dichloromethane (deblock solution) to the support and let it percolate through. Immediately a deep orange color is produced, characteristic of the released dimethoxytrityl cation. Continue to add acid until the effluent is colorless.

(8) Now drain the support using a slight pressure of dry nitrogen or better argon.

(9) Wash the support 8–10 times in a batchwise fashion with acetonitrile using a Teflon wash bottle, removing the supernatant each time with argon pressure.

(10) Just prior to the coupling step, wash the CPG once with very dry acetonitrile containing less than 30 p.p.m. water, flush away with argon pressure, close the tap and stopper the vessel.

(11) Using two gas tight syringes add the desired monomer as a 0.1 M solution in acetonitrile and an equal volume of 0.3 M BTT solution in acetonitrile in a second gas tight syringe to the CPG, stopper the vessel and agitate several times during a period of 6 min.

(12) Whilst the coupling step is in progress clean both syringes thoroughly with acetonitrile and store them in a desiccator.

(13) Flush away the coupling mixture, wash the CPG once with acetonitrile and flush away with argon pressure.

(14) Add a few milliliters of capping mixture to the reaction vessel, stopper and agitate for 1 min and then drain.

(15) Wash the CPG once with acetonitrile and flush away with argon pressure.

(16) Add a few milliliters of oxidation mixture and allow it to slowly percolate through the CPG during 2 min. This step oxidizes the phosphite triester to a phosphate triester.

(17) Drain the CPG and wash once with acetonitrile and drain with argon pressure.

(18) Once again add a few milliliters of capping mixture, agitate for 30 s and then drain using argon pressure.

(19) Now wash the CPG thoroughly with acetonitrile 6 times, draining each time in between using argon pressure.

(20) Repeat Steps 7–19 as many times as necessary until the desired sequence has been reached.

(21) If the RNA is longer than about 25 nt the final trityl group should be left on as a lipophilic purification handle. For RNAs less than 25 nt in length remove the final trityl group as in Step 7 and wash the CPG very thoroughly with acetonitrile.

(22) Finally dry the CPG using a stream of argon.

7.2.1.2 Automated RNA Synthesis

In order to perform automated RNA synthesis follow the instructions for the particular synthesizer plus the programme for the RNA synthesis scale you intend to use. Now the CPG or polystyrene support is placed inside a small plastic cartridge. All the reagents that you will need, including prepacked columns, are commercially available in the correct bottles to fit the various instruments on the market. Activated molecular sieves or trap bags can be added to ensure that reagents stay dry during the synthesis.

7.2.2
Deprotection

In the first deprotection step the succinate linkage connecting the 3′ terminus of the protected RNA to the solid support is cleaved, the β-cyanoethyl phosphate protecting groups are removed by β-elimination and in addition the exocyclic nucleobase protecting groups are cleaved. In our hands this step is best performed with a 1:1 mixture of concentrated aqueous ammonia and 8 M ethanolic methylamine, which prevents premature loss of the TBDMS groups, that would otherwise lead to degradation of the RNA under basic conditions. *Warning, methylamine is not compatible with N^4-benzoyl-protected C.* In the second deprotection step the TBDMS groups are removed using triethylamine *tris*(hydrofluoride) in an appropriate solvent. At this point you will need the following reagents:

(1) High-purity concentrated aqueous ammonium hydroxide. This solution is highly irritating to the eyes and respiratory system and must only be used in a well-ventilated fume cupboard.
(2) Anhydrous 8 M methylamine in ethanol. This compound is also highly irritating to the eyes and respiratory system and must only be used in a well-ventilated fume cupboard.
(3) Anhydrous dimethylsulfoxide (e.g. Fluka, Biotech. Grade).
(4) Triethylamine *tris*(hydrofluoride) (e.g. Aldrich). This compound is hazardous and toxic, and should only be handled wearing full protection and used only in a well-ventilated fume cupboard.
(5) Anhydrous triethylamine.
(6) *N*-Methylpyrrolidone, peptide synthesis grade.
(7) Prop-2-yl trimethylsilyl ether prepared according to Jones [15].
(8) Diethyl ether.

7.2.2.1 Deprotection of Base Labile Protecting Groups

(1) Transfer the support obtained from Section 7.2.1 to a small screw top vial or Duran bottle equipped with a tight fitting screw top.
(2) Add a mixture of concentrated aqueous ammonia and 8 M ethanolic methylamine, 1:1 by volume. A volume of 2 ml is adequate for a 0.2–1-µmol scale synthesis. Otherwise use 2 ml/µmol of support.
(3) Close the vial or bottle tightly and seal further with Parafilm.
(4) Place the vial or bottle in a preheated oven at 65 °C for 20 min for small vials, but 40 min for larger bottles, which take longer to equilibrate thermally.

(5) Allow the vial/bottle to cool completely to room temperature before opening in a fume cupboard.
(6) Carefully remove the supernatant and wash the support several times with a few milliliters of ethanol/sterile water (1:1 by volume).
(7) Combine the supernatant and washings in a Falcon tube and dry in a Speed Vac or, for bigger volumes, evaporate to dryness on a rotary evaporator. Do not use water bath temperatures above 30 °C for trityl-on material.
(8) Dry the residue once by evaporation of absolute ethanol.

7.2.2.2 Desilylation of Trityl-off RNA

(1) Add a 1:1 mixture of dry DMSO and triethylamine *tris*(hydrofluoride) [16], using 600 µl/µmol, to the trityl-off residue in the Falcon tube obtained in section 7.2.2.1 above and sonicate briefly. If you have dried down the oligoribonucleotide in a glass flask, dissolve it in the required volume of dry DMSO using gentle warming of the flask using a hair dryer, transfer the solution to a Falcon tube and add an equal volume of the fluoride reagent.
(2) Close the tube, seal with Parafilm and place it in a preheated oven at 65 °C for 2.5 h.
(3) Cool the tube to room temperature.
(4) Add 2 volumes of isopropyl trimethylsilyl ether [15] to destroy the excess fluoride reagent, close the tube and shake vigorously at intervals during 10 min. At this point a white precipitate appears.
(5) Open the tube carefully and add 5 volumes of diethyl ether, close and agitate vigorously.
(6) Collect the precipitate by centrifugation at 4000 r.p.m. at 4 °C for 5 min.
(7) Remove the supernatant by careful decantation.
(8) Resuspend the pellet in diethyl ether, close the tube, agitate and again collect the precipitate by centrifugation.
(9) Repeat Steps 7 and 8 twice more.
(10) Finally dry the RNA pellet carefully *in vacuo*.

7.2.2.3 Desilylation of Trityl-on RNA

(1) Prepare a solution of N-methylpyrrolidone/triethylamine/triethylamine *tris*(hydrofluoride) (6:3:4 by volume) [17] immediately before use and add to the trityl-on residue obtained in section 7.2.2.1 above using 600 µl/µmol. For material that has been dried down in a glass flask, dissolve the residue in the minimum volume of dry DMSO, transfer the solution to a Falcon tube and add the freshly prepared desilylation solution.
(2) Perform Steps 2–8 as described in Section 7.2.2.2 above.
(3) Finally dry the RNA pellet very briefly using a stream of argon gas, then dissolve it immediately in sterile 0.1 M aqueous ammonium bicarbonate and purify immediately by reversed phase HPLC as the DMTr group has a limited half-life under these conditions.

7.2.3
Purification

This entire section is devoted to the anion-exchange HPLC purification of fully deprotected RNA using gradients of sodium perchlorate [17] or lithium perchlorate [18] as chaotropes, the reversed-phase HPLC purification of trityl-on RNA, detritylation and desalting. In this section you will need the following items:

(1) A biocompatible HPLC system (Amersham Biosciences or other).
(2) A set of anion-exchange HPLC columns, e.g. MonoQ 5/5, Source 15Q 16/10 and/or a FineLINE 35 pilot column packed with Source 15Q (Amersham Biosciences) or Dionex DNAPac PA-100 columns (Dionex, Sunnyvale, CA, USA).
(3) Sodium perchlorate. *Please note that this salt is toxic and corrosive.*
(4) Disodium EDTA.
(5) 1 M sterile Tris–HCl buffer, pH 7.4.
(6) A Hi-Prep 26/10 desalting column (Amersham Biosciences).
(7) Reversed-phase HPLC columns, for instance Hamilton PRP-1, 7 × 305 mm, XTerra™ RP$_8$, 4.6 × 250 mm (Waters), XTerra™ RP$_8$, 19 × 300 mm and/or a FineLINE 35 pilot column packed with Source 15RPC (Amersham Biosciences).
(8) HPLC grade acetonitrile.
(9) High-purity ammonium bicarbonate.
(10) Glacial acetic acid.

7.2.3.1 Anion-exchange HPLC Purification

The purity of oligoribonucleotides less than 25 nt in length, obtained by anion-exchange HPLC as the only purification step is perfectly adequate for most biological applications. It generally results in a purity in the range of 95–98%. Longer RNAs must be purified in the trityl-on mode, see Section 7.2.3.2 below. It is recommended to use a gradient of sodium perchlorate in sterile water/acetonitrile (9:1 v/v) containing 50 mM Tris–HCl buffer pH 7.6 and 50 µM EDTA for anion-exchange HPLC. The reason for adding EDTA is to complex traces of heavy metals that could otherwise lead to cleavage and degradation of the RNA. Recommended columns are the Source 15Q 16/10 columns for purification of 1-µmol scale syntheses with a flow rate of 5 ml/min. For syntheses in the 10–100-µmol scale, purification is best achieved using a FineLINE Pilot 35 column packed with Source 15Q and eluted at 20 ml/min. The low salt or A buffer preferably contains 10 mM sodium perchlorate and the high salt or B buffer contains 600 mM sodium perchlorate. It has been found that a gradient from 10–60% B during 40 min gives good resolution. When not in use the columns should be stored in 20% ethanol in sterile water to prevent microbial growth. For long-term storage of columns it is advisable to add 0.2% sodium azide as an antimicrobial. Prior to using a column that has been stored wash it with several column volumes of sterile water. The column is then equilibrated by washing it with several column volumes of buffer B followed by several column volumes of 90% buffer A plus 10% buffer B before injecting the

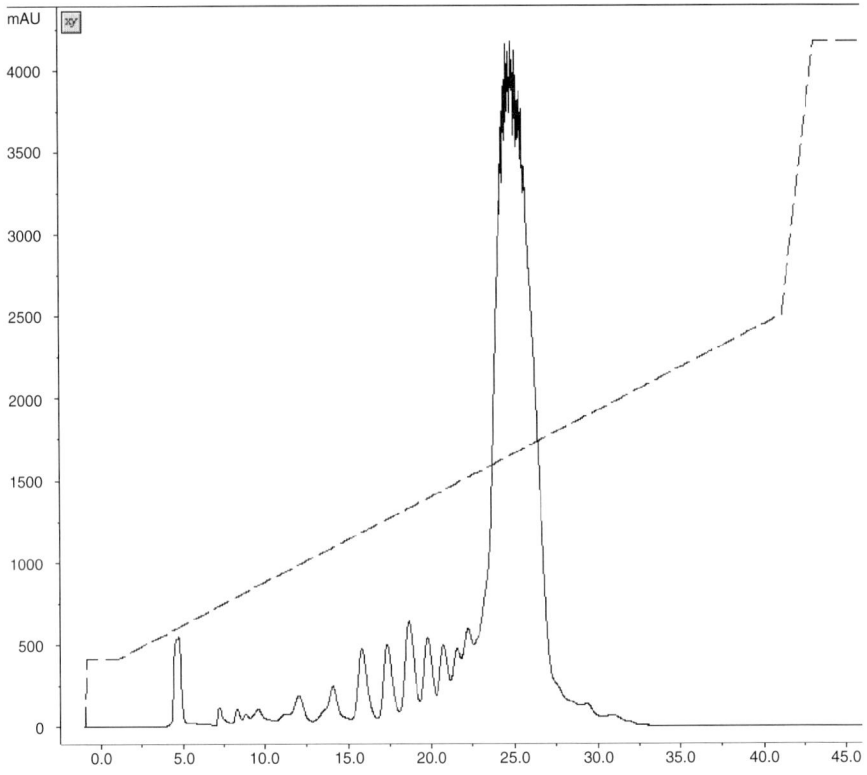

Fig. 7.3. Preparative anion-exchange HPLC trace of a 21mer oligoribonucleotide synthesized manually on 20-µmol scale and purified on Source 15Q packed in a FineLINE Pilot 35 column. The column was eluted with a linear gradient from 10 to 60% B during 40 min at a flow rate of 20 ml/min. Buffer A was 10 mM sodium perchlorate, 50 µM EDTA and 50 mM Tris–HCl, pH 7.6 in sterile water/acetonitrile (9:1 v/v), and buffer B was the same as buffer A except that the sodium perchlorate concentration was 600 mM. Absorbance was monitored at 280 nm. The x-axis is in minutes.

sample of fully deprotected RNA as obtained in Section 7.2.2.2 dissolved in buffer A and running the salt gradient. The desired product peak is the late eluting major component. This material is then desalted as described in Section 7.2.3.4 below. A typical trace of an anion-exchange HPLC purification is shown in Fig. 7.3. In this example the oligomer is a 21mer synthesized manually on a 20-µmol scale and purified on Source 15Q packed in a FineLINE Pilot 35 column. As can be seen the failure peaks are very small compared to the product peak which elutes at about 25 min.

7.2.3.2 Reversed-phase HPLC Purification of Trityl-on RNA

The highly lipophilic dimethoxytrityl group profoundly retards the full-length trityl-on RNA when purification is performed on a reversed phase HPLC column.

Although a better separation of failure peaks from the desired trityl-on product peak is obtained using aqueous triethylammonium acetate/acetonitrile buffers, for ease of salt removal and minimal damage to the RNA the use of ammonium bicarbonate instead of triethylammonium acetate is strongly preferred. However, make up the ammonium bicarbonate buffer fresh, otherwise store it cold, since it has a limited stability at room temperature in contrast to triethylammonium acetate.

Columns recommended for trityl-on purification are the Hamilton PRP-1, 7×305 mm for purifications on the scale of a few micromoles or a 21.5×250 mm column for 10–20-μmol scale purifications. As an alternative for larger-scale purifications, a FineLINE Pilot 35 column packed with Source 15RPC can be used. The buffers required are 0.1 M ammonium bicarbonate prepared in sterile water (buffer A) and 0.1 M aqueous ammonium bicarbonate/acetonitrile (1:1 by volume), which is buffer B. A useful gradient to use is 0–90% B during 40 min. The failure peaks elute early and are well separated from the desired trityl-on product peak which elutes last. The pure product fraction should be collected in a polypropylene Falcon tube and dried down in a Speed Vac. Residual ammonium bicarbonate is then removed by lyophilization of the product, which is then ready for detritylation. A typical trace of a trityl-on RNA purification by reversed phase HPLC is shown in Fig. 7.4. The example shows a trityl-on 34mer oligoribonucleotide synthesized by machine on a 20-μmol scale and purified on a FineLINE Pilot 35 column packed with Source 15RPC. As can be seen the desired product peak elutes at 20–25 min well separated from the trityl-off failure sequences which elute between 7 and 12 min.

7.2.3.3 Detritylation of Trityl-on RNA

(1) Dissolve the HPLC-purified trityl-on RNA, obtained in Section 7.2.3.2 above, in 3% sterile aqueous acetic acid (200 μl/μmol) and keep for 45 min at room temperature. The pH should be about 3.5.
(2) Neutralize the solution by careful addition of solid ammonium bicarbonate until evolution of carbon dioxide ceases. The pH will now be about 7.8.
(3) Dry the sample in a Speed Vac.
(4) Repurify the product by anion-exchange HPLC, as described in Section 7.2.3.1 above, which in addition converts the RNA from the ammonium form into the biologically useful sodium form.
(5) Desalt according to Section 7.2.3.4 below.

As an alternative to Steps 4 and 5 the salt exchange can be performed in a reliable and high yielding fashion by dissolving the residue from Step 3 in sterile 0.3 M aqueous sodium acetate (400 μl/μmol synthesis scale) and adding 2.5 volumes of absolute ethanol. After mixing and storage at $-70\,°C$ for 20 min the precipitated RNA is recovered by centrifugation, washed once with absolute ethanol and then dried carefully *in vacuo*. To ensure a complete exchange of cation from ammonium to sodium the precipitation procedure should be repeated once more. This protocol

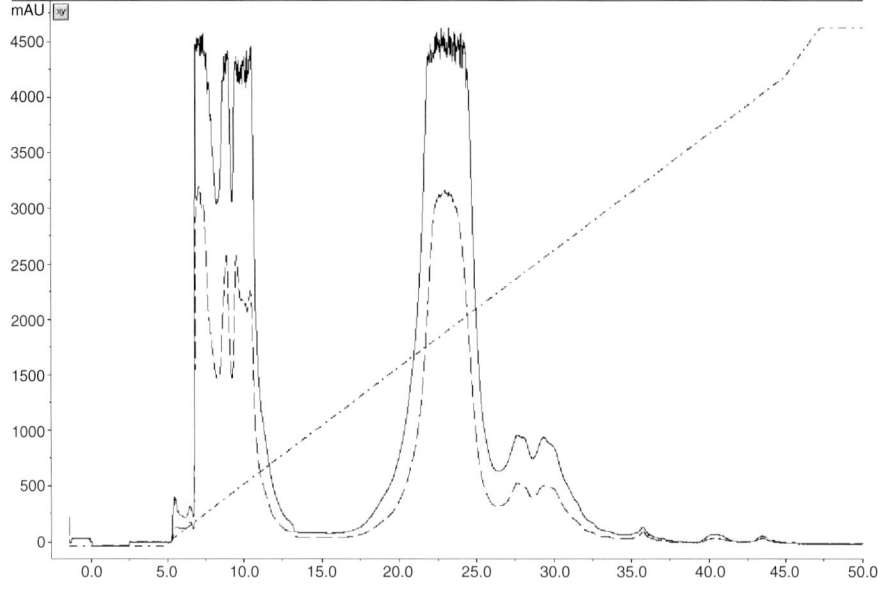

Fig. 7.4. Preparative reversed phase HPLC trace of a trityl-on 34mer oligoribonucleotide synthesized by machine on a 20-µmol scale and purified on Source 15RPC packed in a FineLINE Pilot 35 column. The column was eluted with a linear gradient from 0 to 90% B during 40 min at a flow rate of 25 ml/min. Buffer A was 0.1 M sterile aqueous ammonium bicarbonate and B was 0.1 M aqueous ammonium bicarbonate/acetonitrile (1:1 v/v). The solid line traces absorption at 260 nm, the dashed line that at 280 nm. The x-axis is in minutes.

is convenient if the detritylated RNA is of sufficient purity as determined by analytical anion-exchange HPLC or PAGE.

7.2.3.4 Desalting by HPLC

Oligoribonucleotides purified by anion-exchange HPLC must be free of excess salt, buffer and EDTA regardless of the intended application. This is readily achieved using a desalting column such as the Hi-Prep 26/10 which is filled with Sephadex G-25. The sample should be loaded on in a volume not greater than 15 ml, but preferably less than 10 ml, to obtain a complete separation between the RNA which elutes first and the salt peak which follows it. The column should be stored in and eluted with 20% ethanol in sterile water which prevents microbial growth. For long-term storage it is recommended to add sodium azide as an antibacterial/antimicrobial agent. Careful monitoring of the column effluent by UV and conductivity avoids contamination of the RNA by incompletely removed salt which does not absorb in the UV spectral range.

The desalted RNA in aqueous ethanol is first concentrated in a Speed Vac and finally lyophilized to give the pure RNA in its sodium form as a fluffy white solid.

As an alternative to the HPLC method, fast desalting of micomole-scale RNA samples can be achieved using the NAP columns from Amersham. For very-large-scale desalting, ultrafiltration is the method of choice since in addition it results in sample concentration whereas other desalting methods such as dialysis and gel filtration lead to sample dilution.

7.2.4
Analysis of the Purified RNA

The purity of the RNA purified by the HPLC methods described above should be checked by analytical anion-exchange HPLC using a MonoQ 5/5 or Dionex DNA-Pac PA-100 column and buffers described in Section 7.2.3.1 above. Suitable alternatives are capillary gel electrophoresis and PAGE, both of which operate under denaturing conditions. These methods are not described here.

In addition RNA destined for structural studies such as NMR or X-ray crystallography should also be analyzed in addition by analytical reversed phase HPLC on a high-resolution column such as the 5-μm XTerra™ RP_8, 4.6 × 250 mm column from Waters using a gradient from 0–20% acetonitrile in aqueous ammonium bicarbonate, combined with a flow rate of 1 ml/min. RNA that is deemed not to be of sufficient purity should be further purified by preparative reversed phase HPLC on the 7-μm XTerra™ RP_8, 19 × 300 mm column using a similar gradient to that used for the analytical reversed phase analysis but with a flow rate of 12 ml/min. The material is desalted and exchanged to the sodium form using the procedures described above. As an example the preparative reversed phase HPLC trace of an anion-exchange HPLC purified 27mer oligoribonucleotide, destined for NMR spectroscopy, is illustrated in Fig. 7.5 and an analytical reversed phase HPLC of the now double-purified material is illustrated in Fig. 7.6.

To check product authenticity the molecular weight of the RNA should be determined by mass spectroscopy, either electrospray ionization or matrix-assisted laser desorption ionization (MALDI-TOF) [19]. This is absolutely essential when modified nucleotides are incorporated that contain, for instance, unusual protecting groups. As a final proof of authenticity the RNA can also be sequenced by standard methods to check the absolute order of the monomeric units within the sequence, information which is of course not directly available by mass spectroscopy. These methods are outside the scope of this chapter and the reader is advised to consult the literature.

7.3
Troubleshooting

Many potential sources of problems can be eliminated by taking some simple precautions during the synthesis and purification steps. The following points should

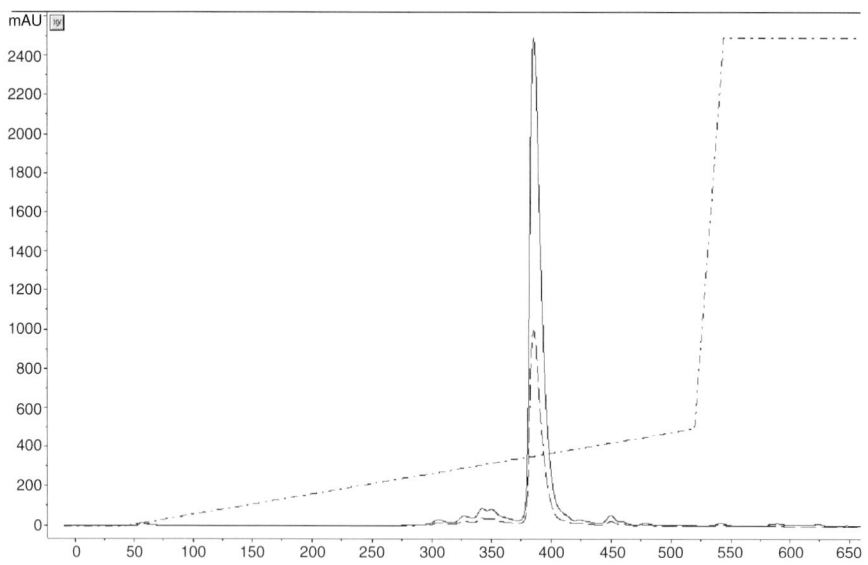

Fig. 7.5. Preparative reversed phase HPLC trace of a 27mer oligoribonucleotide on a 7-μm XTerra™ RP$_8$, 19 × 300 mm column. The compound was initially purified trityl-on by anion-exchange HPLC. The RP$_8$ column was eluted with a linear gradient from 0 to 40% B during 40 min at a flow rate of 12 ml/min. Buffer A was 0.1 M sterile aqueous ammonium bicarbonate and B was acetonitrile. The solid line traces absorption at 260 nm, the dashed line that at 280 nm. The x-axis is in milliliters.

be heeded during the solid-phase synthesis of RNA. The coupling step in the solid-phase synthesis is very sensitive to traces of water and it is essential to use very dry acetonitrile for monomer dissolution. It is also imperative to allow monomers to reach room temperature before opening and weighing out material, otherwise condensation of atmospheric water will occur leading to eventual degradation. The bottle contents should also be put back under argon before sealing and storing again at −20 °C. Addition of activated molecular sieves or trap bags to the monomer and activator solutions ensures that they stay dry during the synthesis. It is also critical to wash away the acid from the detritylation step with copious acetonitrile washes, otherwise residual acid will cause serious problems with the coupling step.

Incomplete oxidation will cause serious problems with the overall synthesis yield and quality as any residual phosphite triester is cleaved at the internucleotide linkage by the acid used in the detritylation step. As mentioned the standard oxidation mixture from several commercial suppliers is only 17 mM in iodine, i.e. 17 μmol/ml, so use enough solution for larger-scale syntheses to ensure that there is an ample excess of reagent and/or increase the iodine concentration to 50 mM. For a 20-μmol scale synthesis use 5 ml of the oxidation mixture. At the end of the solid-phase assembly the CPG or polystyrene bearing trityl-off protected RNA can be stored cold and dry ready for deprotection at an appropriate time; however, CPG

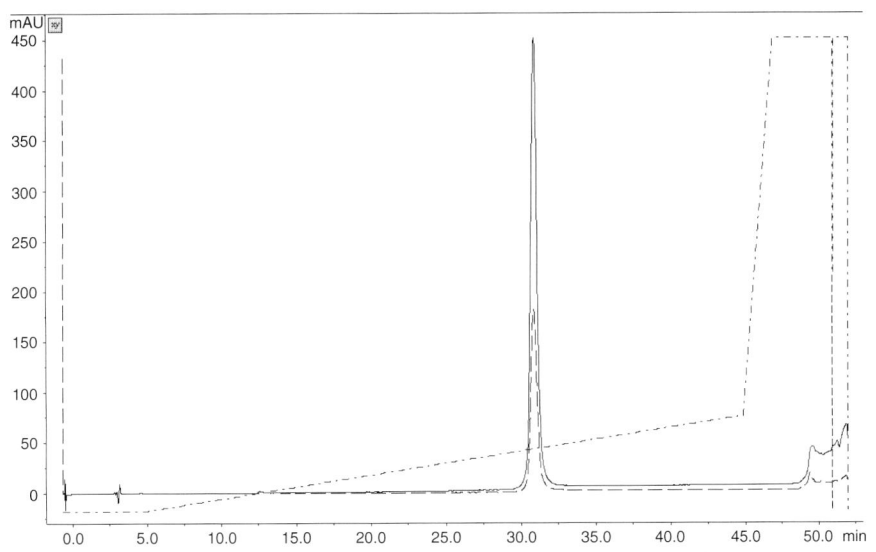

Fig. 7.6. Analytical reversed-phase HPLC trace of double HPLC-purified 27mer oligoribonucleotide run on a 5-μm XTerra™ RP$_8$, 4.6 × 250 mm column. The column was eluted with a linear gradient from 0–40% B during 40 min at a flow rate of 1 ml/min. Buffer A was 0.1 M sterile aqueous ammonium bicarbonate and B was acetonitrile. The solid line traces absorption at 260 nm, the dashed line that at 280 nm. The x-axis is in minutes.

bearing trityl-on RNA must be deprotected and purified immediately upon completion of the synthesis, otherwise there will be partial or complete loss of the trityl group during storage. This is particularly bad for RNAs that terminate at the 5′ end with one or more G residues.

For safety reasons handle all chemicals in a well-ventilated fume cupboard and wear suitable resistant disposable gloves, particularly when handling toxic materials such as triethylamine *tris*(hydrofluoride). In addition, make sure that you have read handling protocols for all chemicals – in particular those with which you are not familiar. Concerning the deprotection step with ammonia/methylamine, avoid too great an air space in the vial or bottle, otherwise most of the ammonia and methylamine will end up in the vapor phase. In the worst case this could lead to an incomplete deprotection. Take care when handling trityl-on RNA, do not over dry or let it get too hot, and purify immediately to avoid partial detritylation that will result in an unnecessary loss of product. This problem seems to be particularly serious with sequences that have one or more Gs at the 5′ end.

To avoid inadvertent degradation of unprotected RNA by RNases use RNase-free salts, sterilize all buffers by autoclaving and sterilize all glassware in a 180 °C oven. In addition, wear disposable gloves and as much as possible use sterile plasticware such as Eppendorf and Falcon tubes. Concerning anion-exchange HPLC purification, oligoribonucleotides that contain four or more consecutive Gs are notoriously difficult to purify since they form tetraplexes and higher aggregates in solution. Such RNAs are best purified using a lithium perchlorate gradient since these structures do not form if lithium ions are present instead of sodium or potassium ions. Of course prior to use in biological experiments the lithium ions must be replaced by sodium ions since lithium ions are toxic in many biological systems. Denaturants such as formamide can also be added to the salt gradient to reduce problems caused by strong secondary structures.

In large-scale purifications, to avoid product shoot through due to the ionic strength of the applied sample solution being too high, it is advisable to apply the crude RNA sample to the FineLINE Pilot 35 column dissolved in a volume of 10–50 ml of 50 mM Tris–HCl buffer, pH 7.4, using a 10- or 50-ml superloop. As an alternative desalt the sample prior to purification.

It is important to note that RNA samples as sodium salts are not suitable for mass spectrometry. Mass spectrometry samples of RNA are best prepared as ammonium salts. This can be done in several ways. One way is to do anion-exchange HPLC using ammonium sulfate for elution, followed by desalting on a small NAP cartridge. In this case store all buffers in plastic bottles and collect the product in an Eppendorf tube. A second way is to take a small aliquot of the RNA in its sodium form and exchange the sodium ions with ammonium ions by using ammonium form Dowex 50 cation exchange resin. A third way is to purify a small sample of RNA by reversed-phase HPLC using the aqueous ammonium bicarbonate/acetonitrile system followed by lyophilization to remove the residual salt. Once in the ammonium form the RNA should not be in contact with glass surfaces, otherwise sodium and potassium ions will be picked up that will severely degrade the quality of the mass spectra. This latter point is of great importance when trying to analyze very long RNAs, e.g. in the 40–70mer range.

Following the protocols and troubleshooting hints given above, the reader should be in a position to synthesize and purify RNA with success.

References

1 G. Ramaswamy, F. J. Slack, *Chem. Biol.* **2002**, *9*, 1053–1055.
2 J. Couzin, *Science* **2002**, *298*, 2296–2297.
3 N. Usman, K. K. Ogilvie, M.-Y. Jiang, R. J. Cedergren, *J. Am. Chem. Soc.* **1987**, *109*, 7845–7854.
4 S. Pitsch, *Helv. Chim. Acta* **1997**, *80*, 2286–2314.
5 S. A. Scaringe, F. E. Wincott, M. H. Caruthers, *J. Am. Chem. Soc.* **1998**, *120*, 11820–11821.
6 S. A. Scaringe, *Methods Enzymol.* **2000**, *317*, 3–18.
7 M. D. Matteucci, M. H. Caruthers, *J. Am. Chem. Soc.* **1981**, *103*, 3185–3191.
8 N. D. Sinha, J. Biernat, J. McManus, H. Köster, *Nucleic Acids Res.* **1984**, *12*, 4539–4557.
9 C. Chaix, D. Molko, R. Téoule, *Tetrahedron Lett.* **1989**, *30*, 71–74.
10 N. D. Sinha, P. Davis, N. Usman, J. Pérez, R. Hodge, J. Kremsky, R. Casale, *Biochimie* **1993**, *75*, 13–23.
11 D. Gasparutto, T. Livache, H. Bazin, A.-M. Duplaa, A. Guy, A. Khorlin, D. Molko, A. Roget, R. Téoule, *Nucleic Acids Res.* **1992**, *20*, 5159–5166.

12 E. Westman, R. Strömberg, *Nucleic Acids Res.* **1994**, *22*, 2430–2431.
13 R. Welz, S. Müller, *Tetrahedron Lett.* **2002**, *43*, 795–797.
14 B. S. Sproat, M. J. Gait, *Oligonucleotide Synthesis: A Practical Approach*, IRL Press, Oxford, **1994**, p. 92.
15 Q. Song, R. A. Jones, *Tetrahedron Lett.* **1999**, *40*, 4653–4654.
16 R. Vinayak, A. Andrus, A. Hampel, *Biomedical Peptides, Proteins & Nucleic Acids* **1995**, *1*, 227–230.
17 F. Wincott, A. DiRenzo, C. Shaffer, S. Grimm, D. Tracz, C. Workman, D. Sweedler, C. Gonzalez, S. Scaringe, N. Usman, *Nucleic Acids Res.* **1995**, *23*, 2677–2684.
18 B. Sproat, F. Colonna, B. Mullah, D. Tsou, A. Andrus, A. Hampel, R. Vinayak, *Nucleosides & Nucleotides* **1995**, *14*, 255–273.
19 U. Pieles, W. Zürcher, M. Schär, H. E. Moser, *Nucleic Acids Res.* **1993**, *21*, 3191–3196.

8
Modified RNAs as Tools in RNA Biochemistry

Thomas E. Edwards and Snorri Th. Sigurdsson

8.1
Introduction

RNA displays a vast variety of functions in that it carries genetic information, regulates gene expression, catalyzes reactions and participates in all facets of protein expression [1]. In addition to the four basic nucleosides (adenosine, guanosine, cytidine and uridine), many RNA molecules contain modified nucleosides essential for function. The fact that these modifications are essential for function in some RNAs, but entirely absent in others, indicates a significant layer of complexity in the hierarchy of RNA structure. With progress in the chemical synthesis of RNA over the last 15 years, modified nucleosides can now be readily incorporated at specific positions in RNA. These advances in solid-phase synthesis have promoted a cornucopia of experiments examining the influence of single-functional-group modification on the biological function of RNA.

Modified nucleosides have also been site-specifically incorporated into RNA as reporter groups for biochemical and biophysical structure–function analysis. There is a large diversity in these approaches. For example, fluorescent probes have been used to report internal changes during RNA folding [2] as well as to measure interhelical distances for determining the global structure of RNA [3–5]. Disulfide crosslinks have been used to restrict RNA helical elements to validate structural models based on other techniques [6]. These are but a few examples of site-specific incorporation of RNA structure–function probes.

The major goals of this chapter are to review the various types of modifications that can be incorporated site-specifically into RNA by chemical synthesis, and to provide a general method for the incorporation of reporter groups into RNA for biochemical and biophysical analysis. This includes comparison of the two central strategies for the incorporation of modified nucleosides into RNA, i.e. the phosphoramidite strategy and post-synthetic labeling. The phosphoramidite strategy utilizes chemical synthesis of a modified nucleoside phosphoramidite in conjunction with solid-phase synthesis, whereas post-synthetic labeling utilizes incorporation by the phosphoramidite method of a convertible nucleoside containing a reac-

Handbook of RNA Biochemistry: Student Edition. Edited by R. K. Hartmann, A. Bindereif, A. Schön, E. Westhof
Copyright © 2009 WILEY-VCH Verlag GmbH & Co. KGaA, Weinheim
ISBN: 978-3-527-32534-4

tive group, which is selectively modified after oligonucleotide synthesis with a labeling reagent. The advantages and disadvantages of each modification strategy will be described. Finally, a general and efficient modification strategy will be presented that utilizes post-synthetic labeling of 2′-amino groups with a wide range of reporter groups through a number of different coupling chemistries.

8.1.1
Modification Strategy: The Phosphoramidite Method

While enzymatic synthesis can be used to prepare uniformly labeled RNA, modified nucleosides can be incorporated site-specifically into RNA by solid-phase chemical synthesis using modified nucleoside phosphoramidites [7]. The main advantage of this method is that it allows for the incorporation of a desired modification or reporter group at a specific position in the RNA. While this is a highly effective and powerful method, it has several disadvantages. In most cases the synthesis of a modified phosphoramidite requires a lengthy and costly synthetic route. Furthermore, the reporter group must be stable to the conditions used in solid-phase oligonucleotide synthesis (e.g. incubation with acid, base and oxidizing solutions) as well as the deprotection conditions. However, phosphoramidites of many desirable modified nucleosides are commercially available, providing rapid, cost-effective access to a variety of modified RNAs.

There are four basic categories of RNA modifications that can be incorporated into RNA via the phosphoramidite method: end (5′ and 3′), base, phosphate and sugar modifications. Figure 8.1 shows selected examples of such modifications, and Table 8.1 lists several of the RNA modifications that are commercially available as phosphoramidites and/or modified RNAs. Of those not commercially available (many of which are reviewed in [8–11]), other notable examples of RNA modification by the phosphoramidite method include the base modifications 2-deoxyribonolactone [12] and 5-ketone pyrimidine derivatives [13]. Internucleotide linkages include boranophosphates [14] and phosphoroselenoates [15]. Sugar modifications

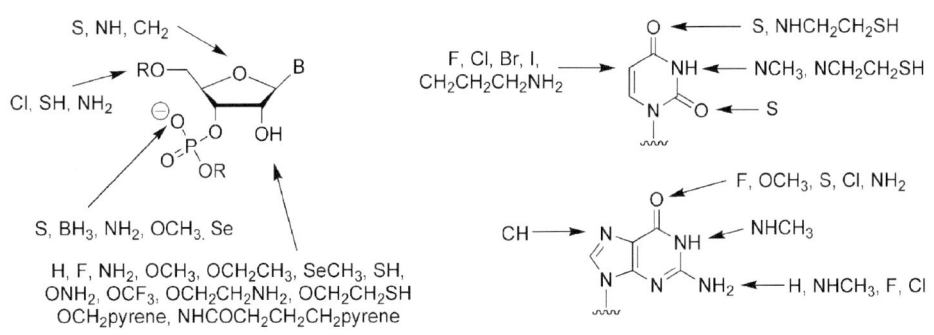

Fig. 8.1. Selected examples of modifications that can be incorporated at the sugar (left), phosphodiester backbone (left) and base (right) using the phosphoramidite method.

Tab. 8.1. Commercially available modifications that can be incorporated into RNA by the phosphoramidite method

Site	Modification	Commercial Source[1]	Reference[2]
End-labeling			reviewed in 74–76
5′ end	fluorescent dyes	CG, Dh, GR	3–5
	amino groups	CG, Dh, GR	77, 78
	biotin	CG, Dh, GR	
	photo-cleavable biotin	Dh, GR	
	5′-thiol	Dh, GR	
	acridine	GR	
3′ end	fluorescent dyes	CG, Dh, GR	
	amino groups	CG, Dh, GR	
	inverted abasic	Dh	
	puromycin	CG, Dh	79
	dideoxy G,C	Dh	
	biotin	CG, GR	
	acridine	GR	
	psoralen	CG, GR	
	cholesterol	CG, GR	
	DNP	CG, GR	
Internucleotide			
	S (non-bridging)	CG, Dh, GR	29–31
	3, 9, 18 atom spacers	CG, Dh, GR	
Sugar			reviewed in 9, 10
1′	abasic	Dh, GR	
2′	NH_2 U,C	CG	37
	F U,C	CG, Dh	
	OCH_3	CG, Dh, GR	
	SCH_3 U	GR	80
	$OCH_2CH_2CH_2NH_2$	CG	65
	$NHCOCH_2CH_2CH_2pyr$ U	Dh	36
	LNA	GR	81
Purine			reviewed in 9, 10
	N^6,N^6-dimethyl A	Dh	
	inosine	CG, Dh, GR	
	purine ribonucleoside	CG, Dh	
	ribavirin	Dh	
	7-deaza A,G	CG	
	2-aminopurine	CG, Dh, GR	2
	2,6-diaminopurine	Dh	
	8-bromo A	CG	
Pyrimidine			reviewed in 9, 10
	N^3-methyl U,rT	CG	
	N^3-thiobenzoyl ethyl U	CG	
	4-triazoylyl U	CG	
	N^4-ethyl C	CG	
	pyridine-2-one	CG	
	pyrrolo-C	GR	

Tab. 8.1. (continued)

Site	Modification	Commercial Source[1]	Reference[2]
	2,2′-anhydro U	CG	
	5-methyl U,C	Dh	
	4-thio uridine	CG, Dh, GR	71
	5-fluoro U	CG, Dh, GR	
	5-bromo U,C	CG, Dh, GR	
	5-iodo U	CG, Dh, GR	
	pseudouridine	CG, Dh, GR	
	5-CHCHCH$_2$NH$_2$ U	Dh	

[1] For commercial sources: CG, ChemGenes; Dh, Dharmacon; GR, Glen Research. Please note that chemical suppliers are subject to change and this list is a representative example at time of publication. Several other companies exist which sell modified RNA and modified RNA phosphoramidites.
[2] References are select examples and may be reviews, applications or synthetic procedures.

include 1′-deutero [16], 2′-modifications (O-(2-thioethyl) [17] and O-(2-aminoethyl) [18]), 5′-modifications (tallo or C-methyl [19, 20], chloro [21], amino [21]) and per-deuterated ribose [22]. Fluorescent labels have also been synthetically attached to the 2′-position via an ether linkage [23], a carbamoyl linker [24], an arabino carbamoyl linker [25] and an amido linkage [26].

The phosphoramidite method has been particularly useful in the investigation of ribozyme cleavage mechanisms. For example, incorporation of a 5′-C-methyl-modified nucleoside near the cleavage site of the hammerhead ribozyme resulted in a kinetically trapped intermediate in a crystal and provided information about a conformational change along the reaction pathway prior to transition state formation [20]. In another example, a crystal structure of the hairpin ribozyme containing a 5′-chloro group at the cleavage site provided structural information for comparison with the non-cleaved state (all RNA) and a vanadyl transition state mimic, providing valuable information about the entire mechanistic pathway [27].

8.1.2
Modification Strategy: Post-synthetic Labeling

Post-synthetic modification of convertible nucleosides enables the site-specific incorporation of a wide variety of reporter groups into RNA. The main advantage of this approach is that once the RNA has been prepared, it enables the rapid and efficient production of a wide variety of modified RNAs. Another advantage is that sensitive reporter groups, which would otherwise be unstable to the conditions of solid-phase oligonucleotide synthesis, can be incorporated into RNA. Possible disadvantages of this strategy are that in some cases additional purification steps are necessary and that it may be necessary to synthesize the convertible nucleoside phosphoramidite if the desired one is not commercially available.

Post-synthetic modification strategies have been developed for attachment of re-

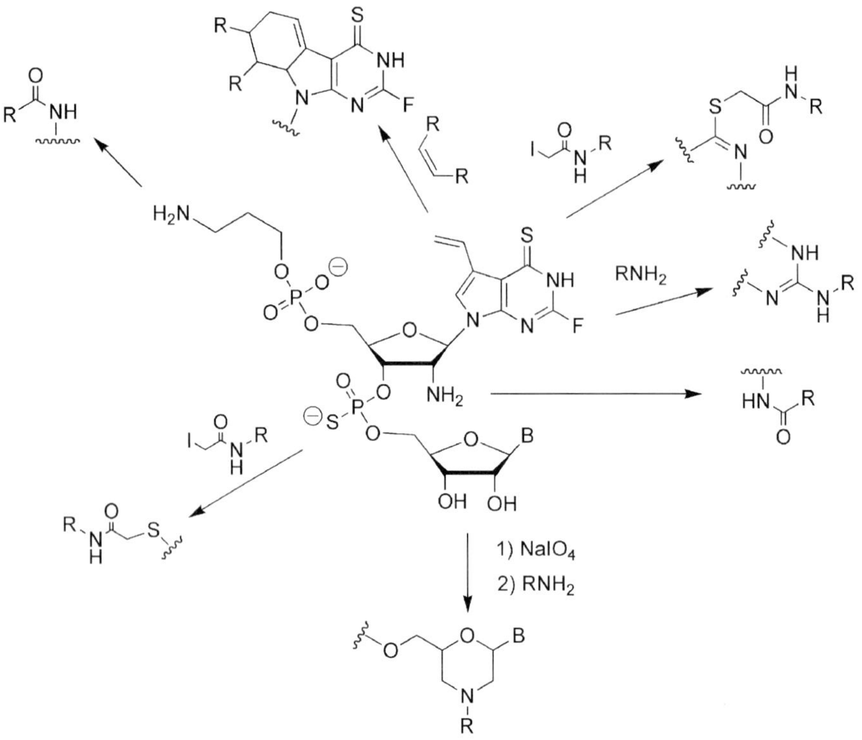

Fig. 8.2. Selected examples of RNA post-synthetic labeling.

porter groups to the 5′ and 3′ ends and at internal sites on the base, phosphate and sugar (Fig. 8.2 and Table 8.2). The main focus of this chapter is the general strategy of post-synthetic labeling of 2′-amino containing RNA and this approach will be described in detail in the next section. In addition to the modifications shown in Fig. 8.2 and Table 8.2, various groups can be attached to the 5-position of pyrimidines via on-column Pd-catalyzed coupling reactions [28]. A variety of molecules have also been attached to the phosphodiester backbone through phosphorothioate [29–31] or phosphoramidate linkages [32]. However, these labeling strategies are problematic for RNA due to the inherent instability of these linkages in the presence of 2′-hydroxyl groups; consequently, this problem is overcome by incorporation of a 2′-deoxy or 2′-methoxy group at the nucleotide 5′ of the modification.

8.2
Description of Methods

8.2.1
Post-synthetic Modification: The 2′-Amino Approach

Post-synthetic labeling of the 2′-amino group (Fig. 8.3a and b) has emerged as an effective approach for the site-specific incorporation of reporter groups into RNA.

Tab. 8.2. Select examples of modifications for post-synthetic RNA derivatization

Modification	Molecular handle	Commercially available?[1]	Labeling reactants	Reference[2]
1	3-amino modifiers	CG, Dh, GR	activated esters	77, 78
2[3]	diene	No	dienophile	82
3	sulfur-containing bases	CG, Dh, GR	iodoacetamides, disulfides	69–71
4	convertible F or ClΦ nucleosides	No	amines	83
5	2′-amino	CG, Dh	isothiocyanates, NHS esters, isocyanates	37
6	2′,3′-diols	All (RNA)	NaIO$_4$, amines	84
7	non-bridging phosphorothioates	CG, Dh, GR	iodoacetamides	29–31

[1] For commercial sources: CG, ChemGenes; Dh, Dharmacon; GR, Glen Research. Please note that chemical suppliers are subject to change and this list is a representative example at time of publication. Several other companies exist which sell modified RNA and modified RNA phosphoramidites.
[2] References are select examples and may be reviews, applications or synthetic procedures.
[3] This modification strategy has only been applied to DNA thus far, but is of select interest.

Several notable examples include the incorporation of disulfide crosslinking reagents for the evaluation of RNA helical orientation [6, 33, 34], the incorporation of fluorescent probes for the study of RNA folding and ligand binding [35, 36] and the incorporation of EPR active probes [37] for the study of RNA internal dynamics [38–40] and for distance measurements [41]. RNAs containing 2′-amino groups at specific pyrimidine nucleotides (Dharmacon) and 2′-amino-modified pyrimidine phosphoramidites (ChemGenes) are now commercially available. Because the 2′-amino group is an aliphatic amine, it is more reactive (i.e. nucleophilic) than the aromatic amines or hydroxyl groups native to RNA, making this method of post-synthetic labeling highly selective. The major advantage of the 2′-amino group over other amino-based modifiers (e.g. 5′- and 3′-amino modifiers, 5-alkylamino-modified pyrimidines) is that it offers a minimal linker length. Several chemical conjugation approaches exist, including reaction with succinimidyl esters (often referred to as NHS esters) to produce amide-modified RNA [33], reaction with aromatic isothiocyanates to form thiourea-linked RNA [6, 42] and reaction with aliphatic isocyanates to prepare urea-tethered RNA [42, 43]. These three methods will be described in detail below, and examples employing these methods to address biochemical and biophysical questions will be provided. Of notable importance for this modification strategy is an alternative 2′ protection approach that has been developed based on a photo-cleavable protecting group in place of the standard 2′-trifluoroacetyl group; after removal of the protecting group, the 2′-amino group may be derivatized on-column, providing many advantages over solution-based post-synthetic modification of deprotected oligonucleotides [44]. Other

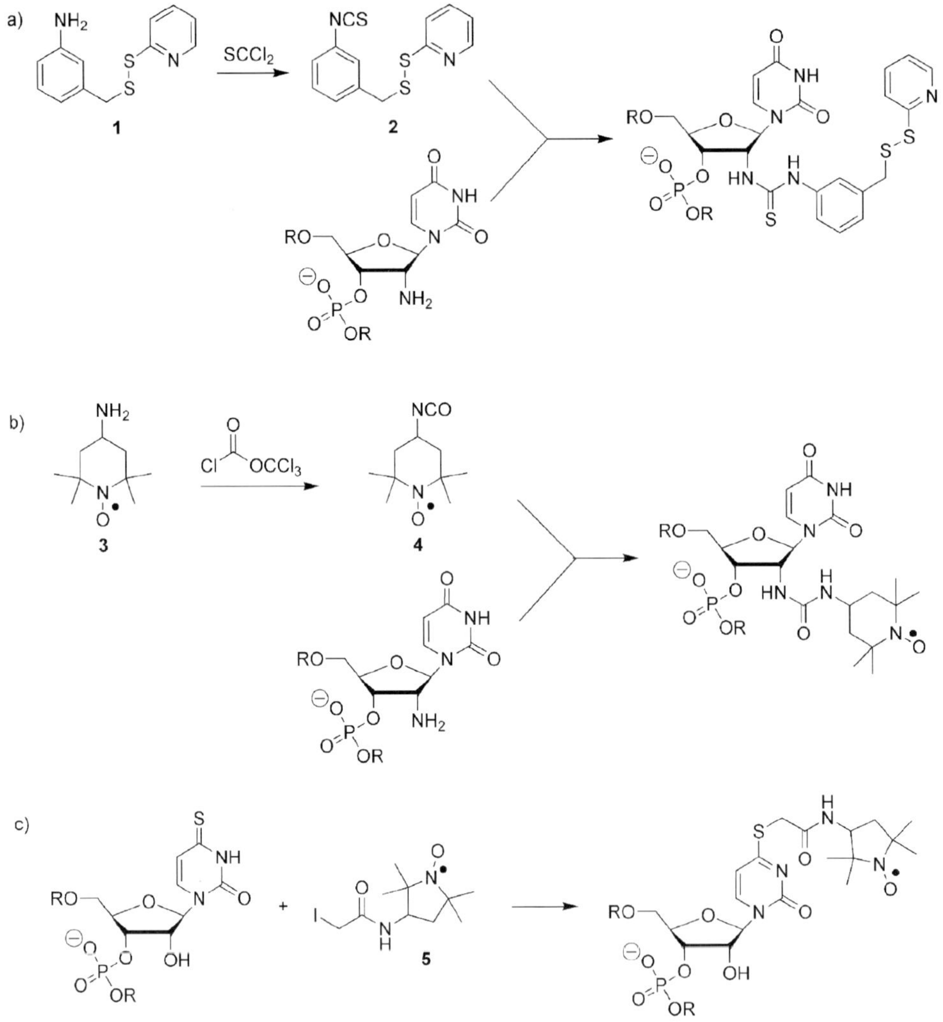

Fig. 8.3. Preparation of an isothiocyanate crosslinking reagent **2** (a) and an EPR spin-labeling reagent 4-isocyanato TEMPO **4** (b) from the corresponding amines using thiophosgene and diphosgene, respectively, and their subsequent incorporation into 2'-amino-modified RNA. (c) Post-synthetic modification of 4-thiouridine by alkylation with spin-label **5**.

approaches for the attachment of reporter groups to the 2'-position of oligonucleotides that will not be addressed in depth include chelation of metal ions such as ruthenium to 2'-amino-modified oligonucleotides [45, 46], incorporation of fluorescamine at a 2'-amino group using a Michael addition and rearrangement reaction [47], reaction of amines with a 2'-O-(acetaldehyde) group [48] and reduction of thiol-containing compounds with 2'-O-(2-thioethyl) to form disulfide-linked modified RNAs [17]. The attachment of sterically hindered compounds to the 2′-amino group may be difficult and can be overcome by use of the 2′-O-(2-aminoethyl) modification [18].

8.2.1.1 Reaction of 2′-Amino Groups with Succinimidyl Esters

The reaction of 2′-amino groups with succinimidyl esters to produce amido-linked modified RNAs has been used to incorporate disulfide crosslinks to evaluate RNA conformational dynamics [33, 49], to convert the hammerhead ribozyme from a ribonuclease to a ligase ribozyme [50], to incorporate photocrosslinking reagents to evaluate RNA tertiary structure [51], to identify base-pair mismatches [52, 53], and to incorporate fluorescent pyrene labels to study RNA folding and ligand binding [36, 54–56]. In conjunction with the isocyanate method described below, this method has been used to probe steric interference in the hammerhead ribozyme [57]. Catalysis of this chemistry by the phosphodiester on the 3′-position adjacent to the 2′-amino-containing nucleoside and/or the 3′-oxygen has been reported [58]. One advantage of this method is that many succinimidyl esters are commercially available (e.g. Molecular Probes and ChemGenes have a wide variety of amine-reactive succinimidyl ester dyes available). The major drawback of this method is that this chemistry often suffers from low labeling efficiency, e.g. the pyrene succinimidyl esters typically coupled with only 20–26% yield after purification [36]. However, in some cases it is possible to overcome this low coupling efficiency by the use of the corresponding carboxylic acid with an activating agent, such as a carbodiimide, which may result in nearly quantitative coupling [18, 44, 49]. Another disadvantage of this modification approach is that 2′-amido modifications destabilize RNA when incorporated at internal positions ($\Delta T_m \sim -5$ to 12 °C per modification) [26, 59, 60]. However, some 2′-amido-linked modifications located at end positions increase RNA stability [26, 61], which is likely a result of these particular modifications that contain large aromatic groups (e.g. fluorescent labels), which may stack onto the end of the helix. Nevertheless, this destabilizing effect of 2′-amido groups at internal sites should be kept in mind when incorporating reporter groups into RNA.

8.2.1.2 Reaction of 2′-Amino Groups with Aromatic Isothiocyanates

The reaction of 2′-amino groups with aromatic isothiocyanates (Fig. 8.3a) has been used to incorporate fluorescent probes [26, 35], disulfide crosslinks [6, 34] and photocrosslinking agents [62] into RNA. The main advantage of this method is the highly efficient chemistry, which has resulted in reported conversion yields in excess of 90% in all cases. In addition, fewer equivalents of the isothiocyanate labeling reagent are required than for the succinimidyl ester chemistry. The main drawback is that although some isothiocyanates are commercially available, most must be prepared from the corresponding amine and thiophosgene; however, this synthetic conversion is relatively straightforward [42, 63]. There is limited available UV thermal stability data for 2′-thioureido modifications and all of the data involves incorporation of large fluorescent probes (fluorescein and rhodamine). Following a similar pattern to that observed for the 2′-amido modifications, these modifications are rather destabilizing at internal positions, but have a stabilizing effect at end positions [26, 35]. Isothiocyanates have also been used to selectively incorporate metal ion chelators at 5-amino-derived pyrimidines [64] and at 2′-(O-propylamino)-derived nucleotides [65].

8.2.1.3 Reaction of 2′-Amino Groups with Aliphatic Isocyanates

The reaction of 2′-amino groups with aliphatic isocyanates (Fig. 8.3b) is a versatile platform for the incorporation of biochemical and biophysical reporter groups into RNA. This method has been used to incorporate disulfide crosslinks [34, 43, 66]; an activated disulfide that can be used to conjugate a wide variety of groups such as cholesterol [67], glutathione and bimane [43]; nitrophenol [44], pyrene [18] and nitroxide spin-labels [37]. Like the isothiocyanate coupling, this chemistry is highly efficient, typically displaying quantitative yields for unhindered isocyanates [43, 44]. Unlike the succinimidyl ester coupling, there is no leaving group for the isocyanate (and isothiocyanate) coupling chemistry, and therefore good yields have been reported for structurally hindered isocyanates (e.g. 90% yields are routinely observed for the secondary isocyanate, 4-isocyanato-TEMPO [37, 38]). Due to the high selectivity and efficiency of this reaction, the crude, deprotected RNA can be labeled directly, allowing for only one purification and therefore high yields. Another advantage is the relatively fast coupling time (15–60 min). In addition, 2′-ureido modifications are not as destabilizing as 2′-amido modifications [43, 66], e.g. incorporation of the EPR spin probe TEMPO through a 2′-urea linkage at internal base-pairing nucleotides resulted in a minor decrease in stability as measured by a small decrease in melting temperature of 1–3 °C [37]. The main drawback is that usually the isocyanate labeling reagent must be prepared from the corresponding amine; however, this chemistry is straightforward and pure isocyanates are obtained in high yields after purification by extraction [42, 43, 68].

8.3
Experimental Protocols

The general experimental protocols for two representative examples of RNA labeling by the 2′-amino approach will be detailed: incorporation of a crosslinking reagent (Fig. 8.3a) for validation of existing structural models [6, 42] and incorporation of an EPR spin-probe (Fig. 8.3b) for biophysical analysis of structure [41] and dynamics of RNA molecular recognition [37–40]. We have also included an example of base-labeling using 4-thiouridine for RNAs that cannot be modified in the 2′-position due to loss of function (Fig. 8.3c).

8.3.1
Synthesis of Aromatic Isothiocyanates and Aliphatic Isocyanates

The 2′-amino post-synthetic labeling approach often requires the synthesis of the desired labeling reagent from the corresponding amine, which may be commercially available. The isothiocyanate crosslinking reagent **2** was prepared from the corresponding amine **1** and thiophosgene (Fig. 8.3a) [6, 42].

(1) Add a solution of amine **1** (for synthesis, see [6]; 8.20 g, 33 mmol) in chloroform (250 ml) drop-wise to a solution of thiophosgene (4.17 g, 36.3 mmol) in chloroform (50 ml) over 10 min at room temperature.
(2) Stir for 1 h at room temperature.
(3) Dilute the mixture with methylene chloride (330 ml).
(4) Wash with NaOH (1 M aq, 165 ml).
(5) Extract the aqueous phase with additional methylene chloride (40 ml).
(6) Combine the organic phases.
(7) Dry the combined organic phases (Na_2SO_4) and filter off the salt.
(8) Remove the solvent *in vacuo*.
(9) Purify the crude product by flash column chromatography (CH_2Cl_2).
(10) This procedure produces an oil (in our hands 8.80 g, 92% yield).

The isocyanate spin-labeling reagent, 4-isocyanato TEMPO **4**, was prepared from 4-amino TEMPO **3** (Acros and Sigma-Aldrich) and diphosgene (Fig. 8.3b) [37].

(1) Pre-cool a solution of **3** (198 mg, 1.15 mmol) in anhydrous CH_2Cl_2 (1.5 ml) in a rock salt/ice water bath at $-8\ °C$.
(2) Separately, pre-cool in the same bath a solution of trichloromethyl chloroformate (diphosgene, 25 μl, 0.21 mmol) in CH_2Cl_2 (1.5 ml).
(3) Rapidly (around 8 s), add the solution of amine under a positive pressure of nitrogen to the stirred solution of trichloromethyl chloroformate at $-8\ °C$.
(4) Remove the cooling bath and allow the reaction to stir for 2 min.
(5) Dilute the crude reaction mixture to 20 ml with CH_2Cl_2.
(6) Wash the organic layer successively with NH_4Cl (1 M aq, 4 × 20 ml) and NaOH (1 M aq, 20 ml).
(7) Dry the organic layer with Na_2SO_4 and filter off the salt.
(8) Remove the solvent *in vacuo*.
(9) This protocol typically yields a peach-colored solid (66 mg, 29% based on starting amine or 87% maximum theoretical yield).
(10) Store the isocyanate desiccated at $-20\ °C$ in CH_2Cl_2 (0.5 mg/50 μl). Small quantities of isocyanates hydrolyze slowly when stored concentrated at $-20\ °C$ (around 30% after 4 weeks) and rapidly when stored in dimethyl formamide (DMF) at $-20\ °C$ [43]. However, isocyanates can be stored in CH_2Cl_2 solutions as described above for several months after preparation.

The syntheses of the isothiocyanate and isocyanate can be readily performed on a scale ranging from 25 mg to several grams. Preparation of isothiocyanates and isocyanates produces acid (HCl), which combines with the starting amine to produce an unreactive ammonium salt. This is particularly problematic for isocyanates where it is only possible to convert one-third of the amine to the corresponding isocyanate using this protocol. Alternatively, the non-nucleophilic base Proton

Sponge® (1,8-bis((dimethyl)amino)naphthylene, 2.5 equivalents; Sigma-Aldrich) can be used, which is especially advantageous if the starting amine is expensive or only available in small quantities. If the light-sensitive Proton Sponge® is used, the reaction should be performed in the dark. After the reaction, Proton Sponge® can be removed by extraction using the protocol described above.

Like most chemicals commonly used in any chemistry laboratory, thiophosgene and diphosgene are harmful, but since they are liquids and used in small quantities, they are relatively simple to handle. However, these reactions should be carried out in a well-ventilated area. Likewise, the isothiocyanates and isocyanates are toxic chemicals, but they are simple to use and require only standard laboratory safety equipment (e.g. nitrile gloves).

8.3.2
Post-synthetic Labeling of 2'-Amino-modified RNA

RNA modified with 2'-amino groups can be purchased from several companies. The standard 2'-trifluoroacetyl protecting group is readily cleaved under standard RNA deprotection conditions and thus no additional deprotection step is necessary. Reaction of 2'-amino-modified RNA with isothiocyanates or isocyanates is typically done under conditions where the RNA is denatured, in aqueous DMF and/or formamide. The organic solvents also act as co-solvents for dissolving the isothiocyanates or isocyanates. It is pertinent that highly pure amine-free anhydrous DMF be used in these reactions, due to the high reactivity of succinimidyl esters, isothiocyanates and isocyanates toward amines. Furthermore, we recommend ethanol precipitation of 2'-amino-modified RNAs, effectively converting ammonium salts of RNA from chemical synthesis into sodium salts, prior to reaction with these amine-reactive compounds as a precaution against unwanted side reactions.

Labeling of 2'-amino-modified RNA with aromatic isothiocyanates

(1) Dissolve the 2'-amino-containing RNA in 5 µl of 50 mM borate buffer, pH 8.6 (RNA concentration around 2 mM).
(2) Add **2** (100 mM in DMF, 5 µl).
(3) Incubate at 37 °C for 28 h (final concentrations: 1 mM 2'-amino RNA, 50 mM isothiocyanate **2**; 50% aqueous DMF, v/v). This reaction proceeds more slowly at room temperature.

Labeling reactions with the aliphatic isocyanates were carried out in a salt/ice water bath (-8 °C) in a cold room (5 °C). If performed at higher temperatures, increased rates of isocyanate hydrolysis result in lower yields. Furthermore, nonspecific labeling has been observed at 37 °C [43]. Analytical-scale reactions can be performed using the following procedure, provided reaction amounts are scaled down in such a way that all concentrations of reactants and buffer remain constant.

Preparative scale reactions of 2′-amino-modified RNA with aliphatic isocyanates

(1) Dissolve the crude (i.e. not yet gel- or HPLC-purified), deprotected 2′-amino-containing RNA (one-quarter of a 1-µmol synthesis) in 100 µl 70 mM boric acid buffer, pH 8.6.
(2) Cool the solution in a salt/ice water bath ($-8\,°C$) in a cold room ($5\,°C$).
(3) Treat the solution sequentially with pre-cooled solutions of formamide (60 µl, $0\,°C$) and freshly prepared isocyanate in anhydrous DMF (75 mM, 40 µl, $-8\,°C$). Final concentrations: 1 mM 2′-amino RNA, 15 mM isocyanate 4; 50% aqueous borate buffer, 30% formamide, 20% DMF, v/v/v.
(4) Incubate for 1 h at $-8\,°C$.
(5) Treat the oligoribonucleotide solution with a second aliquot of freshly prepared isocyanate (40 µl, 75 mM in DMF).
(6) Incubate for 1 h at $-8\,°C$.
(7) Wash the solution with $CHCl_3$ (2×300 µl) at room temperature.
(8) Add sodium acetate (3.0 M, 20 µl, pH 5.3).
(9) Add absolute ethanol ($-20\,°C$, 1.3 ml).
(10) Precipitate the RNA by storage at $-20\,°C$ for 4 h.
(11) Centrifuge the sample (12 000 g, 15 min, $5\,°C$).
(12) Remove the supernatant.
(13) Wash the pellet with cold absolute ethanol (2×50 µl).
(14) Dry the pellet *in vacuo*.
(15) Dissolve the pellet in water (50 µl).
(16) Dilute with aqueous urea (8 M, 150 µl).
(17) Purify the RNA by 20% denaturing PAGE (20-cm gel for short oligos up to 20 nt in length, 20 h at 400–600 V or three-quarters the length of the gel; 40-cm gel for longer oligos up to 50 nt in length, up to 72 h at 600 V or less time if higher voltage, e.g. 1200–1800 V, is used).
(18) Yields typically range from 100–170 nmol for one-quarter of a 1.0-µmol synthesis, depending on the length and quality of the RNA synthesis.

To monitor the extent of labeling:

(1) Remove an aliquot (1.0 µl) of the reaction mixture from step 5 of the above protocol.
(2) Dilute with water (19 µl).
(3) Wash with chloroform (2×75 µl) to remove excess labeling reagent.
(4) Analyze the reaction by one of the following three methods:
  (a) Reversed-phase HPLC on an analytical column (C_{18}, 4.6×250 mm, 5-µm column) at 1.5 ml/min using the following protocol: solvent A, 50 mM Et_3NHOAc (pH 7.0); solvent B, 70% CH_3CN/30% of 50 mM Et_3NHOAc (pH 7.0); 15-min linear gradient from 0 to 23% B, 5-min linear gradient to 100% B, isocratic for 10 min, 3-min linear gradient to initial conditions, 15 min equilibration time between runs. A representative example is given

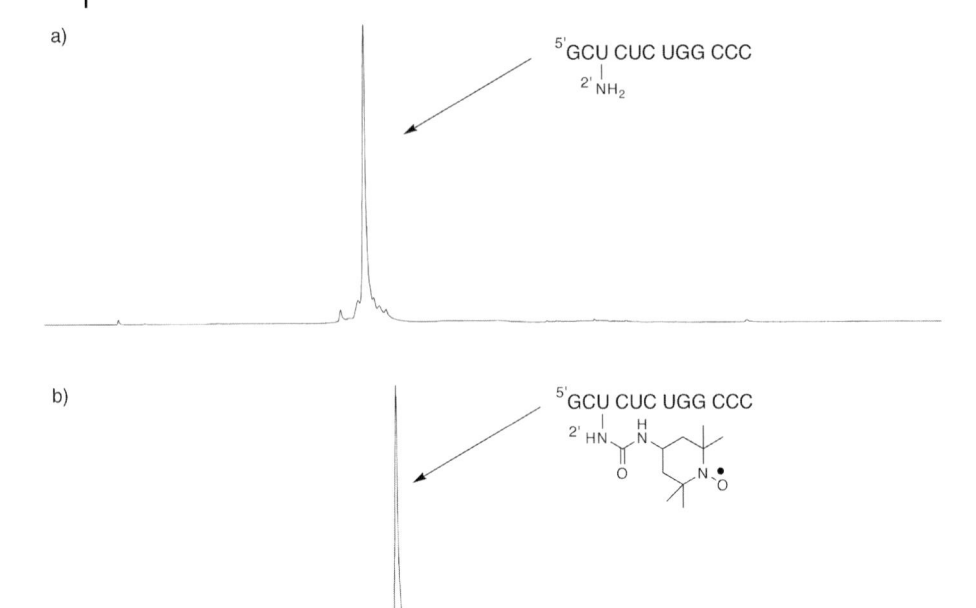

Fig. 8.4. HPLC analysis of 5′-GC(2′-NH$_2$ U) CUC UGG CCC before (a) and after (b) reaction with **4** which shows 95% conversion to the labeled RNA with increased retention time. The asterisks correspond to **4** and its hydrolysis products. HPLC chromatograms were obtained at 260 nm using an analytical column (C18, 4.6 × 250 mm, 5-µm column) run at 1.5 ml/min according to the following protocol: solvent A, 50 mM Et$_3$NHOAc (pH 7.0); solvent B, 70% CH$_3$CN/30% of 50 mM Et$_3$NHOAc (pH 7.0); 15-min linear gradient from 0 to 23% B, 5-min linear gradient to 100% B, isocratic for 10 min, 3-min linear gradient to initial conditions, 15 min equilibration time between runs.

in Fig. 8.4, which shows the reaction of **4** with 5′-GC(2′-NH$_2$ U) CUC UGG CCC.

(b) 20% denaturing PAGE (20-cm gel, 400 V for 3.5 h) by UV shadow visualization.

(c) Analytical ion exchange (IE) HPLC on a Dionex DNA Pac PA-100 4 × 250 mm analytical column heated at 50 °C by a column warmer. Separation will not be achieved without heating the column. Solvent gradients for analytical IE-HPLC were run at 1.0 ml/min as follows: solvent A, 25 mM Tris–HCl, pH 8.0; Solvent B, 1.0 M NaCl, 25 mM Tris–HCl, pH 8.0; 35-min linear gradient from 10% B to 80% B, 5-min linear gradient to 10% B.

Short labeled RNAs (up to 20 nt in length) can also be purified utilizing these HPLC protocols, although we recommend 20% denaturing PAGE purification, since the hydrolysis products of some isocyanates may co-elute with the labeled RNA using RP-HPLC.

8.3.3
Post-synthetic Labeling of 4-Thiouridine-modified RNA

If one knows *a priori* that modification of the 2′-hydroxyl group will likely interfere with biological function (e.g. 2′-OH is involved in an essential hydrogen bond), it may be necessary to label using an alternative post-synthetic labeling strategy. In this case, another simple, straightforward method is the labeling of 4-thiouridine residues with iodoacetamides [69, 70] or sulfur-based compounds [71]. One of the advantages of this method is that the labeling reaction can be followed by monitoring the consumption of UV signal at 320 nm, which corresponds to the thiocarbonyl. This labeling strategy changes the base-pairing properties of this residue. However, UV thermal denaturation melting temperature and hypochromicity data as well as NMR structural data indicate that 4-thiouridine residues can be labeled in this manner without disruption of helical stacking [71]. Labeling of the 4-amine group of cytidine with a crudely analogous modification, however, resulted in severe thermal instability of DNA [72]. Therefore, caution should be exercised when choosing such a labeling strategy for base-pairing residues.

Labeling of 4-thiouridine with the iodoacetamide spin-labeling reagent 3-(2-iodoacetamido)-proxyl (modified procedure from that reported in [70])

(1) Dissolve 4-thiouridine-modified RNA (one-quarter of a 1-µmol synthesis) in 166 µl of buffer (100 mM sodium phosphate, pH 8).
(2) Acquire UV spectrum of an aliquot of the above mixture, monitoring at 260 and 320 nm.
(3) Dissolve 6 mg of 3-(2-iodoacetamido)-proxyl, **5** (Fig. 8.3c, Sigma) into 14 µl of ethanol and 20 µl of anhydrous DMF (0.5 M labeling reagent).
(4) Mix **5** and 4-thiouridine-modified RNA; final concentrations: around 1 mM RNA, 85 mM **5**, 83% phosphate buffer/7% ethanol/10% DMF (v/v/v).
(5) Due to light sensitivity of 4-thiouridine residues, cover samples with aluminum foil.
(6) Vortex vigorously until absorbance at 320 nm disappears (typically 18–28 h).
(7) Once the reaction is complete, precipitate and purify RNA as described above.

8.3.4
Verification of Label Incorporation

Whenever a modification is introduced into RNA, either by solid-phase chemical synthesis using a phosphoramidite or by post-synthetic modification, several steps are necessary to verify that the incorporation was successful. Not all modifications are incorporated as intended. For example, the 5-trifluoromethyl-2′-deoxyuridine phosphoramidite was prepared for the purpose of ^{19}F-NMR spectroscopy of nucleic acids; however, standard oligonucleotide deprotection conditions converted the 5-trifluoromethyl group to a 5-cyano group, prompting the use of alternate mild deprotection conditions [73]. Incorporation of the modified nucleoside should

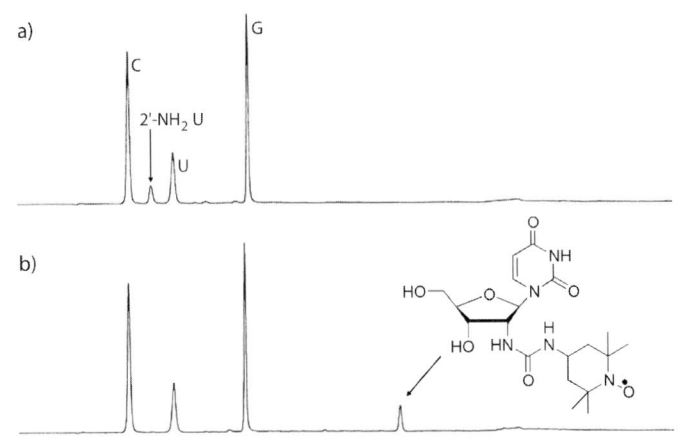

Fig. 8.5. HPLC analysis of enzymatically digested RNA.
(a) Enzymatic digestion of 5′-GCU C(2′-NH$_2$ U)C UGG CCC;
(b) enzymatic digestion of the product of the 2′-NH$_2$-modified oligonucleotide from a after reaction with isocyanate **4**. HPLC chromatograms were obtained as in Fig. 8.4.

be verified by mass spectrometry and enzymatic digestion in conjunction with RP-HPLC analysis. The RNAs (2.0 nmol or around 0.2 OD$_{260}$) should be digested with snake venom phosphodiesterase (0.5 U) and calf intestinal alkaline phosphatase (8 U) at 37 °C for 5 h in 5 mM Tris–HCl, pH 7.4 (20 μl) and then analyzed by analytical RP-HPLC using the same protocol as that listed above for monitoring the extent of labeling. For example, HPLC analysis of the enzymatic digestion of 5′-GCU C(2′-NH$_2$ U)C UGG CCC resulted in peaks corresponding to C, 2′-NH$_2$ U, U and G (Fig. 8.5a), whereas after labeling with the spin-label isocyanate **4** HPLC analysis revealed the absence of the 2′-NH$_2$ U peak and the presence of a new peak (Fig. 8.5b) that co-eluted with the expected modified spin-labeled nucleoside prepared by chemical synthesis [37].

8.3.5
Potential Problems and Troubleshooting

It is always important to determine if the modification interferes (intentionally or unintentionally) with the structure and function of the molecule using a standard structural (e.g. UV thermal denaturation and/or other biophysical spectroscopy or crystallography) and functional (binding or enzymatic) assay. For example, the effect of incorporation of nitroxide spin-labels at the 2′-position on RNA has been investigated by UV thermal denaturation [37], whereas their effect on RNA–protein complex formation was investigated by electrophoretic mobility shift analysis [38].

If the labeling reaction does not work or the yields of the labeling reactions are low, this is generally a result of one of four problems.

(1) The isocyanate may be hydrolyzed or not prepared properly. The quality of the isocyanate can be determined by spectroscopic methods (e.g. NMR) and/or tested by reaction with a simple aliphatic amine such as benzylamine (30 min in CH_2Cl_2 at room temperature).

(2) The 2′-trifluoroacetyl protecting group may not have been fully removed, which may not be readily apparent because 2′-trifluoroacetamido- and 2′-amino-modified RNAs often have similar mobility on HPLC or in gels. However, this can be readily investigated by enzymatic digestion of the RNA, followed by HPLC analysis as described above. For example, if the 2′-trifluoroacetyl group is not fully removed a new peak will be observed by HPLC analysis with a retention time of around 5 min corresponding to the 2′-trifluoroacetamido uridine nucleoside (e.g. in the order of C, 2′-NH_2 U, U, 2′-$NHCOCF_3$ U, G, A).

(3) If the temperature of the reaction is not low enough, the yields are lower, presumably because of the competing hydrolysis of the isocyanate. Therefore, it is important to monitor the temperature of the ice/salt bath with a thermometer.

(4) Lower yields will be obtained if the RNA is not completely dissolved at the beginning of the reaction.

Note added in proof. Recently, a paper published by Pham et al. (Nucleic Acids Res. 2004, 32, 3446–3455) showed that 2′-ureido-modified RNAs are significantly more stable than analogous 2′-amido-modified RNAs.

References

1 R. F. GESTELAND, T. R. CECH, J. F. ATKINS (eds), *The RNA World*, Cold Springs Harbor Laboratory Press, Cold Springs Harbor, NY, **1999**.

2 M. MENGER, F. ECKSTEIN, D. PORSCHKE, *Nucleic Acids Res.* **2000**, *29*, 4428–4434.

3 T. TUSCHL, C. GOHLKE, T. M. JOVIN, E. WESTHOF, F. ECKSTEIN, *Science* **1994**, *266*, 785–788.

4 G. S. BASSI, A. I. H. MURCHIE, F. WALTER, R. M. CLEGG, D. M. J. LILLEY, *EMBO J.* **1997**, *16*, 7481–7489.

5 D. A. LAFONTAINE, D. G. NORMAN, D. M. J. LILLEY, *EMBO J.* **2002**, *21*, 2461–2471.

6 S. T. SIGURDSSON, T. TUSCHL, F. ECKSTEIN, *RNA* **1995**, *1*, 575–583.

7 N. VENKATESAN, S. J. KIM, B. H. KIM, *Curr. Med. Chem.* **2003**, *10*, 1973–1991.

8 B. S. SPROAT, *J. Biotech.* **1995**, *41*, 221–238.

9 D. J. EARNSHAW, M. J. GAIT, *Biopolymers (Nucleic Acid Sci.)* **1998**, *48*, 39–55.

10 S. VERMA, F. ECKSTEIN, *Annu. Rev. Biochem.* **1998**, *67*, 99–134.

11 F. ECKSTEIN, *Biochemie* **2002**, *84*, 841–848.

12 H. J. LENOX, C. P. MCCOY, T. L. SHEPPARD, *Org. Lett.* **2001**, *3*, 2415–2418.

13 S. DEY, T. L. SHEPPARD, *Org. Lett.* **2001**, *3*, 3983–3986.

14 J. S. SUMMERS, B. R. SHAW, *Curr. Med. Chem.* **2001**, *8*, 1147–1155.

15 G. A. HOLLOWAY, C. PAVOT, S. A. SCARINGE, Y. LU, T. B. RAUCHFUSS, *ChemBioChem* **2002**, *3*, 1061–1065.

16 B. CHEN, E. R. JAMIESON, T. D. TULLIUS, *Bioorg. Med. Chem. Lett.* **2002**, *12*, 3093–3096.

17 M. E. DOUGLAS, B. BEIJER, B. S. SPROAT, *Bioorg. Med. Chem. Lett.* **1994**, *4*, 995–1000.

18 J. T. HWANG, M. M. GREENBERG, *J. Org. Chem.* **2001**, *66*, 363–369.

19 L. BEIGELMAN, A. KARPEISKY, N. USMAN, *Nucleosides, Nucleotides* **1995**, *14*, 901–905.

20 J. B. Murray, D. P. Terwey, L. Maloney, A. Karpeisky, N. Usman, L. Beigelman, W. G. Scott, *Cell* **1998**, *92*, 665–673.

21 A. P. Massey, S. T. Sigurdsson, *Nucleic Acids Res.* **2004**, *32*, 2017–2022.

22 A. Földesi, A. Trifonova, Z. Dinya, J. Chattopadhyaya, *J. Org. Chem.* **2001**, *66*, 6560–6570.

23 K. Yamana, R. Iwase, S. Furutani, H. Tschida, H. Zako, T. Yamaoka, A. Murakami, *Nucleic Acids Res.* **1999**, *27*, 2387–2392.

24 V. A. Korshun, D. A. Stetsenko, M. J. Gait, *J. Chem. Soc., Perkin Trans. 1* **2002**, *8*, 1092–1104.

25 N. N. Dioubankova, A. D. Malakhov, D. A. Stetsenko, M. J. Gait, P. E. Volynsky, R. G. Efremov, V. A. Korshun, *ChemBioChem* **2003**, *4*, 841–847.

26 K. Yamana, T. Mitsui, H. Nakano, *Tetrahedron* **1999**, *55*, 9143–9150.

27 P. B. Rupert, A. P. Massey, S. T. Sigurdsson, A. R. Ferré-D'Amaré, *Science* **2002**, *298*, 1421–1424.

28 S. I. Khan, M. W. Grinstaff, *J. Am. Chem. Soc.* **1999**, *121*, 4704–4705.

29 K. Musier-Forsyth, P. Schimmel, *Biochemistry* **1994**, *33*, 773–779.

30 M. M. Konarska, *Methods* **1999**, *18*, 22–28.

31 P. Z. Qin, S. E. Butcher, J. Feigon, W. L. Hubbell, *Biochemistry* **2001**, *40*, 6929–6936.

32 J. A. Fidanza, L. W. McLaughlin, *J. Org. Chem.* **1992**, *57*, 2340–2346.

33 S. B. Cohen, T. R. Cech, *J. Am. Chem. Soc.* **1997**, *119*, 6259–6268.

34 D. J. Earnshaw, B. Masquida, S. Muller, S. T. Sigurdsson, F. Eckstein, E. Westhof, M. J. Gait, *J. Mol. Biol.* **1997**, *274*, 197–212.

35 H. Aurup, T. Tuschl, F. Benseler, J. Ludwig, F. Eckstein, *Nucleic Acids Res.* **1994**, *22*, 20–24.

36 S. K. Silverman, T. R. Cech, *Biochemistry* **1999**, *38*, 14224–14237.

37 T. E. Edwards, T. M. Okonogi, B. H. Robinson, S. T. Sigurdsson, *J. Am. Chem. Soc.* **2001**, *123*, 1527–1528.

38 T. E. Edwards, T. M. Okonogi, S. T. Sigurdsson, *Chem. Biol.* **2002**, *9*, 699–706.

39 T. E. Edwards, S. T. Sigurdsson, *Biochemistry* **2002**, *41*, 14843–14847.

40 T. E. Edwards, S. T. Sigurdsson, *Biochem. Biophys. Res. Commun.* **2003**, *303*, 721–725.

41 O. Schiemann, A. Weber, T. E. Edwards, T. F. Prisner, S. T. Sigurdsson, *J. Am. Chem. Soc.* **2003**, *125*, 3434–3435.

42 S. T. Sigurdsson, *Methods* **1999**, *18*, 71–77.

43 S. T. Sigurdsson, F. Eckstein, *Nucleic Acids Res.* **1996**, *24*, 3129–3133.

44 J. T. Hwang, M. M. Greenberg, *Org. Lett.* **1999**, *1*, 2021–2024.

45 T. J. Meade, J. F. Kayyem, *Angew. Chem. Int. Ed.* **1995**, *34*, 352–353.

46 E. S. Krider, J. E. Miller, T. J. Meade, *Bioconjugate Chem.* **2002**, *13*, 155–162.

47 E. J. Merino, K. M. Weeks, *J. Am. Chem. Soc.* **2003**, *125*, 12370–12371.

48 T. S. Zatsepin, D. A. Stetsenko, A. Arzumanov, E. Romanova, M. J. Gait, T. S. Oretskaya, *Bioconjugate Chem.* **2002**, *13*, 822–830.

49 K. F. Blount, O. C. Uhlenbeck, *Biochemistry* **2002**, *41*, 6834–6841.

50 T. K. Stage-Zimmerman, O. C. Uhlenbeck, *Nat. Struct. Biol.* **2001**, *8*, 863–867.

51 K. L. Buchmueller, B. T. Hill, M. S. Platz, K. M. Weeks, *J. Am. Chem. Soc.* **2003**, *125*, 10850–10861.

52 D. M. John, K. M. Weeks, *Chem. Biol.* **2000**, *7*, 405–410.

53 D. M. John, K. M. Weeks, *Biochemistry* **2002**, *41*, 6866–6874.

54 S. K. Silverman, M. L. Deras, S. A. Woodson, S. A. Scaringe, T. R. Cech, *Biochemistry* **2000**, *39*, 12465–12475.

55 S. K. Silverman, T. R. Cech, *RNA* **2001**, *7*, 161–66.

56 K. F. Blount, Y. Tor, *Nucleic Acids Res.* **2003**, *31*, 5490–5500.

57 K. F. Blount, N. L. Grover, V. Mokler, L. Beigelman, O. C. Uhlenbeck, *Chem. Biol.* **2002**, *9*, 1009–1016.

58 S. I. Chamberlin, E. J. Merino, K.

M. Weeks, *Proc. Natl. Acad. Sci. USA* **2002**, *99*, 14688–14693.
59 C. Hendrix, B. Devreese, J. Rozenski, A. van Aerschot, A. De Bruyn, J. Van Beeumen, P. Herdewijn, *Nucleic Acids Res.* **1995**, *23*, 51–57.
60 C. Hendrix, M. Mahieu, J. Anne, S. V. Calenbergh, A. Van Aerschot, J. Content, P. Herdewijn, *Biochem. Biophys. Res. Comm.* **1995**, *210*, 67–73.
61 O. P. Kryatova, W. H. Connors, C. F. Bleczinski, A. A. Mokhir, C. Richert, *Org. Lett.* **2001**, *3*, 987–990.
62 P. Sergiev, S. Dokudovskaya, E. Romanova, A. Topin, A. Bogdanov, R. Brimacombe, O. Dontsova, *Nucleic Acids Res.* **1998**, *26*, 2519–2525.
63 S. T. Sigurdsson, F. Eckstein, *Trends Biotech.* **1995**, *13*, 286–289.
64 I. Huq, N. Tamilarasu, T. M. Rana, *Nucleic Acids Res.* **1999**, *27*, 1084–1093.
65 L. Huang, L. L. Chappell, O. Iranzo, B. F. Baker, J. R. Morrow, *J. Biol. Inorg. Chem.* **2000**, *5*, 85–92.
66 S. Alefelder, S. T. Sigurdsson, *Bioorg. Med. Chem.* **2000**, *8*, 269–273.
67 B. Bramlage, S. Alefelder, P. Marschall, F. Eckstein, *Nucleic Acids Res.* **1999**, *27*, 3159–3167.
68 S. T. Sigurdsson, B. Seeger, U. Kutzke, F. Eckstein, *J. Org. Chem.* **1996**, *61*, 3883–3884.
69 H. Hara, T. Horiuchi, M. Saneyoshi, S. Nishimura, *Biochem. Biophys. Res. Comm.* **1970**, *38*, 305–311.
70 A. Ramos, G. Varani, *J. Am. Chem. Soc.* **1998**, *120*, 10992–10993.
71 P. Z. Qin, K. Hideg, J. Feigon, W. L. Hubbell, *Biochemistry* **2003**, *42*, 6772–6783.
72 W. Bannwarth, D. Schmidt, *Bioorg. Med. Chem. Lett.* **1994**, *4*, 977–980.
73 J. C. Markley, P. Chirakul, D. Sologub, S. T. Sigurdsson, *Bioorg. Med. Chem. Lett.* **2001**, *11*, 2453–2455.
74 Y. Kinoshita, K. Nishigaki, Y. Husimi, *Nucleic Acids Res.* **1997**, *25*, 3747–3748.
75 J. T. Rodgers, P. Patel, J. L. Hennes, S. L. Bolognia, D. P. Mascotti, *Anal. Biochem.* **2000**, *277*, 254–259.
76 A. J. Harwood, *Methods Mol. Biol.* **2002**, *187*, 17–22.
77 D. L. McMinn, M. M. Greenberg, *J. Am. Chem. Soc.* **1998**, *120*, 3289–3294.
78 P. S. Chockalingam, L. A. Jurado, H. W. Jarrett, *Mol. Biotechnol.* **2001**, *19*, 189–99.
79 R. W. Roberts, J. W. Szostak, *Proc. Natl. Acad. Sci. USA* **1997**, *94*, 12297–12302.
80 M. Teplova, C. J. Wilds, Z. Wawrzak, V. Tereshko, Q. Du, N. Carrasco, Z. Huang, M. Egli, *Biochemie* **2002**, *84*, 849–858.
81 A. A. Koshkin, J. Fensholdt, H. M. Pfundheller, C. Lamholt, *J. Org. Chem.* **2001**, *66*, 8504–8512.
82 A. Okamoto, T. Taiji, K. Tainaka, I. Saito, *Bioorg. Med. Chem. Lett.* **2002**, *12*, 1895–1896.
83 C. R. Allerson, S. L. Chen, G. L. Verdine, *J. Am. Chem. Soc.* **1997**, *119*, 7423–7433.
84 L. Bellon, C. Workman, J. Scherrer, N. Usman, F. Wincott, *J. Am. Chem. Soc.* **1996**, *118*, 3771–3772.

Part II
Structure Determination

II.1
Molecular Biology Methods

9
Direct Determination of RNA Sequence and Modification by Radiolabeling Methods

Olaf Gimple and Astrid Schön

9.1
Introduction

The large numbers of genome sequences now available have allowed the identification of many novel RNA species, simply by deduction from the published DNA sequences [1]. Even though, direct sequencing of RNA molecules is still an indispensable method for a number of reasons. The most important reason is that even nowadays, novel RNA species may be discovered following a "functional assay". If no hint to the sequence can be obtained by genomic data mining, if RNA editing may occur in this organism or, simply, if the RNA is derived from an organism where no genomic sequences are available, the RNA has to be purified prior to direct sequence determination. The second, and probably even more intriguing, rationale is the observation that a large number of RNAs, such as tRNAs, snRNAs, snoRNAs and others, contain modified nucleobases, which in many cases play crucial roles in the function of these RNAs [2–7]. Although in many instances the plain RNA sequence can be extracted from the genomic sequence, the type and position of the modified bases have to be determined by direct analysis of the purified RNA. In this chapter, we will describe methods for the rigorous purification of single small RNA species from the bulk of cellular RNA, their sequence determination and the identification of modified nucleotides by post-labeling methods.

9.2
Methods

It is anticipated that the reader is familiar with standard biochemical and molecular biology practice, such as gel electrophoresis, chromatography and handling of radioactive materials.

In order to avoid RNase contamination, all aqueous solutions and plasticware should be sterilized by autoclaving or prepared from RNase-free [diethylpyrocarbonate (DEPC)-treated] water. Glassware should be baked at 150 °C for 4 h. Centrifugations are performed in a microcentrifuge at 10 000 g, if not stated otherwise.

9.2.1
Isolation of Pure RNA Species from Biological Material

9.2.1.1 Preparation of Size-fractionated RNA

Numerous methods for the preparation of crude RNA from tissues can be found in standard molecular biology reference works, e.g. [8]. They all consist of a cell disruption step under denaturing conditions, followed by separation of the nucleic acids from protein and cell debris. For purification of a single RNA species from this total RNA population, it is preferable to perform a crude size selection prior to further manipulations. The simplest procedure consists of a fractionated precipitation with NaCl, where large RNAs and polysaccharides are separated from the "soluble" (i.e. small) RNAs by centrifugation [9]. A more elaborate scheme for large-scale purification of tRNAs (and other small RNAs) from human and animal tissue has been described by Roe [10]. This procedure can be easily scaled down and adapted to other tissues. The DEAE anion-exchange chromatography described in that paper can be conveniently replaced by ready-to-use columns for small-scale nucleic acid preparations, following the manufacturer's instructions for RNA preparation (e.g. Macherey-Nagel Nucleobond AX). Alternatively, a ribosome-free cell extract (S100) can be used for isolation of non-rRNAs [11]. The RNAs obtained by any of these purification schemes are ready for further purification and functional assays.

9.2.1.2 Isolation of Single Unknown RNA Species Following a Functional Assay

If a functional assay such as aminoacylation, ribozyme activity or similar is available, any RNA of interest can be purified and identified, regardless of prior sequence information. Although chromatographic procedures such as anion-exchange, gel filtration or reverse-phase chromatography can be used in any combination [2, 3], isolation of a single species usually requires preparative separation by one- or two-dimensional gel electrophoresis. To achieve the best possible separation by length, base composition, nucleotide modifications and structure, the first dimension gel should be run at acidic pH under semi-denaturing conditions, and the second dimension at slightly basic pH and fully denaturing conditions as described [12, 13].

Materials for Staining and Elution of RNAs after Electrophoresis

- Staining solution: 0.4% toluidine blue O (w/v) in 50% MeOH (v/v), 10% glacial acetic acid (v/v).
- Destaining solution: 50% MeOH (v/v), 10% glacial acetic acid (v/v).
- Elution buffer: 0.5 M Tris–HCl, pH 7, 0.1% SDS (w/v), 0.1 mM Na_2EDTA, 1 mM $MgCl_2$.

Comments on the Electrophoretic Purification and Elution of RNA Species

To obtain optimal resolution, not more than 50 µg of a pre-fractionated "small RNA" preparation should be loaded onto each 1 cm wide lane of a first dimension gel (10% PAA; 0.5 mm thick; 40 cm total gel length). After electrophoresis, a 16 cm long part of the lane is cut out. The vertical position of this strip depends on the anticipated migration distance of the desired RNA. The strip is then polymerized into the second dimension gel such that the direction of electrophoresis is turned by 90° compared to the first dimension. Following electrophoresis, the RNA species are visualized by toluidine blue staining for 20 min and destaining until a clear background is achieved. The isolated spots are cut out with a scalpel and 150 µl of elution buffer is added to each reaction tube containing a gel piece of 4 mm maximum diameter (if pieces are larger, adjust buffer volume accordingly). After quick-freezing the contents in dry ice, the RNA is eluted by vigorous shaking at room temperature for at least 8 h. The buffer is collected, another 150 µl is added to each tube to wash the gel piece, and the RNA is precipitated from the combined buffer fractions by addition of 750 µl EtOH, overnight incubation at $-20\ °C$ and centrifugation at 10 000 g. After washing with ice-cold 70% ethanol and vacuum drying for 10 min, the RNAs can be dissolved in H_2O or the desired buffer for functional analysis.

9.2.1.3 Isolation of Single RNA Species with Partially Known Sequence

If the primary sequence of an RNA species is known, e.g. from genomic analysis or direct sequencing, a hybrid selection method can be applied to obtain the desired species in sufficient quantities for further studies. This protocol has been optimized following published procedures [14, 15].

Materials for Hybrid Selection of Single RNA Species

Buffers
- $20 \times$ SSC, $6 \times$ SSC, $1 \times$ SSC.
- TE: 10 mM Tris–HCl, pH 7.5, 1 mM Na_2EDTA.
- 2 M NaOAc, pH 5.
- Urea gel loading buffer: 8 M urea, 0.03% (w/v) bromophenol blue (BPB), 0.03% (w/v) xylene cyanol FF (XC), 0.03% (w/v) Sigma brilliant blue (SBB); make up from dye stock (Section 9.2.3.1).

Affinity matrix
- Ultra-Link Streptavidin Plus-Beads (Pierce; 53117); capacity 66.7 pmol biotin/µl according to the manufacturer's information.

Nucleic acids
- 3'-Biotinylated deoxyoligonucleotide, complementary to the desired RNA. Most manufacturers (e.g. IBA, Germany or Eurogentec, Belgium) offer the biotin-coupled oligos at excellent quality. Note that T_m of the oligo should be about 70 °C. If possible, chose a single-stranded variable RNA region as target. The working solution of the oligonucleotide is adjusted to 1 nmol/µl H_2O.
- Crude RNA preparation (see Section 9.2.1.1) in TE or H_2O. Hybrid selection is most efficient with size-fractionated RNA preparations.

Equipment
- Two to three thermostated shakers holding 1.5 ml reaction tubes; alternatively, put standard mixers/shakers in an incubator with the desired temperature. Shaking speed should be adjusted such that the beads are just kept from settling, but do not move vigorously.

Procedure for the Purification of a Single RNA Species from 1 mg Crude Small RNAs

Coupling of biotinylated oligonucleotide to streptavidin beads
The streptavidin beads are supplied as a suspension and tend to settle down fast. Before removing the required amount, shake the suspension well until homogeneous and use a pipette tip with a larger opening (cut with a scalpel) to avoid clogging. Per 1 mg total RNA, take 15 µl of bead suspension. Wash beads twice in 200 µl TE and resuspend in 200 µl TE ("Washing" of beads means: suspend *thoroughly but carefully* by vortexing, collect by centrifugation at 2000 r.p.m. for 5 min in a microcentrifuge, remove supernatant). Add 1 µl biotin–oligonucleotide solution; shake for 15 min at room temperature and wash twice in 6 × SSC.

Hybridization of RNA to coupled oligonucleotides
Pre-heat one shaker at 65 °C and a second one at 90 °C. Adjust RNA volume with TE to 100 µl (final concentration: 10 mg/ml). Denature by heating for 2 min at 90 °C and snap-cool in ice water. Adjust RNA to 6 × SSC by adding 43 µl of 20 × SSC and add to beads; shake for 10 min at 65 °C, turn off heater and continue shaking while the block cools down to room temperature.

Removal of undesired (contaminating) RNAs
After hybridization, collect beads at 2000 r.p.m., wash at room temperature 3 times with 6 × SSC followed by 3 times with 1 × SSC. Save the beads for elution, keep the supernatants in case the hybridization needs further optimization.

Elution of desired RNA species
Pre-heat two shakers at 60 and 75 °C, respectively. Elution of the desired RNA from the beads is achieved in three steps; the supernatant of each step is retained for further analysis. First, the beads are resuspended in 100 µl TE and shaken at room temperature for 5 min; after collecting the beads, the elution is repeated at 60 and 75 °C with the same amount of TE buffer, pre-warmed at the respective temperature. If no RNA is recovered, elution may be repeated at 90 °C.

Electrophoresis of affinity purified RNAs
RNA is precipitated from the fractions by addition of 0.1 volume of 2 M NaOAc (pH 5) and 2.5 volumes of EtOH, followed by incubation at −20 °C for at least 30 min and centrifugation. The precipitate is washed with 100 µl ice-cold 70% EtOH, vacuum-dried and dissolved in 10–20 µl urea gel loading buffer. After denaturation for 2 min at 95 °C, the samples are separated on a denaturing PAA gel (40 cm long, 0.5 mm thick). The gel is stained with toluidine blue (Section 9.2.1.2) and

RNA bands are cut out and eluted (Section 9.2.1.2). Because in some cases, co-purification of RNA species with similar sequence cannot be totally avoided, this step is required to ensure absolute purity of the desired RNA. Alternatively, for the determination of RNA purity on an analytical scale, 2–5 µl of the column eluates are mixed with 4 volumes of loading buffer, electrophoresed as above and detected by silver staining [16].

Comments on electrophoresis and elution of RNAs
Details on preparation-scale electrophoresis of RNA can be found in Chapter 1. The polymer concentration of analytical or preparative gels should be adjusted to the expected size of the RNA of interest, e.g. 15% PAA should be used for an expected length between 75 and 90 nt. In this case, electrophoresis should proceed until XC (the second dye marker) has reached the bottom of the gel. Note that the efficiency of gel elution is also dependent on the gel concentration – whereas a 350 nt long RNA is easily recovered from an 8 or 10% PAA gel, yield is low from a 15% gel. If very small amounts of RNA have to be recovered from large volumes of elution buffer, 1–10 µg glycogen may be added as a co-precipitant if it does not interfere with later analysis.

9.2.2
Radioactive Labeling of RNA Termini

End-labeling of RNA with ^{32}P is a prerequisite for direct sequence analysis and free 5′- or 3′-OH-groups are required for most labeling reactions. The removal of 5′-cap structures using tobacco pyrophosphatase has been described [17]. 5′- and 3′-phosphate residues can be easily removed by calf intestine alkaline phosphatase (CIP) prior to the labeling reaction. Since the different enzymes have different substrate preferences, not all reactions will work equally well with all types of RNA. For example, the "hidden" 5′ end of tRNAs is often hard to dephosphorylate; thus, labeling by phosphate exchange (Section 9.2.2.1) is the preferred method in these cases.

9.2.2.1 5′ Labeling of RNAs

Material Required for 5′-end-labeling of RNAs

Enzymes
- Calf intestine alkaline phosphatase (CIP; 10 mU/µl) and T4 polynucleotide kinase (PNK; 5 U/µl) are from Roche Biochemicals.

Buffers and reagent
- 50 mM nitrilo-tri-acetic acid (NTA), pH 8.
- IMID mix: 250 mM imidazole, 25 mM DTT, 0.5 mM spermine, 0.5 mM Na$_2$EDTA, 50 mM MgCl$_2$, pH 6.6 with HCl.

- MgCl$_2$/spermine solution: 0.2 M MgCl$_2$, 32 mM spermine.
- 100 mM DTT.

Radioactive materials
- [γ-^{32}P]ATP (10 µCi/µl; 3000 Ci/mmol).

5′ Labeling of RNA after Dephosphorylation

For dephosphorylation, 10–100 pmol of purified RNA is vacuum dried, dissolved in 8 µl H$_2$O, denatured for 2 min at 90 °C and snap-cooled in ice water. Then, 1 µl 1 M Tris–HCl, pH 8 and 10 mU CIP are added and the mixture is incubated at 50 °C for 1 h. The enzyme is inactivated by addition of 3.3 µl NTA solution followed by incubation at 50 °C for 20 min.

For radioactive labeling, prepare one 1.5-ml reaction tube containing 200 µCi [γ-^{32}P]ATP and one with 10 µl of urea gel loading buffer (Section 9.2.1.3) and dry in a desiccator. For phosphorylation, the [γ-^{32}P]ATP is dissolved by adding 6.5 µl of the above RNA preparation. Then, 1 µl each of the MgCl$_2$/spermine solution, the 100 mM DTT solution and 1 µl PNK are added, and incubated at 37 °C for 30 min. The reaction is terminated by transferring the whole mixture into the tube containing dry loading buffer. After denaturation (2 min, 90 °C) the labeled RNA is separated on a denaturing PAA gel, localized by autoradiography and recovered by elution (Sections 9.2.1.2). To increase recovery of the labeled RNA during precipitation, 10 µg yeast tRNA per 300 µl elution buffer may be added as a co-precipitant.

5′ Labeling by Phosphate Exchange

Since many small RNAs are highly structured and have a recessed 5′ end difficult to access by the phosphatase, the exchange reaction first introduced by Berkner and Folk [18] is the labeling method of choice for these RNAs.

Between 10 and 100 pmol dry RNA is dissolved in 2 µl IMID mix, 1.25 µl 0.5 mM ADP and 5.75 µl H$_2$O, and transferred into a reaction tube containing 200 µCi dry [γ-^{32}P]ATP. The reaction is initiated by addition of 1 µl PNK, run for 30 min at 37 °C, terminated by pipetting onto loading buffer and separated by electrophoresis as described above.

9.2.2.2 3′ Labeling of RNAs

All intact RNAs possess a 3′-OH end and can thus be directly labeled at this terminus.

The most popular method is the ligation of radioactive pCp to the RNA [19]. The method has been described in detail [12, 20]; an abbreviated and efficient variation including the synthesis of pCp is presented here.

3′ labeling of RNAs is usually more efficient than 5′ labeling and requires less material. However, larger RNAs are often poor substrates for the ligation reaction and may preferably be labeled by poly(A) polymerase, using 3′-deoxyadenosine (Cordycepin) to prevent chain elongation [21]. A special method to label RNAs with (at least partially) known sequence is "splint labeling" of the 3′ end with DNA polymerase [22].

Materials Required for 3′-end-labeling of RNAs

Enzymes
- PNK (5 U/µl) and T4 RNA ligase (1.5–3 U/µl) are from Roche Biochemicals; yeast poly(A) polymerase and T7 DNA polymerase are from United States Biochemicals.

Buffers and reagents
- 1 mM 3′-cytidine monophosphate (3′-Cp).
- pCp mix: 175 mM Tris–HCl, pH 8, 15 mM $MgCl_2$, 12 mM DTT, 2.4 mM spermine.
- D mix: 10 mM HEPES–KOH, pH 8.3 in 33% (v/v) DMSO.
- Ligase mix: 120 mM HEPES–KOH, pH 8.3, 10 mM DTT, 30 mM $MgCl_2$, 30 µg/ml RNase-free BSA or gelatin, 3 mM ATP in 25% (v/v) DMSO. Note that the DMSO should be deionized freshly before preparation of these solutions; aliquots of DMSO or the reaction mixes can be stored frozen for several months.

Radioactive materials
- $[\gamma\text{-}^{32}P]$ATP (10 µCi/µl; 3000 Ci/mmol); $[\alpha\text{-}^{32}P]$Cordycepin triphosphate; $[\alpha\text{-}^{32}P]$dATP (each of highest specific activity available).

3′ Labeling of RNA by Ligation of $[5'\text{-}^{32}P]$pCp

Preparation of $[5'\text{-}^{32}P]$pCp
For one labeling reaction, dry down 100 µCi (3.7 MBq) $[\gamma\text{-}^{32}P]$ATP and dissolve in 2 µl pCp-mix. Add 1 µl 1 mM 3′-Cp and 5 U PNK, incubate for 1 h at 37 °C, and vacuum dry. Alternatively, $[5'\text{-}^{32}P]$pCp can be purchased from several suppliers and used directly for ligation.

Ligation of $[5'\text{-}^{32}P]$pCp to RNA
The dry RNA (4–10 pmol) is dissolved in 2 µl D mix, denatured for 2 min at 90 °C and snap-cooled in ice. Then, 2 µl ligase mix is added and the mixture transferred into the tube containing the dry $[^{32}P]$pCp. The reaction is started by addition of 2 µl T4 RNA ligase and incubated at 4 °C for 16–30 h. Purification is performed by one-dimensional gel electrophoresis as described (Sections 9.2.1.2).

3′ Labeling of RNA with Poly(A) Polymerase and Cordycepin

For this highly sensitive labeling method, 2–10 nmol of RNA 3′ ends is sufficient. The purified RNA is mixed with 2 µl 5 × reaction buffer (supplied by the manufacturer), 2 µl (20 µCi) $[\alpha\text{-}^{32}P]$Cordycepin triphosphate and H_2O to a final volume of 9 µl. Then, 1 µl poly(A) polymerase is added and the reaction incubated at 30 °C for 20 min. Because the enzyme tends to bind to the RNA, a phenol extraction/precipitation step is advisable before proceeding to one-dimensional gel electrophoresis (Section 9.2.1.2).

3′ Labeling of RNA with DNA Polymerase

This "splint labeling" method is a special case because a single RNA species may be labeled specifically within a mixture of other nucleic acids, provided that at least the immediate 3′ sequence of that RNA is known. A DNA oligonucleotide should be obtained which is complementary to the immediate 3′ RNA end (T_m about 40 °C) and has one extra T residue at its 5′ end, providing a "5′-T overhang" after annealing. An RNA preparation containing approximately 10–50 pmol of the desired species is annealed with 50 pmol of this oligonucleotide in a total volume of 17 μl H_2O by incubating for 5 min at 70 °C and cooling down to 50 °C over about 30 min. The annealing process is terminated on ice, 5 μl of 5 × reaction buffer (supplied by the manufacturer), 2 μl (20 μCi) [α-^{32}P]dATP and 1 μl T7 DNA polymerase (about 20 U/μl) are added and the labeling reaction is incubated for 30 min at 37 °C. The reaction products are further analyzed and purified as described (Section 9.2.1.2).

9.2.3
Sequencing of End-labeled RNA

Genomic sequencing has opened the view on a large number of putative novel RNA species. Their existence and primary sequence can easily be verified by a number of indirect techniques, including RT-PCR for the known part of the RNA sequence and variations of the RACE method to determine the initiation and termination points of transcription, or of processing sites during maturation [8]. However, if no hint to the sequence of an interesting functional RNA can be obtained by data mining or if nucleotide modifications are suspected to play a role in RNA function, direct sequence analysis of the RNA should be performed. Base specific enzymatic and chemical cleavages of end-labeled RNA provide hints on the identity of many modified nucleotides [23–28]; detailed working protocols for both methods will be presented here. The enzymatic as well as the chemical sequencing method rely on cleavage reactions that are not completely specific for all of the four major nucleobases; thus, the sequence has to be deduced from a partly ambiguous cleavage pattern in both cases. The advantage of the enzymatic over the chemical method is that sequence can be obtained from either labeled end, that more information on the nature of modifications can be deduced and that the procedure is straightforward and fast. The main disadvantage is that strong secondary structure of an RNA may inhibit cleavage by certain enzymes and that it may be difficult to obtain the required enzymes at sufficient quality. In contrast, the chemical modifications and subsequent cleavage reactions require only a small number of chemicals and are mostly insensitive to secondary structure under the conditions used.

If the exact nature and position of the modification is to be determined, a position-specific nucleotide analysis by thin-layer chromatography (TLC) has to be performed [20, 29]. This post-labeling method avoids end-labeling of the RNA and gives the best information regarding modified nucleotides, but is quite laborious and time-consuming to perform. Also, it is quite often prone to secondary cleavages and thus not absolutely reliable for primary sequence determination. A major disadvantage of all three approaches discussed here is that none of them will allow unambiguous reading of the terminal few nucleotides; the mobility shift method used to solve this problem has been described in detail elsewhere [12, 20, 30] and will not be presented here. If desired, the labeled 5'- or 3'-terminal nucleotide can be determined by cleavage with nuclease P1 or RNase T2, respectively, and subsequent TLC analysis, as described below (Section 9.2.4.1). In conclusion, the inherent advantages and disadvantages of the aforementioned methods demand a careful evaluation of the specific goals of each sequencing project in order to determine in which combination they should be used.

9.2.3.1 Sequencing by Base-specific Enzymatic Hydrolysis of End-labeled RNA

For enzymatic sequencing, either 5'- or 3'-labeled RNA may be used. Although 10 000 c.p.m. per reaction is optimal, as little as 1000 c.p.m. may be used if sufficient exposure time is allowed. To resolve problems arising from secondary structure, all reactions are performed under denaturing conditions (8 M urea, 50 °C).

Materials Required for Enzymatic Sequencing

RNA
- 5'- or 3'-^{32}P-labeled, gel purified RNA (see Sections 9.2.2.1 and 9.2.2.2), minimum amount 10 000 c.p.m. total; 1 µg/µl yeast tRNA (Roche Biochemicals).

Enzymes
- The sequencing nucleases (RNases T1, CL 3, *Staphylococcus aureus* nuclease and RNase U2) are available from Calbiochem, BRL, Roche Biochemicals, Worthington and Pharmacia, respectively. Because the quality differs between production batches, each lot should be tested separately using the protocol below. Working solutions of 100 mU/µl (RNases T1 and U2), 3 U/µl for *Staphylococcus* nuclease and 13 mU/µl for RNase CL3 should be made up fresh in H$_2$O before use.

Reaction mixes
- T1 mix and H$^+$ mix should be made up fresh from appropriate stock solutions; all others may be prepared in advance and stored at −20 °C.
- T1 mix: 20 mM Na citrate, pH 3.5, 1 mM Na$_2$EDTA, 0.03% (v/v) dye stock, 8.1 M urea.
- S7 mix: 20 mM Tris–HCl, pH 7.5, 10 mM CaCl$_2$, 0.03% (v/v) dye stock, 8.1 M urea.
- CL3 mix: 20 mM Tris–HCl, pH 7.5, 0.03% (v/v) dye stock, 8.1 M urea.
- H$^+$ mix: 0.22 N H$_2$SO$_4$, 0.03% (v/v) dye stock, 6.8 M urea.
- Dye stock: 1% (w/v) BPB, 1% (w/v) XC, 2% (w/v) SBB in 10 mM Tris, pH 7.5.

Tab. 9.1. Cleavage reactions for enzymatic sequencing

Reaction	Yeast RNA (μg)	Mix	Enzyme	Incubation	Specificity
–E	1	T1 mix	–	15 min, 50 °C	–
T1			RNase T1		G
U2			RNase U2		A ≫ ms^{2}i^{6}A, G
S7		S7 mix	*Staphylococcus* nuclease		A, U > T, s^{4}U, ms^{2}i^{6}A, m^{2}A
CL3		CL3 mix	RNase CL3		C ≫ A, T
H$^+$	5	H$^+$ mix	–	3 min, 100 °C	All except 2′-O-methyl

Working Procedure for Enzymatic Sequencing

Prepare 10 Eppendorf-type reaction tubes with equal aliquots of your end-labeled, gel-purified RNA (between 1000 and 10 000 c.p.m. per reaction). Label one tube as control (–E), one for the ladder (H$^+$) and two for each of the four enzymes (see Table 9.1). Calculate the amount of carrier RNA (from the elution and precipitation) in each aliquot, and adjust to 1 μg/tube for the control and enzymatic reactions. For the acid ladder adjust carrier to 5 μg. Dry down the contents of all tubes at room temperature (this takes about 30 min in a Speed Vac or 2 h in an evacuated desiccator over fresh desiccant). Prepare an ice box with wet ice and one with finely crushed dry ice, both large enough to hold the 10 tubes deeply immersed. Pre-heat one water bath at 50, 65 and 95–100 °C, respectively. Alternatively, metal-based heating blocks may be used, but heat transfer is faster and more efficient in water.

Add 4 μl of the respective enzyme reaction mix to each labeled tube (except H$^+$; see Table 9.1), spin down shortly, denature for 5 min at 65 °C and quick chill on ice. Add 1 μl of the respective enzyme working solution to the first of your two tubes for the same enzyme (e.g. T1). Mix by pipetting in and out, and transfer exactly 1 μl to the second tube. Immediately put the two tubes in the 50 °C bath and incubate for exactly 15 min; stop the reactions in dry ice. The remaining tubes are treated the same way. For the acid ladder, 4 μl of H$^+$ mix are added to the respective tube, incubated in a boiling water bath for exactly 3 min and quenched in dry ice. The samples may be stored overnight at –80 °C at this stage.

After a short spin, the samples are directly loaded onto a sequencing gel (40 × 20 cm, 0.4 mm thick, 12 lanes per RNA). Details on the composition of RNA sequencing gels may be found in Chapters 10 and 11. For RNAs of 70–90 nt, use 20% PAA; for longer RNAs, use 15% PAA. To read over the whole sequence length, it is advisable to prepare enough material for two runs: a short run (BPB just leaving the 20% gel) and a long run (XC at the edge of the gel). If the sequence should be read up to the labeled end, precipitate the cleavage reactions with EtOH and omit BPB from the loading mix.

Interpretation and Troubleshooting

From the counting ladder (H$^+$) and the highly base specific RNases T1 and U2 (and CL3 if good quality is available) it is straightforward to deduce a large part of the sequence. To read the band pattern created by *Staphylococcus* nuclease, recall that this enzyme cleaves 5' of the respective nucleotide, leading to a band shift (see Fig. 9.1). A gap in the counting ladder (H$^+$) indicates a 2'-ribose methylation of the corresponding nucleotide. Single weak bands may result from base modifi-

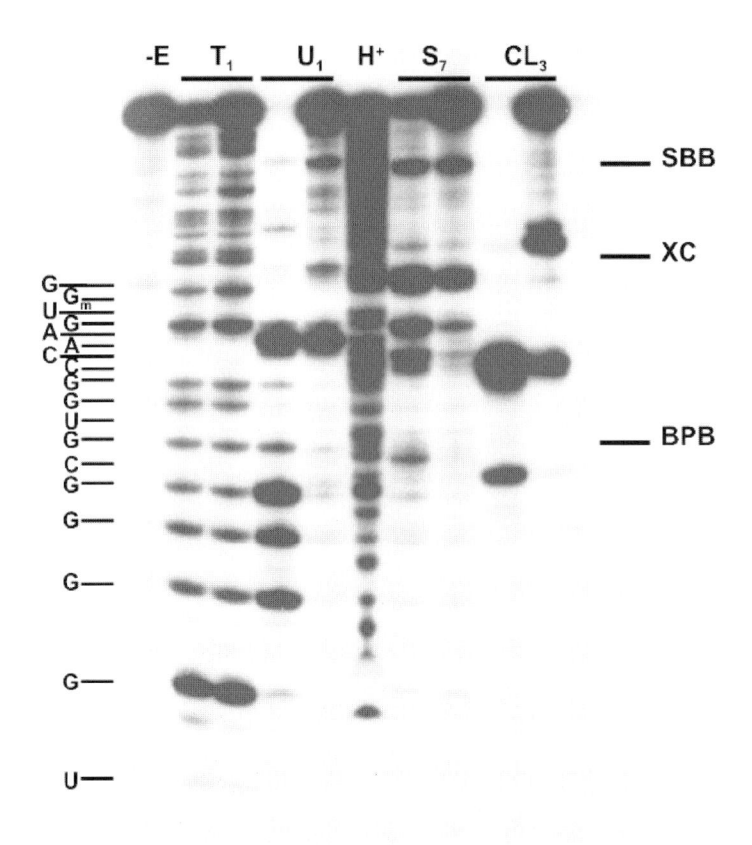

Fig. 9.1. Sequence analysis by enzymatic cleavage of 5'-^{32}P-labeled tRNA^{Gln} from barley chloroplasts. Enzymatic cleavages by RNase T1, U2, CL3 and *Staphylococcus* nuclease (S7), and acid hydrolysis (H$^+$) were performed as described in Section 9.2.3.1; the two lanes with the same specificity differ by a factor of 5 in the amount of enzyme used. A 25% PAA sequencing gel (29:1) was used to allow reading from the second nucleotide (bottom of the gel). The position of the dyes (SBB, XC, BPB) is given on the right, and the sequence of the first 18 nt on the left side of the panel (the terminal U was determined by end group analysis). Note the gap in the ladder (H$^+$) at the position of 2'-O-methyl-guanosine (G$_{m_1}$). For details of the sequence and its interpretation, see [5].

cations; see Table 9.1 for an incomplete overview and [12, 25] for a full discussion of this phenomenon. If a high-quality preparation of RNase CL3 is not commercially available, the enzyme may also be prepared according to the procedure described in [26].

If parts of the gel show weak bands in all lanes, strong secondary structure may hinder efficient cleavage; in this case, denature the RNA at 90 °C and run the reactions at 65 °C (you may have to use more enzyme). If bands are compressed on the gel, insufficient denaturation during electrophoresis is the reason. Make sure that the gel is run at 40 W (for a 20 × 40-cm gel) or use a thermostated electrophoresis apparatus at 60–65 °C.

9.2.3.2 Sequencing by Base-specific Chemical Modification and Cleavage

Chemical sequencing gives clear results only for 3'-labeled RNA, because of the inhomogeneous cleavage products 5' of the attacked nucleotide [24]. The precipitations required to stop the reactions and to remove the aniline prior to electrophoresis lead to some loss of material; thus, a higher amount of radioactive starting material should be used than for enzymatic sequencing. The following protocol follows a simplified and slightly modified version of the original, which should be consulted for full details [24, 27].

Materials Required for Chemical Sequencing

RNA
- 3'-^{32}P-labeled, gel purified RNA (see Section 9.2.2.2), minimum amount 25 000 c.p.m. total; 10 µg/µl yeast tRNA (Roche Biochemicals).

Chemicals
- Hydrazine, DEPC, aniline, NaBH$_4$, dimethylsulfate (DMS) and EtOH should be of the highest purity available and stored dry at 4 °C.

Buffers and reagents
- 50 mM NaOAc, pH 4.
- 1 M aniline acetate, pH 4.5 (mix H$_2$O:HOAc:aniline at a 7:3:1 volume ratio; spin out precipitate, check pH of an aliquot, store frozen in aliquots).
- 0.5 M NaBH$_4$ (make fresh before use).
- 0.3% (w/v) DMS in NaOAc (mix directly before use).
- Hydrazine/H$_2$O: mix equal volumes and keep on ice until use.
- 3 M NaCl in hydrazine: dry NaCl in a 120 °C oven for 2 h, store in a desiccator. Dissolve in hydrazine and keep on ice until use.

Waste disposal
- DMS is a carcinogen; all solutions containing it (e.g. reagents and EtOH supernatants from the first precipitation) should be disposed into a bottle containing 5 M NaOH. Hydrazine waste is inactivated with 3 M FeCl$_3$, aniline, DMS and hydrazine bottles should be opened only under a flow hood.

Tab. 9.2. Working table for chemical sequence analysis of RNA

Specificity	G	A	U	C
Modification		150 µl NaOAc + 1 µl DEPC	10 µl hydrazine/ H_2O	10 µl NaCl/ hydrazine
reagent	10 µl 0.3% DMS in NaOAc	150 µl NaOAc + 1 µl DEPC	10 µl hydrazine/ H_2O	10 µl NaCl/ hydrazine
incubation	40 s, 90 °C	10 min, 90 °C	10 min, 0 °C	10 min, 90 °C
first precipitation	150 µl NaOAc, 650 µl EtOH	400 µl EtOH	150 µl NaOAc, 550 µl EtOH	500 µl 80% EtOH
second precipitation	–	150 µl NaOAc, 450 µl EtOH	150 µl NaOAc, 450 µl EtOH	150 µl NaOAc, 450 µl EtOH
wash	800 µl EtOH	800 µl EtOH	800 µl EtOH	800 µl EtOH
Reduction		–	–	–
reagent	10 µl 0.5 M $NaBH_4$	–	–	–
incubation	10 min, 0 °C (dark)	–	–	–
precipitation	150 µl NaOAc, 650 µl EtOH	–	–	–
wash	800 µl EtOH	–	–	–

Starting material is dried 3′-end-labeled RNA containing 20 µg yeast tRNA per reaction tube. All centrifugations are performed in a microcentrifuge at 4 °C (10 000 g); all precipitations are done in crushed dry ice for 10 min.

Working Procedure for Chemical RNA Sequencing

Prepare five reaction tubes with equal aliquots of your end-labeled, gel purified RNA (between 5000 and 20 000 c.p.m. per reaction) and 20 µg yeast tRNA. Label one tube as control (−E) and for each of the four reactions (A, C, G and U; see Table 9.2). Dry down the contents of all tubes at room temperature (this takes about 30 min in a Speed Vac or 2 h in an evacuated desiccator over fresh desiccant). Prepare an ice box with wet ice, and pre-heat one water bath at 60 and 90 °C. All modifications are done according to the flow sheet (Table 9.2); note that all precipitations are on ice (or at −20 °C) for 5 min, all centrifugations are at 4 °C and 10 000 g for 5 min (precipitations) or 1 min (wash), respectively. Be careful to remove all of the supernatants to avoid a smear on the sequencing gel.

For aniline cleavage, add 10 µl of aniline acetate to all tubes including the control tube. Incubate for 20 min at 60 °C, stop on ice and precipitate with 150 µl NaOAc and 650 µl EtOH. After 2 washes with 800 µl EtOH to completely remove residual aniline, the RNA is dried, dissolved in urea gel loading buffer, denatured for 3 min at 95 °C and analyzed on a sequencing gel (Section 9.2.3.1).

If desired, a counting ladder may be prepared by acid hydrolysis (Section 9.2.3.1) and run in parallel. Note that the resulting banding pattern is shifted about one nucleotide away from the 3′ end if compared to the corresponding band obtained by chemical cleavage.

Interpretation and Troubleshooting
From the counting ladder (H^+) and the highly base-specific cleavages for G, A and U, a large part of the sequence can be easily deduced. Because the C reaction modifies both pyrimidines (although with lower efficiency for U residues), a "subtractive reading" of the U and C lanes is required to unambiguously identify both bases. Some modified bases can be identified very easily: m^7G is sensitive towards aniline without any further modification and thus appears as an extremely strong band in all lanes including the control. All uridine derivatives except pseudouridine are weakly reactive towards the U reaction; ac^4C can be recognized as a band in all lanes, but weaker than the appearance of m^7G. For a more complete overview, see [12, 25]. Band compression due to strong secondary structure of the RNA can be avoided as described above (Section 9.2.3.1).

9.2.4
Determination of Modified Nucleotides by Post-labeling Methods

In many cases, it is desirable to obtain an overview of the modified nucleotides present in a purified RNA species or in an RNA population obtained from a certain organism. If the RNA material can be easily obtained, HPLC analysis is the method of choice because UV spectra provide additional information on the nature of the nucleobase. Coupled HPLC-MS will even identify unknown or novel nucleotides [31]. However, the required apparative infrastructure is not readily available for most laboratories and, even though the sensitivity of the methods is impressive, availability of the biological samples may be limiting. A reliable alternative to determine the nucleotide content of subpicomolar amounts of RNA is the post-labeling of an RNA hydrolysate, followed by two-dimensional TLC analysis of the products [20, 32–34]. The determination of sequence and base modification at the same time has been made possible by the coupling of limited RNA fragmentation and end-group identification of the terminally labeled, separated fragments [29].

9.2.4.1 Analysis of Total Nucleotide Content
The first step of this procedure is the hydrolysis by a mixture of RNases T2 and A. The resulting nucleoside 3′-phosphates are then radioactively labeled at the 5′ end by PNK; these 5′-^{32}P-labeled 3′,5′-nucleoside diphosphates are then converted to the corresponding nucleoside 5′-phosphates by nuclease P1. After elimination of residual ATP by Apyrase, the mixture is subjected to two-dimensional TLC, with an excess of non-labeled nucleoside 5′-phosphates co-migrating as standards.

Materials Required for RNA Nucleotide Analysis

Enzymes
- RNase T2 (Invitrogen); pancreatic ribonuclease (RNase A) and T4 polynucleotide kinase (Roche Biochemicals). Prepare a working solution (T2/A mix) containing 50 mU/μl RNase T2 and 0.1 μg/μl RNase A in H_2O (can be stored frozen).

- Apyrase (5 U/ml; Sigma).
- Nuclease P1 (Gibco/BRL; prepare a working solution of 10 ng/µl in 50 mM ammonium acetate, pH 5.3).

Solvents and plates for TLC
- Solvent A: isobutyric acid:concentrated ammonia:H_2O [57.7:3.8:38.5, (v/v)]; pH 4.3.
- Solvent B: isopropanol:concentrated HCl:H_2O (70:15:15).
- Cellulose TLC plates (plastic or glass backed, non-fluorescent, analytical scale), 20 × 20 cm (Macherey-Nagel or Merck).

Radioactive materials
- [γ-^{32}P]ATP (10 µCi/µl; 3000 Ci/mmol).

Preparation of 5′-^{32}P-labeled Nucleoside Monophosphates
Purified RNA (2–20 pmol) is vacuum dried and dissolved in 2 µl 50 mM ammonium acetate, pH 4.5 and 6 µl H_2O (include a control sample without RNA). 1 µl of RNase T2/A mixture is added and the sample is incubated for 5 h at 37 °C. To the resulting hydrolysate, add 1 µl 10 × concentrated PNK buffer (provided by the manufacturer), 0.5 µl 0.1 mM ATP, 25 µCi [γ-^{32}P]ATP and 5 U PNK; incubate for 30 min at 37 °C. Add 1 µl Apyrase, incubate for another 30 min at 37 °C and proceed with half of the mixture (save the other half at −20 °C). Vacuum-dry this aliquot, add 10 µl of nuclease P1 solution and incubate for 3 h at 37 °C.

Two-dimensional TLC of Nucleoside Monophosphates
For analytical TLC, 1 µl of above hydrolysate is mixed with 1 µl pN marker mix (5 mg/ml each of pA, pG, pC and pU). The start point is marked with a soft pencil in the lower left corner of a cellulose plate, 1.5 cm from each edge. The sample is applied in a small spot (preferably with a drawn-out capillary) and dried. The first dimension is developed in solvent A until the front has reached the top edge; the plate is dried thoroughly under a hood. For chromatography in the second dimension, the plate is turned by 90° compared to the first dimension and developed in solvent B. After drying, the plates can be exposed to X-ray film or a Phospho-Imager. The marker nucleotides are visualized as dark blue spots under a UV lamp at 254 nm; their position is marked as an aid in the identification of unknown nucleotides. If a specific nucleotide has to be prepared for secondary analysis, the whole reaction mix (Section 9.2.4.1) is applied to the plate. The corresponding spot is then localized by positioning the plate on top of the X-ray film, scraped off the plate, and eluted with H_2O [12].

Interpretation and Troubleshooting
The four marker nucleotides should appear under UV as clearly separated spots (see reference patterns in [13, 32–34]; the $^{32}P_i$ (resulting from hydrolysis of unused ATP) should be visible as a prominent spot on the X-ray film in the center of the right edge. If separation of nucleotides is unacceptable, check the pH of the solvents and replace them if necessary; make sure that the sample was dried com-

pletely after application (use an infrared lamp or hot air if necessary) and check that the lids of the tanks close tightly.

If the starting point is streaked out in the first dimension, the problem might be the amount of protein in your sample; try to reduce the amount of enzyme used (extend the incubation times instead). If comparison to the standard pattern reveals that many dinucleotides are present in your sample, you should first analyze an aliquot of your sample before P1 digestion. This reveals whether you should increase the amount of T2/A mix and/or P1 nuclease, or the respective incubation time.

9.2.4.2 Determination of Position and Identity of Modified Nucleotides

In this case, limited RNA hydrolysis (ideally, one cut per RNA molecule) is performed non-enzymatically and the resulting fragments are radioactively labeled [20, 29]. After electrophoretic separation, the 5′-terminal nucleotide of each isolated fragment is determined by TLC.

The material required is mostly identical to that specified in Section 9.2.4.1; in addition, a sterile glass capillary (5 or 10 µl size) and a gas burner is needed.

Generation and Separation of 5′-labeled Random RNA Fragments

In separate reaction tubes, dry down 20–40 pmol purified RNA, 50–100 µCi [γ-^{32}P]ATP and 10 µl of urea gel loading buffer (see Section 9.2.1.3); pre-heat a water bath to 95 °C. Dissolve the RNA in 1.5 µl H$_2$O, transfer it into the center of the capillary by aspiration and seal the ends with the flame. Hydrolysis is performed for exactly 30 s in the boiling water bath and stopped in ice water. Cut open the ends of the capillary, transfer the contents back to the original tube and rinse the capillary with 5 µl H$_2$O. Transfer the whole contents to the tube containing the dry [γ-^{32}P]ATP and proceed with 5′-labeling and electrophoresis on a 15% PAA gel as described in Section 9.2.2.1. For best separation of the RNA fragments, it is advisable to use gels of 60 or 80 cm length and run them until BPB has reached the bottom; if this is not available, a short and long run should be performed on a 40 cm gel. Fragments are localized by autoradiography, cut out and eluted, including 10 µg yeast tRNA per band (see Sections 9.2.1.2).

Identification of the 5′-end Group of the RNA Fragments

Each sample is dissolved in 10 µl of nuclease P1 solution (Section 9.2.4.1), incubated for 2 h at 50 °C, and an aliquot (1–5 µl, depending on labeling efficiency) is removed and dried (the rest may be stored frozen). The dry samples are then dissolved in 2 µl of pN marker mix (Section 9.2.4.1) and equal amounts applied to each of two TLC plates. The cellulose plates are prepared such that 12–16 samples can be applied as thin streaks, 1.5 cm from the bottom edge; they are then analyzed by one-dimensional separation in solvent A and B, respectively (Section 9.2.4.1). The RNA sequence can then be directly read from the TLC plate. If a 2′-O-methylated dinucleotide is detected, an aliquot of the corresponding sample is digested with 1–10 µg of P1 nuclease (5 h at 65 °C) and analyzed as before or by two-dimensional TLC (Section 9.2.4.1).

Interpretation and Troubleshooting

Ideally, the band pattern visible after electrophoresis should show an even distribution, reaching up to the penultimate nucleotide. If the ladder is shifted significantly towards the smaller fragments, try to increase the amount of RNA or reduce the hydrolysis time. A specific problem may arise if a labile modified nucleotide or a C–A bond in a single-stranded region is present in the RNA. In this case, near-quantitative hydrolysis of the corresponding phosphodiester bond may even lead to a complete lack of bands above this point, and the nucleotide 3′ of the cleaved bond will be visible in all other samples [4]. Most problems arising from the TLC systems have been discussed in Section 9.2.4.1. In some cases, it may be necessary to re-analyze specific modified nucleotides in a different solvent system. A two-dimensional chromatography system with slightly different separation properties has been described in [34]. ac^4C and m^5C are not separated in solvents A and B (Section 9.2.4.1), but can be readily distinguished by chromatography on PEI plates [35]. Thionucleotides can be identified after modification with CNBr and separation of the products on Cellulose [36].

9.3
Conclusions and Outlook

The increasing number of genomic sequences has led to the detection of numerous novel RNAs with mostly unknown functions. In many cases, modified nucleotides may play a role in increasing their structural stability, or facilitating specific interactions with proteins or other RNAs; in some cases, editing may even change the primary sequence and coding potential of an RNA. The methods presented here do not only allow the rigorous purification of any desired RNA from biological samples, but also permit the identification of type and position of modified nucleotides. They may thus help in elucidating the function of these RNAs by identifying novel interaction points with other macromolecules. We anticipate an increasing application of direct RNA sequencing methods, specifically in context with the future investigation of novel RNA species.

Acknowledgments

Research on RNase P in my laboratory is supported by grants from the Deutsche Forschungsgemeinschaft (Scho515/7-3).

References

1 W. Filipowicz, *Proc. Natl. Acad. Sci. USA* **2000**, *97*, 14035–14037.
2 H. Beier, M. Barciszewska, G. Krupp, R. Mitnacht, H. J. Gross, *EMBO J.* **1984**, *3*, 351–356.
3 H. Beier, M. Barciszewska, H. D.

Sickinger, *EMBO J.* **1984**, *3*, 1091–1096.
4 A. Schön, G. Krupp, S. Gough, S. Berry-Lowe, C. G. Kannangara, D. Söll, *Nature* **1986**, *322*, 281–284.
5 A. Schön, C. G. Kannangara, S. Gough, D. Söll, *Nature* **1988**, *331*, 187–190.
6 T. Yasukawa, T. Suzuki, N. Ishii, S. Ohta, K. Watanabe, *EMBO J.* **2001**, *20*, 4794–4802.
7 I. Behm-Ansmant, A. Urban, X. Ma, Y. T. Yu, Y. Motorin, C. Branlant, *RNA* **2003**, *9*, 1371–1382.
8 J. Sambrook, E. F. Fritsch, T. Maniatis, *Molecular Cloning: A Laboratory Manual*, Cold Spring Harbor Laboratory Press, Cold Spring Harbor, NY, **2001**.
9 G. Zubay, *J. Mol. Biol.* **1962**, *4*, 247–362.
10 B. Roe, *Nucleic Acids Res.* **1975**, *2*, 21–42.
11 C. Heubeck, A. Schön, *Methods Enzymol.* **2001**, *342*, 118–134.
12 G. Krupp, H. J. Gross, in: *The Modified Nucleotides of Transfer RNA II*, P. F. Agris, R. A. Kopper (eds), Alan R. Liss, New York, **1983**.
13 Y. Kuchino, N. Hanyu, S. Nishimura, *Methods Enzymol.* **1987**, *155*, 379–396.
14 K. Wakita, Y.-I. Watanabe, T. Yokogawa, Y. Kumazawa, S. Nakamura, T. Ueda, K. Watanabe, K. Nishikawa, *Nucleic Acids Res.* **1994**, *22*, 347–353.
15 T. Suzuki, T. Ueda, K. Watanabe, *FEBS Lett.* **1996**, *381*, 195–198.
16 H. Blum, H. Beier, H. J. Gross, *Electrophoresis* **1987**, *8*, 93–99.
17 J. M. d'Alessio, in: *Gel Electrophoresis of Nucleic Acids*, D. Rickwood, B. D. Hames (eds), IRL Press, Oxford, UK, **1982**.
18 L. Berkner, W. R. Folk, *J. Biol. Chem.* **1977**, *252*, 3176–3184.
19 T. E. England, A. G. Bruce, O. C. Uhlenbeck, *Methods Enzymol.* **1980**, *65*, 65–74.
20 H. Beier, H. J. Gross, in: *Essential Molecular Biology II*, T. A. Brown (ed.), IRL Press, Oxford, UK, **1991**.
21 J. Lingner, W. Keller, *Nucleic Acids Res.* **1993**, *21*, 2917–2920.
22 T. P. Hausner, L. M. Giglio, A. M. Weiner, *Genes Dev.* **1990**, *4*, 2146–2156.
23 H. Donis-Keller, A. M. Maxam, W. Gilbert, *Nucleic Acids Res.* **1977**, *4*, 2527–2538.
24 D. A. Peattie, *Proc. Natl. Acad. Sci. USA* **1979**, *76*, 1760–1764.
25 B. Lankat-Buttgereit, H. J. Gross, G. Krupp, *Nucleic Acids Res.* **1987**, *15*, 7649.
26 C. C. Levy, T. P. Karpetsky, *J. Biol. Chem.* **1980**, *255*, 2153–2159.
27 R. Waldmann, H. J. Gross, G. Krupp, *Nucleic Acids Res.* **1987**, *15*, 7209.
28 Y. Kuchino, S. Nishimura, *Methods Enzymol.* **1989**, *180*, 154–163.
29 J. Stanley, S. Vassilenko, *Nature* **1978**, *274*, 87–89.
30 M. Silberklang, A. M. Gillum, U. L. RajBhandary, *Methods Enzymol.* **1979**, *59*, 58–109.
31 J. A. McCloskey, A. B. Whitehill, J. Rozenski, F. Qiu, P. F. Crain, *Nucleosides Nucleotides* **1999**, *18*, 1549–1553.
32 S. Nishimura, in: *Transfer RNA: Structure, Properties and Recognition*, P. Schimmel, D. Söll, J. Abelson (eds), Cold Spring Harbor Laboratory Press, Cold Spring Harbor, NY, **1979**.
33 S. Nishimura, Y. Kuchino, in: *Methods of DNA and RNA Sequencing*, S. Weisman (ed.), Praeger, New York, **1983**.
34 G. Keith, *Biochimie* **1995**, *77*, 142–144.
35 R. C. Gupta, E. Randerath, K. Randerath, *Nucleic Acids Res.* **1976**, *3*, 2915–2921.
36 M. Saneyoshi, S. Nishimura, *Biochim. Biophys. Acta* **1970**, *204*, 389–399.

10
Probing RNA Structures with Enzymes and Chemicals *In Vitro* and *In Vivo*

Eric Huntzinger, Maria Possedko, Flore Winter, Hervé Moine, Chantal Ehresmann and Pascale Romby

10.1
Introduction

A renewal of interest in RNA was brought about the recent discovery of a large number of new regulatory RNA molecules in bacteria and in eukaryotes (for reviews, see [1, 2]). Many studies in bacteria have also now confirmed that mRNA can adopt highly structured domains that serve genetic switches in response to ligand binding, ranging from proteins to RNA and even metabolites (e.g. [3–5]). Thus, the structural features of a RNA most often are of key importance for its biological function and consequently there is an increasing interest in studies on the structure of RNAs either free or in interaction with ligands.

Chemical and enzymatic probing has become one of the most popular approaches for mapping the conformation of RNA molecules of any size under defined experimental conditions. The method maps the reactivity of each nucleotide towards enzymes or chemicals, which reflects its environment within the RNA molecule. The elaboration of a secondary structure model requires the use of probes with different and complementary specificities. For long RNA molecules, the interpretation of the data can be facilitated with the help of computer programs that predict secondary structure from the sequence. One powerful method is to combine energy minimization with co-variation while other programs tend to simulate the kinetics of RNA folding during transcription (for a review, see [6]). Since the probing approach defines unambiguously the unpaired regions of the RNA, they can be given as constraints in the computer folding programs. The resulting secondary structure model can be further validated by a site-directed mutagenesis study coupled to the probing approach to analyze the effect of the mutation on the RNA structure. For instance, compensatory base changes validate the existence of Watson–Crick base pairs and appropriate deletion may help to define independent structural domains.

Probing the structure with chemicals and enzymes may also provide information on the tertiary folding of large RNA molecules. The tertiary structure of large RNAs is composed of stable secondary structure elements that are brought

Handbook of RNA Biochemistry: Student Edition. Edited by R. K. Hartmann, A. Bindereif, A. Schön, E. Westhof
Copyright © 2009 WILEY-VCH Verlag GmbH & Co. KGaA, Weinheim
ISBN: 978-3-527-32534-4

together by long-distance interactions to form compact domains. Crystallographic research on the ribosome has revealed the presence of numerous RNA motifs that have been found in different RNAs ([7] and references therein). These RNA motifs mediate either the specific interactions that induce the compact folding of large RNA or constitute specific ligand-binding sites. They are usually composed of stacked arrays of non-Watson–Crick base pairs, which are characterized by an unusual pattern of chemical reactivity. The correlation between X-ray structure and chemical modification of different RNAs can be used to unravel the existence of particular structural motifs in RNA molecules and certain non-canonical base pairs (e.g. sheared purine base pairs and Hoogsteen reverse AU base pairs). Chemical modifications can also be performed under different experimental conditions (i.e. by varying the temperature or the concentration of divalent ions). Such experiments may provide information on the stability of the different secondary structure domains, but also allow the identification of tertiary elements since these interactions are the first to break in a cooperative manner during the melting process of an RNA structure. Some of the chemicals that react with the phosphate-ribose backbone can be used to probe the inside and outside of large and highly structured RNAs (i.e. *Tetrahymena* ribozyme [8]). Time-resolved small-angle X-ray scattering is a powerful approach to detect transient RNA–RNA interactions and to measure the fast global shape changes of large RNAs under different ionic conditions (e.g. [9, 10]).

One of the main concerns often addressed is how the RNA can be folded in a more complex environment such as in living cells. Ligand binding may indeed change the RNA folding or stabilize a defined conformation. Structure-specific chemical probes are unique tools to map RNA structure *in vivo* under different cell growth conditions. The use of probes is, however, limited by their capability to penetrate the cell wall and membrane due to their size, structure and/or charge. To date, the reagents that have gained widespread use for *in vivo* RNA probing are dimethylsulfate (DMS) (e.g. [11, 12]), to a lesser extent kethoxal [13] and, more recently, lead(II)-induced cleavage [14]. Despite the limited number of probes, which can be used *in vivo*, the comparison between *in vivo* and *in vitro* probing provides complementary data for determining functional RNA structure.

Enzymes and chemicals have been utilized in several other applications [15]. The probes have been extensively used to map the binding site of a specific ligand (antibiotic, RNA, protein, etc.) and to study RNP assembly. Chemicals have also been used for the so-called chemical interference approach. This method defines a set of nucleotides which have lost the capability to interact with a ligand when they are modified by a chemical probe. Finally, chemical probes tethered to protein or RNA can provide topographical information on ligand–RNA complexes by inducing site-specific cleavage of a proximal RNA after binding (e.g. [16, 17]).

The aim of the present chapter is to give an experimental guide of the most commonly used probes for mapping the secondary structure of RNAs *in vitro* and *in vivo*. The mechanism of action and lists of a wide variety of probes can also be found in other reviews [15, 18, 19]. Additional information on probing with lead(II) can be found in Chapter 13.

10.2
The Probes

The probes are used under limited conditions where less than one cleavage or modification occurs per RNA molecule with a statistical distribution. The identification of the cleavages or the modifications can be done by two different methodologies depending on the length of the RNA molecule and on the nature of the modification. The first approach, which uses end-labeled RNA, only detects scissions and is limited to RNA containing less than 300 nt. The second approach, which uses primer extension, detects stops of reverse transcription at modified nucleotides or cleavages, and can be applied to RNAs of any size.

The experimental guide was adapted from the conditions used to probe the structure of the regulatory region of *thrS* mRNA [20] and will be limited to the most commonly used probes. Table 10.1 gives an exhaustive list of the probes for which experimental conditions will be given below. These probes provided complementary data necessary to build the RNA secondary structure model. Other detailed protocols used on different RNAs have been previously described [21–23]. In addition, experimental procedures will be given to map the RNA structure in bacterial cells. Other protocols used for *in vivo* mapping in eukaryotic cells will be given in this handbook.

10.2.1
Enzymes

RNases T1 (specific for unpaired guanines), RNase T2 (with preference for unpaired adenines) and RNase V1 (specific for double-stranded regions) are the most useful enzymes. They are easy to use and help to identify secondary structure RNA elements such as hairpin structures. RNase V1 is the only probe which provides positive evidence for the existence of helical regions. However, due to their size, the RNases are sensitive to steric hindrance. Particular caution has also to be taken since the cleavages may induce conformational rearrangements in RNA that potentially provide new targets (secondary cuts) to the RNase.

10.2.2
Chemical Probes

Base-specific reagents have been largely used to define RNA secondary structure models. The combination of DMS, 1-cyclohexyl-3-(2-morpholinoethyl)carbodiimide metho-*p*-toluene sulfonate (CMCT) and β-ethoxy-α-ketobutyraldehyde (kethoxal) allow to probe the four bases at one of their Watson–Crick positions (Table 10.1). DMS methylates position N^1 of adenines and to a lower extent N^3 of cytosines. CMCT modifies position N^3 of uridine and to a weaker degree N^1 of guanines. Kethoxal reacts with unpaired guanine, giving a cyclic adduct between positions N^1 and N^2 of the guanine and its two carbonyls.

Tab. 10.1. Structure-specific probes for RNA

Probes	MW	Target	Product	Detection			Special considerations Buffers, pH, temperature, etc.
				direct	RT	in vivo	
Chemicals and divalent ion							
DMS	126	$A(N^1)$	N^1–CH_3	–	+	+	reactive at pH ranging from 4.5 to 10 and temperature from 4 to 90 °C; tris buffers should be avoided as DMS reacts with amine groups
		$C(N^3)$	N^3–CH_3	s	+	+	idem
		$G(N^7)$	N^7–CH_3	s	s	+	idem
DEPC	174	$A(N^7)$	N^7–CO_2H_5	s	+	–	reactive at pH ranging from 4.5 to 10 and temperature from 4 to 90 °C; tris buffers should be avoided as DEPC reacts with amine groups
Kethoxal	148	$G(N^1$–$N^2)$	N^1–CHOH | N^2–CROH	+ (a)	+	–	borate ions are required to stabilize the guanine–kethoxal adduct
CMCT	424	$G(N^1)$	N^1–C=N–R | NH–R'	–	+	–	optimal reactivity at pH 8 and over a wide range of temperature; CMCT still soluble up to 300 mg/ml in water
		$U(N^3)$	N^3–C=N–R | NH–R'	–	+	–	idem
Pb(II) acetate	207	specific binding sites dynamic regions	…Np (3'p)	+	+	+	buffers with chlorure ions should be avoided as Pb(II) can form precipitates with it. Pb(II) acetate must be dissolved in water just before use
Biological nucleases							
T1 RNase	11 000	unpaired G	….Gp (3'p)	+	+	–	active under a wide range of conditions (e.g. temperature between 4 and 55 °C, with or without magnesium ion and salt, in urea)
T2 RNase	36 000	unpaired A > C, U, G	….Ap (3'p)	+	+	–	active under a wide range of conditions (e.g. temperature between 4 and 55 °C, with or without magnesium ion and salt)
V1 RNase	15 900	paired or stacked N	pN…. (5'p)	+	+	–	absolutely requires divalent cations; active under a wide range of temperature (4–50 °C)

DMS, dimethylsulfate; DEPC, diethylpyrocarbonate; kethoxal, β-ethoxy-α-ketobutyraldehyde; CMCT, 1-cyclohexyl-3-(2-morpholinoethyl) carbodiimide metho-p-toluene sulfonate. Detection method: (direct) detection of cleavages on end-labeled RNA molecule; (RT) detection by primer extension with reverse transcriptase. (+) the corresponding detection method can be used; (s) a chemical treatment is necessary to cleave the ribose-phosphate chain prior to the detection; (a) RNase T1 hydrolysis can be used after kethoxal modification with end-labeled RNA. Modification of guanine at N1, N2 will prevent RNase T1 hydrolysis [46]. *In vivo* mapping: probes which diffuse efficiently across membranes and walls (+), the other probes can be used only after permeabilization of the cell (–). Molecular weight, specificity, and products generated by the probe action are indicated. The table is adapted from Brunel & Romby [23].

Position N^7 of purines, involved in Hoogsteen or reverse Hoogsteen interactions, can also be probed by diethylpyrocarbonate (DEPC) for adenines and by DMS or nickel complex for guanines. Nickel complex [24] and DEPC [25] are very sensitive to the stacking of the base rings, and therefore N^7 of purines within a helix are never reactive except if the deep groove of the helix is widened.

10.2.3
Lead(II)

Divalent metal ions such as Mg^{2+} are required for stabilization of the RNA structure, but under special conditions can promote cleavage in RNA. This catalytic activity was first discovered with Pb^{2+} ions, and latter with many other di- and tri-valent cations (for a review, see [26]). Strong cleavage was first described as the consequence of a tight divalent metal ion-binding site and of an appropriate stereochemistry of the cleaved phosphodiester bond. Lead(II) is also considered as a single-strand-specific probe since weaker cleavages at several sites occurred mainly in unpaired and flexible regions (interhelical or loop regions and bulged nucleotides). In contrast to RNases, lead(II) is not sensitive to steric hindrance, but detects subtle conformational changes that can occur upon ligand binding. Lead(II) was also successfully used *in vivo* to map the structure of mRNA and regulatory RNAs [14]. In contrast to DMS modification, lead(II) is less sequence dependent, and thus can be used to assess single- and double-stranded regions of RNA. The cleavage patterns obtained on three different RNAs indicated that similar conformations were observed *in vivo* and *in vitro* [14].

10.3
Methods

Probing the conformation of RNAs with different enzymes and chemicals requires the use of defined buffer conditions (pH, ionic strength, magnesium concentration, temperature). Indeed, the optimal conditions vary slightly with the different probes and the possibility exists that subtle conformational changes may occur under different incubation conditions (Table 10.1). The probe:RNA ratio must also be adapted so that the experiments are conducted under limited and statistical conditions. For the first experiment, different concentrations of the probes and a time-scale dependence should be performed. This is also required when the commercial source of the probes has been changed.

10.3.1
Equipment and Reagents

Equipment and Material
Electrophoresis instrument for sequencing gels. Eppendorf tubes, tips and buffers should be sterilized before used.

Chemicals and Enzymes

CMCT and lead(II) acetate can be purchased from Merck; DMS from Acros Organics (ref. 11682-0100); calf intestinal phosphatase and T4 RNA ligase from P-L Biochemicals; avian myeloblastosis virus reverse transcriptase from Q biogene (France) or Life Sciences (USA); T4 polynucleotide kinase, $[\gamma\text{-}^{32}\text{P}]$ATP (3200 Ci/mmol) and $[5'\text{-}^{32}\text{P}]$pCp (3000 Ci/mmol) from Amersham. RNase T1 was from Industrial Research Limited (IRL, New Zealand) or from Fermentas; RNase T2 from Invitrogen (ref. 18031-013) and RNase V1 from Pierce (ref. MB092701).

Safety Rules using Chemicals

Most of the chemical reagents are potential carcinogens and therefore chemical modifications until the removal of the first ethanol supernatant (see below) are carried out under a fume hood while wearing protective gloves. DMS and kethoxal solutions are discarded in 1 M sodium hydroxide waste and CMCT in 10% acetic acid waste.

Buffers

The buffer conditions given here are indicative and can be modulated according to the system used and the nature of the ligand. *buffer N1*: 50 mM sodium HEPES, pH 7.5, 5 mM MgAc, 100 mM KAc; *buffer N2*: 50 mM sodium cacodylate, pH 7.5, 5 mM $MgCl_2$, 100 mM KCl; *buffer D2*: 50 mM sodium cacodylate, pH 7.5, 1 mM EDTA; *buffer N3*: 50 mM sodium borate, pH 8, 5 mM $MgCl_2$, 100 mM KCl; *buffer D3*: 50 mM sodium borate, pH 8, 1 mM EDTA; *buffer N4*: 50 mM sodium borate, pH 7.5, 5 mM $MgCl_2$, 100 mM KCl; *buffer D4*: 50 mM sodium borate, pH 7.5, 1 mM EDTA; *Buffer ΔT1*: 20 mM sodium citrate, pH 4.5, 1 mM EDTA, 7 M urea, 0.02% xylene cyanol, 0.02% bromophenol blue; *Ladder buffer*: Na_2CO_3 0.1 M/$NaHCO_3$ 0.1 M pH 9; *RNA loading buffer*: 0.02% xylene cyanol, 0.02% bromophenol blue in 8 M urea; *DNA loading buffer*: 0.02% xylene cyanol, 0.02% bromophenol blue in formamide; *RTB buffer*: 50 mM Tris–HCl, pH 7.5, 20 mM $MgCl_2$ and 50 mM KCl; *TBE buffer*: 0.09 M Tris–borate, pH 8.3, 1 mM EDTA. All buffers are given 1 × concentrated.

10.3.2
RNA Preparation and Renaturation Step

The RNA is usually transcribed *in vitro* from a DNA template using T7 RNA polymerase. The RNA is then purified from shorter RNA fragments, DNA template and the excess of NTP by using either a gel-filtration column [27], monoQ column [28] or denaturing polyacrylamide–urea gel electrophoresis [29]. More recently, ion-pairing reversed-phase high-performance liquid chromatography (IP-RPLC) was used for the fractionation of short RNA fragments [30].

For 5′-end-labeling, the RNA is first dephosphorylated at its 5′ end, and then labeled using $[\gamma\text{-}^{32}\text{P}]$ATP and T4 polynucleotide kinase [31]. The dephosphorylation step can be avoided if transcription is carried out in the presence of GMP or with ApG. The 3′-end-labeling is performed with $[5'\text{-}^{32}\text{P}]$pCp and T4 RNA ligase [32].

The labeled RNAs are purified by electrophoresis on 8% polyacrylamide (0.5% bis)–8 M urea slab gels. Before each experiment, the RNA is eluted from the gel slice in 100 µl of 500 mM ammonium acetate/1 mM EDTA, precipitated with 2.5 volumes of cold ethanol in the presence of 1 µg of glycogen. After two washing steps with 200 µl of 80% cold ethanol, the pellet is then dissolved in sterile H_2O (to obtain about 50 000 c.p.m./µl).

Since the RNA is often in contact with denaturing reagents during its purification, it is worth spending effort to carry out a renaturation process before the probing experiments. One possible renaturation protocol is as follows: the RNA is pre-incubated 1 min at 90 °C in H_2O, quickly cooled on ice (2 min) and brought back slowly (20 min) at 20 or 37 °C in the appropriate buffer containing 5 mM $MgCl_2$.

10.3.3
Enzymatic and Lead(II)-induced Cleavage Using End-labeled RNA

This direct method which uses end-labeled RNA is restricted to the detection of cleavage in the RNA after RNase hydrolysis or after chemical modifications that allow subsequent strand scission by an appropriate treatment (see Table 10.1).

Enzymatic probing and lead-induced cleavages were adapted for the *thrS* mRNA regulatory region (around 250 nt). Some of the experiments are illustrated in Fig. 10.1(A–C). All reactions were conducted in a total volume of 10 µl. Appropriate dilutions of enzymes and of lead(II) acetate were done in sterile water just before use. Incubation controls in the absence of the probes were performed in order to detect non-specific cleavage in RNA. In these controls, the enzyme or lead(II) was replaced by sterile H_2O. For RNA–ligand footprinting experiments, the complex was pre-formed before the enzymatic or chemical reaction in an appropriate buffer optimal for binding.

Labeled RNA (50 000 c.p.m./µl) sufficient for the planned experiments was first denatured in sterile H_2O at 90 °C for 1 min then cooled on ice for 1 min.

RNase T1
Labeled mRNA (1 µl, 50 000 c.p.m.) was renatured in the presence of 5 µl of H_2O and 2 µl of buffer N1 (5 × concentrated) at 20 °C for 20 min. Then, 1 µl of total tRNA (2 µg/µl) was added and reaction was performed with 1 µl of RNase T1 (0.1 U from IRL, or 0.2 U from Fermentas) for 5 min at 20 °C.

RNase T2
The same protocol as for RNase T1 except that reaction was performed with 1 µl of RNase T2 (0.05 U) for 5 min at 20 °C.

RNase V1
The same protocol as for RNase T1 except that reaction was performed with 1 µl of RNase V1 (0.05 U) for 5 min at 20 °C.

In order to define the best conditions for the hydrolysis, try for the first time three different concentrations of the enzymes: RNase T1 (0.05–0.1–0.5 U from

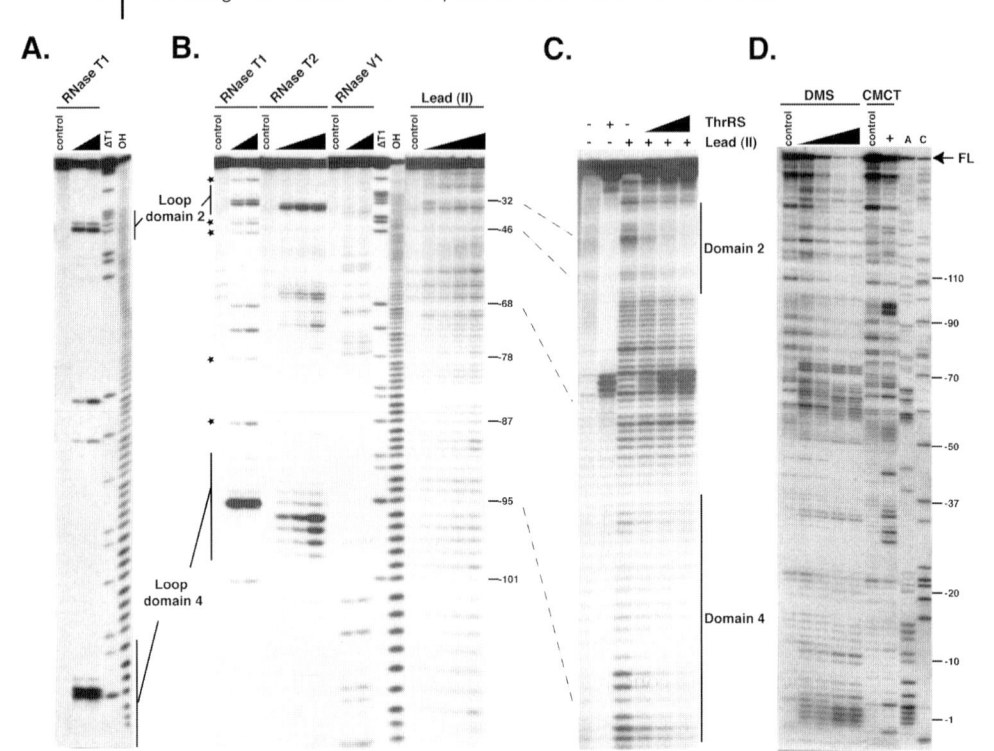

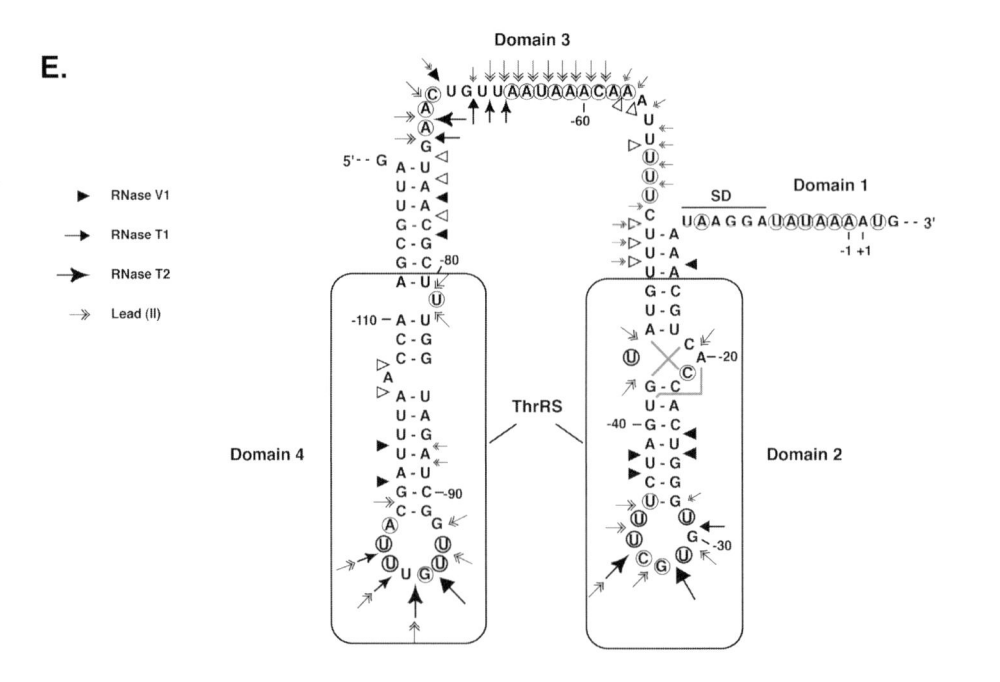

158 | 10 Probing RNA Structures with Enzymes and Chemicals In Vitro and In Vivo

IRL or 0.1–0.2–0.5 U from Fermentas), RNase T2 (0.01–0.05–0.1 U) and RNase V1 (0.01–0.05–0.1 U).

Lead(II) Acetate
Labeled mRNA (1 µl, 50 000 c.p.m.) was renatured in the presence of 3.5 µl of H_2O and 2 µl of buffer N1 (5×) at 20 °C for 20 min. Hydrolysis was initiated with 2.5 µl of different concentrations of lead(II) acetate from 12, 40, 80 to 120 mM for 5 min at 20 °C in the presence of 1 µl of total tRNA (2 µg/µl). Then, 5 µl of 0.1 M EDTA were added to stop the reaction. The best conditions for *thrS* mRNA was 40 mM.

Reaction Stop
Enzymatic hydrolysis were stopped by phenol extraction while the RNA treated with lead(II) was directly precipitated.

- To all samples, 40 µl of 0.3 M sodium acetate, pH 6 and 50 µl of phenol saturated with chloroform (v/v) were added. The samples were mixed for 1 min and then centrifuged 1 min at high speed.
- The aqueous phase was removed carefully and transferred into a new sterile 1.5 ml micro tube and 2.5 volumes of cold ethanol (150 µl) was added to precipitate the RNA. After mixing, the samples were left in a dry ice/ethanol bath for 10 min and centrifuged (13 000 r.p.m. at 4 °C for 15 min).
- The supernatant was discarded and the pellet was washed twice with 200 µl of 80% cold ethanol. After a short centrifugation (13 000 r.p.m. for 5 min at 4 °C), the supernatant was removed and the pellets were vacuum-dried (no more than 5 min) and dissolved in 6 µl of RNA loading buffer.

Fig. 10.1. Enzymatic and chemical probing on *thrS* mRNA. (A–C) Enzymatic hydrolysis and lead(II)-induced cleavages performed on 5′-end-labeled *thrS* mRNA either free (A and B) or in the presence of increasing concentrations of threonyl-tRNA synthetase (C). (A and B) The conditions for RNase and lead(II) concentrations were as follows: (A) RNase T1 from IRL (0.05 and 0.1 U), (B) RNase T1 from IRL (0.2 and 0.5 U), RNase T2 (0.01, 0.05 and 0.1 U) and RNase V1 (0.05 and 0.1 U), and lead(II) (12, 40, 80 and 120 mM). (C) Hydrolysis was performed with 40 mM lead(II). ThrRS concentrations were as follows: 0.01, 0.05 and 0.1 µM. (control) incubation control in the absence of the probes. (ΔT1, OH) RNase T1 under denaturing conditions and alkaline ladder, respectively. RNase T1 cleavages not reproducibly found are noted by an asterisk. (D) Gel electrophoresis fractionation of products resulting from DMS ($N^1A \gg N^3C$) and CMCT ($N^3U \gg N^1G$) modification followed by primer extension analysis. Reactions have been performed on free mRNA under native conditions in the absence (control) or in the presence of increasing concentrations of DMS (see text for details) or in the presence of 4 µl of CMCT 40 mg/ml (+). FL = full-length product. (Lanes A and C) The two sequencing ladders correspond to the RNA sequence. (E) Reactivity of Watson–Crick positions, enzymatic and lead(II)-induced cleavages reported on the secondary structure of the *thrS* mRNA adapted from Caillet et al. [20]. Circled nucleotides are reactive towards DMS (N^1A, N^3C) and CMCT (N^3U, N^1G) modifications. The two domains of the RNA protected by ThrRS are squared.

Fractionation of End-labeled RNA Fragments

RNase T1 and alkaline ladders were required to identify the cleavage positions.

- RNase T1 ladder. Labeled mRNA (0.5 µl, 25 000 c.p.m.) was preincubated at 50 °C for 5 min in 5 µl of buffer ΔT1 containing 1 µg total tRNA. The reaction is then performed at 50 °C for 10 min in the presence of 1 µl of RNase T1 (0.1 U for IRL or 0.5 U for Fermentas).
- Alkaline ladder. Labeled mRNA (2 µl, 100 000 c.p.m.) was incubated at 90 °C for 3 min in the presence of total tRNA (2 µg) in 5 µl of ladder buffer.

The end-labeled RNA fragments were sized by electrophoresis on 12 or 15% polyacrylamide (0.5% bis)–8 M urea slab gels (0.5 mm × 30 cm × 40 cm) in 1 × TBE. Gels should be pre-run (30 min at 75 W) and run warm (75 W) to avoid band compression. The migration conditions must be adapted to the length of the RNA, knowing that on 15% gel, xylene cyanol migrates as a 39-nt RNA and bromophenol blue as 9 nt. The 15% polyacrylamide gel is convenient to collect the data on small-size fragments (1–50 nt RNA fragments). For a 250-nt long RNA, two migrations are necessary to interpret correctly the reactivity of nucleotides of the whole RNA molecule. At the end of the run, the 12% gel is fixed for 5 min in a 10% ethanol/6% acetic acid solution, transferred to Whatman 3 MM paper and dried. The 15% gel was transferred without drying on a plastic support and wrapped with a plastic film. Exposure is done at −80 °C using an intensifying screen.

10.3.4
Chemical Modifications

Examples of chemical modifications performed on *thrS* mRNA regulatory region are shown in Fig. 10.1(D). Reactions were carried out on 1 pmol of unlabeled *thrS* mRNA in a total volume of 20 µl. For enzymatic and lead(II)-induced cleavages, the same experimental conditions could be used as described above except that the end-labeled RNA is replaced by 1 pmol of cold RNA. Control of an unmodified RNA was done in parallel, in order to detect pauses of reverse transcriptase due to stable secondary structures and/or non-specific cleavage. The reactions were conducted either in the presence of mono- and divalent ion ("native conditions") or in the absence of ions ("semi-denaturing" conditions). Unlabeled mRNA was first heated in sterile H_2O at 90 °C for 1 min and then cooled on ice for 1 min.

DMS modification

- Native conditions: 1 µl of mRNA (1 pmol) was first renatured by incubation at 20 °C for 20 min in the presence of 4 µl of buffer N2 (5×) and 13 µl of sterile H_2O. The reaction was performed at 20 °C for 5 min in the presence of 1 µl of tRNA (2 µg/µl) and 1 µl of pure DMS or diluted freshly into ethanol 1:2, 1:5 and 1:10. The optimal chemical modification was obtained with DMS diluted 1:10 (Fig. 10.1D).

- Semi-denaturing conditions: same procedure as for native conditions, but the reaction was performed in buffer D2.

CMCT modification

- Native conditions: 1 µl of mRNA (1 pmol) was first incubated at 20 °C for 20 min in the presence of 4 µl of buffer N3 (5×) and 10 µl of sterile H_2O. The reaction was done at 20 °C during 20 min in the presence of 1 µl of tRNA (2 µg/µl) and 4 µl of CMCT (40 or 60 mg/ml in water just before use). The optimal chemical modification for *thrS* mRNA was obtained with CMCT at 40 mg/ml (Fig. 10.1D).
- Semi-denaturing conditions: same procedure as for native conditions, but in buffer D3.

Kethoxal modifications

- Native conditions: 1 µl of mRNA (1 pmol) was first incubated at 20 °C for 20 min in the presence of 4 µl of buffer N4 (5×) and 12 µl of sterile H_2O. The reaction was done at 20 °C for 10 min in the presence of 1 µl of tRNA (2 µg/µl) and 2 µl of kethoxal (at 10 or 20 mg/ml diluted in 20% ethanol).
- Semi-denaturing conditions: same procedure as for native conditions but reaction was done in buffer D4 for 5 min.

All these steps have to be conducted under a fume hood, and DMS and CMCT solutions should be destroyed in 1 M NaOH and 10% acetic acid waste, respectively.

Reaction stops
All the reactions were stopped by ethanol precipitation of the RNA.

- To all samples, 50 µl of 0.3 M sodium acetate, pH 6 and 250 µl of cold ethanol were added. For RNA–protein footprinting experiments, the protein was removed by phenol extraction. The samples were then mixed, placed in a dry-ice/ethanol bath for 15 min and centrifuged (13 000 r.p.m. at 4 °C for 15 min).
- The supernatants were removed carefully (do not touch the pellet) and 200 µl of 80% cold ethanol added to the pellets. The samples were centrifuged (13 000 r.p.m. at 4 °C for 5 min) and the supernatants were removed with the same caution.
- The pellets were vacuum-dried (no more than 5 min) and resuspended in 4 µl of sterile H_2O.

10.3.5
Primer Extension Analysis

Primer extension with reverse transcriptase was originally developed by HuQu et al. [33] for probing the structure of large RNA molecules. Reverse transcriptase stops its incorporation of dNTP at the residue preceding a cleavage or a modification at a

Watson–Crick position. While carbethoxylation of $A(N^7)$ by DEPC is sufficient to stop reverse transcriptase, a subsequent treatment is necessary to induce a cleavage at $G(N^7)$ after DMS modification (Table 10.1 [23]).

The length of the primer varies usually from 12 to 18 nt. This provides sufficient specificity even if the primers are used on a mixture of RNAs. For long RNA, primers are selected every 200 nt due to the gel resolution limitation. Before probing the RNA structure, assays should be performed to define the best concentration of the RNA, the choice of the primer sequence and the hybridization conditions in order to get an efficient primer extension. For *thrS* mRNA, primer annealing conditions were selected in order to maximize the unfolding of the probed RNA and to minimize RNA degradation. The primer TTACAGCGTGATCGT, complementary to nucleotides +47 to +61 of *thrS* mRNA, was used (Fig. 10.1D).

Hybridization

To the 4 μl of modified mRNA (1 pmol), 1 μl of 5'-end-labeled DNA primer (around 100 000 c.p.m.) was added. The samples were then heated at 90 °C for 1 min and then quickly cooled on ice. Then, 1 μl of 5 × RTB buffer was added and the samples were incubated for 15 min at 20 °C.

Primer Extension

The reaction was done in 15 μl of total volume. To the hybridization mix were added 2 μl of 5 × RTB, 2 μl of dNTP mix (2.5 mM of each dNTP), 4 μl sterile H_2O and 1 μl of reverse transcriptase (2 U/μl diluted freshly into sterile H_2O). Incubate the samples for 30 min at 37 °C.

For kethoxal modification, 2 μl of 50 mM sodium borate, pH 7.0 was added in the extension reaction to stabilize the adduct.

- To all samples, 50 μl of 0.3 M sodium acetate, pH 6 and 200 μl of cold ethanol were added. After precipitation, the pellets were washed twice with 80% ethanol and vacuum-dried as described above. The end-labeled DNA fragments were re-suspended in 6 μl of the DNA loading buffer.
- To improve the resolution of the gels, the RNA template can be subjected to alkaline hydrolysis.
- Just after primer extension, 20 μl of the buffer containing 50 mM Tris–HCl, pH 7.5, 7.5 mM EDTA, 0.5% SDS and 3.5 μl of 3 M KOH were added. The samples were incubated at 90 °C for 3 min then at 37 °C for at least 1 h.
- To all samples, 6 μl of 3 M acetic acid, 100 μl 0.3 M sodium acetate, pH 6 and 300 μl of cold ethanol were added. The following procedure is as described above.

Gel Fractionation

The DNA fragments were denatured by incubation at 90 °C for 3 min and were fractionated on 8% polyacrylamide (0.4% bis)–8 M urea slab gels in 1 × TBE. As described above, gels should be pre-run (30 min at 75 W) and run warm (75 W). The migration conditions must be adapted to the size of the fragments to be analyzed, knowing that on 8% gel, xylene cyanol migrates as 81 nt and bromophenol blue as 19 nt.

The modification or cleavage positions were identified by running in parallel a sequencing reaction. The elongation step was performed as described above except in the presence of one of the didesoxyribonucleotide ddXTP (2.5 µM), the corresponding desoxyribonucleotide dXTP (25 µM) and the three other desoxyribonucleotides (100 µM).

After migration, the gels were dried, and autoradiographed at $-80\ °C$ with an intensifying screen overnight.

10.3.6
In Vivo RNA Structure Mapping

10.3.6.1 *In Vivo* DMS Modification

DMS has been successfully used to probe several RNA species from a variety of cells, including Gram-negative [12, 34] and Gram-positive bacteria [11, 35], yeast [36], protozoa [37], plants [38], and fibroblast cells [39]. This reagent is capable to diffuse efficiently across cell wall and membrane, and to modify unpaired adenines (N^1) and cytosines (N^3). Occasionally, modifications of uridines at N^3 have been identified during *in vivo* DMS modification [13]. Information on the accessibility of guanines at N^7 can also be obtained. The protocol given below was adapted for bacteria and one typical experiment performed on *Staphylococcus aureus* RNAIII is illustrated in Fig. 10.2.

The bacterial strain was grown to mid-log phase and then treated with DMS. As for the *in vitro* experiment, it is important to verify that the reaction occurred under limited conditions such as less than one modification per molecule was statistically induced. Thus, a range of DMS quantities (100 µl of DMS diluted 1:10, 1:5, 1:2 or pure) and time intervals for incubation (2–15 min) were initially tested. After treatment, the reaction was stopped just before the extraction of the total RNA. Sites of DMS modification were detected by primer extension on total RNA extracts (5–20 µg), using end-labeled primers specific for a chosen region of interest of the tested RNA.

- Bacteria (20 ml of culture) were grown in LB medium in a 50-ml sterile tube to mid-logarithmic phase at 37 °C (until an OD_{600} of 0.5 was reached).
- 100 µl of DMS (diluted 1:2 in ethanol) were added and the culture was incubated for an additional 5 min at 37 °C after gentle shaking.
- The reaction was stopped by adding 10 ml of cold stop buffer containing 100 mM Tris–HCl, pH 8, 200 mM β-mercaptoethanol, 5 mM EDTA.
- The cells were then pelleted (3 000 r.p.m. for 15 min at 4 °C), and were resuspended in 1.5 ml of cold buffer 10 mM Tris–HCl, pH 8, 100 mM NaCl, 1 mM EDTA. The cells were transferred in a 1.5-ml micro tube and centrifuged at 13 000 r.p.m. for 15 min at 4 °C.
- The cells were then disrupted by adding 200 µl of buffer containing 50 mM Tris–HCl, pH 8.0, 8% sucrose, 0.5% Triton, 10 mM EDTA, 4 mg/ml lysozyme, and

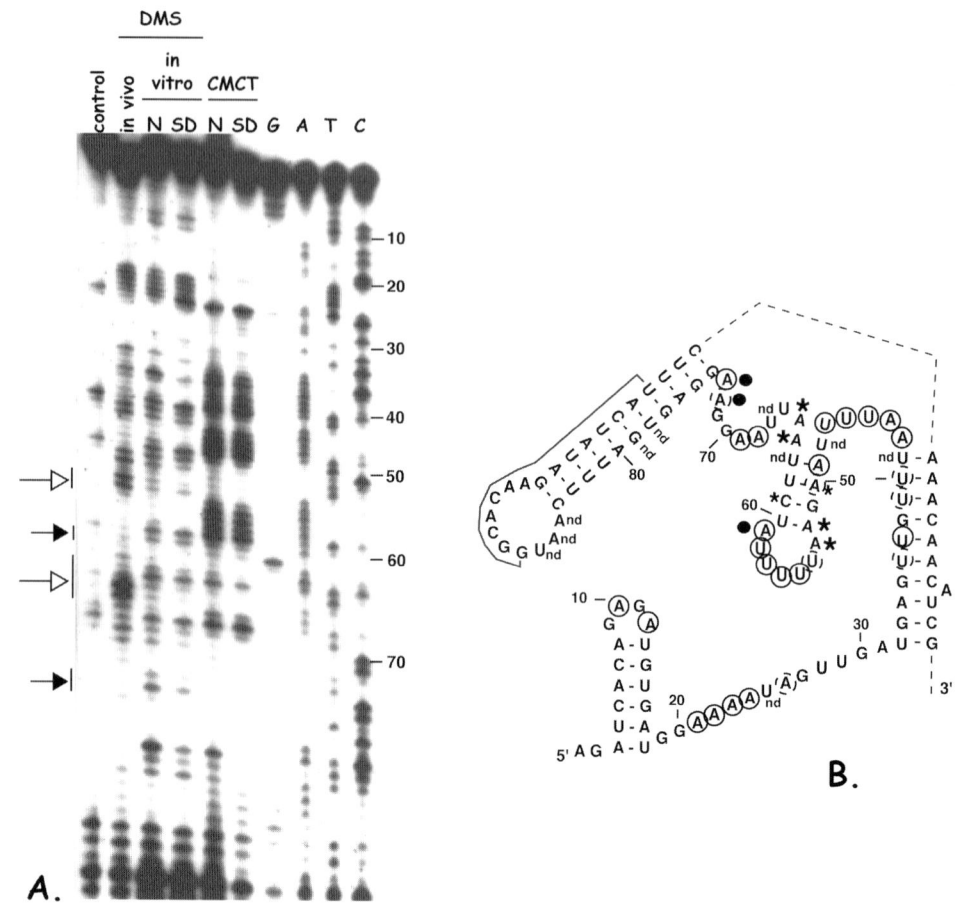

Fig. 10.2. Chemical probing on *S. aureus* RNAIII performed *in vivo* and *in vitro* (A), and part of the secondary structure of *S. aureus* RNAIII (B). (A) DMS and CMCT modifications were performed *in vitro* under native conditions (N) and semi-denaturing conditions (SD), and *in vivo* for DMS. Experimental details are given in the text. (Control) incubation control; (lanes G, A, T, C) DNA sequencing ladders. (B) Circled nucleotides are reactive towards DMS ($N^1A \gg N^3C$) and CMCT ($N^3U \gg N^1G$) modifications. (A and B) Reactivity differences are shown by empty arrows or by stars for nucleotides, which are only reactive *in vivo* and by dark arrows or black circles for nucleotides reactive only *in vitro*. nd = not determined. The position of the primer, complementary to nucleotides G88 to U102, is given. Adapted from Benito et al. [35].

incubated 5 min in ice. For *S. aureus*, the cells were treated with lysostaphine (50 µg/ml) in the presence of 1% SDS.

- Then, 200 µl of phenol saturated with 0.1 M sodium acetate, pH 5.5 and 10 mM EDTA were added. Cells were vortexed for 30 s at high speed. The samples were heated at 65 °C during 15 min and mixed every 5 min.
- The mixture was cooled on ice and centrifuged 10 min at 13 000 r.p.m. The

aqueous phases were carefully collected, and the phenol and interface were re-extracted by vortexing the samples with 100 μl 0.1 M sodium acetate, pH 5.5.
- After centrifugation, the aqueous phases were pooled and extracted once with phenol/chloroform previously saturated in sodium acetate 0.1 M, pH 5.5 and once with chloroform.
- The RNA was then precipitated twice with three volumes of cold ethanol in the presence of 0.3 M sodium acetate (final concentration).
- The pellets were washed twice with 200 μl of 80% ethanol, vacuum-dried (no more than 5 min) and dissolved in a small volume of sterile H_2O. The RNA concentration was measured.
- 10 μg of material was used for primer extension.
- Primer hybridization and elongation by reverse transcriptase were as described above, except that elongation was conducted at 45 °C for 30 min with 5 U of reverse transcriptase. Do not forget the sequencing reactions, which help to identify the position of modifications.
- Incubation control is performed on cells grown and treated in the same conditions as above but in the absence of DMS.
- A stop control was done in order to verify that little or no DMS modification occurred during the RNA extraction. In that control, DMS was added after the addition of the stop buffer.

10.3.6.2 *In Vivo* Lead(II)-induced RNA Cleavages

To avoid secondary cleavages, it is of prime importance to perform lead(II)-induced cleavage under conditions where less than one cleavage per molecule is induced. Thus, a range of lead(II) acetate concentrations (25–200 mM final concentration) and time intervals for hydrolysis (2–15 min) should be tested. After treatment, the reaction is stopped using an excess of EDTA and total RNA can be isolated. Incubation control should be performed under the same experimental conditions, except that lead(II) acetate is avoided. The cleavage positions were detected by primer extension analysis and assigned using in parallel sequencing reactions. The protocol described below was adapted for *Escherichia coli* to map the accessible regions of several non-coding RNAs and of mRNA [14].

- Bacteria (20 ml of culture) were grown in LB medium to mid-logarithmic phase at 37 °C (until an OD_{600} of 0.5 was reached) in a 50 ml conical tube.
- Make up a fresh solution of 1 M lead(II) acetate in sterile water [lead(II) acetate precipitates at high concentration in LB medium]. Then, 2.8 ml of this solution is mixed with 3.2 ml of sterile water and 2 ml of pre-warmed 4 × concentrated LB (at 37 °C) to give 8 ml of lead(II) acetate/LB solution at 350 mM. This step should be done just before use [This step is essential for reproducibility. Some lead(II) acetate precipitation was always observed in LB medium. Consequently, the intracellular concentration must be lower than the nominal concentration in the medium].
- 8 ml of this lead(II) acetate/LB solution (350 mM) was then added to 20 ml of cells at mid-logarithmic phase. This gave a final concentration of lead(II) acetate

100 mM. For the first trials, different concentrations of lead(II) acetate (50, 100, 150 and 200 mM final concentration) should be used.
- Cultures were incubated with gentle shaking for 7 min at 37 °C.
- Reactions were stopped by addition of 10 ml of cold 0.5 M EDTA (1.5-fold molar excess) and immediately put on ice.
- The cells were pelleted (3000 r.p.m. for 15 min at 4 °C) and resuspended in 1.5 ml of cold buffer 10 mM Tris–HCl, pH 8, 100 mM NaCl, 1 mM EDTA. The cells were transferred in a 1.5-ml micro tube and centrifuged at 13000 r.p.m. for 15 min at 4 °C.
- RNA extraction and analysis was carried out as described for the *in vivo* DMS mapping.

Lead(II)-induced cleavage *in vivo* can be performed under different growth conditions. However, lead(II) acetate has some tendency to precipitate, in particular when chloride ions are present within the medium. In that case, it is essential to test different concentrations of the lead(II) acetate. One simple and reliable method for evaluating cleavage conditions is to fractionate total RNA on agarose gels [14]. Upon increases of lead(II) acetate concentration (25–200 mM), the intensities of 16S and 23S rRNA, the predominant RNA species, significantly decreased and optical inspection of the patterns could be used to calibrate conditions.

10.4
Commentary

10.4.1
Critical Parameters

RNA Preparation
The RNA is usually synthesized by *in vitro* transcription using T7 RNA polymerase. Therefore, the 3′ or 5′ end of the RNA might be heterogeneous, and several abortive transcription products might also accumulate. To obtain homogenous size RNA, the method of choice remains the fractionation of the transcript by electrophoresis on denaturing urea gel. For long RNA molecules, electro-elution might help to increase the elution efficiency.

Homogeneous RNA Conformation
Since during the purification, the RNA can be partially denatured, it is essential to design renaturation protocols in order to have a conformationally homogeneous RNA population and to test whether this conformation is biologically relevant (enzymatic activity for ribozyme, efficient ligand binding, etc.). Alternative RNA conformations may co-exist, and can be revealed by the simultaneous presence of single-stranded and double-stranded specific cleavages or modifications at the same position. If the conformers have different electrophoretic mobilities on native

polyacrylamide gel, chemical probing can be used to distinguish them [40]. After chemical modification, the co-existing structures are separated on a native polyacrylamide gel and the modification sites for each conformer are then identified by primer extension.

Chemical and Enzymatic Probing *In Vitro*

The chemical reactions and RNase T1 hydrolysis can be conducted under a variety of experimental conditions. For instance, the influence of divalent ion (such as magnesium) can be tested on the folding of the RNA and by varying the temperature (between 4 and 90 °C) one can follow thermal transition of RNA molecules [41]. It is essential, however, to adapt for each condition the chemical reactions or the enzymatic hydrolysis in order to have less than one cut or modification per molecule, i.e. more than 80% of the RNA should not be modified or cleaved. For example, for DMS modification, reaction at 4 °C is for 20 min in the presence of 1 µl of DMS, whereas at 50 °C, reaction is for 5 min with 1 µl of DMS diluted 1:16.

- Kethoxal might have a partially denaturing effect on RNA structure even if the reaction was not too strong [41]. Concentration of kethoxal or the incubation time should be reduced in order to get only modifications at guanines present in single-stranded regions.
- The RNase cleavages in the RNA molecule can induce conformational rearrangements potentially able to provide new targets for secondary cleavages. Usually these secondary cleavages occur when the reaction is too strong; they also are of weak intensity and are not reproducibly found in all experiments. These cleavages can be distinguished from primary cuts by comparing the hydrolysis patterns obtained from the 5′- or 3′-end-labeled RNA.
- RNase V1 hydrolysis generates RNA fragments which end up with a 3′-OH group in contrast to alkali and most of the RNases. Therefore, 5′-end-labeled fragments generated by alkali will migrate faster than the RNase V1 fragments. This difference is only observed for the shortest RNA fragments (see Fig. 10.1B).
- Appropriate incubation controls are required to identify cleavages that are induced during the incubation treatments and pauses of reverse transcriptase that are due to stable secondary structures. Nucleotides for which strong bands are visible in the control lanes are not considered for the interpretation.
- Each experiment should be repeated several times, and only the reproducible cleavages and modifications will be considered for the interpretation. As mentioned previously, the elaboration of a secondary structure RNA model requires to collect data from enzymes and chemicals with different and complementary specificities. Only this combination will allow to define helical and loop regions. The presence of RNase V1 cleavages and nucleotides not reactive at Watson–Crick position is a strong indication for the existence of a helical region.
- Footprinting assays. The experiments should be done in the presence of increasing concentrations of ligand (Fig. 10.1C). Lead(II)-induced cleavages and hydroxyl radicals are appropriate probes to map the ligand binding site due to their small size and their specificity. Results should be interpreted with care because

decreased reactivity does not necessarily result from a direct shielding effect, but could be due to a steric hindrance effect (particularly observed with the bulky RNases) or to a conformational change of the RNA.

10.4.2
In Vivo Mapping

- Due to the inability to diffuse within the cells, only a few probes have been used to map the RNA structure within the cells. DMS- and lead(II)-induced cleavages are to date the most commonly used probes. Other probes like RNase T1 [42], kethoxal [13] and CMCT [43] have been used *in vivo* after permeabilization of the cells. However, due to this additional treatment, particular caution has to be taken to ensure that the cells remain intact during the time of incubation. It is also essential to verify that the reaction was efficiently stopped before the RNA extraction procedure.
- Alternative to phenol extraction, other protocols used to extract total RNA can be used. Reagents combining phenol and guanidine thiocyanate enable a straightforward isolation of total RNA from samples of human, yeast, bacterial and viral origin.
- Data from *in vivo* probing may be more complex to interpret than the *in vitro* probing. One of the main reasons is that the studied RNA may be involved simultaneously in several complexes (e.g. regulatory RNAs). However, *in vivo* mapping becomes powerful when it is used in a comparative manner. For example, conformational changes of mRNA induced by a *trans*-acting ligand have been identified upon repression or activation of translation (e.g. [11, 34]). DMS- and lead(II)-induced cleavages can also be used to monitor the conformational changes of mRNA under different growth conditions and under different environmental cues such as temperature.

10.5
Troubleshooting

10.5.1
In Vitro Mapping

- RNase T1 cleaves all guanines (Fig. 10.1B). Significant unfolded RNA molecules were present (improve the renaturation protocol) or the hydrolysis was too strong. As shown in Fig. 10.1(A), the cleavage pattern significantly changed by reducing the concentration of RNase T1.
- Compression of bands due to stable secondary structure (in general rich in GC base pairs). Heat the end-labeled RNA samples at 90 °C for 3 min before gel loading. Never heat the alkaline ladder and the RNase T1 ladder.
- Cleavages of end-labeled RNA are doubled: the T7 RNA transcript is not homogenous in size (purify the RNA on polyacrylamide–urea gel).

- Too many bands in the incubation controls of end-labeled RNA: RNase contamination, repurify the RNA and prepare new sterile buffers.
- Aggregation of end-labeled RNA in the gel pockets; only fragments of small size can be visualized. The data cannot be taken into account. The pellets were not correctly dried after ethanol precipitation. Heat the samples before loading on the gel.
- Samples do not migrate correctly during electrophoresis. This is probably due to the presence of salt. Add several washing steps with ethanol 80% at the end of the procedure.
- No full-length RNA after DMS modification (Fig. 10.1D): adapt the conditions by reducing the amount of DMS in order to get more than 80% of RNA molecules unmodified.
- Absence of signal after primer extension: the modified RNA did not efficiently precipitate. Since the modified RNA is not labeled, particular caution should be taken to prevent the loss of the pellet.
- To keep high resolution of the gels, acrylamide, urea solutions and, in particular, ammonium persulfate should be prepared freshly.

10.5.2
In Vivo Probing

- Low yield of total RNA: incomplete homogenization or lysis of samples, degradation of the RNA.
- Strong stops in the control lanes: degradation of RNA, pauses of reverse transcription due to stable secondary structures (perform elongation at higher temperature, increase the concentration of the enzyme and dNTP or change the primer sequence). Many RNA molecules carry post-transcriptional modifications that may interfere with reverse transcriptase elongation (primer should be changed in order to cover the modified base).
- No more full-length RNA product after modification: reaction was too strong (reduce either the quantity of the reagent or/and the time of incubation). Check that the reaction was efficiently stopped before the extraction of total RNA.
- Weak or smearing signal after primer extension: increase the concentration of total RNA, check that the primer hybridization protocol is efficient. Optimal conditions for primer hybridization should be established in a series of pilot experiments, another protocol was described by Sambrook et al. [44]. The optimal temperature for annealing varies from RNA to RNA, depending on the $G+C$ content, the propensity of the RNA to form secondary structure and the length of the primer. The aim is also to minimize the formation of mismatched DNA primer–RNA hybrids.

Acknowledgments

We are grateful to B. Ehresmann for his constant support and for critical reading of the manuscript, and we thank C. Brunel, J. C. Paillart and other members from the laboratory for discussions. This work was supported by the Centre National de Recherche (CNRS), the Ministère délégué à la Recherche et aux Nouvelles Technologies (ACI "Microbiologie et Pathologie Infectieuses"), la Ligue de la Recherche sur le Cancer and Région Alsace (M. P.).

References

1 Storz, G. *Science* 2002, *296*, 1260–1263.
2 Carrington, J. C., V. Ambros. *Science* 2003, *301*, 336–338.
3 Gottesman, S. *Genes Dev.* 2002, *16*, 2829–2842.
4 Johansson, J., P. Cossart. *Trends Microbiol.* 2003, *11*, 280–285.
5 Stormo, G. G. *Mol. Cell* 2003, *11*, 1419–1420.
6 Zuker, M. *Curr. Opin. Struct. Biol.* 2000, *10*, 303–310.
7 Leontis, N. B., E. Westhof. *Curr. Opin. Struct. Biol.* 2003, *13*, 300–308.
8 Latham, J. A., T. R. Cech. *Science* 1989, *245*, 276–282.
9 Ralston, C. Y., B. Sclavi, M. Sullivan, M. L. Deras, S. A. Woodson, M. R. Chance, M. Brenowitz. *Methods Enzymol.* 2000, *317*, 353–358.
10 Das, R., L. W. Kwok, I. S. Millett, Y. Bai, T. T. Mills, J. Jacob, G. S. Maskel, S. Seifert, S. G. Mochrie, P. Thiyagarajan, S. Doniach, L. Pollack, D. Herschlag. *J. Mol. Biol.* 2003, *332*, 311–319.
11 Mayford, M., B. Weisblum. *EMBO J.* 1989, *6*, 4307–4314.
12 Altuvia, S., D. Weinstein-Fischer, A. Zhang, L. Postow, G. Storz. *Cell* 1997, *90*, 43–53.
13 Balzer, M., R. Wagner. *Anal. Biochemistry* 1998, *256*, 240–242.
14 Lindell, M., P. Romby, E. G. H. Wagner. 2002, *RNA 8*, 534–541.
15 Moine, H., B. Ehresmann, C. Ehresmann, P. Romby. Probing the RNA structure and function in solution, in: *RNA structure and function*, Simons, R. W., Grunberg-Manago, M. (eds), Cold Spring Harbor Laboratory Press, Cold Spring Harbor Laboratory, NY, 1998, pp. 77–115.

16 Joseph, S., H. F. Noller. *Methods Enzymol.* 2000, *318*, 175–190.
17 Wilson, K. S., H. F. Noller. *Cell.* 1998, *92*, 131–139.
18 Ehresmann, C., F. Baudin, M. Mougel, P. Romby, J. P. Ebel, B. Ehresmann. *Nucleic Acids Res.* 1987, *15*, 9109–9128.
19 Giegé, R., M. Helm, C. Florentz. *Comp. Natural Prod. Chem.* 1998, *6*, 63–80.
20 Caillet, J., T. Nogueira, B. Masquida, F. Winter, M. Graffe, A. C. Dock-Brégeon, A. Torres-Larios, R. Sankaranarayanan, E. Westhof, D. Moras, B. Ehresmann, C. Ehresmann, P. Romby, M. Springer. *Mol. Microbiol.* 2003, *47*, 961–974.
21 Christiansen, C., J. Egebjerg, N. Larsen, R. A. Garrett. *Methods Enzymol.* 1989, *164*, 456–472.
22 Krol, A., P. Carbon. *Methods Enzymol.* 1989, *180*, 212–227.
23 Brunel, C., P. Romby. *Methods Enzymol.* 2000, *318*, 3–21.
24 Chen, C., S. A. Woodson, C. J. Burrows, S. E. Rokita. *Biochemistry* 1993, *32*, 7610–7616.
25 Weeks, K. M., D. M. Crothers. *Science* 1993, *261*, 1574–1577.
26 Pan, T., D. M. Long, O. C. Uhlenbeck. Divalent metal ions in RNA folding and catalysis, in: *The RNA world*, Gesteland, R. F., Atkins, J. F. (eds), Cold Spring Harbor Laboratory Press, Cold Spring Harbor Laboratory, NY, 1993, pp. 271–302.
27 Romaniuk, P. J., I. L. de Stevenson, H. H. Wong. *Nucleic Acids Res.* 1987, *15*, 2737–2755.
28 Jahn, M. J., D. Jahn, A. M. Kumar, D. Soll. *Nucleic Acids Res.* 1991, *19*, 2786.

29 Milligan, J. F., D. R. Groebe, G. W. Witherell, O. C. Uhlenbeck. *Nucleic Acids Res.* **1987**, *15*, 8783–8798.

30 Gelhaus, S. L., W. R. Lacourse, N. A. Hagan, G. K. Amarasinghe, D. Fabris. *Nucleic Acids Res.* **2003**, *31*, e135–e140.

31 Silberklang, M., A. M. Gillum, U. L. RajBhandary. *Methods Enzymol.* **1979**, *59*, 58–109.

32 England, T. E., A. G. Bruce, O. C. Uhlenbeck. *Methods Enzymol.* **1980**, *65*, 65–74.

33 HuQu, L., B. Michot, J. P. Bachellerie. *Nucleic Acids Res.* **1983**, *11*, 5903.

34 Wulczyn, F. G., R. Kahmann. *Cell* **1991**, *65*, 259–269.

35 Benito, Y., F. A. Kolb, P. Romby, G. Lina, J. Etienne, F. Vandenesch. *RNA* **2000**, *6*, 668–679.

36 Méreau, A., R. Fournier, R. Grégoire, A. Mougin, P. Fabrizio, R. Lührmann, C. Branlant. *J. Mol. Biol.* **1997**, *273*, 552–571.

37 Zaug, A. J., T. R. Cech. *RNA* **1995**, *1*, 363–374.

38 Senecoff, J. F., R. B. Meagher. *Plant Mol. Biol.* **1992**, *18*, 219–234.

39 Granger, S. W., H. Fan. *J. Virol.* **1998**, *72*, 8961–8970.

40 Schröder, A. R. W., T. Baumstark, D. Riesner. *Nucleic Acids Res.* **1998**, *26*, 3449.

41 Jaeger, L., E. Westhof, F. Michel. *J. Mol. Biol.* **1993**, *234*, 331–346.

42 Bertrand, E., M. Fromont-Racine, R. Pictet, T. Grange. *Proc. Natl Acad. Sci. USA* **1993**, *90*, 3496–3500.

43 Drozdz, M., C. Clayton. *RNA* **1999**, *5*, 1632–1644.

44 Sambrook, J., E. F. Fritsch, T. Maniatis. Extraction, purification and analysis of mRNA from eukaryotic cells, in: *Molecular Cloning: A Laboratory Manual*, Cold Spring Harbor Laboratory Press, Cold Spring Harbor, NY, 1989, pp. 7.79–7.81.

45 Swerdlow, H., C. Guthrie. *J. Biol. Chem.* **1984**, *259*, 5197–5207.

11
Study of RNA–Protein Interactions and RNA Structure in Ribonucleoprotein Particles

Virginie Marchand, Annie Mougin, Agnès Méreau and Christiane Branlant

11.1
Introduction

In cells, RNAs almost invariably function in association with proteins and form ribonucleoprotein particles (RNP). For most of the characterized RNP, one or more proteins with RNA-binding properties first associate with the RNA. Subsequently, other protein components may associate to the complex by protein–protein interactions or both protein–protein and RNA–protein interactions. Cellular RNA molecules can be classified into various groups according to their function or localization and different classes of proteins are associated with each of these groups [1, 2]. The RNA-associated proteins have diverse functions. They can stabilize, protect, package or transport RNAs, or participate in their subcellular localization. They can also mediate RNA interactions with other macromolecules or be catalysts [3].

In contrast to DNA, RNA can adopt a large variety of three-dimensional (3-D) structures. RNA–protein interactions may involve a defined nucleotide sequence and/or a specific 2- or 3-D RNA structure. Several protein domains were selected in the course of evolution for their ability to bind peculiar RNA motifs with either a narrow or a broad specificity. One of the best-studied examples is the RNA recognition motif (RRM) [4] that was first discovered in spliceosomal particles. It turned to be a very general RRM present in many different proteins [5]. Members of the RRM family include proteins that bind mRNAs, snRNAs or rRNAs. The KH domain was first identified in the human hnRNP K protein [6] and is also found in a large variety of RNA-binding proteins [1]. RRM and KH domains have well-defined and conserved 3-D structures [2, 7, 8]. Another protein RNA-binding motif with a defined 3-D structure was first described in the ribosomal L30 protein (L30 motif) [9]. It was later found to bind RNAs that form peculiar "K-turn" structures [10, 11]. Several "K-turn" structures were discovered in rRNAs [12, 13], and in small nuclear and nucleolar RNAs (snRNAs and snoRNAs) [10, 11] and they were found to bind "L30 type" protein domains [10, 11, 14–16]. Another characterized RNA-binding domain, which was first found in the *Escherichia coli* RNase III [17], binds specifically double-stranded RNA (dsRBD). This domain is limited to interactions

Handbook of RNA Biochemistry: Student Edition. Edited by R. K. Hartmann, A. Bindereif, A. Schön, E. Westhof
Copyright © 2009 WILEY-VCH Verlag GmbH & Co. KGaA, Weinheim
ISBN: 978-3-527-32534-4

with the A-form RNA helix. Other conserved RNA-binding domains are frequently encountered in proteins that bind RNAs, but their 3-D structures are not well characterized. This is the case for the RGG motif that was initially identified in the hnRNP U protein and is often found in combination with RRM motifs [18]. Other types of RNA-binding motifs have also been described such as zinc fingers, arginine rich and cold shock domains [2, 19–22].

There are numerous approaches to characterize RNA–protein interactions. First of all, one has to identify the proteins which are associated with the studied RNA. This implies the purification to homogeneity of the authentic RNP complex. Classical immunoselection approaches have been used for a long time as a first step in RNP purifications. They were based on the use of antibodies directed against one of the protein component of the complex or against the specific cap structure of the RNA [23, 24]. This first immunoselection step was followed by MonoQ/FPLC chromatography and/or fractionation by glycerol gradient centrifugation. During the last few years, new approaches have been developed for purification of RNP by two successive immunoselection steps [10, 25, 26]. They are based on the expression in cells of tagged components of the RNP (either two proteins or one protein and the RNA). The tagged components are included in the RNA–protein complexes *in vivo*, which allow RNP purification by successive affinity chromatography steps [10, 27, 28]. Identification of protein components of the isolated RNP is done by mass spectrometry.

At this stage the proteins associated with the RNA molecules are identified. The next question is to know which protein(s) bind(s) directly to the RNA and which ones are associated to the complex either only by protein–protein interactions or by protein–protein interactions together with interactions of low stability with the RNA. One way to identify the primary binding proteins is to produce them in a recombinant form in *E. coli*, yeast or animal cells and then to test their *in vitro* capacity to bind RNA by electrophoretic mobility shift assays (EMSA). This approach can produce information on the affinity of the RNA and protein partners, and also on the conditions that favor the interaction (ionic strength, pH, temperature, etc.). Sometimes, it is difficult to produce recombinant protein because of solubility problems. In this case, another approach can be used if specific antibodies directed against the proteins of the RNP are available. This approach is based on the formation of covalent bonds between the RNA and proteins, which are in a very close contact, by UV irradiation at 254 nm. Crosslinking is followed by RNA digestion. The free and crosslinked protein is immunoselected by using antibodies coated on Sepharose beads. After electrophoresis, only the proteins that were covalently linked to the RNA are labeled by the residual crosslinked nucleotides, so that they can be visualized by autoradiography. Measurement of the radioactivity in the gel can give an indication of the affinity of the two partners. Crosslinking experiments may be performed either with cellular extracts or with recombinant proteins. Doing crosslinking experiments in cellular extract is informative because under these conditions all the proteins in the extract are in competition with each other, as is the case in cells. Sometimes, it may happen that two proteins in the extract are in competition with one another for the same site on the RNA. In this case, crosslinking experiments can be performed with the labeled RNA and various relative amounts of the two recombinant proteins. Using this approach, one can obtain important information on the relative affinities of these two proteins for the RNA site.

After identification of the proteins, which directly interact with the RNA, a further step consists of the precise mapping of their binding sites. One possibility is to produce different pieces of the RNA and to test for their ability to bind the proteins. However, pieces of the RNA may fold into structures different from those present in the entire molecule. One more direct method consists of mapping the protein-binding sites by the use of chemical and enzymatic probes in solution. The bound RNA regions are protected against the action of the probes, and are identified by comparison of the cleavages and modifications obtained under the same conditions in the RNP and naked RNA. Such footprinting experiments can be performed on authentic native RNP, and on complexes reconstituted from recombinant proteins and *in vitro* transcribed RNAs or on complexes formed upon incubation of an *in vitro* transcribed RNA in a cellular or nuclear extract. However, as described below, when the experiment is performed in an extract, special digestion and modification conditions have to be used due to the presence of a large amount of protein in the extracts. As sites of cleavages and modifications are identified by primer extension analysis by the use of specific oligodeoxynucleotide primers, the footprinting analyses can be made in the presence of the endogenous RNAs from the extract. Thus, this approach is extremely powerful, since purification of the reconstituted complexes is not required.

Footprinting analysis of RNA–protein complexes formed in cellular extracts can be performed without knowing the identity of the bound proteins. These data allow the delineation of the RNA fragments that are free or bound to proteins within the extract. Then, for rapid identification of the proteins that are bound to the RNA in extracts, without a purification step, supershift experiments can be performed. The principle is to form complexes between the RNA and proteins from the cellular extract, and to incubate the mixture with specific antibodies directed against a protein expected to be bound to the RNA. An EMSA is performed with or without incubation with the antibody. If the antibody binds the protein without dissociation of the complex, the mobility of the RNP complex in the gel is decreased (supershift). If binding of the antibody to the protein dissociates the complex, the RNP band disappears.

If the secondary structure of the studied RNA target is not known, we recommend to study it in parallel with the footprinting experiments, since the same series of chemical and enzymatic probes are used (see Chapter 10). Knowledge of the RNA secondary structure and, if possible, RNA tertiary structure allows a better delineation of the RNA-binding domain. However, the fact that binding of proteins may alter the RNA structure should be taken into consideration for interpretation of these data.

When the binding site of a given protein has been delineated by this approach, confirmation of its functional role can be done *in vitro* and *in vivo*. The absence of protein binding after disruption of the RNA-binding site can be tested *in vitro* by reconstitution and EMSA experiments after site-directed mutagenesis of the RNA [16]. The effect on RNP activity of the disruption of the RNA-binding site can be also studied *in vivo* [16].

More generally, the biological relevance of the *in vitro* probing data can be tested *in vivo*, since one of the chemical probes, dimethylsulfate (DMS), can be used *in vivo* [29, 30] (see Chapter 10). As only a limited number of probes can be used *in vivo*, the best strategy is to perform a deep analysis of the RNP *in vitro* and then, by using DMS as a probe, verify that both the RNA secondary structure and protected areas are identical *in vivo* and *in vitro*.

When an RNA–protein binding site has been identified, the details of the RNA–protein interaction and the mechanism of its formation can be studied at the atomic level. To this end, the 3-D structure of the free RNA and proteins partners and of the complex that they form have to be determined by X-Ray or NMR analysis [11, 31–33].

Finally, for a more precise definition of the RNA-binding specificity of the studied protein, Systematic Evolution of Ligands by EXponential enrichment (SELEX) experiments can be performed [34, 35].

11.2 Methods

11.2.1 RNP Purification

RNP complexes contained in cytoplasmic or nuclear cell extracts are usually purified using immunoaffinity chromatography. The specific antibodies used can be directed against endogenous tagged or untagged proteins expressed from a modified gene (Tap-Tag technique; for experimental details, see [25]). The occurrence of a particular cap structure or the insertion of a tag sequence in the RNA can also be used for RNP immunoselection, for experimental details see [10, 26, 36, 37]. For instance, the presence of a m^3G cap structure at the 5′ extremity of the spliceosomal UsnRNAs and of some snoRNAs was largely used for purification of the spliceosomal UsnRNP and snoRNP using immobilized anti-m^3G cap antibodies [24, 38, 39]. The spliceosomal 25S [U4/U6.U5] tri-snRNP, 20S U5 snRNP, 17S U2 snRNP and 12S U1 snRNP contained in the RNP mixture that is retained on the anti-m^3G antibodies can then be separated by glycerol gradient centrifugation. The importance of the salt concentration in these purification steps is evidenced by the fact that at KCl concentrations above 250 mM, the [U4/U6.U5] tri-snRNP is disrupted into 12S U4/U6 and 20S U5 snRNP, and the 17S U2 snRNP is converted into a 15S or 12S particle [40]. Very powerful methods were recently developed for spliceosomal complex purification [26, 41]. They are based on the addition of an aptamer that binds the tobramycin aminoglycoside at one extremity of the RNA.

11.2.2
RNP Reconstitution

11.2.2.1 Equipment, Materials and Reagents

Equipment

- Electrophoresis instruments for small vertical slab gels. Localization of the RNP complexes in gels is performed either by autoradiography using X-Ray films (Fuji or Kodak) processed in an X-ray film developer or by use of a PhosphorImager.
- Temperature controlled baths (96, 65, 30 and 20 °C).

Materials

Eppendorf tubes, tips, buffers and MilliQ water should be sterilized before use. Wearing gloves is strongly recommended to avoid contamination of the samples by RNases.

Reagents

RNP

Nuclear or cytoplasmic extracts from HeLa cells or other cell lines can be purchased from CilBiotech, Belgium (around 13 mg/ml of total protein) or prepared according to the method developed by Dignam [42]. Before use, the extracts are incubated for 10 min at 30 °C in order to consume the endogenous ATP and kept on ice.

Proteins

RNP proteins can be either extracted from the purified native RNP particles [39] or produced as recombinant proteins in E. coli or using other expression systems [43–45].

Antibodies

Some of the primary antibodies used in the described examples can be purchased from Immuquest.

Chemicals and enzymes

SP6 RNA polymerase, T4 polynucleotide kinase and T4 RNA ligase are purchased from MBI Fermentas (Lithuania); T7 RNA polymerase is from Ambion; calf intestine phosphatase, glycogen (10 mg/ml) and RNase-free DNase I from MBI Fermentas; yeast total tRNA (20 mg/ml) from Roche Diagnostics; heparin sodium salt from porcine intestinal mucosa (H3393) from Sigma; Hybond C nitrocellulose membrane and ECL detection system are purchased from Amersham Pharmacia Biotech.

Radiochemicals
[^{32}P]pCp (3000 Ci/mmol), [γ-^{32}P]ATP (3000 Ci/mmol) and [α-^{32}P]UTP (800 Ci/mmol) are purchased from Amersham Biosciences or ICN.

Buffers
Some of the buffers used here are identical to buffers used in probing experiments.

- 1 × buffer D: 20 mM HEPES–KOH (pH 7.9), 0.2 mM EDTA, 100 mM KCl, 20% glycerol (w/v). Add freshly prepared 0.5 mM dithiothreitol (DTT) and 0.5 mM phenylmethylsulfonyl fluoride (PMSF) (dissolved in 96% ethanol).
- 1 × Tris buffer: 50 mM Tris–HCl (pH 7.5), 100 mM KCl, 2.5 mM $MgCl_2$.
- 1 × binding buffer A: 20 mM HEPES–KOH (pH 7.9), 0.2 mM EDTA, 150 mM KCl, 10% glycerol (w/v), 1.5 mM $MgCl_2$.
- 1 × binding buffer B: 20 mM HEPES–KOH (pH 7.9), 0.2 mM EDTA, 100 mM KCl, 20% glycerol (w/v), 3.125 mM $MgCl_2$.
- CSB loading buffer: 20 mM HEPES–KOH (pH 7.9), 40% glycerol (w/v), 0.05% bromophenol blue, 0.03% xylene cyanol.
- DNA loading buffer: 0.02% bromophenol blue, 0.02% xylene cyanol in formamide.
- SDS–PAGE loading buffer: 62.5 mM Tris–HCl (pH 6.8), 2% SDS, 100 mM β-mercaptoethanol, 10% glycerol, 0.01% bromophenol blue.
- 1 × TBE buffer: 90 mM Tris-borate (pH 8.0), 2 mM Na_2EDTA.
- 1 × elution buffer: 500 mM sodium acetate (pH 5.2), 1 mM EDTA.
- PBS-TM: 1 × PBS containing 0.1% Tween 20 and 5% dry non-fat milk powder.

11.2.2.2 RNA Preparation and Renaturation Step

Production of Labeled and Unlabeled RNAs by *In Vitro* Transcription
RNAs are generated by run-off transcription from a DNA template (usually 0.5–2 pmol of a linearized plasmid or PCR product) using the SP6 or T7 RNA polymerase [45–49]. Efficiency of transcription is usually higher for T7 RNA polymerase than for SP6 RNA polymerase. However, efficient transcription with T7 RNA polymerase requires the presence of at least one G residue at the initiation site. The presence of the GGG, GAG or GGA sequence strongly reinforces the transcription yield [50]. However, addition of these residues at the 5' extremity of the RNA may modify its RNA secondary structure and/or protein-binding capacity. For instance, the presence of a UAGGGA/U sequence at the 5' extremity of the transcript often generates a hnRNP A1 binding site. Noticeably, it is not easy to get small RNA transcripts (less than 50 nt) in high yield. For small RNA production, we recommend the use of the MEGAscript® or MEGAshortscript™ kit provided by Ambion (catalog reference 1330, 1333 or 1354). Several factors affecting the transcription yield must also be taken into account, such as the quantity of DNA template (generally 0.5–2 pmol), the incubation time (2–4 h), the Mg^{2+}/NTP ratio (usually

1/1.75), the pH of NTP stocks, and the preparation in a defined order and at room temperature of the transcription reaction mixture.

Uniformly labeled transcripts are produced by incorporation of an [α-^{32}P]NTP during transcription. After transcription, the DNA template is degraded by RNase-free DNase I (10 U). The RNA transcript is purified from shorter RNA fragments and excess of NTPs by polyacrylamide–urea gel electrophoresis in 1 × TBE buffer (for other protocols, see Chapters 10 and R. Hartmann). RNAs are eluted from the gel slices in 100 µl of 1 × elution buffer and are precipitated by the addition of 3 volumes of 96% EtOH, in the presence of 1 µg of glycogen. After centrifugation, the RNA pellet is washed with 70% EtOH, vacuum-dried and dissolved in MilliQ water.

For 5′-end-labeling, RNA (10–100 pmol) is first dephosphorylated at the 5′ end with the calf intestine phosphatase, and then labeled with [γ-^{32}P]ATP (3000 Ci/mmol) and the T4 polynucleotide kinase [51]. For 3′-end-labeling, [^{32}P]pCp (3000 Ci/mmol) is ligated to the RNA transcript in the presence of the T4 RNA ligase [52]. Labeled RNAs are purified by denaturing PAGE and eluted as described above.

Unlabeled RNA transcripts used RNP reconstitution and 2-D structure analyses are produced by similar methods, except that, in order to improve the transcription efficiency, the concentration of NTPs and Mg^{2+} is higher (up to 5 mM of each NTP).

RNA Transcript Renaturation

A renaturation process is required to produce a homogeneous population of RNA molecules in terms of RNA secondary structure. Before probing of naked RNA or reconstitution of RNP complexes, the RNA transcript dissolved in 1 × buffer D or 1 × Tris buffer is incubated for 10 min at 65 °C and then slowly cooled down to room temperature. Then, addition of divalent ions such as Mg^{2+} (at a concentration between 1.5 and 10 mM) is required to favor RNA 2- and 3-D structure formation and its stabilization during the probing and reconstitution experiments. After Mg^{2+} addition, the RNA is incubated for 10 min at room temperature. RNA should not be heated at 65 °C for a too long time, since phosphodiester bonds can be cleaved under these conditions, especially when Mg^{2+} ions are present in the incubation buffer (1 × Tris buffer) [53]. In spite the frequent use of 1 × Tris buffer described in literature, we recommend the use of buffer D and the addition of Mg^{2+} ions after cooling down to room temperature.

11.2.3
EMSA

EMSA can be used for several purposes. EMSA experiments can be performed with an *in vitro* transcribed RNA and purified proteins or an appropriate cellular extract. For determination of dissociation constant values (K_D), a fixed concentration of labeled RNA and increasing concentrations of the protein or cell extract are used. For estimation of the RNA/protein ratio, which has to be used to form RNP complexes for footprinting analysis, unlabeled RNA is added to the labeled

RNA. Under these conditions the same RNA concentration can be used in the EMSA and probing assays.

To perform EMSA with nuclear extract, the buffer, previously defined for pre-mRNA *in vitro* splicing assays [54] is generally used, except that ATP and creatine phosphate are omitted [48, 55, 56]. We recommend the following incubation mixture: HeLa cell nuclear extract 40–50% of total assay volume (dialyzed against buffer D), 2.5 mM $MgCl_2$ and buffer D. Other conditions can also be used for snRNP reconstitution; however, for the biological relevance of the data, it is important to select *in vitro* conditions, as close as possible, to the *in vivo* conditions.

11.2.3.1 **EMSA Method**

The reactions are performed in a total volume of 10–20 µl.

An amount of labeled RNA between 1 and 50 fmol, with or without 1–10 pmol of cold RNA, can be used for the assay. The use of 3′- or 5′-end-labeled RNA is recommended for K_D determination, since uniformly labeled RNAs give less-defined bands on the EMSA gels, especially in the case of RNAs longer than 60 nt. However, uniformly labeled RNAs (0.2 fmol) mixed with cold RNAs (2–10 pmol) are convenient for verification of RNP reconstitution or determination of RNP formation conditions. Competitor tRNAs (usually total tRNAs from yeast *Saccharomyces cerevisiae*) (10- to 1000-fold mass excess as referred to the tested RNA) can be added to the mixture in order to prevent non-specific RNA–protein interactions.

RNPs are generally reconstituted in the binding buffers (1 × Tris buffer or 1 × buffer D) in the presence of Mg^{2+} ions at concentrations varying from 1 to 10 mM. Nuclear extract, or another appropriate extract, or purified proteins are used as protein sources. A large range of protein concentration can be used: generally from 10 nM to 10 µM for purified proteins and from 1 to 100 µg of total proteins for nuclear extract. Pre-incubation of the protein or nuclear extract with the total tRNA mixture is recommended in order to limit the formation of non-specific RNA–protein interactions. A control experiment is performed in the absence of protein extract or purified protein (replaced by 1 × buffer D).

At the end of the incubation, 7 µl of CSB loading buffer is added. Note that in order to limit the non-specific electrostatic interactions between RNA and proteins, heparin, a negatively charged polyelectrolyte, can be added (at a concentration of 5 µg/µl). Heparin addition is followed by a 10-min incubation at room temperature. The presence of heparin usually improves the selection of specific RNP complexes. Addition of heparin is recommended in the case of nuclear extracts that contain a large amount of positively charged proteins. Depending on the electrostatic properties of the protein, heparin may be omitted when purified proteins are used.

For all types of EMSA, the CSB loading buffer is used. Electrophoresis is performed on 5–10% non-denaturing polyacrylamide gels containing 0.5 × TBE buffer and 5% (v/v) of glycerol (acrylamide:bisacrylamide ratio, 19:1). In order to limit RNP disruption and to obtain reproducible results, electrophoresis is generally performed at 4 °C and 10 V/cm. Conditions of electrophoresis, the acrylamide:bisacrylamide ratio as well as the type of running buffer can be modified if required [57].

At the end of the electrophoresis, the gel is transferred on a sheet of Whatman 3MM paper and dried. Exposure is overnight at −80 °C, using intensifying screens, or at room temperature in a PhosphorImager cassette. Free and bound RNAs are visualized by autoradiography.

For K_D determination, the amounts of radioactivity in the bands of gel are estimated by PhosphorImager measurement and SigmaPlot Software (SPSS Science Software) can be used for K_D calculations using the measured radioactivity [16].

Example: Experimental protocol used for K_D determination (Fig. 11.1)

Figure 11.1 illustrates the determination of the K_D values for the complexes formed between the recombinant S. cerevisiae Snu13 protein and one of its wild-type or mutated target RNA motifs. This target motif is one of the "K-turn" structures present in the nucleolar snoRNA U3 (U3B/C WT). In the variant RNA designated as U3B/C G.C → G.G, 1 bp pair of the motif has been disrupted [33] (Fig. 11.1a). RNA–protein complexes were produced under the following conditions: about 50 fmol of uniformly labeled RNA, mixed with 1 μg of yeast tRNA, was renatured in 6.5 μl of 1 × binding buffer A. The recombinant L7Ae protein was added at various concentrations ranging from 50 to 2000 nM and the mixtures were incubated for 30 min at 4 °C. After addition of 7 μl of CSB loading buffer, the RNA–protein complexes formed were fractionated by electrophoresis on a 6% (19:1) non-denaturing polyacrylamide gel, as described above (Fig. 11.1b).

Example: EMSA protocols used to define optimal conditions for RNP reconstitution in view to probe the RNA structure in the reconstituted RNP (Fig. 11.2)

In Fig. 11.2(a), EMSA was used to define the optimal conditions for formation of an RNP complex. The studied RNA was an HIV-1 BRU RNA fragment designated as SLS2,3, that corresponds to positions 7970–8068 in the entire RNA. It was produced by in vitro transcription with the SP6 RNA polymerase and was incubated in a HeLa cell nuclear extract [45]. In Fig. 11.2(b), the EMSA is used for the same purpose, but in this case the RNP complexes are formed between the SLS2,3 RNA and the recombinant hnRNP A1 protein [45].

The following experimental conditions were used: about 2.5 fmol of the 3′-end-labeled RNA was mixed with 2.5 μg of yeast total tRNA and renatured for 10 min at 65 °C in 6.5 μl of 1 × binding buffer B. After cooling for 10 min, HeLa cell nuclear extract purchased from CilBiotech (26 μg of total protein) (in Fig. 11.2a) or increasing amounts of the purified recombinant hnRNP A1 protein (0, 12.5, 25 and 50 pmol dialyzed against buffer D) [45] (in Fig. 11.2b) were added and the mixture was incubated for 15 min at 4 °C. The RNA–protein complexes formed were subsequently fractionated by electrophoresis at room temperature on a 6% (19:1) non-denaturing polyacrylamide gel for 90 min at 10 V/cm and 4 °C.

11.2.3.2 Supershift Method

Supershift experiments are a variant of EMSA. RNA–protein complex formation is performed in the presence or absence of antibodies. Binding of antibodies to their

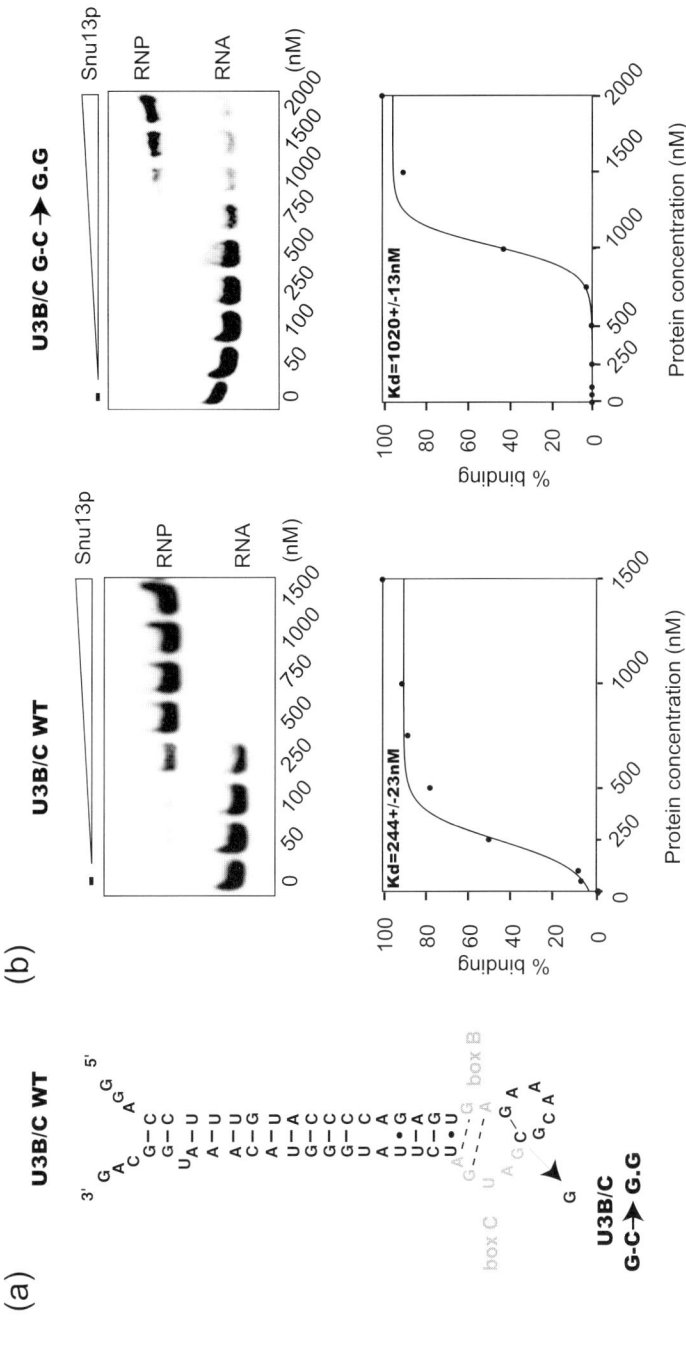

Fig. 11.1. Use of EMSA to study the binding of the recombinant protein Snu13p to the wild-type or mutated B/C motif of yeast U3 snoRNA [33]. The secondary structure of the wild-type U3B/C RNA and position of the C → G mutation in the variant RNA are shown (a). The 3′-end-labeled RNA was incubated in the presence of increasing amounts of the Snu13 protein, as described above. The autoradiograms obtained after gel electrophoresis are shown (b). The protein concentration (in nM) is given below each lane. The K_D values of the RNA–protein complexes (protein concentration for which 50% of the input RNA is shifted to an RNP complex) were calculated for each experiment with the SigmaPlot Software (SPSS Science Software).

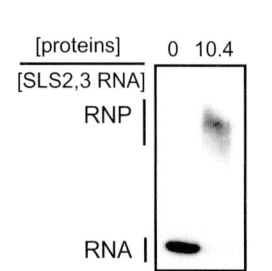

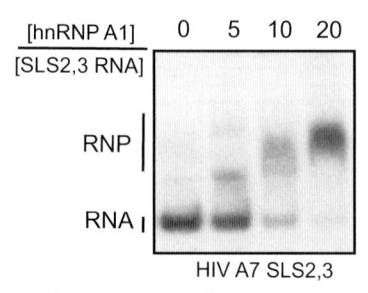

Fig. 11.2. Use of EMSA to study formation of complexes upon incubation of the HIV-1 A7 SLS2,3 RNA with an HeLa cell nuclear extract or the purified recombinant hnRNP A1 protein [45]. (a) and (b) The 3′-end-labeled RNA was incubated in the presence of either a nuclear extract (a) or increasing amounts of the purified recombinant hnRNP A1 protein (b). The [protein] (µg)/[RNA] (pmol) (a) or [hnRNP A1] (pmol)/[RNA] (pmol) (b) ratios (P/R) used in each assays are given above the lanes. Positions of the RNP complexes (RNP) and free RNA (RNA) are indicated on the left of the autoradiograms. The production of complexes with decreasing electrophoretic mobility upon increasing the hnRNP A1 concentration (b) is explained by the multimerization of protein hnRNP A1 along the RNA.

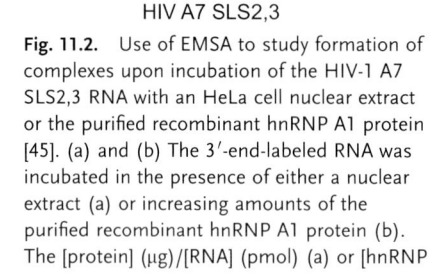

target protein in the RNP complex increases the size of the RNP complex and, thus, lowers its electrophoretic mobility on the gel (so-called "supershift"). In some cases, binding of the antibodies may lead to RNP complex disruption and loss of the shifted RNA band.

For supershift experiments, 0.5–2 µl of antibodies of interest is added before the heparin treatment and incubation is continued for 10 min at 4 °C. The amount of antibody added depends upon the antibody specificity and concentration; both monoclonal and polyclonal antibodies can be used for this type of experiments.

RNP complexes formed with or without antibodies are fractionated in parallel on a non-denaturing polyacrylamide gel (5–10% concentration can be used) and the electrophoresis is performed in 0.5 × TBE buffer containing 5% of glycerol at 4 °C. The gel is dried and the presence of radioactivity is detected as described above.

Example: "Supershift" experimental protocol (Fig. 11.3a)

RNP reconstitution experiments are performed with two different fragments of HIV-1 BRU RNA. The biological components used for the assays are: the SLS2,3 wild-type RNA (positions 7970–8068) described above, the C3 wild-type HIV RNA fragment (positions 5359–5408 in the entire molecule) [58], HeLa cell nuclear extract, and specific anti-hnRNP A1 and hnRNP H antibodies, that were provided by G. Dreyfuss (University of Pennsylvania School of Medicine, Philadelphia, USA) and D. Black (University of California, Los Angeles, USA), respectively. The RNA

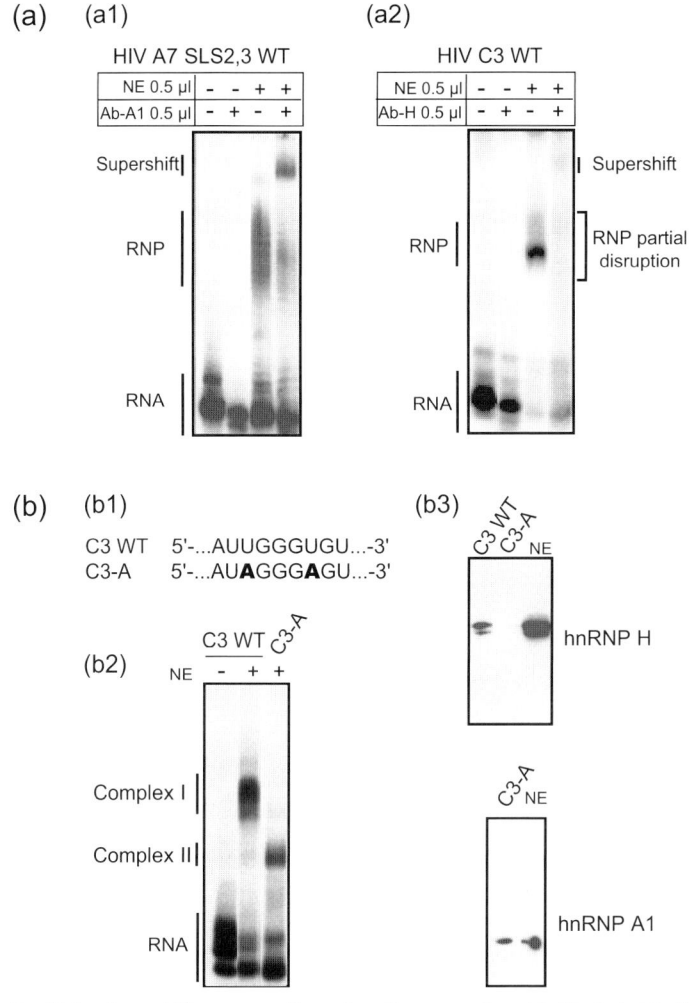

Fig. 11.3. Supershift assays performed on the RNP complexes formed by incubation in a HeLa cell nuclear extract (NE) of the A7 SLS2,3 wild-type (a1) and C3 wild-type (a2) RNAs [45, 58]. (a) RNAs were 3′-end-labeled, and anti-hnRNP A1 (Ab-A1) or anti-hnRNP H (Ab-H) antibodies were used in (a1) and (a2), respectively. Positions of the RNAs, RNP and the supershifted RNP are indicated on the left of the autoradiogram. (b) EMSA experiments coupled with a second gel electrophoresis and Western Blot analysis, performed on complexes formed by the HIV wild-type or mutated C3 RNAs (C3 wild-type and C3-A) in HeLa cell nuclear extract [58]. The mutations present in the C3-A RNA are shown in (b1), the EMSA experiment is illustrated in (b2). The positions of the free RNAs and RNP complexes I and II obtained with the HIV C3 wild-type and C3A RNAs, respectively, are indicated on the left of the autoradiogram. The bands containing complexes I (C3 wild-type) and II (C3 A) were included in an SDS–polyacrylamide gel. After electrophoresis proteins were transferred on a nitrocellulose membrane, that was used for Western blot analysis first with hnRNP H antibodies (upper panel) and then with anti-hnRNP A1 antibodies (lower panel). Proteins from nuclear extract (NE) were loaded on the gel for a control. Complex I contained hnRNP H protein and complex II contained hnRNP A1 protein.

fragments transcribed with SP6 RNA polymerase were 3′-end-labeled. With the HIV A7 SLS2,3 wild-type RNA, a supershift is obtained, demonstrating the presence of protein hnRNP A1 in the complex (Fig. 11.3a1). The Cl complex formed with the C3 wild-type RNA is dissociated by binding of the anti-hnRNP H antibody (Fig. 11.3a2). This suggests the presence of protein hnRNP H in the complex and an essential role of this protein for complex stability.

The following experimental conditions were used: the HIV C3 and A7 SLS2,3 RNAs (2.5 and 6 pmol, respectively) were 3′-end-labeled by using [^{32}P]pCp. Labeled RNAs were incubated at 4 °C in the presence of 5 μg of yeast total tRNA, with 0.5 and 4 μl of nuclear extract, respectively. After a 15-min incubation, 0.5 μl of anti-hnRNP A1 antibodies [59] (Fig. 11.3a1) or 1 μl of the monoclonal anti-hnRNP H antibodies [60] (Fig. 11.3a2) were added, respectively. The incubation was continued for 10 min on ice and was followed by gel electrophoresis. As a control, the two RNAs were incubated under the same conditions in the absence of antibodies.

11.2.3.3 Identification of Proteins Contained in RNP by EMSA Experiments Coupled to a Second Gel Electrophoresis and Western Blot Analysis

This method also allows the identification of proteins present in RNP complexes fractionated on EMSA gels. The EMSA are performed as described above, and the band containing RNP complexes is cut out from the gel, soaked in a SDS–PAGE loading buffer and included in a 5–10% SDS–PAGE. Then, the fractionated proteins are electrotransferred on a nitrocellulose membrane. The search for the presence of defined proteins is done by immunoblotting using specific antibodies directed against these proteins [58].

Example 4: Protocol for an EMSA experiment coupled to a second gel electrophoresis (Fig. 11.3b)

Figure 11.3(b) illustrates the search for the presence of proteins hnRNP H and A1 in RNP complexes I and II by Western blot analysis (Fig. 11.3b). Complexes I and II were formed by incubation in a HeLa cell nuclear extract of the HIV C3 wild-type and C3-A mutant RNAs (Fig. 11.3b2), respectively. The HIV C3-A RNA is a variant of the HIV wild-type C3 RNA (2U were substituted by 2A) (Fig. 11.3b1) [58]. The bands of gel containing each RNP complex were cut out. They were soaked in 10 μl of SDS–PAGE loading buffer for 1.5 h at 37 °C, and the band and buffer were heated for 5 min at 96 °C before their inclusion at the top of a 10% SDS–polyacrylamide gel (1.5 mM thickness). After 2 h electrophoresis at 20 V/cm, the fractionated proteins were electrotransferred onto a Hybond C nitrocellulose membrane (Pharmacia Amersham Biotech) (for 1 h at 100 V). Then, the membrane was blocked with 20 ml of PBS-TM buffer overnight at 4 °C or 2 h at room temperature with gentle shaking. It was then probed with the anti-hnRNP A1 (0.5 μl) or hnRNP H (2 μl) antibodies [58]. The bound antibodies were detected with peroxidase-conjugated anti-mouse and anti-rabbit IgG antibodies, respectively, and visualized by the ECL detection system (Fig. 11.3b3) [58].

11.2.4
Probing of RNA Structure

11.2.4.1 Properties of the Probes Used

Conditions used for RNP probing with chemical reagents or enzymes are chosen as mild as possible in order to preserve the structural integrity of RNP particles: probing reactions are performed in the buffer used for the RNP purification or reconstitution. Incubations are performed at moderate temperature for short times. A defined amount of yeast total tRNA is added, in order to control the [RNA]/[probe] ratio. Modification and enzymatic digestion conditions should be selected in order that less than one modification or cleavage statistically occurs per RNA molecule. The chemical and enzymatic probes used for footprinting are also used for determination of RNA secondary structure in solution (see Chapter 10). When the probing experiments are performed on purified natural complexes or complexes reconstituted between an RNA transcript and recombinant proteins, almost all the probes used for RNA secondary structure analysis can be used. Only the S1 nuclease, which has an optimum pH of action of 4.5, cannot be used. However, when cellular extracts are used for RNP formation without further purification of the complexes, some of the chemical probes, especially 1-cyclohexyl-3(2-morpholinoethyl)carbodiimide metho-p-toluene sulfonate (CMCT), cannot be used. The method employed for the identification of cleavage and modification positions depends on the labeled state of the RNA. For unlabeled RNAs (authentic purified RNP or RNP reconstituted with unlabeled RNAs), primer extension analyses with reverse transcriptase are performed: stops of extension occur at the cleavage site, or one residue before the cleaved (depending on the enzyme used) or modified nucleotide. When 3′- or 5′-end-labeled RNAs are used for RNP reconstitution, only enzymatic probes are used and cleavages are directly localized by gel electrophoresis without the reverse transcriptase step.

Chemical Probes

DMS methylates RNAs at the N^7-G, N^1-A and N^3-C positions of the bases. CMCT alkylates RNAs at the N^3-U and N^1-G positions, and kethoxal reacts at the N^1-G and N^2-G positions. Only N^7-G methylation by DMS can occur in double-stranded RNAs – all the other modifications are impaired.

Enzymes

RNase T1 cleaves the phosphodiester bonds 3′ to G residues, whereas RNase T2 cleaves after any residue. Both enzymes are used in conditions such that they preferentially cleave single-stranded RNA regions. RNase V1 is used to cleave double-stranded or stacked RNA regions. More details on these probes are available in Chapter 10.

To identify the positions that are protected by the proteins in an RNP, the naked RNA and the RNP are subjected to the same chemical and enzymatic treatments, and the reactive positions in RNA and RNP are compared. It should be noticed that in addition to RNA protection, RNP probing may detect some RNA conformational changes occurring upon protein binding [61].

Safety Rules
DMS is a potential carcinogen and special care should be taken when using it (see Chapters 9 and 10 for the safety rules).

11.2.4.2 Equipment, Material and Reagents

Equipment

- Sequencing gels for primer extension analysis.
- Exposure with X-Ray films (Fuji or Kodak) using an X-Ray film developer is recommended. However visualization with a PhosphorImager equipment can also be used.
- Temperature controlled baths (96, 65, 30 and 20 °C).

Reagents

Probes
RNase T1 is purchased from Roche Diagnostics, RNase T2 from Invitrogen, RNase V1 from Ambion, DMS from Aldrich, CMCT from Fluka and kethoxal from Amersham.

Chemicals and enzymes
The avian myeloblastosis virus (AMV) reverse transcriptase is purchased from Q-Bio Gene; glycogen (10 mg/ml), dNTPs (100 mM of each) and ddNTPs are from MBI Fermentas. The yeast total tRNA (20 mg/ml) is from Roche Diagnostics; cacodylic acid and boric acid are from Sigma.

Radiochemicals
$[\gamma\text{-}^{32}P]ATP$ (3000 Ci/mmol) is purchased from Amersham Biosciences or ICN.

Materials
See Section 11.2.2.1.

Buffers
- 1 × buffer D: see Section 11.2.2.1.
- 1 × Tris buffer: see Section 11.2.2.1.
- 1 × TBE: see Section 11.2.2.1.
- 1 × DMS/Ke buffer: 50 mM sodium cacodylate (pH 7.5), 100 mM KCl, 2.5 mM $MgCl_2$.
- 1 × CMCT buffer: 50 mM sodium borate (pH 8.0), 100 mM KCl, 2.5 mM $MgCl_2$.
- DMS stop buffer: 1 M Tris–acetate (pH 7.5), 1.5 M sodium acetate, 1 M β-mercaptoethanol.

- Buffer A: 10 mM Tris–HCl (pH7.5), 10 mM $MgCl_2$, 3 mM $CaCl_2$, 250 mM sucrose, 0.7 M β-mercaptoethanol.
- 0.7 M ice cold β-mercaptoethanol.
- 0.1 M EDTA (pH 8.0).
- 0.5 M potassium borate (pH 7.0).
- 0.5 M sodium cacodylate (pH 7.0).
- 3 M sodium acetate (pH 6.0).
- 10 × RT buffer: 500 mM Tris–HCl (pH 8.3), 60 mM $MgCl_2$, 400 mM KCl (provided with the reverse transcriptase purchased from Q-BioGene).

11.2.4.3 Probing Method

Enzymatic and Chemical Probing of Native Purified RNP Particles

Modifications and cleavages are performed in the purification or storage buffer. To ensure statistical modifications and cleavages, all the reactions are performed in the presence of 1.25 μg of yeast total tRNA.

Chemical Modifications

DMS

DMS modifications are performed for 6 min at 20 °C in 50 μl of 1 × DMS/Ke buffer with 1 μl of a DMS/EtOH solution (1/1, v/v).

Kethoxal

Same protocol as for DMS, except that modifications are performed for 10 min at 0 °C at a kethoxal concentration of 10 mg/ml.

CMCT

CMCT modifications are performed for 6 min at 20 °C in 50 μl of 1 × CMCT buffer and at CMCT concentrations of 30–60 mM.

Enzymatic Cleavages

T1, T2 and V1 RNase cleavages are performed for 6 min at 20 °C in 40 μl of 1 × Tris buffer with 5×10^{-3} U/μl of RNase.

Reaction Stop

DMS modification is quenched by addition of 10 μl of DMS stop buffer (20% of the reaction mixture), followed by phenol extraction. CMCT modification is stopped by phenol extraction, followed by ethanol precipitation. Kethoxal modification is stopped by addition of 0.5 M potassium borate (pH 7.0) to stabilize the kethoxal–guanine adduct (25% of the reaction mixture volume), followed by phenol extraction and ethanol precipitation. RNase V1 digestion is stopped by the addition of 5 μl of 100 mM EDTA (pH 8.0) before phenol extraction. RNase T1 and T2 digestions are stopped by the addition of an excess of yeast total tRNA (10 μg), followed by rapid phenol extraction. To avoid reaction of the enzymes or chemical on the free RNA, phenol extractions should be quickly performed on ice.

Modification and digestion products are ethanol precipitated, washed with 70% (v/v) ethanol, dried and dissolved in MilliQ water (except kethoxal-modified RNA pellets, that are dissolved in potassium borate 25 mM, pH 7.0, in order to stabilize the chemical adducts [62]).

Primer Extension Analysis
For primer extension analysis, 5′-end-labeled primers are annealed to chemically modified or digested RNAs. As the RNA length that can be examined with one primer ranges between 100 and 200 nt, different primers (generally 12–20 nt) complementary to regions that are spaced by 100–200 nt have to be used. As each primer has its own efficiency for reverse priming, preliminary assays should be performed for each primer to define the amount of RNA which is suitable for the analysis. The 5′-end-labeling (with $[\gamma^{-32}P]ATP$, 3000 Ci/mmol) is described in [63]. Extension is achieved with the AMV reverse transcriptase in the presence of the four dNTPs, the conditions are described in [64].

Hybridization
The RNA sample (1–10 pmol in 1 μl) is mixed with the 5′-end-labeled primer (100 000 c.p.m.) and 10 × RT buffer, in a total volume of 2.5 μl. The mixture is incubated for 10 min at 65 °C and quickly cooled on ice for 10 min.

Primer Extension
The primer extension reaction is performed in a final volume of 5 μl. The elongation mixture is prepared as follows: 0.1 μl of dNTP mixture (5 mM of each dNTP), 0.25 μl of 10 × RT buffer, 0.25 μl of reverse transcriptase (2 U/μl extemporaneous dilution) and 1.9 μl of H_2O. The hybridization mixture is mixed with 2.5 μl of elongation mixture and samples are incubated for 45 min at 42 °C. The primer extension is stopped by addition of 4 μl of the DNA loading buffer. To prepare the sequencing ladder, the unmodified RNA is used as a template. The elongation mixture contains a ddNTP at a 0.5 mM concentration and dNTPs with a dNTP:ddNTP ratio of 2.

Gel Fractionation
The elongation mixture is denatured for 2 min at 96 °C and 2 μl aliquots are fractionated on a 7% denaturing (8 M urea) polyacrylamide (19:1 ratio acrylamide:bisacrylamide) sequencing gel in 1 × TBE. The gel is preheated for 30 min at 50 V/cm and electrophoresis is performed at 50 °C using the same voltage. The migration time is adjusted depending on the sequence to be read. After migration, gels are transferred on sheets of Whatman 3MM paper, dried and autoradiographed (X-Ray films from Kodak or Fuji) overnight at −80 °C with an intensifying screen (Amersham, Biosciences).

Example: Native RNP probing protocol (Fig. 11.4)

The yeast U5 snRNA was probed as free RNA (RNA) and in the yeast [U4/U6.U5] tri-snRNP (25S) (Fig. 11.4a). About 200 ng of RNP or 100 ng of free renatured RNA was incubated for 10 min at 20 °C in the presence of 1.25 µg of yeast total tRNA, in 50 µl of the digestion or modification buffer (1 × DMS/Ke buffer or 1 × Tris buffer). For kethoxal, the mixture was then put on ice for 10 min, as the reaction is performed at 0 °C. The reactions with DMS, kethoxal, RNase T2 and V1 were performed and stopped in the conditions described above. Aliquots of the treated RNAs were reverse transcribed with the 5′-labeled specific oligodeoxynucleotide primer O-335, complementary to positions 69–81 of the *S. cerevisiae* U5 snRNA (Fig. 11.4b). The cDNA fragments obtained were fractionated on a sequencing gel, using as a reference a sequencing ladder performed with the unmodified RNA and the same 5′-end-labeled oligodeoxynucleotide (Fig. 11.4a).

Enzymatic and Chemical Probing of RNP Particles Formed in Nuclear Extract or with purified Proteins
As for EMSA experiments, the RNA is renatured in 1 × buffer D, before complex formation, as described in Section 11.2.2.2.

1. Enzymatic Reactions

RNase T1
RNA (200 ng, 1 pmol) is mixed with 1 µl of 62.5 mM $MgCl_2$, 3.6 µl of 1 × buffer D, 5 µg of total tRNA and the final volume is adjusted to 14 µl with water. An adequate amount (based on EMSA) of HeLa cell nuclear extract, as defined by the EMSA experiments, is added and the reaction mixture is incubated at 30 °C for 10 min. The cleavage reaction is performed with 1 µl of RNase T1 (0.025–2 U) at 30 °C for 10 min.

RNase T2
Same protocol as for RNase T1, except that the reaction is performed with 1 µl of RNase T2 (1–3 U) at 30 °C for 10 min.

RNase V1
Same protocol as for RNase T1, except that the reaction is performed with 1 µl of RNase V1 (0.02–0.2 U) at 30 °C for 10 min.

2. Chemical Modification by DMS

The RNA (200 ng, 1 pmol) is mixed with 1 µl of 62.5 mM $MgCl_2$, 3.6 µl of 1 × buffer D, 5 µg of total tRNA are added and the final volume is adjusted with water to 14 µl. An adequate amount of HeLa cell nuclear extract dialyzed against buffer D, as deduced from the EMSA experiment, is added and the reaction mixture is incubated at 30 °C for 10 min. To improve the efficiency of DMS modification, sodium cacodylate at a final concentration of 50 mM (pH 7.5) is added in

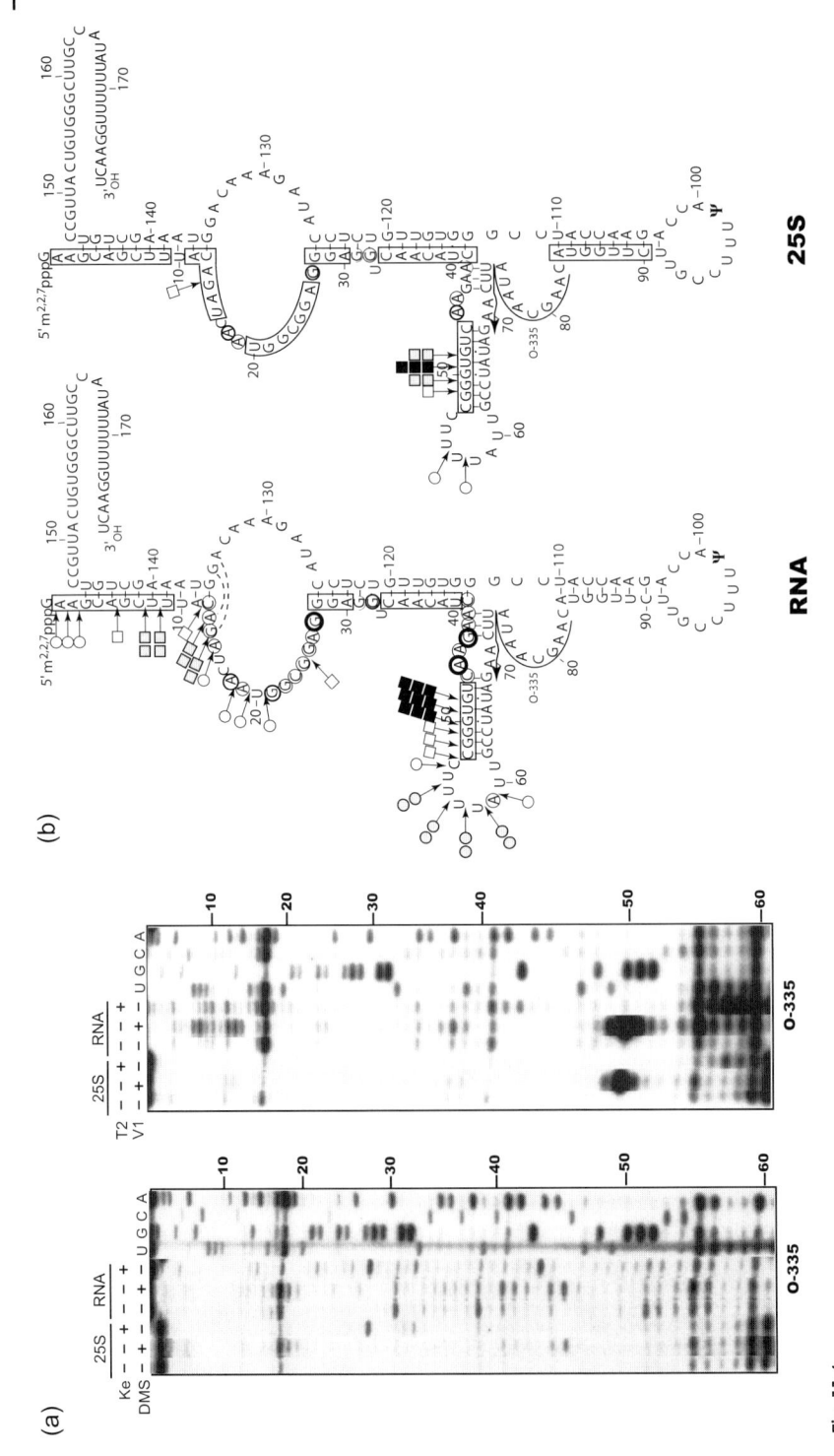

190 | *11 Study of RNA–Protein Interactions and RNA Structure in Ribonucleoprotein Particles*

Fig. 11.4

buffer D before the reaction. The modification reaction is performed at 30 °C for 10 min with between 1 and 5 µl of a 1/1 (v/v) DMS/EtOH solution.

3. Reaction Stop

Same protocol as described in Section 11.2.4.3. The hydrolyses by RNases T1, T2 and V1 are stopped by addition of 20 µg of total yeast tRNA; in addition, 1 µl of 100 mM EDTA is added for the RNase V1. These additions are followed by phenol extraction on ice. DMS modifications are stopped by addition of 10 µl of DMS stop buffer before phenol extraction and ethanol precipitation.

Ethanol precipitation of all samples is done by addition of 10 µl of 3 M sodium acetate (pH 6.0), 1 µg of glycogen and at least 3 volumes of 96% EtOH, followed by a 15 min incubation at −80 °C. After centrifugation at 13 000 r.p.m. for 15 min and at 4 °C, the supernatants are discarded and the RNA pellets are washed with 200 µl of 70% EtOH. A second centrifugation is performed for 5 min at 13 000 r.p.m. and 4 °C. The RNA pellets are vacuum-dried for 2 min and dissolved in 4 µl of MilliQ water before primer extension analysis (see Section 11.2.4.3).

Example: Protocol used for probing of a reconstituted RNP complex (Fig. 11.5)

The complexes formed upon incubation of the HIV-1 A7 SLS1,2,3 RNA fragment (positions 7903–8170) in a HeLa cell nuclear extract or with the purified hnRNP A1 protein were analyzed by chemical and enzymatic probing of the RNA structure and accessibility (Fig. 11.5). The following conditions were used: 1.12 pmol of cold HIV-1 A7 SLS1,2,3 RNA was incubated in the presence of 5 µg of yeast total tRNA with 1 µl of 62.5 mM MgCl$_2$ and 3.6 µl of 1 × buffer D in a total volume of 14 µl. Assays were performed in the presence (+) of 4 µl of nuclear extract ([Protein]/[RNA] (P/R) = 46) (Fig. 11.5a) or 50 fmol of purified hnRNP A1 protein

Fig. 11.4. Probing of U5 snRNA in the spliceosomal [U4/U6.U5] tri-snRNP purified from *S. cerevisiae* by use of chemical and enzymatic probes [49]. The yeast [U4/U6.U5] 25S tri-snRNP (25S, a) and the natural free U5 snRNA (RNA, a) were probed with kethoxal, DMS, RNase T2 and RNase V1 in conditions described in Section 11.2.4.3. (a) Primer extension analyses performed with the primer O-335. For each probe, a control experiment in the absence of the probe was performed (−). Lanes U, G, C and A correspond to the sequencing ladder. Positions of nucleotides in U5 snRNA are indicated on the right side of the panels. (b) The probing data illustrated in (a) are schematically represented on the secondary structure of U5 snRNA (left: 2-D structure results for the naked RNA; right: 2-D structure results for the 25S tri-snRNP). Nucleotides modified by DMS or kethoxal are circled; the thickness of the circles indicates the levels of modification (weak, medium and strong). RNase V1 or T2 cleavages are indicated by arrows linked to squares or circles, respectively. The color and number of symbols indicate the yield of cleavage. Boxed nucleotides are not modified. In the U5 snRNA region that is analyzed with primer O-335, the tri-snRNP components generate a strong protection, except for the lateral stem–loop structure formed by residues 40–75. This stem–loop structure is additional in the yeast U5 snRNA compared to vertebrate U5 snRNA [49].

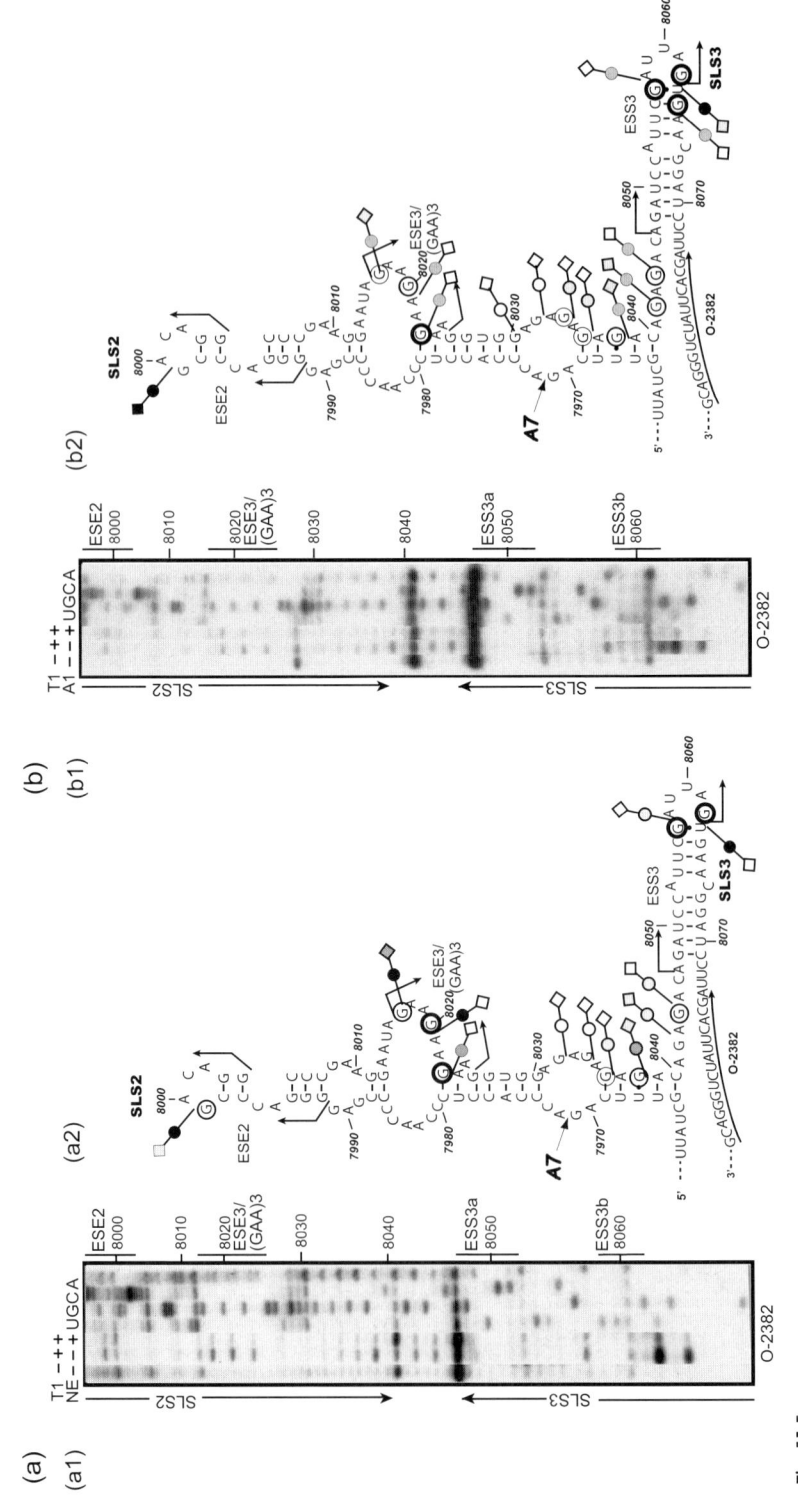

192 | *11 Study of RNA–Protein Interactions and RNA Structure in Ribonucleoprotein Particles*

Fig. 11.5

(P/R = 45) (Fig. 11.5b) or in the absence of added extract or protein (−) (Fig. 11.5a and b). After a 10-min incubation at 30 °C, 0.025 U of RNase T1 was added. For identification of reverse transcriptase pauses which are not due to RNA cleavage, a primer extension was performed on the naked RNA incubated in buffer D without RNase T1. The reactions were stopped by addition of 20 µg of tRNAs, followed by phenol extraction as described in Section 11.2.4.3.

Probing of Yeast RNAs Modified *in vivo* by DMS Treatment

DMS is able to penetrate efficiently in bacterial, yeast and animal cells, and can thus be used to probe RNA structure and accessibility in these cells. However, to get interpretable results the experiments should be performed on RNAs that form well-defined homogeneous particles in cells. Otherwise, information on RNA molecules having different structures and accessibilities would be superimposed in the primer extension analysis. The *in vivo* analysis of the S. cerevisiae U3 snoRNA structure and accessibility in the U3 snoRNP illustrated in Fig. 11.6 was a very successful example of an RNP analysis *in vivo*. The methylation sites detected *in vivo* were compared to those detected *in vitro* by treatment of a partially purified U3 snoRNP and the naked *in vitro* transcribed U3 snoRNA [29]. The results obtained validated the protein binding sites identified by *in vitro* analysis and demonstrated the interaction of U3 snoRNA with the pre-ribosomal RNA [29].

Example: Protocol used for *S. cerevisiae* U3 snoRNP probing by DMS *in vivo* (Fig. 11.6)

Fig. 11.5. Probing of the RNA structure and accessibility in RNP formed by the HIV-1 A7 SLS2,3 RNA and a HeLa cell nuclear extract (NE) or the purified hnRNP A1 protein (A1) [45]. Primer extension analyses of the A7 SLS1,2,3 RNA cleaved by RNase T1 in buffer D in the absence (−) or presence (+) of nuclear extract (a1) or in the presence of the recombinant hnRNP A1 protein (b1) are shown. As a control, a primer extension was performed using the intact RNA transcript incubated without RNase T1 as the template (left lane of the autoradiogram). Lanes U, G, C and A correspond to the sequencing ladder. Positions of nucleotides in the HIV-1 BRU RNA and of the RNA secondary structure elements described for the HIV-1 A7 RNA region are indicated on the right side of the autoradiograms. (a2 and b2) Schematic representations of the probing data illustrated in (a1) and (b1). Positions of RNase T1 cleavages are represented on the RNA secondary structure established for the A7 SLS1,2,3 RNA [45]. Cleavages 3′ to the G residues are indicated by thin lines. They are surmounted with circles when the cleavages occurred in the naked RNA (a2 and b2) Cleavages occurring in the presence of nuclear extract (a2) or hnRNP A1 (b2) are indicated by the presence of a square. The colors of circles and squares indicate the level of cleavage observed in the naked RNA and the RNP complex, respectively (grey, dark gray and black represent increasing intensities of cleavages, respectively). The G residues protected either in the presence of nuclear extract (a2) or with hnRNP A1 protein (b2) are circled; the intensity of the circle corresponds to the yield of the protection. The hybridization site of the oligodeoxynucleotide O-2382 is indicated. The *cis* regulatory elements of splicing, acting at site A7 (ESE2, ESE3/(GAA)3, ESS3) are delimited by two opposite broken arrows and the name of the element is given.

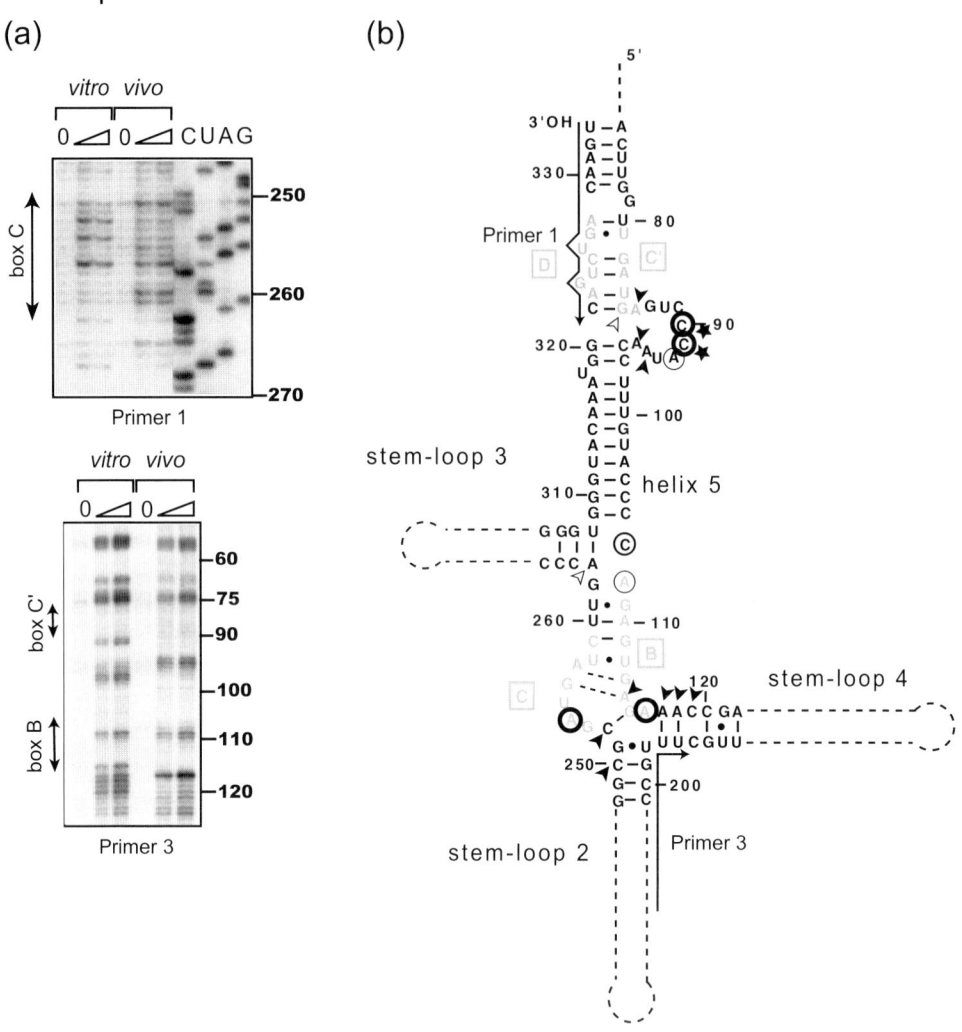

Fig. 11.6. Probing of the structure and the accessibility of the *S. cerevisiae* U3 snoRNA *in vivo* [29]. Primer extension analyses of the U3 snoRNA modified by DMS *in vitro* or *in vivo* are presented in (a). As a control, primer extensions were performed with an untreated U3 snoRNA transcript (lanes 0). Lanes U, G, C and A correspond to the sequencing ladders. Positions of nucleotides in the *S. cerevisiae* U3 snoRNA sequence and the phylogenetically conserved RNA segments of U3 snoRNA [29] are shown on the right and left sides of the autoradiogram, respectively. (b) Experimental data shown in (a) are represented on the secondary structure proposed for the *S. cerevisiae* U3 snoRNA in interaction with the pre-rRNA [16, 29]. Nucleotides modified *in vitro* and protected *in vivo* are indicated by black triangles, nucleotides modified *in vivo* and *in vitro* are circled; the thickness of the circles reflects the yield of modification *in vivo*. Asterisks mark the nucleotides with an increased reactivity *in vivo* compared to *in vitro*.

About 15-ml aliquots of a *S. cerevisiae* culture, grown in YEPD medium at 30 °C to an A_{600} between 0.5 and 1.0, were gently rocked at room temperature for 2 min in the presence of DMS at a concentration between 40 and 160 mM. The reaction was quenched by addition of 0.7 M ice cold β-mercaptoethanol and 5 ml of cold water-saturated isoamyl alcohol, followed by shaking and centrifugation. Cell pellets were washed with Buffer A (10 mM Tris–HCl, pH 7.5, 10 mM $MgCl_2$, 3 mM $CaCl_2$, containing 250 mM sucrose and 0.7 M β-mercaptoethanol). Cells were centrifuged for 10 min at 2500 r.p.m. and 4 °C. Total RNA was extracted by the method described by Domdey et al. [65]. Sites of RNA methylation were mapped by primer extension using the 5′-end-labeled oligodeoxynucleotides complementary to two distinct regions of the yeast U3 snoRNA [29] and 10 μg of total RNA as the template.

11.2.5
UV Crosslinking and Immunoselection

Formation of covalent bonds between RNA and proteins can be established by incorporation of photoactivable residues such as 4-thiouridine in the course of reverse transcription [66]. However, the easiest way to test for a very near proximity between RNA and proteins is to use UV irradiation at 254 nm [45, 58]. Although the yield of crosslinking is low, it is sufficient to detect RNA–protein contacts by using RNA molecules with a high specific radioactive labeling.

11.2.5.1 Equipment, Materials and Reagents

Equipment

- Electrophoresis instruments for small vertical slab gels. Exposure of EMSA gels can be done with X-Ray films (Fuji or Kodak) or a PhosphorImager.
- Temperature controlled baths (65, 50, 30, 20 °C).

Materials
The short wavelength (254 nm) UV lamp and 96-well ELISA plates are purchased from VWR.

Reagents

Antibodies
Some of the primary antibodies used in the described examples can be purchased from Immuquest.

Chemicals and enzymes
Protein G–Sepharose, Protein A–Sepharose, Hybond C nitrocellulose membrane and the ECL detection system are purchased from Amersham Pharmacia Biotech; bovine serum albumin (B8894) is from Sigma; RNase T1 and RNase A are from Roche Diagnostics.

Buffers

- 1 × buffer D: see Section 11.2.2.1.
- SDS–PAGE loading buffer: see Section 11.2.2.1.
- IP buffer: 50 mM Tris–HCl (pH 7.5), 150 mM NaCl, 0.1% (v/v) Nonidet P-40
- PBS-TM: see Section 11.2.2.1

11.2.5.2 UV Crosslinking Method

Coating 96-well ELISA plates
In order to avoid non-specific interactions, each well is coated with 20 µg of bovine serum albumin diluted in 1 × buffer D in a total volume of 25 µl, for 30 min at room temperature. This mixture is then discarded and each well is washed 3 times with 50 µl of 1 × buffer D.

UV Crosslinking
Formation of RNP complexes is performed as described in Section 11.2.2. Note that only RNA transcripts which are uniformly labeled by incorporation of an [α-^{32}P]NTP can be used. Selection of the labeled nucleotide is decided after inspection of the nucleotide sequence of the RNA segment expected to interact with the proteins. For instance, if this fragment contains several uridines, labeling will be done with [α-^{32}P]UTP; if numerous adenines are present, [α-^{32}P]ATP will be used.

Reaction mixtures containing RNP complexes are transferred to a 96-well plate placed on ice and irradiated from the top for 10 min at 254 nm with a UV source placed 1 cm distance from the sample. After irradiation, the RNP complexes are transferred in microcentrifuge tubes, a large excess of RNase T1 (about 50 U/50 fmol of RNA) is added for complete digestion and incubation is done for 1 h at 37 °C. It is followed by 30 min incubation at 50 °C. The crosslinked proteins can then be fractionated by 8–10% SDS–PAGE or they can be purified by immunoselection prior to gel electrophoresis (see below). After electrophoresis, the gel is dried and autoradiographed as previously described.

Example: Crosslinking experimental protocol (Fig. 11.7)

About 50 fmol (about 500 000 c.p.m.) of HIV-1 A7 SLS2 WT RNA uniformly labeled by incorporation of [α-^{32}P]UTP was incubated with 5 µl of nuclear extract or with a mixture of the two recombinant ASF/SF2 and hnRNP A1 proteins, each at a 3 µM final concentration (1:1 ratio). The incubation was performed in conditions established for RNP formation (Fig. 11.2). After a 10-min irradiation at 254 nm, the RNA was digested with 50 U of RNase T1 for 1 h at 37 °C, followed by a 30 min incubation at 50 °C. Then, 8 µl of SDS–PAGE loading buffer are added and the crosslinked products can be fractionated on a 10% SDS–polyacrylamide gel for 1–2 h at 20 V/cm and visualized by autoradiography. Before SDS–PAGE loading buffer addition, one-half of crosslinked products was immunoselected with antibodies bound to Sepharose beads.

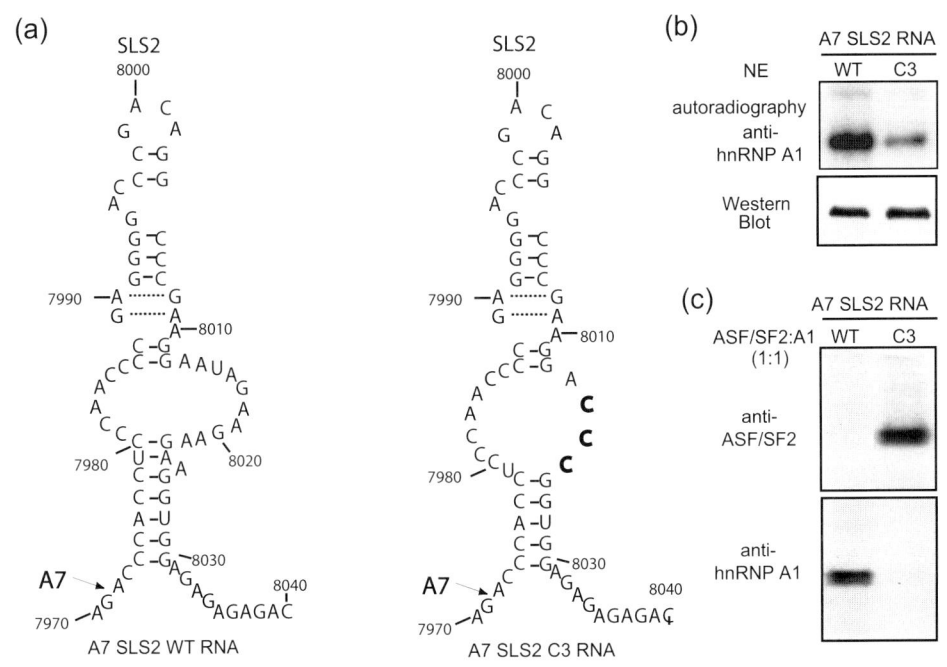

Fig. 11.7. Crosslinking followed by immunoselection of the hnRNP A1 and ASF/SF2 proteins on the wild-type and C3 variant of the HIV-1 A7 SLS2 RNAs (positions 7971 to 8040 of HIV BRU RNA) by irradiation at 254 nm [45]. (a) Secondary structures proposed for the A7 wild-type and C3 variant SLS2 RNAs [45]. Bold characters indicate the CCC sequence, that is substituted for the AUA(GAA)3 sequence in the C3 variant. (b) RNP complexes formed by the wild-type and C3 A7 SLS2 RNAs and proteins of a HeLa cell nuclear extract (NE) were UV irradiated at 254 nm. RNAs were digested with T1 RNase and the hnRNP A1 proteins were immunoselected on Sepharose beads coated with the anti-hnRNP A1 antibodies. One-half of the selected proteins was fractionated by electrophoresis on a 10% SDS–polyacrylamide gel (upper panel). The radiolabeled hnRNP A1 protein was detected by autoradiography. The second half of the selected hnRNP A1 proteins were fractionated on another SDS–polyacrylamide gel used for Western blot analysis (lower panel). HnRNP A1 protein from a HeLa cell nuclear extract was bound to the wild-type RNA and crosslinked at high level; crosslinking was very low with the C3 RNA. RNP complexes were formed by incubation of the wild-type and C3 variant of SLS2 RNAs with equimolar amounts of proteins hnRNP A1 and ASF/SF2 and were UV irradiated at 254 nm. After RNA digestion, the hnRNP A1 and ASF/SF2 protein were immunoselected on Sepharose beads and fractionated on a 10% SDS–polyacrylamide gel. The radiolabeled hnRNP A1 and ASF/SF2 proteins were detected by autoradiography (c). Whereas, protein hnRNP A1 was only crosslinked to the wild-type RNA, ASF/SF2 protein was only crosslinked to the C3 RNA.

Immunoselection

RNase activities are often present in sera and ascite media. It is thus necessary to bind the antibodies to Protein (G or A)–Sepharose beads and to eliminate the possible RNase contaminations by washing the beads. The washing solution contains a large amount of bovine serum albumin that coats the nonspecific protein-binding sites of the beads.

The Protein (G or A)–Sepharose beads are coated with the antibodies. The digested crosslinked products are then incubated with the coated beads in the presence of the immunoselection buffer. Subsequently, the beads are washed with the immunoselection buffer containing 0.25% Nonidet P-40, suspended in the SDS–PAGE loading buffer and heated 5 min at 96 °C for elution of the bound proteins. The proteins are further fractionated by a SDS–polyacrylamide gel (8–10% polyacrylamide concentration).

Each sample of eluted proteins is divided in two parts. One part is fractionated on a gel analyzed by autoradiography. The amount of radiolabeled protein is estimated by a PhosphorImager. The second half of the eluted proteins is fractionated on a gel used for immunoblotting with the antibody that was coated on the beads. This allows an estimation of the total amount of the protein which was retained on the beads. The ratio between the amount of protein estimated by immunoblotting and the radioactivity detected in the gel gives an indication on the level of crosslinking.

Example: Immunoselection protocol (Fig. 11.7)

About 20 µl of a Protein (G or A)–Sepharose beads suspension was coated for 2 h at 4 °C with 1 µl of anti-hnRNP A1 (4B10) or 2 µl anti-ASF/SF2 antibodies provided by G. Dreyfuss (University of Pennsylvania School of Medicine, Philadelphia, USA) and J. Stevenin (IGBMC, Strasbourg, France), respectively. The digested crosslinked products (see Section 11.2.5.2) were incubated with the coated beads by end-over-end rotation for 2 h at 4 °C in 400 µl of immunoselection buffer containing 0.1 mg/ml bovine serum albumin. Subsequently, the beads were washed 3 times by incubation with 500 µl of the immunoselection buffer containing 0.25% Nonidet P-40, followed by centrifugation (2 min at 3000 r.p.m. at room temperature). At this stage, the beads were suspended in 20 µl of SDS–PAGE loading buffer and boiled for 5 min to elute the bound proteins. The proteins are further fractionated by 10% SDS–Polyacrylamide gel for 1–2 h at 20 V/cm.

Each sample of the eluted proteins was divided in two parts: one part was fractionated by electrophoresis and the amount of radiolabeled protein was estimated by PhosphorImager scanning (Fig. 11.7b and c). The other part was fractionated on a gel and subjected to immunoblotting (Fig. 11.7b).

11.3
Commentaries and Pitfalls

11.3.1
RNP Purification and Reconstitution

11.3.1.1   RNA Purification and Renaturation
In order to obtain an RNA transcript of homogeneous size, a step of purification on denaturing PAGE (8 M urea) is advisable. Indeed, minor RNA degradation or abortive transcription products can be generated during transcription.

A homogeneous conformation of the RNA molecules is required. Thus, the RNA renaturation procedure prior to complex formation and probing experiment is a critical step. Alternative conformations lead to the simultaneous detection of single-stranded and double-stranded specific cleavages at some of the positions.

11.3.1.2 EMSA

Some naked RNAs are resolved in two bands in EMSA gels due to the occurrence of two distinct conformations or dimerizations. If the RNA used corresponds to a fragment of a larger RNA, it is advisable in this case to prepare another fragment by choosing other 5′ and 3′ extremities. Templates for production of small RNA may be produced by use of partially complementary DNA oligodeoxynucleotide primers and DNA polymerase to form a full-length double-stranded DNA.

For probing experiments, it is very important to get a complete and homogeneous formation of the RNPs. Otherwise, heterogeneous probing data will be obtained that will be difficult to interpret.

11.3.2
Probing Conditions

The probing conditions (pH, ionic strength, Mg^{2+} ions concentration, temperature, probe concentration) have to be defined by several preliminary tests. The stability of the RNP in these various conditions has to be tested by electrophoresis on native gel [49].

11.3.2.1 Choice of the Probes Used

Some of the widely used probes for RNA 2-D structure investigation [like CMCT, Pb^{2+}, Fe-EDTA(OH) and S1 nuclease] cannot be used when the assays are directly performed in nuclear or cellular extracts. The reasons for this are given below.

- The presence of a nuclear extract inhibits RNA modification by CMCT, probably due to its accelerated hydrolysis or interactions with other components of the extract. Note that CMCT can be used on purified RNP complexes.
- Pb(II) (Pb^{2+}) cannot be used because the presence of chloride ions in the RNP buffers induces precipitation and inhibits the reaction. However, it can be used for purified complex using a suitable buffer and also *in vivo*, using defined conditions (see Chapter 10).
- Fe-EDTA(OH) cannot be used since hydroxylated compounds, such as glycerol present in the nuclear extract and buffer D, are known to be ·OH scavengers and can inhibit ·OH-mediated cleavage. In addition, Tris or HEPES buffers also reduce RNA cleavage by hydroxyl radicals, presumably by acting as free-radical scavengers. Again, Fe-EDTA(OH) can be used on purified RNP.
- S1 nuclease cannot be used for any kind of RNP probing as its optimal pH of action is 4.5, which is generally deleterious for RNP complexes [48].

It should also be taken into account that the overall efficiency of a given enzymatic or chemical probe may be considerably diminished in the cellular extract and this apparent decrease in efficiency of cleavage or modification does not necessarily represent RNA protection due to protein binding. For instance, DMS reacts with both RNA and proteins; thus, DMS has a large number of substrates when modifications are directly performed in extracts, which may decrease its activity on RNA.

To get informative data by RNP probing, the probes should be selected in order that reactive residues or sensitive phosphodiester bonds are present all along the RNA molecule. It is sometimes difficult to fulfill these conditions in the case of highly structured RNAs. The reactivity of the naked RNA with the selected probes should be sufficient to see clear variations upon protein binding. Hence, in order to define conditions suitable for RNP analyses, one has first to test different conditions of modifications and cleavages on the naked RNA. The effects of different parameters can be examined in these preliminary tests, like the [RNA]/[probe] ratio, time of incubation, temperature of incubation and Mg^{2+} concentration. The conditions given in this paper were found to be suitable for analyses of several RNP. However, when they turned out to be unsuitable, they could be use as a starting point to look for other more favorable conditions.

11.3.2.2 Ratio of RNA/Probes

The added exogenous tRNA (usually total yeast tRNA mixture) minimizes the nonspecific interactions between RNA and proteins, and is also used to get a defined ratio of [RNA]/[Probe].

Another difficulty which may be encountered in the course of RNP probing is a very strong protection generated by the proteins. They may mask very large parts of the RNA. This may be the case for probing with extracts containing a large number of proteins or with proteins, like hnRNP A1, that are able to multimerize along the RNA [45]. To get convincing probing data one needs to be sure that the protections observed are not simply due to a global inhibition of the activity of the probes. To this end, it is necessary to obtain nearly unaltered modifications and cleavages, together with strongly diminished ones, in the same experiment. Here, again, conditions of reaction often have to be adapted to obtain such contrasted modifications of the reactivities along the RNA molecule. When proteins of the RNP generate very strong protections we recommend the use of chemical probes which are less sensitive to steric hindrance.

Sometimes, new cleavage sites are observed in RNP compared to naked RNA. This may reflect an RNA conformational change. However, one has to verify that these additional or reinforced cleavages are not due to the presence of an RNase activity present in the extract or the purified protein.

11.3.3
UV Crosslinking

11.3.3.1 Photoreactivity of Individual Amino Acids and Nucleotide Bases

Crosslink formation depends on the photoreactivity of both individual amino acids and nucleotide bases. Pyrimidines residues form covalent bonds with protein more efficiently compared to purine residues. Thus, upon UV irradiation a protein that binds a purine-rich sequence may be undetectable by UV crosslinking.

11.3.3.2 Labeled Nucleotide in RNA

Due to the low level of crosslinking established by UV irradiation, the specific activity of the labeled RNA must be high enough for detection by autoradiography of the crosslinked residues bound to the protein. The choice of the labeled nucleotide and the RNase used for the digestion should be made taking into consideration the nucleotide sequence of the expected binding site.

11.3.4
Immunoprecipitations

The amount of serum, or antibodies, required for complete precipitation of a particular protein has to be determined for each individual batch of serum.

11.3.4.1 Efficiency of Immunoadsorbents for Antibody Binding

Antibodies from humans, rabbits or guinea-pigs have a stronger affinity for Protein A, than those from mouse or rat. Binding to Protein G provides a convenient alternative for the use of mouse and rabbit antibodies [67]. Poor binding of antibodies to Protein A and G can be circumvented by the use of secondary antibodies (e.g. anti-mouse immunoglobulin raised in rabbits) that do bind to Protein A. Alternatively, the secondary antibodies can be directly coupled to CNBr-activated Sepharose. These coupled secondary antibodies will then serve as efficient adsorbents.

11.4
Troubleshooting

11.4.1
RNP Reconstitution

- No RNP complex is observed: first modify the [RNA]/[protein] ratio used for the reconstitution assay, then, if there is still a problem decrease the quantity of competitor tRNA used.
- The RNP complex does not penetrate in the gel: use a lower polyacrylamide concentration.

11.4.2
RNA Probing

- High smearing in the gel: decrease the amount of loaded material or digest the RNA with a DNase-free RNase.
- No elongation stop signal is detected in primer extension analysis of the modified or cleaved RNA: decrease the amount of probe used.

- No protection is observed: check if the RNP complex is formed and stable under the conditions used.
- Too much protections of the RNA: decrease the quantity of purified protein or cellular extract added.

11.4.3
UV Crosslinking

- No crosslinked proteins are obtained: verify that the UV wavelength is correct and that the UV light is still working.
- No crosslinked proteins are immunoselected: if the Western Blot analyses indicate that immunoselection occurred, try to use another labeled nucleotide. If labeled proteins are still not immunoselected, you have to use multiple approaches to understand what are the parameters that govern the binding of the protein to the RNA target.

11.4.4
Immunoprecipitations

- High background in the membrane: it is recommended to use less serum or antibodies.
- No signal is detected on the membrane: check that primary and/or secondary antibodies are still active.

Acknowledgments

S. Jacquenet and A. Clery are acknowledged for providing materials for illustrations of EMSA experiments. Y. Motorin is thanked for helpful discussion and critical reading of the manuscript. V. M. was a fellow of the French Ministère de la Jeunesse, de l'Education Nationale et de la Recherche. C. B., A. M. and A. M. are Staff Scientists from the Centre National de la Recherche Scientifique.

References

1 C. G. BURD, G. DREYFUSS, *Science* **1994**, *265*, 615–621.
2 K. B. HALL, *Curr. Opin. Struct. Biol.* **2002**, *12*, 283–288.
3 T. A. STEITZ, P. B. MOORE, *Trends Biochem. Sci.* **2003**, *28*, 411–418.
4 C. C. QUERY, R. C. BENTLEY, J. D. KEENE, *Cell* **1989**, *57*, 89–101.
5 D. J. KENAN, C. C. QUERY, J. D. KEENE, *Trends Biochem. Sci.* **1991**, *16*, 214–220.
6 H. SIOMI, M. J. MATUNIS, W. M. MICHAEL, G. DREYFUSS, *Nucleic Acids Res.* **1993**, *21*, 1193–1198.
7 C. OUBRIDGE, N. ITO, P. R. EVANS, C. H. TEO, K. NAGAI, *Nature* **1994**, *372*, 432–438.
8 N. V. GRISHIN, *Nucleic Acids Res.* **2001**, *29*, 638–643.
9 H. MAO, S. A. WHITE, J. R. WILLIAMSON, *Nat. Struct. Biol.* **1999**, *6*, 1139–1147.

10 N. J. Watkins, V. Segault, B. Charpentier, S. Nottrott, P. Fabrizio, A. Bachi, M. Wilm, M. Rosbash, C. Branlant, R. Luhrmann, *Cell* **2000**, *103*, 457–466.

11 I. Vidovic, S. Nottrott, K. Hartmuth, R. Luhrmann, R. Ficner, *Mol. Cell* **2000**, *6*, 1331–1342.

12 D. J. Klein, T. M. Schmeing, P. B. Moore, T. A. Steitz, *EMBO J.* **2001**, *20*, 4214–4221.

13 N. Ban, P. Nissen, J. Hansen, P. B. Moore, T. A. Steitz, *Science* **2000**, *289*, 905–920.

14 E. J. Tran, X. Zhang, E. S. Maxwell, *EMBO J.* **2003**, *22*, 3930–3940.

15 T. S. Rozhdestvensky, T. H. Tang, I. V. Tchirkova, J. Brosius, J. P. Bachellerie, A. Huttenhofer, *Nucleic Acids Res.* **2003**, *31*, 869–877.

16 N. Marmier-Gourrier, A. Clery, V. Senty-Segault, B. Charpentier, F. Schlotter, F. Leclerc, R. Fournier, C. Branlant, *RNA* **2003**, *9*, 821–838.

17 I. Fierro-Monti, M. B. Mathews, *Trends Biochem. Sci.* **2000**, *25*, 241–246.

18 M. Kiledjian, G. Dreyfuss, *EMBO J.* **1992**, *11*, 2655–2664.

19 O. Theunissen, F. Rudt, U. Guddat, H. Mentzel, T. Pieler, *Cell* **1992**, *71*, 679–690.

20 J. D. Puglisi, L. Chen, S. Blanchard, A. D. Frankel, *Science* **1995**, *270*, 1200–1203.

21 S. Cusack, *Curr. Opin. Struct. Biol.* **1999**, *9*, 66–73.

22 M. Ranjan, S. R. Tafuri, A. P. Wolffe, *Genes Dev.* **1993**, *7*, 1725–1736.

23 P. Fabrizio, S. Esser, B. Kastner, R. Luhrmann, *Science* **1994**, *264*, 261–265.

24 M. Bach, P. Bringmann, R. Luhrmann, *Methods Enzymol.* **1990**, *181*, 232–257.

25 O. Puig, F. Caspary, G. Rigaut, B. Rutz, E. Bouveret, E. Bragado-Nilsson, M. Wilm, B. Seraphin, *Methods* **2001**, *24*, 218–229.

26 K. Hartmuth, H. Urlaub, H. P. Vornlocher, C. L. Will, M. Gentzel, M. Wilm, R. Luhrmann, *Proc. Natl. Acad. Sci. USA* **2002**, *99*, 16719–16724.

27 G. Neubauer, A. Gottschalk, P. Fabrizio, B. Seraphin, R. Luhrmann, M. Mann, *Proc. Natl. Acad. Sci. USA* **1997**, *94*, 385–390.

28 F. Dragon, J. E. Gallagher, P. A. Compagnone-Post, B. M. Mitchell, K. A. Porwancher, K. A. Wehner, S. Wormsley, R. E. Settlage, J. Shabanowitz, Y. Osheim, A. L. Beyer, D. F. Hunt, S. Baserga, *Nature* **2002**, *417*, 967–970.

29 A. Mereau, R. Fournier, A. Gregoire, A. Mougin, P. Fabrizio, R. Luhrmann, C. Branlant, *J. Mol. Biol.* **1997**, *273*, 552–571.

30 M. Balzer, R. Wagner, *Anal. Biochem.* **1998**, *256*, 240–242.

31 T. Hamma, A. R. Ferre-D'Amare, *Struct.* **2004**, *12*, 893–903.

32 T. Moore, Y. Zhang, M. O. Fenley, H. Li, *Struct.* **2004**, *12*, 807–818.

33 C. Charron, X. Manival, A. Cléry, V. Senty-Ségault, B. Charpentier, N. Marmier-Gourrier, C. Branlant, A. Aubry, *J. Mol. Biol.* **2004**, in press.

34 D. Irvine, C. Tuerk, L. Gold, *J. Mol. Biol.* **1991**, *222*, 739–761.

35 C. Tuerk, L. Gold, *Science* **1990**, *249*, 505–510.

36 H. Urlaub, K. Hartmuth, R. Luhrmann, *Methods* **2002**, *26*, 170–181.

37 M. Bachler, R. Schroeder, U. von Ahsen, *RNA* **1999**, *5*, 1509–1516.

38 C. L. Will, S. E. Behrens, R. Luhrmann, *Mol. Biol. Rep.* **1993**, *18*, 121–126.

39 B. Kastner, R. Luhrmann, *Methods Mol. Biol.* **1999**, *118*, 289–298.

40 A. Gottschalk, G. Neubauer, J. Banroques, M. Mann, R. Luhrmann, P. Fabrizio, *EMBO J.* **1999**, *18*, 4535–4548.

41 K. Hartmuth, H. P. Vornlocher, R. Luhrmann, *Methods Mol. Biol.* **2004**, *257*, 47–64.

42 J. D. Dignam, R. M. Lebovitz, R. G. Roeder, *Nucleic Acids Res.* **1983**, *11*, 1475–1489.

43 F. Baneyx, *Curr. Opin. Biotechnol.* **1999**, *10*, 411–421.

44 C. M. Griffiths, M. J. Page, *Methods Mol. Biol.* **1997**, *75*, 427–440.

45 V. Marchand, A. Mereau, S.

Jacquenet, D. Thomas, A. Mougin, R. Gattoni, J. Stevenin, C. Branlant, *J. Mol. Biol.* **2002**, *323*, 629–652.

46 V. V. Gurevich, I. D. Pokrovskaya, T. A. Obukhova, S. A. Zozulya, *Anal. Biochem.* **1991**, *195*, 207–213.

47 J. F. Milligan, D. R. Groebe, G. W. Witherell, O. C. Uhlenbeck, *Nucleic Acids Res.* **1987**, *15*, 8783–8798.

48 S. Jacquenet, D. Ropers, P. S. Bilodeau, L. Damier, A. Mougin, C. M. Stoltzfus, C. Branlant, *Nucleic Acids Res.* **2001**, *29*, 464–478.

49 A. Mougin, A. Gottschalk, P. Fabrizio, R. Luhrmann, C. Branlant, *J. Mol. Biol.* **2002**, *317*, 631–649.

50 J. F. Milligan, O. C. Uhlenbeck, *Methods Enzymol.* **1989**, *180*, 51–62.

51 M. Silberklang, A. M. Gillum, U. L. RajBhandary, *Methods Enzymol.* **1979**, *59*, 58–109.

52 T. E. England, O. C. Uhlenbeck, *Nature* **1978**, *275*, 560–561.

53 T. Pan, D. M. Long, O. C. Uhlenbeck, in *The RNA World*, R. F. Gesteland, J. F. Atkins (eds), Cold Spring Harbor Laboratory Press, Cold Spring Harbor Laboratory, NY, **1993**.

54 R. Gattoni, P. Schmitt, J. Stevenin, *Nucleic Acids Res.* **1988**, *16*, 2389–2409.

55 A. Mayeda, A. R. Krainer, *Methods Mol. Biol.* **1999**, *118*, 315–321.

56 A. M. Zahler, C. K. Damgaard, J. Kjems, M. Caputi, *J. Biol. Chem.* **2004**, *279*, 10077–10084.

57 M. M. Konarska, *Methods Enzymol.* **1989**, *180*, 442–453.

58 S. Jacquenet, A. Mereau, P. S. Bilodeau, L. Damier, C. M. Stoltzfus, C. Branlant, *J. Biol. Chem.* **2001**, *276*, 40464–40475.

59 S. Pinol-Roma, Y. D. Choi, M. J. Matunis, G. Dreyfuss, *Genes Dev.* **1988**, *2*, 215–227.

60 M. Y. Chou, N. Rooke, C. W. Turck, D. L. Black, *Mol. Cell. Biol.* **1999**, *19*, 69–77.

61 J. R. Williamson, *Nat. Struct. Biol.* **2000**, *7*, 834–837.

62 D. Moazed, H. F. Noller, *Cell* **1986**, *47*, 985–994.

63 J. Sambrook, E. F. Fritsch, T. Maniatis, *Molecular Cloning: A Laboratory Manual*, Cold Spring Harbor Laboratory Press, Cold Spring Harbor, NY, **1989**.

64 A. Mougin, A. Gregoire, J. Banroques, V. Segault, R. Fournier, F. Brule, M. Chevrier-Miller, C. Branlant, *RNA* **1996**, *2*, 1079–1093.

65 H. Domdey, B. Apostol, R. J. Lin, A. Newman, E. Brody, J. Abelson, *Cell* **1984**, *39*, 611–621.

66 A. D. Branch, B. J. Benenfeld, C. P. Paul, H. D. Robertson, *Methods Enzymol.* **1989**, *180*, 418–442.

67 D. Lane, E. Harlow, *Nature* **1982**, *298*, 517.

12
Terbium(III) Footprinting as a Probe of RNA Structure and Metal-binding Sites

Dinari A. Harris and Nils G. Walter

12.1
Introduction

Cations play a pivotal role in RNA structure and function. A functional RNA tertiary structure is stabilized by metal ions that neutralize and, in the case of multivalents, bridge the negatively charged phosphoribose backbone [1]. The current chapter describes the use of the trivalent lanthanide metal ion terbium(III) as a versatile probe of high-affinity metal ion binding sites, as well as of RNA secondary and tertiary structure. Terbium(III) has a similar ion radius (0.92 Å) as magnesium(II) (0.72 Å) and a similar preference for oxygen and nitrogen ligands over softer ones. Thus, terbium(III) binds to similar sites on RNA as magnesium(II); however, with 2–4 orders of magnitude higher affinity. Unlike magnesium(II), the low pK_a of the aqueous terbium(III) complex (around 7.9) generates enough $Tb(OH)_{(aq)}{}^{2+}$ to make it capable of hydrolyzing the RNA backbone around neutral pH via deprotonation of the 2′-hydroxyl group and nucleophilic attack of the resulting oxyanion on the adjacent 3′,5′-phosphodiester to form 2′,3′-cyclic phosphate and 5′-hydroxyl termini [2]. Under physiological conditions, therefore, low (micromolar) concentrations of Tb^{3+} ions readily displace medium (millimolar) concentrations of Mg^{2+} ions from high-affinity binding sites (both specific and nonspecific ones) and promote slow phosphodiester backbone cleavage, revealing their location on the RNA. Higher (millimolar) concentrations of Tb^{3+} ions bind less specifically to RNA and result in backbone cleavage in a sequence-independent manner, preferentially cutting solvent accessible, single-stranded or non-Watson–Crick base-paired regions, thus providing a footprint of the RNA's secondary and tertiary structure at nucleotide resolution.

Terbium(III) can be a very straightforward and useful probe of metal binding and tertiary structure formation in RNA. However, there are several precautions that need to be considered in order to obtain a reliable and reproducible RNA footprinting pattern. Since low (micromolar) concentrations of terbium(III) bind to high-affinity metal binding sites within a folded RNA, while high (millimolar) concentrations produce a footprinting pattern of solvent accessible regions, it is critical to perform terbium(III) induced cleavage reactions over a wide range of Tb^{3+} con-

---

Handbook of RNA Biochemistry: Student Edition. Edited by R. K. Hartmann, A. Bindereif, A. Schön, E. Westhof
Copyright © 2009 WILEY-VCH Verlag GmbH & Co. KGaA, Weinheim
ISBN: 978-3-527-32534-4

centrations. To ensure conformational homogeneity, pre-folding the RNA under optimized buffer conditions and magnesium concentrations is necessary. This is especially important when trying to identify metal-binding sites, since there will be relatively few cleavage events at low Tb^{3+} concentrations. All cleavage reactions should be performed near physiological pH (7.0–7.5) to allow for the accumulation of the cleavage active $Tb(OH)_{(aq)}^{2+}$ species [3]. Insoluble polynuclear hydroxo aggregates of terbium(III) can form at pH 7.5 and above [4, 5], which should be avoided. Another parameter that needs to be empirically optimized is the temperature and duration of the metal-ion-induced cleavage reactions. Higher temperatures result in faster cleavage rates, but also increase the amount of background degradation. Therefore, typical reaction temperatures range from 25 to 45 °C over a period of 0.5–2 h. All of these parameters need to be well established prior to carrying out terbium(III) footprinting experiments in earnest.

12.2
Protocol Description

12.2.1
Materials

Reagents and buffers
Appropriate buffers to fold RNA (usually Tris, MES and/or HEPES of desired pH)
1 M $MgCl_2$ solution.
100 mM $TbCl_3$ in 5 mM sodium cacodylate buffer, pH 5.5 (store in small aliquots at −20 °C).
0.5 M Na_2EDTA, pH 8.0.
Urea loading buffer: 8 M urea, 50 mM EDTA, pH 8.0, 0.01% bromophenol blue, 0.01% xylene cyanol.

Equipment
Heating block at 90 °C.
Water bath.
Phosphor screens and phosphorImager with appropriate software [e.g. PhosphorImager Storm 840 with ImageQuant software (Molecular Dynamics)].

(1) Prior to performing terbium(III) mediated footprinting of an RNA molecule, the RNA should be end-labeled (typically with ^{32}P at either the 5′ or 3′ end), purified by denaturing gel electrophoresis, and stored in water (or an appropriate buffer) at −20 °C.

(2) Prepare a single pool with 250 000–500 000 c.p.m. (typically 0.5–2 pmol) of end-labeled RNA per reaction aliquot under appropriate buffer conditions and heat denature at 90 °C for 2 min. The total pool volume should be sufficient for single or duplicate reaction aliquots at each desired Tb^{3+} concentration.

(3) Pre-fold the RNA, by incubating the pool at an optimized temperature (typically 25–45 °C) for approximately 10 min to ensure structural homogeneity. Some RNAs fold best when a slow-cooling procedure is used or when certain cations are already added at this stage.

(4) To obtain the desired Mg^{2+} concentration, add an aliquot of $MgCl_2$ from an appropriately diluted stock solution and equilibrate at the optimized temperature for an additional 5–10 min. At this point, the total volume of the reaction mixture should be 8 µl per reaction aliquot.

(5) From the 100 mM $TbCl_3$ stock solution, make a serial set of $TbCl_3$ dilutions in water, ranging from micromolar to millimolar concentrations (5× over the final reaction concentration). This wide range of $TbCl_3$ concentrations should be sufficient to probe for both high-affinity metal binding sites and secondary/tertiary structure formation. *Note: The 100 mM $TbCl_3$ stock solution is dissolved in a 5 mM sodium cacodylate buffer at pH 5.5 to prevent formation of terbium(III) hydroxide precipitates at higher pH. The $TbCl_3$ dilutions in water should be made immediately prior to use. A serial set of dilutions is recommended to ensure consistency in cleavage band intensity between gel lanes. Use a fresh aliquot of 100 mM $TbCl_3$ stock solution each time. Final $TbCl_3$ concentrations used in the cleavage reactions should be optimized together with other experimental conditions for the specific RNA and experimental goal.*

(6) To initiate terbium(III) mediated cleavage, mix an 8-µl aliquot from the pool containing the end-labeled RNA, buffer components and Mg^{2+} with 2 µl of an appropriate dilution of $TbCl_3$ to achieve the desired final Tb^{3+} concentration (typically ranging from 5 µM to 5 mM, in addition to a 0 mM Tb^{3+} control). Continue to incubate at the optimized temperature for an optimized amount of time (typically 30 min to 2 h). *Note: The incubation times should be chosen to generate a partial digestion pattern of end-labeled RNA under single-hit conditions. Extended incubation times will increase secondary hits that may reflect structural distortions of the RNA.*

(7) Quench the cleavage reaction by adding EDTA, pH 8.0, to a final concentration of 50 mM (or at least a 2-fold excess over the total concentration of multivalent metal ions in the reaction aliquot).

(8) Perform an ethanol precipitation of the RNA by adding Na(OAc) to a final concentration of 0.3 M and 2–2.5 volumes of 100% ethanol, and precipitate at −20 °C overnight. Centrifuge 30 min at 12 000 g, 4 °C. Decant supernatant, wash with 80% ethanol, decant supernatant, and dry RNA in a Speed Vac evaporator. Re-dissolve samples in 10–20 µl of urea loading buffer.

(9) Partial alkaline hydrolysis and RNase T1 digestion reactions of the same RNA should be performed as calibration standards by incubating the end-labeled RNA in the appropriate buffers.

(10) Heat samples at 90 °C for 5 min and place on ice water to snap cool. Analyze the cleavage products on a high-resolution denaturing (8 M urea) polyacrylamide sequencing gel, using the partial alkaline hydrolysis and RNase T1 digestion reactions as size markers to identify the specific terbium(III) cleavage products at nucleotide resolution. *Note: In the example cited below, a wedged,*

8 M urea, 20% polyacrylamide gel was run at a constant power of 80 W for separating the reaction products of a radiolabeled 39mer RNA. Identical samples can be loaded at different times on the same gel to resolve different regions of longer RNA.

(11) Product bands are directly visualized by exposing the gel to a phosphor screen. *Note: The exposure can take several hours to overnight, depending on the level of radioactivity of the bands in the gel.*

(12) Quantify the full-length RNA and cleavage product bands using a volume count method. (For a more qualitative evaluation, a line scan method can be used.) At every Tb^{3+} concentration, calculate a normalized extent of cleavage (Π) by substituting the peak intensities in the equation:

$$\Pi = \frac{\left(\dfrac{\text{band intensity at nucleotide } x}{\sum_i \text{band intensity at nucleotide } i}\right)_{y[Tb^{3+}]}}{\left(\dfrac{\text{band intensity at nucleotide } x}{\sum_i \text{band intensity at nucleotide } i}\right)_{0 \text{ mM }[Tb^{3+}]}}$$

where y is the terbium(III) concentration in a particular cleavage reaction and x the analyzed nucleotide position of the RNA. 0 mM [Tb^{3+}] signifies a control reaction incubated in the same fashion as the terbium(III) containing ones except that no terbium(III) is added. A Π value of ≥ 2 indicates significant cleavage over background degradation. *Note: By dividing the ratio of a*

Fig. 12.1. Terbium(III) footprinting of the *trans*-acting HDV ribozyme. (A) Synthetic HDV ribozyme construct D1. The ribozyme portion is shown in bold, and consists of two separate RNA strands A and B. 3′ Product (3′P) is shown outlined. The substrate variant S3 contains eight additional nucleotides (gray) 5′ of the cleavage site (arrow). To generate non-cleavable substrate analogs, the 2′-OH of the underlined nucleotide immediately 5′ of the cleavage site was modified to 2′-methoxy and the suffix "nc" added to their name. Dashed lines, tertiary structure hydrogen bonds of C75 and the ribose zipper of A77 and A78 in joiner J4/2. (B) Terbium(III)- and magnesium(II)-mediated footprint of the 5′-^{32}P-labeled HDV ribozyme strand A upon incubation with terbium(III) for 2 h in 40 mM Tris–HCl, pH 7.5, 11 mM MgCl$_2$ at 25 °C. From left to right as indicated: strand A fresh after radiolabeling; incubated in buffer without Tb^{3+}; incubated with excess strand B in buffer without Tb^{3+}; incubated in buffer without Tb^{3+}; incubated with excess strand B and non-cleavable substrate analog ncS3 in buffer without Tb^{3+}; RNase T1 digest; alkali (OH$^-$) ladder; footprint with increasing Tb^{3+} concentrations in the presence of excess strand B and ncS3; incubated in buffer without Tb^{3+}; incubated with excess strand B in buffer without Tb^{3+}; footprint with increasing Tb^{3+} concentrations in the presence of excess strand B and 3′ product (3′P). As the terbium(III) concentration increases, backbone scission becomes more intense. The 5′ and 3′ segments of P1.1 (boxed) footprint very differently in the precursor and product complexes. Far right, magnesium(II)-induced cleavage at pH 9.5 and 37 °C, for comparison; from left to right: precursor (ncS3) complex, control incubated at pH 7.5; precursor complex, footprinted at pH 9.5; product complex, control incubated at pH 7.5; product complex, footprinted at pH 9.5. Reprinted with permission from [6].

Fig. 12.1

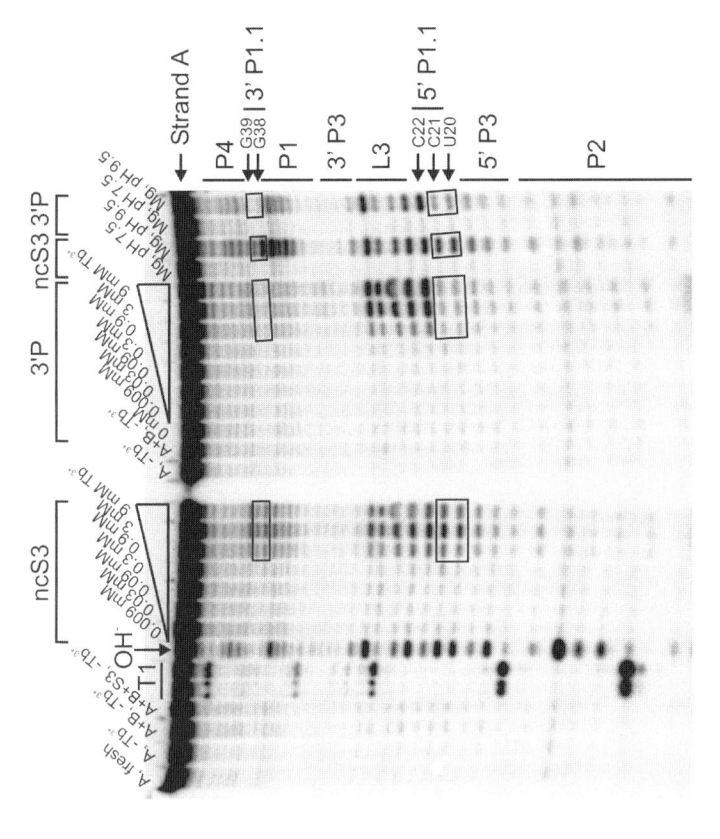

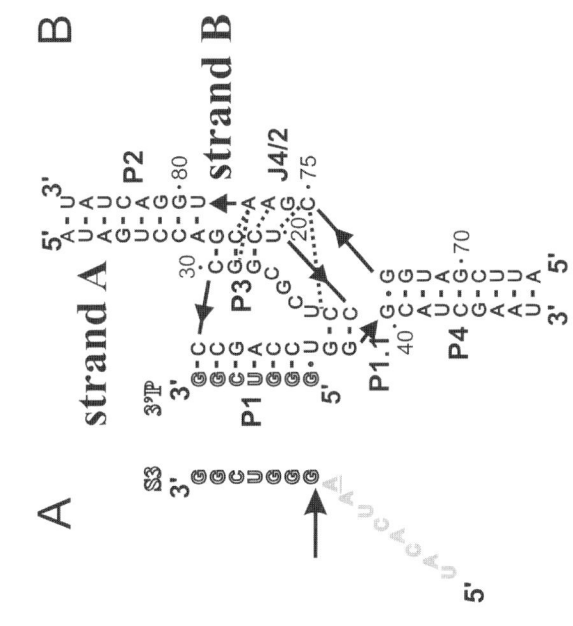

single band intensity over total RNA in the presence of terbium(III) by the ratio of a single band intensity over total RNA in the absence of terbium(III), one normalizes for the effect of non-specific background degradation.

12.3
Application Example

Terbium(III) has been successfully used on a number of RNAs as probe for high-affinity metal-binding sites and tertiary structure formation. For example, taking advantage of its luminescent, as well as RNA footprinting properties, terbium(III) has recently revealed subtle structural differences between the precursor and product forms of the hepatitis delta virus (HDV) ribozyme [6]. The HDV ribozyme is a unique RNA motif found in the human HDV, a satellite of the hepatitis B virus that leads to frequent progression towards liver cirrhosis in millions of patients worldwide [7]. There is a strong interest, both for medical and fundamental reasons, in understanding structure and function relationships in this catalytic RNA. We found that the terbium(III) mediated footprinting pattern of the 3′ product (3′P) complex of a *trans*-acting version of the HDV ribozyme (Fig. 12.1A), obtained in the presence of millimolar Tb^{3+} concentrations, is consistent with a post-cleavage crystal structure. In particular, protection is observed in all five Watson–Crick base-paired stems, P1 through P4 and P1.1, while the backbone of the L3 loop region and that of the J4/2 joining segment are strongly cut (Fig. 12.1) [6]. Cuts in the J4/2 joiner are particularly relevant since it encompasses the catalytic residue C75 and its neighboring G76, and the strong terbium(III) hits implicate it as a region of high negative charge density with high affinity for metal ions.

Strikingly, terbium(III) footprinting reveals the precursor (ncS3) structure as distinct; while P1, P2, P3 and P4 remain protected, both the 5′ and 3′ segments of the P1.1 stem (as well as U20, immediately upstream) are strongly hit, suggesting that this helix in the catalytic core is formed to a lesser extent than in the product complex. In addition, scission in J4/2 extends to A77 and A78, implying that the ribose zipper motif involving these nucleotides may not be fully formed in the precursor complex (Fig. 12.1B). These differences in extent of backbone scission in the precursor versus the 3′ product complexes show that a significant conformational

Fig. 12.2. Sites of backbone scission mediated by 3 mM terbium(III) in 40 mM Tris–HCl, pH 7.5, 11 mM $MgCl_2$ at 25 °C and superimposed onto two-dimensional representations of the precursor and product HDV ribozyme secondary structures. Only the catalytic core residues are explicitly shown. Relative scission intensities were calculated as described in Section 12.2 and are represented by the symbol code. Scission is located 3′ of the indicated nucleotides. Only the product structure is likely to fully form P1.1 and the ribose zipper of A77 and A78 in J4/2, as suggested by solid and dashed lines, respectively. Reprinted with permission from [6].

12.3 Application Example | **211**

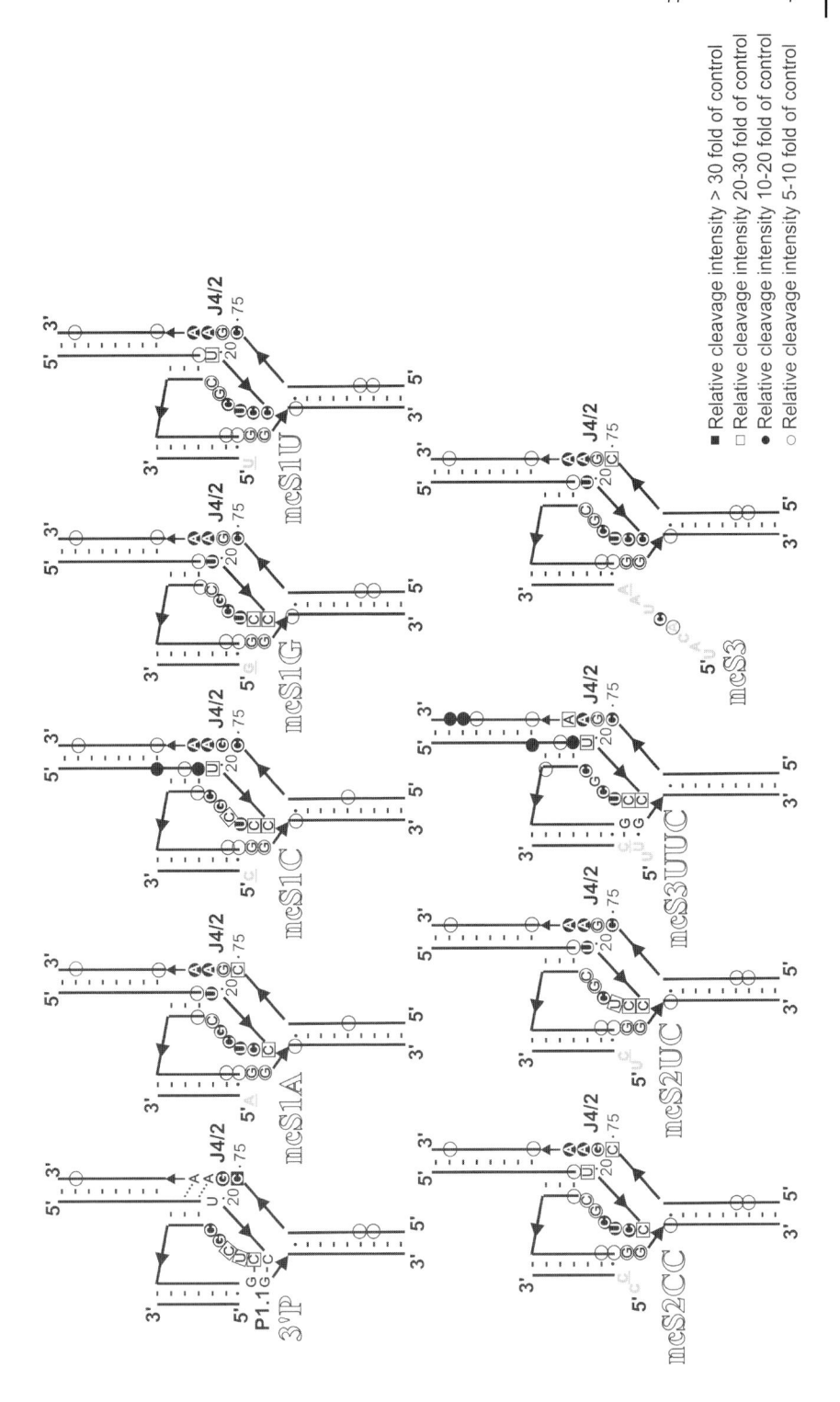

change occurs upon HDV ribozyme catalysis and 5′ product dissociation from the 3′ product [6].

While previous evidence from fluorescence resonance energy transfer [8], 2-aminopurine fluorescence quenching [9] and NMR spectroscopy [10, 11] had already hinted at structural differences between the precursor and 3′ product forms of the *trans*-acting HDV ribozyme, terbium(III)-mediated footprinting complements these techniques by providing specifics of these rearrangements at nucleotide resolution. Particularly intriguing are the differences in the catalytic core structure around C75 and P1.1 that may control access to the cleavage transition state and may therefore explain differences in the catalytic rate constants for substrates with different 5′ sequences (Fig. 12.2) [6]. In fact, the 5′ substrate sequence subtly modulates the terbium(III) footprinting pattern, but all the substrates consistently show strong cuts in the P1.1 stem and the ribose zipper motif of J4/2 (Fig. 12.2). This implies that in the precursor these tertiary structure interactions are not fully formed, in contrast to the 3′P complex. Interestingly, these subtle differences in the catalytic core structure of the various precursor complexes translate into significant changes in fluorescence resonance energy transfer (FRET) efficiency between fluorophores attached to the termini of P4 and P2 stems [6]. Taken together, these results indicate that the various precursor complexes differ in structure both locally (in the catalytic core) and globally (as measured by FRET), providing an explanation for the wide range of catalytic activities of substrates with varying 5′ extensions [6, 12].

Several other labs have also found terbium(III) to be a useful probe of high-affinity metal binding sites and tertiary structure in RNA. Musier-Forsyth and co-workers were able to show that terbium substitutes for several well known metal binding sites in human $tRNA^{Lys,3}$ and works as a sensitive probe of tertiary structure. At low Tb^{3+} concentrations, cleavage of $tRNA^{Lys,3}$ is restricted to nucleotides that were previously identified from X-ray crystallography as specific metal-binding pockets [13]. The use of higher Tb^{3+} concentrations resulted in an overall footprint of the L-shaped tRNA structure, showing increased cleavage in the loop regions (D and anticodon loops). Binding of HIV nucleocapsid protein could then be shown to result in the disruption of the tRNA's metal binding pockets and, at higher concentrations, to induce subtle structural changes in, for example, the tRNA acceptor–TψC stem minihelix [14]. Other RNAs that have similarly been studied by terbium(III)-mediated footprinting include the hammerhead [15], aminoacyl-transferase [16, 17], RNase P [18] and group II intron ribozymes [19].

12.4
Troubleshooting

Initial titration experiments will be necessary to obtain the optimal Tb^{3+} concentration(s) to use for structure probing of any individual RNA [typical terbium(III) and RNA concentrations for determining tertiary structural features are around 1–5 mM and 1 μM, respectively]. The trivalent terbium(III) has been shown to induce slight perturbations in the RNA structure [13], but careful titration will reveal the optimal terbium(III):RNA ratio needed for detecting unbiased secondary and tertiary structure features in a given RNA molecule.

To verify a high-affinity metal-ion-binding site, it is advisable to first decrease the Tb^{3+} concentration until a very narrow cleavage pattern is observed (typically at 10–100 μM Tb^{3+}) and then to perform a competition experiment with increasing concentrations of Mg^{2+}. The intensity (or fraction of RNA cleaved at a particular nucleotide position) should decrease as the Mg^{2+} concentration increases. Quantifying the intensities of cleaved bands at each nucleotide position directly relates to the structure of the RNA. It is critical to keep the extent of total cleavage lower than 20% of the uncleaved or full-length band. This will ensure that the RNA is undergoing a single cleavage event. Finally, it is important to keep in mind that, while terbium(III) footprinting will reveal many high-affinity metal ion binding sites, it may not reveal all. This is due to the fact that there is a steric requirement of Tb^{3+} to bind close to the 2'-hydroxyl group on the ribose for inducing backbone cleavage. This restraint is very unfavorable in A-type RNA helices and, therefore, strong metal sites that occur in RNA helical regions may be underestimated or go undetected by Tb^{3+} cleavage, as may binding sites that are highly specific for a particular metal ion [19].

References

1 Pyle, A. M. *J. Biol. Inorg. Chem.* **2002**, *7*, 679–690.
2 Ciesiolka, J., T. Marciniec, W. Krzyzosiak. *Eur. J. Biochem.* **1989**, *182*, 445–450.
3 Walter, N. G., N. Yang, J. M. Burke. *J. Mol. Biol.* **2000**, *298*, 539–555.
4 Baes, C. F., R. E. Mesmer. *The Hydrolysis of Cations*, Wiley Interscience, New York, **1976**.
5 Matsumura, K., M. Komiyama. *J. Biochem.* **1997**, *122*, 387–394.
6 Jeong, S., J. Sefcikova, R. A. Tinsley, D. Rueda, N. G. Walter. *Biochemistry* **2003**, *42*, 7727–7740.
7 Hadziyannis, S. J. *J. Gastroenterol. Hepatol.* **1997**, *12*, 289–298.
8 Pereira, M. J., D. A. Harris, D. Rueda, N. G. Walter. *Biochemistry* **2002**, *41*, 730–740.
9 Harris, D. A., D. Rueda, N. G. Walter. *Biochemistry* **2002**, *41*, 12051–12061.
10 Luptak, A., A. R. Ferre-D'Amare, K. Zhou, K. W. Zilm, J. A. Doudna. *J. Am. Chem. Soc.* **2001**, *123*, 8447–8452.
11 Tanaka, Y., M. Tagaya, T. Hori, T. Sakamoto, Y. Kurihara, M. Katahira, S. Uesugi. *Genes Cells* **2002**, *7*, 567–579.
12 Shih, I., M. D. Been. *EMBO J.* **2001**, *20*, 4884–4891.
13 Hargittai, M. R., K. Musier-Forsyth. *RNA* **2000**, *6*, 1672–1680.
14 Hargittai, M. R., A. T. Mangla, R. J. Gorelick, K. Musier-Forsyth. *J. Mol. Biol.* **2001**, *312*, 985–997.
15 Feig, A. L., M. Panek, W. D. Horrocks, Jr, O. C. Uhlenbeck. *Chem. Biol.* **1999**, *6*, 801–810.
16 Flynn-Charlebois, A., N. Lee, H. Suga. *Biochemistry* **2001**, *40*, 13623–13632.
17 Vaidya, A., H. Suga. *Biochemistry* **2001**, *40*, 7200–7210.
18 Kaye, N. M., N. H. Zahler, E. L. Christian, M. E. Harris. *J. Mol. Biol.* **2002**, *324*, 429–442.
19 Sigel, R. K., A. Vaidya, A. M. Pyle. *Nat. Struct. Biol.* **2000**, *7*, 1111–1116.

13
Pb^{2+}-induced Cleavage of RNA

Leif A. Kirsebom and Jerzy Ciesiolka

13.1
Introduction

Certain metal ions induce degradation of RNA in a non-oxidative manner, and in some RNA molecules this process is exceptionally efficient and specific. The best-known example, yeast tRNA^{Phe}, undergoes specific fragmentation in the D-loop in the presence of Pb^{2+} [1–3] and other ions, e.g. Eu^{3+} [4, 5], Mn^{2+} [6] and Mg^{2+} [7, 8]. Based on X-ray analysis of yeast tRNA^{Phe} crystals it was suggested that in order to promote cleavage, Pb^{2+} has to be positioned at an optimal distance from the 2′-OH that acts as the nucleophile [9, 10]. These findings gave rise to an experimental approach that uses Pb^{2+} and other ions to localize high-affinity metal ion binding sites as well as to probe the structure of RNA molecules.

Highly efficient and specific Pb^{2+}-induced cleavages are rather rarely observed. The majority of cleavages are weaker and usually comprises several consecutive phosphodiester bonds. Most information on the specificity of Pb^{2+}-induced RNA fragmentation has been obtained from studies on ribosomal 16S RNA [11] and 5S RNAs [12–14]. Cleavages occur preferentially in bulges, loops and other single-stranded RNA regions except those involved in stacking or other higher-order interactions. Double-stranded RNA segments are essentially resistant to breakage. Cleavages are also observed in paired regions destabilized by the presence of non-canonical interactions, bulges or other structural distortions. In general, it seems that flexibility of the polynucleotide chain determines its sensitivity to Pb^{2+}-induced cleavage [11–15].

It has been suggested [16] that the mechanism proposed for the specific, Pb^{2+}-induced fragmentation of yeast tRNA^{Phe} [9, 10, 17] might account for all types of cleavage induced by metal ions. The simplified mechanism shown in Fig. 13.1 is helpful for understanding the relation between RNA structure and sensitivity of a particular RNA region to cleavage.

In the first step, the ionized metal ion hydrate acts as a Brönsted base and abstracts a proton from the 2′-OH group of the ribose. Subsequently, the activated anionic 2′-O$^-$ attacks the phosphorus atom and a penta-coordinated intermediate is formed. The phosphodiester bond is cleaved generating 2′,3′-cyclic phosphate

Handbook of RNA Biochemistry: Student Edition. Edited by R. K. Hartmann, A. Bindereif, A. Schön, E. Westhof
Copyright © 2009 WILEY-VCH Verlag GmbH & Co. KGaA, Weinheim
ISBN: 978-3-527-32534-4

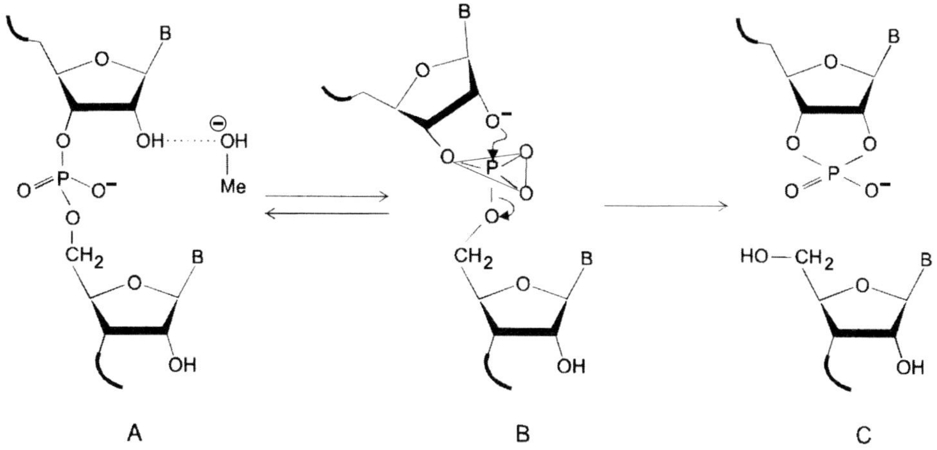

Fig. 13.1. Mechanism of metal ion-induced cleavage of RNA (see text for details).

and 5′-hydroxyl groups as cleavage products. However, based on the discussion about the role of metal ions in, for example, the hammerhead cleavage reaction [18, 19], one has also to consider the possibility that the metal ion acts as a Lewis acid by accepting electrons from the 2′-oxygen, thereby facilitating a nucleophilic attack on the phosphorus atom. Irrespective of mechanism, metal ion interaction with the 2′-OH results in a nucleophilic attack on the phosphorus atom and subsequent cleavage of the phosphodiester bond. Experimentally, the data suggests an inverse correlation between the pK_a values for different metal ion hydrates and cleavage rates: Pb^{2+} with a pK_a of 7.2 induces cleavage more efficiently than, for example, Mg^{2+} ($pK_a = 11.4$). This would be in keeping with the suggestions that the metal ion either acts as a Lewis acid or Brönsted base. For metal ions with higher pK_a values, such as Eu^{3+}, Zn^{2+}, Mn^{2+}, Mg^{2+} and Ca^{2+} ($pK_a = 8.5, 9.6, 10.6, 11.4$ and 12.6, respectively), the reaction pH, time or temperature have to be increased and/or, for example, ethanol has to be added to detect substantial cleavage.

The cleavage efficiency of a particular phosphodiester bond in an RNA molecule depends on: (1) proper localization of the metal ion hydrate facilitating deprotonation of the 2′-OH group (Fig. 13.1, transition A to B), and (2) sufficient conformational flexibility of the analyzed region allowing formation of the penta-coordinated intermediate/transition state and subsequent breaking of the phosphodiester chain (transition B to C). Optimal distance and correct orientation of the bound metal ion hydrate seems to be of primary importance when RNAs undergo efficient and highly specific fragmentation. The cleavage at these sites occurs at relatively low concentrations of Pb^{2+} (below 0.1 mM) – conditions under which breakage of other phosphodiester bonds takes place only at significantly reduced rates. Cleavages with lower efficiencies are most likely induced by ions acting from the solution, from weak binding sites and/or from sites at which the Pb^{2+} ion(s) is positioned suboptimally. Moreover, metal hydrates interact equally well with all accessible 2′-hydroxyl groups. Thus, differences in rigidity/flexibility of the phosphates, hindering or facilitating conformational transitions necessary for the reaction to occur, influence cleavage efficiency at individual phosphodiester bonds [16]. The contribution of rigidity/flexibility to the cleavage reaction is difficult to assess. However, the value of the potential rate enhancement derived from constraining a free RNA linkage to an optimal orientation for nucleophilic attack has recently been estimated not to be greater than 50- to 100-fold [20, 21].

13.2
Pb^{2+}-induced Cleavage to Probe Metal Ion Binding Sites, RNA Structure and RNA–Ligand Interactions

The Pb^{2+} cleavage approach has been used in structural analysis of several RNAs and RNA complexes in various ways as summarized in Table 13.1. The information in the table can be classified into three groups: (1) high-affinity metal ion binding sites, (2) RNA structure and (3) RNA–ligand interactions. Figure 13.2 also shows Pb^{2+}-induced cleavage of RNase P RNA in the presence of various divalent metal ions as a typical example.

13.2.1
Probing High-affinity Metal Ion Binding Sites

A strong, highly specific metal ion-induced cleavage suggests the presence of a tight metal ion binding site in the RNA. Cleavage occurring in a particular RNA region does not implicate, however, the direct involvement of that region in coordination of the metal ion. Also, tightly bound metal ions may not induce cleavage at all due to unfavorable distance constraints and/or high rigidity of the polynucleotide chain. For instance, in yeast tRNA^{Phe} a Pb^{2+} ion induces cleavage in the D-loop, but is bound primarily in the TΨC-loop, while the ion positioned in the anticodon loop does not induce specific cleavage [9, 10, 17]. Furthermore, a metal ion binding pocket can usually accommodate different ions, thereby acting as a "general" metal ion binding site. To probe for a "general" metal ion binding site the following two experimental approaches can be and have been used (see Table 13.1 and Fig. 13.2).

The first approach takes advantage of the fact that metal ion-induced cleavage is suppressed if the reaction is performed in the presence of other ions competing for a common metal ion binding site. Thus, addition of metal ions, such as Mg^{2+}, Mn^{2+} or Ca^{2+}, results in suppression of Pb^{2+}-induced cleavage. Quantitative analysis of the inhibition data can also give the K_D value for binding of Mg^{2+} and information about the relative binding affinity of different metal ions for metal ion binding sites in RNA [22, 23].

The second approach relies on the observation that Pb^{2+}, Mg^{2+}, Mn^{2+} and Eu^{3+} ions, bound in the D–TΨC region of yeast tRNA^{Phe}, induce strong cleavage at the same site (and/or at neighboring sites) in the RNA chain. Thus, it seems likely that

Tab. 13.1. Examples of structural analysis of RNAs and RNA complexes by means of the metal ion-induced cleavage approach[1]

RNA or RNA complex	Structural probe	Type of analysis	References
In vitro selected RNAs, aptamers and model oligonucleotides	Pb^{2+}	ion binding sites RNA structure	24, 41, 54, 70 16, 30, 41, 66, 71–73
tRNAs and mutants, *in vitro* transcripts, and fragments thereof	Pb^{2+} various Me^{2+}	ion binding sites RNA structure ion binding sites RNA structure	2, 3, 17, 22, 37, 50, 74 3, 53, 75, 76 5, 6, 8, 42–45, 77, 78 42–45
HDV ribozyme	various Me^{2+} Pb^{2+}	ion binding sites RNA structure	38, 46, 47 47
4.5S RNA	Pb^{2+}	RNA structure	55
5S rRNA	Pb^{2+}	RNA structure	12–14, 27
U1 snRNA	Pb^{2+}	RNA structure	79
RNase P RNA	Pb^{2+} various Me^{2+}	ion binding sites ion binding sites	28, 29, 51, 52, 59 23, 25, 35
Group I and II intron RNAs	Pb^{2+} various Me^{2+}	ion binding sites ion binding sites	39 26, 48, 49
10Sa RNA (tmRNA)	Pb^{2+}	RNA structure	80
mRNA fragments with trinucleotide repeats	Pb^{2+}	RNA structure	56–58
TfR mRNA fragment	Pb^{2+}	RNA structure	60
SECIS mRNA fragment	Pb^{2+}	RNA structure	81, 82
BRCA1 mRNA fragment	Pb^{2+}	RNA structure	83
CaMV 35S RNA leader	Pb^{2+}	RNA structure	84
16S rRNA fragment in 30S subunit and 70S ribosome	Pb^{2+}	RNA structure RNA–protein interaction RNA–RNA interaction	11
16S and 23S rRNA in 70S ribosome	Pb^{2+} various Me^{2+}	ion binding sites ion binding sites	40 85
RNA aptamer–citrulline complex	Pb^{2+}	RNA–amino acid interaction	65
RNA aptamer–antibiotic complex	Pb^{2+}	RNA–antibiotic interaction	32, 66, 67
HDV ribozyme-antibiotic complex	Pb^{2+}	RNA–antibiotic interaction	38
tRNA–neomycin complex	Pb^{2+}	RNA–antibiotic interaction	31, 69
Phe-tRNA^{Phe}-EF-Tu:GTP complex	Pb^{2+}	RNA–protein interaction	61
4.5S RNA–P48 protein complex	Pb^{2+}	RNA–protein interaction	55
5S rRNA–L18 protein complex	Pb^{2+}	RNA–protein interaction	13
RNase P RNA–neomycin complex	Pb^{2+}	RNA–antibiotic interaction	68
RNase P RNA–tRNA complex	Pb^{2+}	RNA–RNA interaction	52, 63, 64
TfR mRNA fragment–IRP1 complex	Pb^{2+}	RNA–protein interaction	60
3' end of HEV RNA–viral polymerase complex	Pb^{2+}	RNA–protein interaction	62

[1] Reproduced and modified from Table 1 in [36].

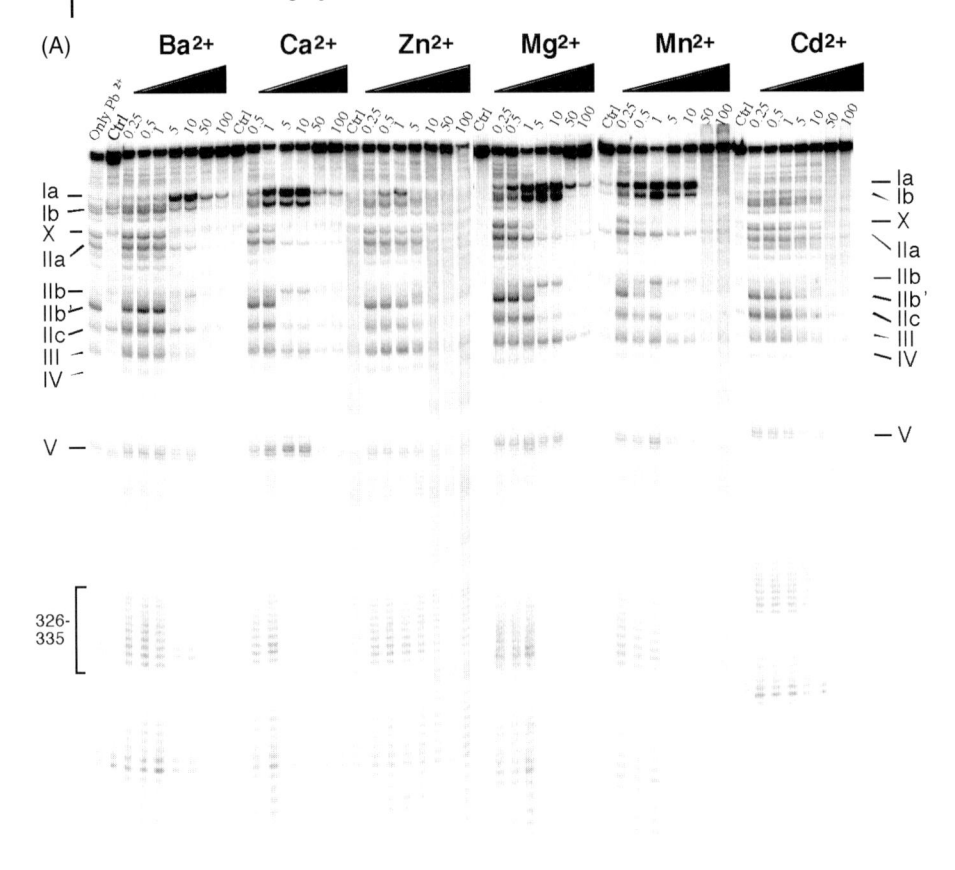

Fig. 13.2. Pb^{2+}-induced cleavage patterns of E. coli RNase P RNA. (A) Pb^{2+}-induced cleavage pattern in the presence of increasing concentrations of Ba^{2+}, Ca^{2+}, Zn^{2+}, Mg^{2+}, Mn^{2+} and Cd^{2+}. Cleavage was performed at specified concentrations and 37 °C as outlined. Lanes: Only Pb^{2+}, incubation in the presence of only 0.5 mM Pb^{2+}; Ctrl, incubation in the absence of Pb^{2+}, but in the presence of Ba^{2+}, Ca^{2+}, Zn^{2+}, Mg^{2+}, Mn^{2+} or Cd^{2+}, 10 mM (final concentration), as indicated; the band denoted X was only observed in the presence of Pb^{2+} and at low concentrations of the other metal ions except Cd^{2+} (see also [23]); reprinted with permission of *Nucleic Acids Research*. (B) Secondary structure of E. coli RNase P RNA; roman numerals refer to the sites of Pb^{2+} cleavage shown in (A). Roman numerals in italic refer to sites where Mg^{2+}-induced cleavage has also been observed [35]. In (A) it is apparent that increasing the concentration of different divalent metal ions results in suppression of Pb^{2+}-induced cleavage at specific sites, although to different degrees (e.g. compare the effects of different metal ions on the cleavage site IIa). This suggests that these metal ions and Pb^{2+} bind to at least overlapping sites (see text for details). Moreover, these data indicate that different divalent metal ions bind with different affinities to RNase P RNA as well as that the conformation of E. coli RNase P RNA is very similar in the presence of Ba^{2+}, Ca^{2+}, Mg^{2+} and Mn^{2+}, while it is changed compared to the Mg^{2+}-induced conformation in the presence of others, e.g. Cd^{2+}. For further details, see Brännvall et al. [23]; region 326–335 represents an example of a flexible single-stranded region with cleavage at several successive phosphodiester bonds [52]. In the case of cleavage sites Ia and Ib, the second divalent metal ion, such as Ba^{2+}, Ca^{2+}, Mg^{2+} and Mn^{2+}, enhances Pb^{2+}-induced cleavage at lower concentrations due to supporting RNA structure formation, but displaces Pb^{2+} at these sites at higher concentrations.

E. coli RNase P RNA

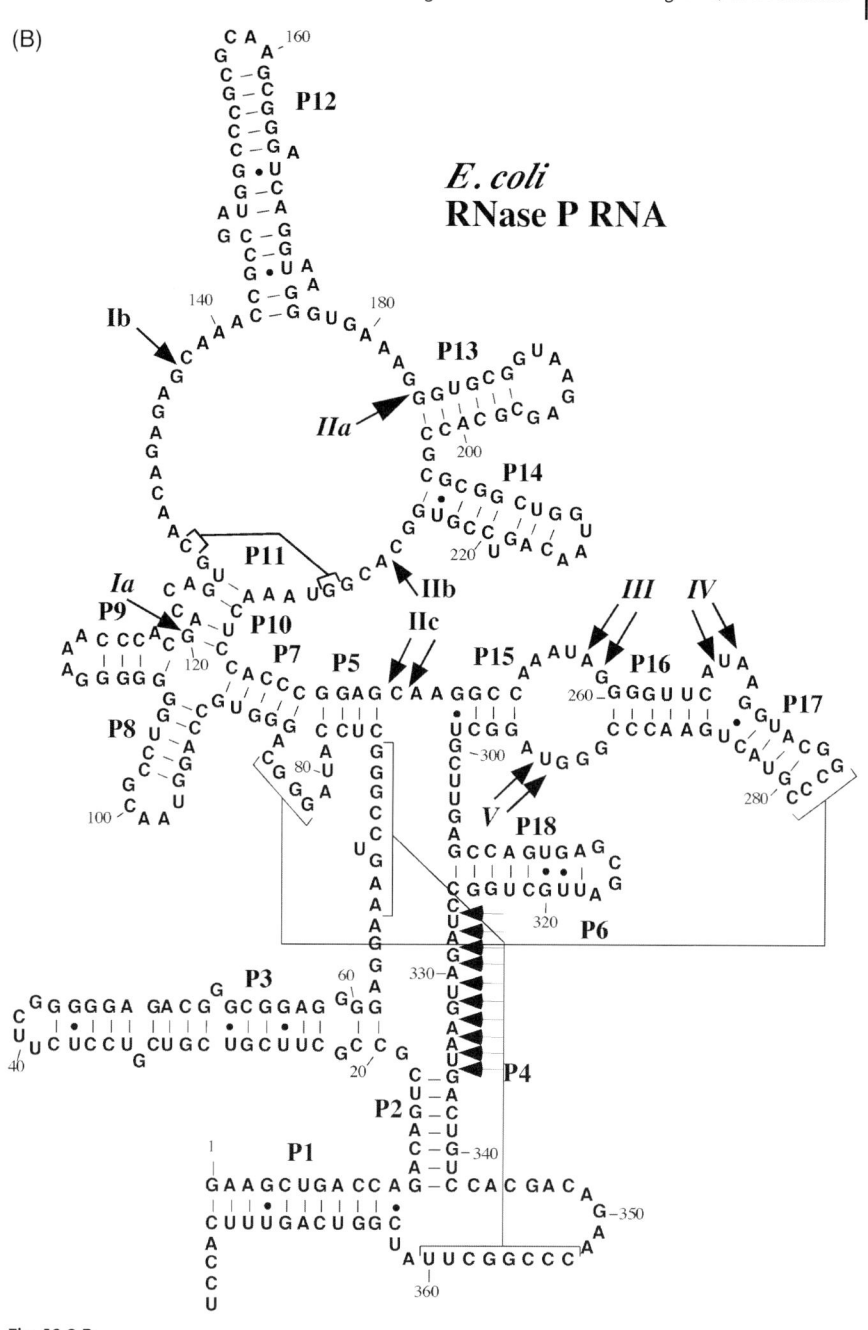

Fig. 13.2.B

different metal ions occupying at least overlapping sites can also induce specific cleavage in the same RNA region in other RNA molecules. However, cleavage induced by different metal ions does not necessarily occur at exactly the same site. Rather, the cleavage sites usually differ by 1 or 2 nt. This is rationalized considering the different coordination preferences and sizes of various metal ions, resulting in a slightly different arrangement of their hydrates in metal ion binding pockets. Note that the cleavage reactions are performed at conditions where the different pK_a values of metal ion hydrates have been taken into account (discussed above, see also below). In addition to Pb^{2+}, typical metal ions used in these experiments are Mg^{2+}, Mn^{2+}, Ca^{2+} and Eu^{3+}.

An additional approach using Tb^{3+}, which has the same coordination geometry as Mg^{2+}, has been used to probe for metal ion binding sites in RNA. Cleavage of RNA with Tb^{3+} gives different cleavage patterns compared to cleavage with, for example, Pb^{2+} [24–26]. Thus, Tb^{3+}-induced cleavage can be used in combination with cleavage induced by other metal ions and thereby more information concerning metal ion binding to RNA can be obtained.

13.2.2
Pb^{2+}-induced Cleavage and RNA Structure

Pb^{2+}-induced cleavage of several RNA molecules has been studied and the cleavage patterns have been used in analysis of RNA structures (see Table 13.1). Moreover, metal ion-induced cleavage allows identification of similarities and differences in related RNA molecules in the regions involved in metal ion binding. However, note that Pb^{2+} patterns do not always correspond precisely to RNA secondary structure models. Experimental results are most consistent with cleavage occurring preferentially in "flexible regions" of an RNA polynucleotide chain. Taking into account that our knowledge of RNA conformational dynamics is still insufficient, the term "flexible regions of RNA" should be used cautiously in the interpretation of experimental data.

Recently, Pb^{2+}-promoted cleavage of several well-defined RNA secondary structure motifs, such as bulges, hairpin loops and single-stranded RNA, has been characterized [16]. These studies show that the cleavage patterns of single nucleotide bulge regions depend on the structural context provided by adjacent base pairs. In general, a pyrimidine flanking the bulged nucleotide, particularly at its 5′ side, facilitates cleavage, while a purine makes the bulge more resistant to cleavage. This effect seems to correlate with the ability of the bulge to form stacking interactions with its neighbors. Cleavage of 2- and 3-nt bulges depends only slightly on their nucleotide composition. In case of terminal loops, cleavage usually increases with the loop size and strongly depends on its nucleotide composition. Particularly resistant are stable tetraloops, most likely due to their high conformational rigidity. Most single-stranded RNA regions are highly susceptible to Pb^{2+}-induced cleavage. However, clusters of G residues and, in most cases, also phosphodiester bonds at the junction of paired and unpaired RNA regions are more resistant. This can be attributed to extensive stacking interactions and increased conformational rigidity

[16]. For some RNAs, however, efficient Pb^{2+}-induced cleavage at the junction between unpaired and double-stranded regions can be seen. It might be that the enhanced reactivity in those cases results from increased "breathing" of the base pair at the junction between single- and double-stranded RNA.

The Pb^{2+} cleavage method is very sensitive for the detection of conformational changes in RNA molecules and useful in determining alternative hairpin structures formed by transcripts composed of trinucleotide repeats (Table 13.1; see also Fig. 13.2 where it is shown that the Pb^{2+}-induced cleavage pattern changes with increasing concentration of other divalent metal ions indicating differences in folding). Several studies have further demonstrated that cleavage of the same structural motifs present in different RNAs results in essentially identical patterns. This raises the interesting possibility to use the Pb^{2+}-induced cleavage approach to identify certain RNA structural motifs in RNA molecules of unknown structure ([16, 27]; Ciesiolka et al., unpublished results).

Lastly, Pb^{2+}-induced cleavage in combination with genetics, i.e. introduction of point mutations, has been used to provide support for the existence of long-range interactions in RNA [3, 50, 51, 59]. Note that point mutations may either result in increased/decreased cleavage at specific positions or in the appearance of cleavage at new positions [28, 29].

13.2.3
Pb^{2+}-induced Cleavage to Study RNA–Ligand Interactions

Remarkable reduction of Pb^{2+}-induced cleavage intensities has been observed upon the formation of RNA complexes with proteins, other RNAs or low-molecular-weight ligands – amino acids, antibiotics and other divalent metal ions (see Table 13.1 and above; also see e.g. [23, 30]).

In RNA–protein complexes, the shielding effect of a bound protein is most likely responsible for changes in cleavage intensities. However, it is not excluded that the RNA changes its conformation due to interaction with protein(s) such that the positioning of Pb^{2+} is altered or that the ion is displaced. In the case of RNA–aminoglycoside interaction, structural studies have provided evidence for displacement of Pb^{2+} as a result of aminoglycoside binding [31]. Furthermore, there are several examples of enhanced Pb^{2+} cleavage upon complex formation (for references, see Table 13.1). Here, moderately enhanced cleavage may suggest increased flexibility of the analyzed RNA regions. The appearance of a very strong cleavage may indicate formation of a new strong metal ion binding site or that a previously inactive metal ion has been repositioned in such a way that efficient metal ion-induced cleavage becomes feasible. In both cases the presence of a tightly bound ion needs to be verified by other methods, since it is conceivable that strand breakage may also occur due to transient, low affinity binding of a metal ion. Needless to say, this mapping method is obviously restricted to RNA regions susceptible to Pb^{2+}-induced cleavage, mainly bulges, loops and other single-stranded RNA stretches.

In some cases, changes in the Pb^{2+} cleavage pattern due to ligand binding may include unexpectedly large regions of the polynucleotide chain [30, 32]. This is probably caused by the loss of flexibility, i.e. formation of a more rigid conformation of a large RNA fragment, which is unstructured in the absence of the ligand.

13.3
Protocols for Metal Ion-induced Cleavage of RNA

The information that can be extracted using the metal ion-induced cleavage approach suggests that a divalent metal ion(s) is positioned close to the metal ion-induced cleavage site, but this does not give any information about how the metal ion is coordinated. However, since the 2′-hydroxyl immediately 5′ of the scissile bond is actively involved in the chemistry of cleavage some structural constraints for the positioning of the divalent metal ion(s) can be derived. These aspects have to be kept in mind when interpreting the data. Here, we describe three protocols used to cleave RNA with divalent metal ions, *in vitro* using Pb^{2+} and Mg^{2+}, and *in vivo* using Pb^{2+} [33]. We will use RNase P RNA, the catalytic subunit of the endoribonuclease P, as an example of an RNA that has been studied using these protocols (see e.g. [34]; unpublished data). Further protocols on Pb^{2+}-induced cleavage of RNA *in vitro* and *in vivo* can be found in Chapter 10.

Protocol 1: Pb^{2+}-induced Cleavage of RNA

(1) The RNA is ^{32}P-labeled at the 3′ end with [5′-^{32}P]pCp or at the 5′ end with [γ-^{32}P]ATP using standard procedures.
(2) The RNA is purified on a denaturing polyacrylamide gel containing 7 M urea in TBE buffer (90 mM Tris–borate, pH 8.5, 2.5 mM EDTA) and eluted (see Chapters 1 and 3). The RNA is renatured by incubation for 5 min at 55 °C in water or buffer of choice.
(3) Pb^{2+}-induced cleavage of, for example, RNase P RNA. Typically, approximately 20 000–30 000 Cerenkov c.p.m. of labeled RNA is mixed with around 2.5 pmol of unlabeled RNA and pre-incubated in 50 mM Tris–HCl, pH 7.5, 100 mM NH_4Cl and 10 mM $MgCl_2$ for 10 min at 37 °C. Note when analyzing metal ion binding and/or structure of an RNA, you have to adjust the conditions such that the RNA adopts a conformation relevant to what you would like to investigate.
(4) Cleavage is initiated by the addition of freshly prepared $Pb(OAc)_2$ to a final concentration of 0.5 mM. Depending on the nature of the experiment you can use other concentrations of $Pb(OAc)_2$, but usually not higher than 2 mM (see also below). Chloride buffer salts can be used, but for higher concentrations of Pb^{2+} (above 2 mM) acetate instead of chloride salts should be used to avoid precipitation of Pb^{2+} ions. The final volume of the reaction is 10 μl.
(5) The reaction is terminated after 10–15 min by the addition of 2 volumes of stop solution (9 M urea, 25 mM EDTA, 0.1% bromophenol blue). The time

of incubation in the presence of Pb^{2+} has to be adjusted experimentally, and depends on the nature of the RNA, cleavage conditions and the question you address.
(6) The cleavage products are separated on denaturing gels, where the percentage depends on the size of the RNA under study (6–12% polyacrylamide gels are generally used).
(7) The Pb^{2+}-induced cleavage sites are mapped using size markers and $[OH^-]$ ladders (see Chapter 10). It is also possible to map cleavage sites by primer extension analysis (see Protocol 4) using primers complementary to specific positions in the RNA. In the case of RNase P RNA, we use 15- to 20-nt primers.

We emphasize that an increase in the concentration of Mg^{2+} (or some other divalent metal ion such as Mn^{2+}) results in suppression of Pb^{2+}-induced cleavage as illustrated for cleavage of RNase P RNA (see Fig. 13.2A), suggesting that Mg^{2+} and Pb^{2+} bind, if not to the same, at least to overlapping sites (see e.g. [23]). In combination with the use of genetics (i.e. by using in our example RNase P RNA variants) or by studying cleavage of the RNase P RNA–substrate complex, it is also possible to use the Pb^{2+}-induced cleavage to probe for structural changes in RNase P RNA. Note that the formation of RNase P RNA–substrate complexes requires a higher concentration of Mg^{2+} (20 mM or above). Therefore, an increased concentration of Pb^{2+} is needed to detect cleavage. This might also apply when other RNA molecules are studied and hence the conditions have to be adjusted accordingly.

Protocol 2: Cleavage of RNA in the presence of Mg^{2+}
RNase P RNA is also cleaved by other divalent metal ions such as Mg^{2+}, first described by Kazakov and Altman [35]. However, Mg^{2+}-induced cleavage of RNase P RNA is less efficient compared to Pb^{2+}-induced cleavage (see above); in order to detect cleavage, the reaction has to be performed at a higher pH and in the presence of 10% ethanol. Hence, Steps 3 and 4 of Protocol 1 are modified.

(3) In our studies of RNase P RNA we have used the following conditions: 50 mM CHES buffer, pH 9.5, 100 mM NH_4Cl, 10 mM $MgCl_2$ (higher concentrations of Mg^{2+} can be used) and 10% ethanol [35].
(4) The reaction is incubated at 37 °C for 6 h.
(5) The reaction is terminated and the cleavage products are separated and characterized as described in Protocol 1.

Protocol 3: Pb^{2+}-induced cleavage of RNA *in vivo*
Here the protocol is adapted to study cleavage in growing bacteria, e.g. *Escherichia coli* [33].

(1) Typically, *E. coli* cells are grown in Luria-Broth media (LB) overnight at 37 °C (or temperature of choice).

(2) The culture is diluted 400-fold in LB and allowed to grow to an $OD_{600} \sim 0.5$ (mid-log phase).
(3) Freshly prepared $Pb(OAc)_2$ solutions of appropriate concentrations are prepared by diluting pre-heated (37 °C) 4 × LB media [3 volumes of $Pb(OAc)_2$ and 1 volume of 4 × LB]. Hence, to give a final (theoretical) concentration of 100 mM in the *E. coli* cell suspension (see Step 4), that typically has been used, the freshly prepared $Pb(OAc)_2$ solution should be 467 mM. For reproducibility mixing has to be performed rapidly. Note that when LB and $Pb(OAc)_2$ are mixed, there is always some precipitation, and hence the final concentration of $Pb(OAc)_2$ in solution is lower. Moreover, replacing LB with minimal media results in substantial precipitation and poor RNA cleavage.
(4) Then 1 volume of the $Pb(OAc)_2$/LB solution ($[Pb(OAc)_2] = 350$ mM) is added to 2.5 volumes of cell culture ($OD_{600} \sim 0.5$, see above) and incubated for 7 min at 37 °C under vigorous shaking (the total volume will be 3.5 volumes = V_{tot}). The final concentration of $Pb(OAc)_2 = 100$ mM (not taking the precipitation into account).
(5) The reaction is stopped by adding excess EDTA to a final concentration of 1.5 times the $Pb(OAc)_2$ concentration, typically 150 mM final concentration of EDTA.
(6) The solution is put on ice.
(7) The cells are harvested by centrifugation and the cell pellet is snap frozen in liquid nitrogen and stored at -70 °C.
(8) The cell pellet is re-suspended in a volume of $0.5 \times V_{tot}$ pre-heated (65 °C) lysis buffer (100 mM Tris–HCl, pH 7.5, 40 mM EDTA, 200 mM NaCl, 0.5% SDS w/v) and incubated at 65 °C for 3–5 min.
(9) This is followed by addition of pre-heated (65 °C) phenol solution (volume: $0.5 \times V_{tot}$). To prepare the phenol solution, 1 volume phenol is saturated with 1 volume 10 mM Tris–HCl, pH 8.0, 10 mM EDTA. The RNA is extracted at 65 °C.
(10) The phenol extraction is followed by chloroform/isoamylalcohol (24:1) extraction at room temperature and ethanol precipitation in the presence of 0.3 M sodium acetate, pH 6.0.
(11) The RNA is dissolved in DNase buffer (40 mM Tris–HCl, pH 7.9, 100 mM NaCl, 60 mM $MgCl_2$, 1 mM $CaCl_2$) and 80 U of DNase I (RNase-free) are added. This mixture is incubated for 15 min at 37 °C followed by standard phenol extraction and ethanol precipitation in the presence of 0.3 M sodium acetate, pH 6.0.
(12) The RNA is stored at -70 °C.
(13) The Pb^{2+}-induced cleavage sites are mapped by primer extension analysis using appropriate oligodeoxyribonucleotides as primers (see Protocol 4).

Note that RNA samples prepared from untreated cells (no Pb^{2+} added) have to be analyzed in parallel to RNA prepared from Pb^{2+}-treated cells. Hence, $Pb(OAc)_2$ is omitted in Step 4 by replacing $Pb(OAc)_2$ with RNase-free water and the RNA is prepared following the same procedure as outlined above. This is an essential control to be able to discriminate signals (stops) in the primer extension analysis that are related to Pb^{2+}-induced cleavage from those that are due to pausing (pre-termination) in the reverse transcription reaction.

Protocol 4: Primer Extension

(1) An appropriate $5'$-^{32}P-end-labeled oligodeoxyribonucleotide is mixed with 10 µg of total cellular RNA from step 11 (Protocol 3) in RNase-free water and incubated for 1 min at 90 °C.
(2) The mixture is put on ice for 1 min followed by warming at 20 °C for 5 min.
(3) The actual primer extension is performed in a total volume of 15 µl in 50 mM Tris–HCl, pH 8.5, 6 mM $MgCl_2$, 40 mM KCl and dNTPs (1.0 mM each) and 200 U of reverse transcriptase (e.g. Superscript II; Life Technologies). The primer extension mixture is incubated at 45 °C for 30 min.
(4) The reaction is terminated by the addition of 20 µl stop buffer (50 mM Tris–HCl, pH 7.5, 0.1% SDS) and 3.5 µl of 3 M KOH. This solution is incubated for 3 min at 90 °C followed by incubation at 37 °C for 3 h.
(5) Add 6 µl of 3 M acetic acid and ethanol precipitate the cDNA in the presence of 0.3 M sodium acetate, pH 6.0. The products are separated on denaturing gels (see Step 6, Protocol 1).

13.4 Troubleshooting

13.4.1 No Pb^{2+}-induced Cleavage Detected

- Old solution of Pb^{2+}. The action is to prepare a new solution.
- Your crystalline $Pb(OAc)_2$ is old or has moistened. The action is to buy a new bottle of solid $Pb(OAc)_2$.
- Cleavage conditions are not optimized with respect to: time of incubation, concentration of other divalent metal ions, e.g. Mg^{2+}, concentration of $Pb(OAc)_2$. The action is to optimize the conditions: increase/decrease concentration of Mg^{2+} and/or Pb^{2+}, increase the time of incubation.
- Poor quality of RNA and/or the RNA solution contains metal ions that interfere with Pb^{2+}-induced cleavage or is contaminated with metal ion chelators, e.g. EDTA. The action is to prepare a new batch of RNA.
- The pH is too low. The action is to increase the pH.

13.4.2 Complete Degradation of the RNA

- Too high concentration of Pb^{2+}. The action is to decrease $[Pb^{2+}]$ and/or time of incubation.

- Time of incubation is too long. The action is to decrease the time and/or to decrease [Pb^{2+}].
- The pH of the reaction is too high. The action is to lower pH.
- Contamination of your solutions with RNase. The action is to change all solutions (from experience, the RNase-free water is usually the candidate that is most often contaminated).

Acknowledgments

We thank our colleagues and, in particular, Mr M. Lindell for critical reading of the section of the chapter concerning Pb^{2+}-induced cleavage *in vivo*. The ongoing research in the laboratories of L. A. K. and J. C. are supported by the Swedish Research Council (L. A. K.), the Wallenberg Foundation (L. A. K.), the Swedish Foundation for Strategic Research (L. A. K.) and the Polish Committee for Scientific Research grant no 6P04B 01720 (J. C.).

References

1 C. WERNER, B. KREBS, G. KEITH, G. DIRHEIMER, *Biochim. Biophys. Acta* **1976**, *432*, 161–175.
2 J. R. SAMPSON, F. X. SULLIVAN, L. S. BEHLEN, A. B. DIRENZO, O. C. UHLENBECK, *Cold Spring Harbor Symp. Quant. Biol.* **1987**, *52*, 267–275.
3 W. J. KRZYŻOSIAK, T. MARCINIEC, M. WIEWIÓROWSKI, P. ROMBY, J. EBEL, R. GIEGÉ, *Biochemistry* **1988**, *27*, 5771–5777.
4 B. F. RORDORF, D. R. KEARNS, *Biopolymers* **1976**, *15*, 1491–1504.
5 J. CIESIOŁKA, T. MARCINIEC, W. J. KRZYŻOSIAK, *Eur. J. Biochem.* **1989**, *182*, 445–450.
6 J. WRZESINSKI, D. MICHAŁOWSKI, J. CIESIOŁKA, W. J. KRZYŻOSIAK, *FEBS Lett.* **1995**, *374*, 62–68.
7 W. WINTERMEYER, G. ZACHAU, *Biochim. Biophys. Acta* **1973**, *299*, 82–90.
8 T. MARCINIEC, J. CIESIOŁKA, J. WRZESINSKI, W. J. KRZYŻOSIAK, *Acta Biochim. Polon.* **1989**, *36*, 115–122.
9 R. S. BROWN, B. E. HINGERTY, J. C. DEWAN, A. KLUG, *Nature* **1983**, *303*, 543–546.
10 J. R. RUBIN, M. SUNDARALINGAM, *J. Biomol. Struct. Dyn.* **1983**, *1*, 639–646.
11 P. GÓRNICKI, F. BAUDIN, P. ROMBY, M. WIEWIÓROWSKI, W. J. KRZYŻOSIAK, J. P. EBEL, C. EHRESMANN, B. EHRESMANN, *J. Biomol. Struct. Dyn.* **1989**, *6*, 971–984.
12 C. BRUNEL, P. ROMBY, E. WESTHOF, C. EHRESMANN, B. EHRESMANN, *J. Mol. Biol.* **1991**, *221*, 293–308.
13 J. CIESIOŁKA, S. LORENZ, V. A. ERDMANN, *Eur. J. Biochem.* **1992**, *204*, 575–581.
14 J. CIESIOŁKA, S. LORENZ, V. A. ERDMANN, *Eur. J. Biochem.* **1992**, *204*, 583–589.
15 L. JOVINE, S. DJORDJEVIC, D. RHODES, *J. Mol. Biol.* **2000**, *301*, 401–414.
16 J. CIESIOŁKA, D. MICHAŁOWSKI, J. WRZESINSKI, J. KRAJEWSKI, W. J. KRZYŻOSIAK, *J. Mol. Biol.* **1998**, *275*, 211–220.
17 R. S. BROWN, J. C. DEWAN, A. KLUG, *Biochemistry* **1985**, *24*, 4785–4801.
18 T. A. STEITZ, J. A. STEITZ, *Proc. Natl. Acad. Sci. USA* **1993**, *90*, 6498–6502.
19 B. W. PONTIUS, W. B. LOTT, P. H. VON HIPPEL, *Proc. Natl. Acad. Sci. USA* **1997**, *94*, 2290–2294.
20 S. MIKKOLA, U. KAUKINEN, H. LÖNNBERG, *Cell Biochem. Biophys.* **2001**, *34*, 95–119.

21 G. M. Emilsson, S. Nakamura, A. Roth, R. R. Breaker, *RNA* **2003**, *9*, 907–918.
22 D. Labuda, K. Nicoghosian, R. Cedergren, *J. Biol. Chem.* **1985**, *260*, 1103–1107.
23 M. Brännvall, N. E. Mikkelsen, L. A. Kirsebom, *Nucleic Acids Res.* **2001**, *29*, 1426–1432.
24 A. Flynn-Charlebois, N. Lee, H. Suga, *Biochemistry* **2001**, *40*, 13623–13632.
25 N. M. Kaye, N. H. Zahler, E. L. Christian, M. E. Harris, *J. Mol. Biol.* **2002**, *324*, 429–442.
26 R. K. Sigel, A. Vaidya, A. M. Pyle, *Nat. Struct. Biol.* **2000**, *7*, 1111–1116.
27 J. Ciesiołka, W. J. Krzyżosiak, *Biochem. Mol. Biol. Int.* **1996**, *39*, 319–328.
28 J. G. Mattsson, S. G. Svärd, L. A. Kirsebom, *J. Mol. Biol.* **1994**, *241*, 1–6.
29 S. G. Svärd, J. G. Mattsson, K. E. Johansson, L. A. Kirsebom, *Mol. Microbiol.* **1994**, *11*, 849–859.
30 J. Ciesiolka, M. Yarus, *RNA* **1996**, *2*, 785–793.
31 N. E. Mikkelsen, K. Johansson, A. Virtanen, L. A. Kirsebom, *Nat. Struct. Biol.* **2001**, *8*, 510–514.
32 S. T. Wallace, R. Schroeder, *RNA* **1998**, *4*, 112–123.
33 M. Lindell, P. Romby, E. G. Wagner, *RNA* **2002**, *8*, 534–541.
34 S. Altman, L. A. Kirsebom, RNase P, in: *The RNA World*, 2nd edn, R. F. Gesteland, T. R. Cech, J. F. Atkins (eds), Cold Spring Harbor Laboratory Press, Cold Spring Harbor, NY, **1999**.
35 S. Kazakov, S. Altman, *Proc. Natl. Acad. Sci. USA* **1991**, *88*, 9193–9197.
36 J. Ciesiołka, Metal ion-induced cleavages in probing of RNA structure, in: *RNA Biochemistry and Biotechnology*, J. Barciszewski, B. F. C. Clark (eds), Kluwer, Dordrecht, **1999**.
37 E. J. Maglott, G. D. Glick, *Nucleic Acids Res.* **1998**, *26*, 301–308.
38 J. Rogers, A. H. Chang, U. von Ahsen, R. Schroeder, J. Davies, *J. Mol. Biol.* **1996**, *259*, 916–925.
39 B. Streicher, U. von Ahsen, R. Schroeder, *Nucleic Acids Res.* **1993**, *21*, 311–317.
40 D. Winter, N. Polacek, I. Halama, B. Streicher, A. Barta, *Nucleic Acids Res.* **1997**, *25*, 1817–1824.
41 J. Ciesiolka, J. Gorski, M. Yarus, *RNA* **1995**, *1*, 538–550.
42 J. Ciesiołka, J. Wrzesinski, P. Górnicki, J. Podkowinski, W. J. Krzyżosiak, *Eur. J. Biochem.* **1989**, *186*, 71–77.
43 T. Marciniec, J. Ciesiołka, J. Wrzesinski, W. J. Krzyżosiak, *FEBS Lett.* **1989**, *243*, 293–298.
44 D. Michałowski, J. Wrzesinski, J. Ciesiołka, W. J. Krzyżosiak, *Biochimie* **1996**, *78*, 131–138.
45 D. Michałowski, J. Wrzesinski, W. J. Krzyżosiak, *Biochemistry* **1996**, *35*, 10727–10734.
46 D. A. Lafontaine, S. Ananvoranich, J. P. Perreault, *Nucleic Acids Res.* **1999**, *27*, 3236–3243.
47 M. Matysiak, J. Wrzesinski, J. Ciesiołka, *J. Mol. Biol.* **1999**, *291*, 282–294.
48 B. Streicher, E. Westhof, R. Schroeder, *EMBO J.* **1996**, *15*, 2556–2564.
49 M. Hertweck, M. W. Mueller, *Eur. J. Biochem.* **2001**, *268*, 4610–4620.
50 L. Behlen, J. R. Sampson, A. B. DiRenzo, O. C. Uhlenbeck, *Biochemistry* **1990**, *29*, 2515–2523.
51 K. Zito, A. Hüttenhofer, N. R. Pace, *Nucleic Acids Res.* **1993**, *21*, 5916–5920.
52 J. Ciesiolka, W.-D. Hardt, J. Schlegl, V. A. Erdmann, R. K. Hartmann, *Eur. J. Biochem.* **1994**, *219*, 49–56.
53 K. N. Nobles, C. S. Yarian, G. Liu, R. H. Guenther, P. F. Agris, *Nucleic Acids Res.* **2002**, *30*, 4751–4760.
54 T. Pan, O. C. Uhlenbeck, *Biochemistry* **1992**, *31*, 3887–3895.
55 G. Lentzen, H. Moine, C. H. Ehresmann, B. Ehresmann, W. Wintermeyer, *RNA* **1996**, *2*, 244–253.
56 M. Napierała, W. J. Krzyżosiak, *J. Biol. Chem.* **1997**, *272*, 31079–31085.
57 A. Jasinska, G. Michlewski, M. de Mezer, K. Sobczak, P. Kozłowski, M. Napierała, W. J. Krzyżosiak, *Nucleic Acids Res.* **2003**, *31*, 5463–5468.
58 K. Sobczak, M. de Mezer, G.

Michlewski, J. Krol, W. J. Krzyżosiak, *Nucleic Acids Res.* **2003**, *31*, 5469–5482.

59 A. Tallsjö, S. G. Svärd, J. Kufel, L. A. Kirsebom, *Nucleic Acids Res.* **1993**, *21*, 3927–3933.

60 J. Schlegl, V. Gegout, B. Schläger, M. W. Hentze, E. Westhof, C. H. Ehresmann, B. Ehresmann, P. Romby, *RNA* **1997**, *3*, 1159–1172.

61 D. E. Otzen, J. Barciszewski, B. F. C. Clark, *Biochem. Mol. Biol. Int.* **1993**, *31*, 95–103.

62 S. Agrawal, D. Gupta, S. K. Panda, *Virology* **2001**, *282*, 87–101.

63 W.-D. Hardt, J. Schlegl, V. A. Erdmann, R. K. Hartmann, *Biochemistry* **1993**, *32*, 13046–13053.

64 W.-D. Hardt, J. Schlegl, V. A. Erdmann, R. K. Hartmann, *J. Mol. Biol.* **1995**, *247*, 161–172.

65 P. Burgstaller, M. Kochoyan, M. Famulok, *Nucleic Acids Res.* **1995**, *23*, 4769–4776.

66 M. G. Wallis, B. Streicher, H. Wank, U. von Ahsen, E. Clodi, S. T. Wallace, M. Famulok, R. Schroeder, *Chem. Biol.* **1997**, *4*, 357–366.

67 C. Berens, A. Thain, R. Schroeder, *Bioorg. Med. Chem.* **2001**, *9*, 2549–2556.

68 N. E. Mikkelsen, M. Brännvall, A. Virtanen, L. A. Kirsebom, *Proc. Natl. Acad. Sci. USA* **1999**, *96*, 6155–6160.

69 S. R. Kirk, Y. Tor, *Bioorg. Med. Chem.* **1999**, *7*, 1979–1991.

70 M. Olejniczak, Z. Gdaniec, A. Fischer, T. Grabarkiewicz, L. Bielecki, R. W. Adamiak, *Nucleic Acids Res.* **2002**, *30*, 4241–4249.

71 Z. Szweykowska-Kulinska, J. Krajewski, K. Wypijewski, *Biochim. Biophys. Acta* **1995**, *1264*, 87–92.

72 I. Majerfeld, M. Yarus, *RNA* **1998**, *4*, 471–478.

73 M. Welch, I. Majerfeld, M. Yarus, *Biochemistry* **1998**, *36*, 6614–6623.

74 H. Y. Deng, J. Termini, *Biochemistry* **1992**, *31*, 10518–10528.

75 C. Baron, E. Westhof, A. Böck, R. Giegé, *J. Mol. Biol.* **1993**, *231*, 247–292.

76 V. M. Perreau, G. Keith, W. M. Holmes, A. Przykorska, M. A. Santos, M. F. Tuite, *J. Mol. Biol.* **1999**, *293*, 1039–1053.

77 J. Ciesiołka, T. Marciniec, P. Dziedzic, W. J. Krzyżosiak, M. Wiewiórowski, in: *Biophosphates and their Analogues: Synthesis, Structure, Metabolism and Activity*, K. S. Bruzik, W. J. Stec (eds), Elsevier, Amsterdam, **1987**.

78 M. Matsuo, T. Yokogawa, K. Nishikawa, K. Watanabe, N. Okada, *J. Biol. Chem.* **1995**, *270*, 10097–10104.

79 E. Zietkiewicz, J. Ciesiołka, W. J. Krzyżosiak, R. Słomski, in: *Nuclear Structure and Function*, J. R. Harris, J. B. Zbarsky (eds), Plenum Press, New York, **1990**.

80 B. Felden, H. Himeno, A. Muto, J. P. McCutcheon, J. F. Atkins, R. F. Gesteland, *RNA* **1997**, *3*, 89–103.

81 R. Walczak, E. Westhof, P. Carbon, A. Krol, *RNA* **1996**, *2*, 367–379.

82 R. Walczak, P. Carbon, A. Krol, *RNA* **1998**, *4*, 74–84.

83 K. Sobczak, W. J. Krzyżosiak, *J. Biol. Chem.* **2002**, *277*, 17349–17358.

84 M. Hemmings-Mieszczak, G. Steger, T. Hohn, *J. Mol. Biol.* **1997**, *267*, 1075–1088.

85 S. Dorner, A. Barta, *Biol. Chem.* **1999**, *380*, 243–251.

14
In Vivo Determination of RNA Structure by Dimethylsulfate

Christina Waldsich and Renée Schroeder

14.1
Introduction

Considerable progress has been made in the past years in elucidating RNA structure, its folding pathways and the functional roles RNA molecules play in cellular processes. Despite the wealth of insights we have obtained about structure and folding, there is a significant drawback as our knowledge is so far predominantly based on, and limited to, biochemical and biophysical analyses of RNA molecules performed *in vitro* [1]. However, the *in vitro* folding conditions significantly contrast the intracellular environment. For example, it is well known that many catalytic RNAs, which usually function at non-physiological reaction conditions *in vitro*, associate with protein cofactors *in vivo*, which are thought to stabilize them [2]. It is therefore essential to extend our understanding of RNA structure and function by studying those molecules within cells [3].

The methodologies useful and suitable for studying RNA structure *in vivo* are limited. Structural probing with dimethylsulfate (DMS), which proved to be a powerful tool *in vitro* [4], has so far been the main chemical used to analyze RNA *in vivo*. DMS is a base-specific probe that modifies, in addition to the N^7 position of guanines, the Watson–Crick positions N^1 of adenines and N^3 of cytidines. The modified sites can be mapped by primer extension and subsequently compiled into a pattern profile of nucleotides protected from or accessible to DMS [4–6]. Notably, certain uridines and guanines are occasionally stabilized in an enol-tautomer due to a specific local environment, and are therefore reactive to DMS at their N^3 or N^1 positions. Also, it has to be kept in mind that naturally modified nucleotides like m^7G in rRNAs could occur in your RNA of interest. A protection from DMS modification can result from base pairing, but also from an interaction with a protein, while accessibility indicates that those residues (at least their N^1, N^3 or N^7 positions) do not participate in any intra- or intermolecular contacts. Thus, DMS modification can be employed to determine RNA structure and folding as well as to study RNA–protein interactions and their associated conformational changes in living cells. As DMS easily and rapidly penetrates cells, this method can be applied

Handbook of RNA Biochemistry: Student Edition. Edited by R. K. Hartmann, A. Bindereif, A. Schön, E. Westhof
Copyright © 2009 WILEY-VCH Verlag GmbH & Co. KGaA, Weinheim
ISBN: 978-3-527-32534-4

to various organisms such as bacteria both Gram-negative [7, 8] and Gram-positive [9], protozoa [10, 11], yeast [12], and plants [13].

However, there are significant limitations to this method that have to be considered to allow a correct interpretation of the results. First, the method hinges on a few out of many functional groups of the RNA nucleotide that can participate in interactions. Thus, you can only determine RNA structure and conformational changes or interactions with proteins that involve those base-functional groups. Additionally, protection from DMS modification observed due to the presence of a protein does not necessarily indicate a physical interaction. Specific binding of a protein to a RNA often leads to a structural stabilization and, as a consequence, to a protection from DMS modification [14]. However, the major "problem" is that the obtained DMS modification pattern is averaged over the entire RNA population and over time. The lack of time resolution is especially problematic when it comes to studying *in vivo* folding and conformational changes, which are time-related events, and thus these questions are rendered more difficult to be addressed within cells. Secondly, your RNA of interest does not necessarily represent a single population, but might be partitioned among distinct species leading to mixed RNA populations. In other words, you have to be aware of what you are looking at. It is important to be able to differentiate, for example, between folded versus unfolded molecules, naked RNA versus RNA–protein particles or spliced versus precursor RNA in order to assign the modification pattern and its concomitant interpretation to a specific population.

In order to prevent fundamental pitfalls it is of great importance to check whether the modified RNA is still functional. Thus, we recommend incorporating the *in vivo* DMS modification step into a well-established experimental procedure that in the end allows you to test the activity of your RNA. For example, we incorporated the modification step into our *in vivo* splicing analysis procedure. As the splicing efficiency and RNA levels were not affected by DMS treatment of cells (no change compared to non-treated cells), we were confident that the RNA we were analyzing was in good condition [15].

14.2
Description of Method

The methodology of *in vivo* DMS structural probing of RNA described in here has been optimized for *Escherichia coli* and has mainly been adjusted from [11].

14.2.1
Cell Growth and *In Vivo* DMS Modification

For the successful application of this method it is important to standardize the way of growing the bacterial cell cultures as well as their DMS treatment [14]. Note that DMS is very toxic (for information, see http://www.state.nj.us/health/eoh/rtkweb/

0768.pdf), therefore you have to take precautions, such as working in a hood as well as wearing gloves and a lab coat, when working with this harmful reagent.

(1) Grow a cell culture of at least 100 ml to an $OD_{600\,nm}$ of 0.2–0.3. Then harvest the cells (2×50 ml) by centrifuging at 5000 r.p.m. in SS34 Sorval tubes at 4 °C for 5 min. Discard the supernatant and resuspend the cell pellet in 1 ml ice cold TM buffer (10 mM Tris–HCl, pH 7.5, 10 mM $MgCl_2$).
(2) Partition the pellet into two equal samples and add DMS to a final concentration of 150 mM to one of the two samples and vortex briefly. Incubate the cells with DMS for one minute and afterwards add β-mercaptoethanol to a final concentration of 0.7 M in order to quench the DMS. Vortex strongly! Centrifuge the tubes immediately at 6000 r.p.m. for 2 min in an Eppendorf tube. After careful removal of the supernatant, freeze the pellet at −80 °C until you proceed with the RNA preparation, but not longer than overnight. As a control, treat the second sample equally but without adding DMS.

After the modification step it is absolutely necessary to get rid of the DMS, because it will interfere with subsequent steps such as RNA extraction (degradation) and it can lead to modification of RNA during its isolation. Making a stop control, in which you add β-mercaptoethanol before DMS, will help to determine whether DMS is sufficiently quenched by β-mercaptoethanol [5]. This will provide confidence that the observed modification did not occur during RNA extraction, but did indeed occur *in vivo*. If you have trouble in removing DMS completely then you can solubilize DMS with isoamylalcohol [5].

14.2.2
RNA Preparation

(1) The frozen cell pellet is resuspended carefully in 157 μl solution A [150 μl TE, 1.5 μl 1 M DTT, 0.75 μl RNasin (35.5 U/μl), 4 μl lysozyme (10 mg/ml), 1 μl ddH_2O]. The cell suspension is frozen in liquid nitrogen and then thawed in a room temperature water bath. These steps (freeze and thaw) have to be done 3 times.
(2) Add 20 μl Solution B [4 μl 1 M $MgOAc_2$, 3.5 μl DNase I (RNase-free) (10 U/μl), 0.1 μl RNasin (40 U/μl), 12.4 μl ddH_2O], mix gently and incubate the samples on ice for 1 h.
(3) Add 20 μl Solution C (10 μl 0.2 M acetic acid, 10 μl 10% SDS), mix gently and incubate the samples at room temperature for 5 min.
(4) Perform phenol extraction: once with 1 volume of phenol, then with PCI (phenol:chloroform:isoamylalcohol = 25:24:1) and finally with CI; for each step vortex the samples and then centrifuge at 15.000 r.p.m. at 4 °C for 5 min. Collect the upper (aqueous) phase and proceed with the next extraction step.
(5) Precipitate the RNA with 1/100 volumes 0.5 M EDTA, pH 8.0, 1/10 volumes 3 M NaOAc, pH 5.0 and 2.5 volumes ethanol abs. Freeze the samples at −80 °C for 1 h and then centrifuge at 15 000 r.p.m. at 4 °C for 30 min. Discard the

supernatant, wash the RNA pellet with 70% ethanol and dry it carefully, but briefly (1 min at 65 °C). Resuspend the pellet in an appropriate volume of ddH$_2$O.

We gained sufficient structural information from analyzing the accessibility of adenine and cytidine residues [15]. However, if you wish to determine whether the N^7 positions of G residues are modified as well, then you need to perform aniline cleavage before reverse transcription [5]. Alternatively, you can study the accessibility of guanine nucleotides using kethoxal *in vivo* [6].

14.2.3
Reverse Transcription

Primer kinase reaction

(1) Set up the reaction in a total volume of 20 μl as follows: 4 μl 5 × PNK buffer (500 mM Tris–HCl, pH 8.0, 50 mM MgCl$_2$, 35 mM DTT), 8 pmol primer with 10 pmol [γ-^{32}P]ATP and 1 μl T4 polynucleotide kinase (10 U/μl).
(2) Incubate the sample at 37 °C for 45 min.
(3) Add 1 μl 500 mM EDTA, pH 8.0. Incubate the sample at 95 °C for 1 min, put the sample immediately on ice.
(4) Precipitate the primer with 1 μl glycogen (10 μg/μl) and 2.5 volumes ethanol p.a. and 1/10 volumes 3 M NaOAc, pH 5.0. Freeze the sample at −20 °C for 30 min and then centrifuge the sample at 4 °C at 15 000 r.p.m. for 30 min. Discard the supernatant and resuspend the washed and dried pellet in an appropriate amount of ddH$_2$O.

Depending on the primer and on the assay it might be necessary to purify the oligonucleotide before labeling it and/or after the labeling reaction with [γ-^{32}P]ATP. In general, newly synthesized oligonucleotides should always be purified prior to use.

Annealing reaction

(1) Combine 2.5 μl of *in vivo* isolated RNA (16 μg/μl) with 1 μl labeled primer (50 000 c.p.m.) and 1 μl 4.5 × hybridization buffer (225 mM K-HEPES, pH 7.0, 450 mM KCl).
(2) Incubate the sample at 90 °C for 1 min.
(3) Subsequently transfer the hot water into another glass box and let the sample cool down to 42 °C.

Extension of the primer

(1) Add 2.2 μl extension mix (0.6 μl 10 × extension buffer (1.3 M Tris–HCl, pH 8.0, 100 mM MgCl$_2$, 100 mM DTT), 0.3 μl nucleotide mix (2.5 mM each dNTP),

1 µl AMV reverse transcriptase (4 U/µl) and 1 µl ddH$_2$O to the 4.5 µl annealing reaction.
(2) Incubate the samples in a water bath at 42 °C for 1 h.
(3) Add 1.5 µl 1 M NaOH in order to degrade the RNA and incubate the sample for another hour at 42 °C. Then add 1.5 µl 1 M HCl in order to neutralize the pH.
(4) Precipitate with 1/10 volumes 500 mM EDTA, pH 8.0, 1 volume 3 M NaOAc, pH 5.0, 3 volumes ethanol p.a. Freeze the sample at −20 °C for 1 h and then centrifuge the sample at 4 °C at full speed for 30 min. Discard the supernatant and resuspend the dried pellet in 10 µl loading buffer [7 M urea, 0.25% bromphenol blue, 0.25% xylene cyanol in 1 × TBE (0.089 M Tris base, 0.089 M boric acid, 2 mM EDTA)]. Separate the extension products on an 8% PAA gel.

For obtaining sequencing ladders proceed as described above, but add in addition to the extension mix 1.5 µl of the appropriate 1 mM ddNTP solution to the reaction. If you have difficulties in generating satisfactory sequencing ladders, optimize the ddNTP concentrations. Usually the A and C lanes are sufficient for orientation along the molecule.

14.3
Evidence for Protein-induced Conformational Changes within RNA *In Vivo*

The *in vivo* DMS modification method described above was used to study the mode of action of the StpA protein in *E. coli* cells. StpA was shown to promote folding of the group I intron containing pre-mRNA of the thymidylate synthase (*td*) gene [16, 17]. First, the modification pattern of the *in vivo* DMS treated intron RNA was used to visualize the *in vivo* folding state of the *td* intron (Fig. 14.1). Importantly, we concluded from this DMS modification pattern that the secondary structure model, which was derived from phylogeny and biochemical data obtained *in vitro*, accurately describes the structure of the *td* group I intron *in vivo* [15]. We then addressed the question how the RNA chaperone StpA rescues folding of the *td* pre-mRNA. For this purpose we determined the DMS modification pattern of *td* RNA in the absence and presence of StpA, and compared it to the *td* RNA in the presence of the specific RNA-binding protein Cyt-18 (Fig. 14.2). In the presence of StpA, residues belonging to tertiary structure elements become more accessible to DMS. In contrast, the presence of Cyt-18 leads to a protection of bases involved in tertiary structure formation. Thus, StpA, a protein with RNA chaperone activity, and Cyt-18, a specific RNA binding protein, have opposite effects on the intron RNA structure *in vivo*.

From these results we concluded that StpA leads to a general loosening of the *td* group I intron RNA structure, while Cyt-18 contributes to the overall compactness of the RNA. In brief, using *in vivo* DMS modification we have been able to provide first evidence for protein induced conformational changes within a catalytic RNA *in vivo* and gained first insights into the mechanism of action of an RNA chaperone.

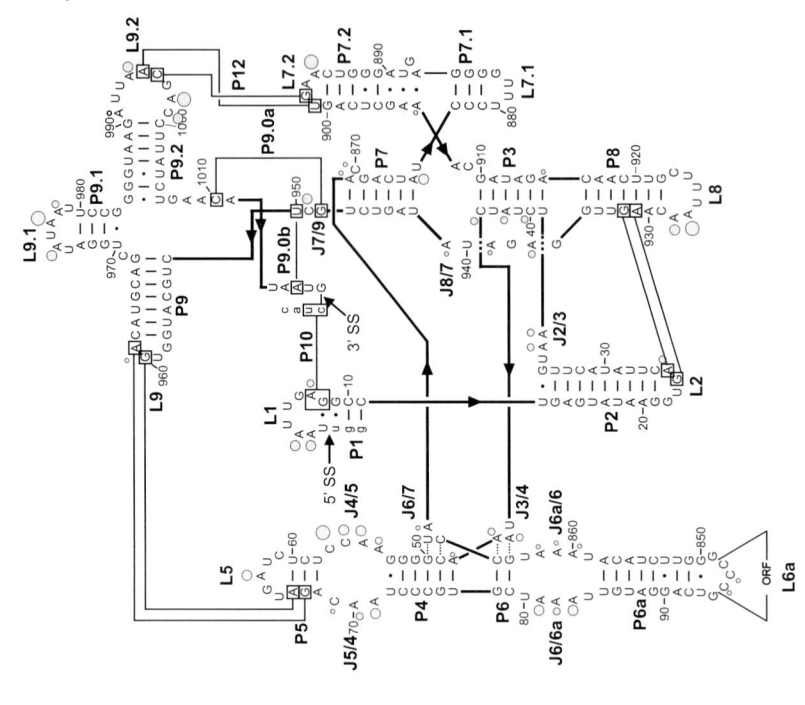

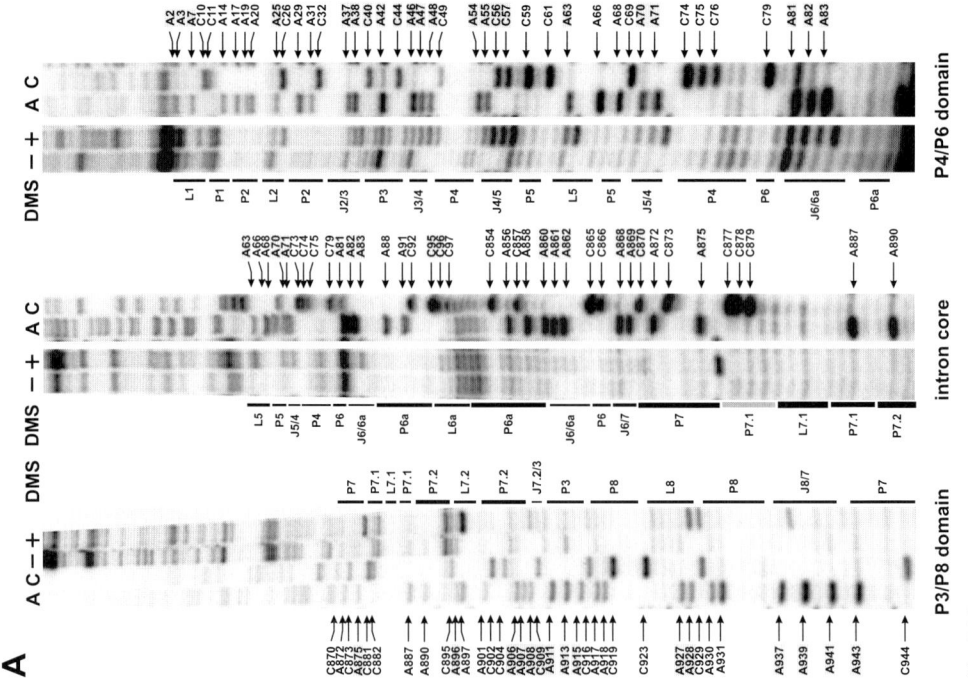

234 | *14 In Vivo Determination of RNA Structure by Dimethylsulfate*

Fig. 14.1

14.4 Troubleshooting

The most likely problems that will occur when performing *in vivo* DMS probing of your RNA of interest is over-modification and RNA degradation. For this reason, we recommend that you carefully determine the optimal concentration of DMS and the incubation time. Both parameters highly depend on the intracellular levels of RNA. The aim is to achieve about one modification per molecule. Thus, reverse transcription from an untreated control as well as from DMS-modified RNA should result in a comparable amount of full-length extension products. If primer extension of DMS-treated RNA runs off a lot earlier than the untreated control, the RNA is probably over-modified. On the other hand it might happen that you observe poor modification. Thus, it is advisable to buy fresh DMS every couple of months (about every 6 months to 1 year). DMS is usually a colorless solution, which becomes more and more yellow the older the solution is. As far as RNA degradation is concerned, there are typically two main reasons for it. First, if DMS is not removed completely before RNA extraction is performed, this results in a very low yield of isolated RNA (proceed as described in the method description). Second, strong stops in the reverse transcription control, which is obtained from untreated RNA, are indicative for contamination with the pancreatic nuclease RNase A that prefers UpA sites for cleavage. In this case you should try to raise the amount of RNase inhibitor. In addition, it is important to note that *in vivo* isolated RNA is not very stable and thus the best results for reverse transcription are obtained within the first 2–3 weeks after the extraction. Potential pitfalls for reverse transcription are the choice of reverse transcriptase, since every reverse transcriptase does not recognize and stop at methylated N^1 of As and N^3 of Cs. We highly recommend using AMV reverse transcriptase. Good and specific priming is typically observed for primers of 18 nt in length. Nevertheless, it is advisable to check the T_m of the primers, which should be above the primer extension temperature (42 °C). If primers do not label or prime efficiently, this might be due to the formation of competing structures within the primer. In that case you should redesign the primer. If the primer extension is not satisfactory, you should try to increase the levels of target RNA and sometimes it is also helpful to optimize the KCl concentration for annealing reaction. A very common phenomenon is that a primer is not significantly extended but there occur very strong reverse transcrip-

Fig. 14.1. DMS modification of the *td* intron *in vivo*. (a) Intron residues accessible to DMS are displayed in these representative gels. Boxed nucleotides correspond to positions within the intron, which are modified by DMS. The P3–P8 domain of the intron core (left panel), the center of the intron core covering the P7 stem as well as the P6–P6a element (middle panel) and the P4–P6 domain of the intron core and the stem–loops P1–P2 (right panel) are shown. A and C denote the sequencing lanes. (B) Summary of the *td* intron residues modified by DMS *in vivo*. Modified sites are indicated by dots. The size of the dots correlates with the relative modification intensities. The largest dot corresponds to the highest modification intensity.

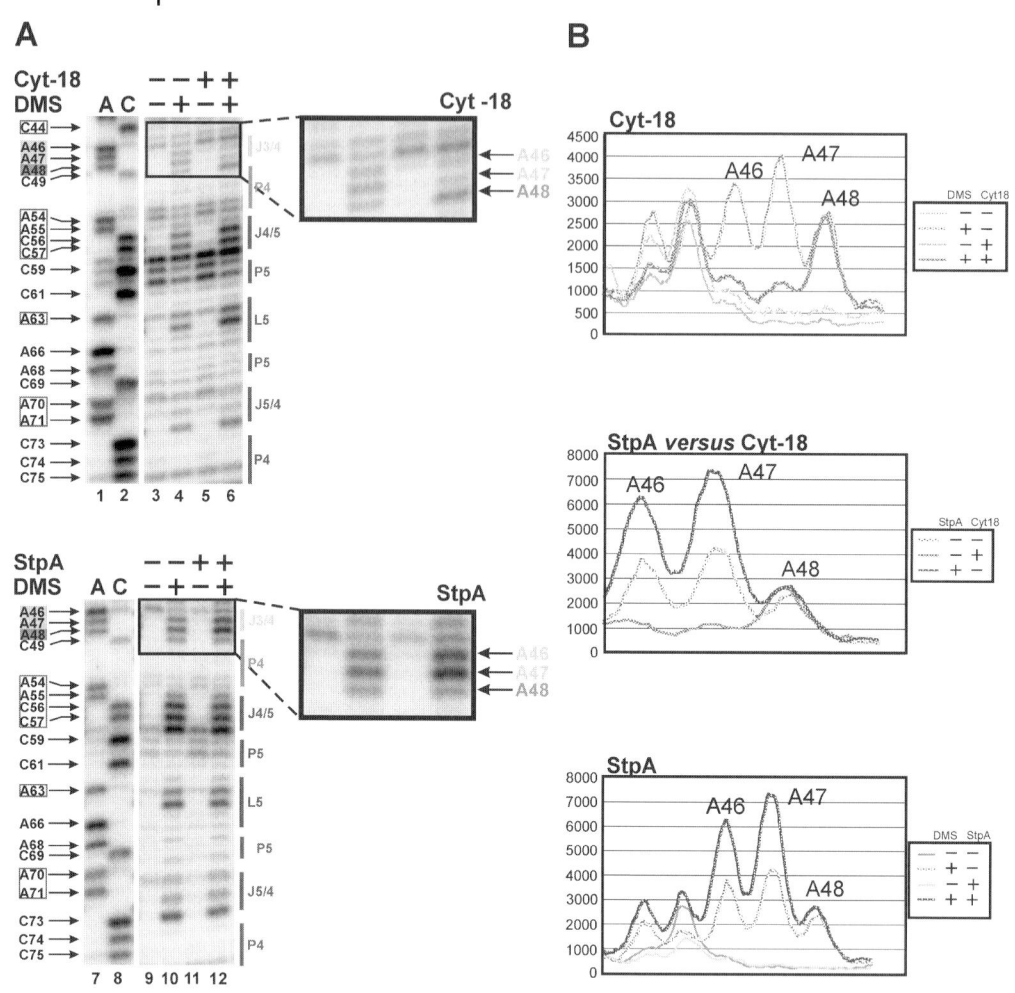

Fig. 14.2. StpA and Cyt-18 induce structural changes in the base triple interactions between adenines in J3/4 and stem P6. Changes in the DMS modification pattern of the td intron in vivo due to the presence of Cyt-18 is shown in the upper panel or due to the presence of StpA in the lower panel. (A) The P4/P5 domain is shown in these representative gels. Numbered nucleotides, which are highlighted by boxes at the left of the gel, are modified by DMS. The gel part boxed in black is outlined to point out the different effect of Cyt-18 versus StpA on the residues A46 and A47 in J3/4. The sequencing lanes are labeled with A and C. In the presence of Cyt-18 the amount of td RNA is increased in the cells as reflected by the increase of non-specific stops in untreated samples (cf. lanes 3 and 5) as well as by the increased modification intensity of residues A55, C56, C57 and A63 in lanes 4 and 6. These differences in the td RNA amount were normalized. (B) PhosphoImager quantification (right panel) of the outlined gel segments in the presence of Cyt-18 or StpA. The opposite effects of these proteins on the accessibility of the two adenines in J3/4 to DMS are summarized in the middle panel.

tion stops at the beginning of the extension. Usually it is sufficient to overcome this problem by setting the primer a few nucleotides more upstream or downstream of the original hybridization site. In summary, the best results for reverse transcription are obtained using clean and freshly labeled primers in conjunction with newly prepared RNA.

References

1 Brion, P., E. Westhof. *Annu. Rev. Biophys. Biomol. Struct.* **1997**, *26*, 113–137.
2 Weeks, K. M. *Curr. Opin. Struct. Biol.* **1997**, *7*, 336–342.
3 Schroeder, R., R. Grossberger, A. Pichler, C. Waldsich. *Curr. Opin. Struct. Biol.* **2002**, *12*, 296–300.
4 Ehresmann, C., F. Baudin, M. Mougel, P. Romby, J. P. Ebel, B. Ehresmann. *Nucleic Acids Res.* **1987**, *15*, 9109–9128.
5 Wells, S. E., J. M. Hughes, A. H. Igel, M. Ares, Jr. *Methods Enzymol.* **2000**, *318*, 479–493.
6 Balzer, M., R. Wagner. *Anal. Biochem.* **1998**, *256*, 240–242.
7 Climie, S. C., J. D. Friesen. *J. Biol. Chem.* **1988**, *263*, 15166–15175.
8 Moazed, D., J. M. Robertson, H. F. Noller. *Nature* **1988**, *334*, 362–364.
9 Mayford, M., B. Weisblum. *EMBO J.* **1989**, *8*, 4307–4314.
10 Harris, K. A., Jr, D. M. Crothers, E. Ullu. *RNA* **1995**, *1*, 351–362.
11 Zaug, A. J., T. R. Cech. *RNA* **1995**, *1*, 363–374.
12 Ares, M., Jr, A. H. Igel. *Genes Dev.* **1990**, *4*, 2132–2145.
13 Senecoff, J. F., R. B. Meagher. *Plant Mol. Biol.* **1992**, *18*, 219–234.
14 Waldsich, C., R. Grossberger, R. Schroeder. *Genes Dev.* **2002**, *16*, 2300–2312.
15 Waldsich, C., B. Masquida, E. Westhof, R. Schroeder. *EMBO J.* **2002**, *19*, 5281–5291.
16 Semrad, K., R. Schroeder. *Genes Dev.* **1998**, *12*, 1327–1337.
17 Clodi, E., K. Semrad, R. Schroeder. *EMBO J.* **1999**, *18*, 3776–3782.

15
Probing Structure and Binding Sites on RNA by Fenton Cleavage

Gesine Bauer and Christian Berens

15.1
Introduction

In the 20 years that have followed the discovery of RNA-based catalysis, many novel biological functions and catalytic activities of RNA have been either discovered *in vivo* or obtained through *in vitro* selection techniques [1]. Roughly, these can be divided into two groups. While the first, exemplified by mRNA, snoRNA, guide RNA and siRNA, utilizes sequence-dependent Watson–Crick interactions, the activities exerted by the second group, containing ribozymes, aptamers and riboswitches, are based on their three-dimensional structures. Knowledge of these structures and how they are formed is a prerequisite to understanding how these RNAs function mechanistically.

As X-ray crystallography of RNA molecules has proven difficult and since many interesting RNA molecules are still too large for NMR analysis, a profusion of RNA probing methods have been developed for structural analysis [2–4]. One very versatile method for analyzing RNA is probing with hydroxyl radicals. They are the smallest molecule species used for chemical probing, cleave nucleic acids with little or no sequence specificity [5, 6], and a significant secondary structure preference has not been observed in radical-induced cleavage of single- and double-stranded forms of RNA and DNA [7].

Hydroxyl radicals are generated physically by radiolysis of water using synchrotron X-ray beams or, more often, chemically by the reduction of peroxo-groups with Fe^{2+} in the so-called Fenton reaction [8]. Like most transition metals, iron has more than one oxidation state besides the ground state and its valence electrons may be unpaired allowing one-electron redox reactions [9]. As such, Fe^{2+} reacts with H_2O_2 (or other peroxo molecules like peroxonitrous acid) to generate short-lived, highly reactive hydroxyl radicals. These cleave the bases of a nucleic acid, its phosphodiester backbone and also peptide bonds in spatial proximity of Fe^{2+}. Sodium ascorbate is often added to the reaction mixture in order to reduce Fe^{3+} to Fe^{2+}, thereby establishing a catalytic cycle and permitting low, micromolar, concentrations of Fe^{2+} to be effective in cleaving the substrates. Consequently,

Handbook of RNA Biochemistry: Student Edition. Edited by R. K. Hartmann, A. Bindereif, A. Schön, E. Westhof
Copyright © 2009 WILEY-VCH Verlag GmbH & Co. KGaA, Weinheim
ISBN: 978-3-527-32534-4

hydroxyl-radical-based probing methods have been widely used for structural analysis of RNA and also as a tool to study interactions of nucleic acids with proteins or other ligands.

The classical interaction study is a footprinting experiment. The presence of the interaction partner protects the nucleic acid at the site bound from cleavage by the hydroxyl radicals. Detailed protocols for footprinting protein–DNA and protein–RNA complexes are given in [10, 11]. However, this method can also be used to identify contact sites of 16S rRNA in 30S subunits with 50S subunits [12] or to determine the structural elements of an internal ribosomal entry site that interact with a 40S ribosomal subunit [13].

The versatility of this approach was greatly extended by tethering Fe^{2+} to defined sites on proteins and RNA using the reagent 1-(*p*-bromoacetamidobenzyl)-EDTA (BABE), originally synthesized by Meares et al. [14, 15]. Hydroxyl radical footprints with Fe^{2+} tethered either to various ribosomal proteins or to rRNA gave important insights into the three-dimensional organization of the ribosome [16–18] that were later confirmed by the crystal structure of the 70S ribosome (reviewed in [19]). Detailed protocols for interaction studies with hydroxyl radicals generated by Fe^{2+} either tethered to proteins or RNA have been published [20, 21].

Hydroxyl radical cleavage is also used for RNA structure analysis. Similar to the interaction analysis method described above, Fe^{2+} can be tethered to RNA to induce intramolecular self-cleavage as was shown by Newcomb and Noller [22] who determined the RNA neighborhoods of specific nucleotides in the rRNA of 70S ribosomes or by Huq et al. [23] who obtained structural information on the three-dimensional fold of the HIV-1 *trans*-activation responsive region RNA.

In addition to tethered Fe^{2+}/EDTA, free Fe^{2+}/EDTA is used to identify solvent-accessible and solvent-excluded sugar moieties and, thus, aids in modeling the three-dimensional structure of an RNA [24]. Protection of tRNA bound to the ribosomal P-site from hydroxyl radical cleavage gave important hints for the mechanism of tRNA-ribosome interaction [25]. Swisher et al. presented hydroxyl radical footprints [26] demonstrating that a group II intron ribozyme has a tightly packed, solvent-inaccessible core like other large ribozymes [27, 28]. In addition, hydroxyl radical footprinting allows us to determine the relative stabilities of individual structural motifs by examining the protection pattern as a function of added Mg^{2+} or urea. Experiments like this have been done with RNase P [29] or the *Tetrahymena* LSU group I ribozyme [30]. Furthermore, synchrotron generated hydroxyl radicals have been employed successfully for time-resolved footprinting of RNA folding [31] (a detailed methods protocol is presented in [32]).

Fe^{2+} is similar to Mg^{2+} in size and coordination geometry [33] and has been used to replace the latter in proteins to map metal ion binding sites [34–36]. Catalytic RNAs either require divalent cations for achieving a stable tertiary structure and for catalysis or their activity is greatly enhanced by the presence of divalent metal ions [37]. The identification of metal ion binding sites is therefore essential for a thorough structure–function analysis of catalytic RNA. In addition to NMR studies (summarized in [38]), hydroxyl-radical-induced cleavage based on limited replacement of Mg^{2+} with Fe^{2+} provides a powerful method for the analysis of RNA–metal ion interactions [39].

Here we present two different protocols for hydroxyl radical probing of RNA. The first describes structural probing of large RNAs. Comparison of cleavage patterns obtained with Fe^{2+} in the presence and in the absence of EDTA allows to distinguish between solvent-exposed and solvent-occluded regions of the RNA and to identify metal-ion-binding sites.

The second protocol describes an interaction study that exploits the ability of Fe^{2+} to replace the Mg^{2+} ion chelated to tetracycline. A subsequent hydroxyl radical digestion can then identify the residues in proximity of the [Fe^{2+}-tetracycline] chelate. This has already been done for the tetracycline proton-antiporter TetA [40] and the tetracycline-dependent regulatory protein TetR [41]. We used hydroxyl radical cleavage of 16S rRNA to identify tetracycline-binding sites in the 70S ribosome [42].

15.2
Description of Methods

15.2.1
Fe^{2+}-mediated Cleavage of Native Group I Intron RNA

The method presented here was used to detect Mg^{2+}-binding sites in the *Tetrahymena* LSU group I intron [39]. For a successful reaction, it is important to prepare the solutions of $FeCl_2$ and H_2O_2 freshly. Sodium ascorbate can be prepared as a 10-fold stock solution and stored at $-20\ °C$. In order to mix the reagents accurately after 1 min, we recommend using a small table Microfuge (Qualitron). The appropriate reagents are added subsequently to the wall of each Eppendorf cap and then mixed simultaneously by briefly applying the centrifuge.

Hydroxyl radical cleavage

- For experiments with native RNA, take 1 µl RNA (5 pmol cold RNA, spiked with approximately 50 000 c.p.m. of the RNA labeled with ^{32}P at either the 5′ or the 3′ end) and add 1 µl of 5 × native cleavage buffer (1 × NCB: 25 mM MOPS–KOH, pH 7.0; 3 mM $MgCl_2$; 400 µM spermidine; 200 mM NaCl). Incubate the RNA for 2 min at 56 °C, followed by 3 min incubation at room temperature.
- Add 1 µl 1.25 mM $FeCl_2$ to the reaction tube, mix by centrifugation and incubate for 1 min before adding 1 µl 12.5 mM sodium ascorbate.
- After 1 min, add 1 µl of 12.5 mM H_2O_2 and mix rapidly to initiate the reaction. The final concentrations are 250 µM for Fe^{2+} and 2.5 mM for both sodium ascorbate and H_2O_2.
- Stop the cleavage reaction after 1 min by adding 1 µl 1 M thio-urea. The RNA is then precipitated with 1 µl glycogen (10 µg/µl), 1 µl sodium acetate (3 M) and 30 µl 96% (v/v) ethanol.

Gel electrophoresis

- After precipitation, resuspend the RNA in gel loading buffer [7 M urea; 0.01% (w/v) bromophenol blue and xylenecyanol each].
- Separate the cleavage products on 6–20% denaturing polyacrylamide sequencing gels.
- For obtaining sequencing ladders, carry out limited hydrolysis with RNase T1 and NaHCO$_3$ [43].

Mg^{2+} competition of Fe^{2+}-mediated cleavage

- Mix equal volumes of a 2.5 mM FeCl$_2$ solution and a Mg^{2+} stock solution. Notice that to achieve the desired Mg^{2+} concentration for competition, you have to take into account the Mg^{2+} already present in the reaction when calculating the Mg^{2+} concentration of the stock solution. For a final MgCl$_2$ concentration of 10 mM, e.g. mix 50 µl of 2.5 mM FeCl$_2$ and 50 µl of 70 mM MgCl$_2$. Then 1 µl of the Fe^{2+}/Mg^{2+} mixture is pipetted into the reaction tube and the cleavage reaction continued as above. The total reaction volume is 5 µl and a combined Mg^{2+} concentration of 10 mM is obtained by 3 mM originating from the native cleavage buffer and 7 mM resulting from the added Fe^{2+}/Mg^{2+} mixture.

Visualization of metal-ion binding sites in group I introns by Fe^{2+}-mediated Fenton cleavage

Cleavage by Fe^{2+} is observed in distinct regions of the group I intron RNA and only with native RNA (*cf.* lanes 7 and 8 in Fig. 15.1). It is competed by Mg^{2+} (*cf.* lanes 8 and 10 in Fig. 15.1) indicating that both ions interact with the same or overlapping binding sites. Lanes 8 and 12 show a comparison of the cleavage sites obtained using Fe^{2+} or Fe^{2+}/EDTA. Most of the sites cleaved by Fe^{2+} are embedded in regions protected from cleavage by Fe^{2+}/EDTA. They are, thus, located in the interior of the ribozyme where they would be expected to be if the metal ions they reflect bury phosphate oxygens [44]. In a three-dimensional model of the bacteriophage T4-derived *td* intron and in the crystal structure of the P4P6 domain of the *Tetrahymena* LSU intron [45], cleavage sites separated in secondary structure come together in three-dimensional space to form several distinct pockets (see figures 8 and 9 in [39]). There is also very good agreement between nucleotides cleaved by Fe^{2+} and nucleotides close to metal ions determined by phosphorothioate substitution [46–48] metal-hydroxyl cleavage [49], or X-ray crystallography [45]. Figure 15.2 shows cleavage sites in the hinge region of the P4P6 domain coincide nicely with a diffusely bound metal ion that was predicted from microenvironment analysis [50], but not observed in the crystal structure [45].

15.2.2
Fe^{2+}-mediated Tetracycline-directed Hydroxyl Radical Cleavage Reactions

This method describes the identification of tetracycline-binding sites on rRNA in 70S ribosomes of *Escherichia coli* [42]. Hydroxyl radical cleavage of the RNA in the

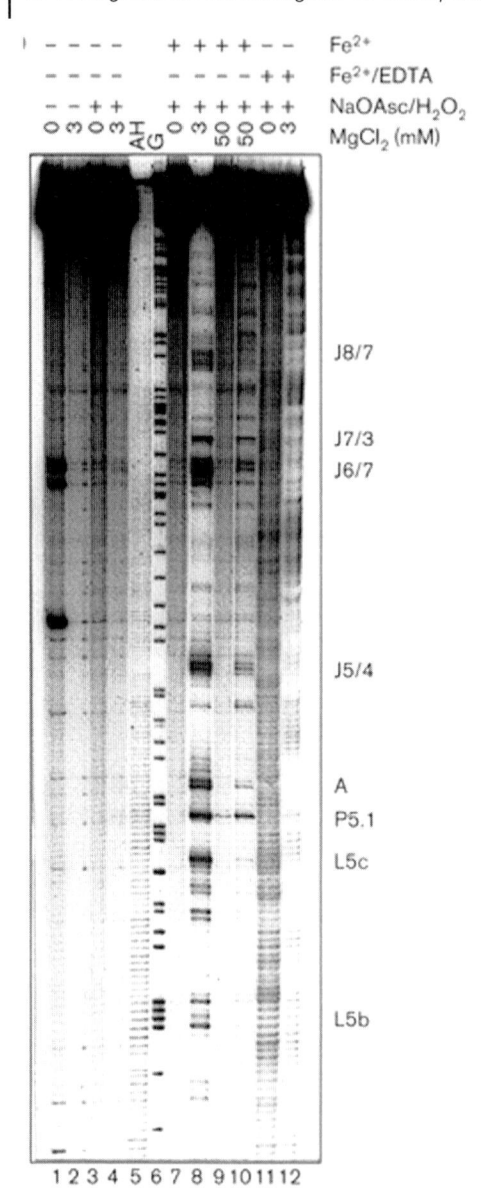

Fig. 15.1. Mapping the Fe^{2+} cleavage sites in the *Tetrahymena* LSU group I intron. Autoradiogram of a 6% denaturing polyacrylamide gel with 5'-end-labeled *Tetrahymena* L-21 RNA cleaved with 10 μM Fe^{2+} (lanes 7 and 9), 250 μM Fe^{2+} (lane 8 and 10) or with 250 μM Fe^{2+}/500 μM EDTA (lanes 11 and 12). Controls with untreated RNA (lanes 1 and 2) and in which Fe^{2+} was omitted (lanes 3 and 4), as well as competition of Fe^{2+} cleavage by 50 mM Mg^{2+} (lanes 9 and 10) are also shown. The respective final concentrations of Mg^{2+}, as well as the presence (+) or absence (−) of Fe^{2+}, sodium ascorbate and H_2O_2 are displayed above each lane. Renatured RNA is in lanes 2, 4, 8, 10 and 12; denatured RNA in lanes 1, 3, 7, 9 and 11. Secondary structure elements cleaved by Fe^{2+} are marked on the right. Sequencing markers are AH (alkaline hydrolysis) and G (RNase T1). Reprinted from [39].

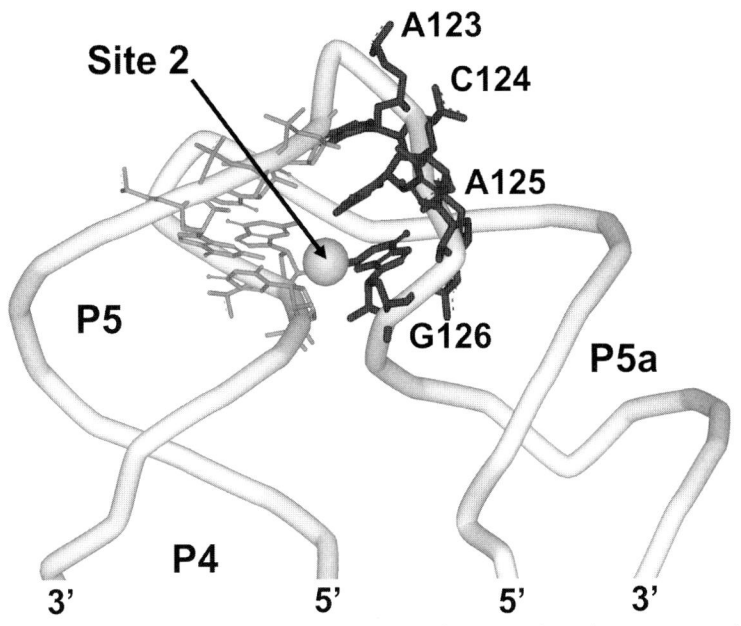

Fig. 15.2. Correlation between a computationally predicted Mg^{2+}-binding site and Fe^{2+} cleavage sites in the *Tetrahymena* P4P6 domain. The phosphodiester backbones of the two RNA strands in the hinge region of the P4P6 domain of the *Tetrahymena* LSU group I intron are shown as closed white ribbons with their polarity and the secondary structure elements indicated. Site 2 (gray sphere) is a potential diffusely bound Mg^{2+} site [50] and possible coordinating residues are displayed as thin gray sticks. Residues that are cleaved by Fe^{2+} are shown as thick black sticks. Coordinates were taken from the RCSB entry 1hr2 [60] and table 3 of [50].

vicinity of bound tetracyclines is detected by primer extension. Fe^{2+}-mediated hydroxyl radical cleavage of the 70S ribosome is carried out similar to the method described above. Tetracycline solutions have to be prepared freshly.

Hydroxyl radical cleavage

- Add 4 µl of a ribosome solution (5 pmol ribosomes in a buffer containing 5 mM $MgCl_2$) to 1 µl of a 10 × tetracycline stock solution of the final tetracycline concentration.
- After addition of 2 µl 5 × cleavage buffer (1 × CB: 25 mM MOPS–KOH, pH 7.0; 3 mM $MgCl_2$; 400 µM spermidine), incubate for 30 min at 37 °C followed by 10 min incubation at room temperature.
- Add 1 µl of 1.25 mM $FeCl_2$ to the reaction tube and mix by centrifugation.
- Incubate for 1 min before adding 1 µl of 6.25 mM sodium ascorbate.
- After 1 min, add 1 µl of 6.25 mM H_2O_2 to initiate the reaction and mix rapidly. The final concentrations are 125 µM for Fe^{2+} and 625 µM for both sodium ascor-

bate and H_2O_2 in the presence of 5 mM Mg^{2+}. The cleavage reaction is stopped after 1 min by the addition of thio-urea to a final concentration of 100 mM. The RNA is precipitated with 1 μl glycogen (10 μg/μl), 5 μl sodium acetate (3 M) and 60 μl 96% (v/v) ethanol.

Mg^{2+} competition of Fe^{2+} cleavage

To assure that Mg^{2+} and Fe^{2+} share the same or overlapping binding sites it is essential to carry out a Mg^{2+} competition experiment. The range of Fe^{2+}:Mg^{2+} ratios necessary for cleavage competition depends on the respective affinities of Fe^{2+} and Mg^{2+} to the appropriate binding site. In case of tetracycline, Fe^{2+} binds 100-fold more tightly than Mg^{2+} [41].

Mg^{2+} competition of Fe^{2+}-mediated cleavage

- Mix equal volumes of a 2.5 mM $FeCl_2$ solution and a Mg^{2+} stock solution. Notice that to achieve the desired Mg^{2+} concentration for competition, you have to take into account the Mg^{2+} already present in the reaction when calculating the Mg^{2+} concentration of the stock solution. In this case, for a final $MgCl_2$ concentration of 10 mM, e.g. mix 50 μl of 2.5 mM $FeCl_2$ and 50 μl of 140 mM $MgCl_2$. Then 1 μl of the Fe^{2+}/Mg^{2+} mixture is pipetted into the reaction tube and the cleavage reaction continued as above. The total reaction volume is 10 μl and a combined Mg^{2+} concentration of 10 mM is obtained by 3 mM originating from the native cleavage buffer and 7 mM resulting from the added Fe^{2+}/Mg^{2+} mixture.

Extraction of rRNA

The rRNA has to be extracted for the following primer extension analysis.

- Resuspend the pellet obtained after the ethanol precipitation in 200 μl ribosomal extraction buffer [REB: 0.3 M sodium acetate; 0.5% (w/v) SDS; 5 mM EDTA] at room temperature.
- In order to remove ribosomal proteins, carry out a phenolization followed by an isoamylalcohol/chloroform (1:24; v/v) treatment. Repeat this procedure twice.
- After an ethanol precipitation, resuspend the RNA in Millipore water and remove residual phenol by diethylether treatment. After a final ethanol precipitation, the RNA is resuspended in 10 μl Millipore water.

Primer extension and gel electrophoresis

Primer extension reaction and gel electrophoresis can be carried out as described in Chapter 14.

Mapping tetracycline binding to ribosomes by drug-directed Fenton cleavage of 16S rRNA

We identified three prominent Fe^{2+}-mediated cleavage sites in the 16S rRNA in the presence of tetracycline. All cleavage sites are in good agreement with published data for tetracycline from genetics [51, 52], biochemistry [53–55] and crystallography [56, 57]. They are located in helices 29 (A1339–U1341) and 34 (C1195–A1197), and in the internal loop of helix 31 (A964–A969) (helical numbering according to [58]). Figure 15.3(B and C) shows sections of denaturing polyacrylamide gels with cleavage sites mapped to tetracycline binding site-1 which is formed by h31 and h34. According to crystal structures of 30S subunits complexed with tetracycline [56, 57], the affected bases overlap with bases within 10 Å distance of tetracycline bound to site-1 (Fig. 15.3A) and tetracycline binding site-4 which is formed by h29 (data not shown).

15.3
Comments and Troubleshooting

- Free Fe^{2+} is present in a large molar excess over the target RNA. Thus, much of the Fe^{2+} will not be bound to RNA, but will be free in solution where it can also generate hydroxyl radicals. Like Fe^{2+}/EDTA-generated hydroxyl radicals, they will cleave the target RNA non-specifically at all surface-exposed positions. This bulk cleavage will reduce the signal to noise ratio, but can be compensated by increasing the amount of target RNA in the reaction mixture. We typically used 500 ng of the group I intron RNA for cleavage with 250 μM Fe^{2+}. For cleavage of yeast $tRNA^{Phe}$, the final cleavage assay contained 10 μg RNA, 10 mM Mg^{2+} and 1 mM Fe^{2+} (C. Berens and R. Schroeder, unpublished). We have not performed experiments with higher RNA amounts or Fe^{2+} concentrations.
- To statistically ensure only a single cleavage event per RNA molecule, about 70% of the population should remain uncleaved after the reaction [59]. We achieve this with Fe^{2+}:Mg^{2+} ratios of 1:10 to 1:20 for the native group I introns or with very low Fe^{2+} concentrations (10 μM) for the non-folded ribozymes in the absence of divalent metal ions. Higher ratios lead to increased unspecific degradation of the RNA, lower ratios to reduced cleavage intensity as a result of the competition between Fe^{2+} and Mg^{2+} for the metal-ion binding sites. Due to the limitations on total RNA and Fe^{2+} that can be added to a reaction assay, this reduces the Mg^{2+} concentrations that can be used for Fenton cleavage to a maximum of 10–20 mM. For RNA molecules that require higher Mg^{2+} concentrations for correct folding, it should be attempted to reduce the divalent cation requirement by substituting spermine and spermidine for Mg^{2+}. These polyamines do not affect the cleavage reaction (C. Berens and R. Schroeder, unpublished).
- Within the limits described above, the native cleavage buffer and the denaturation/renaturation conditions should be adjusted to the requirements of the respective RNA molecule to be probed. For selection of buffer conditions, it is necessary to keep in mind that some buffer additives (EDTA) or reducing reagents (DTT, 2-mercaptoethanol) might scavenge radicals generated by the cleavage reaction.
- Differences in the reaction conditions for cleavage with Fe^{2+} or with Fe^{2+}/ tetracycline indicate that the concentrations of the three chemical species (Fe^{2+},

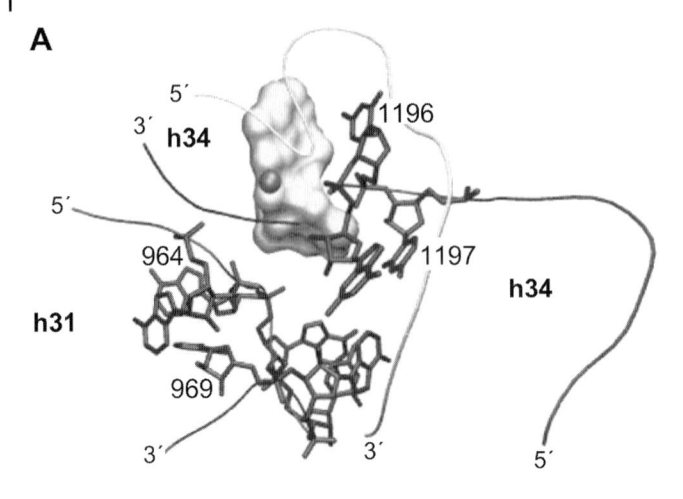

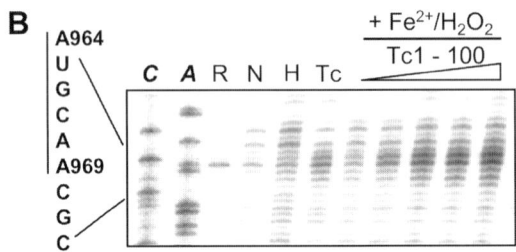

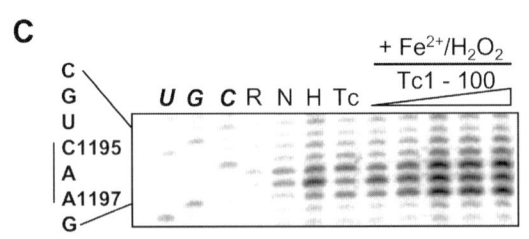

Fig. 15.3. Fe^{2+}-mediated hydroxyl radical cleavage of the 16S rRNA. (A) Surface structure of tetracycline complexed with Mg^{2+} (shown as a grey sphere) bound to the ribosomal binding site-1 which is formed by helix 34 (h34) and the internal loop of helix 31 (h31 [57]). The phosphodiester backbones of RNA strands containing bases that are attacked by hydroxyl radicals are shown as closed black ribbons, the unaffected backbone strand is shown as a light grey ribbon. Residues that are cleaved by Fe^{2+} are shown as black sticks. Bases 964, 969, 1196 and 1197 are marked for orientation. Coordinates were taken from the RCSB entry 1I97 [57]. (B) Autoradiograph of a polyacrylamide gel showing cleavage sites in the internal loop of h31. (C) Cleavage sites in h34. Lanes A, C, G, U: dideoxy sequencing lanes; R: unmodified RNA; N: control in which Fe^{2+} was omitted; H: Fe^{2+}/H_2O_2 cleaved RNA in the absence of antibiotics; Tc: unmodified RNA in the presence of 100 μM Tc and H_2O_2; Tc1–100: Fe^{2+}/H_2O_2 cleavage in the presence of 1, 3, 10, 30 and 100 μM Tc. Lines left of the sequence indicate regions of Fe^{2+}-mediated hydroxyl radical cleavage. E. coli 70S ribosomes were incubated with different amounts of tetracycline and treated with Fe^{2+}/H_2O_2 as described in Section 15.2. Cleavage sites were detected by primer extension and analyzed by electrophoresis on a denaturating 10% polyacrylamide gel.

sodium ascorbate and H_2O_2) might have to be varied to optimize the generation of hydroxyl radicals. We generally recommend using equimolar amounts of sodium ascorbate and H_2O_2. For naked RNA, it is not necessary to precipitate the 5'- or 3'-labeled RNA for initial evaluation and optimization of the experimental protocol. Addition of an equal volume of denaturing loading buffer and immediate electrophoresis gives data of sufficient quality. However, a purification step is still recommended for final mapping of the cleavage sites and for quantification of cleavage intensity.

- An incubation time of 1 min before the addition of sodium ascorbate and H_2O_2 was the shortest period of time in which the six reaction tubes that fit into the tabletop microcentrifuge could be manipulated easily. Extending the incubation time for Fe^{2+} and sodium ascorbate to 10 min did not lead to different cleavage patterns. The extension of the reaction time after the addition of H_2O_2 might increase cleavage intensities. However, we recommend changing the $FeCl_2$ concentration if the extent of cleavage intensity should be altered.
- Do not pre-mix the sodium ascorbate and $FeCl_2$ solutions before adding them to the RNA, as they will form a chelate-complex that effectively titrates the Fe^{2+} in the reaction mixture.
- It is necessary to perform control experiments with Fe^{2+} and H_2O_2 in the absence of tetracycline, in order to distinguish between tetracycline-mediated cleavage and cleavage caused by metal ions (either bound to specific binding pockets or diffusely associated with the surface of the RNA). Tetracycline is known to bind RNA unspecifically. This may lead to additional, unspecific cleavage sites at high concentrations. It is therefore advisable to titrate the reaction with tetracycline. In addition, one control should contain the highest amount of tetracycline used in the titration in the absence of Fe^{2+} and H_2O_2 since binding of tetracycline to RNA could cause a stop of reverse transcription. For all controls, the compound omitted is substituted by Millipore or double-distilled water.

References

1 J. A. DOUDNA, T. R. CECH, *Nature* **2002**, *418*, 222–228.
2 J. A. PICCIRILLI, J. S. VYLE, M. H. CARUTHERS, T. R. CECH, *Nature* **1993**, *361*, 85–88.
3 S. P. RYDER, S. A. STROBEL, *J. Mol. Biol.* **1999**, *291*, 295–311.
4 C. BRUNEL, P. ROMBY, *Methods Enzymol.* **2000**, *318*, 3–21.
5 T. D. TULLIUS, B. A. DOMBROSKI, M. E. CHURCHILL, L. KAM, *Methods Enzymol.* **1987**, *155*, 537–558.
6 B. BALASUBRAMANIAN, W. K. POGOZELSKI, T. D. TULLIUS, *Proc. Natl. Acad. Sci. USA* **1998**, *95*, 9738–9743.
7 D. W. CELANDER, T. R. CECH, *Biochemistry* **1990**, *29*, 1355–1361.
8 S. UDENFRIED, C. T. CLARK, J. AXELROD, R. B. BRODIE, *J. Biol. Chem.* **1954**, *208*, 731–739.
9 F. A. COTTON, G. WILKINSON, C. A. MURILLO, M. BOCHMANN, R. N. GRIMES, *Advanced Inorganic Chemistry*. 6th edn, Wiley, New York, **1999**.
10 P. ANSEL-MCKINNEY, L. GEHRKE, in: *Analysis of mRNA Formation and Function*, J. D. RICHTER (ed.), Academic Press, New York, **1997**, pp. 285–303.
11 W. J. DIXON, J. J. HAYES, J. R. LEVIN,

M. F. Weidner, B. A. Dombroski, T. D. Tullius, *Methods Enzymol.* **1991**, *208*, 380–413.

12 C. Merryman, D. Moazed, G. Daubresse, H. F. Noller, *J. Mol. Biol.* **1999**, *285*, 107–113.

13 T. Nishiyama, H. Yamamoto, N. Shibuya, Y. Hatakeyama, A. Hachimori, T. Uchiumi, N. Nakashima, *Nucleic Acids Res.* **2003**, *31*, 2434–2442.

14 T. M. Rana, C. F. Meares, *J. Am. Chem. Soc.* **1990**, *112*, 2457–2458.

15 D. P. Greiner, R. Miyake, J. K. Moran, A. D. Jones, T. Negishi, A. Ishihama, C. F. Meares, *Bioconjug. Chem* **1997**, *8*, 44–48.

16 G. M. Heilek, R. Marusak, C. F. Meares, H. F. Noller, *Proc. Natl. Acad. Sci. USA* **1995**, *92*, 1113–1116.

17 T. Powers, H. F. Noller, *RNA* **1995**, *1*, 194–209.

18 G. M. Culver, J. H. Cate, G. Z. Yusupova, M. M. Yusupov, H. F. Noller, *Science* **1999**, *285*, 2133–2136.

19 V. Ramakrishnan, *Cell* **2002**, *108*, 557–572.

20 G. M. Culver, H. F. Noller, *Methods Enzymol.* **2000**, *318*, 461–475.

21 S. Joseph, H. F. Noller, *Methods Enzymol.* **2000**, *318*, 175–190.

22 L. F. Newcomb, H. F. Noller, *Biochemistry* **1999**, *38*, 945–951.

23 I. Huq, N. Tamilarasu, T. M. Rana, *Nucleic Acids Res.* **1999**, *27*, 1084–1093.

24 J. A. Latham, T. R. Cech, *Science* **1989**, *245*, 276–282.

25 A. Hüttenhofer, H. F. Noller, *Proc. Natl. Acad. Sci. USA* **1992**, *89*, 7851–7855.

26 J. Swisher, C. M. Duarte, L. J. Su, A. M. Pyle, *EMBO J.* **2001**, *20*, 2051–2061.

27 T. S. Heuer, P. S. Chandry, M. Belfort, D. W. Celander, T. R. Cech, *Proc. Natl. Acad. Sci. USA* **1991**, *88*, 11105–11109.

28 T. Pan, *Biochemistry* **1995**, *34*, 902–909.

29 X. Fang, T. Pan, T. R. Sosnick, *Biochemistry* **1999**, *38*, 16840–16846.

30 C. Y. Ralston, Q. He, M. Brenowitz, M. R. Chance, *Nat. Struct. Biol.* **2000**, *7*, 371–374.

31 B. Sclavi, M. Sullivan, M. R. Chance, M. Brenowitz, S. A. Woodson, *Science* **1998**, *279*, 1940–1943.

32 C. Y. Ralston, B. Sclavi, M. Sullivan, M. L. Deras, S. A. Woodson, M. R. Chance, M. Brenowitz, *Methods Enzymol.* **2000**, *317*, 353–368.

33 I. B. Brown, *Acta Crystallogr. B* **1988**, *44*, 545–553.

34 J. M. Farber, R. L. Levine, *J. Biol. Chem.* **1986**, *261*, 4574–4578.

35 C. H. Wei, W. Y. Chou, S. M. Huang, C. C. Lin, G. G. Chang, *Biochemistry* **1994**, *33*, 7931–7936.

36 J. J. Hlavaty, J. S. Benner, L. J. Hornstra, I. Schildkraut, *Biochemistry* **2000**, *39*, 3097–3105.

37 A. L. Feig, O. C. Uhlenbeck, in: *The RNA World*, 2nd edn, R. F. Gesteland, T. R. Cech, J. F. Atkins (eds), Cold Spring Harbor Laboratory Press, Cold Spring Harbor, NY, **1999**, pp. 287–319.

38 R. L. Gonzalez, I. Tinoco, *Methods Enzymol.* **2001**, *338*, 421–443.

39 C. Berens, B. Streicher, R. Schroeder, W. Hillen, *Chem. Biol.* **1998**, *5*, 163–175.

40 L. M. McMurry, M. L. Aldema-Ramos, S. B. Levy, *J. Bacteriol.* **2002**, *184*, 5113–5120.

41 N. Ettner, J. W. Metzger, T. Lederer, J. D. Hulmes, C. Kisker, W. Hinrichs, G. A. Ellestad, W. Hillen, *Biochemistry* **1995**, *34*, 22–31.

42 G. Bauer, C. Berens, S. J. Projan, W. Hillen, *J. Antimicrob. Chemother.* **2004**, *53*, 592–599.

43 H. Donis-Keller, A. M. Maxam, W. Gilbert, *Nucleic Acids Res.* **1977**, *4*, 2527–2538.

44 J. H. Cate, R. L. Hanna, J. A. Doudna, *Nat. Struct. Biol.* **1997**, *4*, 553–558.

45 J. H. Cate, J. A. Doudna, *Structure* **1996**, *4*, 1221–1229.

46 E. L. Christian, M. Yarus, *J. Mol. Biol.* **1992**, *228*, 743–758.

47 E. L. Christian, M. Yarus, *Biochemistry* **1993**, *32*, 4475–4480.

48 M. Lindqvist, K. Sandström, V. Liepins, R. Strömberg, A. Gräslund, *RNA* **2001**, *7*, 1115–1125.

49 B. Streicher, U. von Ahsen, R. Schroeder, *Nucleic Acids Res.* **1993**, *21*, 311–317.
50 D. R. Banatao, R. B. Altman, T. E. Klein, *Nucleic Acids Res.* **2003**, *31*, 4450–4460.
51 J. I. Ross, E. A. Eady, J. H. Cove, W. J. Cunliffe, *Antimicrob. Agents Chemother.* **1998**, *42*, 1702–1705.
52 C. A. Trieber, D. E. Taylor, *J. Bacteriol.* **2002**, *184*, 2131–2140.
53 D. Moazed, H. F. Noller, *Nature* **1987**, *327*, 389–394.
54 R. Oehler, N. Polacek, G. Steiner, A. Barta, *Nucleic Acids Res.* **1997**, *25*, 1219–1224.
55 J. W. Noah, M. A. Dolan, P. Babin, P. Wollenzien, *J. Biol. Chem.* **1999**, *274*, 16576–16581.
56 D. E. Brodersen, W. M. Clemons, A. P. Carter, R. J. Morgan-Warren, B. T. Wimberly, V. Ramakrishnan, *Cell* **2000**, *103*, 1143–1154.
57 M. Pioletti, F. Schlünzen, J. Harms, R. Zarivach, M. Glühmann, H. Avila, A. Bashan, H. Bartels, T. Auerbach, C. Jacobi, T. Hartsch, A. Yonath, F. Franceschi, *EMBO J.* **2001**, *20*, 1829–1839.
58 F. Mueller, R. Brimacombe, *J. Mol. Biol.* **1997**, *271*, 524–544.
59 M. Brenowitz, D. F. Senear, M. A. Shea, G. K. Ackers, *Methods Enzymol.* **1986**, *130*, 132–181.
60 K. Juneau, E. Podell, D. J. Harrington, T. R. Cech, *Structure* **2001**, *9*, 221–231.

16
Measuring the Stoichiometry of Magnesium Ions Bound to RNA

A. J. Andrews and Carol Fierke

16.1
Introduction

RNAs are large polyanions containing negative charges on the many backbone phosphodiester groups that interact with positively charged ions in the solution by charge–charge, or Coulombic, interactions [1]. This Coulombic field of negative charge attracts large numbers of positively charged counter-ions. These counter-ions typically consist of both monovalent ions, such as potassium or sodium, and divalent ions, such as magnesium. These ions can both loosely associate with the phosphodiester backbone of RNA mainly by electrostatic interactions, forming an "ionic cloud", or form specific interactions with the RNA backbone and bases to bind in a unique position [2]. Specific binding sites can include direct coordination of the cation by the RNA (inner-sphere interaction) or a hydrogen bond contact via a coordinated water molecule (outer-sphere interaction).

In this chapter we present a method to measure the total number of divalent magnesium ions, including ions that are either specifically or electrostatically bound, that interact with an RNA molecule under a given set of conditions. This method can be useful for determining how the composition of the electrostatic cloud is affected by solution conditions, and whether changes in the RNA structure or the addition of protein cofactors affect the number and composition of interacting ions [3]. For instance, alteration of a nucleotide in RNA or addition of a protein component may reduce the total ionic charge of the RNA and, therefore, the number of associated magnesium ions. Quantification of the number of ions that bind or interact with polyanions, such as RNA or DNA, becomes more difficult as the size of the polyanion, and therefore the number of bound ions, increases. In some cases, ions bound to the RNA can be distinguished from free ions using spectroscopic analysis, e.g. electron paramagnetic resonance spectroscopy has been used to measure the binding of manganese ions to RNA [4]. However, a more general approach is to physically separate the bound and free metal ions and then determine the concentration of ions in both fractions.

The main methods for separating ions bound to large RNAs are gel-filtration chromatography, equilibrium dialysis and forced equilibrium dialysis [5]. Forced

Handbook of RNA Biochemistry: Student Edition. Edited by R. K. Hartmann, A. Bindereif, A. Schön, E. Westhof
Copyright © 2009 WILEY-VCH Verlag GmbH & Co. KGaA, Weinheim
ISBN: 978-3-527-32534-4

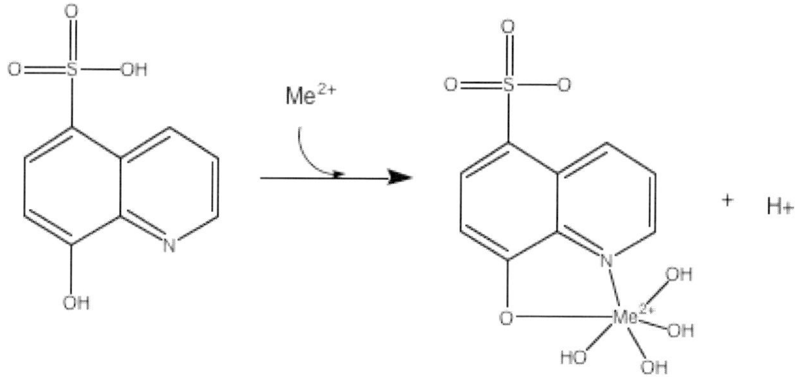

Fig. 16.1. The chemical structure of HQS binding to metal as seen in the nickel crystal structure [10]. The magnesium–HQS complex is assumed to have a similar structure.

dialysis has given the most consistent results and is the easiest to perform; therefore this method will be the main focus of this chapter. Once the bound and free ions are separated, the total concentration of ions in each sample can be measured using a number of techniques [inductively coupled plasma emission mass spectrometry (ICP-MS), atomic absorption or fluorescent dye binding], but most require expensive equipment and in-depth training, except for the fluorescent methods. A number of fluorophores have been used to detect metal ions including 8-hydroxyquinoline-5-sulfonic acid (HQS). HQS at neutral pH is minimally fluorescent, but becomes highly fluorescent upon binding to magnesium and forming a magnesium-8-quinolinol complex (Fig. 16.1). Here, we describe in detail the use of the fluorophore HQS for determining magnesium concentrations following the separation of bound and free magnesium ions.

16.2 Separation of Free Magnesium from RNA-bound Magnesium

Equilibrium dialysis is still the best way to separate small molecules from ones associated with larger molecules under equilibrium conditions. While numerous methods of dialysis exist, from the traditionnal simple dialysis tubing to fully automated machines, the deciding factors are the concentration and volume of RNA required, and the amount of time needed for the dialysis experiment to reach equilibrium. Long equilibration times and large volumes often limit traditional equilibrium dialysis experiments. The ideal method would use a small amount of sample, have a high degree of precision and be rapid. Forced dialysis is an equilibrium method with shorter equilibration times and smaller volumes than traditional dialysis methods. While new and more advanced methods for the dialysis of small samples are available, the use of ultrafiltration devices provides a quick and cheap alternative.

16.3
Forced Dialysis is the Preferred Method for Separating Bound and Free Magnesium Ions

The main advantages of forced dialysis compared to conventional dialysis are: (1) equilibration occurs in the absence of a dividing membrane, which significantly decreases the amount of time needed to reach equilibrium, (2) small volumes (approximately 100 µl) of sample are required, and (3) separation of the free and bound molecules is rapid. The forced dialysis method is accomplished by first allowing the sample to equilibrate at a given set of experimental conditions in the absence of a separating membrane. Separation is then carried out after equilibrium is reached. A simple method for carrying out this separation is to use a Microcon centrifugal filter device (Millipore, Billerica, MA; www.millipore.com) or other similar products. Microcon filter devices are manufactured with membranes that limit the size of the nucleic acid that can pass through the pores, with limitation sizes of 10–300 single-stranded nucleotides. Liquid is forced through the membrane by centrifugation (up to 14 000 g) which separates unbound small molecules from those associated with the larger nucleic acid (Fig. 16.2). Furthermore, equilibrium is maintained throughout the experiment as the concentration of unbound ligand remains constant. Two important notes are that Microcon devices, as manufactured, contain a small amount of glycerin in the filter and the filter has roughly 10 µl retention volume. Before beginning any dialysis experiment, it is necessary to confirm that your ligand of interest can pass through the membrane easily and that neither component will preferentially bind to the dialysis membrane.

The success of this experiment also depends significantly on the purity of the

Forced dialysis method

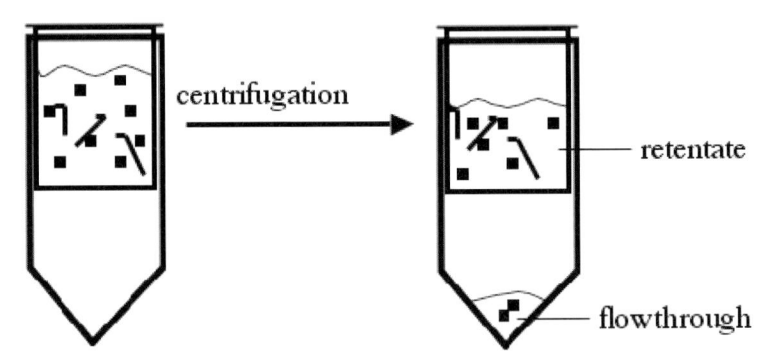

Fig. 16.2. A cartoon illustrating the forced dialysis method. Squares represent magnesium ions and larger lines are RNA molecules. After enough time has elapsed to reach equilibrium, the device is centrifuged so that a small amount of liquid is forced through the dialysis membrane. The retentate contains the bound magnesium plus the free magnesium and the flow-through contains only the free magnesium ions.

RNA and the reagents. RNA samples should be as clean as possible, with both contaminating metals (i.e. magnesium) and chelators (i.e. EDTA) removed. All solutions should be prepared using metal-free tubes and pipette tips. Buffers should be as close to physiological or assay conditions as possible, with close to neutral pH, low monovalent salt concentrations and varying concentrations of magnesium ions. All small molecules with high affinity for magnesium, such as EDTA, should be removed. Excess metal ions can be removed from microfuge tubes by soaking in 100 mM EDTA overnight followed by extensive washing with metal-free water. To remove the glycerin found in the membranes, the device should be washed by centrifugation of at least 500 µl of metal-free water followed by the same amount of buffer. The RNA should be prepared and refolded as usual, although the magnesium concentration should be kept as low as possible.

The minimum concentration of RNA needed for this experiment is dependent on both the binding affinity for magnesium and the stoichiometry for magnesium. There must be sufficient RNA to bind enough magnesium ions such that the total magnesium concentration ($[Mg^{2+}]_{tot} = [Mg^{2+}]_{bound} + [Mg^{2+}]_{free}$) is greater than the free magnesium concentration alone. The following equation (1) demonstrates that for a stoichiometry (n) of 1 and $K_{1/2}$ of 1 mM, the concentration of RNA would need to be 1 mM to see a 2-fold difference between the bound and free fractions:

$$\frac{[RNA \cdot Mg^{2+}]}{[RNA]_{total}} = \frac{n[Mg^{2+}]}{([Mg^{2+}] + K_{1/2})} \tag{1}$$

However, most RNAs have a stoichiometry for bound Mg^{2+} that is much higher than 1, which lowers the required concentration of RNA. For instance, if $n = 100$ and $K_{1/2} = 1$ mM, then only 20 µM RNA is required to achieve equal concentrations of free and bound magnesium. Therefore, as the size of the RNA and the resulting Coulombic field increases, the concentration of RNA required to see a measurable difference in fluorescence decreases. The number of metal ions binding to RNA can be estimated to be between 0.3 and 0.7 M^{2+}/nt [3, 6].

To initiate the experiment, the RNA is diluted into buffer containing magnesium. A recommended buffer volume for this experiment is 100 µl, but this depends on the size of the filter apparatus. This allows for the removal of sufficient volume for the analysis of free magnesium without changing the RNA concentration by a large amount. The half-time for equilibration can be estimated from the K_D and a reasonable guess as to the association rate constant (k_a), assuming a simple two-step binding reaction, as shown in Eq. (2):

$$t_{1/2} = \frac{\ln 2}{k_{obs}} = \frac{0.693}{k_a([Mg^{2+}] + K_D)} \tag{2}$$

Even assuming a value of K_D of 1 µM and a slow second-order association rate constant of 1×10^5 M^{-1} s^{-1}, the calculated $t_{1/2}$ is 3 s suggesting that a 30-s equilibrium time should be sufficient. In practice, the best way to test whether the incubation time is adequate to achieve equilibrium is to demonstrate that doubling (or halving) the equilibration time does not alter the final result.

Once equilibrium is reached, the sample can be added to the pre-washed Microcon device. After adding the RNA to the device, centrifuge the Microcon at 7000 g until a small volume has passed through (around 50 μl). The flow-through can then either be added back to the Microcon or replaced with an equal volume of the original magnesium buffer. Allow the reaction mixture to re-equilibrate (i.e. 15 min at 25 °C) then spin again at 7000 g to collect the flow-through containing only the free magnesium. This sample should be a small percentage (20% or less) of the total volume. This $[Mg^{2+}]_{Free}$ sample is ready for analysis and the $[Mg^{2+}]_{total}$ sample can be obtained directly from the solution that is retained in the top of the Microcon.

16.4
Alternative Methods for Separating Free and Bound Magnesium Ions

Size exclusion chromatography and equilibrium dialysis are alternative methods that can be used to separate free from bound magnesium ions. In all cases, care needs to be taken to make sure all solutions have no other divalent metals beside magnesium and no magnesium chelators.

Gel filtration columns using size-exclusion matrices such as Sephadex G-50 and others can be used to separate small molecules from large RNAs. However, even small columns take on the order of 5 min to run which allows ample time for re-equilibration of the metal ions during the separation. An alternative rapid separation method is gel-filtration spin columns [7, 8]. Using the gel-filtration spin columns, separation can be accomplished on the order of seconds, which greatly reduces the likelihood of re-equilibration during separation. However, for micromolar binding constants, the dissociation rate constant for metal ions from RNA could be on the order of $1-10 \text{ s}^{-1}$ ($t_{1/2} = 0.1-1$ s) indicating that re-equilibration can occur on the same time scale as the separation by spin column method. Therefore, this technique is only applicable for very tight or slowly equilibrating magnesium-binding sites.

Equilibrium dialysis is probably the most well-known and common method of determining the stoichiometry and affinity of ligand binding sites. Equilibrium dialysis is often quite slow, taking many hours to complete since movement through the membrane can be slow [7]. Therefore, it is important to test both the magnesium equilibration time and the RNA stability. As before, great care should be taken to make sure that all solutions and tubes are free from metals or chelators that could interfere with the experiments as well as contaminating RNase. Therefore, the equipment used should be autoclaved if possible and then soaked in 70% ethanol followed by a 100% ethanol wash. After choosing the correct dialysis membrane, the next important step is to determine the equilibration time. This can be estimated from measuring the time required for a magnesium solution to reach equal concentrations in both chambers of the equilibrium dialysis apparatus in the absence of RNA. After the dialysis experiment is complete, the RNA should

16.5
Determining the Concentration of Free Magnesium in the Flow-through

Magnesium standards should be made in the same buffer concentrations as the RNA being tested. The magnesium standards will be used to make a standard curve to determine the concentration of magnesium in the sample. Three very important factors need to be addressed in the use of HQS to measure bound magnesium. (1) The RNA needs to be denatured to prevent high affinity sites from competing with HQS for binding Mg^{2+}. Therefore, guanidine–HCl is included in the assay buffer (5 M guanidine). (2) The pH should be near neutral since, at high pH, HQS ionizes which increases the background fluorescence and decreases the sensitivity of the metal analysis. (3) HQS should be in high enough concentration to completely bind the available magnesium ions. Therefore, the HQS concentration should be much larger than both the magnesium dissociation constant (K_D) and the $[Mg^{2+}]$. In summary, the recommended assay conditions are 5 M guanidine–HCl, 0.1 M Tris, pH 8 and 2 mM HQS. A small volume of the sample to be tested (20 µl or less) is mixed with 150 µl of the HQS solution and the fluorescence is then measured in a 120-µl cuvette at 25 °C. The excitation wavelength is set at 397 nm and the emission wavelength is 502 nm. At this point, the experimentally determined fluorescence can be compared to the standard curve of fluorescence as a function of magnesium concentration under the exact same conditions to calculate the concentrations of magnesium ions in the experimental samples. See Fig 16.3.

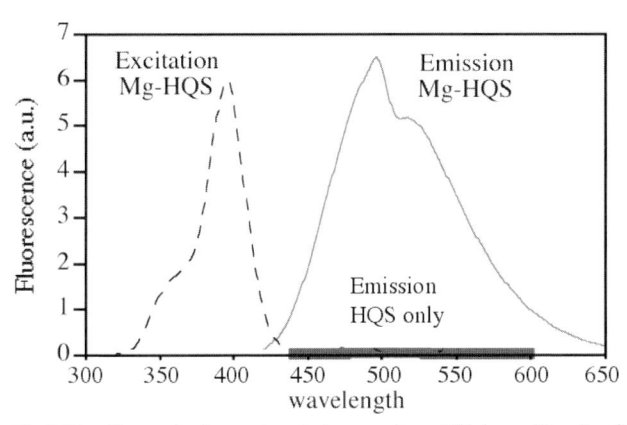

Fig. 16.3. The excitation and emission spectra of Mg^{2+}–HQS with the emission spectra of HQS shown as a reference. The spectra were collected on an Amico-Bowman Series 2 spectrometer (ThermoSpectronic, Rochester, NY) by exciting the flourophore at 397 nm. The slit width for both excitation and emission spectra was kept at 8 nm. The sample was placed in 0.1 M Tris, pH 8, 2 mM HQS and with or without 1 mM magnesium at 25 °C.

16.6
How to Determine the Concentration of Magnesium Bound to the RNA and the Number of Binding Sites on the RNA

In the forced dialysis method, the concentration of bound magnesium ions $[Mg^{2+}]_{bound}$ can be calculated by subtracting the concentration of free magnesium $[Mg^{2+}]_{free}$ determined from the flow-through from the total magnesium concentration $[Mg^{2+}]_{total}$ determined from the retentate. Similarly, in a standard equilibrium dialysis experiment the concentration of bound magnesium can be determined by measuring the difference between the magnesium concentration on the two sides of the dialysis membrane. The $[Mg^{2+}]_{free}$ is measured on the side without the RNA, while the concentration on the side with RNA equals the combination of RNA–magnesium and free magnesium ($[Mg^{2+}]_{bound} + [Mg^{2+}]_{free}$). Therefore the concentration of bound magnesium ions can be determined by subtracting the magnesium concentration on the "free" side from the concentration on the side with the RNA:

$$[Mg^{2+}]_{bound} = ([Mg^{2+}]_{bound} + [Mg^{2+}]_{free}) - [Mg^{2+}]_{free} \tag{3}$$

As stated before, important control experiments include demonstrating that the membrane does not bind significant magnesium ions and that the RNA of interest does not affect the determination of the magnesium concentration. If the RNA competes with HSQ for binding magnesium, you can either: (1) decrease the magnesium affinity of the RNA by adding RNase, (2) measure the standard curve for magnesium in the presence of a known concentration of the competing RNA or (3) estimate the bound magnesium solely from the free magnesium by subtracting two times the free magnesium $[Mg^{2+}]_{free}$ from the total magnesium added:

$$[Mg^{2+}]_{bound} = [Mg^{2+}]_{total} - 2[Mg^{2+}]_{free} \tag{4}$$

Once the concentrations of bound and free magnesium are determined at different magnesium concentrations, the $K_{1/2}$ for magnesium binding and the stoichiometry, n, can be determined by fitting Eq. (5) to these data (Fig. 16.4) [3, 7]. This analysis assumes that the magnesium is binding to the RNA via multiple, independent binding sites [5], where v equals magnesium bound divided by total RNA added. This assumption can be tested by making a Scatchard plot where $v/[Mg^{2+}]_{free}$ is plotted versus v; the slope of a linear fit equals $1/K_{1/2}$ (Eq. 6). If only one type of site is observed, the Scatchard plot will be linear. If multiple types of binding sites are observed, the Scatchard plots will not be linear [5]. In the case of most large RNAs, the ratio of specifically bound ions to loosely bound or interacting magnesium ions is so small that only one class of ions will be seen.

$$v = \frac{\left(\dfrac{n[Mg^{2+}]_{free}}{K_{1/2}}\right)}{\left(1 + \dfrac{[Mg^{2+}]_{free}}{K_{1/2}}\right)} \tag{5}$$

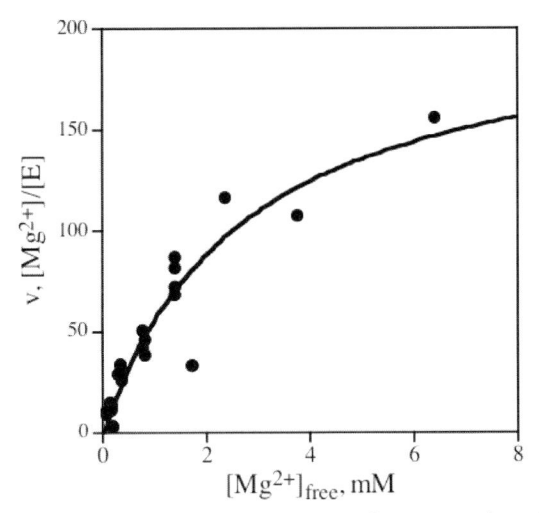

Fig. 16.4. A plot showing the amount of magnesium bound to the RNase P holoenzyme at various concentrations of magnesium ions determined using the forced dialysis method [3]. The stoichiometry of magnesium bound to the holoenzyme is 160 Mg^{2+} per enzyme and the $K_{1/2}$ is 1.5 mM [3].

The data can also be plotted as a Scatchard plot which results in a linear correlation:

$$\frac{v}{[Mg^{2+}]_{free}} = \frac{n}{K_{1/2}} - \frac{v}{K_{1/2}} \tag{6}$$

In this analysis we have made the assumption that magnesium ions will bind to RNA in a non-cooperative fashion (Eq. 1). However, if the Mg^{2-} ions bind cooperatively, the fraction of magnesium bound to RNA will be described by Eq. (7), so that the binding can be measured and analyzed in a manner similar to what has been described for non-cooperative binding.

$$\frac{[RNA \cdot Mg^{2+}]}{[RNA]_{total}} = \frac{[Mg^{2+}]}{([Mg^{2+}]^n + K_{1/2})} \tag{7}$$

16.7
Conclusion

This method is a straightforward way to quantitate the number of metal ions directly interacting with RNA by using common laboratory equipment. The important step in this method that is missing in other systems is the separation of free and bound magnesium ions. Other systems add HQS directly to the RNA and measure competition of the two ligands for the magnesium ions [9]. While these methods may work for one or two binding sites, the analysis of this type of competition experiment becomes extremely complicated as the number of binding sites increases. The measurement of the number of metal ions interacting with an RNA molecule under various conditions will advance our knowledge of RNA–metal interactions and will be useful for testing the validity of RNA modeling techniques.

16.8
Troubleshooting

(1) If you are using Microcons, make sure that you have removed the glycerol from the filter and you have not exceeded the maximum g force.
(2) Confirm that RNA, but not magnesium, is retained by the membrane and that neither RNA nor magnesium sticks to the membrane.
(3) Make sure that all of the solutions, tips and tubes are free from contaminating metals, chelators and RNases.
(4) Determine that the RNA is folded and stable throughout the experiment.
(5) Demonstrate that the reaction was incubated for sufficient time to reach equilibrium. Make sure that the magnesium in the experimental samples is determined under exactly the same conditions as the magnesium standard curve (buffer, temperature, wavelengths, voltage, etc.).

References

1 G. S. MANNING, J. RAY, J. Biomol. Struct. Dyn. **1998**, 16, 461–476.
2 V. K. MISRA, D. E. DRAPER, Biopolymers **1998**, 48, 113–135.
3 J. C. KURZ, C. A. FIERKE, Biochemistry **2002**, 41, 9545–9558.
4 V. J. DEROSE, Curr. Opin. Struct. Biol. **2003**, 13, 317–324.
5 C. R. CANTOR, P. R. SCHIMMEL, Biophysical Chemistry, Freeman, New York, 1980.
6 A. L. FEIG, O. C. UHLENBECK, The role of metal ions in RNA biochemistry, in The RNA World, 2nd edn, R. F. GESTELAND, T. R. CECH, J. F. ATKINS, (eds), Cold Spring Harbor Laboratory Press, Cold Spring Harbor, NY, **1999**, pp. 287–319.
7 J. A. BEEBE, J. C. KURZ, C. A. FIERKE, Biochemistry **1996**, 35, 10493–10505.
8 J. J. GARCIA, A. GOMEZ-PUYOU, E. MALDONADO, M. TUENA DE GOMEZ-PUYOU, Eur. J. Biochem. **1997**, 249, 622–629.
9 V. SEREBROV, R. J. CLARKE, H. J. GROSS, L. KISSELEV, Biochemistry **2001**, 40, 6688–6698.
10 S. B. RAJ, P. T. MUTHIAH, G. BOCELLI, L. RIGHI, Acta Crystallogr. **2001**, E57, m591–m594.

17
Nucleotide Analog Interference Mapping and Suppression: Specific Applications in Studies of RNA Tertiary Structure, Dynamic Helicase Mechanism and RNA–Protein Interactions

Olga Fedorova, Marc Boudvillain, Jane Kawaoka and Anna Marie Pyle

17.1
Background

17.1.1
The Role of Biochemical Methods in Structural Studies

Recent advances in structure determination of RNA and RNA–protein complexes by diffraction and NMR methods have radically expanded our understanding of RNA architecture [1–5]. Due to the complexity and resolution of these structures, the role of accompanying biochemical studies for construct design, testing of function and, ultimately, for interpretation of high-resolution structural data has never been greater. Classical methods such as photo-crosslinking, footprinting and chemical modification interference remain powerful tools [6]. However, a new set of biochemical tools for high-resolution structure determination and testing is now available in the form of Nucleotide Analog Interference Mapping (NAIM) and Nucleotide Analog Interference Suppression (NAIS) [7–10]. These methods are based on the selection for functionally active RNA molecules. Structural constraints provided by NAIM/NAIS experiments are uniquely powerful because they reflect the active RNA conformation. Therefore data from NAIM/NAIS experiments are particularly informative for meaningful structure–function analysis. NAIM and NAIS also provide biochemical data at an unprecedented level of resolution, as they interrogate individual atoms, and predict specific hydrogen bonds and RNA interaction motifs [9–12].

Even if it were possible to solve the crystal structure of every interesting RNA, the classical and NAIM/NAIS methods would remain essential, complementary approaches. The biochemical approaches can help distinguish functionally relevant structural information from crystal packing artifacts. NAIM and NAIS can probe whether hydrogen bonds predicted from crystallographic studies are actually important for molecular function or activity [5]. When a crystal structure provides important clues about catalytic mechanism, chemical details and models can be probed through biochemical methods. Finally, the very success of crystallographic

Handbook of RNA Biochemistry: Student Edition. Edited by R. K. Hartmann, A. Bindereif, A. Schön, E. Westhof
Copyright © 2009 WILEY-VCH Verlag GmbH & Co. KGaA, Weinheim
ISBN: 978-3-527-32534-4

studies often rests on the ability to design a functional construct that is capable of crystallizing. Biochemical methods guide in the identification of stable RNA structural domains and provide tests for activity of the resultant RNA. Thus, at all stages of investigation, high-resolution RNA structure determination is aided by a diverse arsenal of biochemical methods.

There remain many important RNA molecules that have eluded high-resolution analysis by diffraction and NMR. Even if one form of a structure is known, there is often no information about alternative molecular conformations that are important for function. High-resolution information is often found for small, stable subdomains of larger molecules which, in their intact form, have not been possible to crystallize [13]. Thus, there remains ample room for the application of biochemical methods in structure determination and modeling. A diversity of methods have historically been used to provide distance constraints and to elucidate long-range tertiary interactions [14–16]. This type of information has facilitated the construction of three-dimensional (3-D) models for RNA molecules. While methods such as UV crosslinking and footprinting have been valuable tools in modeling efforts, they are unable to differentiate active from inactive conformations of a molecule and they are not applicable for testing the role of specific RNA functional groups. Thus, there has been a need for biochemical approaches that provide functional information at high resolution.

Mutational analysis has traditionally been a powerful tool for elucidating the contribution of functional groups in an RNA [17]. In these studies, the identity of an entire nucleobase is changed or, in a more precise adaptation, a single functional group is altered. The latter modifications are incorporated through chemical synthesis of RNA (if the RNA is no longer than around 50 nt) or by a combination of chemical and enzymatic syntheses (for longer RNAs). The modified species are then analyzed for function in parallel with unmodified RNA (i.e. to evaluate binding or a reaction of interest). The relative importance of the functional group is then determined from the difference in activity between modified and unmodified molecules. This approach necessitates the synthesis of many different modified RNA molecules, in which important atoms are changed one at a time. The systematic screening of functionalities by this method requires an enormous synthetic effort. The approach has been successfully applied in many cases, including the determination of tertiary interactions between the group I intron core and its substrate helix [18], and the identification of catalytically important functional groups on domain 5 of a group II intron [17]. However, the method is simply inapplicable for screening the importance of functional groups throughout large RNA molecules such as group II introns, which are often around 1000 nt in length.

The problem of identifying specific RNA atoms that are essential to function is solved by NAIM. In a single experiment, this method screens a combinatorial library of modified RNA molecules for atoms that are important for function. The RNA of interest is prepared by *in vitro* transcription in the presence of a nucleoside analog thiotriphosphate that contains desired sugar or base modifications. Because it is susceptible to cleavage by iodine, the phosphorothioate linkage serves as a tag that reflects the location of an important modified residue. As the first application

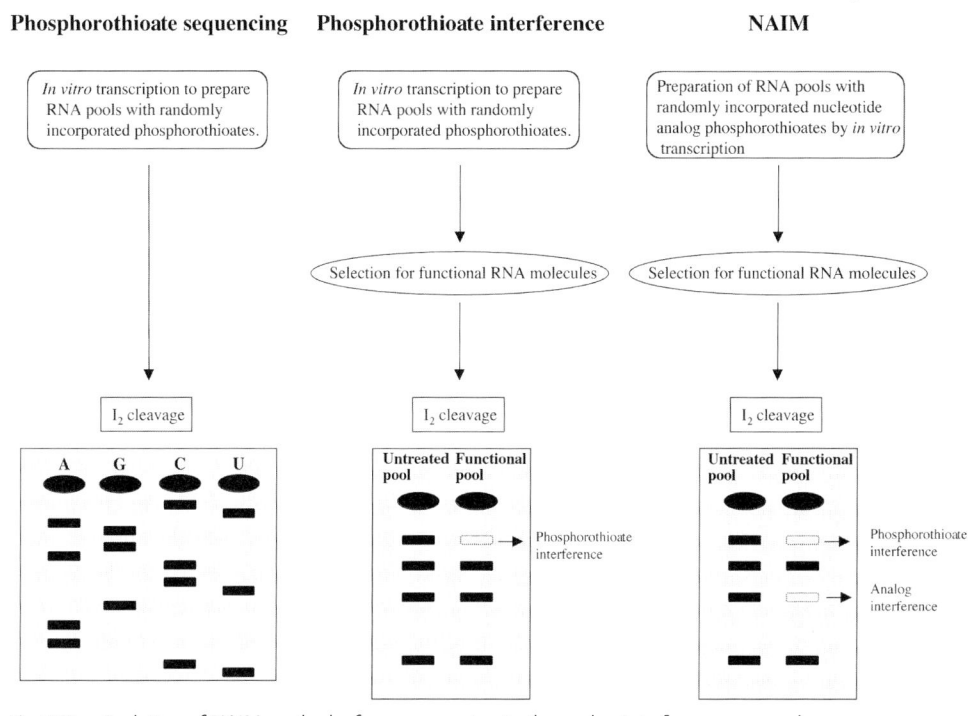

Fig. 17.1. Evolution of NAIM methods: from sequencing to the analog interference approach.

of phosphorothioate tagging, Gish and Eckstein reported a technique for RNA and DNA sequencing that was based on selective susceptibility of a phosphorothioate linkage to iodine cleavage (Fig. 17.1) [19]. The method originally employed iodoethane as an activating agent, which alkylated the sulfur, thus making the phosphorus more susceptible to a nucleophilic attack by an adjacent 2'-OH group followed by strand scission [19]. Iodoethane has since been replaced by ethanolic iodine solution. The exact mechanism of iodine interaction with the phosphorothioate linkage has never been studied in detail, contrary to iodoethane cleavage where all the possible pathways and products have been thoroughly investigated [19]. The putative mechanism of iodine cleavage, assuming that it is similar to iodoethane cleavage, is shown in Fig. 17.2.

Although not discussed further here, it is notable that these investigators also proposed a similar method for chemical probing of DNA. In that case, iodoethanol replaces iodine as the activating electrophile. While the nucleophile for strand scission of RNA is the adjacent 2'-OH group (Fig. 17.2), in DNA it is the hydroxyl group of the ethanol moiety [20].

The phosphorothioate tagging method quickly evolved when several research groups realized that it could be coupled with a selection to identify functionally important phosphate groups in an RNA (Fig. 17.1) [21–24]. The approach was taken

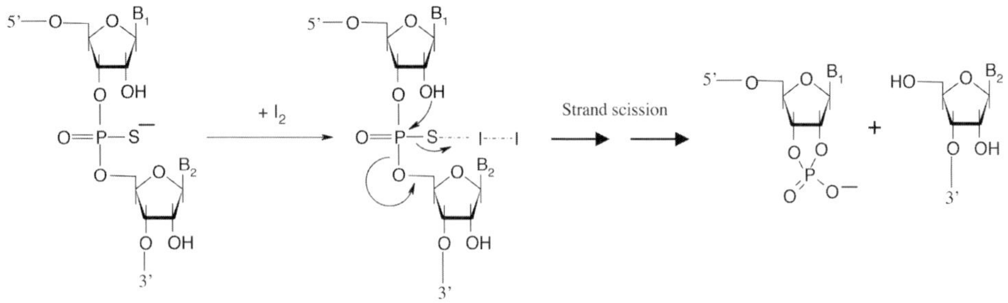

Fig. 17.2. The putative mechanism of cleavage of phosphorothioate-modified RNA by iodine. Iodine electrophilically attacks the sulfur, thus making phosphorus more susceptible to the nucleophilic attack by the adjacent 2′-hydroxyl. Sulfur (path a) or oxygen (path b) can be the leaving group. Products of interest, which make this reaction applicable for NAIM/NAIS, are formed as a result of strand scission. The actual reaction mechanism is still under discussion [19, 24].

yet to another level when it was expanded to include other nucleotide analogs that contained sugar and base modifications [11, 25–30]. In this approach, i.e. NAIM, the phosphorothioate linkage was still used as a tag, but it was incorporated along with a second modification of interest on either the base or the sugar. After a selection step, the phosphorothioate linkage reported not only the positions of important phosphate atoms, but also of other nucleotide atoms that influence function (Fig. 17.1). Recent studies have applied NAIM to the analysis of RNA structural motifs [11, 31], metal binding sites [32, 33] and mechanisms of ribozyme catalysis [34–36]. NAIM subsequently evolved into NAIS, which permits the precise identification of tertiary interaction partners [9, 12, 28]. In combination, the two methods now represent the most powerful biochemical method for RNA structure determination and analysis.

17.1.2
NAIM: A Combinatorial Approach for RNA Structure–Function Analysis

17.1.2.1 Description of the Method
NAIM involves the following steps [7].

(1) Preparation, by *in vitro* transcription, of an RNA library that is doped with a small population of modified phosphorothioate nucleotide analog.
(2) Selection and separation – a process of interest (i.e. binding, conformational change or catalytic reaction) is used to separate active from inactive pools of RNA.
(3) Visualization of interferences – iodine cleavage is used to detect the locations of modified atoms that affect function (Fig. 17.3A). In order to detect cleaved molecules, the RNA pool must be labeled at either the 3′ or 5′ terminus. In some cases, precursor molecules are labeled after transcription [7, 8], while in

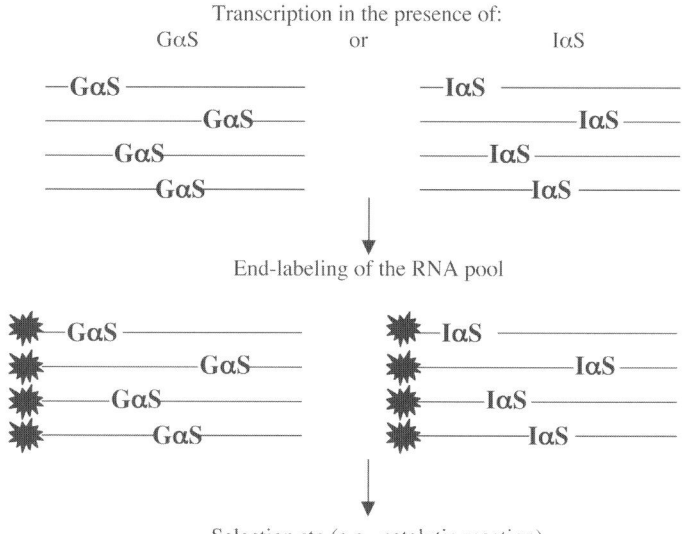

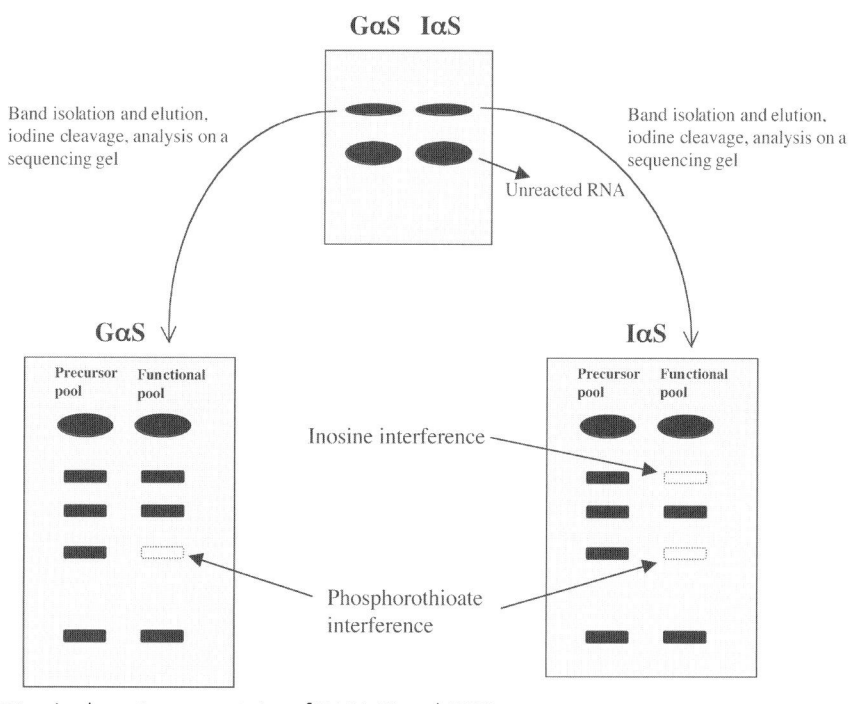

Fig. 17.3. A schematic representation of NAIM (A) and NAIS (B) assays. Note that in (B) the term "mutant RNA 1" is used not to describe a classical base mutation, but a single-atom substitution (2'-OH is changed to 2'-H). In NAIS literature this is frequently done for the purpose of simplicity.

264 17 Nucleotide Analog Interference Mapping and Suppression

B

The selection step is carried out in the presence of

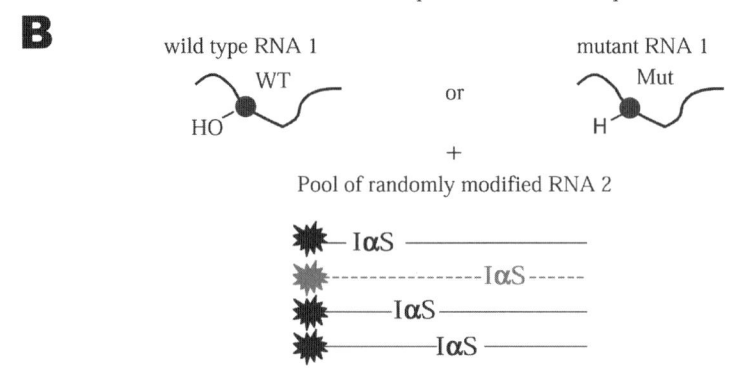

wild type RNA 1   or   mutant RNA 1

WT   Mut

HO⁻   H

+

Pool of randomly modified RNA 2

— IαS
-------------- IαS ------
——— IαS ———
————— IαS ———

A functionally important tertiary interaction:

WT

Precursor / Functional pool

→ interference

is disrupted for molecules in RNA 2 pool with an inosine modification at this position

Mut

Precursor / Functional pool

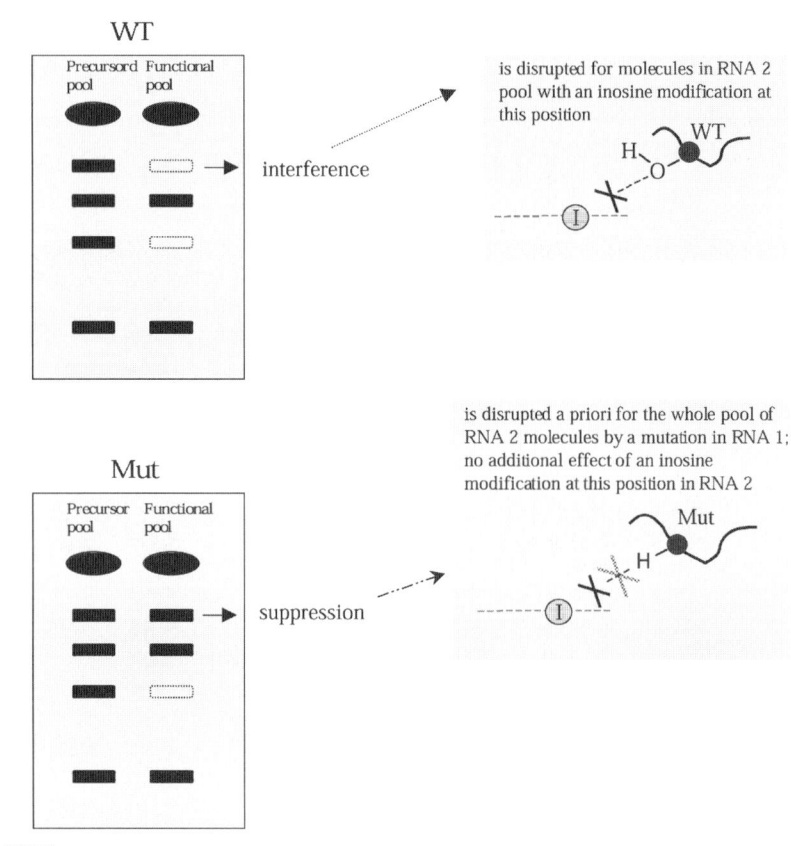

→ suppression

is disrupted a priori for the whole pool of RNA 2 molecules by a mutation in RNA 1; no additional effect of an inosine modification at this position in RNA 2

Fig. 17.3.B

other studies, functional molecules are radiolabeled during the selection step [28].

In preparing the pool of randomly modified RNAs, the ratio of a thiotriphosphate analog to a corresponding unmodified nucleoside triphosphate in the transcription mixture is adjusted so that the analog is incorporated with an efficiency of one or two modifications per molecule. Incorporation levels of up to 5% are considered satisfactory [8]. For optimal efficiency of the polymerase and for ease of analysis, only one modified analog is usually included at a time in a given transcription reaction [7, 8].

A typical NAIM experiment is exemplified as follows. To determine the functional importance of guanosine exocyclic amino groups in an RNA, it is transcribed in the presence of inosine phosphorothioate nucleotide triphosphates. The resulting pool of RNA molecules has a statistical probability of including a single inosine substitution at each guanosine position in the RNA (Fig. 17.3A). The RNA pool is then bound to another molecule or it undergoes a catalytic reaction of interest. If the inosine modification is located at a position where it interferes with binding or reaction, this RNA molecule will be depleted from the pool of functional molecules. The fraction of reacted molecules is then separated from unreacted molecules using electrophoresis or an affinity column, although separation methods vary widely depending on the application of interest. Then the reacted fraction (functional pool) as well as the precursor RNA pool and/or the unreacted fraction are treated with iodine and analyzed on a denaturing polyacrylamide gel (Fig. 17.3A). If the modification at a certain position interferes with function, the corresponding band will be missing or underrepresented in the lane that represents the pool of active RNAs (Fig. 17.3A). Since the phosphorothioate linkage itself can interfere with activity, all NAIM assays contain a control experiment where only phosphorothioate is incorporated and its effects on function are analyzed in parallel.

17.1.2.2 Applications

Identifying Atoms and Functional Groups Critical for Ribozyme Activity
NAIM has been widely used to map chemical groups that are important for the catalytic function of ribozymes such as the group I intron [11, 26, 27, 37], group II intron ribozymes [28] and the hairpin ribozyme [29]. Applications to the study of RNase P are discussed in Chapter 18.

Generally, the first insight into structural organization and 3-D architecture of a large ribozyme is through phylogenetic analysis, which allows identification of highly conserved nucleotides and, in some cases, tertiary interactions [38–40]. NAIM takes this type of approach to the next level and obtains information with atomic resolution. Since most tertiary interactions involve a network of contacts among sugar-phosphate backbone residues, traditional phylogenetic approaches often fail to identify critical tertiary contacts. However, an alternative form of phylogenetic information can be obtained by comparing the NAIM profiles of ribo-

NAIM "Signatures" for Elucidation of RNA Structural Motifs

Although NAIM provides a map of chemical groups that are important for RNA function, it remains challenging to translate this into concrete information on RNA tertiary structure. One can distill the code of interference patterns into probable types of structural motifs through judicious choice of modified nucleotides and careful comparison with known structures. For example, if an adenosine residue exhibits interference with 7-deazaAαS, m^{6}AαS, PurαS and 2APαS, it is likely to form hydrogen bonds from its Hoogsteen face [11, 41]. Similarly, if N^2-methyl guanosine or inosine substitutions interfere with a specific guanosine, its minor groove is likely to be involved in a tertiary interaction [27]. Further examples of analogs are provided in Fig. 17.4 and in [8].

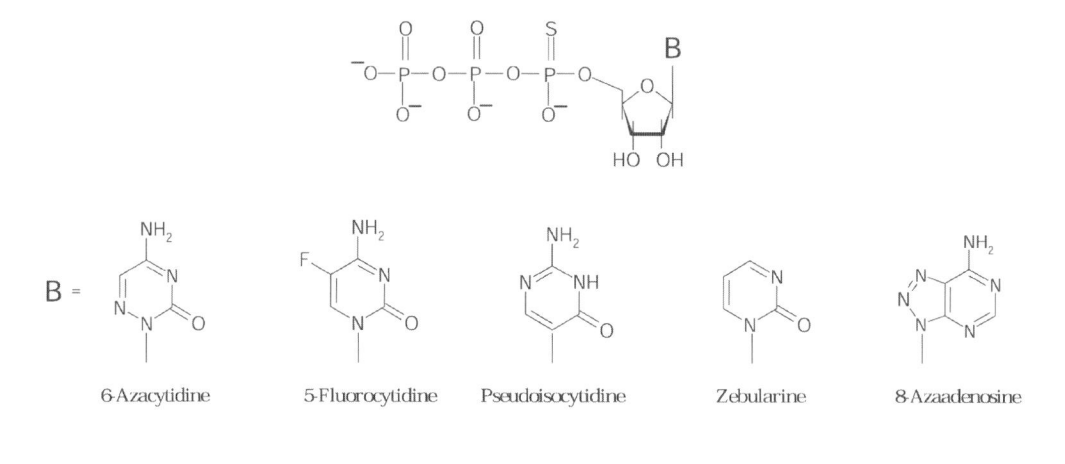

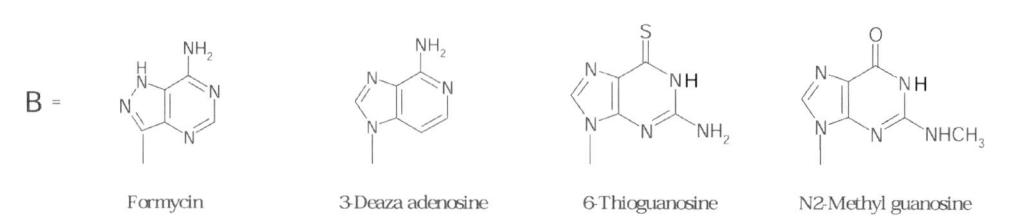

Fig. 17.4. Nucleotide analog α-thiotriphosphates synthesized by the Strobel group. These analogs are not commercially available, but can be synthesized using published protocols [8, 64, 66]. Commercially available analogs can be found in the Glen Research catalog (www.glenres.com).

The growing number of RNA crystal and NMR structures provide an invaluable library of known RNA structural motifs. From the hydrogen bonding patterns in these motifs, one can predict a pattern of interferences with certain nucleotide analogs and then screen an RNA of interest for such patterns. For example, using the crystal structures of two different ribozymes [2, 42], one can predict that a GAAA tetraloop involved in an interaction with a cognate tetraloop receptor will exhibit 2′-deoxy and 7-deaza adenosine interferences at the second and third adenosine positions of the loop. Such a pattern has been observed for a GAAA tetraloop on domain 5 of a group II intron, confirming its interaction with a cognate receptor in domain 1 [28]. Another very common RNA packing motif is the A-minor motif, which involves the interaction of an adenosine's N^1, N^3 and/or 2′-OH functional groups with 2′-OH groups and the minor groove edges of a GC or a GU base pair [43]. Using a combination of 2′-deoxy AαS and 3-deazaAαS analogs, Strobel et al. have identified A-minor motifs in the active site of the group I ribozyme [10, 31].

Identifying Nucleobases Involved in Ribozyme Catalysis

Proton transfer represents an important catalytic strategy of many protein enzymes. General acid or base catalysis usually requires ionization of specific functionalities, such as a histidine side chain. In some cases, the function of an important side chain requires perturbation of its normal pK_a [44]. An increasing number of ribozymes have now been shown to effect catalysis via proton transfer [45–47] and, in certain cases, this activity appears to be divalent metal-ion independent [48–50]. This suggests that nucleobases themselves may be directly involved in proton transfer during catalysis, particularly if they undergo a shift in pK_a value. The adenosine N^1/N^3 and cytosine N^3 functionalities have most typically been implicated, as they are most easily protonated. Strobel et al. have established specialized NAIM assays and analogs for identifying nucleotides with shifted pK_a values [34–36]. For this purpose, a series of adenosine and cytosine analogs have been designed that probe pK_a values at N^3 (Fig. 17.4). The resulting interference patterns have then been studied as a function of pH [34–36]. Nucleotide positions that show pH-dependent interferences with the new analogs are good candidates for participation in catalysis via proton transfer. However, unambiguous conclusions about direct involvement in catalysis cannot be made on the basis of NAIM alone. There is always the possibility that an ionized nucleobase participates in a pH-dependent tertiary interaction or folding event that is not directly involved in catalytic activity.

Locating Metal-binding Sites in RNA

The incorporation of single phosphorothioates (Rp or Sp isomers) at certain positions, followed by kinetic analysis in the presence of magnesium versus different thiophilic metal ions, has long been instrumental in establishing and characterizing catalytically important metal-binding sites in RNA [51–55]. When phosphorothioate substitution results in a drastic decrease in catalytic activity that is rescued by the addition of a thiophilic metal ion, such as Mn^{2+} or Cd^{2+}, then the corresponding Rp or Sp oxygen is implicated in direct coordination with Mg^{2+}. Random incorporation of phosphorothioate analogs through NAIM can be used to screen for potential metal-binding sites. However, the experiment is limited to Rp-phosphorothioates due to stereospecific incorporation of Sp-α phosphorothioate NTPs by T7 RNA polymerase. This NAIM approach has been explored in a variety of ribozyme systems [56–58], including the *Tetrahymena* group I intron [59]. Basu and Strobel revisited this system and found that metal ion rescue of 2′-deoxy or phosphorothioate interference effects may not always have been due to Mg^{2+} coordination [33], as metal binding at these sites was not confirmed by X-ray crystallography. However, they found that reduction of Mn^{2+} concentration to 0.3 mM eliminates all non-specific rescue artifacts and produces data that were in a good agreement with crystallographic results. Other thiophilic metal ions are sometimes used in rescue experiments [22, 52, 53]. However, in some cases they can inhibit ribozyme activity [60] or produce incomplete subsets of metal-binding sites [33]. The possibility of artifacts in metal rescue experiments underscores the need for proper controls and cautious interpretation of metal rescue data.

NAIM is not limited to the detection of divalent metal ion binding sites. Strongly associated monovalent ions can also be detected if they bind to a site through interaction with guanosine and uracil ketone groups. To probe for the presence of site-bound potassium ion, RNA is transcribed in the presence of 6-thioguanosine thiotriphosphate (Fig. 17.4) [32, 61]. If interference is observed, a metal ion rescue experiment is attempted using thiophilic thallium ion, which has properties similar to those of potassium. This approach resulted in the identification of a unique potassium-binding site located below the AA platform of a GNRA tetraloop receptor [32].

Determining the Free Energy Contributions of Individual Functional Groups on RNA

To identify RNA functional groups that are important for interacting with other molecules, one can set up an NAIM experiment in which binding is used for selection. A free energy profile can be constructed for each site of interference by measuring the magnitude of NAIM effects as a function of ligand concentration [62]. The efficacy of this approach has been successfully demonstrated on the signal recognition particle (SRP) system where the results obtained from NAIM experiments were in good agreement with mutational analysis [62].

17.1.3
NAIS: A Chemogenetic Tool for Identifying RNA Tertiary Contacts and Interaction Interfaces

17.1.3.1 **General Concepts**

Although NAIM is useful for identifying the locations of important atoms in an RNA molecule, it does not provide information on how these atoms interact with each other and it does not reveal tertiary interaction partners. However, by comparing the NAIM profiles of a wild-type molecule with that of a mutant, one can discern patterns of energetic communication and tertiary contact in a system. This variation of the NAIM experiment, which is called NAIS, is generally carried out after all interference (NAIM) data has been collected. The original NAIM patterns are used as a guide for the design of mutants with a critical nucleobase or a highly important atom altered, which are then produced by transcription or chemical synthesis. The resulting mutants are usually much less reactive than wild-type RNA, as important tertiary interactions are disrupted by introduced mutations.

In a two-part system for probing tertiary contacts along an interaction interface between RNA molecules (Fig. 17.3B), one of the RNA molecules (RNA 1) is typically mutated. The other RNA molecule (RNA 2) is transcribed in the presence of nucleotide analogs. The latter RNA is then screened for changes in its interference pattern upon reaction with either the mutant or the wild-type versions of RNA 1. Assume, for example, that the 2'-OH group of a nucleotide A in RNA 1 interacts with the exo-cyclic amine of the guanosine residue at position B in RNA 2 (Fig. 17.3B) and that this interaction is critical for a selectable function. Upon reaction with wild-type RNA 1, RNA 2 will exhibit strong interference with an inosine and/or 2-methyl guanosine analog at position B. However, when RNA 2 reacts with a mutant RNA 1 (under conditions altered to result in the same extent of reaction as with wild-type RNA 1) that contains a single 2'-deoxynucleotide at position A, the interaction between nucleotides A and B will have already been disrupted and the penalty for this disruption will have already been paid in the form of lesser reactivity by mutant RNA 1. Therefore, when RNA 2 interacts with the specified mutant RNA 1, there will be no interference observed at position B (Fig. 17.3B). This effect is generally referred to as "interference suppression". Note that this is not a "rescue", because the disappearance of the interference effect is not attributed to a restored interaction, but rather to elimination of an interaction partner.

17.1.3.2 Applications: Elucidating Tertiary Contacts in Group I and Group II Ribozymes

NAIS as a method to identify long-range tertiary contacts in large ribozymes was first demonstrated during studies on group I intron ribozymes. There it was used to predict an extended minor groove triple helix between the P1 helix and the single-stranded J8/7 region, in which each triple appeared to be mediated by at least one 2'-hydroxyl group [9]. In the group II intron ribozymes, for which little structural information exists, NAIS identified two tertiary interactions, $\kappa-\kappa'$ and $\lambda-\lambda'$, that are essential for visualizing active-site architecture in the group II intron core (Fig. 17.5) [12, 28]. Hydrogen bonding contacts elucidated by NAIS have been extensively used as constraints for building models of group I and II ribozyme active sites [9, 63].

In subsequent Sections 17.2 and 17.3, we will illustrate applications of NAIM and NAIS, respectively, by using three diverse systems as examples: (1) ribozyme studies on a group II intron, (2) studies of dynamic interactions between protein and RNA during the NPH-II helicase reaction, and (3) studies of RNA–protein interactions that are involved in transcription termination. Detailed protocols will be provided for each NAIM/NAIS step as it applies to these systems, although they can be tailored to meet the specific needs of the reader. To avoid repetition, Section

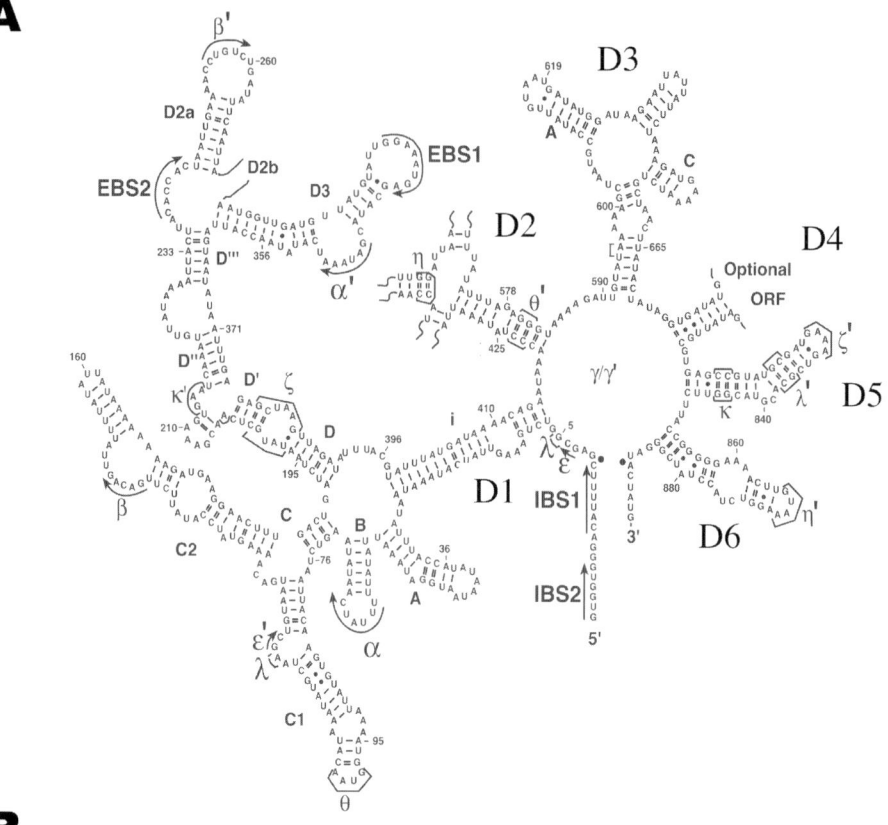

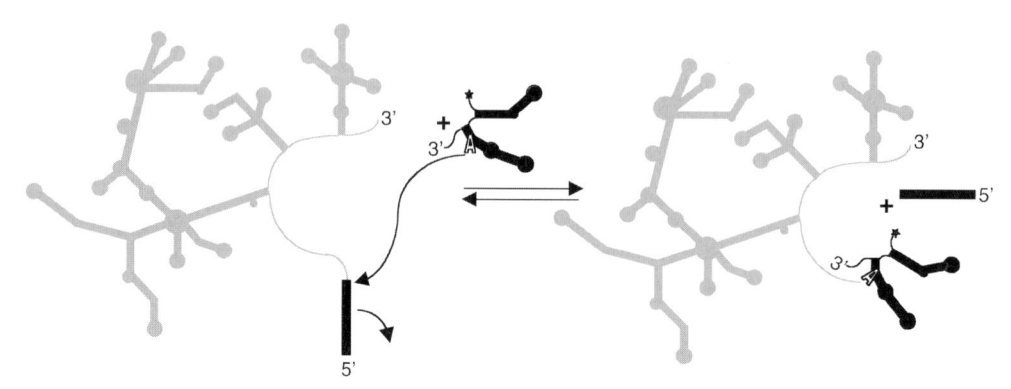

Fig. 17.5. (A) Secondary structure of the ai5γ group II intron. (B) Schematic representation of the *trans*-branching reaction used as a selection step in NAIM and NAIS studies of the ai5γ group II ribozyme.

17.2 begins with the basic protocols and methods that are used for all applications of NAIM. These are followed by specific examples, which differ primarily in the selection method used for NAIM. NAIM protocols will then be followed by NAIS methodologies that were developed to study group II intron tertiary structure (Section 17.3).

17.2 Experimental Protocols for NAIM

17.2.1 Nucleoside Analog Thiotriphosphates

Many phosphorothioate nucleotide analogs are commercially available from Glen Research (for European suppliers, see Chapter 18). Strobel et al. have expanded the existing collection by developing strategies and protocols for the efficient synthesis of other analogs (Fig. 17.4) [8]. Most analogs can be synthesized from unprotected nucleosides [generally available from Sigma, R.I. Chemical (Orange, CA) or ChemGenes (Wilmington, MA)] in two steps by adapting the procedure first described by Arabshahi and Frey [64]. The first step involves reaction of a nucleoside with thiophosphoryl chloride, resulting in the formation of 1,1-dichloro-1-thionucleoside phosphate, which is then converted to a nucleotide thiotriphosphate by reaction with tributylammonium pyrophosphate [8]. The synthetic procedures described above are not suitable for some cytidine analogs, which require the use of a salicyl phosphoramidite approach [65]. The synthesis of these analogs is described in detail by Oyelere and Strobel [66].

17.2.2 Preparation of Transcripts Containing Phosphorothioate Analogs

RNA molecules that contain randomly incorporated phosphorothioate analogs are prepared by *in vitro* transcription from a double-stranded DNA template (usually either a PCR product or a restriction digest of plasmid DNA). Transcription efficiency and fidelity tend to be low, however, unless the DNA template contains blunt or 5′-overhanging ends. Thus, restriction enzymes must be chosen accordingly [67].

To optimize transcription efficiency, only one phosphorothioate NTP analog at a time is usually added to the transcription reaction. Since in many cases nucleotide analog α-thiotriphosphates are not incorporated into the transcript with the same efficiency as unmodified triphosphates, the ratio of NTPαS to NTP needs to be optimized in order to achieve the desired level of analog incorporation (generally 5%). For commercially available analogs, such an optimization has already been performed and they are sold as 10× solutions at a concentration that is adjusted for 5% incorporation. For example, while conventional NTPs are added to the transcription at 1 mM final concentration, the modified analogs IαS and m^6AαS

Tab. 17.1. Concentrations (mM) of most common NTPαS analogs (and parental rNTP) required for around 5% NαS incorporations per transcript[1,2].

Incorporation	[NTPαS]	[rNTP]	RNAP
AαS, CαS, GαS, UαS	0.05	1	WT
dAαS, dCαS	1.5	1	Y639F
dGαS, dUαS	0.5	1	Y639F
IαS, m^{6}AαS	0.4	1	WT
C^7-AαS, C^7-GαS	0.1	1	WT
PurαS, 2APαS	2	0.5	WT
DAPαS	0.025	1	WT
m^{2}GαS	1.5	0.5	Y639F

[1] Further details on *in vitro* transcription with the Y639F mutant or wild-type (WT) T7 RNAP are described in the text.
[2] Adapted from [8].

are often added at a concentration of 0.4 mM. An optimal NTP/NTPαS ratio for many nucleotide analogs has been optimized and reported elsewhere (Table 17.1) [8]. Wild-type T7 RNA polymerase is used for the incorporation of most analogs, except for those containing minor groove substitutions. The N^2-methyl guanosine, 3-deaza adenosine and sugar modifications (2′-deoxy, 2′-fluoro, 2′-O-methyl) are usually incorporated using mutant Y639F polymerase [68], which has an enhanced tolerance for minor groove modifications [7].

For most constructs employed in ribozyme and helicase studies, the optimal yield of RNA can be obtained using the following transcription protocol.

(1) Prepare the reaction mixture containing 40 mM Tris–HCl, pH 7.5, 15 mM MgCl$_2$, 2 mM Spermidine, 5 mM DTT, 0.02 μg/μl DNA template and 0.08 μg/μl of wild-type T7 RNA polymerase or its mutant Y639F. In our hands, 1 ml transcription of exD123 or D56 was sufficient to carry out several NAIM/NAIS experiments.
(2) Incubate the transcription mixture at 37 °C for 3–4 h.
(3) Precipitate the transcript by the addition of 0.25 M NaCl, 0.02 M EDTA and 3 volumes of ethanol followed by incubation at −80 °C for 1 h and centrifugation.
(4) Re-suspend the pellet in 10 mM MOPS (pH 6.0), 1 mM EDTA (M$_{10}$E$_1$ buffer) that is supplemented with an equal volume of denaturing loading buffer (8 M urea, 0.05% each of xylene cyanol and bromophenol blue, 17% sucrose, 83 mM Tris, pH 7.5, and 1.7 mM EDTA) and purify on a denaturing polyacrylamide gel. Visualize the band corresponding to the desired transcript by UV shadowing (for a detailed description, see Chapter 3), excise from the gel and elute in M$_{10}$E$_1$ buffer at 4 °C for 1–3 h.
(5) Precipitate the RNA as above; re-suspend the pellet in M$_{10}$E$_1$ buffer and store at −80 °C.

Tips and troubleshooting

For the best yield of RNA, the transcription conditions always need to be optimized for each particular construct. One can vary, for example, the reaction time, the concentrations of DNA template, T7 RNA polymerase, $MgCl_2$ in the reaction buffer and triphosphates. The addition of 0.1% Triton X-100 (final concentration), extra DTT or RNase inhibitor can also help increase the transcription efficiency (see Chapter 1). If it is necessary to optimize the incorporation level for a nucleotide analog, the transcription is performed using different NTP:NTPαS ratios.

17.2.3
Radioactive Labeling of the RNA Pool

In order to employ the NAIM assay, one component of the reaction must be radiolabeled. This can be achieved by either labeling the entire RNA pool after transcription and purification or by reaction of the unlabeled RNA pool with a radioactively labeled molecule during the selection step. For example, prior to the selection that is based on helicase unwinding of duplex RNA, one strand of the duplex in the RNA pool is end-labeled with ^{32}P. The selection step for reaction by a group II intron ribozyme is performed by reacting a labeled pool of one RNA with an unlabeled pool of a second RNA or *vice versa*.

Radiolabeling of an NAIM transcript can be performed either at the 5′ end (kinase reaction) or at the 3′ end (via templated addition of a single α-^{32}P-labeled dAMP residue by Klenow polymerase) [69]. In order to 5′-end label an RNA transcript, one has to follow the protocol below.

(1) Prior to 5′-end-labeling, dephosphorylate transcribed RNA (around 60 pmol) with calf intestinal phosphatase (CIP, 40 U) in a final volume of 50 μl using the reaction buffer supplied by the manufacturer (Roche). Incubate the reaction mixture at 37 °C for 30 min, then phenol-extract and ethanol-precipitate the RNA.
(2) For 5′-end-labeling, incubate 10 pmol of dephosphorylated RNA with 1 μl of [γ-^{32}P]ATP (6000 Ci/mmol) and 10 U of T4 polynucleotide kinase (PNK) in a final volume of 10 μl at 37 °C for 1 h.
(3) Purify the labeled RNA on a denaturing polyacrylamide gel.

17.2.4
The Selection Step of NAIM: Three Applications for Studies of RNA Function

17.2.4.1 Group II Intron Ribozyme Activity: Selection through Transesterification

General aspects
Group II introns are complex ribozymes with a diverse catalytic repertoire [40, 70]. Self-splicing of these ribozymes occurs in two steps, with the first step occurring via competing parallel pathways: hydrolysis and branching [71, 72]. Despite the lack of primary sequence conservation, group II intron secondary structure is very

conserved. It consists of six domains, each of which has a distinct catalytic function (Fig. 17.5A) [38, 40, 70]. The largest, domain 1 (D1), folds independently [73], serves as a scaffold for binding other intronic domains as well as the 5′-exonic substrate, and contributes key elements of the intron active site. Domain 5 (D5) contains many active site residues and is absolutely critical for any catalytic reaction performed by the intron. Domain 3 (D3) is a catalytic effector that accelerates every reaction catalyzed by the intron. Domain 6 (D6) contains the bulged branch-point adenosine, which is essential for the transesterification pathway of splicing. Domain 2 (D2) mediates conformational rearrangements along the splicing pathway and Domain 4 (D4) often contains an open reading frame that encodes a protein cofactor.

One of the most remarkable features of group II intron ribozymes is their modularity, which allows reconstitution of an active ribozyme from two or more separately transcribed components. This modularity provides an opportunity to dissect the entire splicing pathway into separate reactions, which can be studied independently. One of the most informative constructs for studying group II intron catalysis is the *trans*-branching system, which divides the intron into two critical parts: exD123 (comprised of the 5′ exon and intronic domains 1–3) and D56 (containing domains 5 and 6) (Fig. 17.5B). When these two RNA components are combined in the presence of Mg^{2+}, the 2′-OH of the D6 branchpoint (in the D56 construct) attacks the 5′ splice site in exD123, resulting in the formation of a covalent linkage. The resulting 2′–3′–5′ branched D56/D123 molecule is highly stable and can be isolated. A pool of these branched molecules represents the population of species that were capable of undergoing the first step of group II intron self-splicing. Thus, the *trans*-branching assay represents a useful selection for NAIM and NAIS studies on the first step of splicing. NAIM assays that employ this *trans*-branching system have provided information on functionally important residues in both RNAs (exD123 and D56) (Fig. 17.6) [28] and NAIS applications have identified a number of critical tertiary interactions (Fig. 17.7) [12, 28].

Experimental procedures
The following protocol is used for identification of functional groups in D56 RNA that are important for branching.

(1) Transcribe D56 in the presence of nucleotide analogs and 5′-end-label the transcript as described above.
(2) Denature trace amounts of modified, labeled D56 (1–10 nM) and unlabeled, unmodified exD123 (1.5 µM) separately at 95 °C for 1 min in 40 mM MOPS pH 6.0. Cool down to about 42 °C by leaving the tubes at room temperature for around 1 min and then mix the two RNA fractions with the simultaneous addition of salts to final concentrations of 100 mM $MgCl_2$, 2 mM $Mn(OAc)_2$ and 0.5 M $(NH_4)_2SO_4$. Incubate the mixture at 42 °C until the fraction of branched product reaches around 20%.
(3) Separate unreacted D56 and branched product on a 5% polyacrylamide denaturing gel, recover both species from the gel and subject to iodine cleavage as described below.

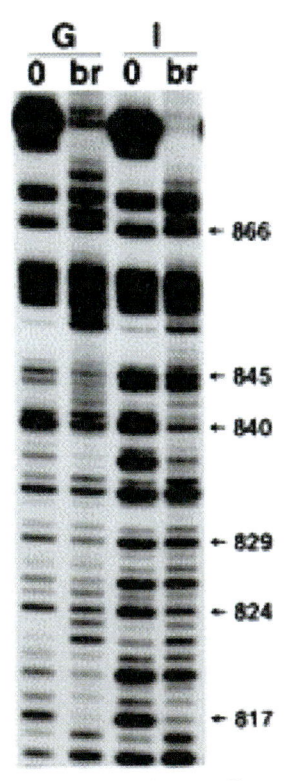

Fig. 17.6. Interference effects in the ai5γ group II intron. (A) A high-resolution sequencing gel summarizes the effects of GTPαS and inosine substitutions (G and I lanes, respectively) in D56 after iodine cleavage of the unreacted (0) and branched (br) fractions. A weak phosphorothioate effect at G840 could not be verified in subsequent experiments. Reprinted with permission from *The EMBO Journal*. (B) Summary of NAIM effects in D5 of the group II intron.

The experimental setup is essentially the same for mapping the 5'-half of D123 in the exD123 construct. In this case, exD123 is transcribed with nucleotide analogs and D56 is unmodified. The reaction is carried out according to the above protocol, and the branched product is purified in a 5% denaturing gel and subjected to sequencing by iodine. The cleavage pattern is compared to that of precursor exD123.

In order to map the 3'-half of exD123, the procedure is modified as follows.

(1) Transcribe exD123 in the presence of the nucleotide analogs of interest and label at the 3' end [69].

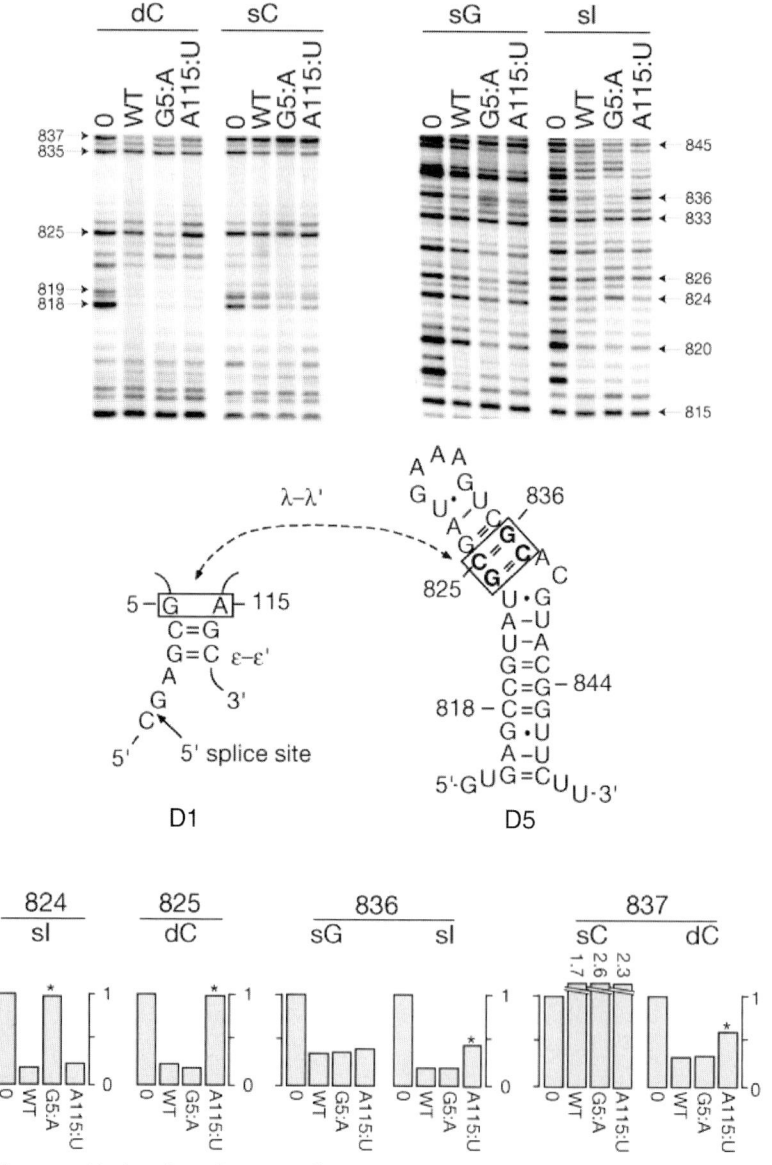

Fig. 17.7. Nucleotide analogue interference suppressions within D5 and D1 of the ai5γ group II intron. Autoradiographs correspond to iodine-cleaved transcripts modified with deoxy-CαS (dC), cytosineαS (sC), guanosineαS (sG) or inosineαS (sI). Comparative data are shown for NαS-containing D56 that was unreacted (0) or branched to wild-type exD123 (WT), exD123(G5:A) and exD123(A115:U) mutants. Band intensities were quantitated and corrected for background phosphorothioate effects. The normalized intensities shown in the bar diagram at the bottom represent mean values, each having a maximum variance of 20%, from two to four independent experiments. Asterisks indicate positions where NAIS effects were observed (loss or alleviation of interference compared with the wild-type).

(2) React trace amounts of radiolabeled exD123 (1–10 nM) with 1.5 µM unlabeled, unmodified D56 according to the same protocol as described above.
(3) Separate the branched product from the unreacted exD123 on a 5% denaturing gel; subject both RNAs to iodine cleavage.

Since the 3′ terminus of exD123 is labeled in this approach, one can also observe the formation of both branched and hydrolysis products, both of which can be analyzed for interference effects.

17.2.4.2 Reactivity of RNA Helicases: Selection by RNA Unwinding

General aspects
DExH/D proteins are involved in nearly every aspect of RNA metabolism and represent the largest group of RNA helicases [74, 75]. NPH-II is a prototypical member of the DExH/D family [76], and it was the first one shown to unwind RNA [77]. NPH-II has robust ATPase activity that is partially stimulated by nucleic acids [77]. NPH-II readily unwinds RNA substrates that contain a 3′ overhang [78] and it has a high degree of processivity during unwinding of long RNA duplexes [79]. The overall unwinding reaction is limited by the rate constant for unwinding initiation, which is 1–3 min^{-1}. NPH-II is both a robust motor for RNA strand displacement [79] and for the removal of proteins that are bound to RNA sites [80].

Currently there are about one dozen solved helicase structures. While these provide invaluable mechanistic clues, they are limited to providing a static picture, or "freeze-frames", of unwinding events. Mechanistic models for unwinding by an RNA helicase of the DExH/D family have not been well developed. NAIM has been used to probe the dynamic interactions that occur between NPH-II and its substrate during translocation (J. Kawaoka and A. M. Pyle, manuscript in preparation). NPH-II has been shown to specifically recognize a cluster of 2′-OH groups and a critical Rp-phosphate at the initiation site of unwinding (located in the first 4 bp of the duplex substrate). After initiation, the enzyme recognizes phased, periodic clusters of 2′-OH groups every 10 nt along the bottom strand of its substrate (J. Kawaoka and A. M. Pyle, manuscript in preparation). This periodic phasing suggests a defined translocation step size of discrete magnitude and is consistent with an inchworm-like model, as suggested for various helicases [81] including RecBC [82].

For suppressions at positions 824 and 825, significant phosphorothioate effects were not observed, so corresponding bars are not shown. The band intensity observed for unreacted material was arbitrarily set to one. For branched products, the size of the bars is related to 1/(interference effect), i.e. the smaller the bar, the stronger the interference. The values of band intensities over 1 are shown above the corresponding bars. In the middle of the figure, the intra-domain 1 ε–ε' interaction (base pairs G3–C117 and C4–G116) as well as the tertiary interaction (λ–λ') with D5 are illustrated; nucleotides G5 and A115, which were mutated, are shown boxed on the left. Reprinted with permission from Nature.

Substrate design

Substrates used in unwinding assays normally consist of two strands that are annealed through base-pairing. For most SF2 helicases like NPH-II, the typical substrate for *in vitro* studies is a duplex RNA that is flanked by a 3′-overhang. The overhanging strand (on which NPH-II tracks; J. Kawaoka and A. M. Pyle, manuscript in preparation) is called the "bottom or loading strand". The "top strand" is stripped away during the unwinding reaction. To evaluate unwinding, the two strands are synthesized independently, one of them is end-labeled with ^{32}P, then the two strands are annealed and the resultant duplex is purified on a native gel. For NAIM analysis of RNA functional groups that contribute to unwinding, the substrate strands were designed using mfold [91] to minimize intra-strand secondary structure. A typical top strand RNA used in these studies is 5′-CUG UGG CAU GUC CUA GCG UCG UAU CGA UCU GGU CGU CUCC-3′, which anneals to the complementary bottom strand with the following sequence: 5′-GGA GAC GAC CAG AUC GAU ACG ACG CUA GGA CAU GCC ACA GAC GUA CUA ACA GCA UCA AUG ACA UCA AUGA-3′ (nucleotides overlapping with top strand underlined).

Previous studies demonstrated that functional groups on the top strand do not contribute to NPH-II recognition during unwinding. Therefore, NAIM has only been used to determine the location of important functional groups on the bottom strand. A pool of modified bottom strands is created as follows.

(1) Transcribe bottom strands with phosphorothioate ribonucleotides (NTPαS) or phosphorothioate deoxyribonucleotides (dNTPαS) statistically incorporated to a level of 5% as described in Section 17.2.2. Add trace amounts of [α-^{32}P]UTP to the transcription mixture to label the transcripts internally, so that reaction products could be easily identified and cut from the gel.
(2) Gel-purify, dephosphorylate and 5′-end label transcripts as described in Sections 17.2.2 and 17.2.3.
(3) Test incorporation of analogs by treating around 30 000 c.p.m. of each modified transcript with 3 mM iodine (prepared freshly in EtOH) at 25 °C for 5 min. Analyze samples on a denaturing polyacrylamide sequencing gel and visualize by radioanalytic imaging.

NAIM selection – defining optimal unwinding conditions

Previous optimization experiments served to define the reaction conditions with the strongest influence on NPH-II unwinding [79]. Because it was not known how single deoxy (or phosphorothioate) modifications would affect NPH-II helicase activity, the NAIM assay was tested under a variety of conditions. Variables that were tested included extent of unwinding (5–50%), concentration of NaCl in the reaction (10–120 mM) and enzyme turnover (single versus multiple cycle kinetics). Nearly all interference effects were found to disappear under low salt conditions (NaCl < 30 mM), even when total unwinding proceeded to only 10%. In addition, interference was not seen under multiple cycle conditions (i.e. when helicase is allowed to rebind in the absence of a trap molecule), regardless of the extent of total unwinding (as low as 10% of the total population).

Under single cycle conditions, when the [NaCl] is around 70 mM, interference effects are strong. Also smaller degrees of total unwinding (below 20%) tend to amplify the weaker interference effects (around $I = 1.5$, as defined below). Therefore, these unwinding conditions [single cycle, low extent of unwinding (below 20%) and 70 mM NaCl] were selected for NAIM because they provided good selectivity while maintaining a high level of unwinding activity.

An NAIM helicase assay

(1) Prior to initiation of unwinding, pre-incubate NPH-II protein (15 nM) with duplex substrate (1–2 nM, labeled with ^{32}P at the 5' end of the bottom strand) in 40 mM Tris, pH 8.0, 4 mM Mg(OAc)$_2$, 70 mM NaCl for 3 min at room temperature.
(2) Initiate reactions by the simultaneous addition of 3.5 mM ATP and 400 nM duplex RNA trap (which prevents helicase rebinding and causes the reaction to be under "single cycle" conditions). Total reaction volume can be 35–100 µl.
(3) Quench the unwinding reaction upon attaining a reaction extent of around 15%.
(4) Separate duplex and unwound fractions by native gel electrophoresis.
(5) Visualize the duplex and unwound species and isolate them from the gel.
(6) In order to make sure that the difference in the iodine cleavage pattern of the duplex and unwound fractions is due to the analog effect and not to the difference in accessibility to the iodine, isolate the bottom strand from the duplex fraction by denaturing PAGE before the cleavage procedure.
(7) Re-suspended all samples in a solution of 10 mM MOPS (pH 5.0), 1 mM EDTA, and subject to iodine cleavage (see below) and analysis on a denaturing polyacrylamide gel.
(8) Quantify and analyze interference effects using the same methods as for other NAIM applications (see below).

17.2.4.3 RNA–Protein Interactions: A One-pot Reaction for Studying Transcription Termination

General aspects

During the processive phase of transcription in bacteria, the fast-moving transcription elongation complex (TEC) is held together by an intricate network of cooperative interactions between the DNA template, RNA polymerase and the RNA product ([83] and references therein). To induce TEC dissociation, many transcription termination signals (termed intrinsic or rho-independent terminators) rely on the formation of a specific stem–loop structure within the nascent transcript that is directly upstream from a short U-rich 3' end. These RNA functional elements largely contribute to the disruption of TEC-stabilizing interactions, albeit by a mechanism that is not totally understood. To investigate the role of terminator components in the highly dynamic context of transcription elongation, NAIM experiments were applied [30]. At present, these experiments have been restricted to the study of RNA polymerases (RNAPs) that efficiently utilize NTPαS analogs, which include certain mutants of bacteriophage RNAPs (prototypes are the Y639F and Y639F/H784A mutants of T7 RNAP [68, 84]; M. Boudvillain, unpublished results). However, efforts to evolve suitable mutants of multi-subunit RNAPs are also underway, and these may broaden the scope of NAIM/NAIS-based transcription experiments in the near future (M. Boudvillain and R. Rahmouni, unpublished results).

The implementation of an NAIM strategy for studying transcription termination has resulted in a simple "one-pot" reaction that combines the first two steps of NAIM (preparation of the RNA pool *and* transcript selection) [30]. In this approach, linear DNA templates containing a terminator sequence downstream from a phage promoter are transcribed with an appropriate RNAP variant (here we will only consider the case of the Y639F mutant of T7 RNAP) and in the presence of a NTPαS analog. Due to the incomplete efficiency of most termination signals, only a fraction of the TECs are released at the termination points, whereas other TECs continue transcription to the end of the template (Fig. 17.8A). Purification of both types of transcript products (i.e. the NαS-modified transcripts that are released at the terminator, and those that are released at the template end) allows one to compare the functional groups that are important to the formation of each. By ^{32}P-labeling the two types of products and then treating them with iodine, one reveals the RNA atoms and functional groups that are important for transcription termination (Fig. 17.8C). Because high-resolution crystal structures of the TEC now exist for various RNAPs ([83] and references therein), it is possible to link some of the NαS effects with specific known interactions between the RNAP sidechains and the transcript [30].

Experimental procedure

(1) In a 0.5 ml tube, mix 1.1 pmol of DNA template (either linearized plasmid or purified PCR fragment) with 8 μl of 5 × transcription buffer (30 mM MgCl$_2$,

Fig. 17.8. NAIM analysis of transcription termination. (A) *In vitro* transcription of a linear DNA template (schematically depicted on the left) containing the sequence of the *rrnB* T1 terminator from *E. coli* (boxed) downstream from a T7 promoter (arrow). As shown on the gel (right), termination (T) and runoff (RO) transcripts are formed in comparable amounts during transcriptions with the Y639F and wild-type (WT) T7 RNAPs. (B) Selection of NAIM interference effects on *rrnB* T1 transcription termination [30] using either κ (or 1/κ) values (top) or λ discrimination factors (bottom). Broken lines correspond to standard interference thresholds (see text). (C) A representative gel showing Rp-phosphorothioate (GαS) and 2′-deoxy (dGαS) interference effects at G positions during intrinsic termination of transcription at the major site (T) of the *rrnB* T1 terminator; R = runoff transcripts. Positions of Rp-phosphorothioate (triangles) and 2′-deoxy (circles) modifications that favor (open symbols) or are detrimental (filled symbols) to transcript release at the T site are illustrated in the context of the terminator secondary structure.

17.2 Experimental Protocols for NAIM | 281

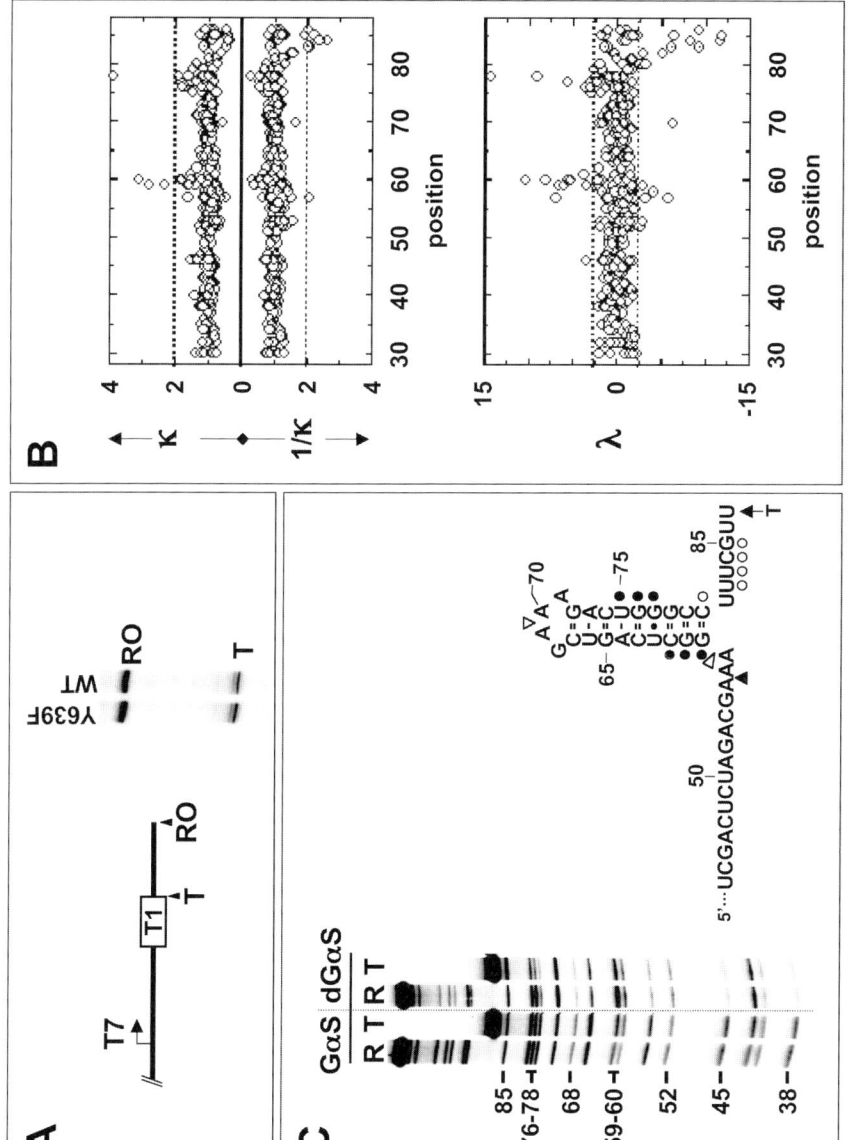

Fig. 17.8

50 mM NaCl, 100 mM HEPES, pH 7.5, 30 mM DTT, 0.05% Triton X-100, 10 mM spermidine), appropriate amounts of NTPαS analog and parental rNTP (Table 17.1), 20 nmol of each of the other rNTPs and 3 pmol of the Y639F RNAP. The final reaction volume should be 40 µl, which usually yields sufficient quantity of transcripts for subsequent NAIM analysis.

(2) Incubate transcription mixtures for 15–30 min at 37 °C before addition of single-stranded M13 DNA (1 µg) and KCl (250 mM) to prevent non-specific association between free RNAPs and transcripts [85].

(3) Quickly load the mixtures onto Microcon columns (100-kDa cut-off for around 100-nt transcripts) that had been saturated with BSA as described by the manufacturer (Millipore). After centrifugation of the columns for 2 min at 10 000 r.p.m. in a desktop centrifuge, collect the filtrates, which should then be free of any unwanted high-molecular-weight species [86]. Unwanted salts and nucleotides are also eliminated by gel filtration on a 1–2 ml Sephadex G-50 (Sigma) column.

(4) To remove triphosphate moieties at the 5' ends of transcripts, mix the column eluate with 2 µl of calf intestinal alkaline phosphatase (CIAP) (Roche Diagnostics) and incubate for 30 min at 37 °C. The enzyme efficiently dephosphorylates transcripts under these conditions, no additional buffer or salts are required.

(5) Add sodium acetate to 0.3 M final concentration and extract twice with one volume of a phenol:chloroform:isoamyl alcohol mix (25:24:1; Amresco) and twice with 1 volume of chloroform buffered with 10 mM MOPS, pH 7. Then, add 3 volumes of ethanol and precipitate the RNA for 2 h at −20 °C.

(6) Centrifuge for 30 min at 5000 g and 4 °C, discard the supernatant, briefly dry the pellet in a Speed Vac apparatus and re-suspend in 14 µl of $M_{10}E_1$.

(7) Add 3 µl of [γ-^{32}P]ATP (3000 Ci/mmol; Amersham Biosciences), 2 µl of 10 × kinase buffer (100 mM $MgCl_2$, 50 mM DTT, 700 mM Tris–Cl, pH 7.5) and 1 µl of T4 polynucleotide kinase (New England Biolabs). Incubate for 30 min at 37 °C, then phenol-extract the sample and precipitate it with ethanol as described above.

(8) Re-suspend the pellet in 10 µl of denaturing buffer (95% formamide, 5 mM EDTA, 0.1% xylene cyanol) and incubate for 1 min at 95 °C before loading on a denaturing polyacrylamide gel (8–10% gels are adequate for around 100-nt long transcripts and should also contain 7 M urea and 30% formamide to ensure strong denaturing conditions) that had been pre-heated to 60 °C. After gel migration (about 2 h at 40 W for adequate separation of around 100-nt transcripts on an 8% acrylamide, 20 × 40-cm gel), recover termination and runoff transcripts from gel slices (see Section 17.2.2) and then analyze through iodine sequencing as described below.

Tips and troubleshooting

Terminator stem–loop structures usually contain many GC pairs. In some instances, complete unfolding of the structure does not occur within the polyacrylamide gel, resulting in anomalous electrophoretic migration (e.g. [85]) even under the harsh denaturing conditions described above. This may result in improper partitioning of the transcription products and in an inextricable mixing of NαS effects on termination and unrelated RNA structural features (A. Schwartz, R. Rahmouni and M. Boudvillain, unpublished results). For similar reasons, results implicating NαS modifications that strengthen RNA base pairs (such as DAPαS) should be analyzed with great caution. In general, it is best to use terminators with no more than three or four consecutive GC pairs in the hairpin stem such as in the *rrn*B T1 terminator of *Escherichia coli* [30]. The DNA template should also be designed to yield transcripts as short as possible (70–100 nt should leave enough positions of neutral NαS incorporation [for calibration of NAIM effects; see Section 17.2.6] between the initiation and termination regions); the terminator sequence should also be introduced sufficiently upstream from the template end to yield termination and runoff transcripts with significantly different lengths. Appropriate DNA templates can be easily prepared by sub-cloning oligonucleotides within commercial vectors that bear a multi-cloning site surrounded by phage promoters such as the pSP73 plasmid (Promega).

In order to calibrate the assay, it is useful to compare NAIM signals obtained with both the Y639F and wild-type T7 RNAPs and analogs that are good substrates for the two enzymes (such as ITPαS or C7-RTPαS analogs). With the *rrn*B T1 terminator, we did not observe significant differences (M. Boudvillain, unpublished observations).

Transcripts of various lengths are usually released in the termination window albeit with different efficiencies (Fig. 17.8A). It is usually best to isolate and analyze those termination species separately as significant differences in NAIM patterns are likely to be observed [30].

The efficiency of transcript release by a given terminator may be affected by modifications of the experimental conditions [87]. This lack of control on the termination reaction precludes an easy adjustment of the detection threshold of NAIM effects (see also Section 17.2.6). For this reason, the use of λ discrimination factors in place of κ interference values (see Section 17.2.6) usually facilitates the identification of weak and moderate NAIM effects (Fig. 17.8B; [30]; M. Boudvillain, unpublished results). Although recent results suggest that, at least in some cases, the assay sensitivity may be controlled through variations of the RNA polymerase/DNA template ratio [88], we still prefer to rely on statistical discrimination (λ values) for detection of NAIM effects on transcription termination.

17.2.5
Iodine Cleavage of RNA Pools

The preceding selection methods result in three different pools of RNA, all of which are examined by iodine cleavage in order to assay function: (A) the selected pool of functional RNA molecules, (B) the selected pool of nonfunctional RNA molecules and (C) the unreacted starting pool of RNA molecules. By comparing

Pool A to Pool B or Pool C (or both B and C), one can deduce the role of specific atoms on function.

Experimental procedure

(1) Re-suspend RNAs in 10 μl of a solution containing 5 μl $M_{10}E_1$ buffer (10 mM MOPS, pH 6.0, 1 mM EDTA) and 5 μl of formamide, and denature at 95 °C for 1 min, followed by chilling on ice.
(2) Add a freshly prepared solution of iodine [1 μl of a 10 mM iodine (Sigma) solution dissolved in ethanol] and incubate the reaction mixture at 37 °C for 3 min.
(3) Precipitate the RNA by adding 240 μl of 0.3 M sodium acetate (pH 5.0), 1 μg tRNA carrier and 750 μl of ethanol.
(4) Analyze samples on a denaturing polyacrylamide gel. Depending on the application and the length of the RNA transcript, the percentage of acrylamide varies from 4 to 20%. It is essential to load samples that have *not* been treated with iodine on the same gel, to provide controls for non-specific RNA degradation. In many cases NAIM gels in published articles do not show lanes of iodine-untreated samples in order to preserve space; however, it is always assumed that the researcher has performed this important control. In order to facilitate the comparison of iodine cleavage patterns in lanes corresponding to precursor RNA and selection product, it is advisable to determine the amount of radioactive material in each sample and load equal amounts of radiolabeled RNA onto the gel. In this case, the 2-fold or greater interference effects will be easily detectable after radioanalytic imaging, even prior to quantitation.

Tips and troubleshooting
The stated iodine cleavage conditions may require optimization, depending on the RNA sample. The final concentration of iodine can vary from 0.1 to 1 mM, and often iodine cleavage reactions are carried out at room temperature. Incubation times can vary from 1 to 5 min. It is also possible to load the samples onto the gel directly without precipitation, thereby sparing additional sample handling. Frequently, however, this results in a salt front on the gel, which can adversely affect migration of the samples.

17.2.6
Analysis and Interpretation of NAIM Results

17.2.6.1 **Quantification of Interference Effects**
Dried polyacrylamide gels are subjected to radioanalytic imaging [using, for example, a PhosphorImager (Molecular Dynamics)]. For most applications, the PhosphorImager Imagequant program (Molecular Dynamics) is used to quantify the radioactive intensity of each band in precursor (pre) and product (pro) lanes for both unmodified (NαS) and modified (N'αS) nucleoside phosphorothioate-containing

RNAs. Then the interference at each position is calculated using the following equation [8, 11]:

$$\text{Int} = [N\alpha S_{\text{pro}}/N'\alpha S_{\text{pro}}]/[N\alpha S_{\text{pre}}/N'\alpha S_{\text{pre}}]$$

This allows one to normalize the interference value for the effect of the phosphorothioate itself as well as for the incorporation differences between unmodified and modified nucleotides. After interferences have been calculated for each nucleotide position, their values are adjusted to correct for variable sample loading on the gel, as previously described [8] (see Table 17.2): an average interference value is calculated for all positions that are within two standard deviations of the mean. Each interference value is then divided by this average. Adjusted interference values (usually referred to as κ values) that are higher than 2 or less than 0.5 are interpreted as interference or enhancement effects, respectively.

Utilization of λ discrimination factors

When the stringency of the NAIM assay cannot be easily modulated, a statistical filter, such as the λ discrimination factor [30], may help to discriminate weak and moderate interference effects (Table 17.2 and Fig. 17.8B). The λ discrimination factors are derived from κ values [8] as follows: κ values deviating by more than two standard deviations from the mean of a homogeneous data set (inosine effects, for instance) are not included in a new calculation of the standard deviation (SD'). Then for every κ value of the data set, the discrimination factor is defined as:

$$\lambda = (\kappa - 1)/\text{SD}' \quad \text{if } \kappa > 1$$
$$\lambda = (1 - 1/\kappa)/\text{SD}' \quad \text{if } \kappa < 1$$

Note that the sign of λ is completely arbitrary so that the above formulas can be formatted to suit an inverted reference system of 'negative' and 'positive' effects (as in [30]).

The use of λ factors presents several advantages. It normalizes NAIM signals for varying experimental quality and population extent among the different data sets and provides identical intensity scales for favorable ($\lambda < 0$) and detrimental ($\lambda > 0$) NAIM effects. Moreover, if one assumes that SD' adequately reflects the standard deviation of random (i.e. no interference) NAIM signals, then z tables, which are found in most statistics textbooks, may be used to select a confidence interval. For instance, an interference threshold of $|\lambda| \sim z = 2.5$ would correspond to a confidence interval for random NAIM signals above 98% (the probability, p, that a value of $|\lambda| > 2.5$ is indicative of a random signal rather than interference would be less than 0.0124). Of course, this would also depend on a number of assumptions such as signals being truly random, following a normal distribution, not being interdependent, etc. In any case, the sample size should be sufficiently large (at least over 20 random signals for every data set) and the κ values deter-

Tab. 17.2. Analysis of NAIM data obtained upon selection of reactive molecules (pro) from a pool of AαS-containing transcripts (pre).

Imager counts[1]		Int[2]	κ[3]	λ[4]
AαS$_{pre}$	AαS$_{pro}$			
53385	49431	1.08	0.96	−0.47
82981	66920	1.24	1.11	1.25
69918	97108	0.68	0.61	*−7.10*
62386	58855	1.06	0.95	−0.60
60916	59142	1.03	0.92	−0.99
51981	49982	1.04	0.93	−0.85
39672	20991	1.89	1.69	*7.71*
53039	44200	1.20	1.07	0.79
56126	41885	1.34	1.20	2.27
52541	53072	0.99	0.88	−1.55
16793	17313	0.97	0.87	−1.70
24755	23576	1.05	0.94	−0.72
28757	23571	1.22	1.09	1.02
18172	16225	1.12	1.00	0.00
35408	35057	1.01	0.90	−1.26
32444	29765	1.09	0.97	−0.35
39831	36210	1.10	0.98	−0.23
28660	26537	1.08	0.96	−0.47
27706	25654	1.08	0.96	−0.47
26325	21578	1.22	1.09	1.02
26404	21467	1.23	1.10	1.13
31698	26197	1.21	1.08	0.91
29349	24458	1.20	1.07	0.79
SD[5]		0.21	0.18	
Limits[6]				
$1 + 2 \times SD$		1.42	1.36	
$1/(1 + 2 \times SD)$		0.70	0.74	
Mean[7]		1.12	1.00	
SD′ [5,7]			0.09	

[1] Virtual imager counts were created to simulate a NAIM experiment. Each line corresponds to a single RNA position.
[2] For parental nucleotides such as AαS, Int values are determined with the formula: Int = NαS$_{pre}$/NαS$_{pro}$.
[3] The κ values are obtained by dividing the corresponding Int values by the mean (1.12), thereby correcting for differences in loading of the gel lanes.
[4] The λ discrimination factors were determined with the formulas given in the text. Statistically significant interference effects (threshold ± 2.5) are shown in italics. Such effects would not have been revealed by a standard analysis of the κ values (0.5 < κ < 2).
[5] Standard deviations were determined with Microsoft Excel; here, values marked by grey boxes were included.
[6] The formula $L_{low} = 1 - 2 \times SD$, which may also be used to calculate the lower limits, does not account for the inherently skewed distributions of Int and κ values.
[7] Data not included within limits (grey boxes) were excluded for mean and SD′ calculations.

mined with enough accuracy (in three to four independent experiments with an experimental error below 20%) for the λ values to reflect statistically significant interference effects ([30]; M. Boudvillain, unpublished results).

17.3 Experimental Protocols for NAIS

In this section we will describe experimental procedures specific for the NAIS assay on the group II intron system. Some of the protocols used in this method, such as *in vitro* transcription, end-labeling and iodine treatment are shared with a general NAIM assay (see Section 17.2).

17.3.1 Design and Creation of Mutant Constructs

General considerations

NAIS experiments are basically NAIM experiments on mutant and wild-type molecules that are performed in a side-by-side manner. It is generally thought that NAIS experiments are more meaningful when mutants contain single-atom changes. However, when studying molecules for which there is little structural information, it may be reasonable to initially perform NAIS using mutant constructs that contain full-base mutations. Based on the results in these coarse experiments, one can then refine the system by making single-atom mutants. This approach was essential for identifying critical tertiary interactions in the group II intron active site (such as κ–κ' and λ–λ') (Fig. 17.7) [12, 28].

When NAIM was first performed on the ai5γ group II intron, clusters of interference effects were observed in several regions of D1, including the very beginning of the intron (nt 1–5) and an asymmetric bulge in the C1 stem (Fig. 17.5A) [28]. When full-base mutations were introduced in these regions at positions 5 (G5:A) and 115 (A115:U), respectively, the resulting mutants exhibited suppression of specific interferences in D5 (Fig. 17.7), suggesting a complex tertiary interaction. This interaction (λ–λ') was then studied at higher resolution by NAIS that employed a series of D56 constructs containing single-atom changes [12]. Notably, when there is no appropriate analog available for studying certain atoms or functional groups by NAIM/NAIS (e.g. N^1 of adenosine), NAIM/NAIS can be easily complemented by chemical modification interference (e.g. using dimethylsulfate as a modifying reagent) [12].

Preparation of RNA molecules containing single-atom substitutions

RNAs that contain single-atom substitutions can be either purchased commercially or synthesized on an automated DNA–RNA synthesizer via standard solid-phase phosphoramidite method (Chapter 7). Although the variety of modified RNA oligonucleotides offered commercially is growing, one still has more options when synthesizing them in-house on an automated synthesizer (such as Applied

Biosystems, Pharmacia, etc). A variety of modified phosphoramidites with single-atom or functional group modifications can be purchased commercially (from Glen Research, ChemGenes, etc). When choosing a modified phosphoramidite for incorporation into an RNA oligonucleotide, it is important to ensure that base and/or 2′-OH protecting groups on the modified monomer are compatible with the synthetic cycle and can be easily and quantitatively removed by using standard RNA-deprotecting protocols. For example, if the desired modification is 2′-deoxy cytidine and the RNA oligo is to be deprotected by the Wincott procedure [89], it is advisable to use the phosphoramidite with an acetyl protecting group on the base, which is most compatible with the RNA deprotection protocol.

In our hands, the Wincott deprotection protocol is the most effective of all existing RNA deprotection procedures. Our protocol is similar but somewhat simpler than the procedure described in Chapter 7. Nevertheless, it allows one to synthesize the catalytic domain D5 of the group II intron which is as active as the transcribed D5 RNA.

(1) Our base-deprotection protocol is essentially the same as described in Chapter 7 with the following exceptions.
   (a) We use 40% methylamine in water instead of 8 M ethanolic methylamine in our base-deprotection mixture. Our mixture consists of concentrated (28–30%) ammonium hydroxide: methylamine (40% in water) (1:1 by volume).
   (b) We add 4 ml (not 2 ml) of base-deprotecting mixture to the polymer support and incubate it at 65 °C for 10 min (not 20–40 min) with occasional stirring.
(2) We generally synthesize trityl-off RNA and our 2′-OH deprotection protocol also differs from the one described in Chapter 7. After base deprotection, we use the following procedure.
   (a) Separate the supernatant from the support, aliquot into six to eight Eppendorf tubes and dry in the Speed Vac.
   (b) Re-suspended pellets in 250 µl of triethylamine:triethylamine *tris*(hydrofluoride):N-methylpyrrolidinone mix (0.75:1:1.5) and incubate 1.5 h at 65 °C.
   (c) Precipitate the deprotected RNA by adding 25 µl of 3 M KOAc and 1 ml of butanol to each tube, followed by incubation at −20 °C overnight.
   (d) Purify the RNA on a denaturing polyacrylamide gel. In order to facilitate gel purification and prevent gel overloading with salts, one may consider desalting the oligo on a C_{18} disposable cartridge (Sep-Pak, Waters; or OPC, Applied Biosystems).

For NAIS experiments on the group II intron system, D56 molecules containing single-atom substitutions were prepared by chemical synthesis. The D56 construct is 80 nucleotides long, so it was synthesized in two pieces using the procedure described above. The pieces were joined by splint-directed ligation using T4 DNA ligase (see Chapter 3), which has been particularly successful for preparing functionally active D56 molecules using the procedure described below.

(1) Combine 5′-phosphorylated downstream RNA fragment, the upstream fragment and a DNA splint oligonucleotide (60 nt) in equimolar ratio, denature in water at 95 °C for 1 min and slowly cool down to 30 °C.
(2) Supplement the mixture with 10 × buffer [500 mM Tris–HCl, 100 mM $MgCl_2$, 100 mM DTT, 10 mM ATP, 250 µg/ml BSA, pH 7.5 (25 °C); provided by the manufacturer (New England Biolabs)], RNase Inhibitor and T4 DNA ligase and incubate at 30 °C for 10–12 h.
(3) Ethanol-precipitate and gel-purify the ligated RNA.

In this method, the RNA concentration, reaction volume, and amounts of RNase inhibitor and ligase must be optimized and conditions are highly dependent on the system. For the preparative ligation we generally use 3′- and 5′-fragments of the D56 construct at 20 µM each in the total reaction volume of 100 µl containing 8 µl of RNase Inhibitor (40 U/µl) and 8 µl of T4 DNA ligase (400 U/µl).

17.3.2
Functional Analysis of Mutants for NAIS Experiments

Before using mutant molecules in an NAIS experiment, it is essential to evaluate activity using the same functional assay as for the NAIS selection step. If the mutation does not affect activity, then it is unlikely to disrupt important interactions and therefore useless for NAIS analysis. By contrast, if the mutant retains only traces of activity, it becomes very difficult to harvest sufficient material for subsequent iodine treatment. In our hands, the optimal reduction in mutant activity is between 3- and 20-fold relative to wild-type. The *trans*-branching reaction was used both for selection and for analysis of D56 mutant activity (Fig. 17.5B) [28].

17.3.3
The Selection Step for NAIS

The selection step for NAIS studies was carried out by following essentially the same protocol as described for the NAIM experiments on this system (see Section 17.2). However, the following considerations were incorporated into the experimental design.

(1) In order to compare the interference pattern of the wild-type and mutant RNAs, the corresponding reactions must be carried out under exactly the same conditions (buffer, pH, ionic conditions).
(2) The extent of the reaction for wild-type and mutant constructs must be the same (usually about 20% of product formation). This is generally achieved by varying the reaction time, which can be, for example, 20 min for the wild-type and 4 h for the mutant.
(3) If the reaction conditions must be varied to make the mutant more reactive (for example, the monovalent concentration is raised to 1 M instead of 0.5 M), also the wild-type has to be tested under these changed conditions to deter-

mine the reaction time at 20% product formation, and, most importantly, to make sure that the wild-type still shows the interferences of interest.

(4) Iodine treatment of the samples is performed according to the same protocol as described in Section 17.2. It is always necessary to have lanes of iodine-untreated reaction products for both wild-type and mutant RNAs on the gel next to iodine-treated samples, to make sure that the band attributed to the interference suppression in the presence of the mutant is not simply a degradation product due to the prolonged incubation time.

(5) In order to demonstrate that the tertiary interaction identified by NAIS is specific, it is helpful to carry out a reverse NAIS experiment. In this setup, the RNAs that contain the single-atom modification and the pool of analog-modified transcripts are switched. Ideally, a mutation or single-atom modification is introduced at the position where suppression was found in the previous experiment, and the other RNA is screened for interferences. If such an interference is observed, there is strong evidence that a specific interaction exists between the two RNAs. The validity of this approach has been successfully demonstrated on a group II intron system [12].

17.3.4
Data Analysis and Presentation

NAIS results are analyzed and quantified essentially in the same manner as NAIM data (see Section 17.2.6). The presentation of results usually includes a gel clearly showing analog interference with the wild-type RNA and suppression of this interference with the mutant RNA, and a bar graph with nucleotide analog effects for mutant molecules in comparison with the wild-type [9, 12] (Fig. 17.7). It is important to ensure that the mutation or a single-atom change results in a specific suppression of the analog interference at one or two positions with the rest of interference effects remaining unchanged, and not in five or more suppression effects at various positions throughout the molecule. The latter indicates a problem with the selection step, i.e. wild-type and mutant not treated under the same conditions or not reacted to the same extent. If one mutation causes suppression of interference at more than one position, reverse NAIS (see above) may allow one to distinguish between a complex multi-component tertiary interaction, e.g. $\lambda-\lambda'$ in the group II intron [12], and an artifact or an indirect effect.

Exacerbation of interferences caused by a mutation is more difficult to interpret. The presence of a mutation sometimes makes a system more susceptible to disturbance by modifications and can result in non-specific exacerbation of interferences with various analogs at different positions. At the same time, if a certain function is supported by a multi-component interaction involving some redundant elements, a mutation at one of these elements can make the system more sensitive to modifications at other components of this interaction. In this case the appearance of additional interferences compared to the wild-type is selectively caused by a specific mutation. This type of specific exacerbation has also been observed during studies of the ai5γ group II intron [90]. While wild-type RNA does not exhibit inosine interference at the θ' tetraloop receptor in D2 (Fig. 17.5A), it is observed upon incorporation of specific mutations in D3 (A627:G), suggesting a functional connection between this nucleotide and the θ–θ' tertiary contact.

Acknowledgments

We thank S. Hamill for contributing data used in creating Fig. 17.6(B). A. M. P. and O. F. thank the Howard Hughes Medical Institute for the financial support of this work. Research on transcription termination (M. B.) has been supported in part by the Ligue Contre le Cancer (Comité du Loiret). This work is also supported by grants from the National Institutes of Health (NIH GM50313 and GM60620 to A.M.P.).

References

1 A. FERRE-D'AMARE, J. A. DOUDNA, Annu. Rev. Biophys. Biomol. Struct. 1999, 28, 57–73.
2 J. H. CATE, A. R. GOODING, E. PODELL, K. ZHOU, B. L. GOLDEN, C. E. KUNDROT, T. R. CECH, J. A. DOUDNA, Science 1996, 273, 1678–1685.
3 N. BAN, P. NISSEN, J. HANSEN, P. B. MOORE, T. A. STEITZ, Science 2000, 289, 905–920.
4 P. NISSEN, J. HANSEN, N. BAN, P. B. MOORE, T. A. STEITZ, Science 2000, 289, 920–930.
5 R. BATEY, R. P. RAMBO, L. LUCAST, B. RHA, J. A. DOUDNA, Science 2000, 287, 1232–1239.
6 J.-L. CHEN, J. M. NOLAN, M. E. HARRIS, N. R. PACE, EMBO J. 1998, 17, 1515–1525.
7 S. RYDER, L. ORTOLEVA-DONNELLY, A. B. KOSEK, S. A. STROBEL, Methods Enzymol. 2000, 317, 92–109.
8 S. RYDER, S. A. STROBEL, Methods 1999, 18, 38–50.
9 A. A. SZEWCZAK, L. ORTOLEVA-DONNELLY, S. P. RYDER, E. MONCOEUR, S. A. STROBEL, Nat. Struct. Biol. 1999, 5, 1037–1042.
10 J. SOUKUP, N. MINAKAWA, A. MATSUDA, S. A. STROBEL, Biochemistry 2002, 41, 10426–10438.
11 L. ORTOLEVA-DONNELLY, A. A. SZEWCZAK, R. R. GUTELL, S. A. STROBEL, RNA 1998, 4, 498–519.
12 M. BOUDVILLAIN, A. DELENCASTRE, A. M. PYLE, Nature 2000, 406, 315–318.
13 L. ZHANG, J. A. DOUDNA, Science 2002, 295, 2084–2088.
14 A. BURGIN, N. R. PACE, EMBO J. 1990, 9, 4111–4118.
15 M. HARRIS, J. M. NOLAN, A. MALHOTRA, J. W. BROWN, S. C. HARVEY, N. R. PACE, EMBO J. 1994, 13, 3953–3963.
16 S. JOSEPH, M. L. WHIRL, D. KONDO, H. F. NOLLER, R. B. ALTMAN, RNA 2000, 6, 220–232.
17 D. L. ABRAMOVITZ, R. A. FRIEDMAN, A. M. PYLE, Science 1996, 271, 1410–1413.
18 A. M. PYLE, F. L. MURPHY, T. R. CECH, Nature 1992, 358, 123–128.
19 G. GISH, F. ECKSTEIN, Science 1988, 240, 1520–1522.
20 K. L. NAKAMAYE, G. GISH, F. ECKSTEIN, H. P. VOSBERG, Nucleic Acids Res. 1988, 16, 9947–9959.
21 R. B. WARING, Nucleic Acids Res. 1989, 17, 10281–10293.
22 D. E. RUFFNER, O. C. UHLENBECK, Nucleic Acids Res. 1990, 18, 6025–6029.
23 K. L. MASCHHOFF, R. A. PADGETT, Nucleic Acids Res. 1992, 20, 1949–1957.
24 D. SCHATZ, R. LEBERMAN, F. ECKSTEIN, Proc. Natl. Acad. Sci. USA 1991, 88, 6132–6136.

25 F. Conrad, A. Hanne, R. K. Gaur, G. Krupp, Nucleic Acids Res. **1995**, 23, 1845–1853.
26 S. A. Strobel, K. Shetty, Proc. Natl. Acad. Sci. USA **1997**, 94, 2903–2908.
27 L. Ortoleva-Donnelly, M. Kronman, S. A. Strobel, Biochemistry **1998**, 37, 12933–12942.
28 M. Boudvillain, A. M. Pyle, EMBO J. **1998**, 17, 7091–7104.
29 S. Ryder, S. A. Strobel, J. Mol. Biol. **1999**, 291, 295–311.
30 A. Schwartz, A. R. Rahmouni, M. Boudvillain, EMBO J. **2003**, 22, 3385–3394.
31 S. Strobel, Biochem. Soc. Trans. **2002**, 30, 1126–1131.
32 S. Basu, R. P. Rambo, J. Strauss-Soukup, J. H. Cate, A. R. Ferre-D'Amare, S. A. Strobel, J. A. Doudna, Nat. Struct. Biol. **1998**, 5, 986–992.
33 S. Basu, S. A. Strobel, RNA **1999**, 5, 1399–1407.
34 S. Ryder, A. K. Oyelere, J. L. Padilla, D. Klostermeier, D. P. Millar, S. A. Strobel, RNA **2001**, 7, 1454–1463.
35 A. Oyelere, J. R. Kardon, S. A. Strobel, Biochemistry **2002**, 41, 3667–3675.
36 F. Jones, S. A. Strobel, Biochemistry **2003**, 42, 4265–4276.
37 J. K. Strauss-Soukup, S. A. Strobel, J. Mol. Biol. **2000**, 302, 339–358.
38 F. Michel, K. Umesono, H. Ozeki, Gene **1989**, 82, 5–30.
39 F. Michel, E. Westhof, J. Mol. Biol. **1990**, 216, 585–610.
40 F. Michel, J.-L. Ferat, Annu. Rev. Biochem. **1995**, 64, 435–461.
41 A. Szewczak, L. Ortoleva-Donnelly, M. Zivarts, A. K. Oyelere, A. V. Kazantsev, S. A. Strobel, Proc. Natl. Acad. Sci. USA **1999**, 96, 11183–11188.
42 H. M. Pley, K. M. Flaherty, D. B. McKay, Nature **1994**, 372, 111–113.
43 P. Nissen, J. A. Ippolito, N. Ban, P. B. Moore, T. A. Steitz, Proc. Natl. Acad. Sci. USA **2001**, 98, 4899–4903.
44 A. Fersht, Structure and Mechanism in Protein Science, Freeman, New York, **1999**.
45 S. Nakano, D. M. Chadalavada, P. C. Bevilacqua, Science **2000**, 287, 1493–1497.
46 A. Ferre-D'Amare, K. Zhou, J. A. Doudna, Nature **1998**, 395, 567–574.
47 A. Perrotta, I. Shih, M. D. Been, Science **1999**, 286, 123–126.
48 A. Nesbitt, L. A. Hegg, M. J. Fedor, Chem. Biol. **1997**, 4, 619–630.
49 D. J. Earnshaw, M. J. Gait, Nucleic Acids Res. **1998**, 26, 5551–5561.
50 J. B. Murray, A. A. Seyhan, N. G. Walter, J. M. Burke, W. G. Scott, Chem. Biol. **1998**, 5, 587–595.
51 P. Gordon, E. Sontheimer, J. Piccirilli, Biochemistry **2000**, 39, 12939–12952.
52 P. Gordon, J. A. Piccirilli, Nat. Struct. Biol. **2001**, 8, 893–898.
53 J. M. Warnecke, J. P. Furste, W.-D. Hardt, V. A. Erdmann, R. K. Hartmann, Proc. Natl. Acad. Sci. USA **1996**, 93, 8924–8928.
54 J. M. Warnecke, R. Held, S. Busch, R. K. Hartmann, J. Mol. Biol. **1999**, 290, 433–445.
55 A. Pyle, J. Biol. Inorg. Chem. **2002**, 7, 679–690.
56 E. C. Christian, M. Yarus, Biochemistry **1993**, 32, 4475–4480.
57 G. Chanfreau, A. Jacquier, Science **1994**, 266, 1383–1387.
58 V. Sood, T. L. Beattie, R. A. Collins, J. Mol. Biol. **1998**, 282, 741–750.
59 J. H. Cate, R. L. Hanna, J. A. Doudna, Nat. Struct. Biol. **1997**, 4, 553–558.
60 T. Pfeiffer, A. Tekos, J. M. Warnecke, D. Drainas, D. R. Engelke, B. Seraphin, R. K. Hartmann, J. Mol. Biol. **2000**, 298, 559–565.
61 S. Basu, S. A. Strobel, Methods, **2001**, 23, 264–275.
62 J. Cochrane, R. T. Batey, S. A. Strobel, RNA **2003**, 9, 1282–1289.
63 J. Swisher, C. M. Duarte, L. J. Su, A. M. Pyle, EMBO J. **2001**, 20, 2051–2061.
64 A. Arabshahi, P. A. Frey, Biochem. Biophys. Res. Commun. **1994**, 204, 150–155.
65 J. Ludwig, F. Eckstein, J. Org. Chem. **1989**, 54, 631–635.
66 A. Oyelere, S. A. Strobel, J. Am. Chem. Soc. **2000**, 122, 10259–10267.

67 E. Schenborn, R. C. Mierendorf, *Nucleic Acids Res.* **1985**, *13*, 6223–6236.
68 R. Sousa, R. Padilla, *EMBO J.* **1995**, *14*, 4609–4621.
69 Z. Huang, J. W. Szostak, *Nucleic Acids Res.* **1996**, *24*, 4360–4361.
70 A. M. Pyle, *Nucleic Acids and Molecular Biology*, F. Eckstein and D. M. J. Lilley (eds), Springer Verlag, New York, **1996**, pp. 75–107.
71 K. A. Jarrell, C. L. Peebles, R. C. Dietrich, S. L. Romiti, P. S. Perlman, *J. Biol. Chem.* **1988**, *263*, 3432–3439.
72 D. Daniels, W. J. Michels, A. M. Pyle, *J. Mol. Biol.* **1996**, *256*, 31–49.
73 P. Z. Qin, A. M. Pyle, *Biochemistry* **1997**, *36*, 4718–4730.
74 J. de la Cruz, D. Kressler, P. Linder, *Trends Biochem. Sci.* **1999**, *24*, 192–198.
75 E. Silverman, G. Edwalds-Gilbert, R. J. Lin, *Gene*, **2003**, *312*, 1–16.
76 A. Gorbalenya, E. V. Koonin, *Curr. Opin. Struct. Biol.* **1993**, *3*, 419–429.
77 S. Shuman, *Proc. Natl. Acad. Sci. USA* **1992**, *89*, 10935–10939.
78 C. H. Gross, S. Shuman, *J. Virol.* **1996**, *70*, 2615–2619.
79 E. Jankowsky, C. H. Gross, S. Shuman, A. M. Pyle, *Nature* **2000**, *403*, 447–451.
80 E. Jankowsky, C. H. Gross, S. Shuman, A. M. Pyle, *Science* **2001**, *291*, 121–125.
81 S. S. Velankar, P. Soultanas, M. S. Dillingham, H. S. Subramanya, D. B. Wigley, *Cell* **1999**, *97*, 75–84.
82 P. R. Bianco, S. C. Kowalczykowski, *Nature* **2000**, *405*, 368–372.
83 K. S. Murakami, S. A. Darst, *Curr. Opin. Struct. Biol.* **2003**, *13*, 31–39.
84 R. Padilla, R. Sousa, *Nucleic Acids Res.* **2002**, *30*, e138.
85 M. Kashlev, N. Komissarova, *J. Biol. Chem.* **2002**, *277*, 14501–14508.
86 V. Gopal, L. G. Brieba, R. Guajardo, W. T. McAllister, R. Sousa, *J. Mol. Biol.* **1999**, *290*, 411–431.
87 L. E. Macdonald, Y. Zhou, W. T. McAllister, *J. Mol. Biol.* **1993**, *232*, 1030–1047.
88 V. Epshtein, F. Toulme, A. R. Rahmouni, S. Borukhov, E. Nudler, *EMBO J.* **2003**, *22*, 4719–4727.
89 F. Wincott, A. DiRenzo, C. Shaffer, S. Grimm, D. Tracz, C. Workman, D. Sweedler, C. Gonzalez, S. Scaringe, N. Usman, *Nucleic Acids Res.* **1995**, *23*, 2677–2684.
90 O. Fedorova, T. Mitros, A. M. Pyle, *J. Mol. Biol.* **2003**, *330*, 197–209.
91 M. Zuker, *Nucleic Acids Res.* **2003**, *31*, 3406–3415.

18
Nucleotide Analog Interference Mapping: Application to the RNase P System

Simona Cuzic and Roland K. Hartmann

18.1
Introduction

18.1.1
Nucleotide Analog Interference Mapping (NAIM) – The Approach

In classical mutational studies, only the base moiety of a nucleotide can be replaced with one of the three natural alternatives. Even a simple C → U transition affects more than one functional group and exchange of the 4-amino for a keto group represents a rather radical chemical change that can have profound effects on RNA functionality. A more specific and versatile chemogenetic approach is NAIM, which allows us to probe the functional consequences of changes as minor as single-atom substitutions in the base, sugar or phosphate moiety. For example, in the case of a guanosine to inosine modification, the chemical alteration is restricted to deletion of the 2-amino group without additionally replacing the 6-keto with an amino group as in G → A mutations.

At the onset of NAIM studies, a pool of RNA molecules with limited numbers of randomly distributed nucleotide analogs is synthesized. Such a pool of RNAs is then subjected to a selection procedure to separate active variants from those with impaired function due to modification at a particular location. Subsequent comparative analysis of the distribution of modifications in the active RNA fraction and a reference fraction (e.g. the fraction of molecules with impaired function or the original unselected pool) reveals positions critical for function. The salient feature of the method is that all incorporated nucleotide analogs additionally carry a phosphorothioate modification (one non-bridging phosphate oxygen replaced with sulfur), which permits to specifically cleave the nucleic acid chain by iodine [1] exclusively at the sites of analog incorporation (for details, see Chapter 17). Iodine treatment thus results in A-, C-, G- or U-specific sequence ladders on denaturing polyacrylamide gels.

The partial modification of RNA is achieved by the presence of nucleoside α-thiotriphosphate analogs during *in vitro* transcription by T7 RNA polymerase, resulting in the aforementioned pool of RNA molecules, each carrying a low number of randomly distributed modifications. The elegance of the method lies in the capacity to simultaneously screen for the functional contribution of a particular chemical group at almost every, for example, A residue in an RNA chain. There are two major limitations: (1) residues that show a strong phosphorothioate interference effect *per se* will be insensitive to the effect of the additional modification, such as a 2′-deoxy substitution in case of 2′-deoxy NTPαS analogs and (2) phage RNA polymerases do not accept all kinds of nucleotide analogs as substrates (see Section 18.1.2 and Chapter 1). Chemical RNA synthesis expands the scope of possible modifications (e.g. introduction of *Sp*- in addition to *Rp*-phosphorothioates [2]), but such approaches are usually more tedious and require equipment for chemical RNA synthesis as well as special protocols in order to introduce a low level of randomly distributed modifications. Also, chemical RNA synthesis is practically limited to an RNA chain length of about 50 nt, thus excluding directly screening of larger RNAs, such as self-splicing introns or RNase P RNA.

The analogs available for NAIM studies can be divided into three categories [3], according to their main attributes: (1) if they primarily change the chemical properties of the substitutent, (2) delete a functional group or (3) introduce a bulky substituent. Depending on the type of modification introduced, NAIM experiments have the potential to reveal the following information:

- An *Rp*-phosphorothioate modification *per se* (AMPαS, GMPαS, CMPαS, UMPαS) may identify crucial coordination sites for Mg^{2+} ions. Substitution of sulfur for a non-bridging phosphate oxygen essentially abolishes inner-sphere coordination to Mg^{2+}, because Mg^{2+}, a "hard" Lewis acid, prefers to bind oxygen, a "hard" Lewis base, relative to the much more polarizable and thus "softer" sulfur [4]. However, addition of more thiophilic metal ions ("softer" Lewis acids) such as Mn^{2+} or Cd^{2+} may restore, to varying extent, metal ion binding to the thiophosphate, leading to a (partial) rescue of the functional defect [5].
- C^7-deaza purine analogs (c7-AMPαS, c7-GMPαS) are employed to reveal N^7 positions involved in hydrogen binding or metal ion coordination. The latter aspect may be particularly relevant if RNA structure and function is probed in the presence of transition metal ions, such as Mn^{2+} or Zn^{2+}, which form inner-sphere complexes with the N^7 of purines [6, 7].
- Ribose 2′-deoxy modifications allow to probe for 2′-hydroxyls involved in tertiary contacts.
- IMPαS, incorporated by T7 RNA polymerase instead of G nucleotides, is suited to probe the role of guanine exocyclic amino groups in hydrogen bonding. For *Escherichia coli* RNase P RNA, relatively few inosine interference effects were detected in regular helices, suggesting that helix destabilization by this modification is of minor importance for the function of the RNase P ribozyme [8]. However, destabilization of secondary structure can become important if inosines are part of a short intermolecular hybrid helix required to bind the substrate to the ribozyme, particularly under conditions where ribozyme molecules compete for a limited amount of substrate RNAs [9]. For such cases, combined analysis of IMPαS and N^2-methyl-GMPαS interference patterns was reported as a strategy

to differentiate between helix destabilization and loss of important tertiary interactions as the cause of interference [3, 9]. The N^2-methyl group can still form a hydrogen bond with the O^2 of cytosine in Watson–Crick base pairs, but has lost its capacity to participate in a bifurcated hydrogen bonding frequently observed in tertiary contacts that involve the 2-amino group of G residues [8–11].
- Analogs incorporated at A positions by T7 RNA polymerase, such as purine, N^6-methyl-adenosine, 2-aminopurine and 2,6-diaminopurine, all of which are commercially available, probe the N^6 position in terms of chemical properties and steric constraints and the tolerance for an additional 2-amino group on the minor groove edge of the base, respectively [12].

18.1.2
Critical Aspects of the Method

18.1.2.1 Analog Incorporation

T7 RNA polymerase incorporates Sp-NTPαS analogs, yielding Rp-phosphorothioate-modified RNAs due to inversion of configuration at the phosphorus atom during polymerization [13]. It has previously been documented that Sp-NTPαS analogs are incorporated with essentially the same efficiency as normal NTP substrates by T7 RNA polymerase [14]. However, for other phosphorothioate-tagged analogs, which carry an additional modification at the base or sugar moiety, incorporation efficiency is in most cases lower. One exception is ITPαS which is accurately incorporated in place of guanosine and with comparable efficiency [8, 15]. Many analogs (such as those with 2'-ribose modifications) are better accepted by the Y639F mutant T7 RNA polymerase which shows a greater tolerance toward changes of functional groups in the minor groove [16, 17]. The reader is referred here to Chapter 17 and the publication by Ryder and Strobel [16] for more details on individual analogs.

The extent of analog incorporation is adjusted to 2.5–10% (usually 5%) in NAIM studies, although incorporation efficiency may not exceed 1–2% in the case of some analogs [16]. A modification extent of 5% permits good detection and quantitation of iodine hydrolysis bands, but avoids two problems associated with higher modification extents: (1) RNA inactivation due to phosphorothioate-tagged analog incorporation at multiple sites per molecule, as seen with fully (100%) AMPαS-, CMPαS- or GMPαS-modified E. coli RNase P RNA [18] and (2) an increased probability that each RNA molecule carries a modification at a site of strong interference in addition to weakly interfering modifications; as a consequence, the sites of strong interference will, by themselves, fully determine the deficiency status of an RNA molecule irrespective of additional weaker interferences, thus masking the latter.

One should also be aware that analogs are not incorporated to the same extent at all transcript positions, the incorporation pattern being largely specific for the analog and the individual RNA under investigation. One observation is the lack or reduction of analog incorporation at homo-di- or homo-oligonucleotide stretches [16].

We usually follow a simple strategy to adjust the extent of analog incorporation, based on the roughly equal incorporation efficiency of NTP and Sp-NTPαS substrates. For example, to assess ITPαS incorporation efficiency, we transcribed RNase P RNA in the presence of 2.5, 5.0 and 10% ITPαS; in parallel, we performed transcriptions in the presence of 5% GTPαS and 95% GTP as the reference. Gel analysis of samples after iodine hydrolysis then revealed that transcripts with 2.5% IMPαS modification resulted in iodine hydrolysis band intensities comparable with those of transcripts carrying 5% GMPαS modifications. For other analogs, however, this relation is reversed due to the less efficient incorporation of analog versus its NTPαS reference substrate.

18.1.2.2 Functional Assays

The second crucial step in NAIM studies, aside from analog incorporation, is the selection assay that partitions functional and defective RNA molecules. We will mention here examples of such assays that have been used in NAIM analyses of RNase P RNA (examples of selection assays developed for other ribozyme systems are described in Chapter 17):

- Separation of modified RNase P RNA pools into precursor tRNA-binding and non-binding fractions via adsorption of biotinylated precursor tRNA to streptavidin–agarose beads [12].
- Separation of modified RNase P RNA pools into RNase P protein-binding and non-binding fractions via adsorption of the His-tagged protein to Ni-NTA agarose beads [19].
- Partitioning of self-cleaving RNase P RNA ribozyme–substrate conjugates into reacted and unreacted fractions and separation by denaturing PAGE [20–22].
- Separation of modified RNase P RNA pools into tRNA-binding and non-binding fractions by gel retardation [8, 18, 23, 24].

18.1.2.3 Factors Influencing the Outcome of NAIM Studies

Factors that influence the outcome of NAIM studies include:

- The functional aspect selected for. As an example, Boudvillain and Pyle [25] reported their *trans*-branching assay for group II intron self-splicing to be more sensitive to modification interference and thus perturbations of tertiary structure than *cis*-splicing assays. They argued that the *trans*-branching approach is so effective because the stabilization energy resulting from essential tertiary interactions has to counterbalance the entropic penalty inherent to the assembly of a two- or multi-component system.
- Reaction conditions of the selection assay (nature and concentration of mono- and divalent cations, pH and temperature). High salt conditions, for example, suppressed weaker interference effects in group II intron *cis*- and *trans*-splicing assays [26]. Variations in pH were shown to alleviate or exacerbate interference effects [20, 21], attributable to changes in rate-limiting steps of ribozyme-catalyzed reactions [24]. Several modifications that interfere with tRNA binding

to RNase P RNA were suppressed at higher RNA concentrations [18]. Likewise, all factors that alter the dissociation constant of complexes in *trans*-binding assays (such as the temperature) will, to some extent, change the pattern and strength of interference effects.
- Inefficient analog incorporation at certain positions in the RNA chain (e.g. at homonucleotide stretches) and gel artifacts, such as band compressions at a string of G residues [8], which can limit the information content of NAIM experiments. The reader should further be aware that fluctuations in the strength of interference effects between individual experiments are considerable. The extent of such fluctuations is expected to rise with (1) an increasing number of experimental steps involved throughout the entire procedure, (2) difficulties to kinetically control a reaction catalyzed by an enzyme or (3) difficulties to define and control the proportion of functional protein in the case of RNA–protein-binding studies (e.g. if a His-tagged protein is coupled to a Ni-NTA affinity resin; [19]).

18.1.3
Interpretation of Results

The *Rp*-phosphorothioate modification itself can cause interference effects that are due to disruption of hydrogen bonding or inner-sphere coordination of Mg^{2+} normally involving the *pro*-Rp oxygen at this location. At most sites, the sulfur substitution has little or no effect on RNA structure; in rare cases, however, it may substantially change local structure [27] and, as a consequence, could induce global conformational changes. When NAIM is used to probe, for example, RNA–ligand interactions, a phosphorothioate modification could thus affect ligand binding even if the site of modification is at considerable distance to the binding interface. The resulting interference effect will then be hard to interpret solely on the basis of NAIM data. Along the same lines, phosphorothioate interference effects are often found to be clustered in densely packed RNA core structures which play a key role in RNA architecture [18, 23]. Such RNA regions are apparently very sensitive even to minor structural and chemical perturbations caused by the sulfur substitution, and thus many interference effects are likely to reflect changes of global structure rather than direct contacts to an RNA ligand.

Partial suppression of phosphorothioate interference in the presence of Mn^{2+} is often considered to indicate direct metal ion coordination to the respective phosphate oxygen [28]. The ability of Mn^{2+}, but the inability of Mg^{2+}, to coordinate to the sulfur is made responsible for this effect. However, although Mg^{2+} and Mn^{2+} ions can occupy basically the same metal ion binding pockets, they may coordinate in a slightly different way, as originally observed in tRNA crystals [29]. Likewise, Mg^{2+} and Mn^{2+} interacted differently with the phosphate of A9 in two hammerhead ribozyme crystal structures [11, 30]. Nevertheless, Mn^{2+} rescue of phosphorothioate interferences in the central P4 helix of RNase P RNA [18, 20] provided the basis for detailed studies of metal ion coordination in this region using RNase P RNA constructs with single-site phosphorothioate modifications that were analyzed for kinetic defects as well as Mn^{2+} and Cd^{2+} rescue effects [31, 32].

Generally, interpretation of NAIM results remains ambiguous without further information from other approaches. For example, if NAIM is performed to identify functional groups that are involved in binding of a protein or another RNA, it is impossible to differentiate whether interference effects represent direct contact sites or indirect effects due to destabilization of the global RNA fold. However, additional information often permits to interpret interference data, as illustrated by two examples. For the *Tetrahymena* ribozyme, the crystal structure of the P4–P6 domain [10] revealed a hydrogen bonding interaction between the exocyclic amine of G212 in helix P4 and A184 in the A-rich bulge motif, which bridges the core helix P4 and the three-way junction of helices P5a, b and c. NAIM identified N^2-methyl-GMPαS, but not IMPαS, interference at G212 [9]. The authors concluded that deletion of the hydrogen bond due to inosine modification was not sufficient to significantly impair intron splicing, while introduction of the bulky N^2-methyl group in the P4 minor groove substantially destabilized this tightly packed region of the RNA. Another example comes from the RNase P RNA system. Comparative sequence analysis and results from biochemical studies in combination with computer-aided derivation of RNase P RNA architecture led to the proposal of the loop–helix tertiary interactions L18–P8 and L8–P4 [33, 34]. Further, Easterwood and Harvey [35] proposed a model of tRNA 3′-CCA end binding to the P15 loop of *E. coli* RNase P RNA, which involves formation of two consecutive base triples. A variety of nucleotides in the corresponding structural elements of RNase P RNA showed Rp-phosphorothioate, 2′-deoxy, inosine and/or c7-deaza interference effects in a tRNA binding assay [8, 18, 23, 24]. The interference data went surprisingly well with the above-mentioned interactions and allowed us to confirm and refine their atomic details [8, 24].

On the other hand, NAIM approaches can be an important tool in cases where it is unclear if a ribozyme crystal structure represents the catalytically competent conformer. A study combining point mutations and NAIM showed formation of a base triple in the core of the *Tetrahymena* ribozyme, which contributes to substrate helix docking and stabilization of active site conformation [36]. The formation of this base triple indicated that a crystallized form of the ribozyme [37] would require a substantial rearrangement to adopt an active conformation.

For *E. coli* RNase P RNA, results from a series of NAIM experiments permit to extract the following conclusions:

(1) Although largely different functional assays and assay conditions were employed (self-cleavage of ribozyme–substrate conjugates at 3 M Na$^+$ or NH$_4^+$ and 1–25 mM Mg^{2+}, gel-resolvable binding of mature tRNA to RNase P RNA in the presence of 0.1 M NH$_4^+$ and 0.1 M Mg^{2+}, RNase P RNA binding to immobilized precursor tRNA in the presence of 1 M NH$_4^+$ and 25 mM Ca^{2+}), a substantial number of identical interference effects has been observed ([24] and references therein; [22]). The conclusion is that modifications at these vulnerable sites destabilize the tertiary fold, thus affecting substrate and product binding as well as catalysis. In fact, the vast majority of interference effects likely reflects perturbation of RNA tertiary structure rather than representing

functional groups directly involved in catalysis or the binding of mature or precursor tRNA.

(2) Interference effects cluster in regions which (i) organize the conserved catalytic core of RNase P RNA (such as P4, J2/4, J3/4, J18/2 and J5/15), which (ii) are tightly packed with the help of multiple metal ions (the P2–P3/P1–P4 four helix junction [32]) and which (iii) organize the metal ion-rich core of the specificity domain (J11/12, P11 [8, 12, 18, 23, 24, 38]).

18.1.4
Nucleotide Analog Interference Suppression (NAIS)

NAIS is an extension of NAIM that permits to overcome the uncertainty of NAIM experiments which leave open if an interference effect reflects a direct contact between two interacting macromolecules. In NAIS, the partially analog-modified RNA pool (RNA 1) is analyzed for interference effects in two parallel setups: one employing the interacting RNA (RNA 2) in its wild-type form, as in NAIM, and the second using RNA 2 with a point mutation or a single chemical group altered at a location suspected to interact directly with RNA 1. If a functional group in RNA 1 normally interacts with the functional group in RNA 2 that has been changed by the aforementioned mutation or modification, then the interference effect observed in RNA 1 for disrupting this contact should disappear, because the contact has already been abolished by the mutation or modification in RNA 2. A critical point of such NAIS experiments is to adapt the functional assay with the mutant or modified RNA 2 in such a way that the extent of reaction or complex formation is the same as in the reference assay with wild-type RNA 2. For a more detailed description and illustration of NAIS, see Chapter 17.

18.2
NAIM Analysis of *Cis*-cleaving RNase P RNA–tRNA Conjugates

18.2.1
Characterization of a *Cis*-cleaving *E. coli* RNase P RNA–tRNA Conjugate

Cis-cleaving RNase P RNA (P RNA)–tRNA conjugates of the type shown in Fig. 18.1(A) open up the perspective to identify functional groups that are crucial to cleavage chemistry, either applying NAIM or NAIS. In previous related approaches, the tRNA substrate was tethered to internal positions of *E. coli* P RNA [20–22]. However, these constructs required 3 M monovalent salt for efficient self-cleavage and we were concerned that such conditions may suppress informative interference effects to an unwanted extent. We therefore pursued a somewhat different strategy, based on a transcript consisting of *E. coli* P RNA, a linker region and the 5′ half of a bacterial tRNA^{Gly} which is annealed to the tRNA 3′ half (added in excess over P RNA–tRNA 5′ half), creating a self-cleaving P RNA–tRNA conjugate (Fig. 18.1A). This bipartite system prevents self-cleavage at the tRNA 5′ end already during RNA preparation and permits to use tRNA 3′ halves with single-site modifications for NAIS experiments. Before conducting NAIM experiments, it was essential to analyze the kinetic behavior of the conjugate in order to determine the conditions under which *cis*-cleavage only occurs. Moreover, we had to clarify what limits the reaction rate, for example cleavage chemistry or a refolding step. Our finding that the cleavage rate remained constant in the tested range of P RNA–tRNA 5′ half concentrations (Fig. 18.1C) demonstrated that this type of ribozyme acts only *in cis* and not *in trans* under dilute conditions. Exclusion of *trans*-cleavage is essential for the separation of active ribozymes from less active and inactive RNA variants. The data points for the time course of the *cis*-cleavage reaction were best described by the sum of two first-order reactions (Fig. 18.1B), suggesting that there are at least two populations of ribozymes reacting with different velocities: the correctly folded population (around 20%, see Limit 1 in Fig. 18.1B, panel II) reacts fast, while the fraction of slower reacting conjugates either has to change its conformation before *cis*-cleavage can occur or uses an alternative, albeit slower reaction pathway. The linear relationship for log k_1 versus pH with a slope of about 1 in the range of pH 5.2–6.5 (Fig. 18.1D, panel I) indicates that the chemical step is rate-limiting in the initial phase of the reaction [39]. In contrast, the cleavage rate constant k_2 describing the slow turnover (Fig. 18.1D, panel II) was independent of pH, supporting the idea that k_2 reflects the rate of a slow refolding step. The rate constant k_1 was further shown to be independent of the linker length in the range of 45–130 nt (Fig. 18.1E); also, tRNA 3′ half concentrations exceeding that of P RNA–tRNA 5′ half (0.1 nM) by a factor of 800 did not further increase k_1 (Fig. 18.1F).

18.2.2
Application Example

In the following section we will describe, as an application example, IMPαS modification experiments using the *E. coli* P RNA–tRNA conjugate with a 53-nt long linker connecting the P RNA and tRNA moieties. Usually, one compares IMPαS with GMPαS interference patterns to be able to ascribe interference effects to the thioate and/or additional base modification (see Chapter 17). However, our initial intention was to determine to which extent IMPαS interference patterns obtained with the *cis*-cleaving conjugate overlap with those observed in a gel retardation assay selecting for high affinity tRNA binding to *E. coli* P RNA [8]. The protocol is tailored to NAIM analysis of the 3′ half of P RNA. Usually, analysis from the 5′ end is conducted in parallel in order to fully resolve interference effects for an RNA of the size of P RNA (around 400 nt). Corresponding analysis from the P RNA 5′ end requires two major changes in the procedure outlined below and illustrated in Fig. 18.2: (i) in Step 1, the procedure is started exclusively with 5′-^{32}P-end-labeled pool RNA (instead of adding some 5′-end-labeled to predominantly unlabeled RNA), using about 15 times more radioactivity than specified in Protocol 5, and (ii) 3′-end-labeling (Step 6) is omitted. In addition, the simplest strategy for this setup is to directly excise the cleaved and uncleaved fractions from the same gel lane (no need to perform Step 4b), since 5′-end-labeled bands derived from

18 Nucleotide Analog Interference Mapping: Application to the RNase P System

A E. coli P RNA

linker sequence — CCA

5' half | 3' half

tRNA

B

I single exponential fit

Variable	Value	Std. err.
Limit	0.6	0.04
k	0.2	0.04

II double exponential fit

Variable	Value	Std. err.
Limit 1	0.20	0.03
k_1	1.55	0.48
Limit 2	0.49	0.03
k_2	0.07	0.01

C concentration of P RNA-tRNA 5' half:

0.5 nM	2 nM	10 nM	20 nM	40 nM

% cleavage product: 0 43 55 63 | 0 44 56 67 | 0 47 60 68 | 0 45 56 67 | 0 49 58 71

D k_1 (min^{-1}) vs pH (I); k_2 (min^{-1}) vs pH (II)

E k_1 (min^{-1}) vs linker length (nt)

F k_1 (min^{-1}) vs [tRNA 3' half] / [P RNA - tRNA 5' half]

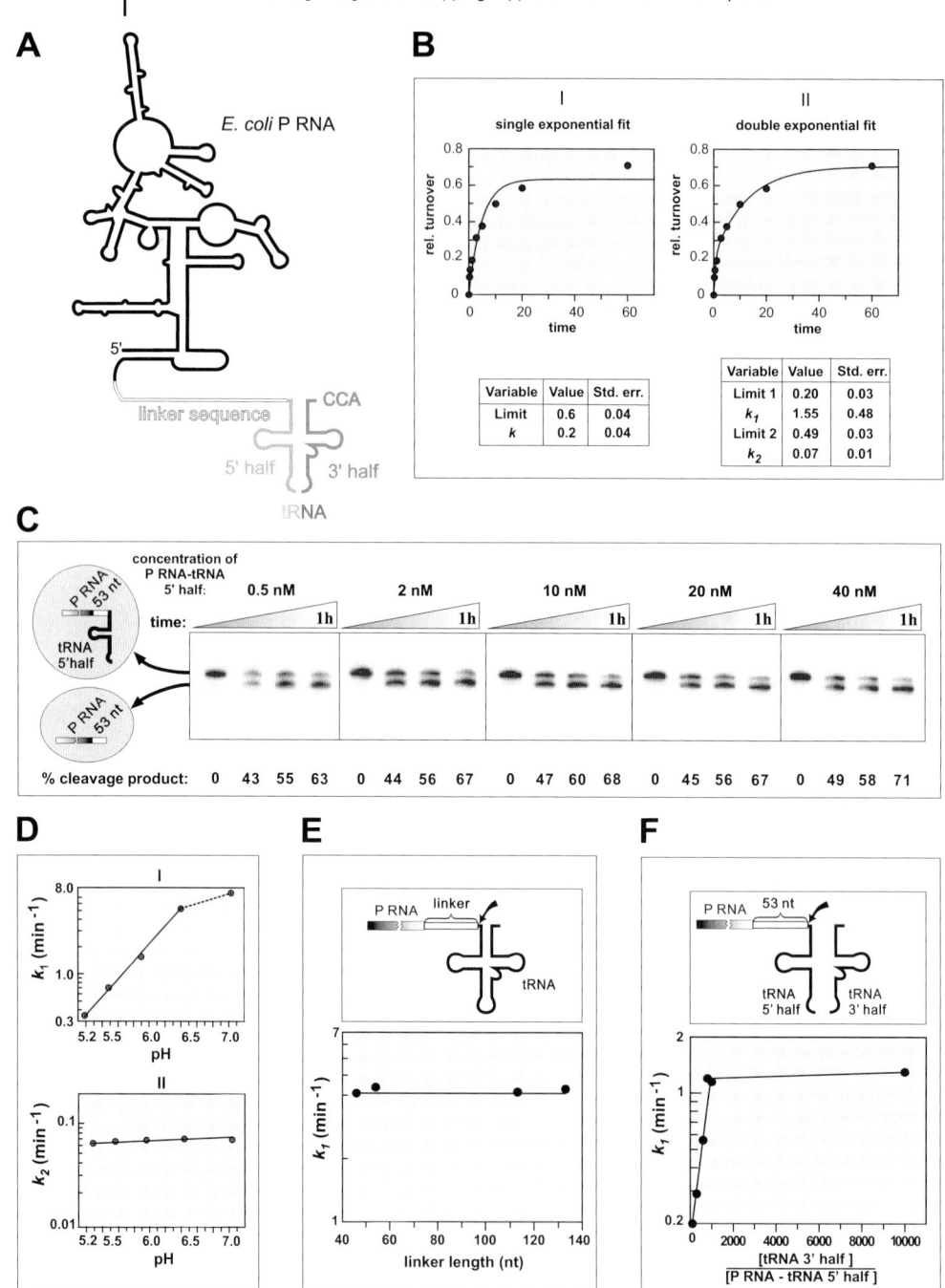

iodine hydrolysis within the P RNA moiety will co-migrate for the cleaved and uncleaved fraction in the final gel analysis (Step 7).

The experimental procedure involved the following steps, illustrated in Fig. 18.2:

(1) Transcription of P RNA–tRNA 5′ half carrying a low degree (2.5%) of randomly distributed IMPαS modifications, 5′-^{32}P-end-labeling of an aliquot of the RNA pool for the detection of cleaved and uncleaved molecules on denaturing PAA gels, and transcription of tRNA 3′ half (unmodified for NAIM). The concept for NAIS experiments is illustrated as well. For example, with the goal to identify functional groups in P RNA that interact with 2-hydroxyls in the T arm, two tRNA 3′ halves will be utilized: an all-ribose 3′ half and a variant thereof with a single 2′-deoxyribose (indicated by a filled triangle); interference patterns obtained with the two tRNA 3′ halves will then be compared in order to identify interference suppression effects when using the tRNA 3′ half carrying a single 2′-deoxy modification in the T arm. Experimental steps for NAIS will be identical to those of NAIM, except that the reaction conditions have to be adapted for the 2′-deoxy-modified 3′ half to give the same extent of reaction as for the all-ribose tRNA 3′ half.

Fig. 18.1. Characterization of a model *E. coli* P RNA–tRNA conjugate. (A) Schematic representation of the construct consisting of the P RNA moiety (black), a linker sequence (thin double lines) and the tRNA 5′ half (light grey); the complete tRNA substrate is reconstituted by annealing the tRNA 3′ half (dark grey). (B) Self-cleavage of the conjugate shown in (A), equipped with a 53-nt long linker. Assays were performed as follows. The P RNA–tRNA^{Gly} complex was formed by annealing the tRNA 3′ half to 5′-end-labeled P RNA–tRNA 5′ half in the presence of 100 mM NH$_4$Cl and 5 mM CaCl$_2$ to avoid uncontrolled self-cleavage. The P RNA–tRNA 5′ half concentration was 0.3 nM and that of tRNA 3′ half 300 nM. The annealing mixture was heated to 95 °C for 2 min, then transferred to and cooled down in a heating block adjusted to 50 °C and pre-incubated for 30 min at 50 °C. After pre-incubation, the mixture was diluted to 0.1 nM P RNA–tRNA 5′ half and 100 nM tRNA 3′ half by adding NH$_4$Cl, urea and CaCl$_2$ to a final concentration of 100, 100 and 5 mM, respectively. The reaction was started by adding MgCl$_2$ to a final concentration of 36 mM. Samples were withdrawn at different time points and analyzed by 8% denaturing PAGE. The time course was fit to either a single first-order (panel I) or two consecutive first-order reactions (panel II). (C) This panel documents that no significant *trans*-cleavage occurred under the conditions tested, since the cleavage rate was constant at different concentrations of P RNA–tRNA 5′ half (0.5–40 nM final concentration in the *cis*-cleavage assay). (D) pH dependence of k_1 (measuring the rate of the fast initial phase of the reaction) and k_2 (measuring the rate of the second slower phase of the reaction; see panel B II); the linear relationship of log k_1 versus pH with a slope of about 1 in the range of pH 5.2–6.5 suggests that the chemical step is rate-limiting in the initial phase of the reaction. (E) Influence of the linker length on the cleavage rate constant k_1. The linker length is defined as the number of nucleotides that separate the P RNA 3′-end from the tRNA 5′-end. No significant differences were observed in the cleavage rate constant k_1 among complexes with linker lengths between 45 and 133 nt. (F) To find the saturation limit of the tRNA 3′ half, its concentration was varied up to a 10,000-fold excess over P RNA–tRNA 5′ half (53-nt linker) whose concentration was 0.1 nM; tRNA 3′ half concentrations exceeding that of P RNA–tRNA 5′ half by more than a factor of 800 did not further increase the cleavage rate constant k_1. For reaction conditions in (D)–(F), see legend to (B).

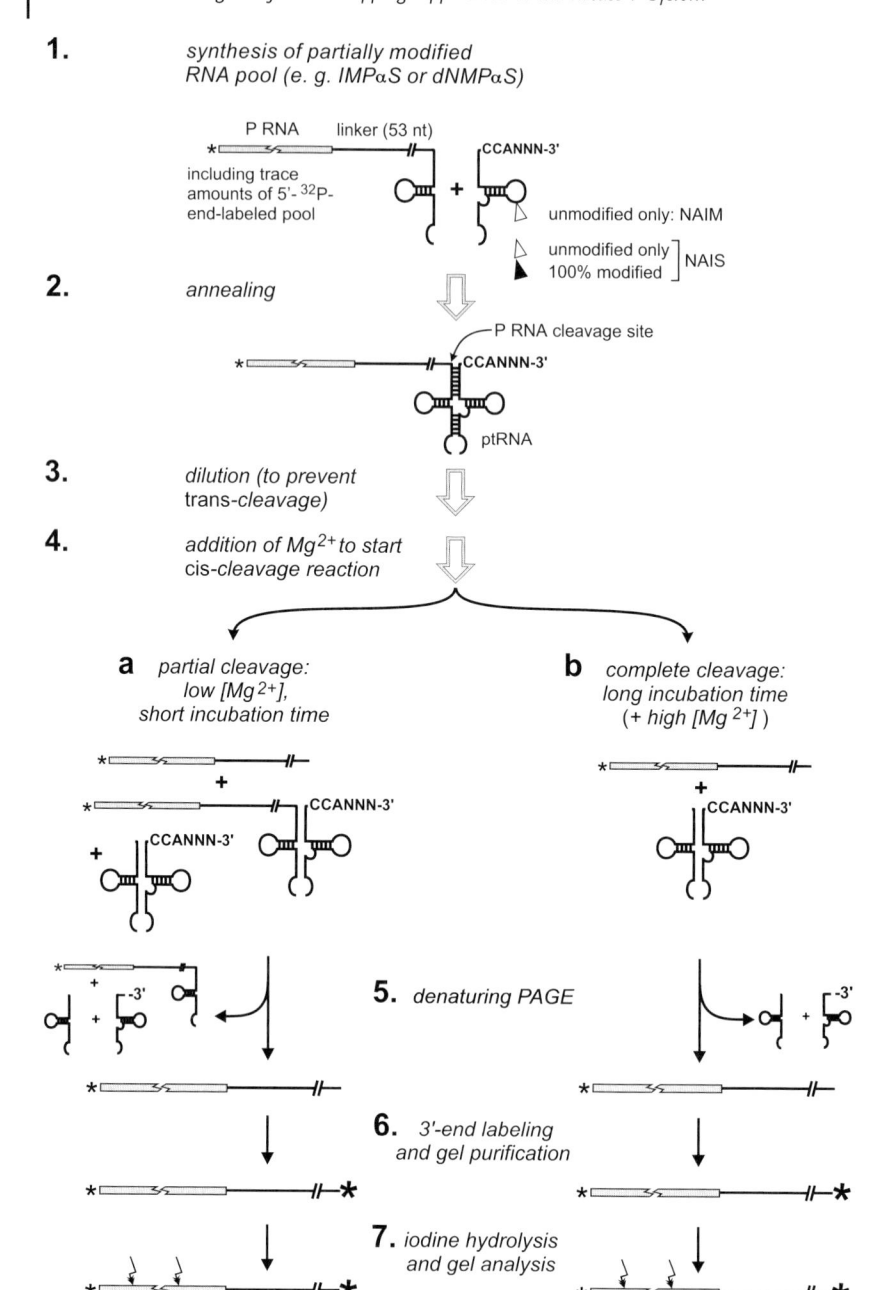

Fig. 18.2. Flow scheme for NAIM (NAIS) analysis of a cis-cleaving *E. coli* RNase P RNA–tRNA conjugate. For details, see Sections 18.2.2 and 18.2.4.

(2) Formation of the P RNA–tRNA^{Gly} complex by annealing the tRNA 3′ half (8.8 µM) to the P RNA–tRNA 5′ half (91 nM, including trace amounts of 5′-end-labeled material) in the presence of 100 mM NH$_4$Cl and 5 mM CaCl$_2$ at pH 5.9, conditions that prevent uncontrolled self-cleavage; the annealing mix is heated to 95 °C for 2 min, then transferred to and cooled down in a heating block adjusted to 50 °C and pre-incubated for 30 min at 50 °C.

(3) 3.8-fold dilution of the annealed mix to 24 nM P RNA–tRNA 5′ half and 2.3 µM tRNA 3′ half by addition of NH$_4$Cl, urea and CaCl$_2$ to final concentrations of 100, 100 and 5 mM, respectively.

(4) Starting the *cis*-cleavage reaction by addition of MgCl$_2$ to a final concentration of 36 mM.

  (a) Stopping the reaction after 2 min at 50 °C, resulting in 20–30% product formation corresponding to the fast phase of the reaction (Fig. 18.1B, panel II).

  (b) A parallel reaction was incubated for 2 h, resulting in essentially complete substrate conversion, serving as the reference RNA pool for NAIM analysis. The reason for taking this sample as the reference pool, and not the original, untreated RNA pool, is that the same length species as in the short incubation is generated (P RNA plus linker). One potential drawback is that modification at some positions may entirely block *cis*-cleavage, resulting in the absence of an iodine hydrolysis band for the 2-min fraction as well as the reference pool, with the effect that strong interference effects would escape notice. A strategy to circumvent this problem is to perform the 2-h incubation for the reference pool under *trans*-cleavage assay conditions by elevating the Mg^{2+} concentration. Yet another option is to first load the iodine-hydrolyzed starting pool (P RNA–tRNA 5′ half), or possibly the uncleaved fraction from Step 4a, onto the gel (Step 7 below), let it run for some time, and then load the RNA fraction that was cleaved within 2 min to compensate for its reduced length due to the absence of the tRNA 5′ half (Fig. 18.2). In any case, it is advisable to compare the iodine hydrolysis patterns of alternative reference pools to address the potential ambiguity mentioned above.

(5) Denaturing PAGE and elution of 5′-end-labeled cleavage products (P RNA plus linker).

(6) 3′-End-labeling of eluted RNAs, such that the radioactivity of the 3′-^{32}P-label exceeds that of the 5′-label by a factor of around 100.

(7) Iodine hydrolysis, denaturing PAGE analysis, phosphoimaging and quantification of band intensities.

18.2.3
Materials

Sp-ITPαS was custom-synthesized and purified by IBA (Göttingen, Germany; http://www.iba-go.com/). A variety of other nucleoside α-thiotriphosphates, mainly those with 2′-ribose or adenine base modifications in addition to ATPαS, CTPαS, GTPαS, ITPαS and UTPαS, are available from Glen Research and can be found in their catalogue under the keyword NAIM (http://www.glenres.com/index.html). The catalogue, which can be downloaded as a pdf, also lists their authorized distributors outside the US (e.g. Eurogentec in France). Likewise, IBA (see above) offers numerous nucleoside α-thiotriphosphates including several halogen-derivatized base analogs. For additional analogs, see Chapter 17.

18.2.4
Protocols

Protocol 1: Transcription of P RNA–tRNA 5′ half carrying randomly distributed IMPαS modifications

Transcripts can be initiated with the nucleoside guanosine that introduces a 5′-terminal hydroxyl group to permit direct 5′-end-labeling (Fig. 18.2, Step 1). Alternatively, one may perform a standard transcription initiating RNA chains with 5′-guanosine triphosphate, followed by a dephosphorylation step with alkaline phosphatase prior to 5′-end-labeling, as described in Chapters 3 and 17. With the protocol outlined below, the average yield of P RNA–tRNA 5′ half was in the range of 400–1000 µg (2.4–6 nmol) per 500 µl transcription mix.

(1) Before starting to prepare the reaction mix, incubate the guanosine stock solution at 75 °C in a thermoshaker until the solution becomes clear; then stop shaking, but leave the solution at 75 °C.
(2) Prepare the reaction mix – except for guanosine and T7 RNAP – at room temperature; add the components in the order they are presented in the table below, and afterwards pre-warm to 37 °C before addition of guanosine. Add the preheated guanosine solution rapidly to the reaction mix and vortex to avoid guanosine precipitation; start the reaction by addition of enzyme.

Transcription reaction, 500 µl:		Final concentration
RNase-free water	83.5 µl	
HEPES, pH 8.0, 1 M	40 µl	80 mM
DTT 100 mM	75 µl	15 mM
MgCl$_2$ 3 M	5.5 µl	33 mM
spermidine 100 mM	5 µl	1 mM
rNTP mix (25 mM each)	75 µl	3.75 mM (each)
ITPαS 4.33 mM	11 µl	0.095 mM
Template (linearized plasmid 3.2 kb) 0.5 µg/µl	40 µl	40 µg/ml
Pyrophosphatase 200 U/ml	5 µl	2 U/ml
• Pre-warm mixture to 37 °C, then add:		
Guanosine (30 mM, kept at 75 °C)	150 µl	9 mM
T7 RNAP 200 U/µl	10 µl	4000 U/ml

(3) Incubate for 4 h at 37 °C.
(4) Extract RNA once with an equal volume phenol (saturated with 10 mM Tris–HCl, 1 mM EDTA, pH 7.5–8.0) and twice with equal volumes of chloroform.
(5) Precipitate by adding 1/5 volumes 2 M NH$_4$Ac (pH 7.0) and 2.7 volumes

of ethanol. Mix and keep at −20 °C for at least 1 h; centrifuge at 4 °C and 16 000 g for 1 h in a desktop centrifuge.
(6) Dissolve the pellet in 50 µl RNase-free water and 50 µl gel loading buffer (see Protocol 2a).
(7) Purify the RNA by denaturing PAGE as described in Protocol 2a.

Protocol 2a: Purification of analog-modified P RNA–tRNA 5′ half pools by denaturing PAGE

- Gel loading buffer: 2.7 M urea; 1 × TBE; 67% (v/v) formamide; 0.01% (w/v) each bromophenol blue and xylene cyanol.
- RNA elution buffer 1: 200 mM Tris; 1 mM EDTA; 0.1% SDS; pH 7.0 at room temperature (this buffer may also be prepared without SDS, e.g. when the eluted RNA is afterwards subjected to 3′-end-labeling).
- RNA elution buffer 2: 1 M NH_4OAc; pH 7.0.

Prepare a 5% polyacrylamide (24:1 acrylamide:N,N'-methylene bisacrylamide) gel in 8 M urea and 1 × TBE, 15-cm wide, 35-cm long and 1- or 0.5-mm thick. The pocket size depends on the amount of RNA that has to be purified (about 10-cm pocket width for a 500-µl preparative transcription in case of 1 mm gel thickness).

(1) Load the RNA sample from Protocol 1, Step 6, onto the gel immediately after extensive rinsing of the pocket with a syringe to remove urea solution that has diffused from the gel matrix into the pocket; run the gel at 20–25 mA for about 3 h until the xylene cyanol has reached the bottom of the gel.
(2) Separate the glass plates and place the gel between two sheets of kitchen wrapping film.
(3) Visualize RNA band(s) by UV shadowing (for details, see Chapter 3). The exposure should be minimized to avoid UV-induced damage of the RNA. Mark the band of interest with a pen or marker and excise it with a sterile scalpel under normal light. Check correct excision by UV shadowing.
(4) Elute the RNA in the appropriate volume of RNA elution buffer 1 or 2 (3 ml for RNA purified from a 500-µl transcription assay) overnight at 4 °C with shaking.
(5) Collect the supernatant.
(6) For ethanol precipitation, add 1/5 volumes 2 M NH_4OAc (pH 7.0) and 2.7 volumes of ethanol (omit NH_4OAc when using elution buffer 2). Mix and keep at −20 °C for at least 1 h and centrifuge at 4 °C and 16 000 g for 1 h in a desktop centrifuge.
(7) Wash the pellet with 100–200 µl ice-cold 70% ethanol, centrifuge at 4 °C and 16 000 g for 5 min; air-dry the pellet and resuspend in 200 µl RNase-free water for RNA derived from a 500-µl transcription assay (the RNA concentration should be 2–5 µg/µl).
(8) Measure the RNA concentration by UV spectroscopy (see Chapter 4 and Appendix).

One may repeat the elution (Steps 4–8) to recover higher amounts of RNA.

Protocol 2b: Purification of aliquots of analog-modified pool RNA after ^{32}P-end-labeling
Follow Protocol 2a, with the following alterations:

- Load the radiolabeled RNA (5–20 pmol) into a 0.5-cm wide (1 mm gel thickness) or 1.3-cm wide (0.5 mm gel thickness) gel pocket.
- In Step 3, visualize the radiolabeled RNA with a phosphoimager (instead of UV shadowing) after an image plate has been exposed to the gel for 1–20 min, depending on the amount of radioactivity loaded on the gel.
- In Step 4, elute the RNA in 500–1000 µl elution buffer 1 or 2.
- Resuspend the RNA pellet after elution and ethanol precipitation in 10–20 µl RNase-free water.

Protocol 3: 5′-End-labeling of analog-modified pool RNA

(1) Prepare the reaction mix by adding the components in the order as they are presented in the table below; vortex, spin down and incubate at 37 °C for 60–120 min.
  - 10 × T4 polynucleotide kinase (T4 PNK) buffer (forward reaction): 500 mM Tris–HCl, pH 7.6; 100 mM MgCl$_2$; 50 mM DTT; 1 mM spermidine; 1 mM EDTA

Labeling reaction, 15 µl:		Final concentration
10 × T4 PNK buffer (forward reaction)	1.5 µl	1×
25 mM DTT	1.5 µl	2.5 mM
RNase-free water	6 to 7 µl	
Pool RNA purified according to Protocol 2a	1–2 µl (10–20 pmol)	0.66–1.33 µM
[γ-^{32}P]ATP (3000 Ci/mmol, 10 µCi/µl, 3.3 µM)	3.0 µl	0.66 µM
10 U/µl T4 PNK	1 µl	0.66 U/µl

(2) After incubation, add 35 µl of RNase-free water and vortex; add 7 µl 2 M NH$_4$OAc and 94.5 µl ethanol for RNA precipitation; proceed as described in Protocol 2a, Step 6. Wash the pellet with 100 µl ice-cold 70% ethanol, centrifuge at 4 °C and 16 000 g for 5 min and air-dry the pellet.
(3) Resuspend the pellet in 10 µl gel loading buffer (Protocol 2a) and purify the radiolabeled RNA by 5% denaturing PAGE as described in Protocols 2a and b.
(4) After gel elution and ethanol precipitation resuspend the RNA pellet in 10–20 µl RNase-free water and determine the overall yield of labeled RNA by measuring 1 µl using a scintillation counter.

Protocol 4: 3′-End-labeling of analog-modified pool RNA
- 10 × T4 RNA ligase buffer: 500 mM HEPES–NaOH, pH 8.0; 100 mM MgCl$_2$; 100 mM DTT

18.2 NAIM Analysis of *Cis*-cleaving RNase P RNA–tRNA Conjugates

3′-end-labeling reaction, 6 µl:		Final concentration
Cleaved RNA after Step 5 of Fig. 18.2 (air-dried pellet)	5–10 pmol	0.83–1.66 µM
10 × T4 RNA ligase buffer	0.6 µl	1×
1.5 mM ATP	0.33 µl	82.5 µM
[5-^{32}P]pCp (3000 Ci/mmol, 10 µCi/µl)	4 µl	2.2 µM
10 U/µl T4 RNA ligase	1 µl	1.67 U/µl

(1) Prepare the reaction mix by adding the components in the order given in the table above; vortex, spin down and incubate at 4 °C overnight.
(2) Add 10 µl gel loading buffer and purify the radiolabeled RNA by 5% denaturing PAGE as described in Protocols 2a and b.
(3) After gel elution and ethanol precipitation according to Protocols 2a and b, resuspend the RNA pellet in 10–20 µl RNase-free water and determine the overall yield of labeled RNA by measuring 1 µl using a scintillation counter.

Protocol 5: Selection for *cis*-cleavage of P RNA–substrate conjugates

The analog-modified pool RNA (P RNA conjugated to tRNA 5′ half, Fig. 18.1A) is first annealed to the tRNA 3′ half in order to reconstitute the full-length substrate, followed by dilution and concomitant addition of Mg^{2+} to start the *cis*-cleavage reaction (Fig. 18.2).

- 4 × annealing buffer: 400 mM MES–NaOH, pH 5.9; 400 mM NH$_4$Cl.
(1) Prepare the annealing master mix:

Annealing mix 660 µl:		Final concentration
4 × annealing buffer	165 µl	1×
30 mM CaCl$_2$	110 µl	5 mM
24 µM IMPαS-modified P RNA–tRNA 5′ half	2.5 µl	91 nM
26 000 c.p.m./µl radiolabeled IMPαS-modified P RNA–tRNA 5′ half	5 µl	197 c.p.m./µl
166 µM tRNA 3′ half	35 µl	8.8 µM
RNase-free water	342.5 µl	

(2) Distribute in 132-µl aliquots to five different tubes.
(3) Heat to 95 °C for 2 min, then transfer to and cool down in a heating block adjusted to 50 °C and pre-incubate for 30 min at 50 °C.
(4) Add to each aliquot of annealed mix the components listed in the table below; mix thoroughly and incubate for 2 min at 50 °C.

Self-cleavage reaction 506 µl:		Final concentration
Annealed mix	132 µl	0.26 × annealing buffer
300 mM CaCl$_2$	6.23 µl	5 mM
4 × annealing buffer	93.5 µl	0.74 × annealing buffer
5 M urea	10 µl	99 mM
RNase free water	258.2 µl	

(5) Then start the *cis*-cleavage reaction by adding 6.1 μl 3 M MgCl$_2$ to each of the five mixes (final concentration 36 mM).
(6) For four tubes, stop the reaction after 2 min (resulting in 20–30% product formation, representing the fraction of functional RNA) by placing them immediately on ice, followed by addition of 100 μl 2 M NH$_4$OAc, 2 μl of 20 μg/μl glycogen as carrier and 1.2 ml ethanol.
(7) Mix vigorously and store at −20 °C overnight.
(8) Keep the fifth tube at 50 °C for 2 h to allow the *cis*-cleavage reaction to proceed to quasi completion (represents the endpoint of the reaction, where essentially all RNA molecules of the original RNA pool have been cleaved); prepare ethanol precipitation as in Steps 6 and 7.
(9) Centrifuge all five samples for 30–60 min at 4 °C and 16 000 g, wash the pellets with 100 μl 70% ethanol, briefly centrifuge and air-dry the pellets; resuspend (combine) the pellets of tubes 1–4 in 15 μl gel loading buffer; resuspend the pellet of the fifth tube separately in 15 μl gel loading buffer; run the two samples on a 5% PAA/8 M urea gel (0.5-mm thick, pocket width 1.3 cm) as described in Protocol 2b.
(10) Expose an image plate to the gel for 10 min and visualize the bands using a phosphoimager. Excise the cleaved product band from the two lanes and elute each in 500–1000 μl elution buffer 2. Do not perform elution in buffer 1 because residual SDS may disturb the next step (3′-end-labeling). Ethanol-precipitate the eluted RNA and air-dry the pellets as in Protocol 2a.
(11) 3′-end-label the eluted RNA fractions according to Protocol 4.

Protocol 6: Iodine-induced hydrolysis of analog-modified RNA fractions after functional selection

(1) Prepare a fresh I$_2$ solution as in the table below:

I$_2$ solution, 50 μl:		Final concentration
10 mg/ml I$_2$ solution in ethanol	5 μl	1 mg/ml
Ethanol	5 μl	10%
RNase-free water	40 μl	

(2) Prepare the iodine hydrolysis reaction mix as follows:

Iodine-induced hydrolysis reaction, 50 μl:		Final concentration
3′ (or 5′)-end-labeled RNA	1–10 μl (50 000 c.p.m.)	1000 c.p.m./μl
100 mM HEPES pH 7.5	5 μl	10 mM
1 mg/ml I$_2$ solution from step 1	5 μl	0.1 mg/ml
RNase-free water	30–39 μl	

(3) Incubate the reaction mix for 10–20 min at 37 °C.

(4) Add 150 μl RNase-free water, and ethanol-precipitate by addition of 40 μl of 2 M NH$_4$OAc, 2 μl glycogen (20 mg/ml) and 540 μl ethanol.
(5) Resuspend the pellet in 10 μl gel loading buffer and apply to a 10% PAA gel containing 8 M urea (gel thickness: 0.5 mm; length: 35 cm; pocket width: 0.6–1.3 cm). Run the gel at 10 mA for 3–4 h until the distance of the xylene cyanol dye is 5 cm to the gel bottom. To also resolve longer iodine hydrolysis fragments, and thus a larger portion of the RNA molecule, the 10% denaturing PAA gel may be run for an extended period, or a lower gel percentage (e.g. 5%) may be used. However, according to our experience, longer runs or the use of 5% gels have the drawback of usually resulting in more diffuse bands.
(6) Remove the glass plates. Place the gel between one layer of gel drying (Whatman) paper and one layer of kitchen wrapping film.
(7) Dry the gel for 30 min at 70 °C in a gel dryer under vacuum. Switch off the heating and leave under vacuum for another 30 min.
(8) Expose an image plate to the dried gel overnight.
(9) Scan the image plate with a phosphoimager; encircle each band, either by a rectangle, an ellipsoid or by following the individual contours of the band; quantify the image quants therein and evaluate interference data as outlined in Section 18.2.5.

18.2.5
Data Evaluation

Iodine hydrolysis bands were visualized and quantified using a phosphoimager, in our case a Bio-Imaging Analyzer BAS-1000 or FLA 3000-2R (Fuji Film) and the analysis software PCBAS/AIDA (Raytest). An application example is illustrated in Fig. 18.3, representing two experiments run in parallel (lanes 1–4 and 5–8). The quantification boxes positioned with the program AIDA are drawn with thin black or white lines (white when black lines are masked by high band intensities). In the case of insufficient separation of individual bands (e.g. boxes 1 and 2), two or more bands are enclosed in a single box. Some bands, such as those of boxes 31, are hardly visible in Fig. 18.3 due to loss of resolution, but can be clearly differentiated from the background within the analysis program AIDA. The next step is to determine the normalization factor κ to compensate for differences in total radioactivity in lanes A versus B, where A represents the cleaved conjugate at the endpoint (here after 2 h) and B the fraction of conjugates *cis*-cleaved in the initial fast phase of the reaction (within 2 min). A' and B' represent data from another individual experiment; data from a third experiment (A" and B") were also included for the calculation of the average R value, but are not shown in Fig. 18.3 and Table 18.1. The normalization factor κ is then calculated from the ratio of the sum of all band intensities in lane A versus B ($\Sigma I_A / \Sigma I_B$). Interference and enhancement effects for individual bands (or two or more bands if quantified as one due to low gel resolution) are then determined by calculating the ratio $R = (\kappa \times I_B)/I_A$. Interference effects are associated with R values below 1.0, whereas enhancement effects will result in R values above 1.0. In a previous study [19], only R values below 0.82

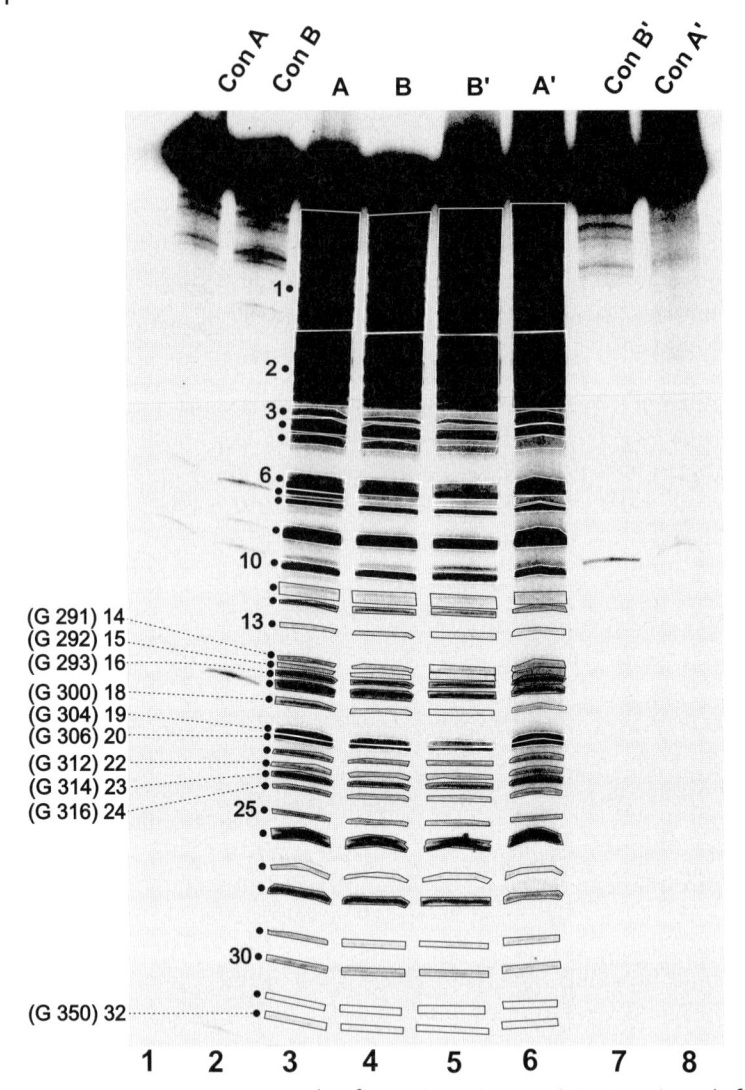

Fig. 18.3. NAIM experiment to identify IMPαS modifications that interfere with *cis*-cleavage of an *E. coli* P RNA–tRNA conjugate. 3′-end-labeled RNA samples, treated with iodine (lanes A, A′, B, B′) or not (lanes Con A, A′, B, B′) were loaded on a 10% PAA/8 M urea gel and separated by electrophoresis until xylene cyanol reached the bottom of the gel. Radioactive bands were visualized as described in Section 18.2.5. Lane A: pool of RNA molecules after 2 h of incubation, representing the endpoint of the reaction (Fig. 18.2, Step 4b); lane B: fraction of conjugates *cis*-cleaved in the initial fast phase of the reaction (within 2 min; Fig. 18.2, Step 4a). Lanes A′ and B′: same as lanes A and B, respectively, but representing a second individual experiment.

and above 1.2 were considered significant, but these cutoff values are arbitrary and depend on the quality of the data, such as the number of and fluctuations between individual experiments. In the example of Table 18.1, the data are based on three individual experiments, which we consider to be the absolute minimum for such studies. Figure 18.4(A) shows a graphical representation of the mean R values (including errors). We also evaluated the data following the calculation procedure described in Chapter 17 (Fig. 18.4B). Both evaluation procedures revealed G300 as a site of IMPαS interference, while several other weaker interferences are only identified according to the evaluation procedure presented in Fig. 18.4(A).

The gain of knowledge derived from NAIM experiments can be largely extended when comparing NAIM results obtained for the same system but with different functional assays. For example, comparison of the interference results shown in Fig. 18.4(A) with those observed in a gel retardation assay selecting for high affinity tRNA binding to E. coli P RNA [8] reveals substantial overlap. In the region of nucleotides 291–350, IMPαS modifications at G291–293, G300, G304, G306 and G314 caused interference in both functional assays, although with different amplitudes. It should be noted that, in the tRNA binding assay [8], interference at G300 was predominantly a phosphorothioate effect, already observable with the GMPαS modification alone, which, however, was not analyzed in the *cis*-cleavage assay (Fig. 18.3). Modifications at G291–293 directly weaken the interaction with the 3'-CCA terminus of tRNA [8, 40], explaining why interferences at these positions are detected in both functional assays. Modifications at G300, G304, G306 and G314 apparently also destabilize substrate binding to the ribozyme, either directly or by inducing conformational changes of ribozyme structure. The tRNA binding assay revealed additional IMPαS interferences at G329 and G356, but no effect at G312, G316 and G350 [8]. G312 and G316 are borderline cases due to the weakness of interference effects (Fig. 18.4A). However, G350 may represent a position whose interference is specific to the catalytic step, as evidence was provided that G350 contributes to the binding of catalytically important Mg^{2+} near the active site of RNase P RNA [41].

18.3 Troubleshooting

RNA transcription reaction did not work

- pH too low: check if all the reaction components were added in the required quantities, particularly the HEPES buffer; check the pH of the reaction mixture (should be in the range of 7.5–8.0).
- For further troubleshooting, see Chapter 1.

RNA degradation

- RNase contamination: prepare all solutions freshly using RNase-free water.

Tab. 18.1. Analysis of NAIM data obtained upon selection of fast-cleaving (within 2 min) E. coli P RNA–tRNA conjugates (fractions B and B′) compared with the IMPαS-modified pool of RNA molecules after 2 h of incubation, representing the endpoint of the reaction (fractions A and A′).

No.	Position	Fraction A I_A	Fraction B I_B	$I_B \times \kappa$	Fraction B′ $I_{A'}$	Fraction A′ $I_{B'}$	$I_{B'} \times \kappa'$	R	R′	R″	∅R	SD
1		267325.7	268508.1	285370.41	205632.4	247931.9	229510.56	1.07	1.16	1.02	1.07	0.03
2		99128.7	97563	103689.96	93011.7	96683.2	89499.64	1.06	0.96	0.98	0.10	0.03
3		5103.5	4019.6	4272.03	8152.5	3743.8	3465.64	0.84	0.43	1.08	0.78	0.19
4		11505.4	11559.2	12285.12	8920.8	9722.3	8999.93	1.07	1.01	1.04	1.04	0.02
5		6894.3	6143.4	6529.21	6028.7	5262.1	4871.13	0.95	0.81	0.99	0.92	0.06
6		16869.3	15114.1	16063.27	11072	13779.9	12756.05	0.96	1.15	1.11	1.07	0.06
7		5823.9	3511.6	3732.13	3464.9	2963.2	2743.03	0.65	0.79	0.96	0.8	0.09
8		6918.6	5624.5	5977.72	5825.2	5363.9	4965.36	0.86	0.85	0.96	0.89	0.03
9		20596.9	16283.7	17306.32	17227.6	14217.7	13161.32	0.85	0.76	1.15	0.92	0.12
10	271 + 270	9981.1	8461.5	8992.88	6623.5	8808	8153.57	0.90	1.23	1.21	1.11	0.11
11	275 + 276	4298.9	4396.8	4672.92	4542.9	4306.1	3986.16	1.09	0.88	1.22	1.06	0.10
12	280	3472.3	3499.8	3719.59	2768	3338	3089.99	1.07	1.12	0.93	1.04	0.06
13	285	1660.9	1402.6	1490.68	1699.1	1788.3	1655.43	0.90	0.97	1.38	1.09	0.15
14	291	2395.1	1871.8	1989.35	3471.2	2678.9	2479.84	0.83	0.71	0.86	0.80	0.04
15	292	2225.7	1535	1631.40	3451.9	2099.5	1943.51	0.73	0.56	0.84	0.71	0.08
16	293	5352.9	3555.6	3778.89	5085.5	4225.7	3911.73	0.71	0.77	0.80	0.76	0.027
17	296 + 297	8472.7	7253.3	7708.81	5296.3	5313.3	4918.52	0.91	0.93	1.36	1.07	0.15
18	300	3822.9	1686.3	1792.20	2301.7	1650.6	1527.96	0.47	0.66	0.63	0.59	0.06
19	304	6238.6	4291.2	4560.69	4714.8	3548.7	3285.03	0.73	0.70	0.88	0.77	0.06
20	306	7388.6	5131.3	5453.55	5342.3	4166.4	3856.84	0.74	0.72	0.81	0.76	0.03
21	310	3425.2	2531	2689.95	3154.1	2732.1	2529.10	0.79	0.80	0.95	0.85	0.05
22	312	2973	2091.6	2222.95	3304	2428.8	2248.34	0.75	0.68	0.97	0.80	0.09
23	314	5925.1	3661	3890.91	4617.2	4307.2	3987.17	0.66	0.86	0.81	0.78	0.06
24	316	2856.3	2112.4	2245.06	2818	1905.8	1764.20	0.79	0.63	1.02	0.81	0.12

25	320	2667.7	2265.8	2408.09	2014.7	2228.7	2063.11	0.90	1.02	0.81	0.91	0.06
26	323 + 324	10050.2	9544.4	10143.79	10304.4	10058.5	9311.15	1.01	0.90	0.93	0.95	0.03
27	329	1942.9	1449.4	1540.42	1867.6	1696.1	1570.08	0.79	0.84	0.90	0.85	0.03
28	332	6697.3	6058.8	6439.29	3723.5	5509.8	5100.42	0.96	1.37	0.80	1.04	0.17
29	336	2334.2	2167.8	2303.94	2261.5	1788.6	1655.71	0.99	0.73	0.83	0.85	0.07
30	340	2966	2978.4	3165.44	2224.3	2556.2	2366.27	1.07	1.06	0.71	0.95	0.12
31	346	1264.6	804.2	854.70	702.1	632.2	585.23	0.68	0.83	1.14	0.88	0.14
32	350	1944.3	1486.7	1580.06	1461.1	1236.6	1144.72	0.81	0.78	0.68	0.76	0.04
Sum		540522.8	508563.9		443085.5	478672.1						

$\kappa = 1.06284146$; $\kappa' = 0.92565558$.
I = intensity; κ = normalization factor (see text).
Position: according to *E. coli* P RNA numbering system (see secondary structure in Fig. 13.2B of Chapter 13).
$R = I_B \times \kappa / I_A$; $R' = I_{B'} \times \kappa' / I_{A'}$; $R'' = I_{B''} \times \kappa'' / I_{A''}$, primary data ($I_{A''}$; $I_{B''}$; $I_{B''} \times \kappa$) not included in the table; $\varnothing R$ = mean of R, R' and R'';
SD = standard deviation of the mean.

R. K. Hartmann, *RNA* **1999**, *5*, 102–116.

9 L. Ortoleva-Donnelly, M. Kronman, S. A. Strobel, *Biochemistry* **1998**, *37*, 12933–12942.

10 J. H. Cate, A. R. Gooding, E. Podell, K. Zhou, B. L. Golden, C. E. Kundrot, T. R. Cech, J. A. Doudna, *Science* **1996**, *273*, 1678–1685.

11 H. W. Pley, K. M. Flaherty, D. B. McKay, *Nature* **1994**, *372*, 111–113.

12 D. Siew, N. H. Zahler, A. G. Cassano, S. A. Strobel, M. E. Harris, *Biochemistry* **1999**, *38*, 1873–1883.

13 A. D. Griffiths, B. V. Potter, I. C. Eperon, *Nucleic Acids Res.* **1987**, *15*, 4145–4162.

14 E. L. Christian, M. Yarus, *J. Mol. Biol.* **1992**, *228*, 743–758.

15 S. A. Strobel, K. Shetty, *Proc. Natl. Acad. Sci. USA* **1997**, *94*, 2903–2908.

16 S. P. Ryder, S. A. Strobel, *Methods* **1999**, *18*, 38–50.

17 Y. Huang, F. Eckstein, R. Padilla, R. Sousa, *Biochemistry* **1997**, *36*, 8231–8242.

18 W.-D. Hardt, J. M. Warnecke, V. A. Erdmann, R. K. Hartmann, *EMBO J.* **1995**, *14*, 2935–2944.

19 C. Rox, R. Feltens, T. Pfeiffer, R. K. Hartmann, *J. Mol. Biol.* **2002**, *315*, 551–560.

20 M. E. Harris, N. R. Pace, *RNA* **1995**, *1*, 210–218.

21 A. V. Kazantsev, N. R. Pace, *RNA* **1998**, *4*, 937–947.

22 N. M. Kaye, E. L. Christian, M. E. Harris, *Biochemistry* **2002**, *41*, 4533–4545.

23 W.-D. Hardt, V. A. Erdmann, R. K. Hartmann, *RNA* **1996**, *2*, 1189–1198.

24 C. Heide, R. Feltens, R. K. Hartmann, *RNA* **2001**, *7*, 958–968.

25 M. Boudvillain, A. M. Pyle, *EMBO J.* **1998**, *17*, 7091–7104.

26 G. Chanfreau, A. Jacquier, *Science* **1994**, *266*, 1383–1387.

27 J. S. Smith, E. P. Nikonowicz, *Biochemistry* **2000**, *39*, 5642–5652.

28 E. L. Christian, M. Yarus, *Biochemistry* **1993**, *32*, 4475–4480.

29 A. Jack, J. E. Ladner, D. Rhodes, R. S. Brown, A. Klug, *J. Mol. Biol.* **1977**, *111*, 315–328.

30 W. G. Scott, J. T. Finch, A. Klug, *Cell* **1995**, *81*, 991–1002.

31 E. L. Christian, N. M. Kaye, M. E. Harris, *RNA* **2000**, *6*, 511–519.

32 E. L. Christian, N. M. Kaye, M. E. Harris, *EMBO J.* **2002**, *21*, 2253–2262.

33 J. W. Brown, J. M. Nolan, E. S. Haas, M. A. Rubio, F. Major, N. R. Pace, *Proc. Natl. Acad. Sci. USA* **1996**, *93*, 3001–3006.

34 C. Massire, L. Jaeger, E. Westhof, *J. Mol. Biol.* **1998**, *279*, 773–793.

35 T. R. Easterwood, S. C. Harvey, *RNA* **1997**, *3*, 577–585.

36 A. A. Szewczak, L. Ortoleva-Donnelly, M. V. Zivarts, A. K. Oyelere, A. V. Kazantsev, S. A. Strobel, *Proc. Natl. Acad. Sci. USA* **1999**, *96*, 11183–11188.

37 B. L. Golden, A. R. Gooding, E. R. Podell, T. R. Cech, *Science* **1998**, *282*, 259–264.

38 A. S. Krasilnikov, X. Yang, T. Pan, A. Mondragon, *Nature* **2003**, *421*, 760–764.

39 T. Persson, S. Cuzic, R. K. Hartmann, *J. Biol. Chem.* **2003**, *278*, 43394–43401.

40 S. Busch, L. A. Kirsebom, H. Notbohm, R. K. Hartmann, *J. Mol. Biol.* **2000**, *299*, 941–951.

41 T. A. Rasmussen, J. M. Nolan, *Gene* **2002**, *294*, 177–185.

19
Identification and Characterization of Metal Ion Binding by Thiophilic Metal Ion Rescue

Eric L. Christian

19.1
Introduction

One of the most important characteristics of RNA is its ability to fold into complex three-dimensional structures that participate both directly and indirectly in a wide range of biochemical functions. These closely packed structures result in a high degree of unfavorable electrostatic repulsion that must be offset by interactions with positively charged monovalent and divalent ions [1, 2]. While RNA secondary structure and some tertiary structure can form in the presence of monovalent ions alone, divalent ions are required by large RNAs to adopt their active conformations [3–5]. The large majority of divalent metal ion interactions are thought to be weak, non-specific and to exchange rapidly between positions of elevated negative charge density, precluding their identification by high-resolution methods such as X-ray crystallography or nuclear magnetic resonance (NMR) [2, 6–10]. In contrast, a relatively small fraction of divalent metal ions form tight and specific interactions that can be resolved by high-resolution methods [2, 7, 8, 11–16].

Metal ions have been shown to both direct RNA folding and to stabilize specific RNA structures [17]. In addition, metal ions in some RNA catalysts (ribozymes) can interact directly with substrate phosphates to catalyze phosphodiester bond hydrolysis or transesterfication reactions [18–25]. While several catalytic RNAs have been shown to have bases positioned in the appropriate structural environment to play this role, the pK_as of RNA functional groups generally lie far from neutral pH, inhibiting their participation in acid–base catalysis under physiological conditions [17]. Metal ions thus provide a crucial addition to the RNA functional group repertoire, and their identification is often key to our understanding of folding and catalytic mechanism. Locating specific metal ions within structural or catalytic RNAs, however, is experimentally challenging and thus only a few metal ion interactions have been linked to a specific aspect of RNA function.

The central role metal ions play in RNA structure and catalysis creates at least two distinct challenges for defining their specific biological function. First, unlike the analysis of a single or small number of metal ions associated with proteins, there are often many metal ions associated with RNA, (hundreds in the case of

Handbook of RNA Biochemistry: Student Edition. Edited by R. K. Hartmann, A. Bindereif, A. Schön, E. Westhof
Copyright © 2009 WILEY-VCH Verlag GmbH & Co. KGaA, Weinheim
ISBN: 978-3-527-32534-4

the larger ribozymes). Thus there is significant difficulty in distinguishing an individual metal ion from what has often been described as the "sea" of metal ions associated with an RNA molecule [5, 6, 13–15, 26–34]. Second, individual metal ions may have multiple roles in structure or catalysis, making quantification of their contribution to specific functions difficult to deconvolute. In addition, traditional methods of structural analysis such as crystallography and NMR are unlikely to completely define the functional properties of divalent metal ion binding. Different numbers of divalent metal ions have been defined in different structural studies for the same enzyme [35, 36] and different catalytic roles have been proposed for closely spaced active site metal ion candidates [37, 38]. Moreover, metal ion interactions important to the reaction transition state may exhibit only weak or transitory binding in the ground state, precluding their detection by high-resolution structural methods.

Despite these complications significant progress has been made in our understanding of the influence of metal ions on RNA folding using a range of biochemical approaches including footprinting [39, 40], thermal denaturation [41, 42], fluorescence energy transfer [43], UV absorbance and circular dichroism [44], small-angle X-ray scattering [45, 46], and single-molecule florescence [47, 48]. Notably, the information these experimental approaches provide reflects the combined influence of metal ions and does not address the role of individual metal ions at a specific location. Significant progress has also been made in our understanding of the role of metal ions in RNA catalysis using well-studied RNA enzymes and traditional enzyme kinetics [32, 49–52]. These studies have examined enzymatic RNA cleavage as a function of metal type and concentration and have provided evidence for catalytic mechanisms involving one or more metal ions. The usefulness of this approach, however, is limited as it does not provide a distinction between direct and indirect metal ion effects on catalysis, and it does not examine the functional role of an individual metal ion. Sites of metal ion binding have been examined by metal-dependent RNA cleavage studies, but this approach provides only the approximate location of the subset of metal ions able to cleave the RNA backbone rather than specific functional groups involved in metal coordination and does not necessarily reflect the position of the native magnesium ions (Mg^{2+}) [53, 54]. Thus, despite this arsenal of experimental approaches it remains difficult to conclusively address many of the most pressing questions regarding the role of divalent metal ions in RNA. These structure and function questions include identifying the number of metal ions in a folding domain or an RNA active site, the ligands involved in metal ion coordination, the energetic contribution of an individual ion to folding or catalysis, the step in the folding or catalytic pathway where specific metal binding exerts its effect, and the fundamental mechanisms by which an individual metal ion can direct RNA folding and catalysis.

Many of these questions can be addressed using the method of thiophilic metal ion rescue (or metal ion specificity switch) experiment (Fig. 19.1). This approach involves the atomic substitution of an individual oxygen atom in an RNA molecule with sulfur or nitrogen at a position that produces significant changes in RNA folding or catalytic rate (Fig. 19.1A and B) [55]. The majority of these experiments have

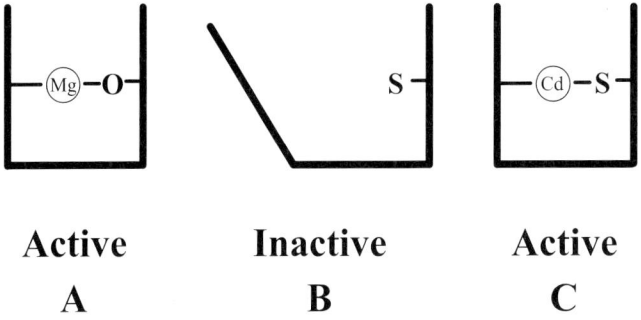

Fig. 19.1. Schematic of thiophilic metal ion rescue. (A) RNA activity dependent on specific Mg^{2+} coordination. (B) Loss of RNA activity due to the inability of Mg^{2+} to coordinate site-specific sulfur substitution. (C) Rescue of RNA activity by the addition of a thiophilic metal (e.g. Cd^{2+}) capable of direct coordination of sulfur as well as oxygen.

used phosphorothioate nucleotide analogs, which contain a sulfur substitution in place of a bridging or non-bridging phosphate oxygen. Such substitution at positions involved in metal ion binding, disrupts direct inner-sphere coordination of Mg^{2+}, as sulfur is a poor ligand for Mg^{2+} relative to oxygen (Fig. 19.1B, note: Mg^{2+} binding to oxygen is approximately 31 000-fold greater than that for sulfur [59]). Consistent with the direct disruption of metal ion coordination, and the central role that metal ions can play in RNA structure and function, phosphorothioate inhibition can produce substantial (greater the 10^4-fold) changes in the activity of ribozymes, rivaling the magnitude of effects observed in experiments that directly alter or delete essential active site elements in both RNA and protein enzymes [18, 20–22, 56, 57]. Moreover, these effects parallel the decrease in affinity of Mg^{2+} for sulfur relative to oxygen in model complexes [58, 59] and are consistent with the predicted energetic penalty for a change in ligand coordination of a Mg^{2+} ion [24]. In contrast to Mg^{2+}, "thiophilic" metal ions such as manganese (Mn^{2+}), cadmium (Cd^{2+}) or zinc (Zn^{2+}) can more readily accept both oxygen, nitrogen or sulfur in their inner coordination sphere and can thus alleviate deleterious effects of specific oxygen to sulfur or nitrogen substitutions in RNA that perturb the binding of Mg^{2+} (Fig. 19.1C) [55, 58–62]. The ability of low concentrations of such thiophilic metal ions to "rescue" deleterious effects of site-specific sulfur substitutions provides strong evidence for direct metal ion coordination [58].

Thiophilic metal ion rescue experiments have been used to identify likely metal contacts in the conserved structural domains and at the reactive phosphates of all three of the large ribozymes as well as metal coordination sites within small ribozymes and spliceosomal RNAs [18–22, 24, 25, 34, 52, 63–83], including interactions linked to the RNA active site [18–22, 24, 66, 67, 70, 73] and RNA folding [84, 85]. As will be discussed in detail in Section 19.2, the metal rescue approach has also been applied to determine the apparent affinity of a specific metal ion interaction by measuring the reaction rate over a range of metal ion concentrations [24, 25, 34, 72, 81, 86]. This quantitative characterization of individual metal ion interactions allows both the direct comparison of metal ion binding at different positions and the determination of the relative energetic contribution to the ribozyme reaction. These measurements are central to tests of specific structural and mechanistic proposals for the role of individual metal ions in enzyme function.

Metal binding sites identified by thiophilic metal ion rescue have generally been confirmed in the small (but growing) number of cases where alternative methodologies have been applied. These include independent confirmation of metal ion binding by X-ray crystallography [87–91], by ^{15}N- and ^{31}P-NMR [92–94], and by electron paramagnetic resonance [95], Q-band electron nuclear double resonance [95], and electron spin echo envelope modulation [96] spectroscopies. While some discrepancies have been observed in the detection of metal ion binding by the methods above [92, 97], the preponderance of independent analyses suggests that metal ion interactions detected by thiophilic metal ion rescue reflect genuine metal ion interactions within the native RNA structure.

It is important to underscore, however, that the method of thiophilic metal ion rescue is by no means definitive and has its own ambiguities and limitations that can lead to concluding something is a ligand when it is not or missing a functional group that is a ligand. Uncertainty in interpreting thiophilic metal ion rescue studies can stem from a wide range of potential indirect effects that may be induced by the phosphorothioate analog, the rescuing metal or the combination of both. Although the single atomic substitution of sulfur or nitrogen represents a minimal perturbation of enzymatic structure, changes in bond length, ionic radius, and local charge distribution can have significant effects beyond altering metal ion affinity [98–100]. For example, steric constraints imposed by the larger sulfur atom can alter the geometry of metal ion interactions to preclude metal ion rescue [101]. Thiophilic metal ions can also bind to positions other than the site of interest to indirectly affect the ribozyme's structure and reactivity [78, 102]. Moreover, different metals have different coordination properties that may be an issue in binding sites where precise geometry is important. In addition, while it is generally assumed that phosphorothioate modifications do not significantly perturb the native structure or induce *de novo* metal ion binding sites, it is difficult to exclude the possibility entirely.

The main limitations of thiophilic metal ion rescue are the small number of analogs available for this type of analysis and the inability to examine outer sphere interactions involved in metal ion coordination. This fact and the likelihood of geometric constraints can significantly restrict both the number of metal ions that can be identified as well as the number of contacts that can be identified in an individual coordination sphere. Moreover, even if the position of metal ion binding can be established, the determination of its biological role requires a detailed understanding of the reaction kinetics of the experimental system and the ability to analyze quantitatively an individual step in RNA function such as folding, substrate binding or enzymatic cleavage. Finally, while the general theory of thiophilic metal ion rescue is straightforward, well-controlled experimental design and interpretation of findings is not. Indeed, the absence of any one of a number of experimental considerations can substantially limit both the significance and the interpretability a great deal of experimental effort. Nevertheless, it is important to point out that many of the limitations and ambiguities noted above are not unique to thiophilic metal ion rescue and must be placed in the equally important context of the limitations and ambiguities of RNA (and protein) structural analysis in general, and in the significant difficulty in demonstrating that any given structure is functionally relevant.

Details for the proper application of thiophilic metal ion rescue will vary depending on the RNA being studied, the reaction conditions used, and the specific experimental question being addressed (e.g. whether the metal ions being examined are linked to folding, substrate binding, or catalysis). Thus, it is not possible in the scope of this short review chapter to cover the rationale or experimental detail for each of these circumstances. The study of ribozyme catalysis, however, provides an excellent illustration of the sensitivity of this probe for functional metal ions and for the measurement of energetic consequences of metal ion–RNA interactions. This review chapter highlights general experimental considerations with specific examples from the group I and group II introns, the hammerhead ribozyme, and bacterial ribonuclease P (RNase P) where the majority of thiophilic metal rescue studies have been performed. These experimental considerations, however, should also be central to other experimental systems and biological questions where thiophilic metal ion rescue can be applied.

19.2
General Considerations of Experimental Conditions

19.2.1
Metal Ion Stocks and Conditions

Highly purified preparations (above 99.99%) of $MgCl_2$, $MnCl_2$, $CdCl_2$, $ZnCl_2$ and other metal salts are available commercially (Aldrich), and should be stored tightly sealed to minimize absorption of water. Note that the relatively high pK_a of Mg^{2+} (11.4) makes stock solutions of this metal ion stable in water and most buffer conditions at or below pH 9. In contrast, the lower pK_as of thiophilic metal ions (10.6, 9.0 and 9.0 for $MnCl_2$, $CdCl_2$ and $ZnCl_2$, respectively) results in the formation of insoluble metal hydroxides at concentrations in the millimolar range above neutral pH [103, 104]. Thus solutions of thiophilic metal ions must be used immediately after preparation or made as concentrated, acidic stocks (pH 2) and diluted into buffer immediately prior to use.

Although optimal concentrations vary between experimental systems, the study of metal ions required for catalysis generally requires a baseline concentration Mg^{2+} to completely fold the RNA being studied and to minimize the effect of thiophilic metal ion binding at sites other than that associated with the phosphorothioate modification. Background levels of Mg^{2+} used to minimize non-specific effects in thiophilic metal ion rescue studies in group I, group II, bacterial RNase P and the hammerhead ribozyme, for example, are generally in the range of 5–

10 mM [34, 66, 72, 86]. The upper limit of total divalent metal ion concentration is also an important experimental consideration since elevated levels of divalent metal ion can both alter RNA structure and the rate-limiting step of the reaction, and produce significant non-specific cleavage of the RNA backbone. The range over which experiments can be conducted can restrict both the level and types of metal ions that can be used, and the ability to detect or completely rescue individual phosphorothioate positions. The concentration and types of metal ions that can be used in thiophilic metal ion rescue are those that can be shown by independent experiments to maintain the same rate limiting step of reaction (e.g. catalysis) as that of the native enzyme (see below).

Two additional factors can influence the concentration and type of divalent ion present in thiophilic metal ion rescue. First, elevated levels of RNA used to achieve single turnover conditions can have significant effects on the concentration of free metal ion in solution. At 10 µM an RNA of 400 nt such as RNase P will have 400 negative charges on its backbone that can bind several hundred Mg^{2+} ions or 2 mM of the free divalent metal ion. Changes in the concentration of free metal ion in the millimolar range may thus produce indirect effects if the total background magnesium concentration is not in sufficient excess to insure RNA folding. Second, while EDTA may be useful to prevent degradation due to metal-dependent hydrolysis, it preferentially binds thiophilic metal ions even in an excess of Mg^{2+}, which can lead to a significant overestimate of the amount of thiophilic metal ion required for rescue [105]. EDTA can generally be omitted from solutions without consequence, thus removing this potential complication. However, since EDTA chelates most thiophilic divalent ions five to eight orders-of-magnitude more tightly than Mg^{2+}, it is routinely added to control reactions done in the presence of Mg^{2+} to demonstrate that the observed effect is not due to the presence of contaminating thiophilic metal ions [105] (see below).

19.2.2
Consideration of Buffers and Monovalent Salt

Although a wide range of buffers can be used to examine metal ion rescue, it is important to avoid the use of buffers that chelate metal ions (e.g. citrate, phosphate or buffers containing acetate). Buffers used in thiophilic metal ion rescue studies include MES, MOPS, EPPS, HEPES, PIPES, Tris–HCl and BisTris–propane–HCl (Sigma, molecular biology grade) and are commonly used at a final concentration of 50 mM. Side by side comparisons of these buffers at a given pH generally do not produce large buffer-specific changes in the level of phosphorothioate inhibition or thiophilic metal ion rescue (usually less than 2-fold), allowing direct comparison of metal rescue studies under different buffer conditions. However, control experiments should be done to show that this is, in fact, the case.

In contrast, thiophilic metal ion rescue can be strongly influenced by both the concentration and type of monovalent salt. This observation is not surprising given that monovalent ions can compete directly with divalent ions for binding to the vast majority of charged positions, and thus are an important determinant of the final folded structure. The effect of monovalent salt becomes particularly important when the optimal concentration of monovalent salt is in excess of the concentration of divalent metal ion (e.g. 1 M M^+ versus 25 mM M^{2+} for *Escherichia coli* RNase P). While thiophilic metal ion rescue should be done under the same monovalent in which the kinetic pathway of the system of interest has been defined, changes in monovalent conditions can sometimes be advantageous under some experimental circumstances. Elevated levels of monovalent ion can act to both compensate for perturbations in diffuse metal ion binding and significantly dampen indirect effects due to changes in ionic strength upon the addition of thiophilic metal ions. Moreover, changes in monovalent ion species can significantly alter the level of metal ion rescue. Conditions that produce the best rescue, however, are not necessarily comparable to the behavior observed under a standard set of conditions. Comparisons of this type should be made with caution and with evidence that the rate-limiting step of the reaction has not been altered (see below). Careful consideration should also be given in using monovalent salts containing acetate, as elevated concentrations required for some systems may significantly alter the level of free divalent metal ion.

19.2.3
Incorporation of Phosphorothioate Analogs

The most commonly used phosphorothioate analogs contain sulfur substitutions at the non-bridging phosphate oxygens and are available commercially (Glen Research). These analogs are stored and handled in a manner analogous to that of other ribonucleotides (-80 °C in dH_2O). As noted above, sulfur substitutions have also been introduced at bridging phosphate oxygen positions [71, 73, 75]. However, these analogs are not available commercially and must therefore be chemically synthesized [106]. Phosphorothioate analogs are generally incorporated in one of two ways, by *in vitro* transcription or by solid phase chemical synthesis and ligation. These methods are covered in detail in Chapters 5 and 2.6.1, and thus only a brief overview and discussion of their relative merits with respect to thiophilic metal ion rescue will be presented here.

The incorporation of phosphorothioates by transcription generally involves the addition of low levels of nucleotide triphosphate analogs (ATPαS, CTPαS, GTPαS or UTPαS) to a standard transcription mixture to yield a population of randomly modified RNA that usually contain no more than one analog per molecule [107]. This mixture of randomly modified RNAs can be quickly analyzed for the effects of phosphorothioate substitution as well as for thiophilic metal ion rescue using the method of nucleotide analog interference modification (NAIM, see Chapter 17). This approach offers the advantage of being able to analyze all modified positions in a given molecule in a single experiment, a feature particularly important in the initial analysis of large RNAs. However, there are two important limitations to this approach that restrict both detection and thorough characterization of metal ion interactions. First, RNA polymerase is somewhat limited in the number of different analogs that can serve as substrates in transcription. For example, phosphorothioate analogs leading to the modification of the *pro*-Sp position cannot be incorporated by RNA polymerase, thus excluding half of the non-bridging phosphate oxygens that are likely to be involved in metal ion interactions. Second, the use of a mixed population of randomly modified RNAs make it particularly difficult to characterize the effect of an individual phosphorothioate modification and associated metal ion rescue beyond qualitative changes in folding, reaction rate or substrate binding. Thus, while random incorporation is useful in the initial survey of putative positions of metal ion binding, a uniform population of molecules with a site-specific substitution is ultimately necessary to utilize the power and quantitative potential of standard enzyme kinetics to confirm and characterize interactions involved in thiophilic metal ion rescue.

Site-specific phosphorothioate modifications are incorporated by chemical synthesis into short oligonucleotide fragments of approximately 10–20 nt (see Chapter 7). *Pro*-Sp and *pro*-Rp stereoisomers are separated by reverse-phase HPLC [108], with the *pro*-Rp peak usually emerging before *pro*-Sp. The identity of individual stereoisomers is verified by digestion with RNase T1 and snake venom phosphodiesterase, which cleave the *pro*-Sp isomer more slowly than *pro*-Rp, and P1 nuclease, which shows the opposite preference [109]. Note, however, that it can become increasingly difficult to separate the individual stereoisomers by HPLC for fragments greater than approximately 13 nt. The exact length will vary with sequence but separation can be enhanced by placing the phosphorothioate away from the center of the fragment toward the 5′ or 3′ end. Larger oligonucleotides containing racemic mixtures of the two stereoisomers can still be informative if the resulting reaction profile produces distinct kinetic phases (see below) and can be compared to unmodified and, optimally, purely *pro*-Rp isomers produced by transcription.

Isolated oligonucleotides are often 5′-end-labeled using $[\gamma\text{-}^{32}P]$ATP and T4 polynucleotide kinase and further purified by denaturing or non-denaturing PAGE prior to their use as substrates for joining to other RNA fragments. Purified oligonucleotides are generally ligated by the method of Moore and Sharp [110] (Chapter 3) to synthesized or transcribed RNA fragments containing the remaining sequence to yield a complete RNA. Oligonucleotide concentration and ligation efficiency can be assessed using UV absorbance or the specific activities of the radioactive RNAs.

Because ligation efficiency can be significantly reduced in reactions that contain more than a single ligation junction, oligonucleotide fragments are generally added to the 5′ or 3′ end of the larger RNA fragment rather than in the interior of the molecule as part of a reaction involving the joining of three or more fragments. Ligation efficiency is an important consideration in phosphorothioate studies since some ligation reactions are limited by structural constraints that reduce efficiency to only a few percent. Thus it may be difficult to produce enzyme or substrate in the quantities that are necessary for certain experimental conditions (e.g. single turnover conditions, see below). The insertion of functional groups in the interior of large RNAs such as the RNase P ribozymes can be accomplished by moving the native 5′ and 3′ ends of the RNA near the site of modification by circular permutation [111] to allow a simple two-part ligation. Note, however, that any modification of RNA structure must be kinetically characterized to show that it retains the functional properties of the native enzyme. Purified oligonucleotide fragments can also be positioned in the substrate or ribozyme by simple annealing using a protocol analogous to that described below for general folding of enzyme or substrate prior to enzymatic reactions. While this has been shown to be useful in the study of smaller RNAs, such as the pre-tRNA substrate for bacterial RNase P [112, 113], it is not clear whether this approach is well suited for studying the interiors of more complex RNAs where saturating levels of oligonucleotide binding may be difficult to obtain.

19.2.4
Enzyme–Substrate Concentration

The amount of RNA used in thiophilic metal ion rescue studies varies greatly depending on the phenomenon being studied (e.g. folding, binding or catalysis) and whether the experimental system is intra- or inter-molecular. However, under all circumstances, it is imperative that the relative RNA concentrations and experimental conditions isolate or uniquely reflect the phenomenon of interest. For the current discussion, enzyme reactions must be limited by the rate of catalysis rather than other kinetic events such as folding, substrate binding or product release. Because multiple turnover reactions for ribozyme systems in which thiophilic metal ion rescue has been applied are not generally limited by chemistry, single turnover conditions [e.g. the relative concentrations of enzyme and substrate (E:S) > 5:1] have been used to isolate the catalytic step. Control reactions at higher enzyme concentrations (e.g. E:S = 10:1 or 20:1) are generally required to demonstrate that the reaction is not dependent on enzyme concentration and that the substrate is completely bound in all cases. Note, however, that a wide range of other ratios of enzyme and substrate has been used to address different aspects of the kinetic profile [24].

Other methods to insure the reaction is dependent on the rate of cleavage involve the use of modifications at or adjacent to the scissile phosphate that inhibit the reaction [114, 115], as well as mutants that produce strong catalytic defects within the ribozyme itself [74]. In addition, enzymatic catalysis can be analyzed in the context of self-cleaving RNAs in which both ribozyme and substrate are contained in the same RNA fragment [74]. These enzymatic systems offer the advantage of dramatically increasing the effective local substrate concentration associated with the ribozyme to allow the analysis of cleavage of the enzyme substrate complex with radiochemical amounts ($\ll 1$ pM) of ligated RNA as opposed to the significantly higher enzyme concentrations (e.g. 1–10 µM) generally required for intermolecular single turnover reactions. Constructs that tether ribozyme and substrate require different temperature, monovalent and divalent metal ion conditions than the native *trans* reaction, however, and can differ in both the position and magnitude of the metal ion effects observed, making direct comparison problematic in the absence of additional controls [74, 82].

19.2.5
General Kinetic Methods

As noted above, analysis of metal ions important for catalytic function is generally done under single turnover conditions with ribozyme in excess of a 5′-end-labeled substrate. Although protocols vary, ribozyme and substrate are usually denatured by heating (e.g. 95 °C for 2–3 min) in reaction buffer without divalent metal ions and cooled stepwise or gradually to the reaction temperature. Reactions in which ribozyme and substrates are renatured together to allow proper annealing (as in the case of the hammerhead and group I ribozymes) are typically initiated by the addition of divalent ions. In reactions where ribozyme and substrate are folded independently, divalent metal ions are added during or after the cooling phase (to minimize metal dependent hydrolysis) and reactions are initiated by mixing of enzyme and substrate. Renaturation and subsequent incubation of the cleavage reaction can be done in a heat block, water bath, or PCR machine, although the latter is often convenient to insure reproducibility of the reaction conditions.

Reaction volumes usually vary between 10 and 60 μl and aliquots (1–2 μl) taken at specified times are added to solutions (around 2–10 μl) containing formamide (above 80%) and an excess of EDTA (generally greater than 2-fold) to terminate the reaction. Termination solutions also often contain buffer (e.g. 1 mM Tris, pH 7.5) and 0.005% xylene cyanol and bromophenol blue when products will be separated by gel electrophoresis. Long time points (lasting several hours to several weeks) should be taken from reactions covered with mineral oil (50–100 μl) and kept in an incubator or PCR machine with a heated lid to minimize changes in reaction volume due to evaporation. Substrate and products are typically resolved by polyacrylamide/7 M urea gel electrophoresis and quantified on a PhosphoImager (Molecular Dynamics).

Data are fit to the appropriate kinetic equation using KaleidaGraph (Synergy Software) or SigmaPlot (Jandel Scientific), preferably with a sufficient number of time points (usually six to 10) to give non-linear least-squares fits with $R^2 > 0.98$. Reaction time courses are typically fit to a single exponential:

$$\frac{[P]}{[E \cdot S]_{total}} = A - Be^{-k_{obs}t} \qquad (1)$$

where $[P]$ is the amount of product formed at time t, $[E \cdot S]_{total}$ is the initial concentration of bound substrate, A represents the maximal extent of reaction and B is the amplitude of the exponential fit. Two independent exponentials are generally used if the reaction appears to have distinct (e.g. fast and slow) phases which may arise from the existence of multiple enzyme–substrate (E · S) complexes with different observed rate constants (see Eq. 2 where k_1 and B, and k_2 and C represent the observed rate and amplitude of independent reaction phases, respectively).

$$\frac{[P]}{[E \cdot S]_{total}} = A - (Be^{-k_1 t} + Ce^{-k_2 t}) \qquad (2)$$

Very slow reactions (longer than 10 days) often only allow the measurement of the initial phase of the reaction and are typically fit to a linear equation or to an exponential equation if an endpoint can accurately be measured by another means [72]. Note that experimental error can be significant for large changes in the observed reaction rate. This is due to the inherent difficulty in obtaining accurate measurements of very slow rates ($\ll 10^{-4}$ min^{-1}) where the reactions cannot be followed to completion; caution should be exercised in comparing the magnitude of inhibitory effects due to site-specific substitution or inefficient thiophilic metal ion rescue. Independent experiments should also be conducted to show that the ribozyme remains fully active after extended reaction times.

The accurate measurement of the level of thiophilic metal ion rescue of a specific modification requires that it be distinguished from effects due to thiophilic metal ion binding at other positions. This requirement is based on the fact that changes in both the concentration and type of metal ion present can have significant effects on folding and activity of the unmodified ribozyme. Consequently, the observed rate of both the modified ribozyme or substrate (k_S) and the unmodified RNA control (k_O) must be measured in the presence and absence of added thiophilic metal ion [34]. For example, in the analysis of Cd^{2+} rescue of a specific phosphorothioate substitution, the observed rate of the modified RNA in the presence and absence of Cd^{2+} would be expressed by the terms k_S^{Cd} and k_S^{Mg}, while the observed reaction rate of the unmodified ribozyme under the same conditions would be expressed by the terms k_O^{Cd} and k_O^{Mg}. The relative rate of rescue (k_{rel}) of thiophilic metal ion binding at a specific site of phosphorothioate modification is determined by the fold rate enhancement of the modified ribozyme or substrate in the presence of the thiophilic metal ion over that observed in its absence normalized to the rate enhancement (or inhibition) observed in the absence of the phosphorothioate modification. For example, using the four rate constants in the example immediately above, $k_{rel} = (k_S^{Cd}/k_S^{Mg})/(k_O^{Cd}/k_O^{Mg})$ [34]. Rescue observed at thiophilic metal ion concentrations that show comparatively little or no effect on an RNA lacking a specific modification provides strong evidence for thiophilic metal ion binding to the specific modification itself. It must be emphasized, however, that the ability to interpret k_{rel} in this way requires independent evidence that the reactions being compared share the same rate-limiting step, that the effect of other metal ion sites is the same for both reactions and there is no direct effect of the rescuing metal ion on the "normal" reaction (see below).

19.2.6
Measurement of Apparent Metal Ion Affinity

As noted in the introduction, measurement of the dependence of reaction rate on the concentration of rescuing metal ion(s) allows the determination of the apparent metal ion affinity for binding to the sulfur substitution (Fig. 19.2) [34, 72].

The dependence of the relative rate (k_{obs}^{rel}) on the concentration of added thiophilic metal ion (M_S) is determined by the relative fraction of RNA with site-specifically bound thiophilic metal ion (Eq. 3) which is a function of the thiophilic

$$E \cdot S + M_s \underset{}{\overset{K_d^{M_s}}{\rightleftharpoons}} E \cdot S \cdot M_s$$

$$\downarrow k^o \qquad\qquad \downarrow k'$$

Products Products

A

B

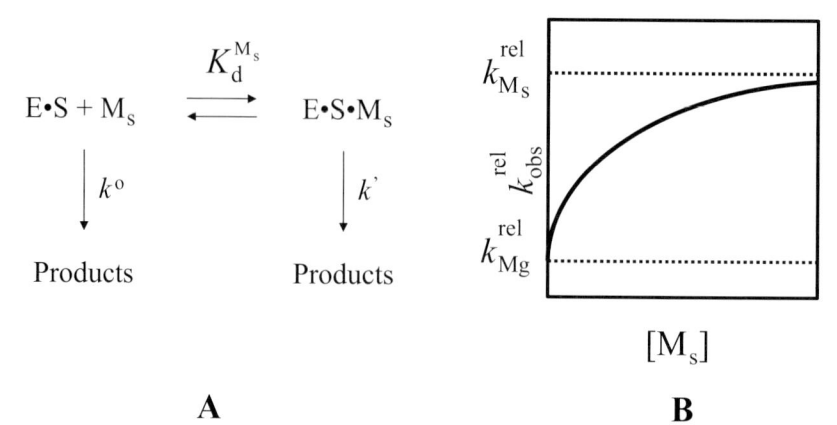

Fig. 19.2. Dependence of the reaction rate on thiophilic metal ion binding. (A) Kinetic scheme showing thiophilic metal ion binding ($K_d^{M_s}$) to an enzyme–substrate complex that produces an observed reaction rate (k') that is distinct from that observed in the absence of thiophilic metal ion binding (k^o). (B) Plot of the observed relative rate (k_{obs}^{rel}) as a function of thiophilic metal ion concentration ($[M_S]$). Note that the relative contribution of the individual rates (k_{Mg}^{rel} and $k_{M_s}^{rel}$) to k_{obs}^{rel} is directly proportional to the fraction of RNA bound to the thiophilic metal ion. Determination of apparent thiophilic metal ion affinity is derived from equations described in the text.

metal ion's affinity ($K_d^{M_s}$) (Eq. 4) [34]:

$$k_{obs}^{rel} = k_{Mg}^{rel} \times \{\text{fraction }^{Mg}E\} + k_{M_s}^{rel} \times \{\text{fraction }^{M_s}E\} \qquad (3)$$

$$k_{obs}^{rel} = k_{Mg}^{rel} \times \frac{K_d^{M_s}}{[M_S] + K_d^{M_s}} + k_{M_s}^{rel} \times \frac{[M_S]}{[M_S] + K_d^{M_s}} \qquad (4)$$

Note that under conditions where background (e.g. Mg^{2+}) metal ion competes with rescuing thiophilic metal ion (as is often seen) the observed dissociation constants, $K_d^{M_s}$, are "apparent" values. Under conditions where added thiophilic metal ions produce non-specific inhibition (e.g. inhibition of unmodified RNA), $K_d^{M_s}$ has been measured using a model analogous to that described by Equation 4 but includes the inhibitory binding of an additional thiophilic metal ion ($K_i^{M_s}$) [72]:

$$k_{obs}^{rel} = k_{Mg}^{rel} \times \frac{[M_S]}{\{[M_S] + K_d^{M_s}\}\left\{1 + \frac{[M_S]}{K_i^{M_s}}\right\}} \qquad (5)$$

Observed reaction rates are often normalized to facilitate comparison of metal ion dependence between reactions that show large differences in the absolute change in reaction rate, such as those involving reactions where thiophilic metal ion rescue is incomplete [72]. Under conditions where saturating concentrations of thiophilic metal ion cannot be obtained, lower limits for $K_d^{M_S}$ can be estimated from curve fits if partial saturation is observed, or the lower limit set as the highest concentration of thiophilic metal ion tested when significant levels of saturation are not evident [72, 86]. Since the level of rescue is proportional to the fraction of ribozyme bound by thiophilic metal ion, comparisons of relative affinity can also be made in the absence of complete thiophilic metal ion saturation by comparing the concentration of added thiophilic metal ion required to produce the same level of rate enhancement [25]. Note that variation in the individual observed reaction rates from experiment to experiment can be significant, and that greater precision can be achieved by comparing values of relative rate or affinity obtained in side-by-side experiments.

Because of the large range in the observed reaction rate and thiophilic metal ion concentrations used in these experiments it is often advantageous for the interpretation of the data to plot k_{obs}^{rel} vs. $[M_S]$ on a log-log scale (Fig. 19.3 [72]):

$$\log k_{obs}^{rel} = \log \left\{ k_{Mg}^{rel} \times \frac{K_d^{M_S}}{[M_S] + K_d^{M_S}} + k_{M_S}^{rel} \times \frac{[M_S]}{[M_S] + K_d^{M_S}} \right\} \qquad (6)$$

Single metal ion binding is consistent with plots in which k_{obs}^{rel} increases linearly with added thiophilic metal ion at concentrations below $K_d^{M_S}$ (Fig. 19.3A) [72]:

$$\log k_{obs}^{rel} = \log \left\{ k_{Mg}^{rel} + k_{M_S}^{rel} \times \frac{[M_S]}{[M_S] + K_d^{M_S}} \right\}; \quad [M_S] \ll K_d^{M_S} \qquad (7)$$

Note that the apparent non-linearity of plots at thiophilic metal ion concentrations well below $K_d^{M_S}$ is due to the contribution of enzymatic activity in the absence of thiophilic metal ion (k_{Mg}^{rel}). Log–log plots that exhibit a steeper dependence of the observed reaction rate on added thiophilic metal ion concentration when compared to that for single metal ion binding indicate the binding of multiple metal ions. In this case experimental data are generally fit to a non-linear form of the Hill equation:

$$k_{obs}^{rel} = \frac{n K_d^{M_S} k_{max} [M_S]^n}{1 + K_d^{M_S} [M_S]^n} \qquad (8)$$

where n is the Hill coefficient that gauges the cooperative dependence of thiophilic metal ion stoichiometry on ribozyme function ($n = 1$ reflects no cooperativity, $n = 2$ reflects the cooperative effect of *at least* two metal ions) and $K_d^{M_S}$ represents the thiophilic metal ion concentration required to attain half the maximal reaction rate (k_{max}) (Fig. 19.3A) [72]. Note, however, that the Hill equation does not directly imply cooperative *binding* when reactivity is used as the experimental signal since "cooperative" effects on function can also be observed if metal ions bind independently. In addition, it is important to remember that the Hill constant reflects the *minimal* number of metal ions. In the current example, $n_{Hill} = 2$ indicates the involvement of 2 *or more* metal ions in rescue.

constraints to determine RNA (and in particular active site) structure and specific features of the catalytic mechanism itself [25, 72]. In general, the finding of similar affinities (e.g. within experimental error) is consistent with rescue by a common ion, while the observation of distinctly different affinities (above 10-fold) provides strong evidence for rescue by different metal ions (for additional considerations see below).

In order to *directly* compare metal ion affinities one must be able to measure the affinity of a rescuing thiophilic metal ion that is *not* perturbed by a given modification (for a detailed discussion, see [34]). That is, although a given modification is required for thiophilic metal ion rescue, reaction conditions must be set such that the affinity of the rescuing metal ion is not altered by the presence of sulfur or nitrogen substitution. Perturbation of metal ion affinity by the sulfur or nitrogen modification is highly dependent on local RNA structure and the specific combination of modification and thiophilic metal ion involved, and is thus not necessarily indicative of the native metal ion interaction or an accurate basis for *direct* comparison of kinetic and thermodynamic values. In the group I ribozyme from *Tetrahymena*, the measurement of metal ion affinity unperturbed by site-specific substitution has been met by examining thiophilic rescue by several distinct kinetic methods. One particularly robust approach that is unique to this system takes advantage of the observation that oligonucleotide substrates in this system bind in two steps, first as an "open complex" by simple base pairing with the ribozyme, and second by formation of a "closed complex" which involves docking of the substrate helix into the active site via tertiary interactions [34]. Introduction of 2′-deoxy or 2-methoxy modifications at or upstream of the cleavage site results in ground state binding predominantly in the open complex with no effect on subsequent reaction steps. Although rescue occurs through the interaction of the thiophilic metal ion in the transition state, the transition state itself is transient and does not affect the observed metal ion affinity. The measurement of thiophilic metal ion affinity (K_d^{Ms}) using substrate bound in the open complex thus reflects the apparent metal ion affinity to the unmodified enzyme.

K_d^{Ms} can also be measured under conditions where the concentration of enzyme is subsaturating with respect to modified substrate and well below K_d^{Ms} [34]. In this case the active site is left sufficiently unoccupied such that thiophilic metal ion affinity is not perturbed by atomic (e.g. sulfur) substitution. In addition, thiophilic metal ion affinities independent of modification can be calculated from (1) the measurement of K_d^{Ms} under conditions where substrate is bound and (2) the independent determination of the effect of thiophilic metal ion on the affinity of modified and unmodified substrates, using simple thermodynamic cycles involving alternative pathways of binding and catalysis [34]. Note, however, that the ability to measure thiophilic metal ion binding independent of the phosphorothioate modification is made possible because of the detailed kinetic understanding of the *Tetrahymena* group I ribozyme and that it may be difficult to meet this criteria in other systems where the kinetic mechanism is less defined.

Although thiophilic metal ion binding to sulfur or nitrogen does not necessarily reflect the metal ion interaction in the native context, quantitative measurement of such an affinity along with other characteristics can nevertheless provide experimental evidence for the number of metal ions involved in binding and for the coordination of individual ligands to the same or different metal ions. Experiments often involve comparative analysis of the effect of additional modifications on n_{Hill} or K_d^{Ms} of thiophilic metal ion binding. Such comparisons can be made in the context of a second phosphorothioate modification or additional modification(s) of the backbone or base. The most common form of alternative (non-phosphorothioate) modifications are changes in ring nitrogens (e.g. N^7-deaza) and 2' hydroxyl positions (e.g. 2'-H, -NH$_2$) because of the propensity of metal ions to interact with these functional groups [34, 72, 78, 81, 86]. Additional mutations that result in a shift to weaker or stronger metal ion binding indicate that the modified functional groups are thermodynamically-linked. This result is consistent with a model in which the functional groups in question are common ligands to the same metal ion (e.g. Fig. 19.3B). In contrast, additional mutations that produce no change in affinity are more consistent with independent interactions [72, 81, 86]. Similarly, second-site modifications that alter affinity but do not perturb the single metal ion binding characteristic ($n_{Hill} = 1$) of the original sulfur or nitrogen modification are consistent with the binding of both ligands to the same metal ion, while combined mutations that alter the level of apparent cooperativity (e.g. from $n_{Hill} = 1$ to $n_{Hill} = 2$) are indicative of independent metal ion binding (Fig. 19.3A) [25, 34]. In addition, second-site modification can allow detection of metal ion interaction to the original sulfur or nitrogen substitution by providing a ligand or structural environment more favorable to binding, positioning, or specificity of a particular thiophilic metal ion [25, 72, 73, 86] (see below). Note that it is important to show that the observed change in signal is specific to the functional group(s) in question, with other sites having no significant effect [72, 86].

The binding of metal ions to individual functional groups can also be compared by measuring the competing effect of background metal ions (such as Ca^{2+} or Mg^{2+}) on the affinity of the rescuing thiophilic metal ion [25]. Such experiments are most easily interpreted using metal ions that do not directly contribute to the observed rescue of the reaction rate, but rather appear to weaken the apparent binding of thiophilic metal ion by direct competition. Increasing concentrations of competing metal ion should result in a decrease in the apparent affinity of thiophilic metal ion binding without changing the shape or apparent cooperativity of the binding curve (e.g. Fig. 19.3B). A shift in rescuing thiophilic metal ion binding should only be observed at concentrations of competing metal ion in the range of that observed for K_d^{Ms}. Therefore, the presence or absence of a shift in K_d^{Ms} at different concentrations of competing metal ion can be used to discern whether thiophilic metal ion affinity at one position is similar or distinct from that previously observed at other positions [25]. This approach is particularly useful under conditions where saturating concentrations of thiophilic metal ion cannot be obtained and apparent affinity must be measured indirectly from comparing the amount of thiophilic metal ion required to achieve the same enhancement of the observed rate for modified and control RNAs (see above). In addition, the observation of a shift in apparent $K_d{}^{Ms}$ at increasing concentrations of background metal ion suggests (but does not prove) that the background metal ion occupies the binding site rescued by thiophilic metal ion and reflects a metal binding site in the native ribozyme [24]. As with many of the experiments above, it is advantageous to identify analogous shifts in $K_d{}^{Ms}$ with different competing metal ions to demonstrate the generality of the observation.

19.2.8
Further Tests of Metal Ion Cooperativity

Although the observed metal ion dependence can appear to be consistent with a model for single metal ion binding, there are a number of circumstances where the binding of two metal ions can produce the same results. For example, thiophilic metal ion dependence consistent with single metal ion binding would also be observed if a modification allowed the first of two separate metal ions to bind tightly and did not produce a rate defect until a second metal ion was bound at a lower affinity site [72, 86]. The presence of a higher affinity site can be probed by examining the extent of rescue at different concentrations of ribozyme–substrate complex [72]. Because the model for tight metal binding requires the binding of a much lower affinity ion to observe rescue, significant rescue can only be observed if the concentration of thiophilic metal ion exceeds that of the ribozyme substrate complex. In contrast, the observed rescue of a single metal ion is not affected by enzyme–substrate concentration.

Thiophilic metal ion dependence consistent with single metal ion binding would also be observed for independent or cooperative binding of multiple metal ions with similar affinities, but with anti-cooperative effects on catalysis in which binding of one ion reduces the stimulatory effect of the other [72]. Moreover, affinities differing by less than 10-fold are difficult to resolve as independent elements of a binding curve. These models for multiple metal ion binding, as well as that above involving the tight binding of a second metal ion can be addressed using base deletions or functional group modifications that perturb the observed thiophilic metal ion dependence. These modifications are not expected to produce identical changes in affinity of two distinct metal ions and can reveal biphasic binding predicted for models involving two metal ions [72, 116]. Distinct metal binding sites also have the potential for distinct affinities for different thiophilic metal ion species and thus may produce biphasic dependence when different thiophilic metal ions are used. Note, however, that different ligands to the same metal ion may have distinct effects on binding activity due to changes in the steric or electrostatic environment engendered by the phosphorothioate modification, as has been observed in protein enzymes [101]. Thus, the observation of single metal ion dependence, even in the presence of modifications that perturb metal ion affinity or in the presence of different thiophilic metal ions that support rescue of the observed rate, can provide evidence that suggests, but does prove, the involvement of only a single metal ion [72].

19.3
Additional Considerations

19.3.1
Verification of k_{rel}

Because the interpretation of thiophilic metal ion rescue studies is dependent on the extent to which k_{rel} reflects the effect of an individual functional group on the native reaction, it is crucial to provide independent evidence that that this is in fact the case. As noted above the ability to interpret k_{rel} in this way requires: (1) the reactions being compared share the same rate-limiting step, (2) the effect of other metal ion sites is the same for both reactions and (3) there is no direct effect of the rescuing metal ion on the "normal" reaction. First, one cannot overstate the importance of isolating the step of interest and providing evidence that the step of the reaction studied under conditions of metal ion rescue is the same as that for the native ribozyme. The establishment of this condition is particularly vital since a shift in rate-limiting step (which is likely in cases where the perturbation of the observed rate is large or the rescue is incomplete) can result in apparent (i.e. false) rescue and thus an inaccurate and misleading interpretation [78]. However, this control is a difficult criterion to fulfill given the small number of tests that can be used to identify an individual step in a ribozyme reaction and the fact that even these tests are not definitive. While the kinetic conditions required to isolate an individual step such as catalysis are system dependent and beyond the scope of this chapter, it can generally be said that the insight that can be gained from thiophilic metal ion rescue studies is directly proportional to the degree to which the kinetic scheme of a ribozyme has been defined. Indeed the ability to examine the effect of metal ion binding on different steps of a given reaction lies at the heart of defining the metal ion's functional role.

In the kinetically defined ribozyme systems from the Hammerhead, the group I and group II ribozymes and bacterial RNase P, single turnover conditions have been defined in which the observed rate of cleavage shows a log-linear dependence on pH [49, 75, 104, 117, 118]. Although the mechanisms of cleavage differ, the log-linear dependence of observed rate on pH is nevertheless considered consistent with rate-limiting catalysis, and not reflective of upstream events such as substrate binding or a conformation change prior to catalysis. Thus changes in reaction conditions (e.g. changes introduced into the RNA structure or in the type or concentration of divalent metal ion) that provide evidence of a change in rate limiting step of the reaction preclude meaningful comparison with the wild-type enzyme and interpretation of thiophilic metal rescue studies. Conversely, modification-induced changes in RNA structure that do not alter the apparent pH dependence of the observed reaction rate provide evidence (but do not prove) that the rate-limiting step of the reaction has not been perturbed, and allow the effects of modification and thiophilic metal ion rescue to be more directly and quantitatively compared. Similarly, perturbations from base deletion, mutation, or additional functional group modification that produce analogous effects for all RNA species and conditions being compared in thiophilic rescue also provide evidence (but do not prove) that the ribozyme–substrate complexes of the RNAs being compared follow the same mechanism and involve the same molecular interactions [72, 75, 119, 120].

For the study of catalysis, additional care should also be taken to make sure that steps upstream of the cleavage step are in rapid equilibrium. Support for the presence of rapid equilibrium can be obtained (1) by showing that the rate of the reaction is not affected by either the time of incubation with thiophilic metal ion before the initiation of the reaction or by the order of addition of thiophilic metal ion and other reaction components, (2) by showing that reactions follow good first order reaction kinetics without a lag phase, or (3) by showing that pulse chase studies do not perturb the extent and apparent rate of reaction [24]. Although the methods for accurately measuring a given rate-limiting step will differ in different systems, determination of the thiophilic metal ion rescue over a range of conditions provides an important gauge of the robustness and generality of the observed effect and the extent to which the effect can be reliably measured.

The second criterion that the effect of other metal ion sites is the same for both reactions reflects the need for equivalent indirect effects of metal ion binding in order to justify normalizing or canceling-out of these effects when the observed rate of one RNA is divided by the other to produce k_{rel}. Differences in indirect effects of metal ion binding can produce significant under or over estimates of apparent metal ion affinity or cooperativity and should be suspected in plots of k_{rel} versus thiophilic metal ion concentration that cannot be fit to a standard binding isotherm. The likelihood of differences in indirect metal ion binding effects can be assessed by comparing the observed rate dependence on metal ion concentration for thiophilic metal ions that do not rescue, for the native metal ion Mg^{2+}, and for alternative background metal ions such as Ca^{2+}.

The third criterion that there be no direct effect of the rescuing metal ion on the "normal" reaction addresses the issue of whether the binding of thiophilic metal ion alters the chemical mechanism and is thus not representative of the native reaction. Changes in the reaction mechanism can result form changes in local geometry, alterative coordination, or the introduction of novel metal sites within the active site or elsewhere. The likelihood of such changes can be assessed by examining individual reaction characteristics (e.g. k_{obs}, n_{Hill}, pH dependence) over the full range of metal ion conditions tested with particular attention to changes in an individual parameter being greater reaction than expected for a small change in thiophilic metal ion concentration. In this respect, clear interpretation of thiophilic metal ion concentrations that produce a thousand fold enhancement in k_{obs} for unmodified ribozyme is problematic while changes of 10-fold or less are unlikely to reflect changes in reaction mechanism.

19.3.2
Contributions to Complexity of Reaction Kinetics

Reaction kinetics in thiophilic metal ion rescue studies are often complex (containing two or more phases). This complexity is typically the result of differences in the level or type of molecular substitution or heterogeneity in RNA folding, any of which can significantly affect the interpretation of the experiment results. Stereoisomers of individual phosphorothioates can have distinct effects on reaction rate; consequently incomplete separation during purification can lead to biphasic kinetics. Cross-contamination of the individual stereoisomers can be verified by testing whether the observed rate and amplitude of each kinetic fraction show a reciprocal correlation between the pro-R_p and pro-S_p purified fractions and by comparing the observed rates to that of unmodified RNA [66].

Reaction kinetics of RNAs containing inhibitory phosphorothioate substitutions can reveal a small burst phase of approximately 1–3% (but can be much higher), which reflects the level of contaminating unmodified nucleotide analog and must be excluded from calculations of rate defects due to the sulfur substitution itself. The presence of unmodified phosphate is likely the result of problems in synthesis or purification, but may also arise from desulfurization during the course of the reaction [121]. Short (10 or less) oligonucleotides containing either phosphate or phosphorothioate often migrate differently on higher percentage (20% or higher) denaturing polyacrylamide gels, allowing the relative level of oxygen contamination to be measured throughout purification and enzymatic analysis from a comparison of modified and unmodified oligonucleotides or fragments derived from RNase T1 nuclease cleavage. Note that contamination with unmodified RNA or a stereoisomer that is not inhibitory can result in a significant underestimation of the observed level of a specific phosphorothioate effect if substrate dissociation is fast relative to cleavage even under single turnover conditions [66].

Distinct kinetic phases can also result from disruption of structural elements important for RNA folding to produce alternative conformations that may be inactive and exchange slowly with the native structure [122, 123]. The presence of alternative conformations may be monitored by non-denaturing gel electrophoresis [82], and the fraction of correctly folded RNA can be determined from active site titration and burst kinetics [124, 125]. Although some additional insight into the nature of structural complexity may be gained through chemical modification, crosslinking or metal-dependent cleavage, the resolution and conclusions that can be drawn from these techniques is somewhat limited. Finally, it is possible that complex reaction kinetics is due to the formation of multimeric complexes. This possibility can be excluded by demonstrating that the rates and relative amplitudes of the individual kinetic phases are not changed upon dilution of the ribozyme–substrate complex by more than 100-fold [66].

19.3.3
Size and Significance of Observed Effects

Because effects on the observed rate from atomic substitution and thiophilic metal ion rescue vary over a broad range and can be significantly larger than 1000-fold, it is tempting to consider the size of the effect as a measure of its biological importance. However, as noted above, there are many factors that contribute to the observed effect that may mask the energetic contributions of specific metal ion interactions and the nature of metal ion binding in the native state. In particular, geometric constraints present in the native metal binding site or imposed by atomic substitution or binding of non-native metal species preclude prediction of the presence or magnitude of thiophilic rescue based simply on the affinity of a metal ion for a given functional group [98–100]. Both sulfur and thiophilic metal ions are significantly larger than their oxygen and Mg^{2+} counterparts, and significantly alter the bond to phosphate, which in turn can alter the position of the backbone as well as the position or occupancy by the metal ion itself [126–128]. It is also important to note that while Mn^{2+} has a greater thiophilicity than Mg^{2+}, it nevertheless retains a strong preference for oxygen that may mute the magnitude of rescue, particularly in cases where more than one metal ion is involved in coordinating a single sulfur substitution and contributes more than one unfavorable interaction [25, 58, 129]. The combination of sulfur and thiophilic metal ion may also lead to an interaction not present in the native structure such as a change from an outer-sphere to an inner-sphere interaction [5]. In addition, the ligands for Mg^{2+} and thiophilic ions may not be equivalent and may reflect distinct binding sites that are mutually exclusive, either because of electrostatic repulsion or conformational rearrangements [78]. Electrostatic or conformational rearrangements may also allow thiophilic metal ions to bind better than Mg^{2+} in the active conformation of the ribozyme or produce alternative structures capable of activity [5, 72, 78]. Finally, thiophilic metal ions may be bound or better positioned in the transition state compared to the ground state of the reaction [66, 72]. The significance of incomplete rescue should therefore be interpreted with caution and the absence of rescue interpreted simply as a negative result. Nevertheless, rescue effects less than 10-fold fall close to the uncertainty from experimental error. Thus, while this does not mean that smaller effects are not indicative of functionally important interactions, it does increase the burden of distinguishing direct from indirect effects. The finding of similar effects from phosphorothioate substitution or thiophilic metal ion rescue in a structurally distinct but evolutionarily related system, however, can help to provide evidence for the significance of more subtle effects [81, 130]. While phylogenetic comparison bares the burden of finding two or more systems that can meet the criteria stated above, it can provide a powerful control to help rule out experimental artifacts and to isolate structural features central to function rather than idiosyncratic to a particular RNA species.

19.4
Conclusion

As can be seen from the considerations outlined above, application of thiophilic metal ion rescue involves careful design of the experiment within a well-defined kinetic scheme to control for numerous sources of indirect effects on experimental signal. However, its proper application has been invaluable in defining the biochemical role of metal ions in RNA structure and enzymatic catalysis, and as an independent and essential approach to validating high-resolution studies.

Acknowledgments

Sincere thanks go to Drs. Jonatha Gott, Frank Campbell, Nathan Zahler, Mike Harris, Joe Piccirilli and Dan Herschlag for helpful discussions, comments and careful reading of the manuscript.

References

1 MISRA, V. K., DRAPER, D. E., *Biopolymers* **1998**, *48*, 113–135.
2 MISRA, V. K., SHIMAN, R., DRAPER, D. E., *Biopolymers* **2003**, *69*, 118–136.
3 SHELTON, V. M., SOSNICK, T. R., PAN, T., *Biochemistry* **2001**, *40*, 3629–3638.
4 WADKINS, T. S., SHIH, I., PERROTTA, A. T., BEEN, M. D., *J. Mol. Biol.* **2001**, *305*, 1045–1055.
5 DEROSE, V. J., *Curr. Opin. Struct. Biol.* **2003**, *13*, 317–324.
6 GLUICK, T. C., WILLS, N. M., GESTELAND, R. F., DRAPER, D. E., *Biochemistry* **1997**, *36*, 16173–16186.
7 HORTON, T. E., CLARDY, D. R., DEROSE, V. J., *Biochemistry* **1998**, *37*, 18094–18101.
8 SCHREIER, A. A., SCHIMMEL, P. R., *J. Mol. Biol.* **1974**, *86*, 601–620.
9 YAMADA, A., AKASAKA, K., HATANO, H., *Biopolymers* **1976**, *15*, 1315–1331.
10 ANDERSON, C. F., RECORD, M. T., JR, *Annu. Rev. Phys. Chem.* **1995**, *46*, 657–700.
11 OTT, G., ARNOLD, L., LIMMER, S., *Nucleic Acids Res.* **1993**, *21*, 5859–5864.
12 CATE, J. H., GOODING, A. R., PODELL, E., ZHOU, K., GOLDEN, B. L., KUNDROT, C. E., CECH, T. R., DOUDNA, J. A., *Science* **1996**, *273*, 1678–1685.
13 CATE, J. H., DOUDNA, J. A., *Structure* **1996**, *4*, 1221–1229.
14 CATE, J. H., HANNA, R. L., DOUDNA, J. A., *Nat. Struct. Biol.* **1997**, *4*, 553–558.
15 WALTER, F., MURCHIE, A. I., THOMSON, J. B., LILLEY, D. M., *Biochemistry* **1998**, *37*, 14195–14203.
16 PYLE, A. M., *J. Biol. Inorg. Chem.* **2002**, *7*, 679–690.
17 HANNA, R., DOUDNA, J. A., *Curr. Opin. Chem. Biol.* **2000**, *4*, 166–170.
18 PICCIRILLI, J. A., VYLE, J. S., CARUTHERS, M. H., CECH, T. R., *Nature* **1993**, *361*, 85–88.
19 SONTHEIMER, E. J., SUN, S., PICCIRILLI, J. A., *Nature* **1997**, *388*, 801–805.
20 WARNECKE, J. M., FURSTE, J. P., HARDT, W. D., ERDMANN, V. A., HARTMANN, R. K., *Proc. Natl. Acad. Sci. USA* **1996**, *93*, 8924–8928.
21 WEINSTEIN, L. B., JONES, B. C., COSSTICK, R., CECH, T. R., *Nature* **1997**, *388*, 805–808.
22 CHEN, Y., LI, X., GEGENHEIMER, P., *Biochemistry* **1997**, *36*, 2425–2438.
23 NARLIKAR, G. J., HERSCHLAG, D., *Annu. Rev. Biochem.* **1997**, *66*, 19–59.
24 SHAN, S. O., HERSCHLAG, D., *Biochemistry* **1999**, *38*, 10958–10975.
25 SHAN, S., KRAVCHUK, A. V., PICCIRILLI, J. A., HERSCHLAG, D., *Biochemistry* **2001**, *40*, 5161–5171.
26 BUJALOWSKI, W., GRAESER, E., MCLAUGHLIN, L. W., PROSCHKE, D., *Biochemistry* **1986**, *25*, 6365–6371.
27 QUIGLEY, G. J., TEETER, M. M., RICH, A., *Proc. Natl. Acad. Sci. USA* **1978**, *75*, 64–68.
28 CELANDER, D. W., CECH, T. R., *Science* **1991**, *251*, 401–407.
29 PAN, T., LONG, D. M., UHLENBECK, O. C., *Divalent Metal Ions in RNA Folding and Catalysis*, Cold Spring Harbor Laboratory Press, Cold Spring Harbor, NY, **1993**.
30 PAN, T., *Biochemistry* **1995**, *34*, 902–909.
31 BASSI, G. S., MURCHIE, A. I., LILLEY, D. M., *RNA* **1996**, *2*, 756–768.
32 BEEBE, J. A., KURZ, J. C., FIERKE, C. A., *Biochemistry* **1996**, *35*, 10493–10505.

33 Draper, D. E., *Trends Biochem. Sci.* **1996**, *21*, 145–149.

34 Shan, S., Yoshida, A., Sun, S., Piccirilli, J. A., Herschlag, D., *Proc. Natl. Acad. Sci. USA* **1999**, *96*, 12299–12304.

35 Zhang, Y., Liang, J. Y., Huang, S., Ke, H., Lipscomb, W. N., *Biochemistry* **1993**, *32*, 1844–1857.

36 Bone, R., Frank, L., Springer, J. P., Atack, J. R., *Biochemistry* **1994**, *33*, 9468–9476.

37 Wilcox, D. E., *Chem. Rev.* **1996**, *96*, 2435–2458.

38 Cowan, J. A., Ohyama, T., Howard, K., Rausch, J. W., Cowan, S. M., Le Grice, S. F., *J. Biol. Inorg. Chem.* **2000**, *5*, 67–74.

39 Rangan, P., Masquida, B., Westhof, E., Woodson, S. A., *Proc. Natl. Acad. Sci. USA* **2003**, *100*, 1574–1579.

40 Hampel, K. J., Burke, J. M., *Biochemistry* **2003**, *42*, 4421–4429.

41 Misra, V. K., Draper, D. E., *J. Mol. Biol.* **2002**, *317*, 507–521.

42 Nakano, S., Cerrone, A. L., Bevilacqua, P. C., *Biochemistry* **2003**, *42*, 2982–2994.

43 Hammann, C., Lilley, D. M., *ChemBioChem* **2002**, *3*, 690–700.

44 Pan, T., Sosnick, T. R., *Nat. Struct. Biol.* **1997**, *4*, 931–938.

45 Fang, X. W., Thiyagarajan, P., Sosnick, T. R., Pan, T., *Proc. Natl. Acad. Sci. USA* **2002**, *99*, 8518–8523.

46 Russell, R., Millett, I. S., Tate, M. W., Kwok, L. W., Nakatani, B., Gruner, S. M., Mochrie, S. G., Pande, V., Doniach, S., Herschlag, D., Pollack, L., *Proc. Natl. Acad. Sci. USA* **2002**, *99*, 4266–4271.

47 Zhuang, X., Rief, M., *Curr. Opin. Struct. Biol.* **2003**, *13*, 88–97.

48 Sosnick, T. R., Pan, T., *Curr. Opin. Struct. Biol.* **2003**, *13*, 309–316.

49 Smith, D., Pace, N. R., *Biochemistry* **1993**, *32*, 5273–5281.

50 McConnell, T. S., Herschlag, D., Cech, T. R., *Biochemistry* **1997**, *36*, 8293–8303.

51 Lott, W. B., Pontius, B. W., von Hippel, P. H., *Proc. Natl. Acad. Sci. USA* **1998**, *95*, 542–547.

52 Warnecke, J. M., Held, R., Busch, S., Hartmann, R. K., *J. Mol. Biol.* **1999**, *290*, 433–445.

53 Brannvall, M., Mikkelsen, N. E., Kirsebom, L. A., *Nucleic Acids Res.* **2001**, *29*, 1426–1432.

54 Sigel, R. K., Vaidya, A., Pyle, A. M., *Nat. Struct. Biol.* **2000**, *7*, 1111–1116.

55 Eckstein, F., Gish, G., *Trends Biochem. Sci.* **1989**, *14*, 97–100.

56 Peracchi, A., Beigelman, L., Usman, N., Herschlag, D., *Proc. Natl. Acad. Sci. USA* **1996**, *93*, 11522–11527.

57 McKay, D. B., *RNA* **1996**, *2*, 395–403.

58 Jaffe, E. K., Cohn, M., *J. Biol. Chem.* **1978**, *253*, 4823–4825.

59 Pecoraro, V. L., Hermes, J. D., Cleland, W. W., *Biochemistry* **1984**, *23*, 5262–5271.

60 Burgers, P. M., Eckstein, F., *J. Biol. Chem.* **1979**, *254*, 6889–6893.

61 Burgers, P. M., Eckstein, F., *J. Biol. Chem.* **1980**, *255*, 8229–8233.

62 Connolly, B. A., Eckstein, F., *J. Biol. Chem.* **1981**, *256*, 9450–9456.

63 Christian, E. L., Yarus, M., *Biochemistry* **1993**, *32*, 4475–4480.

64 Chanfreau, G., Jacquier, A., *Science* **1994**, *266*, 1383–1387.

65 Hardt, W. D., Warnecke, J. M., Erdmann, V. A., Hartmann, R. K., *EMBO J.* **1995**, *14*, 2935–2944.

66 Peracchi, A., Beigelman, L., Scott, E. C., Uhlenbeck, O. C., Herschlag, D., *J. Biol. Chem.* **1997**, *272*, 26822–26826.

67 Sjogren, A. S., Pettersson, E., Sjoberg, B. M., Stromberg, R., *Nucleic Acids Res.* **1997**, *25*, 648–653.

68 Ortoleva-Donnelly, L., Szewczak, A. A., Gutell, R. R., Strobel, S. A., *RNA* **1998**, *4*, 498–519.

69 Sood, V. D., Beattie, T. L., Collins, R. A., *J. Mol. Biol.* **1998**, *282*, 741–750.

70 Scott, E. C., Uhlenbeck, O. C., *Nucleic Acids Res.* **1999**, *27*, 479–484.

71 Sontheimer, E. J., Gordon, P. M., Piccirilli, J. A., *Genes Dev.* **1999**, *13*, 1729–1741.

72 Wang, S., Karbstein, K., Peracchi, A., Beigelman, L., Herschlag, D., *Biochemistry* **1999**, *38*, 14363–14378.

73 Yoshida, A., Sun, S., Piccirilli, J. A., *Nat. Struct. Biol.* **1999**, *6*, 318–321.

74 Christian, E. L., Kaye, N. M., Harris, M. E., *RNA* **2000**, *6*, 511–519.
75 Gordon, P. M., Sontheimer, E. J., Piccirilli, J. A., *Biochemistry* **2000**, *39*, 12939–12952.
76 Gordon, P. M., Sontheimer, E. J., Piccirilli, J. A., *RNA* **2000**, *6*, 199–205.
77 Pfeiffer, T., Tekos, A., Warnecke, J. M., Drainas, D., Engelke, D. R., Seraphin, B., Hartmann, R. K., *J. Mol. Biol.* **2000**, *298*, 559–565.
78 Shan, S. O., Herschlag, D., *RNA* **2000**, *6*, 795–813.
79 Yean, S. L., Wuenschell, G., Termini, J., Lin, R. J., *Nature* **2000**, *408*, 881–884.
80 Warnecke, J. M., Sontheimer, E. J., Piccirilli, J. A., Hartmann, R. K., *Nucleic Acids Res.* **2000**, *28*, 720–727.
81 Gordon, P. M., Piccirilli, J. A., *Nat. Struct. Biol.* **2001**, *8*, 893–898.
82 Crary, S. M., Kurz, J. C., Fierke, C. A., *RNA* **2002**, *8*, 933–947.
83 Szewczak, A. A., Kosek, A. B., Piccirilli, J. A., Strobel, S. A., *Biochemistry* **2002**, *41*, 2516–2525.
84 Shan, S. O., Herschlag, D., *RNA* **2002**, *8*, 861–872.
85 Lindqvist, M., Sandstrom, K., Liepins, V., Stromberg, R., Graslund, A., *RNA* **2001**, *7*, 1115–1125.
86 Christian, E. L., Kaye, N. M., Harris, M. E., *EMBO J.* **2002**, *21*, 2253–2262.
87 Pley, H. W., Flaherty, K. M., McKay, D. B., *Nature* **1994**, *372*, 68–74.
88 Scott, W. G., Murray, J. B., Arnold, J. R., Stoddard, B. L., Klug, A., *Science* **1996**, *274*, 2065–2069.
89 Curley, J. F., Joyce, C. M., Piccirilli, J. A., *J. Am. Chem. Soc.* **1997**, *119*, 12691–12692.
90 Brautigam, C. A., Sun, S., Piccirilli, J. A., Steitz, T. A., *Biochemistry* **1999**, *38*, 696–704.
91 Murray, J. B., Dunham, C. M., Scott, W. G., *J. Mol. Biol.* **2002**, *315*, 121–130.
92 Maderia, M., Hunsicker, L. M., DeRose, V. J., *Biochemistry* **2000**, *39*, 12113–12120.
93 Tanaka, Y., Kojima, C., Morita, E. H., Kasai, Y., Yamasaki, K., Ono, A., Kainosho, M., Taira, K., *J. Am. Chem. Soc.* **2002**, *124*, 4595–4601.
94 Huppler, A., Nikstad, L. J., Allmann, A. M., Brow, D. A., Butcher, S. E., *Nat. Struct. Biol.* **2002**, *9*, 431–435.
95 Morrissey, S. R., Horton, T. E., DeRose, V. J., *J. Am. Chem. Soc.* **2000**, *122*, 3473–3481.
96 Morrissey, S. R., Horton, T. E., Grant, C. V., Hoogstratten, C. G., Britt, R. D., DeRose, V. J., *J. Am. Chem. Soc.* **1999**, *121*, 9215–9218.
97 Suzumura, K., Yoshinari, K., Tanaka, Y., Takagi, Y., Kasai, Y., Warashina, M., Kuwabara, T., Orita, M., Taira, K., *J. Am. Chem. Soc.* **2002**, *124*, 8230–8236.
98 Smith, J. S., Nikonowicz, E. P., *Biochemistry* **2000**, *39*, 5642–5652.
99 Maderia, M., Horton, T. E., DeRose, V. J., *Biochemistry* **2000**, *39*, 8193–8200.
100 Horton, T. E., Maderia, M., DeRose, V. J., *Biochemistry* **2000**, *39*, 8201–8207.
101 Brautigam, C. A., Sun, S., Piccirilli, J. A., Steitz, T. A., *Biochemistry* **1999**, *38*, 696–704.
102 Basu, S., Strobel, S. A., *RNA* **1999**, *5*, 1399–1407.
103 Krageten, J. *Atlas of Metal–Ligand Equilibria in Aqueous Solution*, Halsted Press, New York, **1978**.
104 Dahm, S. C., Derrick, W. B., Uhlenbeck, O. C., *Biochemistry* **1993**, *32*, 13040–13045.
105 Anderegg, G., in: *Comprehensive Coordination Chemistry: The Synthesis, Reactions, Properties, Applications of Coordination Compounds*. Wilkinson, G., Gillard, R. D., McCleverty, J. A. (eds), Pergamon Press, Oxford, **1987**, pp. 777–792.
106 Sun, S., Yoshida, A., Piccirilli, J. A., *RNA* **1997**, *3*, 1352–1363.
107 Christian, E. L., Yarus, M., *J. Mol. Biol.* **1992**, *228*, 743–758.
108 Slim, G., Gait, M. J., *Nucleic Acids Res.* **1991**, *19*, 1183–1188.
109 Burgers, P. M., Eckstein, F., *Proc. Natl. Acad. Sci. USA* **1978**, *75*, 4798–4800.

110 MOORE, M. J., SHARP, P. A., *Science* **1992**, *256*, 992–997.
111 HARRIS, M. E., CHRISTIAN, E. L., *Methods* **1999**, *18*, 51–59.
112 HANSEN, A., PFEIFFER, T., ZULEEG, T., LIMMER, S., CIESIOLKA, J., FELTENS, R., HARTMANN, R. K., *Mol. Microbiol.* **2001**, *41*, 131–143.
113 ZAHLER, N. H., CHRISTIAN, E. L., HARRIS, M. E., *RNA* **2003**, *9*, 734–745.
114 HERSCHLAG, D., ECKSTEIN, F., CECH, T. R., *Biochemistry* **1993**, *32*, 8312–8321.
115 YOSHIDA, A., SHAN, S., HERSCHLAG, D., PICCIRILLI, J. A., *Chem. Biol.* **2000**, *7*, 85–96.
116 CHRISTIAN, E. L., ZAHLER, N. H., KAYE, N. M., HARRIS, M. E., *Methods* **2002**, *28*, 307–322.
117 HERSCHLAG, D., KHOSLA, M., *Biochemistry* **1994**, *33*, 5291–52287.
118 PYLE, A. M., GREEN, J. B., *Biochemistry* **1994**, *33*, 2716–2725.
119 HERSCHLAG, D., PICCIRILLI, J. A., CECH, T. R., *Biochemistry* **1991**, *30*, 4844–4854.
120 MICHELS, W. J., JR, PYLE, A. M., *Biochemistry* **1995**, *34*, 2965–2977.
121 HAMM, M. L., NIKOLIC, D., VAN BREEMEN, R. B., PICCIRILLI, J. A., *J. Am. Chem. Soc.* **2000**, *122*, 12069–12078.
122 CLOUET-D'ORVAL, B., UHLENBECK, O. C., *RNA* **1996**, *2*, 483–491.
123 RUFFNER, D. E., DAHM, S. C., UHLENBECK, O. C., *Gene* **1989**, *82*, 31–41.
124 BEEBE, J. A., FIERKE, C. A., *Biochemistry* **1994**, *33*, 10294–10304.
125 CORNISH-BOWDEN, A. J., *Fundamentals of Enzyme Kinetics*, Portland Press, London, **1995**.
126 WEAST, R. C., *Handbook of Physics and Chemistry*, CRC Press, Boca Raton, FL, **1989**.
127 SHANNON, R. D., *Acta Crystallogr. A* **1976**, *32*, 751–767.
128 LIANG, C. X., ALLEN, L. C., *J. Am. Chem. Soc.* **1987**, *109*, 6449–6453.
129 STROBEL, S. A., ORTOLEVA-DONNELLY, L., *Chem. Biol.* **1999**, *6*, 153–165.
130 KUO, L. Y., PICCIRILLI, J. A., *Biochim. Biophys. Acta* **2001**, *1522*, 158–166.

20
Identification of Divalent Metal Ion Binding Sites in RNA/DNA-metabolizing Enzymes by Fe(II)-mediated Hydroxyl Radical Cleavage

Yan-Guo Ren, Niklas Henriksson and Anders Virtanen

20.1
Introduction

The presence and requirement for divalent metal ions in the active sites of nucleic acid metabolizing enzymes which participate in phospho(di)ester formation and breakage has emerged as a common theme (reviewed in [1, 2]). One of the best-studied active sites involved in the cleavage of a phosphodiester bond is the 3′ exonucleolytic site of *Escherichia coli* DNA polymerase (Pol) I, which has been characterized by a combination of genetics, biochemistry and structural techniques (see [3–5] and references therein). Most importantly, crystallographic studies provided direct evidence that divalent metal ions are coordinated in this active site directly or via bridging water to oxygens in four acidic amino acid residues as well as to a non-bridging oxygen at the scissile phosphodiester. The divalent metal ions in the 3′-exonucleolytic site of DNA Pol I play a critical role during catalysis, and it has been proposed that the nucleophile (water or hydroxide ion) attacking the scissile phosphate during cleavage is oriented by one metal ion and two amino acid residues. After cleavage, a second divalent metal ion stabilizes the negative charge on the leaving group. A similar mechanism, where one divalent metal ion activates the initially attacking nucleophile while the other stabilizes the leaving group, is used by several other enzymes involved in breaking and forming phospho(di)ester bonds, such as endo- or exonucleases, kinases, phosphatases and polymerases (reviewed in [1, 2, 6]). Taken together, it has become apparent that a very fruitful strategy to study the active site of any enzyme participating in phospho(di)ester formation and breakage is to identify and characterize its divalent metal ion binding sites.

Here we will describe protocols that we have used to characterize and map divalent metal ion binding sites in the active sites of the Klenow fragment of *E. coli* DNA Pol I and human poly(A)-specific ribonuclease (PARN) [7, 8]. PARN is a 3′ exonuclease that efficiently degrades mRNA poly(A) tails [9–17] and belongs to the RNase D family of nucleases [8, 10, 18], of which the 3′ exonuclease domain of *E. coli* DNA Pol I is one of the best-studied examples. The method of Fe(II)-mediated hydroxyl radical cleavage [8] described here has been applied to map di-

Handbook of RNA Biochemistry: Student Edition. Edited by R. K. Hartmann, A. Bindereif, A. Schön, E. Westhof
Copyright © 2009 WILEY-VCH Verlag GmbH & Co. KGaA, Weinheim
ISBN: 978-3-527-32534-4

valent metal ion binding sites in a large variety of metalloenzymes (see, e.g. [19–25] and references therein). We want to emphasize, however, that this method needs to be combined with several other approaches before a complete picture of a divalent metal ion binding site can be drawn. In our case, the Fe(II)-mediated hydroxyl radical cleavage assays of PARN were preceded by two important steps: (1) bioinformatic identification of amino acids potentially located in the active site of the enzyme and (2) site-directed mutagenesis of amino acids expected to be important for catalysis. Several protocols and descriptions of these two steps are available elsewhere (see, e.g. [26, 27]) and thus will not be given here.

20.2
Probing Divalent Metal Ion Binding Sites

One of the most important prerequisites for the successful analysis of any enzyme is the availability of sufficiently large and pure preparations of the enzyme. It is therefore important to spend some time to determine a simple and efficient protocol for the expression and purification of a recombinant form of the enzyme of interest. A large variety of expression systems are commercially available and we have successfully used several of them (e.g. the pET system from Novagen or the pCAL system from Stratagene). When choosing your expression system it is important to investigate if the recombinant form of the enzyme has the same key properties as the non-recombinant one, which is often not the case. For example, we observed when studying human poly(A) polymerases that the position, N- or C-terminally, of the tag used for affinity purification significantly affected the K_m parameter of the enzyme (our unpublished observation).

20.2.1
Fe(II)-mediated Hydroxyl Radical Cleavage

The induction of hydroxyl radicals through the Fenton reaction [28] (Fig. 20.1) in the vicinity of Fe(II) ion binding sites has become a powerful tool to identify divalent metal ion binding sites in protein and RNA enzymes. In the presence of reductants, such as ascorbic acid or DTT, Fe(II) generates hydroxyl radicals which efficiently cleave the polypeptide or nucleic acid backbone in the vicinity of the Fe(II) binding site.

Before performing the Fe(II)-mediated hydroxyl radical cleavage assay, it is advisable to investigate if the enzyme under study is active in the presence of Fe(II) [29].

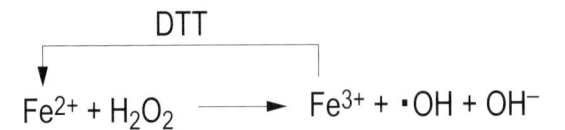

$$Fe^{2+} + H_2O_2 \xrightarrow{DTT} Fe^{3+} + \cdot OH + OH^-$$

Fig. 20.1. The Fenton reaction.

For PARN we could readily detect enzymatic activity when we replaced the essential divalent metal ion Mg(II) with Fe(II) [8]. A positive result from this simple assay argues immediately that Fe(II) functions catalytically, which implies that some of the Fe(II) ion binding sites overlap with binding sites for Mg(II) ions. It is important to remember that the latter statement is one of the key assumptions of your analysis since you will argue that some of the binding sites for Fe(II) ions that you eventually will identify correspond to binding sites for the natural metal cofactors in the active site of the enzyme.

Technically, the Fe(II)-mediated hydroxyl radical cleavage assay is easy to perform. The enzyme (0.5–10 µM) is incubated in a buffer containing 50–100 mM HEPES, pH 7.0, 10 mM DTT and 0.2–20 µM $Fe(NH_4)_2SO_4$. The exact reaction volume, incubation time and temperature as well as the amount of enzyme incubated have to be determined empirically in order to resolve visible and distinct cleavage products by SDS–PAGE. As a rule, the reaction volume is about 10 µl, and incubation times and temperatures are between 2 and 30 min and 0 and 37 °C. Protocol 1 describes the conditions we used for PARN [8], while Protocol 2 describes our conditions for the Klenow Pol fragment [7]. Often, a small amount of H_2O_2 (approximately 0.1–0.2% v/v) has to be added, as well as a small amount of NaCl. In the case of Fe(II)-mediated cleavage of PARN and Klenow Pol we could omit H_2O_2, while we included 5 mM NaCl in reactions containing the Klenow Pol fragment. The presence of substrate can also influence the cleavage pattern or the efficiency of cleavage. For PARN, the inclusion of the substrate poly(A_{50}) improved the cleavage reaction significantly, while the addition of DNA to Klenow Pol had a minor effect only.

The reaction is terminated by the addition of one reaction volume of $2 \times$ SDS loading buffer and directly fractionated by SDS–PAGE. Subsequently, the cleavage products are visualized by silver staining if a non-radioactive polypeptide was reacted, or by autoradiography if the polypeptide was radioactively labeled. Figure 20.2 shows a typical result obtained when Klenow Pol is subjected to Fe(II)-mediated cleavage. Please note the importance of the control lanes 2, 3, 5 and 6. Lane 2 controls for the dependence on Fe(II), lane 3 demonstrates the essential role of the reducing agent DTT, lane 6, including the chelator EDTA, documents the requirement for Fe(II) and/or possibly traces of other divalent metal ions brought in with the enzyme preparation, and lane 5 in comparison with lane 4 suggests that Fe(II) and Mg(II) occupy overlapping binding sites since addition of Mg(II) suppresses the appearance of the main cleavage product. Note that the reactions in lanes 2, 3, 5 and 6 also control for the presence or appearance of non-specific cleavage products.

20.2.2
How to Map Divalent Metal Ion Binding Sites

The Fe(II)-induced cleavage site(s) will be in the vicinity, within a few Ångstroms, of the binding site(s) for Fe(II). Thus, a major effort is dedicated to localizing the cleavage site. For this purpose, radioactive labeling of the N or C terminus of the

348 | *20 Identification of Divalent Metal Ion Binding Sites*

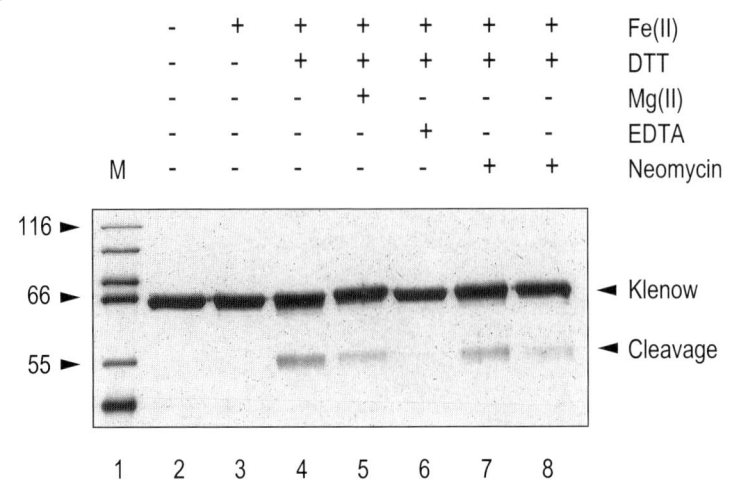

Fig. 20.2. Fe(II)-mediated cleavage of recombinant Klenow Pol fragment. A sample of 2 μg of Klenow Pol fragment was incubated as described in Protocol 2. The resulting cleavage products were analyzed by SDS–PAGE followed by silver staining. Presence (+) or absence (−) of indicated reagent; Fe(II) (lanes 3–8), 20 μM Fe(NH$_4$)$_2$(SO$_4$)$_2$; DTT (lanes 4–8), 10 mM DTT; Mg(II) (lane 5), 10 mM MgCl$_2$; EDTA (lane 6), 10 mM EDTA·Na$_2$; neomycin (lanes 7 and 8), 2 and 10 mM neomycin B. The molecular size marker was fractionated in lane 1. The arrowheads on the right mark the position of Klenow Pol fragment and its cleavage product; arrowheads on the left depict selected size markers with their indicated molecular weight (in kDa).

enzyme and size fractionation of cleavage fragments by SDS–PAGE is often used [30–34]. A good assignment of the cleavage position then depends on an accurate correlation of molecular size and migrational distance. Another advantage of using radioactively labeled polypeptides is the possibility to quantify the cleavage reaction. Protocol 3 describes our procedure for radioactive labeling of PARN at the N terminus [8]. In this particular case, we have made use of a protein kinase recognition motif that was present in the N-terminal tag encoded in the commercially available expression plasmid pET33 (Novagen).

However, due to the tertiary structure of proteins, induced cleavages and Fe(II) binding sites are not always close to each other in the primary sequence. As a matter of fact, accurate mapping of the cleavage sites is not always required. This is very well exemplified by our studies of the active site of PARN. Here we could apply a different strategy instead, since bioinformatic characterization followed by site-directed mutagenesis had already revealed amino acids presumably located in the active site of the enzyme. Thus, we simply investigated if any of these introduced mutations affected the appearance of the Fe(II)-induced cleavage products in comparison with the wild-type enzyme. An altered cleavage pattern for the mutant polypeptide then indicated that the mutated amino acid is required for Fe(II)-mediated cleavage. The effects caused by the mutations that we have observed ranged from minute decreases to complete disappearance of cleavage product. An excellent way to quantify such effects is to determine an apparent K_D (appK_D) for the Fe(II) ions causing the cleavages [22]. For this purpose, an increasing amount of Fe(II) is added to the individual reactions and the cleavage product at each concentration of Fe(II) is quantified. Finally, an appK_D is calculated, e.g. using Lineweaver–Burk formalism. It is worthwhile mentioning that a difference in the calculated appK_D for two or more cleavage sites in the same polypeptide implies that (1) the cleavages are induced by different Fe(II) ions and, thus, (2) multiple binding sites for Fe(II) ions have been identified.

20.2.3
How to Use Aminoglycosides as Functional and Structural Probes

Aminoglycosides bind frequently to negatively charged binding pockets present in both protein enzymes and RNA (see [7, 35–37] and references therein). Often these binding sites overlap with binding sites for divalent metal ions, and experimental evidence suggests that aminoglycosides displace functionally important divalent metal ions upon binding and thereby perturb the function of RNA and protein metalloenzymes (e.g. [7, 36, 37]). Aminoglycosides have therefore turned out to be convenient probes in studies of divalent metal ion binding and function. For application of the strategy we used in our studies of PARN, Klenow Pol and poly(A) polymerase ([7] and unpublished data), one may follow the experimental scheme outlined below (see [7] for a detailed description):

(1) Investigate if aminoglycosides inhibit the enzymatic activity by simply adding increasing amounts of aminoglycoside to the reaction. The chemical properties of the aminoglycoside will, of course, influence how efficient it inhibits enzyme activity and one should therefore investigate a repertoire of commercially available aminoglycosides. As a rule of thumb: the higher its number of positively charged amino groups, the more efficiently the aminoglycoside will inhibit the enzyme. The interaction is highly electrostatic; the pH of the reaction therefore plays a decisive role and should usually be below 7.0. The inhibition constants are often in the micromolar range.
(2) Once conditions for inhibition have been established, one should investigate if the aminoglycoside perturbs the Fe(II)-mediated cleavage reaction. For this purpose, include increasing amounts of aminoglycoside in the Fe(II)-mediated cleavage reaction, followed by SDS–PAGE. This is illustrated in Fig. 20.2 (lanes 7 and 8) for neomycin B and the Klenow Pol fragment.
(3) Finally, you should investigate if increasing amounts of a second divalent metal ion, such as Mg^{2+}, relieve the inhibition.

Provided certain aminoglycosides bind to the metalloenzyme of interest with reasonable affinity and specifically displace active site metal ion(s) as inferred from suppression of Fe(II)-mediated cleavage, one has established an elegant experimental platform to investigate the structural and functional role of divalent metal ions in much detail.

20.3 Protocols

Protocol 1: Fe(II)-mediated cleavage of PARN

(1) Label PARN with ^{32}P at the N-terminus using [γ-^{32}P]ATP and bovine heart protein kinase (see Protocol 3).
(2) Prepare the following stock solutions: 0.5 M HEPES–KOH, pH 7.0, 20 µM Fe(NH$_4$)$_2$(SO$_4$)$_2$ [Fe(NH$_4$)$_2$(SO$_4$)$_2$·6 H$_2$O; Sigma F 3754], 50 mM DTT, 0.3 µM poly(A$_{50}$) and 2 × SDS–PAGE loading buffer (0.25 mM Tris–HCl, pH 6.8 at room temperature, 20% glycerol, 2% SDS, 0.025% bromophenol blue) supplemented with 0.2 M DTT.
(3) For each reaction, mix on ice 5 pmol of ^{32}P-labeled PARN, 1 µl 0.5 M HEPES–KOH, pH 7.0, 1 µl 20 µM Fe(NH$_4$)$_2$(SO$_4$)$_2$ and adjust the volume to 8 µl with H$_2$O.
(4) Start the reaction by the addition of 2 µl 50 mM DTT. Mix by gently flicking the tube with your finger and transfer to 37 °C. Incubate for 15–30 min.
(5) Stop the reaction by the addition of 10 µl of 2 × SDS–PAGE loading buffer supplemented with 0.2 M DTT.
(6) Boil the sample for 3 min.
(7) Load on a 10% SDS–polyacrylamide gel with a 4% stacking gel.
(8) Run the gel until the dye reaches the bottom.
(9) Fix and dry the gel, and expose an X-ray film or phosphoimage screen.

Protocol 2: Fe(II)-mediated cleavage of Klenow Pol

(1) Prepare a pure preparation of Klenow Pol at approximately 1–2 mg/ml in 20 mM HEPES–KOH, pH 7.0, and 5 mM NaCl.
(2) Prepare the following stock solutions: 0.5 M HEPES–KOH, pH 7.0, 50 mM NaCl, 100 µM Fe(NH$_4$)$_2$(SO$_4$)$_2$, 50 mM DTT and 2 × SDS–PAGE loading buffer (see Protocol 1) supplemented with 0.2 M DTT.
(3) For each reaction, mix on ice 2–4 µg Klenow Pol, 2 µl 0.5 M HEPES–KOH, pH 7.0, 1 µl 50 mM NaCl, 2 µl 100 µM Fe(NH$_4$)$_2$(SO$_4$)$_2$ and adjust the volume to 8 µl with H$_2$O.
(4) Start the reaction by the addition of 2 µl 50 mM DTT. Mix by gently flicking the tube with your finger and transfer to 37 °C. Incubate for 15–30 min.
(5) Stop the reaction by the addition of 10 µl of 2 × SDS–PAGE loading buffer supplemented with 0.2 M DTT.
(6) Boil the sample for 3 min.
(7) Load on a 10% SDS–polyacrylamide gel with a 4% stacking gel.
(8) Run the gel until the dye reaches the bottom.

(9) Fix and silver stain the gel (Sigma ProteoSilver Silver Staining Kit or Amersham Biosciences PlusOne Silver Staining Kit).

Protocol 3: Radioactive labeling of recombinant polypeptide
The recombinant polypeptide should contain a protein kinase recognition motif, either at the N or C terminus. A number of recombinant protein expression systems (e.g. pET33; Novagen) will provide such a motif in-frame with the affinity tag and placed in the multiple cloning site. We have successfully labeled polypeptides expressed by the bacterial pET33 expression system.

(1) Apply 30 µl of the purified recombinant polypeptide at 10 nM onto a G-50 spin column (MicroSpin G-50 columns; Amersham Biosciences) pre-equilibrated with 20 mM Tris–HCl, pH 7.5, 0.1 M NaCl and 12 mM $MgCl_2$. Spin the column at 2000 g for 1 min.
(2) Mix the eluate with 1 µl [γ-^{32}P]ATP (10 mCi/ml, 3000 Ci/mmol), 1 µl 50 µM ATP and 3 µl (10 U/µl) of a freshly dissolved batch of bovine heart protein kinase A catalytic subunit (Sigma P 2645, supplied as lyophilized powder).
(3) Incubate for 30 min on ice.
(4) Apply the labeling mixture onto a G-50 spin column equilibrated in 25 mM HEPES–KOH, pH 7.0, and 100 mM NaCl. Spin the column at 2000 g for 1 min. Collect the eluate and check the efficiency of labeling by fractionating a small sample by SDS–PAGE.

20.4
Notes and Troubleshooting

No Fe(II)-mediated cleavage detected

(1) Use fresh Fe(II) and DTT solutions.
(2) Check if the pH of the reaction is altered. Usually, a higher pH (>7) facilitates Fe(II)-mediated cleavage while a low pH (<6) abolishes cleavage.
(3) Make sure that there is no metal ion-chelating compound such as EDTA present or check if contaminating divalent metal ions compete with Fe(II) in the reaction.
(4) Try titrating H_2O_2 into the reaction.

No distinct cleavage product(s) detected by SDS–PAGE after Fe(II)-mediated cleavage

(1) The concentration of Fe(II) is too high. Optimize the concentration of Fe(II) in the reaction.
(2) If H_2O_2 is used, optimize the concentration of H_2O_2 or omit it.
(3) Shorten the incubation time and/or lower the temperature.
(4) Try addition of enzyme substrate or product.

Low efficiency of ^{32}P-incorporation after kinase labeling reaction

(1) Be sure to use a freshly dissolved batch of protein kinase (see Protocol 3).
(2) Check if the protein kinase recognition sequence tag is intact and has not been removed by protein degradation.

References

1 T. A. Steitz, J. A. Steitz, *Proc. Natl. Acad. Sci. USA* **1993**, *90*, 6498–6502.
2 C. M. Joyce, T. A. Steitz, *J. Bacteriol.* **1995**, *177*, 6321–6329.
3 A. Bernad, L. Blanco, J. M. Lazaro, G. Martin, M. Salas, *Cell* **1989**, *59*, 219–228.
4 C. A. Brautigam, T. A. Steitz, *J. Mol. Biol.* **1998**, *277*, 363–377.
5 C. M. Joyce, T. A. Steitz, *Annu. Rev. Biochem.* **1994**, *63*, 777–822.
6 T. A. Steitz, *Nature* **1998**, *391*, 231–232.
7 Y. G. Ren, J. Martinez, L. A. Kirsebom, A. Virtanen, *RNA* **2002**, *8*, 1393–1400.
8 Y. G. Ren, J. Martinez, A. Virtanen, *J. Biol. Chem.* **2002**, *277*, 5982–5987.
9 C. G. Körner, E. Wahle, *J. Biol. Chem.* **1997**, *272*, 10448–10456.
10 C. G. Körner, M. Wormington, M. Muckenthaler, S. Schneider, E. Dehlin, E. Wahle, *EMBO J.* **1998**, *17*, 5427–5437.
11 J. Martinez, Y. G. Ren, P. Nilsson, M. Ehrenberg, A. Virtanen, *J. Biol. Chem.* **2001**, *276*, 27923–27929.
12 J. Martinez, Y. G. Ren, A. C. Thuresson, U. Hellman, J. Astrom, A. Virtanen, *J. Biol. Chem.* **2000**, *275*, 24222–24230.
13 J. Åström, A. Åström, A. Virtanen, *EMBO J.* **1991**, *10*, 3067–3071.
14 J. Åström, A. Åström, A. Virtanen, *J. Biol. Chem.* **1992**, *267*, 18154–18159.
15 E. Dehlin, M. Wormington, C. G. Körner, E. Wahle, *EMBO J.* **2000**, *19*, 1079–1086.
16 P. R. Copeland, M. Wormington, *RNA* **2001**, *7*, 875–886.
17 M. Gao, D. T. Fritz, L. P. Ford, J. Wilusz, *Mol. Cell* **2000**, *5*, 479–488.
18 I. S. Mian, *Nucleic Acids Res.* **1997**, *25*, 3187–3195.
19 J. M. Farber, R. L. Levine, *J. Biol. Chem.* **1986**, *261*, 4574–4578.
20 M. R. Ermacora, J. M. Delfino, B. Cuenoud, A. Schepartz, R. O. Fox, *Proc. Natl. Acad. Sci. USA* **1992**, *89*, 6383–6387.
21 N. B. Grodsky, S. Soundar, R. F. Colman, *Biochemistry* **2000**, *39*, 2193–2200.
22 A. Mustaev, M. Kozlov, V. Markovtsov, E. Zaychikov, L. Denissova, A. Goldfarb, *Proc. Natl. Acad. Sci. USA* **1997**, *94*, 6641–6645.
23 J. Lykke-Andersen, R. A. Garrett, J. Kjems, *EMBO J.* **1997**, *16*, 3272–3281.
24 G. N. Godson, J. Schoenich, W. Sun, A. A. Mustaev, *Biochemistry* **2000**, *39*, 332–339.
25 J. J. Hlavaty, J. S. Benner, L. J. Hornstra, I. Schildkraut, *Biochemistry* **2000**, *39*, 3097–3105.
26 F. M. Ausubel, R. Brent, R. E. Kingston, D. D. Moore, J. G. Seidman, J. A. Smith, K. Struhl (eds), *Current Protocols in Molecular Biology*, Wiley, New York, **2004**.
27 J. Sambrook, D. Russel, *Molecular Cloning: A Laboratory Manual*, 3rd edn, Cold Spring Harbor Laboratory Press, Cold Spring Harbor, NY, **2001**.
28 H. J. H. Fenton, *Proc. Chem. Soc.* **1893**, *9*, 113.
29 I. B. Brown, *Acta Crystallogr. B* **1988**, *44*, 545–553.
30 T. H. Jensen, A. Jensen, J. Kjems, *Gene* **1995**, *162*, 235–237.
31 T. H. Jensen, A. Jensen, A. M. Szilvay, J. Kjems, *FEBS Lett.* **1997**, *414*, 50–54.

32 T. H. Jensen, H. Leffers, J. Kjems, *J. Biol. Chem.* **1995**, *270*, 13777–13784.
33 R. Hori, S. Pyo, M. Carey, *Proc. Natl. Acad. Sci. USA* **1995**, *92*, 6047–6051.
34 M. Zhong, L. Lin, N. R. Kallenbach, *Proc. Natl. Acad. Sci. USA* **1995**, *92*, 2111–2115.
35 T. Hermann, E. Westhof, *J. Mol. Biol.* **1998**, *276*, 903–912.
36 N. E. Mikkelsen, M. Brännvall, A. Virtanen, L. Kirsebom, *Proc. Natl. Acad. Sci. USA* **1999**, *96*, 6155–6160.
37 N. E. Mikkelsen, K. Johansson, A. Virtanen, L. A. Kirsebom, *Nat. Struct. Biol.* **2001**, *8*, 510–514.

21
Protein–RNA Crosslinking in Native Ribonucleoprotein Particles

Henning Urlaub, Klaus Hartmuth and Reinhard Lührmann

21.1
Introduction

Protein–RNA interactions lie at the structural and functional heart of ribonucleoprotein (RNP) particles. They govern such fundamental cellular processes as pre-mRNA processing, rRNA maturation, post-transcriptional control (mRNA stability), RNA export, translation and translational control. In this chapter, we present a method we have developed in recent years that allows us to characterize sites of direct protein–RNA contact in native particles, after the contacts have been made permanent by UV crosslinking [1–4].

Our method is especially suitable in situations where the objects of investigation are native RNP particles for which the RNA and the protein compositions are known, while little or no information is available on which proteins are in contact with RNA or where such contacts take place. The method has further been proven to be of value for the identification of direct RNA–protein contact sites in RNP particles reconstituted *in vitro* in which several proteins interact with the RNA component.

In the protocols listed, we refer to isolated U snRNP particles from HeLa cells [5, 6] involved in pre-mRNA processing (for review, see [7]). Importantly, we would like to note that the entire approach can be regarded as a general one, so that the protocols can be easily adapted to investigations of other native RNP particles, or of RNP particles reconstituted *in vitro*.

21.2
Overall Strategy

The overall experimental strategy that we have used for the identification of protein–RNA contact sites in native RNP particles comprises crosslinking of RNP particles by UV irradiation at 254 nm, which fixes protein–RNA interactions covalently by generating a zero-length crosslink, followed by analytical procedures to identify the exact nucleotide(s) on the RNA where the crosslink occurred and to identify the crosslinked polypeptide.

Handbook of RNA Biochemistry: Student Edition. Edited by R. K. Hartmann, A. Bindereif, A. Schön, E. Westhof
Copyright © 2009 WILEY-VCH Verlag GmbH & Co. KGaA, Weinheim
ISBN: 978-3-527-32534-4

The approach is primarily a primer extension analysis of the crosslinked RNA derived from the UV-irradiated native particles. The correct assignment of putative protein–RNA crosslinking sites on the RNA requires parallel analysis of UV-irradiated "naked" (protein-free) RNA and of non-irradiated naked RNA. Comparison of the reverse transcriptase patterns obtained in these three experiments leads to the identification of the RNA bases at which proteins are crosslinked. This first set of experiments gives an excellent overview if a certain protein – or several proteins of multiprotein complexes – is/are in direct contact with the RNA, but it yields no information about which protein of the RNA is crosslinked.

The identification of the corresponding crosslinked protein is achieved by immunoprecipitation combined with primer extension analysis. Thereby, one can define which protein of the RNP is crosslinked to the bases of the RNA that have been identified in the first set of experiments. It is obvious that this type of identification depends on the availability of antibodies against the different proteins and upon the efficiency with which each antibody precipitates its corresponding protein, especially under mild denaturing conditions (for details, see below). The advantage of the method is that it can reveal multiple crosslinks between one protein and its cognate RNA in native particles. For example, we found in this manner that in U1 snRNP particles the U1 70K protein is in contact with two nucleotides in the loop of stem I of the U1 snRNA [1]. Another example is the U4/U6-specific protein 61K: this was found in contact with two distinct sites on the U4 snRNA in native tri-snRNP particles, i.e. the loop in the 5′ stem–loop of the U4 snRNA and nucleotides upstream of the 5′ stem–loop [4].

21.3
UV Crosslinking

UV crosslinking of RNP particles is a straightforward technique. UV crosslinking at 254 nm generates a covalent bond between an amino acid side chain of a protein and a base of the RNA, whenever the relative position of the two components is favorable. In earlier studies we found that UV irradiation of native complexes at 254 nm leads to crosslinking of the side chains of the following amino acids: methionine, tyrosine, histidine, leucine, phenylalanine, and cysteine ([1–4, 8, 9] and our unpublished observations). On the basis of work with halopyrimidine-substituted RNAs, Koch et al. [10] suggested two possible mechanisms for UV-induced protein–RNA crosslinking events: (1) UV-induced electron transfer from the amino acid residue to the halopyrimidine followed by a loss of halide and subsequent radical combination or (2) UV-induced homolysis of the carbon–halogen bond followed by radical addition to the aromatic ring of the amino acid residue. The fact that we have also found highly specific amino acid-RNA crosslinks in non-substituted RNAs of native complexes strongly supports the first mechanism.

Our approach is highly specific, but it has some limitations. The crosslinking yield is relatively low when compared with that of crosslinking in particles reconstituted *in vitro* that carry an RNA species site-specifically labeled with a crosslinking moiety [11–19]. Furthermore, not all proteins that are tightly bound by RNA can also be directly crosslinked by UV irradiation. Examples of this are the U1A protein bound to the U1 snRNA particle ([20] and our unpublished observations) and the human 15.5K protein bound to the human U4 snRNA ([21] and our unpublished observations). On the other hand, every direct UV-induced protein–RNA crosslink found, in particular in native RNP particles, must reflect a "real" interaction because of the short distance between the crosslinked entities. For that reason such crosslinks are referred to as "zero-length". Moreover, work with particles reconstituted *in vitro* that carry a site-specifically labeled RNA is dependent on the efficiency of reconstitution and may even produce false-positive results if heterogeneous populations are generated as a result of incomplete assembly.

Protocol 1 describes the UV irradiation procedure. Some critical points have to be considered when one performs UV crosslinking experiments with native RNP particles:

(1) Concentration of the RNP particles. For UV crosslinking, purified native RNP particles are typically adjusted to a concentration of not more than about 0.1 mg/ml. The final concentration of native particles in solution as such is not critical, as native particles are fully assembled, and inter-particle crosslinking events in native RNP particles are highly unlikely. This item becomes much more of a problem when particles reconstituted *in vitro* are studied. Because an excess of protein over RNA has to be used for the efficient *in vitro* reconstitution, non-specific crosslinks due to the excess of protein may pose a problem [see Troubleshooting (1) for details]. For RNP particles reconstituted *in vitro*, an RNA concentration of 0.1 pmol/µl is in general well sufficient.

(2) Choice of buffer. First of all, the buffer should not contain a high concentration of reagents that are known to scavenge radicals, for example glycerol. UV crosslinking is a UV-induced radical reaction that generates a new covalent bond between the side chain of an amino acid and a base of the RNA. Thus, radical scavengers drastically reduce the crosslinking yield and, for example, glycerol concentrations should be kept as low as possible. Further, since samples are irradiated in small droplets (see below) any detergents in the buffer must be avoided, as the droplets will start to spread out over the sample plate. Finally, if treatment with proteinase K is necessary (see Protocol 2) the buffer should not contain potassium ions.

(3) Crosslinking conditions. These include the choice of UV lamp, the distance between the lamp(s) and the sample and the irradiation time. Our laboratory uses a specially constructed device for UV irradiation at 254 nm (see Fig. 21.1 and Protocol 1). Alternatively, other commercially available devices can be used (e.g. a UV Stratalinker 2400, Stratagene, La Jolla, USA); however, the conditions of UV irradiation, in particular the irradiation time as a function of the power of the UV source and the distance of the lamp(s) to the sample, have to be adjusted accordingly (see below). The samples can be irradiated in different ways, in droplets on a glass slide or Parafilm (Pechiney Plastic Packing, Menasha, USA), or in open plastic tubes. In cases of high sample volumes, custom-made larger glass dishes (4–12 cm in diameter) with a planar surface can be used. Pre-cooling of the samples (4 °C) and the glassware is essential. In our

21.4 Identification of UV-induced Protein–RNA Crosslinking Sites by Primer Extension Analysis

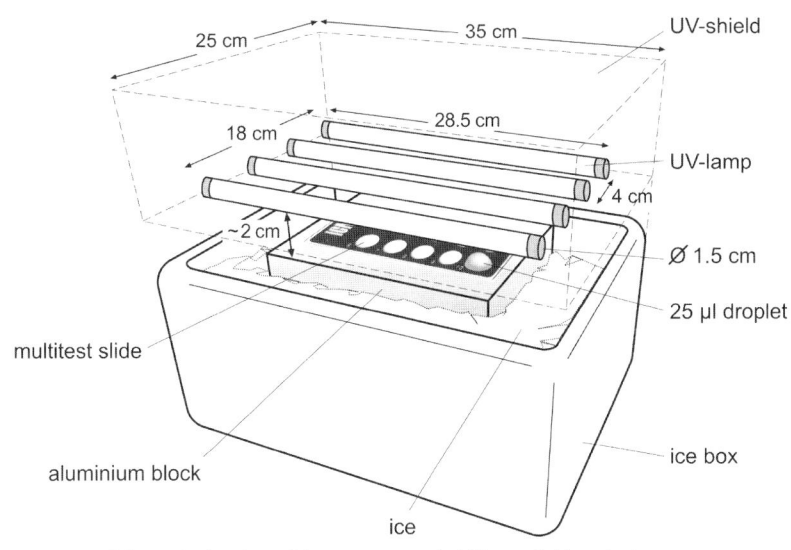

Fig. 21.1. Schematic drawing of the custom made UV crosslinking device.

hands, 25-µl droplets on a pre-cooled 10-well multitest slide (see Protocol 1) work best. For smaller volumes (e.g. 10 µl) we use Eppendorf tubes (Eppendorf AG, Germany) mounted directly under the UV source. We irradiate the samples at a distance of 2 cm from the source (corresponding to the height of a tube mounted directly under an 8-W lamp). The glass slide with the samples is put on top of an aluminium block placed in ice. In addition to these items, the most critical point is the duration of direct UV irradiation. We have observed that the maximum yield of crosslinks under our conditions is obtained after 2 min. Longer irradiation (3 min) does not increase the crosslinking yield significantly and further extended irradiation times lead to substantial loss of particles. On the other hand, when one is working with more rigid RNP particles such as ribosomes [8, 9], UV irradiation times may be prolonged. In any case, as a starting point we recommend performing Protocol 1 with different durations of UV irradiation.

21.4
Identification of UV-induced Protein–RNA Crosslinking Sites by Primer Extension Analysis

After exhaustive hydrolysis of the protein moiety of crosslinked RNP particles by proteinase K treatment, a few amino acids remain covalently attached to the RNA at the sites of crosslinking. In a primer extension reaction, RNA is primed with a 5'-^{32}P-labeled DNA oligonucleotide complementary to a chosen region on the RNA. The reverse transcriptase enzyme then adds dNTPs, which are complementary to the nucleotides of the RNA, to the 3' end of the labeled DNA primer and thus generates radioactively labeled DNA molecules that are complementary to the entire RNA sequence. At a nucleotide on the RNA that has been covalently modified (i.e. by crosslinking), no complementary DNA nucleotide can be added by the reverse transcriptase, owing to either incomplete Watson–Crick base pairing or to steric hindrance due to the presence of crosslinked amino acids. Thus, this nucleotide will cause a stop, or at least a "stuttering", of the reverse transcriptase. It should be noted that the reverse transcriptase stops one nucleotide before the actual crosslinking site. The complementary DNA generated has a certain length and the stop sites (i.e. the length of the generated DNA) can be deduced from a sequencing gel when analyzed next to a marker of complementary DNA that has been generated with the help of dideoxynucleotides [22].

Crosslinking induced by UV irradiation at short wavelengths can also cause intra-RNA crosslinks or induce strand breaks in the RNA; both of these also lead to stops or stuttering of the reverse transcriptase. Therefore, the reverse transcriptase patterns from three RNAs must be compared: (1) RNA from UV-irradiated RNP particles, (2) UV-irradiated naked RNA and (3) non-irradiated naked RNA.

Comparison of the primer extension reaction from these sets of experiments on a high-resolution sequencing gel leads to the identification of putative protein–RNA crosslinking sites. Figure 21.2(A) illustrates the principle of the three experiments necessary for the identification of protein–RNA crosslinking sites and Fig. 21.2(B) gives an example of identified protein–RNA crosslinking sites in native UV-irradiated U1 snRNPs [1, 2].

Protocol 2 describes the purification of RNA derived from UV-irradiated RNP particles. Protocol 3 describes the experimental steps that are required in order to generate naked UV-irradiated and non-irradiated RNAs for controls. Protocol 4 gives a detailed description of the primer extension reaction (including DNA primer purification and labeling) and the subsequent gel electrophoresis that are needed for the reproducible visualization and identification of the crosslinking sites on the RNAs.

For a correct assignment of protein–RNA crosslinking sites in native particles by this method we would like to emphasize several important points:

(1) The first set of experiments probes putative crosslinking sites in samples that contain an excess of non-crosslinked (or unmodified) RNA and in which only a small percentage of the RNA is modified by the UV irradiation (depending on the crosslinking yield and on the degree of UV-induced damage). Therefore, signals corresponding to the full-length RNA will be strongest, as seen by autoradiography of a sequencing gel. To systematically compare crosslinks on RNA from UV-irradiated particles with those on UV-irradiated naked RNA, the amount of RNA probed by primer extension must be the same. In order to ensure that similar amounts of crosslinked material were loaded, the signals corresponding to full-length RNA should be of comparable intensity in both sample preparations. In those cases where UV-induced crosslinks significantly reduce the signal intensity of the full-length transcript, the intensity of the naturally occurring stops on the RNA before the crosslinks should be of comparable

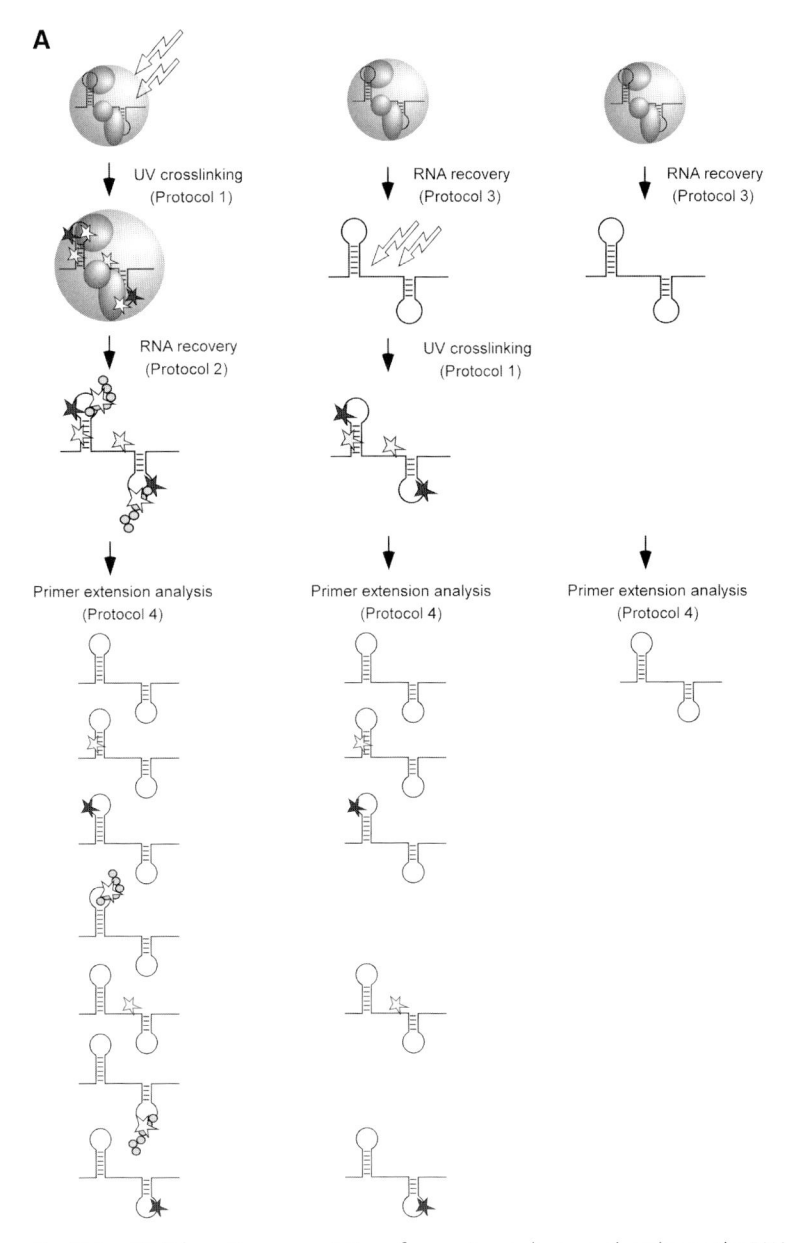

Fig. 21.2. (A) Schematic representation of the initial experiments necessary for the identification of putative protein–RNA crosslinking sites in native particles. RNA derived either from UV-irradiated particles, or RNA that was stripped of proteins before UV irradiation, or non-irradiated "naked" RNA is analyzed by primer extension analysis. White stars indicate nucleotides on the RNA that are covalently modified by crosslinked proteins, grey stars indicated intra-RNA crosslinks and black stars indicate UV-induced strand breaks on the RNA. Grey balls indicate those amino acids that remain covalently attached to the sites of crosslinking after digestion of the RNP particles with proteinase K.

This approach results in the precipitation of most of the non-crosslinked protein together with the portion of the same protein that is crosslinked to RNA. After digestion of the entire protein moiety with proteinase K and extraction of the RNA, the only RNA molecules isolated are those crosslinked to the precipitated protein. Importantly, non-crosslinked RNA molecules are not precipitated. Consequently, in a subsequent primer extension analysis, the only transcripts detected are those with the reverse transcriptase stop at the sites of protein–RNA crosslinking (Fig. 21.3A). Full-length transcripts should not be visible. In practice, however, additional background stops or stops due to UV-induced RNA damage or intra-RNA crosslinking events (Fig. 21.3B) are usually visible. However, the bands due to the stops at the crosslinking sites of the precipitated protein have significantly greater intensities. Figure 21.3(A) illustrates the principle of the immunoprecipitation combined with primer extension analysis of the co-precipitated RNA and Fig. 21.3(B) summarizes results that we have obtained from our crosslinking experiments with native U1 snRNPs [1–3].

This type of analysis is dependent on the number of antibodies available for the different proteins of the particle and on how efficiently each antibody precipitates its corresponding protein, in particular under semi-denaturing conditions that disrupt protein–protein interaction but still preserve the reactivity of the antibodies during immunoprecipitation. As a control, the performance of similar experiments with the pre-immune sera of the corresponding antibodies is highly recommended.

Protocol 5 describes in detail all the steps necessary for this analysis. The most critical steps during the analysis are the procedures for dissociation of the RNP particles before immunoprecipitation and for washing of the beads after immunoprecipitation to remove non-specifically bound material. In our hands, the dissociation of the particles before immunoprecipitation usually works best in the presence of 1% SDS and is improved by subsequent heating of the samples to 70 °C. However, in several cases we observed that higher concentrations of SDS are required to dissociate the particles completely [2, 3]. Washing of the Protein A–Sepharose beads after precipitation should include an additional washing step in a new tube. Furthermore, we observed that washing the samples with buffer containing detergent (e.g. Nonidet P-40) leads to a dramatic increase of non-specific background signals in the subsequent primer extension. We have no explanation for this.

21.6
Troubleshooting

(1) The above procedures can be applied to the analysis of RNP particles assembled *in vitro* from a known number of defined components. The RNA is most easily synthesized *in vitro* by phage RNA polymerases (see Chapter 1) or by chemical synthesis (see Chapter 7). The protein(s) can be produced in either

Escherichia coli or insect cells, or they can be purified from a readily available biological source [22]. When embarking on such a project, a number of considerations should be borne in mind at the outset. The major problem is directly related to the efficiency of RNP assembly *in vitro*. It would be difficult to discern artificial crosslinking events due to incomplete or non-specific assembly from genuine crosslinks. Similar problems will arise if the protein preparation is contaminated by interfering proteins or by eubacterial RNA. This is in particular an obstacle when bacterially expressed RNA-binding proteins are being studied.

(2) Technically speaking, the most demanding aspect of the procedures outlined is that the interpretability of the final result is heavily dependent on the recovery of RNP and RNA in a large number of consecutive experimental manipulations. Great care has to be taken that the ethanol precipitations are quantitative. Similarly, recovery of RNA or RNP from the glass plate after UV irradiation may pose a problem. Also, care has to be taken that all steps requiring the resuspension of a dry RNA pellet in buffer are performed with the necessary patience.

(3) The numerous manipulations required to achieve the aims of the experiments are also possible entry points for contaminations by RNases. Standard precautions have to be taken at the outset. The most important ones are: gloves must be worn at all times; when preparing solutions, only double-distilled or Millipore Q water should be used; all solutions should be sterilized by filtering through 0.2-µm nitrocellulose filters.

(4) The primer extension itself should be performed at least 3 times with RNA obtained from independent experiments and it may be necessary to use the RNA from one particular experiment with two different primers.

(5) The most critical point of the entire analysis is the immunoprecipitation combined with primer extension analysis. False-positive results are obtained if immunoprecipitation is performed under conditions where the RNP complexes are not fully disrupted. For example, the U5 snRNP-specific proteins 40K, 116K, 200K and 220K form a remarkably stable heteromeric protein complex [23]. Using immunoprecipitation combined with primer extension analysis, we demonstrated that the U5 220K protein crosslinks to loop 1 of U5 snRNA [1]. However, immunoprecipitation under less stringent conditions compared to Protocol 5 [e.g. dissociation in 0.05% (v/v) SDS; see Fig. 21.4] resulted in co-precipitation of crosslinked RNA with antibodies against each of the proteins and thus a comparable reverse transcriptase pattern for the four proteins (data not shown). A similar situation was observed when crosslinking sites in the highly conserved Sm site of U1 snRNPs were analyzed. The seven U snRNP-specific Sm proteins form a highly stable heteromeric ring-like structure that interacts with the Sm site [24]. Immunoprecipitation combined with primer extension analysis under our standard conditions [1% SDS (v/v), 5% (v/v) Triton X-100] revealed a similar crosslinking pattern of the SmF and SmG protein to U128. To demonstrate that only SmG is in contact with U128 in the

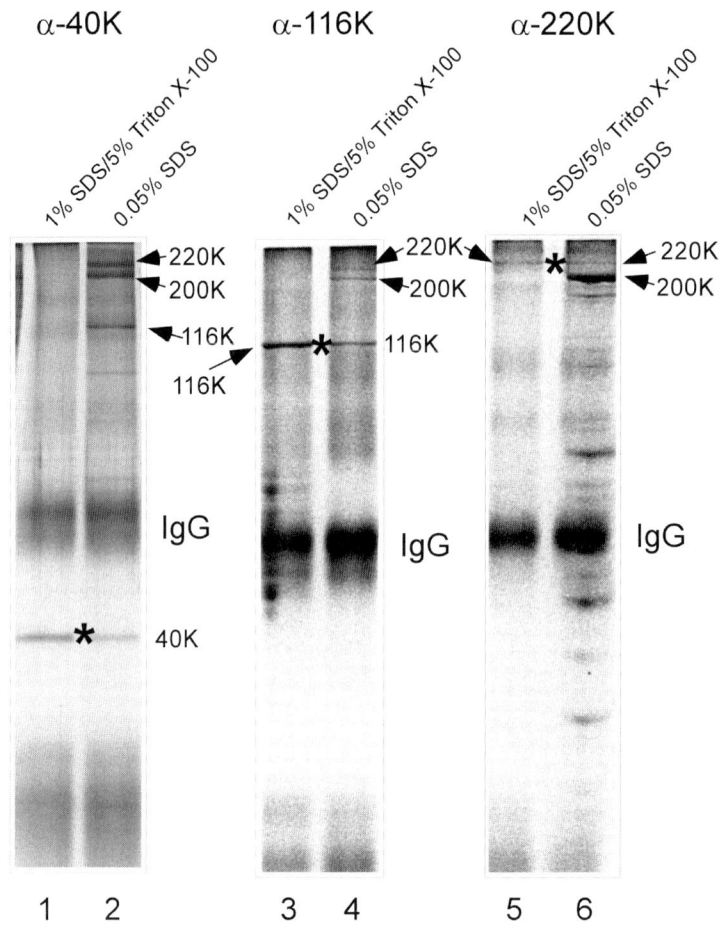

Fig. 21.4. Immunoprecipitation of single U5 snRNP-specific proteins from the U5 snRNP through dissociation of the U5 snRNPs by SDS/Triton X-100. U5 snRNPs were dissociated in 1% SDS and 5% Triton X-100 (lanes 1, 3 and 5, see Protocol 5.3) or in the presence of 0.05% (v/v) SDS only (lanes 2, 4 and 6). The immunoprecipitation was performed with covalently coupled antibodies (see Protocol 5.2) against the U5 snRNP specific proteins 40K (α-40K), 116K (α-116K) and 220K (α-220K). Proteins were visualized by silver staining. IgG: residual antibody released from the beads. Proteins precipitated under stringent conditions are marked with an asterisk.

U1 snRNA, we had to raise the SDS concentration to 2% (v/v) to allow complete dissociation of SmF and SmG proteins [2, 3]. The high specificity of our immunoprecipitation procedure is demonstrated in Fig. 21.4 for the precipitation of the above-mentioned U5 snRNP-specific proteins 40K, 116K and 220K. The silver-stained SDS–polyacrylamide gels of the precipitated proteins shows that a single protein can only be precipitated upon denaturation in the presence of 1% SDS with subsequent addition of Triton X-100 to 5%.

(6) Another important aspect is the reliability of the reactivity of the antibodies under these harsh conditions. To exclude the possibility that negative results (i.e. failure to detect protein-RNA crosslinks) are due to a poor reactivity of the antibodies, each antibody can be tested for its capability to precipitate a single protein from the RNPs (see above). Protocol 5 describes all steps necessary to visualize the precipitated proteins on a silver-stained SDS–polyacrylamide gel. Here, it is essential to couple the antibodies covalently to the beads in order to minimize the IgG background (Protocol 5.2).

21.7 Protocols

Protocol 1: UV irradiation of RNPs

(1) Starting materials are purified snRNP particles [5, 6], spliceosomal complexes [25–29] or reconstituted RNP particles [4]. The sample concentration is adjusted to approximately 0.1 mg/ml. Any buffer is suitable, provided it conforms to the criteria as stated above [Section 21.3 (2)]. The particles that we analyze are in 20 mM Tris–HCl (pH 7.0), 370 mM NaCl, 1.5 mM $MgCl_2$, 0.5 mM DTT (in case of U1 snRNPs) or in 20 mM HEPES–KOH (pH 7.9), 1.5 mM $MgCl_2$, 250 mM NaCl, 0.5 mM DTT, 0.2 mM EDTA (in case of 25S[U4/U6.U5] tri-snRNPs).
(2) Divide the sample into droplets of 25 µl and place the droplets carefully onto pre-cooled 10-well multitest slides (ICN Biomedical, USA). Ensure that the droplets stay intact and do not spread over the slide.
(3) Irradiate for 2 min at 254 nm at a distance of 4 cm from the UV source. We use a custom-made holder with four 8-W germicidal lamps (G8T5, Herolab, Germany) mounted in parallel, 4 cm apart (Fig. 21.1).
(4) Carefully pipette the droplets from the glass slide back into a new tube. It is essential that recovery of the sample from the slide is as complete as possible.
(5) Recover the RNA from crosslinked complexes as outlined in Protocol 2 or perform the immunoprecipitation (Protocol 5) before.

Protocol 2: RNA recovery from UV-irradiated RNPs

(1) To 50 µl of the pooled irradiated samples obtained in Protocol 1, add 40 µl of the buffer in which the native particles were initially purified.
(2) Add SDS to a final concentration of 1% (v/v), by adding 10 µl 10% SDS (v/v) to the above volume.
(3) Incubate for 10 min at 70 °C with gentle agitation and then allow the sample to cool down to room temperature over a period of 5 min.
(4) First add EDTA to 7.5 mM (1.5 µl of 0.5 M EDTA to above volume), and then add proteinase K to around 1 mg/ml (w/v) [10 µl proteinase K (10 mg/ml; Roche, Germany) to above volume]. Incubate the samples for a minimum of 30 min at 37 °C.

(5) Extract the RNA by adding 100 µl phenol/chloroform/isoamyl alcohol (PCI; Roth, Germany) and subsequent vigorous shaking for 2 min. Centrifuge (5 min, 13 000 r.p.m., 10 000 g) and transfer the aqueous phase containing the RNA to a new tube.

(6) Add 20 µg glycogen (Roche, Germany) and 1/10 of the sample volume 3 M sodium acetate, pH 5.3, and precipitate in 3 sample volumes of ethanol (p.a. grade; Merck, Germany) for a minimum of 2 h at −20 °C.

(7) Collect the RNA by centrifugation (20 min, 13 000 r.p.m. at 4 °C) and discard the supernatant. RNA recovery is monitored by inspection of the glycogen pellet, which must be clearly visible.

(8) Dissolve the RNA in 100 µl 0.3 M NaOAc, pH 5.3, and precipitate once more in 3 volumes ethanol for a minimum of 2 h at −20 °C. Collect the RNA by centrifugation (see Step 7) and dry the sample for 3 min in a Speed Vac.

(9) Dissolve the RNA in 6.5 µl CE buffer (10 mM cacodylic acid–KOH, pH 7.0, 0.2 mM EDTA, pH 8.0) with shaking for 10 min. The RNA is stored at −20 °C.

Protocol 3: UV irradiation of naked RNA

(1) As starting material, use twice as much naked RNA as that contained in the corresponding RNP particle employed in Protocol 1 (Step 1) to compensate for loss of RNA on the glass slides during UV irradiation.

(2) Perform the proteinase K digestion and RNA extraction essentially as described in Protocol 2, Steps 2–8, except that glycogen is omitted.

(3) Dissolve the RNA in 50 µl of the buffer used for the RNP (see Protocol 1, step 1).

(4) Perform UV irradiation and sample recovery exactly as described in Protocol 1, Steps 2–5.

(5) Further processing of the samples is as described in Protocol 2, Steps 6–9. In those cases where the starting material was not doubled, the RNA is resuspended in 3.5 µl instead of 6.5 µl CE buffer.

(6) Primer extension is performed as outlined in Protocol 4.

Protocol 4: Primer extension analysis

Protocol 4.1: Purification of the primer
The primer is obtained from any commercial source. It must be gel-purified for reproducibly clean primer extension reactions. Approximately 5 nmol of the primer is first dissolved in 100 µl of 80% formamide, 0.5 × TBE, 0.001% xylene cyanol and 0.001% bromophenol blue, and denatured at 96 °C for 3 min. After cooling to room temperature, it is loaded onto a 25-mm wide slot of a 1 mm thick, 15–20-cm long, denaturing 20% polyacrylamide/8.3 M urea gel and electrophoresed at approximately 1.5 W/cm. Electrophoresis time depends on the primer length (2 h for a 24mer). The region of the gel, which contains the primer, is identified by UV shadowing and excised. The gel slice is wetted with elution buffer (20 mM Tris–HCl, pH 7.5, 0.2 mM EDTA, pH 8, 0.15 M NaCl, 0.5% SDS) and cut into small cubes, which are subsequently transferred to a tube and overlaid with 300–500 µl

elution buffer. Elution is performed by diffusion (16 h at 30 °C). The eluate is recovered, extracted with phenol/chloroform, and precipitated twice essentially as described for the RNA extraction in Protocol 2, Steps 5–9, except that 300 µl PCI is used (Step 5), and that the primer is dissolved at 5 pmol/µl CE (Step 9).

Protocol 4.2: $5'$-^{32}P-labeling of the primer

(1) For one 10-µl reaction, the following components are mixed: 2 µl (10 pmol) of purified DNA oligonucleotide, 3 µl CE buffer, 1 µl 10 × T4 polynucleotide kinase (PNK) buffer (0.7 M Tris–HCl, pH 7.6, 0.1 M MgCl$_2$, 0.05 M DTT), 1 µl T4 PNK (New England Biolabs, USA), and 6 µl [γ-^{32}P]ATP (6000 Ci/mmol; Amersham Biosciences, UK). Incubate for 40 min at 37 °C.
(2) 50 µl CE buffer are added to the reaction and unincorporated nucleotides are removed by G-50 or G-25 Sephadex spin column chromatography (Amersham Biosciences, UK) according to the manufacturer's instructions. The volume is adjusted to 100 µl with CE and the extent of incorporation is determined (usually about 0.8 to 1.0×10^6 c.p.m. per pmol of primer).
(3) Residual protein and other impurities are removed essentially as described in protocol 2, Steps 5–9 with the following changes: (i) 10 µg glycogen is used in Step 6; (ii) the labeled primer is resuspended in 40 µl CE buffer (Step 9).

Protocol 4.3: Primer extension reaction

The following different samples are probed by primer extension analysis: (i) UV-irradiated RNAs from Protocols 2 and 3; (ii) non-irradiated RNA; (iii) crosslinked RNA isolated after immunoprecipitation; (iv–vii) template RNAs (either native RNA isolated according to Protocol 2 or RNA transcribed *in vitro* with bacteriophage RNA polymerases from an appropriate plasmid template) for the sequencing reactions used as markers. Template RNAs for marker synthesis should have a concentration of 0.2 pmol/µl. The experimental procedure for the primer extension closely follows that described in [30].

(1) For each RNA sample to be analyzed, 1.5 µl of a hybridization mix (HY) is required. It is composed of 0.25 µl 10 × hybridization buffer (0.5 M Tris–HCl, pH 8.4, 0.6 M NaCl, 0.1 M DTT), 0.5 µl ^{32}P-labeled DNA oligonucleotide, and 0.75 µl H$_2$O. Enough HY mix for the number of samples to be processed must be prepared.
(2) To anneal the primer, 1 µl of the RNA is first mixed with 1.5 µl of the HY mix, then heated for 60 s at 96 °C, and allowed to cool at room temperature for 5 min. Samples are briefly centrifuged.
(3) 1 µl of ddNTP is added to each of the four marker RNA samples (0.5 mM ddGTP, ddATP, ddTTP or ddCTP; Amersham Biosciences, UK).
(4) For each RNA sample, 2.5 µl of a reverse transcriptase mix (RT) is now required. It is composed of 0.25 µl 10 × reverse transcriptase buffer (0.5 M Tris–HCl, pH 8.4, 0.1 M MgCl$_2$, 0.6 M NaCl, 0.1 M DTT), 0.1 µl dNTPs (5 mM

each dGTP, dATP, dTTP and dCTP; Amersham Biosciences, UK), 0.08 µl (about 2 U) reverse transcriptase (30 U/µl, Seikagaku, Japan) and 2.07 µl H_2O. Enough RT mix for the number of samples to be processed must be prepared. 2.5 µl RT mix is added per sample, mixed and incubated for 45 min at 42.5 °C. A hybridization oven is recommended to avoid condensation at the lid of the tube.

(5) 6.5 µl loading buffer [8.3 M urea, 0.5 × TBE, 0.001% (w/v) bromophenol blue, 0.001% (w/v) xylene cyanol] is added to all samples, except for the markers, which receive 10 µl. Samples can be stored at −20 °C for at least 1 week.

Protocol 4.4: Gel electrophoresis

The transcribed cDNA products are analyzed on a 9.6% polyacrylamide (acrylamide:bisacrylamide, 19:1)/8.3 M urea gel in 1 × TBE in a Gibco/BRL Model S2 apparatus (0.5-mm thick gel) with 1 × TBE as electrophoresis buffer. Pre-electrophoresis is for 30 min at 65 W. Electrophoresis is at 65 W for a time depending on the length of the primer (approximately 2 h for a 24mer). For autoradiography, the sequencing gels are first transferred to a used X-ray film for support and covered with kitchen wrapping film. Alternatively, sequencing gels can be fixed in 40% methanol/10% acetic acid, transferred to Whatman 3MM paper and dried under vacuum (Bio-Rad model 583 gel dryer). A BioMax film (Kodak) is exposed to the gel at −70 °C for 1–10 days in the presence of intensifying screens. The long exposure times are required when performing the immunoprecipitation experiments combined with primer extension analysis, because of the inherently low yields of immunoprecipitation.

Protocol 5: Immunoprecipitation of the RNA-protein crosslinks

Protocol 5.1: Non-covalent coupling of antibodies to Protein A–Sepharose

Immunoprecipitation was found to be optimal with per assay 15 µl packed matrix volume of Protein A–Sepharose beads (Amersham Biosciences, UK) coupled with antiserum. Depending on the number of samples that are assayed, proportionally more bead slurry can be coupled with correspondingly increased amounts of antiserum. The coupled beads can be distributed afterward between the different tubes.

(1) For coupling of the antibody to beads an amount of slurry (30 µl) corresponding to 15 µl of beads (packed volume) is taken and washed 3 times with 500 µl aliquots of PBS (20 mM Na_2HPO_4, pH 8.0, 130 mM NaCl).

(2) The antiserum is diluted with PBS to 500 µl and added to the washed beads. Normally, 50 µl of antiserum is sufficient for one immunoprecipitation, but this volume may have to be adjusted, depending on the titer of the antiserum. Coupling is performed overnight by head-over-tail rotation at 4 °C.

(3) After coupling, beads are washed 3 times with 500 µl PBS. Tubes are changed by transferring the beads with the last washing aliquot to a new tube using a plastic pipette tip with a cut-off end. The washed beads with the coupled antibody are then overlaid with 15 µl PBS and kept on ice until use.

Protocol 5.2: Covalent coupling of antibodies to Protein A–Sepharose

Covalent coupling of antibodies is recommended when the capability of the antibodies to selectively precipitate a single protein from an RNP under semi-denaturing conditions (see Protocol 5.3) is tested, in order to exclude the possibility that negative results are due to the poor reactivity of the antibodies under these conditions [see also Troubleshooting (6)].

For covalent coupling it is recommended to increase the total amount of Protein A–Sepharose beads and the amount of antiserum is usually twice the volume of the beads.

(1) 30 µl Protein A–Sepharose beads are washed with PBS (Protocol 5.1) and then incubated with 60 µl of antiserum in a final volume of 500 µl PBS, 0.05% (v/v) Nonidet P-40 (NP-40) with head-over-tail rotation overnight at 4 °C.
(2) The beads are washed 5 times with 500 µl PBS, 0.05% NP-40 at 4 °C.
(3) The antibody-coupled beads are equilibrated 2 times with 300 µl 200 mM Na borate (pH 9.0) at room temperature.
(4) Crosslinking of the antibodies to the beads is achieved by incubation with 500 µl DMP (dimethyl pimelinidate dihydrochloride; Sigma, USA) at a final concentration of 5.2 mg/ml in 200 mM Na borate for 1 h at room temperature with head-over-tail rotation. Note that the pH of the solution must be above 8.3.
(5) The supernatant is removed as completely as possible and the reaction is stopped by addition of 300 µl 0.2 M ethanolamine–HCl, pH 8.0, to the beads and further incubation with head-over-tail rotation for 1 h.
(6) The beads are then washed with PBS (Protocol 5.1) and residual non-crosslinked antibodies are removed by three additional washes with 500 µl 0.1 M glycine–HCl, pH 2.7.
(7) After a final wash with PBS containing 0.02% NaN_3, the slurry can be stored for at least 6 months at 4 °C.
(8) The covalently coupled antibody beads are now used in the immunoprecipitation exactly as described in Protocol 5.3. After the final wash beads are incubated with an appropriate volume of SDS sample buffer [125 mM Tris–HCl, pH 6.8, 1% (v/v) SDS, 5% (v/v) glycerol, 10 mM DTT, 0.005% (w/v) bromophenol blue] and heated for 5 min at 70 °C. Beads are spun down with maximum speed (10 000 g) and the supernatant is loaded onto an SDS–PAA gel [31].
(9) Proteins are visualized by silver staining according to [32].

Protocol 5.3: Dissociation of RNP particles and immunoprecipitation

For the immunoprecipitation experiments native or reconstituted RNP particles in a volume of 50 µl of appropriate buffer (for buffer conditions see Protocol 1, Step 1) are used. For the reliable assignment of crosslinks it is essential to include a sample that was not UV-irradiated, but otherwise treated in an identical manner.

(1) Add SDS to a final concentration of 1% (v/v) to the samples from Protocol 1 and incubate for 10 min at 70 °C on a shaker. Use 2% SDS (v/v) in those cases

where the protein–protein interactions are known or were found to be extremely strong (see above).

(2) Allow the samples to cool at room temperature for 5 min. Then, add Triton X-100 (density 1.06 g/l, molecular biology grade; Sigma, USA) to a final concentration of 5% (v/v). Use of the concentrated Triton X-100 stock solution is necessary to keep the final volume as low as possible. Gently mixing is necessary to completely dissolve the added Triton X-100, which initially forms a separate phase at the bottom of the tube.

(3) Adjust the sample volume to 350 µl with PBS and add the mixture to the prepared antibody-coupled beads (Protocol 5.1). Incubate with head-over-tail rotation for 1–1.5 h at 4 °C.

(4) Wash the samples 4 times with 500 µl aliquots of PBS and transfer the slurry into a new tube at the fourth washing step. Wash the beads once more with 500 µl PBS. Carefully check recovery of the beads during the washing procedure by inspecting the amount of beads visible in the tube after each step; any loss of material must be avoided.

(5) Remove the supernatant as completely as possible, then add 90 µl of buffer (see Protocol 1, Step 1) and proceed with proteinase K digestion and RNA recovery essentially as described in Protocol 2, Steps 2–9, except that shaking is for 5 min (Step 5) and that the RNA is dissolved in 3.5 µl (Step 9).

(6) Proceed with the primer extension as outlined in Protocol 4.

Acknowledgments

We thank our colleagues for their excellent technical assistance at the time when we established our method, in particular Peter Kempkes for HeLa cell fermentation, Axel Baduin and Winfried Lorenz for the snRNP preparations, and Irene Öchsner for her expertise with the snRNP-specific antibodies. This work is supported by a grant from the BMBF (031U251B), Fonds der Chemischen Industrie and DFG Forschergruppe (LU 294/12-1) to R. L.

References

1 H. Urlaub, K. Hartmuth, S. Kostka, G. Grelle, R. Lührmann, *J. Biol. Chem.* 2000, 275, 41458–41468.
2 H. Urlaub, V. Raker, S. Kostka, R. Lührmann, *EMBO J.* 2001, 20, 187–196.
3 H. Urlaub, K. Hartmuth, R, Lührmann, *Methods* 2002, 26, 170–181.
4 S. Nottrott, H. Urlaub, R. Lührmann, *EMBO J.* 2002, 21, 5527–5538.
5 C. L. Will, B. Kastner, R. Lührmann, Analysis of ribonucleoprotein interactions in: *RNA Processing, Vol. I, A Practical Approach*, S. J. Higgins, B. D. Hames (eds), IRL Press, Oxford, 1994.
6 B. Kastner, Purification and electron microscopy of spliceosomal snRNPs, in: *RNP Particles, Splicing and Autoimmune Diseases*, J. Schenkel (ed.), Springer, Berlin, 1998.
7 C. B. Burge, T. Tuschl, P. A. Sharp,

Splicing of precursors to mRNA by the spliceosome, in: *The RNA World*, R. F. GESTELAND, T. R. CECH, J. F. ATKINS (eds), Cold Spring Harbor Laboratory Press, Cold Spring Harbor, NY, **1999**.

8 H. URLAUB, V. KRUFT, O. BISCHOF, E. C. MÜLLER, B. WITTMANN-LIEBOLD, *EMBO J*. **1995**, *14*, 4578–4588.

9 H. URLAUB, B. THIEDE, E. C. MÜLLER, R. BRIMACOMBE, B. WITTMANN-LIEBOLD, *J. Biol. Chem*. **1997**, *272*, 14547–14555.

10 K. M. MEISENHEIMER, P. L. MEISENHEIMER, T. H. KOCH, *Methods Enzymol*. **2000**, *318*, 88–104.

11 R. BRIMACOMBE, W. STIEGE, A. KYRIATSOULIS, P. MALY, *Methods Enzymol*. **1988**, *164*, 287–309.

12 B. S. COOPERMAN, *Methods Enzymol*. **1988**, *164*, 341–361.

13 M. J. MOORE, P. A. SHARP, *Science* **1992**, *256*, 992–997.

14 J. R. WYATT, E. J. SONTHEIMER, J. A. STEITZ, *Genes Dev*. **1992**, *6*, 2542–2553.

15 I. DIX, C. S. RUSSELL, R. T. O'KEEFE, A. J. NEWMAN, J. D. BEGGS, *RNA* **1998**, *4*, 1675–1686.

16 R. REED, M. D. CHIARA, *Methods* **1999**, *18*, 3–12.

17 M. M. KONARSKA, *Methods* **1999**, *18*, 22–28.

18 Y. T. YU, *Methods Enzymol*. **2000**, *318*, 71–88.

19 B. RHODE, K. HARTMUTH, H. URLAUB, R. LÜHRMANN, *RNA* **2003**, *9*, 1542–1551.

20 C. OUBRIDGE, N. ITO, P. R. EVANS, C. H. TEO, K. NAGAI, *Nature* **1994**, *372*, 432–438.

21 S. NOTTROTT, K. HARTMUTH, P. FABRIZIO, H. URLAUB, I. VIDOVIC, R. FICNER, R. LÜHRMANN, *EMBO J*, **1999**, *18*, 6119–6123.

22 J. SAMBROOK, E. F. FRITSCH, T. MANIATIS, *Molecular Cloning: A laboratory Manual*, 2nd edn, Cold Spring Harbor Laboratory Press, Cold Spring Harbor, NY, **1989**.

23 T. ACHSEL, K. AHRENS, H. BRAHMS, S. TEIGELKAMP, R. LÜHRMANN, *Mol. Cell. Biol*. **1998**, *18*, 6756–6766.

24 C. KAMBACH, S. WALKE, R. YOUNG, J. M. AVIS, E. DE LA FORTELLE, V. A. RAKER, R. LÜHRMANN, J. LI, K. NAGAI, *Cell* **1999**, *96*, 375–387.

25 R. REED, L. PALANDIJAN, Spliceosome assembly, in: *Eukaryotic mRNA Processing*, A. R. KRAINER (ed.), IRL Press, Oxford, **1997**.

26 M. S. JURICA, M. J. MOORE, *Methods* **2002**, *28*, 336–345.

27 E. M. MAKAROV, O. V. MAKAROVA, H. URLAUB, M. GENTZEL, C. L. WILL, M. WILM, R. LÜHRMANN, *Science* **2002**, *298*, 2205–2208.

28 K. HARTMUTH, H. URLAUB, H.-P. VORNLOCHER, C. L. WILL, M. GENTZEL, M. WILM, R. LÜHRMANN, *Proc. Natl. Acad. Sci. USA* **2002**, *99*, 16719–16724.

29 K. HARTMUTH, H.-P. VORNLOCHER, R. LÜHRMANN, *Methods Mol. Biol*. **2004**, *257*, 47–64.

30 A. J. ZAUG, T. R. CECH, *RNA* **1995**, *1*, 363–374.

31 U. K. LAEMMLI, *Nature* **1979**, *227*, 680–685.

32 H. BLUM, H. BEIER, H. J. GROSS, *Electrophoresis* **1987**, *8*, 93–99.

22
Probing RNA Structure by Photoaffinity Crosslinking with 4-Thiouridine and 6-Thioguanosine

Michael E. Harris and Eric L. Christian

22.1
Introduction

Chemical crosslinking, including photoaffinity crosslinking, has been widely used to gain insight into structures associated with the biological function of large, structurally complex RNAs and ribonucleoproteins (RNPs). Examples include analysis of catalytic RNAs and the major cellular RNPs, the ribosome [1–4] and the spliceosome [5–9]. Combined with continuing improvements in the ability to generate RNAs with site-specific modifications, crosslinking continues to be a key analytical method for investigating structure–function relationships. If carried out with due care, crosslinking experiments can establish that specific residues are (or were) proximal when the crosslinking reaction occurred. Thus, when applied in a targeted way this information together with kinetic and thermodynamic studies of structure variants can be used to reveal residues involved in catalysis and molecular recognition. If sufficient information is available from other biochemical and comparative analyses, it can be possible to use the information gained from crosslinking as constraints for molecular modeling [1, 10–14]. Although the resolution of structures obtained this way is necessarily low (generally of the order of ± 10 Å), they present an explicit context for designing new structure–function experiments and for interpreting structural information.

Although a wide variety of chemical and photo-crosslinking reagents are available, 4-thiouridine and 6-thioguanosine are excellent choices due to their simple molecular structure, relative stability and high reactivity (Fig. 22.1) [15–20]. s^4U and s^6G introduce only minimal perturbations of the native structure since they differ from their corresponding "parent" nucleoside by a single atomic substitution, the replacement of a nucleobase oxygen by sulfur. This substitution renders the reagent sensitive to UV light and exposure yields reactive sulfur radical that can react efficiently with functional groups that are in proximity. Crosslinking reactions involving these reagents can be very efficient, making it an easier task to isolate sufficient quantities of crosslinked species for mapping of crosslinked nucleotides and assessment of retention of biological activity. Additionally, these reagents are advantageous in that they are relatively short range (around 3 Å), and thus in

Handbook of RNA Biochemistry: Student Edition. Edited by R. K. Hartmann, A. Bindereif, A. Schön, E. Westhof
Copyright © 2009 WILEY-VCH Verlag GmbH & Co. KGaA, Weinheim
ISBN: 978-3-527-32534-4

6-thioguanidine (s⁶G)   4-thiouracil (s⁴U)

Fig. 22.1. Structures of 6-thioguanidine and 4-thiouracil.

principle provide spatial information that is higher resolution than, for example, azido derivatives that generally introduce a linker between the RNA and photoagent that can be 10 Å or greater.

There are several excellent and up to date reviews available that describe methods for generation and incorporation of these and other photoaffinity reagents including chapters in this work [2, 21–25]. Because the choice of crosslinking reagent and method of incorporation will depend on the specific experimental application, the reader is referred to these important resources. Here, we will focus on simple procedures and considerations for generating and isolating crosslinked RNAs, and for primer extension mapping of crosslinked nucleotides. In the examples given below, crosslinking is applied to identify active site components within the RNase P ribozyme–substrate complex [16, 26]. The description is designed to be sufficiently general in order to be of maximum use as a guideline for an experimenter at least at the graduate level who is considering the application of photocrosslinking of RNA in their research. However, a basic understanding of techniques for handling nucleic acids is assumed.

It is important to note that the descriptions included here are by necessity brief and only a starting point because of the significant condition dependence of the crosslinking reactions and variability in the physical behavior of different RNAs. It cannot be overemphasized that achieving efficient crosslinking and obtaining clean and convincing primer extension mapping data will require significant effort toward optimization of different experimental parameters. In this section we attempt to describe the logic behind the choice of the basic experimental parameters we have used, and to illustrate the experimental constraints and controls necessary for interpretation of crosslinking data in terms of biological function.

As with any experimental approach it is important to first consider the difficulties inherent to its application and limitations to interpretation of the data. Despite its conceptual simplicity, successfully applying any crosslinking approach can be sometimes difficult and time consuming. Despite one's best efforts at optimization, the crosslinking reaction itself can be inefficient due to inherently unfavorable geometry or the chemical environment at the site of photoagent incorporation. Although methods using radioactive labeling that have good sensitivity are used to map crosslink sites, the clearest and best results are obtained when nanogram quantities of the crosslinked species can be obtained. Similarly, it can sometimes be hard to generate high-quality primary data since several manipulations of RNA are required (i.e. photoagent modification, crosslinking, gel purification, etc.). As described in more detail below, interpretation of primer extension–termination mapping of crosslinked sites can be difficult and great care must be taken to insure that the data truly reflect the formation of novel crosslinks. Much of the ambiguity can be resolved with the appropriate controls, and the most important considerations in this regard are outlined below.

Because of these issues, it is critical to test in the most direct way possible that the crosslinking data reflects the functional form or native folded structure of the RNA and to consider what experimental evidence can be brought to bear to establish the functional relevance of the data set. Optimally, this goal can be achieved by assaying directly whether the crosslinked RNA retains biological activity. However, this can be a problem when probing the functional core of an RNA since the crosslink itself can alter chemical groups important for biological function. Alternatively, the proximity data from crosslinking can be considered in light of other structural constraints from, for example, phylogenetic comparative studies, chemical and enzymatic probing and high-resolution structures of homologous molecules.

Important new insights into the validity as well as the limitations of crosslinking as an approach for exploring RNA structure comes from the comparison of the recent three-dimensional structure of the ribosome and the extensive collection of biochemical structure probing data [1]. Overall a large percentage of the crosslinking data were consistent with the structure from X-ray crystallography; however, the resolution of the structural information was less than expected given the chemical structure and size of the different crosslinking reagents used. Furthermore, no individual crosslinking reagent appeared to be superior with respect to validity of the data; however, the method of detection did have an important impact since most of the lower quality data was obtained by primer extension mapping. Most likely this limited accuracy is due to misidentification of non-specific terminations as crosslink sites. The highest quality data was obtained by direct physical mapping of the crosslinked nucleotides, underscoring the importance of optimization of the crosslinking procedure. Despite this track record, primer extension mapping is still a convenient method due to its sensitivity and flexibility; however, obtaining more direct data such as gel mobility, RNase H mapping or optimally by fingerprinting is obviously desirable.

Despite these limitations and considerations, crosslinking approaches can provide important structural information in those numerous instances when it is impossible to obtain material in adequate amounts or in sufficient purity for high-resolution structural analysis. Often it is desirable to probe structure in a context such as within cell extracts where high-resolution studies are impractical. In principle, crosslinking reports on the structure or structures as they occur in solution and in instances when conditions can be found to favor one conformation over another, it can be possible to use crosslinking to define the characteristic structural features of these different states. Perhaps the greatest advantage is sensitivity, since relatively small amounts of crosslinked material are needed for mapping. Once the sites of crosslinking are defined, the formation of a specific crosslink can be used analytically, again with high sensitivity using radiolabeled RNA.

22.2 Description

22.2.1 General Considerations: Reaction Conditions and Concentrations of Interacting Species

It is important to take into consideration that RNA structure, and thus its biological activity, can be highly dependent on reaction conditions. Individual RNAs often adopt multiple conformations and obviously it is necessary that the crosslinking experiment be performed under conditions that favor the correct structural form or the structure of interest [27–29]. Therefore, it must be considered how to optimally fold the RNA sample prior to initiating the crosslinking reaction. Thus, it is important to have as detailed an understanding as possible about the influence of mono- and divalent ion concentrations and identity as well as pH on the biological activity and RNA structure. It is also useful to examine the effect of these parameters on the crosslinking reaction as well, since gaining the highest efficiency possible is important for subsequent identification of crosslinked nucleotides and analysis of the retention of biological activity of the purified crosslinked species.

Crosslinking is very useful for initial analysis of intermolecular interactions, and can be used to define the potential interface between two RNAs or between RNA and a specific protein. Because of the aforementioned penitent for misfolding and condition dependence, RNAs can self-associate or bind in non-productive ways. Similarly, even specific RNA-binding proteins can interact weakly with RNA in a non-specific fashion. Thus, it is important to consider the relative concentrations of the interacting species in the reaction to minimize the potential for formation of non-specific complexes. Examining the effect of macromolecular concentration on the crosslinking reaction can provide insight into whether the information gained accurately reflects formation of high affinity or biologically active complexes.

In the following example the interaction between the RNase P ribozyme and its substrate were examined using intermolecular crosslinking with s^4U- and s^6G-modified tRNA precursors (pre-tRNA) (Fig. 22.2) [16, 26]. RNase P is a widespread and essential ribonucleoprotein enzyme that generates the 5′ end of mature tRNAs via a site-specific phosphodiester bond hydrolysis reaction [30–32]. In bacteria, RNase P enzymes are heterodimers composed of a small, but essential, protein subunit and a larger RNA component that is the catalytic subunit of the enzyme. Whereas most ribozymes catalyze self-cleavage or self-splicing reactions and have to be engineered to work in *trans*, for RNase P RNA catalysis of a multiple turnover reaction is intrinsic to its biological function. Although RNase P, like many other RNA processing RNPs, can recognize a broad spectrum of substrates, the mechanistic basis for its multiple substrate recognition properties is not clearly defined.

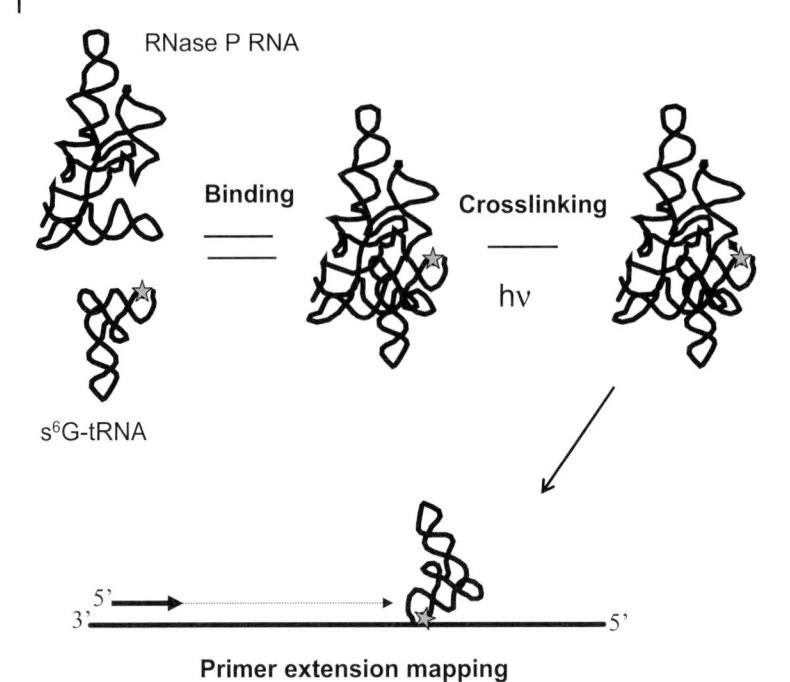

Fig. 22.2. Overview of photoaffinity crosslinking and primer extension mapping. The RNase P RNA is represented in this example as a black ribbon diagram. The photoagent modified pre-tRNA substrate is shown in grey. The position of the photoagent is indicated by a star. As described in the text, the two RNAs are allowed to bind (Binding) and the photoagent is activated by exposure to the appropriate wavelength of UV light (Crosslinking). Subsequently, the appropriate crosslinked RNA species are isolated, generally by gel purification and the sites of crosslinking determined by primer extension mapping. The radiolabeled primer used in the reaction is indicated by an arrow.

Thus, the interactions between RNase P RNA and pre-tRNA substrates, in particular those that underlie specificity, continue to be the subject of considerable interest [33].

To identify residues in the RNase P ribozyme that are proximal to the substrate cleavage site we positioned s^4U and s^6G on either side of the reactive phosphodiester bond in a model tRNA precursor. Kinetic and thermodynamic studies demonstrated that the inclusion of the photoagent at the substrate cleavage site did not interfere with high affinity binding and that the modified substrate was processed at a rate that was essentially identical to the unmodified substrate. To insure proper folding of the two RNAs and efficient formation of the enzyme–substrate complex the following procedure was used. The RNAs are resuspended separately in reaction buffer, in this case (2 M ammonium acetate; 50 mM Tris–HCl, pH 8.0) for refolding. The RNA-containing solutions are heated to 90 °C for 1 min in a programmable heating block (MJ Research) and then cooled to room temperature using a standard water bath over a period of approximately 20 min. Divalent metal ions, in this example 25 mM CaCl$_2$, are added and the RNAs incubated at 37 °C for 15–30 min to insure as much of the RNA as possible has attained the native, folded form. Equal volumes of substrate and enzyme RNA are mixed and incubated for 2 min. In this instance Ca^{2+} is used to replace the optimal metal ion for the reaction, Mg^{2+}, in order to slow the rate of catalysis and permit the assessment of the binding affinity of the substrate [34].

Preparative intermolecular crosslinking reactions generally contained 100 nM photoagent-containing pre-tRNA and 1 µM RNase P ribozyme in order to insure that the majority of the photoagent-modified substrate was bound to the ribozyme. Importantly, it could be demonstrated that formation of crosslinks was dependent on the presence of the ribozyme and occurred in a concentration-dependent manner over a broad range of concentrations (Fig. 22.3). Additionally, the same

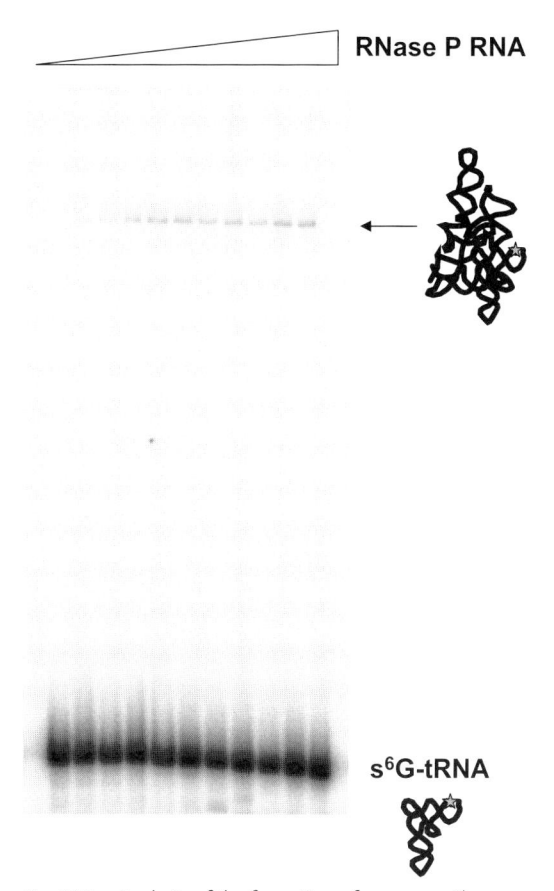

Fig. 22.3. Analysis of the formation of crosslinked species by gel electrophoresis. In this example, radiolabeled and photoagent modified pre-tRNA (s^{6}G-tRNA) was incubated with increasing concentrations of RNase P ribozyme and the reactions exposed to UV light. The formation of a single crosslinked species that requires the presence of the ribozyme and is dependent on its concentration is indicated by the arrow.

crosslinked species were detected at both high and low concentrations of the ribozyme. The concentration dependence clearly demonstrates that the crosslinks are intermolecular in nature and reflect the structure of high-affinity complexes between the two RNAs.

22.2.2
Generation and Isolation of Crosslinked RNAs

Once the conditions and concentrations of the reaction are set or optimized crosslinking is easily initiated by irradiation with the appropriate wavelength of light. Subsequently, the reactions are analyzed for the formation of new crosslinked species. Identification and isolation is almost always accomplished by taking advantage of the altered mobility of the crosslinked RNAs relative to uncrosslinked RNA on denaturing polyacrylamide gels. The crosslinked RNAs are subsequently eluted from the gel and recovered by ethanol precipitation using standard methods.

For analytical reactions in which the photoagent-modified pre-tRNA substrate was also radioactively labeled, aliquots of 12 µl were transferred to a parafilm covered aluminum block. A convenient source is the block from a standard dry-bath incubator, pre-cooled in ice for at least 1 h prior to the experiment. We found that crosslinking occurred optimally at 4 °C. Parafilm and samples were placed on the block just before irradiation to minimize dilution or contamination by condensation. The samples were irradiated for 5–15 min at 366 nm at a distance of 3 cm using a model UVGL-58 ultraviolet lamp from UVP, Upland, CA. A standard (3–4 mm) thick glass plate was placed between the lamp and the sample to help filter out shorter wavelengths of UV light that can damage the RNA sample. Aliquots were recovered from the block, diluted to 200 µl with 10 mM Tris–HCl, pH 8.0, 0.5 mM EDTA, 0.3 M sodium acetate, then extracted twice with 50/50% phenol/chloroform and once with chloroform alone and precipitated by addition of 3 volumes of ethanol.

Because both inter- and intramolecular crosslinking alter the linear topology of the targeted RNA, one can generally identify and isolate crosslinked species based on their slower mobility in denaturing acrylamide gels (Fig. 22.3). Appropriate controls should be run in parallel in which the photoagent is omitted from the reaction in order to demonstrate that the formation of the more slowly migrating species depends on presence of the crosslinking reagent and not from adventitious crosslinking due to ambient UV light. Similarly, control samples that are not irradiated must also be compared since crosslinking can occur during sample workup that may not necessarily reflect the functional structure. Additionally, for intermolecular crosslinking it is essential to demonstrate that formation of the crosslinked species requires the presence of the interacting partner RNA or protein and that its formation is concentration dependent.

Once the specificity of the crosslinking reaction is established, the next step is to isolate sufficient quantities of the individual crosslinked species to map the crosslink sites. Keeping in mind that picomole amounts of material will be optimal for primer extension mapping it is necessary to scale the crosslinking reaction up accordingly. We have had good success in simply "spiking" preparative reactions with a small quantity of radiolabeled RNA to use a marker for gel purification. For electrophoresis, the sample can be loaded in a continuous well across the top of the gel and electrophoresis conditions optimized using analytic reactions to achieve the best degree of separation between crosslinked and uncrosslinked RNA. Standard methods are acceptable for location of bands by autoradiography, excision of the appropriate gel slices and elution and recovery of the RNA. We have found that addition of 0.01 µg/µl glycogen as a carrier greatly improves recovery from larger volumes of gel elution buffer and does not interfere in subsequent primer extension mapping experiments.

22.2.3
Primer Extension Mapping of Crosslinked Nucleotides

The general principle behind primer extension mapping of crosslinked nucleotides is that polymerase will continue to synthesize a DNA stand up to, but not beyond the site of crosslinking due to chemical disruption of the template strand [15]. These specific terminations observed in reactions containing crosslinked RNA as a template and not observed in control, uncrosslinked RNA samples are interpreted as sites whether the photoagent has formed a new covalent bond (Fig. 22.2). Terminations are interpreted as occurring one nucleotide 5' to the site of crosslinking. A key advantage is that only relatively small amounts of template RNA are required (1–0.1 pmol); however, best results are obtained when at least picomolar amounts are available. Additionally, the technique is relatively easy, rapid and requires reagents and equipment that are widely available.

However, the biggest problem here is that reverse transcriptase will pause at specific sites on virtually any RNA template. Therefore, it is essential to distinguish between a termination that is due to a crosslink and one that is due to RNA structure, non-specific radiation damage or degradation. To control for these phenomena it is important to perform a parallel analysis of RNA taken through the protocol but not subjected to irradiation, as well as a control RNA sample irradiated in the absence of ligand, or with a ligand population that has not been modified with the photoagent. Additionally, corroborative information from RNase H mapping [22] or the mobility of the crosslinked species in denaturing gels [10] can be used to gain information on where crosslinks are located and this information can be important for resolving any potential ambiguities in the primer extension termination results.

For primer extension analysis of intermolecular crosslinks between photoagent-modified pre-tRNA and the RNase P ribozyme 0.2 pmol of $5'$-^{32}P-end-labeled primer are annealed to 0.05–0.2 pmol of gel-purified crosslinked RNA in a total volume of 5 µl. The annealing solution is composed of 50 mM Tris–HCl, pH 8.3, 15 mM NaCl and 10 mM dithiothreitol. Individual samples are heated to 65 °C for 3 min and then set immediately on dry ice. The annealed samples are thawed on ice and 1 µl of 30 mM $MgCl_2$ is added followed by addition of the four deoxynucleotide triphosphates to a final concentration of 400 µM. These reactions (8 µl) are initiated by the addition of 2 U (in 2 µl) of AMV reverse transcriptase (Boehringer Mannheim) and then incubated at 47 °C for 5 min. Reactions are then quenched by the addition of an equal volume of 0.5 M NaCl, 20 mM EDTA and 0.5 µg glycogen, and the extension products recovered by ethanol precipitation.

Primer extension products are then resolved next to dideoxy sequencing standards on denaturing polyacrylamide gels. The concentration of dideoxynucleotide added can be varied from 5 to 100 µM to detect nucleotides from less than 10 nt to more than several hundred away from the primer binding site. After recovery by ethanol precipitation and washing with 80% ethanol to remove excess salt the radiolabeled products are resuspended in a small volume (2–5 µl) of formamide loading buffer (95% formamide, 150 mM Tris–HCl, pH 8.0, 15 mM EDTA, and trace amounts of bromophenol blue and xylene cyanol FF). Only a fraction of the reaction (e.g. 2 µl) is loaded in an individual lane such that the sample just covers the bottom of an individual well in order to generate the sharpest banding pattern possible.

Obtaining clean primer extension results necessarily requires significant attention to optimization of the reaction parameters. We find that there can be large differences in the quality of data and the pattern of non-specific terminations with different primers, and some attempt to compare two or more sites of primer binding was necessary in some cases. Additionally, the optimal primer concentration is often also idiosyncratic to individual oligonucleotide sequences. An additional area in which the procedure can be optimized is in the annealing procedure where it is useful to compare slow cooling to rapid cooling. Additionally, we have found that increasing the reaction temperature can result in fewer non-specific transcription terminations, but temperatures in excess of 50 °C result in enzyme denaturation or inhibition. Despite the number of parameters that can be varied, generally a few days spent optimizing these few aspects of the procedure using control, uncrosslinked RNA will be time very well spent, since the payoff will be in obtaining clearer and therefore more convincing primary data. See Fig. 22.4.

22.3
Troubleshooting

The key problems associated with these particular methods are inefficient formation of crosslinked species and ambiguous or unclear primer extension mapping results. In the case of the former, locating the photoagent to a nearby region of the molecule may overcome unfavorable geometric constraints. Additionally, as mentioned above, it is usually important to optimize the folding of the RNAs of interest in order to insure that the maximum fraction of the sample is in the correctly folded and biologically active form. If the particular site of photoagent attachment is sufficiently interesting then trying additional, longer-range crosslinking agents such as phenylazides [23] can be used to increase crosslinking efficiency.

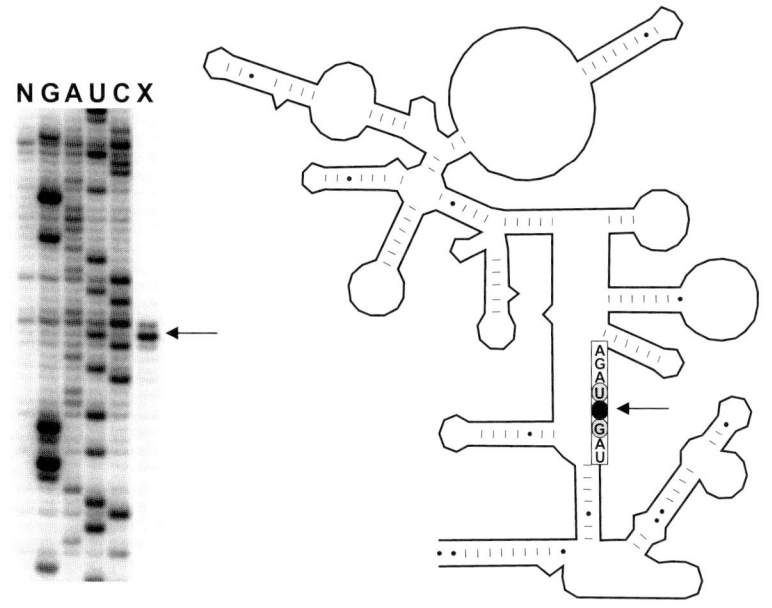

Fig. 22.4. Primer extension mapping. Products from reactions containing gel-purified crosslinked RNA (X) as well as control RNA from a control reaction that was not irradiated were resolved by gel electrophoresis (N). Lanes G, A, U and C contain products from sequencing reactions containing the appropriate dideoxynucleotide. An arrow indicates the position of the termination due to crosslink formation. A diagram of the secondary structure of *Bacillus subtilis* RNase P is shown on the right with the position of the crosslink indicated by an arrow.

One of the key difficulties that we have encountered is in protecting the photoagent modified RNA from ambient UV light. Although it can be somewhat awkward, working up the photoagent modified RNA and performing the crosslinking reactions in a darkened laboratory environment is important for obtaining the cleanest possible results.

In our experience, by far the most challenging aspect is to generate appropriately clean primer extension mapping data. As with all procedures involving RNA, making sure that the sample is not exposed to heating in the presence of metal ions is essential. Degradation of the RNA sample is apparent as an intense background of terminations in the control RNA primer extensions. Beyond repetition of the experiment to confirm the reproducibility of the results, good experimental technique and careful handling of the RNA is implicit. We have found that longer incubation times for the reverse transcriptase reaction can increase the background and thus it is important to consider assessing the effects of varying the reaction time. Also, some pilot experiments addressing what conditions are most appropriate for annealing of the radiolabeled primer will also pay off in the long run.

References

1 M. Whirl-Carrillo, I. S. Gabashvili, M. Bada, D. R. Banatao, R. B. Altman, *RNA* **2002**, *8*, 279.
2 D. I. Juzumiene, P. Wollenzien, *RNA* **2001**, *7*, 71.
3 P. V. Baranov, S. S. Dokudovskaya, T. S. Oretskaya, O. A. Dontsova, A. A. Bogdanov, R. Brimacombe, *Nucleic Acids Res.* **1997**, *25*, 2266.
4 T. Kao, D. L. Miller, M. Abo, J. Ofengand, *J. Mol. Biol.* **1983**, *166*, 383.
5 P. A. Maroney, Y. T. Yu, M. Jankowska, T. W. Nilsen, *RNA* **1996**, *2*, 735.
6 A. J. Newman, S. Teigelkamp, J. D. Beggs, *RNA* **1995**, *1*, 968.
7 D. A. Wassarman, J. A. Steitz, *Science* **1992**, *257*, 1918.
8 C. H. Kim, J. Abelson, *RNA* **1996**, *2*, 995.
9 Y. T. Yu, J. A. Steitz, *Proc. Natl Acad. Sci. USA* **1997**, *94*, 6030.
10 M. E. Harris, A. V. Kazantsev, J. L. Chen, N. R. Pace, *RNA* **1997**, *3*, 561.
11 J. L. Chen, J. M. Nolan, M. E. Harris, N. R. Pace, *EMBO J.* **1998**, *17*, 1515.
12 R. Pinard, D. Lambert, J. E. Heckman, J. A. Esteban, C. W. t. Gundlach, K. J. Hampel, G. D. Glick, N. G. Walter, F. Major, J. M. Burke, *J. Mol. Biol.* **2001**, *307*, 51.
13 A. Malhotra, S. C. Harvey, *J. Mol. Biol.* **1994**, *240*, 308.
14 F. Mueller, H. Stark, M. van Heel, J. Rinke-Appel, R. Brimacombe, *J. Mol. Biol.* **1997**, *271*, 566.
15 M. E. Harris, E. L. Christian, *Methods* **1999**, *18*, 51.
16 E. L. Christian, D. S. McPheeters, M. E. Harris, *Biochemistry* **1998**, *37*, 17618.
17 P. V. Sergiev, I. N. Lavrik, V. A. Wlasoff, S. S. Dokudovskaya, O. A. Dontsova, A. A. Bogdanov, R. Brimacombe, *RNA* **1997**, *3*, 464.
18 E. J. Sontheimer, *Mol. Biol. Rep.* **1994**, *20*, 35.
19 Y. L. Dubreuil, A. Expert-Bezancon, A. Favre, *Nucleic Acids Res.* **1991**, *19*, 3653.
20 A. Favre, C. Saintome, J. L. Fourrey, P. Clivio, P. Laugaa, *J. Photochem. Photobiol. B* **1998**, *42*, 109.
21 B. S. Cooperman, R. W. Alexander, Y. Bukhtiyarov, S. N. Vladimirov, Z. Druzina, R. Wang, N. Zuno, *Methods Enzymol.* **2000**, *318*, 118.
22 Y. T. Yu, *Methods Enzymol.* **2000**, *318*, 71.
23 A. B. Burgin, N. R. Pace, *EMBO J.* **1990**, *9*, 4111.
24 M. M. Konarska, *Methods* **1999**, *18*, 22.
25 B. C. Thomas, A. V. Kazantsev, J. L. Chen, N. R. Pace, *Methods Enzymol.* **2000**, *318*, 136.
26 E. L. Christian, M. E. Harris, *Biochemistry* **1999**, *38*, 12629.
27 O. C. Uhlenbeck, *RNA*, **1995**, *1*, 4.
28 D. K. Treiber, J. R. Williamson, *Curr. Opin. Struct. Biol.* **2001**, *11*, 309.
29 D. K. Treiber, J. R. Williamson, D. Thirumalai, N. Lee, S. A. Woodson, D. Klimov, *Curr. Opin. Struct. Biol* **2001**, *11*, 309.
30 J. Hsieh, A. J. Andrews, C. A. Fierke, *Biopolymers* **2004**, *73*, 79.
31 M. E. Harris, E. L. Christian, *Curr. Opin. Struct. Biol.* **2003**, *13*, 325.
32 V. Gopalan, A. Vioque, S. Altman, *J. Biol. Chem.* **2002**, *277*, 6759.
33 E. L. Christian, N. H. Zahler, N. M. Kaye, M. E. Harris, *Methods* **2002**, *28*, 307.
34 D. Smith, A. B. Burgin, E. S. Haas, N. R. Pace, *J. Biol. Chem.* **1992**, *267*, 2429.

II.2
Biophysical Methods

23
Structural Analysis of RNA and RNA–Protein Complexes by Small-angle X-ray Scattering

Tao Pan and Tobin R. Sosnick

23.1
Introduction

Small-angle X-ray scattering (SAXS) is a solution technique that measures the size and shape of an individual or a complex of macromolecules. The method is well suited for the analysis of RNA structure, folding and association with proteins. For example, partially folded states that are not readily measured using other biophysical methods such as NMR or crystallography can be readily studied using SAXS.

In the typical SAXS experiment, X-rays are scattered by the sample and the scattering profile is measured at very low angles θ (Fig. 23.1A). The radius of gyration, R_g, of the particle can be determined from the width of the scattering profile. For globular objects, the profile can be approximated as a Gaussian $I(Q) = I_0 e^{-Q^2 R_g^2/3}$, where $Q = 2\pi \sin\theta/\lambda$ and λ is the X-ray wavelength. R_g is the root-mean-square of the distances of all regions to the center of mass of the particle weighted by their (excess) electron density. Typically, R_g is obtained from the slope of the Guinier plot of $\ln I(Q)$ versus Q^2.

The entire scattering profile can be used to obtain the R_g as well as other shape information such as the maximum distance in the particle, d_{max}, and the pair-distribution function $P(r)$ [1]. $P(r)$ is the probability distribution of distances between scattering atoms within the macromolecule (Fig. 23.1B). It has a maximum at the most probable distance in the object (e.g. slightly larger than the radius for a sphere) and goes to zero at d_{max} (e.g. the diameter). The R_g value can also be calculated from the second moment of the $P(r)$ distribution.

SAXS can be used to test the consistency of either high- or low-resolution structural models obtained experimentally or from modeling. The R_g value and the $P(r)$ function measured in a SAXS experiment can be compared with the predicted values [2]. As SAXS is a solution technique, this capability is particularly useful for determining whether a structure undergoes a conformational change upon crystallization or electron microscopy preparation. Multi-domain RNAs or RNA–protein complexes can be modeled using the crystal or NMR structures of the individual components. Generally, constructing a unique three-dimensional model from one-

Handbook of RNA Biochemistry: Student Edition. Edited by R. K. Hartmann, A. Bindereif, A. Schön, E. Westhof
Copyright © 2009 WILEY-VCH Verlag GmbH & Co. KGaA, Weinheim
ISBN: 978-3-527-32534-4

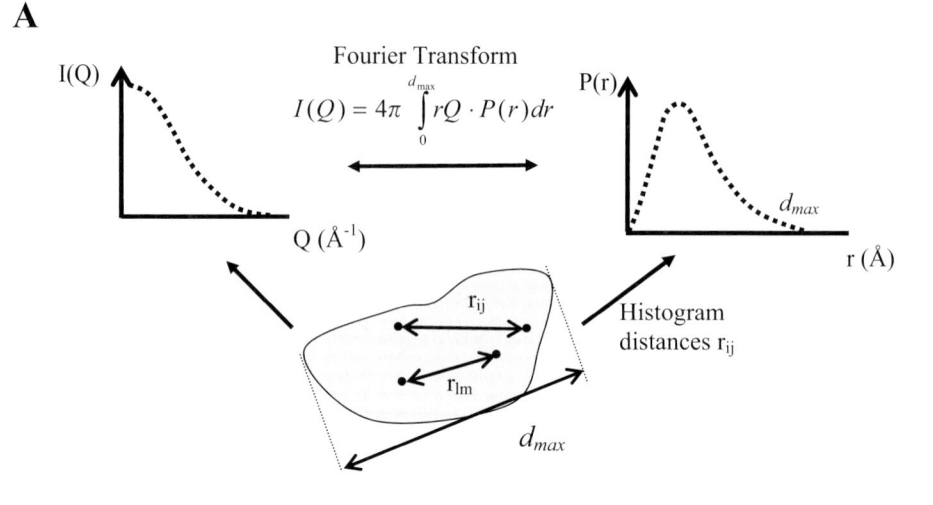

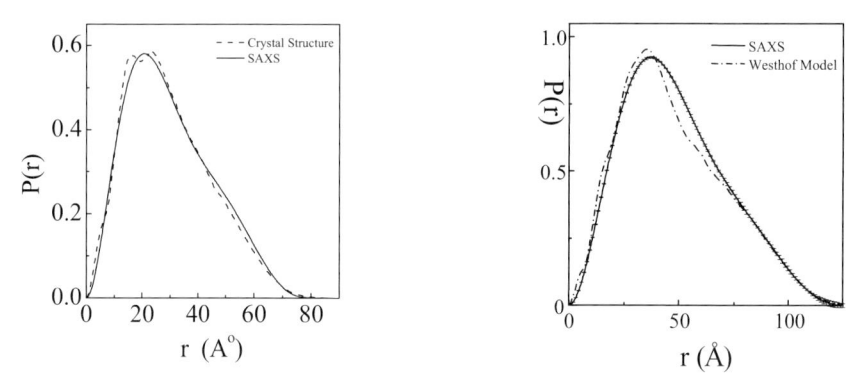

Fig. 23.1. (A) Relationship between $I(Q)$, $P(r)$ and the macromolecule of interest. (B) Comparing the $P(r)$ function of the crystal structure of yeast tRNA^{Phe} (left) and the Westhof model of the catalytic domain of *B. subtilis* RNase P RNA (right) with the SAXS data.

dimensional scattering data is difficult in the absence of other information, although significant progress has been made in this area [3].

Another useful parameter from SAXS measurements is the absolute scattering intensity at zero angle, I_0. For a monodispersed system, I_0 is related to the molecular weight of the scattering species (MW), macromolecular weight concentration (C, in mg/ml) and the electron density (ρ) according to [4]:

$$I_0 \propto (\rho_{macromolecule} - \rho_{solvent})^2 \times MW \times C \qquad (1)$$

This proportionality allows the determination of the oligomerization state of the RNA or RNA–protein complex when a suitable RNA mass standard is used such as the yeast tRNA^{Phe}. Because the crystal structure is known for this tRNA, its monodispersity can be confirmed from the measured R_g and $P(r)$ values with those calculated from the crystal structure. Therefore, the I_0 value of the tRNA^{Phe} can be used as a molecular weight and size standard for the studies of RNA or RNA–protein complexes of interest [5].

An advantage of SAXS in studying RNA–protein complexes is that RNA scatters more strongly than protein due to its higher electron density. Generally, the relative scattering power of the RNA to the protein is $(\rho_{RNA} - \rho_{solvent})^2 / (\rho_{protein} - \rho_{solvent})^2 \sim$ 5-fold [6]. This property allows the size and the shape of the RNA to be analyzed in the presence of moderate amounts of protein – a feature particularly useful for identifying conformational changes in RNA upon protein binding.

The downside of the sensitivity of SAXS method to size and to oligomeric state is that even mild amounts of aggregation can lead to spurious $P(r)$ and R_g values. Before synchrotron facilities were available, a typical SAXS experiment at a home source required relatively high RNA concentrations, e.g. 10 mg/ml. A SAXS experiment at a synchrotron such as the Advanced Photon Source (APS) at the Argonne National Laboratory, however, only requires RNA samples in the range of 0.1–1 mg/ml (4–40 µM yeast tRNA^{Phe}) and proportionally lower for larger RNAs. Data acquisition takes just a few seconds, which further reduces the chances of aggregation and sample degradation. In addition, lower concentrations can be critical for samples that are difficult to obtain or permit measurements to be conducted over a variety of conditions.

23.2
Description of the Method

23.2.1
General Requirements

Biological SAXS experiments require dedicated instruments designed for low angle measurements and optimized for low background levels. As mentioned, synchrotron-based measurements have a significant number of advantages (Fig. 23.2). In addition, a programmable titrator in conjunction with a flow-through sample cell enables an entire titration series (e.g. varying the ion or denaturant concentration) to be expediently carried out with a single sample in an hour or less.

The following steps are recommended for an SAXS experiment:

(1) Find a synchrotron-based beam-line with personnel who have experience with biological samples. As biological samples scatter weakly, their signal can be orders of magnitude weaker than background levels. Hence, minimizing background levels at the outset of an experiment is highly advantageous in terms of

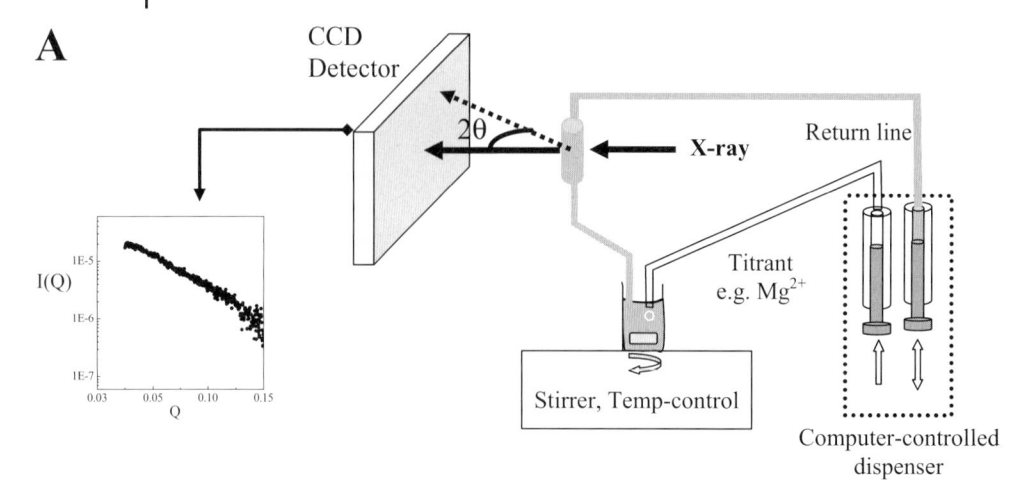

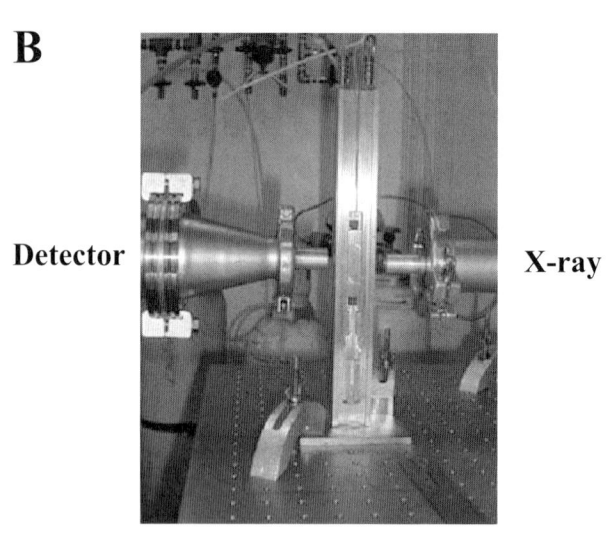

Fig. 23.2. (A) Experimental setup of SAXS at a synchrotron facility. (B) The sample holder used in our SAXS studies, sandwiched between the X-ray beam on the right and the CCD detector on the left.

sample requirements and reproducibility. Frequent interactions are necessary between the user and the scientists; fortunately, many facilities now have established user programs which provide expert support.

(2) Sample requirement. Approximately 200 μl of around 0.3 mg/ml RNA of about 50 kDa is needed for a single SAXS measurement. Large RNAs scatter more strongly than small RNAs ($I_0 \propto MW$), so the operational concentration can be

proportionally lower for large RNAs. For a titration series of 10–20 data points, at least 1 ml of sample is recommended.

(3) Time requirement. The most time-consuming component is the instrumental setup which is done by the staff at the synchrotron facility. A single SAXS measurement takes about 10 min, most of which is spent on sample manipulation. At the BioCat and BSSERC ID-12 beam-lines at the APS, the actual data acquisition takes just a few seconds or less. A titration series can take approximately 1 h, including the background (buffer) measurements.

(4) Data analysis. To confirm that measurements are generating useful information, preliminary data analysis should be conducted during the course of the experiment. This analysis generally includes background subtractions and Guinier analysis to obtain I_0 and R_g. These quantities enable the user to confirm that background levels are minimized, the signal is reproducible and the sample is monodispersed – three critical elements of a high-quality experiment.

23.2.2
An Example for the Application of SAXS

RNase P is an essential enzyme required for the 5′ maturation of all tRNAs and is conserved in all three kingdoms of life [7, 8]. In bacteria, RNase P is composed of one RNA subunit of 330–420 nt and a small protein subunit of 13–15 kDa. The bacterial RNase P RNA subunit (P RNA) alone is catalytically active, but the protein subunit is required for full activity under physiological conditions. We used SAXS to address the following questions for the RNase P ribozyme from *Bacillus subtilis* (for more details see [9, 10]):

(1) What is the oligomerization state of the P RNA alone and complexed with the RNase P protein (P protein) at varying monovalent ion concentration?
(2) What is the overall shape of the P RNA alone and complexed with the P protein?
(3) What is the size and shape of the ribozyme–substrate complexes?

23.3
General Information

RNase P RNA from various organisms was obtained by standard transcription using T7 RNA polymerase and purified using denaturing gels. The P protein was prepared from an overexpression clone. The concentration of the P RNA and the P protein was determined by UV absorbance using previously determined extinction coefficients.

The P RNA was renatured as follows: (1) heat the RNA in the buffer alone at

90 °C for 2 min, (2) incubate at room temperature for 3 min, (3) add $MgCl_2$ to 10 mM final concentration, (4) incubate at 50 °C for 5 min and (5) add NH_4Cl or KCl to the desired concentration.

To reconstitute the RNase P holoenzyme, an equal amount of P protein was added to the renatured P RNA, followed by the incubation at 37 °C for 5 min.

SAXS experiments were carried out at the SAXS instrument on the BESSRC ID-12 beam-line of the APS at the Argonne National Laboratory located in Illinois, USA [11]. Data were collected using a nine-element mosaic CCD area detector (15 cm × 15 cm) and exposure times were 1–6 s for each measurement. Sample–detector distance was 3 m; the energy of X-ray radiation was set to 13.5 keV. Computer-controlled Hamilton syringes injected sample into a thermostated flow cell made of 1.5-mm diameter cylindrical quartz capillary (Fig. 23.2). The background scattering was from a buffer solution in the identical configuration. Samples were measured under constant flow conditions in order to reduce the possibility of radiation damage.

23.4
Question 1: The Oligomerization State of P RNA and the RNase P Holoenzyme

I_0 and $P(r)$ function are used to deduce the oligomerization state of P RNA in the absence and presence of the P protein. Without the P protein, the I_0 ratio of P RNA and yeast $tRNA^{Phe}$ at the same weight concentration (0.1–1 mg/ml) is proportional to their molecular weight ratio (5.6 ± 0.6 versus 5.4). Yeast $tRNA^{Phe}$ has been shown conclusively to be a monomer under these conditions [5], and experimentally derived $P(r)$ and R_g for this tRNA agree well with the crystal structure. Hence, this result shows that P RNA is a monomer without the P protein.

The *B. subtilis* holoenzyme contains two P RNA molecules as indicated by SAXS (Fig. 23.3A). The I_0 of the P RNA–P protein complex reconstituted at 1:1 RNA–protein is twice the I_0 of P RNA alone. The higher scattering signal from RNA and the significantly larger size of the P RNA over P protein ensure that the observed I_0 is almost entirely derived from the scattering from P RNA. Consistent with the dimer formation, the $P(r)$ function of the holoenzyme has two times more distance pairs compared to those of the P RNA alone. As the stoichiometry of the P RNA and P protein is 1:1 in the holoenzyme, SAXS results show that the *B. subtilis* holoenzyme contains two P RNA and two P protein subunits.

The scattering profile of P RNA without the P protein shows considerable variation in the absence and the presence of 0.1 M NH_4Cl or 0.1 M KCl. The native structure of P RNA is composed of two independently folding domains. The variation in the scattering data of the P RNA monomer at different solution conditions may be explained by a difference in the relative orientation of the two domains.

The fraction of the holoenzyme dimer and monomer as a function of monovalent salt also is determined by SAXS (Fig. 23.3B). Changes in the I_0 and the R_g values are used to deduce the dimer fraction. Comparing 0.8 versus 0.1 M NH_4Cl, I_0 is 2-fold at 0.8 M NH_4Cl and the R_g value is also reduced from 57 to 50 Å at

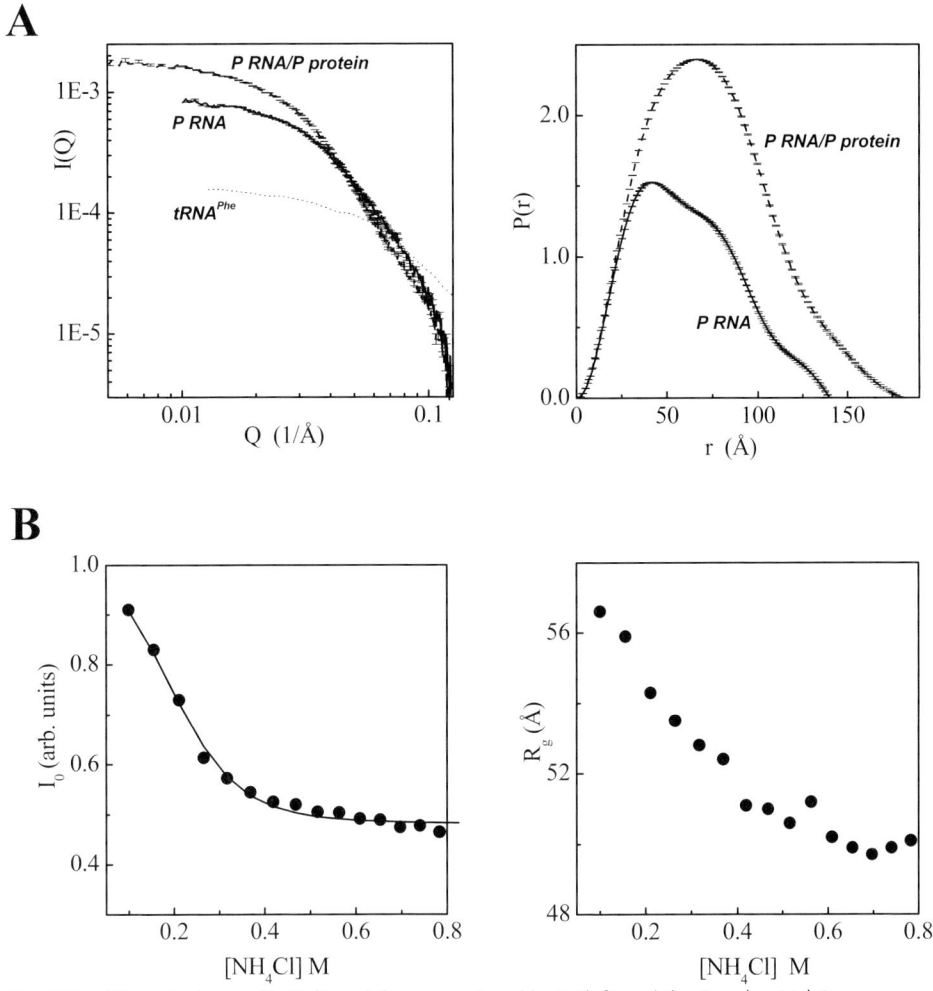

Fig. 23.3. (A) Scattering profile (left) and the P(r) function (right) of the P RNA alone and complexed with the P protein. (B) The dependence of the holoenzyme dimer:monomer equilibrium on the concentration of NH$_4$Cl, monitored by I_0 (left) and the R_g value (right). Changes in the I_0 value can be fit (solid line) to obtain a fifth-power dependence of the dimerization constant on the NH$_4$Cl concentration.

0.8 M NH$_4$Cl. These results show that at the holoenzyme is predominantly a dimer at 0.1 M NH$_4$Cl, but is a monomer at 0.8 M NH$_4$Cl. The holoenzyme is in a dimer–monomer equilibrium at the intermediate NH$_4$Cl concentrations.

The dimer–monomer equilibrium has a fifth-power dependence on the NH$_4$Cl concentration according to the SAXS data, i.e. the dimerization constant, K_D, is reduced by 2^5-fold when the NH$_4$Cl concentration is increased by 2-fold. At 0.1 M NH$_4$Cl, the K_D value is around 50 nM.

23.5
Question 2: The Overall Shape

The shape of the P RNA monomer and the holoenzyme is modeled and compared to the R_g and $P(r)$ function from SAXS. The modeling is performed in two steps. First, the P RNA model from Westhof et al. [12] is modified to allow a better fit to the SAXS data. Second, two P RNAs are brought together and the dimer model compared to the SAXS results.

The R_g of the P RNA monomer derived from the Westhof model is smaller than those determined by SAXS (43 versus 47–55 Å). The $P(r)$ function of the Westhof model also has a narrower distribution of mass pairs than that derived from the SAXS data. The P RNA in the Westhof model, however, was constructed in the context of a bound tRNA substrate. As discussed in the next section, the presence of tRNA may affect the conformation of the P RNA.

The Westhof model was modified to determine whether a change in the domain orientation alone could explain the difference in the R_g and $P(r)$ (Fig. 23.4A). In our case, the structures of both domains are kept the same as those in the Westhof model. However, one of the two domain–domain connections is used as a hinge and the other connection is extended to allow one domain to be rotated away from the other domain. Rotation of the inter-domain angle from around 30° in the Westhof model to around 60° and 90° changes the R_g for the new P RNA models from 43 to 47–53 Å, respectively, much closer to the measured R_g with and without 0.1 M salt. Similarly, the $P(r)$ functions of the models with altered domain orientations agree better with the SAXS data.

Two P RNA molecules with modified inter-domain orientations are brought together with the catalytic domain of one P RNA proximal to the specificity domain of the other P RNA and vice versa to generate the holoenzyme model (Fig. 23.4B). To obtain a better fit to the SAXS data, the precise angle between the domains in the holoenzyme is similar, but not identical, to that in either P RNA monomer model. Changing the angle between the domains should be feasible because P protein binding could easily compensate for any potential energetic cost of altering the domain orientation. The model has similar $P(r)$ and R_g (56 Å) to the experimentally measured $P(r)$ and R_g (57 Å).

23.6
Question 3: The Holoenzyme–Substrate Complexes

Although the pre-tRNA substrates used in almost all biochemical studies contain a single tRNA, the cellular substrate is more diverse. *B. subtilis* has a total of 86 tRNA genes, only eight of which produce single tRNA transcripts [13]. The remaining 78 tRNAs are arranged in 13 operons that produce 2–21 tRNAs per transcript. At least in *B. subtilis*, the holoenzyme is likely to encounter tRNA transcripts containing two or more tRNAs. Therefore, two types of substrate are used in the

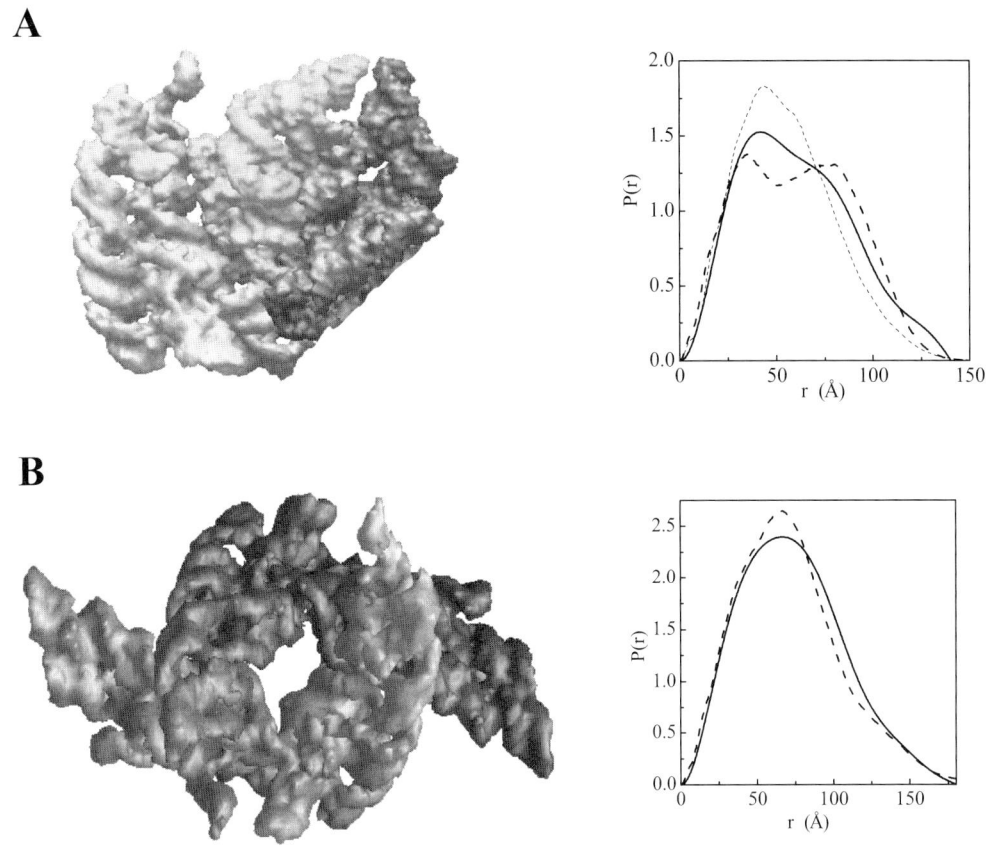

Fig. 23.4. (A) Models of the P RNA (left) and the $P(r)$ functions calculated from these models compared to the SAXS data (right). The Westhof model of the *B. subtilis* P RNA is shown in light gray. The modified model has a different inter-domain angle and is shown in dark gray. The $P(r)$ from the Westhof model is shown as a thin dashed line, from the modified model as a thick dashed line and from the SAXS data as a solid line. (B) A model of the holoenzyme dimer (left) and the $P(r)$ functions (right) calculated from the model (dashed line) and derived from the SAXS data (solid line).

SAXS study – one containing a single tRNA precursor and the other containing two tRNA precursors.

Substrate binding of the holoenzyme is again analyzed by SAXS (Fig. 23.5). Upon substrate addition and complex formation, the I_0 value is proportional to the second power of the molecular weight ratio of the ES complex and the holoenzyme alone, i.e. $I_0^{ES}/I_0^E = (MW^{ES}/MW^E)^2$. Changes in the R_g value also provide information on the structure of the ribozyme–substrate complexes.

The formation of the holoenzyme–substrate complex is monitored at 0.1 and 0.8 M NH_4Cl upon varying the molar enzyme:substrate ratio from 0 to 2. The

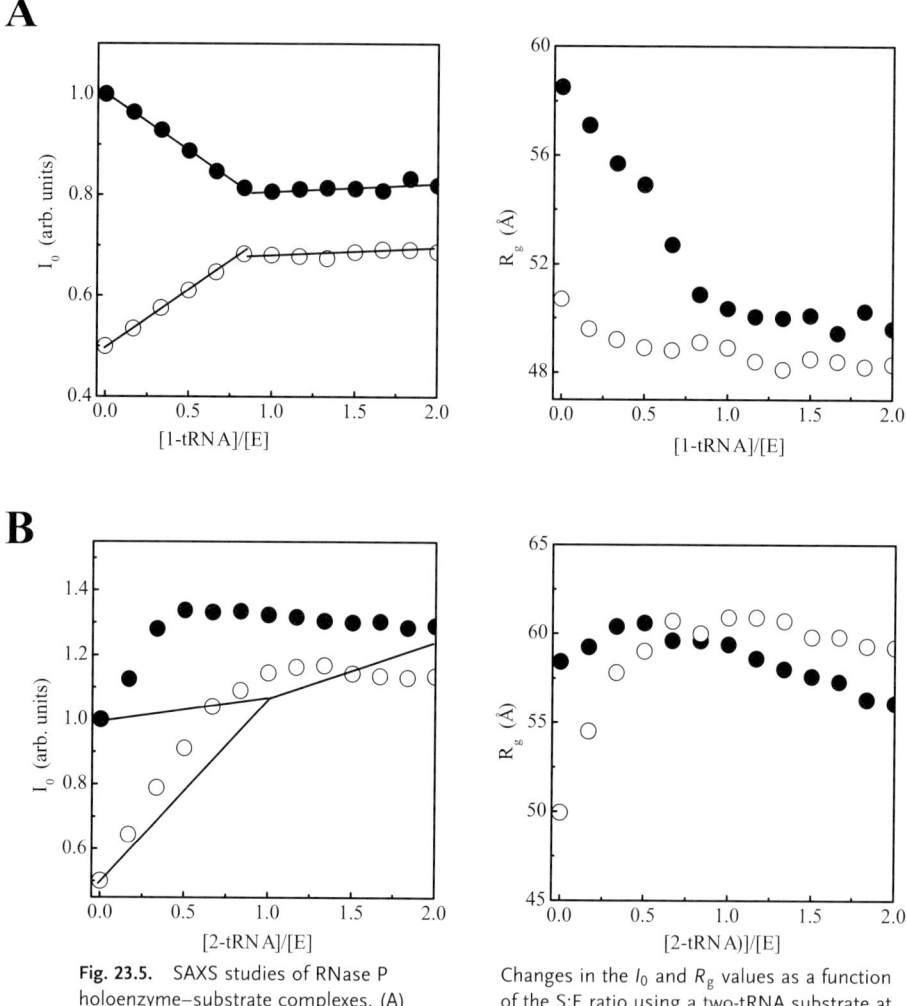

Fig. 23.5. SAXS studies of RNase P holoenzyme–substrate complexes. (A) Changes in the I_0 and R_g values as a function of the S:E ratio using a one-tRNA substrate at 0.1 M NH_4Cl (filled circles) and 0.8 M NH_4Cl (open circles). Only around 83% of this substrate is capable of binding to the enzyme under this condition, resulting in the non-identical I_0 and R_g value at high S:E ratio. (B) Changes in the I_0 and R_g values as a function of the S:E ratio using a two-tRNA substrate at 0.1 M NH_4Cl (filled circles) and 0.8 M NH_4Cl (open circles). Only around 60% of this substrate is capable of binding to the enzyme under this condition, resulting in the non-ideal I_0 and R_g value at higher S:E ratio. The solid lines show the predicted I_0 changes if all ES complexes are monomers.

holoenzyme concentration is kept constant at 2.4 μM to ensure that the initial holoenzyme is a dimer at 0.1 M NH_4Cl or a monomer at 0.8 M NH_4Cl. Under these conditions, the affinity of P RNA for a tRNA substrate is strong enough so that all properly folded substrates are bound to the P RNA when the stoichiometry of substrate to ribozyme is less than 1.

When the holoenzyme initially is a dimer at 0.1 M NH_4Cl, the I_0 value decreases upon the addition of the one-tRNA substrate until the molar ratio of the substrate to the total RNase P holoenzyme is approximately 1. Further addition of pre-tRNA^{Phe} produces only a very slight increase in the I_0 value, accounted for by the presence of the small, uncomplexed pre-tRNA substrate. When the holoenzyme is initially a monomer at 0.8 M NH_4Cl, the I_0 value increases upon the addition of pre-tRNA^{Phe} until the molar ratio of the substrate to the total holoenzyme is approximately one. These results show that the ES complex with a one-tRNA substrate is a monomer under all conditions.

Different ES complexes are observed when the two-tRNA substrate is bound to the holoenzyme. The ES complex is almost exclusively a dimer at 0.1 M NH_4Cl when the holoenzyme is in molar excess over the substrate, taking into account that only around 60% of this substrate is properly folded. Upon addition of more two-tRNA substrate, monomeric ES complex begins to form, presumably due to the holoenzyme binding to just one of the two tRNAs in this substrate. Now, both dimeric and monomeric substrate can exist at the same time. Both monomeric and dimeric ES complexes still form at 0.8 M NH_4Cl, but the fraction of the dimeric complex is significantly lower. These results show that the ES complex with a two-tRNA substrate is a mixture of dimer and monomer under very different conditions.

23.7 Troubleshooting

23.7.1 Problem 1: Radiation Damage and Aggregation

Due to the extremely high flux at a synchrotron source, radiation damage is a serious concern. The intense X-ray generates hydroxyl radicals that react with the RNA and proteins in the sample to result in aggregation. This problem can often be visualized upon comparing the scattering profiles of the same sample between multiple exposures.

There are two simple remedies to deal with the aggregation problem. First, the Tris and other organic buffers scavenge free radicals, so that their presence can significantly reduce radiation damage. Phosphate or other inorganic buffers are not recommended unless supplemented by a radical scavenger. Second, the measurement should be carried out under constant flow conditions, so that a fresh batch of molecules is exposed to X-ray at all times during measurement. Constant flow can be achieved using the sample-handling device depicted in this article.

23.7.2 Problem 2: High Scattering Background

Beyond optimizing beam-line performance, RNA samples purified by denaturing gel electrophoresis sometimes contain minute amounts of polyacrylamide particles that also scatter X-rays. Because these particles may be much bigger than the molecules of interest, the scattering profile may contain signals derived from these particles. Passing the RNA samples through a 0.22-μm filter often alleviates this problem.

23.7.3
Problem 3: Scattering Results cannot be Fit to Simple Models

Before jumping to a complicated conclusion, it is often advisable to determine the fractional folding or activity of the RNA and proteins present in the sample by another method. Even though a single RNA species is present, it may exist in two or more conformational populations at significant fractions. This subject is particularly important for interpreting titration experiments where the concentration of one or more components is varied.

Another trivial explanation for a complicated result is the loss of integrity of the RNA or the proteins during the experiment. Ideally, samples after exposure to X-ray should be examined by other methods after the SAXS measurements to ensure that the majority of the RNA is still intact.

The relatively high concentrations of RNA used in SAXS measurements can reduce the concentration of free cations. This reduction can result in an apparent concentration shift in the divalent cation requirement in the studies of RNA folding.

23.8
Conclusions/Outlook

The advance of synchrotron technologies has given this old biophysical method new life. In addition to the RNase P work described here, SAXS has also been applied recently to RNA folding studies [5, 14–19]. The two advantages of the SAXS method are the determination of the shape and size change in real-time and the accommodation of SAXS data to structural models. We anticipate more broad applications of SAXS to the understanding of other, even more complex biological systems in the near future.

Acknowledgments

All SAXS experiments were carried out at the Argonne National Laboratory in collaboration with Dr P. Thiyagarajan. Dr Xing-wang Fang designed the sample holder and carried out all SAXS experiments described here. We thank the NIH and the US Department of Energy for their financial support.

References

1 Moore, P. B., *J. Appl. Crystallogr.* **1980**, *13*, 168–175.
2 Svergun, D., C. Barberato, M. H. Koch, *J. Appl. Crystallgr.* **1995**, *28*, 768–773.
3 Svergun, D. I., V. V. Volkov, M. B. Kozin, H. B. Stuhrmann, *Acta Crystallogr. A* **1996**, *52*, 419–426.
4 Glatter, O., O. Kratky, *Small Angle X-ray Scattering*. Academic Press, London, **1982**.
5 Fang, X., K. Littrell, X. Yang, S. J. Henderson, S. Siefert, P. Thiyagarajan, T. Pan, T. R. Sosnick, *Biochemistry* **2000**, *39*, 11107–11113.
6 Cantor, C., P. Schimmel, *Biophysical Chemistry: Part II*. Freeman, New York, **1980**.
7 Altman, S., L. Kirsebom, in *The RNA World*, 2nd edn, Gesteland, R. F., Cech, T. R., Atkins, J. F. (eds) Cold Spring Harbor Laboratory Press, Cold Spring Harbor, NY, **1999**, pp. 351–380.
8 Frank, D. N., N. R. Pace, *Annu. Rev. Biochem.* **1998**, *67*, 153–180.
9 Fang, X. W., X. J. Yang, K. Littrell, S. Niranjanakumari, P. Thiyagarajan, C. A. Fierke, T. R. Sosnick, T. Pan, *RNA* **2001**, *7*, 233–241.
10 Barrera, A., X. Fang, J. Jacob, E. Casey, P. Thiyagarajan, T. Pan, *Biochemistry* **2002**, *41*, 12986–12994.
11 Seifert, S., R. E. Winans, D. M. Tiede, P. Thiyagarajan, *J. Appl. Crystallogr.* **2000**, *33*, 782–784.
12 Massire, C., L. Jaeger, E. Westhof, *J. Mol. Biol.* **1998**, *279*, 773–793.
13 Kunst, F., N. Ogasawara, I. Moszer, A. M. Albertini, G. Alloni, V. Azevedo, M. G. Bertero, P. Bessieres, et al. *Nature* **1997**, *390*, 249–256.
14 Russell, R., I. S. Millett, S. Doniach, D. Herschlag, *Nat. Struct. Biol.* **2000**, *7*, 367–370.
15 Fang, X. W., B. L. Golden, K. Littrell, V. Shelton, P. Thiyagarajan, T. Pan, T. R. Sosnick, *Proc. Natl. Acad. Sci. USA* **2001**, *98*, 4355–4360.
16 Fang, X. W., P. Thiyagarajan, T. R. Sosnick, T. Pan, *Proc. Natl. Acad. Sci. USA* **2002**, *99*, 8518–8523.
17 Russell, R., X. Zhuang, H. P. Babcock, I. S. Millett, S. Doniach, S. Chu, D. Herschlag, *Proc. Natl. Acad. Sci. USA* **2002**, *99*, 155–160.
18 Russell, R., I. S. Millett, M. W. Tate, L. W. Kwok, B. Nakatani, S. M. Gruner, S. G. Mochrie, V. Pande, S. Doniach, D. Herschlag, L. Pollack, *Proc. Natl. Acad. Sci. USA* **2002**, *99*, 4266–4271.
19 Das, R., L. W. Kwok, I. S. Millett, Y. Bai, T. T. Mills, J. Jacob, G. S. Maskel, S. Seifert, S. G. J. Mochrie, P. Thiyagarajan, S. Doniach, L. Pollack, D. Herschlag, *J. Mol. Biol.* **2003**, *332*, 311–319.

24
Temperature-Gradient Gel Electrophoresis of RNA

Detlev Riesner and Gerhard Steger

24.1
Introduction

As expected from the name of the method, temperature-gradient gel electrophoresis (TGGE) is a combination of a thermodynamic method and an analytical separation technique. The electrophoretic mobility of one or several RNA molecules is analyzed in dependence upon the temperature. Quite different parameters of an RNA affect this temperature-dependent mobility, i.e. size, sequence, secondary and tertiary structure, structural stability, hydrodynamic flexibility, and electrical properties like counterion condensation. Since different electrophoretic mobilities result in well-separated bands in a gel, RNA molecules of different sizes and/or with differences in the other parameters mentioned above can be analyzed in one and the same experiment. Although this feature might appear trivial, one should keep in mind that in other well-established physical methods like spectroscopy, in particular optical melting curves or most hydrodynamic methods, a superimposition of the parameters of different molecules or different conformations is always measured, and can hardly be deconvoluted into the individual parameters.

TGGE can be applied to a wide variety of nucleic acids, DNA and RNA, single-stranded and double-stranded, from oligonucleotides to the size limit of polyacrylamide gel electrophoretic resolution, i.e. a few thousand bases; the most relevant range is between 100 and 1000 bases. Also, different staining protocols can be used for detection of the nucleic acid; the most common is silver staining, but very specific methods like hybridization of a particular sequence in a crude nucleic acid preparation may also be used. The present chapter is restricted to RNA analysis, predominantly single-stranded RNA, and a few more specialized examples of RNA–RNA complexes and RNA–protein complexes. For a detailed description of other examples and application of TGGE to DNA and protein analysis, the reader may refer to several chapters in textbooks [1–3].

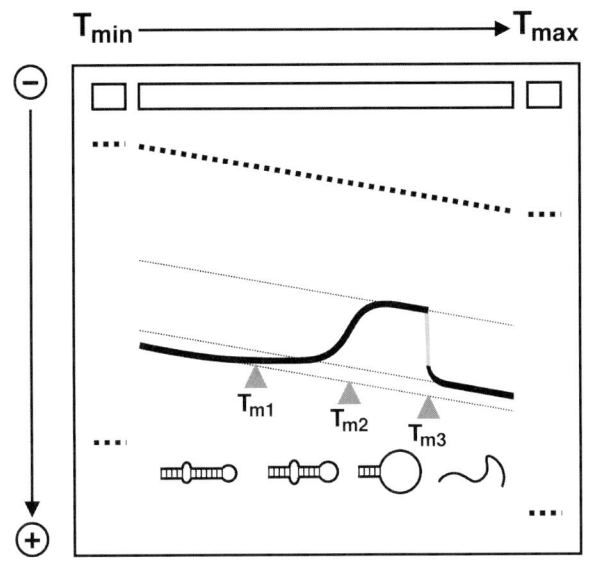

Fig. 24.1. Principle of TGGE. A linear temperature gradient is applied perpendicular to the electric field. The sample is applied to the long slot; small marker slots are on the left and right side of the gel. The mobility of an RNA with secondary structure is slightly decreased after a cooperative, reversible transition at T_{m1}, it is drastically decreased after a further transition at T_{m2} and it is increased after the irreversible transition at T_{m3} from a partially denatured molecule to a state without any base pairs. The secondary structure drawings below the transition curve symbolize the major structures before and after the different transitions. The dashed lines show the migration behavior of double-stranded nucleic acids that do not undergo any transition in the chosen temperature range; their increase in mobility is due to a decrease in viscosity with increasing temperature.

24.2 Method

24.2.1 Principle

As shown in Fig. 24.1, a nucleic acid sample is applied to a slab polyacrylamide gel in a broad slot that extends over nearly the whole width of the gel at the side of the negative electrode. A linear temperature gradient is established perpendicular to the direction of the electric field. The molecules at the left side of the gel migrate at low temperatures, the molecules at the right side at high temperatures and those at positions in between at corresponding intermediate temperatures. Each individual molecule, however, migrates during the whole electrophoretic run at constant temperature. In the example of Fig. 24.1, an RNA is analyzed which undergoes three cooperative conformational transitions: one from a native, highly base-paired state to an altered conformation, a second to a partially denatured state and a third transition to the totally denatured state. The transitions occur at well-defined denaturation temperatures $T_{m1} < T_{m2} < T_{m3}$. The molecules at $T > T_{m1}$ migrate slower than those at $T < T_{m1}$; the molecules at $T > T_{m2}$ migrate much slower than those at $T < T_{m2}$. In the narrow temperature range of the first transition the molecules switch between two states reversibly and assume a mobility averaged according to the degree of transition. The same holds for the second transition. Therefore, after staining the nucleic acid the band represents a transition curve. The third transition is an example for an irreversible step, i.e. there is no thermodynamic equilibrium between the paired structure below T_{m3} with the completely denatured state above T_{m3} under the low salt condition of the gel electrophoresis. The RNA denatures at T_{m3}, but renaturation is not possible, which leads to the jump in migration behavior without a continuous band in between the two states.

In addition to the gradient perpendicular to the electric field, one in parallel to the electric field can also be applied. In that case, the samples are applied to narrow slots and run from low to high temperature. This mode of TGGE is applied mostly to mutant analysis, when heteroduplices of wild-type and mutant sequences form mismatches, and are retarded due to partial melting at a lower temperature as compared to homoduplices [1, 2].

24.2.2
Instruments

The original instruments were home-made instruments, in which the gradient was established by thermostating the two edges of a thermostating block by liquids from two thermostats [4]. Although these instruments worked quite well and were available for some years commercially, they are no longer available. Therefore, the presently available system from Biometra (Göttingen, Germany) will be described.

The Biometra TGGE system is a well-constructed commercial instrument, which is developed for routine use as well as for research use. The microprocessor-driven gradient block based on Peltier elements allows well-defined temperature gradients with good resolution and reproducibility. The Biometra TGGE system is available in two formats. The standard TGGE "mini" system operates small gels and is therefore ideally suited for fast, "first-slot" experiments. The TGGE "maxi" system provides a large separation distance, allows high electrophoretic resolution and is well suited for systematic TGGE analysis of RNA. High parallel sample throughput can be achieved for DNA mutation analysis.

24.2.3
Handling

Handling and protocols are intimately connected with the TGGE instruments. Thus, we refer to the detailed manual (Version 3.02 TGGE MAXI System) from Biometra. The single steps are:

(1) Casting of gels (assembly of the gel cuvette, preparing gel and buffer solutions, pouring gels).

(2) Electrophoresis (electrophoresis conditions, sample loading, setup electrophoresis unit, gel electrophoresis run).
(3) Running conditions (voltage, temperature gradient, electrophoresis time).
(4) Staining.

24.3
Optimization of Experimental Conditions

TGGE relies on the fact that the electrophoretic mobility of a biopolymer is altered due to a conformational change. The extent of the alteration critically depends on several parameters. For a systematic study, it is essential to optimize these parameters first in order to obtain a large alteration. Furthermore, reversibility of a transition may be achieved only by choosing appropriate conditions. Therefore, the variables outlined below should be optimized, which has to be done primarily according to empirical rules.

24.3.1
Attribution of Secondary Structures to Transition Curves in TGGE

In order to attribute secondary structures to transition curves from TGGE, thermodynamic features as well as gel-electrophoretic mobilities have to be taken into account. Whereas the transition temperatures may be calculated quite accurately, the interpretation of gel-electrophoretic mobilities has to rely more on qualitative arguments.

(1) Branched structures migrate slower than extended structures. This effect is known from the denaturation of double-stranded nucleic acid, which leads to drastic retardation as long as the denaturation is incomplete [5]. The effect has also been described with dimeric transcripts of potato spindle tuber viroid (PSTVd) RNA [6].
(2) Structures with large loops migrate extremely slowly. The low mobility of denatured circular viroids and the lower mobility of plasmids in the form of relaxed circles as compared to supercoils are examples of this tendency.
(3) Because of their higher molecular weight and their usually high degree of bifurcations, bimolecular complexes migrate much slower than the corresponding uncomplexed molecules.
(4) Most RNA transitions appear as smooth curves; this is based on the reversible nature of the transitions. Irreversible transitions are recognized by their stepwise behavior; irreversibility is mostly due to a low chance of renaturation in low ionic strengths conditions of gel electrophoresis.

Attempts have been made to derive quantitative relationships from the general rules (see [7]; Mundt and Steger, unpublished). The mobility of partly denatured double-stranded nucleic acid has been calculated with fair success [8].

24.3.2
Pore Size of the Gel Matrix

The pore size of the gel matrix can be varied by the concentration of polyacrylamide and the ratio acrylamide-N,N'-methylenebisacrylamide (Bis). A compromise has to be found between a large change in electrophoretic mobility and an acceptable migration velocity. For example, a concentration of 6% polyacrylamide might be too low because of only small changes, whereas an optimum pore size might be reached at 8% acrylamide. The polyacrylamide concentration may be varied between 4% for larger RNA and up to 20% for short oligonucleotides, and the ratio of acrylamide:bisacrylamide between 20:1 and 40:1, respectively.

24.3.3
Electric Field

The electric field tends to stretch the molecules. These effects are very sensitive to the charge distribution, the conformation and the flexibility of the molecule. The mobility changes of two conformational transitions in one and the same molecule can exhibit different dependencies upon the electric field. Evidently, such a dependence cannot be predicted. An excessive electric field, although inducing large changes, may be disadvantageous because of the large electric current.

24.3.4
Ionic Strength and Urea

A variation of the ionic strength is always connected with a variation of T_m of the conformational transition. Therefore, this parameter cannot be varied independently from other features of TGGE. For nucleic acids one may keep in mind as a general rule that changes in electrophoretic mobility are larger in low ionic strength, which means, however, also at lower temperature. The effect was particularly large with circular RNA such as viroids (cf. Fig. 24.2), where the change in mobility was reduced to 40% if the ionic strength was raised from 8.9 to 89 mM Tris–borate. With nucleic acids, high ionic strength always improves the reversibility of a transition, and reversible transitions may be evaluated more easily and more accurately than irreversible or discontinuous transitions. Some examples shown below will prove this. The increase in the transition temperature due to high ionic strength may be compensated by addition of urea. Thus, addition of 5–10 mM NaCl and 4–6 M urea to the standard buffer (8.9 mM Tris, 8.9 mM boric acid) was found advantageous.

24.4
Examples

We will describe four different examples that should show the potency of the TGGE method. All examples were produced with an apparatus which was commercially available from Qiagen (Hilden, Germany) up to about 1995 and which was very similar to the home-made instrument mentioned above [4]; the gel length in this apparatus was 190 mm. However, after adaptation of the applied voltage to the length of a different gel system, the examples should be reproducible on any TGGE apparatus.

Most of the following examples were obtained with PSTVd RNA (for reviews, see [9, 10]). PSTVd is a covalently closed RNA molecule of 359 nt, has no protein-coding capacity, but is able to infect certain plants. It is replicated by DNA-dependent RNA polymerase of the host plant in an asymmetric rolling circle; multimeric (+)-stranded intermediates are processed by plant RNases/ligases to the mature circular molecule.

24.4.1
Analysis of Different RNA Molecules in a Single TGGE

Optical melting curves of a nucleic acid are based on the structure-dependent extinction coefficients; consequently, any impurity of the sample as well as different conformations of the sample at the same temperature add up to the signal and end in a non-interpretable transition curve. In TGGE, molecules are separated according to their hydrodynamic shape, as in any standard gel-electrophoretic method, i.e. nucleic acid molecules of different length are separated according to their size and molecules of identical length are separated according to their different conformation. Thus, sample impurities like abortive transcripts from T7 transcription or DNA templates are not prohibitive for analysis of denaturations in TGGE. For visualization, all gel-staining methods might be applied; of course, silver staining detects all nucleic acids, whereas after blotting, specific RNAs can be detected by hybridization or double-stranded RNAs by double-strand-specific immunostaining.

An example with silver staining is shown in Fig. 24.2. A crude RNA extract, consisting of at least circular and linear PSTVd (359 nt), 7S RNAs (about 305 nt), 5S rRNA (120 nt), and tRNAs, is separated on a TGGE, and the RNA bands are stained by silver. At the lowest temperature of the gel, circular (cPSTVd) as well as linear PSTVd (lPSTVd) have a rod-shaped secondary structure, which leads to a relatively high mobility. At a certain temperature, about 57 °C under the gel-electrophoretic conditions (see arrow in Fig. 24.2), the native structure denatures completely and three extra-stable hairpins are formed; the nucleotide regions forming these hairpins are given in bold characters in the schematic drawing of PSTVd's structure (see Fig. 24.2). The main transition is highly cooperative, leading to the sharp change in mobility at this temperature – the mobility is decreasing because the branched hairpin-containing structure is much bulkier than the rod-like native structure. At higher temperature the most stable hairpin II (HPII) denatures (see arrow in Fig. 24.2); because less base pairs are involved, this transition takes place over a broader temperature range. The completely denatured structure is an expanded circle, which is strongly retarded. Linear PSTVd molecules, which are either replication intermediates or are created by RNases during preparation of the RNA extract, migrate proportional to their length after full denaturation,

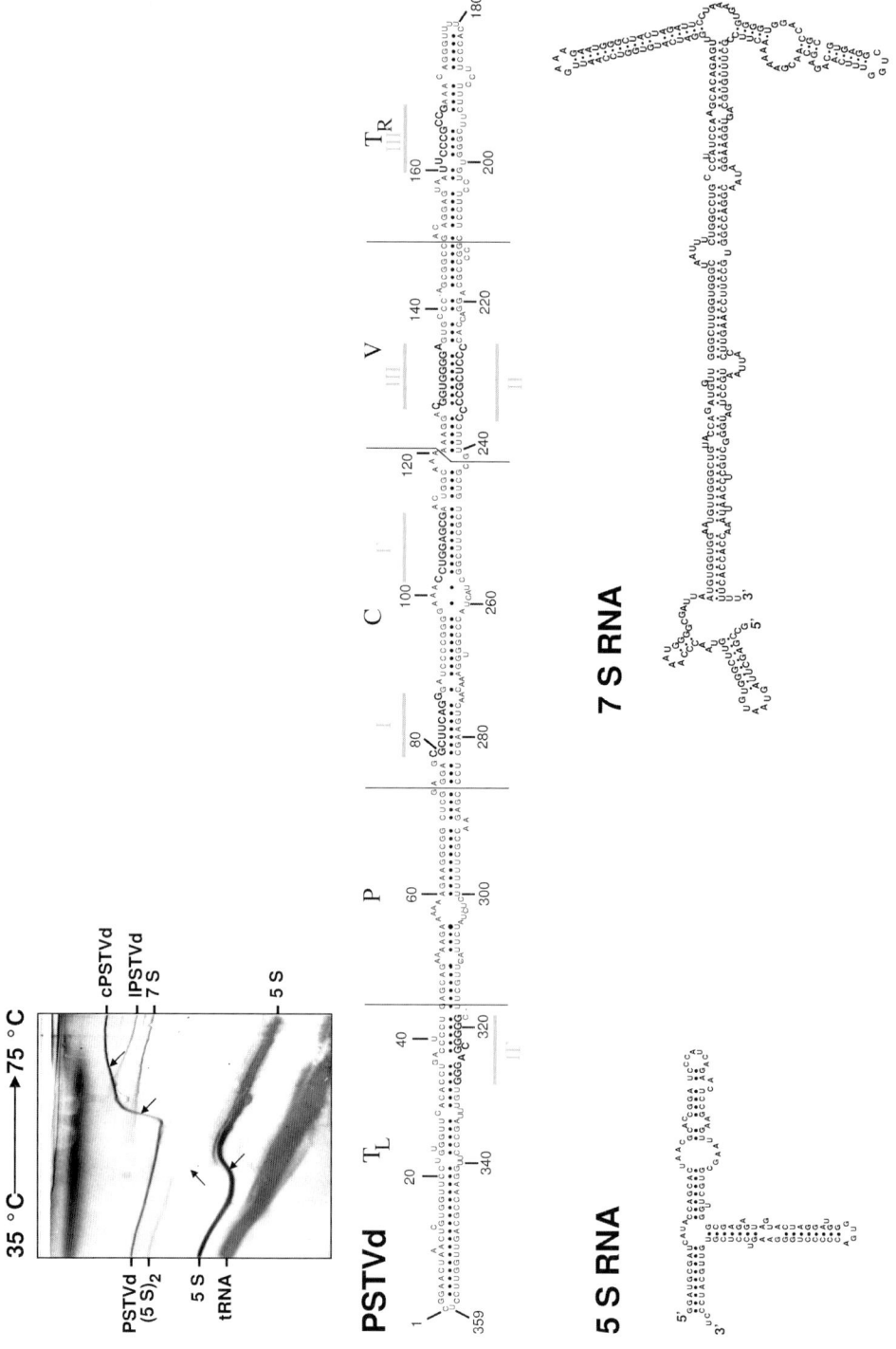

404 *24 Temperature-Gradient Gel Electrophoresis of RNA*

Fig. 24.2

i.e. lPSTVd migrates faster than cPSTVd after full denaturation. The temperature of the main transition of lPSTVd depends on the point of linearization; due to the low concentration of individual lPSTVd molecules, these transitions are not visible on the gel.

Similarly, at low temperature, a band for the 7S RNAs is not visible; after full denaturation, all 7S RNAs co-migrate. The 5S rRNA shows a single transition (see arrow). At the denaturation temperature of 5S RNA, the dimeric complex of 5S RNA also dissociates. At high temperature, the bands derived from monomeric and dimeric 5S RNA do not co-migrate; the small migration difference corresponds to the migration distance of dimeric and monomeric 5S RNA during the pre-electrophoresis step at low temperature before the temperature-gradient was applied.

Note that all bands show an increase in mobility with increasing temperature. This general effect is not based on structural rearrangements, but on the decreasing viscosity with increasing temperature.

24.4.2
Analysis of Structure Distributions of a Single RNA – Detection of Specific Structures by Oligonucleotide Labeling

The bases for structure formation of nucleic acids are hydrogen bonds between bases (Watson–Crick and wobble base pairs) and stacking interactions between neighboring bases and base pairs, which are thermodynamically favorable, i.e. structure formation is a chemical reaction and many different structures are in a thermodynamic equilibrium. Whereas only highly specialized sequences favor a dominant structure, most sequences exhibit a structure distribution. In the case of "RNA switches", an RNA molecule might even be evolutionarily optimized to exhibit more than a single structure. A further possibility for structure distributions is not based on thermodynamics, but on the kinetics of folding, and is called "sequential folding": in most cases the formation of RNA structural elements is faster than the rate of synthesis; therefore, the partial, still elongating RNA chain already

Fig. 24.2. Analysis of a crude RNA extract from tomato plants infected by PSTVd. Conditions of electrophoresis: 0.2 × TAE (8 mM Tris–HCl, 20 mM NaOAc, 0.2 mM EDTA, pH 8.4), 5% (w/v) acrylamide, 0.125% (w/v) bisacrylamide (acrylamide/bisacrylamide 40:1), 0.1% (v/v) N,N,N',N'-tetra-methylethylene-diamine (TEMED) and 0.04% (w/v) ammonium peroxodisulfate for starting the polymerization; 10 min pre-electrophoresis at 25 °C, 10 min equilibration for the 35–75 °C temperature gradient, 300 V for 90 min; gel size: 180 × 190 × 0.9 mm; slot size: 130 × 4 mm; silver staining. cPSTVd and lPSTVd: circular and linear forms of PSTVd, respectively; 5 S: 5S RNA; (5 S)$_2$: dimeric complex of 5S RNA; 7 S: 7S RNA. Transitions described in the text are marked by arrows. In the schematic drawing of the PSTVd structure, the following regions are marked: T$_L$ and T$_R$, terminal left and terminal right region, respectively (cf. Fig. 24.4); P, pathogenicity-related region; C, central conserved region; V, variable region; grey lines marked I/I' and II/II' form extra-stable hairpins I and II, respectively, at temperatures above the main transition (cf. Fig. 24.3).

folds during synthesis into structures which might be thermodynamically suboptimal for the full-length molecule. After finishing the synthesis, the structures generated during synthesis can rearrange if sufficient activation energy is available. Such structure distributions or co-existing structures, either based on thermodynamics or on kinetics, are one of the many problems of chemical and enzymatic mapping for structure determination [11]. TGGE allows for separation of structure distributions, at least when the different structures do not interconvert fast or co-migrate.

Figure 24.3 shows the analysis of sequential folding during synthesis of a (−)-stranded PSTVd transcript [11, 12]. Under thermodynamic equilibrium conditions, which can be established by complete denaturation and slow renaturation under low salt conditions to avoid formation of bimolecular complexes, the transcript forms a single dominant structure (see Fig. 24.3). The native conformation is rod-

---

Fig. 24.3. Detection of specific structures of a synthetic PSTVd transcript (linearized at a *Sty*I site, nt 337) by oligonucleotide labeling. (A) TGGE of the transcript under equilibrium conditions. The transcript in 0.2 × TBE (17.8 mM Tris, 17.8 mM borate, pH 8.4, 2.5 mM EDTA) was denatured at 70 °C for 15 min and slowly renatured to room temperature at about 0.1 °C/min. Conditions of electrophoresis: 0.2 × TBE, 4% (w/v) acrylamide, 0.13% (w/v) bisacrylamide (acrylamide/bisacrylamide 30:1), 0.1% (v/v) TEMED and 0.04% (w/v) ammonium peroxodisulfate for starting the polymerization; 15 min pre-electrophoresis at 15 °C with 30 V/cm, 10 min equilibration for the 20–55 °C temperature gradient, 30 V/cm for 120 min; gel size: 180 × 190 × 0.2 mm; slot size: 130 × 4 mm; gel was stained with silver. (B and C) TGGE analysis of the transcript after different times of transcription and incubation. T7 transcription assays with cNTP = 1 mM at 25 °C, corresponding to an elongation rate of about 130 nt/s, were stopped after 1 h (B) or after 30 s (C). Conditions of electrophoresis as in (A); detection of transcripts by staining with NBT/BCIP (B) or by autoradiography (C). In (B), the uppermost band is the DNA template. For further details, see [12]. (D) After transcription, (−)-stranded PSTVd exists in different structural conformations due to sequential folding; one of these conformations, which is metastable, contains a long G:C-rich hairpin (HPII) that is thought to be critical for transcription to (+)-strands *in vivo*. In the native conformation the two halves of the helix-stem of HPII are involved in base pairings at two distant positions. The oligonucleotide 27AB (5′-CUUACUUGCUUCCUUUGCGCUGUCGCU-3′), complementary to PSTVd (−)-sequence 318–307/237–251, is designed to hybridize with its full length to the HPII loop of the HPII-containing conformation, whereas binding to the native conformation is possible only with a part of its sequence. (E) Complexes were formed by incubating 200 ng of the *in vitro* transcript for 20 min in buffer (500 mM NaCl, 4 M urea, 1 mM sodium cacodylate, 0.1 mM EDTA, pH 7.0) with 105 c.p.m. oligonucleotide 27AB. After subsequent dialysis against 0.2 × TBE buffer, TGGE analysis was performed. Conditions of electrophoresis: 0.2 × TBE, 5% (w/v) acrylamide, 0.17% (w/v) bisacrylamide (acrylamide/bisacrylamide 30:1), 0.1% (v/v) TEMED and 0.04% (w/v) ammonium peroxodisulfate for starting the polymerization; 15 min pre-electrophoresis at 15 °C with 30 V/cm, 10 min equilibration for the 20–55 °C temperature gradient, 30 V/cm for 90 min; gel size: 180 × 190 × 0.2 mm; slot size: 130 × 4 mm; gel was stained with silver and exposed to X-ray film (Kodak Xomat AR) for detection of the radioactive transcript. On the silver-stained gel (left) several bands are detectable (S, M, Q/P, R), which represent different conformations of the same transcript; conformation S behaves identical to the transcript in the native conformation. On the autoradiograph of the silver-stained gel (right) only the RNA species Q/P and R are visible due to complexing with RNA oligonucleotide 27AB. For further details, see [11].

Fig. 24.3

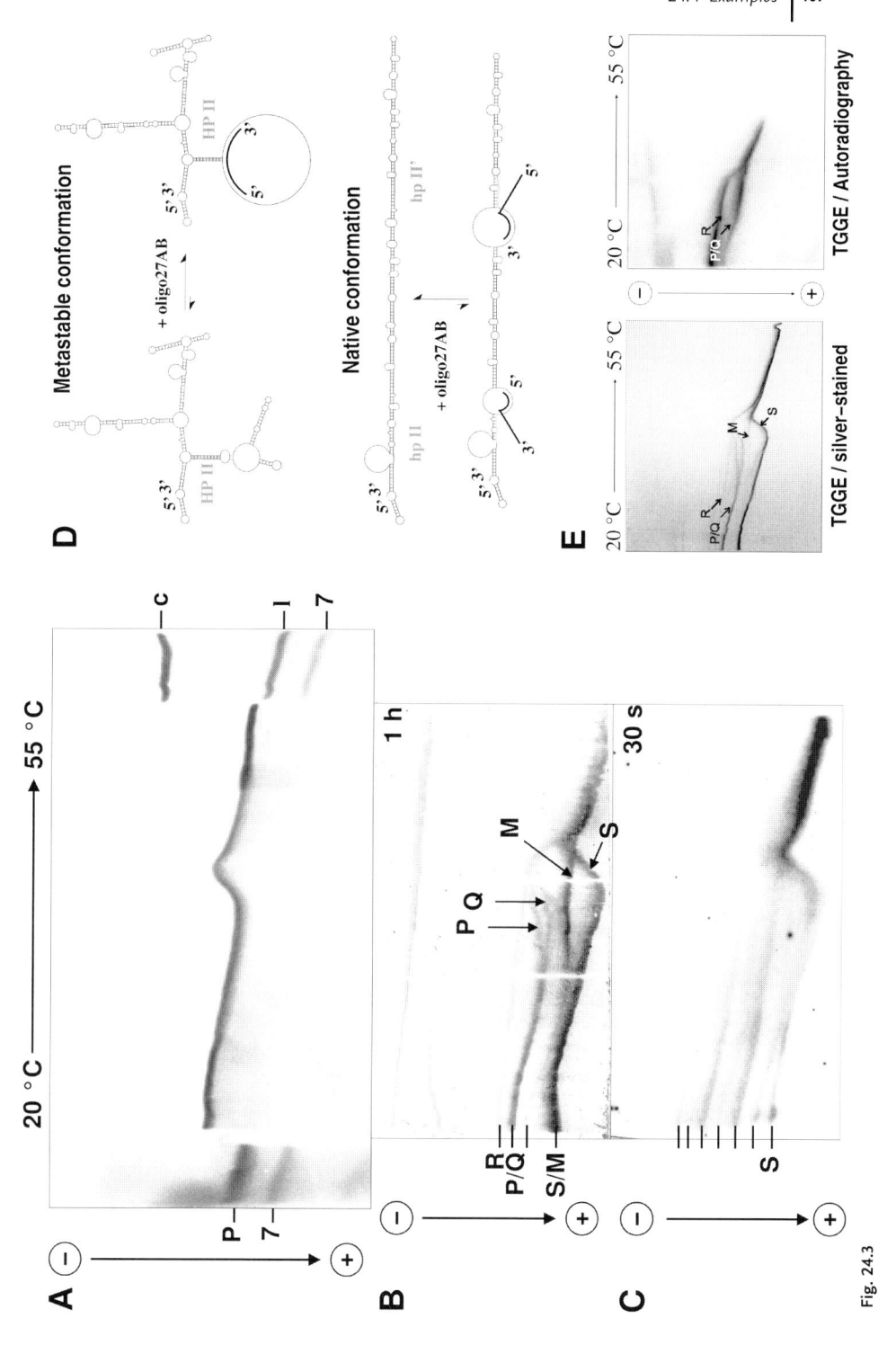

like (see Fig. 24.3D, bottom), and exhibits a single main transition in which the rod-like structure denatures and a particularly stable hairpin HPII is formed. The structure containing HPII is bulky; denaturation of additional helices and finally HPII leads to an increase in mobility at high temperature.

Transcripts from *in vitro* T7 polymerization can also exhibit several different structures (see Fig. 24.3B and C). The band of highest mobility, marked by "S", resembles the band from thermodynamic equilibrium conditions as shown in Fig. 24.3A. If analyzed directly after 30 s of transcription (*cf.* Fig. 24.3C), however, the thermodynamic equilibrium is not established and the rod-like structure is only a minor one. The other bands show lower mobility and therefore represent bulkier structures; they contain many thermodynamically instable helices and the slower bands from the more bulky structures denature at lower temperatures irreversibly. After slower synthesis and incubation under the high salt conditions of synthesis, most structures are of higher mobility and stability in comparison to those after a fast synthesis (*cf.* Fig. 24.3B with C). Note that a single band in TGGE does not ensure a single structure; the bands marked "P/Q" and "S/M" co-migrate at low temperature, but separate prior to the temperature of the main transition.

The migration behavior of certain bands, i.e. steepness of transitions, acceleration or retardation transitions, temperature of transitions, might be correlated with the results from structural calculations. The programs mfold [13] or rnafold [14], which are based on thermodynamic equilibrium calculations, or programs, which take into account kinetics and sequential folding [15, 16], were applied; in both cases the ionic strength difference in gel electrophoresis and calculations had been corrected for. A more direct, experimental confirmation of a structure model is possible, e.g. by "oligonucleotide mapping" – an oligonucleotide designed to hybridize only to a certain structural element is used to mark bands containing that structural element. An example is given in Fig. 24.3(D and E). Those oligonucleotides were used to detect structures of (−)-stranded PSTV transcripts which contain HPII; it should be noted that HPII is thought to be critical for synthesis of (+)-stranded replication intermediates *in vivo*. An oligonucleotide was designed that pairs to both regions neighboring the HPII helix (see Fig. 24.3D, top). With structures containing HPII, the oligonucleotide pairs over its full length; whereas with structures without HPII, e.g. a rod-like structure (see Fig. 24.3D, bottom), the oligonucleotide is able to pair only with half its length. Furthermore, the length of the oligonucleotide was carefully chosen not to shift the concentration ratio between existing structures towards HPII-containing structures. In Fig. 24.3(E), a hybridization analysis of the structure distribution of the transcript (as in Fig. 24.3B) with the radioactively labeled oligonucleotide is shown. By comparison of the silver-stained gel and its autoradiograph, it is obvious that only structures in bands "P/Q" and "R" are marked by the oligonucleotide; these bands are visible in the autoradiograph up to a temperature at which the oligonucleotide dissociates. These bands, however, are not visible in the silver-stained gel due to the low concentration of the oligonucleotide–transcript complex; the assignment is made on the nearly identical migration behavior of the respective complexes from the autoradiograph and the uncomplexed bands after silver staining. Note that the migration behavior of a complex may not be identical to the uncomplexed structure; the molecular weight of the complex is increased and the hydrodynamic shape might be altered.

24.4.3
Analysis of Mutants

Single nucleotide mutations might alter the structure and the structural stability of an RNA [18]. Visualization of those alterations in TGGE can be done either by analysis of a mixture of a mutant and the wild-type RNA or by analysis of the different RNAs after loading them sequentially to the gel. The first method allows for detection of even subtle differences in hydrodynamic migration and/or thermodynamic stability, whereas the second makes it easier to identify the individual species.

An analysis with sequential loading of three slightly different RNAs is shown in Fig. 24.4. The terminal left region (T_L) of PSTVd contains two repeats (marked in light and dark grey, respectively, in Fig. 24.4, right), which are partially complementary to each other. This allows for two different structural arrangements of the T_L region: a rod-like conformation, which is by far the dominant conformation according to calculations, and a branched conformation. For an easy comparison of the different structures, the wild-type "native" structure was modified by 2-bp changes that favor drastically either the rod-like or the branched structure, respectively. In the "rod" mutant, an A_{344}:U_{18} pair is changed to a G:C pair; as seen from Fig. 24.4, this change stabilizes the rod conformation and destabilizes the branched conformation. In the "branched" mutant, an A_5:U_{18} pair is changed to a G:C pair which stabilizes the branched conformation and destabilizes the rod-like conformation. These "designer mutants" could be analyzed experimentally in the silver-stained TGGE (Fig. 24.4, left), as well as in optical melting curves and NMR [17]. The stability of the "rod" structure is increased in comparison to the "native" structure; note the increase in melting temperature and the slight decrease in half-width of the transition, which is based on the increase in $\Delta H°$. In contrast, the "branched" structure shows a much broader transition at relatively low temperature, which might be due to opening of one of the branch helices, overlaid with a second transition at higher temperature (see open arrowhead in Fig. 24.4, left). From all experiments – TGGE, optical melting curves and NMR – it was concluded that the conformations of "native" and "rod" coincide, so that the native conformation of PSTVd is indeed rod-like.

24.4.4
Retardation Gel Electrophoresis in a Temperature Gradient for Detection of Protein–RNA Complexes

A relatively simple analysis of RNA–protein interaction is possible via gel-retardation experiments. In contrast to DNA–protein interaction, RNA–protein interaction is quite often not sequence dependent, but a function of the RNA struc-

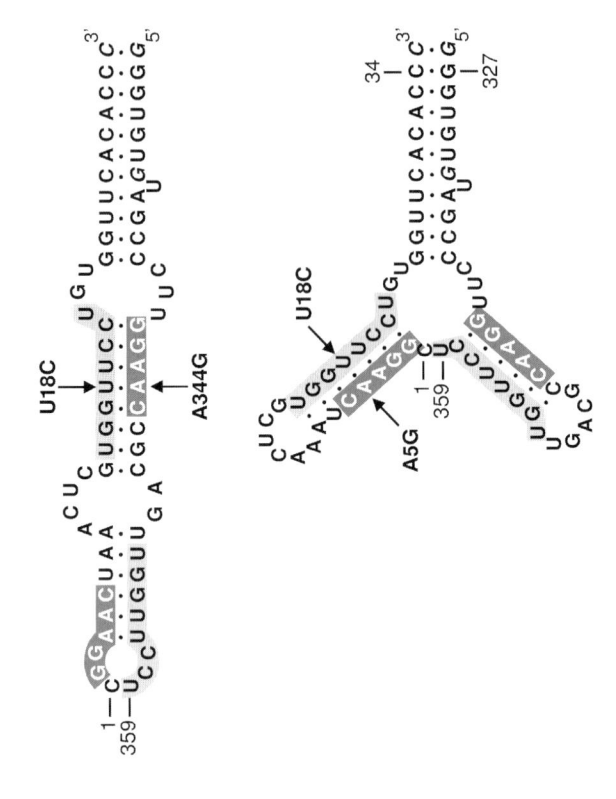

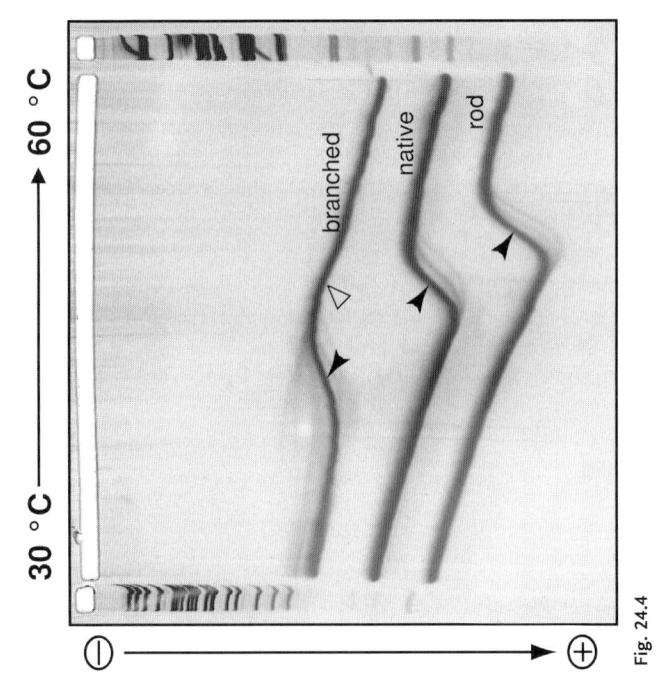

410 | *24 Temperature-Gradient Gel Electrophoresis of RNA*

Fig. 24.4

ture. Thus, a combination of gel-retardation for analysis of the interaction and of TGGE for the simultaneous analysis of RNA structure might be helpful.

A corresponding analysis [7] of the interaction of the 5′-untranslated region of spinach chloroplast *psbA* mRNA, which encodes the D1 protein of photosystem II, and a stromal protein extract is shown in Fig. 24.5. To establish complex formation, samples were incubated at 25 °C in the presence of heparin and tRNA to avoid unspecific binding. In a gel-shift assay at 25 °C with 0.5 mg/ml protein extract, 50% of the radioactively labeled RNA (0.5 pmol/ml) is retarded and in the presence of 1 mg/ml protein extract, the complete RNA is shifted into the complexed form (see Fig. 24.5A). Because the fraction of binding proteins in the extract is not known, no binding constant could be estimated. The percentages of bound RNA, however, are drastically dependent on the temperature of incubation and electrophoresis, as shown in Fig. 24.5(B): incubation and analysis of 1 mg/ml protein extract with 0.5 pmol RNA under the same conditions as above, but at 10 °C did not result in any RNA retardation. To analyze the temperature dependence of complex formation in more detail, 67 fmol RNA was incubated with 260 µg protein in 400 µl directly in the long slot of a gel in the presence of a temperature gradient from 10 to 40 °C, then separated by electrophoresis and the RNA visualized by autoradiography (see Fig. 24.5C). The gel shows a distinct temperature range in which retardation of the RNA and thereby complex formation with proteins can be observed. Retardation is observed between about 18 °C up to about 35 °C with a maximum at 22–25 °C. With additional experiments it could be verified that neither the pH gradient, produced by the temperature gradient, nor a

---

Fig. 24.4. Analysis of transcripts with a sequence derived from the T_L domain of PSTVd. The two possible structural conformations of the "native" sequence are shown on the right; numbering is according to circular PSTVd; the terminal helix was stabilized by addition of a terminal G:C base pair and a mutation $U_{332}G$ in comparison to PSTVd's T_L domain (*cf.* Fig. 24.2). In the mutant transcript "rod", the rod-like conformation was stabilized by the two mutations $U_{18}C$ and $A_{344}G$; in the mutant transcript "branched", the branched conformation was stabilized by the two mutations A_5G and $U_{18}C$. Samples were loaded sequentially onto the gel in the order of elongated rod, native and branched oligomers. The very weak additional bands observed during the transitions are due to $n+1$ transcripts that could not be removed completely during the purification procedure. The black arrowheads denote the main melting transitions; the open arrowhead denotes an additional transition in the "branched" mutant. Conditions of electrophoresis: 0.2 × (v/v) TBE (17.8 mM Tris, 17.8 mM boric acid, 0.4 mM EDTA, pH 7.5), 5% (w/v) polyacrylamide, 0.17% (w/v) bisacrylamide (acrylamide/bisacrylamide 30:1), 0.1% (v/v) TEMED and 0.04% (w/v) ammonium peroxodisulfate for starting the polymerization. The RNA samples (300 ng) were applied to the broad sample slot (130 × 4 mm), while the small slots (5 × 4 mm) at both sides were used for appropriate size marker DNA (pBR322 digested with *MspI*). Upon applying 500 V at a uniform temperature of 20 °C for 25 min, the RNA sample migrates a few millimeters into the gel matrix. This step was repeated for the loading of the second and third RNA samples, except the third sample electrophoresis period was only 10 min. Size marker DNA was loaded at the same point as the second RNA loading. Electrophoresis was paused for 15 min for the equilibration of the 30–60 °C gradient and the electrophoresis continued for 75 min at 500 V. The gel was stained by silver. For further details, see [17].

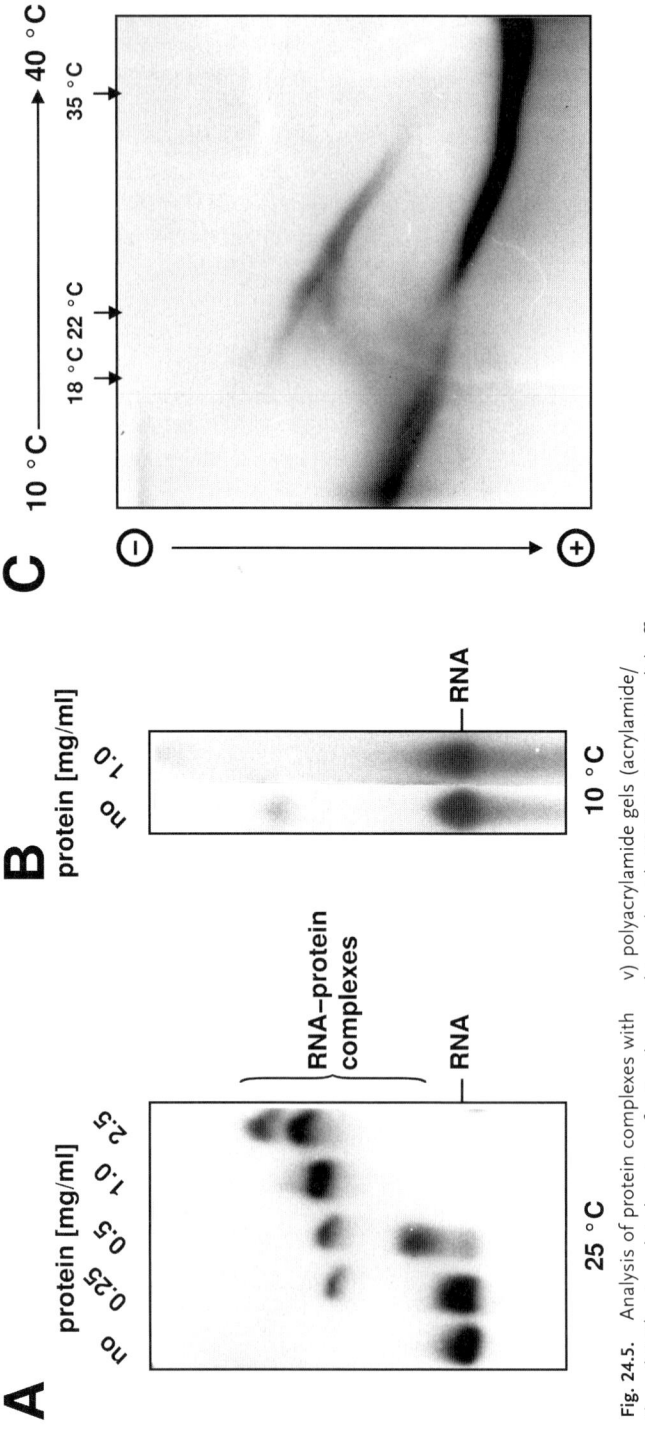

Fig. 24.5. Analysis of protein complexes with the *psbA* 5′-untranslated region of spinach. (A and B) Effect of temperature on protein binding. In a volume of 10 μl, 5 fmol of radiolabeled RNA was incubated for 15 min at 22 °C (in 20 mM Tris–HCl, pH 8.5, 20 mM KCl, 10 mM MgCl$_2$, 5 mM DTT, 2 mg/ml heparin, 0.5 mg/ml tRNA) with varying amounts of stromal protein as indicated. Free and complexed RNA was separated on 8% (w/v) polyacrylamide gels (acrylamide/bisacrylamide 20:1) in 2 × Laemmli buffer (50 mM Tris, 0.38 M glycine) at 25 (A) and 10 °C (B), respectively, at 200 V for 75 min. (C) Radiolabeled transcripts (67 fmol) were incubated with 260 μg of protein in 400 μl in the gel slot (130 × 4 mm) in the presence of a temperature gradient from 10 to 40 °C and then separated on a gel (other conditions as in A and B). For further details, see [7].

protein conformational change (aggregation, precipitation) in the low temperature range, but RNA conformational changes near 22 and 33 °C form basis for the temperature-dependent complex formation. One has to conclude that only the RNA conformation dominant between 22 and 33 °C is responsible for protein binding, whereas the structures dominant below 22 and above 33 °C, respectively, are not adapted for protein binding [7]. In the temperature range above 20 °C, binding of protein to the RNA decreases with increasing temperature; thus, the binding seems to be driven by enthalpy.

24.4.5
Outlook

In the examples we have concentrated on experiments with viroid RNA from our own work. TGGE is not limited, however, to the analysis of viroids, but has to be viewed more generally as a replacement or at least a supporting method for thermodynamic analysis of RNA structure by UV melting curves with the additional advantage to allow for analysis of RNA mixtures.

The group of Bevilacqua [19, 20] has used TGGE to select for and isolate thermodynamically more stable loop variants from combinatorial libraries of small RNA hairpins. They have mainly used a version of TGGE with the temperature gradient in parallel to the direction of RNA migration. This allowed for an easy excision of molecules with structures more stable than the bulk of molecules, because the already denatured molecules run slower than the still structured hairpins at a certain temperature.

TGGE may be used also for the analysis of tertiary structures. Take note, however, that the use of Mg^{2+} ions is necessary for stabilization of tertiary interactions in most cases, but this also catalyzes the degradation of RNA at elevated temperatures similar to basic conditions. The groups of R. Schroeder and E. Westhof monitored the tertiary structure transitions of the *td* intron of bacteriophage T4 and several mutants of this group I intron by TGGE [21]. With two mutant RNAs applied to one gel, similar to the experiment shown in Fig. 24.4, small stability differences of the mutants could be detected while simultaneously checking for deviating conformations and formation of intermolecular dimers.

Guo and Cech [22] searched for *Tetrahymena* ribozymes with an enhanced activity at elevated temperature by *in vitro* evolution. They selected for the thermodynamically most stable tertiary structure variants in the first step and for activity in the second step. In contrast to Bevilacqua's selection procedure [19, 20], they used temperature gradients perpendicular to the migration direction and excised small rectangular regions near to the unfolding transition from the fully folded state including tertiary interactions into a state with mainly secondary structure of the RNA. During eight rounds of selection, the temperature of the tertiary structure unfolding increased from 45 to 52 °C in a buffer containing about 0.4 mM free Mg^{2+} ions and in five subsequent rounds from 35 to 40 °C in about 0.1 mM free Mg^{2+} ions. Indeed the final variants contained up to 11 mutations, which mainly strengthened the active conformation through tertiary interactions and had an increase of the maximum active temperature by 10 °C in comparison to the starting variant.

The last examples demonstrate that TGGE is a favorable method to test for mutations and altered conformation and/or stability in the same analysis. We expect more of these applications in the future.

References

1 D. Riesner, K. Henco, G. Steger, in: *Advances in Electrophoresis 4*, A. Chrambach, M. J. Dunn, B. J. Radola (eds), VCH, Weinheim, **1991**.
2 K. Henco, H. Harders, U. Wiese, D. Riesner, in: *Methods in Molecular Biology*, J. M. Walker (ed.), Humana Press, Clifton, NJ, **1994**.
3 D. Riesner, in: *Antisense Technology: A Practical Approach*, C. Lichtenstein, W. Nellen (eds), Oxford University Press, Oxford, **1997**.
4 V. Rosenbaum, D. Riesner, *Biophys. Chem.* **1987**, *26*, 235–246.
5 G. Steger, P. Tien, J. Kaper, D. Riesner, *Nucleic Acids Res.* **1987**, *15*, 5085–5103.
6 R. Hecker, W. Zhi-min, G. Steger, D. Riesner, *Gene* **1988**, *72*, 59–74.
7 P. Klaff, S. Mundt, G. Steger, *RNA* **1997**, *3*, 1480–1485.
8 L. S. Lerman, S. G. Fischer, I. Hurley, K. Silverstein, N. Lumelsky, *Annu. Rev. Biophys. Bioeng.* **1984**, *13*, 399–423.
9 M. Tabler, M. Tsagris, *Trends Plant Sci.* **2004**, *9*, 339–348.
10 A. Hadidi, R. Flores, J. W. Randles, J. S. Semancik (eds), *Viroids*. CSIRO, Melbourne, **2003**.
11 A. R. W. Schröder, D. Riesner, *Nucleic Acids Res.* **2002**, *30*, 3349–3359.
12 D. Repsilber, U. Wiese, M. Rachen, A. R. Schröder, D. Riesner, G. Steger, *RNA* **1999**, *5*, 574–584
13 M. Zuker, *Nucleic Acids Res.* **2003**, *31*, 3406–3425.
14 I. L. Hofacker, *Nucleic Acids Res.* **2003**, *31*, 3429–3431.
15 M. Schmitz, G. Steger, *J. Mol. Biol.* **1996**, *255*, 254–266.
16 C. Flamm, W. Fontana, I. L. Hofacker, P. Schuster, *RNA* **2000**, *6*, 325–338.
17 A. J. Dingley, G. Steger, B. Esters, D. Riesner, S. Grzesiek, *J. Mol. Biol.* **2003**, *334*, 751–767.
18 P. Schuster, W. Fontana, P. F. Stadler, I. L. Hofacker, *Proc. R. Soc. Lond. B Biol. Sci.* **1994**, *255*, 279–284.
19 J. M. Bevilacqua, P. C. Bevilacqua, *Biochemistry* **1998**, *37*, 15877–15884.
20 D. J. Proctor, J. E. Schaak, J. M. Bevilacqua, C. J. Falzone, P. C. Bevilacqua, *Biochemistry* **2002**, *41*, 12062–12075.
21 P. Brion, F. Michel, R. Schroeder, E. Westhof, *Nucleic Acids Res.* **1999**, *27*, 2494–2502.
22 F. Guo, T. R. Cech, *Nat. Struct. Biol.* **2002**, *9*, 855–861.

25
UV Melting, Native Gels and RNA Conformation

Andreas Werner

25.1
Monitoring RNA Folding in Solution

Due to its conformational versatility, RNA is a performant catalyst and involved in specific recognition. Therefore, it is essential for any biochemical or structural study to monitor the correct formation of its three-dimensional shape. The larger the RNA, the more rugged the free energy landscape and thus the possibility that folding intermediates or alternative conformations persist [1]. Moreover, depending on renaturation protocol and RNA concentration, intermolecular annealing may be preferred over intramolecular folding, as in the case of hairpin versus duplex [2]. UV melting curves cannot only provide a means of controlling these phenomena, but also allow determination of equilibrium constants and energy parameters.

By heating and cooling under quasi-stationary conditions, the RNA can be reversibly forced through the folding process between an ordered, native and a disordered, denatured state. UV absorbance provides a convenient means of monitoring this transition in solution and with little material. This is due to the phenomenon of hypochromicity of stacked bases, which provides a exquisitely sensitive signal as temperature increases and RNA unfolds (Fig. 25.1A). The temperature at the midpoint of this transition is called T_m. When plotting the derivative of the absorbance as a function of temperature (a melting profile), each dA/dT peak corresponds to a partial transition (Fig. 25.1B).

Non-covalent interactions in RNA are commonly classified as secondary or tertiary structure. Secondary structure consists of Watson–Crick base pairs forming helical stacks linked by bulges, loops and junctions. Tertiary structure subsumes all other intra- or intermolecular interactions based on canonical and non-canonical base pairs, pseudoknots, water bridges, and interactions with metal ions. Translated into melting curves, the first peak in the low temperature range usually corresponds to the opening of the tertiary structure. Higher temperatures are required to unzip stacked bases and thus unfold secondary structure helices. Given a largely modular RNA structure, each independently folding element called a thermosome has its own T_m [3]. The derivative melting curve will be a convolution of normally

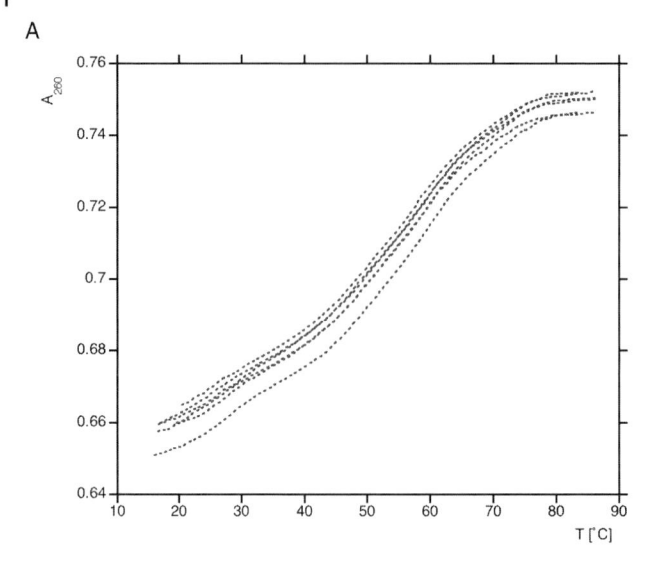

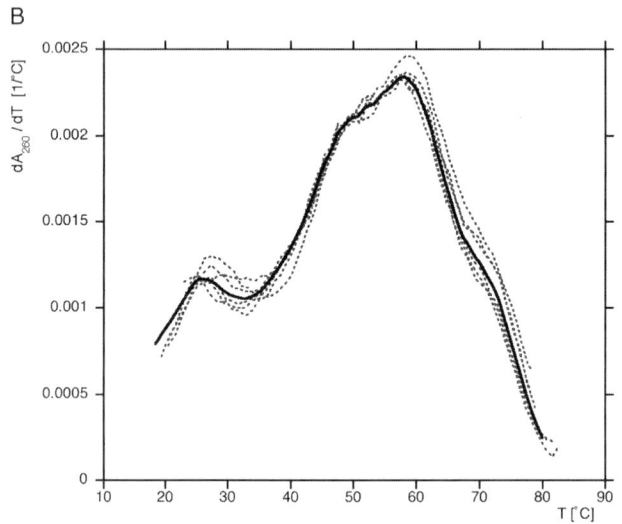

Fig. 25.1. UV melting profile of an RNase P RNA annealed in 10 mM sodium cacodylate pH 7, 10 mM NaCl and 0.5 mM MgCl$_2$. (A) Absorbance melting curve, three subsequent runs. (B) Derivative dA_{260}/dT of (A), the curve showing evaporation has been excluded. Black: average. Transitions visible at 26, 49, 58 and 70 °C.

distributed peaks around these T_ms. Many closely spaced thermosomes give rise to a complex melting profile. In simpler systems (less than 100 nt), it has been possible to deconvolute these thermosomes and assign them to individual transitions [4]. However, interpretation requires additional data through corroborating calorimetric and biochemical methods. For example, native polyacrylamide gels can provide complementary data to understand how the RNA collapses to the native state. In particular cases, a combination of melting and native gels could be applied even to thermostable RNA–protein complexes [5].

25.2
Methods

Most standard UV spectrophotometers can be equipped to measure RNA melting curves. We have used a double-beam Uvikon XL spectrophotometer attached to a Biotek temperature-control unit and a microcomputer running the Biotek Life-Power T_M Junior software (Fig. 25.2A). The instrument offers the advantage that it can hold up to 12 cuvettes in an automatic cell changer. Therefore, up to nine samples plus two standards and a temperature reference can be measured during one experimental run (Fig. 25.2B). Both racks are thermoregulated with a Peltier element and a liquid cooling circuit that allows precise heating/cooling rates of up to 5 °C/min. However, experimental heating rates will be typically in the range of 0.1–0.5 °C/min. On the one hand, a slow heating rate increases the risk of sample degradation by depurination and hydrolysis at high temperatures. On the other hand, the heating ramp should be slow enough to exclude kinetic effects in the processes being observed. For example, the hairpin–duplex transition can be quite

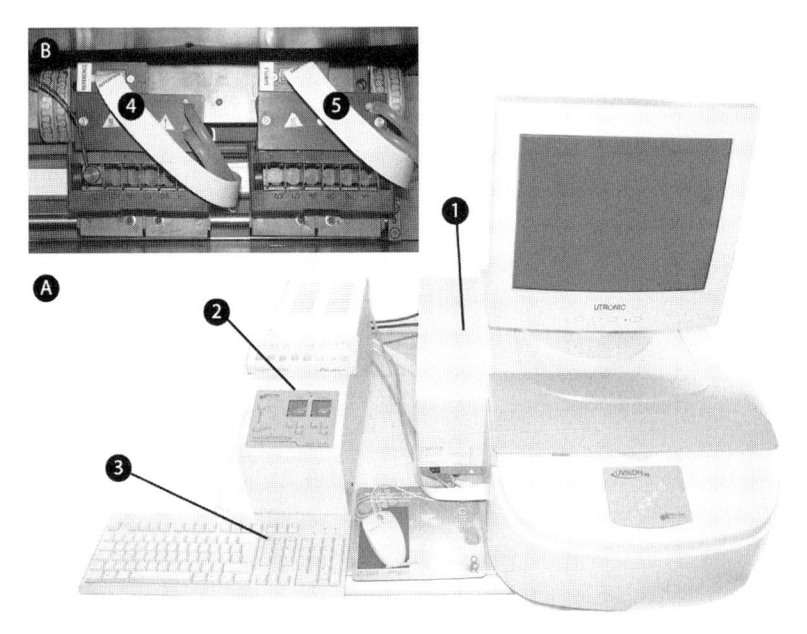

Fig. 25.2. (A) Setup for UV melting studies: UV spectrophotometer Uvikon XL (1), Temperature control unit (2) and computer interface (3). (B) Two independent cuvette racks (4,5).

slow and some group I introns fold on a timescale of minutes. RNA can be annealed prior to the melting run (5 min at 80 °C, then slow cooling to ambient temperature), but preferably in a more controlled way directly in the cuvettes, using the first heating/cooling cycle. Three or more heating/cooling cycles can be run overnight to ensure that results are reproducible. Data is collected from 10 to 90 °C, with data points every 0.2–0.5 °C. Samples would start boiling at higher temperatures and atmospheric pressure, and lower temperatures result in problems with condensation.

RNA can be prepared by transcription or chemical synthesis. A check of purity is advisable prior to spectroscopy. For annealing involving two or more strands, it is important to determine the concentration of each monomer precisely. Concentration is usually determined based on nearest-neighbor calculations of nucleotide extinction coefficients ε. However, depending on base composition, the calculated ε may deviate from the empirical value by up to 20% [6]. If necessary, the ε for any native structure can be determined experimentally by hydrolyzing the RNA enzymatically and measuring the A_{260} of the resulting nucleotides (see Appendix). Different cell path lengths can be used to allow measurements at different concentrations. We have used a set of custom-made 1-cm path length cuvettes (cat.-no. 115; Hellma, Germany) holding 500 µl of sample volume. Shorter path lengths are desirable for low sample concentrations and allow faster heat transfer.

Only little material is required – the hypochromic RNA must be adjusted to an absorbance of 0.1–0.2, corresponding to concentrations of 0.1–100 µM. To remain within the linearity range of the instrument, the maximum signal should not exceed an absorbance of 1.5. Measurements are normally done at the UV absorption maximum of AT base pairs at 260 nm. However, depending on base composition [7] and helix structure, measurements at a different wavelength may yield a better signal difference [8]. This suggests an initial wavelength run at lowest and highest temperature to determine the wavelength with the greatest difference in hypochromicity. Alternatively, the transition can also be monitored by observing circular dichroism as a function of temperature [5]. For model fitting, it may be necessary to obtain absorbance data at two different wavelengths simultaneously. This is because the hypochromicity of A-U and G-C basepairs is the same at 260 nm, but different at 280 nm, which can help to deconvolute two overlapping transitions.

Any ideal buffer for melting curves should not absorb any UV light and its pK_a should not vary with temperature. The standard buffer typically used is sodium cacodylate pH 7. Phosphate buffers, while being temperature insensitive, have the drawback of a high binding constant to divalents. Tris and HEPES show a too high pH variation with temperature. The pH of a Tris-buffered solution drops considerably when increasing the temperature from 35 to 90 °C (Fig. 25.3). For salt ions subject to deprotonation, the buffering capacity should also be considered: 100 mM NH_4^+ buffered in 50 mM Tris leads to a considerable drop in pH by about 1 unit, while the same solution in 10 mM sodium cacodylate (pH 7.0) buffer shows slight acidification (0.3 pH units at 100 mM NH_4). Depending on the salt, a higher buffer concentration may be advisable.

Choice of salt is also critical. Because divalent ions can cause hydrolysis of RNA at high temperatures, many experimenters include 0.1–1 mM EDTA. The major

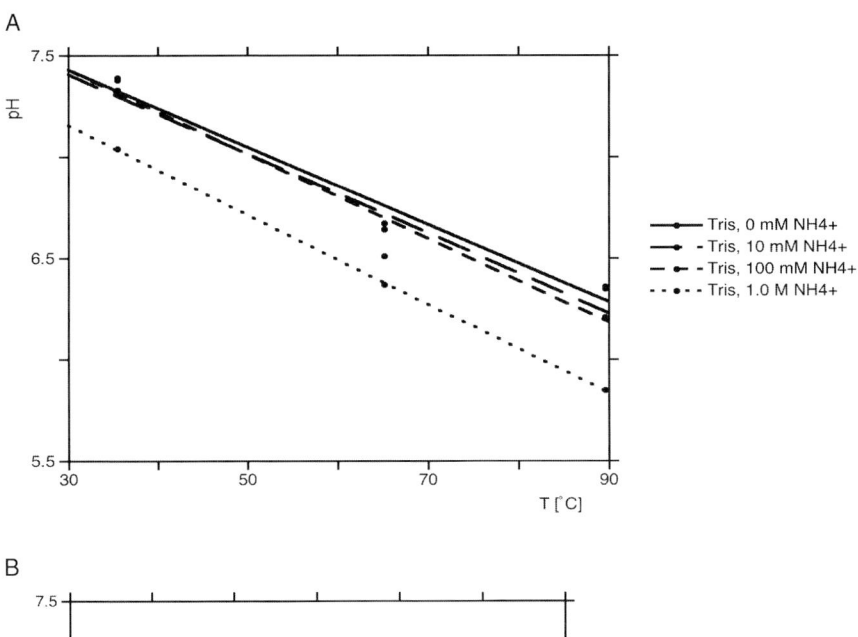

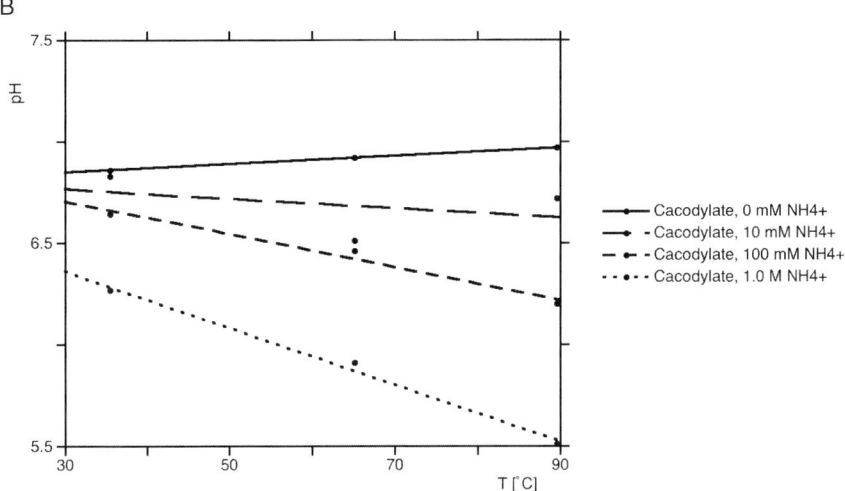

Fig. 25.3. (A) Temperature dependency of pH in Tris- or cacodylate-buffered solutions of NH$_4$Cl. Conditions: 50 mM Tris, pH 7.5 or 10 mM sodium cacodylate, pH 7.0 (100 mM NaCl, 0.5 mM EDTA).

role of Mg^{2+} is electrostatic, minimizing electrostatic repulsion between nucleic acid strands. The collapse to a compact, native-like state can be induced alternatively by low concentrations of divalents or high concentrations of monovalents, thereby avoiding the problem of hydrolysis. Roughly, each additional charge on the counterion decreases the required ion concentration by 2 log units [9]. The ion concentration should be varied to determine which transitions are visible in the experimentally accessible range (10–80 °C).

After each completed melting/annealing cycle, comparison between initial and final absorbance allows us to determine whether any evaporation or hydrolysis has occurred. This is a frequently encountered problem and some precautions can help to avoid it. Small path lengths seem to favor bubble formation, in this case the solution in each cuvette should be degassed prior to use by bubbling through an inert gas (N_2 or argon) or by applying a vacuum. The surface is sealed with 5–6 drops of PCR-quality mineral oil (Acros Organics, NJ). When applying the oil, care has to be taken not to touch the etched glass surface that should make a tight contact with the Teflon stopper or else the oil will spread and the sample will evaporate. The stopper is sealed with plumber's Teflon tape. For determination of derivative melting curves, minor evaporation losses are not critical.

It is possible to recover and precipitate the RNA after the measurements and separate it on a polyacrylamide gel to check for any degradation. This may be especially important if hydrolysis in the presence of a large concentration of divalent ions is a concern. Before precipitation, the oil can be removed from the aqueous phase by rolling the solution on a Teflon dish or by freezing-out. Separation of our RNA on denaturing 5% polyacrylamide gels containing 8 M urea followed by staining with toluidine blue O failed to detect any RNA degradation after three heating cycles.

25.3
Data Analysis

As stacked bases are hypochromic, thermal unfolding of RNA is accompanied by an increase in absorbance. By assuming a linear temperature dependence for absorbance coefficients of both folded and unfolded RNA, the transition between the two indicates the fraction of unfolded molecules in equilibrium at any given temperature (Fig. 25.4A). A lower (B_0) and upper (B_1) baseline has to be fit to each unfolding transition. In a graphical determination of T_m, the bisector between both baselines intersects the absorbance curve at the midpoint of transition where $\Delta G = 0$. A normalized representation as the fraction of folded molecules θ takes into account the baselines (Fig. 25.4B):

$$\theta = (B_0 - A_{260})/(B_0 - B_1)$$

This is the only way to determine T_m precisely. The data can be fitted with a sigmoidal function, where the peak of the derivative $d\theta/dT$ (or $d\theta/dT^{-1}$) indicates T_m. For intramolecular folding, it can be shown that $d\theta/dT^{-1}$ gives the precise value of the midpoint, while systematically overestimating the T_m of bimolecular annealing. However, most melting profiles with multiple transitions do not allow fitting of baselines with precision. Instead, we use the locally weighted least-square error algorithm provided with the data analysis software to obtain the derivative dA_{260}/dT (Fig. 25.1B). The estimated difference to the real T_m is in most cases no larger than ± 1–$2\,°C$.

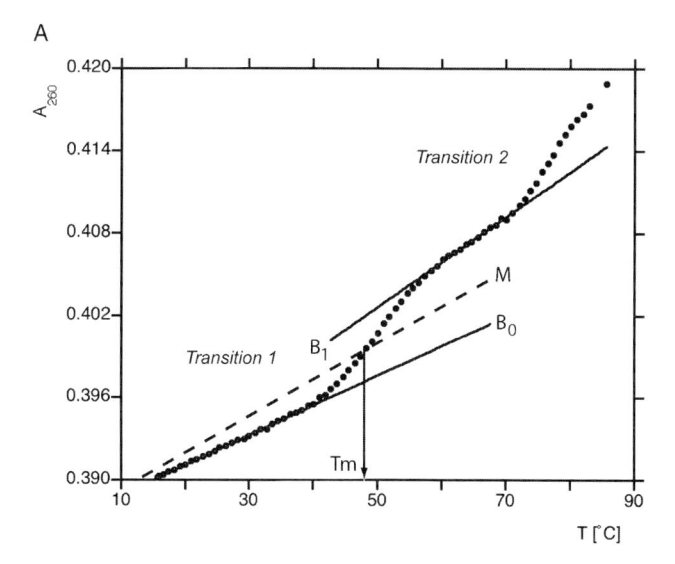

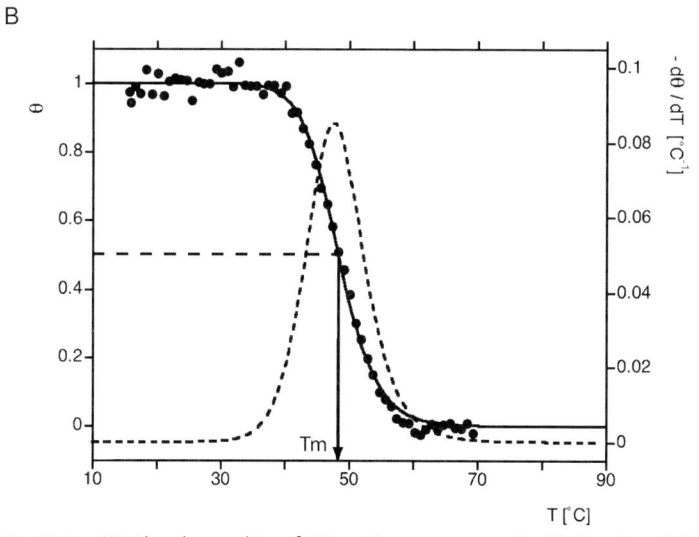

Fig. 25.4. (A) Absorbance data of RNase P RNA annealed in 10 mM sodium cacodylate, pH 7, 100 mM NaCl and 0.5 mM EDTA. The midpoint of transition T_m is defined by the bisector M between upper B_1 and lower tangents B_0. (B) Fraction of dissociated molecules θ calculated from baselines of transition 1 in (A) (left ordinate) and derivative $-d\theta/dT$ (right ordinate, dotted line).

A different approach is based on statistical mechanics. Using the tabulated nearest-neighbor enthalpies for Watson-Crick and non-canonical basepairs ("Turner rules"), the sequential unfolding of every basepair is considered independently to determine the corresponding partition function [4]. Additional fitting parameters are then added to account for the effect of tertiary interactions and make the theoretical curve coincide with the experimental one. Measuring at two different wave-

Tab. 25.1. Calculation of K_a.

Equation	$U \Leftrightarrow N$	$U_1 + U_2 \Leftrightarrow N$	$2U \Leftrightarrow N$
Reaction	monomolecular	bimolecular	bimolecular
Equilibrium constant $K_a =$	$\dfrac{\theta}{(1-\theta)^2}$	$\dfrac{\theta}{[U]_0(1-\theta)^2}$	$\dfrac{\theta}{2[U]_0(1-\theta)^2}$
Examples	Hairpin	Complementary oligos ($[U_1] = [U_2]$)	Autocomplementary oligos

U: unfolded RNA, N: folded RNA. Similar calculations are possible for equilibria involving n molecules [10].

lengths simultaneously allows to determine the minimum number of transitions that have to be considered [13]. Since secondary structures can also be predicted using nearest-neighbor free energy calculations, the peaks can be deconvoluted and, in the case of simple systems, assigned to secondary versus tertiary structure transitions. A number of software packages are available to help performing this kind of analysis, e.g. MeltWin (Windows, http://www.meltwin.com/), and Global-MeltFit (Macintosh, [4]).

25.4
Energy Calculations and Limitations

The spectrophotometric data for each sample consists of a single column of absorbance versus temperature. Sometimes, hysteresis is observed between the denaturation and the annealing curve. This is normally a kinetic effect and an indicator that the heating rate was too high to allow complete equilibration. Using a script, one can extract ascending and descending parts and put them into a table separately. Next, the file is imported into KaleidaGraph (Synergy Software, PA). A plot of absorbance versus temperature (Fig. 25.1A) is fit with the built-in smoothing function (window size 5). Data points of the smoothing function are then differentiated with respect to T and averaged over several runs. Since the derivative is calculated over a small data window (3–5 °C), very sharp transitions are more accurately reproduced by second-order polynomials, as performed by the Savitsky-Golay algorithm included in some software packages. The resulting derivative melting curve is plotted versus temperature (Fig. 25.1B) and contains a peak for each partial transition.

If T_m has been accurately determined, it can be used to calculate the energy parameters [10]. The first step consists in determining if the system can be approximated by any two-state model and accordingly which expression from mass action law is used to calculate the affinity constant K_a (Table 25.1). Next, the natural logarithm of K_a is plotted versus the inverse temperature T^{-1} (in Kelvin) following:

$$\Delta G^\circ = -RT \ln K_a = \Delta H^\circ - T\Delta S^\circ$$

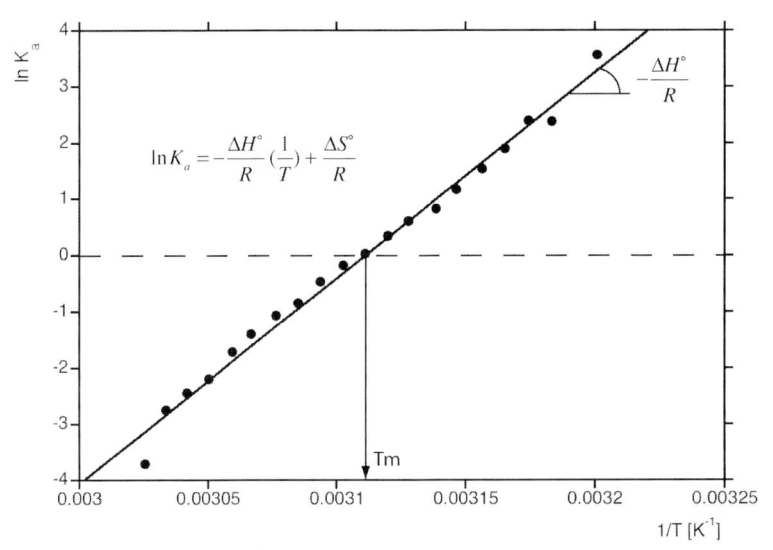

Fig. 25.5. Van t'Hoff plot of data in Fig. 25.4 (see text). K_a: affinity constant, $\Delta H°$: standard enthalpy, $\Delta S°$: entropy, R: molar gas constant.

where $\Delta G°$ is the free Gibbs enthalpy, $\Delta H°$ is the standard enthalpy, $\Delta S°$ is the entropy and R is the molar gas constant. Here, $\Delta H°$ also describes the temperature dependence of the equilibrium. By rearranging, one obtains the van t'Hoff expression:

$$\ln K_a = -\frac{\Delta H°}{R}\left(\frac{1}{T}\right) + \frac{\Delta S°}{R}$$

which allows us to determine $\Delta H°$ from the slope and $\Delta S°$ from the ordinate intersection (Fig. 25.5). In general, the energies determined by this method are less reliable than those determined by calorimetric methods, because they depend on several assumptions: (1) baselines have been determined correctly, (2) a two-state model is valid, (3) the system is in perfect equilibrium at all temperatures (absence of kinetic effects), and (4) $\Delta H°$ is temperature-independent and therefore $\Delta C_p = 0$ [11]. For example, for DNA duplex formation, ΔC_p has been shown to be quite negative [12].

It can be important to determine ΔC_p, which gives information about the exposed surface in folded and unfolded states. In that case, or if precise energy parameters are to be known, it is preferable to rely upon differential scanning calorimetry (DSC) or isothermal calorimetry (ITC) at varied temperatures. These methods provide necessary additional constraints to fit parameters to sequential models consisting of a succession of two-state transitions [4]. Measuring at two different wavelengths simultaneously allows to us determine the minimum number of transitions that have to be considered [13]. Since secondary structures can also be predicted using nearest-neighbor free energy calculations, the peaks can be deconvoluted and, in the case of simple systems, assigned to secondary versus tertiary structure transitions. Because calorimetric methods require large quantities of RNA, it can be useful to narrow down the number of buffer and salt conditions initially by UV melting, before carrying out a small number of calorimetric experiments [14].

25.5
RNA Concentration

For intramolecular annealing only, the T_m is independent of RNA concentration. In all other cases involving secondary and tertiary interactions between molecules, an increase in RNA concentration will facilitate association. Therefore, finding a concentration dependence of the melting profile is a clear indicator of dimerization, as is often observed with small hairpins. The dimerization will raise the T_m, which allows us to determine the enthalpy $\Delta H°$ of the association reaction. Data from a series of melting curves at different RNA concentrations is plotted as $1/T_m$ versus $\ln[RNA]_0$. Again provided that $\Delta C_p = 0$, the reaction enthalpy for the association of n molecules can be directly determined from the slope, $(n-1)R/\Delta H°$. In the case of a bimolecular association:

$$\frac{1}{T_m} = \frac{\Delta S°}{\Delta H°} + \frac{R}{\Delta H°} \ln\left(\frac{1}{2}[RNA]_0\right)$$

which follows from the equilibrium constant given in Table 25.1. Any precise determination supposes that $[RNA]_0$ can be varied over a large range while the absorbance remains within the linearity limits of the instrument.

25.6
Salt and pH Dependence

Increasing the ionic strength of the solution directly increases the melting temperature of the RNA (Fig. 25.6A). From polyelectrolyte theory it has been predicted that the T_m of a secondary structure transition will follow a linear dependence on the logarithm of ion concentration [15]. As for monovalent ions, the effect of divalent ions in solution can be explained by purely electrostatic effects. Through charge screening of the backbone phosphates, an increase in concentration allows the compaction of the structure. Taking into account specific Mg^{2+}-binding sites is only necessary at high background concentrations of monovalents [16] and in very specific cases of large RNAs, where Mg^{2+} locks the already compact, folded structure into the native state.

For helix melting, the T_m will vary by 16–17 °C for every log unit over a wide range of salt concentrations (Fig. 25.6B), but levels off at higher ionic strength. Thus the slope of salt dependence of T_m provides a clue to distinguish secondary from tertiary structure transitions. Moreover, the melting of a tertiary structure is

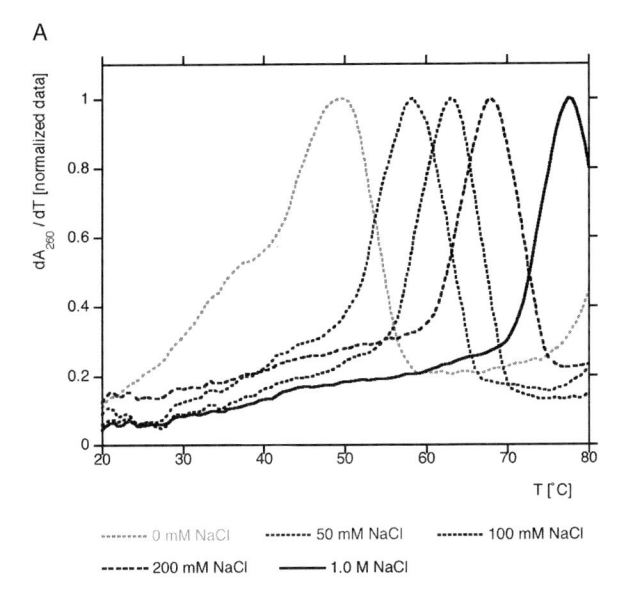

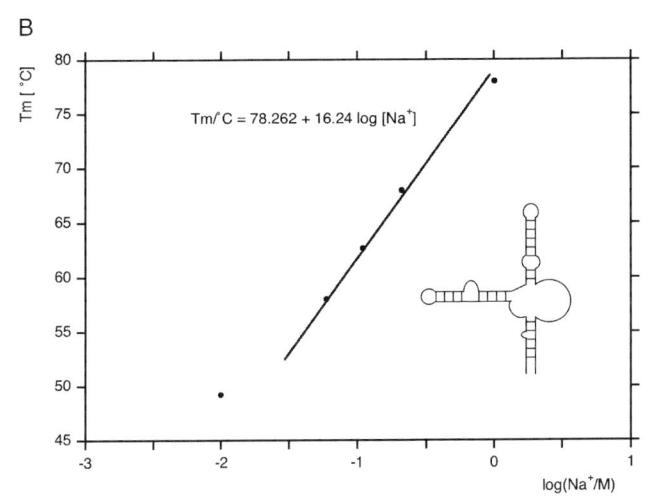

Fig. 25.6. Salt dependence of UV melting curves. (A) RNA annealed in 10 mM sodium cacodylate, pH 7, 0.5 mM EDTA and 0–1.0 M NaCl added. (B) plot of T_m versus $\log[Na^+]$.

much more dependent on the ionic radius than secondary structure [17]. Therefore, it can be useful to measure melting profiles in the presence of a series of cations with different radii. The pH can also greatly affect the melting temperature, depending on base composition. This is especially true for certain cytosines protonated on N^3. Extensive studies have been performed on the pH dependence of the T_m of pseudoknot motifs [18].

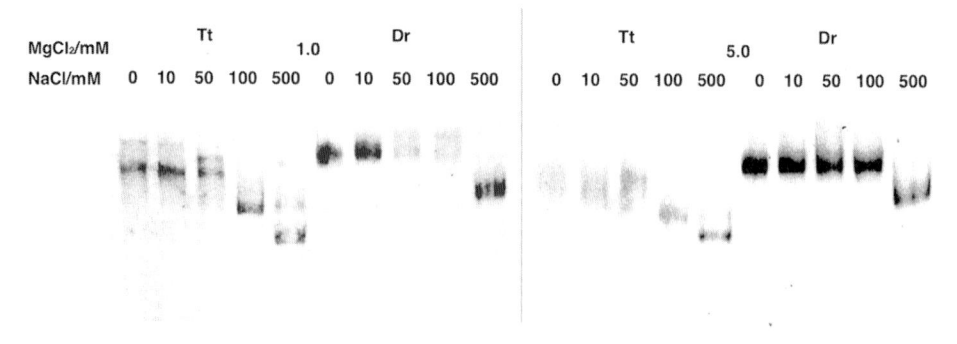

Fig. 25.7. Native gel electrophoresis (5% polyacrylamide). Tt and Dr RNase P RNA has been annealed in 10 mM sodium cacodylate, pH 7, 0–500 mM NaCl before addition of 1.0 (left) or 5.0 mM (right) $MgCl_2$.

25.7
Native Gels

A useful visual control of RNA folding is provided by native gels (Fig. 25.7). Consistent with what has been said for melting curves, the compaction of RNA structure is globally achieved by electrostatic effects. A 100× increase in concentration approximately compensates for the loss of one charge when passing from trivalent to divalent to monovalent ions [9]. Only final rearrangements may depend on the presence of Mg^{2+} at specific binding sites. Here we provide a protocol that can be adapted to the particular conditions of the RNA under investigation. Working under standard RNase-free conditions, RNA is heated to 90 °C for 5 min and annealed by slow cooling in the same buffer used for UV melting studies (10 mM sodium cacodylate, pH 7.0, 10–300 mM NaCl and/or 0–5 mM $MgCl_2$). $MgCl_2$ is added after cooling down to approximately 40 °C. RNA is either radiolabeled to approximately 30 000 c.p.m./μl or stained with the toluidine blue method.

Polyacrylamide gels are adjusted to correct separation range for the given RNA [19] and cast in 1 × TB running buffer (89 mM Tris, pH 8.3, 89 mM sodium borate). It is also possible to include 5% glycerol or low concentrations of Mg^{2+} in the gel matrix. For example, for a 5% gel, we mix 8.33 ml 30% acrylamide:bisacrylamide solution (Roth), 10 ml 5 × buffer, 31.5 ml H_2O, 150 μl 1 M $MgCl_2$, 500 μl 10% ammonium persulfate and 50 μl TEMED. To keep the RNA in its conformation, it is important to work quickly and in the cold room. The samples are mixed with 6 × loading buffer directly prior to charging (50% glycerol, 6 × TB buffer, 0.3% bromophenol blue, 0.3% xylene cyanol FF). The gel is run at 15 W and 4 °C. Alternatively, gel units for room temperature exist that are constantly cooled to 4 °C by a circulating water bath.

Finally, for concentrated RNA deposits (below 100 ng per band) the gel stain can be stained for 1 h with toluidine blue solution [0.5 g toluidine blue O (Sigma, MO) in 200 ml EtOH, 295 ml H$_2$O, 5 ml glacial acetic acid], followed by destaining (H$_2$O) and drying on a gel dryer. Alternatively, radioactively marked RNA is revealed by autoradiography or phosphoimaging.

Acknowledgments

A. W. acknowledges support from the European contract HPRN-CT2002-00190 (CARBONA).

References

1. Woodson, S. A. *Nat. Struct. Biol.* **2000**, *7*, 349–352.
2. Holbrook, S. R., C. Cheong, I. Tinoco, Jr, S. H. Kim. *Nature* **1991**, *353*, 579–581.
3. Gabarro, J. *Anal. Biochem.* **1978**, *91*, 309–322.
4. Draper, D. E., T. C. Gluick. *Methods Enzymol.* **1995**, *259*, 281–305.
5. Brescia, C. C., P. J. Mikulecky, A. L. Feig, D. D. Sledjeski. *RNA* **2003**, *9*, 33–43.
6. Kallansrud, G., B. Ward. *Anal. Biochem.* **1996**, *236*, 134–138.
7. Puglisi, J. D., I. Tinoco Jr. *Methods Enzymol.* **1989**, *180*, 304–325.
8. Mergny, J. L., A. T. Phan, L. Lacroix. *FEBS Lett.* **1998**, *435*, 74–78.
9. Heilman-Miller, S. L., D. Thirumalai, S. A. Woodson. *J. Mol. Biol.* **2001**, *306*, 1157–1166.
10. Marky, L. A., K. J. Breslauer. *Biopolymers* **1987**, *26*, 1601–1620.
11. Chaires, J. B. *Biophys. Chem.* **1997**, *64*, 15–23.
12. Rouzina, I., V. A. Bloomfield. *Biophys. J.* **1999**, *77*, 3242–3251.
13. Theimer, C. A., Y. Wang, D. W. Hoffman, H. M. Krisch, D. P. Giedroc. *J. Mol. Biol.* **1998**, *279*, 545–564.
14. Gluick, T. C., N. M. Wills, R. F. Gesteland, D. E. Draper. *Biochemistry* **1997**, *36*, 16173–16186.
15. Record, M. T., Jr. *Biopolymers* **1967**, *5*, 975–992.
16. Manning, G. S. *Biophys. Chem.* **1977**, *7*, 141–145.
17. Heerschap, A., J. A. Walters, C. W. Hilbers. *Biophys. Chem.* **1985**, *22*, 205–217.
18. Nixon, P. L., D. P. Giedroc. *J. Mol. Biol.* **2000**, *296*, 659–671.
19. Sambrook, J., D. W. Russell, J. Sambrook. *Molecular Cloning: A Laboratory Manual*, Cold Spring Harbor Laboratory Press, Cold Spring Harbor, NY, **2001**.

- Phenol/chloroform/isoamylalcohol (25:24:1), saturated with TE buffer (10 mM Tris–HCl, pH 8.0, 1 mM EDTA).
- Buffer D (20 mM HEPES–KOH, pH 8.0, 100 mM KCl, 0.2 mM EDTA, pH 8.0, 1.5 mM $MgCl_2$, 20% glycerol, in DEPC-H_2O, add fresh 0.5 mM PMSF and 0.5 mM DTT).
- 10 × buffer G (200 mM HEPES–KOH, pH 8.0, 1.5 M KCl, 15 mM $MgCl_2$).
- 10% glycerol solution (1 × buffer G, 10% glycerol, in DEPC-H_2O, add fresh per 100 ml 100 µl leupeptin, 50 µl 1 M DTT and 500 µl 0.1 M PMSF solution, filter through a 0.45-µm filter).
- 30% glycerol solution (same as 10% glycerol solution, but with 30% glycerol).
- SDS–PAGE protein sample buffer (2% SDS, 10% glycerol, 50 mM Tris–HCl, pH 6.8, 0.005% bromophenol blue).
- 5 × agarose gel loading buffer (5 × TBE, 10% glycerol, 0.025% bromophenol blue).
- Leupeptin (4 mg/ml in DEPC-H_2O).
- 16S and 23S rRNAs from *Escherichia coli* (Roche; 0206938).

26.2.3
Method

26.2.3.1 **Preparation of the Glycerol Gradient**

- Clean the gradient mixer with 100 and 50% ethanol, close valves, and rinse the chambers with 30 and 10% glycerol solution, respectively; make sure that no air bubbles become trapped in the connections.
- Place the gradient mixer on a magnetic stirrer, fill the chambers (valves closed) with 5.5 ml of 10 (chamber with the outlet pipe) and 30% glycerol solution without trapping air bubbles, add sterile magnetic stir bar to the chamber with the outlet pipe (Fig. 26.1).
- Place centrifugation tube on ice, fix sterile glass capillary to the pipe of the mixer and place it in the tube so that it touches the bottom.
- Start magnetic stirrer and open the valves, so that the centrifugation tube is slowly filled with the glycerol solution, the 10% solution being underlayed with the denser solution.
- Before pouring another gradient rinse the mixer again with 30% and 10% glycerol solution, respectively.
- Place the gradients at 4 °C for 1 h for equilibration.

26.2.3.2 **Sample Preparation and Centrifugation**

- Thaw 400 µl of HeLa nuclear extract slowly on ice, clear it of aggregates by a short spin, and mix it with 1.1 ml of 1 × buffer G freshly supplemented with 2 mM DTT and 0.5 mM PMSF.
- Carefully overlay the prepared gradients with the solution and balance the tubes together with the centrifugation buckets with 1 × buffer G.

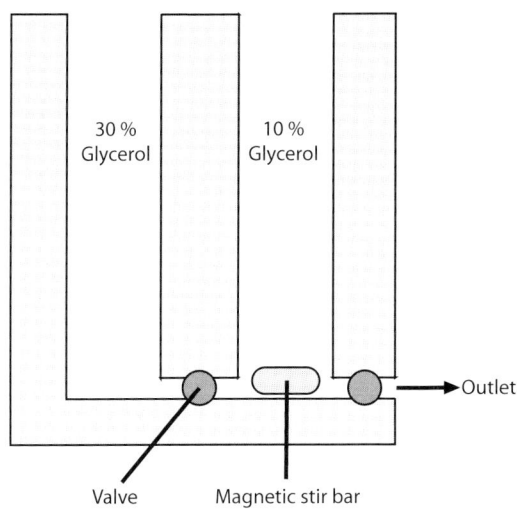

Fig. 26.1. Schematic display of the Hoefer Gradient Mixer for formation of glycerol gradients. For each chamber the concentration of the glycerol stock solution for preparation of a linear 10–30% gradient is given; the outlet pipe is indicated by an arrow.

- Centrifuge for 17 h at 32 000 r.p.m. (corresponding to 130 000 g) and at 4 °C in a precooled SW-40 rotor.
- Carefully fractionate the gradient from top to bottom by taking off 25 fractions of 500 μl each and resuspend the pellet in the last fraction.

26.2.3.3 Preparation of RNA from Gradient Fractions

- Split the fraction in two for easier handling during preparation of the RNA.
- Add 250 μl of phenol/chloroform/isoamylalcohol to each of the fractions, vortex thoroughly and centrifuge for 10 min at 14 000 r.p.m.
- Transfer upper aqueous phase to fresh Eppendorf tubes, and add 20 μl 3 M sodium acetate, 2 μl glycogen (20 mg/ml) and 550 μl 100% ethanol to each of the tubes and mix by inversion.
- Precipitate nucleic acids for 10 min at −70 °C, centrifuge for 20 min at 14 000 r.p.m. and 4 °C, remove supernatant from pellets, and wash at room temperature with 500 μl 70% ethanol.
- Take up pellets in DEPC-H_2O and combine the two aliquots from one fraction.
- Analyze RNA by denaturing PAGE.

26.2.3.4 Simultaneous Preparation of RNA and Proteins

In parallel to the isolation of RNA, proteins can be prepared from the same fractions. Because of the high concentration of proteins in some fractions an initial dilution step of the gradient fractions (1:1 with DEPC-H_2O) is recommended, when crude cellular extracts are fractionated.

- After phenolization and removal of the upper aqueous phase, add 5 volumes of acetone to the phenol phase for precipitating proteins.
- Mix and store samples for at least 1 h at −20 °C.
- Centrifuge for 30 min at 14 000 r.p.m., remove supernatant from pellet and wash with 500 µl of 80% ethanol.
- Dissolve pellet in a small amount (10–20 µl) of SDS–PAGE protein sample buffer, boil for 10 min and analyze by SDS–PAGE.

26.2.3.5 Control Gradient with Sedimentation Markers

Since 5S ribosomal RNA is no longer commercially available, it is recommended to analyze a silver-stained denaturing polyacrylamide gel of a gradient containing nuclear extract to detect the peak of free 5S rRNA (see Fig. 26.2). For the 16S and 23S sedimentation markers proceed as follows:

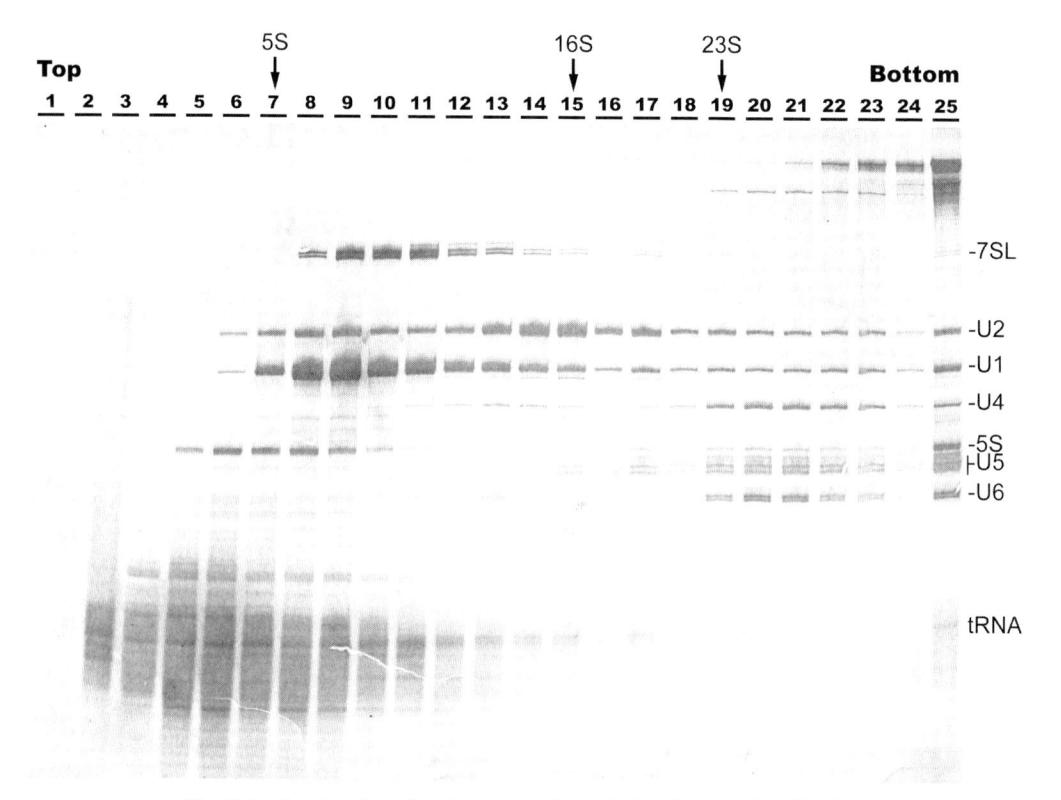

Fig. 26.2. Fractionation of nuclear extract by glycerol gradient centrifugation: RNA analysis. After ultracentrifugation of 200 µl of HeLa nuclear extract through an 11-ml linear 10–30% glycerol gradient (see Section 26.2.3.2 for conditions), 25 fractions of 500 µl each were taken from the top to the bottom of the gradient. RNA from 50 µl of each fraction was isolated and analyzed by denaturing PAGE and silver staining. Marker sizes (in nucleotides, Roche DIG V) are shown on the left, the identities of the RNAs are indicated on the right (U1, U2, U4, U5 and U6 snRNAs, 5S rRNA, 7SL RNA, and tRNAs). The positions of the 5S, 16S and 23S sedimentation markers are given on the top.

- Prepare gradient as described above.
- Mix 10 µl 16S/23S rRNA (4 µg/µl, Roche; 0206938) with 390 µl buffer D (or the buffer corresponding to the sample loaded on the analytical gradients).
- Add 1.1 ml 1 × buffer G and mix.
- Load onto gradient and run as described above.
- Fractionate gradient.
- Mix 12 µl of every second fraction with 3 µl agarose gel sample buffer and analyze on a 0.6% agarose gel in 0.5 × TBE.
- Visualize RNAs by ethidium bromide staining.

26.2.3.6 Notes and Troubleshooting

(1) Depending on the size range and properties of the RNPs to be separated, the running time (17 h), rotor type (SW-40) and conditions (32 000 r.p.m.; 4 °C) described in the example above (see Section 26.2.3.2) may be adjusted. Run the control gradient (see Section 26.2.3.5) under exactly the same conditions.

(2) Instead of glycerol, as described here, gradient sedimentation can also be performed in sucrose gradients (e.g. 15–30% [4] and 15–45%, [8]).

(3) Handle gradient fractions and prepared RNAs always on ice to minimize degradation. During RNA preparation make sure that only the aqueous phase is transferred to a fresh Eppendorf tube; do not touch the interface or the phenol phase which will lead to poor results upon analysis of the RNA due to contaminations. Before taking up the RNA in DEPC-H_2O make sure the pellet is completely dried and that no residual ethanol is left inside the Eppendorf tube. On the other hand, do not overdry the pellet (3–5 min at room temperature are normally sufficient), since otherwise dissolving it later in DEPC-H_2O may become difficult.

(4) Instead of using the Hoefer Gradient Mixer to prepare the gradients (see Section 26.2.3.1), the BioComp Gradient Master, a programmable, gradient-forming instrument (Frederickton, NB) is recommended for higher reproducibility and for faster preparation of the gradients.

(5) Make sure that the samples to be subjected to glycerol gradient centrifugation do not contain too much glycerol – here more than 10% – so that they do not sink below the surface of the prepared gradient upon loading. If necessary, dilute or dialyze against a suitable buffer to reduce the glycerol concentration.

(6) If samples obtained from the glycerol gradient centrifugation are to be subjected to immunoaffinity purification or other methods, we also recommend dilution or dialysis (e.g. with NET-100 buffer: 50 mM Tris–HCl, pH 8.0, 100 mM NaCl, 0.05% Nonidet P-40, 0.5 mM DTT) in order to reduce the high glycerol concentration.

(7) For visualization of RNAs in agarose gels by ethidium bromide staining (see Section 26.2.3.5) the ethidium bromide should be added directly to the agarose solution (final concentration of 500 ng/ml) before casting the gel; if the gel is stained after the run, degradation of the RNA may occur.

26.3
Fractionation of RNPs by Cesium Chloride Density Gradient Centrifugation

When applying isopycnic ultracentrifugation for the analysis of proteins or RNA–protein complexes, one has to consider that the high ionic strength may destabilize and dissociate the complexes, resulting in denaturation and precipitation of the proteins. Spliceosomal core snRNPs, however, are stable under high-salt conditions in the presence of 15 mM $MgCl_2$. This allows us to separate the snRNPs from free proteins that stay at the top of the gradient and from free RNAs that are pelleted. The sedimentation behavior of the snRNPs yields additional information on their protein:RNA ratio, since that determines their buoyant density ρ (see Section 26.3.3.4). As a typical example we describe here the fractionation of nuclear extract by cesium chloride density gradient ultracentrifugation.

26.3.1
Equipment

- Beckman Optima TLX benchtop ultracentrifuge.
- Beckman TLA 120.2 rotor with thick-walled polycarbonate centrifugation tubes (11 × 34 mM, Beckman 343778).
- Sterile, RNase-free 1.5-ml Eppendorf tubes.
- Cooling microcentrifuge.

26.3.2
Reagents

- DEPC-H_2O (see Section 26.2.2).
- 1 M $MgCl_2$ in DEPC-H_2O.
- 3 M sodium acetate, pH 5.2, in DEPC-H_2O.
- 20 mg/ml glycogen in DEPC-H_2O.
- 1 M DTT in DEPC-H_2O.
- 0.1 M PMSF in ethanol.
- Phenol/chloroform/isoamylalcohol (25:24:1) saturated with TE buffer (10 mM Tris–HCl, pH 8.0, 1mM EDTA).
- 10% SDS.
- 70 and 100% ethanol.
- Buffer D/$MgCl_2$ (see Section 26.2.2, but with 15 mM $MgCl_2$).
- CsCl stock solution (dissolve CsCl in buffer D/$MgCl_2$ to a final density of 1.55 g/ml, the easiest way to achieve this is to mix two buffer D/$MgCl_2$ solutions, one containing no CsCl, the other containing approximately 1.8 g/ml CsCl; adjust the CsCl density of 1.55 g/ml precisely, since this is critical for the reproducibility of the gradients. The amount x of CsCl (expressed in grams) needed to prepare 1 ml of a solution with the density ρ can also be calculated using the following formula [9]: $x = (\rho - 1)/0.92$ (ρ indicating the numerical value of the density expressed in g/ml). Note that it is not necessary to add PMSF and DTT to buffer D/$MgCl_2$ at this stage, store at room temperature, see Section 26.3.3.4).

26.3.3
Method

26.3.3.1 Preparation of the Gradient and Ultracentrifugation

- Thaw nuclear extract carefully on ice, clear it of precipitates by a short spin (1 min at 14 000 r.p.m. and 4 °C) and add $MgCl_2$ solution to a final concentration of 15 mM.
- Take an aliquot of the extract as input control.
- Supplement 3 ml of the CsCl stock solution freshly with 0.5 mM DTT and 0.5 mM PMSF.
- Mix 200 µl of the extract with 300 µl of the prepared CsCl solution.
- Pipette 500 µl CsCl stock solution supplemented with PMSF and DTT into a pre-cooled 1-ml polycarbonate tube (11 × 34 mM; part no. 343778) and overlay it carefully with the prepared extract CsCl mixture.
- Balance the tubes carefully with buffer D.
- Centrifuge at 90 000 r.p.m. for 20 h at 4 °C in a Beckman TLX tabletop ultracentrifuge, using a precooled TLA 120.2 rotor.
- Carefully fractionate the gradient from top to the bottom by taking off 10 100-µl fractions; in the 10th tube collect the pellet in the residual gradient solution by resuspension.

26.3.3.2 Preparation of RNA from the Gradient Fractions

- Add 300 µl of DEPC-H_2O and 40 µl of 10% SDS to each fraction, then add 400 µl phenol/chloroform/isoamylalcohol, vortex thoroughly and separate phases by centrifugation (10 min, 4 °C, 14 000 r.p.m.).
- Transfer upper aqueous phase to a new Eppendorf tube and add 40 µl of 3 M sodium acetate, pH 5.2, 2 µl glycogen (20 mg/ml) and 1 ml ethanol.
- Mix solution by inversion of the tube and incubate for 10 min at −70 °C.
- Precipitate nucleic acids by centrifugation for 20 min at 14 000 r.p.m. and 4 °C, remove supernatant from pellets and wash with 500 µl of 70% ethanol (room temperature).
- Analyze RNA by denaturing PAGE (Fig. 26.3).

26.3.3.3 Control Gradient for Density Calculation

- Prepare a gradient as described, replacing the nuclear extract by buffer D.
- Run gradient and fractionate as described above.
- Precisely weigh an aliquot of each fraction (e.g. 50 µl), which gives the density distribution across the gradient.

26.3.3.4 Notes and Troubleshooting

(1) By measuring the buoyant density of a particle the approximate percentage of protein mass in the complex can be calculated using the following empirical

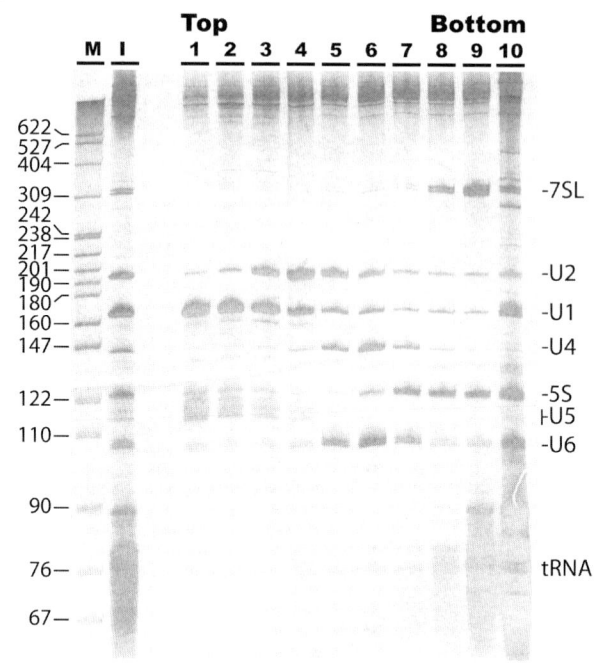

Fig. 26.3. Fractionation of nuclear extract by CsCl density gradient ultracentrifugation: RNA analysis. After ultracentrifugation of 200 μl of HeLa nuclear extract in a 1-ml CsCl density gradient (see Section 26.3.3.1 for conditions), 10 fractions of 100 μl each were taken from the top to the bottom of the gradient. RNA was prepared from each fraction (1–10, 10 including the pellet) and analyzed by denaturing polyacrylamide gel electrophoresis, followed by silver staining. Marker sizes (pBR322 DNA digested with HpaII) are shown on the left, RNAs are indicated on the right. For example, the U4/U6 snRNP with base-paired U4 and U6 snRNAs peaks in fractions 5–7, free tRNAs are found in the pellet fraction (10). For comparison, lane I contains RNA prepared from 10 μl of HeLa nuclear extract corresponding to 10% of the input.

formula [9, 10]: % protein $= (1.85 - \rho)/0.006$ (ρ indicating the numerical value of the density expressed in g/ml). For example, densities of 1.36 and 1.51 g/ml, respectively, were observed for the *Trypanosoma brucei* U5 and U4/U6 core snRNPs; these values correspond to protein ratios of 82 (U5 core snRNP) and 57% (U4/U6 core snRNP). Taking the known masses of the RNA components into account, this results in total protein masses of 93 kDa per U5 core complex and 89 kDa per U4/U6 core complex [11].

(2) Instead of the standard CsCl, Cs_2SO_4 has also been used in isopycnic density gradient centrifugation (see, e.g. [4, 12, 13]); note that this appears to result in different stringencies for the RNPs (as discussed for the trypanosomal U2 snRNP [14]).

(3) For the preparation of the CsCl stock solution it is not recommended to add PMSF and DTT to buffer D/$MgCl_2$, since they are degraded during storage. In-

stead, PMSF and DTT should be added to the CsCl stock solution immediately prior to gradient preparation (see Section 26.3.3.1).

(4) Due to the high ionic strength of the fractions prepared from the gradient it is recommended to dilute or to dialyze the fractions (e.g. against buffer D) before subjecting them to immunoaffinity purification or other methods (see, e.g. [15]). The high ionic strength also interferes with degradation of the RNAs, nevertheless it is recommended to handle the obtained samples on ice, especially after removal of the salts by precipitation and washing. During phenol/chloroform extraction make sure that only the upper, aqueous phase is transferred to a new Eppendorf tube without touching the interface or the phenol phase, as this may lead to poor results during analysis of the RNA due to contaminations.

(5) Always completely remove ethanol from the RNA pellet by drying (which normally takes 3–5 min at room temperature) before taking it up in water. Avoid overdrying the pellet, since this may make dissolving it in DEPC-H_2O difficult.

Acknowledgments

This work was supported by grants from the Deutsche Forschungsgemeinschaft (Bi 316/9-2 and 10-3).

References

1 C. L. WILL, R. LÜHRMANN, Curr. Opin. Cell Biol. 2001, 13, 290–301.
2 C. BRUNEL, J. S. WIDADA, M. N. LELAY, P. JEANTEUR, J. P. LIAUTARD, Nucleic Acids Res. 1981, 9, 815–830.
3 C. BRUNEL, G. CATHALA, Methods Enzymol. 1990, 181, 264–273.
4 T. PEDERSON, J. Mol. Biol. 1974, 83, 163–183.
5 C. HEUBECK, A. SCHÖN, Methods Enzymol. 2001, 342, 118–134.
6 J. D. DIGNAM, R. M. LEBOVITZ, R. G. ROEDER, Nucleic Acids Res. 1983, 11, 1475–1489.
7 E. M. MAKAROV, O. V. MAKAROVA, H. URLAUB, M. GENTZEL, C. L. WILL, M. WILM, R. LÜHRMANN, Science 2002, 298, 2205–2208.
8 G. AST, D. GOLDBLATT, D. OFFEN, J. SPERLING, R. SPERLING, EMBO J. 1991, 10, 425–432.
9 M. G. HAMILTON, Methods Enzymol. 1971, 20, 512–521.
10 A. S. SPIRIN, Eur. J. Biochem. 1969, 10, 20–35.
11 S. LÜCKE, T. KLÖCKNER, Z. PALFI, M. BOSHART, A. BINDEREIF, EMBO J. 1997, 16, 4433–4440.
12 S. MICHAELI, T. G. ROBERTS, K. P. WATKINS, N. AGABIAN, J. Biol. Chem. 1990, 265, 10582–10588.
13 W. SZYBALSKI, Methods Enzymol 1968, 12, 330–360.
14 M. CROSS, A. GÜNZL, Z. PALFI, A. BINDEREIF, Mol. Cell. Biol. 1991, 11, 5516–5526.
15 Z. PALFI, G.-L. XU, A. BINDEREIF, J. Biol. Chem. 1994, 269, 30620–30625.

27
Preparation and Handling of RNA Crystals

Boris François, Aurélie Lescoute-Phillips, Andreas Werner and Benoît Masquida

27.1
Introduction

The crystal structures of prokaryotic ribosomal subunits [1–3], as well as sequence analysis coupled to molecular modeling, have demonstrated that RNA structure is modular [4, 5]; in other words, it can be decomposed in individual building blocks (modules), recurrently found in various RNA molecules, that are assembled together to form the overall RNA fold.

Because ribosomal subunits are the most abundant native particles of a growing cell, there is no need to use *in vitro* transcription or chemical synthesis to overproduce them. This is crucial since it 'reduces' molecular handling to biochemical purification of native particles. However, in most cases, overproduction techniques are needed and the biochemist faces subsequent RNA folding or protein–ligand/RNA association problems, unless the RNA is rather short. This partly explains why small RNA structures can be solved fairly quickly, whereas RNAs beyond 100 nt require time-consuming biochemical characterization before successful crystallization. In fact, except ribosomal subunits, only three RNA structures over 100 nt have been solved to date [6–8]. Thus, the folding of individual RNA motifs is apparently easier to control than the whole assemblies they are part of. Consequently, the structure of these motifs can be studied individually. Here, we focus on intermediate-size RNA motifs either produced by *in vitro* transcription or chemical synthesis (below 50 nt) [9–12].

The motifs of interest are extracted from a large RNA assembly and sequences are designed in order to favor a unique secondary structure. This can be assessed by using *in silico* folding programs [13–15] and native PAGE techniques (see Chapter 25). When the RNA has a substrate, further biochemical characterization might be required. Then the fragments are either cloned and *in vitro* transcribed or chemically synthesized. In this chapter we describe experimental procedures routinely used in the laboratory to obtain highly pure RNA molecules suitable for crystallization studies. Once the RNA has been synthesized, it has to be purified and concentrated. The correct length product is separated from contaminants by gel elec-

Handbook of RNA Biochemistry: Student Edition. Edited by R. K. Hartmann, A. Bindereif, A. Schön, E. Westhof
Copyright © 2009 WILEY-VCH Verlag GmbH & Co. KGaA, Weinheim
ISBN: 978-3-527-32534-4

trophoresis and/or chromatography. The RNA is then eluted, concentrated and desalted. Folding assays are performed prior to crystal screening.

27.2
Design of Short RNA Constructs

RNA structures can be seen as assemblies built from a construction set consisting of building blocks of various shape and complexity which obey conservation rules at the level of sequence and structure. In order to understand RNA architecture, it is therefore necessary to elucidate the structure of these building blocks. To achieve this goal, the motifs are analyzed in their wild-type contexts and, after alignment of the sequences, designed so as to ensure that they will conserve their original structure. It is worth to note that the best situation is when the secondary structure is well supported by biochemical data and that the edges of the motif (5' and 3' ends) are well located. Attention should be given to the design process in order to increase the probability for the structure to adopt the wild-type conformation. The stability of the constructs can be evaluated using UV-melting techniques under various ionic strengths (see Chapter 25).

Design of short RNA constructs is greatly helped by *in silico* folding programs [13–16]. They allow for testing modifications of the secondary structure upon modifications of the various base pairs. This step is crucial because apparently insignificant events such as the reversal of a GC pair can sometimes have major implications. A second point to think about is that the RNA motif may not reveal any propensity to pack and consequently to grow crystals. To address this problem it is advised to design constructs exhibiting various edges producing different situations regarding length and sequence that would eventually be more favorable for crystal growth.

27.3
RNA Purification

RNA molecules can be purified either by PAGE or liquid chromatography (HPLC or FPLC). These methods can even be coupled to improve the results. While PAGE is applicable to any RNA length, HPLC is dedicated to RNA up to about 35 nt long, but the latter method is always useful to cleanse RNA preparations purified on gels. Routine techniques mentioned in this chapter are described in more detail in [17].

27.3.1
HPLC Purification

When the RNA oligonucleotide is shorter than around 35 nt, it can be purified using FPLC or HPLC techniques. The best results are obtained using salt gradients

Tab. 27.1.

	Buffer A	Buffer B
MES	20 mM pH 6.2	20 mM pH 6.2
Urea	4 M	4 M
NaClO$_4$	1 mM	400 mM

A sodium perchlorate gradient is run over 70 min from 15 to 70% buffer B with a 1 ml/min flow rate [MES: 2-(N-morpholino)ethanesulfonic acid]. In the lab we run these gradient on a Dionex system equipped with a Nucleopac-PA-100 (0.45 × 25 × π mm^3) column. After a wash step at 90% buffer B, the column is re-equilibrated in 15% of buffer A to prepare the next run.

on anion exchange columns bearing quaternary amines like mono-Q matrices. HPLC presents the advantage that the column can be heated in an oven to temperatures up to 90 °C, thus promoting the unfolding of the RNA and increasing the retention time on the column for a better separation. The addition of chaotropic agents like urea or formamide enhances the effect of heating the sample. However, formamide should be used with caution in the presence of RNA. Heat leads to formamide decomposition into carbon monoxide and ammonia – the latter can very quickly hydrolyze the RNA preparation. In such denaturing conditions, the RNA mix is fractionated according to the size of the present species, close to what can be achieved using gel electrophoresis. A typical protocol is described in Table 27.1. Other HPLC purification procedures for RNA have been described elsewhere [18].

Several pitfalls should nonetheless be avoided. If the RNA has been produced by *in vitro* transcription, proteins should be removed by phenol/chloroform extraction. Otherwise the column bed may get coated with proteins and the column will lose its loading capacity over time. The sample should be assayed for precipitation by mixing with the highest salt buffer that is going to be used for separation to avoid clogging the HPLC. Since the sample is going to be heated, divalent ion contamination should be avoided. Hence, the pK_a value of the buffer should be in the acidic range so as to minimize spontaneous hydrolysis of phosphodiester bonds in the case of a contamination with divalent cations. To achieve this, we recommend to use PEEK (poly-ether-ether-ketone)-coated pumps as well as peek tubing.

27.3.2
Gel Electrophoresis

RNAs of any size (up to 500 nt) can be purified efficiently using PAGE under denaturing, semi-denaturing or native conditions. Various urea concentrations can be tried at an analytical scale before going to preparative scale in order to identify the most appropriate protocol. A sequencing electrophoresis apparatus with an aluminum back to homogenize the glass plates' temperature allowing the use of 30 × 40 cm^2 gel plates is recommended. The running temperature is usually set between

50 and 60 °C. Of course, the gel thickness has to be significantly increased when going to preparative scale (use at least 1.5-mm thick spacers) to separate precisely the RNAs of different length in the sample. The volume of gel to be prepared is thus around 250 ml.

The following equipment is needed to set up the experimental procedure:

- PAGE equipment: electrophoresis apparatus, siliconized glass plates (around 30×40 cm^2), comb and spacers (at least 1.5-mm thick).
- Acrylamide 20%/urea 8 M stock solution (made from 500 ml acrylamide-bisacrylamide 38% (w/v) acrylamide, 2% (w/v) N,N'-methylene bisacrylamide mixed with 480.5 g urea and Millipore water to give a volume of 1 l).
- $10 \times$ TBE (Tris–borate–EDTA buffer [17]).
- 8 M Urea solution.

Once a gel of the appropriate acrylamide percentage has been prepared (see [17]), the RNA solution to be fractionated is mixed with one volume of 8 M urea and then loaded onto the gel. The preparative gel electrophoresis usually requires a power value around 25 W. Progress of the migration is followed by the course of the bromophenol and xylene cyanol dyes loaded in a lane containing no RNA. At the end of the migration, glass plates are removed, the gel is wrapped in plastic film, and the RNA bands are visualized by UV shadowing using a UV lamp and a silica plate as a screen (see Chapter 3 for details).

27.3.3
RNA Recovery

The RNA contained in the visualized bands has to be eluted from the gel and concentrated before subsequent experiments. To achieve this goal, the bands are first delineated with an indelible marker on the plastic wrap. Then, they are cut out of the gel using a sterile scalpel blade.

27.3.3.1 Elution of the RNA from the Gel
The oligoribonucleotides are recovered by passive elution at 4 °C in Millipore water. The gel is crushed in a mill (A11 basic analysis mill; IKA) and poured in a 50 ml polypropylene conical tube with water (around 30 ml of gel under preparative conditions). The RNA-containing tube is placed in a "rock & roll" stirrer at 4 °C overnight. Finally, the eluted RNA solution is filtered on a 0.22-µm sterile filtration unit (Nalgene) to get rid of the acrylamide particles.

27.3.3.2 Concentration and Desalting
Whatever the technique employed to purify the RNA, it is necessary to desalt and concentrate it prior to use in crystallization trials. A very efficient way of achieving this is to use reverse-phase Sep-Pak columns that can be used on the bench (Waters Sep-Pak C$_{18}$ Classic short-body). These are operated by gravity or using a syringe.

A classical protocol consists of the following steps:

(1) Plug the column inlet to the luer of a 10 ml syringe and fix it to a bench stand.
(2) Equilibrate the column using 10 ml of methanol.
(3) Pass through 10 ml of Millipore water.
(4) Load the sample.
(5) Wash the sample with 10 ml of Millipore water.
(6) Elute the sample with 5 ml of water/acetonitrile (1:4) in 1 ml fractions.

Three facts should be kept in mind when using Sep-Pak cartridges. The pH of the sample should not exceed 7 to guarantee efficient binding to the column bed. The loading step should not exceed 10 min to minimize loss of material due to driving by the mobile phase. If loading would take longer, the sample should be fractionated on more than one column. The column should never run dry to prevent the loss of the sample. Hence, the syringe luer should be removed with caution in intermediate steps. The next solution should be added when there is still a small volume (100 µl) of the previous phase in the syringe.

The RNA-containing water/acetonitrile solution is then evaporated to dryness in a Speed Vac. The pellet can be resuspended in the solution of choice for further studies.

27.4
Setting Crystal Screens for RNA

After purification of a sufficient amount of RNA, conditions for crystallization have to be found. An economical screening method should use the least possible amount of RNA. Therefore, it is recommended to start by screening a large number of combinations, and then switch to other methods to optimize crystal shape and size. For a broad general screen, specific crystallization sparse matrices for proteins or nucleic acids have been published [19–22] and some are commercially available (www.hamptonresearch.com, www.decode.com, www.nextalbiotech.com). These have been designed based on extensive mining of previously published crystal-yielding conditions. Although the general considerations for crystal screens of proteins equally apply to RNA, some particularities have been identified. If no crystals have been obtained during the first trials, it is often more promising to vary the sequence instead of sampling a larger variety of combinations. In the crystallization process, sequence and shape of the molecules will drive the nucleation and subsequent crystal growth through a network of packing interactions mediated among symmetry-related molecules. Considering RNA, these factors have even more drastic effect than for proteins since the former are usually less globular in shape than the latter. Thus, various RNA constructs with different sequences and helix length, in other words with different shapes, should be tried when no crystals appear for a given construct [21, 23]. Chemical synthesis makes this process relatively straightforward for small RNAs (below 30 nt). Higher crystallization temper-

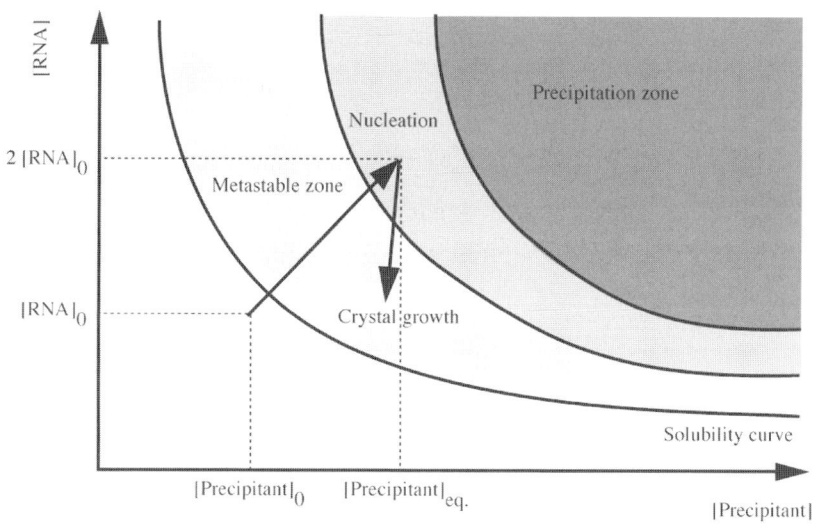

Fig. 27.1. Solubility diagram. During equilibration, the concentration of both precipitant and macromolecule increase until precipitation occurs. The formation of crystal nuclei reduces the amount of solvated macromolecule and allows the system to remain in the metastable zone where crystals can grow.

atures (37 °C) also seem to favor formation of RNA crystals. Finally, the choice of crystallization method (vapor diffusion or batch crystallization) may also influence the success.

Once suitable starting conditions have been found, the strategy consists of a rational variation of conditions. The crystallization process, as visualized in the phase diagram (Fig. 27.1), is influenced by numerous variables $x_1, x_2, \ldots$ (called *factors*), i.e. RNA sequence, crystallization temperature, buffer and pH, and kind and percentage of precipitant and salts. Each of these factors can be adjusted to different *levels* (for example, factor [LiCl] to 150 mM and factor [MPD] to 25%). In order to quantitate each observation (clear drop, precipitate, spherulites, microcrystals or crystals), an arbitrary score (*response*) is assigned to it and represented on a multi-dimensional *response surface* $f(x_1, x_2, \ldots)$. The aim of the crystallization screen is to explore this surface (Fig. 27.2) where the expected summit would yield the optimal result (best crystals). Yet, in practice, only a limited number of all possible factor combinations can be tested. The simplest approach would be varying a single factor at a time, while keeping all others constant. While it is possible to reach the optimum by sheer luck, the response surface shows that the score will more likely converge to a plateau or local maximum. Furthermore, the results of each testing series cannot be generalized and interactions between different factors are neglected.

These problems are avoided using *experimental design*, where multiple factors are varied simultaneously between different crystallization trials. Each experiment n represents a combination $Cn = (x_1, x_2, \ldots, x_k)$ of k factors. For each combination,

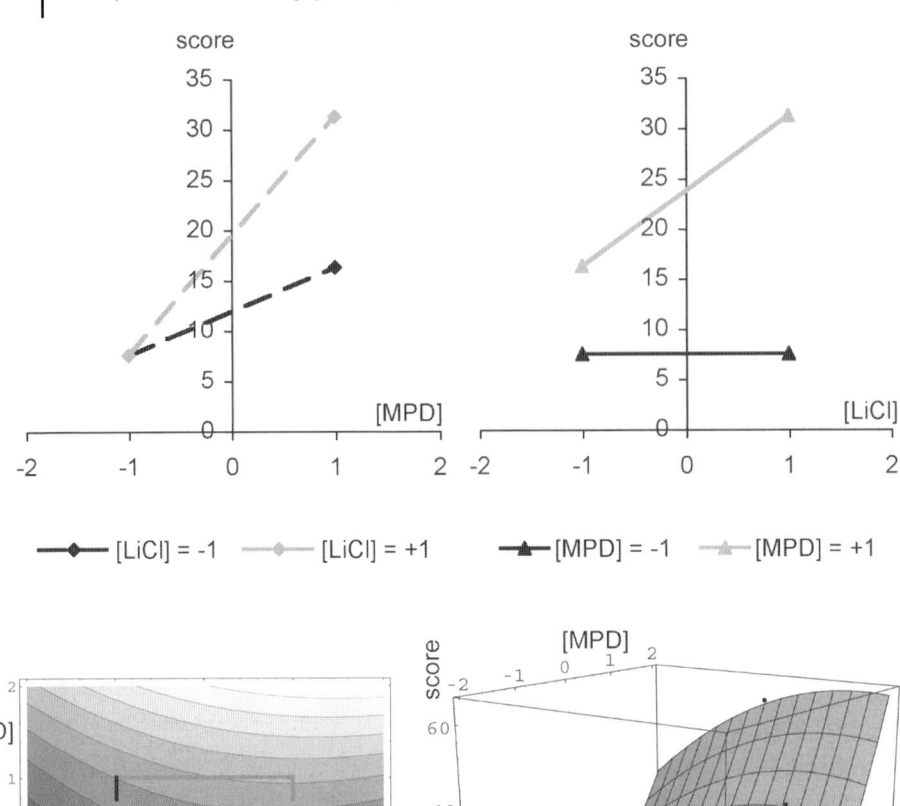

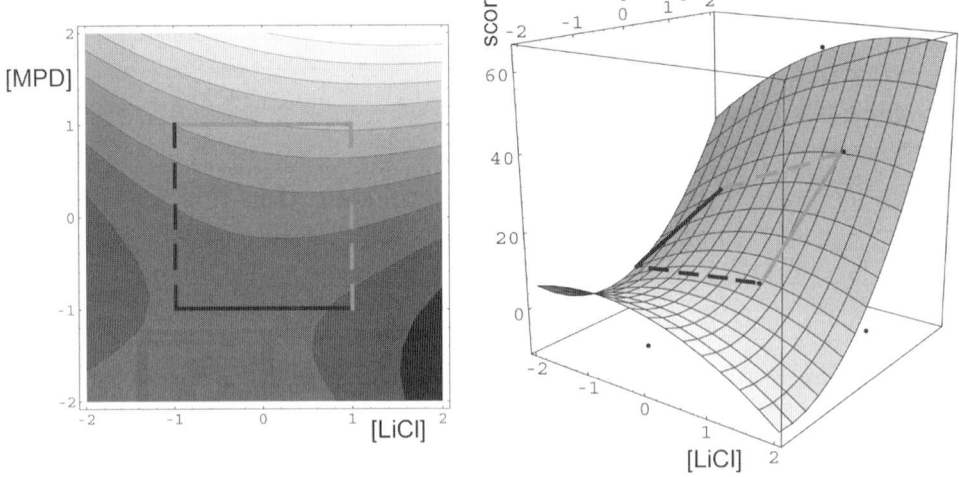

Fig. 27.2. (Top) Scoring of influence of two factors on crystal growth (see text, score in arbitrary units). (Bottom) response surface fitted to result, top view (left) and three-dimensional view (right).

multiple factors are changed simultaneously according to a predefined plan. A detailed exploration of this powerful technique is beyond the scope of this text and instead can be found in books on engineering statistics, or see references [24] and [25] for its application to crystallization. In general, at least two levels are defined for each factor at a chosen, equal distance around a central point, staking out a well-balanced experimental space in which the response of the system (the quality of crystals) is noted. Rather than just trying out some extreme values, it is important to choose reasonable values within a well-known range to avoid veering towards some jagged region, or close to an asymptote. The different levels of each factor are coded in coefficients. In our two-factor example ($k = 2$), both 50 mM/250 mM [LiCl] and 10%/40% MPD would be listed in an experimental design matrix and coded as $+1/-1$. Following scoring, the response surface spread out over all combinations indicates the direction where the best score is to be expected (Fig. 27.2). If k is not too large, one can also take into account possible interactions between factors. If factor x influences factor y, then the response surface $f(x_1, x_2)$ is not only a polynomial of $c_1 \cdot x_1$ and $c_2 \cdot y_2$, respectively, but the additional interaction term $c_3 \cdot x_1 x_2$ also has to be considered. To help with the design process, several computer programs have been made available [26]. A number of different designs have been coined with the common goal of reducing the number of experiments without compromising the well-balanced exploration of the experimental space.

Initial screens can be distinguished between those used to determine what factors are most important, and follow-up screens that allow optimization and improvement of crystal quality (Table 27.2). In experimental design, this is known

Tab. 27.2. Application of various experimental designs in crystallization.

Field of application	Experimental design	Factors and levels	No. of experiments	Comments
Initial screen: which factors are most important?	full 2-level factorial design	k factors at 2 levels	2^k	accounts for interactions between factors, but too many experiments necessary if $k > 5$ factors
	incomplete 2-level factorial design		2^{k-p}	p factors are confounded, effect of interactions cannot be evaluated, but less experiments necessary
	Plackett–Burman design		$k+1$	greatly reduced number of experiments; interaction bias neglected
Optimization	steepest ascent	k factors at 2 levels		follow-up on 2-level design: reduction of step size when approaching maximum
	central composite Box–Behnken; Hardin–Sloane	k factors at 5 levels		follow-up on steepest ascent design. quadratic model, approach to optimum
	Randomized block designs	3–5 factors at 3–4 levels		one single factor of primary interest interaction bias neglected
	Simplex matrix	k factors at 3 levels	initially $k+1$	iterative triangulation towards optimum

as the "Box–Wilson strategy" [27]. The first group of screens is generally based on a so-called *factorial plan* which determines the polynomial coefficients of a function with k variables (factors) fitted to the response surface. It can be shown that the number of necessary experiments n increases with 2^k if all interactions are taken into account. Instead of running an unrealistic large number of initial experiments, the full factorial matrix can be advantageously replaced by a fractional factorial matrix or a Plackett–Burman design [28]. Here, interactions between factors are partially or completely neglected. For example, if a multiplicative effect of salt concentration and MPD can be ruled out, the interaction between these factors can be neglected, thereby reducing the number of necessary experiments.

Based on the response surface obtained, a second round of optimization follows, using the *steepest ascent method* where the direction of the steepest slope indicates the position of the optimum. Alternatively, a quadratic model can be fitted around a region known to contain the optimum somewhere in the middle. This so-called *central composite design* contains an imbedded factorial design with center points, points at $-1/+1$ and an additional group of outlying "star points" α as upper and lower limits, which allows an estimation of the curvature (see Fig. 27.3 for an example with combinations of three factors). There are alternative designs, if the number of factors is small and optimization is the main goal. *Randomized block designs* (Latin Squares and Greco-Latin Squares) are useful if there is one main factor to consider. The design helps to separate it from the influence of *nuisance factors* that may affect the measured result, but are not of primary interest.

Finally, the simplex design has also been adopted for crystallization purposes [29]. This is an iterative approach starting with one more combination than factors under investigation. In an example with three factors at three equally spaced levels 0, p and q, the first set consists of combinations C0 (0, 0, 0), C1 (p, q, q), C2 (q, p, q) and C3 (q, q, p) (Fig. 27.3). Combination C2 giving the worst result, it is replaced in the following round by combination C4 with coordinates exactly opposite to C2, where the mirror plane is defined by C0, C1 and C3. Comparing these

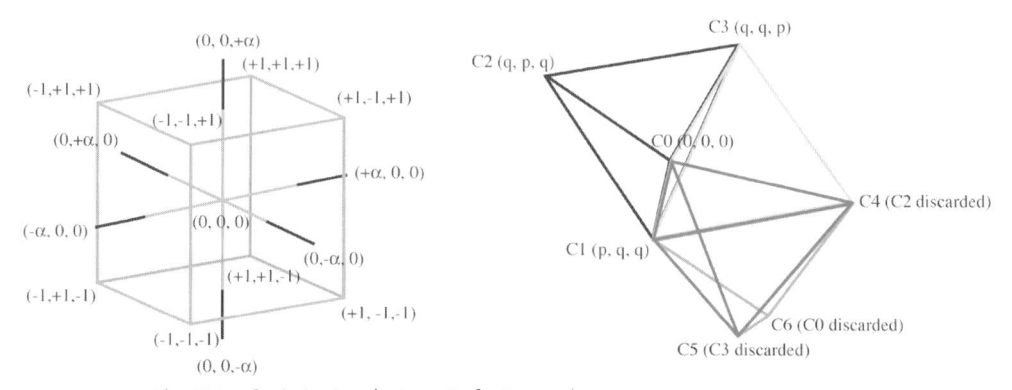

Fig. 27.3. Optimization designs. (Left) Centered composite design with three factors. (Right) Three factors optimized in four steps using a simplex design.

points with C4, the worst result is now C3 and therefore replaced by mirror point C5, and so forth. After several rounds of triangulation in the experimental space, the optimum is reached when no further improvements are observed. While multiple rounds of optimization are required, this extremely economical approach is especially useful when too little sample is available for extended factorial plans.

27.4.1
Renaturing the RNA

Prior to setting up crystallization experiments, the concentrated RNA has to be properly folded in the native state. This is performed by a heating step in a heating-block for 1 min at 70–85 °C (depending on the melting temperature) in the presence of monovalent salts only. Then, the solution is left in the switched-off heating-block to cool down slowly until room temperature is reached. In order to avoid self-cleavage of the RNA, the pH is usually chosen slightly acidic and divalent cations are added only around 35 °C.

27.4.2
Setting-up Crystal Screens

The main technique employed to setup crystal screens is the vapor diffusion method either in the hanging drop or sitting drop setup. This method is based on slowly concentrating the droplet solution against a reservoir solution of infinite volume (milliliter scale) compared to the volume of the droplet (microlitre scale, see Fig. 27.1). The choice between the various plastic-ware commercially available will be driven by the amount of RNA sample and the number of crystallization conditions to be tested. Nowadays, more and more laboratories have the opportunity to use crystallization robots that permit to decrease the drop volume to hundreds of nanoliters and, with the same amount of material, to set up thousands of trays on very short time scales.

The different crystallization screens are set up by adding the biomolecule solution to the crystallization solution. Once a first hit has been obtained, one needs to optimize the conditions. In the example of the aminoglycoside/ribosomal A site complexes (see below), crystals were optimized using different crystallization solutions to test various glycerol/MPD ratios (see Table 27.3). All trials are done at 37 °C using the hanging drop vapor diffusion method: 1 µl of RNA-antibiotic complex solution is added to 1 µl crystallization solution and equilibrated over a reservoir containing 500 µl of 40% MPD.

27.4.3
Forming Complexes with Organic Ligands: The Example of Aminoglycosides

RNA molecules bind various organic ligands. Different RNA fragments based on the *Escherichia coli* A site located in the penultimate helix of the 16S ribosomal RNA [10–12, 30] have been tested in the presence of their natural ligands, anti-

Tab. 27.3. Crystallization conditions testing various glycerol/MPD ratios.

Crystallization condition (reservoir)	Stock	1	2	3	4	5	6
MPD (%)	60	1	2	2	2	1.5	2
Glycerol (%)	100	5	4	2	1.5	1	1
Na cacodylate pH 6.4 (M)	0.85	0.05					
KCl (M)	3	0.15					
Glycerol/MPD ratio		5	2	1	0.75	0.67	0.5

biotics of the aminoglycoside family. The RNA construct was designed as a self-complementary oligonucleotide so as to incorporate two A sites in a head-to-head manner (Fig. 27.4). This choice eliminates two drawbacks. First, since the internal loop is asymmetric, one would otherwise need to synthesize, purify and mix 1:1 two different RNA strands in order to obtain a single site. Second, one could also use a single site capped by a stable hairpin of the GNRA family, for example. However, in such cases, it is frequently observed that the crystallized structure reveals a

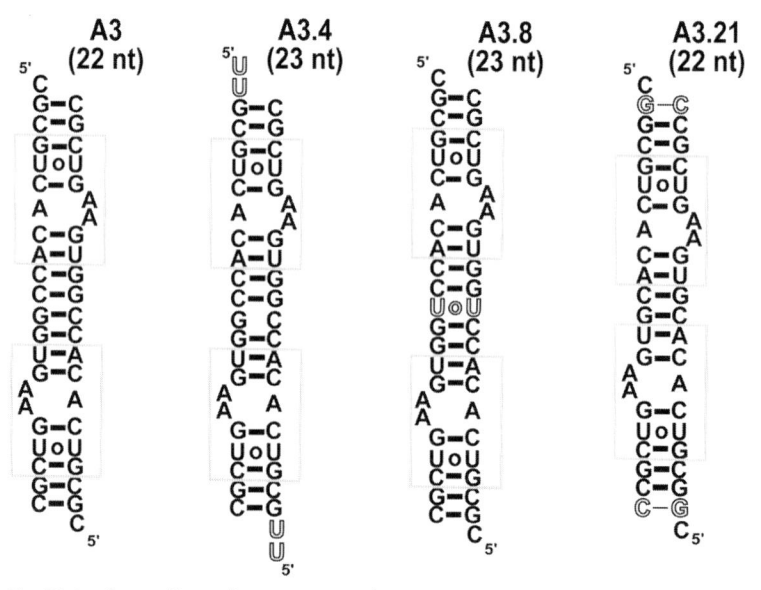

Fig. 27.4. Four self-complementary RNA fragments containing a tandem array of two E. coli 16S ribosomal A site modules [10–12].

full duplex with several non-Watson–Crick pairs [31]. In order to monitor the effect of sequence variations for the crystallization of these complexes, various modifications have been performed such as addition of a 5′ UU overhang, insertion of a UU pair or moving the two A sites closer to one another (Fig. 27.4).

Routinely, the purified oligoribonucleotides are solubilized in a solution containing 2 mM RNA, 25 mM NaCl, 5 mM MgSO$_4$ and 100 mM sodium cacodylate buffer, pH 6.4. This solution is first heated to 85 °C for 2 min and then slowly cooled until a temperature of 37 °C is reached. One volume of a 4 mM aminoglycoside solution is added to the RNA solution and incubated for 2 h at 21 °C. The two solutions should be at the same temperature. Since the RNA fragments contain two antibiotic-binding sites, the aminoglycoside concentration is twice the RNA concentration. A general rule is that the organic ligand concentration should be 100 times higher than its dissociation constant (K_D) to ensure binding site saturation which is usually easily achieved in the millimolar range.

In the example of the aminoglycoside/A Site complexes, different crystallization solutions were prepared to test various glycerol/MPD ratios: 5, 2, 1, 0.75, 0.67 and 0.5 (Table 27.3). All trials are performed at the optimal temperature of 37 °C using the vapor diffusion method in the hanging drop setup: 1 μl RNA-antibiotic complex solution is added to 1 μl crystallization solution and equilibrated over a 40% MPD reservoir.

27.4.4
Evaluate Screening Results

Since first crystallization attempts will not automatically yield crystals or they may be of too poor quality for X-ray diffraction experiments, evaluating screening results is required prior to proceed to crystallization optimization. This is performed by using a binocular microscope hooked up to a digital camera to record observations. A numerical scoring value describing the content of the droplet (Fig. 27.5) is reported on a paper scoring sheet (Table 27.4).

Two weeks are enough for droplets of about 3 μl to equilibrate under any conditions [32]. During this period, droplets should be inspected daily to follow up the appearance of crystals. Crystals may still form after 2 weeks, but this is less likely in the case of oligonucleotides. Crystals can then be cryo-protected and frozen or capillary mounted to be tested.

27.4.5
The Optimization Process

Here are provided non-exhaustive guidelines to interpret the droplet content of crystallization screenings (Table 27.4) and possible ways to optimize positive hits. See also [33] for more details.

- *Clear drops* – indicates that the RNA supersaturation state has not been reached, the RNA concentration is outside the nucleation zone (Fig. 27.1). These experi-

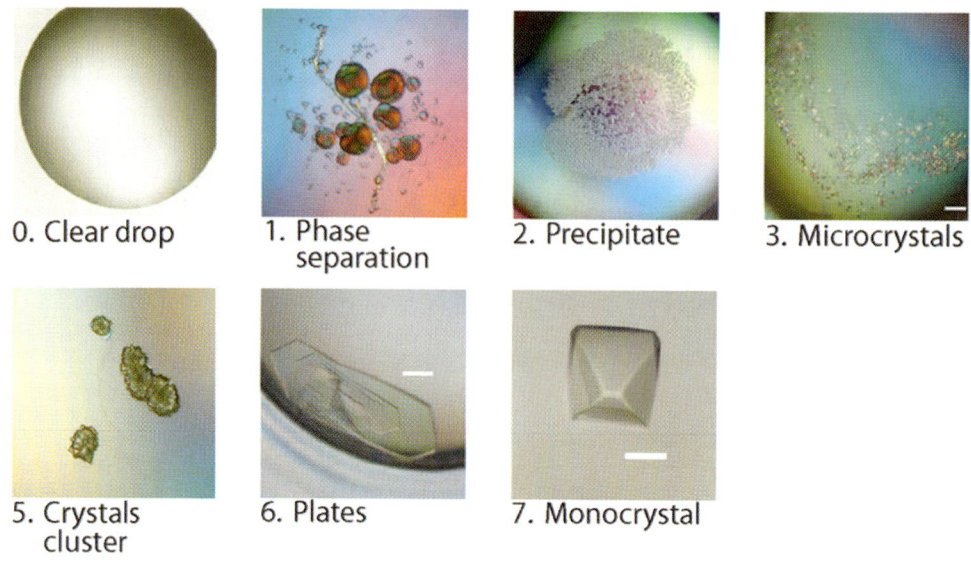

0. Clear drop **1. Phase separation** **2. Precipitate** **3. Microcrystals**

5. Crystals cluster **6. Plates** **7. Monocrystal**

Fig. 27.5. Numerical scoring terms.

Tab. 27.4. Scoring sheet for rows A and B from a 24-well LINBRO crystallization plate.

| CRYSTALLIZATION PLATE NUMBER: |
| SAMPLE & ADDITIVES: |
| DATE OF SETTING: |
| TEMPERATURE: |

	DATE										
A	1										
	2										
	3										
	4										
	5										
	6										
B	1										
	2										
	3										
	4										
	5										
	6										

ments must be repeated with higher sample and/or salt concentrations. The temperature can also be lowered.
- *Phase separation* – indicates a need to increase the monovalent salt concentration and/or to test a smaller precipitant concentration (MPD, PEG) to make the RNA sample more soluble.
- *Light precipitates* – indicates that the relative supersaturation between sample and reagent is too high. Prepare new tests with a decreased RNA and/or precipitant concentration or dilute the droplet by vapor diffusion by adding water into the reservoir.
- *Strong precipitates* – indicates that the sample has been partially denatured. The sample must be tested at a lower concentration or less salt should be used. Note that a fresh test should be prepared in this case.
- *Small precipitates* – must be carefully inspected using polarized light because it may contain a microcrystalline shower. A microscope with a magnification factor greater than 100-fold can be useful in this case.
- *Cluster of homogenous crystals* – try to slow down the nucleation; test the conditions at lower temperature, cover the reservoir solution with oil to slow down the water diffusion.

Different parameters can be tested to optimize the growth of single monocrystals: salt type, additives, temperature. Finally, if no crystal can be obtained with a given construct, new RNA sequences have to be designed so as to provide new potential scaffoldings to help crystal packing.

27.5 Conclusions

Protocols to purify and concentrate large amounts of RNA under controlled buffer and salt conditions for crystallization experiments have been described. The crucial points to preserve the RNA sample are: the use of slightly acidic pH and the avoidance of divalent cations. Usually RNases, feared by most RNA scientists, are introduced into the solution by an upstream experiment such as plasmid and protein preparations (e.g. T7 RNA polymerase). It is, thus, strictly recommended to assess the RNase activity of a solution before using it on the whole RNA sample by incubating an aliquot in the presence of the RNA for few hours and check the extent of the digestion by PAGE. Then strategies to crystallize RNA oligonucleotides in the presence or absence of ligands have been presented. A peculiarity of crystallization experiments is the absence of a negative control. To circumvent this, we advise to attempt to crystallize simultaneously several related RNA sequences in the same well. Only one or few RNAs, if any, will crystallize if any, leading to the conclusion that RNA crystals instead of salt crystals have been obtained. In cases where no crystal is observed, it is recommended to design a new set of oligonucleotides bearing slight sequence changes in order to enhance interactions between symmetry-related molecules that lead to regular crystal packing interactions.

the detected fluorescence originates from only one molecule. In order to fulfill both requirements, the measuring volume has to be minimized by shrinking the excitation and/or the detection volume. Depending on the goal of the study, different techniques are applied to realize small measuring volumes: optical near-field excitation through pointed fibers (scanning near-field microscope), total internal reflection (TIR) at a glass–water interface and the confocal microscopy technique (CM) where a laser beam is focused into the sample and a special optical arrangement restricts the excitation as well as the detection volume.

Applying TIR, the molecules under investigation are usually fixed to the surface, which has the advantage of long observation times, so that even slow dynamics are accessible. However, care has to be taken regarding the possibility that the biomolecule can be strongly influenced by interaction with the surface. To observe single molecules on a surface, their density should not be more than one per 10 µm^2 to avoid an overlap of the single fluorophores images.

The CM setup gives detection volumes in the femtoliter range, which corresponds to the size of bacteria. Compared to the TIF technique, the CM setup has an inferior signal-to-noise ratio, but measurements on freely diffusing molecules are possible, avoiding the risk of surface artifacts. To guarantee that mostly only one fluorophore will diffuse through the focus of the CM setup at a time, the fluorophore concentration has to be in the range of 100 pM, leaving approximately one fluorophore in the detection volume.

A major point which has to be considered in single molecule detection is photodestruction of the fluorophore. A fluorophore can typically do 10^6–10^7 absorption–emission cycles on average before it is likely to be destroyed by chemically reacting in the excited state, mainly with oxygen. Thus, the total number of photons that can be detected from one fluorophore is limited. They can be detected either in a short time, when exciting the fluorophore with a higher intensity, getting a high time resolution to observe fast dynamics, but only for a short observation time, or, when exciting with a lower intensity, one can achieve a long observation times up to minutes, but will miss fast dynamic events.

A prerequisite for single molecule fluorescence detection is the existence of a suitable fluorophore. In biomolecules, however, only a few intrinsic fluorophores like the flavin-adenine dinucleotide (FAD) are suitable for single molecule spectroscopy [3]. In nucleic acids, one can substitute bases by some fluorescent analogs, e.g. 2-aminopurine or ethenoadenosine. Unfortunately they all suffer from low photostability and are therefore not suitable for single molecule measurements. Therefore, in most cases efficient fluorescent dyes are covalently coupled to the sample. These fluorophores are used as probes to test the molecular properties of the biomolecules.

In fluorescence microscopy, the static distribution of fluorescently labeled molecules in cells is directly observed. Their dynamic transport between different cell compartments can only be observed when it is triggered either by the injection of labeled molecules or by bleaching them in certain areas and observing their redistribution, called fluorescence recovery after photobleaching (FRAP). From these methods only average diffusion coefficients can be calculated and the mobility of the single molecules is still hidden behind the average.

The full dynamic behavior of the molecules is only accessible when the movement of individuals is observed. Today, digital cameras are available that are sensitive and fast enough to take video sequences of single fluorescent molecules. This allows tracking the pathway of individual molecules through the cell [4], to observe their diffusion in membranes [5] or their directed transport in cells.

The motion of a labeled molecule depends on its mass and, thus, the analysis of dynamic behavior gives information about the change of mass due to molecular complexation. Another technique able to analyze the mobility of molecules is fluorescence correlation spectroscopy (FCS). In this technique, the intensity fluctuations in the fluorescence signal are observed that are caused by single molecules diffusing in and out of a confocal detection volume. The analysis of the signal fluctuations gives the typical diffusion times at the defined position of the detection volume [6–8].

Several physical properties of the fluorophores are influenced by their surroundings and by that provide functional information about the sample. Close contact of the fluorophore to certain molecules can lead to quenching and thus to the emission of less fluorescence photons. In ensemble assays this decrease of fluorescence intensity can be used to monitor binding events. In addition to the fluorescence intensity, the fluorescence lifetime changes when the fluorophore is quenched and thus also can indicate molecular interaction.

A further important parameter of fluorescence is anisotropy. It can give information about the mobility of the fluorophore. When exciting fluorophores are in solution with linear polarized light, the fluorescence emission is not isotropic, i.e. has an anisotropy, even if the fluorophores have a random orientation. If the fluorophore rotates within its lifetime, i.e. in the time between excitation and fluorescence emission, this leads to a decrease of anisotropic emission. This can be used to determine the molecular rotational time [9] and, thereby, the binding events due to the change in mass of the molecular complexes.

In addition to these one-label techniques that allow the analysis of position and mobility of one molecule or molecular complex, techniques using two different labels allow us to analyze the proximity of different molecules.

With a mixture of molecules labeled with two different fluorophores, one can visualize the two fluorophores separately by taking images in the spectral ranges of only one or the other fluorophore. Superposition of those two images allows one to determine the co-localization with an accuracy of up to tenths of nanometers, far below the wavelength of light, depending only on the number of photons one can detect per molecule [10, 11].

Using two different labels in fluorescence correlation analysis, interactions can be monitored much better than via the diffusion times. So-called two-color cross-correlation analysis detects only fluctuations that occur simultaneously in two different detection channels, each sensitive for only one of the two fluorophores.

Thus, only that part of the intensity fluctuations is detected which originates from both fluorophores moving together bound to one complex [12–14].

28.2
Fluorescence Resonance Energy Transfer (FRET)

Co-localization studies can only measure distances down to tenths of nanometers. To measure inter- or intramolecular distances below 10 nm, one can take advantage of a process called FRET. Instead of emitting a photon, an excited fluorophore (called a donor) can transfer its energy to another fluorophore (called an acceptor) which instead emits a photon. The acceptor has to be in a vicinity of 1–10 nm and its absorption spectrum has to overlap with the emission spectrum of the donor. From this decrease of the donor fluorescence and the increase of acceptor fluorescence the efficiency of the energy transfer and the distance between the two fluorophores can be calculated. In addition to the donor and the acceptor fluorescence intensity, the donor lifetime is also reduced if energy transfer occurs and thus the measured change in lifetime can also be used to calculate the distance.

FRET is probably the most versatile fluorescence tool for single molecule measurements. It allows one to get both dynamic and structural information at the single molecule level. If the donor and acceptor are bound to two different segments of an RNA structure, all structural changes that affect the distance between the segments will be visible as a change in the energy transfer. Conformational changes can be observed at a time range from milliseconds up to minutes. While dwell times or time constants of fluctuations can be directly calculated from the FRET signal, the measurement of exact distances is more complicated. Different parameters have to be taken into account when calculating distances from the measured fluorescence intensities or the measured lifetime. Here, single molecule measurements are a great advantage, since many artifacts that occur in ensemble measurements due to averaging can be eliminated, which allows a more accurate calculation of distances.

28.2.1
Measurement of Distances via FRET

According to Förster's theory [15, 16] the efficiency E of energy transfer from the donor fluorophore to the acceptor fluorophore depends on their distance R:

$$E = R_0^6 / (R_0^6 + R^6)$$

Here R_0 is called the Förster distance. At that distance the transfer efficiency is 50%. R_0 is a typical parameter for each donor–acceptor pair.

R_0 depends on several parameters:

$$R_0^6 = 8.8 \times 10^{-28} J \kappa^2 \Phi_D n^{-4}$$

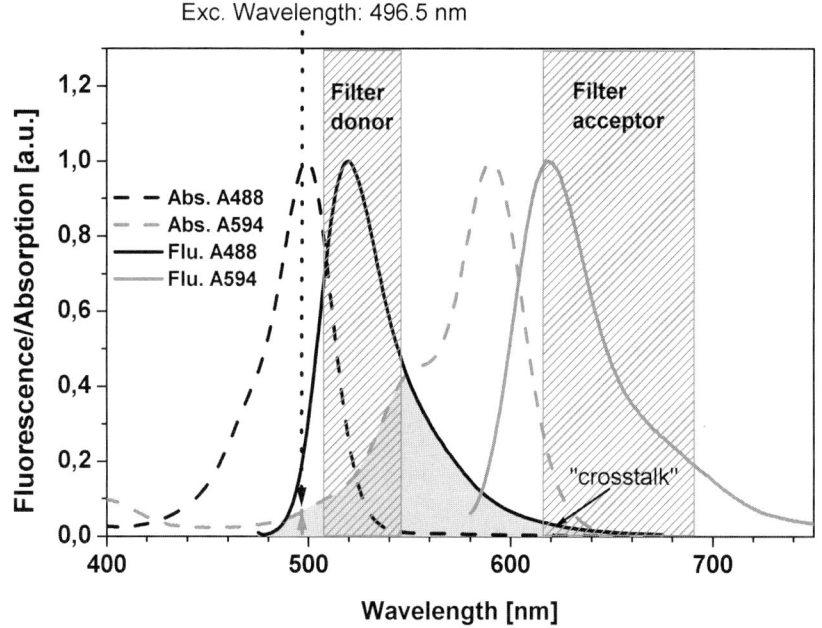

Fig. 28.1. Donor (Alexa 488) and acceptor (Alexa 594) spectra are shown. FRET occurs if the donor emission overlaps with the acceptor absorption; illustrated here by the grey area. Fluorescence detection with different filters allows detection of donor and acceptor fluorescence separately. The fluorescence intensities have to be corrected for direct excitation (indicated by the grey arrow at 496.5 nm) of the acceptor and for crosstalk of the donor fluorescence into the acceptor channel.

Here J is the spectral overlap between donor emission and acceptor absorption. κ^2 accounts for the relative orientation of the two dyes. If both dyes are free to rotate on a time scale faster than their fluorescence lifetimes, an averaging of all orientations results in $\kappa^2 = 2/3$. Φ_D is the fluorescence quantum yield of the donor fluorophore and n the optical refraction index of the medium. For the most used donor–acceptor pairs, R_0 is in the range between 2 and 6 nm. See Fig. 28.1.

The transfer efficiency needed to calculate distances can be determined in two independent ways. Since the lifetime of the donor fluorophore is shortened by the energy transfer, the ratio of the donor lifetimes in the presence ($\tau_{D(A)}$) and absence (τ_D) of the acceptor is thus a measure of the transfer efficiency:

$$E = 1 - (\tau_{D(A)}/\tau_D)$$

On the other hand, the transfer efficiency can also be determined by measuring the fluorescence intensities of the donor (F_D) and the acceptor (F_A):

$$E = (1 + F_D \Phi_A / F_A \Phi_D)^{-1}$$

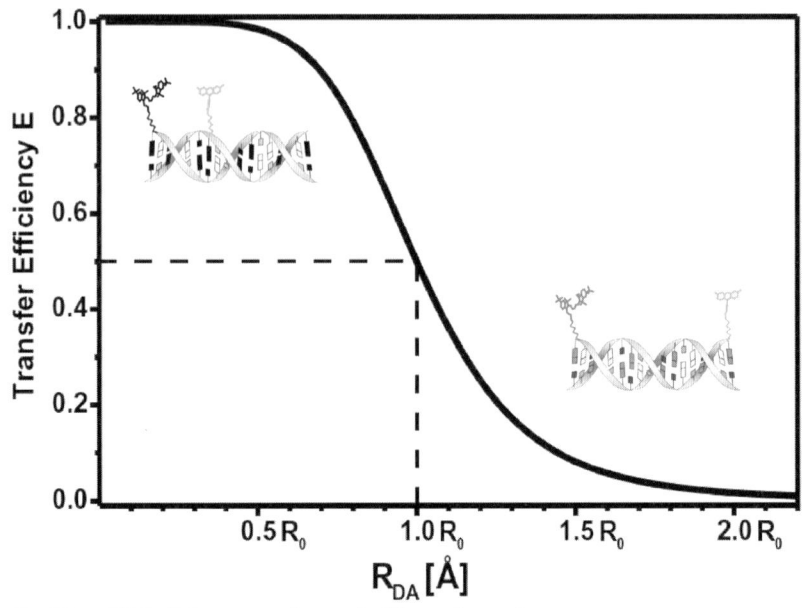

Fig. 28.2. FRET efficiency according to the Förster theory. At small distances the energy transfer approaches 100% and decreases for longer distances. At R_0 the FRET efficiency is 50%. It is there that FRET is most sensitive to distance changes.

Here Φ_D and Φ_A are the fluorescence quantum yields for donor and acceptor. The simultaneous measurement of the fluorescence lifetime as well as the fluorescence intensity for the donor and acceptor fluorescence reduces statistical errors and helps to identify systematic ones.

To test the accuracy of different single molecule or ensemble techniques that measure distances via FRET, several groups used the DNA helix with its well-known structure as a molecular ruler [9, 17–19]. See Fig. 28.2.

28.3
Questions that can be Addressed by Single Molecule Fluorescence

Very good work is published about the dynamics of RNA structures measured in detail by single molecule fluorescence. Conformational fluctuations of three- and four-way junctions as well as protein-induced structural changes have been investigated *in vitro*. Folding pathways of complex RNA structures like ribozymes were analyzed [20–24]. Single molecule measurements in living cells promise detailed insight in molecular distribution and dynamic processes. The simultaneous acquisition of lifetime information, FRET efficiencies and anisotropies can reveal molecular conformation and its interaction with other molecules. Thus, one could get a functional image of specific molecules in a cell. The autofluorescence of cells makes single molecule detection difficult, but single molecule imaging in cells has nonetheless already been shown.

Meanwhile, the advantages of single molecule detection are also used in different techniques for nucleic acid analysis. Single molecule detection is applied especially in sequencing and fragment sizing in order to increase the throughput and reduce the amount of sample needed.

28.3.1
RNA Structure and Dynamics

The FRET technique was used by Kim et al. [20] to study Mg^{2+}-facilitated conformational changes in a three-helix junction – a ribosomal junction that initiates the folding of the 30S ribosomal subunit. The junctions were bound to a surface and labeled with donor and acceptor at the ends of two helical arms. Structural changes of the junction caused a change in the angle between the helical arms and thus could be observed as a change in the energy transfer. The authors analyzed the fluctuation of the single molecule signals visible in the measured time traces by correlation analysis. In contrast to solution FCS measurements, fluorescence correlation of surface-bound single molecules is not limited by diffusion times. The experiments showed that for this particular junction the structural fluctuations were Mg^{2+} and Na^+ dependent, but did not result from binding and unbinding of the ions. Just the intrinsic conformational fluctuations were altered by the uptake of ions.

Ha et al. used FRET to measure structural changes induced by protein binding [21]. They observed the conformational change of the 16S rRNA three-way junction induced by the binding of ribosomal protein S15. In ribosomal assembly the binding of the S15 protein nucleates the assembly of the central domain of the 30S ribosomal subunit by folding the RNA such that distant sites are brought together, allowing the binding of subsequent proteins in the assembly cascade. This distance change was monitored by FRET. In two-color images of surface-bound junctions, those with no protein bound could easily be distinguished from the ones that formed a complex with a S15 protein. The protein-binding-induced structural change reduced the distance between donor and acceptor, leading to high energy transfer and high red acceptor fluorescence. In the same image, protein-free junctions are distinguished by their high green donor fluorescence.

Zhuang et al. demonstrated the potential of FRET for the study of folding of RNA structures. They measured the complex dynamics of the hairpin ribozyme in its minimal form [22]. There it consists of two helix–loop–helix segments. These associate non-coaxially in the active folded structure in a way that brings catalytically important loop nucleotides into close proximity. Donor and acceptor bound to the ends of the two segments allowed direct observation of the enzyme opening and closing. The enzyme–substrate complex exists in either docked (active) or undocked (inactive) conformations. The authors found complex structural dynamics with several docked states of distinct stabilities. With the complex structural dynamics they could quantitatively explain the heterogeneous cleavage kinetics common to many catalytic RNAs.

However, in the natural form where the ribozyme assembles in the context of a four-way helical junction, the folding pathway includes an additional intermediate state. Tan et al. [23] found that this intermediate step originates from the four-way junction and is obligatory for the folding process. This intermediate step, which could not be discovered by ensemble measurements, brings the two loop elements in close vicinity. It increases the probability of their interaction and accelerates the folding by nearly three orders of magnitude, allowing the ribozyme to fold rapidly in physiological conditions.

Zhuang et al. [24] used the single molecule FRET technique to analyze much bigger RNA structures. They observed folding steps of individual *Tetrahymena* group I intron ribozymes. The analysis of time trajectories allowed them to identify a rarely populated state, which was not measured by ensemble measurements before. Intermediate folding states and multiple pathways were observed. As well as previously established pathways, a new folding pathway could also be observed.

28.3.2
Single Molecule Fluorescence in Cells

The spatial and temporal distribution of RNA species in living cells has been studied successfully in many systems. The mechanisms of RNA localization and pathways of cellular transport in transcription, splicing and translation processes have been investigated.

When observing single molecules in cells one has to take into account the autofluorescence background. To reduce the background fluorescence, one should select the fluorophores emission spectrum according to a minimal overlap with the autofluorescence spectrum. Also, several reagents have been described that reduce autofluorescence, but they cannot remove it completely [25].

28.3.2.1 **Techniques used for Fluorescent Labeling RNA in Cells**
Several techniques have been established to introduce fluorophores into living cells. Direct microinjection of labeled RNA allows us to use photostable fluorophores with high quantum efficiency [26–28]. After microinjection one can easily follow the temporal and spatial distribution by live cell microscopy. Transport between cytoplasm and nucleus and the assembly of RNA in different foci was observed by several groups [29–31]. Microinjection also makes it possible to control the number of molecules injected to a single cell. Knemeyer et al. showed that one can count the number of injected molecules by positioning a confocal detection volume at the end of the micropipette [28]. However, it must be taken in account that the injected labeled nucleotides may follow different routes and show different kinetics due to the fluorescent modifications. When working with excess RNA, cellular transport and processing routes may also become saturated.

Another approach uses RNA-binding domains that are fused to Green Fluorescent Protein (GFP) [32]. This approach is useful for organisms that cannot be microinjected or do not allow the penetration of macromolecules. Due to unfavorable photophysical properties and low photostability of all fluorescent proteins, the detection of single RNA molecules is only achieved if multiple binding is possible.

Instead of directly labeling RNA, the hybridization of labeled oligonucleotide probes to endogenous RNA, called fluorescence *in vivo* hybridization (FIVH), is often used to label RNA specifically in live cells [33, 34]. Different groups used nucleotide probes complementary to polyadenylated RNA, spliceosomal small nuclear RNA or ribosomal rRNA to detect their *in vivo* distribution dynamics [28, 35]. In the cytoplasm ribonucleic probes are degraded by RNases. To circumvent this, nuclease resistant 2'-O-methyl oligoribonucleotides with high affinity to complementary RNA or DNA strands are used. A major problem in FIVH is to distinguish between specifically hybridized probes and background due to non-bound fluorescent probes.

The concept of molecular beacons, developed for the detection of nucleic acid amplification products, is used to localize specific nucleic acids in fixed cells. They can also be applied to live cells. Molecular beacons are double-labeled oligonucleotides with a fluorophore at one end and a quencher at the opposite end. The molecular beacons are designed to have a target-specific probe sequence positioned centrally between two short self-complementary segments. These special sequences form a structure that consists of a loop and a stem, so that the dye and quencher are in close vicinity. Upon hybridization to the target, the stem opens and the fluorophore starts to fluoresce. This technique has been used to detect different mRNAs in the cytoplasm of living cells [36, 37]. Unfortunately it is reported that molecular beacons, especially in the nucleus, often do not show better results than linear oligonucleotide probes, meaning that molecular beacons often already open before hybridization to the target sequence due to non-specific interaction with nuclear proteins or RNAs.

Another promising approach uses FRET to detect only specific hybridized probes. In this case the acceptor and donor are bound to two different nucleotide probes, which hybridize with the target sequence such that donor and acceptor are placed close enough for energy transfer to occur [37–40]. FRET occurs only when donor and acceptor probes are hybridized to the target sequence simultaneously. However, the probability to assemble a FRET probe is much smaller than to hybridize one single fluorophore probe alone.

The detection of singly fluorescently labeled oligo(dT) probes hybridized to mRNA was demonstrated [28, 41]. This was possible due to the use of spectrally resolved fluorescent lifetime microscopy. The combination of spectral and lifetime information allows one to distinguish clearly between autofluorescence background and the used probes. Oligo(dT) nucleotides were microinjected in 3T3 mouse fibroblast cells and hybridized to polyadenylated mRNA. From single molecule imaging Knemeyer et al. estimated the fraction of immobile and mobile mRNA. This technique showed that imaging single molecules in living cells is possible. Thus, the toolbox of single molecule multiparameter fluorescence detection allows us to analyze structural and dynamic properties of single molecules and molecular complexes in living cells.

28.3.2.2 Intracellular Mobility

Several attempts have been made to detect the detailed individual mobility of single particles or even single molecules. Different groups use fast CCD cameras that are sensitive enough to detect few or single fluorophores. Time series of those single particle images reveal particle movement in cells (time-lapse microscopy). The trajectories of single viruses and spliceosomal particles have been analyzed, revealing diffusive movement and active transport [4, 42, 43]. A different approach is under development; it uses a confocal scanning system, which allows tracking the movement of single particles [44, 45]. There the detection focus is moved repetitively in a circular path, allowing us to determine the position of a fluorophore in the plane of movement by the intensity distribution along the circular scanning path.

28.3.3
Single Molecule Detection in Nucleic Acid Analysis

In nucleic acid analysis, single molecule detection is an advantage due to the small amount of sample needed. It also allows the development of fast assay formats with high throughput [46].

28.3.3.1 Fragment Sizing

To determine the size of restriction fragments, one can take advantage of nucleic acid staining fluorophores that intercalate or bind in the helical grooves. Fluorophores are available for single-stranded as well as for double-stranded oligonucleotides. Several fluorophores can bind to a single DNA fragment. The number of fluorophores depends on the length of the fragment and thus the fluorescent brightness of the labeled fragment can be used as a measure of its length. Since single molecule detection counts the number of fragments, an accurate calculation of sample concentration is easily possible. This is a big advantage over gel electrophoresis, where the brightness of the bands depends on the number of molecules as well as on the length of the fragments. To analyze the labeled fragments individually, microfluidic systems similar to flow cytometric instruments are used. Capillary systems lead the molecules to an excitation and detection volume. For the detection of the individual fluorescent molecules, imaging with a sensitive CCD camera or detection in a confocal setup is usually used [47, 48].

28.3.3.2 Single Molecule Sequencing

With the availability of biological sequences new applications emerge and the need for sequence information grows exponentially. Single molecule sequencing has the potential to produce faster and cheaper sequences by facilitating a high parallelization to reduce the time and amount of material needed. Sequencing then enables also fast analytical purposes such as single nucleotide polymorphism (SNP) detection or microbial typing [49, 50].

In the search for a robust technique for single molecule sequencing, two different approaches are under development – the sequential degrading of fluorescently labeled nucleic acid strands and the successive incorporation of labeled bases.

To observe sequential degrading, the labeled nucleic acid is placed in an ultrasensitive flow cytometric setup. When the exonuclease degrades the nucleotide, the flow drags the cleaved nucleotides to the detection volume. There the labeled nucleotides are excited with a focused laser beam and detected mostly by a CCD camera. If all four bases are labeled with different fluorophores, the sequence can be read directly. Since the efficiency of enzymatic processing of nucleic acids is reduced by labeled nucleotides, it is useful not to label all nucleotides. Different groups recently demonstrated this technique for DNA sequences containing labeled nucleotides at each T position [51], and two distinguishable fluorophores at the U and C positions, respectively [52]. Instead of labeling all nucleotides, it is enough to label only two different base types at a time. That will result in a two-base sequence. Measuring all six possible two-base sequences allows the assembly of the full sequence.

The other approach uses the possibility to image single fluorophore labeled molecules on a surface with a CCD camera. Therefore, primers are hybridized to nucleic acid strands that are bound to the surface. Then a primer extension is performed by a DNA polymerase that incorporates a fluorescently labeled nucleotide. In principle, when using distinguishable fluorophores for the four base types, this would allow us to read the sequence while the extension takes place. Mainly due to unspecific adsorption to the surface and high background fluorescence caused by the high amount of labeled nucleotides needed for the polymerase reaction, the construction of an experimental setup is more complicated. It was demonstrated that with the use of FRET between a fluorescently labeled primer and the newly incorporated fluorophore, it is possible to detect clearly specific incorporation and to identify the sequence [53].

This technology is still under development and recent progress, especially in the field of background reduction [54], allows to hope that in near future it will lead to fast sequencing that can be highly parallelized, and needs only the tiniest amounts of material and probe. Direct sequencing of DNA as well as RNA without previous amplification should be possible. mRNA sequencing can then be performed by using either the reverse transcriptase or DNA polymerase after a DNA strand is synthesized.

28.4
Equipment for Single Molecule FRET Measurements

28.4.1.1 Excitation of the Fluorophores

For any fluorescence experiment an excitation source is mandatory. Modern single molecule FRET measurements are performed using a laser with suitable wavelength to excite the donor dye, usually in the range from 450 to 550 nm. For most applications, a diode laser operated in the in the continuous wave (c.w.) mode is sufficient. For experiments were fluorescence lifetime information is needed, the laser has to shoot short flashes of light (pulsed excitation).

A special excitation technique is the two-photon excitation. Here, excitation is induced by simultaneous absorption of two photons that have twice the wavelength used in one-photon excitation. Since the two photons have to hit the dye at almost the same time, two-photon excitation requires a high instantaneous photon flux and therefore bleaching is a major problem. The advantage of this technique is a higher spatial resolution compared to one-photon excitation, since the intensity and thus the probability that two photons are absorbed at the same time is sufficiently high only in a very small area. The long wavelength of the exciting laser reduces autofluorescence significantly, which is especially useful for measurements in cells.

The light is mostly focused on the sample via an objective. Total internal reflection fluorescence (TIRF) experiments use oil immersion objectives with a high numerical aperture ($NA = 1.4$) to create an evanescent field at the glass surface where the sample is bound. For experiments where the sample molecules diffuse free in solution, water immersion objectives ($NA = 1.2$) are used.

In particular, y for experiments on immobilized molecules in combination with confocal detection, it is necessary to scan the sample. Therefore a piezo-driven scanning stage is needed, which allows precise movement of the e sample precisely the x-, y- and z-directions.

In many setups, the laser excitation light as well as the fluorescence passes the same objective. In this case a dichroic mirror has to be implemented to separate the fluorescence from excitation light (laser is reflected and fluorescence passes or vice versa, see Fig. 28.3).

28.4.1.2 Fluorescence Detection

In a setup for a single molecule FRET measurement, the fluorescence originated by the donor dye is separated from acceptor fluorescence either by another dichroic and/or by suitable band pass filters. The fluorescence is detected by an avalanche photodiode or a photomultiplier tube to collect single photons. To get an image, the sample is scanned or a CCD camera is used for wide-field detection.

Thus, a laser, an objective, two dichroics, two lenses and two detectors are needed for the most simple setup which allows single molecule FRET experiments. This "minimal" setup allows the registration of the intensities for the donor and acceptor fluorescence, and permits the calculation of the FRET efficiency for every single molecule event. If a polarizing beam splitter is implemented and the donor fluorophore is excited with pulsed excitation, the fluorescence lifetime as well as the anisotropy for the donor and acceptor dye becomes additionally accessible. Employing this supplementary information, the precision of single molecule FRET measurements can be enhanced and possible artifacts, like restricted motion (κ^2 effects) or local quenching of the dyes, can be recognized and taken into account for further analysis [55, 56]. Therefore this kind of setup, shown in

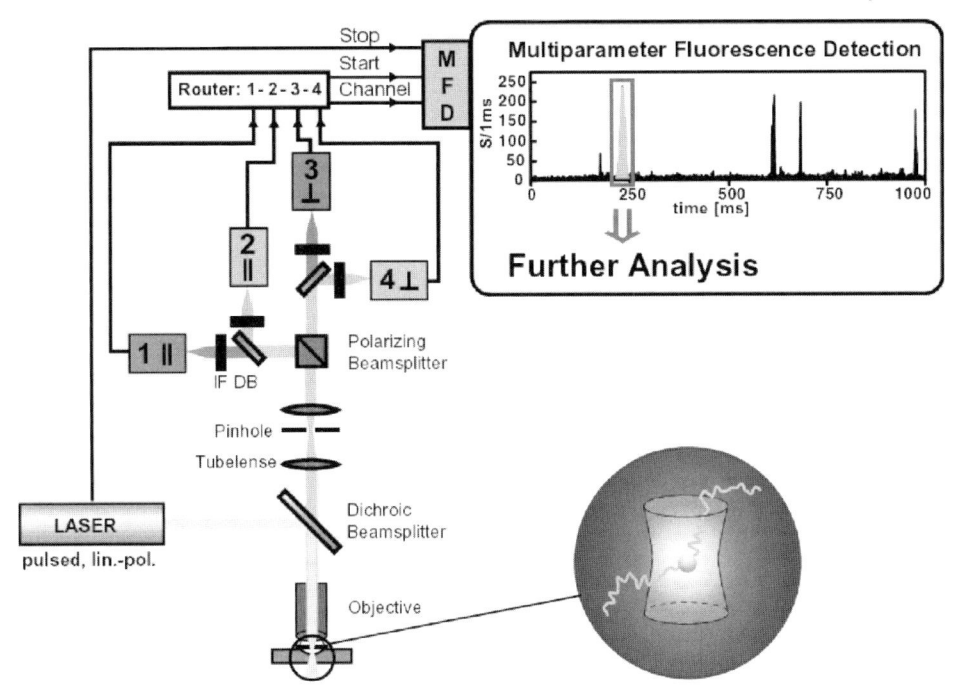

Fig. 28.3. A confocal microscopy arrangement for multiparameter fluorescence detection (MFD). Freely diffusing molecules are excited by a pulsed, linear polarized laser. After reflection by a dichroic mirror the laser beam is focused by a microscope objective. The fluorescence is collected and then refocused by the same objective to the image plane, where a pinhole is placed, which ensures that only fluorescence from the focal plane can pass. After the collected light is separated according to its polarization by a polarizing beam splitter, the fluorescence is divided by a dichroic mirror into the fractions originating from the acceptor and donor fluorophores. Subsequently, band pass filters are used to discriminate fluorescence from scattered laser light. Finally, the fluorescence light is detected by four detectors: two for the donor (2 + 4) and two the acceptor fluorescence (1 + 3).

Fig. 28.3, is especially useful to investigate processes where the expected FRET changes are small or structural information should be gained.

28.4.1.3 Data Analysis

Even more challenging than building a setup capable of detecting single fluorophores, is the data processing of single molecule experiments. Using a confocal setup with time-correlated single photon counting, the amount of data gained in a single molecule experiment is tremendous. However, since commercial computers with fast processors and a memory with a high capacity are available, data collection is rarely a problem anymore.

Most of the collected data originate from the background due to dark counts of the detector and remaining Raman, respectively, Rayleigh scattering. After a single molecule solution experiment all intensity peaks, called fluorescence bursts, that originate from single fluorophores diffusing through the focus can be identified and further evaluations of the spectroscopic properties can be restricted to the selected events.

28.5
Sample Preparation

28.5.1
Fluorophore–Nucleic Acid Interaction

The interaction between the fluorophore and the biomolecule under investigation should be minimized or well known. The fluorophores should not introduce any perturbation to the biomolecules' structure or dynamics. When labeling RNA, care has to be taken so that the interaction of the bases is not disturbed, especially in complex non-helical structures. Also, the fluorophores' characteristics should not be changed by the biomolecule [57].

For FRET measurements it is desirable that the dyes rotate fast and freely, so one can assume an average κ^2 of 2/3. Therefore, great care has to be taken so that the reporter dyes do not stick to the nucleic acid under investigation. Steric hindrance of the fluorophores can introduce a reasonable uncertainty [58].

On the other hand, a sticky fluorophore can be used to calculate the characteristic rotation of the oligonucleotide from its fluorescence anisotropy.

However, the interaction with the bases can lead to quenching of the fluorophores, mainly due to electron transfer from the fluorophore to guanosine bases [59, 60]. Then the fluorophore is dark in the bound state and bright if it is free. Thus fluctuations between the bound and unbound state become visible as intensity fluctuations that could be misinterpreted as dynamics of the nucleic acid structure. Fluorophore–nucleic acid interactions can be reduced if the fluorophores are negatively charged and thus rejected by the negatively charged backbone.

28.5.2
RNA Labeling

28.5.2.1 Fluorophores for Single Molecule Fluorescence Detection
Fluorophores used for single molecule fluorescence have to fulfill several criteria.

- They need high photostability to allow long observation times, a high probability to absorb photons for excitation (high extinction coefficient) and a high fluorescence quantum yield.
- They have to be excitable with one of the commercially available lasers in the visible range.
- They have to be water soluble and should not be pH insensitive.

Several fluorophores have been developed that fulfill these conditions. Many of them belong to the rhodamine or cyanine dye family. They are available under trade names like Alexa, Cy, Atto and others.

28.5.2.2 Fluorophores used for FRET Experiments

For FRET applications the selected donor–acceptor pair should have large spectral separation to minimize leakage of donor fluorescence into the acceptor channels and direct excitation of the acceptor, but an overlap of the donor emission and acceptor absorption spectrum still big enough to guarantee energy transfer.

Most publications report the use of Cy3–Cy5 [23, 61] or Alexa488–Cy5 [9, 56] as FRET pairs for single molecule FRET studies. With their Förster distances R_0 of 5.8 nm (Cy3–Cy5) [62] and 5.1 nm (Alexa488–Cy5) [63] they allow distance measurements between 2 and 8 nm with highest sensitivity around 5 nm.

In spite of the fact that Cy5 is widely used, it has some drawbacks. Often up to 50% of the fluorophores are found to be inactive after coupling to biomolecules, probably due to pre-bleaching. However, in single molecule experiments this is not a severe problem, since these can be separated from the ones carrying active acceptors. Cy5 also undergoes photo-induced isomerization between a highly fluorescent *trans* and a weakly fluorescent *cis* state. This leads to a loss of almost 50% of the fluorophores' capacity and to intensity fluctuations that interfere with correlation analysis of other fluctuations [64]. Alternatively, the Alexa594 dye is sometimes used as acceptor. This dye from the rhodamine family has no dark states, but direct excitation of the acceptor is higher due to its blue-shifted fluorescence spectrum.

The constant need for brighter and more photostable fluorophores ensure the continuing rapid development of fluorophores. Thus, for single molecule experiments it is always worth looking for the best fluorophore available at the time.

28.5.2.3 Attaching Fluorophores to RNA

Intercalating or groove-binding dyes can be used to stain DNA, but these dyes do not bind in a sequence-specific manner to the DNA. For structural investigations on DNA or RNA, one has to attach the dyes to specific positions within the oligonucleotide.

To attach fluorophores to specific positions in RNA, modified bases can be incorporated in oligo synthesis to which then fluorophores are attached in a second step. Mostly amino modifications are used. All standard fluorophores are available as activated amine-reactive derivatives. There are four major classes of commonly used reagents to label amines: succinimidyl esters, isothiocyanates, sulfonylchlorides and tetrafluorophenyl esters. Of the four, tetrafluorophenyl esters are the preferred chemistry for conjugations. Similar to the succinimidyl esters, they produce stable carboxamide bonds. They are less susceptible to hydrolysis than succinimidyl esters and therefore can provide more reaction time in aqueous-based reactions. However, the succinimidyl ester is still by far the most commonly used amine-reactive group. Oligos with 3'- or 5'-end amino-modifications or end-labeled with fluorophores are commercially available almost everywhere. Here, the dyes are usually attached as phosphoramidites during synthesis.

For internal labeling the modification should be located at the 5 position of the base U or C. The fluorophore is then positioned in the major groove of the double strand. This position minimizes interference to protein–RNA interactions which mostly take place in the minor groove. RNA oligos with internal modification are available from a few companies.

5-Aminoallyl U and 5-aminoallyl C are commercially available as triphosphates. Some companies also offer fluorophores already bound to UTP. Since this modification is compatible with enzymatic processing, primer extension and ligation can be used to incorporate the modifications at the desired position. For enzymatic incorporation or successive attachment of fluorophores we refer to the methods recommended by the respective provider of the modified bases or fluorophores.

For FRET experiments, donors and acceptors are mostly bound to the complementary strands. Double labeling of one RNA strand is much more complicated and mostly leads to loss of sample. Ligation of different labeled strands might introduce pre-bleaching while handling. Using different RNA modifications and coupling strategies for donor and acceptor is often limited by the availability of the respective modifications. However, some commercial suppliers do offer double-labeled oligonucleotides.

28.5.2.4 Linkers

The linker used to couple the fluorophore to the RNA has to be considered carefully. To avoid orientational effects the linker has to be long enough to give the fluorophore enough mobility to test all orientations within its fluorescence lifetime. Usually a carbon C_6 linker serves this purpose well enough. Care has to be taken when calculating absolute distances for shorter linkers. The value of the orientation factor κ^2 might deviate from its average value of $2/3$, introducing uncertainties in the calculated distances. When using C_6 linkers it always has to be taken into account that the mean position of the fluorophores due to the linker length is not at the base where it is attached to. The distance between the Base and the mean position of the fluorophore mostly is between 1 and 2 nm. The position of the fluorophore relative to the anchoring point either has to be determined experimentally, e.g. from systematic distance measurements on a linear nucleic acid helix (then being valid only in systems where the fluorophore is bound to a helical segment) or has to be estimated by molecular modeling.

28.5.3
Fluorescence Background

28.5.3.1 Raman Scattered Light

The laser light used for excitation produces Raman light that is shifted to longer wavelengths and thus can interfere with the fluorescence. Intensity and spectral distribution depend on the solvent used and the excitation wavelength. This can be efficiently suppressed with the right selection of excitation wavelength, filters and fluorophores.

28.5.3.2 Cleaning Buffers

Using picomolar concentrations of host molecules, background fluorescence of the solvents becomes a big issue and particles scattering the excitation light as well as background fluorescence of ingredients of the used buffer have to be carefully avoided. Impurities in the sample are less critical, since they also get diluted when preparing the right sample dilution for single molecule experiments. Fluorescent impurities of the buffers can be efficiently removed by adding activated charcoal granula, mixing and subsequently filtration with standard sterile filters [65].

28.5.3.3 Clean Surfaces

In single molecule measurements at a surface, two sources of background signal always have to be taken into account – impurities adsorbed to the surface, which can be reduced by cleaning, and autofluorescence of the surfaces material itself.

Two procedures are commonly used for cleaning – sonicating in a series of different solvents [66] or flaming of the surface [20].

The following protocol for sonication is given in [66]:

(1) 30% detergent solution 1 h
(2) Distilled water 5 min
(3) Acetone 15 min
(4) Distilled water 5 min
(5) 1 M KOH 15 min
(6) Ethanol 15 min
(7) 1 M KOH 15 min
(8) Distilled water 15 min

Additionally, or alternatively, the surface can be flamed for a few seconds in a propane torch.

As well cleaning, the sonication procedure serves the purpose of providing a chemically uniform surface for subsequent functionalization. The autofluorescence of the surface substrate cannot be removed, but instead of glass, a quartz glass can be used which has significantly less autofluorescence.

28.5.4
Surface Modification

Surface modification serves two purposes – it has to provide specific attachment of the molecules and at the same time prevents unspecific adsorption.

28.5.4.1 Coupling Single Molecules to Surfaces

The biotin–streptavidin system is mostly used for specific coupling of RNA to surfaces [22, 66]. Biotinylated bovine serum albumin (BSA) is adsorbed to the surface and subsequently streptavidin is attached. Then biotinylated RNA is anchored to the surface. The following protocol for RNA coupling is given in [20]:

(1) Biotinylated BSA 1 mg/ml
(2) Streptavidin 0.2 mg/ml
(3) 50 pM biotinylated RNA

Each step lasts 5 min followed by washing with buffer (10 mM Tris/50 mM NaCl, pH 8).

For covalent surface modification, aminosilanization with subsequent binding of PEG is often used in DNA chip production. For most single molecule experiments there is no need for covalent modification since no long-term stability is needed, but quick and simple sample preparation is desired.

28.5.4.2 Surface Passivation

Unspecific adsorption of the sample to the surface has to be prevented since it produces strong artifacts or leads to loss of sample. BSA provides already good passivation against RNA adsorption. Unfortunately, this procedure does not apply to proteins. A dense layer of polyethylene glycol (PEG) is usually used to prevent reject protein adsorption [66].

28.5.5
Preventing Photodestruction

Photobleaching of the fluorophores is an issue in any single molecule experiment. Even the most photostable organic fluorophores like rhodamine or cyanine dyes bleach sooner or later. Since the presence of oxygen is known to reduce the photostabiltity of almost all dyes used in single molecule experiments, the reduction of the oxygen content helps to prevent photobleaching and longer observation times are achievable. So far, most groups use an enzymatic oxygen-scavenging system, where oxygen is consumed in an enzymatic (glucose oxidase/catalase) catalyzed oxidation of glucose. The use of approximately 400 µM ascorbic acid is another possibility to remove the oxygen from the sample [11]. However, one has to keep in mind that oxygen also serves as a triplet quencher for many dyes. Consequently, if the oxygen is removed completely, the fluorescence intensity drops because the dyes are in the triplet state most of the time and do not fluoresce [66].

Thus, there are no universally valid conditions to avoid photobleaching in a single molecule experiment; in fact, the excitation power as well as the oxygen content has to be adjusted for every experiment individually.

28.6
Troubleshooting

Here we want to discuss some typical problems that can occur in sample preparation or in the measurement.

28.6.1.1 Orientation Effects
After labeling of the sample, unbound fluorophores have to be removed. This is mostly done by PAGE. If the fluorophores could not be removed thoroughly, they have to be distinguished from the host molecules that carry only the donor (e.g. a

RNA single strand) in the single molecule measurements. They, understandably, reduce the relative amount of host molecules. In solution experiments free fluorophores can be distinguished by their diffusion times either in single molecule events or in FCS.

Since the orientation factor κ^2 is important for the calculation of absolute distances, it has to be ensured that the fluorophore is freely mobile so that the average value for κ^2 may be taken. Also, in experiments that concentrate on dynamic aspects, signal fluctuations might arise from changes in the way the fluorophore interacts with the host molecule. As a control, the measurement of the anisotropy is recommended either in an ensemble or single molecule experiment. If the fluorophore is nearly freely mobile, the measured anisotropy should be close to zero, if the host molecules rotation is also slow.

If single molecules bound to a glass surface are observed, interaction with this surface may also reduce the mobility of the fluorophore. Here the polarization response of the immobilized molecules should be analyzed. Again, if the fluorophore is nearly freely mobile, the polarization should be close to zero.

28.6.1.2 Dissociation of Molecular Complexes

To measure single molecules in solution, the sample has to be diluted to subnanomolar concentrations. Thereby experiments on molecular complexes may be limited due to their low affinity constant, leading to dissociation upon dilution. To prevent dissociation, some components may be added in higher concentrations. Normally the component carrying the acceptor can be used in up to 10-fold higher concentrations compared to the donor, before the background gets to high due to the increased direct excitation. If the experiment is designed such that the donor and acceptor are bound to the same molecule, the other unlabeled binding partners can be added in much higher concentrations as long as no unspecific complexation is induced or their autofluorescence becomes significant.

28.6.1.3 Adsorption to the Surface

In solution experiments, the low concentration of host molecules makes the sample highly sensitive to loss due to adsorption to the surface. If the measured mean count rate drops during the measurement, this is mainly due to adsorption. Bleaching also leads to depopulation of the sample, but this can be neglected in typical sample volumes around 50 µl and gets important only in sub-microliter volumes.

28.6.1.4 Diffusion Limited Observation Times

Diffusion normally limits observation times of biomolecules of around 100 kDa to a few milliseconds. For longer observation times, molecules are mostly bound to surfaces at the cost of possible sample–surface interaction. A good alternative is the reduction of diffusion time by attaching the host molecule to a bigger mass that has a low diffusion time. In nucleic acid analysis, a long DNA sequence fragment with a sticky end can easily serve this purpose. Another alternative sometimes proposed is changing the viscosity of the used solvent. Great care has to be taken when using organic solvents, which can produce strong Raman light that interferes with the detected fluorescence.

28.6.1.5 Intensity Fluctuations

Since most dyes are not ideal emitters, when analyzing fluctuations in FRET signals one has to take into account that some fluorophores show intrinsic fluctuations. Fluctuations of the donor between the bright and dark state can be disregarded, since it results in total annihilation of donor and acceptor fluorescence, and thus just no information is available during dark states. In contrast, dark states of the acceptor lead to a reduction of the energy transfer and thus to an increase of the donor fluorescence. This can be misinterpreted as big conformational changes where the dyes are brought totally out of the range of energy transfer. Photo-induced transitions between the dark and bright state, as found in the case of Cy5, are intensity dependent. Thus, an intensity series can decide whether the fluctuations are due to conformational changes of the host molecule or due to intrinsic properties of the fluorophore.

References

1. Moerner, W. E., Kador, L., *Phys. Rev. Lett.* **1989**, *62*, 2535–2538.
2. Orrit, M., Bernard, J., *Phys. Rev. Lett.* **1990**, *65*, 2716–2719.
3. Lu, H. P., Xun, L., Xie, X. S., *Science* **1998**, *282*, 1877–1882.
4. Kues, T., Dickmanns, A., Luhrmann, R., Peters, R., Kubitscheck, U., *Proc. Natl. Acad. Sci. USA* **2001**, *98*, 12021–12026.
5. Sako, Y., Minoguchi, S., Yanagida, T., *Nat. Cell Biol.* **2000**, *2*, 168–172.
6. Schwille, P., Korlach, J., Webb, W. W., *Cytometry* **1999**, *36*, 176–182.
7. Schwille, P., Haupts, U., Maiti, S., Webb, W. W., *Biophys. J.* **1999**, *77*, 2251–2265.
8. Köhler, R. H., Schwille, P., Webb, W. W., Hanson, M. R., *J. Cell Sci.* **2000**, *113*, 3921–3930.
9. Widengren, J., Schweinberger, E., Berger, S., Seidel, C. A. M., *J. Phys. Chem. A* **2001**, *105*, 6851–6866.
10. Enderle, T., Ha, T., Ogletree, D. F., Chemla, D. S., Magowan, C., Weiss, S., *Proc. Natl. Acad. Sci. USA* **1997**, *94*, 520–525.
11. Yildiz, A., Forkey, J. N., McKinney, S. A., Ha, T., Goldman, Y. E., Selvin, P. R., *Science* **2003**, *300*, 2061–2065.
12. Schwille, P., Meyer-Almes, F. J., Rigler, R., *Biophys. J.* **1997**, *72*, 1878–1886.
13. Kettling, U., Koltermann, A., Schwille, P., Eigen, M., *Proc. Natl. Acad. Sci. USA* **1998**, *95*, 1416–1420.
14. Heinze, K. G., Koltermann, A., Schwille, P., *Proc. Natl. Acad. Sci. USA* **2000**, *97*, 10377–10382.
15. Förster, T., *Ann. Phys.* **1948**, *2*, 55–75.
16. Stryer, L., Haugland, R. P., *Proc. Natl. Acad. Sci. USA* **1967**, *58*, 719–726.
17. Deniz, A. A., Dahan, M., Grunwell, J. R., Ha, T. J., Faulhaber, A. E., Chemla, D. S., Weiss, S., Schultz, P. G., *Proc. Natl. Acad. Sci. USA* **1999**, *96*, 3670–3675.
18. Jares-Erijman, E. A., Jovin, T. M., *J. Mol. Biol.* **1996**, *257*, 597–617.
19. Clegg, R. M., Murchie, A. I. H., Zechel, A., Lilley, D. M., *Proc. Natl. Acad. Sci. USA* **1993**, *90*, 2294–2298.
20. Kim, H. D., Nienhaus, G. U., Ha, T., Orr, J. W., Williamson, J. R., Chu,

S., *Proc. Natl. Acad. Sci. USA* **2002**, *99*, 4284–4289.

21 Ha, T., Zhuang, X. W., Kim, H. D., Orr, J. W., Williamson, J. R., Chu, S., *Proc. Natl. Acad. Sci. USA* **1999**, *96*, 9077–9082.

22 Zhuang, X., Kim, H., Pereira, M. J. B., Babcock, H. P., Walter, N. G., Chu, S., *Science* **2002**, *296*, 1473–1476.

23 Tan, E., Wilson, T. J., Nahas, M. K., Clegg, R. M., Lilley, D. M., Ha, T., *Proc. Natl. Acad. Sci.* **2003**, *100*, 9308–9313.

24 Zhuang, X., Bartley, L. E., Babcock, H. P., Russell, R., Ha, T., Herschlag, D., Chu, S., *Science* **2000**, *288*, 2048–2051.

25 Andersson, H., Baechi, T., Hoechl, M., Richter, C., *J. Microsc.* **1998**, *191*, 1–7.

26 Ainger, K., Avossa, D., Diana, A. S., Barry, C., Barbarese, E., Carson, J. H., *J. Cell Biol.* **1997**, *138*, 1077–1087.

27 Wang, J., Cao, L. G., Wang, Y. L., Pederson, T., *Proc. Natl. Acad. Sci. USA* **1991**, *88*, 7391–7395.

28 Knemeyer, J.-P., Herten, D.-P., Sauer, M., *Anal. Chem.* **2003**, *75*, 2147–2153.

29 Jacobson, M. R., Pederson, T., *Proc. Natl. Acad. Sci. USA* **1998**, *95*, 7981–7986.

30 Jacobson, M. R., Cao, L. G., Wang, Y. L., Pederson, T., *J. Cell Biol.* **1995**, *131*, 1649–1658.

31 Lange, T. S., Gerbi, S. A., *Mol. Biol. Cell* **2000**, *11*, 2419–2428.

32 Bertrand, E., Chartrand, P., Schaefer, M., Shenoy, S. M., Singer, R. H., Long, R. M., *Mol. Cell* **1998**, *2*, 437–445.

33 Politz, J. C., Tuft, R. A., Pederson, T., Singer, R. H., *Curr. Biol.* **1999**, *9*, 285–291.

34 Politz, J. C., Browne, E. S., Wolf, D. E., Pederson, T., *Proc. Natl. Acad. Sci. USA* **1998**, *95*, 6043–6048.

35 Dirks, R. W., Molenaar, C., Tanke, H. J., *Histochem. Cell Biol.* **2001**, *115*, 3–11.

36 Perlette, J., Tan, W., *Anal. Chem.* **2001**, *73*, 5544–5550.

37 Sokol, D. L., Zhang, X., Lu, P., Gewirtz, A. M., *Proc. Natl. Acad. Sci. USA* **1998**, *95*, 11538–11543.

38 Tsuji, A., Sato, Y., Hirano, M., Suga, T., Koshimoto, H., Taguchi, T., Ohsuka, S., *Biophys. J.* **2001**, *81*, 501–515.

39 Tsuji, A., Koshimoto, H., Sato, Y., Hirano, M., Sei–Iida, Y., Kondo, S., Ishibashi, K., *Biophys. J.* **2000**, *78*, 3260–3274.

40 Matsuo, T., *Biochim. Biophys. Acta* **1998**, *1379*, 178–184.

41 Sako, Y., Hibino, K., Miyauchi, T., Miyamoto, Y., Ueda, M., Yanagida, T., *Single Mol.* **2000**, *1*, 159–163.

42 Seisenberger, G., Ried, M. U., Endreß, T., Büning, H., Hallek, M., Bräuchle, C., *Science* **2001**, *294*, 1929–1932.

43 Kues, T., Peters, R., Kubitscheck, U., *Biophs. J.* **2001**, *80*, 2954–2967.

44 Dahan, M., Levi, S., Luccardini, C., Rostaing, P., Riveau, B., Triller, A., *Science* **2003**, *302*, 442–445.

45 Levi, V., Ruan, Q., Kis-Petikova, K., Gratton, E., *Biochem. Soc. Trans.* **2003**, *31*, 997–1000.

46 Keller, R. A., Ambrose, W. P., Arias, A. A., Cai, H., Emory, S. R., Goodwin, P. M., Jett, J. H., *Anal. Chem.* **2002**, *74*, 316A–324A.

47 van Orden, A., Keller, R. A., Ambrose, W. P., *Anal. Chem.* **2000**, *72*, 37–41.

48 Gao, Q., Shi, Y., Liu, S., *Fresenius. J. Anal. Chem.* **2001**, *371*, 137–145.

49 Ronaghi, M., Elahi, E., *J. Chromatogr. B* **2002**, *782*, 67–72.

50 Pourmand, N., Elahi, E., Davis, R. W., Ronaghi, M., *Nucleic Acids Res.* **2002**, *30*, e31.

51 Werner, J. H., Cai, H., Jett, J. H., Reha-Krantz, L., Keller, R. A., Goodwin, P. M., *J. Biotechnol.* **2003**, *102*, 1–14.

52 Sauer, M., Angerer, B., Ankenbauer, W., Foldes-Papp, Z., Gobel, F., Han, K. T., Rigler, R., Schulz, A., Wolfrum, J., Zander, C., *J. Biotechnol.* **2001**, *86*, 181–201.

53 Braslavsky, I., Hebert, B., Kartalov, E., Quake, S. R., *Proc. Natl. Acad. Sci. USA* **2003**, *100*, 3960–3964.

54 Levene, M. J., Korlach, J., Turner, S. W., Foquet, M., Craighead, H. G., Webb, W. W., *Science* **2003**, *299*, 682–686.

55 Margittai, M., Widengren, J., Schweinberger, E., Schroder, G. F., Felekyan, S., Haustein, E., Konig, M., Fasshauer, D., Grubmuller, H., Jahn, R. et al., *Proc. Natl. Acad. Sci. USA* **2003**, *100*, 15516–15521.

56 Rothwell, P. J., Berger, S., Kensch, O., Felekyan, S., Antonik, M., Wöhrl, B. M., Restle, T., Goody, R. S., Seidel, C. A. M., *Proc. Natl. Acad. Sci. USA* **2003**, *100*, 1655–1660.

57 Gaiko, N., Hillisch, A., Berger, S., Seidel, C. A. M., Diekmann, S., Griesinger, C. **2002**, in preparation.

58 Ha, T., Glass, J., Enderle, T., Chemla, D. S., Weiss, S., *Phys. Rev. Lett.* **1998**, *80*, 2093–2096.

59 Seidel, C. A. M., Schulz, A., Sauer, M. H. M., *J. Phys. Chem.* **1996**, *100*, 5541–5553.

60 Seidel, C., *Proc. SPIE Int. Soc. Opt. Eng.* **1991**, *1432*, 91–104.

61 McKinney, S. A., Declais, A. C., Lilley, D. M., Ha, T., *Nat. Struct. Biol.* **2003**, *10*, 93–97.

62 Ishii, Y., Yoshida, T., Funatsu, T., Wazawa, T., Yanagida, T., *Chem. Phys.* **1999**, *247*, 163–173.

63 Widengren, J., Schweinberger, E., Berger, S., Seidel, C. A. M., *J. Phys. Chem. A* **2001**, *105*, 6851–6866.

64 Widengren, J., Schwille, P., *J. Phys. Chem. A* **2000**, *104*, 6416–6428.

65 Borsch, M., Diez, M., Zimmermann, B., Reuter, R., Graber, P., *FEBS Lett.* **2002**, *527*, 147–152.

66 Ha, T., *Methods* **2001**, *25*, 78–86.

29
Scanning Force Microscopy and Scanning Force Spectroscopy of RNA

Wolfgang Nellen

29.1
Introduction

RNA research has significantly advanced by using sophisticated methods to investigate secondary and tertiary structures, and RNA–RNA as well as RNA–protein interactions. Most biochemical and biophysical methods rely on large numbers of molecules – the average reaction or deduced structures are interpreted to reconstruct the behavior or shape of a single molecule. This "top-down approach" can answer many fundamental questions to understand RNA function. It is contrasted by the "bottom-up approach" where single molecules are investigated and a large number of individual observations are used to statistically determine the average behavior or structure of the molecule(s). (It should be kept in mind that biochemists and biophysicists, on the one hand, and single molecule investigators, on the other, are approximately 17 orders of magnitude apart when they speak of "large numbers of molecules"!).

One of the highly promising methods in single molecule research is scanning probe microscopy (SPM), based on the scanning force microscopy (SFM) developed by Binnig et al. [1]. This allows us to record three-dimensional (3-D) topographic maps of biological samples with a resolution of a few nanometers.

The basic principle of SFM (Fig. 29.1) is scanning of a sample with a tip a few nanometers in diameter. The tip is attached to a flexible cantilever arm fixed at one end in a holder. A laser beam is focused on the cantilever and reflected to the center of a four-quadrant diode. The sample is mounted on a stage that can be moved by piezo elements in the x-, y- and z-directions. When, during scanning in the x-direction, the tip hits an obstacle, i.e. the molecule of interest, the cantilever arm will bend and the laser beam is deflected off the center of the diode. An electronic feedback will re-adjust the stage in the z-direction and the signal is recorded. Every scan line thus presents a height profile of the sample. Usually, 512 x-profiles adjacent to each other in the y-direction create a topographic image.

In contrast to electron microscopy, images are taken directly from "live" molecules and not indirectly from metal or carbon coatings of the molecules. Since

Handbook of RNA Biochemistry: Student Edition. Edited by R. K. Hartmann, A. Bindereif, A. Schön, E. Westhof
Copyright © 2009 WILEY-VCH Verlag GmbH & Co. KGaA, Weinheim
ISBN: 978-3-527-32534-4

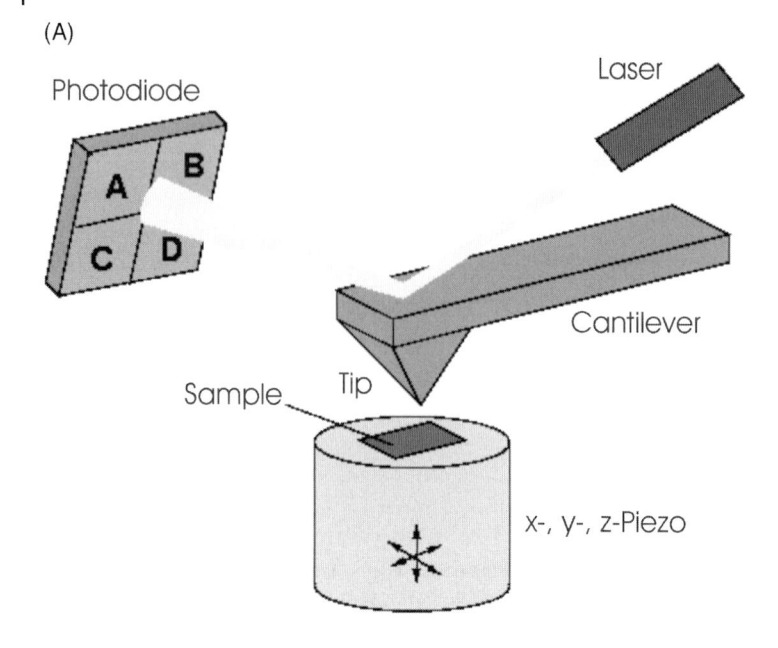

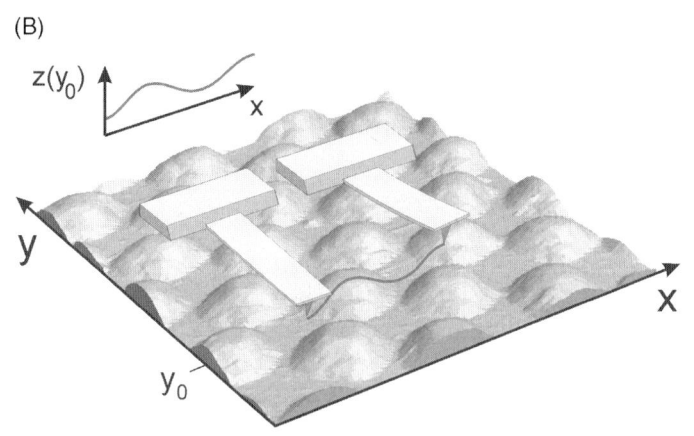

Fig. 29.1. (A) The principle of the scanning force microscope: see text for details. (B) Scanning of a 3-D surface. The cantilever is moving in the x-direction across the sample, and is shown in the start (left) and end (right) position based on the diagram above. Note the bending of the cantilever arm (right position) when it encounters an obstacle.

samples are physically scanned under ambient conditions, biomolecules may also be investigated under near physiological conditions, i.e. in aqueous solutions.

In the "contact mode", the rigid cantilevers are in direct contact with the sample and "scrape" the surface. Thus, shearing forces are exerted on the sample and soft material may be scratched or cut. Therefore, the "tapping mode" is usually employed to minimize contact with the sample. The cantilever oscillates and only briefly touches the sample and lateral forces are therefore minimized. Cantilevers used for these measurements usually oscillate with a resonance frequency of 250–400 kHz and a free amplitude of about 100 nm. The set point is adjusted to 10–30% below the free amplitude to obtain clear images with minimal damage to the sample.

29.2
Questions that could be Addressed by SFM

At present, SPM is mainly within the realm of physicists who, with the support of biologists and biochemists, developed methods to provide a "proof of principle". The number of publications presenting new, additional information in biology and biochemistry is still limited. However, the relative simplicity of sample preparation and use of the instrument should result in a rapid increase in applications in the near future. RNA research by SFM started out with the visualization of molecules [2, 3]. Imaging of mRNA circularization by interaction of poly(A)-binding proteins with EIF4E was one of the major contributions of SFM to the understanding of translation complexes [4]. The detection of preferential binding sites of an antibody directed against double-stranded RNA provided evidence that RNA–protein interactions could be demonstrated that had previously escaped biochemical analysis [5] (Fig. 29.2).

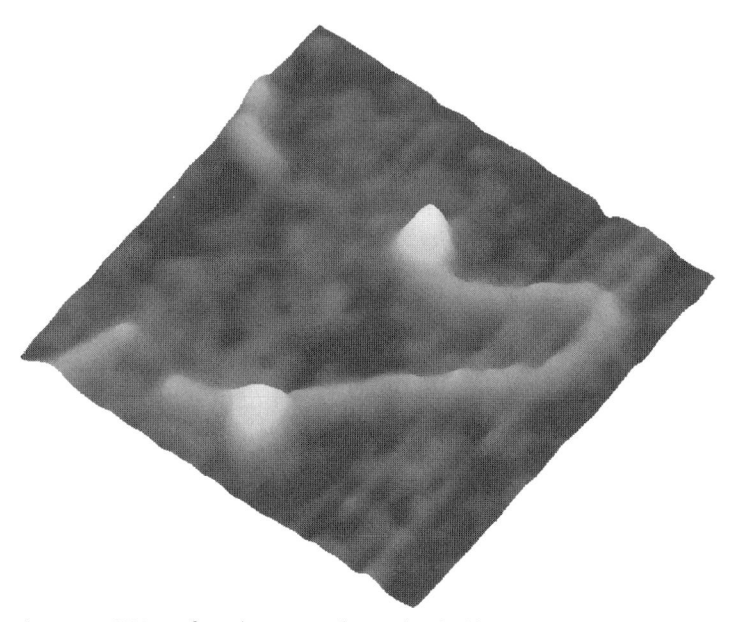

Fig. 29.2. SFM: surface plot image of a 700-bp double-stranded RNA with two proteins bound. Image size is 125 × 125 nm.

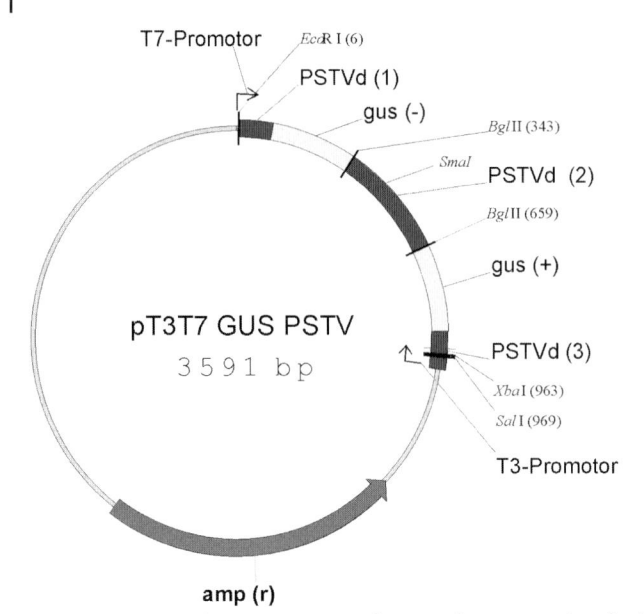

Fig. 29.3. Plasmid for the generation of double-stranded RNA with inserted protein binding sites. The fragment to be transcribed by T7 or T3 RNA polymerase consists of two gus segments that are complementary and form a complete double helix; in addition, three PstVd segments form imperfect double strands flanking the gus hybrid. The single SmaI site within PSTVd (2) is used to insert a sequence of interest.

A more general approach to study RNA–protein interactions was initiated by the construction of a versatile vector that could be used to generate a fold-back RNA into which any potential interaction site of interest could be inserted (Fig. 29.3). The backbone RNA contained one completely double-stranded and two partially double-stranded regions derived from a coding mRNA and the potato spindle tuber viroid (PstVd), respectively. The RNA of 980 nt that forms a rigid rod-like structure thus contains, in addition to the interaction site of interest, a multitude of non-specific competitor sequences and structures. This backbone with the appropriate sequences inserted allows for the determination of protein-binding specificity in comparison to other sites and for large-scale structural changes (like bending and kinking) upon binding of a partner. With the insertion of a second site for binding of a different molecule, protein–protein interactions could be investigated as long as the distance between both sites is large enough to allow for sufficient flexibility [6].

RNA-modifying enzymes like adenosine deaminase that acts on RNA (ADAR) have a general, promiscuous binding affinity to and modifying activity on double-stranded RNA *in vitro*, but modify only specific sites *in vivo*. By SFM, the binding frequency to specific and non-specific sites could be determined with a precision of approximately 40 bp (the sequence covered by one protein molecule). In parallel biochemical experiments, editing efficiency was measured at the respective sites and could be directly compared to binding. These experiments provided new, unexpected insights into the mechanisms of RNA editing [7]. It should, however, be noted that the SFM experiments cannot be correlated to association constants, but rather give quantitative values of a less-well-defined "binding frequency" in comparison to competing sites.

SFM measurements as described above can be easily done on dried samples in air – they visualize the situation after a reaction has been carried out in solution and stopped by spreading the molecules on the mica surface. More sophisticated experiments can be done in liquid using a "fluid cell" that allows for imaging under close to physiological conditions. While measurements in air are usually done at ambient temperature, reaction conditions in a fluid cell can be controlled and reaction partners can be sequentially added by a delivery system. Technical problems arise because of insufficient adherence of molecules to the surface and consequently displacement by the lateral forces of the cantilever during scanning. Between 1 and 3 mM $NiCl_2$ has been used to improve binding of RNA to the substrate. Fay et al. [8] have used these conditions to visualize hairpin ribozymes in solution and, by using mutants, were able to distinguish between different conformational states during the self-cleaving process. Obviously, adjusting salt conditions for optimal binding of nucleic acids to the surface may interfere with molecular interactions with other reaction partners.

The resolution of imaging depends on the variable scanning field (usually 10×10 μm) and the scanning speed (0.3–2 Hz). Images of 512×512 pixels are set up within 4–25 min. More rapid scanning may damage the sample and decrease the resolution. Even though some attempts have been made to monitor dynamic changes in biological samples [9–11], the compromise between resolution and speed of commercial cantilevers is not yet favorable for real-time recording of molecular movements. However, new, very small cantilevers of 10–20 μm length, 2–5 μm width and 100–140 nm thickness with resonance frequencies up to 650 kHz (in water) have now been developed [12] and permit rapid scanning. Images of 100×100 pixels can be captured within 80 ms [13], thus allowing for the observation of biomolecules in motion.

"Pseudo-dynamic" measurements may be done when a reaction or interaction can be started, for example, by adding ATP. By spreading and drying the sample in the mica surface, the reaction is stopped at different times and the evaluation of the reaction intermediates could provide insight into consecutive structural alterations with time [14].

SFM of biological samples is limited by several parameters. To avoid background structures, extremely flat surfaces are required for loading the sample. Freshly cleaved mica (Goodfellow, Cambridge) has proven functional in most cases of nucleic acid and protein analysis. For the specific, oriented attachment of molecules (see below), gold-coated chips have frequently been used.

Even though the tip forces on the sample have been strongly reduced with the development of the "tapping mode", molecules are still subjected to mechanical insult that may change the integrity of a molecule or a complex. In fact, cantilever forces can be adjusted so that single molecules may be "nano-dissected" in a defined way [15, 16]. Lateral forces exerted by the cantilever can dislocate the sample during the scanning process and result in blurred images. On the other hand, strong binding of the sample to the surface could interfere with structural flexibility and thus biochemical properties. The "non-specific" interaction forces of nucleic acids and proteins with mica are not defined, but they are sufficiently strong in air. In liquid, however, nucleic acid molecules are more loosely attached or are only bound at a few sites along the chain. Therefore, specific precautions have to be taken to avoid displacement by the cantilever forces (see above).

Tip diameters present another limitation of SFM: commercially available tips have a radius of 1–10 nm and thus result in an apparent broadening of the sample. Double-stranded RNA with a width of 2.5 nm therefore appears 5–10 nm wide depending on the quality of the tip. In contrast, the height of a sample is frequently underestimated because the mechanical contact of the tip with the soft material causes flattening. Heights of 1.2–1.7 nm are usually measured instead of the expected 2.5 nm for double-stranded RNA. Due to these errors and to the fact that scanning only visualizes the top view of an object (but not possible indentations at the bottom), volumes of a sample can hardly be determined and the analysis of 3-D fine structures is also limited.

Double-stranded DNA and double-stranded RNA are easily visualized since they form rather rigid rods. The investigation of, for example, protein-binding sites is, however, hampered by the fact that the orientation of the nucleic acid cannot be determined. For double-stranded RNA, the constructs to produce transcripts can be modified by including defined asymmetric secondary structures of at least 30 bp (equivalent to 9 nm). Branches of this size are easily visible by SFM, and unambiguously identify the left and right of the molecule.

Loops, bulges and longer stretches of unpaired bases may contribute to the length of the molecules but so far in a rather unpredictable way. An unpaired loop of 60 nt appeared as a bulge of approximately 10 nm length and 15 nm width; in addition, the height of this structure was significantly increased in comparison to completely base-paired double-stranded RNA. Presumably, a predominant 3-D structure was adopted by the single-stranded RNA, but the resolution of the method was insufficient to obtain any more detailed information (Bonin and Nellen, unpublished).

Many RNA-binding proteins display high affinities to double-stranded RNA ends. Substrates generated by *in vitro* transcription of sense and antisense strands followed by hybridization generate open termini that are "sticky" for some proteins. Limited digestion with RNase A to remove single-stranded overhangs did not abolish the problem. A general backbone construct largely reduced the problem of non-specific end binding: *in vitro* transcripts are generated from an inverted repeat that folds back to a double-stranded and partially double-stranded molecule with the 5′ end, and the 3′ end embedded in a secondary structure [6].

Visualization of single-stranded RNA has only been possible in some special cases. This is due to extensive inter- and intramolecular secondary structures that are rapidly formed in solution or even in the remaining liquid on mica before drying. Single strands may become visible when they are fixed by a protein on one side and stabilized by a double-stranded region on the other side [14]. For some applications, single-stranded RNA can be incubated with the interacting protein and subsequently parts of the RNA are "stretched out" by annealing DNA oligos [4].

29.3
Statistics

Biochemical analysis provides an average of the reactivity of 10^{17}–10^{20} molecules, but does not address the detailed behavior of single molecules. On the other hand, SPM, observes single molecules and every non-typical reactivity gains a much higher weight than in bulk analysis. Non-typical behavior could be due to rare but significant reactions, but also to damaged molecules, mistakes in synthesis of the components or other artifacts. To distinguish between artificial and rare, but specific, interactions, a large number of molecules have to be inspected. Overview images rapidly supply a first general, although subjective, impression of molecular interactions. To measure sizes, distances and heights of several 100 molecules in a reasonable time and to gather sufficient data for solid statistics, the appropriate software is usually supplied with the instrument. With measuring 100–500 events, even minor interactions that may escape biochemical analysis could be detected.

29.4
Scanning Force Spectroscopy (SFS)

The scanning force microscope can also be employed to determine inter- and intramolecular binding forces. The basic principle is that one binding partner is attached to the cantilever and the other to the surface. Upon approaching the cantilever to the surface in the z direction, two molecules eventually interact. When the cantilever is then retracted from the sample, binding forces bend the arm until the retracting force equals and finally exceeds the binding force and the interaction is disrupted. The force required for a conformational change or disruption of binding is calculated from the recorded bending of the cantilever and the spring constant – a value describing the flexibility of an individual cantilever.

Conventional scanning force microscopes may be used for SFS but a separate z-piezo has to be integrated to individually control the approach and retract movements. Specific force microscopes are available that do not allow for imaging, but provide optimal hardware and software for interaction measurements. Hybrid instruments have now been developed and become especially interesting for biological applications: a sample can be scanned to provide a topological image and then a specific molecule for force measurements can be approached. It has to be noted that a functionalized cantilever, i.e. a tip with a biomolecule attached, may well be used for scanning and subsequent force measurements.

Figure 29.4 displays a typical force–distance curve taken from double-stranded DNA. In this case, non-specific binding forces between the nucleic acid and the mica surface, on one side, and between the DNA and the silicon nitrate tip, on

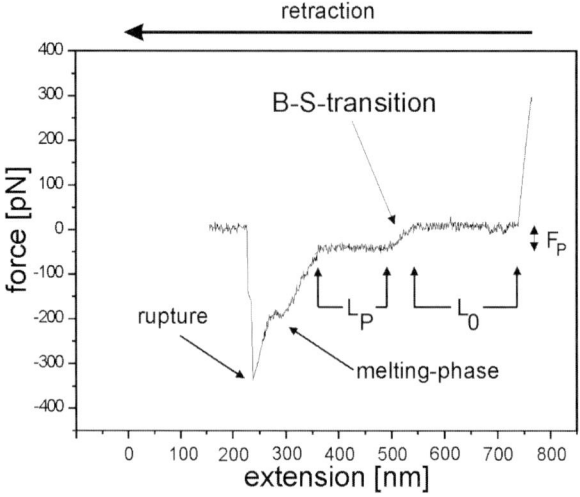

Fig. 29.4. SFS: force–distance curve taken from double-stranded DNA. See text for explanation.

the other side, are sufficiently strong to measure binding forces within the molecule. The graph shows the bending of the cantilever while the z-piezo moves away from the tip: on the far right, tip and substrate are in contact. Overcoming nonspecific adhesion to the substrate, the tip jumps to the zero line. A DNA molecule attached to the surface and the tip is lifted up without applying additional force until it is extended to its entire free length (here approximately 200 nm = L_0). The plateau force (F_p) of 65 pN is then required to convert the B-form DNA to the stretched S-form. After the B–S transition, S-form DNA can be further extended without additional force by a factor of 0.6 (S factor) of the original B-form length. This is shown by the plateau length L_p. Then, a strong force depending on various parameters (attachment site to tip and surface, $G + C$ content, length of molecule) has to be applied to reach the melting phase and a further force finally leads to either disruption of the two strands, disruption of the bond to the tip or disruption of the bond to the surface. After this rupture, the cantilever swings back to the zero line [17, 18].

This example demonstrates the principle of SFS – pulling at a complex of two molecules may lead to one or more defined structural conversions that require defined forces and a final disruption of the interaction caused by an additional force. As for SFM, multiple measurements are made to obtain statistically solid data. When the x–y position of the cantilever is not changed, the same complex may be recorded multiple times. This is useful to demonstrate reversibility and reproducibility, but different complexes at other locations should also be measured. Attempts have been made to detect secondary structures in RNA molecules by SFS, but the results are so far difficult to interpret [19] or have only confirmed relatively simple structures that were already known.

DNA and double-stranded RNA are easy samples since they make strong, non-specific bonds to the mica surface. The exact nature of these bonds is not known and also the position of attachment to the surface within a long nucleic acid is variable. With a 1000-bp double-stranded RNA (or DNA) molecule L_0 values from close to zero up to a maximum of 300 nm can be obtained. In the latter case, the attachment to the mica surface would be close to one end while the molecule is picked up by the cantilever close to the other end. For most other experiments, both or at least one of the interacting partners should be covalently attached to the substrate or the cantilever. The most common method for RNA is to use thiolated oligos that can be bound to a gold surface. Alternatively, thiolated DNA oligos may be used as an anchor and the RNA of interest is transcribed *in vitro* with a tail complementary to the oligo. The hydrogen bonds of the hybrid are usually stronger than, for example, the protein–RNA interactions to be measured. However, one has to consider that the B–S transition of the hybrid may be superimposed on the forces disrupting the interaction of interest.

The measured unbinding forces depend on the velocity of cantilever retraction (loading rate = retraction velocity × elasticity of molecule). With low loading rates, thermal fluctuation contributes significantly to overcoming the energy barrier to separate the molecules and therefore lower forces are measured. With higher loading rates, higher forces are usually required to achieve unbinding. Different protein binding characteristics to target nucleic acids may only be detected in the high loading rate regime [20].

In many cases, it is advisable to attach RNA and/or protein via a flexible linker either to the surface or the cantilever or both. Polyethylene glycol (PEG) linkers of defined length have been successfully used [21].

29.5
Questions that may be Addressed by SFS

SFS can address binding forces between different molecules as well as intramolecular forces like secondary structures of proteins [22] or the forces that hold a protein in a membrane [23]. In combination with molecular genetics, the influence of defined mutations on interactions can be determined. Although association kinetics cannot so far be determined, thermal off-rates can be derived from experiments using different loading rate regimes [20]. At least in some cases this has provided insights into binding mechanisms that could not be obtained by conventional molecular biology approaches.

29.6
Protocols

Protocol: SFM studies on RNA–protein interactions

- To provide internal controls and to reduce the frequently observed preferential end binding of proteins to RNA, cloning of the (putative) protein interaction site into a vector like pT3T7–gus–PstVd is recommended. A single *Sma*I site within the PstVd segment (Fig. 29.3) is used to insert the sequence of interest (usually 30–100 bp). mfold analysis [24] of the expected transcript is carried out to confirm that the inserted sequence is exposed and does not disturb the structure of the backbone molecule. RNA is synthesized by *in vitro* transcription with T7 or T3 RNA polymerase, the reaction mixture is extracted with phenol/chloroform, precipitated with ethanol and washed with 70% ethanol. RNA is taken up in an appropriate volume of buffer (see below) and the concentration is determined by $OD_{260/280}$. It is recommended to evaluate the RNA by native as well as by denaturing gel electrophoresis. In native agarose gels, the approximately 1000-nt gus–PstVd backbone transcripts display a mobility similar to a 500-bp DNA fragment. Recombinant protein is purified by affinity chromatography under native conditions.

- Mica plates are mounted with double-adhesive tape to 1-cm diameter metal disks for easier handling in the SFM. Mica is freshly cleaved by removing the top layer with adhesive tape and is then activated by exposure for 1 min to an air plasma at 0.2 mbar, 600 V and 20 kHz.
- Reactions are carried out under the appropriate buffer conditions in solution. Buffers should contain 3–15 mM $MgCl_2$. Tris and KCl should be avoided since they form salt precipitates on the mica when dried. Instead, HEPES and NH_4Cl are recommended. The concentrations of the reaction partners have to be tested to identify the optimal density for molecules in the microscope. Depending on intrinsic properties of the reactants (like affinity to the surface) and the properties of the mica, concentrations in the range of 3–30 nM are appropriate. Too high concentrations may result in the formation of aggregates during or after drying. Overcrowding of the reaction mixture may not be directly obvious since large aggregates may be preferentially washed off in the following step leaving only a few molecules on the surface. Aliquots of 10 µl of the reaction are placed on the mica surface for 2–10 min and then washed off with 1 ml of water. The surface is blow-dried with nitrogen or argon and ready for microscopy. If the reaction conditions are not compatible with SFM (high salt), a more concentrated assay may be diluted 10-fold in 5 mM $MgCl_2$ and then spread on mica.
- All water and buffers used for SFM are set up with MilliQ purified water, autoclaved and passed through a 0.45-µm sterile filter unit.
- The scanning force microscope is usually placed on a heavy stone slab supported by bungee cords. Alternatively, a dynamic vibration isolation system is used to grant stable measurements.
- SFM conditions depend very much on the microscope that is used. Scanners with a range of $125 \times 125 \times 5$ µm (x, y, z) or $10 \times 10 \times 2.5$ µm are employed. Silicon cantilevers with a nominal resonance frequency of 200–400 kHz (in air), a spring constant of 10–50 N/m and a point diameter of 1–5 nm provide good resolution in tapping mode scanning. High-quality images are taken at 0.5–1 Hz (5–10 min per image). Appropriate software to process the images is provided by the suppliers of the microscope. Additional software for length measurements may be required. Height profiles can be calculated by integrated software for any straight line drawn across the sample.

Protocol: SFS studies on protein–nucleic acids interactions

To our knowledge, measurements of RNA–protein interaction forces have so far not been reported, but can be carried out according to established protocols for DNA–protein interaction.

Functionalized tips and surfaces

- Si_3N_4 tips may be prepared for binding of ligands by different ways.
  (1) Tips are activated by brief incubation in concentrated nitric acid and then silanized for 2 h in 2% aminopropyltriethoxysilane in dry toluene and extensively washed in toluene. They are then incubated in 0.1 mM potassium phosphate pH 8 containing 1 mM N-hydroxysuccinimide–PEG–maleimide for 30 min. Cantilevers are washed with potassium phosphate buffer and are ready for binding the nucleic acid. 5'-SH modified RNA oligonucleotides (10 ng/µl) are incubated with the prepared tips in 50 mM Tris, 100 mM NaCl, 0.1 mM $NiCl_2$ at pH 8.3 at 4 °C for 10 h. After washing with binding buffer, the tips are ready for spectroscopy measurements or may be stored at 4 °C for about 1 week [20].
  (2) An alternative is the use of defined length PEG spacers with an amine-reactive and a thiol-reactive end. Silicon nitride cantilevers are activated for 10 min in chloroform and then for 30 min in H_2SO_4/H_2O_2 (70:30), washed in sterile water and baked for 2 h at 180 °C. Tips are then incubated in 55% ethanolamine chloride in dimethylsulfoxide overnight at 100 °C together with 0.3-µm molecular sieve beads under vacuum with H_2O trapping [21]. PEG spacers are coupled to the amine groups as described by Haselgrübler et al. [25]. Thiolated RNA oligos are then coupled in an oriented way to the thiol-reactive end as described above.
- Several cantilevers (five to 10) should be derivatized simultaneously. The concentration of protein or nucleic acid to be bound to the tip is so low that statistically only a few molecules will bind to the tip. Cantilevers with no ligand or too many ligands have to be identified experimentally and will be discarded.
- For derivatization of surfaces, mica is freshly cleaved, baked at 180 °C and processed as described under (2) for nucleic acid or protein binding.
- For protein binding, the surface is silanized with aminopropyltriethoxysilane in an exsiccator and then incubated for 1 h at 4 °C with the purified protein of interest (5 µM) in 0.1 M potassium phosphate buffer (pH 7.5) containing 20 µM bis(sulfosuccinimidyl)suberate-sodium salt [20], this couples the N-terminus of the protein covalently to the derivatized surface.
- Another method to functionalize surfaces is using gold chips or gold-coated mica to anchor thiolated oligos or proteins covalently [26].

29.7
Troubleshooting

The most sensitive component for imaging, but also for spectroscopy, are the tips. It is very difficult to predict the lifetime of a cantilever: they may be good for measuring 2–3 days or they may become blunt within one measurement. The manufacturing process and quality control is becoming much more reliable, but 10% of rejects are still to be expected. A broadening of the image indicates that a cantilever is blunt. A frequent observation is a "shadow image", a weaker duplication of an image in the same frame. This indicates a double tip either due to the original cantilever, to some damage during scanning or to contamination picked up during scanning of the sample. The shadow image may appear 100 nm or more from the original image and not be obvious at the first glance. Since double tips usually have a different distance to the substrate, the weaker and often thinner shadow image may be misinterpreted as single-stranded DNA or RNA.

As with electron microscopy, the interpretation of an image may be problematic since contaminations may be similar to the expected shape of the sample. A famous example is the steps in carbon surfaces that resembled in shape structure and dimension DNA double strands.

High concentrations of nucleic acids result in aggregates – they cannot be interpreted and may be regarded as contamination only and a complete failure of the experiment. High concentration of protein can cause the same effect; in addition, non-specific multimers may give the impression that the sample is inhomogenous with many different sizes of protein. Since only a "top view" of the molecules is possible, even specific dimers are frequently not observed as particles with double the volume or double the surface of a monomer. To reduce the concentration of the sample is a simple remedy for this problem. However, especially for RNA–protein interactions measured in air, concentrations in the reaction tube are not necessarily reflected in the image because RNA, protein and complexes may adhere with different efficiency to the surface.

Low contrast could be due to set point adjustment too close to the free amplitude of the cantilever. Samples measured in air are usually robust under ambient conditions. High humidity of the air may, however, decrease contrast because a water film will form on the sample. Repeated drying with argon or nitrogen will temporarily solve the problem but water will re-accumulate after 20–40 min.

To discuss specific problems and applications the SFM/SFS forum at http://spm.di.com/listinfo.html is recommended.

29.8
Conclusions

SFM and SFS have advanced beyond the realm of physicists and of providing proof-of-principle for nucleic acids applications. The current generation of instruments is user friendly and designed for biologists who wish to approach biological problems, and not necessarily improve the technology. In combination with biochemistry and molecular biology, SFM and SFS have proven to provide additional information and substantial insight into molecular mechanisms.

Acknowledgments

I thank F. W. Bartels (University of Bielefeld) and I. Sagi (Weizmann Institute) for making data available prior to publication. Nils Anspach and C. Hammann are acknowledged for critical reading of the manuscript. K. Pross (Veeco) is acknowledged for contributing an extensive compilation of the relevant literature.

References

1 G. Binnig, C. F. Quate, C. Gerber, Phys. Rev. Lett. **1986**, 56, 930–933.
2 Y. Lyubchenko, L. Shlyakhtenko, R. Harrington, P. Oden, S. Lindsay, Proc. Natl. Acad. Sci. USA **1993**, 90, 2137–2140.
3 A. Y. Lushnikov, A. Bogdanov, Y. L. Lyubchenko, J. Biol. Chem. **2003**, 278, 43130–43134.
4 S. E. Wells, P. E. Hillner, R. D. Vale, A. B. Sachs, Mol. Cell **1998**, 2, 135–140.
5 M. Bonin, J. Oberstrass, N. Lukacs, K. Ewert, E. Oesterschulze, R. Kassing, W. Nellen, RNA **2000**, 6, 563–570.
6 M. Bonin, J. Oberstrass, U. Vogt, M. Wassenegger, W. Nellen, Biol. Chem. **2001**, 382, 1157–1162.
7 Y. Klaue, A. M. Kallman, M. Bonin, W. Nellen, M. Ohman, RNA **2003**, 9, 839–846.
8 M. J. Fay, N. G. Walter, J. M. Burke, RNA, **2001**, 7, 887–895.
9 M. Argaman, R. Golan, N. H. Thomson, H.-G. Hansma, Nucleic Acids Res. **1997**, 25, 4379–4384.
10 M. Guthold, M. Bezanilla, D. A. Erie, B. Jenkins, H. G. Hansma, C. Bustamante, Proc. Natl. Acad. Sci. USA **1994**, 91, 12927–12931.
11 M. Guthold, X. Zhu, C. Rivetti, G. Yang, N. H. Thomson, S. Kasas, H. G. Hansma, B. Smith, P. K. Hansma, C. Bustamante, Biophys. J. **1999**, 77, 2284–2294.
12 M. B. Viani, T. E. Schäffer, A. Chand, M. Rief, H. E. Gaub, P. K. Hansma, J. Appl. Phys. **1999**, 86, 2258–2262.
13 T. Ando, N. Kodera, E. Takai, D. Maruyama, K. Saito, A. Toda, Proc. Natl. Acad. Sci. USA **2001**, 98, 12468–12472.
14 A. Henn, O. Medalia, S. P. Shi, M. Steinberg, F. Franceschi, I. Sagi, Proc. Natl. Acad. Sci. USA **2001**, 98, 5007–5012.
15 S. Iwabuchii, T. Mori, K. Ogawa, K. Sato, M. Saito, Y. Morita, T. Ushiki, E. Tamiya, Arch. Histol. Cytol. **2002**, 65, 473–479.
16 S. Thalhammer, R. W. Stark, S. Muller, J. Wienberg, W. M. Heckl, J. Struct. Biol. **1997**, 119, 232–237.
17 H. Clausen-Schaumann, M. Rief, C. Tolksdorf, H. E. Gaub, Biophys. J. **2000**, 78, 1997–2007.
18 M. Rief, H. Clausen-Schaumann, H. E. Gaub, Nat. Struct. Biol. **1999**, 6, 346–349.
19 M. Bonin, R. Zhu, Y. Klaue, J. Oberstrass, E. Oesterschulze, W. Nellen, Nucleic Acids Res. **2002**, 30, e81.
20 F. W. Bartels, B. Baumgarth, D. Anselmetti, R. Ros, A. Becker, J. Struct. Biol. **2003**, 143, 145–152.
21 P. Hinterdorfer, W. Baumgartner, H. J. Gruber, K. Schilcher, H. Schindler, Proc. Natl. Acad. Sci. USA **1996**, 93, 3477–3481.
22 M. Rief, M. Gautel, F. Oesterhelt, J. M. Fernandez, H. E. Gaub, Science **1997**, 276, 1109–1112.
23 D. J. Muller, J. B. Heymann, F. Oesterhelt, C. Moller, H. Gaub, G. Buldt, A. Engel, Biochim. Biophys. Acta **2000**, 1460, 27–38.
24 M. Zuker, Nucleic Acids Res. **2003**, 31, 3406–3415.
25 T. Haselgrubler, A. Amerstorfer, H. Schindler, H. J. Gruber, Bioconjug. Chem. **1995**, 6, 242–248.
26 O. Medalia, J. Englander, R. Guckenberger, J. Sperling, Ultramicroscopy **2001**, 90, 103–112.

Part III
RNA Genomics and Bioinformatics

30
Comparative Analysis of RNA Secondary Structure: 6S RNA

James W. Brown and J. Christopher Ellis

30.1
Introduction

The purpose of this chapter is to provide a "primer" on the comparative analysis of RNA secondary structure. The emphasis here is on the initial stages of the analysis; in other words, how one goes about creating a working model of the secondary structure *de novo* using the comparative approach. This is a common scenario; you, a student or coworker have discovered that an RNA is involved in a biological system under investigation. The sequence of the RNA is determined, usually either from cDNA or from the gene. Or perhaps it is discovered that a region of a messenger RNA or viral RNA is important in some process and it is suspected that the structure of this region is critical for that function. You are interested, then, it obtaining information about the structure of this RNA in order to help guide experiments and to organize data about the RNA. The determination of the three-dimensional (3-D) structure of the RNA is unlikely to be cost-effective or feasible (certainly not as a first step), but you correctly realize that the single most thermodynamically favorable predicted secondary structure is not going to suffice. How, then, to proceed? Usually, the answer is by creating a secondary structure model based on comparative sequence analysis. The detailed analysis of very-high-resolution secondary structure, the identification and evaluation of tertiary interactions, and the construction of 3-D models based on comparative analysis will not be considered here; these aspects of comparative analysis of RNA structure require specialized experience. Comparative analysis, like X-ray crystallography, is as much art as science, but the creation of a basic secondary structure is well within the range of a newcomer to the "RNA World", the target audience for this chapter.

The approach taken here is to follow the construction of a basic secondary structure of an example RNA: 6S RNA. The 6S RNA was discovered in *Escherichia coli* in 1971 [1], but its function remained unknown until very recently [2]. 6S RNA is not essential for viability [3], but accumulates during the stationary phase, binds directly and specifically to RNA polymerase, and regulates RNA polymerase function in a growth stage-specific manner [2]. The secondary structure of the 6S RNA has not been examined in any detail; the existing secondary structure proposed for this

Handbook of RNA Biochemistry: Student Edition. Edited by R. K. Hartmann, A. Bindereif, A. Schön, E. Westhof
Copyright © 2009 WILEY-VCH Verlag GmbH & Co. KGaA, Weinheim
ISBN: 978-3-527-32534-4

RNA was based on a comparison of only the *E. coli* and *Pseudomonas aeruginosa* sequences [4, 5].

30.1.1
RNA Secondary Structure

What *is* an RNA secondary structure? Although most researchers would agree on simple definitions of primary structure (sequence) and tertiary structure (3-D coordinates), there is a surprising extent of disagreement about *exactly* what RNA secondary structure is, even (perhaps especially!) among established RNA researchers that work with secondary structures on a daily basis [6]. At its most basic, however, a secondary structure is a list of adjacent, antiparallel Watson–Crick (or G∘U) base pairs in an RNA chain; these are the pairings for which the rules are clear and that are readily predicted by comparative sequence analysis. Uncertainty about what exactly is "secondary structure" deal primarily with the distinction between secondary and tertiary interactions. For example, are non-Watson–Crick base pairs other than G∘U included? Are isolated base pairs included? What about helical stacks? In the case of a pseudo-knot, which helices are considered secondary and which, if any, are tertiary? All of these are subject to some level of disagreement. It is also worth remembering that "secondary" does not mean 2-D; secondary structures contain a plethora of 3-D information, beginning with the presumption that the helices are generally A-form in structure. However, in the comparative analysis of secondary structure of an RNA, the basic definition of secondary structure is generally most useful.

Secondary structure can be represented in a variety of ways, but is most often presented as a string of letters, the sequence, twisted around on a page (i.e. in two dimensions) such that these antiparallel adjacent interactions can be shown as dashes between each pair of bases. By formal convention, G∘U pairs are shown with a hollow dot instead of a dash and non-Watson–Crick pairings with a closed dot, such as G•A [7]. Typically (tRNA is the exception, here) structures are drawn to flow generally clockwise 5' to 3'. A convenient way to specify whether or not there is specific evidence for a base pairing is to only put in the dash (or dot or circle) if there is such evidence.

RNA secondary structure is very specific and highly defined; secondary structure is the central organizing principle in RNA structure. This is a fundamental difference between protein and RNA (and DNA secondary structure, of course). Pragmatically, experiments are almost always developed and results represented in the context of the secondary structure of an RNA.

30.1.2
Comparative Sequence Analysis

Comparative sequence analysis is the process of extracting information about a macromolecule (in this case RNA) from the similarities and differences between different, but homologous, sequences (for review, see [8–11]). The underlying assumption is that the higher order structure of the molecule is more highly conserved than is the sequence; in other words, the sequence is free to change during evolution as long as the 3-D structure is generally maintained. In terms of secondary structure, this means that changes in the identity of a base involved in a pairing should generally be allowed by a compensatory change in its pairing partner so that the ability of the two to form isosteric base pairs is retained. The two bases that pair then vary together, or *covary*. The work involved in the construction of a secondary structure of an RNA by comparative analysis is primarily the search for these sequence *covariations*. If sufficient numbers of sequences are available, these covariations can be identified statistically directly from a sequence alignment [12, 13]. Comparative analysis, then, is an iterative process in which improvements in the alignment result in additional structural information, which can be used in turn to improve the alignment. Although attempts have been made to automate this process (see, e.g. [14–16]), with varying levels of success, in practice this is generally still a manual process.

30.1.3
Strengths and Weakness of Comparative Analysis

Comparative analysis is the "Gold Standard" method for determining secondary structure of RNAs; computational methods for predicting secondary structure are typically validated by comparison with "true" secondary structures as determined by comparative analysis (see, e.g. [17, 18]). However, other methods for determining secondary structure can be very useful supplements to comparative analysis or serve as last resort alternatives if comparative analysis is not feasible, e.g. if few or only one sequence is available for analysis.

A particularly useful supplement to comparative analysis is the genetic analysis of mutation and second-site compensatory mutation; in fact, these methods are formally equivalent, the difference being whether you create the variations or observe them in nature. This method is typically laborious and so has not been used generally as an alternative to comparative analysis, but can be especially useful either to confirm the presence of a particular feature of secondary structure (see, e.g. [19]) or to probe secondary structure than cannot be assessed by the comparative method, such as pairings involving invariant sequences. For instance, the 6S RNA secondary structure used as an example of comparative analysis in this chapter contains a stem–loop in which none of the base pairing are specifically supported by sequence covariations; the paired sequences are invariant among the sequences available. An alternative to obtaining additional 6S RNA sequences in hope of finding covariations in this potential stem–loop would be to make point mutations in this region of the RNA in *E. coli* that affect the function of the molecule and then make the compensatory change. If the RNA with two substitutions, such that the potential base pairing is maintained, functions better than the RNA with either single substitution that disrupts the potential pairing, then the pairing is presumed to be legitimate. Genetic analysis has also been used in the absence of comparative data in cases where only a single instance of the functional RNA is known, such as the delta virus ribozyme [20].

Another useful supplement to comparative analysis, as we will see, is the prediction of structures thermodynamically. This is, in reality, where secondary structure modeling usually begins. These predictions are steadily improving, especially with the ability to predict a variety of structures near the minimum free energy and assess the frequency that particular base pairings are predicted in these collections of structures [21–23]. Thermodynamic predictions are routinely used to predict the structures of idiosyncratic elements of structure that appear as insertions in specific instances of an RNA. The danger of thermodynamic prediction is the tendency to consider these structures endings rather than beginnings. A measure of the success of thermodynamic prediction is that the predicted lowest free energy structures contain, on average, about 73% of base pairs that would exist in a "true" secondary structure determined by comparative analysis [17].

The last commonly used method for assessing secondary structure in RNA is chemical and enzymatic probing. Although these methods have been used extensively in attempt to determine structure, their utility is mostly in the examination of *changes* or *differences* in structure that result from mutation, binding to other molecules, and the like. Chemical and enzymatic probing data are notoriously difficult to judge directly in terms of the secondary structure of the RNA.

30.1.4
Comparison with Other Methods

Comparative analysis is similar to, but more sensitive than, genetic experiments because natural selective pressure is more sensitive than our biochemical or genetic methods. Comparative analysis past the initial stages is objective, quantitative and conceptually automatable. Given sufficient numbers of variable sequences, a secondary structure can be very high resolution, in which every base pair is assessed individually. Only biologically relevant base pairings are identified by comparative analysis. Nevertheless, there are limitations to comparative analysis. The most important of these is that no structure can be assessed in the absence of sequence variation; as a result, the most important aspects of structure, those comprised of the most highly conserved sequences, are the most difficult to prove by comparative analysis. The initial stage of a comparative analysis, the subject of this chapter, is basically a manual process. No specific information is provided about unique sequences that cannot be meaningfully aligned. Although tertiary interactions can also be detected by comparative analysis (although this typically required large collections of sequences), only base-base interactions in which more than one isosteric possibility is structurally acceptable will be detected. Nevertheless, comparative analysis is certainly the method of choice whenever possible. The list of structures determined definitively by comparative analysis is nearly as long as the list of known RNA types: large and small subunit ribosomal RNAs [24], transfer RNAs [25], RNase P [26] and MRP [27] RNAs, SRP RNA [28], tmRNA [29], group I [30] and II [31] introns, nuclear splicing RNAs (e.g. [32]), H/ACA [33] and box C/D [34] snoRNAs, telomerase RNA [35], etc.

30.2
Description

30.2.1
Collecting Sequence Data

The raw material needed to determine the secondary structure of an RNA by comparative analysis is sequence data; more specifically, what is needed is a collection of different, but functional and homologous sequences. There are two ready sources for sequences: nature and GenBank [36, 37]. The first step, then, is to mine the available databases for homologous sequences. Very often there are sufficient numbers of suitable sequences available for the generation of at least an initial secondary structure. If this is not the case or if a higher-resolution secondary structure is desired, it will be necessary to obtain additional sequences experimentally.

A variety of approaches are needed to identify as many homologous sequences in GenBank as possible. A good starting point is to search the GenBank using BLAST [38] with your sequence of interest. In our example, the *E. coli* 6S RNA sequence (X01238) returned a number of other sequence records containing the *E. coli* 6S RNA:

```
AE016766.1 Escherichia coli CFT073 section 12 of 18 of the
complete genome
X01238.1 E. coli 6S ribosomal RNA
AE005521 Escherichia coli O157:H7 EDL933 genome, contig 3
of 3, section 140 of
AE000374 Escherichia coli K12 MG1655 section 264 of 400 of
the complete genome
U28377.1 Escherichia coli K-12 genome; approximately 65 to
68 minutes
M12965.1 E.coli ssr gene encoding 6S RNA
AP002563.1 Escherichia coli O157:H7 DNA, complete genome,
section 14/20
```

These are all identical to the original sequence and so of no use to us. Note that this need not be the case; for some RNAs there may be useful variants in different strains of the same species. Other sequences obtained in this search were:

```
AE016988.1 Shigella flexneri 2a str. 2457T section 11 of 16
of the complete genome
```

```
AE015303.1 Shigella flexneri 2a str. 301 section 266 of 412
of the complete
AE016844.1 Salmonella enterica subsp. enterica serovar
Typhi Ty2, section 11 of
AL627277.1 Salmonella enterica serovar Typhi (Salmonella
typhi) strain CT18
AE008840.1 Salmonella typhimurium LT2, section 144 of 220
of the complete genome
AE008841.1 Salmonella typhimurium LT2, section 145 of 220
of the complete genome
AE013931.1 Yersinia pestis KIM section 331 of 415 of the
complete genome
AJ414145.1 Yersinia pestis strain CO92 complete genome;
segment 5/20
```

The sequences from the two *Shigella flexneri* strains are identical, as are those of the two strains of *Salmonella enterica* and *Yersinia pestis*. The *Salmonella typhimurium* sequences represent the same sequence from the genome sequence, split in two by the separation of sections 144 and 145 of the genome record. This is frequently the case for RNA-encoding genes because genome sequences are divided into sections with an eye toward larger "intergenic" regions (spaces between open reading frames) that often turn out to be RNA-encoding genes. It is simply a matter of extracting the two fragments of sequence and merging them. For most of the other sequences, the entire sequence can be extracted simply by cutting and pasting from the BLAST results page. Sometimes, however, it is necessary to go to the original sequence record. For example, in the case of the *Y. pestis* sequence, the 3' end of the sequence is different enough from that of *E. coli* that it was not returned in the BLAST alignment and had to be retrieved from the original.

Additional sequences can often be identified by repeating the BLAST searches with the sequences identified in the initial search. In the case of our example, however, a search using the most disparate sequence identified so far, that of *Yersinia pestis*, yielded the same list of sequences.

Another obvious approach is to search using the name of the RNA, but unlike protein-encoding genes, RNA-encoding genes (except those of rRNAs and tRNAs) are often not annotated even in genome sequences. Using "6S RNA" as the search term for our example locates the sequence from *E. coli* (X01238), all of the sequences listed above and that of *P. aeruginosa* (Y00334). However, already we see one of the weaknesses of relying on sequence annotations; the *E. coli* 6S RNA is misannotated as the "*E. coli* 6S ribosomal RNA"! In addition, a number of "6S RNA" sequence annotations are typographical errors where "16S RNA" was meant. A number of other matches are spurious because of the presence of the term "6S" in strain or clone names, enzyme name (e.g. the "6Fe–6S prismane cluster-containing protein") or other RNAs with the same names (it seems there are different "6S RNAs" in vertebrates and in λ). Annotations must always be scrutinized critically.

Nevertheless, the identification of the annotated 6S RNA sequence from *P. aeruginosa* provides a fresh avenue for the search; a BLAST search using this sequence identified a homolog in the *Pseudomonas syringae* genome (AE016875) (as well as several instances of the *P. aeruginosa* sequence, of course). In addition, a weak match to the *P. aeruginosa* 6S RNA sequence was found in the *Pasteurella multocida* genome (AE006208); this region of the genome sequence was extracted and used, in turn, in a BLAST search that identified a homologous sequence is *Haemophilus influenzae* Rd (U32767). In the cases of both of these sequences, the ends of the RNA are not obvious from sequence similarity and so a generous amount for sequence was taken from either end.

The 6S RNA is encoded by the *ssrS* gene and has in the past been referred to as the "ssr RNA" [3]; a search of the GenBank using these terms did not identify any additional sequences.

A number of complete genome sequences are available for organisms that are related to those for which 6S RNA sequences had been identified, but in which homologous sequences had not been found in general BLAST searches of the GenBank. The genomes of all of the γ proteobacteria in which the 6S RNA had not yet been found were then searched individually from the NCBI genome-specific web pages in the hope of extracting additional sequences. (Phylogenetic information about these organisms can be found on the Taxonomy Browser at http://www.ncbi.nlm.nih.gov/Taxonomy/ [36]). The *Pseudomonas putida* KT2440 (NC_002947) 6S RNA sequence was identified in a search using the *P. aeruginosa* sequence as the query; it is perhaps surprising that this sequence failed to be identified in the original search of the entire GenBank, but this is not unusual. More surprising still is that no 6S-like sequences could be identified in the complete genomes of *Vibrio cholerae*, *Vibrio parahaemolyticus* or *Vibrio vulnificus* using the sequence from the closely related *E. coli* as the query. Nor could a 6S-like sequence be identified in the *Haemophilus ducreyi* complete genome sequence using the *H. influenzae* sequence as query.

Another source of sequences are secondary databases, such as, in this case, the Small RNA Database (http://mbcr.bcm.tmc.edu/smallRNA/) [39], the Noncoding RNAs Database (http://biobases.ibch.poznan.pl/ncRNA/) [40] or the Washington University Rfam Database (http://rfam.wustl.edu/index.html) [41]. The first two of these include only the *E. coli* and *P. aeruginosa* sequences, but the Washington University Rfam site contains an alignment of 6S-like sequences from a number of bacterial genomes, including three that were not found in our previous searches: *Shewanella onedensis* (AE015522), *V. vulnificus* (AE016802) and *V. cholera* (AE004317). Using these sequences in turn to search the global GenBank and individual genome sequences using BLAST yielded a sequence from the *Vibrio parahaemolyticus* genome using the *V. vulnificus* (but surprisingly not the *V. cholerae*) sequence as the query.

At this point, 14 presumptive 6S RNA sequences have been identified and extracted from the GenBank. These sequences range from nearly identical (those of *E. coli* and *S. flexneri* differ by only 1 nt) to less than 50% identical; a reasonable collection to begin a comparative analysis. It is important to have a wide range of sequence variation. The closely related sequences are useful because they are readily aligned and allow the initial identification of structure in the most variable parts of the RNA, but provide no useful information in the conservative regions of the sequence. The distantly related sequences are needed (often at later stages of the analysis) as a source of sequence variation for the analysis of the conservative (and therefore most important) regions of the RNA.

If additional sequences are needed, either because homologous sequences cannot be found by mining sequence databases or to increase the resolution of a secondary structure based on available sequences, they will have to be obtained experimentally. PCR amplification is typically used to obtain these sequences, but because the sequences flanking the gene are unlikely to be conserved, primers for amplification most often are within the gene itself and so only partial sequences are obtained. Although partial sequences have been very useful in comparative analyses of RNA structure, the entire sequence can usually be obtained using a variety of technologies available in "kit" form. It is important to note that the primer target sequences at either end of a PCR product should *not* be used in a comparative analysis; these sequences are derived from the primers, not the target. A particularly useful approach to collecting large numbers of sequences quickly has been the use of PCR amplification from DNA extracted from complex microbial natural populations, rather than pure cultures [42]. The amplification products are populations of sequences, and so must be separated by cloning, but hundreds of sequences can be obtained in a single experiment. The species from which any particular sequence originates is unknown, but this information is unnecessary for the purposes of comparative analysis; all that matters is that the sequence is a valid sequence. In any case, the phylotype of the sequence itself can be determined after the fact by the construction of phylogenetic trees based on the final sequence alignment.

30.2.2
Thermodynamic Predictions

It is useful, early in the process, to have the thermodynamic predictions of the structures of all of the RNA sequences in the collection. These are generated using mfold, most conveniently using the Mfold Web Server [43] at http://www.bioinfo.rpi.edu/applications/mfold/old/rna/form1.cgi. For the purposes of initial comparative analysis, the default settings should suffice for most RNAs of reasonable

Fig. 30.1. Potential structures of the *E. coli* 6S RNA predicted thermodynamically. These are all of the structures predicted by mfold using the default parameters; in particular, only structures within 5% of the minimum free energy were allowed and a window parameter (which defines how dissimilar two structures must be to be considered distinct) of 10 was used. Structures from left to right are from most to least favorable, respectively. Any base pairings in the helices identified in the comparative analysis (see Fig. 30.4) are boxed. Structures were downloaded from the mfold server [43] as connect (.ct) files and displayed using LoopDloop [55].

Fig. 30.1

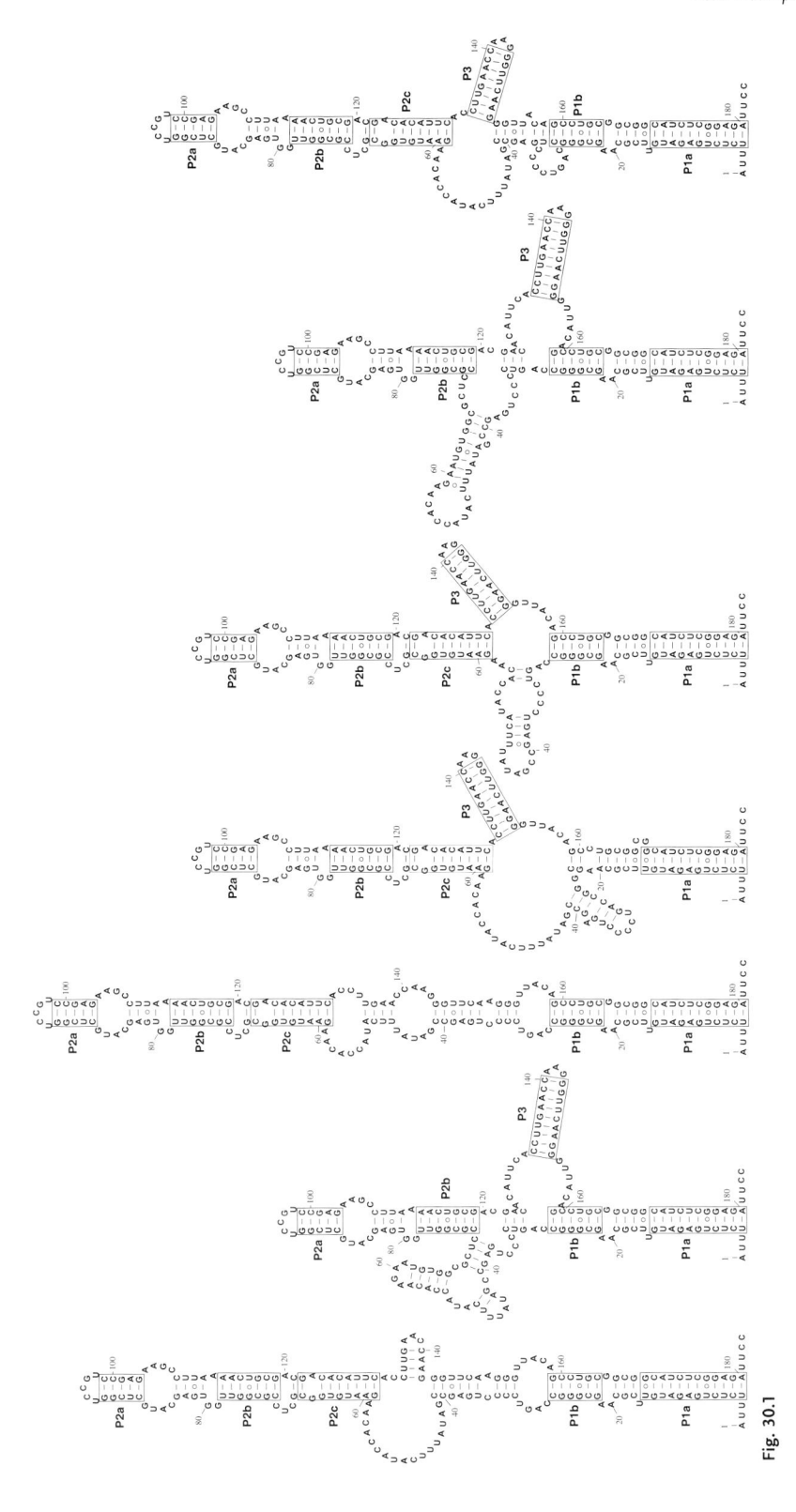

length. If an unmanageable number of structures are predicted, the window parameter can be increased. If only one or two structures are generated, increase the percent suboptimality parameter to 10. The predicted structures can be downloaded as images for printing, but also download and print the energy table; this represents all of the predicted suboptimal foldings. Consistencies in these folding predictions among the different RNA sequences provide a starting point for comparative analysis.

In the case of the 6S RNAs, mfold consistently predicts pairing of the middle regions (roughly bases 60–130) of the RNAs in a stem–loop, and the two ends (the first and last around 20 nt) of the RNA as a terminal helix (see Figs 30.1 and 30.2 for the predicted *E. coli* 6S RNA structures). The interior of this extended stem–loop structure is less consistently predicted. The most common alternatives for the central region of the RNA (between the consistently predicted terminal stem and the medial stem–loop) are base-pairing across this internal region such that the entire RNA would form an extended irregular hairpin or the presence of local stem–loops on either side of the "conserved" central stem–loop. A stem–loop on the 3' side (position around 130–150) is predicted more frequently and the placement of this predicted stem–loop is more consistent than predictions on the 5' side.

30.2.3
Initial Alignment

A comparative analysis requires that the homologous sequences be aligned; in fact, it is the continuous building and refinement of the alignment that drives the structure analysis. Comparative analysis is an iterative process; additions to or improvements in the alignment result in additional structural information that, in turn, allows the alignment to be refined and provides insight required to add increasingly distantly related sequences to the analysis.

The first step, of course, is to collect all of these sequences into a sequence alignment editor. A variety of alignment editors are available for various computer platforms and many of them are freely available from the authors. For Windows/PC computers, a particularly useful alignment editor and analysis program, available at no cost, is BioEdit (http://www.mbio.ncsu.edu/BioEdit/). Most commercial DNA manipulation and analysis software packages include an alignment editor. Because you will most often be adding sequences by extracting them from larger (often much larger) sequence records, it is usually most convenient to move them to the alignment editor by simply cutting and pasting. Retyping sequences manually, although it might seem to be a small task, is a last resort; any hand-typed sequences will need to be painstakingly checked and rechecked for errors.

Once the sequences are all added to the alignment editor, they will need to be aligned preliminarily. If the sequences are all fairly closely related, this might be easily done "by eye", but generally one would use an automated method, CLUSTAL [44] being the most common method incorporated into most alignment editors. Note that this is your *initial* alignment, not your *final* alignment! Much of

```
boxplot_ng by D. Stewart and M. Zuker
©2003 Washington University
```

Fold of Eco 6S RNA at 37 C.

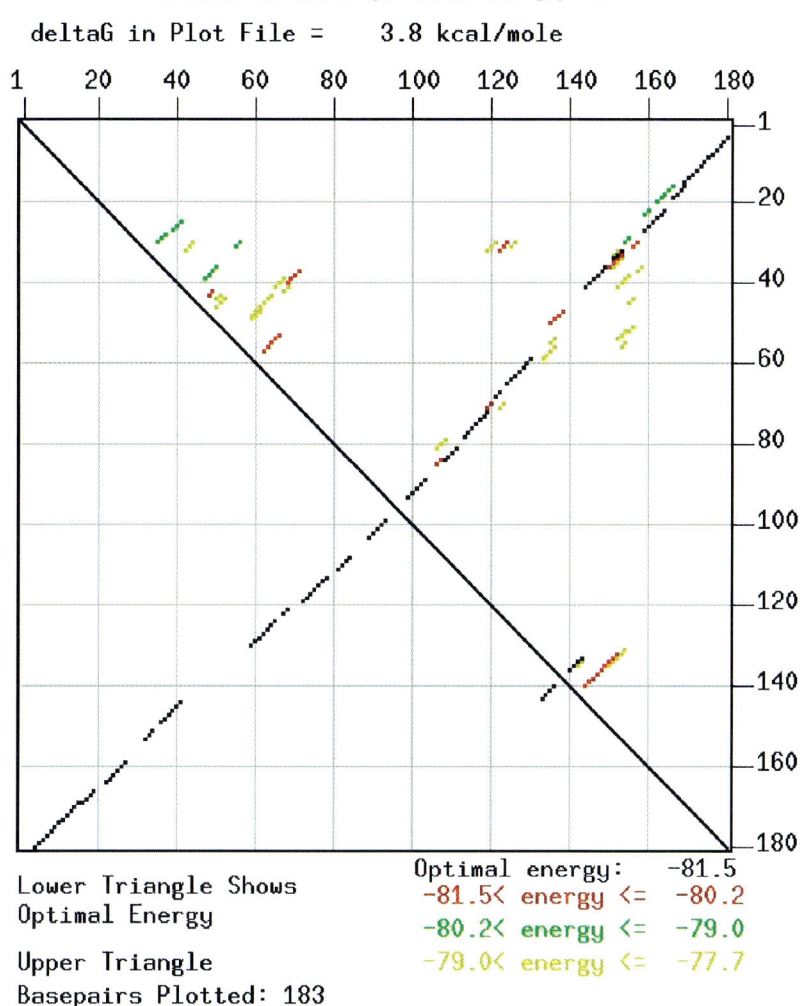

deltaG in Plot File =   3.8 kcal/mole

Lower Triangle Shows
Optimal Energy

Upper Triangle
Basepairs Plotted: 183

Optimal energy: −81.5
−81.5< energy <= −80.2
−80.2< energy <= −79.0
−79.0< energy <= −77.7

Fig. 30.2. Energy table of the *E. coli* 6S RNA from mfold. This represents the same structures shown in Fig. 30.1. The *x*- and *y*-axis each represent the sequence in the 5′ to 3′ direction. Each "dot" indicates a predicted base pairing; the single best predicted structure (the left-hand-most structure in Fig. 30.1) is shown below the diagonal, all of the predicted structures are show above the diagonal. Helices identified in the comparative analysis (see Fig. 30.4) are boxed above the diagonal. This energy table was generated by the mfold server [22, 43].

the work of a comparative analysis is the iterative improvement of the alignment. Even a novice can usually scan through a preliminary CLUSTAL alignment and find room for improvement. There is a fundamental difference between protein sequence alignments, which are generally based only on some maximizing measure of similarity between all pairs of sequences, and RNA alignments, that are based on the higher-order structure of the molecules. Ultimately, of course, the goal of any sequence alignment is to have homologous residues in alignment, but protein alignments attempt to achieve this by maximizing sequence similarity, because the richness of amino acid variation provides substantially more information on which to base an alignment than do the 4 bases in nucleic acid alignments. On the other hand, protein secondary structure is less informative than the highly organized secondary structures of RNA, which are based on one-to-one interactions between bases, and so RNA alignments are more easily based directly on higher-order structure.

Before proceeding further, it is important to arrange the sequences phylogenetically within the alignment (see Fig. 30.3). The NCBI Taxonomy Browser web site (http://www.ncbi.nlm.nih.gov/Taxonomy/) is a useful guide to general phylogenetic relationships. In our example, the 6S RNAs of *E. coli* and *S. flexneri* are nearly identical; they should therefore be adjacent in the alignment, as should the two sequences from *Salmonella* species (*S. typhimurium* and *S. enterica*), *Vibrio* species, and *Pseudomonas* species. *E. coli*, *S. flexneri* and *Salmonella* species form a larger cluster, and so should be brought together; likewise, all of the sequences from the enteric bacteria (the species just mentioned and *Vibrio* species) should be clustered. *Haemophilus* and *Pasteurella* are relatives, and so these two sequences belong together as well. It is convenient to have our prototype sequence, that of *E. coli*, at the top of the alignment, with increasingly distant sequences arranged downwards.

30.2.4
Terminal Helix (P1a)

Pairing of the sequences near the 5′ and 3′ ends to form a terminal helix is a common element of RNA structure, and is a good starting point in the construction of the secondary structure of an RNA. Assuming that the ends of at least one example of the RNA of interest has been determined experimentally, the identification of a terminal helix allows the prediction of the location of the ends of the remaining RNAs in the alignment. In the case of our example, the 6S RNA, a terminal helix is also consistently predicted thermodynamically (Figs 30.1 and 30.2). In fact, all of

Fig. 30.3. Alignment of 6S RNA sequences following comparative analysis. Sequences are ordered phylogenetically (see text) and the alignment (*not* any particular sequence) is numbered at the top. The base pairing identified by comparative analysis are defined using parentheses in the last line. The structure is also shown diagrammatically at the bottom; the upstream (5′) and downstream (3′) nucleotides in each helix are shown with arrows. Four regions of absolutely conserved sequence longer than 5 nt are labeled "Conserved Region" (CR) I–IV, as used for the RNase P RNA [56].

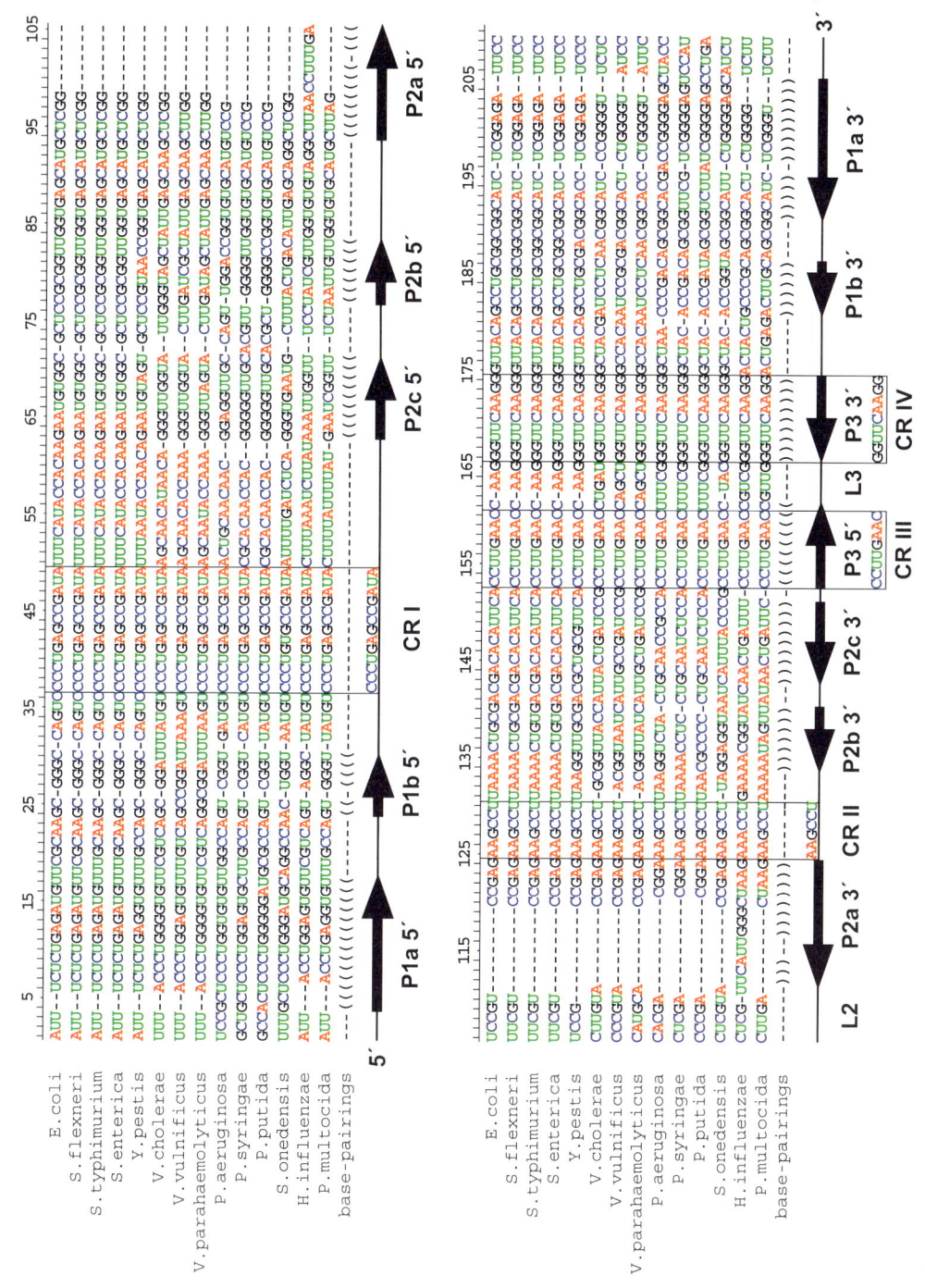

Fig. 30.3

the sequences in the collection are complementary near the ends, but the length of that complementarity varies somewhat and two sequences contain a bulged "A" interrupting this helix. Alignment of the nucleotides on either strand of this helix is straightforward, however, on the basis of sequence conservation; only minor alteration of the CLUSTAL alignment is required to bring the bases in each position of the helix into the same columns (Fig. 30.3). When aligned on the basis of sequence similarity, it becomes clear that the variation in helix length results from the addition of 2 bases to the distal ends of the helix (i.e. then end of the helix that contains the 5′ and 3′ tails) in *Pseudomonas* and *S. onedensis*. If the alignment is correct, there is a one to one correspondence in pairing partners in columns of the alignment; notice how the helix is opened by one column to accommodate the bulged "A" in two of the *Pseudomonas* sequences in the 3′ strand of this helix. To solidify the specific base pairs and their homology among the sequences, a new line in the sequence alignment is added to hold right- and left-facing parenthesis to specify pairing partners (see Fig. 30.3). Additional lines can be added to the alignment for annotations to make it easier to visualize the helices. Once the alignment of this terminal helix (if present) is finalized, the alignment can be trimmed at the ends to match the native ends of any RNAs in which the ends have been determined experimentally. In our example, the ends of the *P. multocida* and *H. influenzae* sequences could not be clearly defined on the basis of sequence conservation relative to either the *E. coli* or *P. aeruginosa* sequences (in which the native ends are known [1, 4]), but these ends can now be predicted on the basis of the structure, and the alignment trimmed to match. The predicted terminal helix in these organisms is 1 bp shorter (at the distal end) than the other sequences other than those of *Pseudomonas* species and *S. onedensis*. Following the nomenclature used for group I introns and RNase P [45, 46], we will call this helix "P1a"; P for "pairing", "1" because it is the first helix counting from the 5′ end and "a" because, as we will see, P1 continues after an interruption.

Before moving on, it is important to evaluate in detail the evidence supporting the existence of this helix. The basic bits of evidence upon which secondary structures are built are sequence *covariations*. Two positions in an alignment are said to covary (for the purposes of secondary structure analysis) if *both* positions vary while maintaining the ability to form A–U, G–C or G∘U base pairs. Covariation of 2 bp in a potential helix is generally accepted as proof that the helix exists. In our example, the presence of the terminal helix P1a is supported by sequence covariation in most of the base pairs of the helix with only a few discrepancies. The ends of a helix can be harder to define; what is needed at each end of the helix is sequence covariations supporting pairing on the terminal base pair and clear failure of the adjacent 2 bases to covary arguing against their pairing. Ultimately, one would like to have evidence supporting the pairing of every base pair shown in a secondary structure; one useful way to denote how close you are to this is to only draw the line (or open dot, in the case of G∘U pairs, by convention) connecting base pairs in a secondary structure if these positions in the alignment covary, i.e. if that individual base pair is supported by sequence covariation (see Fig. 30.4). In our example, we have covariation supporting the terminal base pair of the con-

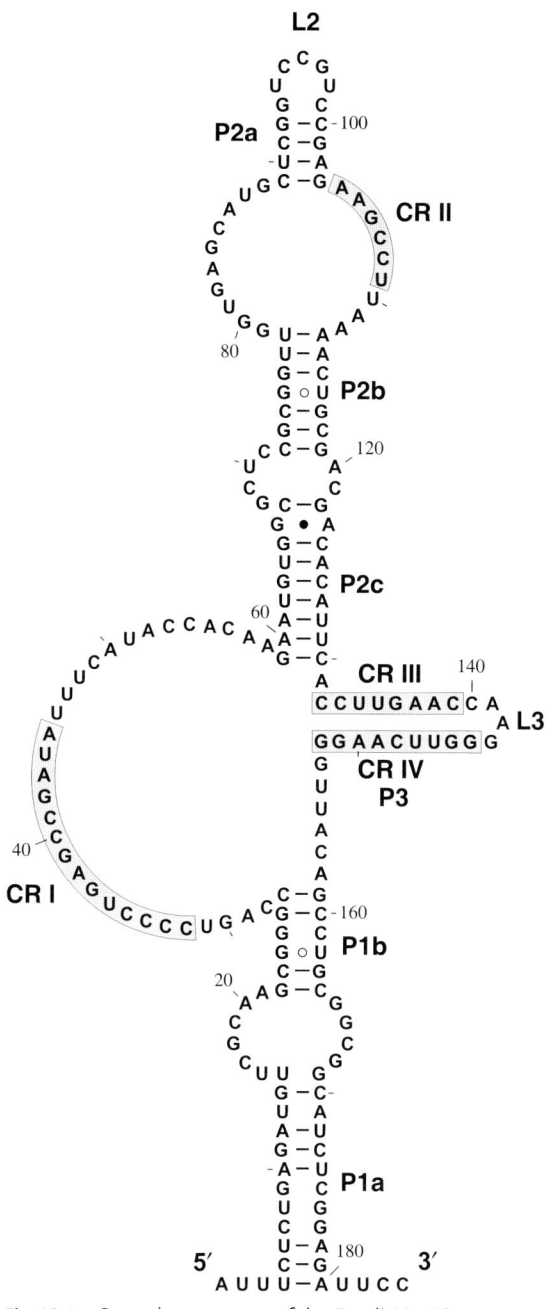

Fig. 30.4. Secondary structure of the *E. coli* 6S RNA. Helices are labeled as described in the text. Base pairings supported individually by sequence covariation are indicated by the connecting lines or dots; unsupported pairings lack these markers. The sequence is numbered 5′ to 3′ every 20 nt, with a tick mark every 10 nt. Four regions of absolutely conserved sequence longer than 5 nt are boxed and labeled "Conserved Region" (CR) I–IV as in Fig. 30.3. This structure was generated in connect (.ct) format directly from the alignment in Fig. 30.3 using a Hypertalk script and displayed using LoopDloop.

served structure and the consistent inability of the three 5' and four 3' nucleotides to pair, so this end of the helix is well defined, *except* in *Pseudomonas* and *S. onedensis*, in which this helix is potentially lengthened at this end by an additional base pairs and two flanking unpaired nucleotides (Fig. 30.3). Whether or not this extra potential base pair is really paired is not clear, because the 5' base is a C and the 3' base always G in these sequences; in the absence of sequence variation, comparative analysis provides no evidence for or against the pairing. At the proximal end of this helix, the last base pair is likewise uncertain; the 5' base is U or C, the 3' base is an invariant G. This is consistent with their pairing, but does not constitute specific evidence for it. The penultimate base pair, on the other hand, is supported by covariation; this is typically G=C, but is A–U in *P. putida*. The G∘U at this position in *P. syringae* constitutes a covariation neither with G=C nor with A–U and so is not evidence for or against this pairing. The flanking two bases are often U and G, and so might be thought to pair, but these fail to covary and so should not be included as part of this helix; the 3' G is invariant, and the 5' base, although U in most sequences, is a G in *P. putida* and an A in *S. onedensis*. There are three adjacent unsupported base pairs in the interior of this helix, but they are given provisional acceptance given that they have the potential to base pair and are flanked on both sides by well-supported pairings.

30.2.5
Subterminal Helix (P1b)

The sequence immediately interior to the 11–13 base pairings that make up the terminal helix P1a of the 6S RNA cannot pair, but complementarity resumes after only a few bases on either side. The pairing of these sequences is predicted in most of the mfold structures from all of the sequences, although there are usually some idiosyncratic alternatives (see Figs 30.1 and 30.2). Adjustment of the alignment of the 5' region of this potential helix is needed to accommodate an extra nucleotide present only in *V. vulnificus* and *V. parahemolyticus*; assignment of homology is straightforward if you keep conserved purines (G or A) or pyrimidines (U or C) aligned (Fig. 30.3). There is reasonable covariation of all but one of the positions in the potential six base pair helix, which we will call "P1b", with an occasional mismatch and a bulged nucleotide representing the extra nucleotide in two *Vibrio* sequences.

30.2.6
Apical Helix (P2a)

In addition to the consistently predicted terminal helix, all of the 6S RNA sequences are predicted by mfold to have a stem–loop in the middle of the RNA containing some conservative sequence elements: CUCGG on the 5' side, and CCGAG on the 3' side (Figs 30.1 and 30.2). Attention is also draw to the potential pairing of these sequences because of the presence in most of the 6S RNAs of the conserved tetraloop sequences UNCG or CUYG (GNRA is the other conserved tetraloop motif [47]), although these would be unusual stem–loops in that the tetraloop sequences are in this case followed (except in *Y. pestis*) by 1 or 2 extraneous nucleotides before the 3′ strand of the stem (Fig. 30.3). Nevertheless, the alignment of these sequences is straightforward based on sequence conservation. The minor exception is the RNA of *H. influenzae*, which contains extra nucleotides between the tetraloop sequence (CUCG) and the conserved complementary sequences on both sides; these extra nucleotides are generally complementary and so would presumably create a terminal extension to the stem–loop. There are sequence covariations confirming all of the base pairs of the helix, which we will call "P2a", with the exception of the terminal pairing, which is always G=C except in the *H. influenza* sequence in which this is an A·C mismatch in the middle of the extension of this stem.

30.2.7
Subapical Helices (P2b and P2c)

Flanking the apical helix P1a are highly conserved sequences that are not complementary and what sequence variation that does occur does not support the specific interaction of these sequences. However, flanking these conserved sequences in turn are variable sequences that are generally complementary (Fig. 30.3). Variation in these regions makes them difficult to assign definitive homologies solely on the basis of sequence, but they are readily aligned, by default, simply based on their conserved distance from the flanking conserved sequences; the only adjustment necessary is the addition of a gap downstream of the apical helix corresponding to an obviously absent U in the *Vibrio* sequences. A similar gap is required in the *S. onedensis* sequence, although the location of this gap is questionable. This helix is 1 bp shorter in the *Pseudomonas* sequences than in the others; this deletion seems to be from the proximal end (furthest from the apex). There are covariations supporting all of the base pairings in this helix, which will be designated "P2b". One RNA, that of *S. onedensis*, has two non-Watson–Crick mispairs in P2b. Although these might seem to argue against the pairing of these sequences, the remaining sequences covary cleanly. More importantly, the mispairs in *S. onedensis* are adjacent G·A/A·G pairs, a 3-D motif that is sometimes is seen as an alternative to Watson–Crick pairs in helices and is known not to interrupt the flanking A-form helical structure [48, 49]. Otherwise, there is only a single instance of a mispair between these two sequences (a C·C pairing in *P. putida*). The internal loop between P2b and P2a is nearly symmetrical, and is comprised of highly conserved sequences, suggesting an important functional role.

Closely flanking P2b are additional complementary sequences. In this case, these sequences are so divergent from one group of sequences to another that alignment based on sequence is impossible (Fig. 30.3). Furthermore, there is significant variation in the spacing between these complementary sequences and the surrounding conserved sequences and helix. Nevertheless, because of the conservation of general complementarity, the presence of this helix in the favorable structures predicted by mfold, and that fact that there is no apparent covariation between either of these sequence regions and anything else in the RNA, we will include this potential helix "P2c" in the secondary structure on a preliminary basis. After aligning these sequences based on this complementarity, there is good covariation of all of the potential base pairings except that a G·A mispairing is present in the penultimate position of the helix in *E. coli* and its closest relatives (*S. flexneri* and *Salmonella* species).

30.2.8
Potential Interior Stem–loop (P3)

The secondary structure of the 6S RNA, as we understand it at this point, is an extended terminal helix P1 and an extended apical helix P2 flanking a central loop of as-yet undefined structure. Very highly conserved sequences on either side of this loop have the potential to pair to form a 9-bp helix with a single bulged A. mfold includes this helix in many of the structures it produced. However, the predictions in this region are *not* consistent; a number of equivalently favorable structures are predicted for each sequence, and the pairings predicted are idiosyncratic for each sequence (Figs 30.1 and 30.2). Only three sequence variants exist in this region; all three of these changes (an A to U in *S. onedensis* and a G to A in *H. influenzae* and *P. multocida*) disrupt the complementarity between these sequences, arguing against their pairing (Fig. 30.3).

A search for frequently occurring tetraloop sequences (UNCG, CUYG and GNRA) [47] flanked by complementary sequences, however, reveals an alternative; a potential stem–loop on the 3′ strand of this interior loop. This 9-bp stem would be composed of very highly conserved sequences; the only sequence variation among the potentially paired nucleotides is the change of a conserved C to U in the terminal base pair in the *Pseudomonas* sequences (Figs 30.3 and 30.4). This is consistent with the pairing of this nucleotide with the conserved G opposite, but does not constitute specific evidence for that pairing. In the absence of evidence for any of the pairings in this helix, the conservative approach would be not to propose this helix until additional sequences with variation in this region can be obtained or a genetic experiment performed. However, two aspects of the potential loop sequences argue for the provisional acceptance of this stem–loop, "P3". First, in the *Pseudomonas* sequence, this loop is a UUCG tetraloop motif, implying a stem–loop structure for at least these RNAs and, by extension, the others as well given the conservation of this region. Second is the observation that all of the sequence and length variation in this region is very specifically located in this potential loop (and the closing base pair in *Pseudomonas*), and that variation is consistent with loop structure.

30.2.9
Is There Anything Else?

At some point in the initial analysis of the secondary structure of your RNA, you will reach a point where no additional structure is obvious. What else can you do in attempt to find structure? Perhaps the most useful approach is to draw all of the RNAs in the secondary structure as it is at this point and compare them with an eye towards common potential helices; how useful this is likely to be depends on how much structure you have already gleamed; the more you already know the better. Some will find it convenient to draw a single "reference" structure and then annotate this with sequence variants. Another fruitful approach is to go back to mfold and generate another round of structures, using the structure you already know as constraints, i.e. force the pairing of all of the helices you are sure of and see what structures are predicted in this context. These structures would then be scrutinized from sequence to sequence in search of commonalities, as before. If you have already identified a large part of the secondary structure, there are likely to be only a small number of favorable structures generated. Another useful approach is to generate a sequence logo from the alignment (a web server for this can be found at http://weblogo.berkeley.edu/logo.cgi) [50]. This allows you to consider potential pairing in the context of sequence variation; sequences are expected to pair with other sequences with similar extents of variation. In the case of the 6S RNA, these methods failed to provide any additional insight into the structure, perhaps because so much structure has already been identified.

30.2.10
Where To Go From Here

Once you are satisfied that you have extracted all of the secondary structure information you can out of your sequences, you have a useful "working model" for the structure of the RNA. Have you identified all of the base pairing in the RNA? Not likely. Do you have extraneous base pairs in the structure? Probably. What direction do you take from here? In the words of the Cheshire Cat, "That depends a good deal on where you want to get to" [51]. If this working model is sufficient for your needs, then you are finished. If you wish to learn more about the secondary structure of the RNA or identify tertiary interactions, then you will need to continue the comparative analysis after additional sequences are obtained.

The choice of where to get additional sequences depends on what you want to learn about the RNA and how well the initial secondary structure analysis went. Most likely, you will want to know more about both the variable and most highly conserved regions of the RNA, and so you will want to obtain sequences that are closely and distantly related to those already in hand. Sequences similar to those already in hand are typically easy to obtain experimentally, and can often be obtained in large numbers. Distantly related sequence are much harder to obtain, but are needed to provide details about the regions of the RNA that are most important for function and so are very highly conserved. Thermophiles are a good source of useful sequences; these RNAs typically contain the fewer irregularities than those of mesophiles, and are much better fodder for thermodynamic prediction [21, 52, 53]. With new sequences in hand, of course, you have the opportunity to mine the sequence databases again and potentially identify sequences that were there all along but remained unrecognized.

30.3
Troubleshooting

Comparative analysis is a straightforward process, but as with any approach, it is possible to run into trouble. Below are listed some of the common problems that arise and how you might try to get around them.

- *But I only have one sequence!* This *is* a major problem; you cannot do a *comparative* analysis with only one sequence. However, this is the usual starting point; you are interested in the structure of a specific RNA from a specific organism and that is the only one you have in hand. Usually you will be able to get at least one or two additional sequences from the genomes of related species. If data mining fails to yield the sequences you need, you have no choice but to get the sequences experimentally.
- *I just do not get it – how do I get started once I have some sequences?* If you are having trouble getting started and you cannot seem to get a handle on the alignment, then reduce the problem by starting with a smaller collection of very similar sequences that you can align easily by eye. Look at every difference in the sequences – can you find one change that corresponds to another change that means sequences could remain complementary? Again, start out, as we did in our example, by looking to see if the two ends of the RNA might form a helix. Also be on the lookout for common tetraloop sequences (UUCG, UCCG, GAGA, GAAA, GUGA, GUAA, GCGA, GCAA and CUUG [9]) flanked by complementary sequences – these are very likely to form stem–loops. Another approach is to start with the single best structure predicted by mfold and then pick through each helix by comparative analysis to prove or disprove each one. You could then move on to the unique helices predicted in the less favorable structures or the structures of the other sequences.
- *Some of my sequences have the most highly conserved sequences, but otherwise cannot be aligned.* It is common to have sequences that you can only align to others in the conservative regions of the molecules, at least at first. In our example of the 6S RNAs, the *H. influenzae* and *P. multocida* sequences do not align well to the others at first. This is especially a problem for sequences that are quite different in length that the others; sometimes localizing the sites of the insertions or deletions can be difficult. This is usually best dealt with from both directions: aligning those regions you can to identify structure in common and dividing the alignment into smaller groups of similar sequences to identify structure in the regions unique to each group. Once some insight on both is obtained, the alignments can be merged on the basis of structure rather than just sequence.
- *PCR or sequencing artifacts.* It is critically important that the sequences that go into a comparative analysis be valid. If you must enter sequences manually, check them very carefully for errors. The qualities of sequences are only as good as the abilities of the person or machine that did the sequence determination; there is *always* a chance that the sequence is incorrect. Genome sequences are usually reliable, but even here errors occasionally arise. In some cases, it might even be worth your effort to confirm an unusual sequence experimentally. As

was seen above, sequence annotations are imperfect; if a sequence does not look like what you expect, it probably is not what you want, and even if it is you will not (yet) be able to use it. Generally speaking, the more recent the sequence, the less likely it is to contain errors. A common source of problematic sequences is PCR amplification and there are two commonly seen types of these errors: point "mutations" and chimeras [54]. Point mutations are a problem, but a limited one. These changes will most often appear in the analysis as the occasional mismatch or idiosyncrasy. Chimeric sequences are sequences that have been artificially spliced together during the amplification process; this is a common problem when amplifying genes from DNA extracted from microbial populations rather than pure cultures. Two aspects of chimeric RNAs usually reveal their nature; their failure to conform to long-range structure that is well-maintained among the remaining sequences or their similarity to one sequence at one end of the RNA, but a very different sequence at the other end. Suspected chimeric sequences should, of course, be removed from the analysis.

Acknowledgments

Research in the authors laboratory is supported by NIH grant GM52894 to J. W. B. and is dedicated to the memory of Dr Elizabeth Suzanne Haas (1957–2002).

References

1 BROWNLEE, G. G., *Nat. New Biol.* **1971**, *229*, 147–149.
2 WASSERMAN, K. M., STORZ, G., *Cell* **2000**, *101*, 613–623.
3 LEE, C. A., FOURNIER, M. J., BECKWITH, J., *J. Bacteriol.* **1985**, *161*, 1156–1161.
4 VOGEL, D. W., HARTMANN, R. K., STRUCK, J. C., ULBRICH, N., ERDMANN, V. A., *Nucleic Acids Res.* **1987**, *15*, 4583–4591.
5 WASSERMAN, K. M., ZHANG, A., STORZ, G., *Trends Microbiol.* **1999**, *7*, 37–45.
6 WAUGH, A., GENDRON, P., ALTMAN, R., BROWN, J. W., CASE, D., GAUTHERET, D., HARVEY, S. C., LEONTIS, N., WESTBROOK, J., WESTHOF, E., ZUKER, M., MAJOR, F., *RNA* **2002**, *8*, 707–717.
7 LEONTIS, N. B., WESTHOF, E., *RNA* **2001**, *7*, 499–512.
8 JAMES, B. D., OLSEN, G. J., PACE, N. R., *Methods Enzymol.* **1989**, *180*, 227–239.
9 WOESE, C. R., PACE, N. R., Probing RNA structure, function and history by comparative analysis, in: *The RNA World*, R. F. GESTELAND, J. F. ATKINS (eds), Cold Spring Harbor Laboratory Press, Cold Spring Harbor, NY, **1993**, pp. 91–117.
10 MICHEL, F., COSTA, M., Inferring RNA structure by phylogenetic, genetic analysis, in: *RNA Structure and Function*, R. W. SIMONS, M. GRUNBERG-MANAGO (eds), Cold Spring Harbor Laboratory Press, Cold Spring Harbor, NY, **1998**, pp. 175–202.
11 MICHEL, F., COSTA, M., MASSIRE, C., WESTHOF, E., *Methods Enzymol.* **2000**, *317*, 491–510.
12 WINKLER, S., OVERBEEK, R., WOESE, C. R., OLSEN, G. J., PFLUGER, N., *Comput. Appl. Biosci.* **1990**, *6*, 7–18.
13 CHIU, D. K., KOLODZIEJCZAK, T., *Comput. Appl. Biosci.* **1991**, *7*, 347–352.
14 PARSCH, J., BRAVERMAN, J. M., STEPHAN, W., *Genetics* **2000**, *154*, 909–921.
15 JUAN, V., WILSON, C., *J. Mol. Biol.* **1999**, *289*, 935–947.

16 Han, K., Kim, H. J., *Nucleic Acids Res.* **1993**, *21*, 1251–1257.
17 Zuker, M., Jaeger, J. A., Turner, D. H., *Nucleic Acids Res.* **1991**, *19*, 2707–2714.
18 Fields, D. S., Gutell, R. R., *Fold Des.* **1996**, *1*, 419–430.
19 Haas, E. S., Morse, D. P., Brown, J. W., Schmidt, F. J., Pace, N. R., *Science* **1991**, *254*, 853–856.
20 Perrotta, A. T., Been, M. D., *Nature* **1991**, *350*, 434–436.
21 Jaeger, J. A., Turner, D. H., Zuker, M., *Methods Enzymol.* **1989**, *183*, 281–306.
22 Jacobson, A. B., Zuker, M., *J. Mol. Biol.* **1993**, *221*, 403–420.
23 Zuker, M., Jacobson, A. B., *RNA* **1998**, *4*, 699–679.
24 Gutell, R. R., Larsen, N., Woese, C. R., *Microbiol. Rev.* **1994**, *58*, 10–26.
25 Zachau, H. G., Dutting, D., Feldmann, H., Melchers, F., Karau, W., *Cold Spring Harb. Symp. Quant. Biol.* **1966**, *31*, 417–424.
26 Harris, J. K., Haas, E. S., Williams, D., Frank, D. N., Brown, J. W., *RNA* **2001**, *7*, 220–232.
27 Schmitt, M. F., Bennett, J. L., Dairaghi, D. J., Clayton, D. A., *FASEB J.* **1993**, *7*, 208–213.
28 Larsen, N., Zwieb, C., *Nucleic Acids Res.* **1991**, *19*, 209–215.
29 Kelley, S. T., Harris, J. K., Pace, N. R., *RNA* **2001**, *7*, 1310–1316.
30 Michel, F., Westhof, E., *J. Mol. Biol.* **1990**, *216*, 585–610.
31 Michel, F., Umesono, K., Ozeki, H., *Gene* **1989**, *82*, 5–30.
32 Frank, D. N., Roiha, H., Guthrie, C., *Mol. Cell Biol.* **1994**, *14*, 2180–2190.
33 Ganot, P., Caizergues-Ferrer, M., Kiss, T., *Genes Dev.* **1997**, *11*, 941–956.
34 Ni, J., Tien, A. L., Fournier, M. J., *Cell* **1997**, *89*, 391–400.
35 Chen, J. L., Blasco, M. A., Greider, C. W., *Cell* **2000**, *100*, 503–514.
36 Wheeler, D. L., Chappey, C., Lash, A. E., Leipe, D. D., Madden, T. L., Schuler, G. D., Tatusova, T. A., Rapp, B. A., *Nucleic Acids Res.* **2000**, *28*, 10–41.
37 Benson, D. A., Karsch-Mizrachi, I., Lipman, D. J., Ostell, J., Rapp, B. A., Wheeler, D. L., *Nucleic Acids Res.* **2000**, *28*, 15–18.
38 Altschul, S. F., Gish, W., Miller, W., Myers, E. W., Lipman, D. J., *J. Mol. Biol.* **1990**, *215*, 403–410.
39 Gu, J., Chen, Y., Reddy, R., *Nucleic Acids Res.* **1998**, *26*, 160–162.
40 Szmanski, M., Erdmann, V. A., Barciszewski, J., *Nucleic Acids Res.* **2003**, *31*, 429–431.
41 Griffiths-Jones, S., Bateman, A., Marshall, M., Khanna, A., Eddy, S. R., *Nucleic Acids Res.* **2003**, *31*, 439–441.
42 Brown, J. W., Nolan, J. M., Haas, E. S., Rubio, M. A., Major, F., Pace, N. R., *Proc. Natl. Acad Sci. USA* **1996**, *93*, 3001–3006.
43 Zuker, M., *Nucleic Acids Res.* **1994**, *31*, 3406–3415.
44 Cheena, R., Sugawara, H., Koike, T., Lopez, R., Gibson, T. J., Higgins, D. G., Thompson, J. D., *Nucleic Acids Res.* **2003**, *31*, 3497–3500.
45 Burke, J. M., Belfort, M., Cech, T. R., Davies, R. W., Schweyen, R. J., Shub, D. A., Szostak, J. W., Tabak, H. F., *Nucleic Acids Res.* **1987**, *15*, 7217–7221.
46 Haas, E. S., Brown, J. W., Pitulle, C., Pace, N. R., *Proc. Natl. Acad. Sci. USA* **1994**, *91*, 2527–2531.
47 Woese, C. R., Gutell, R. R., *Proc. Natl. Acad. Sci. USA* **1990**, *87*, 8467–8471.
48 Gautheret, D., Damberger, S. H., Gutell, R. R., *Nucleic Acids Res.* **1994**, *242*, 1–8.
49 Wu, M., Turner, D. H., *Biochemistry* **1996**, *35*, 9677–9689.
50 Schneider, T. D., Stephens, R. M., *Nucleic Acids Res.* **1990**, *18*, 6097–6100.
51 Carroll, L., *Alice's Adventures in Wonderland*, available from the Gutenberg Project at http://www.cs.indiana.edu/metastuff/wonder/wonderdir.html, **1865**.
52 Pace, N. R., Smith, D. K., Olsen, G. J., James, B. D., *Gene* **1989**, *82*, 65–75.
53 Brown, J. W., Haas, E. S., Pace, N. R., *Nucleic Acids Res.* **1993**, *21*, 671–679.
54 Wang, G. C., Wang, Y., *Microbiology* **1996**, *142*, 1107–1114.
55 Gilbert, D., *Methods Mol. Biol.* **2000**, *132*, 149–184.
56 Chen, J. L., Pace, N. R., *RNA* **1997**, *3*, 557–560.

31
Secondary Structure Prediction

Gerhard Steger

31.1
Introduction

The biological activity of RNA is quite diverse: messenger RNA codes mainly by its sequence for proteins, but structural elements in the non-transcribed regions or even in the open reading frame may be of importance for translation; for most other RNA (e.g. ribosomal RNA, natural antisense RNA, small interfering RNA, ribozymes, etc.), which are summarized by the terms structural or non-coding (nc) RNAs, the two- and/or three-dimensional structure is of utmost importance for their biological function. In this chapter we will concentrate on the prediction of such structures given the RNA sequence. Prediction of RNA secondary structure is much easier than prediction of protein structure, because the stability of RNA secondary structure is dominated by nearest-neighbor interaction (base pair stacking) of Watson–Crick and wobble base pairs, for which thermodynamic parameters are known. This thermodynamic background is introduced in Section 31.2. Given a complete set of thermodynamic parameters and the sequence, the problem to predict the structure of minimum of free energy (mfe structure) or the structural equilibrium is an optimization problem that is solved by computational methods (see Section 31.3). Both background information should help the reader to make use of the two web services and the underlying programs mfold (see Section 31.4) and RNAfold (see Section 31.5) as well as to interpret the results more easily. The next step after prediction of a secondary structure – and its experimental verification – is computation and modeling of the tertiary structure, which will be mentioned briefly in the final section. This is possible in principle because the tertiary interactions contribute much less in energy than the basic secondary structure; in other words, nature builds RNA's tertiary structure on top of one of the best secondary structures.

31.2
Thermodynamics

Structure formation – from an unfolded, random coil structure C into the folded structure S – is a standard equilibrium reaction described best in an all-or-none manner by a temperature-dependent equilibrium constant K:

$$C \rightleftharpoons S$$

$$K = [S]/[C]$$

$$\Delta G_T^\circ = -RT \ln K = \Delta H^\circ - T \Delta S^\circ$$

At the denaturation temperature $T_m = \Delta H^\circ / \Delta S^\circ$ (melting temperature or midpoint of transition) the folded structure S has the same concentration as the unfolded structure ($K = 1$; $\Delta G_{T_m}^\circ = 0$). This is only true if the structure S denatures in an all-or-none transition. In most cases, however, structural rearrangements and/or partial denaturation take place depending on the temperature prior to the complete denaturation.

The number of possible secondary structures grows exponentially ($\sim 1.8^N$) with the sequence length N [1]. Accordingly, all possible structures S_i of a single sequence coexist in solution with concentrations dependent on their free energies ΔG_i°, i.e. each structure is present as a fraction:

$$f_S = \exp\left(\frac{\Delta G_T^\circ(S)}{RT}\right) \bigg/ Q$$

based on a partition function Q for the ensemble of all possible structures:

$$Q = \sum_{\text{all structures } S} \exp\left(-\frac{\Delta G_T^\circ(S)}{RT}\right) \quad (1)$$

The structure of lowest free energy is called the optimal structure or structure of minimum free energy (mfe). However, quite different structures with identical energies might exist for a single sequence. This is of special biological relevance for RNA switches [2]. Thus, one should not assume that an RNA folds into a single, static structure.

The free energy ΔG_T° of a structure S is a sum over the free energy contributions of all structural elements of S:

$$\Delta G_T^\circ = \sum_i (\Delta H_{\text{stack}}^\circ - T \cdot \Delta S_{\text{stack}}^\circ) + \sum_j (\Delta H_{\text{loop}}^\circ - T \cdot \Delta S_{\text{loop}}^\circ)$$

Energetic contributions of base pairs are favorable ($\Delta G^\circ < 0$) due to stacking of the base pairs on top of each other, thus forming regular helices. Formation of loops is unfavorable ($\Delta G^\circ > 0$); exact values depend on loop type, nucleotides neighboring the loop-closing base pair(s) as well as on the exact sequence of the loops. Loop types are classified according to the number of loop-closing base pair(s): a single base pair closes hairpin loops, bulge loops (with no nucleotides in one strand) and interior loops (with symmetric or asymmetric numbers of nucleotides in both strands) are closed by 2 bp, and multiloops (bifurcations, junctions) connect more

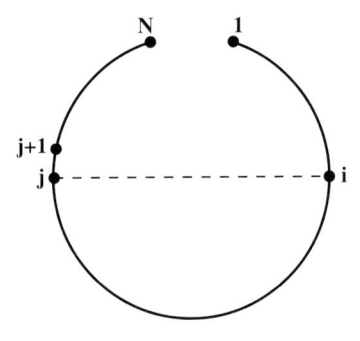

Fig. 31.1. Principle of base pair maximization. The graphs symbolize a sequence of length N with given indices $1 \leq i \leq k < j < N$. A dotted line limits a sequence segment of which the maximum number of pairs is already known; the continuous line marks a base pair. (Left) The maximum number of base pairs for the sequence range i, j is $M_{i,j}$; for each subsequence of that range the structures with maximum number of base pairs are also known. (Right) The sequence range was increased to $i, (j+1)$ and a base pair $(j+1){:}k$ with $i \leq k \leq (j-3)$ should exist.

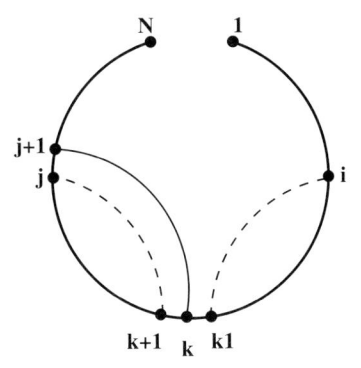

than two helices (see, e.g. Fig. 31.3). Note that for a certain loop type of n nucleotides total length, there are (including the possibilities for the closing base pairs) up to $6 \times 6 \times n^4$ different sequence combinations with different energetic contributions. A major set of these parameters was measured by the group of D. Turner [3]. Parameters are known only within certain error limits; because these errors are smallest near $T = 37\,°C$, $\Delta G°_{37\,°C}$ values are mostly reported.

A loop should not be thought of as a floppy structural element; in many cases loop nucleotides have fixed orientations due to stacking and/or non-Watson–Crick interactions with other loop nucleotides. Famous examples are loop E of eukaryotic 5S rRNA and the multiloop of tRNA. The loop E is an internal loop of four and five bases in its parts; all nucleotides are involved in non-Watson–Crick interactions, including one triple-strand interaction. In tRNA, the stacking of multiloop-closing base pairs across the multiloop is a major energetic contribution to the stability of the cloverleaf and is critical for the formation of the tRNA's tertiary structure.

Compensation of the negatively charged phosphate backbone of nucleic acids by positively charged counterions M^+ leads to stabilization of structural elements according to:

$$C + n \cdot M^+ \rightleftharpoons S$$

$$K = [S]/([C] \cdot [M]^n)$$

From this a logarithmic dependence between denaturation temperature T_m and salt concentration (ionic strength) follows:

$$\frac{dT_m}{d\ln[M]} = -n\frac{RT_m^2}{\Delta H°}$$

All thermodynamic parameters for RNA structure formation were determined in 1 M NaCl. This is not far from the ion conditions in cells except when specific interactions with divalent cations play a role. If necessary, however, values for the ionic strength dependence of a structure or a structural element may be found in the literature, including functions for G:C content of the RNA, or depending on various types of buffers (e.g. Tris–borate) and co-solvents like formamide or urea.

31.3
Formal Background

A structure of an RNA sequence R consists of base stacks and loops. It is defined as:

$$R = r_1, r_2, \ldots, r_N$$

with the indices $1 \leq i \leq N$ numbering the nucleotides $r_i \in \{A, U, G, C\}$ in the 5′ → 3′ direction. Base pairs are denoted by $r_i{:}r_j$ or, for short, $i{:}j$ with

$$1 \leq i < j \leq N.$$

Allowed base pairs are A:U, U:A, G:C, C:G, G:U and U:G. Formation of base pairs of a single secondary structure is restricted by:

$j - i \geq 4$ which gives the minimum size of an hairpin loop, and the order of two base pairs $i{:}j$ and $k{:}l$ has to be

$i = k$ and $j = l$ or

$i < j < k < l$ or

$i < k < l < j$

The second condition allows neighboring base pairs, but disallows any triple-strand formation. The third condition allows formation of several hairpin loops in a structure. The fourth condition disallows explicitly any tertiary interaction.

The search for an optimal secondary structure was initially solved by maximizing the number of base pairs for a given sequence [4, 5] using graph theory and dynamic programming [6]. The algorithm uses the possibility to dissect the problem into shorter and easier subproblems: if the maximum number of base pairs $M_{i,j}$ in the sequence region i, j ($1 \leq i < j < N$) and also the maximum number of all subsequences thereof are known, the following recursion does hold (cf. Fig. 31.1):

$$M_{i,j+1} = \max\left\{ M_{i,j},\ \max_{i \leq k \leq (j-3)} [M_{i,k-1} + 1 + M_{k+1,j}] \cdot p(r_k, r_{j+1}) \right\} \quad (2)$$

with:

$$r_k, r_{j+1} \in \{A, U, G, C\}$$

and:

$$p(r_k, r_{j+1}) = \begin{cases} 1 & \text{if } r_k \text{ and } r_{j+1} \text{ may form a base pair} \\ 0 & \text{otherwise} \end{cases} \tag{3}$$

The procedure starts with $i = 1$ and $j = 4$; no sequence range of this size may contain base pairs. The maximum number of base pairs $M_{i, j+1}$ for an increased sequence range $i, (j+1)$ is the maximum of $M_{i, j+1}$, which means a single-stranded nucleotide $(j+1)$, and the maximum of all possibilities with a base pair $(j+1)$:k, $i \leq k < (j-3)$. The maximum number of base pairs for the complete sequence is $M_{1, N}$. The corresponding structure is found by "backtracking", i.e. starting from the value $M_{1, N}$ the structure is reconstructed from the known values of the matrix M. The calculation cost for the recursion (Eq. 2) is $\mathcal{O}(N^3)$; the memory cost is $\mathcal{O}(N^2/2)$ because $1 \leq i < j < N$.

With a matrix of size $N^2/2$ it is only possible to start a backtrack from $M_{1, N}$ towards the main diagonal. Structures with a lower, suboptimal number of base pairs are not found by the recursion Eq. (2) because suboptimal solutions are overwritten by the optimal solution. The "trick" for not overwriting them is to use a virtually doubled sequence and a matrix V of size N^2 instead of the matrix M:

- Each base pair i:j divides a structure into two halves, called the "included fragment" from r_i to r_j and the "excluded fragment" from r_j to r_i. The optimal structure for the total sequence $r_1, \ldots, r_N$ is obtained by joining of the two partial structures, which are optimal for the subsequences $r_i, \ldots, r_j$ and $r_j, \ldots, r_i$, respectively, from the two parts of matrix V [7, 8]: $V_{\max} = \max\{V_{i, j} + V_{j, i}\}$. This argument allows also us to compute structures for circular sequences. See Fig. 31.2.
- To allow for multiloops, a further matrix W has to be used. In V, at position i, j structures are stored that are optimal and do contain the base pair i:j; in W, structures are stored that are optimal for the sequence range $r_i, \ldots, r_j$ but are not forced to have the base pair i:j.

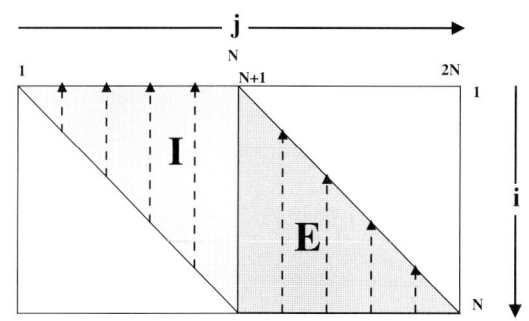

Fig. 31.2. Memory allocation for matrices V and W in the algorithm for prediction of suboptimal structures. The parts of the matrix filled grey are used. Part I of the matrix contains values for the "included fragment" from r_i to r_j. Part E of the matrix contains values for the "excluded fragment" from r_j to r_N to $r_{(N+i)} = r_i$. The matrix is filled during the recursion in direction of the arrows.

- mfold uses an algorithm similar to that sketched above. For energy minimization, however, the ρ values from Eq. (3) are replaced by energies according to the loop type (or base pair stack) under consideration. Additional cost in terms of memory and CPU time is needed because the energies depend on the type of loop nucleotides under consideration. This is especially critical for multiloops. Here a simple linear dependence of energy upon loop size is used during dynamic programming; only after the backtrack are corrections applied that take into account the possibility for base pair stacking across the multiloop.
  mfold allows us to find the mfe structure [9] as well as structures suboptimal in energy [8], which are optimal for any single, admissible base pair. Note that the calculable suboptimal structures do not cover all possible structures – another algorithm has to be used for this [10, 11].
- In RNAfold [12] two algorithms are used. The first computes the mfe structure only and produces the same structure as mfold when using identical energy parameters. The second algorithm computes the partition function (Eq. 1) as described by McCaskill [13], thus allowing us to calculate the probability for the formation of each base pair instead of individual structures.

31.4
mfold

The mfold server (version 3.1) for detailed structural analysis of a single RNA sequence is located at the Rensselaer–Wadsworth Bioinformatics Center, Troy, NY. More specialized variants of the main server – for folding many sequences at once (quikfold server), prediction of minimum folding energy only, prediction of two-state melting temperatures (T_m server) and hybridization server – will not be mentioned further. Authors who make use of the mfold service should cite the excellent description by the main developer [14]; more detailed information is found in [8, 15] and references therein. The free energy parameters of mfold 3.1 are only valid at 37 °C; in case other temperatures are of importance, e.g. for RNAs from plants or thermophile organisms, one should use the mfold 2.3 server that uses a less complete set of parameters, of which, however, entropies for extrapolation to other temperatures are available.

31.4.1
Input to the mfold Server

31.4.1.1 Sequence Name
A sequence name of maximum 40 characters may be given; otherwise a date-based name is used. A few characters (", <, >, ', \) are modified or deleted.

31.4.1.2 Sequence
All characters except "A–Z" and "a–z" are removed from the text area box, i.e. numbers and spaces are deleted but not headings like those from a GenBank-formatted sequence. Instead use sequence format conversion tools (see Table

31.1) like readseq or sreformat from the SQUID library. Lower case characters are converted to upper case; T is converted to U. Certain characters have a special meaning, which does not coincide with the IUPAC ambiguous nucleic acid character convention:

- B, D, H and V denote A, C, G and U/T, respectively. These nucleotides may pair only if their 3′ neighbor is unpaired.
- W, X, Y and Z denote again A, C, G and U/T, respectively. These nucleotides may pair only at the end of a helix.
- N means any nucleotide but is not allowed to pair.

These conventions may be of help for example in denoting experimental knowledge from nuclease mappings.

31.4.1.3 Constraints

Constraints forcing and/or prohibiting base pair(s) have to be given in a text area using a rigid format. Per line only one of the following constraints may be given:

- Forcing a base pair or a helix:
  F i j k
  will force the formation of a helix of k bp (or a single base pair with $k = 1$)

  $$5' - r_i - r_{i+1} - \cdots - r_{i+k-2} - r_{i+k-1} - 3'$$
  $$ | | \phantom{r_{i+1} - \cdots -} | \phantom{r_{i+k-2} -} |$$
  $$3' - r_j - r_{j-1} - \cdots - r_{j-k+2} - r_{j-k+1} - 5'$$

  The first base pair of the forced helix is $r_i{:}r_j$ followed by $k - 1$ consecutive base pairs.
- Forcing a string of consecutive bases to pair (or a single base with $k = 1$):
  F i 0 k
  will force nucleotides $r_i, r_{i+1}, \ldots, r_{i+k-1}$ to pair; the third character of the input string is "zero".
- Prohibiting a base pair or a helix:
  P i j k
  will prevent the formation of a helix of k bp (or a single base pair with $k = 1$); the base pairs $r_{i+h}{:}r_{j-h}$, $0 \leq h < k$ are not allowed to form.
- Prohibiting a string of consecutive bases from pairing (or a single base with $k = 1$):
  P i 0 k
  will prevent nucleotides $r_i, r_{i+1}, \ldots, r_{i+k-1}$ from pairing; the third character of the input string is "zero".
- Prohibiting a string of consecutive bases from pairing with a string of consecutive bases:
  P i j k l
  will prevent nucleotides $r_i, r_{i+1}, \ldots, r_j$ from pairing with nucleotides $r_k, r_{k+1}, \ldots, r_l$ ($i \leq j, k \leq l$). If $i = k$ and $j = l$, then any base pair within the string $r_i, \ldots, r_j$ is prohibited.

Tab. 31.1. Internet addresses of programs and manuscripts mentioned in the text.

Program/manuscript	URL	Reference/authors
mfold	http://www.bioinfo.rpi.edu/applications/mfold	14
mfold manual	http://bioinfo.math.rpi.edu/~zukerm/export/mfold-3.0-manual.ps	15
	http://bioinfo.math.rpi.edu/~zukerm/export/mfold-3.0-manual.pdf	
RNAfold server	http://rna.tbi.univie.ac.at/cgi-bin/RNAfold.cgi	I. L. Hofacker
Vienna package	http://www.tbi.univie.ac.at/~ivo/RNA/	I. L. Hofacker
	http://www.tbi.univie.ac.at/papers/Abstracts/93-07-07.ps.gz	12
Sequence conversion		
readseq	ftp://ftp.bio.indiana.edu/molbio/readseq/	D. G. Gilbert
	http://packages.debian.org/stable/science/readseq.html	
SQUID library	ftp://ftp.genetics.wustl.edu/pub/eddy/software/squid.tar.gz	S. R. Eddy
Graphics formats		
png	http://www.libpng.org/	G. Roelofs
PostScript®	http://www.adobe.com/products/postscript/main.html	Adobe®
SVG®	http://www.w3.org/TR/SVG/	W3C®
Structure drawing		
RNAMLview	http://ndbserver.rutgers.edu/services/	17
RnaViz	http://rrna.uia.ac.be/rnaviz/	20
GCG	http://www.accelrys.com/about/gcg.html	Accelrys Inc.
XRNA	http://rna.ucsc.edu/rnacenter/xrna/xrna.html	B. Weiser and H. F. Noller
naview	http://www.bioinfo.rpi.edu/~zukerm/export/naview.tar.Z	21
loopdloop (Mac ct)	http://iubio.bio.indiana.edu/IUBio-Software+Data/molbio/loopdloop/	D. Gilbert
SStructView	http://smi-web.stanford.edu/projects/helix/sstructview/home.html	25
Consensus structure prediction		
ConStruct	http://www.biophys.uni-duesseldorf.de/local/ConStruct/ConStruct.html	31
SLASH server	http://www.bioinf.au.dk/slash/	32
ESSA	http://www.inra.fr/bia/T/essa/Doc/essa_home.html	33
alifold	http://rna.tbi.univie.ac.at/cgi-bin/alifold.cgi	34
structure logo	http://www.cbs.dtu.dk/~gorodkin/appl/slogo.html	36
Pseudoknot prediction		
pknots	http://bibiserv.techfak.uni-bielefeld.de/pknotsrg/	39

Be careful with these constraints. For example, forcing a single base pair $r_i:r_j$ by the constraint F i j 1 will fail when either no neighboring base pair is possible or this base pair itself cannot exist. Forcing too many base pairs, prohibiting too many nucleotides to be paired or an N nucleotide to pair will cause failure of the job.

31.4.1.4 Further Parameters

- RNA sequences may be linear or circular. This is of importance because, for example, in a linear structure the first and last nucleotides of the sequence may pair with each other without dangling ends; in a circular sequence the first and last nucleotides of the linear input string may not base pair with each other, but have to form a hairpin loop (or a different, energetically more favorable structure).
- The percent suboptimality number P. This parameter controls the free energy range $\Delta G°_{m.f.e.} + \delta\delta G°$ in which structures will be computed and their base pairs plotted in the energy dot plot (see below), i.e. if you would like to see more suboptimal structures in the dot plot, increase the parameter and *vice versa*. Per default, $\delta\delta G° = |\Delta G°_{m.f.e.}| \cdot P/100$ with $P = 5$, but $\delta\delta G°$ is rounded up to 1 kcal/mol or down to 12 kcal/mol if outside this range. Take note, that a suboptimal structure with $\Delta G° = (\Delta G°_{m.f.e.} + 1 \text{ kcal/mol})$ at 37 °C will be a factor of 5 lower in concentration than the mfe structure and an energy difference of 10 kcal/mol will result in a factor of about 10^7!
- The upper bound on the number of computed structures. This number determines how many structural drawings are generated. It has no influence on the number of structures in the energy dot plot.
- The window parameter W. This parameter controls how many structures will be automatically computed and how different they will be from one another. A smaller value of this parameter will result in more computed structures that may be quite similar to one another. A larger value will result in fewer structures that are very different from one another. More precisely, a distance between two base pairs $r_i:r_j$ and $r_k:r_l$ is defined as $W = \max\{|i-k|, |j-l|\}$. If $m-1$ structures have already been predicted by mfold, the mth folding must have at least W base pairs that are at least a distance W from any of the base pairs in the first $m-1$ structures.
- The maximum size of a bulge or interior loop/the maximum asymmetry of a bulge or interior loop. The first parameter is set per default to 30 nt. For most sequences at $T = 37$ °C this is more than sufficient; larger values increase the computing time. The second parameter limits the maximum size of a bulge loop. For interior loops with i and j single-stranded nucleotides on their two sides, the parameter limits the size difference to $|i-j|$. Again, for most sequences at $T = 37$ °C this is more than sufficient and larger values increase the computing time. Take note, however, that even in natural sequences long stretches of unpaired nucleotides do appear and large loops may be present also at elevated temperatures prior to complete denaturation. In such cases, it is worthwhile to set these parameters to the sequence length.

- The maximum distance between paired bases. If this parameter M is given, then any base pair $r_i:r_j$ with $i < j$ must satisfy $j - i \leq M$ in a structure of a linear sequence and $\min\{j - i, N + i - j\}$ for a circular sequence of length N. Thus a lower value of M favors more local pairings, whereas $M = N - 1$ allows for a pairing $r_1:r_N$.
- Further parameters or options concern the types of output that is generated; most parameters may be given or modified also on the result page (see below). The structure drawing in Fig. 31.3 was produced by default parameters.
  - Images are available in different image sizes and resolutions as .png (portable network graphics format with lossless compression) and .jp(e)g (Joint Photographic Experts Group format with not-lossless compression) files. Default (regular) values for size and resolution are 612×792 and 72×72 pixels per square inch, respectively. All output is available also in PostScript format, which may be scaled to any size without loss in quality. Be aware, however, that the .png and .jpg hyperlinks of the output page lead to clickable images, which allow one to get additional information (see below).
  - In structure drawings, nucleotides are shown as characters ("bases") or not ("outline"); in "automatic" mode characters are drawn for sequences up to a length of 800 nt.
  - Nucleotides will be numbered automatically according to the length of the sequence or at user-selectable distances. A numbering frequency of 0 turns off numbering.
  - The rotation angle, with positive values for counter-clockwise rotation, modifies orientation of structure drawings.
  - Structure drawings may be annotated in different ways; for "p-num" and "ss-count" annotation see below. The "high-light" mode allows to enter region(s) of nucleotides that are colored differently; regions may be given as a comma-separated list of positions (e.g. "10–20,30–40").

31.4.1.5 Immediate versus Batch Jobs

Calculations for sequences of up to 800 nt are completed in several seconds, depending on the server load, and presented to the user; results will be deleted after

Fig. 31.3. Major output produced by mfold. In both parts, certain structural elements are named in grey. The originally colored plots are redrawn in grey scale; in the text, original colors are given in brackets. (Top) Energy dot plot; the lower left triangle shows the optimal (mfe) secondary structure; the upper triangle shows the mfe structure as black dots, and suboptimal structures – in order of decreasing stability – in dark grey (red), grey (green) and light grey (yellow) dots; the free energy ranges for these are given below the plot. In this example the best suboptimal structures [dark grey (red)] are only variants of the mfe structure, whereas structures with more positive energy [light grey (yellow)] deviate greatly from the mfe structure. (Bottom) Drawing of optimal structure by naview; base pairs are connected by dots [black (red) for G:C and grey (blue) for A:U and G:U pairs]. Parameters: mfold 3.1, $T = 37$ °C, $P = 15$, $W = 1$; AC: L27168.

31.4 mfold

Fold of HALRRAH at 37 C.

δG in Plot File = 6.1 kcal/mole

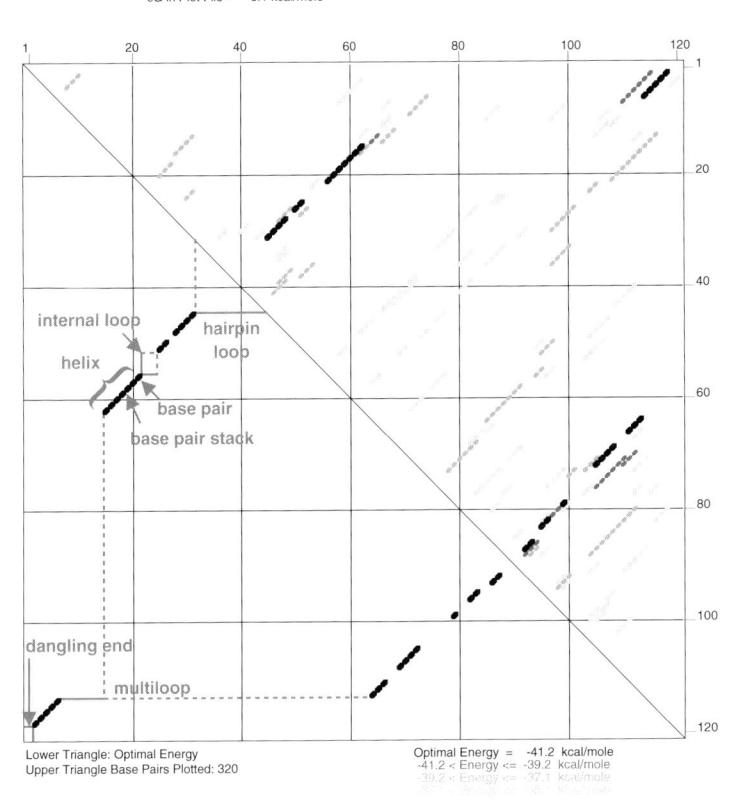

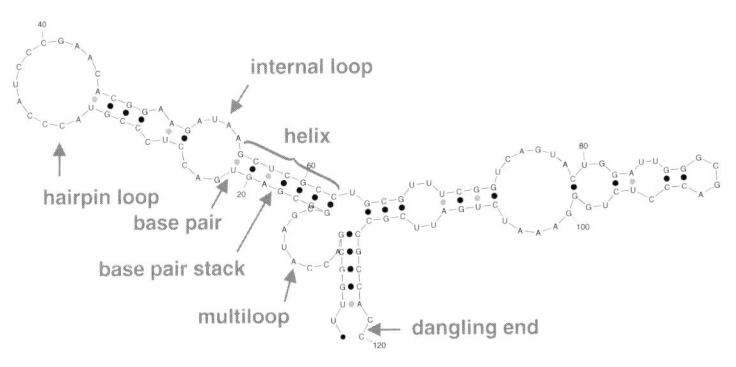

Lower Triangle: Optimal Energy
Upper Triangle Base Pairs Plotted: 320

Optimal Energy = -41.2 kcal/mole
-41.2 < Energy <= -39.2 kcal/mole
-39.2 < Energy <= -37.1 kcal/mole

dG = -40.46 [initially -41.2]   HALRRAH

Fig. 31.3

about a day. For sequences up to 6000 nt, an E-mail address should be given and a "batch job" has to be started, which produces a page showing the URL for the results and the E-mail address to which a notification will be sent on completion of the calculations, which needs mostly a few hours. Results will be available for about 2 days after submission.

31.4.2
Output from the mfold Server

Computed results may be viewed by following links from the initial results page.

31.4.2.1 Energy Dot Plot

In the lower left triangular region of the plot (see, e.g. Fig. 31.3) the mfe structure is presented; a dot in row i and column j represents a base pair between the ith and jth nucleotide. The upper triangular region of the plot contains the superposition of all computed structures within $P\%$ of the minimum energy, where P is the percent suboptimality parameter (see Section 31.4.1.4). Black dots again represent the mfe structure. Colored dots represent superpositions of computed suboptimal structures. The color ranges are red, blue and yellow, representing base pairs that are in structures with energies $\Delta G°$ in the ranges $\Delta G°_{m.f.e.} < \Delta G° \leqslant \Delta G°_{m.f.e.} + \delta\delta G°/3$, $\Delta G°_{m.f.e.} + \delta\delta G°/3 < \Delta G° \leqslant \Delta G°_{m.f.e.} + 2\delta\delta G°/3$ and $\Delta G°_{m.f.e.} + 2\delta\delta G°/3 < \Delta G° \leqslant \Delta G°_{m.f.e.} + \delta\delta G°$, respectively. Thus the yellow dots represent base pairs that are least likely to form.

The "text" format of the energy dot plot gives the basic information on position of helices, the energy of the corresponding structure and its energy range.

The .png and .jpg hyperlinks each lead to a new page with a .png or .jpg image of the energy dot plot and buttons that allow the user to get the position $i:j$ of a base pair and the energy of the optimal structure with this base pair, to magnify the dot plot, to modify the number of energy ranges or to limit the lower size of depicted helices ("filter").

31.4.2.2 RNAML (RNA Markup Language) Syntax

Computed structures are available in the RNAML format, which is the first standardized format for information exchange on RNA structure [16]. Up to now, only RNAMLview [17] is able to read this format and allows the user to perform dynamic editing of the two-dimensional structure drawing.

31.4.2.3 Extra Files

These are raw text files which make available the "p-num" values, also used for coloring structure representations, and "h-num" values [18, 19]. Note that p-num and h-num values depend on the chosen percent suboptimality parameter P.

P-$num(i)$ is a measure on how likely is the pairing of nucleotide i with any other nucleotide:

$$\delta(a) = \begin{cases} 1 & \text{if expression } a \text{ is true} \\ 0 & \text{otherwise} \end{cases}$$

$$P\text{-}num(i) = \sum_{k<i} \delta(\Delta G^\circ_{k,i} \leq \Delta G^\circ_{\text{m.f.e.}} + \delta\delta G^\circ) + \sum_{i<j} \delta(\Delta G^\circ_{i,j} \leq \Delta G^\circ_{\text{m.f.e.}} + \delta\delta G^\circ) \quad (4)$$

That is, the $P\text{-}num(i)$ value depends on the total number of dots in the ith row and ith column of the unfiltered upper triangle of the energy dot plot. Low values indicate "well-defined" nucleotides; in particular, values of 0 or 1 indicate that a nucleotide is single stranded or paired with a defined partner, respectively, in all structures depicted in the energy dot plot.

$H\text{-}num(i, j)$ is a measure on the tendency of a nucleotide i to pair with a certain other nucleotide j:

$$H\text{-}num(i, j) = [P\text{-}num(i) + P\text{-}num(j) - 1]/2$$

Helices with lower h-num values are believed to be predicted more reliably.

31.4.2.4 Download All Foldings

All computed structures may be downloaded in a single zipped file (extension .zip) or a compressed tar file (.tar.Z). Images are available in PostScript, .png or .jpg format. Different formats useful for further processing by other programs (e.g. RnaViz [20], GCG or XRNA; see Table 31.1) are available.

31.4.2.5 View ss-count Information

The ss-count file contains statistics on single strandedness of each nucleotide in all of the computed structures: $ss\text{-}count(i)$ is the number of structures in which the nucleotide i is single stranded divided by the number of all computed structures. A corresponding plot is available, for which an "averaging window" parameter determines the nucleotide range over which the $ss\text{-}count(i)$ values are averaged (see, e.g. Fig. 31.4D). The plot is useful for comparison with mapping data.

31.4.2.6 View Individual Structures

- Clicking to "Structure i" gives an .html file with the structure i in a simple, easily recognizable text format.
- The "thermodynamic details" lists type, position and free energy contribution of each atomic structural element in structure i.
- The "PostScript" option gives a static plot drawn in a manner as asked for on the input .html page.
- The .png and .jpg hyperlinks each lead to a new page with a .png or .jpg image of structure i generated by naview [21]. Here the structure may be magnified, differently annotated (for a description of the colors with $P\text{-}num(i)$ or $ss\text{-}count(i)$

526 | *31 Secondary Structure Prediction*

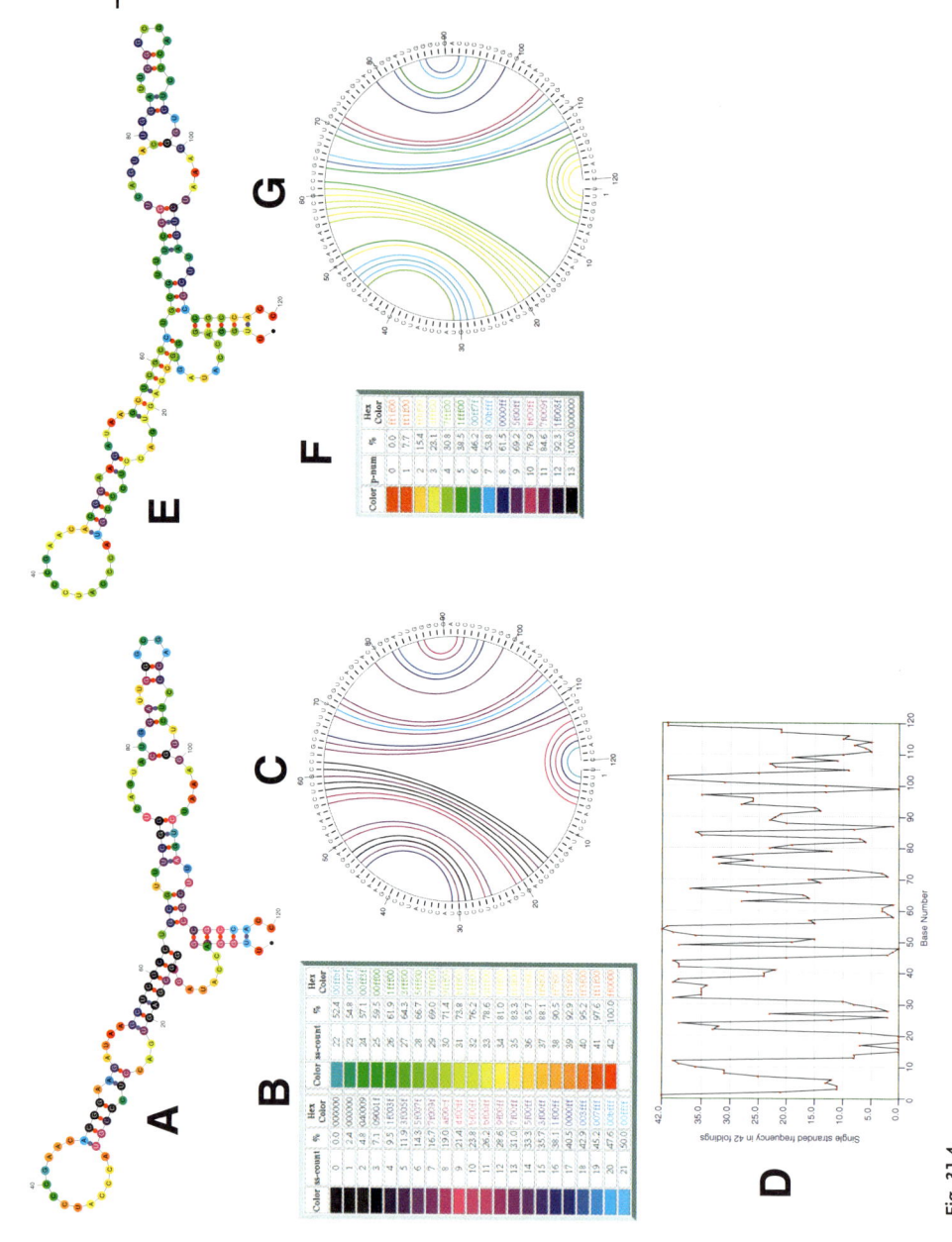

Fig. 31.4.

annotation see in the top right edge of the final html page), and labeled with dots and/or characters for the nucleotides (see, e.g. Fig. 31.4).
- The "new" options as well as the "new structure viewing options" link to structure output generated by sir_graph [14], which is slightly different from that produced by naview. Additional options do exist for representation of the external loop (connecting 5' and 3' end), three versions for drawing loops are available. In addition, sir_graph allows one to produce circle graphs for structure representation (see, e.g. Fig. 31.4C and G); this representation is quite useful for comparison of different structures, but fails for longer sequences.
- ct (connect format) [21], RnaViz ct [20], Mac ct, GCG connect and XRNA are text formats useful for input into respective programs and further processing/refinement/labeling as wanted by the user (see Table 31.3).

31.4.2.7 Dot Plot Folding Comparisons

This option plots several computed structures, selectable by the user, in a dot plot similar to the "energy dot plot option" for an easy recognition of those parts that are different between the depicted structures. Base pairs that occur in all selected structures are colored black, those that occur in two or more are colored grey and base pairs unique to a structure are depicted in other colors. With the "Multicolor Overlap" option the dots (squares) otherwise colored grey are depicted in different colors as well as different trapezoids to allow for an easier discrimination between the different structures. For a comparison of three structures, in the lower left triangle base pairs are plotted that occur in structures 1 and 2 but not 3, 1 and 3 but not 2, and 2 and 3 but not 1 in three different colors; base pairs common to all three structures are again colored black.

31.5
RNAfold

The RNAfold server (see Table 31.1) is part of the Vienna package (see Table 31.1), which contains additional programs/services related to RNA structure prediction and comparison like:

- RNAeval for calculation of free energy of a given RNA secondary structure.

Fig. 31.4. Structure output and annotation by mfold. (Left) Annotations by ss-count(i). (Right) Annotations by P-num(i). (A and E) Drawings by naview; nucleotides are depicted as characters and dots, and base pairs by dots. (B and F) Coloring schemes for ss-count and p-num coloring, respectively. Note that the ss-count coloring (A–C) follows the reverse scheme of p-num coloring (E–G), i.e. in both cases red is most likely to be single stranded (high value of ss-count but low value of p-num). (C and G) Circle drawings by sir_graph; base pairs are connected by arcs. (D) ss-count information useful for comparison with mapping data. Parameters: mfold 3.1, $T = 37\,°C$, $P = 15\%$, $W = 1$; AC: L27168.

- RNAheat for calculation of specific heat of an RNA sequence/structure distribution.
- RNAinverse for inverse folding and design of sequences with a predefined structure.
- RNAdistance for comparison of secondary structures.
- RNApdist for comparison of base pair probability plots.
- RNAsubopt for complete suboptimal folding.
- RNAlfold for scanning very long sequences for locally stable structures.
- alifold for prediction of a consensus structure for a set of aligned RNA sequences.

The server is hosted by the Theoretical Biochemistry Group at the University of Vienna. The package is free software and can be downloaded as C source code that compiles easily on almost any flavor of Unix and Linux. Authors who make use of the RNAfold service should cite the description by Hofacker [22], for further details see references therein.

31.5.1
Input to the RNAfold Server

In the following we will concentrate on the web server; additional options, however, are available for a locally installed program, which will be mentioned also.

31.5.1.1 Sequence and Constraints

On the server all non-alphabetic characters are removed from the text area box, i.e. numbers and spaces are deleted, but not headings like those from a GenBank-formatted sequence. Instead use sequence format conversion tools (see Table 31.1) like readseq, which is also part of the Vienna package, or sreformat from the SQUID library.

Lower-case characters are converted to upper case; T is converted to U. Any symbols except AUCGTXKI will be interpreted as non-bonding bases.

- N means any nucleotide but is not allowed to pair.
- I can pair with A and U.
- K and X can only pair with each other.

With respect to the thermodynamic parameters, X is equivalent to G and K to C. A:I and I:U pairs use A:U parameters; a non-paired I is equivalent to N. A reasonable use of these nucleotides is rare, but they may help in giving certain constraints.

The locally installed program reads sequences from stdin. The allowed sequence format is similar to FastA format, i.e. the first line has to consist only of:

⟩ name

where "name" is a single string of up to 12 characters. The full sequence has to be given on the next line. Further sequences may follow, up to either a "@" or EOF (end of file). Constraints may be given by special characters (see also Section 31.5.2.2) in a third line:

- A dot "." is an unconstrained nucleotide.
- A pair of matching brackets "()" force that base pair.
- A pipe symbol "|" forces this nucleotide to be paired.
- Angle brackets "⟨" or "⟩" force pairing of the nucleotide with an upstream or downstream nucleotide, respectively.
- An "x" prohibits pairing of this nucleotide.

This constraint or structure string has to have the same length as the sequence string. In partition function folding symbols except "(", ")" and "x" are ignored. To allow for reading of such a structure string, the option "–C" has to be given, i.e. a possible command for calculation of the partition function with reading from a file "mysequence.vie" with content:

```
> dummy
AGAGGGUUUCCCUCU
.x.(......)....
```

and writing the text output to a file "mysequence.log" would be:

RNAfold –C –p ⟨ mysequence.vie ⟩ mysequence.log

Because constrained folding works by assigning bonus energies to all structures complying with the constraints, single bracket pairs or a single angle bracket is sometimes not sufficient to get the desired output.

31.5.1.2 Further Parameters

In the following the options are given that may be activated by appropriate buttons on the server or by command line options. A few additional options are available only on the command line; for these check the manual page.

- Fold algorithm. RNAfold uses two algorithms as mentioned in Section 31.3. The first computes the optimal (mfe) structure only. The second, activated with option "–p", computes the partition function and pair probabilities. The mfe structure as well as the pair probabilities are depicted in a probability dot plot (see below), which is similar to the energy dot plot of mfold.
- Temperature "–T". The only possible temperature for thermodynamic parameters from Mathews et al. [3] is 37 °C (as in mfold 3.1). When choosing other temperatures, parameters, which include ΔS values (see Section 31.2), from Walter et al. [23] are used (as in mfold 2.3).
- No special tetraloops "–4". Certain hairpin-loop sequences (especially UNCG

and GNRA with N for any nucleotide and R for purine) are thermodynamically more stable than other sequences [24]. This option discards the appropriate tetraloop parameters, so that the influence of such sequences and parameters on a structure distribution can be recognized.
- No dangling end energies "–d[0|1|2|3]". This option has influence on the treatment of dangling ends (exterior loop at 5'/3' end) and of multiloops: do nucleotides at free ends stack onto a neighboring base pair and is there coaxial stacking of helices in multiloops?
  - Adding this option, which is equivalent to "–d0" or "–d", ignores a possible influence of dangling ends altogether.
  - With "–d1", which is the default for calculation of the mfe structure, only unpaired bases can participate in at most one dangling end. Partition-function folding will work as with "–d2".
  - With "–d2", which is the default for partition-function folding, the check from "–d1" is ignored. This treatment of dangling ends gives more favorable energies to helices directly adjacent to one another, which can be beneficial since such helices often do engage in stabilizing interactions through co-axial stacking.
  - With "–d3" mfe folding will allow co-axial stacking of adjacent helices in multi-loops; partition-function folding will work as with "–d2".
  Note that by default partition function and mfe folding treat dangling ends differently, use "–d2" in addition to "–p" to ensure that both algorithms use the same energy model.
- No GU pairs at the end of helices "–noCloseGU" and no GU pairs "–noGU". Quite outdated options.
- Avoid isolated base pairs "–noLP". Avoid structures with lonely pairs (helices of length 1). For partition-function folding this disallows only pairs that can only occur isolated. Other pairs may still occur occasionally.
- Mountain plot. A mountain representation of the mfe structure shows each base pair $i{:}j$ with $\Delta y = 1$ and $\Delta y = -1$ at positions $x = i$ and $x = j$, respectively, and $\Delta y = 0$ for loops. This results in a mountain-like graph where helices are slopes, loops are plateaus and hairpin loops are peaks. If the partition function is computed, the Δy values correspond to the pairing probability.
- View mfe structure as SVG (scalable vector graphics) image. Produces a structure drawing in SVG, i.e. you need a SVG-plugin for your browser (see Table 31.1). The drawing is identical to the default drawing (in PostScript) but allows one to remove the nucleotide characters.
- View mfe structure via the SStructView Java applet. The drawing is identical to the default drawing (in PostScript), but allows one to see individual nucleotide numberings and to zoom into the drawing. The Java applet does not seem to work with recent browser/Java combinations.

31.5.1.3 Immediate versus Batch Jobs

Calculations for sequences of up to 300 nt are completed in several seconds (depending on the server load) and presented to the user. For sequences up to

4000 nt, an E-mail address has to be given for notification on completion of the calculations.

31.5.2
Output from the RNAfold Server

31.5.2.1 Probability Dot Plot

If partition function folding ("–p") was selected, the major output of RNAfold is a dot plot (see, e.g. Fig. 31.5B) that depicts the base pairs of the mfe structure in the lower left triangle. All probable base pairs from the partition function are depicted in the upper right triangular region; the area of a dot in row i and column j is proportional to the probability of a base pair between the ith and jth nucleotide.

31.5.2.2 Text Output of Secondary Structure

The following text output (see, e.g. Fig. 31.5A) is available only in parts from the web server.

The first output is the input's name line. The second line is the sequence in uppercase with Ts converted to Us if necessary. The third line contains the optimal (mfe) structure in bracket-dot notation, i.e. at positions of unpaired nucleotides is a dot and base pairs are denoted by pairs of brackets. At the end of the line is given the structure's free energy. If the partition function was computed, a fourth and fifth line is given: the fourth shows a representation of the partition folding with symbols denoting nucleotides that are essentially unpaired ("."), weakly paired (","), strongly paired without preference ("|"), weakly upstream (downstream) paired ("{", "}") or strongly upstream (downstream) paired ("[", "]"), respectively, followed by the free energy of the structure distribution; the fifth line gives the relative concentration of the mfe structure in the structure distribution. For thermodynamic reasons the free energy of the distribution has to be lower (more negative) than that of the mfe structure. The relative concentration of the mfe structure may be very low when there are many structural alternatives.

31.5.2.3 Graphical Output of Secondary Structure

Here the optimal (mfe) secondary structure is shown in up to three different versions. The standard representation is available via a hyperlink as a PostScript plot (see Fig. 30.5C). The second uses the SStructView Java applet and the third is the SVG image (see Section 31.5.1.2). Note that all structures are given by default in counter-clockwise orientation; for mirroring and/or rotating the structures the Vienna package contains a small Perl script. For annotation of a secondary structure by "reliability information", the Vienna package contains a small Perl script that calculates the positional entropy $S(i)$ by:

$$S(i) = -\Sigma p_{i,j} \cdot \log(p_{i,j})$$

and encodes it as color hue, ranging from red (low entropy, well-defined) via green

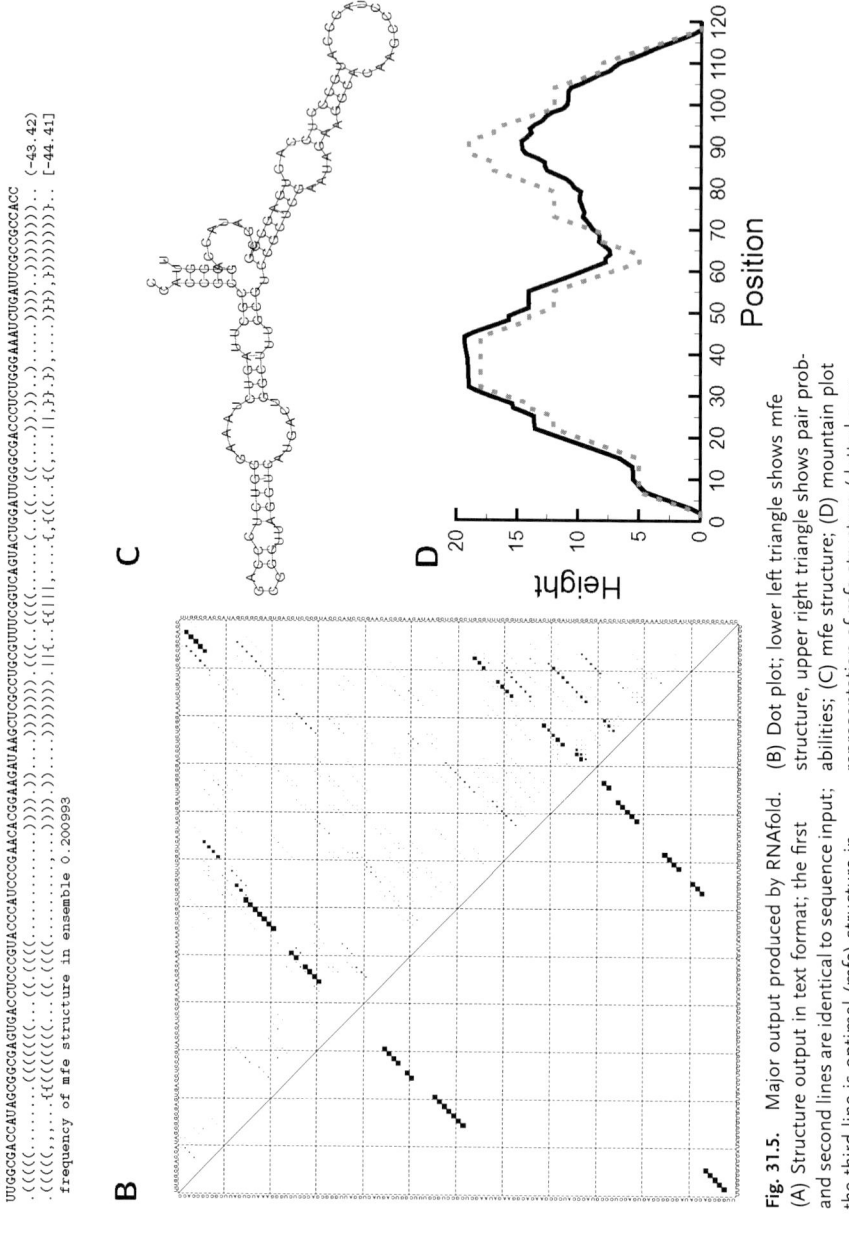

A

```
> HSRRAH
UUGGCGACCAUAGCGGAGUGACCUCCGUACCCAUCCCGACACGGAAGAGAUAAGCUCGCCUGCCUGGGUCAGUACUGGAUUGGGCGACCCUCUGGAAAUCUGAUUCGCCGGCACC
.((((.........(((((((...((.(((((............))))).))......)))).)).....(((....((.(((..........))).)).....))).........    (-43.42)
.(((((,,,.{{(((((((((.....((.(((((............))))).))......))))))))....{(....((.(((..........))).))......}}.,}}}))))).    [-44.41]
frequency of mfe structure in ensemble 0.200993
```

B

C

D

Fig. 31.5. Major output produced by RNAfold. (A) Structure output in text format; the first and second lines are identical to sequence input; the third line is optimal (mfe) structure in bracket-dot notation; the fourth line is a crude representation of the structure distribution. (B) Dot plot; lower left triangle shows mfe structure, upper right triangle shows pair probabilities; (C) mfe structure; (D) mountain plot representation of mfe structure (dotted grey line) and pair probabilities (black). Parameters: "–p –d3"; AC: L27168.

to blue and violet (high entropy, ill-defined). This is similar to p-num annotation by mfold [see Eq. (4) and Fig. 31.4(E)].

31.5.2.4 Mountain Plot

The Vienna server produces a .png image with mountain representation of the mfe structure as well as a representation of the partition folding; a PostScript image with same information is available via a Perl script from the Vienna package (see, e.g. Fig. 31.5D).

31.6 Troubleshooting

My major advice with any structure prediction program is to get some experience. For example, fetch two, relatively short, similar sequences from GenBank or EMBL data bank; either a consensus structure or an experimentally confirmed structure should be known for the sequence family from the literature. The example used here is an archaeal 5S rRNA (AC: L27168); its predicted mfe structure is near the 5S rRNA consensus structure. A second RNA could be another 5S rRNA (e.g. AC: AE010834, nt 66–200), for which the programs predict a deviating mfe structure. Next, reformat the sequences into a suitable format and then play with the programs.

- Do not expect that a "good" structure has more base pairs than a "bad" structure; even for a random RNA sequence about two-thirds of its nucleotides are involved in base pairs.
- The quality of structure predictions decreases with increasing sequence length [26]. On the one hand, this may reflect uncertainties in the basic thermodynamic parameters: errors, especially due to multiloops or other special loops, may sum up with increasing sequence length. On the other hand, kinetics may increasingly influence the folding with increasing sequence length. The programs mentioned here are based on thermodynamics only; for programs that take into account the kinetics of folding, see [27–29].
- In case data from *in vitro* experiments like chemical mapping are available, take into account the experimental conditions. For example, most gel buffers have a very low ionic strength (e.g. 8.9 mM Tris–borate has about 3 mM ionic strength), but temperature conditions may also be very different from 37 °C.
- Be very careful using mapping data in constraining any computations. For example, having more than a single structure in solution, mapping data would lead one to expect more nucleotides to be single-stranded than are possible in a single structure. In my opinion it is easier to compare mapping data with pairing probabilities or ss-num information; if this does not fit, than add few constraints to favor one of the suboptimal structures.
- In case your sequence is one of a family of sequences, this might be extremely helpful because all sequences should form a similar structure. Dynalign [30] pre-

dicts the secondary structure common to two RNA sequences. For more than two sequences the tools ConStruct [31], the SLASH server [32], ESSA [33] or alifold [34] are useful (see Table 31.1). Visualization of the additional information from RNA structural alignments is possible via MatrixPlot [35] or structure logo [36] (see Table 31.1).

- A few programs are available that allow to compute tertiary interactions like pseudoknots [37–39]; be careful, however, with the necessary computing resources and the predicted output. For further modeling of three-dimensional structure, refer to [40].
- The energy units used within the programs mfold and RNAfold are kcal/mol; this unit should be replaced by the standard IUPAC unit of kJ/mol (1 kcal = 4.1868 kJ).

References

1 M. S. Waterman, *Introduction to Computational Biology. Maps, Sequences and Genomes*, Chapman & Hall, London, **1995**.
2 J. H. Nagel, C. W. Pleij, *Biochimie* **2002**, *84*, 913–923.
3 D. H. Mathews, J. Sabina, M. Zuker, D. H. Turner, *J. Mol. Biol.* **1999**, *288*, 911–940.
4 R. Nussinov, G. Pieczenik, J. R. Griggs, D. J. Kleitman, *SIAM J. Appl. Math.* **1978**, *35*, 68–82.
5 M. S. Waterman, T. F. Smith, *Math. Biosci.* **1978**, *42*, 257–266.
6 R. Bellman, R. Kalaba, *SIAM J. Appl. Math.* **1960**, *8*, 582–588.
7 G. Steger, H. Hofmann, J. Förtsch, H. J. Gross, J. W. Randles, H. L. Sänger, D. Riesner, *J. Biomol. Struct. Dyn.* **1984**, *2*, 543–571.
8 M. Zuker, *Science* **1989**, *244*, 48–52.
9 M. Zuker, P. Stiegler, *Nucleic Acids Res.* **1981**, *9*, 133–148.
10 M. Waterman, T. Byers, *Math. Biosci* **1985**, *77*, 179–188.
11 S. Wuchty, W. Fontana, I. L. Hofacker, P. Schuster, *Biopolymers* **1999**, *49*, 145–165.
12 I. L. Hofacker, W. Fontana, P. F. Stadler, S. Bonhoeffer, M. Tacker, P. Schuster, *Monatsh. Chem.* **1994**, *125*, 167–188.
13 J. S. M. McCaskill, *Biopolymers* **1990**, *29*, 1105–1119.
14 M. Zuker, *Nucleic Acids Res.* **2003**, *31*, 3406–3415.
15 M. Zuker, D. H. Mathews, D. H. Turner, Algorithms and thermodynamics for RNA secondary structure prediction: A practical guide, in: *RNA Biochemistry and Biotechnology*, J. Barciszewski, B. F. C. Clark (eds), NATO ASI Series, Kluwer, Dordrecht, **1999**.
16 A. Waugh, P. Gendron, R. Altman, J. W. Brown, D. Case, D. Gautheret, S. C. Harvey, N. Leontis, J. Westbrook, E. Westhof, M. Zuker, F. Major, *RNA* **2002**, *8*, 707–717.
17 H. Yang, F. Jossinet, N. Leontis, L. Chen, J. Westbrook, H. Berman, E. Westhof, *Nucleic Acids Res.* **2003**, *31*, 3450–3460.
18 M. Zuker, A. B. Jacobson, *Nucleic Acids Res.* **1995**, *23*, 2791–2798.
19 M. Zuker, A. B. Jacobson, *RNA* **1998**, *4*, 669–679.
20 P. De Rijk, J. Wuyts, R. De Wachter, *Bioinformatics* **2003**, *19*, 299–300.
21 R. E. Bruccoleri, G. Heinrich, *Comp. Appl. Biosci.* **1988**, *4*, 167–173.
22 I. L. Hofacker, *Nucleic Acids Res.* **2003**, *31*, 3429–3431.
23 A. E. Walter, D. H. Turner, J. Kim, M. H. Lyttle, P. Muller, D. H. Mathews, M. Zuker, *Proc. Natl. Acad. Sci. USA* **1994**, *91*, 9218–9222.
24 G. Varani, *Annu. Rev. Biophys. Biomol. Struct.* **1995**, *24*, 379–404.

25 R. M. Felciano, R. O. Chen, R. B. Altman, *Gene* **1997**, *190*, GC59–70.
26 M. Zuker, J. A. Jaeger, D. H. Turner, *Nucleic Acids Res.* **1991**, *19*, 2707–2714.
27 A. P. Gultyaev, F. H. van Batenburg, C. W. Pleij, *J. Mol. Biol.* **1995**, *250*, 37–51.
28 M. Schmitz, G. Steger, *J. Mol. Biol.* **1996**, *255*, 254–266.
29 C. Flamm, W. Fontana, I. L. Hofacker, P. Schuster, *RNA* **2000**, *6*, 325–338.
30 D. H. Mathews, D. H. Turner, *J. Mol. Biol.* **2002**, *317*, 191–203.
31 R. Lück, S. Gräf, G. Steger, *Nucleic Acids Res.* **1999**, *27*, 4208–4217.
32 J. Gorodkin, S. L. Stricklin, G. D. Stormo, *Nucleic Acids Res.* **2001**, *29*, 2135–2144.
33 F. Chetouani, P. Monestié, P. Thébault, C. Gaspin, B. Michot, *Nucleic Acids Res.* **1997**, *25*, 3514–3522.
34 I. L. Hofacker, P. F. Stadler, *Comp. Chem.* **1999**, *23*, 401–414.
35 J. Gorodkin, H. H. Stærfeldt, O. Lund, S. Brunak, *Bioinformatics* **1999**, *15*, 769–770.
36 J. Gorodkin, Heyer L. J., S. Brunak, G. D. Stormo, *Comp. Appl. Biosci./Bioinformatics* **1997**, *13*, 583–586.
37 F. H. van Batenburg, A. P. Gultyaev, C. W. Pleij, *J. Theor. Biol.* **1995**, *174*, 269–280.
38 E. Rivas, S. R. Eddy, *J. Mol. Biol.* **1999**, *285*, 2053–2068.
39 J. Reeder, R. Giegerich, BMC Bioinformatics **2004**, *5*, 104.
40 F. Michel, M. Costa, C. Massire, E. Westhof, *Methods Enzymol.* **2000**, *317*, 491–510.

32
Modeling the Architecture of Structured RNAs within a Modular and Hierarchical Framework

Benoît Masquida and Eric Westhof

32.1
Introduction

Although, as in proteins, the structural information of an RNA molecule is intrinsically hidden in the sequence of residues, it is encoded in different chemical ways and exploits different physicochemical forces. In proteins, amino and carbonyl groups involved in secondary structure formation belong to the backbone, and, thus, are common to all amino acids. The native three-dimensional (3-D) folding of a protein results from the assembly of those secondary structure elements via tertiary interactions and tight packing between amino acid side-chains. On the contrary, specific interactions in RNA, such as Watson–Crick (WC) pairs, are mediated by the chemical groups of the bases which confer their identity to each nucleotide. Furthermore, although both modularity and hierarchy are central to protein and RNA folding, they appear and are exploited more clearly in RNA folding. Upon synthesis of an RNA strand, A-form helices build up and constitute the secondary structure. The energy content of the secondary structure can become rather high since it increases with the number of base pairs in helices. This is why it is often possible to observe an RNA molecule with only secondary structure helices and no tertiary fold. In contrast, and as in proteins, the free energy of the native 3-D state is not very negative (around -10 kcal/mol). The scaffoldings provided by RNA helices yield a population of loose structures that promote the adoption of the native state, which is achieved when tertiary interactions, mainly through non-canonical pairs, form depending on ionic conditions or on the presence of interacting partners. This view of RNA folding is biologically relevant since RNA will have to adopt various conformations in the course of chemical reactions or to interact with other molecules to fulfill their role.

Secondary structure diagrams provide a framework summarizing sequence and biochemical data without providing a full structural basis. The secondary structure diagrams, ideally with the accompanying sequence alignment, constitute the starting point for 3-D modeling (see below). Being closer to reality, 3-D molecular models offer a framework to rationalize the data, with 3-D coherence respectful of the sugar-phosphate backbone stereochemistry, molecular geometry and non-covalent interactions. Thus, the whole modeling process consists of extracting the structural information from a set of sequences belonging to organisms which scan the phylogeny, and providing a model of architecture compatible with sequence covariations and biochemical observations.

After early success in the late 1960s with the prediction of the tRNA anticodon loop by Fuller [1] and the beautiful work of Michael Levitt on the tRNA structure [2], the RNA modeling field emerged at the end of the 1980s. In the meantime, molecular biology techniques led to the harvest of numerous RNA sequences and to the discovery that the functions of RNA molecules are not restricted to their roles in translation. Ribozymes were discovered by Cech [3] and Altman [4]. The first model of a large RNA, the core of the group I introns [5], was built *ab initio* in a context where the body of structural knowledge consisted of the well-known WC helix and the tRNA crystal structures [6–8]. Non-canonical base interactions observed in tRNA crystal structures gave a snapshot of the potential plasticity of these molecules. However, the observed interactions could not always be unambiguously derived from sequence analysis. The rules governing the interaction schemes in non-canonical regions could only be formulated in terms of daring assumptions that needed to be experimentally validated.

Then, the outcome of modeling studies led to the topology and architecture of the large RNAs, as in the studies on the group I intron [9]. Nowadays, there is an increasing availability of RNA-containing crystal structures such as ribozymes [10–12], and ribosomal particles and subunits [13–15]. In this chapter, we will emphasize how crystal structures of folded RNA fragments, together with systematic sequence comparisons, can be used to understand better the overall architecture of large RNAs. An overview of the computer programs used to generate RNA 3-D models will be presented.

32.2
Modeling Large RNA Assemblies

The starting point for modeling an RNA molecule is the secondary structure which represents, on a planar drawing, the WC paired helical regions linked by the non-helical segments: hairpin loops, internal bulges and multiple junctions. The folded architecture results from the assembly of the helical framework through interactions between the helices and the single-stranded regions. There are three main categories of tertiary structure interactions: those between two double-stranded helices, those between a helix and a single strand and those between two single-stranded regions. Sequence analyses together with the growing number of RNA crystal structures have shown that RNA architectures are assembled from modules in a hierarchical manner. The modeling process developed in the laboratory is based on this principle of natural folding processes [16]. Modules can be either architectural (forming a bend, e.g. kink–turn motif [17], or a reorientation within a helix or between helices) or anchors for association (e.g. tetraloop–tetraloop receptor interaction [18]). Modules generally contain RNA motifs with an ordered assembly of non-WC base pairs. Modules should possess the following properties: (1) their borders and interfaces should be clearly identifiable, (2) if they interact with other modules, they should do so via defined protocols, and (3) they should be able to undergo modifications and evolution independently of each other. The properties of the protocols are such that they define the rules of association between modules in a robust manner capable of sustaining biological evolution. The modules are limited in number and are recurrent, as well as some of the rules of association between them [19]. Thus, the motifs can show sequence variability without impairing their ability to adopt a structure close to the archetype that preserves the borders and interfaces as well as protocols of interaction. For example, the loop E motif varies from one organism to another, but the geometry of the base-pairing ensemble of the motif is preserved [20, 21]. It sustains RNA stabilization and RNA–protein interactions in the ribosomal 50S subunit [22] and was extended to RNA–RNA interactions in the group I intron from the *td* gene of bacteriophage T4 [23].

Because of the underlying sequence variations within the motifs (the rules of which are still unknown), it is crucial to perform extensive sequence alignment and secondary structure analysis prior to any 3-D modeling in order to identify with confidence the nature of the various motifs. The secondary structure can then be parsed into modules based on elementary motifs, the 3-D coordinates of which can be generated later using appropriate programs (see below). These modules can be afterwards assembled to form the RNA architecture. Finally, the geometry of the model is regularized using least-square refinement. In this process, hydrogen bonds between nucleotides are used as explicit constraints.

One of the most difficult tasks is the arrangement of the multiple way junctions. Usually, a 2-D structure is a set of hairpins linked together by various single-stranded regions and junctions. The main problem is then the proper choice of helices which stack on each other. If an internal loop lies within a helical stem, does it favor co-axial stacking or does it make a kink that prevents co-axially stacking between the helical parts of the stem? In the case of three- or four-way junctions, which helices do stack, if any? A thorough analysis of the sequence of the junctions should be performed to exhibit features homologous to junctions for which a crystal structure is known. If no convincing homology is found, assumptions should be made that eventually lead to the building of several models [16]. Then tertiary interactions coupled to biochemical data may help to discriminate between the various possibilities.

32.2.1
The Modeling Process

The suite of in-house modeling programs runs on SGI graphical stations under the IRIX 6.2 system or beyond. Figure 32.1 summarizes the overall process. The manip [24] package contains programs dedicated to generate 3-D coordinates from the sequence information and is available at URL http://www-ibmc.u-strasbg.fr/upr9002/westhof/. Coseq [25] will soon be made available.

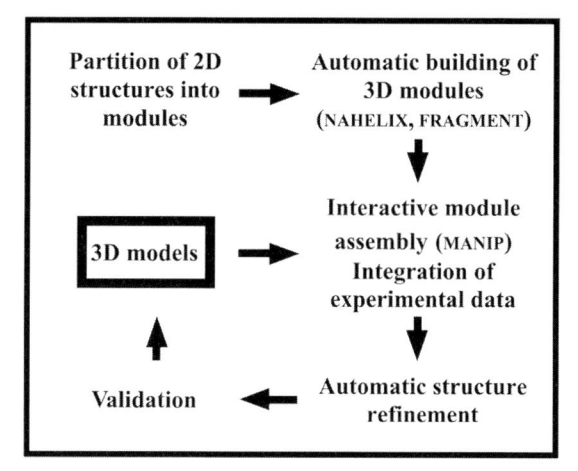

Fig. 32.1. Diagram of the modeling process. From an accurate sequence alignment, the secondary (2-D) structure can be deduced. Then, the first step consists of partitioning the 2-D structure into modules for which 3-D coordinates can be generated using adequate computer programs. These modules can then be assembled interactively on a graphic computer. When all the modules are placed next to each other while respecting the covalent linkages of the RNA chain, refinement can be performed. Validation of the model proceeds through collating it to biochemical data. Additional steps of interactive modeling/refinement can be performed until the model is fully satisfactory.

32.2.1.1 Getting the Right Secondary Structure

The best way to build a reasonable secondary structure for a given RNA is to have at hand enough sequences to perform statistical analysis of sequence variations. When this criterion is not met, *in vitro* evolution methods can be used to generate sequence datasets for relatively short RNAs [26]. Chemical probing techniques can compensate for a lack of sequences and help discriminate between various possibilities [27, 28]. With experience, secondary structure prediction programs such as mfold can be very outstanding [29–31]. As a first step of the molecular modeling process, sequences are roughly aligned according to fully conserved nucleotide stretches, if present. Then, helices can be identified by checking for sequence covariations obeying WC rules. Within this scenario, conserved WC base pairs cannot be proven and, thus, should be regarded as non-proven or with extreme caution. Indeed, conservation of WC pairs can reveal either tertiary contacts or an alternative pairing geometry. This can be the case for AU pairs when they form Hoogsteen–WC interactions (pair U135–A187 in the crystal structure of the P4–P6 domain of a group I intron [18]). Further, it is even likely that apparent conserved bases at helix edges do not interact with each other, as seen in the 16S ribosomal binding site of protein S15 (G587 and C754 [32]). Finally, the RNA parts which do not follow the WC rules can be suspected to form specific tertiary motifs that can be scrutinized for sequence homologies to motifs of known structure. Once such a motif is recognized in an existing crystal structure, the interacting nucleotides' chemical groups are identified, which allows the prediction of covariations. Thus, compatibility between a new sequence and the interaction scheme provided by the crystal structure can be checked. This process enables unraveling non-WC interaction rules and is thus of considerable importance to RNA modeling.

The full process of sequence analysis can be performed using the in-house developed program coseq [25] or BioEdit (http://www.mbio.ncsu.edu/BioEdit/bioedit.html). This program allows the display and editing of sequence alignment files (.gb: GenBank format) and statistical analysis of covariations. First, a GenBank format file should be created containing all sequences of interest. Then, this file can be edited and interactively modified in the alignment window of coseq. Once the alignment is satisfactory, the statistics window allows us to monitor covariations. To do so, two boxes (X and Y) can be filled in by the user with position numbers and then computed (stretches of nucleotides can be input in both boxes). The program outputs ij matrices $[(1, 1) \le (i, j) \le (n, m)]$ containing percentages or sequence counts for every combination of nucleotides as terms (Fig. 32.2). When the output is too complex, the amount of information can be filtered out using simplification masks.

32.2.1.2 Extrusion of the Secondary Structure in 3-D

Two complementary programs help extruding secondary structure elements into three dimensions and have been described elsewhere [33].

- Nahelix [33] is dedicated to building nucleic acid helices and is run using either a graphical interface or interactive mode in a shell window. The program reads the sequence length, numbering and sequence as input, and outputs a coordinate file for all the RNA atoms of the described sequence. In the case non-WC helices are built, the sequence should be given for both strands, 5' to 3' for the first one and 3' to 5' for the second one. The file is written in the Wayne Hendrickson format (.hd: Hendrickson data) and can be transformed to .pdb format using the hd2pdb program (pdb2hd exists in order to perform the reverse action).
- Fragment [33] is dedicated to mutate sequences of known structure. A given .pdb file can be transformed to a fragment file (.fgm) using the makefgm routine (reads in .hd files). An output from makefgm contains no sequence information, but only important conformational parameters for maintaining the sugar-phosphate backbone characteristic of the motif. Then, a polynucleotide backbone with any base sequence can be threaded onto that of the starting folded polynucleotide, which considerably facilitates the building of variants of a given motif.

32.2.1.3 Interactive Molecular Modeling

Once all secondary structure elements have been built in 3-D, they can be assembled interactively using the manip program. manip is a user-friendly program operated by pop-up menus in the main window. It allows definition of zones that

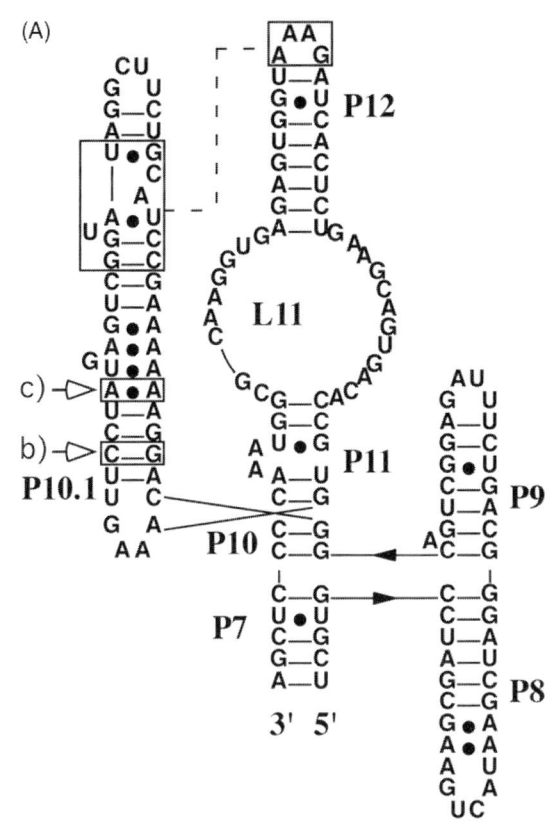

Fig. 32.2. Examples of covariation matrices taken from stem P10.1 in a subset of the S domain RNase P sequences related to the *Bacillus subtilis* type [35]. (a) Secondary structure of the S domain. Arrows point towards boxed base pairs that have been analyzed using the program coseq. For a base pair analysis coseq typically outputs 5 × 5 covariation matrices corresponding to all nucleotide and gap combinations [(ACGU−) × (ACGU−)]. The numbers in parentheses stand for the statistically expected occurrences of each pair calculated on the basis of the number of each type of base (A, G, C, U) at the appropriate positions in the structurally aligned sequences. (b) In the case of a WC base pair, terms corresponding to WC covariations (G=C, C=G, A–U, U–A) and wobble pairs predominate. (c) On the other hand, in non-canonical base pairs, represented combinations reflect other rules. Here, A139 can be mutated to G without impairing the interaction. The WC population results from a subset of sequences in which the canonical motif (and thus the non-WC pairing) is not present.

can be independently moved using the **move** command in order to generate a 3-D model. The other main modeling function is **tor** which activates all the dihedral angles of a selected nucleotide including $O3'_{n-1} - P_n$. A switch allows the user to quickly activate dihedral angles from previous or next residues all along the RNA chain. RNA fragments can thus be moved like tentacles and efficiently be linked to each other.

(B)

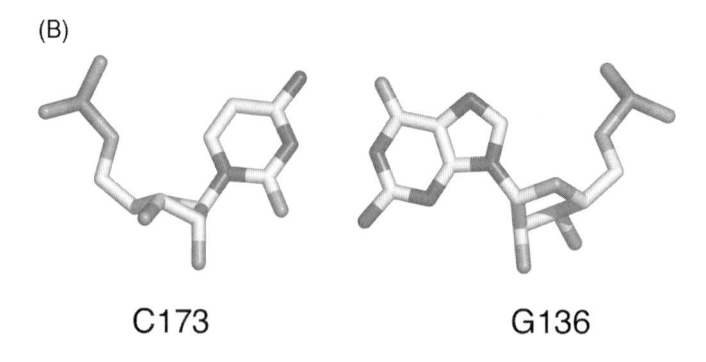

C173   G136

C173=G136	A	C	G	U	-	All
A	2.6 (2.0)	0.0 (0.2)	0.0 (3.7)	**5.1 (1.0)**	0.0 (0.8)	7.7
C	0.0 (11.8)	0.0 (1.2)	**46.2 (22.5)**	0.0 (5.9)	0.0 (4.7)	46.2
G	2.6 (3.3)	**2.6 (0.3)**	0.0 (6.2)	7.7 (1.6)	0.0 (1.3)	12.8
U	**20.5 (5.9)**	0.0 (0.6)	2.6 (11.2)	0.0 (3.0)	0.0 (2.4)	23.1
-	0.0 (2.6)	0.0 (0.3)	0.0 (5.0)	0.0 (1.3)	**10.3 (1.1)**	10.3
All	25.6	2.6	48.7	12.8	10.3	100.0

Fig. 32.2B

32.2.1.4 Refinement of the Model

Refinement of the model is achieved by geometrical least squares using the Konnert–Hendrickson algorithm [34] implemented in the program nuclin/nuclsq [8]. The algorithm takes into account bond lengths, valence angles, dihedrals and has an anti-bump restraint. The resulting function is minimized against a dictionary of values that have been observed in high-resolution crystal structures of nucleotides and oligonucleotides. Since the refinement program optimizes the geometry along the steepest descent, the conformation of the starting model should not present extremely distorted regions to avoid refinement failure. The refined model can then be collated to the data and the process of interactive modeling and least-squares refinement can be looped until the model is satisfactory.

A subset of the data should be used as a blind test during the building of the

(C)

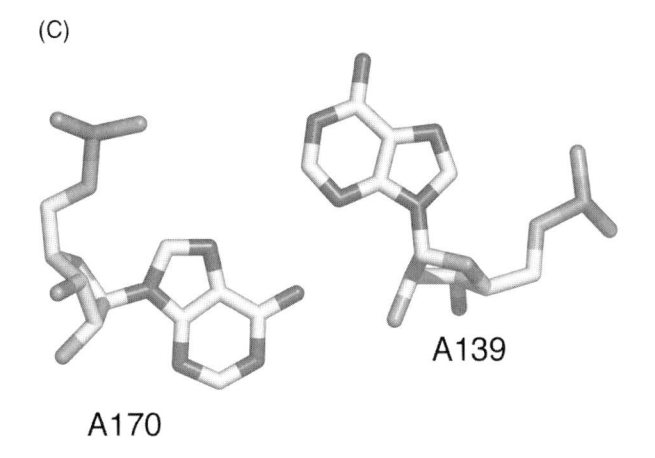

A170∴A139	A	C	G	U	-	All
A	**22.5** (16.9)	2.5 (8.4)	**37.5** (28.7)	5.0 (5.1)	0.0 (8.4)	67.5
C	0.0 (1.2)	0.0 (0.6)	5.0 (2.1)	0.0 (0.4)	0.0 (0.6)	5.0
G	0.0 (2.5)	10.0 (1.2)	0.0 (4.2)	0.0 (0.8)	0.0 (1.2)	10.0
U	2.5 (1.2)	0.0 (0.6)	0.0 (2.1)	2.5 (0.4)	0.0 (0.6)	5.0
-	0.0 (3.1)	0.0 (1.6)	0.0 (5.3)	0.0 (0.9)	12.5 (1.6)	12.5
All	25.0	12.5	42.5	7.5	12.5	100.0

Fig. 32.2C

model so as to help validating the model. Further steps of interactive modeling may then be required until a satisfactory solution is reached.

32.3
Conclusions

This review updates RNA molecular modeling techniques that had been described [3, 24, 33]. The first major improvement to the overall process is the availability of the program coseq or BioEdit (see chapter 30) which allows for sequence alignment and covariation analysis. It is clear that with the availability of genome sequences (www.tigr.org), comparative sequence analysis to elucidate RNA structure has already undergone wide use. The second major improvement is the increasing number of available RNA crystal structures which provide a structural basis to understand sequence variations. Sequences and crystal structures can be merged together in order to decipher the rules governing nucleotide interactions in complex RNA motifs and, hence, provide new tools to unravel RNA structures.

References

1. W. Fuller, A. Hodgson, *Nature* **1967**, *215*, 817–821.
2. M. Levitt, *Nature* **1969**, *224*, 759–763.
3. T. R. Cech, A. J. Zaug, P. J. Grabowski, *Cell* **1981**, *27*, 487–496.
4. C. Guerrier-Takada, K. Gardiner, T. Marsh, N. Pace, S. Altman, *Cell* **1983**, *35*, 849–857.
5. F. Michel, E. Westhof, *J. Mol. Biol.* **1990**, *216*, 585–610.
6. J. L. Sussman, S. R. Holbrook, R. W. Warrant, G. M. Church, S.-H. Kim, *J. Mol. Biol.* **1978**, *123*, 607–630.
7. J. E. Ladner, A. Jack, J. D. Robertus, R. S. Brown, D. Rhodes, B. F. Clark, A. Klug, *Proc. Natl. Acad. Sci. USA* **1975**, *72*, 4414–4418.
8. E. Westhof, P. Dumas, D. Moras, *J. Mol. Biol.* **1985**, *184*, 119–145.
9. V. Lehnert, L. Jaeger, F. Michel, E. Westhof, *Chem. Biol.* **1996**, *3*, 993–1009.
10. H. W. Pley, K. M. Flaherty, D. B. McKay, *Nature* **1994**, *372*, 68–74.
11. A. R. Ferré-D'Amaré, K. H. Zhou, J. A. Doudna, *Nature* **1998**, *395*, 567–574.
12. P. B. Rupert, A. R. Ferre-D'Amaré, *Nature* **2001**, *410*, 780–786.
13. W. M. Clemons, Jr, D. E. Brodersen, J. P. McCutcheon, J. L. May, A. P. Carter, R. J. Morgan-Warren, B. T. Wimberly, V. Ramakrishnan, *J. Mol. Biol.* **2001**, *310*, 827–843.
14. N. Ban, P. Nissen, J. Hansen, P. B. Moore, T. A. Steitz, *Science* **2000**, *289*, 905–920.
15. H. F. Noller, M. M. Yusupov, G. Z. Yusupova, A. Baucom, K. Lieberman, L. Lancaster, A. Dallas, K. Fredrick, T. N. Earnest, J. H. Cate, *Cold Spring Harbor Symp. Quant. Biol.* **2001**, *66*, 57–66.
16. E. Westhof, B. Masquida, L. Jaeger, *Fold. Des.* **1996**, *1*, R78–R88.
17. D. J. Klein, T. M. Schmeing, P. B. Moore, T. A. Steitz, *EMBO J.* **2001**, *20*, 4214–4221.
18. J. H. Cate, A. R. Gooding, E. Podell, K. Zhou, B. L. Golden, C. E. Kundrot, T. R. Cech, J. A. Doudna, *Science* **1996**, *273*, 1678–1684.
19. N. B. Leontis, E. Westhof, *Curr. Opin. Struct. Biol.* **2003**, *13*, 300–308.
20. N. B. Leontis, E. Westhof, *RNA* **1998**, *4*, 1134–1153.
21. N. B. Leontis, J. Stombaugh, E. Westhof, *Biochimie* **2002**, *84*, 961–973.
22. M. Lu, T. A. Steitz, *Proc. Natl. Acad. Sci. USA* **2000**, *97*, 2023–2028.
23. C. Waldsich, B. Masquida, E. Westhof, R. Schroeder, *EMBO J.* **2002**, *21*, 5281–5291.
24. C. Massire, E. Westhof, *J. Mol. Graph. Model.* **1998**, *16*, 197–205, 255–197.
25. C. Massire, E. Westhof, in preparation.
26. A. Berzal-Herranz, S. Joseph, B. M. Chowrira, S. E. Butcher, J. M. Burke, *EMBO J.* **1993**, *12*, 2567–2574.
27. T. Powers, H. F. Noller, *EMBO J.* **1991**, *10*, 2203–2214.
28. C. Isel, C. Ehresmann, B. Ehresmann, R. Marquet, *Pharm. Acta Helv.* **1996**, *71*, 11–19.

29 A. B. Jacobson, M. Zuker, *J. Mol. Biol.* **1993**, *233*, 261–269.
30 H. Isambert, E. D. Siggia, *Proc. Natl Acad. Sci. USA* **2000**, *97*, 6515–6520.
31 Y. Ding, C. E. Lawrence, *Nucleic Acids Res.* **2003**, *31*, 7280–7301.
32 A. Nikulin, A. Serganov, E. Ennifar, S. Tishchenko, N. Nevskaya, W. Shepard, C. Portier, M. Garber, B. Ehresmann, C. Ehresmann, et al. *Nat. Struct. Biol.* **2000**, *7*, 273–277.
33 E. Westhof, *J. Mol. Struct. Dyn.* **1993**, *286*, 203–210.
34 J. H. Konnert, W. A. Hendrickson, *Acta Crystallogr. A* **1980**, *36*, 344–349.
35 C. Massire, L. Jaeger, E. Westhof, *J. Mol. Biol.* **1998**, *279*, 773–793.

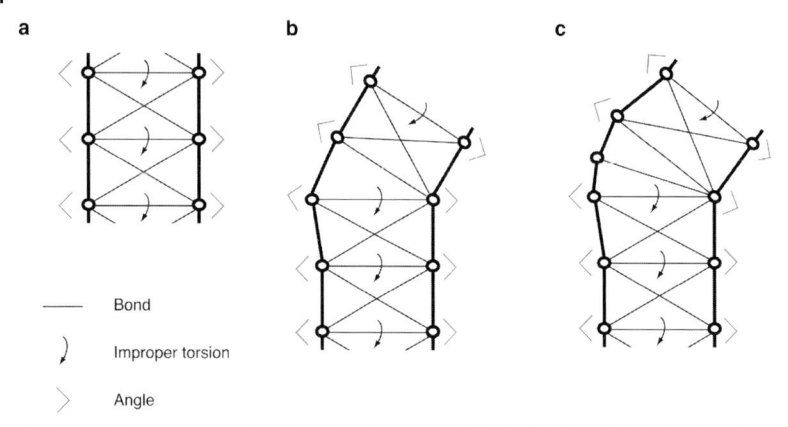

Fig. 33.2. Common restraints in reduced representation models. In our simulations, we use harmonic restraints to hold nucleotides in positions that are consistent with A-form RNA helices. (a) The harmonic restraints include bonds, angles and improper torsions to insure that the P-atoms accurately represent the double helix. (b and c) Bulged nucleotides can induce kinks in the RNA helices because of the opposing strand asymmetry. The severity of kinking is dependent on the number of unpaired nucleotides in the bulge: (b) 1-nt kink and (c) 2-nt kink.

tion of P- and X-atoms will provide interhelix volume exclusions consistent with the crystallographic database of RNA structures.

33.2.2
Implementing RNA Secondary Structure

In a very simple form of reduced representation modeling, secondary structure can be implemented by representing each helical element as a single pseudo-atom. If we use an all-P model, the complexity requires more restraints on the atoms to form a helix. In order to create an A-form helix, a set of five bonds, two angles and one pseudo-torsion are used for each base pair in the helix stem (Fig. 33.2a). X-atoms are then anchored at the geometric center of each base pair in the helix stem. Helical stacking can be introduced such that two helices are essentially treated as an intact helix, with the same restraints found in a canonical helix.

Many helical elements have bulged nucleotides that may alter the geometry of a helix by imparting flexibility. All-P models for regular helices are modified at the bulge site to introduce a small kink in the helical axis for bulges with one nucleotide and to introduce flexibility for larger bulged loops (Fig. 33.2b and c).

Hairpin loops in RNA structure are very diverse in both size (number of nucleotides) and arrangement. There are a few common motifs including the U-turn found in the tRNA anticodon [1], the GNRA tetraloop [2, 19] and others. In order to decrease computational complexity, it may be worthwhile to constrain the loops into conformations that are consistent with loops determined in X-ray crystallography studies. Internal loops caused by mismatched base pairs are another important feature of RNA structure. As with the helical loops, it is generally advantageous to use internal loop structures that are consistent with crystallography data. With the wealth of data available from the ribosome crystal structures, there are many types of internal and helical loops that can be used as a library for modeling these regions in reduced representation models.

Single-stranded regions of RNA are usually left relatively unstructured in our models and bonds are used to maintain connectivity along the chain. The phosphate–phosphate distance along an RNA backbone depends on the backbone and sugar conformation, but on average is roughly 6 Å. This distance can be used for the pseudo-bonds between P-atoms in single-stranded regions of the RNA chain. Flexibility about an individual nucleotide would be exaggerated if only bond constraints are used, because the sphere that it represents allows the chain to kink much more than is physically possible. Backbone angles in an RNA chain have well-defined ranges and this restraint can be implemented as a semi-harmonic energy penalty when P–P–P bond angles fall below a certain value. This energy term can be restrictive or flexible and is easily incorporated into the input files of the *yammp* system.

33.2.3
Protein Components

Before the crystal structures of the ribosomal subunits had been determined at a reasonably high resolution, there was low-resolution data about the orientation of proteins in the subunits available from neutron scattering experiments [20] and also from chemical crosslinks. Therefore, it was appropriate to treat proteins as individual spheres with radii appropriate for their size in our original ribosome models [10, 11]. Now that the structures of the proteins associated with the ribosome have been determined, the proteins can be incorporated as either all-atom models or pseudo-atom models. The pseudo-atom detail can vary and we have used a protocol where proteins are represented as a collection of pseudo-atoms centered on the C_α atom of each amino acid residue. Pseudo-bonds connect consecutive residues in the protein sequence and a web of pseudo-bonds is used to retain the proteins conformation. Increasing the complexity of protein representation allows for incorporation of specific interactions between amino acids and nucleic acids.

33.2.4
Implementing Tertiary Structural Information

Chemical crosslinks are capable of determining the interactions between RNA and protein or different segments within an RNA sequence. Such tertiary information is incorporated into our models as pseudo-bonds, bringing crosslinked elements together. The lengths of the bonds are consistent with the type of crosslinking agent used in the experiment and the sum of the radii of the reactive nucleotides/

proteins. The force constant of the bond is also proportional to the quality of the experimental data, so the force constants for tertiary data are generally much weaker than those for the secondary structure motifs.

Footprinting studies examine differences in RNA accessibility upon the binding of proteins. These changes in reactivity at certain sites can be attributed directly to protein binding, or to conformational changes that are an indirect consequence of binding. These data are usually less reliable than the direct crosslinking studies, but they provide an extensive set of contacts that can be incorporated into a structural model. As with crosslinks, the interactions determined by footprinting are treated as pseudo-bonds in the model.

Although there are only a few crystal structures of large RNP complexes, they provide a wealth of structural data, including direct interactions. Since the data are very precise, a stronger bond potential is used to restrain the proteins and RNA when incorporating specific interactions into a model of the ribosome. Interactions determined by X-ray crystallography allow us to model dynamic processes associated with RNP complexes. Specifically, protein–RNA or RNA–RNA contact information can be implemented in the model to imitate various states of RNP assembly, as will be presented later.

Other tertiary data can be used to further restrain the model. Results from all types of experiments can be incorporated into models as distance constraints (pseudo-bonds) with force constants consistent with the precision of the data.

33.2.5
Modeling Protocol

It is important to emphasize at the outset that all terms in the force field are empirically derived for these succinct models. This is different from the force field parameters in traditional all-atom simulations, which are derived in a manner that is intended to reproduce the physics underlying all the interactions as closely as possible. In an all-atom model, the inter-atomic potential for any pair of non-covalently bound atoms should be equivalent to the appropriate potential of mean force between them. In our models, pseudo-atoms either interact (in which case they are held together by appropriate terms in the force field) or they do not (in which case the only term for them in the force field is a pure non-bonded repulsion, to prevent steric overlaps). Here, the phrase "pseudo-atoms interact" means they are part of a well-defined structural subunit such as an RNA double helix or an individual protein, or that there is a piece of experimental data – such as a crosslink – that provides information on the distance between them.

Harmonic potential functions are used for all the constraints in our models, including bonds, angles, and improper torsions:

$$E_{\beta i} = k_{\beta i}(\beta_i - \beta_{io})^2 \quad \text{for bonds} \tag{1}$$

$$E_{\alpha i} = k_{\alpha i}(\alpha_i - \alpha_{io})^2 \quad \text{for angles} \tag{2}$$

$$E_{\tau i} = k_{\tau i}(\tau_i - \tau_{io})^2 \quad \text{for improper torsions} \tag{3}$$

where $E_{\beta i}$, $E_{\alpha i}$ and $E_{\tau i}$ denote energy of the ith bond, ith angle and ith torsion, respectively; $k_{\beta i}$, $k_{\alpha i}$ and $k_{\tau i}$ are the force constants for the ith bond, ith angle and ith torsion; β_i, α_i and τ_i are the ith bond, ith angle and ith torsion; and β_{io}, α_{io} and τ_{io} are the corresponding equilibrium or ideal values. Harmonic potentials are easy to minimize, have a unique minimum and the potential function is non-negative, regardless of the conformation of the model. One major advantage of this formulation is that zero energy indicates that all experimental constraints are satisfied. Any other value can be used as a rough indicator of the quality of a model. Force constants are chosen to reflect the uncertainties associated with a specific bond, angle or torsion. As a concrete example, we note that a harmonic bond potential function is equivalent to the Gaussian distribution of bond β_i about the equilibrium value β_{io} with variance equal to $RT/2k_{\beta i}$. This allows the direct calculation of the appropriate force constant from the experimentally estimated variance in the distance. Therefore, low-resolution tertiary constraints are more flexible than constraints determined from X-ray crystal structures. Experimentally determined tertiary interactions are treated as semi-harmonic interactions:

$$E_{\beta i} = k_{\beta i}(\beta_i - \beta_{io})^2 \qquad \text{if } \beta_i \geq \beta_{io} \qquad (4)$$

$$E_{\beta i} = 0 \qquad \text{if } \beta_i < \beta_{io} \qquad (5)$$

By using the semi-harmonic function, a pseudo-bond is not penalized for being too small, which is advantageous for low-resolution structural data. An energetic penalty is only introduced if the bond distance is too large, thereby forcing the two constituents to remain close to one another. If the tertiary data comes from a more precise method, the force constant is increased and a normal harmonic pseudo-bond is used.

Non-bond interactions are used to provide volume exclusion for all pairs of pseudo-atoms (both P- and X-atoms). We use semi-harmonic terms for non-bond repulsions:

$$E_{ij} = k_{ij}(r_{ij} - r_{ijo})^2 \qquad \text{if } r_{ij} \leq r_{ijo} \qquad (6)$$

$$E_{ij} = 0 \qquad \text{if } r_{ij} > r_{ijo} \qquad (7)$$

where E_{ij} is the non-bond interaction energy between atoms i and j, k_{ij} is the non-bond force constant for the atom pair ij, r_{ij} is the distance between atoms i and j, and r_{ijo} is the minimum distance allowed between the two atoms (usually the sum of their radii). Atoms far apart do not sense one another, but as atoms approach each other in conformational space, they repel with a force determined by k_{ij}. By using the semi-harmonic term, the computational calculations are more efficient since they do not need to consider the interactions between pseudo-atoms that are far apart. Only those atoms close enough to sense one another are important for the calculations. It is also essential to exclude non-bond interactions for atoms that are connected by bonds, angles or pseudo-torsions. We also exclude non-bond interactions for all pairs of atoms in a given double helix. On occasion, it may be worthwhile to obtain a larger sampling of conformational space by allowing RNA chains to pass through one another. In this case, the non-bond force constant can be "softened" or completely disregarded during initial refinement, then turned on for final refinement.

The modeling protocol is implemented using *yammp*, an in-house molecular modeling package. In the past, a script file was used to describe an RNA chain or RNP. The script file could be converted into a file describing the molecular topology and force field parameters (called a descriptor file in *yammp*) with a program called *mksnad* (make succinct nucleic acids descriptor). Such an example is described in our earlier review using a tRNA molecule as an example [21]. However, a new force field assembler is currently being developed in the lab using the Python scripting language. We are developing a newer version of *yammp* (termed YUP, for *yammp under python*). A force field assembler (FFA) instantiates a particular model by dynamically linking appropriate objects in the model as fragments of code. One benefit of organizing a model in this way is that it allows for true multiscale modeling. Therefore, some regions of the model can be treated at the all-atom level while others are represented at a lower resolution. Another is that the program structure provides dynamic access to all the parameters of the model, allowing a level of user control not available in conventional modeling programs.

The models are refined using energy minimization and simulated annealing with Monte Carlo. Because all of the energy terms are harmonic, the lowest energy value that can be reached is zero. In larger RNP systems, like the ribosome, all constraints are unlikely to be simultaneously satisfied. Therefore, the energy term is a measure of the quality of the model. This method generates several models corresponding to many possible conformers. Then it is possible to evaluate which regions of a particular structure are better resolved based on experimental data incorporated in the model. Increased conformational freedom in a specific region correlates with fewer restraints.

33.3
Application of Modeling Large RNA Assemblies

33.3.1
Modeling the Ribosome Structure at Low Resolution

Before the crystal structures of the ribosomal subunits were determined, they were a great mystery. Several groups were building models of the small, 30S, subunit manually. Using reduced representation methods described in this chapter, it was possible to incorporate low-resolution experimental data into an automated RNA folding procedure that could generate several models for the ribosomal subunit [10, 11].

The secondary structure of the 16S rRNA was well established from comparative sequence analysis [22, 23] and was used to generate the fold of the RNA helices in the model. Some RNA tertiary interactions had been determined using compar-

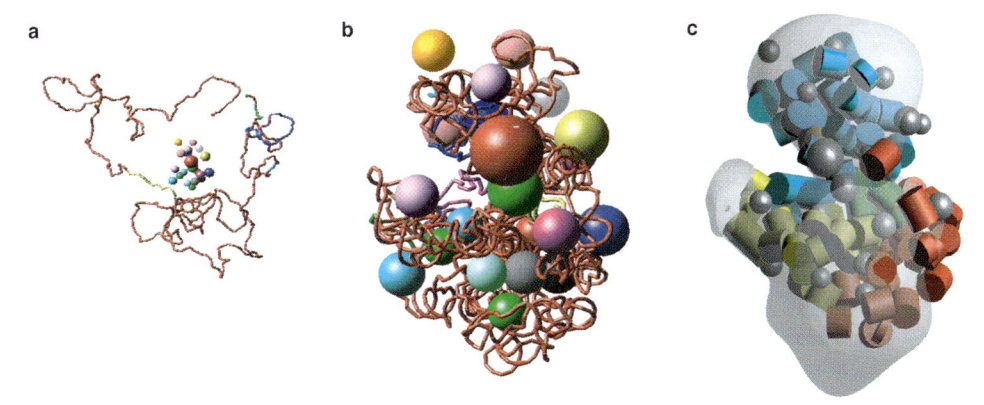

Fig. 33.3. Low-resolution model of the 30S subunit of the ribosome. Studies aimed at determining the structural relationships between different domains of the small subunit of the ribosome started with a randomized RNA chain around the ribosomal proteins position based on data from neutron diffraction (a). Using a multitude of tertiary restraints, several models could be generated to explain experimental data, including the model shown in (b). (c) A model of the 30S subunit was built using the refinement protocol that included shape data from early electron microscopy data. Cylindrical helices are colored by domain and ribosomal proteins are represented as grey spheres.

ative studies [24] and were also incorporated into the model. Crosslinking and footprinting information was introduced using harmonic pseudo-bonds for both RNA–RNA and RNA–protein interactions. Proteins were tethered to their respective locations as determined by neutron scattering based on the average of the standard error in the protein coordinates [20]. Shape information from electron microscopy was also incorporated into the model, but that protocol will not be addressed in this chapter [11, 21].

The modeling protocol involved the evolution of structural complexity in RNA representations. To begin, helices were treated as single pseudo-atoms (called 1H), and several random walks of the RNA chain were generated at this resolution (Fig. 33.3a). These chains were optimized using the energy refinement of the potential terms that enforce the structural constraints of the models. The refined 1H models were then extrapolated to a resolution where five pseudo-atoms are used to represent each double-stranded RNA stem (called 5H). The refinement procedure was repeated and the model was further extrapolated to an all-P resolution for a final round of energy refinement.

Not all random walk chains refined by the protocol led to acceptable structures (low energies) and those that did not were discarded. Of the remaining structures, visual and energetic analysis could determine which had folds that were more acceptable because of helix stacking and the overall arrangement of ribosomal RNA and proteins (Fig. 33.3b). The overall energies never reached a value of zero (all constraints were not satisfied), which was not surprising considering the size of the system and the resolution of the experimental data. This led us to identify conflicts in the data that were reported to some of the experimental groups that had generated the data.

The models generated in this study explored the range of different arrangements that were compatible with the experimental data that was available at that time. The best representation of the models generated is seen in Fig. 33.3(c). There were many similarities with the manually built models, but several helices had a range of alternate positions because of the lack of structural data. The models generated with the refinement protocol examined alternative sets of experimental data and were useful for designing future experiments. It was easy to incorporate new information into the models for future refinements.

33.3.2
Modeling Dynamic Assembly of the Ribosome with Reduced Representation

Once the crystal structures for the small subunit had been determined [4, 5], we were interested in using our molecular mechanics protocol to rationalize the order of protein binding observed in ribosome assembly in terms of ribosome folding [12]. The order in which proteins bind to the 16S ribosomal RNA, the assembly map, was determined by Nomura et al. in the early 1970s [25], but precise interactions between RNA and protein could not be determined until the crystal structures were available. The obvious questions were whether we could imitate ribosome assembly by adding the proteins to the ribosomal RNA in the order that they bind in the assembly map and what we might learn from such an exercise. Using the crystallographic distances between amino acids and nucleotides, harmonic restraints were introduced to tether each protein to its binding site on the RNA chain. The RNA was represented using the all-P model with X-atoms centered in the helices, and proteins had pseudo-atoms centered on every C_α in the amino acid backbone.

During the simulations, the 16S rRNA helices and proteins were treated as rigid bodies. The single-stranded regions of the RNA chain were flexible and the 16S rRNA was randomized in the absence of proteins to generate a starting conformation. The force field for the simulation consisted of harmonic bonds and semi-harmonic non-bonds. Rigid-body Monte Carlo with simulated annealing was used to generate an ensemble of feasible structures at each stage of assembly.

Using only protein–RNA contacts as restraints, we were able to fold the 30S subunit of the ribosome into a conformation that was very similar to the crystal structure. The major difference was in the orientation of the major structural domains relative to one another (Fig. 33.4d and e). This may have been corrected if tertiary RNA–RNA contacts or interactions between different proteins had been incorporated, but the individual domains match the crystal structures very well considering the lower resolution of the model (Fig. 33.4a–c). Therefore, in the absence of tertiary RNA or inter-protein contacts, protein–RNA interactions are able to organize the RNA helices in positions that are relatively close to the native structure observed in the X-ray crystal structure. A similar situation is observed using NMR refinement techniques. Close-range interactions provide many local restraints resulting in good prediction of the local structure. However, it is difficult to determine global structure using local restraints.

The ensemble of structures from the simulated RNA in the absence of proteins indicates the flexibility of the protein-binding sites on the RNA chain at the initial stage of ribosome assembly. Protein-binding sites were more ordered for proteins that bound early in the assembly pathway. As individual proteins were added, the constraints that were imposed changed the topology of the RNA such that the binding sites for later binding proteins became more ordered. In this way, early protein binding events shape the RNA to allow consecutive proteins to bind to their respective binding pocket on the RNA chain.

33.4
Conclusion

Reduced representation models provide a useful tool for modeling large RNA assemblies because they provide efficient methods for compiling experimental data into a coherent 3-D model. They can also suggest the functional relevance of dynamic processes that cannot be seen in the low energy structures determined by X-ray crystallography and NMR. The highest resolution in our examples of reduced models use one pseudo-atom to represent each nucleotide and amino acid. With the new version of *yammp* (*YUP*), the user can treat certain regions at the all-atom level while keeping other regions at a reduced level. The power of this utility will be to examine local, atomic interactions within a large assembly while still accounting for long-range effects.

The protocol described here is for large RNA and RNP assemblies. It has also been used to model smaller RNA structures associated with *trans*-translation [26] and HIV initiation [27]. Similar methods can be used to refine or simulate low-resolution models of any macromolecular system for which a body of tertiary data is available.

33.5
Troubleshooting

An essential routine in building reduced representation models of RNA is to measure the energy of the model with secondary structure restraints alone. As mentioned earlier, when tertiary interactions are implemented as restraints in a large system like the ribosome, the overall energy for the system will be greater than zero because of conflicts in low-resolution structural data. However, the energy should be zero when only secondary structure restraints are incorporated into the model (meaning all restraints are satisfied) because RNA helices, loops and bulges should convert from the secondary structure into a 3-D model. If the energy cannot reach zero, the model should be examined to determine where the conflict(s) occurs and then be modified appropriately. The refinement method may not find

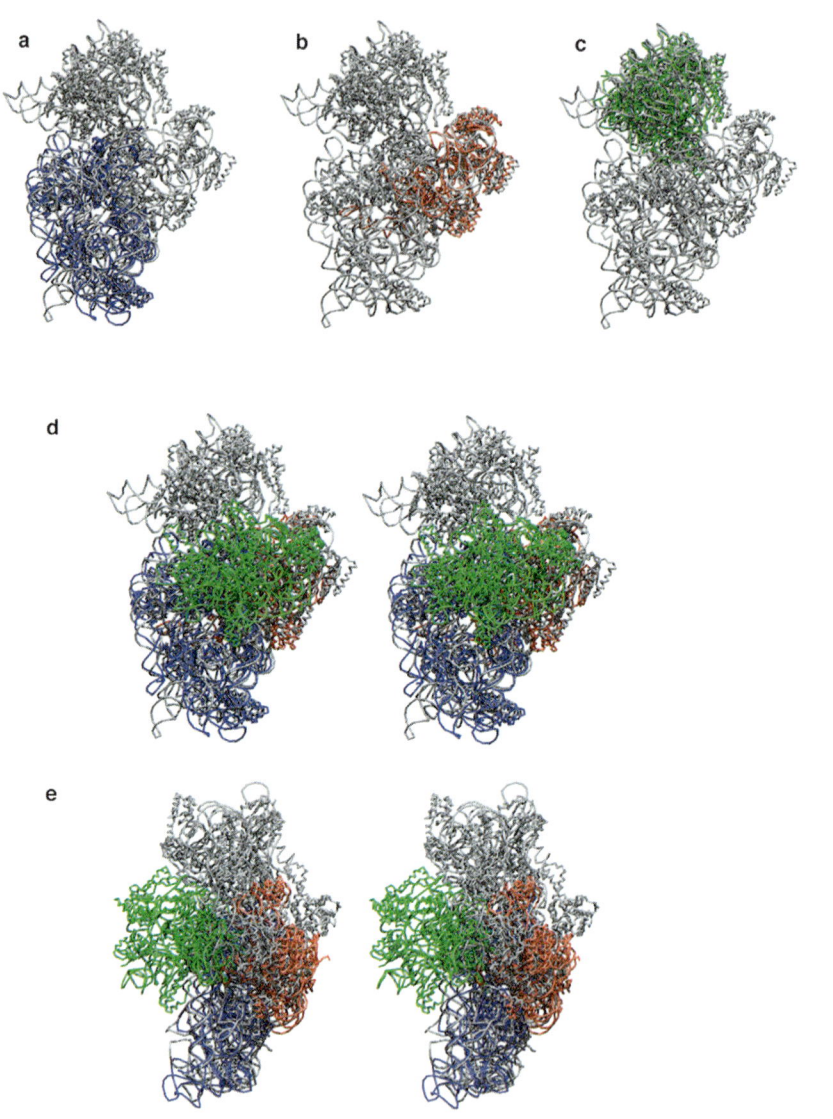

Fig. 33.4. Modeling the assembly of the 30S subunit of the ribosome. Using reduced representation models of the small subunit of the ribosome, we were able to refold the randomized 16S rRNA chain using only protein-RNA interactions. (a–c) The individual domains superpose very well onto the crystal structure, indicating that protein-RNA contacts are sufficient for the assembly of the separate domains of the 30S subunit [the body is blue (a), the platform is red (b) and the head is green (c)]. (d) However, the domain orientation is not the same as was found in the native X-ray crystal structure (gray). The view from the interface side of the subunit shows that the head (green) is rotated roughly 90° relative to the other domains. (e) The side view of the subunit provides a different view.

the lowest energy conformation(s) if (1) the restraints were not properly incorporated into the model, (2) there are errors in the experimentally determined RNA helices, loops and bulges or (3) the refinement does not search a large enough sample of conformational space. Condition (1) reflects an error on the part of the modeler. Condition (2) calls the experimental secondary structure model into question. The user must be certain that condition (3) is not the case before concluding that condition (1) or (2) applies. Once the secondary structure data is resolved in the model, tertiary constraints can be implemented to further refine the structure. This protocol then allows for critical evaluation of experimentally determined tertiary interactions.

References

1 KIM, S. H., et al., *Science* **1973**, *179*, 285–288.
2 CATE, J. H., et al., *Science* **1996**, *273*, 1678–1685.
3 BAN, N., et al., *Science* **2000**, *289*, 905–920.
4 WIMBERLY, B. T., et al., *Nature* **2000**, *407*, 327–339.
5 SCHLUENZEN, F., et al., *Cell* **2000**, *102*, 615–623.
6 AGRAWAL, R. K., et al., *J. Cell Biol.* **2000**, *150*, 447–460.
7 AGRAWAL, R. K., et al., *Science* **1996**, *271*, 1000–1002.
8 GAO, H., et al., *Cell* **2003**, *113*, 789–801.
9 TAN, R. K.-Z., S. C. HARVEY, *J. Comput. Chem.* **1993**, *14*, 455–470.
10 MALHOTRA, A., R. K.-Z. TAN, S. C. HARVEY, *Proc. Natl. Acad. Sci. USA* **1990**, *87*, 1950–1954.
11 MALHOTRA, A., S. C. HARVEY, *J. Mol. Biol.* **1994**, *240*, 308–340.
12 STAGG, S. M., J. A. MEARS, S. C. HARVEY, *J. Mol. Biol.* **2003**, *328*, 49–61.
13 WATSON, J. D., F. H. CRICK, *Nature* **1953**, *171*, 737–738.
14 OLSON, W. K., P. J. FLORY, *Biopolymers* **1972**, *11*, 1–23.
15 SCHELLMAN, J. A., *Biopolymers* **1974**, *13*, 217–226.
16 LEVITT, M., A. WARSHEL, *Nature* **1975**, *253*, 694–698.
17 LEVITT, M., *J. Mol. Biol.* **1976**, *104*, 59–107.
18 TAN, R. K.-Z., S. C. HARVEY, *J. Mol. Biol.* **1989**, *205*, 573–591.
19 WOESE, C. R., S. WINKER, R. R. GUTELL, *Proc. Natl. Acad. Sci. USA* **1990**, *87*, 8467–8471.
20 CAPEL, M. S., et al., *Science* **1987**, *238*, 1403–1406.
21 MALHOTRA, A., R. K.-Z. TAN, S. C. HARVEY, *Biophys J.* **1994**, *66*, 1777–1795.
22 GUTELL, R. R., et al., *Prog. Nucleic Acid Res. Mol. Biol.* **1985**, *32*, 155–216.
23 WOESE, C. R., et al., *Microbiol Rev.* **1983**, *47*, 621–669.
24 GUTELL, R. R., et al., *Nucleic Acids Res.* **1992**, *20*, 5785–5795.
25 HELD, W. A., et al., *J. Biol. Chem.* **1974**, *249*, 3103–3111.
26 STAGG, S. M., et al., *J. Mol. Biol.* **2001**, *309*, 727–735.
27 ELGAVISH, T., M. S. VANLOOCK, S. C. HARVEY, *J. Mol. Biol.* **1999**, *285*, 449–453.

34
Molecular Dynamics Simulations of RNA Systems

Pascal Auffinger and Andrea C. Vaiana

34.1
Introduction

As a result of important methodological advances, the number of molecular dynamics (MD) simulations related to RNA systems has become significant, and simulations on more than 30 different RNA, RNA–protein and RNA–ligand systems have been reported (Table 34.1). However, the simulation of the dynamics of such systems presents specific problems associated with the complex three-dimensional folds adopted by their highly charged polyanionic backbone and by the large variety of naturally modified nucleotides they comprise [1]. It is, thus, timely to present the current state of the art of MD simulations of RNA systems in a comprehensive and practically oriented manner. Several reviews on nucleic acid MD simulations [2–10] already address important methodological issues.

Here, we will focus on how to setup a MD simulation of an RNA system by using explicit representation of solvent and Ewald summation methods for the treatment of the long-range electrostatic interactions. Specific features related to MD simulations of RNA systems using implicit solvation models can be found in [5]. The simulation protocols that will be described in the following refer mainly to our own experience with the AMBER program with which we are most familiar and which is used in many laboratories working on nucleic acids. They should be easily transposable to other MD packages.

34.2
MD Methods

Common force field-based simulation methods make use of an empirical potential energy function and parameter set to describe the physical properties of the system to be studied. In MD simulations, the "possible" time evolution of the system starting from an initial set of coordinates and velocities is calculated by integrating Newton's equations of motion over a certain period of time [11]. Descriptions of the integration algorithms associated with such programs can be found in [12–

Handbook of RNA Biochemistry: Student Edition. Edited by R. K. Hartmann, A. Bindereif, A. Schön, E. Westhof
Copyright © 2009 WILEY-VCH Verlag GmbH & Co. KGaA, Weinheim
ISBN: 978-3-527-32534-4

Tab. 34.1. List of recent MD simulations of RNA systems (up to December 2003) using an explicit representation of the solvent and Ewald summation methods for the treatment of the long-range electrostatic interactions (simulations using truncation [107–110] or density functional methods [39] are not listed in the table).

Starting structures	nt	Length (ns)	Ions	Modified nt	References
Model					
PNA.RNA duplex	12	2.5	Na$^+$	PNA	111
HNA.RNA duplex	16	1.1	Na$^+$	HNA	112
Model helix r{(ApU)$_{12}$}$_2$	48	2.4	~0.2 M KCl		20
Model helix r{(CpG)$_{12}$}$_2$	48	2.4	~0.2 M KCl		19, 20, 22, 85
Model helix r{(CmpGm)$_{12}$}$_2$	48	4.4	~0.2 M KCl	Cm; Gm	22
r(CCAACGUUGG)$_2$	20	2.0	Na$^+$		21, 83
r(CCAACGUUGG)$_2$	20	1.3	Na$^+$	MOE	113
r(CGCGCG)$_2$	12	0.7	Na$^+$		23
r(GAGUACUC)$_2$	16	5.0	~0.3 and ~1.0 M NaCl		114
r(GCGAGUACUCGC)$_2$	24				
r(CGCGAUCGCG)$_2$	20				
r(CCUUUCGAAAGG)$_2$	24				
tetraloop hairpins	26	3.0	~0.1 M NaCl		26
	10	1.4	Na$^+$	I	115
U1A hairpin	21	5.0	NaCl		116
Human U4 snRNA	62	3.0	K$^+$		117
X-ray					
ApU and GpC steps (in crystal)	8	2.0	Na$^+$		44
r(CmGmCmGmCmGm)$_2$	12	0.7	Na$^+$	Cm; Gm	23
r(GGACUUCGGUCC)$_2$	24	4.0	~0.1 M NaCl		118
r(UAAGGAGGUGAU)$_2$	24	5.0	~0.3 and ~1.0 M NaCl		114
tRNA^{Asp}	76	0.5	NH$_4^+$	D; Ψ; m^{1}G; m^{5}C; m^{5}U	90, 119
tRNA^{Asp} anticodon hairpin	17	0.5	NH$_4^+$	Ψ; m^{1}G	38, 89, 91–94, 119
5S rRNA loop E	24	10.0	Na$^+$ & Mg^{2+}		50
	24	11.5	~0.2 and ~1.0 M KCl and Mg^{2+}		38, 48, 49
HIV kissing loop complexes	46	7.5	Na$^+$ and Mg^{2+}		52
16S rRNA	81	5.5	~0.1 M NaCl		120
Pseudoknot	26	5.0	Na$^+$	C$^+$	34
Hammerhead ribozyme	41	1.1	Na$^+$ and Mg^{2+}		121, 122
	41	0.8	~0.1 M NaCl and Mg^{2+}		51, 123
RNA–ligand (X-ray)					
16S rRNA site A/neomycin B	21	10.0	Na$^+$		124

Tab. 34.1 (continued)

Starting structures	nt	Length (ns)	Ions	Modified nt	References
RNA–protein (X-ray)					
U1A RNA–protein complex	21	1.0	Na$^+$		125
	21	1.8	~0.1 and ~1.0 M NaCl		126
	21	5.0	Na$^+$ and Cl$^-$		116
U2 snRNA–protein complex	23	2.2	K$^+$		127
NMR					
GCAA tetraloop hairpin	12	0.2	Na$^+$		128
UUCG tetraloop hairpin	12	2.0	Na$^+$		129
tRNA^{Ala} acceptor stem hairpin	22	2.5	NH$_4^+$		45
	22	2.0	Na$^+$	I; 2AA; IsoC; dU; Z; M; 7DAA	24, 25
HIV-1(lai) SL1 hairpin	23	10.0	Na$^+$		130, 131
HIV-1 TAR RNA hairpin	29	1.6	Na$^+$	2AP	132
H3 kissing loop complex	28	16.0	Na$^+$		52
TAR–aptamer complex	29	3.0	Na$^+$		133
RNA/ligand (NMR)					
FMN aptamer	35	1.7	Na$^+$		134
BIV tat–TAR complex	28	1.2	Na$^+$		135
RNA/protein (NMR)					
U1A RNA–protein complex	30	1.0	Na$^+$		125
dsRBD/dsRNA	30	2.0	Na$^+$ and Cl$^-$		136

15]. These are common to all current simulation packages and will not be further discussed here. Various methods having the goal of pushing back currently accessible timescale limits by using multiple time-step integrators are currently under development [12, 13, 16–18].

34.3
Simulation Setups

34.3.1
Choosing the Starting Structure

Starting structures are mainly derived from (1) model built systems, (2) low- to high-resolution X-ray data and (3) NMR data.

34.3.1.1 Model Built Structures
Such structures can be used to study precise issues related to (1) the conformation and solvation of simple RNA building blocks such as Watson–Crick base pairs and base pair steps [19, 20], (2) the differences between RNA and DNA structures of similar sequence [19–21] or (3) the effects associated with the introduction of natural [22, 23] and non-natural modified nucleotides [24, 25]. In order to generate larger systems, model built fragments can also be assembled with motifs derived from X-ray or NMR structures [26].

34.3.1.2 X-ray Structures
When derived from moderate- to high-resolution data, X-ray experiments certainly provide the most accurate source of starting structures since, besides precise coordinates for the solute, they also supply important information related to its hydration and ionic environment [27, 28]. The main sources for X-ray structures are the NDB (Nucleic acid Database; http://ndbserver.rutgers.edu/ [29]) and the PDB (Protein Data Bank; http://www.rcsb.org/pdb/ [30]).

34.3.1.3 NMR Structures
NMR structures are similarly deposited in the NDB and PDB. Due to the lower amount of detail contained in these structures compared to those derived from X-ray data, they are generally chosen only when no X-ray data for a given system are available. Furthermore, NMR cannot provide data related to the hydration and ionic environment of the solute.

34.3.2
Checking the Starting Structure

The imprecision associated with model built and NMR structures is usually well characterized. Whereas X-ray structures are generally very precise, these may nonetheless contain local errors which, if not noticed and corrected, may seriously impair the quality of the calculated trajectories which greatly depends on the correctness of the starting structure.

34.3.2.1 Conformational Checks
For proteins, the precise orientation of several residues (essentially His, Asn and Glu) cannot be correctly assessed by a single examination of the electron density maps obtained at resolutions above around 1.0 Å. Some groups have addressed this problem [31] and have proposed various automated solutions based on an examination of "contact maps" (see, e.g. MolProbity; http://kinemage.biochem.duke.edu/ [32]) for detecting incorrect or suspicious orientations of amino acid residues in existing experimental structures. Checking the orientation of such residues is therefore essential if one desires to perform accurate simulations of protein–RNA complexes. In nucleic acids, the orientation of the natural nucleotides can generally be determined without ambiguity even at low resolution. However, this is not

the case for the orientation of the pseudouridine or Ψ residue which is a naturally modified nucleotide [33] and possibly other side-chains in modified nucleotides.

34.3.2.2 Protonation Issues

It is now recognized that some nucleic acid residues such as adenines or cytosines can be found in their protonated state [34]. Yet, X-ray structures rarely provide direct evidence for such protonation states. These have to be derived from chemical intuition by inspecting the structural environment of the involved residues. In cases where it is impossible to decide which protonation state is the most probable, it is advisable to perform simulations of both the charged and uncharged states [34]. Evidently, protonation states of charged amino acids and of drugs such as aminoglycosides [35, 36] have also to be clearly defined before starting a simulation.

34.3.2.3 Solvent

The interpretation of solvent density maps derived from X-ray experiments is often difficult. A recent survey of the NDB has revealed that it is not unusual for water molecules to be assigned to sites with very large electron densities (above those expected for a water molecule); this may be indicative of the presence of a metal ion at those locations [37]. It has also recently been emphasized that some density spots resulting from the presence of anions (SO_4^{2-}, Cl^-, etc.) in the vicinity of nucleic acids have been wrongly assigned to divalent cations (Mg^{2+} or Mn^{2+}) [38]. Incorrectly assigned metal ions may represent an important source of errors in MD simulations.

34.3.3
Adding Hydrogen Atoms

Most of the structures deposited in the NDB or PDB are devoid of hydrogen atoms. Hence, before starting a simulation, it is essential to add them either by using the tools provided by the main MD packages or by using external programs such as MolProbity (http://kinemage.biochem.duke.edu/ [32]).

34.3.4
Choosing the Environment (Crystal, Liquid) and Ions

Two types of environment may be chosen: the crystal and the liquid phase. Simulations of the crystal phase are rare for nucleic acids [39–43] and only the r(CpG) and the r(ApU) dinucleotides have been simulated in this environment [39, 44] even though such simulations probably provide the most accurate confrontation between experimental and theoretical data [3]. Simulations of RNA systems in the aqueous phase are the most common. The ions used to neutralize the nucleic acid can be of various types (Na^+, K^+, NH_4^+, Mg^{2+}) and different ionic conditions ranging from a number of ions sufficient to neutralize the charge carried by the nucleotidic backbone to a number of ions that would appropriately represent a solution containing 0.1–1.0 M of added salt can be selected (Table 34.1). In this last case, co-ions (Cl^-, SO_4^{2-}, etc.) must be placed in the simulation box as well.

34.3.5
Setting the Box Size and Placing the Ions

34.3.5.1 **Box Size**
For simulations of the crystal phase, the box size is determined by the size of the crystal cell. In solution, the box size is usually allowed to extend by approximately 8–12 Å around the solute, thus ensuring three to four hydration layers in each direction.

34.3.5.2 **Monovalent Ions**
Monovalent ions are then placed in the simulation box by replacing the water molecules with the lowest (for cations) and highest (for anions) electrostatic potential [19, 45]. If one wants to avoid biases due to the initial placement of ions, it is also essential to set limits ensuring that no ions are closer than a certain distance, usually 5–6 Å, to any nucleic acid atom or to another ion. In this manner, no ions are initially found in the nucleic acid grooves, in direct proximity of the phosphate groups or forming contact ion pairs.

34.3.5.3 **Divalent Ions**
The limited length of the simulations does not allow reproduction of dehydration processes of Mg^{2+} ions and formation of direct $Mg^{2+}\cdots O$ contacts with the RNA which take microseconds to occur [46, 47]. Thus, divalent ions cannot be placed in the simulation box as described above. Such systems can only be realistically simulated if the Mg^{2+} ions occupy their crystallographic positions already at the beginning of the MD run [48–52]. It has to be noted that, although the Mg^{2+} ions occupy their crystallographic positions, the water molecules located in the first hydration shell of the ions find their equilibrium positions during the equilibration phase or, in other words, the crystallographic position of these water molecules is not taken into account.

34.3.6
Choosing the Program and Force Field

34.3.6.1 **Programs**
Two very widely used MD simulation packages are AMBER (Assisted Model Building with Energy Refinement; http://amber.scripps.edu/ [53]) and CHARMM (Chemistry at HARward Molecular Mechanics; http://yuri.harvard.edu/). Each of these is associated with its own force field. Other more recent and non-commercial MD packages such as GROMOS (http://www.igc.ethz.ch/gromos/ [54]) and NAMD (http://www.ks.uiuc.edu/Research/namd/ [55]) are also available. Package choice is a very subjective matter and is generally dictated by the lab history. However, one

should be attentive to the fact that some desired features may not be present in all of these packages.

34.3.6.2 Force Fields

AMBER [56] and CHARMM [57, 58] are the most commonly used force fields for simulating the dynamics of nucleic acids and proteins. They co-exist with the less frequently used BMS [59] and OPLS force fields [60]. However, it is essential to be aware of their limitations. Although they are all able to reproduce quite well most of the important structural and dynamical features of biological systems, they do differ in some specific aspects. Comparisons of the AMBER and CHARMM force fields performed on DNA duplexes [61, 62] have led to the conclusion that in some instances a "force field dependent polymorphism" [3] is observed. This polymorphism contributes to an artifactual drift of the duplexes toward an A- or a B-form that is independent of the starting conformation. Some of these problems have been corrected in recent versions of the AMBER and CHARMM force fields [63, 64]. Hence, since these force fields are constantly evolving, it is recommended to check for the latest available updates. Nevertheless, it can be concluded that up to now, none of the widely used force fields is able to reproduce fully all of the structural features experimentally observed for nucleic acids in solution [8, 64]. New, polarizable force fields [65] such as those recently implemented in AMBER are expected to bring significant improvement in the informative and predictive power of MD simulations (although at the expense of the use of more computer resources). These force fields are still at an initial stage of development and will have first to be thoroughly tested on small systems before using them for simulating solvated macromolecules.

34.3.6.3 Parameterization of Modified Nucleotides and Ligands

Most empirical force fields used in common MD packages are equipped with parameters for modeling all canonical nucleotides and amino acids, but not for modified nucleotides and ligands. For simulation of such systems, new parameter sets must be developed. Although some general sets of procedures for automated parameterization based on fitting to both experimental and quantum chemical data have been proposed, force field parameterization remains a matter for experts. A classical review on force field parameterization has been written by Dinur and Hagler [66]. Fortunately some tools and methods for generating reasonable to high-quality force field parameters are available [67, 68]. Among them it is worth mentioning the AMBER module Antechamber which greatly simplifies the generation of new parameters for the AMBER force field and the Automated Frequency Matching Methodology (AFMM) developed for the CHARMM force field but extendible to all other atom based force fields [69]. AFMM allows development of parameter sets for small to medium sized molecules using high quality quantum chemical calculations as reference data. A new set of scripts (RED) has also been made available to facilitate the derivation of partial charge sets from electrostatic potential calculations (http://www.u-picardie.fr/labo/lbpd/RED/). In addition to those for nucleic acids and amino acids, some force fields have been developed for specific classes of molecules, such as carbohydrates (OPLS-AA [70]; GLYCAM http://glycam.ccrc.uga.edu/), sulfates and sulfamates [71] or polyphosphates [72]. Parameters for the ions which are compatible with the chosen force field can be easily retrieved from the literature.

34.3.6.4 Water Models

Among the available water models, the TIP3P (Transferable Interatomic Potential – 3 Point [73]) model is generally associated with the AMBER force field. The more computationally "expensive" four (TIP4P [73]) and five (TIP5P [74]) point models have been developed and tested on various systems. SPC (Single Point Charge [75]) and SPC/E (SPC/Extended [76]) models are also very popular. For DNA, it has been shown that the use of TIP3P and SPC/E water models resulted in comparable hydration patterns although the water densities associated with the water model with the highest diffusion rate (TIP3P) appear to be more blurred than those calculated with the SPC/E model that has calculated diffusion rates in better agreement with experimental data [10]. Besides, a MD simulation of a protein in a crystal environment has led to the conclusion that results obtained with the SPC/E model are in much better agreement with neutron-scattering data than those collected with the TIP3P model [77]. Thus, a particular water model can significantly alter the results from MD simulations and more studies are needed to precisely evaluate their influence.

34.3.7
Treatment of Electrostatic Interactions

The ensemble of parameters in the above referenced force fields is usually considered to be the most important factor in determining the quality of a MD trajectory. Yet, if the empirical equation that models in a condensed way the intermolecular forces at play in biomolecular systems is not accurate enough or if one of its terms is not correctly evaluated then even the most precisely developed set of parameters cannot realistically account for the dynamics of these systems. Indeed, numerous studies have revealed that severe artifacts emerge when the long-range electrostatic contributions are neglected. In other words, when only a part of the Coulomb term $q_i q_j / 4\pi\varepsilon_0 r_{ij}$ is estimated [78–80]. It has been shown that an accurate treatment of the long-range electrostatic interactions is of paramount importance for generating realistic trajectories of nucleic acid systems. The particle mesh Ewald (PME) summation method [81, 82] has proven to lead to very stable nucleic acid trajectories [80, 83].

The program default parameters are generally adequate for using the PME method. These parameters are such that a cubic interpolation scheme and a 10^{-5} Å tolerance for the direct space sum cutoff is chosen. To speed up the fast Fourier transform in the calculation of the reciprocal sum, the size of the PME charge grid is chosen to be an integer power of 2, 3 and 5, and to be slightly larger than the size of the periodic box. This leads to a grid spacing of around 1 Å or less.

34.3.8
Other Simulation Parameters

34.3.8.1 Thermodynamic Ensemble

MD simulations can be conducted in a microcanonical ensemble or (N,V,E) for constant number of particles, volume and energy. Yet, the isothermal–isobaric or (N,P,T) ensemble for constant number of particles, pressure and temperature is more commonly chosen. Working at constant pressure is particularly important during the equilibration stage since during the building stage of the system, the water molecule positions are not optimized at the solute/solvent interface. This can lead to the generation of holes in he simulation box when constant volume options are selected.

34.3.8.2 Temperature and Pressure

In order to maintain a constant temperature and pressure, the Berendsen temperature coupling scheme and an isotropic molecule based constant pressure scaling with a time constant of 0.2–0.5 ps for both are often used [84], although other coupling schemes and time constants are found in the literature.

The target temperature and pressure values are usually set to 1 atm and 298 K or 25 °C, also called "room temperature". At this point, it is interesting to note that the choice of this target temperature is dictated by early studies which used experimental data obtained at room temperature in order to parameterize the force fields. This is close to the temperature at which many *in vitro* experiments are performed in a laboratory. Nonetheless, simulations using different target temperatures (e.g. 5, 25 and 37 °C) have been used to reveal some differences in the hydration of RNA systems [85].

34.3.8.3 Shake, Time Steps and Update of the Non-bonded Pair List

A tolerance of 0.0005 Å is generally used for the SHAKE algorithm [86] that allows to "safely" utilize a 2-ps time step (instead of 1 ps) by artificially freezing the most rapid vibration motions (C–H, O–H, N–H, etc., elongations) which can be observed in biomolecular systems. A strong tolerance value has been found to reduce the so-called "flying ice cube" phenomena described below [87, 88]. Another time saving device is associated with the use of a pair list for calculating non-bonded interactions between all atoms in the system. This pair list is not calculated at each time step, which would be computationally expensive, but only once every nth step (usually every 10 steps).

34.3.8.4 The Flying Ice Cube Problem

Since the non-bonded pair list is not updated every step (in order to save CPU resources) and constant pressure algorithms as well as uniform scaling of velocities by the Berendsen coupling scheme are used, some small energy drain during the simulation can occur and the center of mass velocity can slowly grow. Therefore, periodically in the simulation this center of mass velocity has to be removed otherwise the "flying ice cube" syndrome may appear [87, 88]. It is also advised to update the non-bonded pair list more frequently [87, 88].

34.3.9
Equilibration

The main aim of equilibration procedures is to alleviate tensions created in the system during the early building stages. These tensions would, if neglected, lead to unrealistic trajectories. In the following, a typical equilibration protocol used by us will be described.

First, 500 steps of steepest descent minimization are applied to the entire system with periodic boundary conditions in the (N,P,T) ensemble. This is followed by 25 ps of simulation in the same thermodynamic ensemble where only the water molecules and the hydrogen atoms of the solute are allowed to move (the heavy atoms of the solute and the ions are frozen by using the BELLY option in AMBER). During the next 25 ps, the constraints placed on the monovalent ions are released so that they can start to equilibrate around the RNA system while the BELLY option is still used to freeze the heavy atoms of the solute. Then, several rounds of 50-ps MD are performed during which positional constraints of 10, 5, 2, 1, 0.5, 0.1, 0.01 and 0.001 kcal/mol Å^2 are applied to the heavy atoms of the solute, yielding a partially constrained 450 ps MD trajectory. Then the production phase can take place. However, in order to allow a better sampling of the conformational space by the solute and the ions, the next 0.5–1.0 ns are generally included in the equilibration phase [19].

The length of the equilibration phase can be extended or shortened at will. However, it has to be noted that the phase during which constraints are applied ensures that the solvent is appropriately equilibrated in the vicinity of the solute, but also throughout the simulation box. The part during which no constraints are applied allows then for the solute to relax in the presence of relatively well-equilibrated solvent environment. Other equilibration protocols start at low temperature (50 K); then, the temperature is raised to the target temperature at discrete time intervals [51]. An apparent advantage of the former procedure is that the sampling of the conformational space by the solvent is conducted at the target temperature and, hence, probably more efficient.

34.3.10
Sampling

34.3.10.1 How Long Should a Simulation Be?
Ideally, MD simulations should be long enough to sample all the conformational transitions occurring in a particular biomolecular system. Given present and predictable computational means, this goal will certainly not be achieved in the next decade. It is, thus, impossible to address in the near future biochemical processes with very long relaxation times by using classical MD simulations with explicit solvent representations. Up to now, the longest MD simulations of RNA systems have barely overcome the 10-ns timescale (Table 34.1). Nevertheless, many processes associated with limited conformational perturbations lie within reach of present methods. Among them, those that involve the binding of the smallest ligands to nucleic acids (monovalent ions and water molecules [19, 20, 48]) or the orientation

of the 2′-OH hydroxyl groups [89] can trustfully be evaluated on the current nanosecond timescale.

34.3.10.2 When to Stop a Simulation

Another important and often overlooked issue is: *when to stop a MD simulation or when does it start to be a waste of time to continue sampling?* Indisputably, a very large number of approximations are included in MD simulations. These approximations are, among others, related to: (1) imprecision in the force field parameters, (2) incomplete evaluation of the intermolecular forces, (3) neglect of polarization and charge transfer effects and (4) an incomplete knowledge of the starting conditions. This last point by itself can easily lead to the generation of partially or totally unrealistic simulations. For example, an instability of the tertiary core structure of the yeast tRNA^{Asp} molecule characterized by a reordering of several base triples has been observed during a 500-ps MD simulation [90]. This has first been attributed to the absence of Mg^{2+} ions in the model. Later on, it was proposed that an adenine involved in the above-mentioned base triples is protonated [34]. Probably both effects led to the calculated transitions and further sampling would, in this case, not have contributed to a better understanding of the structure and the dynamics of this molecule. Hence, there is a certain risk for long MD simulations to oversample regions of the "configurational space" attached to the current force field and MD package that are not overlapping with the "true" configurational space explored by the investigated system.

34.3.10.3 Multiple MD (MMD) Simulations

An alternative to long MD simulations is to generate an ensemble of several "shorter" trajectories by using the MMD simulation technique. This method introduces slight perturbations (such as different initial velocities) in the starting conditions and, thus, exploits the chaotic nature of MD simulations in order to generate several uncorrelated trajectories [91–97]. It has been stated that several trajectories generated from similar but slightly different starting conditions may provide, on a statistical basis, more information than a single long trajectory. A parallel view developed by other authors consists of generating and comparing several trajectories obtained by starting from different initial configurations [96]. In short, MMD methods allow a statistical evaluation of MD simulation that cannot be achieved from the analysis of single trajectories [80].

34.4 Analysis

34.4.1 Evaluating the Quality of the Trajectories

Before trying to extract information from MD simulations, it is imperative to evaluate the quality of the generated trajectories by checking their internal consistency and confronting them to all available experimental data [98].

34.4.1.1 Consistency Checks

Internal consistency checks are related to the detection of unphysical behaviors such as those resulting from the use of truncation methods for the evaluation of the electrostatic interactions [78, 79] or "flying ice cube" syndromes. When such issues come to the foreground, it is clear that the simulation can and should no longer be used to derive biologically relevant information but that the focus of the study must shift toward the correction of the detected artifacts. By doing so, the importance of the neglected contributions is generally brought to light. For example, the stabilizing role of "hydration forces" could be assessed by comparing trajectories issued from simulations including or discarding long-range electrostatic interactions [3, 80, 92].

34.4.1.2 Comparison with Experimental Data

Confrontation with available experimental data (that are most frequently of a structural type) is in all cases mandatory. For instance, if important tertiary interactions (Watson–Crick base pairing, etc.) present in the starting structure break during the simulation over short timescales it is probable that the simulation protocols are not adequate and the reasons underlying such behavior must be uncovered [3, 79]. Early simulations performed in *"in vacuo"* conditions or with truncation of the electrostatic interactions where disruptions of important tertiary interactions were observed clearly illustrate this point [79, 99]. Moreover, if the structure of the investigated system diverges significantly from the initial structure, in part or entirely, during the time course of the simulation, it must be considered that this observation may result from force field inaccuracies. Only comparisons with experimental data on the same or related systems can help to resolve such issues.

34.4.1.3 Visualization

Visual checks are also mandatory and can be performed by using the programs that are delivered with each main MD package or by using programs such as VMD (Visual Molecular Dynamics; http://www.ks.uiuc.edu/Research/vmd/) that can read most of the available MD outputs. In rare cases, the Open BABEL utility (http://openbabel.sourceforge.net/) is needed in order to convert data written in different formats.

34.4.2
Convergence Issues

Afterwards, the convergence of the simulation with respect of the investigated properties has to be assessed. For this, it is necessary for the simulation length to exceed the average relaxation time of the investigated property which is not necessarily the longest relaxation time of the system that can largely exceed the currently accessible nanosecond timescales [98]. For example, it is possible to estimate on a statistical basis the residence times of water molecules (between 10 and 500 ps) around a regular RNA helix from nanosecond MD simulations, although the statistics will obviously be less good for the water molecules with the longest residence times [19, 20]. For less regular RNA systems, water molecules may be trapped into specific pockets for much longer times and only lower bounds for the residence times can be obtained [100]. A similar problem is associated with the estimation of the binding properties of monovalent and divalent ions. While it is feasible to estimate the exchange time of monovalent ions located in RNA grooves from nanosecond MD simulations [49], it is impossible on such timescales to estimate the same values for divalent ions [48].

34.4.3
Conformational Parameters

Most of the analysis programs delivered with major MD packages (such as CARNAL for AMBER) can monitor nearly all of the main conformational parameters such as interatomic distances, angles and dihedral angles, and also parameters that are specific to nucleic acids such as sugar puckers or user defined variables. Other programs such as CURVES (http://www.ibpc.fr/UPR9080/Curindex.html) [101] or 3DNA (http://rutchem.rutgers.edu/%7Exiangjun/3DNA/) [102] that allow to estimate parameters specific to nucleic acids such as bending, shift, slide, rise, tilt, roll, twist, etc., can be also very useful in the analysis process. The majority of these parameters are now attached to the experimental structures deposited at the NDB.

34.4.4
Solvent Analysis

Solvent analysis is an important aspect of the evaluation of MD trajectories. It is related to the characterization of water and ion binding sites and to the estimation of their residence times. Several reviews have already addressed these issues for nucleic acid systems [100, 103–106] and we will not discuss them in further details.

34.5
Perspectives

The current simulation protocols are now able to deliver stable trajectories of RNA systems ranging from small duplexes to large RNA–protein complexes over nanosecond timescales. Important conformational transition and solvation processes can be, thus, studied in detail. The current trend is to head toward longer timescales. This would allow us to address, among others, folding and unfolding issues. Nevertheless, one should be aware that by extending the currently accessible timescales, new artifacts that remain up to now hidden will appear and necessitate the development of more refined force fields and simulation methods that will most evidently incorporate polarizability and charge transfer effects. The detection and correction of artifacts that will manifest themselves on the longer timescales must be regarded as an important duty of modelers.

Acknowledgments

A. C. V. acknowledges support from the European contract HPRN-CT2002-00190 (CARBONA).

References

1 H. Grosjean, R. Benne, *Modification and Editing of RNA*, ASM Press, Washington, DC, **1998**.
2 P. Auffinger, E. Westhof, in: *Encyclopedia of Computational Chemistry*, P. v. R. Schleyer, N. L. Allinger, T. Clark, J. Gasteiger, P. A. Kollman, H. F. Schaefer III, P. R. Schreiner (eds), Wiley, Chichester, **1998**, pp. 1629–1640.
3 P. Auffinger, E. Westhof, *Curr. Opin. Struct. Biol.* **1998**, *8*, 227–236.
4 P. Auffinger, E. Westhof, in: *Perspective in Structural Biology*, M. Vijayan, N. Yathindra, A. S. Kolaskar (eds), University Press, Hyderabad, **1999**, pp. 545–555.
5 M. Zacharias, *Curr. Opin. Struct. Biol.* **2000**, *10*, 311–317.
6 J. Norberg, L. Nilsson, *Acc. Chem. Res.* **2002**, *35*, 465–472.
7 E. Giudice, R. Lavery, *Acc. Chem. Res.* **2002**, *35*, 350–357.
8 T. E. Cheatham, 3rd, M. A. Young, *Biopolymers* **2000**, *56*, 232–256.
9 T. E. Cheatham, P. A. Kollman, *Annu. Rev. Phys. Chem.* **2000**, *51*, 435–471.
10 T. E. Cheatham, P. A. Kollman. *Structure, Motion, Interaction and Expression of Biological Macromolecules: Proceedings of the Tenth Conversation*, R. H. Sarma, M. H. Sarma (eds), Adenine Press, New York, **1998**, pp. 99–116.
11 W. F. van Gunsteren, H. J. C. Berendsen, *Angew. Chem. Int. Ed.* **1990**, *29*, 992–1023.
12 M. P. Allen, D. J. Tildesley, *Computer Simulation of Liquids*, Clarendon Press, Oxford, **1987**.
13 D. C. Rapaport, *The Art of Molecular Dynamics Simulation*, Cambridge University Press, Cambridge, **1995**.
14 J. A. McCammon, S. C. Harvey, in: *Dynamics of Proteins and Nucleic Acids*, J. A. McCammon, S. C. Harvey (eds), Cambridge University Press, Cambridge, **1987**.
15 A. R. Leach, *Molecular Modelling: Principles and Applications*, Pearson Education, Harlow, **2001**.
16 T. Schlick, E. Barth, M. Mandziuk, *Annu. Rev. Biophys. Biomol. Struct.* **1997**, *26*, 181–222.
17 P. F. Batcho, D. A. Case, T. Schlick, *J. Chem. Phys.* **2001**, *115*, 4003–4018.
18 T. Schlick, *Structure* **2001**, *9*, R45–R53.
19 P. Auffinger, E. Westhof, *J. Mol. Biol.* **2000**, *300*, 1113–1131.
20 P. Auffinger, E. Westhof, *J. Mol. Biol.* **2001**, *305*, 1057–1072.
21 T. E. Cheatham, P. A. Kollman, *J. Am. Chem. Soc.* **1997**, *119*, 4805–4825.
22 P. Auffinger, E. Westhof, *Angew. Chem. Int Ed.* **2001**, *40*, 4648–4650.
23 K. Kulinska, T. Kulinski, A. Lyubartsev, A. Laaksonen, R. W. Adamiak, *Comp. Chem.* **2000**, *24*, 451–457.
24 M. C. Nagan, P. Beuning, K. Musier-Forsyth, C. J. Cramer, *Nucleic Acids Res.* **2000**, *28*, 2527–2534.
25 M. C. Nagan, S. S. Kerimo, K. Musierforsyth, C. J. Cramer, *J. Am. Chem. Soc.* **1999**, *121*, 7310–7317.
26 C. Schneider, J. Suhnel, *J. Biomol. Struct. Dyn.* **2000**, *18*, 345–352.
27 M. Sundaralingam, B. Pan, *Biophys. Chem.* **2002**, *95*, 273–282.
28 M. Egli, *Chem. Biol.* **2002**, *9*, 277–286.
29 H. M. Berman, J. Westbrook, Z. Feng, L. Iype, B. Schneider, C. Zardecki, *Acta Crystallogr. D* **2002**, *58*, 889–898.
30 H. M. Berman, T. Battistuz, T. N. Bhat, W. F. Bluhm, P. E. Bourne,

K. Burkhardt, Z. Feng, G. L. Gilliland, L. Iype, S. Jain et al., *Acta Crystallogr. D* **2002**, *58*, 899–907.

31 K. McDonald, J. M. Thornton, *Prot. Eng.* **1995**, *8*, 217–224.

32 S. C. Lovell, I. W. Davis, W. B. Arendall, 3rd, P. I. de Bakker, J. M. Word, M. G. Prisant, J. S. Richardson, D. C. Richardson, *Proteins* **2003**, *50*, 437–450.

33 P. Auffinger, E. Westhof, in: *Modification and Editing of RNA*, H. Grosjean, R. Benne (eds), American Society for Microbiology, Washington, DC, **1998**, pp. 103–112.

34 Csaszar, N. Spackova, R. Stefl, J. Sponer, N. B. Leontis, *J. Mol. Biol.* **2001**, *313*, 1073–1091.

35 M. Kaul, C. M. Barbieri, J. E. Kerrigan, D. S. Pilch, *J. Mol. Biol.* **2003**, *326*, 1373–1387.

36 D. S. Pilch, M. Kaul, C. M. Barbieri, J. E. Kerrigan, *Biopolymers* **2003**, *70*, 58–79.

37 U. Das, S. Chen, M. Fuxreiter, A. A. Vaguine, J. Richelle, H. M. Berman, S. J. Wodak, *Acta Crystallogr. D* **2001**, *57*, 813–828.

38 P. Auffinger, L. Bielecki, E. Westhof, *Structure* **2004**, *12*, 379–388.

39 Hutter, P. Carloni, M. Parrinello, *J. Am. Chem. Soc.* **1996**, *118*, 8710–8712.

40 D. R. Bevan, L. Li, L. G. Pedersen, T. A. Darden, *Biophys. J.* **2000**, *78*, 668–682.

41 H. Lee, T. A. Darden, L. G. Pedersen, *J. Chem. Phys.* **1995**, *102*, 3830–3834.

42 D. M. York, W. Yang, H. Lee, T. Darden, L. G. Pedersen, *J. Am. Chem. Soc.* **1995**, *117*, 5001–5002.

43 N. Korolev, A. P. Lyubartsev, A. Laaksonen, L. Nordenskiold, *Biophys. J.* **2002**, *82*, 2860–2875.

44 H. Lee, T. Darden, L. Pedersen, *Chem. Phys. Lett.* **1995**, *243*, 229–235.

45 Nina, T. Simonson, *J. Phys. Chem. B* **2002**, *106*, 3696–3705.

46 H. Ohtaki, *Monatsh. Chem.* **2001**, *132*, 1237–1268.

47 H. Ohtaki, T. Radnai, *Chem. Rev.* **1993**, *93*, 1157–1204.

48 P. Auffinger, L. Bielecki, E. Westhof, *J. Mol. Biol.* **2004**, *335*, 555–571.

49 P. Auffinger, L. Bielecki, E. Westhof, *Chem. Biol.* **2003**, *10*, 551–561.

50 K. Reblova, N. Spackova, R. Stefl, K. Csaszar, J. Koca, N. B. Leontis, J. Sponer, *Biophys. J.* **2003**, *84*, 3564–3582.

51 T. Hermann, P. Auffinger, E. Westhof, *Eur. J. Biophys.* **1998**, *27*, 153–165.

52 K. Reblova, N. Spackova, J. E. Sponer, J. Koca, J. Sponer, *Nucleic Acids Res* **2003**, *31*, 6942–6952.

53 D. A. Pearlman, D. A. Case, J. W. Caldwell, W. S. Ross, T. E. Cheatham, S. DeBolt, D. Ferguson, G. Seibel, P. Kollman, *Comp. Phys. Commun.* **1995**, *91*, 1–41.

54 W. R. P. Scott, P. H. Hünenberger, I. G. Tironi, A. E. Mark, S. R. Billeter, J. Fennen, A. E. Torda, T. Huber, P. Krüger, W. F. van Gunsteren, *J. Phys. Chem. A* **1999**, *103*, 3596–3607.

55 Kale, R. Skeel, M. Bhandarkar, R. Brunner, A. Gursoy, N. Krawetz, J. Phillips, A. Shinozaki, K. Varadarajan, K. Schulten, *J. Comput. Phys.* **1999**, *151*, 283–312.

56 W. D. Cornell, P. Cieplak, C. I. Bayly, I. R. Gould, K. M. Merz, D. M. Ferguson, D. C. Spellmeyer, T. Fox, J. W. Caldwel, P. A. Kollman, *J. Am. Chem. Soc.* **1995**, *117*, 5179–5197.

57 Foloppe, A. D. MacKerell, *J. Comput. Chem.* **2000**, *21*, 88–104.

58 A. D. MacKerell, N. Banavali, *J. Comput. Chem.* **2000**, *21*, 105–120.

59 D. R. Langley, *J. Biomol. Struct. Dyn.* **1998**, *16*, 487–509.

60 G. A. Kaminski, R. A. Friesner, J. Tirado-Rives, W. L. Jorgensen, *J. Phys. Chem. B* **2001**, *105*, 6474–6487.

61 M. Feig, B. M. Pettitt, *J. Chem. Phys. B* **1997**, *101*, 7361–7363.

62 M. Feig, B. M. Pettitt, *Biophys. J.* **1998**, *75*, 134–149.

63 T. E. Cheatham, P. Cieplak, P. A. Kollman, *J. Biomol. Struct. Dyn.* **1999**, *16*, 845–862.

64 S. Y. Reddy, F. Leclerc, M. Karplus, *Biophys. J.* **2003**, *84*, 1421–1449.

65 T. A. Halgren, W. Damm, *Curr. Opin. Struct. Biol.* **2001**, *11*, 236–242.
66 U. Dinur, A. T. Hagler, in: *Reviews in Computational Chemistry*, K. B. Lipkowitz, D. B. Boyd (eds), VCH, New York, **1991**, pp. 99–164.
67 J. Wang, P. Kollman, *J. Comput. Chem.* **2001**, *22*, 1219–1228.
68 S. L. Njo, W. F. van Gunsteren, F. Muller-Plathe, *J. Chem. Phys.* **1995**, *102*, 6199–6207.
69 A. C. Vaiana, A. Schulz, J. Wolfrum, M. Sauer, J. C. Smith, *J. Comput. Chem.* **2003**, *24*, 632–639.
70 D. Kony, W. Damm, S. Stoll, W. F. Van Gunsteren, *J. Comput. Chem.* **2002**, *23*, 1416–1429.
71 C. J. M. Huige, C. Altona, *J. Comput. Chem.* **1995**, *16*, 56–79.
72 K. L. Meagher, L. T. Redman, H. A. Carlson, *J. Comput. Chem.* **2003**, *24*, 1016–1025.
73 W. L. Jorgensen, J. Chandrasekhar, J. D. Madura, R. W. Impey, M. L. Klein, *J. Chem. Phys.* **1983**, *79*, 926–936.
74 M. W. Mahoney, W. L. Jorgensen, *J. Chem. Phys.* **2000**, *112*, 8910–8922.
75 H. J. C. Berendsen, J. P. M. Postma, W. F. van Gunsteren, J. Hermans, in: *Intermolecular Forces*, B. Pullman (ed.), Reidel, Dordrecht, **1981**, pp. 331–342.
76 H. J. C. Berendsen, J. R. Grigera, T. P. Straatsma, *J. Phys. Chem.* **1987**, *97*, 6269–6271.
77 M. Tarek, D. J. Tobias, *Biophys. J.* **2000**, *79*, 3244–3257.
78 Auffinger, D. L. Beveridge, *Chem. Phys. Lett.* **1995**, *234*, 413–415.
79 G. Ravishanker, P. Auffinger, D. R. Langley, B. Jayaram, M. A. Young, D. L. Beveridge, in: *Reviews in Computational Chemistry*, K. B. Lipkowitz, D. B. Boyd (eds), VCH, New York, **1997**, pp. 317–372.
80 S. Louise-May, P. Auffinger, E. Westhof, *Curr. Opin. Struct. Biol.* **1996**, *6*, 289–298.
81 C. Sagui, T. A. Darden, *Annu. Rev. Biophys. Struct.* **1999**, *28*, 155–179.
82 T. Darden, L. Perera, P. Li, L. Pedersen, *Structure* **1999**, *7*, R55–R60.
83 T. E. Cheatham, J. L. Miller, T. Fox, T. A. Darden, P. A. Kollman, *J. Am. Chem. Soc.* **1995**, *117*, 4193–4194.
84 H. J. C. Berendsen, J. P. M. Postma, W. F. van Gunsteren, A. DiNola, *J. Chem. Phys.* **1984**, *81*, 3684–3690.
85 P. Auffinger, E. Westhof, *Biophys. Chem.* **2002**, *95*, 203–210.
86 J. P. Ryckaert, G. Ciccotti, H. J. C. Berendsen, *J. Comput. Phys.* **1977**, *23*, 327–336.
87 S. C. Harvey, R. K. Z. Tan, T. E. Cheatham, *J. Comput. Chem.* **1998**, *19*, 726–740.
88 S. W. Chiu, M. Clark, S. Subramaniam, E. Jakobsson, *J. Computat. Chem.* **2000**, *21*, 121–131.
89 P. Auffinger, E. Westhof, *J. Mol. Biol.* **1997**, *274*, 54–63.
90 P. Auffinger, S. Louise-May, E. Westhof, *Biophys. J.* **1999**, *76*, 50–64.
91 P. Auffinger, S. Louise-May, E. Westhof, *J. Am. Chem. Soc.* **1995**, *117*, 6720–6726.
92 P. Auffinger, S. Louise-May, E. Westhof, *J. Am. Chem. Soc.* **1996**, *118*, 1181–1189.
93 P. Auffinger, E. Westhof, *Biophys. J.* **1996**, *71*, 940–954.
94 P. Auffinger, E. Westhof, *J. Mol. Biol.* **1997**, *269*, 326–341.
95 M. Braxenthaler, R. Unger, D. Auerbach, J. A. Given, J. Moult, *Proteins* **1997**, *29*, 417–425.
96 L. S. D. Caves, J. D. Evanseck, M. Karplus, *Prot. Sci.* **1998**, *7*, 649–666.
97 H.-B. Zhou, L. Wang, *J. Phys. Chem.* **1996**, *100*, 8101–8105.
98 W. F. van Gunsteren, A. E. Mark, *J. Chem. Phys.* **1998**, *108*, 6109–6116.
99 Swaminathan, G. Ravishanker, D. L. Beveridge, *J. Am. Chem. Soc.* **1991**, *111*, 5027–5040.
100 P. Auffinger, E. Westhof, *Biopolymers* **2000**, *56*, 266–274.
101 Lavery, H. Sklenar, *J. Biomol. Struct. Dyn.* **1989**, *6*, 655–667.
102 X. J. Lu, W. K. Olson, *Nucleic Acids Res.* **2003**, *31*, 5108–5121.
103 P. Auffinger, E. Westhof, in: *Water Management in the Design and Distribution of Quality Food*, Y. H. Ross, R. B. Leslie, P. J. Lillford (eds), Technomic, Basel, **1999**, pp. 165–198.

104 M. Feig, B. M. Pettitt, *Biopolymers* **1998**, *48*, 199–209.
105 V. A. Makarov, B. M. Pettitt, M. Feig, *Acc. Chem. Res.* **2002**, *35*, 376–384.
106 P. Auffinger, B. Masquida, E. Westhof, in: *Computational Methods for Macromolecules: Challenges and Applications*, T. Schlick, H. H. Gan (eds), Springer Verlag, Heidelberg, **2002**, pp. 61–70.
107 Y. Tang, L. Nilsson, *Biophys. J.* **1999**, *77*, 1284–1305.
108 A. Lahiri, L. Nilsson, *Biophys. J.* **2000**, *79*, 2276–2289.
109 J. Sarzynska, T. Kulinski, L. Nilsson, *Biophys. J.* **2000**, *79*, 1213–1227.
110 J. Norberg, L. Nilsson, *J. Chem. Phys.* **1996**, *104*, 6052–6057.
111 R. Soliva, E. Sherer, F. J. Luque, C. A. Laughton, M. Orozco, *J. Am. Chem. Soc.* **2000**, *122*, 5997–6008.
112 H. De Winter, E. Lescrinier, A. Van Aerschot, P. Herdewijn, *J. Am. Chem. Soc.* **1998**, *120*, 5381–5394.
113 K. E. Lind, V. Mohan, M. Manoharan, D. M. Ferguson, *Nucleic Acids Res.* **1998**, *26*, 3694–3799.
114 Y. Pan, A. D. MacKerell, *Nucleic Acids Res.* **2003**, *31*, 7131–7140.
115 J. Sarzynska, L. Nilsson, T. Kulinski, *Biophys J* **2003**, *85*, 3445–3459.
116 F. Pitici, D. L. Beveridge, A. M. Baranger, *Biopolymers* **2002**, *65*, 424–435.
117 J. Guo, I. Daizadeh, W. H. Gmeiner, *J. Biomol. Struct. Dyn.* **2000**, *18*, 335–344.
118 C. Schneider, M. Brandl, J. Sühnel, *J. Mol. Biol.* **2001**, *305*, 659–667.
119 P. Auffinger, S. Louise-May, E. Westhof, *Faraday Discuss.* **1996**, *103*, 151–174.
120 W. Li, B. Ma, B. A. Shapiro, *Nucleic Acids Res.* **2003**, *31*, 629–638.
121 R. A. Torres, T. C. Bruice, *Proc. Natl. Acad. Sci. USA* **1998**, *95*, 11077–11082.
122 R. A. Torres, T. C. Bruice, *J. Am Chem. Soc.* **2000**, *122*, 781–791.
123 Hermann, P. Auffinger, W. G. Scott, E. Westhof, *Nucleic Acids Res.* **1997**, *25*, 3421–3427.
124 J. L. Asensio, A. Hidalgo, I. Cuesta, C. Gonzalez, J. Canada, C. Vicent, J. L. Chiara, G. Cuevas, J. Jimenez-Barbero, *Chemistry* **2002**, *8*, 5228–5240.
125 C. M. Reyes, P. A. Kollman, *RNA* **1999**, *5*, 235–244.
126 Hermann, E. Westhof, *Nat. Struct. Biol.* **1999**, *6*, 540–544.
127 J. Guo, W. H. Gmeiner, *Biophys. J.* **2001**, *81*, 630–642.
128 D. A. Zichi, *J. Am. Chem. Soc.* **1995**, *117*, 2957–2969.
129 J. Miller, P. A. Kollman, *J. Mol. Biol.* **1997**, *270*, 436–450.
130 F. Kieken, E. Arnoult, F. Barbault, F. Paquet, T. Huynh-Dinh, J. Paoletti, D. Genest, G. Lancelot, *Eur. Biophys. J.* **2002**, *31*, 521–531.
131 G. La Penna, D. Genest, A. Perico, *Biopolymers* **2003**, *69*, 1–14.
132 T. Kulinski, M. Olejniczak, H. Huthoff, L. Bielecki, K. Pachulska-Wieczorek, A. T. Das, B. Berkhout, R. W. Adamiak, *J. Biol. Chem.* **2003**, *278*, 38892–38901.
133 F. Beaurain, C. Di Primo, J. J. Toulme, M. Laguerre, *Nucleic Acids Res.* **2003**, *31*, 4275–4284.
134 C. Schneider, J. Sühnel, *Biopolymers* **1999**, *50*, 287–302.
135 C. M. Reyes, R. Nifosi, A. D. Frankel, P. A. Kollman, *Biophys. J.* **2001**, *80*, 2833–2842.
136 T. Castrignano, G. Chillemi, G. Varani, A. Desideri, *Biophys. J.* **2002**, *83*, 3542–3552.

35
Seeking RNA Motifs in Genomic Sequences

Matthieu Legendre and Daniel Gautheret

35.1
Introduction

RNA molecules display a considerable functional diversity, ranging from genetic data storage to gene regulation, sensing, transport or targeting. The last few years have seen a continuous stream of novel RNA genes being reported, the most striking of which were microRNAs, involved in the regulation of essential events such as cell differentiation and cell death. These discoveries have enticed biologists into the systematic scrutiny of genomic and expressed sequences for yet other functional, non-coding RNAs (ncRNA), either in the form of independent genes or of structural motifs in the untranslated regions of transcripts. Part of this effort is carried out experimentally, using "RNomics" [1], a general term for amplification techniques specifically targeted at short, non-polyadenylated transcripts. However, bioinformatics now emerges as an inexpensive, yet efficient alternative to experiment. Indeed, with the growth of sequence databases and the development of specific algorithms for RNA detection, bioinformatics has become a practical option to consider in most RNA search situations. In this chapter, we will present some of the computational tools available for the identification of RNA genes or motifs in sequence databases. Our focus here is the detection of *known* genes or motifs (i.e. for which some structural or sequence information is already available). We will not cover the identification of unknown, potential ncRNA genes, which is best carried out using comparative genomics [2]. First, we will outline the difficulties associated to RNA search and explain what tools are best suited depending on the available sequence information. We will then show how to organize RNA sequence data, evaluate search results and, finally, use search software for actual genome scans.

We will use as an example a relatively simple motif: the Signal Recognition Particle (SRP) RNA. The SRP is an RNA–protein complex present in all organisms, involved in the targeting of proteins to the plasma membrane or endoplasmic reticulum. The RNA component of SRP is a molecule of 200–300 nt with a generally variable structure, except for highly conserved region of about 45 nt called domain IV [3], which is usually considered as a signature for the identification of SRP

Handbook of RNA Biochemistry: Student Edition. Edited by R. K. Hartmann, A. Bindereif, A. Schön, E. Westhof
Copyright © 2009 WILEY-VCH Verlag GmbH & Co. KGaA, Weinheim
ISBN: 978-3-527-32534-4

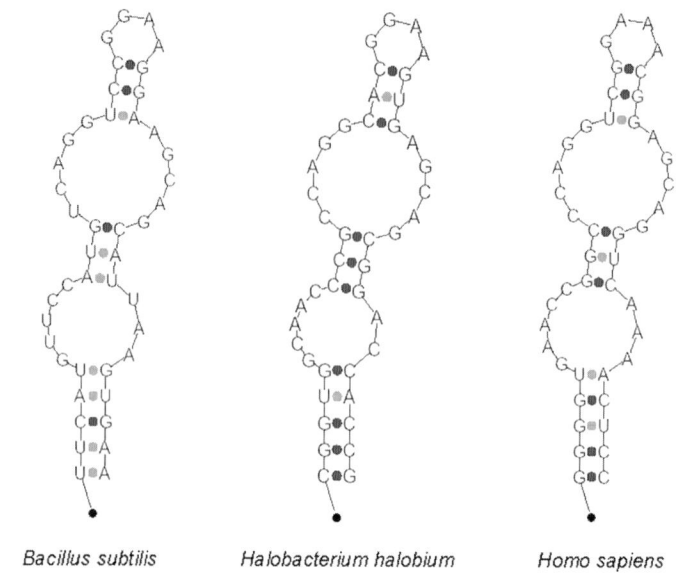

Bacillus subtilis *Halobacterium halobium* *Homo sapiens*

Fig. 35.1. The sequence and secondary structure of SRP RNA domain IV from representatives of the major phylogenetic domains: Bacteria, Archaea and Eukarya.

genes. This structure is shown in Fig. 35.1 for three representative species. Our goal will be to identify SRP genes in complete genomes, using two RNA search programs based on very different computational approaches – RNAMOTIF [4] and ERPIN [5].

35.2
Choosing the Right Search Software: Limitations and Caveats

Biologists studying RNA motifs sometimes wish they were working on protein motifs instead. Protein comparison benefits from powerful amino acid substitution models, such as the PAM and BLOSUM matrices that, used in conjunction with sequence alignment programs, allow for a reliable detection of protein signatures, even after millions of years of divergent evolution. RNA motifs cannot be detected in this way, on the sole basis of sequence comparison. This is due in part to the poor functional information carried by nucleotide bases compared to amino acid residues and also to the particular structure of RNA molecules, defined by distant interactions as well as by linear sequences. Therefore, usual sequence alignment tools cannot be applied to RNA detection, except in a few rare cases where RNA sequences are large and conserved enough, such as in ribosomal RNA.

All successful RNA detection programs developed so far involve, in one way or another, scanning target sequences for distant base-paired fragments. To achieve

Tab. 35.1. Pros and Cons of RNAMOTIF and other descriptor-based software.

Pros	Cons
Draft descriptors can be quickly sketched and tested	Requires a good prior knowledge of secondary structure and sequence constraints
No alignment is required, although it is very helpful to have one	Requires basic computer skills to translate biological constraints into the descriptor language
Biologists decide what features are important or not (see also Cons!)	Biologists have the responsibility of correctly weighting each important feature

this, algorithms need to scan the sequence back and forth, implying lots of extra computation compared to a linear sequence search. As a consequence, all RNA search programs are much slower and more complicated to handle than basic homology search programs.

Currently available methods for RNA motif identification can be classified into three categories. In the first category are programs especially developed for the detection of a specific RNA molecule, such as tRNA or C/D box snoRNA. Few such programs exist [6–8] and it is unlikely that one has already been developed for your favorite RNA motif. In the next category are descriptor-based programs, allowing biologists to describe virtually any type of RNA motif using a special descriptor language. Several languages and search engines have been developed over the years, including RNAMOT, PALINGOL, PATSEARCH, PATSCAN and RNAMOTIF [4, 9–12]. All provide convenient syntactic structures for describing the usual RNA building blocks, i.e. helices and single-stranded elements, with optional specification of strand lengths, conserved sequences, base-pair mismatches, etc. Generally, these programs do not compute statistics: they either reject or accept solutions based on compliance with motif constraints. The advantages and limitations of descriptor-based software are summarized in Table 35.1. We recommend using such software when structure/sequence information is scarce and yet some basic structural features are determined, e.g. when a new class of RNA is identified with only a couple of known members and a conserved secondary structure is predicted. Descriptor-based programs are also able to express complex logical constraints that no other program would capture, such as "when helix H is longer than x bp, then strand S does not exist".

The third category is that of training set-based programs. These programs use a collection of known RNA sequences as a training set, from which they extract a statistical model that is then used for database search. The major advantage of these approaches is that they infer a scoring system automatically from the training set, relieving biologists from the responsibility of weighting each important feature. The classical training set-based software for RNA motifs is COVE [13], which uses the model of Stochastic Context Free Grammars (SCFG). SCFGs can be viewed as a comprehensive statistical description of sequence and secondary structure constraints in the training set. These constraints are translated into "production rules"

Tab. 35.2. Pros and Cons of ERPIN.

Pros	Cons
All constraints in the training set are efficiently exploited, resulting in highly specific detections	Alignment and secondary structure constraints must be accurate
After alignments and secondary structures are created, no further programming is needed	Helices of variable length need to be reduced to their shortest consensus
Scoring system is defined automatically	Program will not depart from initial alignment in terms of motif size
E-values are provided for each hit	Users still have to decide on search order and masked elements

representing the successive structural elements with their associated probability of occurrence. A search algorithm is then used to discover sites in database sequences where these rules are satisfied. Computational demands are very high with this complete probabilistic approach and, as a result, combinations of programs are required to pre-filter databases and reduce run-time. In addition, current SCFGs do not support pseudo-knots, which may be a hindrance in some cases. These practical limitations have limited the adoption of SCFGs outside the community of bioinformatics experts.

Another training-set based software, ERPIN [5] uses weight matrices to represent RNA sequences and secondary structures. Weight matrices are simple statistical representations that basically translate a sequence alignment into a scoring system. ERPIN creates a weight matrix for each secondary structure element (helix or single strand) in the alignment. A dynamic programming algorithm then detects sequences matching these matrices above a certain score threshold. Since version 3.9 of the software, an E-value is calculated for each occurrence. The E-value tells us how many hits could be obtained by chance with a given score, in a given database. This is a very useful hint at the biological significance of a motif. The pros and cons of the ERPIN approach are shown in Table 35.2. We recommend using ERPIN when a significant (more than 10 sequences) and reliable alignment is available.

RNA motif identification is still a very active research field, with several valuable approaches presently in development. Our software selection is of course based on our own experience with tools we have either used routinely in our lab or contributed to develop. Whatever method is used, there are many pitfalls in the quest for RNA motifs. First, software will easily spin out of control (seemingly run forever) when large multi-helix motifs are entered without a careful search strategy being implemented. Typically, this means you have to identify the most specific fragments in the RNA motif and restrict search to these fragments, at least in a first approximation. Second, whatever scoring system is adopted, reaching a correct balance between specificity and sensitivity always remains under your responsibility. Undertaking an RNA motif search will thus require departing from the "push-

Tab. 35.3. Computer programs required in our examples, along with their download location and current version.

Program	Download location	Version used
ClustalW	ftp://ftp.ebi.ac.uk/pub/software/unix/clustalw/	1.83
RNAalifold	http://www.tbi.univie.ac.at/~ivo/RNA/ViennaRNA-1.5beta.tar.gz	1.5 beta
Pfold	http://www.daimi.au.dk/~compbio/rnafold/	N/A
Shuffle	ftp://ftp.genetics.wustl.edu/pub/eddy/software/squid.tar.gz	1.9 g
RNAMOTIF	ftp://ftp.scripps.edu/case/macke/	3.0.0
ERPIN	http://tagc.univ-mrs.fr/erpin/	3.9.9

button'' approach by a large extent. Although laborious, this work will get you to learn a lot about structure–function relationships in your RNA molecule and will eventually make this enterprise very rewarding.

35.3
Retrieving Programs and Sequence Databases

Table 35.3 presents the www or ftp sites for downloading the programs used in our protocols (ftp servers can be accessed using any www browser). Whenever program implementations for different Operating Systems (OS) were available, we chose the Linux version. Linux (or other flavors of Unix) is by far the most common OS for bioinformatics and we strongly recommend using it for RNA motif search. The program command lines are shown as they should be typed under Linux. All tests were performed on a 2.4-GHz Pentium computer with 1 Gb of memory.

Genome sequence databases were retrieved from the EBI ftp site ftp://ftp.ebi.ac.uk/pub/databases/genomes/. For RNA motif search, files must be in the FASTA format (usually with extension .fa, .fna or .fasta). Here we retrieved all archaeal and bacterial genomes available on the EBI server (17 archaeal genomes – 373 Mb and 128 bacterial genomes – 410 Mb) and one eukaryotic genome (*Drosophila melanogaster*, 117 Mb).

35.4
Organizing RNA Motif Information

Independently of the type of algorithm used for motif search, RNA sequence and structure information has to be organized in such way that key features, such as stem and loop lengths or conserved sequences, can be accurately defined. The best way to achieve this is to perform a structure-based alignment of the RNA sequences. Training set-based algorithms require these alignments as input, but descriptor-based approaches will also largely benefit from them. High-quality structure-based alignments are already available for many RNA genes and motifs.

Several of them are listed in the "RNA World" website (http://www.imb-jena.de/RNA.html). If you are lucky enough and find an expert-built alignment for your target RNA, just use it. Otherwise, you are poised to build you own, as explained below.

RNA sequence alignments should be structure based. This means the secondary structure, not the primary sequence, is the major constraint. Although they can be useful as a first approximation, regular sequence alignment programs cannot be relied on here. RNA sequence alignment is an iterative process in which the conserved secondary structure is progressively inferred by seeking covarying base pairs – distant positions where mutations occur in a concerted fashion, for instance G:C to C:G – and realigning the sequences to optimize both covariations and conserved sequences. This lengthy process, which may also involve using a phylogenetic tree, is out of the scope of this presentation (see the description by Gutell [14]). Several algorithms have been developed to perform simultaneous RNA alignment and secondary structure prediction. Unfortunately, they require a lot of computing power and patience, and many base pairs turn out incorrectly folded, especially when sequence sets contain outliers.

A more pragmatic approach to RNA alignment is to use a regular multiple alignment program such as CLUSTAL [15] in a first approximation, keeping in mind that these programs are solely based on sequence similarity and do not know about base-pairing, and, in a second step, infer secondary structure using a program that does not try to realign sequences, such as RNAalifold [16]. RNAalifold performs a thermodynamic folding prediction and then uses sequence covariation to predict a consensus secondary structure for the alignment. Other programs like Pfold [17] can do the same using SCFGs. Once this first cycle is completed, one may try to manually improve the alignment based on secondary structure.

As an expert alignment was available for SRP RNA, we could use the somewhat simplified protocol below. The initial SRP RNA alignment of 172 sequences was downloaded in the FASTA format from SRPdb [18] (http://psyche.uthct.edu/dbs/SRPDB/SRPDB.html) and the domain IV region (about 50 nt long including gaps) was extracted using a text editor. We then wanted to use RNAalifold to obtain a consensus secondary structure for the domain IV alignment. Since RNAalifold requires alignment files in the CLUSTAL format, we need to perform a format conversion first. This can be done with CLUSTAL, by typing:

```
clustalw SRPaligned.fasta -convert
```

Here, SRPaligned.fasta is the domain IV part of the multiple alignment obtained from SRPdb. This automatically produces a CLUSTAL format file called SRPaligned.aln. An extract of this alignment is shown in Fig. 35.2, with representative sequences from the three phylogenetic domains. At this stage, secondary structure annotation is absent from the alignment. We now use RNAalifold to compute the consensus secondary structure:

```
RNAalifold SRPaligned.aln
```

```
                            (((   ((   .......   (((   ....   (((   ......   )))   ....   )))   ....   ))   .   )))
                            000   00   0000000   000   0000   000   000000   000   0000   000   1111   00   1   000
                            222   44   3333333   666   5555   888   777777   888   9999   666   1111   44   3   222
         ┌ Escherichia coli         UCU  GU  UUACC-A  GGU  CAGG  UCC  GGA--A  GGA  AGCA  GCC  AA-G  GC  -  AGA
         │ Thermus thermophilus     GGC  GU  GAACC-G  GGU  CAGG  UCC  GGA--A  GGA  AGCA  GCC  CUAA  GC  -  GCC
Bacteria │ Clostridium perfringens  CCU  GU  GAACC-U  CGU  CAGG  UCC  GGA--A  GGA  AGCA  GCG  AUAA  GC  -  AGU
         │ Bacillus subtilis        UUC  AU  GAACC-A  UGU  CAGG  UCC  GGA--A  GGA  AGCA  GCA  UUAA  GU  -  GAA
         │ Chlorobium tepidum       UGC  CC  A-ACC-A  UGU  CAGG  UCC  GGA--A  GGA  AGCA  GCA  U-CC  GG  U  AAU
         └ Rickettsia prowazekii    CUU  GC  UUAGU-U  GGU  CAGG  UCU  GAA--A  AGA  AGCA  GCC  AG-G  GU  -  AAG

        ┌ Archaeoglobus fulgidus    CGG  GG  GGAAC-G  GCC  CAGG  CCC  GGA--A  GGG  AGCA  GGC  UA-A  CC  -  CCG
Archaea │ Pyrococcus abyssi         GCC  CC  AAACC-C  CGC  AAGG  CCC  GGA--A  GGG  AGCA  GCG  GU-A  GG  -  GGC
        └ Thermococcus celer        CCG  CC  GAACC-C  CGU  CAGG  CCC  GGA--A  GGG  AGCA  GCG  GU-A  GG  -  CGG

           ┌ Arabidopsis thaliana        GGA  GG  GUAAU-G  CGU  GAGG  CUG  GCUUCA  CAG  AGCA  GCG  ACUA  CU  -  UCC
           │ Oryza sativa                GGC  AG  GCACA-G  CGU  GAGG  CUG  GCUUCA  CAG  AGCA  GCG  AUCA  CU  -  GCC
           │ Triticum aestivum           GGC  AG  GCACA-G  CGU  GAGG  CUG  GCUUCA  CAG  AGCA  GCG  ACAA  CU  -  GCC
Eukaryotes │ Homo sapiens                GGG  GU  GAACC-G  GCC  CAGG  UCG  GAA--A  CGG  AGCA  GGU  CAAA  AC  -  UCC
           │ Drosophila melanogaster     GGG  AU  GAACC-G  GGC  CAGG  GGU  GAA--A  ACC  AGCA  GCC  AAGA  GU  -  UCC
           │ Caenorhabditis elegans      GUC  GU  GGAU--G  GUU  CAGG  ACC  GAA--A  GGU  AGCA  GAC  AAAA  GC  -  GAC
           │ Lycopersicon esculentum     GGG  GC  GGACC-G  CAU  GAGG  CUG  GCUUCA  CAG  AGCA  GUG  AA-C  GC  -  UCC
           └ Leptomonas collosoma        UAG  AG  GAACU-G  GGU  CAGG  CCG  GCA--A  CGG  AGCA  GCC  CA--  CC  -  UCG
```

Fig. 35.2. An extract of the SRP RNA alignment obtained from SRPdb [18], along with secondary structure annotation. Top line: parenthesized structure as generated by the RNAalifold program [16]; lines 2 and 3: secondary structure element numbering as used by the ERPIN software. For clarity, we inserted spacers between each secondary structure element in the alignment.

RNAalifold displays on the screen a consensus primary sequence, consensus secondary structure in the parenthesis notation and postscript drawing of the consensus structure. The parentheses are shown on top of Fig. 35.2. This structure-annotated alignment can be fed with little change to the ERPIN program or used directly for the design of an RNAMOTIF descriptor.

35.5 Evaluating Search Results

Before we run the search programs, we need to prepare a consistent protocol to evaluate search results. Like other predictive methods, motif search is generally evaluated using two parameters: specificity and sensitivity. Specificity is the fraction of predicted motifs that are true motifs. Sensitivity is the fraction of true motifs in the search space that are indeed detected by the program. These two parameters can be computed as follows:

Sensitivity: $SN = TP/(TP + FN)$

Specificity: $SP = TP/(TP + FP)$

Where TP (True Positives) is the number of true motifs that are identified as such, FN (False Negatives) is the number of true motifs that are missed and FP (False Positives) is the number of predicted motifs that are not true. TP and FN can be measured easily using a sequence database where known motifs are present: the positive control set. This can be either the training set itself or sequences where the motif has previously been identified at specific positions. FP is more difficult to assess, since any new hit in a biological sequence can be either a misprediction (FP) or a true, previously unreported motif (TP). A possible workaround is to count FPs in a randomized database where any prediction can safely be regarded as a misprediction (the negative control set).

As a positive control set, we will use the SRP RNA sequences present in our expert alignment. Because there are enough sequences in this alignment ($n = 172$), we can afford the rigorous approach of splitting the alignment into two equal parts and use one part as a training set (in the case of training set-based programs) and the second part as a positive control. This ensures we are really measuring our method's capacity to discover new sequences, and not only known ones. For a robust TP and FN estimation, this operation of random database splitting can be repeated several times. Here, we will produce 100 random control sets of 86 sequences, and re-compute TP and FN 100 times. This will provide our sensitivity measure.

As mentioned earlier, specificity is more problematic as it also depends on the density of true sites in the sequence under study. This density affects the TP to FP ratio, independently of program performance. Hence, it is common to use only a relative number of FP (e.g. FP per megabase) as a measure of specificity. This is what we will do here, using a random sequence as a negative control set. Random sequences with a uniform nucleotide composition (25% each) are not advised, since background compositions (single- and dinucleotide frequencies) strongly affect the number of chance hits, e.g. an AU-rich motif is more likely to occur in an AU-rich sequence. It is therefore important to reproduce in our negative control set the overall compositional biases of the target database. Several programs can do this. We will use shuffle, by S. Eddy (Table 35.3), which randomizes an input sequence while preserving the single- and dinucleotide frequencies. If specificity has to be usefully quantified, the negative set should be large enough to find at least a few FPs. For the SRP motif, trial and error experience has shown that we need at least 100 Mb. The next step is to select the sequence to be shuffled. If our goal was to analyze a specific genome (say, E. coli), we would just randomize this genome. However, here we intend to perform searches in all types of genomes, with very different nucleotide compositions. In this context, a good negative control is the randomized training set. Because its composition is, by definition, close to that of SRP motifs, the randomized training set more likely to give rise to false positives than any other sequence. Therefore it can be considered a rigorous and conservative control. As the shuffle program produces a sequence of same size as the input sequence, the shuffling procedure must be iterated to produce a random sequence of the desired size. Here, we need to perform 13225 iterations in order to obtain a 100 Mb negative control sequence set:

```
shuffle -d -n 13225 SRP.fasta
```

As this command prints output on the screen, it is more convenient to use a redirection and send output it directly into a file:

```
shuffle -d -n 13225 SRP.fasta > negative.fasta
```

Here, the random sequence is stored into file negative.fasta. As most motif search programs are not able to swallow such a large sequence in a single chunk, (the size limit for a single sequence is 30 Mb for RNAMOTIF and 300 Mb for ERPIN) we should either split this large sequence or generate several shorter sequences. In any case, make sure that each control sequence in the set is at least two orders of magnitude larger than the RNA motif under study, to minimize the so-called "border effects".

35.6
Using the RNAMOTIF Program

RNAMOTIF requires that we write a descriptor, such as the one shown in Fig. 35.3. Let us take a quick look at this new language. The first section (`parms`), says that G:U pairs are allowed as well as Watson–Crick pairs. The next section (`descr`) is the actual descriptor, listing all secondary structure elements. There are three types of elements: `h5` stands for the 5′ strand of a helix, `h3` for the 3′ strand of a helix and `ss` for a single strand. When no other indication is provided, RNAMOTIF assumes pairing between `h3` and `h5` elements in a purely nested fashion. That is to say, each `h3` pairs to the last unpaired `h5` encountered. Non-

```
parms
   wc += gu;
descr
   h5(minlen=4, maxlen=6)
        ss(minlen =5, maxlen=7)
        h5(minlen=3, maxlen=4)
                ss(len=4, seq="AGR$")                      }
                h5(len=3)                                  } A
                        ss(tag='hp', minlen=4, maxlen=6)   }
                h3                                         }
        ss(len=4, seq="AGCA")                              } B
        h3
        ss(minlen=2, maxlen=4)
   h3

score
   {
        if (( ss(tag='hp') !~ "^GNRA$") && ( ss(tag='hp') !~ "GYUUCA")) {
                REJECT;
        }
   }
```

Fig. 35.3. RNAMOTIF descriptor for SRP RNA domain IV: (A) first version: "rnamotif A" and (B) second version: "rnamotif B".

nested helices (i.e. pseudo-knots) require a special tag that modifies this default behavior. Associated to each `h5`, `h3` or `ss` element is a list of optional parameters that specify minimum and maximum length, conserved sequences, conserved base pairs, etc. By default, RNAMOTIF reports all sequences matching the structure in the `descr` section. Sometimes, however, additional filters are required that cannot be applied at the `descr` level. The optional `score` section is used for this purpose. In the `score` section, tests can be performed using a C-like computer language, and matches are rejected if the tests fail. Now let us see how our knowledge of the SRP RNA can be turned into an efficient descriptor. Our example will use only the most basic features of the RNAMOTIF language. Many other functions are available to define complex motifs. For an in-depth explanation of RNAMOTIF descriptors, refer to the detailed manual included with the program.

We have outlined in Fig. 35.2 the main base-pair stems, obtained by extending manually the consensus helices generated by RNAalifold. We can observe three distinct stems. In our alignment, the outermost stem is interrupted in its 3' strand by a bulge. RNAMOTIF does not support bulge-containing stems, which have to be represented using two stems separated by an asymmetrical single-strand. As this bulge can be, in some cases, replaced by a regular but shifted helix (see the *Chlorobium tepidum* sequence), we decided to neglect it and thus simplify our descriptor for this tutorial. Then the outermost stem has 4–7 bp, the central stem has 3–4 bp and the apical stem has 3 bp. In the descriptor shown in Fig. 35.3(A), this corresponds to the three `h5` and `h3` statements, for the 5' and 3' parts of these helices, respectively. Stems are separated by single strands noted `ss`. The `minlen=` and `maxlen=` options specify the size boundaries for each element. When a helix or single strand has a fixed size, the `len=` option is used instead, as in the third and fourth single strand, each of size 4. When a helix or single strand includes a conserved sequence, this constraint can be specified using the `seq=` option. As the fourth single strand was a conserved AGCA, we typed `seq="AGCA"` in the corresponding `ss` line. RNAMOTIF offers a lot of flexibility in the description of conserved sequences. The usual IUPAC code for ambiguous nucleotides (R, Y, W, etc.) can be used, as well as some of the special characters used in Unix regular expressions. The third single strand ends with a conserved AGG in our alignment or, more accurately, AGR if one considered all available SRP sequences. In Fig. 35.3(A), we specified this using `seq="AGR$"`. As the $ sign stands for the end of the strand, this expression means the strand ends with sequence AGR. Alternatively, we could have used `seq="NAGR"` to specify just the same constraint in a 4-nt strand. The option `tag='hp'` in the apical stem (giving it a name) is dispensable at this point; it will be used later on in the more elaborate version of the descriptor. Let us now test this first descriptor using the positive and negative test sets defined in the previous section. Assuming the descriptor is saved in the srp1.descr file, the command line to run the search should be:

```
rnamotif -descr srp1.descr database.fasta
```

Where database.fasta should be replaced either by the positive control set, the neg-

Tab. 35.4. Search results obtained with three different RNAMOTIF descriptors and four different ERPIN search parameters.

	Rnamotif A	Rnamotif B	Rnamotif NAR2001	Erpin −8,8	Erpin −6,6	Erpin −2,2	Erpin optimal
TP	67	65	79	82	79	81	80
FN	19	21	7	4	7	5	6
FP (100 Mb)	204	2	2	673196	10	1	1
E-value (100 Mb)	–	–	–	737000	10.7	1.13	1.06
SN	78%	75%	92%	95%	92%	94%	93%
SP	25%	97%	98%	0%	89%	99%	99%
Time (100 Mb)	1 min 12 s	1 min 12 s	52 min 50 s	2 min 45 s	3 min 51 s	21 min 52 s	1 min 35 s

Descriptors Rnamotif A and Rnamotif B are those shown in Fig. 35.3. Descriptor Rnamotif NAR2001 refers to that used in [4]. TP, FN and FP are defined in Section 35.5.

ative control set, or the database against which to do the actual search. RNAMOTIF results are displayed on the screen, which can be impractical when too many solutions are produced. We recommend to use a redirection as explained above. Using the positive control set, this command finds 456 solutions. This is obviously too large, as there are only 86 SRP sequences in this set. Due to variations in helix sizes and positions, RNAMOTIF tends to generate several overlapping solutions for each actual motif. A utility program, rmprune, is provided to filter out these overlapping solutions and retain only the highest scoring one at each site. This program is executed in combination with RNAMOTIF, as follows:

```
rnamotif -descr srp1.descr database.fasta | rmprune
```

From this point, we will use the rmprune program at each run. Results are shown under "Rnamotif A" in Table 35.4. RNAMOTIF now finds only 67 solutions in the positive control set, more in line with the ideal value of 86. This amounts to a sensitivity of 78%. Our descriptor is thus a bit too restrictive. Indeed, a closer look at the complete alignment shows that there are several exceptions to our secondary structure definition. For instance, some C:A base pairs occur in the alignment at the innermost position of the middle helix, which we did not take into account. This alone would have captured a dozen more SRPs in the training set. But consider how our descriptor performs against the negative set. In the 100-Mb random sequence, RNAMOTIF identifies 204 hits, amounting to a specificity of only 25%. Obviously, we should tackle this aspect first, and tighten our descriptor accordingly. An important constraint we have not yet exploited is the sequence of the apical loop. Apical loops are either 6 nt long with a GYUUCA sequence or 4 nt long with a GNRA sequence. This is a strong constraint that we should certainly benefit from, but version 3.0 of RNAMOTIF does not permit to express a constraint of type "contains either ... or ..." in the `descr` section. We have to use the `score` section to do this. Our "Rnamotif B" descriptor (Fig. 35.3B) adds such a section to the previous descriptor. The `score` section contains a program that will be automatically executed for each solution. Some knowledge of C or Perl programming is required to write such a program. We cannot present here the complete syntax and possibilities of this language, which are covered in detail in the documentation, so we will limit ourselves to this particular example. The `tag=` option is used to assign a name to a given helix or single strand in the descriptor. Here, we used the option `tag='hp'` to designate the apical loop in the `descr` section, so that we can access this element again in the `score` section and perform a test on it. Our test verifies that the single strand of `tag='hp'` does not contain (!~ operator) either the string ^GNRA$ or the string GYUUCA. As the ^ and $ signs designate the beginning and end of a strand, respectively, this expression means that the GNRA sequence cannot be a part of a larger loop sequence. A passed test indicates that neither of the required sequence was found in the apical loop and we therefore want to reject the proposed solution. This is done using the `REJECT;` command. If this command is not encountered, the solution is retained. Now that we have added this test on the apical loop sequence, let us see how our extended descriptor performs. RNAMOTIF was run using the same data and command lines as above. Results are shown in Table 35.4, under "rnamotif B". True Positives are down to 65, indicating that two true SRP sequences are lost due to the additional constraint. However, this 3% loss in sensitivity is a reasonable price to pay for the concomitant 100-fold drop in false positive rate, down to 2/100 Mb, a considerable gain in specificity (up to 97%).

At this point, our analysis of the alignment and descriptor design are still relatively superficial. One could significantly improve search sensitivity by relaxing constraints in the `descr` section (notably helix length) and writing additional tests for rejecting unfit solutions in the `score` section. Eventually, very efficient RNAMOTIF descriptors can be obtained for the SRP RNA. Sampath et al. [4] have designed an SRP descriptor that is 91% sensitive and 84% specific by our tests (Table 35.4). This descriptor is rather too complex to be presented in length in this chapter, but it can be understood quite easily using the simple syntax elements we just presented.

In Table 35.5, we show the results of actual "real world" database searches performed using the "rnamotif B" descriptor in bacteria, archaea and *Drosophila* ge-

Tab. 35.5. Whole genome scan results (the eukaryotic genome is that of *D. melanogaster*).

Domain (number of genomes)	Total database size (nt)	RNAMOTIF		ERPIN		
		Hits	FP	Hits	FP	E-value ≤ 0.01
Bacteria (128)	410431203	91	8	125	4	122
Archaea (17)	372925507	12	7	17	4	16
Eukarya (1)	116781562	2	2	3	1	3

nomes. We estimated the number of FP for each genome on the basis of the number of FP per 100 Mb, as computed above (Table 35.4). In doing this, we assumed that all genomes had the same composition and hence the same number of FP/100 Mb, which is not true. A more accurate control would imply creating a shuffled negative test set for each genome under study. Nonetheless, with the current controls, False Positives would remain acceptable for bacterial searches (8 FPs for 91 hits) but would be too high in Archaea (7 FPs for 12 hits) or *Drosophila* (2 FPs for 2 hits). It is probably safer to refrain from publishing these *Drosophila* hits!

35.7 Using the ERPIN Program

Now let us address the SRP identification problem using the ERPIN software. As explained earlier, ERPIN extracts motif information directly from an alignment, thus circumventing the descriptor design problem. When both accurate sequence alignment and secondary structure information are available, we will reap important benefits from such a statistical approach. First, we need to convert the alignment into the ERPIN format. Here is how the first lines of our SRP alignment should look like:

```
>structure
00000000000000000000000000000000000000001111001000
22244333333366655558887777778889999666111443222
>AQU.AEO.
AGGGUGAACU-CCCCCAGGCCCGAA—AGGGAGCAAGGGUAAGC-CCG
>THE.THE.
GGCGUGAACC-GGGUCAGGUCCGGA—AGGAAGCAGCCCUAAGC-GCC
```

The first three lines provide secondary structure information. In this example, the secondary structure is the same as the parenthesized structure in Fig. 35.2, but each structural element (single strand or helix) is assigned a different number, that is read vertically. When segments at different positions have the same number (e.g. 04 in this example), the corresponding element is understood as a helix, otherwise, it is a single strand. These secondary structure lines can be created manually using a text editor, or automatically from a parenthesized secondary structure file using the b2epn script provided in the ERPIN distribution. The following lines contain the aligned sequences in the classical FASTA format. Our ERPIN formatted alignment is saved as srp.epn.

We are almost ready to launch an ERPIN search, but there is a question about search efficiency that we need to resolve first: should the whole motif or just a specific element be used for searching? To answer this, let us have another look at the alignment in Fig. 35.2 which contains 14 elements (the ERPIN numbering is shown on top of the figure). Note that the bulged helix now produces two helices, noted 02 and 04, as ERPIN does not support bulges or gaps within helices. A direct search for a 14-element region could be prohibitive in terms of CPU and memory usage, especially when elements vary in size. In our case, there are few gaps (hence not much size variation) and a direct complete search may be possible. However, we will start cautiously and use a short motif first. The apical hairpin (elements 7 and 8) has a relatively well-defined loop sequence. To find out how good a search would be using those elements alone, let us limit the search to this region, by using:

```
erpin srp.epn sequence.fasta -8,8 -nomask
```

The first two arguments of this ERPIN command line are the training set and the target sequence database. The -8,8 argument defines the boundaries of the selected query region. A minus sign is required to differentiate 5' from 3' strands of helices. Here the search region goes from the 5' strand of helix 8 to the 3' strand of helix 8. The -nomask option means that we use the whole region between these two elements (helix 8 and strand 7) for the search. We could use only the helical part ignoring strand 7 by using -mask 7 instead. We could also use the -cutoff option to set the cutoff score above which solutions are retained. The default cutoff score is that of the lowest scoring sequence in the training set.

Let us see in Table 35.4 how this search performs. Running on our positive control set, we are able to catch 82 out of 86 true SRP molecules (sensitivity = 95%). The 100-Mb negative control set is scanned in only 2 min 45 s, but yields a huge number of hits ($n = 673\,196$). Obviously, this single stem–loop region does not carry enough information to yield a specific search. ERPIN does provide a useful clue to assess the reliability of solutions. For any hit with a score S, ERPIN computes an E-value, i.e. the number of hits of same or higher score that can be expected by chance in the same database. Here, ERPIN tells us that the E-value for the default cutoff is 737 000. This means 737 000 expected hits for any 100-Mb database! When doing an actual genome scan, E-value information can help you evaluate search specificity without going through tedious control runs. To increase specificity, we need to use a larger region of the alignment. Let us try the -6,6 region:

```
erpin srp.epn sequence.fasta -6,6 -nomask
```

The widened search region includes single strands 5 and 9, which not only are highly conserved, but are also deprived of gaps. Gapless regions are particularly interesting in that they require little CPU time. As we can see from Table 35.1, the program still runs fast enough (3 min 51 s) and sensitivity remains at a reasonable 92%. More importantly, FPs are now down to 10/100 Mb. A most welcomed gain indeed! However, this acceptable specificity level for a bacterial genome remains problematic for large eukaryotic genomes: it amounts to 15 spurious hits in the *Drosophila* genome or 300 in the human genome. Can we further enhance specificity without loosing too much in sensitivity? Since the program runs acceptably fast to this point, let us be bold and just give it the whole domain IV structure:

`erpin srp.epn sequence.fasta -2,2 -nomask`

The region from element −2 to 2 encompasses the whole alignment. Sensitivity remains at a satisfying 94% and specificity is down to a single FP for 100 Mb, which is compatible with the scanning of a small eukaryotic genome. However, search time rose sharply to about 22 min for 100 Mb. Using the whole structure thus remains tractable in this case, but it will not be so for other RNA alignments that contain more gaps or fewer conserved regions than SRP domain IV. Just keep in mind that more gaps (in strands, helix cannot contain gaps) and more variation in the alignment means longer search times and a reduced specificity.

It is relatively easy to decrease CPU time by implementing a search strategy. The main idea is to perform a stepwise search using masks. In the first step, one should mask most of the selected region and retain only a few key elements. ERPIN will disregard all masked elements and restrict the search to the unmasked part. The best parts to unmask during the first step are those conserved elements that occur rarely in the database and are close spatially so that they are quickly identified. Then, at each successive step, one unmasks additional elements for ERPIN to look for. ERPIN will consider these latter elements only after elements at the first round have been identified at the required cutoff score. Any number of search steps can be specified on the same command line. For our SRP motif, let us try the following two-step strategy:

`erpin srp.epn sequence.fasta -2,2 -umask 5 9 -nomask`

In the first step (`-umask 5 9`), we seek elements 5 and 9, which are well conserved and close together. In the second step, the whole region is unmasked. As shown in Table 35.3, sensitivity and specificity are basically unaffected (93 and 99%), but search time decreases from 22 min to only 1 min 35 s.

Results of real genome scans for the ERPIN program are shown in Table 35.5, using the last command. As a measure of hit quality, we show the expected number of FP based on the negative test set as above. However, ERPIN E-values are a much better way to measure a hit quality as they evaluate the significance of each solution. When presenting your own ERPIN search results, you should select hits based on their E-value. An E-value of 0.01 or less is significant, but higher E-values may also be interesting in specific biological contexts. Here, we expect one FP in the *Drosophila* genome at the default cutoff (Table 35.4), but all three hits in this genome have an E-value of less than 0.01. This means all three hits are significant. Finally, E-values depend on the database size. Therefore, when a genome is available in separate sequence files (e.g. with one file per chromosome), be sure to save these sequences into a single file so that E-values reflect the complete genome search.

35.8
Troubleshooting

35.8.1
RNAMOTIF

35.8.1.1 Too Many Solutions
A common problem with RNAMOTIF is overlapping solutions, i.e. multiple hits occurring around a single site. Most of these overlapping solutions can be eliminated using the rmprune utility. To filter RNAMOTIF output with rmprune, use the following command line:

```
rnamotif -descr descriptor-file database-file | rmprune
```

Note that this filter is not 100% efficient and may still retain a minority of overlapping hits.

35.8.1.2 Program Too Slow
Like other RNA search programs, RNAMOTIF is highly sensitive to size variations in gap-containing regions. When program execution is too slow, the first thing to do is inspect the alignment for regions with large numbers of gaps, and then identify a few helical and single-strand elements that do not encompass or fall within these regions. Design a first descriptor that includes only these elements. If search time is satisfying, then progressively enlarge this descriptor.

35.8.2
ERPIN

35.8.2.1 Too Many Solutions
As with the RNAMOTIF program (albeit to a lesser extent), ERPIN is likely to find several overlapping solutions in lieu of a single motif. Yet another source of output inflation is the presence of Ns in database sequences. Due to the underlying statistical model in ERPIN, sequences containing multiple Ns often end up with reasonable scores that pass the default cutoff. A utility Perl script, readerpin.pl, can be used to filter out both overlapping and N-containing solutions. This script can be combined with an ERPIN command as below:

```
erpin srp.epn sequence.fasta -6,6 -nomask | readerpin.pl
```

The readerpin.pl utility can also be used to filter out solutions with low scores or high *E*-values. For instance:

```
erpin srp.epn sequence.fasta -6,6 -nomask | readerpin.pl -e 0.0001
```

will display only those solutions with an *E*-value lower than 0.0001.

35.8.2.2 Program Too Slow

If ERPIN run time becomes excessive, one should either use a smaller search region or implement a stepwise search strategy. A good starting point for a quick search is usually an apical stem–loop. Find in your alignment an apical stem–loop, ideally with some conserved nucleotides in it, and run the search on this region alone. In the previous section about ERPIN usage, the search region for the SRP RNA was progressively enlarged from an initial -8,8 apical stem-loop to the larger -2,2 region. As search times increased with the -2,2 region, we used a stepwise strategy, with two masking levels. More complex, multistep strategies can be devised for larger motifs and ERPIN offers different kinds of masking commands to facilitate this.

A key in understanding performance issues with ERPIN is the number of configurations. This is the number of possible combinations of helix positions in a given search region, induced by size variations in single strands. The more configurations, the slower the program. A utility program cfgs serves to evaluate the number of configuration for a given search. Here is an example using our SRP.epn training set:

```
cfgs SRP.epn 2,2 -nomask
```

The output tells us there are 54 possible configurations for the whole -2,2 region. If we now use strand 5 and 9 only:

```
cfgs SRP.epn -2,2 -umask 5 9
```

The output tells us there are only three configurations for elements 5 and 9. Therefore, search will run much faster.

Acknowledgments

We thank Dr Pascal Hingamp for his careful reading of the manuscript.

References

1. A. Huttenhofer, J. Brosius, J. P. Bachellerie, *Curr. Opin. Chem. Biol.* **2002**, *6*, 835–843.
2. K. M. Wassarman, F. Repoila, C. Rosenow, G. Storz, S. Gottesman, *Genes Dev.* **2001**, *15*, 1637–1651.
3. K. Nagai, C. Oubridge, A. Kuglstatter, E. Menichelli, C. Isel, L. Jovine, *EMBO J.* **2003**, *22*, 3479–3485.
4. T. J. Macke, D. J. Ecker, R. R. Gutell, D. Gautheret, D. A. Case, R. Sampath, *Nucleic Acids Res.* **2001**, *29*, 4724–4735.
5. D. Gautheret, A. Lambert, *J. Mol. Biol.* **2001**, *313*, 1003–1011.
6. G. A. Fichant, C. Burks, *J. Mol. Biol.* **1991**, *220*, 659–671.
7. T. M. Lowe, S. R. Eddy, *Science* **1999**, *283*, 1168–1171.

8 L. P. Lim, N. C. Lau, E. G. Weinstein, A. Abdelhakim, S. Yekta, M. W. Rhoades, C. B. Burge, D. P. Bartel, *Genes Dev.* **2003**, *17*, 991–1008.

9 D. Gautheret, F. Major, R. Cedergren, *Comput. Appl. Biosci.* **1990**, *6*, 325–331.

10 B. Billoud, M. Kontic, A. Viari, *Nucleic Acids Res.* **1996**, *24*, 1395–1403.

11 G. Pesole, S. Liuni, M. D'Souza, *Bioinformatics* **2000**, *16*, 439–450.

12 M. Dsouza, N. Larsen, R. Overbeek, *Trends Genet.* **1997**, *13*, 497–498.

13 S. R. Eddy, R. Durbin, *Nucleic Acids Res.* **1994**, *22*, 2079–2088.

14 R. Gutell, *Curr. Opin. Struct. Biol.* **1993**, *3*, 313–322.

15 J. D. Thompson, D. G. Higgins, T. J. Gibson, *Comput. Appl. Biosci.* **1994**, *10*, 19–29.

16 I. L. Hofacker, M. Fekete, P. F. Stadler, *J. Mol. Biol.* **2002**, *319*, 1059–1066.

17 B. Knudsen, J. Hein, *Nucleic Acids Res.* **2003**, *31*, 3423–3428.

18 M. A. Rosenblad, J. Gorodkin, B. Knudsen, C. Zwieb, T. Samuelsson, *Nucleic Acids Res.* **2003**, *31*, 363–364.

36
Approaches to Identify Novel Non-messenger RNAs in Bacteria and to Investigate their Biological Functions: RNA Mining

Jörg Vogel and E. Gerhart H. Wagner

> *Prior to the dawn of systematic screens, coming across a new non-coding RNA was often considered a serendipitous surprise. The English word "Serendipity" derives from the name that Muslim traders once gave Sri Lanka, calling it Serendib, the "island of gems". With the wonders of the RNA world in mind, one is tempted to liken the mining for new RNAs to the mining for gems.*

36.1
Introduction

In the traditional view, the RNA population of a cell is categorized in three major classes with distinct cellular roles: mRNAs, including the genetic material of some viruses, function in genetic information transfer, while ribosomal RNAs and tRNA adaptors are essential players in protein synthesis. Small, untranslated RNAs (sRNAs) that do not fall into these major classes have been known for more than 30 years, but despite a growing appreciation both of their specific biological roles, and of the catalytic and regulatory potential of RNA in general, they were mostly regarded as a minor addition. The year 2001 brought about a startling change in the way scientists perceived the roles of small, non-coding RNAs, when systematic searches in the genomes of prokaryotic and eukaryotic model organisms showed these molecules to be far more prevalent than previously assumed.

Mining for novel non-messenger RNAs represents a rapidly expanding field of research, making use of a variety of individual approaches. This chapter presents an overview of the experimental and bioinformatic strategies that have been used to identify sRNAs over the years (summarized in Table 36.1). Somewhat biased by the authors' main field of research, we will emphasize searches for these RNAs in prokaryotic genomes, and point out advantages and disadvantages of the individual approaches. The methodology of searching for sRNAs in the various kingdoms of life overlaps significantly, and some of the techniques listed here have been successfully applied to eukaryotic and archaeal sRNAs as well.

It is worth noting that small RNA terminology is sometimes confusing. Many terms have been proposed and are often used interchangeably, e.g. ncRNA (non-

Handbook of RNA Biochemistry: Student Edition. Edited by R. K. Hartmann, A. Bindereif, A. Schön, E. Westhof
Copyright © 2009 WILEY-VCH Verlag GmbH & Co. KGaA, Weinheim
ISBN: 978-3-527-32534-4

Tab. 36.1. Overview of the sRNA screening approaches discussed.

Screening type	Advantages	Disadvantages	Selected references
Direct labeling and cloning	Detects most abundant sRNAs and/or sRNAs with highest synthesis rate under a given set of conditions Independent of prior knowledge of sRNA and sRNA gene features Can identify the entire sRNA biochemically (and processed versions thereof) Detection of species-specific sRNAs	Involves handling of highly radiolabeled bacterial cultures (orthophosphate labeling) Does not distinguish between sRNAs or abundant processed fragments of other RNAs	3, 7–9, 11–12, 14–16
Functional genetic screens	Gives immediate clues about functional involvement Can utilize the entire strain collection/method arsenal of a genetics lab	Labor-intensive Difficult if sRNA is essential sRNAs acting under special conditions hard to identify	17, 20–22, 25–29, 31, 32
Biocomputational screens	Often rapid, generates many candidates Very fruitful when phylogenetic comparisons are incorporated	Requires prior knowledge Requires validation of many candidates	13, 33, 34, 36–39
Microarray detection	Can give transcriptional profiles for all genes Fast profiling of condition-dependent sRNA expression patterns Detection of species-specific sRNAs	Expensive; requires microarrays that cover intergenic regions Frequently inconsistent sRNA detection results as compared to Northern blot signals	13, 44, 45, 46
Shotgun cloning (RNomics)	Does not built on prior knowledge Can be performed in highly automated manner Detection of processed and species-specific sRNAs; also permits detection of primary transcripts	Cost-intensive (sequencing) cDNA synthesis may be biased against highly structured sRNAs	47–58

coding RNA), snmRNA (small non-messenger RNA), fRNA (functional RNA), eRNA (effector RNA), regulatory RNA and riboregulator. Each of these denotations has its problem and will be inappropriate for at least some known RNAs, or may subsume additional RNAs (mRNAs, tRNAs, etc.). We will here refer to the bacterial RNAs as sRNAs (small RNAs).

36.2
Searching for Small, Untranslated RNAs

36.2.1
Introduction

Quite different from animals and plants, in which only a minor fraction of the genome encodes proteins, prokaryotic genomes are dominated by mRNA-encoding regions. This may be illustrated by the 4.3-Mbp chromosome of the model bacterium *Escherichia coli*, of which 87.8% account for protein-coding genes, and 0.8% is occupied by rRNA and tRNA genes [1]. Thus, sRNAs could either originate from within the remaining 11% of the genome or be encoded opposite to mRNA genes, thus possessing full complementarity to the transcript(s) from the opposite DNA strand. Some considerations regarding the latter type, *bona fide* antisense RNAs, are found in Section 36.2.8.

Of primary interest here are sRNAs that are encoded by separate transcription units in "empty" intergenic regions (IGRs) between known open reading frames (ORFs) or other genes. Most of the 10 chromosomal *E. coli* sRNAs known prior to 2001 are encoded in such regions [2]. Consequently, many of the following systematic screens were based on the assumption that conservation of promoters, transcription terminators, and primary sequence among closely related species – within IGRs – would predict new sRNA genes (see biocomputation-aided and microarray-based approaches, Sections 36.2.4 and 36.2.5). As a result of these screens, the number of chromosomally encoded sRNAs now known in *E. coli* exceeds 60, with around 1000 additional predicted candidates still awaiting validation. Even though the majority of sRNAs originate from IGRs, functional sRNAs might not necessarily be transcribed from single genes flanked by own promoters and terminators. 6S RNA of *E. coli* is an example of a regulatory sRNA which is part of a larger transcription unit. This RNA was discovered by *in vivo* RNA labeling experiments [3] and was subsequently sequenced by enzymatic digestion [4]. Accumulating to around 10 000 molecules per cell (in stationary phase), 6S RNA is one of the most abundant RNAs in *E. coli*. A function in regulating RNA polymerase has been proposed by Wassarman and Storz [5]. Different from many other stable sRNAs, 6S RNA lacks a strong Rho-independent terminator and requires processing from a dicistronic transcript that includes the downstream *ygfA* gene [6]. Since the 3′ end of 6S RNA and the start codon of *ygfA* are only around 70 nt apart, this important sRNA could have easily been dismissed as a processed leader fragment in predictions based on "classical" sRNA features.

36.2.2
Direct Labeling and Direct Cloning

The identification of bacterial sRNAs other than tRNAs and 5S rRNA was pioneered by the fractionation of ^{32}P-labeled total cellular RNA about 40 years ago [3, 7–9]. Many of the sRNAs discovered in these studies have since been assigned important housekeeping or regulatory functions, such as M1 RNA of RNase P, tmRNA, 4.5S RNA, 6S RNA and Spot 42 RNA.

Total cellular RNA was labeled by treating *E. coli* cells with ^{32}PO$_4^{3-}$ (orthophosphate), which is readily taken up and incorporated into newly synthesized nucleic acids. Isolated RNA was separated on polyacrylamide gels and autoradiographed, and selected bands were extracted for analysis by nuclease digest (fingerprinting). An advantage of this method lies in an almost direct correlation of the signal for a given RNA with its abundance in the cell, provided a labeling time long enough to compensate for the slow turnover of particularly stable sRNAs. In principle, experiments could as well be designed to specifically detect those RNAs with the highest synthesis rate under a given condition, i.e. by using short labeling times (pulse labeling). Although the interest in direct labeling approaches has ceased, some of the problems initially associated with this method can now be handled easily. (1) Improved electrophoresis techniques, including pre-fractionation of sRNAs on columns (to reduce background signal), would improve separation already in the first separation step. (2) Sequence determination of isolated RNAs by nuclease fingerprinting is cumbersome and time consuming, but rapid cDNA cloning of extracted RNAs is carried out routinely today. (3) Potential health risks associated with the handling of highly radiolabeled bacterial cultures are reasons for concern. However, better RNA isolation techniques and higher detection sensitivity should allow for a significant reduction in the radiolabeled orthophosphate required. Thus, with these improvements, direct labeling may re-emerge as a valuable approach to catch an initial glimpse of the most abundant or most actively synthesized sRNAs in bacteria other than *E. coli*. This method could also be useful to identify highly upregulated sRNAs under certain growth or stress conditions. For example, OxyS RNA, a regulator of oxidative stress-related genes, is barely detectable in *E. coli* in normal growth. However, after induction by H$_2$O$_2$ treatment, this 109-nt RNA accumulates to around 4500 molecules per cell, compared to 13 000 tmRNA copies [10], and thus should become detectable as a strongly labeled band.

As an alternative to *in vivo* labeling by ^{32}P-orthophosphate, *extracted* RNAs can be post-labeled *in vitro* at their 5′ or 3′ termini. However, it needs to be noted that secondary structure of RNA can affect the efficiency of labeling at the 5′ end (with T4 polynucleotide kinase and [γ-^{32}P]ATP) as well as at the 3′ end (with T4 RNA ligase and [^{32}P]pCp) in an unpredictable manner, thus skewing the representation of individual sRNAs when visualized after electrophoresis. An impressive example of how dramatically labeling efficiencies can differ between the two termini is given by Watanabe et al. [11]: the patterns of labeled bands from the size fraction between 16S rRNA and 4S RNA from the cyanobacterium *Synechococcus* were almost mutually exclusive for the 5′ and 3′ reactions. In *E. coli*, however, the sRNA profiles are comparable after 5′ or 3′ labeling (Karen M. Wassarman, personal communication).

In principle, the chemical group at the 5′ end of sRNAs could be a reason for concern. Recent 5′ RACE (rapid amplification of cDNA ends) experiments suggest that many sRNAs (those that are primary transcripts) retain a 5′-triphosphate which, unless removed, will result in inefficient labeling. Processed sRNAs are less of a problem in this respect, since all bacterial ribonucleases that are involved in sRNA processing create 5′-phosphate and 3′-hydroxyl ends (RNases E, P and III; *cf*. Table 1 in [2]).

We are not aware of any study that has exploited *in vitro* labeling to identify new sRNAs at a larger scale. In the afore-mentioned *Synechococcus* sRNA study, nine of the 11 most strongly 5′-labeled bands were identified as rRNA fragments. The remaining two originated from the tmRNA homolog of this organism [12] and an abundant 185-nt RNA (6Sa RNA) of as yet unknown function [11]. Neither of these two RNAs was efficiently labeled at its 3′ terminus. However, 3′ labeling was successful in visualizing sRNAs co-immunoprecipitated with the Hfq protein in *E. coli* [5, 13].

In a few cases, radioactive labeling was not required; some abundant sRNAs were detectable by staining of total RNA after separation on polyacrylamide gels. Two sRNAs of *Bacillus subtilis*, BS190 and BS203, were stained equally well with ethidium bromide, as was 4.5S RNA [14, 15]. MP200 RNA(s) from *Mycoplasma pneumoniae* and an MP170 RNA homolog from *M. genitalium* were first detected by silver staining [16].

Different approaches were taken to identify the sequences, and genes, of the abundant RNAs described above. Watanabe et al. [11, 12] determined the sequences of gel-extracted, 5′-labeled RNA fragments by enzymatic sequencing and then derived oligonucleotide probes to screen a *Synechococcus* genomic phage library by plaque hybridization. Göhlmann et al. [16] gel-extracted the silver-stained MP200 RNAs, conducted 5′ labeling with T4 polynucleotide kinase and probed on Southern blots against a cosmid library of *M. pneumoniae*. The sequences of BS190 and BS203 RNA were determined by cDNA cloning of gel-purified fragments and gene location was deduced by searching the available genomic sequence of *B. subtilis* [14, 15].

36.2.3
Functional Screens

Some sRNAs in *E. coli* were discovered as cloned genomic fragments in genetic analyses of certain loci or in genetic screens aimed at finding factors that modulated certain activities. MicF RNA of *E. coli*, the first *trans*-encoded antisense RNA described in bacteria, was found serendipitously when studying the genetic basis for regulation of the two outer membrane proteins OmpC and OmpF [17]. Upon characterization of the *ompC* promoter it was found that a promoter-proximal 300-bp DNA segment, when present on a multicopy plasmid, caused depletion of OmpF protein. Inspection of this region identified a short gene, *micF*, lacking protein-coding potential. The sequence of the putative *micF* transcript showed a significant complementarity to the translation initiation region of the *ompF* mRNA, thus suggesting MicF to act by an antisense mechanism. The concept of a regulatory antisense RNA that is *trans*-encoded (in contrast to many *cis*-encoded antisense RNAs in plasmids [18, 19]) was novel, but later analyses uncovered several similar cases (reviewed in [20]).

The 87-nt regulatory RNA DsrA of E. coli was discovered in studies of regulatory elements involved in capsular synthesis. Overproduction of capsule, for example when stimulated by the positive regulator RcsA, results in a mucoid phenotype. The *rcsA* gene is normally silenced by the histone-like protein, H-NS, but becomes upregulated when a region immediately downstream of *rcsA* is present on multicopy plasmids. Subcloning of this region on pACYC184 (p15A origin of replication) resulted in the isolation of the gene encoding DsrA, which downregulates the *hns* message by an antisense mechanism ([21, 22] and references therein).

In addition to its role as an antisilencer of capsular polysaccharide synthesis, DsrA was later also shown to activate translation of *rpoS* (encoding stationary phase sigma factor RpoS [23, 24]) at low temperature. The *rpoS* coding region in the mRNA is preceded by a long, structured leader to which DsrA base-pairs. Binding results in an alternative, translation-competent structure. In a screen for factors that can activate RpoS translation in the absence of DsrA, the Gottesman group isolated yet another regulatory sRNA, RprA [25]. Here, a pBR322-based E. coli library with inserts ranging from 1.5 to 5 kb ([26]; Nadim Majdalani, personal communication) was introduced into a Δ*dsrA*, *rpoS–lacZ* (translational fusion) strain. Of the 12 plasmids isolated from red colonies on MacConkey lactose plates (red colony color indicates β-galactosidase fusion protein activity; 25 000 colonies were screened in total), eight mapped to the genomic region that contained the *rprA* gene. The other four plasmids contained other genomic segments. The smallest *rprA*-linked fragment supported a 6-fold upregulation of *rpoS–lacZ* expression in a Δ*dsrA* background, as compared to only 2-fold in the presence of a chromosomal *drsA* copy. This illustrates the need to inactivate one sRNA gene when screening for additional sRNAs that might regulate the same target.

UptR RNA (92 nt) was found in a screen for suppressors of export toxicity – an effect associated with proteins that fail to adopt appropriate structure when passing the membrane [27]. The authors made use of "unfoldable" DsbA′–PhoA hybrid proteins which, when their genes are carried by pBR322 plasmids, confer lethality to a protease-deficient E. coli strain (K10 Δ*degP*). Cells collectively carrying a random library of host genome fragments cloned on phage MudII were transformed with a plasmid encoding the toxic fusion protein, and viable colonies were selected for. Of 18 phagemids (carrying putative suppressor genes), 11 carried the *degP* gene which restored the protease deficiency and five contained *recG* which decreases the copy number of the toxic pBR322 plasmids. The remaining two plasmids carried a DNA region from which the *uptR* locus (unfolded protein toxicity-relieving factor) was subsequently isolated through subcloning. As a facet, this region is not present on the host chromosome but on a resident F plasmid.

Two small RNAs of E. coli, CsrB and CsrC, are antagonists of the regulatory protein CsrA which modulates (usually inhibits) translation of certain target mRNAs. CsrC was identified in a functional screen. Inserts cloned in a low-copy plasmid, pGL339, were used to identify genes that affected glucan biosynthesis [28], Two insert-carrying plasmids (genetically distinct) increased expression of a *glgC'–'lacZ* fusion gene. Deletion analysis identified a 360-bp region that was sufficient to confer activation. This resulted in identification of the *csrC* gene, encoding a 245-nt RNA [29]. Note that, at that time, it was already established that *glgC* was primarily regulated by another small RNA, CsrB, which alleviates the inhibitory effect of CsrA on *glgC* translation (reviewed in [30]). CsrB had previously been identified by direct cloning of the major RNA species that binds the CsrA protein (see Section 36.2.7).

Homologs of the CsrA/CsrB system are found in a wide range of bacteria (see also Section 36.2.7). PrrB RNA, a functional homolog of CsrB, was found in a multicopy plasmid screen in the biocontrol strain *Pseudomonas fluorescens* F113 [31]. This screen was carried out to identify genes that could restore secondary metabolite production in *gacS/gacA*-deficient (two-component system) mutant strains. A single plasmid carrying a 5.4-kb fragment from a genomic library suppressed the mutant phenotype. Further analysis of a subfragment in which all putative ORFs lacked identifiable ribosome binding sites revealed the *prrB* gene encoding a 132-nt RNA.

Interestingly, none of the multicopy screens listed above was designed to specifically identify new sRNA genes. In all cases, libraries would (and often did) contain DNA fragments long enough to code for even large proteins. For specific sRNA gene searches it might be useful to construct libraries that consist of small-sized DNA fragments (by using frequently cutting restriction endonucleases or mechanical shearing of genomic DNA followed by size fractionation). For example, the minimal fragment capable of expressing *uptR* (see above) was 195 bp and contained around 100 bp upstream of the *uptR* transcription start site. Considering that the majority of known sRNAs are at most 200 nt long, DNA fragments ranging from 300 to 400 bp should be appropriate for finding new sRNA in functional screens. For statistical considerations, a significantly higher number of colonies must then be screened for adequate genome coverage.

Some sRNA genes, in particular those involved in stress responses, may not be highly expressed even under multicopy conditions (e.g. due to autoregulation or repression under the screening condition employed) and thus may fail to give a measurable phenotype. This may be overcome by cloning random DNA fragments downstream of a constitutive or regulatable promoter. Such an approach does not always work: some *E. coli* sRNAs are highly toxic, or tend to acquire mutations when expressed from a strong promoter (e.g. P_L from phage λ) and when present on high copy number plasmids (own unpublished results). Constructing the same library on both low and high copy number plasmids may be worth the effort.

Apart from functional multicopy screens, gene activation by random genome insertion mutagenesis can be useful to identify sRNA genes that are essential for a certain phenotype or whose absence allows survival under a given screening condition. Of the best-characterized *E. coli* sRNAs, only M1 RNA of RNase P and 4.5S

RNA are essential under normal growth conditions [2]. However, other regulatory sRNAs may be conditionally required, so that essentiality becomes manifest only under initially unknown stress conditions. A problem of insertional mutagenesis is associated with the comparatively small size of sRNA genes. Given an average ORF size of around 960 nt (in *E. coli* [1]), insertions in protein-coding genes are expected to occur 5- to 10-fold more often than in sRNA genes. Moreover, most sRNAs are encoded by autonomous transcription units, whereas proteins are often translated from polycistronic messages. Thus, inactivation of upstream genes in an operon often compromises expression of the downstream genes (polarity). In spite of a statistically unfavorable bias, transposon insertion mutagenesis identified a 213-nt sRNA, encoded by the *sra* gene of *Bradyrhizobium japonicum*, that is essential for symbiotic nodule development [32].

36.2.4
Biocomputational Screens

Following many years of finding new sRNAs by and large serendipitously, biocomputation-based approaches have lately led to a methodological breakthrough, and to a rapid increase in the number of sRNA genes discovered in bacteria. Even though all of the studies discussed below concerned *E. coli*, the developed methods are generally applicable to many other bacteria.

Three pioneering studies quadrupled the number of known *E. coli* sRNAs in 2001 [13, 33, 34]. Conceptual similarity lies in that the prediction of sRNA genes was based on the conservation of IGRs between related enterobacteria. The complete DNA sequences of *E. coli* and closely related *Salmonella* species were known; partial genome sequences were also available for more distant relatives (*Yersinia pestis* and *Klebsiella pneumoniae*). Argaman et al. [33] extracted the sequences of IGRs from *E. coli* and subjected them to a predictive scheme that built on four criteria: (1) transcription initiation signals (promoters), (2) Rho-independent termination signals, (3) sequence conservation and (4) candidate gene orientation. Sequences in which the distance between a predicted promoter (+1 position) and a terminator (run of Us downstream of stem–loop) was 50–400 bp were compared to other genomes by BLAST searches. The upper limit of 400 bp was based on the size of the longest sRNAs known at that time (M1 RNA, 377 nt; CsrB, 369 nt; tmRNA, 363 nt). Regions of significant conservation were selected as candidates and further inspected with regard to the orientation of the putative sRNA relative to that of the adjoining genes; sRNA candidates orientated oppositely to both neighboring genes were considered particularly "safe", that is they could not correspond to mRNA leaders or trailers. Incorporation of the different criteria resulted in a list of 24 putative sRNA genes (Fig. 36.1). The promoter algorithm, based on a σ^{70} promoter consensus, gave redundant results and predicted several putative promoters for each sRNA, the majority of which was not supported by subsequent biochemical analysis. Hence, the final candidate list was assembled by giving higher weight to the other three criteria. Twenty-three of these candidates were tested experimentally by Northern probing, which led to the discovery of 14 new sRNA

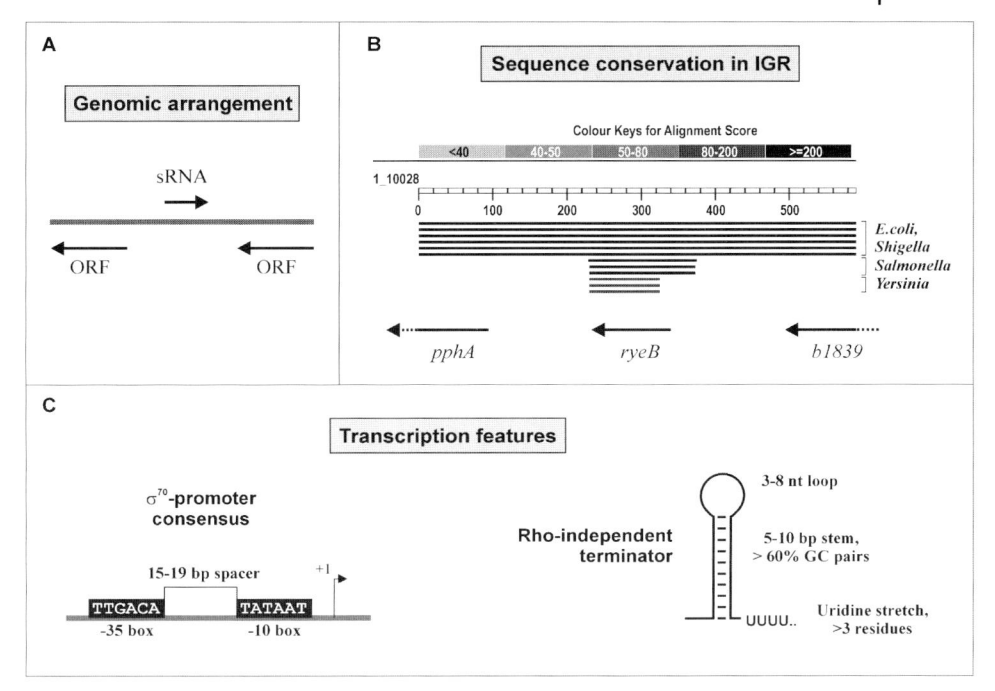

Fig. 36.1. Individual criteria of the predictive scheme by Argaman et al. [33]. (A) sRNA candidates in intergenic regions that were orientated oppositely to both neighboring genes were considered particularly "safe", i.e. they could not correspond to mRNA leaders or trailers. (B) BLASTN search result for the intergenic region (including 100 bp of upstream and downstream coding sequence) that harbors the small RNA gene *ryeB*. Conservation of the entire region from *E. coli* MG1655 is limited to the genomes of pathogenic *E. coli* and *Shigella* strains. However, the *ryeB* sequence itself displays high homology values in related enterobacteria such as *Salmonella* and *Yersinia* species. (C) Transcription features shared by many *E. coli* sRNAs that served as input to search for promoter and terminator sequences in intergenic regions.

genes. In retrospect, the high success rate is attributable to stringent application of prediction criteria or, in other words, to keeping the candidate list short and free of seemingly "weak" candidates. Moreover, most of the candidates predicted bore the hallmark of known sRNAs, that is a Rho-independent terminator; sRNAs with Rho-dependent terminators or with 3′ ends generated by processing, such as 6S RNA, would have been difficult to predict using this approach.

Wassarman et al. [13] took a somewhat similar route by extracting the 1087 *E. coli* IGR sequences of a length exceeding 180 bp, rating their conservation in *Salmonella* and *Klebsiella* species, and evaluating them further with respect to transcription signals and gene orientation. This study also involved an independent set of experiments in which putative sRNA transcripts were detected on *E. coli* microarrays. Finally, they assembled a list of 59 sRNA candidates (partly supported by the microarray experiments; see also Section 36.2.5), of which 23 were detected on Northern blots. Of these, 17 represented new sRNA genes and six were reclassified as putative mRNAs, since their conservation and translation potential suggested the presence of small protein-coding regions.

A conceptually different prediction algorithm was developed by Rivas et al. [34], based on conservation of potential RNA secondary structure elements rather than of primary sequence. It has previously been noted that secondary structure alone is not sufficient to predict non-coding RNAs [35]. However, by combining structure prediction with comparative analysis of *E. coli*, *Salmonella* and *Klebsiella* genomes, the different conservation biases acting either on protein-coding regions or on RNA structure-encoding regions was exploited. This algorithm was called QRNA [36] and predicted 556 candidate sRNA loci from *E. coli* IGRs, of which 275 loci remained after removal of known regulatory and repetitive elements. When 49 of the final candidate genes were assayed experimentally, 11 were found to express sRNA transcripts. In addition, several of the new sRNA genes that had been confirmed shortly before [13, 33] were on the list of the 275 candidates. In this study, the authors cautiously point out that the high number of sRNA candidates is preliminary and requires validation. Northern analysis was limited to RNA from a single growth condition, exponential phase [34]. In contrast, the results of the other two screens, using ten [33] or three [13] different time points and/or growth conditions, showed that many of the new sRNAs were upregulated only upon entry into stationary phase. Concerning detection methods, Wassarman et al. [13] used RNA antisense probes (riboprobes) for either strand of the candidate regions. This provides higher sensitivity than the use of single-stranded DNA oligonucleotide probes (used in the other two screens) and permitted the identification of the only example of overlapping sRNAs expressed from both strands of an IGR (RyeA/SraC, RyeB). Recently, we have confirmed another high-scoring candidate, *psrA19*, from the list published by Argaman et al. [33]. In this case, several DNA oligonucleotides targeted to various regions of this 75 nt RNA, now denoted IstR, were insufficient as probes, whereas a riboprobe permitted its detection (J. Vogel, L. Argaman, E. G. H. Wagner and S. Altuvia, unpublished results).

Subsequently, two additional genome-wide screens of *E. coli* IGRs were published. Carter et al. [37] developed a machine learning approach to extract features shared by known sRNAs in order to predict new candidates. The output of this screen, 562 predicted genes, contains many redundant candidates; frequently, several candidates were predicted in the same IGR and/or on both strands. No attempt was made to experimentally verify hitherto unpredicted sRNAs. Nevertheless, like QRNA, this approach seems to be less dependent on prior knowledge of specific RNA gene features in a given organism. In another study, Chen et al. [38] confined their sRNA search strategy to predicting solely σ^{70} promoter/Rho-independent terminator pairs (same orientation) with a distance range of 45–350 bp. Of 227 initial sRNA candidates, 51 were removed as putative short mRNAs (approximately half of these ORFs had not been annotated, whereas the other half were annotated in some but not all *E. coli* databases). Another 32 were filtered as orphan tRNA genes, short leaders, tRNA/rRNA operon fragments, or known sRNA genes. Importantly, of about 40 sRNA genes known at the time of publication, only 10 were recognized by the search algorithm. Of 144 final candidates, eight were experimentally tested (Northern blots) and seven confirmed. Although this screen is reminiscent of the one carried out by Argaman et al. [33], Chen et al. did detect new sRNA genes, most of which had not been predicted before. The omission of conservation aspects may have resulted in a stronger bias towards predicting sRNA genes specific to, in this case, *E. coli*. For instance, most of the seven newly identified sRNAs lack clear homologs outside the range of some pathogenic *E. coli* strains and/or the closely related bacterium *Shigella flexneri* [39]. As a note of caution – without support from phylogenetic conservation, experimental validation is even more important as a criterion for whether species-specific sRNAs indeed are functional.

An extension of a prediction scheme based on promoters could involve parallel analysis of putative binding sites of stress-related activator or repressor proteins in the vicinity of the predicted sRNA promoters (see also Chapter 37). This might identify sRNA genes that are part of known regulons, thus facilitating a subsequent functional characterization. Consistent with the presence of a binding site of Fur (a repressor of Fe^{2+}-regulated genes) overlapping the promoter of the 90-nt RyhB RNA, also known as SraI [13, 33], this sRNA was shown to be regulated by iron [40]. In this case, the Fur site was found by genetic analysis prior to discovery of the *ryhB* gene [41].

The rapidly growing list of completed bacterial genomes will certainly close the phylogenetic gaps between distantly related organisms, thereby facilitating the discovery of sRNA homologs by bioinformatics using conservation criteria. As yet, obstacles for homology searches lie in a limited conservation of sRNAs at the primary sequence level. Here, it should be further taken into account that longer sRNA genes can become split or permuted during evolution and even be encoded by two separate loci. Circularly permuted tmRNA genes have recently been reported in *Caulobacter* and cyanobacteria. In these cases, a segment normally found at the 3' end of tmRNA genes is located upstream of the segment normally at the 5' end. The two conserved tmRNA segments are linked by a short, non-conserved sequence [42, 43] which is excised post-transcriptionally to yield the standard tRNA acceptor stem termini, thus resulting in a two-piece mature tmRNA.

36.2.5
Microarray Detection

Microarrays have become a valuable tool to monitor mRNA expression at a genome-wide scale and should, in principle, also be useful to detect new transcripts such as sRNAs. However, most microarrays available for bacterial genomes carry only probes specific for protein-coding regions, whereas most sRNA genes are found in IGRs. An exception to the "standard" microarray is the high-density oligonucleotide probe array for *E. coli* introduced by Selinger et al. [44]. It carries strand-specific oligodeoxyribonucleotide probes for all ORFs, tRNAs and rRNAs, but additionally covers both strands in IGRs at least 40 bp in length and here with one probe every six bases. In the initial study, focusing primarily on development of this technology, a few sRNAs and antisense RNAs were detected *en passant*. Employing this array type in independent experiments later bolstered the sRNA prediction by comparative genome analysis in the study by Wassarman et al. [13] and came into full play in a study by Tjaden et al. [45], in which microarray signals suggested nine new intergenic sRNA candidates (supported by significant homology in *S. typhimurium*) in addition to the around 45 sRNA genes known at the time. These two latter studies report and discuss several problems associated with reliable detection of small structured RNAs on microarrays, in particular with respect to probe preparation. Representative examples of sRNA expression profiles are given in Fig. 2 in Wassarman et al. [13]. Frequently, only a subset of the oligos within the sRNA transcript region gave a clear signal although the same sRNAs were often visible as strong bands on Northern blots; under a single growth condition, about one-third of the sRNAs identified in this study gave signals that paralleled Northern results. Curiously, Tjaden et al. [45] occasionally observed sRNA signals on the strand opposite to the one experimentally found to be transcribed. Whether these signals account for yet unknown antisense RNAs or might represent experimental noise is currently unclear.

Notwithstanding these problems, it can be expected that technical progress (and decreasing costs) will make microarrays a standard tool for the identification as well as for the expressional profiling of bacterial sRNAs. Microarrays with IGR coverage are now available for bacteria other than *E. coli*, such as *B. subtilis*, *Pseudomonas aeruginosa* and *Staphylococcus aureus*, though limiting IGR coverage to those of 150 bp or larger (Carsten Rosenow, Affymetrix, personal communication). Recently, microarray approaches have also been used to detect *E. coli* sRNAs that were selectively co-immunoprecipitated with the RNA-binding protein Hfq ([46]; see also Section 36.2.7). In this study, the authors used a novel detection method employing antibodies specific for RNA:DNA hybrids, which was reported to improve the sensitivity of detection and to circumvent the problems associated with labeling of small structured RNAs (however, note that as of the time of writing, these antibodies are not commercially available).

36.2.6
Shotgun Cloning (RNomics)

The cDNA cloning of RNAs in the size range of 50–500 nt aims at identifying sRNAs that are expressed in a given genome under a given set of conditions, irrespective of whether they represent primary transcripts or processing products. This approach, "Experimental RNomics", is described in detail in Chapter 38 and has been successfully employed to discover numerous non-coding RNAs in several eukaryotic organisms and in the archaeon *Archaeoglobus fulgidus* [47–53]. As for eubacteria, it has so far been applied only to *E. coli* [54]. RNomics is in some respects similar to the early approaches that identified sRNAs by orthophospate labeling and biochemical isolation (see Section 36.2.2). The main difference is that reverse transcription followed by cloning has since revolutionized isolation and sequence determination of many sRNAs in parallel. As an additional example, small-sized

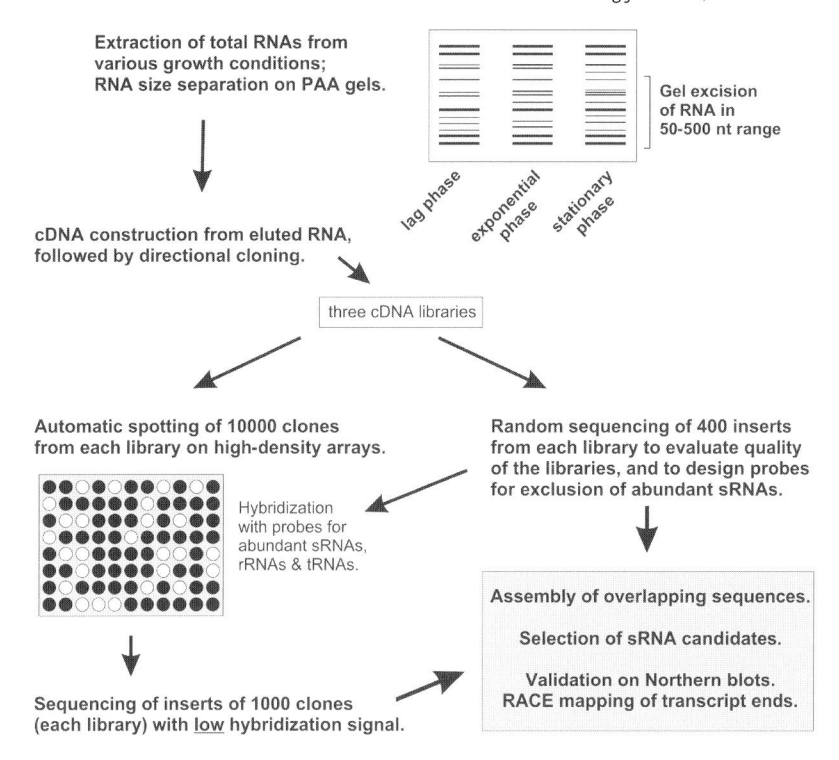

Fig. 36.2. Flow chart of the RNomics screen in *E. coli* as published by Vogel et al. [54]. See text for a more detailed explanation of the individual steps. For experimental details on RNomics, see Chapter 38.

(around 22 nt) subclasses of eukaryotic sRNAs, such as miRNAs and siRNAs, were identified by similar protocols [55–58].

Vogel et al. [54] constructed different cDNA libraries from total RNA representing lag phase, exponential/log phase and stationary phase, respectively, from *E. coli* grown under standard laboratory conditions. (For a flow chart of the *E. coli* RNomics screen, see Fig. 36.2) The construction of three individual libraries was taken as a precaution since growth rate-specific expression had already been observed for many *E. coli* sRNAs (e.g. [13, 33]). Indeed, the number of individual clones obtained for a given sRNA generally matched signal intensities on Northern blots run with RNA from different growth phases. In contrast, some well-characterized sRNAs that are specifically stress-induced were not detectable during normal growth, e.g. OxyS [10] was not found in the libraries.

Total RNA was fractionated on denaturing polyacrylamide gels and extracted (size range about 50–500 nt). The size-selected RNA pool was C-tailed with poly(A) polymerase, reverse transcribed and the cDNA libraries were constructed by directional cloning in *E. coli*. Since cDNA fragments were cloned (i.e. promoterless sequences), we did not anticipate difficulties with cloning on high-copy plasmids. Then, 10 000 individual clones were spotted on high-density filters, which were hybridized with probes for rRNA and most of the tRNA genes. In total, 1000 cDNA clones from each growth phase-specific library, exhibiting the *lowest* hybridization scores (i.e. those that did not hybridize to rRNA or tRNA probes), were sequenced. Overlapping sequences were automatically sorted and assembled into *contigs* (groups of identical and/or overlapping sequences) and mapped to the *E. coli* genome sequence by BLASTN searches.

Although RNAs extracted from polyacrylamide gels should have a lower size limit of around 50 nt, the libraries contained a high proportion of smaller fragments, often 20 nt or less. In the subsequent analysis, all fragments of 15 bp or less were disregarded, since they frequently gave multiple hits in BLASTN searches and thus could not be mapped unequivocally to the *E. coli* genome.

The final set of 451 *contigs* represented about half of the initial sequences. Of these, 78% were derived from within coding regions of known mRNA genes and their leaders, as well as from ORFs of unknown function. Directional cloning unambiguously specified the orientation of the cloned RNA fragments. About 5% of the final sequences had matches within a gene or ORF, but in antisense orientation. These antisense fragments were not analyzed further. Hence, it remains unclear whether they represent distinct antisense RNAs or resulted from global antisense transcription (see also Section 36.2.8). Fragments from intergenic or, more precisely, intercoding regions accounted for 17%. Since only less than 11% of the *E. coli* genome belongs to this category [1], the sequences obtained from the library exhibited a slight bias towards IGRs.

In this study, the number of cloned sRNA-specific sequences per contig varied greatly. Also, the length and position of 5′ and 3′ ends of individual contig members differed. Most sRNAs were represented by one to five individual sequences, but this number increased to 59 sequences for RyeB (derived from the RNA pool obtained under growth conditions at which this sRNA is strongly induced).

To summarize the results of this screen, about half of the sRNAs known at that time were present in the libraries. Some of the newly found sRNAs from independent genes were specific to *E. coli*. Particularly, several mRNA leader and trailer fragments were found to accumulate as distinct sRNA species; two of these, SroA and SroG, contain aptamer-like binding sites known as riboswitch elements [59]. It is unlikely that such distinct, processed sRNA species would have been predicted by other methods.

Comparing methods, the underlying principles of RNomics are related to those of a microarray-based screen and so are its technical problems. RNA secondary structure certainly imposes technical limitations, since sRNAs may differ dramatically in their efficiency of C-tailing, thus skewing the representation of individual sRNA in the cDNA library. For example, 6S RNA was poorly represented even in the stationary phase library despite its high abundance in this growth phase. In principle, one might introduce changes to the individual enzymatic steps, such as to carry out cDNA synthesis at elevated temperature (using a thermostable reverse transcriptase). Further improvement could be expected by raising the sampling number, employing a higher stringency of the pre-screen (for tRNA and rRNA fragments) and by decreasing the fraction of very small fragments through a repeated gel extraction of the initially excised RNA (or pre-fractionating RNA on columns or sucrose gradients).

Many sRNA transcripts retain a triphosphate at their 5′ end [33, 54]. This suggests that, if cDNA library construction included a 5′-linker ligation step, the 5′-triphosphates should be converted to 5′-monophosphates to avoid depletion of primary sRNA transcripts. This can be achieved by treatment of the initial RNA pool with tobacco acid pyrophosphatase (TAP), which specifically converts RNA 5′-triphosphates to 5′-monophosphates.

36.2.7
Co-purification with Proteins or Target RNAs

Some sRNAs require proteins for activity or stability, some others in turn act on and modify the activity of proteins. In either case, they will interact with these proteins, suggesting that their isolation and purification from sRNA–protein complexes should be feasible. The RNA CsrB of *E. coli* was discovered in studies of the RNA-binding protein CsrA, which promotes decay of a number of mRNAs. A recombinant HIS-tagged CsrA protein from *E. coli* was found to purify as a globular ribonucleoprotein complex that, apart from 18 CsrA subunits, contained a 366-nt RNA [60]. The corresponding *csrB* gene was then identified by poly(A)-tailing of the isolated RNA, subsequent cDNA cloning, and BLASTN searches. A small RNA, RsmZ, that binds to RsmA, the CsrA homolog of *P. fluorescens* CHA0, was identified similarly [61]. Interestingly, the *rsmZ* gene shows no significant sequence similarity to the *csrB* homologs in *E. coli*, *Salmonella* and *Erwinia* species, even though all these RNAs share some overall secondary structure elements. However, they are functionally equivalent; these RNAs interact with CsrA/RsmA homologs which appear to recognize short sequence motifs (GGA [60, 62]) in the sRNAs. Hence, cloning of co-purified RNAs can be helpful in cases where sRNA sequences have diverged to a point that does not permit identification based on similarity, but where the protein component can be identified by BLAST searches.

Binding of helper proteins may stabilize sRNAs by protecting them from RNases, and/or aid annealing to target mRNAs (in antisense cases). Proteins known to bind RNAs are, for example, the Sm-like protein Hfq (also known as HF-I), the RNA chaperone StpA, the histone-like protein HU, and the transcriptional regulator H-NS (see references in [18]). These proteins are very abundant, ranging from 20 000–30 000 (StpA, H-NS) to around 50 000 (Hfq, HU) molecules per cell during logarithmic growth [63]. Of these, Hfq is the best-characterized sRNA binder. For example, RprA, DsrA, RhyB, OxyS and Spot 42 are dependent on this protein for regulatory activity (references in [18]). Wassarman et al. [13] found that seven newly identified sRNAs could be co-immunoprecipitated with Hfq. Subsequently, co-immunoprecipitation with anti-Hfq antibodies, combined with sRNA detection on microarrays, was employed as a new screening approach (see Section 36.2.5). We expect that tagging of RNA binding proteins by affinity tags (e.g. FLAG or 6 × His), followed by immunoprecipitation with a tag-specific antibody, could be a general strategy to identify associated sRNAs, provided that the affinity tags do not interfere with RNA binding.

Surprisingly, antibodies against *eukaryotic* snRNP (small nuclear ribonucleoprotein) also immunoprecipitated small bacterial RNAs of 140–240 nt. These RNAs were detected in the cyanobacterium *Synechococcus leopoliensis* and in *B. subtilis*, but not in *E. coli* [64]. To the best of our knowledge, this interesting observation has not been followed up nor have their sequences been determined.

36.2.8
Screens for *Cis*-encoded Antisense RNAs

Cis-encoded antisense RNAs appear to be mainly confined to plasmids, transposons, and bacteriophages (for comprehensive reviews, see [19, 65]). Indeed, very few antisense RNA genes of this type have so far been found in bacterial chromosomes. However, significant transcription antisense to protein-encoding genes was detected by high-density microarrays that covered both strands of the *E. coli* genome [44]. Unfortunately, this study did not address a possible biological significance of the frequent occurrence of antisense transcripts. Are some of these functional antisense RNAs? In the same organism, a substantial number of antisense RNA transcripts, or fragments thereof, were also identified in the shotgun-cloning approach (RNomics) described above ([54]; see Section 36.2.6). Although these fragments accounted for more than 5% of the total clones, their further evaluation awaits functional tests. At this point, the presence of antisense transcripts is interesting, but inconclusive. In most cases, one can assume that fortuitous transcription events have been sampled and hence an assignment of *bona fide* sRNAs must require either the detection of distinct antisense transcripts (of fairly unique size) – at least under some conditions – and/or a suggested functional involvement in some biological response, e.g. inferred from genetic experiments. Thus, in spite of the interesting possibility of ORF-internal antisense RNAs, there is little data that indicates whether bacteria use such RNAs in regulatory pathways. If a researcher suspects an sRNA to be encoded opposite a protein-encoding gene, it is advisable to conduct initial experiments in *rnc* mutant strains (deficient for RNase III), since a putative target interaction is expected to result in cleavage of the antisense/target RNA duplex. It is also advisable to search for promoters (bioinformatically or by promoter–reporter gene fusion experiments) to substantiate the claim that a new sRNA has been found.

36.3
Conclusions

The strategies discussed above have all been successful in identifying novel sRNAs in *E. coli* as well as in other bacteria. Although this in itself is of importance – since it emphasizes that sRNAs are far more prevalent than previously assumed – the key question is: what are they doing? Functional characterizations and searches for biological roles are the subject of Chapter 37.

A few conclusions can be drawn from a survey of the various detection strategies. Computational screens, transcriptional profiling (microarrays) and RNomics (shotgun cloning) approaches usually give no immediate clues as to what function a newly discovered sRNA may carry out. However, these approaches, when followed up by, for example, Northern analyses under relevant conditions, at least validate that a distinct sRNA has been found. In contrast, strategies that use affinity to certain proteins as the main search criterion may immediately identify functionally related sRNAs. If indeed common mechanisms – involving proteins such as Hfq or others – apply, the elucidation of one sRNA with respect to its function could facilitate the subsequent study of other sRNAs' roles by proxy. Conversely, there might be a widely used machinery for RNA–RNA interactions, which is utilized for entirely different purposes. Thus, *finding* a new sRNA is only the beginning of an exciting journey into the biology of small RNAs.

In the last few years, many new sRNAs have been discovered in bacteria as well as in organisms of all kingdoms of life. Clearly, the number of searches in *E. coli* alone bear witness to the growing interest in the roles that these RNAs may play in various life processes and responses. As shown by the discussion of different strategies used, there is no single superior method for sRNA discovery. Each method has its strengths and weaknesses. What has become clear – we believe – is that common denominators are rarely found. The *E. coli* sRNAs being investigated at present do not exhibit shared primary or secondary structures [39], they display a great range of metabolic stabilities [54] and they can act by different mechanisms (protein sequestration, antisense mechanisms or others). Thus, the obvious next step is to carry out a search for function. In many, possibly most of the cases, this will involve a search for targets of sRNAs and/or a search for traits and responses that are affected by the expression of single sRNAs.

Acknowledgments

The authors gratefully acknowledge funding from HFSP (Human Frontier Science Program), VR (The Swedish Research Council), Wallenberg Foundation (all to E. G. H. W.) and from EMBO (long-term fellowship, J. V.).

References

1 F. R. BLATTNER, G. PLUNKETT, C. A. BLOCH, N. T. PERNA, V. BURLAND, M. RILEY, J. COLLADO-VIDES et al., *Science* **1997**, *277*, 1453–1474.

2 K. M. WASSARMAN, A. ZHANG, G. STORZ, *Trends Microbiol.* **1999**, *7*, 37–45.

3 J. HINDLEY, *J. Mol. Biol.* **1967**, *30*, 125–136.

4 G. G. Brownlee, *Nat. New Biol.* **1971**, *229*, 147–149.

5 K. M. Wassarman, G. Storz, *Cell* **2000**, *101*, 613–623.

6 L. M. Hsu, J. Zagorski, Z. Wang, M. J. Fournier, *J. Bacteriol.* **1985**, *161*, 1162–1170.

7 T. Ikemura, J. E. Dahlberg, *J. Biol. Chem.* **1973**, *248*, 5033–5041.

8 T. Ikemura, J. E. Dahlberg, *J. Biol. Chem.* **1973**, *248*, 5024–5032.

9 B. E. Griffin, *FEBS Lett.* **1971**, *15*, 165–168.

10 S. Altuvia, D. Weinstein-Fischer, A. Zhang, L. Postow, G. Storz, *Cell* **1997**, *90*, 43–53.

11 T. Watanabe, M. Sugiura, M. Sugita, *FEBS Lett.* **1997**, *416*, 302–306.

12 T. Watanabe, M. Sugiura, M. Sugita, *Biochim. Biophys. Acta* **1998**, *1396*, 97–104.

13 K. M. Wassarman, F. Repoila, C. Rosenow, G. Storz, S. Gottesman, *Genes Dev.* **2001**, *15*, 1637–1651.

14 S. Suzuma, S. Asari, K. Bunai, K. Yoshino, Y. Ando, H. Kakeshita, M. Fujita, K. Nakamura, K. Yamane, *Microbiology* **2002**, *148*, 2591–2598.

15 Y. Ando, S. Asari, S. Suzuma, K. Yamane, K. Nakamura, *FEMS Microbiol. Lett.* **2002**, *207*, 29–33.

16 H. W. Göhlmann, J. Weiner, A. Schön, R. Herrmann, *J. Bacteriol.* **2000**, *182*, 3281–3284.

17 T. Mizuno, M. Y. Chou, M. Inouye, *Proc. Natl. Acad. Sci. USA* **1984**, *81*, 1966–1970.

18 E. G. H. Wagner, J. Vogel, in: *Noncoding RNAs*, J. Barciszewski, V. Erdmann (eds), Landes, Austin, TX, **2003**, 243–259.

19 E. G. H. Wagner, R. W. Simons, *Annu. Rev. Microbiol.* **1994**, *48*, 713–742.

20 N. Delihas, *Mol. Microbiol.* **1995**, *15*, 411–414.

21 D. Sledjeski, S. Gottesman, *Proc. Natl. Acad. Sci. USA* **1995**, *92*, 2003–2007.

22 R. A. Lease, M. E. Cusick, M. Belfort, *Proc. Natl. Acad. Sci. USA* **1998**, *95*, 12456–12461.

23 D. D. Sledjeski, A. Gupta, S. Gottesman, *EMBO J.* **1996**, *15*, 3993–4000.

24 N. Majdalani, C. Cunning, D. Sledjeski, T. Elliott, S. Gottesman, *Proc. Natl. Acad. Sci. USA* **1998**, *95*, 12462–12467.

25 N. Majdalani, S. Chen, J. Murrow, K. St John, S. Gottesman, *Mol. Microbiol.* **2001**, *39*, 1382–1394.

26 N. D. Ulbrandt, J. A. Newitt, H. D. Bernstein, *Cell* **1997**, *88*, 187–196.

27 A. Guigueno, J. Dassa, P. Belin, P. L. Boquet, *J. Bacteriol.* **2001**, *183*, 1147–1158.

28 T. Romeo, J. Moore, J. Smith, *Gene* **1991**, *108*, 23–29.

29 T. Weilbacher, K. Suzuki, A. K. Dubey, X. Wang, S. Gudapaty, I. Morozov, C. S. Baker, D. Georgellis, P. Babitzke, T. Romeo, *Mol. Microbiol.* **2003**, *48*, 657–670.

30 T. Romeo, *Mol. Microbiol.* **1998**, *29*, 1321–1330.

31 S. Aarons, A. Abbas, C. Adams, A. Fenton, F. O'Gara, *J. Bacteriol.* **2000**, *182*, 3913–3919.

32 S. Ebeling, C. Kundig, H. Hennecke, *J. Bacteriol.* **1991**, *173*, 6373–6382.

33 L. Argaman, R. Hershberg, J. Vogel, G. Bejerano, E. G. Wagner, H. Margalit, S. Altuvia, *Curr. Biol.* **2001**, *11*, 941–950.

34 E. Rivas, R. J. Klein, T. A. Jones, S. R. Eddy, *Curr. Biol.* **2001**, *11*, 1369–1373.

35 E. Rivas, S. R. Eddy, *Bioinformatics* **2000**, *16*, 583–605.

36 E. Rivas, S. R. Eddy, *BMC Bioinformatics*, **2001**, *2*, 8.

37 R. J. Carter, I. Dubchak, S. R. Holbrook, *Nucleic Acids Res.* **2001**, *29*, 3928–3938.

38 S. Chen, E. A. Lesnik, T. A. Hall, R. Sampath, R. H. Griffey, D. J. Ecker, L. B. Blyn, *Biosystems* **2002**, *65*, 157–177.

39 R. Hershberg, S. Altuvia, H. Margalit, *Nucleic Acids Res.* **2003**, *31*, 1813–1820.

40 E. Massé, S. Gottesman, *Proc. Natl. Acad. Sci. USA* **2002**, *99*, 4620–4625.

41 N. Vassinova, D. Kozyrev, *Microbiology* **2000**, *146*, 3171–3182.
42 K. C. Keiler, L. Shapiro, K. P. Williams, *Proc. Natl. Acad. Sci. USA* **2000**, *97*, 7778–7783.
43 C. Gaudin, X. Zhou, K. P. Williams, B. Felden, *Nucleic Acids Res.* **2002**, *30*, 2018–2024.
44 D. W. Selinger, K. J. Cheung, R. Mei, E. M. Johansson, C. S. Richmond, F. R. Blattner, D. J. Lockhart, G. M. Church, *Nat. Biotechnol.* **2000**, *18*, 1262–1268.
45 B. Tjaden, R. M. Saxena, S. Stolyar, D. R. Haynor, E. Kolker, C. Rosenow, *Nucleic Acids Res.* **2002**, *30*, 3732–3738.
46 A. Zhang, K. M. Wassarman, C. Rosenow, B. C. Tjaden, G. Storz, S. Gottesman, *Mol. Microbiol.* **2003**, *50*, 1111–1124.
47 J. Cavaille, K. Buiting, M. Kiefmann, M. Lalande, C. I. Brannan, B. Horsthemke, J. P. Bachellerie, J. Brosius, A. Hüttenhofer, *Proc. Natl. Acad. Sci. USA* **2000**, *97*, 14311–14316.
48 A. Hüttenhofer, J. Brosius, J. P. Bachellerie, *Curr. Opin. Chem. Biol.* **2002**, *6*, 835–843.
49 A. Hüttenhofer, M. Kiefmann, S. Meier-Ewert, J. O'Brien, H. Lehrach, J. P. Bachellerie, J. Brosius, *EMBO J.* **2001**, *20*, 2943–2953.
50 C. Marker, A. Zemann, T. Terhorst, M. Kiefmann, J. P. Kastenmayer, P. Green, J. P. Bachellerie, J. Brosius, A. Hüttenhofer, *Curr. Biol.* **2002**, *12*, 2002–2013.
51 T. H. Tang, J. P. Bachellarie, T. Rozhdetvensky, M. L. Bortolin, H. Huber, M. Drungowski, T. Elge, J. Brosius, A. Hüttenhofer, *Proc. Natl. Acad. Sci. USA* **2002**, *99*, 7536–7541.
52 T. H. Tang, T. S. Rozhdestvensky, B. C. d'Orval, M. L. Bortolin, H. Huber, B. Charpentier, C. Branlant, J. P. Bachellerie, J. Brosius, A. Hüttenhofer, *Nucleic Acids Res.* **2002**, *30*, 921–930.
53 G. Yuan, C. Klambt, J. P. Bachellerie, J. Brosius, A. Hüttenhofer, *Nucleic Acids Res.* **2003**, *31*, 2495–2507.
54 J. Vogel, V. Bartels, T. H. Tang, G. Churakov, J. G. Slagter-Jager, A. Hüttenhofer, E. G. Wagner, *Nucleic Acids Res.* **2003**, *31*, 6435–6443.
55 S. M. Elbashir, W. Lendeckel, T. Tuschl, *Genes Dev.* **2001**, *15*, 188–200.
56 M. Lagos-Quintana, R. Rauhut, W. Lendeckel, T. Tuschl, *Science* **2001**, *294*, 853–858.
57 N. C. Lau, L. P. Lim, E. G. Weinstein, D. P. Bartel, *Science* **2001**, *294*, 858–862.
58 R. C. Lee, V. Ambros, *Science* **2001**, *294*, 862–864.
59 E. Nudler, A. S. Mironov, *Trends Biochem. Sci.* **2004**, *29*, 11–17.
60 M. Y. Liu, G. Gui, B. Wei, J. F. Preston, L. Oakford, U. Yuksel, D. P. Giedroc, T. Romeo, *J. Biol. Chem.* **1997**, *272*, 17502–17510.
61 S. Heeb, C. Blumer, D. Haas, *J. Bacteriol.* **2002**, *184*, 1046–1056.
62 C. Valverde, M. Lindell, E. G. Wagner, D. Haas, *J. Biol. Chem.* **2004**, *279*, 25066–25074.
63 T. Ali Azam, A. Iwata, A. Nishimura, S. Ueda, A. Ishihama, *J. Bacteriol.* **1999**, *181*, 6361–6370.
64 S. A. Kovacs, J. O'Neil, J. Watcharapijam, C. Moe-Kirvan, S. Vijay, V. Silva, *J. Bacteriol.* **1993**, *175*, 1871–1878.
65 E. G. H. Wagner, S. Altuvia, P. Romby, *Adv. Genet.* **2002**, *46*, 361–398.

37
Approaches to Identify Novel Non-messenger RNAs in Bacteria and to Investigate their Biological Functions: Functional Analysis of Identified Non-mRNAs

E. Gerhart H. Wagner and Jörg Vogel

37.1
Introduction

Small non-coding/non-messenger RNAs have been discovered at a staggering rate in bacteria (primarily *Escherichia coli* [1–6]) as well as in archaea and eukaryotes ([7] and accompanying Chapter 36). To date, many more than 1000 non-coding RNAs are known to be encoded by various cells and organisms. The functions of the vast majority of them are unknown. How can we find out about their biological roles? In this chapter, possible approaches are discussed. We disregard two abundantly characterized classes of non-messenger RNAs, tRNAs and rRNAs, while concentrating on the heterogeneous class of small, non-coding RNAs of unknown function. As will become apparent, we start from the unproven yet circumstantially supported conviction that most of these new non-coding RNAs will turn out to have regulatory roles and only a few will support housekeeping functions.

In eukaryotes, many non-coding RNAs appear to fall into categories or classes that suggest related functions in certain cellular processes. For instance, snRNAs are mostly involved in RNA processing (splicing). The snoRNAs (small nucleolar RNAs, classified as H/ACA and C/D box snoRNAs) guide the site-specific pseudouridylation or methylation of, in particular, rRNA in eukarya and archaea. However, shotgun cloning (often referred to as RNomics) approaches have turned up many more snoRNAs than those required to account for modification activities. It has been suggested that these orphan snoRNAs may act on other – so far unknown – targets [7]. The microRNAs (miRNAs) are a very heterogeneous class of about 22-nt long RNAs derived from longer precursors by successive enzymatic action of two RNase III-like enzymes, Drosha and Dicer [8–11]. Many of them have been shown to act as developmental regulators in animals and plants. The small interfering RNAs (siRNAs) of about 22 nt are the effectors of the RNA interference pathways and can – at least in some cases – also direct heterochromatin formation [12, 13]. Thus, even though the functional significance of *individual* members of each of the above classes of small RNAs is mostly unknown, there appear to be unifying themes that can aid functional assignment.

Small, non-coding RNAs (here denoted as sRNAs) in bacteria are heterogeneous. At present there are no easily discernible criteria such as size, genomic organization, common structures or structural motifs, etc., that would sort them into meaningful functional categories [14–16]. This, in turn, makes it difficult to "guess" at their roles and motivates an attempt to outline approaches that can be used to elucidate their biological functions.

As indicated in the preceding Chapter 36, more than 60 sRNAs are present in *E. coli*, the majority of so-far unknown function. In other bacteria, much fewer sRNAs have been found to date [17–19], although this may be attributed to a lack of systematic searches. So far, the absence of obvious classes of structurally related prokaryotic sRNAs indicates that function must be assigned on a case-by-case basis.

The nature of the problem addressed in this chapter – identifying biological roles for novel sRNAs – implies that feasible and appropriate strategies are considered more important than detailed methods (although methods can be found in the papers cited). Below, we will describe both large- and small-scale approaches that may be useful in the process of determining sRNA function.

37.2
Approaches for Elucidation of Bacterial sRNA Function

The number of identified genes encoding bacterial sRNAs of unknown function has increased dramatically in recent years and consequently there are many RNAs to work on. However – at this point – most often it has only been ascertained that a certain sRNA is encoded at a specific location within a given genome. Additional information about expression patterns (e.g. by Northern analysis) may or may not be available. To elucidate what function is associated with an sRNA, large-scale approaches could be the first choice. Alternatively, leads derived from determined expression patterns or bioinformatics-based information can be used to approach a biological role more directly. A few examples will be given below.

37.2.1
Large-scale Screening for Function

Given sufficient manpower and/or access to core facilities, large-scale approaches can be fruitful. The availability of genomic sequences is a prerequisite for comprehensive screenings for sRNA-dependent changes in the transcriptome, the proteome, the metabolome, or any other '-ome' that may be defined in the years to come.

37.2.2
Preparing for Subsequent Experiments: Strains and Plasmids

Whether one chooses larger- or smaller-scale analyses, a number of strains and plasmids should be constructed. For any analysis that aims at the function of an sRNA, one should create a strain with a deletion of the gene in question (knockout strain), which then can be complemented by an sRNA-expressing plasmid.

Mutations in bacterial strains can be obtained in a variety of ways: chemical-induced mutagenesis, transposon mutagenesis, etc. In terms of ease and precision, genetically engineered replacement of the chromosomal gene by a marker gene is often superior. The Court lab [20] has devised an efficient method that permits deletion/gene replacement in Gram-negative bacteria such as *E. coli*, making use of λ phage recombination functions. As an example, an sRNA gene is PCR-amplified from chromosomal DNA with primers designed to include its upstream and downstream DNA regions, and – in some cases – cloned in a plasmid of choice. An inner segment of the sRNA gene is replaced by a selectable marker gene (often an antibiotic resistance gene bordered by terminator sequences to prevent transcriptional read-through/read-out). After validation of such a construct, the engineered sRNA gene region is PCR-amplified and, as a linear fragment, introduced into a strain (see [20] for details) in which a temperature shift promotes homologous recombination. Alternatively, the antibiotic resistance cassette can, without prior cloning, be directly amplified using primers that introduce the flanking sequences of the sRNA gene to be deleted. As little as 30–50 bp of authentic flanking sequences suffice to exchange the chromosomal copy of the sRNA gene by the deletion allele. Once stable recombinants have been obtained, the marker-tagged sRNA deletion allele will have to be moved into the appropriate strain background by conventional P1 phage transduction [21], thus creating an isogenic sRNA-null mutant strain for subsequent analysis.

What problems can one anticipate? If an sRNA gene is essential, a deletion strain cannot be constructed directly by linear transformation. Based on available data, this is an unlikely outcome; the sRNA gene deletions analyzed so far have not resulted in lethality under "normal" growth conditions, except for the RNA subunit of RNase P, 4.5S RNA and tmRNA – in some or all of the bacterial species analyzed. If one suspects an essential role (i.e. if linear transformation failed repeatedly), a chromosomal deletion can be introduced in the presence of a plasmid carrying the intact sRNA gene (supplying the "essential" function *in trans*). A second problem can arise from a "tight" location: an sRNA gene may overlap DNA sequences that are required for proper expression or control of neighboring genes. It is worth noting that intergenic location, as defined by most researchers, only considers the position relative to neighboring open reading frames (ORFs), not the genes *per se*. Therefore, the promoter of a neighboring gene, or its regulatory sites, may lie within the DNA segment which has been removed and replaced as described above. If so, obtained phenotypic effects can be due to either the absence of the sRNA, a defect in expression of a neighboring gene, or both.

The next step in obtaining a tool kit for functional analysis is the construction of a conditional sRNA overexpressor strain. The easiest approach involves cloning of the sRNA sequence (as DNA) in a plasmid downstream of its own or a suitable heterologous promoter. The choice of plasmid may be important: naturally occurring plasmids and commercial vectors come in many flavors, in particular with respect to copy number, from close to single-copy plasmids like (in *E. coli*) F and P1 at about 1 copy per chromosome, over pSC101* plasmids at about 3, over p15A plasmids at about 15, to high copy ColE1/pUC vector plasmids at 50–300 copies.

In addition, many different cloning vectors exist, some of which have an extended host range (i.e. they can be introduced into/maintained in different bacterial species), some carry stable maintenance functions (particularly important for low-copy number vectors which otherwise are easily lost from the population) and some carry useful promoter/operator combinations. Since it is beyond the scope of this article to list examples representing all cases, the reader is referred to [22–29].

The second consideration is the choice of promoter that drives expression of the sRNA gene. Regulatable promoters are mostly derived from controlled bacterial operons (p*lac*, p*tac* – a hybrid *trp/lac* promoter – and pBAD [30–32]) or from bacteriophages (λ P_L, phage T7 promoter [33, 34]). These promoters have often been modified to enhance transcription rates and regulatory elements, such as operators, have been introduced for improvement with respect to control. The *lac* and *tac* promoters are strong, but prone to leakiness when under the control of a chromosomally encoded *lacI* (repressor) gene; induction is accomplished by addition of IPTG. Using a chromosomally or plasmid-encoded *lacI*^{q} allele (a repressor variant that is more highly expressed) control becomes tighter. In this case, however, higher concentrations of IPTG are required for full induction. The pBAD promoter which is derived from the *ara* operon is controlled by the AraC protein. This promoter has the beneficial property of being tightly controllable by arabinose, such that strong induction (−glucose, +arabinose) but also tight repression (+glucose, −arabinose) can be obtained. To avoid significant fluctuations in gene expression in individual cells, *E. coli* strains carrying mutations in transport functions are useful (for details, see [32]). In general, if pBAD-containing constructs are to be used, host strains must be chosen in which arabinose addition is non-toxic. Many other promoters can be adapted to an sRNA-overexpressor plasmid (e.g. T7 polymerase promoters which, however, require plasmid- or chromosome-encoded T7 RNA polymerase). A widely used promoter is P_L from phage λ, which normally is controlled by the cI repressor. Temperature-sensitive alleles (*cI857*) are available and permit transcription at non-permissive temperature. Alternatively, control can be exerted by other regulators after insertion of the operator sequence of choice. A number of very useful plasmids with a set of different tightly controlled promoters, origins of replications and antibiotic markers can be found in [34].

Based on what is known about regulatory sRNAs in bacteria, secondary structures are often crucial to the functionality of the RNA [14, 35]. Thus, if a promoter is simply placed upstream of a DNA segment that contains an sRNA gene, transcripts will usually carry additional sequences 5′ of those normally present in the authentic sRNA. This can result in a different folding of the RNA which may compromise its activity. We advise to map the transcription start site of the sRNA by methods like 5′-RACE (rapid amplification of cDNA 5′ ends [1, 36], see also Section 37.5, Protocol B) or primer extension analysis [37, 38] and then to introduce the chosen promoter so that its +1 position coincides with the one determined for the sRNA.

At this point, the researcher will have constructed a basic set of strains for further analysis: an isogenic strain pair, wild-type or deleted for the sRNA gene, a second set of strains, deleted for the sRNA in question, carrying either a plasmid with the sRNA gene under the control of a chosen promoter, or a control plasmid (lacking the insert or carrying an unrelated gene segment). Naturally, additional strain backgrounds/plasmids may be desirable, and can be chosen on a case-by-case basis.

Some considerations

There are occasional obstacles. As mentioned above, deletion of an sRNA gene may affect the expression of neighboring genes if promoter or regulatory elements are removed, thus complicating interpretations. A second aspect, which here only can be addressed in general terms, concerns regulatory circuits. If an sRNA is suspected to play a role in a global regulatory circuit (e.g. some stress response) or requires for its function/mechanism of action certain cellular proteins (Hfq, RNase III etc.), it will be beneficial to create additional isogenic strain pairs with lesions in any of the suspected key genes. Since most often the roles of sRNAs are not immediately apparent, this may be a useful step in later analysis.

37.2.3
Experimental Approaches

With a suitable set of knock-out, wild-type and overexpressor strains at hand, several global characteristics of bacteria can be compared, scoring for changes that are attributable to the presence or absence of the sRNA. A fashionable description of many of these large-scale approaches is "functional genomics".

37.2.4
Physiological Phenotypes (Lethality, Growth Defects, etc.)

If – as can be expected in most cases – the deletion of an sRNA gene is not lethal, obvious first experiments can be carried out with the repertoire of the traditional bacteriologist. By growing strains on rich and poor media, in the presence or absence of different carbon sources, at various temperatures, aerobically or anaerobically, etc., one may observe conditional lethality (i.e. cells die under certain conditions if they lack or overexpress the sRNA). As for recent examples from *E. coli*, deletion of the *istR* locus cannot be obtained in a strain in which the SOS response is constitutively turned on (Vogel et al., submitted), since IstR is an antisense RNA acting on an SOS-induced toxin target mRNA; overexpressed RyhB/SraI prevents growth on succinate as the sole carbon source [6, 39], since RyhB targets the *sdhCDAB* mRNA for degradation.

Also, growth rate measurements under different conditions may reveal growth retardation in the presence or absence of the sRNA. The outcome of such experiments cannot be predicted, but one may obtain leads that aid in the subsequent assignment of function when complemented with other methods.

A recently described method can be used to screen strains for growth under many different nutritional conditions. These so-called phenotypic arrays are carried out in microtiter plates [40]. A color change is used as a read-out that signifies growth deficiency, e.g. when utilizing a certain sugar as carbon source. An advantage of this method lies in its speed and high throughput of bacterial samples; its disadvantage is its high cost.

37.2.5
Analyzing sRNA Effects on Specific mRNA Levels by Microarrays

Microarrays are used for transcriptional profiling and determination of RNA levels under a variety of conditions. For many bacterial genomes, microarrays are commercially available or can be custom-made. Transcriptional profiling is now routine and has already been used in many bacteria [41–55]. Technically, microarrays can differ in several ways. Mostly, glass slides are spotted either with PCR-generated entire ORF-DNA, or with gene-specific oligodeoxyribonucleotides (often 35–70 nt in length) of either sense or antisense orientation (relative to known or predicted ORFs). In some cases, only one oligodeoxyribonucleotide per gene is used. In other cases, arrays have coverage of both DNA strands, with multiple probes per gene strung together at very short intervals. Such chips are useful since they potentially also detect transcripts from intergenic regions (where most sRNAs appear to be encoded) and leaders/trailers of mRNAs [4, 53, 56].

Depending on the DNA immobilized on the chip, an appropriate labeling method must be chosen. RNA is extracted from strains (wild-type and deletion strain, with control or sRNA gene-carrying plasmid). If chips carry antisense DNA sequences, total RNA can be chemically labeled after preparation with fluorescent dyes (usually Cy3 or Cy5). If sense DNA is immobilized, fluorescent cDNA probes can be generated by reverse transcription of RNA using random priming with a pool of short oligonucleotides. A recently described detection method relies on antibodies specific for RNA:DNA hybrids. This method therefore does not require labeling of RNA or cDNA [53]. Several other ways of generating probes have been published, but are not covered here.

Usually, the same chip is used for a simultaneous hybridization reaction with two differently labeled RNA preparations (e.g. cells + sRNA, Cy3; cells − sRNA, Cy5). Most spots will, upon detection with an appropriate reader device, display a relatively constant ratio between Cy3 and Cy5 signals. Spots that significantly deviate from this (much higher or lower ratio) identify candidate genes that are either up- or downregulated by the sRNA. The reader is advised to consult the expert literature for pitfalls, important controls and appropriate statistical treatment of data sets (see references above).

Considerations

If some spots show altered Cy3:Cy5 signal ratios, this does not necessarily imply that a candidate target gene (regulated by a given sRNA) has been identified. Since microarrays measure relative RNA abundance only, transcriptional downregulation of a gene cannot be distinguished from increased RNA degradation. Thus, microarrays may primarily detect downstream effects, rather than direct effects on an sRNA target gene. For example, if an sRNA inhibits translation of an mRNA that encodes a transcriptional regulator, this particular mRNA may even remain unchanged in abundance, whereas the levels of mRNAs encoded by downstream targets of the regulatory protein are expected to change. Putative secondary targets can, however, provide important leads to deduce a primary target upstream in a regulatory cascade.

37.2.6
Analyzing sRNA Effects by Proteomics

As stated above, sRNAs that regulate mRNA expression may not affect the abundance of their primary target RNA. A microarray experiment then cannot directly pinpoint the target, whereas techniques that monitor the proteins encoded can. Proteomics, the analysis of the total protein complement of a cell, provides such information.

Instead of analyzing total RNA from cells, total protein is extracted and subjected to two-dimensional (2-D) gel electrophoresis. In the first dimension, proteins are separated by isoelectric point; in the second dimension (presence of SDS), by size. A number of variations of the basic experiment with respect to labeling (in vivo labeling by [^{35}S]methionine, labeling of extracted proteins by fluorescent dyes, etc.), and also to downstream processing and analysis of gels, have been described. For details on methods, the reader is referred to company web sites (e.g. http://www.amershambiosciences.com, http://expressionproteomics.com, http://proteome.incyte.com, http://www.genomicsolutions.com and many others), articles in the journal *Proteomics* (http://www3.interscience.wiley.com/cgi-bin/jhome/76510741) as well as lab protocols available on the web. Similar to microarray experiments, differences in signal intensities of proteins (instead of RNA) between sRNA-proficient and -deficient strains are monitored via radioactive or fluorescent labels. Proteomics analyses can identify direct targets of an sRNA – in cases in which a target mRNA is translationally inhibited without major effects on mRNA abundance – and also downstream targets. A drawback is that only a fraction of the total protein complement can be resolved on 2-D gels. This reflects abundance as well as electrophoretic and solubility properties. That is, minor proteins may be difficult to detect reliably, some proteins may co-migrate with other more abundant proteins, and some may entirely fail to be solubilized and thus are not resolved in the analysis.

If a spot with significantly altered signal intensity ratio has been found, the identity of the protein can be revealed by excision, extraction and subsequent mass spectrometric sequence determination. Even partial amino acid sequences are usually sufficient to deduce the candidate target gene by database searches (if the genomic sequence of the bacterium in question is available). Proteomics analyses have been performed for several bacterial species, most often to identify regulatory pathways or to address pathogenesis-related questions [57–63]. In some cases, even 1-D gel electrophoresis can identify significant changes (see, e.g. the case of OxyS [64]). Problems with the solubility of membrane proteins are discussed in [65].

37.2.7
Analyzing sRNA Effects by Metabolomics

A third large-scale screening strategy of a related kind is sometimes referred to as metabolomics. Up to now, metabolomics has primarily been used in plant research and medicine/toxicology/pharmacology [66–69]. In general, this approach involves the analysis and quantification of as many metabolites as possible, produced by a given cell type or culture. So far, this methodology has – to our knowledge – not yet been employed in searches for sRNA function. In principle, it should be feasible to obtain information on metabolites whose production depends on the presence of an sRNA. This in turn could suggest enzymes involved in production of the metabolites and thus provide candidates for sRNA targets.

37.2.8
Finding Targets by Reporter Gene Approaches

A labor-intensive, but often fruitful, approach involves read-out reporter systems for sRNA-mediated regulation. These can be used for large-scale screening of effects, but are even more useful in experiments directed at confirming or falsifying candidate target genes.

Reporter genes can be engineered to be located downstream of a promoter, permitting a transcription activity read-out, or fused in-frame to provide a measurement of translation (as well as transcription). For shortness, we will here only describe one such system which may serve as an example. The *lacZ* (β-galactosidase) gene has frequently been used as a reporter, since cells expressing this gene form blue colonies on plates, due to the color change upon cleavage of a substrate, X-Gal, by this enzyme. In addition, enzyme production can be quantified by a simple colorimetric assay that works well even in crude cellular extracts [21]. Other reporters, each with their particular benefits, are bacterial (*Vibrio harveii*; *luxAB*) or insect (firefly; *luc*) luciferases [70–72] and green fluorescent protein from the jellyfish *Aequorea victoria* (*gfp*) [73].

For large-scale, single-copy monitoring of the expression of adjacent genes, the *lacZ* gene has been engineered into a deficient μ phage (Mud) such that it will be located close to the site the phage utilizes for integration into the host genome [74–76]. Usually, a promoterless *lacZ* gene is oriented "inwards" so that a neighboring chromosomal promoter, if present, can drive its expression after phage integration, resulting in a transcriptional fusion. In addition, these phages carry selectable marker genes (e.g. conferring ampicillin and kanamycin resistance phenotypes). Integration of these so-called *Mud–lac* phages is random, resulting in a library of *Mud–lac* insertions in a given bacterial strain. Each individual cell – in principle – will carry a different insertion, permitting a read-out in the form of blue colonies (for all cells that carry an active promoter near the site of insertion). Thus, most studies have employed *Mud–lac* fusions to monitor effects on the regulation of neighboring promoters. For studies of sRNA effects this is not expected to be very fruitful since one expects mostly interference with translation and/or tar-

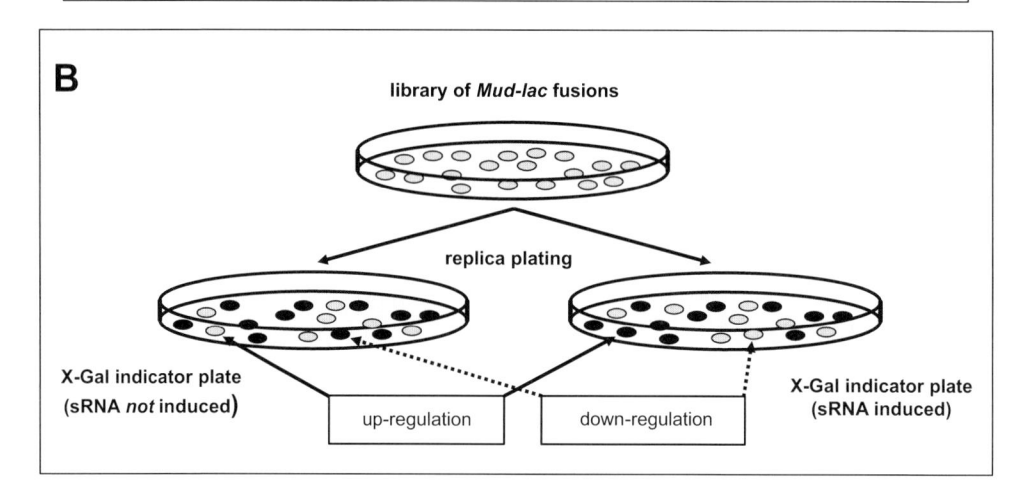

Fig. 37.1. Strategy for the identification of sRNA targets by reporter gene fusion. (A) A putative protein target gene (arbitrarily termed gene Y) of an sRNA is shown at the top, including its promoter (P), the 5′-untranslated region (5′-UTR) of its mRNA, the ATG start codon (open circle) and the coding region (hatched). One strategy to identify such an sRNA target gene is the use of engineered Mud–lac phages that carry a promoterless lacZ-coding region lacking a ribosome-binding site and start codon immediately adjacent to one end of the phage sequence, oriented "inwards". Random Mud–lac chromosomal insertion events which have resulted in in-frame fusion of a gene Y 5′ segment and the lacZ-coding region (A, lower panel) then permit to monitor sRNA-mediated effects on transcription, translation and stability of the target Y mRNA. (B) Analysis of sRNA effects using a library of transcriptional/translational Mud–lac fusion insertions; the strain in this example is deleted for the gene of the sRNA under investigation. The analysis is carried out on X-Gal plates, such that in-frame fusions may result in blue colonies, with color intensity dependent on expression level (dark blue indicated as black, light blue indicated as white) and out-of frame fusions in white colonies. Since only few genes (exemplified by target gene Y in panel A) are likely to be a target of a particular sRNA, few bacterial clones within a library of random Mud–lac fusion insertions will be affected by induction of a plasmid-borne sRNA gene, as visible by a change in their color. The latter then likely represent clones with insertions in potential target genes, since induction of sRNA expression affected lacZ expression (up-regulation causing an increase in blue color intensity, downregulation a decrease), probably by an effect on translation (or mRNA abundance) of the truncated gene upstream of the insertion point.

get mRNA stability. Variants of the *Mud–lac*/mini-*Mu-lac* phages [77–79] can, however, also be engineered to identify translationally and possibly stability-controlled target RNAs. Here, the phage carries a promoterless *lacZ* lacking a translation initiation site. Hence, insertions in target genes that are in-frame with *lacZ* (statistically 1/6 in correct frame and orientation) will generate an active fusion enzyme that provides a transcriptional and translational/stability readout. Figure 37.1 illustrates an idealized experiment. Let us assume that we have created a library of (transcriptional/translational) *Mud–lac* fusion insertions in an sRNA deletion strain. Since only few genes are likely to be a target of a particular sRNA, few bacterial clones within a library of random *Mud–lac* fusion insertions will be affected by induction of the plasmid-borne sRNA gene. Thus, on X-Gal replica plates, induction of the sRNA will not change the blue color intensity of the vast majority of the colonies, but may alter that of a few. The latter then likely represent clones with insertions in potential target genes, since induction of sRNA expression affected *lacZ* expression, probably by an effect on translation (or mRNA abundance) of the truncated gene upstream of the insertion point. The gene in question can then easily be identified by cloning of the region flanking the inserted phage, i.e. by making use of the antibiotic resistance gene nearby, or more directly by inverse PCR [80].

37.2.9
Bioinformatics-aided Approaches

Since the number of sequenced prokaryotic and eukaryotic genomes has increased dramatically in recent years, bioinformatics-aided approaches have emerged as invaluable tools. The power of bioinformatics has already been demonstrated by the identification of many novel sRNAs (accompanying Chapter 36 and references therein). The availability of genomic sequences and the development of efficient search algorithms provide methods for the identification of sequence as well as structure motifs and similarities, interaction sites or other searchable features. If we consider the task at hand, how could bioinformatics help in identifying sRNA targets or give us functional leads?

37.2.10
Prediction of Regulatory Sequences in the Vicinity of sRNA Gene Promoters

Most sRNAs appear not to be constitutively expressed, but accumulate under specific (often stress) conditions. Changed abundance of an sRNA can be caused by up- or downregulation at the transcriptional level or by changes in transcript stability. If transcriptional regulation applies, binding sites for regulatory proteins (repressors, activators) should be present in the vicinity of the predicted or experimentally verified promoter of the sRNA gene.

Searchable databases are available in which consensus binding sites of many known transcriptional regulators are catalogued (e.g. in *E. coli*: http://arep.med.harvard.edu/ecoli_matrices/; http://tula.cifn.unam.mx/%7Emadisonp/

E.colipredictions.html). Thus, knowing the exact location of the sRNA gene of interest and its promoter within an annotated genome, one may sometimes be able to identify known motifs near the site of interest "by eye". However, programs can more quickly screen for known motifs within a sequence window defined individually (e.g. from 300 bp upstream to 100 bp downstream of the transcription start site).

By carrying out such a search, significant information can be derived. The presence of a Fur binding site (Fur is an iron-dependent repressor) upstream of an sRNA gene denoted *ryhB/sraI* suggested an involvement in iron metabolism. Subsequent work showed that *ryhB* was indeed repressed by Fur. Under de-repressed conditions RyhB inhibits, post-transcriptionally by an antisense mechanism, at least six genes encoding iron-containing or iron-storing proteins [39].

A limitation of a search for regulatory elements lies in the quality of prior knowledge (sometimes only one or two binding sites for a given regulator are validated experimentally and thus the "consensus" sequence is based on little data) and in degeneracy (some consensus sites score numerous candidates that are not experimentally supported). Nevertheless, information derived from such searches can sometimes pinpoint involvement in putative regulatory circuits and therefore narrow down the number of likely targets.

Another consideration is phylogenetic conservation of genetic context of a given sRNA gene, since neighboring genes may be part of the same regulatory pathway. For example, the antisense RNA gene *oxyS* (regulator of many genes in oxidative stress) is located next to *oxyR* (encoding a master regulator of the oxidative stress response) in *E. coli* as well as in its relatives [64]. Similarly, the gene encoding a regulatory sRNA, IstR-1, is located next to an operon, *tisAB*, in close relatives to *E. coli*; recent work has shown that the *tisAB* mRNA is an antisense target for IstR-1 (Vogel et al., submitted).

37.2.11
Finding Interacting Sites (Complementarity/Antisense)

Most functionally characterized known sRNAs in *E. coli*, but also RNAIII (a regulatory RNA that controls virulence genes in *Staphylococcus aureus*) are *trans*-encoded antisense RNAs [14, 81], i.e. they base-pair to target RNAs resulting in inhibition (or, less often, activation) of the target RNA. Irrespective of whether the mechanism involves translational block or induced RNA degradation, the recognition of the target by the sRNA involves the interaction of complementary sequences in the two RNAs. Thus, simple BLAST searches could potentially identify sites complementary to sequences within the sRNA.

In a few cases, this has indeed been borne out (usually, however, some additional information helped to select promising candidates from many sequences). Thus, targets for RyhB (see above) and Spot42 (an sRNA that regulates *galK*) were found by straightforward complementarity searches [39, 82]. As a general strategy, this is not expected to be feasible for most sRNAs, because a survey of the known examples shows that base-paired regions can (1) vary considerably in length, (2) are often non-contiguous (in case of OxyS and its target *fhlA*, two separate stretches of 7 and 9 bp, respectively), (3) may contain bulges and internal loops (as for DsrA and its target *rpoS*) and (4) may form non-canonical base pairs, mainly G–U [14]. Thus, algorithms that take all these variables into account tend to generate numerous putative targets, making subsequent validation a labor-intensive task.

Including further criteria might improve the outcome of the search. We have incorporated genomic target location and phylogenetic conservation as additional scores (Reimegård and Wagner, unpublished). Most (antisense) sRNAs appear to target sequences in the 5' portion of an mRNA or sometimes the region overlapping the 5' end of a coding frame within a polycistronic mRNA [39, 82] and thus higher scores can be given to target candidates that meet these criteria. Furthermore, if sRNA homologs are available in closely related bacteria, searches can be reiterated in these related genomes. If an sRNA from, for example, *Salmonella* species carries nucleotide changes compared to that in *E. coli*, and compensatory changes are encountered in a predicted target in the heterologous host (compared to its homolog in *E. coli*), this increases the likelihood that a correct target has been identified. At present, published programs or web sites for bacterial sRNA target identification are not available.

Whether or not structural features (such as recognition loops, motifs like U-turns, predicted unstructured regions, etc. [35]) in sRNAs and putative target RNAs have predictive value is a question that, at this point, cannot be answered, but could be adequately addressed when more and more cases are understood in mechanistic detail.

37.3
Additional Methods Towards Functional and Mechanistic Characterizations

No matter if one arrives at candidate targets and/or suspected biological roles of a given sRNA by the large-scale screening procedures outlined above or works from a "hunch", detailed studies are needed to discard or confirm candidates. Below, we will suggest only a few lines of experiments as incomplete guidance.

37.3.1
Finding sRNA-associated Proteins

Most RNAs are associated with proteins in living cells. Interactions may be fortuitous (many proteins will associate with RNAs due to non-specific electrostatic interactions) or functionally important. Yet, some RNAs require interactions with certain proteins for their activity or stability. Hence, if one works with an sRNA that is functionally ill defined, it may sometimes be of interest to identify protein partners.

High affinity of proteins to certain RNAs can be utilized in, for example, immunoprecipitation experiments. The hexameric Sm-like protein Hfq is known to be required for the *in vivo* activity of several *trans*-encoded antisense RNAs (RyhB, OxyS, Spot42, etc.) [39, 83, 84]. Accordingly, immunoprecipitation by anti-Hfq antibodies has identified additional RNAs of this "family" [53]. In this analysis, RNA was isolated from *E. coli* cells, immunoprecipitated with anti-Hfq antibodies and analyzed on high-density Affimetrix microarray chips.

Reversed strategies could also be useful, although – to our knowledge – they have never been employed in search of sRNA function, but may be promising for sRNAs suspected to act by sequestration of proteins. For example, whole-cell lysates can be prepared under non-denaturing conditions, followed by affinity purification of sRNA–protein complexes. Biotinylated oligodeoxyribonucleotides complementary to a segment of the sRNA in question can be incubated with extract. Using streptavidin-coated matrices (or coated magnetic beads), the specific sRNA–protein complexes can be isolated and separated from the bulk material, and specifically bound protein(s) identified by mass spectrometry (or other sequencing techniques). Variants of such a protocol would most likely have identified CsrA (bound to the regulatory RNA CsrB) [85] and SmpB (bound to the housekeeping tmRNA, also known as SsrA) [86].

Related strategies make use of aptamer modules that can be genetically engineered into sRNA genes. Thus, the *in vivo* generated sRNAs will contain additional sequences which fold into structural motifs that permit high-affinity purification of the sRNA variant (streptomycin [87] or tobramycin [88] affinity tag). Consequently, these aptamer–sRNA hybrid RNAs will be recovered as ribonucleoprotein particles from cell extracts, permitting the identification of specifically associated proteins.

37.3.2
Regulation of the Target RNA – Use of Reporter Gene Fusions

As outlined above (Section 37.2.8), reporter gene fusions can help in large-scale screening for putative targets. If, based on this or other approaches, a manageable number of candidate targets for a given sRNA has been selected, fusion gene experiments are highly recommended. Many plasmid-borne *lacZ* cassettes are available, and some of these can be converted into single copy, i.e. by subsequent integration into a chromosomal locus. Strains used should carry a mutation or deletion of the *lacZ* gene. In the favored type of fusion construct, a promoterless *lacZ* coding region lacking its own ribosome binding site and start codon is used. Putative target 5' gene segments have to be inserted in-frame with *lacZ*. Such transcriptional/translational fusions require the target gene promoter to be active, and translation initiated at the target gene to continue throughout *lacZ*, generating the active fusion protein (see Fig. 37.1A). Such fusion constructs permit monitoring of sRNA-mediated effects on transcription, translation and stability of the target RNA.

Plasmid-borne or chromosomal insertion of, in particular, target gene–*lacZ* translational fusions can be used to confirm, and quantitate, the regulatory effect of an sRNA. For example, blue colonies on X-Gal plates indicate that the fusion protein is synthesized and active. If the correct target is present in this construct, expression of the cloned sRNA gene should result in light blue or white colonies. This is expected in all cases in which the sRNA acts by an antisense RNA mechanism on translation of the target RNA or triggers its degradation. The efficiency of inhibition can be further assessed by conducting β-galactosidase assays in cell extracts [21].

Moreover, if regulation has been observed, the fusion constructs and sRNA-overexpressing plasmids immediately provide tools for more detailed mechanistic studies. If two RNAs are suspected to interact functionally by base-pairing, mutations can be introduced at selected sites and the effect of mismatched antisense/sense combinations evaluated by the same assay. If unknown, interaction sites may be found by random mutagenesis of the target-fusion, scoring for maintained activity but loss of control. The most convincing experimental strategy usually involves mismatched mutant/wild-type and wild-type/mutant combinations, both showing loss of control, and combinations with compensatory mutations which (partially or entirely) restore a wild-type phenotype, e.g. inhibition of *lacZ* expression. As mentioned in Section 37.2.8, reporter gene fusions other than *lacZ* can be used as well.

37.3.3
Northern Analyses

Analyses of sRNA expression patterns are instructive and may give clues to possible biological functions. An sRNA that is abundantly present only after onset of a response, such as SOS or oxidative stress, is likely to be involved in the response in question (or its regulation). Northern analyses, in addition to giving a measure of presence and abundance, may also reveal size alterations under different physiological conditions. For example, if an sRNA is almost perfectly complementary to a (known or yet unknown) target RNA such that a helix of two or more turns is formed, RNase III is expected to cleave both RNAs. That is, observation of shorter species of an sRNA (or a target RNA, if known) is indicative of an antisense mechanism. Confirmatory experiments are, however, required, since RNA processing is also carried out by single-strand-specific RNases, such as RNase E.

Northern blot experiments have also been very useful in monitoring changes in abundance and/or processing of either sRNA or target RNAs as part of detailed mechanistic studies. For example, if a suspected antisense/target RNA pair is to be analyzed, the sRNA can be overexpressed, and the fate of the target RNA in terms of processing and degradation can be followed. A protocol for Northern analysis that has worked well in our hands is provided as Protocol A (Section 37.5).

37.3.4
Analysis of sRNAs – RACE and Primer Extensions

Northern blots provide approximate sizes, but fail to identify the position and exact nature of RNA 5′ and 3′ ends. Such information is required if one wants to generate the authentic sRNA *in vitro* for subsequent analyses (see below). Additionally, 5′ ends may represent either +1 (transcription initiation) sites or be generated by endonucleolytic processing; if two or more sRNA species are detected, one cannot *a priori* be certain that they are generated from two promoters. Furthermore, metabolic stability often depends on polyadenylation and hence 3′ ends may carry short A-tails.

5′-End mapping can conveniently be carried out by reverse transcription of RNA extracted from cells, using an sRNA-specific, 5′-end-labeled primer (e.g. [89]). However, this technique does not distinguish between 5′-triphosphate ends (transcriptional initiation sites) and 5′-monophosphates generated by processing of primary transcripts. This information can be obtained by 5′-RACE experiments using a protocol originally devised by Bensing et al. [37], which uses the enzyme tobacco acid pyrophosphatase (TAP; Protocol B, Section 37.5).

37.3.5
Structures of sRNAs and Target RNAs

To gain insight into the mechanism of sRNA/target RNA interaction, structural analyses are required. For RNAs that are short (less than around 120 nt or so), secondary structure prediction programs can often give fairly reliable information (e.g. mfold [90] and RNAStar [91]). Structures of much longer RNAs, such as most mRNAs, are difficult to predict, because many energetically favorable folding schemes may be possible. In addition, the sRNA target region within a long RNA may be unknown. As an additional complication, an antisense RNA *in vivo* will generally interact with a "moving target", i.e. a population of nascent mRNAs of different lengths and folding states. Finding out which folding intermediates are effective targets is non-trivial. Nevertheless, *in vitro* structure mapping has been shown to be a powerful tool to determine secondary structures of sRNAs as well as selected target RNA segments. In contrast, high-resolution NMR or X-ray structures are certainly desirable, but can most often not be quickly obtained.

To conduct *in vitro* structure mapping, one needs transcription templates. PCR fragments are created such that the RNA-encoding DNA is preceded by a suitable promoter sequence (e.g. a T7 phage promoter) introduced by the upstream PCR primer. RNAs are generated in conventional *in vitro* transcription reactions using T7 RNA polymerase (Chapter 1) and then subjected to partial cleavage using nucleases, such as RNase T1, A, T2 and V1, or modification and chemical cleavage by dimethylsulfate (DMS), 1-cyclohexyl-3-(2-morpholinoethyl)carbodiimide metho-*p*-toluene sulfate (CMCT), lead(II) acetate, kethoxal and hydroxyl radicals, to name the most popular chemical probes. Each of these chemicals has distinct structure and/or base specificities that can be useful to obtain significant information [92–94], as detailed in Chapters 10–15. Depending on the sizes of RNAs to be analyzed, two alternative methods of analysis can be chosen. Short RNAs can be 5′- or 3′-end-labeled (see Chapters 3 and 9). For long RNAs and for any RNA that has been modified rather than cleaved, reverse transcription is used instead. Particularly when a long RNA is to be mapped, several different 5′-end-labeled primers can be employed to cover, in conjunction, essentially all regions of the RNA. Examples of such analyses can be found in [89, 95, 96]. Information derived from mapping experiments can then be used to refine results of prediction algorithms.

With purified RNAs at hand, RNA–RNA interactions can be assessed. Binding kinetics of antisense-target RNA pairs can be determined by gel-shift assays using denaturing or non-denaturing gels, depending on the expected length, and the GC content, of the base-paired regions [97, 98]. This can be carried out in the presence of specific proteins if protein involvement is suspected [83, 84]. In cases in which an antisense RNA is not fully complementary to its target, but forms only a limited number of base pairs (such as OxyS, MicF, IstR-1, etc.), comparisons of partial lead(II) cleavage patterns of end-labeled target RNA in the presence and absence of the sRNA can be used to footprint the binding site (Darfeuille and Wagner, unpublished).

Several *in vitro* methods can be used to map protein-binding domains in RNA. It exceeds the frame of this chapter to cover this subject area adequately. However, for high-resolution information on contacts between a protein ligand and specific bases in an RNA, NAIM (nucleotide analog interference mapping; see Chapters 17 and 18, as well as [99]) is a very useful strategy.

Finally, it is worth noting that at least two chemicals, DMS and lead(II) acetate, have been used successfully to obtain RNA secondary structure information *in vivo*, and even to assess protection by proteins (see Chapters 9 and 14, and [64, 100, 101]).

37.4
Conclusions

In this chapter, we have provided an overview of some strategies that can be used when attempting to identify the biological function of an uncharacterized sRNA. It will not have escaped the attention of an observant reader that some techniques, approaches and strategies have not (or only incompletely) been covered. The reason for this lies partly in a bias towards approaches that we have found useful or others have successfully used and partly in space limitations. A second bias is the focus on target identification, which reflects our conviction that most sRNAs of so far unknown function will turn out to act as regulators – either acting on RNAs (as antisense RNAs) or on proteins (by sequestration or by modulating their activities).

Finally, in our view, the emerging picture regarding sRNAs in bacteria suggests a greater variety in targets, functions, and mechanisms than, for instance, seen in the case of miRNAs in eukaryotes. This, if true, implies that standard recipes for sRNA function assignment may not resolve the questions asked. Instead, each case will require a substantial portion of creativity in designing the appropriate experiments that will lead to an understanding of the function of the sRNA in question.

37.5
Protocols

Protocol A: Northern blot protocol

We routinely use the RNA extraction part of the protocol provided below for preparation of larger quantities of RNA (100 µg to 1 mg) from *E. coli* and other enterobacteria. If smaller amounts of RNA are needed for small-scale analysis, the more expensive TRIzol reagent is recommended (see Protocol B).

RNA extraction

(1) Dilute overnight bacterial culture in 50–100 volumes of growth medium.
(2) Grow to the OD_{600} of choice.
(3) For RNA extraction, withdraw the required aliquot of culture volume and stop growth by adding to 1/5 volumes of pre-chilled STOP solution [ethanol/5% phenol (v/v)].
(4) Mix thoroughly.
(5) Pellet bacterial culture by centrifugation (approximately 3500 g, 10 min, 4 °C) and discard supernatant. (At this point, samples can be stored at -20 °C after snap-freezing in liquid nitrogen.)
(6) Resuspend pellet in 1/2 volumes of RNA lysis buffer (100 mM Tris–HCl, pH 7.5, 40 mM EDTA, 200 mM NaCl, 0.5% SDS) pre-heated to 65 °C by gently pipetting up and down. For example, if you started with a 10-ml culture, 5 ml of RNA lysis buffer is added. For samples of $OD_{600} > 1.5$, use 1 volume of RNA lysis buffer.
(7) Immediately place at 65 °C for 5 min.
(8) Add an equal volume of water- or buffer-saturated phenol pre-heated to 65 °C and vortex thoroughly.
(9) Centrifuge (approximately 10 000 g, 30 min, 4 °C) and transfer the aqueous phase to fresh RNase-free tube.
(10) Add an equal volume of chloroform and vortex thoroughly.
(11) Centrifuge (approximately 10 000 g, 30 min, 4 °C) and transfer the aqueous phase to fresh tube.
(12) Add 1/10 volumes of 3 M sodium acetate (pH 6.5) and 2.5 volumes ice-cold EtOH.
(13) Precipitate at -20 °C for 1–3 h.
(14) Pellet RNA by centrifugation (at least 12 000 g, 30 min, 4 °C), discard EtOH, wash pellet with ice-cold 75% EtOH and centrifuge as above but for 10 min. Remove supernatant and air-dry RNA pellet at room temperature. Resuspend pellet in, for example, 500 µl of RNase-free water when RNA has been isolated from 50 ml of an *E. coli* culture grown to 1.0 OD_{600}.

DNase I treatment

(1) Treat RNA samples with RNase-free DNase I for a maximum of 1 h according to the manufacturer's protocol (see also Protocol B, Step 4).
(2) Repeat phenol extraction as above, but use phenol at room temperature.

Gel separation and Northern blot

(1) Quantify RNA using a spectrophotometer measuring absorbance at 260 and 280 nm.
(2) Analyze a small aliquot of each RNA sample on a 1% RNase-free agarose gel with $1 \times$ TBE as electrophoresis buffer. This permits one to roughly assess

RNA integrity. The three ribosomal RNA bands (5S, 16S and 23S rRNA) should be clearly visible and without substantial smear. Comet-like tails from the 16S down to the 5S rRNA bands are indicative of degradation and, at this stage, one should consider repeating the RNA preparation.

(3) Prepare a 5–10% denaturing polyacrylamide gel (7–8 M urea, $1 \times$ TBE), depending on the size of your RNA of interest.

(4) Pre-run your gel at a sufficiently high voltage, such that the gel temperature is perceived as hot (50–60 °C). This ensures a continuously denatured state of your RNAs during electrophoresis.

(5) Load onto the gel 10 µg of RNA per sample diluted at a 1:1 ratio with RNA loading dye [95% (v/v) formamide, 0.1% (w/v) bromophenol blue, 2 mM EDTA]. Also load a lane with an appropriate DNA or RNA ladder that has been end-labeled with [γ-^{32}P]ATP. The Fermentas pUC mix 8 is recommended, including diverse size markers ranging from 19 to 1116 bp. RNA size markers are expensive and may not be worth the more precise size prediction.

(6) Cut out six sheets of 3MM paper of the size of your gel region of interest; pre-wet them in $1 \times$ TBE.

(7) Cut blotting membrane of same size (Amersham BioSciences Hybond N+ recommended); pre-wet in $1 \times$ TBE.

(8) After the gel run, detach one of the glass plates and apply one of the pre-wetted 3MM papers to the desired region of the gel. Start from one end and avoid trapping of bubbles between paper and gel; stack two more 3MM paper sheets on the first. Peel the gel off the second glass plate and turn around such that the naked gel surface is facing up.

(9) Carefully place the membrane on the naked gel surface. Once in contact with the gel, *avoid any lateral movement!* Using a Pasteur pipette, gently roll out any bubbles that may have been trapped between gel and membrane.

(10) Finish building the sandwich by placing the last three sheets of 3MM paper on top of the membrane, rolling out any bubbles during this process.

(11) Transfer the RNA to the membrane by either electroblotting (preferred) or capillary blotting techniques. Consult the manual of your blotting apparatus for an appropriate protocol. If a radiolabeled size marker is used, its successful transfer to the membrane can be evaluated with a hand-held β-radiation detector.

(12) When transfer is completed, disassemble the sandwich and crosslink the RNA to the membrane either by baking at 80 °C for 1 h or preferably by using a UV crosslinker (for energy settings and crosslinking times, see manual of your UV crosslinker). After crosslinking, rinse the membrane with RNase-free water.

(13) Place the membrane in a hybridization tube and wet it with hybridization buffer (20 ml for a membrane of 20×15 cm; hybridization buffer: 0.5 M phosphate buffer, pH 7.2, 7% SDS, 10 mM EDTA). Pre-hybridize at the chosen hybridization temperature for at least 30 min while preparing your radioactive probe. Note that the ends of your rolled membrane should not overlap in the tube.

(14) Denature probe by heating to 95 °C for 5 min. For single-stranded DNA oligonucleotide probes, 5–50 pmol of a 5'-end-labeled oligonucleotide should be used. Although DNA may suffice to detect most sRNAs, single-stranded antisense RNA probes are recommended if higher sensitivity is required. Such riboprobes can be prepared by *in vitro* transcription with T7 RNA polymerase in the presence of labeled rUTP from PCR templates (created such that the sRNA region is transcribed in antisense direction from a T7 phage promoter introduced by the downstream PCR primer). Approximately 10^6 c.p.m. of riboprobe/ml hybridization buffer should be used. To decrease background signal, both labeled DNA oligonucleotides and riboprobes should be freed from unincorporated nucleotides by column purification (e.g. Amersham MicroSpin™ G25).

(15) Add denatured probe to the tube and hybridize for at least 6 h at the temperature of choice. We generally hybridize DNA oligonucleotide probes at 42 °C, and riboprobes at 60–70 °C in a hybridization buffer containing 50% deionized formamide, 7% SDS, 250 mM NaCl, 120 mM sodium phosphate buffer, pH 7.2.

(16) Discard the radioactive probe-containing liquid appropriately or save for another membrane.

(17) Rinse the membrane briefly in a small volume of 2 × SSC, 0.1% SDS.

(18) Wash the membrane at the hybridization temperature in three subsequent 15-min washing steps, using at least 100 ml for each step (first wash: 2 × SSC/0.1% SDS, second wash: 1 × SSC/0.1% SDS and third wash: 0.1 × SSC/0.1% SDS).

(19) Aspirate any liquid from the membrane. Seal the membrane in plastic and expose it to a PhosphorImager screen overnight.

Protocol B: 5'-RACE protocol for mapping of transcription start sites and processed 5' ends

Although primer extension analysis is commonly employed for 5' end mapping of RNA, it is less well appreciated that this method fails to distinguish between primary 5' ends and those generated by processing. To identify the start of transcription initiation and to distinguish it from processed 5' ends of a small RNA, a 5'-RACE protocol that includes treatment with TAP is more suitable. Additional advantages of 5'-RACE are the small amount of RNA required and that it obviates the use of radiolabeled oligonucleotides. Primary transcripts in bacteria carry a 5'-triphosphate, which is hydrolyzed by TAP specifically between the α- and β-phosphate groups. Via the resulting 5'-monophosphate, these RNAs are subsequently ligated to the 3'-hydroxyl group of an RNA oligonucleotide (5' adaptor), which is followed by reverse transcription with a gene-specific deoxyoligonucleotide, and subsequent PCR amplification using a 5'-adaptor primer and a nested gene-specific primer. TAP treatment is expected to yield a specific or at least strongly enhanced signal for primary transcripts in the amplification step, as compared to untreated RNA samples. However, 5' ends resulting from processing (retaining a 5′-monophosphate) will also be amplified and are analyzed in parallel (Fig. 37.2).

The protocol below is the one commonly used in our laboratories [1]. Given are conditions and primers for 5′-end analysis of the small SraC RNA of *E. coli*, which may serve as a control when establishing the method for other sRNAs. The *sraC* locus yields four different RNAs (Fig. 37.2A) that vary in size and abundance depending on growth conditions, as detected by detected by Northern hybridization [1]. Our RACE analysis revealed that only one of these four species represents a primary transcript (band 5a in Fig. 37.2B). Essentially, this protocol is based on a publication by Bensing et al. [37], but has been optimized for sRNAs. For example, we use a thermostable reverse transcriptase enzyme to perform cDNA synthesis at an elevated temperature, which should significantly increase the yield of full-length cDNAs with authentic 5′ ends, particularly in the case of highly structured RNAs. If 5′ ends of several sRNAs need to be determined in parallel, it is possible to prime cDNA synthesis with several gene-specific oligonucleotides in the same reaction tube (see Step 7 below). However, we do not recommend the use of random, e.g. hexamer, oligonucleotides since this usually results in a lower yield of specific 5′-RACE products.

Required reagents

- STOP mix: ethanol, 5% phenol (v/v).
- TRIzol reagent (Invitrogen, The Netherlands).
- RNase-free DNase I (Roche Diagnostics, Germany; 20 U/µl).
- 10 × DNase I digestion buffer: 100 mM sodium acetate (pH 5.2), 5 mM $MgSO_4$.
- RNase inhibitor SUPERase•In (20 U/µl; Ambion, USA).
- P:C:I (water-saturated phenol/chloroform/isoamylalcohol mixture, 25:24:1, v/v).
- Ethanol containing 0.3 M sodium acetate (pH 5.7).
- TAP (Epicentre Technologies, USA; 10 U/µl). The enzyme is provided with 10 × TAP digestion buffer: 500 mM sodium acetate (pH 6.0), 10 mM EDTA, 1% β-mercaptoethanol, 0.1% Triton X-100.
- T4 RNA ligase (New England Biolabs, USA; 20 U/µl).
- 10 × RNA ligation buffer: 500 mM Tris–HCl (pH 7.9), 100 mM $MgCl_2$, 40 mM DTT, 150 µM ATP.
- DMSO.
- ThermoScript RT-PCR System (Invitrogen, The Netherlands); components required here: ThermoScript reverse transcriptase (15 U/µl), 5 × cDNA synthesis buffer (250 mM Tris–acetate, pH 8.4, 375 mM potassium acetate, 40 mM magnesium acetate, stabilizer), 0.1 M DTT, 10 mM dNTP mix, RNaseOUT RNase inhibitor (40 U/µl), DEPC-treated ddH_2O, *E. coli* RNase H (2 U/µl).
- HotStar *Taq* polymerase (Eurogentec, Belgium; 20 U/µl; provided with 10 × PCR buffer and 25 mM $MgCl_2$ solution). The use of hot start PCR enzymes is highly recommended as it will significantly decrease the background resulting from unspecific amplification products.
- Small-size DNA marker (e.g. pUC19 DNA/*Msp*I marker from Fermentas, Lithuania).

A)

exponential phase RNA
stationary phase RNA

M
- 331
- 242
- 190
- 147

1.5 2 4 6 8 10

B)

exponential phase
T- T+

stationary phase
T- T+

404
331
242
190
147
110
64

5a
5b 5c
5d

C)

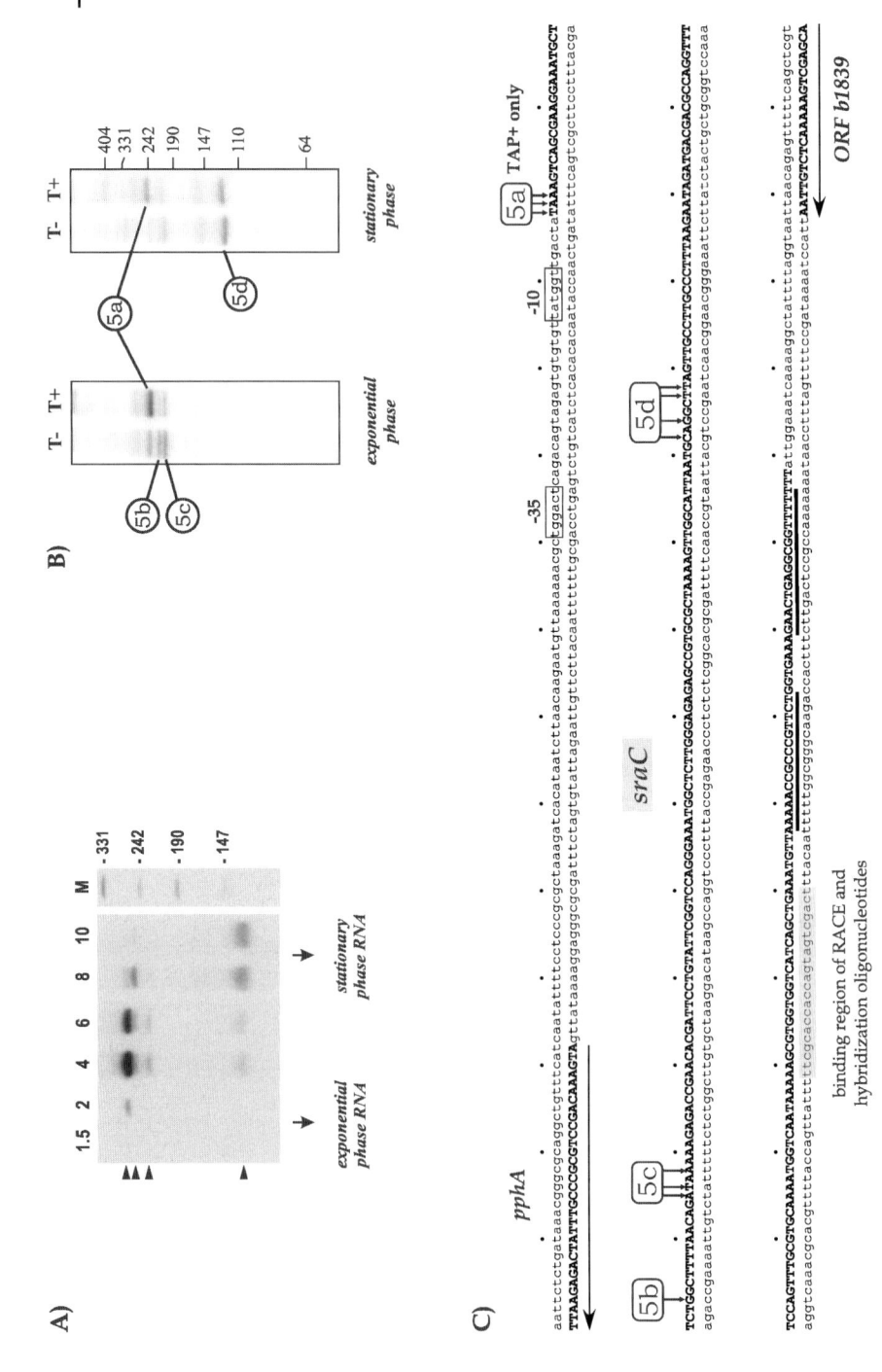

pphA

aattctctgataaacgcgcaggctgtttcatcatattttcctccccgctaaagtcacatatttttaaacagaagatgttaaaaaacgc
TTAAGACTATTTGCCCCGACCAAAGTAgttataaaaggaggcgcgcattcttagtgattgttcttacaattttttgcgaccgagtcgt

5b

sraC

-35
cagacagtagagtgtgttatgtgactaTAAAGTCAGCGAAGGAAATGCT

5a TAP+ only

-10

TCTGGCTTTTAACAGATAAAAAGAGACCGAACACGATTCCTGTATTCGGTCCAGGAAATGCTCTTGGAAGAGACCGTGCGTAAAAGTTGGCATTAATGCAGCCTTAGTTGCCTTGCCCTTTAAGAATAGATGACGACGCCAGGTTT
agaccgaaaattgtctattttctcggctgtgctaaggacataagccaggtcccttacgcagttcctcgagaaccctctcggcacgcgatttcaaccgtaattacgtccgaatcaaccgaacgggaaattcttacctactgtcgtccaaa

5c

TCCAGTTGCGTGCAAAATGCTCAATAAAAAGCGTGGTGGTCATCAGCTGAAATGTTAAAAACCGCCGTTCTGTGAAGAACTGAGCGGCGGTTTTTTattggaaatcaaaagctatttaggtaattaacagagttttcagtcgt
aggtcaaacgcacgttttaccagtcattttcgcaccacagtagctgacttacaattttgcgcgcaggaccacttcttgactccgccaaaaaaatccaatcctttagttttccgataaaatccat**AATTGTCTCAAAAGTCGAGCA**

5d
ORF *b1839*

binding region of RACE and hybridization oligonucleotides

Fig. 37.2

- TOPO TA cloning kit: pCR2.1-TOPO vector cloning kit with competent TOP10F' cells (Invitrogen, The Netherlands).
- Regular *Taq* polymerase (5 U/µl), 10 × PCR buffer, 25 mM $MgCl_2$ solution.
- RNA oligonucleotide A3 (5'-AUAUGCGCGAAUUCCUGUAGAACGAACACUA-GAAGAAA-3'). Any other RNA oligonucleotide may be used as long as the 5'-adapter-specific DNA oligonucleotide for subsequent PCR amplification is designed accordingly. Note, however, that for efficient ligation with T4 RNA ligase, the 3' end of the RNA oligonucleotide should be purine-rich.
- Gene-specific DNA oligonucleotide jb-101 for reverse transcription (5'-CAGCT-GATGACCACCA-3').
- Gene-specific DNA oligonucleotide jb-101-B for PCR amplification (5'-TGAT-GACCACCACGCTT-3').
- 5'-Adapter-specific DNA oligonucleotide B6 for PCR amplification (5'-GCGCGAATTCCTGTAGA-3').
- DNA oligonucleotide Seq-UNI-61 for colony PCR (5'-ACGACGTTGTAAAAC-GACGG-3').
- DNA oligonucleotide Seq-REV for colony PCR (5'-TTCACACAGGAAACAGC-TATGAC-3').
- DNA oligonucleotide Seq-REV-50 for sequencing (5'-CACAGGAAACAGCTAT-GAC-3').
- Phase Lock Gel (PLG) tubes, PLG Heavy, 0.5 ml and 2 ml (Eppendorf, Germany). These tubes are recommended to facilitate phase separation, but ordinary Eppen-

Fig. 37.2. (A) Northern blot with an SraC-specific oligonucleotide probe and RNA sampled from cells of different growth phases (adapted from [1]). RNA was prepared 1.5, 2, 4, 6, 8 and 10 h after dilution of an overnight culture into fresh LB medium. Four distinct RNA species are observed, indicated by arrowheads on the left. Approximate growth points at which RNA was sampled for 5'-RACE experiments (exponential and stationary phase sample) are indicated by arrows below the blot picture. (B) RACE mapping of SraC 5' ends at the two growth points indicated. Total *E. coli* RNA was linked to a 5'-adaptor RNA without or after treatment with TAP (lanes T− and T+, respectively). TAP converts the 5'-triphosphate of an RNA to a monophosphate. A comparison can therefore distinguish between primary transcript 5' ends and internal 5'-processing sites (carrying 5'-monophosphates). Subsequently, cDNA was made and the 5'-end sequence of the respective sRNA was amplified using gene- and adaptor-specific primers. PCR products were then separated on a native 5% polyacrylamide gel, and bands of interest were excised, cloned and sequenced. Only one out of four 5' ends, i.e. 5a, was strongly enhanced after TAP treatment, thus identifying the transcription initiation site of the *sraC* gene. In contrast, no significant difference was found upon TAP treatment for bands 5b–5d, indicating that the RNAs contained in these bands are processing products of the primary *sraC* transcripts. Note that bands on Northern blots and amplified cDNAs differ in size due to only partial amplification of the *sraC* transcripts as well as the presence of additional nucleotides that stem from the 5' adapter RNA oligonucleotide. (C) Sequence of the *sraC* locus of *E. coli* (*sraC* gene on the "+" strand) in the intergenic region between *pphA* and ORF b1839 (both on the "−" strand). Coding sequences of the adjoining genes and the *sraC* gene are capitalized and in bold face. Horizontal lines between the "+/−" strand sequence show the position of the *sraC* terminator. Labeled vertical arrows denote the 5' ends obtained by cloning and sequencing of the four 5'-RACE products (5a–5d) from above. Putative −10 and −35 sequence elements of the *sraC* promoter are boxed.

dorf tubes may alternatively be used as well. Keep in mind that PLG tubes must not be vortexed. Shake samples vigorously by hand instead.
- Glass milk-based gel purification kit (e.g. QIAEX II; Qiagen, Germany).

Step 1: Growth of bacterial cultures
Dilute a fresh overnight culture of *E. coli* K12 cells 1/100 into 25 ml LB medium in a 100-ml flask and keep shaking at 37 °C at 200 r.p.m. until cells have reached the desired OD_{600} value.

Step 2: Harvest of bacteria
Harvesting bacterial cultures as described below usually yields 20–30 µg of total RNA.

- *For RNA preparation from exponential phase cells ($OD_{600} = 0.3$)*. Transfer 15 ml of culture into a 50-ml plastic tube. Add 3 ml (20% v/v) of STOP mix. Mix thoroughly and immediately freeze in liquid nitrogen. Thaw frozen mixture on ice with occasional mixing. Once completely thawed, pellet bacteria by centrifugation for 10 min at 3500 g and 4 °C. Remove supernatant and any traces thereof.
- *For RNA preparation from stationary phase cells ($OD_{600} = 2$)*. Remove 2 ml of culture into a 2-ml Eppendorf tube. Immediately collect cells by centrifugation in a table-top microcentrifuge for 1 min at maximum speed. Remove supernatant and any traces of it.

Step 3: RNA isolation
Immediately dissolve pellet in 1 ml of TRIzol reagent by pipetting up and down. Incubate at room temperature for 5 min. Transfer to a 2-ml PLG tube; it is important to collect the gel at the bottom of the PLG tube by centrifugation in a table-top microcentrifuge for 30 s at maximum speed immediately prior to use. Add 200 µl chloroform, shake vigorously for 30 s and incubate for 2 min at room temperature. Separate organic and aqueous phase by centrifugation in a microcentrifuge at 10 000 g or higher for 15 min at room temperature (it is important not to centrifuge PLG tubes at low temperature since this will interfere with proper placement of the phase lock gel between the two phases). Transfer 500 µl of the aqueous (upper) phase to a fresh Eppendorf tube. Add 400 µl of isopropyl alcohol, mix and incubate sample at room temperature for 10 min. Centrifuge (approximately 12 000 g, 10 min, 4 °C) and remove supernatant. Wash the RNA pellet once by adding 500 µl of 75% ethanol and vortex briefly. Pellet RNA by centrifugation (approximately 7500 g, 5 min, 4 °C). Remove supernatant, and air-dry pellet at room temperature. Dissolve RNA pellet in 133 µl RNase-free ddH_2O. Incubate for 5 min at 65 °C with occasional vortexing. Centrifuge briefly to collect RNA solution at the bottom of the tube.

Step 4: Removal of DNA
Add to tube:

- 15 µl of 10 × DNase I digestion buffer
- 0.5 µl RNase inhibitor
- 2 µl DNase I

Mix and incubate for 20 min at 37 °C. Transfer to a 0.5-ml PLG tube and add 150 µl of P:C:I. Shake vigorously for 30 s and separate phases by centrifugation in a table-top centrifuge at maximum speed for 15 min at room temperature. Transfer aqueous phase (150 µl) to a fresh Eppendorf tube and add 450 µl of ethanol containing 0.3 M sodium acetate (pH 5.7). Precipitate for 30 min on ice and pellet RNA by centrifugation (at least 12 000 g, 30 min, 4 °C). Wash the pellet once with 75% ethanol (see Step 3) and finally dissolve in 50 µl RNase-free ddH$_2$O. Determine the RNA concentration in a UV spectrophotometer.

Step 5: Treatment with TAP

Bring 12 µg of dissolved RNA to a volume of 87.5 µl by adding the required volume of RNase-free ddH$_2$O. Add 10 µl of 10 × TAP digestion buffer and 0.5 µl RNase inhibitor. Mix and split into two tubes (49 µl each). Add to one tube 1 µl of TAP enzyme (treated sample) and incubate both samples at 37 °C for 30 min. At the end of incubation, add 500 pmol of RNA oligonucleotide A3 and 100 µl RNase-free ddH$_2$O to each tube. Remove enzyme and buffer by P:C:I extraction followed by ethanol precipitation as in Step 4.

Step 6: 5′ RNA adapter ligation

Dissolve RNA pellets in 14 µl RNase-free ddH$_2$O. Heat to 90 °C for 5 min. Immediately place on ice for 5 min. Spin briefly and to each tube add:

- 2 µl of 10 × RNA ligation buffer
- 2 µl DMSO
- 2 µl T4 RNA ligase pre-mixed 9:1 with RNase inhibitor

Incubate overnight at 17 °C. At the end of incubation, add 2 pmol of the gene-specific DNA oligonucleotide for reverse transcription, here oligonucleotide jb-101, and RNase-free ddH$_2$O to bring the total volume in each tube to 150 µl. Remove enzyme and buffer by P:C:I extraction followed by ethanol precipitation as in Step 4.

Step 7: Reverse transcription

Dissolve RNA pellets from Step 6 in 20 µl RNase-free H$_2$O$_{bidest}$. Transfer 10 µl to a 0.2-ml PCR tube and place in the PCR block. The following steps can be programmed in a single routine, but it is important to add the required reagents and enzymes on time.

Denature RNA and oligonucleotide for 5 min at 65 °C. Place on ice. Add 10 µl of reverse transcriptase enzyme/buffer mix (20 µl pre-mixed for two reactions: 8 µl of 5 × reverse transcription buffer, 4 µl of 10 mM dNTP mix, 3 µl RNase-free ddH$_2$O, 2 µl 0.1 M DTT, 2 µl Thermoscript reverse transcriptase, 1 µl RNase inhibitor).

Mix and incubate for 5 min at 42 °C, followed by three subsequent 20-min incubation steps at 55, 60 and 65 °C, respectively. Terminate reverse transcriptase reaction by heating for 5 min at 85 °C. Cool down to 37 °C and add 1 μl RNase H. Incubate at 37 °C for 20 min. Then keep on ice or at −20 °C until PCR amplification.

Step 8: PCR amplification of cDNA
Set up the following PCR reactions (50 μl final volume per reaction). As template, use cDNAs from untreated and TAP-treated samples, and a 50% dilution of the ligated RNA that was taken for reverse transcription (remaining 10 μl from Step 9) as negative control.

- 5 μl   10 × PCR buffer without Mg^{2+}
- 3 μl   25 mM $MgCl_2$
- 1 μl   10 mM dNTP mix
- 2.5 μl   oligo jb-101-B (10 pmol/μl)
- 2.5 μl   oligo B6 (10 pmol/μl)
- 2 μl   template (cDNA or RNA)
- 0.25 μl HotStar *Taq* DNA polymerase
- 34 μl ddH_2O

Run PCR with the following cycles:

- 95 °C/10 min
- 35 × (95 °C/40 s, 58 °C/40 s, 72 °C/40 s)
- 72 °C/7 min

Add 10 μl of agarose gel loading buffer (6×) to each reaction and run 20 μl on a 3% Nusieve Agarose gel in 1 × TAE buffer, along with an appropriate DNA size marker.

Step 9: Cloning of amplified cDNAs and determination of 5′ ends
Excise bands of interest from the agarose gel and purify with a gel extraction kit. Consult the kit manual for lower cutoff size of DNA fragments. This is because the size of sRNA frequently will give you RACE products of 100 bp or less, which most column-based kits fail to purify efficiently. We have routinely used the QIAEX II Gel Extraction Kit (Qiagen, Germany), which is based on DNA binding to glass milk. As a slight alteration of QIAgen's protocol, we use 7 μl of glass milk per gel piece (instead of the recommended volume) and elute the purified DNA in only 15 μl Tris buffer (10 mM, pH 8). It is important to ensure that the eluate is free of glass milk since the latter may interfere with subsequent ligation into a cloning vector.

For the eluted DNA, 4 μl is used in a TOPO TA cloning kit standard reaction. Approximately 100 clones are usually obtained when using the pCR2.1-TOPO vector and chemically competent TOP10F′ cells, and more than 95% of the clones will contain an insert. Keep in mind that the cloned fragments are present in very high copy numbers (several hundred per cell). Hence, we do not advise to use blue/white screening with X-gal to select positive clones, since this implies transcription of the cloned sRNA fragment, which in some cases may be lethal to the bacteria at this copy number. Instead, use colony PCR to identify positive inserts, products of which can then be sequenced, thereby obviating isolation of the plasmid. A 25-μl colony PCR reaction is usually sufficient for both gel visualization and subsequent sequencing. We recommend assaying six colonies from each ligation.

Set up the following PCR reaction mixture (150 μl final volume):

- 15 μl   10 × PCR buffer without Mg^{2+}
- 9 μl   25 mM $MgCl_2$
- 3 μl   10 mM dNTP mix
- 3 μl   oligonucleotide Seq-UNI-61 (10 pmol/μl; for ligation into plasmid pCR2.1)
- 3 μl   oligonucleotide Seq-REV (10 pmol/μl; for ligation into plasmid pCR2.1)
- 1 μl   *Taq* DNA polymerase
- 116 μl   ddH_2O

Split into six PCR tubes (6 × 25 μl) and place them in a PCR block set to 4 °C. Slightly touch a colony with a pipette tip (10- to 200-μl tip) and then swirl the tip at the bottom of the PCR tube. Avoid transferring too much of the colony since this will inhibit amplification or result in high background. Once all PCR tubes are provided with 'template', start PCR with the following cycles:

- 95 °C/10 min
- 30 × (95 °C/40 s, 65 °C/40 s, 72 °C/40 sec)
- 72 °C/7 min

Add 5 μl of 6 × agarose gel loading buffer [10 mM Tris-HCl, pH 7.6, 0.03% (w/v) each bromophenol blue and xylene cyanol, 60% (v/v) glycerol, 60 mM EDTA] to each reaction, and analyze for inserts by running 4 μl on a 2% agarose gel. The size of colony PCR products from religated vector, i.e. without insert, is approximately 220 bp. Hence, positive inserts should have a size of 220 bp plus the size of the DNA fragment that was cloned. Purify and sequence PCR products according to the standard protocol of your sequencing service. For sequencing, use oligo Seq-REV-50, which has a lower annealing temperature than oligonucleotide Seq-REV. It is highly recommended to determine the sequence of at least four inserts per candidate since both primary and processed 5′ ends may vary by a few nucleotides (see Fig. 37.2C). For primary transcripts (band intensity increased after TAP treatment), the most upstream 5′ nucleotide is regarded as the transcription initiation site.

Acknowledgments

The authors gratefully acknowledge funding from HFSP (Human Frontier Science Program), VR (The Swedish Research Council), Wallenberg Foundation (all to E. G. H. W.) and from EMBO (long-term fellowship, J. V.).

References

1 L. Argaman, R. Hershberg, J. Vogel, G. Bejerano, E. G. H. Wagner, H. Margalit, S. Altuvia, *Curr Biol.* **2001**, *11*, 941–950.
2 S. Chen, E. A. Lesnik, T. A. Hall, R. Sampath, R. H. Griffey, D. J. Ecker, L. B. Blyn, *Biosystems* **2002**, *65*, 157–177.
3 E. Rivas, R. J. Klein, T. A. Jones, S. R. Eddy, *Curr Biol.* **2001**, *11*, 1369–1373.
4 B. Tjaden, D. R. Haynor, S. Stolyar, C. Rosenow, E. Kolker, *Bioinformatics* **2002**, *18* (Suppl. 1), S337–S344.
5 J. Vogel, V. Bartels, T. H. Tang, G. Churakov, J. G. Slagter–Jäger, A. Hüttenhofer, E. G. H. Wagner, *Nucleic Acids Res.* **2003**, *31*, 6435–6443.
6 K. M. Wassarman, F. Repoila, C. Rosenow, G. Storz, S. Gottesman, *Genes Dev.* **2001**, *15*, 1637–1651.
7 A. Hüttenhofer, J. Brosius, J. P. Bachellerie, *Curr. Opin. Chem. Biol.* **2002**, *6*, 835–843.
8 M. Lagos-Quintana, R. Rauhut, J. Meyer, A. Borkhardt, T. Tuschl, *RNA* **2003**, *9*, 175–179.
9 N. C. Lau, L. P. Lim, E. G. Weinstein, D. P. Bartel, *Science* **2001**, *294*, 858–862.
10 M. Lagos-Quintana, R. Rauhut, W. Lendeckel, T. Tuschl, *Science* **2001**, *294*, 853–858.
11 R. C. Lee, V. Ambros, *Science* **2001**, *294*, 862–864.
12 A. J. Hamilton, D. C. Baulcombe, *Science* **1999**, *286*, 950–952.
13 T. A. Volpe, C. Kidner, I. M. Hall, G. Teng, S. I. Grewal, R. A. Martienssen, *Science* **2002**, *297*, 1833–1837.
14 E. G. H. Wagner, J. Vogel, in: *Noncoding RNAs*, J. Barciszewski, V. Erdmann (eds), Landes, Austin, TX, **2003**, 243–259.
15 K. M. Wassarman, A. Zhang, G. Storz, *Trends Microbiol.* **1999**, *7*, 37–45.
16 G. Storz, *Science* **2002**, *296*, 1260–1263.
17 R. J. Klein, Z. Misulovin, S. R. Eddy, *Proc. Natl. Acad. Sci. USA* **2002**, *99*, 7542–7547.
18 S. Suzuma, S. Asari, K. Bunai, K. Yoshino, Y. Ando, H. Kakeshita, M. Fujita, K. Nakamura, K. Yamane, *Microbiology* **2002**, *148*, 2591–2598.
19 Y. Ando, S. Asari, S. Suzuma, K. Yamane, K. Nakamura, *FEMS Microbiol. Lett.* **2002**, *207*, 29–33.
20 D. Yu, H. M. Ellis, E. C. Lee, N. A. Jenkins, N. G. Copeland, D. L. Court, *Proc. Natl. Acad. Sci. USA* **2000**, *97*, 5978–5983.
21 J. H. Miller, *A Short Course in Bacterial Genetics: A Laboratory Manual and Handbook for* Escherichia coli *and Related Bacteria.* Cold Spring Harbor Laboratory Press, Cold Spring Harbor, NY, **1992**.
22 K. Nordström, B. E. Uhlin, *Biotechnology (NY)* **1992**, *10*, 661–666.
23 M. Hirano, K. Shigesada, M. Imai, *Gene* **1987**, *57*, 89–99.
24 M. J. Casadaban, S. N. Cohen, *J. Mol. Biol.* **1980**, *138*, 179–207.
25 G. Churchward, D. Belin, Y. Nagamine, *Gene* **1984**, *31*, 165–171.
26 F. Bolivar, R. L. Rodriguez, P. J. Greene, M. C. Betlach, H. L. Heyneker, H. W. Boyer, *Gene* **1977**, *2*, 95–113.
27 S. Austin, F. Hart, A. Abeles, N. Sternberg, *J. Bacteriol.* **1982**, *152*, 63–71.
28 D. R. Helinski, *CRC Crit. Rev. Biochem.* **1979**, *7*, 83–101.
29 J. Shi, D. P. Biek, *Gene* **1995**, *164*, 55–58.
30 E. Amann, J. Brosius, M. Ptashne, *Gene* **1983**, *25*, 167–178.
31 L. M. Guzman, D. Belin, M. J. Carson, J. Beckwith, *J. Bacteriol.* **1995**, *177*, 4121–4130.
32 R. M. Morgan-Kiss, C. Wadler, J. E. Cronan, Jr, *Proc. Natl. Acad. Sci. USA* **2002**, *99*, 7373–7377.
33 F. W. Studier, B. A. Moffatt, *J. Mol. Biol.* **1986**, *189*, 113–130.
34 R. Lutz, H. Bujard, *Nucleic Acids Res.* **1997**, *25*, 1203–1210.

35 E. G. H. Wagner, S. Altuvia, P. Romby, *Adv. Genet.* **2002**, *46*, 361–398.
36 C. Clepet, I. Le Clainche, M. Caboche, *Nucleic Acids Res.* **2004**, *32*, e6.
37 B. A. Bensing, B. J. Meyer, G. M. Dunny, *Proc. Natl. Acad. Sci. USA* **1996**, *93*, 7794–7799.
38 C. G. Simpson, J. W. Brown, *Methods Mol. Biol.* **1995**, *49*, 249–256.
39 E. Massé, S. Gottesman, *Proc. Natl. Acad. Sci. USA* **2002**, *99*, 4620–4625.
40 B. R. Bochner, P. Gadzinski, E. Panomitros, *Genome Res.* **2001**, *11*, 1246–1255.
41 R. A. Britton, P. Eichenberger, J. E. Gonzalez-Pastor, P. Fawcett, R. Monson, R. Losick, A. D. Grossman, *J. Bacteriol.* **2002**, *184*, 4881–4890.
42 K. J. Cheung, V. Badarinarayana, D. W. Selinger, D. Janse, G. M. Church, *Genome Res.* **2003**, *13*, 206–215.
43 P. M. Dunman, E. Murphy, S. Haney, D. Palacios, G. Tucker-Kellogg, S. Wu, E. L. Brown, R. J. Zagursky, D. Shlaes, S. J. Projan, *J. Bacteriol.* **2001**, *183*, 7341–7353.
44 R. Grifantini, S. Sebastian, E. Frigimelica, M. Draghi, E. Bartolini, A. Muzzi, R. Rappuoli, G. Grandi, C. A. Genco, *Proc. Natl. Acad. Sci. USA* **2003**, *100*, 9542–9547.
45 J. D. Helmann, M. F. Wu, P. A. Kobel, F. J. Gamo, M. Wilson, M. M. Morshedi, M. Navre, C. Paddon, *J. Bacteriol.* **2001**, *183*, 7318–7328.
46 Y. Hihara, K. Sonoike, M. Kanehisa, M. Ikeuchi, *J. Bacteriol.* **2003**, *185*, 1719–1725.
47 J. M. Lee, S. Zhang, S. Saha, S. Santa Anna, C. Jiang, J. Perkins, *J. Bacteriol.* **2001**, *183*, 7371–7380.
48 A. Løbner-Olesen, M. G. Marinus, F. G. Hansen, *Proc. Natl. Acad. Sci. USA* **2003**, *100*, 4672–4677.
49 S. Minagawa, H. Ogasawara, A. Kato, K. Yamamoto, Y. Eguchi, T. Oshima, H. Mori, A. Ishihama, R. Utsumi, *J. Bacteriol.* **2003**, *185*, 3696–3702.
50 M. A. Schembri, K. Kjaergaard, P. Klemm, *Mol. Microbiol.* **2003**, *48*, 253–267.
51 V. E. Wagner, D. Bushnell, L. Passador, A. I. Brooks, B. H. Iglewski, *J. Bacteriol.* **2003**, *185*, 2080–2095.
52 M. Wilson, J. DeRisi, H. H. Kristensen, P. Imboden, S. Rane, P. O. Brown, G. K. Schoolnik, *Proc. Natl. Acad. Sci. USA* **1999**, *96*, 12833–12838.
53 A. Zhang, K. M. Wassarman, C. Rosenow, B. C. Tjaden, G. Storz, S. Gottesman, *Mol. Microbiol.* **2003**, *50*, 1111–1124.
54 L. Zhou, X. H. Lei, B. R. Bochner, B. L. Wanner, *J. Bacteriol.* **2003**, *185*, 4956–4972.
55 M. Zheng, X. Wang, L. J. Templeton, D. R. Smulski, R. A. LaRossa, G. Storz, *J. Bacteriol.* **2001**, *183*, 4562–4570.
56 B. Tjaden, R. M. Saxena, S. Stolyar, D. R. Haynor, E. Kolker, C. Rosenow, *Nucleic Acids Res.* **2002**, *30*, 3732–3738.
57 C. Arevalo-Ferro, M. Hentzer, G. Reil, A. Gorg, S. Kjelleberg, M. Givskov, K. Riedel, L. Eberl, *Environ. Microbiol.* **2003**, *5*, 1350–1369.
58 S. Balasubramanian, T. Schneider, M. Gerstein, L. Regan, *Nucleic Acids Res.* **2000**, *28*, 3075–3082.
59 K. Buttner, J. Bernhardt, C. Scharf, R. Schmid, U. Mader, C. Eymann, H. Antelmann, A. Volker, U. Volker, M. Hecker, *Electrophoresis* **2001**, *22*, 2908–2935.
60 S. Fulda, F. Huang, F. Nilsson, M. Hagemann, B. Norling, *Eur. J. Biochem.* **2000**, *267*, 5900–5907.
61 P. R. Jungblut, *Microbes Infect.* **2001**, *3*, 831–840.
62 F. Tremoulet, O. Duche, A. Namane, B. Martinie, J. C. Labadie, *FEMS Microbiol. Lett.* **2002**, *215*, 7–14.
63 V. C. Wasinger, I. Humphery-Smith, *FEMS Microbiol. Lett.* **1998**, *169*, 375–382.
64 S. Altuvia, D. Weinstein-Fischer, A. Zhang, L. Postow, G. Storz, *Cell* **1997**, *90*, 43–53.
65 B. Vecerek, I. Moll, T. Afonyushkin, V. Kaberdin, U. Bläsi, *Mol. Microbiol.* **2003**, *50*, 897–909.

66 O. Fiehn, *Plant Mol. Biol.* **2002**, *48*, 155–171.
67 L. W. Sumner, P. Mendes, R. A. Dixon, *Phytochemistry* **2003**, *62*, 817–836.
68 S. M. Watkins, J. B. German, *Curr. Opin. Biotechnol.* **2002**, *13*, 512–516.
69 W. Weckwerth, *Annu. Rev. Plant Biol.* **2003**, *54*, 669–689.
70 I. Bronstein, J. Fortin, P. E. Stanley, G. S. Stewart, L. J. Kricka, *Anal. Biochem.* **1994**, *219*, 169–181.
71 S. J. Gould, S. Subramani, *Anal. Biochem.* **1988**, *175*, 5–13.
72 L. F. Greer, 3rd, A. A. Szalay, *Luminescence* **2002**, *17*, 43–74.
73 S. R. Kain, M. Adams, A. Kondepudi, T. T. Yang, W. W. Ward, P. Kitts, *Biotechniques* **1995**, *19*, 650–655.
74 J. L. Slonczewski, T. N. Gonzalez, F. M. Bartholomew, N. J. Holt, *J. Bacteriol.* **1987**, *169*, 3001–3006.
75 B. L. Wanner, S. Wieder, R. McSharry, *J. Bacteriol.* **1981**, *146*, 93–101.
76 Y. Wei, C. G. Miller, *J. Bacteriol.* **1999**, *181*, 6092–6097.
77 B. A. Castilho, P. Olfson, M. J. Casadaban, *J. Bacteriol.* **1984**, *158*, 488–495.
78 P. Ratet, J. Schell, F. J. de Bruijn, *Gene* **1988**, *63*, 41–52.
79 B. L. Wanner, *J. Bacteriol.* **1987**, *169*, 2026–2030.
80 D. L. Hartl, H. Ochman, *Methods Mol. Biol.* **1996**, *58*, 293–301.
81 E. G. H. Wagner, R. W. Simons, *Annu. Rev. Microbiol.* **1994**, *48*, 713–742.
82 T. Møller, T. Franch, C. Udesen, K. Gerdes, P. Valentin-Hansen, *Genes Dev.* **2002**, *16*, 1696–1706.
83 T. Møller, T. Franch, P. Hojrup, D. R. Keene, H. P. Bachinger, R. G. Brennan, P. Valentin-Hansen, *Mol. Cell* **2002**, *9*, 23–30.
84 A. Zhang, K. M. Wassarman, J. Ortega, A. C. Steven, G. Storz, *Mol. Cell* **2002**, *9*, 11–22.
85 M. Y. Liu, G. Gui, B. Wei, J. F. Preston, 3rd, L. Oakford, U. Yuksel, D. P. Giedroc, T. Romeo, *J. Biol. Chem.* **1997**, *272*, 17502–17510.
86 A. W. Karzai, M. M. Susskind, R. T. Sauer, *EMBO J.* **1999**, *18*, 3793–3799.
87 M. Bachler, R. Schroeder, U. von Ahsen, *RNA* **1999**, *5*, 1509–1516.
88 K. Hartmuth, H. Urlaub, H. P. Vornlocher, C. L. Will, M. Gentzel, M. Wilm, R. Lührmann, *Proc. Natl. Acad. Sci. USA* **2002**, *99*, 16719–16724.
89 F. A. Kolb, C. Malmgren, E. Westhof, C. Ehresmann, B. Ehresmann, E. G. H. Wagner, P. Romby, *RNA* **2000**, *6*, 311–324.
90 M. Zuker, *Nucleic Acids Res.* **2003**, *31*, 3406–3415.
91 F. H. van Batenburg, A. P. Gultyaev, C. W. Pleij, *J. Theor. Biol.* **1995**, *174*, 269–280.
92 C. Ehresmann, F. Baudin, M. Mougel, P. Romby, J. P. Ebel, B. Ehresmann, *Nucleic Acids Res.* **1987**, *15*, 9109–9128.
93 C. Brunel, P. Romby, *Methods Enzymol.* **2000**, *318*, 3–21.
94 H. Moine, C. Ehresmann, B. Ehresmann, P. Romby, in: *RNA Structure and Function*, R. W. Simons, M. Grunberg-Manago (eds), Cold Spring Harbor Laboratory Press, Cold Spring Harbor, NY, **1997**, pp. 77–115.
95 F. A. Kolb, H. M. Engdahl, J. G. Slagter-Jäger, B. Ehresmann, C. Ehresmann, E. Westhof, E. G. H. Wagner, P. Romby, *EMBO J.* **2000**, *19*, 5905–5915.
96 F. A. Kolb, E. Westhof, B. Ehresmann, C. Ehresmann, E. G. H. Wagner, P. Romby, *J. Mol. Biol.* **2001**, *309*, 605–614.
97 C. Persson, E. G. H. Wagner, K. Nordström, *EMBO J.* **1988**, *7*, 3279–3288.
98 C. Persson, E. G. H. Wagner, K. Nordström, *EMBO J.* **1990**, *9*, 3777–3785.
99 C. Rox, R. Feltens, T. Pfeiffer, R. K. Hartmann, *J. Mol. Biol.* **2002**, *315*, 551–560.
100 C. Valverde, M. Lindell, E. G. H. Wagner, D. Haas, *J. Biol. Chem.* **2004**, *279*, 25066–25074.
101 M. Lindell, P. Romby, E. G. H. Wagner, *RNA* **2002**, *8*, 534–541.

38
Experimental RNomics: A Global Approach to Identify Non-coding RNAs in Model Organisms

Alexander Hüttenhofer

38.1
Introduction

Non-coding RNAs (ncRNAs) do not encode proteins, but function directly on the level of RNA in the cell. Over the last few years, the importance of this surprisingly diverse class of molecules has been widely recognized [1–4]. ncRNAs have been identified in unexpectedly large numbers, with present estimates in the range of thousands per eukaryotic genome. They play key roles in a variety of fundamental processes in all three domains of life, i.e. Eukarya, Bacteria and Archaea. Their functions include DNA replication and chromosome maintenance, regulation of transcription, RNA processing (not only RNA cleavage and religation, but also RNA modification and editing), translation and stability of mRNAs, and even regulation of stability and translocation of proteins [4–15]. Many of them have been discovered fortuitously, suggesting they merely represent the tip of the iceberg. The vast majority of ncRNAs are relatively small, much shorter than most mRNAs; hence, their generic denomination as small non-messenger or non-coding RNAs. The highly specific roles of ncRNAs reflect in most cases their ability to selectively bind a small set of proteins as well as their potential to specifically recognize definite RNA targets via regions of sequence complementarity.

Recently, systematic searches for ncRNAs, dubbed experimental RNomics, have been initiated in various model organisms, by screening cDNA libraries constructed from sized RNA fractions [16–18]. Since a large number of ncRNAs can be grouped into specific RNA classes (based on size, structural signatures, protein partners, RNA targets or subcellular location), experimental RNomics can be focused on a particular ncRNA type, using tailor-made cDNA libraries and/or appropriate screening procedures. First, we will describe procedures for the construction and screening of a general purpose cDNA library from sized RNA fractions. To illustrate a more specialized RNomics strategy, we will next use as a paradigm one of the two large families of small nucleolar RNAs (snoRNAs), both associated with a small set of specific proteins. The two snoRNA subtypes, termed C/D and H/ACA snoRNAs, guide two prevalent types of nucleotide modification, 2′-*O*-methylation and pseudouridylation, respectively, by forming a specific duplex around each target RNA modification site. Cellular RNAs targeted by guide sno-RNAs include rRNAs in Eukarya and Archaea, small nuclear RNAs and archaeal tRNAs, as well as probably a range of still unidentified RNAs [4, 10, 11].

38.2
Materials

38.2.1
Oligonucleotide Primers

- Gibco I primer-adapter: 5′-TCG CGA GCG GCC GCG GGG GGG GGG GGG GG-3′ (*Not*I site underlined).
- M13 fsp: 5′-GCT ATT ACG CCA GCT GGC GAA AGG GGG ATG TG-3′.
- M13 rsp: 5′-CCC CAG GCT TTA CAC TTT ATG CTT CCG GCT CG-3′.
- Primer 1: 5′-ATAAAGCGGCCGCGGATCCAA-3′.
- Primer 2: 5′-TTGGATCCGCGGCCGCTTTATTNNNNTCAG-3′.
- Primer 3: 5′-AATAAAGCGGCCGCGGATCCAANNNNNRTGATGA-3′.

38.2.2
Enzymes

- Poly(A) polymerase (MBI Fermentas).
- T4 polynucleotide kinase (New England Biolabs).
- T4 RNA ligase (New England Biolabs).
- *Taq* polymerase I (Appligene).
- Other enzymes used but not listed here were supplied with the "SuperScript™ Plasmid System for cDNA Synthesis and Plasmid Cloning" kit (Invitrogen).

38.2.3
Buffers

- Electrophoresis buffer: $1 \times$ TBE (90 mM Tris–HCl, pH 8.3, 90 mM boric acid, 2.5 mM EDTA).
- $1 \times$ Tailing buffer: 50 mM Tris–HCl, pH 8.0, 200 mM NaCl, 10 mM $MgCl_2$, 2 mM $MnCl_2$, 0.4 mM EDTA, 1 mM DTT, 2 mM CTP.
- $5 \times$ First strand buffer: 250 mM Tris–HCl, pH 8.3, 375 mM KCl, 15 mM $MgCl_2$.
- $5 \times$ Second strand buffer: 100 mM Tris–HCl, pH 6.9, 450 mM KCl, 23 mM $MgCl_2$, 0.75 mM β-NAD^+, 50 mM $(NH_4)_2SO_4$.
- $5 \times$ T4 DNA ligase buffer: 250 mM Tris–HCl, pH 7.6, 50 mM $MgCl_2$, 5 mM DTT, 5 mM ATP, 25% (w/v) PEG 8000.
- $10 \times$ *Not*I buffer: 500 mM Tris–HCl, pH 8.0, 100 mM $MgCl_2$, 1 M NaCl.
- Hybridization buffer 1: 0.5 M sodium phosphate, pH 7.2, 7% SDS, 1 mM EDTA.
- Hybridization buffer 2: 1 M sodium phosphate buffer, pH 6.2, 7% SDS.
- Pre-hybridization buffer = Hybridization buffer 2.

- Wash buffer 1: 2 × SSC (20 mM sodium phosphate, pH 7.4, 0.3 M NaCl, 2 mM EDTA), 0.1% SDS.
- Wash buffer 2: 0.1 × SSC (1 mM sodium phosphate, pH 7.4, 15 mM NaCl, 0.1 mM EDTA), 0.5% SDS.
- 1 × RNA ligation buffer: 50 mM Tris–HCl, pH 7.8, 10 mM $MgCl_2$, 10 mM DTT, 1 mM ATP.
- 1 × Kinase buffer: 70 mM Tris–HCl, pH 7.6, 10 mM $MgCl_2$, 5 mM DTT, 1 mM ATP.
- 1 × Taq I buffer: 10 mM Tris–HCl, pH 9.0, 50 mM KCl, 0.1% Triton X-100.
- NET-150 buffer: 50 mM Tris–HCl, pH 7.4, 150 mM NaCl, 0.05% Nonidet P-40.

38.2.4
Reagents, Kits, Vectors and Bacterial Cells

- TRIzol® Reagent (Invitrogen).
- Phenol/chloroform: water-saturated phenol/chloroform (1:1).
- cDNA synthesis and cloning kit: SuperScript™ Plasmid System for cDNA Synthesis and Plasmid Cloning with Gateway® Technology (Invitrogen; pSPORT 1 vector included).
- *Escherichia coli* strain DH10B™ competent cells (Invitrogen).
- BigDye terminator cycle sequencing reaction kit (PE Applied Biosystems).
- Protein A–Sepharose (Sigma).
- Glycogen (Promega): 20 mg/ml.
- [γ-$^{32/33}$P]ATP.
- Nylon membranes (Qiabrane Nylon Plus; Qiagen, Germany, or Qbiogene, USA).
- pKS vector (Stratagene).

38.3
Protocols for Library Construction and Analysis

The methods described below outline the construction and screening of a cDNA library from sized RNA fractions (Protocol 1) and the construction of a more specialized cDNA library (Protocol 2), either starting from an RNA pool enriched in the ncRNA subtype of interest and/or using specialized oligonucleotide primers for preferential cDNA synthesis of this ncRNA subtype.

As for the construction of the two different cDNA libraries (Protocols 1 and 2, respectively), two different methods for reverse transcription of ncRNAs into cDNAs are applied (Fig. 38.1). In the first method (Protocol 1), 3′-tailing of ncRNAs by CTP and poly(A) polymerase is followed by reverse transcription of ncRNAs by an oligo-d(G) primer (see below). As an alternative to C-tailing of RNAs, we also used a "linker-ligation" method for cloning small RNA molecules (Protocol 2, see Fig. 38.1): both ends of ncRNAs are joined to a short 5′-phosphorylated DNA or RNA linker by T4 RNA ligase. Subsequently, ligated RNAs are transcribed into cDNA by RT-PCR using primers derived from the linker sequence. The disadvantage of this

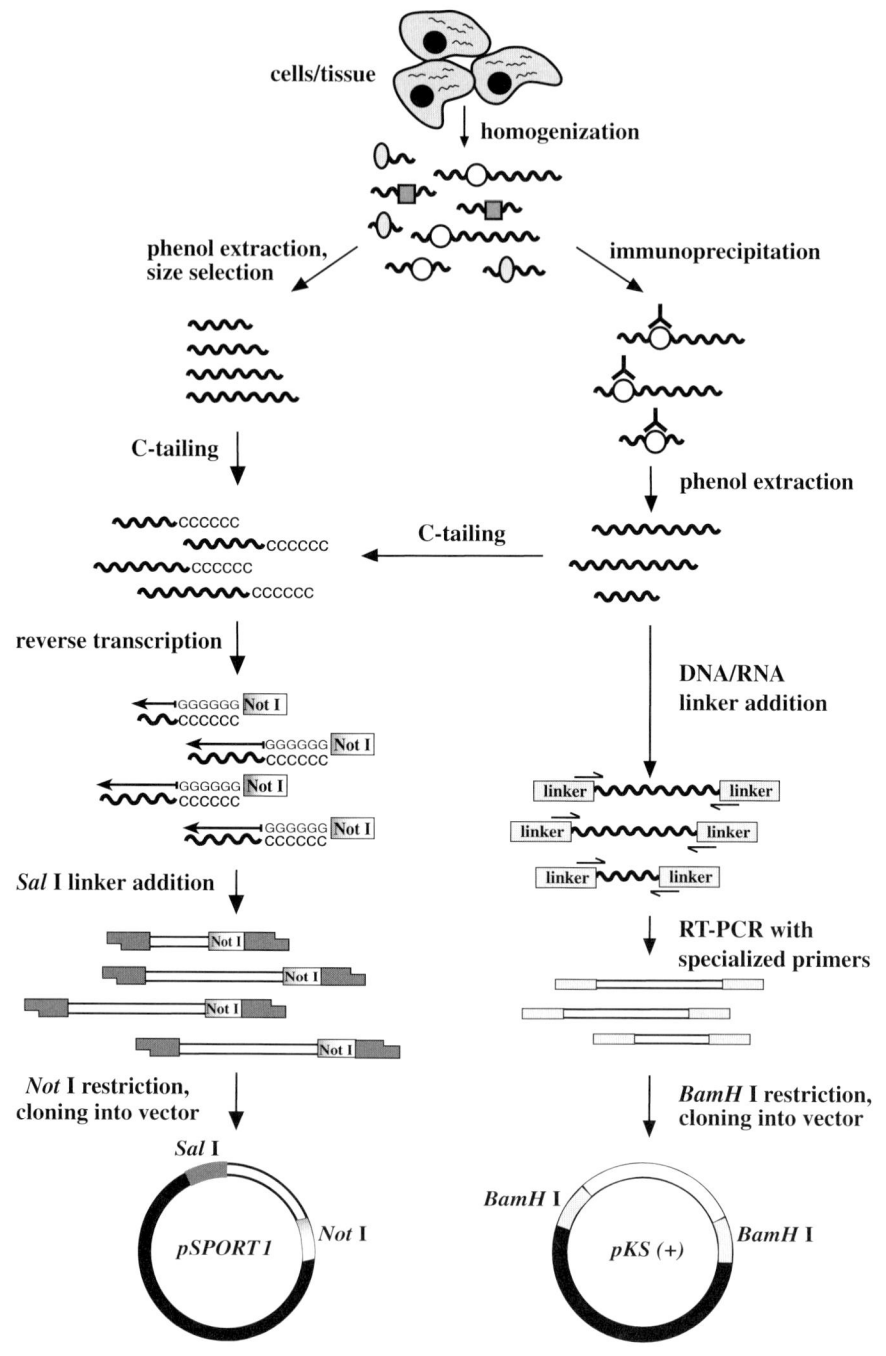

Fig. 38.1

method is, however, the possibility of oligomerization of linker oligonucleotides and the limitation of linker attachment to RNAs with 5′-monophosphate and 3′-hydroxyl ends, excluding for example primary transcripts with 5′-triphosphate termini. The advantage of this compared with the C-tailing method is that full-length cDNA clones can be obtained since in the C-tailing method the second strand synthesis by an RNase H/DNA polymerase I approach results in truncated 5′ termini.

Protocol 1: cDNA library construction and screening

This section describes the isolation of total RNA from the organism of interest, the purification and size selection and the reverse transcription of ncRNAs into cDNA (Fig. 38.1). Since ncRNAs do not contain a poly(A) tail (as do mRNAs), ncRNAs are, in a first step, tailed with CTP and poly(A) polymerase (which also uses CTP as a substrate). After the addition of C-tails, ncRNAs are reverse transcribed employing an oligo-d(G) primer and the SuperScript™ II Reverse Transcriptase. This procedure also tags small poly(A)-tailed messenger RNAs that co-fractionate with the ncRNA pool, but their cDNAs will be identified by the feature of an A-tail followed by a C-tail.

Protocol 1.1: Construction of the cDNA library encoding ncRNAs

(1) Freeze 1–2 mg of tissue in liquid nitrogen and prepare total RNA with the TRIzol reagent (Invitrogen) according to the protocol provided by the manufacturer.
(2) Fractionate about 200 µg of total RNA on a denaturing 8% polyacrylamide gel (7 M urea, 1 × TBE buffer); run the gel (20 cm wide, 30 cm long, 2 mm thick) at 400 V for 1.5 h.
(3) Excise RNA from the gel corresponding to the desired size range (e.g. 50–500 nt), passively elute the RNA by vigorous shaking in 0.3 M sodium acetate (pH 5.2), 0.1% SDS, 0.2% phenol at 4 °C overnight, extract with phenol/chloroform and precipitate the RNA with ethanol.
(4) Tail 5 µg of RNA with CTP using poly(A) polymerase. The C-tailing reaction is carried out in a volume of 50 µl containing 1 × tailing buffer and 1.5 U of poly(A) polymerase. Incubate the reaction mix at 37 °C for 1.5 h. After incubation, add 20 µl 3 M sodium acetate (pH 5.2) and 130 µl double-distilled water

Fig. 38.1. Construction of an unbiased, general purpose cDNA library (upper left) of ncRNAs or a more specialized library encoding RNAs of a certain ncRNA subclass, e.g. snoRNAs (upper right). Here, ncRNA subclass enrichment was achieved by immuno-precipitation of a cell lysate with an antibody directed against an RNA-binding protein specific to this subclass (indicated by open circles; grey ovals and squares represent other RNA-binding proteins). Two different ways to generate a cDNA library are illustrated in the bottom part. For the general purpose cDNA library, the C-tailing approach is the method of choice (left path); for the specialized cDNA library, a "linker ligation" protocol (right path) represents a preferable alternative in cases where primers specific to the RNA subclass of interest can be used in RT-PCR reactions (for details, see Fig. 38.2).

(ddH$_2$O). Extract the RNA with phenol/chloroform, precipitate by ethanol and dissolve in 5 μl of RNase-free ddH$_2$O.

(5) First and second cDNA strand synthesis. All enzymes, reagents, the SalI adapter and the pSPORT1 vector used in Steps 5–7 are supplied with the Invitrogen SuperScript™ Plasmid System (see Section 38.2.4). However, it is important not to use the NotI primer-adapter included in this kit, since it contains an oligo-d(T) stretch suitable for reverse transcription of mRNAs only, but to use the Gibco I primer-adapter oligonucleotide (custom-made, Section 38.2.1) instead. Before first-strand cDNA synthesis, combine 5 μl C-tailed RNA (obtained in Step 4) with 2 μl Gibco I primer-adapter oligonucleotide (0.5 μg/μl). Denature the mixture at 90 °C for 1 min, immediately transfer to 37 °C and incubate for 10 min. Subsequently, add 4 μl of 5 × first strand buffer, 2 μl of 0.1 M DTT, 1 μl of 10 mM dNTP mix, and 1 μl of ddH$_2$O, and incubate at 37 °C for 2 min. Start first strand synthesis by addition of 5 μl of SuperScript™ II (AMV) reverse transcriptase (200 U/μl) and let the reaction proceed for 1.5 h at 45 °C. For second strand DNA synthesis, add to the first strand cDNA mix 30 μl of 5 × second strand buffer, 1 μl of E. coli DNA ligase (10 U/μl), 4 μl of E. coli DNA polymerase I (10 U/μl), 1 μl of E. coli RNase H (2 U/μl), 3 μl of 10 mM dNTPs and 91 μl ddH$_2$O to give a final volume of 150 μl. Gently mix and incubate at 16 °C for 2 h. Add 2 μl of T4 DNA polymerase (5 U/μl) to the reaction mix, continue incubation at 16 °C for another 5 min and place the mixture on ice for 5 min. Stop the reaction by adding 10 μl of 0.5 M EDTA (pH 8.0), followed by one phenol/chloroform extraction and ethanol precipitation. Subsequently, dissolve the pellet in 25 μl of RNase-free ddH$_2$O.

(6) Addition of SalI adapters. Before cloning into the pSPORT1 vector, add SalI adapters to the 5′ and 3′ ends of cDNAs as follows: combine the 25 μl of cDNA from Step 5 with 10 μl of 5 × T4 DNA ligase buffer, 10 μl of SalI adapter (1 μg/μl) and 5 μl of T4 DNA ligase (1 U/μl), resulting in a total volume of 50 μl. Incubate at 16 °C for 16 h. Extract the DNA with phenol/chloroform, precipitate by ethanol and dissolve in 41 μl of RNase-free ddH$_2$O. Subsequently, digest the DNA with NotI in a total volume of 50 μl containing 5 μl of 10 × NotI buffer and 4 μl NotI (10 U/μl) for 2 h at 37 °C. The restriction fragments are separated on a native 6% polyacrylamide/1 × TBE gel. The cDNA is visualized by staining with ethidium bromide, subsequently excised from the gel and passively eluted in 0.3 M sodium acetate buffer, pH 7.2, at 4 °C overnight. The DNA is extracted with phenol/chloroform, precipitated with ethanol and dissolved in 15 μl of RNase-free ddH$_2$O.

(7) Cloning of cDNAs into the pSPORT 1 vector employing the Invitrogen SuperScript™ Plasmid System. Of the 15 μl of NotI-digested cDNA prepared in Step 6, use 1.5 μl (about 100 ng) for the ligation reaction, containing 1 × T4 DNA ligase buffer, 1 U of T4 DNA ligase and 50 ng pSPORT 1 vector in a total volume of 20 μl; incubate at room temperature for 3 h. Extract the ligation mixes with phenol/chloroform, precipitate the DNA with ethanol and redissolve it in 10 μl ddH$_2$O; 3 μl are used for transformation of E. coli strain DH10B™ competent cells.

(8) Amplify cDNA inserts from individual plasmid clones by PCR using primers M13 fsp and M13 rsp (see Section 38.2.1) and spot PCR products on filters, if available with robots in high-density arrays [19].

Protocol 1.2: Analysis of cDNA clones: sequencing and sequence analysis

(1) Sequence a subset of cDNA clones using the M13 rsp primer and, for example, the BigDye terminator cycle sequencing reaction kit on an ABI Prism 3700 (Perkin Elmer) sequencer (see Section 38.5, Note 3).
(2) Based on the sequence data obtained in Step 1, design oligonucleotide probes to identify clones of abundant known ncRNAs, such as fragments of ribosomal RNA, by filter hybridization screening (see Protocol 1.3). Such clones are excluded from the second round of sequencing.
(3) Then sequence the remaining cDNA clones and compare them to each other using, for example, the Lasergene Seqman II program package to identify clones with identical or overlapping cDNA sequences.
(4) Search the GenBank database (NCBI) for all novel sequences using the BLASTN program. All sequences which have not been annotated in this database previously can be treated as potential candidates for novel ncRNAs.

Protocol 1.3: Exclusion of cDNA clones for the most abundant, known RNA species

(1) End-label DNA oligonucleotides (20–30 nt in length) derived from abundant sequences identified in Protocol 1.2, Step 1, with $[\gamma\text{-}^{33}P]ATP$ and T4 polynucleotide kinase using standard molecular biology techniques [20].
(2) Hybridize the labeled oligonucleotides to cDNA arrays spotted on filters (see above). Prior to hybridization, pre-treat the membrane by rinsing twice with pre-hybridization buffer. Dot-blot hybridization is performed in hybridization buffer 1 at 53 °C for 12 h.
(3) Wash the filters twice for 15 min in wash buffer 1 at room temperature, followed by another washing step in wash buffer 2 for 1 min at 53 °C.
(4) Expose a phosphor-imaging screen and rank intensities of hybridization signals by computer-aided analysis as described [21].

Protocol 1.4: Verification of the expression of cloned ncRNAs by Northern blot analysis

(1) Fractionate total RNA prepared from the organism or tissue of interest using the TRIzol method described above on an 8% denaturing polyacrylamide gel (7 M urea, 1 × TBE buffer).
(2) Transfer onto nylon membranes, e.g. using a Bio-Rad semi-dry blotting apparatus (Trans-blot SD, Bio-Rad, Germany), and immobilize the RNAs (e.g. for 1.5 min in a Stratagene UV crosslinker at 1200 J/cm^2).
(3) Incubate nylon membranes for 1 h in pre-hybridization buffer.
(4) End-label 22- to 24-nt long oligonucleotides complementary to potential novel RNA species with $[\gamma\text{-}^{32}P]ATP$ and T4 polynucleotide kinase according to stan-

dard procedures and hybridize the nylon membranes for 12 h at 58 °C in hybridization buffer 2.
(5) Wash blots twice at room temperature for 15 min in wash buffer 1 and subsequently at 58 °C for 1 min in wash buffer 2.
(6) Expose Kodak MS-1 film to membranes for 1 h to 5 days or use a PhosphorImager.

Protocol 2: Construction of a specialized cDNA library
A specialized cDNA library is generated by use of RNA isolated from a defined subcellular fraction [22] or RNA enriched in the ncRNA type of interest due to pre-fractionation of a total cellular extract (Fig. 38.1, Protocol 2.1). Alternatively, or in addition, enrichment can be achieved at the level of cDNA synthesis (see Protocol 2.2).

Protocol 2.1: Using an RNA sample enriched through immunoprecipitation
The approach is illustrated for a cDNA library specific to C/D snoRNAs. Enrichment was achieved by using antibodies against fibrillarin, a protein which binds specifically all members of this large ncRNA family. The monoclonal antibody 72B9 (provided by M. Pollard) has been successfully employed for this purpose.

(1) Rapidly cut freshly prepared or frozen tissues (stored at −80 °C) into small pieces using a sterile cutter and resuspend in 8 ml of cold NET-150 buffer.
(2) Homogenize the cellular sample, for example by using a Homogenizer One Shot Model (Constant Systems LTD) at 2.0 kbar and clarify the extract by centrifugation (12 000 g, 10 min, at 4 °C).
(3) Immobilization of IgG antibodies on Sepharose beads. In siliconized RNase-free Eppendorf tubes, incubate 10–50 μl of antibody solution (here a supernatant of a monoclonal culture for antibody 72B9) with 2.5 mg of swollen Protein A–Sepharose (PAS) in 0.5 ml NET-150 buffer for 2 h at room temperature with gentle agitation. Pellet the immunoglobulin-coated PAS (PAS–Ig) matrix by brief centrifugation (at 500 g for 20 s) and wash the beads twice with 1 ml NET-150 buffer before resuspension in 0.5 ml NET-150.
(4) Add 0.5 ml of cell extract supernatant from Step 2 to the 0.5 ml of resuspended PAS–Ig material. After 60 min at 4 °C with gentle agitation, re-pellet and wash the PAS–Ig material 8 times as in Step 3.
(5) To the washed pellet, add 300 μl NET-150, 1 μl glycogen (20 mg/ml), 30 μl SDS (10%) and 300 μl phenol/chloroform. Incubate for 15 min at 37 °C with occasional vortexing.
(6) After centrifugation (16 000 g, 10 min at 4 °C) recover the aqueous phase, add 40 μl 3 M sodium acetate, pH 5.3, and 1 ml of cold 100% ethanol, and chill at −20 °C for 30 min. Centrifuge at 4 °C and 16 000 g for 30 min; wash the pellet with 1 ml 70% ethanol, briefly centrifuge and redissolve the air-dried pellet in 10 μl RNase-free water. Store at −20 °C.
(7) Then proceed as described in Protocol 1.1 (starting from Step 4) or in Protocol 2.2.

Protocol 2.2: Using an RT-PCR procedure with specialized primers

Extracted RNAs can be oligonucleotide-tagged at both ends by T4 RNA ligase and amplified by RT-PCR [22]. Note that the reaction catalyzed by T4 RNA ligase requires RNA substrates with 5'-monophosphate and 3'-OH termini. In a variation of this approach, the RT-PCR primers are designed in such a way as to preferentially amplify RNA molecules of the selected ncRNA subtype. This can be performed whenever the ncRNA type of interest exhibits sequence signatures at a short, fixed distance from the 5' and/or 3' terminus. This is illustrated below for the C/D snoRNAs, which contain the box C and D motifs only a few nucleotides away from their 5' and 3' termini, respectively [23]. Oligonucleotide-tagged RNA molecules are then submitted to RT-PCR using a pair of primers matching the tag as well as the adjacent terminal nucleotides of a typical C/D snoRNA specimen (Fig. 38.2). The 3'-terminal sequence of both primers, termed 2 and 3, matches the

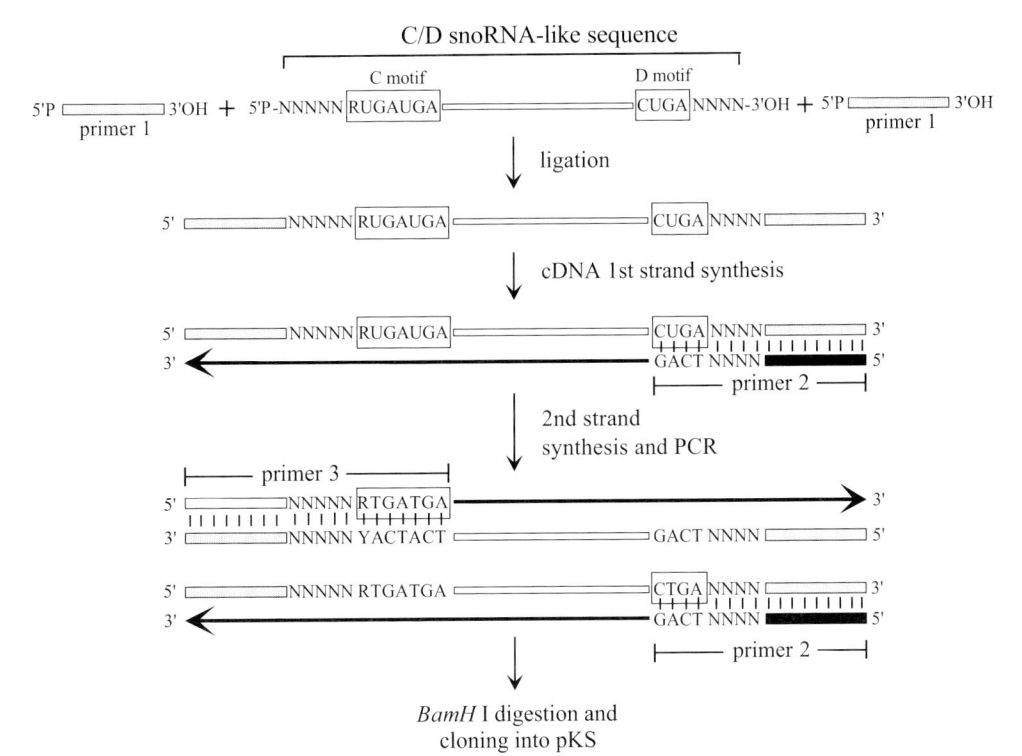

Fig. 38.2. Preferential amplification of a specific RNA subset by an RT-PCR procedure with specialized primers. C/D snoRNAs immunoprecipitated by the anti-fibrillarin antibody are oligonucleotide-tagged at both ends. They are then amplified by RT-PCR using primers 2 and 3 which are designed to amplify only RNA molecules with the prevalent terminal sequences of C/D snoRNAs, i.e. the RUGAUGA (box C) and CUGA (box D) positioned at 5 nt from their 5' end and 4 nt from their 3' end, respectively. Accordingly, the two primers contain 5 and 4 degenerated positions, respectively, to cover all sequence variants in the region upstream of box C and downstream of box D.

conserved box D and C motifs, respectively, thereby ensuring the preferential amplification of C/D snoRNAs.

(1) Primer 1 (preferentially RNA or DNA with a 3′-terminal ribonucleotide; see Chapter 4) must be phosphorylated at its 5′ end. For the kinase reaction, incubate 60 pmol of primer 1 with 10 U of T4 polynucleotide kinase in 15 µl 1 × kinase buffer for 60 min at 37 °C. Dilute with 200 µl of RNase-free ddH$_2$O, extract with an equal volume of phenol/chloroform and precipitate with ethanol, using 1 µl of glycogen (20 mg/ml) as a carrier. The pellet is washed with 70% ethanol, air-dried and redissolved in 10 µl RNase-free ddH$_2$O.

(2) Ligate the extracted RNAs (from Protocol 2.1, Step 6) to phosphorylated primer 1 (see also Section 38.5, Note 4). The reaction is carried out in 20 µl of 1 × RNA ligation buffer, with 20 pmol of phosphorylated primer 1 (3.3 µl from Step 1) and 20 U of T4 RNA ligase. Incubate overnight at 4 °C. Then dilute with 200 µl of RNase-free ddH$_2$O and proceed with a phenol/chloroform extraction and ethanol precipitation as in Step 1 above. Redissolve in 10 µl RNase-free ddH$_2$O.

(3) The ligation product is then reverse transcribed using primer 2, in a final reaction volume of 20 µl. For this purpose, add 6 µl of RNase-free ddH$_2$O to 5 µl of the oligonucleotide-tagged RNA from Step 2 and 1 µl of primer 2 (0.5 µg/µl). Heat the mixture at 70 °C for 5 min and quickly chill on ice. Immediately add 4 µl of 5 × first strand buffer, 2 µl of 0.1 M DTT and 1 µl of 10 mM dNTP (10 mM each). Mix and incubate 2 min at 42 °C. Add 200 U of SuperScript™ II reverse transcriptase and incubate for 2 h at 42 °C.

(4) Use the resulting cDNA product as template for PCR by *Taq* polymerase I with primers 2 and 3. The PCR reaction is carried out in a final volume of 100 µl in 1 × *Taq* I buffer, after addition of 6 µl MgCl$_2$ (25 mM), 2 µl dNTPs mix (10 mM each), 2 µl of the reverse transcription reaction mix from Step 3, 1 µl each of primer 2 and 3 solutions (each 0.5 µg/µl) and 2.5 U of *Taq* I DNA polymerase. Perform 30 cycles, with 1 min denaturation step at 90 °C, 1 min hybridization step at 55 °C and 1 min elongation step at 72 °C. Extract with an equal volume of phenol/chloroform and precipitate with ethanol.

(5) Digest the PCR product overnight at 37 °C by *Bam*HI (the PCR amplification primers are designed to contain a *Bam*HI site) and clone in pKS vector by standard recombinant DNA methods [20]. Individual clones can be manually sequenced (T7 sequenase version 2.0; US Biochemical Corp.). About 70–80% of them correspond to C/D snoRNA-like sequences.

38.4
Computational Analysis of ncRNA Sequences

The cloning strategy described in Protocol 1.1 will result in 5′ truncated cDNAs that are significantly shorter than the corresponding ncRNAs detected by Northern blot analysis (see above). For organisms with a completely sequenced genome, the missing 5′-terminal sequence of the ncRNA can be inferred from the genomic sequence, taking into account the size determined by the Northern analysis. All ncRNA sequences are then searched for the presence of known sequence or structure motifs typical of a particular ncRNA subtype, especially the two widespread C/D and H/ACA snoRNA families. Motif searches can be carried out with the DNAMAN sequence analysis program. Folding of ncRNA structures can be predicted using the mfold version 3.0 by Zuker [24].

38.5
Troubleshooting

(1) After isolation of the RNA of choice by the TRIzol method, we find it useful to additionally perform one or two phenol/chloroform extractions to remove residual amounts of proteins still attached to RNAs.
(2) When running total RNA on denaturing polyacrylamide gels for size separation, try to minimize the running distance of RNA into the gel because otherwise you will have too much gel material from which the RNA has to be passively eluted: the more gel material you end up with, the larger is the elution volume and hence the more difficult it will be to quantitatively precipitate the size-selected RNA.
(3) When employing the C-tailing method, sequence analysis of clones may be severely impeded when a primer downstream from the C-tail is used. Indeed, according to our experience sequencing through the C-tail to determine the ncRNA gene of interest gives unsatisfactory results. Hence, for sequence analysis a primer upstream from the ncRNA gene of interest is recommended.
(4) As the ligation reaction is not restrained in terms of primer 5′ and 3′ end reactivity (Protocol 2.2, Step 2), this step can lead to the production of various forms of primer–RNA concatemers, which may result in the cloning of inserts spanning two (or more) C/D snoRNA sequences.

Acknowledgments

This work is supported by grant HU 467/5-1 of the Deutsche Forschungsgemeinschaft (DFG) and grant FWF 171370 of the Österreichischer Fonds für Wissenschaft und Forschung (FWF) to A. H.

References

1 J. COUZIN, Science 2002, 298, 2296–2297.
2 C. DENNIS, Nature 2002, 420, 732.
3 C. DENNIS, Nature 2002, 418, 122–124.
4 A. HÜTTENHOFER, J. BROSIUS, J. P. BACHELLERIE, Curr. Opin. Chem. Biol. 2002, 6, 835–843.
5 G. STORZ, Science 2002, 296, 1260–1263.

6 S. Gottesman, *Genes Dev.* **2002**, *16*, 2829–2842.
7 T. Tuschl, *Nat. Biotechnol.* **2002**, *20*, 446–448.
8 T. Tuschl, *Nature* **2003**, *421*, 220–221.
9 V. Ambros, *Cell* **2001**, *107*, 823–826.
10 J. P. Bachellerie, J. Cavaille, A. Hüttenhofer, *Biochimie* **2002**, *84*, 775–790.
11 T. Kiss, *Cell* **2002**, *109*, 145–148.
12 N. C. Lau, L. P. Lim, E. G. Weinstein, D. P. Bartel, *Science* **2001**, *294*, 858–862.
13 R. C. Lee, V. Ambros, *Science* **2001**, *294*, 862–864.
14 M. W. Rhoades, B. J. Reinhart, L. P. Lim, C. B. Burge, B. Bartel, D. P. Bartel, *Cell* **2002**, *110*, 513–520.
15 T. A. Volpe, C. Kidner, I. M. Hall, G. Teng, S. I. Grewal, R. A. Martienssen, *Science* **2002**, *297*, 1833–1837.
16 A. Hüttenhofer, M. Kiefmann, S. Meier-Ewert, J. O'Brien, H. Lehrach, J. P. Bachellerie, J. Brosius, *EMBO J.* **2001**, *20*, 2943–2953.
17 C. Marker, A. Zemann, T. Terhorst, M. Kiefmann, J. P. Kastenmayer, P. Green, J. P. Bachellerie, J. Brosius, A. Hüttenhofer, *Curr. Biol.* **2002**, *12*, 2002–2013.
18 T. H. Tang, J. P. Bachellerie, T. Rozhdestvensky, M. L. Bortolin, H. Huber, M. Drungowski, T. Elge, J. Brosius, A. Hüttenhofer, *Proc. Natl. Acad. Sci. USA* **2002**, *99*, 7536–7541.
19 A. O. Schmitt, R. Herwig, S. Meier-Ewert, H. Lehrach, High density cDNA grids for hybridization fingerprinting experiments, in: *PCR Applications: Protocols for Functional Genomics*, M. A. Innis, D. H. Gelfand, J. J. Sninsky (eds), Academic Press, San Diego, **1999**, pp. 457–472.
20 J. Sambrook, E. F. Fritsch, T. Maniatis, *Molecular Cloning, A Laboratory Manual*, 2nd edn, Cold Spring Harbor Laboratory Press, Cold Spring Harbor, NY, **1989**.
21 E. Maier, S. Meier-Ewert, A. R. Ahmadi, J. Curtis, H. Lehrach, *J. Biotechnol.* **1994**, *35*, 191–203.
22 Z. Kiss-Laszlo, Y. Henry, J. P. Bachellerie, M. Caizergues-Ferrer, T. Kiss, *Cell* **1996**, *85*, 1077–1088.
23 J. Cavaille, P. Vitali, E. Basyuk, A. Hüttenhofer, J. P. Bachellerie, *J. Biol. Chem.* **2001**, *276*, 26374–26383.
24 M. Zuker, *Methods Mol. Biol.* **1994**, *25*, 267–294.

39
Large-scale Analysis of mRNA Splice Variants by Microarray

Young-Soo Kwon, Hai-Ri Li and Xiang-Dong Fu

39.1
Introduction

Alternative splicing plays an important role in regulating gene expression and increasing the complexity of the proteome in higher eukaryotes. Alternative splicing is now a rule rather than an exception for most genes in humans as up to 60% of human genes are alternatively spliced to give rise to multiple mRNA isoforms [1]. Different mRNA isoforms may encode functionally distinct proteins and/or follow different pathways for nuclear export, stability and translation. Aberrant splicing has also been increasingly linked to diseases [2, 3]. To decipher the function of mRNA isoforms and understand how genes are regulated at the splicing level, methods for detecting and quantifying mRNA isoforms in large scale would be highly desirable.

Two microarray approaches have been developed to meet the challenge of detecting mRNA isoforms in high throughput. One is based on oligonucleotides to detect exon–exon junctions [4, 5]. The other, which is described here, is a methodology called RASL (RNA-mediated Annealing, Selection and Ligation) jointly developed by our group in collaboration with Illumina [6]. RASL technology uses a pair of oligonucleotides linked to a unique address sequence (or index) to detect a specific mRNA splicing event (see below). Although RASL technology was developed based on a sophisticated fiber-optic array platform, and is still in its development and optimization phases, we describe here a simplified version so that individual laboratories may explore its applications on a small-to-intermediate scale. The limitations of RASL technology and various potential pitfalls are discussed in Troubleshooting (Section 39.4).

39.2
Overview of RASL Technology

RASL technology takes advantage of exon–exon junction information. The assay uses total RNA from cells without cDNA conversion or signal amplification, and

Handbook of RNA Biochemistry: Student Edition. Edited by R. K. Hartmann, A. Bindereif, A. Schön, E. Westhof
Copyright © 2009 WILEY-VCH Verlag GmbH & Co. KGaA, Weinheim
ISBN: 978-3-527-32534-4

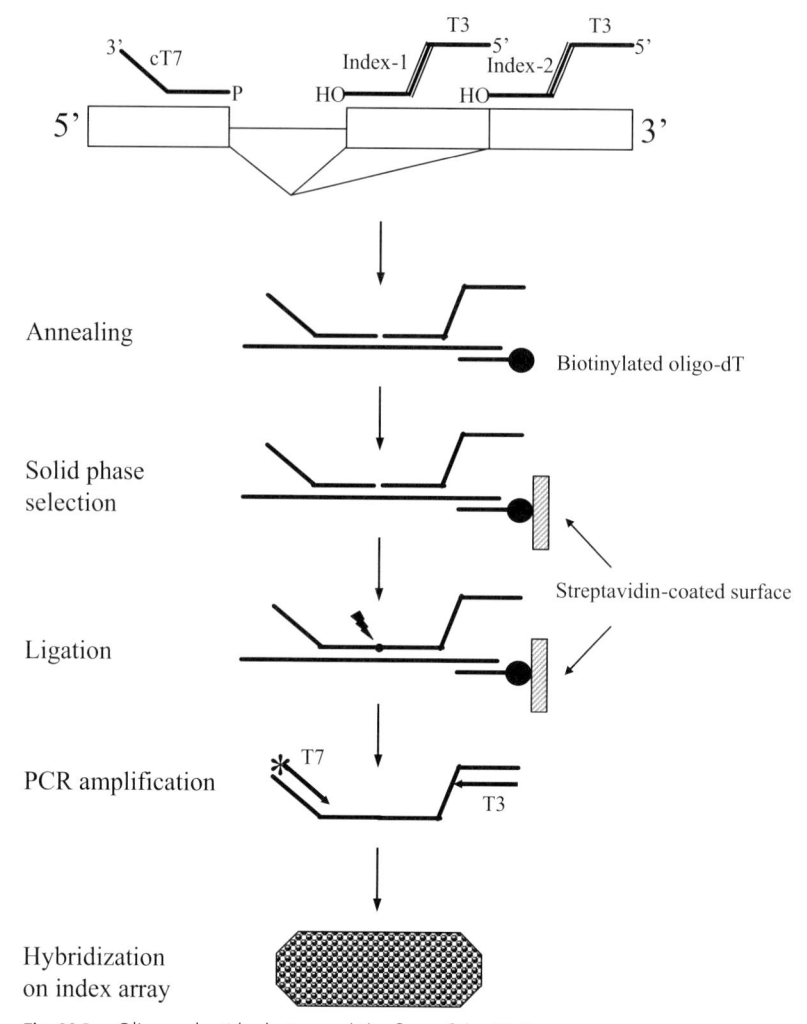

39 Large-scale Analysis of mRNA Splice Variants by Microarray

Fig. 39.1. Oligonucleotide design and the flow of the RASL assay.

aims to minimize potential distortion of mRNA isoform representation. As shown in Fig. 39.1, a pair of oligonucleotides is used to detect a specific mRNA isoform, one hybridizing to the upstream exon and the other to the downstream exon. Unique index sequences (20–25mers) are each linked to the oligonucleotides targeting the exonic sequences of choice during splicing (i.e. only alternative splice sites are indexed). In addition, a universal primer-landing site (23mer complementary T7 sequence) is linked to all oligonucleotides targeting the upstream exons and another universal primer landing site (23mer T3 sequence) linked to all oligonucleotides targeting downstream exons. All oligonucleotides targeting upstream exons are required to carry the 5′-phosphate, which may be pooled and kinased with T4 kinase. Subsequently, all oligonucleotides are pooled for the RASL assay.

The RASL procedure can be divided into five steps:

(1) Annealing. Pooled oligonucleotides are mixed with total RNA along with biotinylated oligo-dT$_{(25)}$, which are treated under denaturing and annealing conditions.
(2) Solid phase poly(A)$^+$ selection. The annealed mix is transferred to a streptavidin-coated tube to allow mRNA and annealed oligonucleotides to be trapped on the tube surface through the interaction between the poly(A) tail and immobilized oligo-dT$_{(25)}$. Excess oligonucleotides are then washed away.
(3) Ligation. T4 DNA ligase is used to seal the nick between correctly juxtaposed oligonucleotides, which converts aligned oligonucleotide pairs from half amplicoms to complete amplicoms for PCR.
(4) Signal amplification. Ligated oligonucleotides are released from the solid phase and used as templates for PCR with universal T7 and T3 primers. T7 is 5′-end-labeled with a fluorescence dye or biotin.
(5) Hybridization on a universal index array. The RASL-PCR products are detected and quantified on an array of probes consisting of index sequences. In the following section, detailed instructions to each of these steps are provided.

39.3
Description of Methods

39.3.1
Preparation of Index Arrays

RASL technology was co-developed with Illumina, where the universal index arrays are manufactured on both fiber-optic bundles and more conventionally on glass slides. These products should be commercially available soon. In the meantime, one can contact the company (www.illumina.com) to obtain the list of about 1500 index sequences (20–25mers), which are designed to avoid significant homology to any genomic sequences in databases and are characterized based on similar behavior during hybridization. Individual oligonucleotides are amino-derived during chemical synthesis and spotted on conventional glass slides. We use the 3D CodeLink^{TM} slides (Amersham Biosciences) to prepare index arrays for profiling splicing. Slide printing is in accordance with the instructions of slide manufacturer. Briefly, probe oligonucleotides are diluted to 10 µM in 1 × Printing Buffer (150 mM sodium phosphate, pH 8.5) and printed with a pin-type microarray printer (Cartesian Technologies). The probes are covalently crosslinked to the slide by overnight incubation in a NaCl-saturated chamber.

39.3.2
Annotation of Alternative Splicing

In order to obtain precise information on exon junction sequences, a database of alternative splicing would greatly assist RASL technology. For individual laboratories, the information on genes of interest may be extracted from a number of published alternative splicing databases coupled with a literature search. Since none of the published databases has tabulated information for microarray purposes, we have designed a database containing manually annotated alternative splicing events or MAASE ([7]; http://splice.sdsc.edu). The MAASE system actually contains two components. One is a computer-aided annotation system to streamline and catalyze the manual annotation process. The second component is a curated database. The system is designed for the splicing community to both utilize its content as well as contribute to its expansion. The full content of the database will be available soon for public use.

39.3.3
Target Design

The design for targeting oligonucleotides is according to the configuration shown in Fig. 39.1. Since only around 1500 index sequences are available, we generally design a set of 1200 targets as a pool and set aside a few hundred index sequences for future addition or inclusion of missed isoforms after initial profiling experiments. To pair a specific targeting oligonucleotide to a particular index sequence, we have developed software to aid target design that is now coupled with MAASE. The program will hypothetically pair each targeting oligonucleotide to all index sequences and calculate T_m, G/C content and the potential secondary structure for each potential pair. A specific pair is chosen in order for each pair to have similar T_m and G/C content with minimal secondary structure. All information is then tabulated in a spreadsheet for oligonucleotide synthesis.

39.3.4
Preparation of Target Pool

(1) Resuspend oligos in RNase-free water at 100 µM (in terms of each oligo).
(2) Mix oligos targeting upstream and downstream exonic sequences separately.
(3) Phosphorylate the upstream pool in the following reaction:
   (a) 5 µl 10 × kinase buffer (0.7 M Tris–HCl, pH 7.6, 0.1 M $MgCl_2$, 50 mM DTT)
   (b) 2.5 µl oligo pool
   (c) 0.5 µl 100 mM ATP
   (d) 41 µl RNase-free H_2O
   (e) 10 U of T4 DNA kinase (NEB)
(4) Incubate the reaction at 37 °C for 30 min.
(5) Inactivate the kinase at 65 °C for 20 min (no further purification required).

(6) Store both kinased and unkinased oligo pools at −20 °C.
(7) Pool both kinased and unkinased oligos and dilute to 100 nM before use.

39.3.5
The RASL Assay Protocol

(1) Mix the following in a test tube:
  (a) 0.1 µg total RNA extracted with TRIzol
  (b) 1 µl oligo pool (100 fmol/µl)
  (c) 1 µl biotin–(dT)$_{25}$ (2 pmol/µl)
  (d) 1 µg carrier yeast tRNA
(2) Add 20 µl 2 × lysis/binding buffer (40 mM Tris–HCl, pH 7.6, 1 M NaCl, 2 mM EDTA, 0.2% SDS).
(3) Add RNase-free H$_2$O to give a total volume of 40 µl.
(4) Heat the mixture to 94 °C for 2 min, cool for 1 min to 45 °C and incubate for a further 5 min (this and the following steps may be carried out in a PCR machine).
(5) Transfer the mixture to a streptavidin-coated tube (Roche, Indianapolis) and incubate at 45 °C for 1 h. To increase the capture sensitivity, streptavidin-coated magnetic beads may be used. We use the Sera-Mag^{TM} streptavidin-coated magnetic beads (Seradyn, Indianapolis).
(6) Remove the liquid by aspiration and wash the tube 3 times with 200 µl wash buffer (20 mM Tris–HCl, pH 7.6, 0.1 M NaCl, 1 mM EDTA, 0.01% Triton X-100), preheated to 45 °C.
(7) Wash the tube once with 200 µl 1 × ligation buffer (50 mM Tris–HCl, pH 7.6, 10 mM MgCl$_2$, 1 mM ATP, 1 mM DTT, 5% polyethylene glycol-8000).
(8) Add 40 µl of 1 × ligase buffer containing 10 U of T4 DNA ligase (BRL, high concentration) and incubate for 1 h at 37 °C.
(9) Remove the ligation solution and wash the tube twice with 200 µl wash buffer.
(10) Elute ligated oligos by incubating with 40 µl H$_2$O at 65 °C for 5 min.

39.3.6
PCR Amplification

(1) Universal PCR primer sequences are as follows:
  T7 primer sequence (23mer): 5'-TAATACGACTCACTATAGGGAGA-3'
  T3 primer sequence (23mer): 5'-AATTAACCCTCACTAAAGGGAGA-3'
(2) The 5' end of T7 primer is labeled with a fluorescent dye. To carry out a comparison experiment (treated versus untreated, normal versus mutant, etc.), we perform two-color experiments using T7 labeled separately with Cy3 and Alexa 647 (Molecular Probes; the dye choice is based on their similar signal strength and stability). There are two options to prepare dye-labeled T7 primer: (1) order dye-labeled primers directly from a commercial vendor or (2) use a dye-labeling kit from Molecular Probes. The labeled primer can be purified in a sequencing

gel or by HPLC. Both T3 and dye-labeled T7 primers are adjusted to a final concentration of 10 pmol/μl.
(3) Assemble the following in a test tube for PCR:
   (a) 4 μl RASL products
   (b) 5 μl 10 × AmpliTaq buffer
   (c) 3 μl 25 mM MgCl$_2$
   (d) 1 μl dNTP (10 mM each)
   (e) 3 μl T3 primer
   (f) 3 μl dye-labeled T7 primer
   (g) 0.3 μl AmpliTaq Gold
   (h) Add H$_2$O to 50 μl
(4) Activate the *Taq* polymerase at 93 °C for 10 min.
(5) Run PCR for 35 cycles (30 s at 94 °C, 30 s at 54 °C and 30 s at 72 °C).
(6) Check 5 μl in a 2% agarose gel. A single band of around 110 bp should be visible.

39.3.7
Hybridization on Index Array

(1) Before hybridization, slides are placed in a pre-warmed blocking solution (1 M Tris–HCl, 50 mM ethanolamine, pH 9.0, 0.1% SDS) at 50 °C for 15 min to block residual reactive groups.
(2) Rinse twice with deionized H$_2$O and wash in pre-warmed 4 × SSC with 0.1% SDS at 50 °C for 60 min. The slides are further rinsed twice with deionized H$_2$O and individually centrifuged at 800 r.p.m. for 3 min in a 50-ml conical tube in a table-top centrifuge.
(3) 10 μl of RASL-PCR products from Cy3 and Alexa reactions are combined and diluted into the final 100-μl hybridization mixture (5 × SSC, 0.1% SDS, 0.1 mg/ml sonicated salmon sperm DNA, 0.3 μM each of T7 and T3 primers, which are used to block potential background hybridization through the T7 and T3 sequences in PCR products).
(4) Heat the hybridization mixture in a boiling water bath for 2 min followed by a brief spin (about 1 min) in a microfuge.
(5) Apply 50 μl of the mixture onto an index array slide and place the slide in a Corning Slide Hybridization Chamber (ref. no. 2551), followed by laying a coverslip on top of the solution.
(6) Incubate the chamber in a water bath at 48 °C for 4 h.
(7) Remove the slide from the chamber and wash twice with pre-warmed 2 × SSC with 0.1% SSC at 48 °C for 5 min, once with 0.2 × SSC and then with 0.1 × SSC at room temperature for 1 min.
(8) Dry the slide by spinning at 800 r.p.m. for 3 min in a conical tube in a table-top centrifuge.
(9) Scan the slide in a standard slide scanner (we use the Axon GenePix 4000B scanner with photomultiplier tube voltage set at 600 for Cy3 and 700 for Alexa 647; the voltage should be adjusted so that most housekeeping transcripts give rise to yellow signals). The scan resolution is set at 10 μm.

39.3.8
Data Analysis

The analysis of microarray data needs special attention. There are several approaches that one can use to process the data and there is no single approach that is universally applicable to all situations. Discussing the methods used in the analysis in detail is beyond the scope of this chapter and the reader is directed to reviews published specifically on this topic [8, 9]. However, it is worthwhile to touch upon a few essential points. The very first transformation that is applied to array data is referred to as normalization. This is a process wherein the systematic variations that are contributed by several factors during the experiment are removed. This is an essential step that will facilitate comparison of (1) different dyes (used to label experimental sample versus control), (2) different genes on the same array and (3) results from different arrays. Different methods of normalizing microarray data are discussed in detail in several recent articles [10, 11]. After the data are normalized, they can be further analyzed by determining statistically significant changes in expression. One convenient way to approach this is to use the SAM package (significance analysis of microarrays; http://www-stat.standford.edu/~tibs/SAM) [12].

We generally carry out three independent experiments using Cy3 to label one sample and Alexa to label the other to be compared. Then we conduct another three dye-swap experiments to minimize the dye effect. Any pair of repeats should reach $R^2 > 0.85$. Representative results from SAM analysis are then individually confirmed by conventional RT-PCR.

39.4
Troubleshooting

Although RASL technology has many advantages in terms of its specificity and sensitivity, there are shortcomings, many of which are still waiting to be solved. The following section describes a number of system limitations we have realized during the development of the technology as well as several potential experimental problems we have frequently encountered during the implementation of the technology.

39.4.1
System Limitation and Pitfalls

(1) Limitation of the RASL technology. First, the density of the microarray system is currently not high enough to allow genome scans because of the limited number of the index sequences available. Although it is possible to generate and test additional index sequences, the question remains regarding the feasibility of genome scans in a single experiment when considering a huge number of potential mRNA isoforms expressed in the cell. In our opinion, the system is more suitable for applications at a small-to-intermediate scale. The

system may best be served by coupling with other genome scan methods based on high-density exon junction arrays. In other words, once a set of potential alternative splicing events of interest is identified, RASL technology may be used for additional experiments to take advantage of its high sensitivity and specificity.

(2) Variation introduced by index sequences. Although the index sequences currently being used behave similarly in hybridization, they exhibit quite distinct hybridization kinetics when coupled with specific targeting oligonucleotides. Thus, it is only possible to compare the difference of a specific mRNA isoform in different biological samples, rather than deduce the relative abundance of individual isoforms from the same pre-mRNA in a given biological sample.

(3) The requirement for prior knowledge of alternative splicing. Based on its design, RASL technology does not permit *de novo* discovery of unknown mRNA isoforms. Fortunately, the information from existing EST databases provides a wealthy resource to deduce potential alternative splicing events to be built into the assay. In fact, the number of potential mRNA isoforms deduced from EST databases would surpass the capacity of any of the existing microarray technologies.

(4) Compound effects of transcription and splicing on data analysis. This is an ongoing challenge to data analysis because a change in the expression of a specific isoform detected may result from alteration at the level of transcription or splicing, or both. In principle, one can include an oligonucleotide pair that targets an invariant region in a given mRNA in order to simultaneously deduce the level of total transcript and individual isoforms. We are currently testing this approach.

(5) The likelihood of highly abundant transcripts to obscure low abundant ones in the assay. PCR amplification of ligated oligonucleotides will eventually reach saturation and thus hinder the detection of low-abundance transcripts. Although this issue is not unique to the RASL assay, we feel that this problem may be overcome to some extent by grouping transcripts according to their relative abundance. For example, based on initial experiments, transcripts may be subdivided into low-, intermediate- and high-abundance categories to be reassayed. The versatility of RASL technology allows implementation of this idea in high-throughput applications, which cannot be approached by other microarray platforms based on conversion of total RNA to hybridization targets.

39.4.2
Potential Experimental Problems

(1) Quality of targeting oligonucleotides. T4 DNA ligase can only seal the gap between two juxtaposed oligonucleotides on an RNA template. Therefore, high-quality oligonucleotides are essential for efficient ligation.

(2) RNA quality. A partially degraded RNA sample is obviously bad for the assay. We prefer TRIzol-extracted RNA.

(3) The capture efficiency. As described in the protocol, streptavidin-coated tubes

are convenient, but have a limited surface area. The mRNA capture efficiency can be significantly improved by using streptavidin-coated magnetic beads.

(4) Contamination problems. The PCR amplification step is highly susceptible to contamination by RASL amplicons in the environment. To avoid the problem, we always include a blank control (using H_2O in place of RNA). In addition, we conduct the RASL assay in one room and PCR amplification in a separate room.

(5) Stochastic PCR variation. It is well known that the founder effect of PCR makes it impossible to amplify low-concentration targets quantitatively. Thus, if the concentration of a specific ligated RASL product is low, stochastic fluctuation may aggravate its detection after PCR amplification. We found that carrying out three to eight independent PCR reactions using the same ligated products after RASL and applying combined PCR products in hybridization dramatically increases the reproducibility of the RASL assay.

References

1 E. S. LANDER et al. (International Human Genome Sequencing Consortium), *Nature* **2001**, *409*, 860–921.
2 M. NISSIM-RAFINIA, B. KEREM, *Trends Genet.* **2002**, *18*, 123–127.
3 N. A. FAUSTINO, T. A. COOPER, *Genes Dev.* **2003**, *17*, 419–437.
4 T. A. CLARK, C. W. SUGNET, M. ARES JR, *Science* **2002**, *296*, 907–910.
5 J. M. JOHNSON, J. CASTLE, P. GARRETT-ENGELE, Z. KAN, P, M. LOERCH, C. D. ARMOUR, R. SANTOS, E. E. SCHADT, R. STOUGHTON, D. D. SHOEMAKER, *Science* **2003**, *302*, 2141–2144.
6 J. M. YEAKLEY, J.-B. FAN, D. DOUCET, L. LUO, E. WICKHAM, Z. YE, M. S. CHEE, X.-D. FU, *Nat. Biotechnol.* **2002**, *20*, 353–358.
7 C. L. ZHENG, T. M. NAIR, M. GRIBSKOV, Y.-S. KWON, H-R. LI, X.-D. FU, *Pacific Symposium on Biocomputing 2004*, World Scientific, Singapore, **2004**, pp. 78–88.
8 J. QUACKENBUSH, *Nat. Genet. Suppl.* **2002**, *32*, 496–501.
9 S. DRAGHICI, *Data Analysis Tools for DNA Microarrays*, CRC Press, Boca Raton, FL, **2003**.
10 S. DUDOIT, Y. H. YANG, M. T. P. SPEED, J. CALLOW, *Statistica Sinica* **2002**, *12*, 111–139.
11 Y. H. YANG, S. DUDOIT, P. LU, D. M. LIN, V. PENG, J. NGAI, T. P. SPEED, *Nucleic Acids Res.* **2002**, *30*, e15.
12 V. G. TUSHER, R. TIBSHIRANI, G. CHU, *Proc. Natl. Acad. Sci. USA* **2001**, *98*, 5116–5121.

Part IV
Analysis of RNA Function

IV.1
RNA–Protein Interactions in vitro

40
Use of RNA Affinity Matrices for the Isolation of RNA-binding Proteins

Steffen Schiffer, Sylvia Rösch, Bettina Späth, Markus Englert, Hildburg Beier and Anita Marchfelder

40.1
Introduction

All living organisms use a variety of RNA–protein interactions either in stable ribonucleoprotein (RNP) complexes (e.g. the ribosome or the spliceosome) or in transient RNA–protein interactions. Transcripts have to be capped, modified, edited, spliced, processed, transported and, finally, degraded. Thus, RNA-binding proteins play an essential role in key processes in the cell. To understand their function and to investigate them in detail, it is often required to purify and isolate these proteins. Several different approaches can be employed for the isolation of RNases. Purifications generally start with the isolation of a soluble protein extract which is initially fractionated using precipitation and ion-exchange chromatography. Subsequently group-specific affinity matrices (e.g. Cibacron blue or Heparin) are applied, which result in good purification and yield. The final and most efficient purification step can be achieved for RNA-binding proteins by using ligand-specific affinity matrices. Since affinity chromatography exploits the highly specific interaction between molecules, it is highly selective, and offers high yield and purity. Sequence-specific RNAs have been successfully used to purify RNA binding proteins [1, 2].

Here the application of an RNA affinity column to purify a tRNA-processing enzyme is described. tRNA molecules are transcribed as precursors containing 5′ and 3′ additional sequences which have to be removed to yield functional tRNAs [3]. Accurate tRNA 3′ end maturation is essential for aminoacylation and thus for protein synthesis in all organisms. In archaea and the majority of eukaryotes the enzyme catalyzing the removal of the 3′ trailer sequence is the endonuclease RNase Z [4–6]. Only recently it was shown that in *Bacillus subtilis* an RNase Z homolog is responsible for the 3′ end maturation of a certain set of tRNAs [7]. Knock down of the RNase Z homolog in *Drosophila melanogaster* using RNAi resulted in accumulation of nuclear and mitochondrial pre-tRNAs containing 3′ trailers [8].

To purify the RNase Z from wheat germ we used a tRNA affinity column. Although the tRNA is not a substrate, but the product of the reaction catalyzed by RNase Z, RNase Z shows a high affinity to tRNAs as can be shown by EMSA

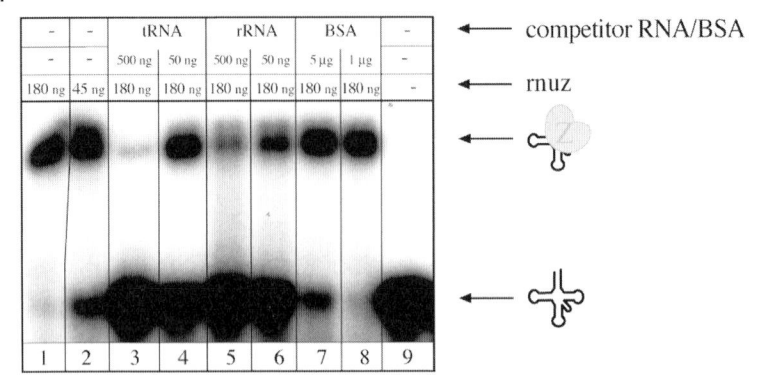

Fig. 40.1. The tRNA 3′ processing endonuclease RNase Z has a high affinity for tRNAs. Although the 3′-end-processed tRNA is a product of the reaction catalyzed by the endonuclease RNase Z, RNase Z still has a high affinity for tRNA molecules as shown by electrophoretic mobility shift assays (EMSA). Recombinant RNase Z from *A. thaliana* (rnuz) [4] was incubated with wheat tRNA which was 3′-end-labeled with [^{32}P]pCp (in lanes 1 and 3–8 180 ng rnuz was used, in lane 2 only 45 ng rnuz was used). In addition either unlabeled wheat tRNA (lanes 3 and 4) or unlabeled rRNA (lanes 5 and 6) was added as competitor RNA (50 or 500 ng) to the reaction. Lanes 7 and 8 show reactions to which bovine serum albumin (BSA) was added (5 or 1 µg). Lane 9 shows a control incubation of labeled wheat tRNA without the addition of protein. The tRNA and the tRNA–RNase Z complex are shown at the right schematically. Incubation of 180 ng rnuz with the labeled wheat tRNA resulted in an almost complete shift of the tRNA (lane 1). Addition of 500 ng unlabeled wheat tRNA as competitor RNA resulted in a drastic reduction of shifted tRNA molecules (lane 3). The addition of unlabelled rRNA as competitor reduced only slightly the amount of shifted tRNA (lane 5), showing that RNase Z binds specifically to wheat tRNAs. Addition of 1 µg BSA did not interfere with the shift; addition of 5 µg BSA slightly reduces the RNA–protein complex formation.

(Fig. 40.1). The use of a NHS–tRNA affinity column subsequent to a number of other chromatographic steps likewise turned out to be a powerful tool for the purification of tRNA ligase from wheat germ (M. Englert and H. Beier, unpublished results). In addition, tRNA affinity columns have previously been successfully used to purify the tRNA 5′ processing enzyme RNase P [9–11].

40.2
Materials

40.2.1
CNBr-activated Sepharose 4B Affinity Column

CNBr-activated Sepharose 4B (Amersham), 1 mM HCl, coupling buffer 1 (0.5 M NaCl, 0.1 M NaHCO$_3$, pH 8.3), 0.2 M glycine, pH 8.0, 0.1 M NaOAc, pH 4.5 containing 0.5 M NaCl, 0.1 M Tris–HCl buffer, pH 8.0, containing 0.5 M NaCl, sintered glass filter, tRNA (e.g. wheat tRNA V; Sigma), an empty column (e.g. C10/10; Amersham).

40.2.2
NHS-activated HiTrap Columns

5 ml NHS-activated HiTrap Column (Amersham), coupling buffer 2 (0.5 M NaCl, 0.2 M NaHCO$_3$, pH 8.3), 1 mM HCl, buffer A (0.5 M ethanolamine, 0.5 M NaCl, pH 8.3), buffer B (0.1 M acetate, 0.5 M NaCl, pH 4.0), tRNA (e.g. wheat tRNA V; Sigma), 5- or 10-ml syringe.

40.3
Methods

For the purification of the wheat RNase Z we initially used the CNBr–tRNA affinity column, which yielded good purification. Subsequently we used a NHS–tRNA affinity column, which resulted in an even better purification of the wheat RNase Z. That could be due to the fact that in the CNBr–tRNA affinity column the ligand (tRNA) is only separated by a single C atom from the gel matrix, whereas in the NHS–tRNA affinity column the tRNA is separated by 10 C atoms from the gel matrix, giving the RNase Z more space to bind to the tRNA (Fig. 40.2).

40.3.1
Coupling of tRNAs to CNBr-activated Sepharose 4B

Preparing the gel

(1) Put 2.5 g CNBr-activated Sepharose 4B (1 g powder gives about 3.5 ml final volume of gel) in 25 ml 1 mM HCl into a sintered glass filter and mix carefully using a spatula. (Do not let the CNBr-activated Sepharose 4B powder stand at room temperature for too long since it takes up air moisture.) Let the powder swell at room temperature for 2 min.
(2) Wash with 4 × 100 ml 1 mM HCl, each wash for about 4 min, do not let the powder run dry.
(3) Wash with 100 ml coupling buffer 1 (0.5 M NaCl, 0.1 M NaHCO$_3$, pH 8.3).

Coupling the ligand

(1) Dissolve 100 mg tRNA in 40 ml coupling buffer 1 (0.5 M NaCl, 0.1 M NaHCO$_3$, pH 8.3). Cool to 4 °C.
(2) All solutions should be cold (4 °C) and this step should be carried out at 4 °C to avoid degradation of the tRNA and to prevent the coupling reaction from pro-

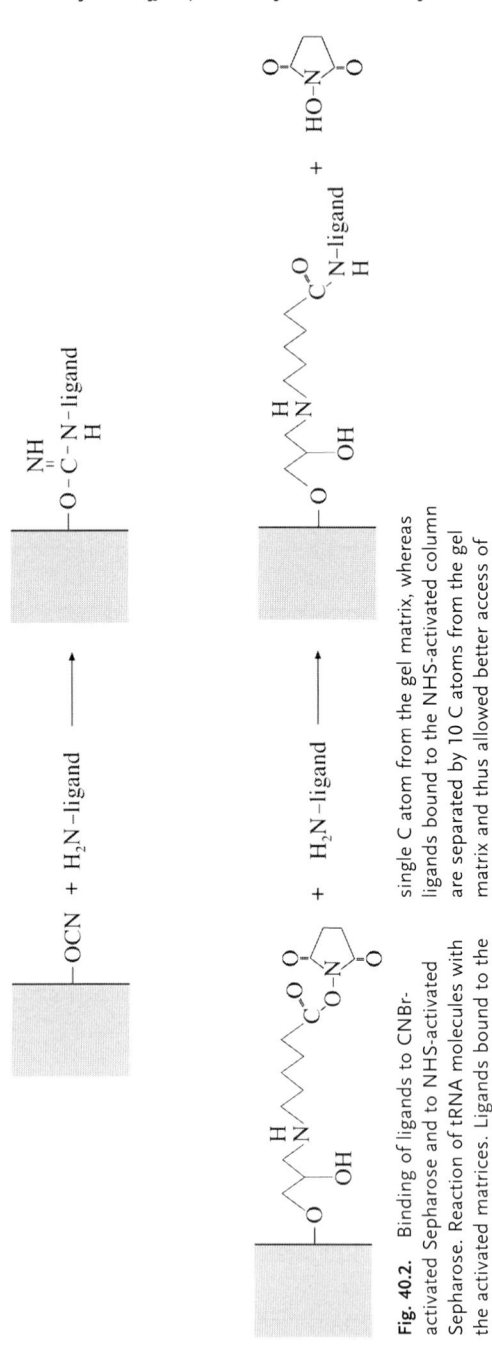

Fig. 40.2. Binding of ligands to CNBr-activated Sepharose and to NHS-activated Sepharose. Reaction of tRNA molecules with the activated matrices. Ligands bound to the CNBr-activated Sepharose are separated by a single C atom from the gel matrix, whereas ligands bound to the NHS-activated column are separated by 10 C atoms from the gel matrix and thus allowed better access of interacting molecules.

gressing too fast. This would result in binding of only a single tRNA to several CNBr. Mix the coupling solution containing the ligand with the gel in a stoppered vessel. Rotate the mixture end-over-end overnight (16–20 h) at 4 °C. Other gentle stirring methods may be employed alternatively. *Do not use a magnetic stirrer.*

(3) Pour gel onto a sintered glass filter and suck the solution off (but do not let the gel run dry, keep it moist). Wash away excess ligand with 50 ml coupling buffer.
(4) Block any remaining active groups by adding 40 ml 0.2 M glycine pH 8.0. Incubate for 2 h at room temperature and over night at 4 °C.
(5) Remove the solution above the gel matrix (but do not let the gel run dry, keep it moist). Wash the gel with 50 ml 0.1 M NaOAc, pH 4.5 containing 0.5 M NaCl.
(6) Wash the gel with 50 ml 0.1 M Tris–HCl buffer, pH 8.0 containing 0.5 M NaCl.
(7) Repeat Steps 6 and 7 twice.
(8) The gel is ready to be packed into a column (e.g. a C10/10; Amersham) yielding a 5-ml tRNA affinity column.

40.3.2
Coupling of tRNAs to a 5-ml NHS-activated HiTrap Column

(1) Dissolve 20 mg tRNA in 5 ml coupling buffer 2 (0.5 M NaCl, 0.2 M $NaHCO_3$, pH 8.3). Cool to 4 °C.
(2) Remove the top cap of the column and apply a drop of ice-cold 1 mM HCl to the top of the column to avoid air bubbles.
(3) Connect the HiTrap luer adaptor to the top of the column. Remove the twist-off end.

Coupling the ligand

(1) Wash out the isopropanol using a 5- or 10-ml syringe with ice-cold 1 mM HCl (3 × 10 ml) at a flow rate of 1 ml/min (1/2 drop per second). Do not exceed the flow rate or the gel can be irreversibly compressed.
(2) Immediately inject 5 ml of ligand solution into the column.
(3) Seal the column and let it stand for 4 h at 4 °C.
(4) Inject 3 × 10 ml buffer A (0.5 M ethanolamine, 0.5 M NaCl, pH 8.3).
(5) Inject 3 × 10 ml buffer B (0.1 M acetate, 0.5 M NaCl, pH 4.0).
(6) Repeat Step 4. Let the column stand for 30 min.
(7) Inject 3 × 10 ml buffer B.
(8) Inject 3 × 10 ml buffer A.
(9) Repeat Step 7.
(10) Inject a buffer with neutral pH to adjust the pH. The column is now ready to use.

We usually load 29 mg wheat extract purified through five steps onto the 5-ml NHS–tRNA affinity column.

40.4
Application

40.4.1
Purification of the Nuclear RNase Z from Wheat Germ

Earlier experiments had shown that the tRNA 3' processing activity in wheat germ is present in very low amounts, similar to other tRNA processing enzymes such as RNase P and tRNA splicing endonuclease [11, 12]. For the isolation of the wheat tRNA 3' endonuclease a lot of material was thus required, although highly efficient fractionation steps had been worked out in earlier experiments [13]. Briefly, a soluble protein fraction (S30) was extracted from 2.9 kg of wheat germ and purified through six purification steps (Fig. 40.3A–D), the most efficient purification step provided by a tRNA affinity column, to which RNase Z bound tightly. Initially we used a CNBr–tRNA affinity column [13], which already resulted in a high purification factor. Since an NHS–tRNA affinity column showed an even higher purification than the CNBr–tRNA affinity column [4], it was subsequently used. The tRNA molecules are separated from the gel matrix in the CNBr–tRNA affinity column by only one C atom leaving only little space between the gel matrix and the tRNA, whereas in the NHS–tRNA affinity column the tRNAs are separated by 10 C atoms, giving the RNase Z enzyme ample space to bind to the tRNA (Fig. 40.2).

The 5-ml NHS–tRNA affinity column was made with wheat tRNA (Sigma), which was coupled to HiTrap NHS-activated Sepharose (Amersham) as described in Section 40.3.2. The column was equilibrated with buffer A (40 mM Tris–HCl, pH 8, 5 mM $MgCl_2$, 5% glycerol, 2 mM DTT, 0.5 mM PMSF) and loaded with 29 mg protein of the RNase Z active fraction (0.6–0.8 M KCl) from the preceding Blue column, using a flow rate of 200 µl/min. Proteins were eluted with a step gradient (0.2, 0.5, 1.0 and 2.0 M KCl), using a flow rate of 500 µl/min. Then, 1-ml fractions were collected, and fractions from one gradient step were pooled, concentrated and dialyzed using Centriplus 10 concentrators (Millipore). Pooled fractions were subsequently analyzed for tRNA 3' processing activity and RNase Z activity was recovered in the 0.5 M KCl fraction. RNase Z purified with this tRNA affinity column showed an apparent molecular mass of 64 kDa on the subsequent gel filtration column. After the gel filtration column, RNase Z active fractions were analyzed by SDS–PAGE, where only a 43-kDa protein correlated with the activity (Fig. 40.3C). Since gel filtration analysis indicated a molecular mass of 64 kDa for the tRNA 3' processing activity, the active enzyme might be present as homodimer.

Approximately 0.5 µg of the 43-kDa protein were excised from SDS–PAGE and subjected to tryptic digestion and MS/MS analysis. Subsequent database searches using the algorithm SEQUEST [14] and programs developed at the Harvard Microchemistry Facility [15] did not identify the corresponding gene. Therefore, peptides were separated by HPLC and sequenced using Edman degradation. Four peptide sequences were obtained, and database searches revealed a wheat cDNA sequence and two open reading frames in the *Arabidopsis thaliana* genome with high sequence similarity to these peptides. The wheat cDNA sequence translates into a

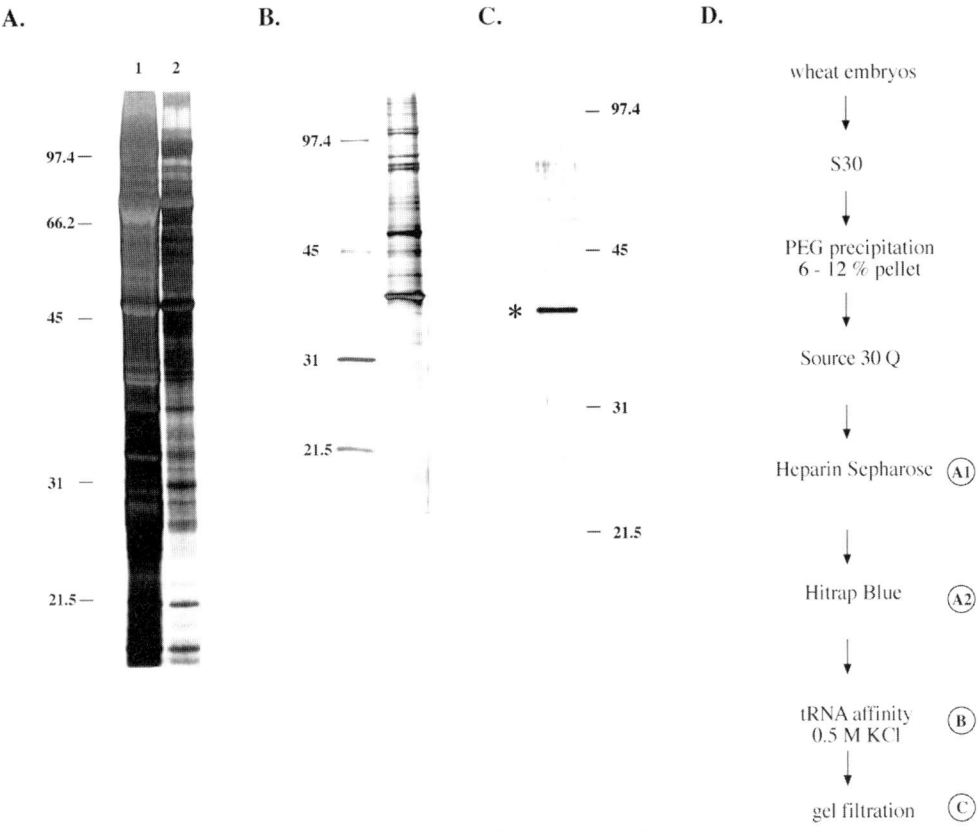

Fig. 40.3. Isolation of the nuclear RNase Z from wheat. (A) SDS–PAGE of RNase Z active fractions from two purification steps. Aliquots of RNase Z active fractions from the two purification steps A1 and A2 (see purification scheme in D) were loaded onto SDS gels; lane 1: 57 µg of the 0.25 M Heparin fraction, lane 2: 10 µg of the 0.6 M KCl Blue fraction. Protein molecular weight markers are given in kDa. (B) Since little protein was left after the last purification step, protein concentrations could not be determined and therefore 10% of the tRNA affinity fraction was loaded. At the left a protein size standard is given in kDa. (C) SDS–PAGE of the RNase Z-active fraction after the last purification step. Again 10% of the RNase Z-active gel filtration fraction were loaded onto the gel. A protein size standard is given at the right in kilodaltons. (D) Purification scheme. RNase Z was purified from the soluble protein fraction from wheat germ (S30) in six fractionation steps.

protein 100% identical to one of the sequenced wheat peptides. Since the wheat sequence is only a partial cDNA sequence, the other three peptides are outside of this sequence and consequently show no match to the cDNA. The corresponding wheat genomic sequence could not be identified in the few wheat sequences available in public databases. However the full-length cDNA and genomic sequence were found in *Arabidopsis*, and the encoded proteins were termed Nuz for nuclear RNase Z and Cpz for chloroplast RNase Z. These plant protein sequences were the source sequence to identify homologues of this protein in all kingdoms.

40.5
Notes

RNA affinity columns are a powerful method to purify RNA-binding proteins. Since specific affinity columns usually are difficult to prepare, they are smaller and do not have such a high capacity as ion-exchange columns or group-specific affinity columns. In contrast to DNA affinity columns [16], RNA affinity columns are always sensitive to degradation especially from RNases and are therefore less stable than, for example, DNA affinity columns. Thus only extracts, which already have been purified through several steps and are highly enriched for the desired protein, should be loaded onto such columns. The preceding purification steps must be sufficient to remove ribonucleases, which can degrade the RNA target attached to the column. Since specific RNA affinity matrices usually do not have high capacities, the prior purification steps should in addition remove proteins binding unspecifically to the RNA matrix and thus blocking binding sites for the protein being purified.

The NHS-activated Sepharose was originally developed for the coupling of proteins via freely accessible primary amino groups. Since in highly structured RNA molecules like transfer RNAs only few amino groups are accessible, the coupling efficiency is low. Since the coupled tRNAs are presumably attached by only one covalent linkage, the free accessibility of the ligand to the RNA-binding protein is ensured.

Recently, new methods to prepare RNA columns with higher capacity and better stability have been developed, which might circumvent some of the above-mentioned problems [17]. Updated methods for RNA affinity columns are published frequently describing the development of specific RNA affinity tags or aptamers with high affinity [18, 19].

References

1 H. Li, H. P. Zassenhaus, *Biochem. Biophys. Res. Commun.* **1999**, *261*, 740–745.
2 R. Trippe, H. Richly, B. J. Benecke, *Eur. J. Biochem.* **2003**, *270*, 971–980.
3 M. Mörl, A. Marchfelder, *EMBO Rep.* **2001**, *2*, 17–20.
4 S. Schiffer, S. Rösch, A. Marchfelder, *EMBO J.* **2002**, *21*, 2769–2777.
5 K. Kruszka, F. Barneche, R. Guyot, J. Ailhas, I. Meneau, S. Schiffer, A. Marchfelder, M. Echeverría, *EMBO J.* **2003**, *22*, 621–632.
6 K. Schierling, S. Rösch, R. Rupprecht, S. Schiffer, A. Marchfelder, *J. Mol. Biol.* **2002**, *316*, 895–902.
7 O. Pellegrini, J. Nezzar, A. Marchfelder, H. Putzer, C. Condon, *EMBO J.* **2003**, *22*, 4534–4543.
8 E. B. Dubrovsky, V. A. Dubrovskaya, L. Levinger, S. Schiffer, A. Marchfelder, *Nucleic Acids Res.* **2004**, *32*, 255–262.
9 Y. C. Lee, B. J. Lee, D. S. Hwang, H. S. Kang, *Eur. J. Biochem.* **1996**, *235*, 289–296.
10 H. L. True, D. W. Celander, *J. Biol. Chem.* **1996**, *271*, 16559–16566.
11 S. Zimmerly, D. Drainas, L. A.

Sylvers, D. Söll, *Eur. J. Biochem.* **1993**, *217*, 501–507.

12 K. Kleman-Leyer, D. W. Armbruster, C. J. Daniels, *Cell* **1997**, *89*, 839–847.

13 M. Mayer, S. Schiffer, A. Marchfelder, *Biochemistry* **2000**, *39*, 2096–2105.

14 J. K. Eng, A. McCormick, J. R. Yates, III, *J. Am. Soc. Mass Spectrom.* **1994**, *5*, 976–989.

15 H. S. Chittum, W. S. Lane, B. A. Carlson, P. P. Roller, F. D. Lung, B. J. Lee, D. L. Hatfield, *Biochemistry* **1998**, *37*, 10866–10870.

16 H. Gadgil, S. A. Oak, H. W. Jarrett, *J. Biochem. Biophys. Methods* **2001**, *49*, 607–624.

17 C. R. Allerson, A. Martinez, E. Yikilmaz, T. A. Rouault, *RNA* **2003**, *9*, 364–374.

18 C. Srisawat, D. R. Engelke, *Methods* **2002**, *26*, 156–161.

19 K. Hartmuth, H. Urlaub, H. P. Vornlocher, C. L. Will, M. Gentzel, M. Wilm, R. Lührmann, *Proc. Natl. Acad. Sci. USA* **2002**, *99*, 16719–16724.

41
Biotin-based Affinity Purification of RNA–Protein Complexes

Zsofia Palfi, Jingyi Hui and Albrecht Bindereif

41.1
Introduction

Many cellular functions are carried out by proteins that are components of large complexes. The identification of proteins present in biological complexes requires their purification, which can be achieved through affinity procedures. Often the protein components are then analyzed and identified by a combination of mass spectrometry and database search. This chapter will focus on ribonucleoprotein (RNP) complexes, for which affinity purification techniques are particularly suitable, since either a protein- or RNA-specific affinity tag can be incorporated in the complex. The use of protein affinity tags has become routine for the purification and detection of protein complexes. A variety of affinity tags are available for labeling proteins of interest [1, 2]. On the other hand, RNA–protein complexes can also be affinity-purified through their RNA component, using RNA aptamers as affinity tags, e.g. the streptavidin-binding S1 RNA aptamer [3], the D8 Sephadex-binding RNA minimal motif [4] or the tobramycin aptamer [5]. Alternatively, a well-known RNA–protein interaction can be exploited for this purpose, by incorporation binding sites for the MS2 coat protein into the RNA of interest and using an MS2 coat protein–maltose binding protein (MS2–MBP) fusion as an affinity tag for purification on an amylose resin. The bound RNP complex can subsequently be eluted from the resin under native conditions with free maltose [6, 7]. A recent example for this approach was the isolation of human spliceosomes in a functional form, resulting in their comprehensive proteomic analysis [8].

In the following we will concentrate on the affinity purification of RNA–protein complexes through their RNA component, using biotin-based methods. Biotin can be introduced into the RNA (1) directly and internally in the form of biotinylated nucleotides during *in vitro* SP6 or T7 transcription, or (2) into an antisense oligonucleotide against the RNA of interest during its chemical synthesis. Subsequently, affinity selections make use of the tight recognition of biotin residues by avidin or streptavidin.

Examples for the first approach come from the initial identification of spliceosomes, when biotinylated nucleotides were randomly incorporated into pre-mRNA substrate during *in vitro* transcription. Spliceosomes were then assembled

Handbook of RNA Biochemistry: Student Edition. Edited by R. K. Hartmann, A. Bindereif, A. Schön, E. Westhof
Copyright © 2009 WILEY-VCH Verlag GmbH & Co. KGaA, Weinheim
ISBN: 978-3-527-32534-4

on such biotinylated splicing substrates in extract, followed by binding to streptavidin resin under native conditions [9–11]. This was the way the set of spliceosomal snRNAs U1, U2, U4, U5 and U6 had first been identified. Alternatively, spliceosomes assembled on unmodified pre-mRNAs can be captured on streptavidin resin with biotinylated antisense oligonucleotides [12].

The use of biotinylated antisense 2′-O-methyl RNA (2′OMe) oligonucleotides for the specific selection of RNP complexes was first introduced by Lamond, Sproat and co-workers [12–14, reviewed 15]. They found that 2′OMe RNA oligonucleotides have several advantages over other types of oligonucleotides, which can be exploited for the affinity selection procedure. Most importantly, 2′OMe RNA oligonucleotides are not only nuclease resistant, but also possess – in comparison to DNA oligonucleotides – a higher affinity for RNA, increased specificity, faster hybridization kinetics and a superior ability to bind to structured RNA targets [16–18]. The antisense affinity technology using biotinylated 2′OMe RNA oligonucleotides is very versatile and allows the rapid purification of any RNA–protein complex in which the RNA component is accessible to oligonucleotide binding [19–23]. Purified complexes can be directly processed for protein identification by mass spectrometric analysis and for analysis of their RNA components. Both protein and RNA analyses can be performed on the same samples (see Section 41.3.1.3).

However, because of the extremely stable biotin–avidin interaction, this standard procedure requires denaturation of the affinity-purified complex. To overcome this limitation, an important variation of the original protocol was suggested by Lingner and Cech [24], allowing the elution of purified RNP complexes from the streptavidin matrix and their further functional analysis. This so-called displacement strategy results in the release from the affinity matrix under native conditions and is based on disrupting the 2′OMe RNA–target RNA hybrid by an excess of a DNA oligonucleotide directed against the 2′OMe RNA oligonucleotide. This strategy has been applied for the affinity purification and functional assays of different RNPs [25–28].

Here we will describe in experimental detail procedures how RNPs can be affinity-purified on the basis of the biotin–streptavidin interaction, focusing on two major principles used: (1) using biotinylated 2′OMe RNA oligonucleotides that are antisense to the RNA component of the RNP and (2) using biotin-labeled RNAs. Specific examples will be presented for each of the two principles and potential problems will be discussed.

41.2 Materials

41.2.1 Oligonucleotides

The following types of biotinylated oligonucleotides are used for affinity selections: (1) 2′OMe RNA oligonucleotides complementary in sequence to the targeted RNAs [12, 20], (2) chimeric 2′OMe RNA/DNA oligonucleotides in combination with displacement DNA oligonucleotides [24] and (3) biotinylated RNAs [9, 29, 30]. For oligonucleotide design, a good starting point is the secondary structure model of the target RNA. As a general rule, single-stranded regions are chosen and their accessibility within intact RNPs can be experimentally confirmed by oligodeoxynucleotide-directed RNase H digestion [31]. Most importantly, the affinity oligonucleotide should be specific for the target RNA. The oligonucleotide used for selection generally comprises at least 10–12 bases and contains one or more biotin residues for immobilization on affinity matrices. Accessibility of the biotin residue(s) to the affinity matrix is important for the efficiency of selection and can be influenced by the 3'- or 5'-terminal position of the biotin residue(s) on the oligonucleotide. Incorporation of biotin at both termini might improve binding efficiency [32]. All biotin reagents used for biotinylation of oligonucleotides or RNAs should contain a spacer arm, at least six C atoms in length, to reduce steric hindrance.

41.2.2
Affinity Matrices

Different types of affinity matrices are available for the immobilization of biotinylated oligonucleotides. Most frequently used are streptavidin–agarose (SAg; Sigma S-1638) or immunopure immobilized streptavidin (Pierce 20349). Streptavidin is a roughly 60-kDa protein composed of four identical subunits, each of which has a binding site for biotin. Streptavidin is immobilized on agarose beads, supplied as a 1:1 suspension in aqueous solution. It shows relatively low levels of non-specific binding. Neutravidin–agarose (NAg; Pierce 29200) is a chemically modified (deglycosylated) version of streptavidin, and shows very low non-specific binding compared to streptavidin and high binding capacity for biotinylated molecules. The binding affinity of immobilized monomeric avidin (Pierce 20228) is several orders of magnitude lower than tetrameric avidin. Immobilized monomeric avidin can be used for reversible binding of biotinylated probes – bound biotinylated molecules can be specifically eluted by ligand competition using a biotin-containing buffer. Streptavidin-coated magnetic beads (e.g. Dynabeads M-270 Streptavidin; Dynal Biotech 353.02 or Strepavidin MagneSphere Paramagnetic Particles; Promega Z5482) are paramagnetic polystyrene beads with streptavidin covalently attached to the beads surface. They carry lower binding capacity for biotinylated molecules than SAg or NAg, but negligible non-specific binding for negatively charged molecules, and show less aggregation in high-salt buffers. They are handled fast and conveniently by a magnetic particle concentrator. These magnetic beads are not recommended for selection of abundant RNPs or for using with unfractionated cell lysates [15]. Another disadvantage is that heating causes release of iron, which may degrade RNA.

41.2.3
Cell Extracts

Extracts from cells, tissues, or organisms can be prepared by a variety of procedures. The choice of the appropriate procedure will depend on the RNA–protein complex to be purified and on prior experience. In general, it will be necessary to check whether the applied extraction procedure is efficient and if the targeted RNP particle is stable under those conditions. Particularly relevant to the success of the affinity selection is the ionic strength of the extracts, the presence of non-ionic detergents [Nonidet P-40 (NP-40) or Triton X-100], and of RNase and protease inhibitors. Cell lysates are best used freshly for affinity selections. In case of long-term storage, extracts are supplemented with 10% glycerol, aliquoted, flesh-frozen in liquid nitrogen and kept at $-80\,°C$. In the example given below, total cell lysates were prepared from *Trypanosoma brucei* cells as described [33]. Cell fractionation can facilitate the affinity purification by enriching for the RNP. The preparation of nuclear extracts from human cells has been described by Dignam et al. [34]; modifications for efficient affinity selections have been introduced by Blencowe and Lamond [15]; the extract preparation from small quantities of cells has been developed by Lee et al. [35]. In case of selecting RNPs of low abundance, chromatographic fractionation methods can also be applied to enrich for certain RNP particles (e.g. DEAE–Sepharose chromatography in the case of snRNPs [31, 33]).

41.2.4
Buffers and Solutions

Buffers should be prepared freshly in diethylpyrocarbonate (DEPC)-treated water and kept at 4 °C (except when otherwise stated). Note that Tris buffers cannot be DEPC-treated, because Tris reacts with DEPC. To make a Tris buffer, the water should therefore be DEPC-treated first and autoclaved before adding Tris. After addition of Tris, the solution should be autoclaved again. Dithiothreitol (DTT) and phenylmethylsulfonyl fluoride (PMSF) or complete protease inhibitor cocktail tablets (e.g. Roche 10266500) are added to the buffers just before use.

- SAg blocking buffer (SAgBB): 1 mg of RNase-free BSA, 200 µg glycogen and 200 µg yeast tRNA/ml low-salt wash buffer.
- Low-salt wash buffer (LS-WB): 20 mM HEPES, pH 8.0, 100 mM KCl 10 mM $MgCl_2$, 0.01% NP-40, 1 mM DTT.
- High-salt wash buffer (HS-WB): same as LS-WB, except that the KCl concentration is 300 mM or higher.
- 6 and 9 M urea solutions in DEPC-treated water containing 1 mM DTT and 0.01% NP-40.
- 2 × PK buffer: 0.2 M Tris, pH 7.5, 0.3 M NaCl, 2% SDS, 25 mM EDTA.
- 10% NP-40 in DEPC-treated water
- Phenol/chloroform/isoamylalcohol (25:24:1) saturated with TE buffer (10 mM Tris, pH 8.0, 1 mM EDTA).
- 70 and 100% ethanol, 100% acetone.

41.2.5
Additional Materials

- Sterile, RNase-free 1.5-ml Eppendorf tubes, 15- and 50-ml Falcon tubes, pipette tips.

- Microcentrifuge.
- Rotating wheel.
- Thermoshaker for Eppendorf tubes.
- Magnetic particle concentrator (when using streptavidin magnetic particles).

41.3
Methods

In the following we describe (1) how biotinylated antisense 2′OMe oligoribonucleotides are used for affinity purification of RNP complexes and (2) how specific RNA-binding proteins can be isolated through unmodified biotinylated RNAs from cell extracts.

41.3.1
Affinity Purification of RNPs

To select the optimal sequence for the biotinylated 2′OMe RNA oligonucleotide, regions of the target RNP should be determined that are accessible for oligonucleotide binding. This can be achieved, for example, by RNase H cleavage with oligonucleotides directed to different regions of the target RNA. When an RNP particle should be purified under native conditions for functional assays, chimeric 2′OMe RNA/DNA oligonucleotide in combination with a displacement DNA oligonucleotide can be used (see Section 41.3.1.4). In case of low-abundance RNPs it might be important to enrich the RNPs of interest by chromatographic fractionation or by glycerol gradient centrifugation. As a control, mock selection (without oligonucleotide or with an unrelated oligonucleotide) must always be run in parallel. In case when the mock control shows background binding, pre-clearing of the cell extract might be helpful (see Section 41.3.1.1). SAg should be pre-blocked in blocking buffer before each selection to saturate non-specific binding sites. As a specific example, the affinity purification of snRNPs from *T. brucei* cell lysate will be described in detail in the following.

41.3.1.1 Depletion of Total Cell Lysate from SAg-binding Material (Pre-clearing)
All steps should be carried out at 4 °C. SAg should be handled with care – never use high-speed centrifugation to collect the beads, since this may damage the structure of the agarose particles, and do not vortex SAg-containing suspensions. For mixing, use a rotating wheel or a shaker adjusted to low speed.

(1) To pre-clear 10 ml extract (protein concentration around 10 mg/ml) resuspend SAg in the commercial bottle and transfer 200 μl slurry into two 1.5-ml Eppendorf tubes (The amount of affinity matrix used to pre-clear crude cell lysates should be determined experimentally, but as a rough guide we recommend to use approximately 10–20 μl packed SAg beads per 10 mg of protein in the lysate). Centrifuge the suspension at 1500 g for 1 min to remove the buffer (contains preservative). This gives 100 μl packed beads in each tube.

(2) Add 1 ml LS-WB to the pelleted beads and mix the suspension on a rotating wheel for 5 min.
(3) Centrifuge the beads to remove the buffer (1 min at 1500 g).
(4) Repeat washing the beads in HS-WB (Steps 2 and 3).
(5) Take the necessary amount of fresh cell extract (depending on whether analytical or preparative-scale affinity selections are planned) and supplement with NP-40 (final concentration: 0.01%). Frozen aliquots of cell lysates should be thawed quickly by warming in a 30 °C water bath, cooled on ice, and then centrifuged at 10 000 g for 10 min to remove large aggregates or precipitates (which would otherwise contaminate everything). Transfer the supernatant into fresh tubes.
(6) Transfer one aliquot of the pre-washed, pelleted SAg beads into the cell extract by resuspending the beads in a small volume of extract.
(7) Rotate the tube slowly on a turning wheel for 1 h at 4 °C.
(8) Centrifuge the suspension for 2 min at 1500 g.
(9) Take the supernatant into a new tube and repeat Steps 6–8 with the second aliquot of pre-washed SAg beads.
(10) Repeat centrifugation of the supernatant once more to be sure that all SAg beads are removed from the extract.

41.3.1.2 Pre-blocking SAg Beads

(1) Transfer the required amount of SAg slurry into a 1.5-ml Eppendorf tube. Centrifuge the suspension at 1500 g for 1 min, discard the supernatant.
(2) Resuspend the SAg (100 µl packed beads) in 1 ml cold LS-WB and rotate the tube on a turning wheel for 5 min.
(3) Collect the beads by a short centrifugation (1500 g; 1 min).
(4) Repeat Steps 2 and 3.
(5) Resuspend the beads in SAgBB, using 1 ml buffer for blocking 100 µl packed beads.
(6) Rotate the tube slowly on a turning wheel for 1 h at 4 °C.
(7) Collect the beads by a 1-min centrifugation at 1500 g and discard the supernatant.
(8) Mix the beads in 1 ml cold HS-WB containing 300 mM KCl (5 min on a turning wheel).
(9) Collect the beads again by a short centrifugation.
(10) Repeat Steps 8 and 9 twice. The pre-blocked SAg beads can be stored for 1–2 days at 4 °C.

41.3.1.3 Affinity Selection of RNPs for Structural Studies

In principle this can be done in two different ways: (1) by pre-binding the biotinylated antisense 2′OMe RNA oligonucleotide to SAg beads, then reacting the immobilized oligonucleotide with the target RNP in the extract, and (2) by binding the oligonucleotide to the target RNP in solution (extract), followed by immobilizing the complex on SAg beads. Which version is better? To decide, first carry out small-scale selections. The efficiency of selection may be influenced by the order of these steps. Below we describe a specific example for an affinity selection of snRNPs, demonstrating how low-abundance RNP particles can be purified.

Example: Affinity purification of *T. brucei* U2, U4/U6 and SL RNPs

In this case the cell lysate is first enriched for snRNPs by DEAE–Sepharose chromatography [33]. The biotinylated antisense 2'OMe RNA oligonucleotides are immobilized first on SAg, then incubated with the DEAE fraction. The following 2'OMe RNA oligonucleotides were used for affinity selection: Tb-U2-5', Tb-U4-int and Tb-SL-int2 [20]. An unrelated oligonucleotide, HuU4-5' [36], served as a mock control. The result of this affinity selection procedure is shown in Fig. 41.1.

Procedure

(1) Take 200 µl pre-blocked SAg beads (1:1 slurry in HS-WB) for each selection into a 1.5-ml Eppendorf tube (for four selections: 4 × 200 µl).
(2) Add 800 µl HS-WB containing 10 µg of biotinylated 2'OMe RNA oligonucleotide (around 2 nM of a 15-mer oligonucleotide) into each tube. For each affinity selection use the corresponding oligonucleotide. Mock selection can be also done by leaving out the oligonucleotide. Bind the oligonucleotides to SAg during a 2-h incubation at 4 °C, slowly rotating the tubes on a turning wheel.
(3) Collect the SAg beads by a short centrifugation (1500 g, 1 min), discard the supernatants.
(4) Resuspend the pelleted beads in 1 ml HS-WB and incubate the samples for 5 min at 4 °C by slow rotation.
(5) Spin down the beads (1500 g, 1 min) and repeat washing in 1 ml HS-WB 3 times (Steps 4 and 5).
(6) Transfer the 100 µl packed SAg beads each carrying 10 µg oligonucleotide into 2-ml Eppendorf tubes by using 200 µl HS-WB for the transfer.
(7) Add 2 ml pooled DEAE fraction (1 ml DEAE fraction corresponds to around 5×10^{10} cell equivalents) into each tube (for four selections: 4 × 2 ml) and mix the suspension on a turning wheel for 1 h by slowly rotating the tubes.
(8) Collect the SAg beads by a short spin (1500 g, 1 min). Save the supernatant and resuspend the pellet in 1 ml HS-WB. Transfer the suspension into a fresh 1.5-ml Eppendorf tube. The supernatant can be used to check for the efficiency of depletion of the targeted RNP.
(9) Centrifuge the SAg in a microfuge at 1500 g for 1 min. Remove the supernatant.
(10) Add 1 ml HS-WB to the pellet and rotate the samples for 5 min.
(11) Repeat Steps 9 and 10 five times.

Analysis of the affinity-selected material: protein and RNA components

Both proteins and RNAs of the selected RNPs are recovered from the same sample.
First, *proteins* are recovered by the following procedure:

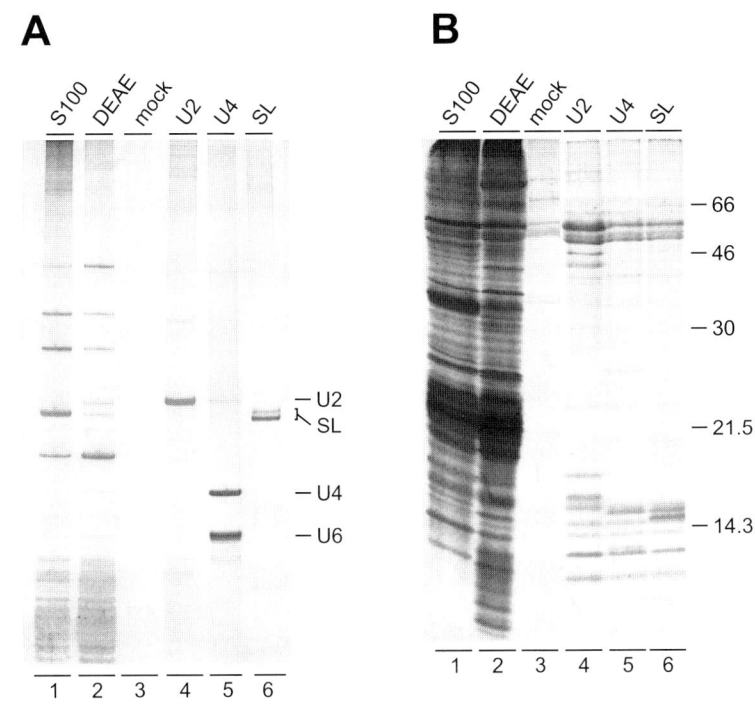

Fig. 41.1. Affinity purification of *T. brucei* U2, U4/U6 and SL RNPs using biotinylated 2'OMe RNA oligonucleotides. (A) RNA analysis of affinity-selected snRNPs. After affinity selections from 2-ml DEAE fractions with 2'OMe RNA oligonucleotides specific for U2, U4 and SL RNAs (lanes 4–6), and subsequent protein dissociation, bound RNA was released, separated by 10% denaturing gel and visualized by silver staining. As a mock control, RNA was selected in the presence of an unrelated oligonucleotide (lane 3). For comparison, RNA purified from 25 μl of *T. brucei* S100 (lane 1) and 25 μl of the DEAE fraction (lane 2) is also shown. The positions of U2, SL, U4 and U6 RNAs are indicated on the right. (B) Protein analysis of affinity-selected snRNPs. Following affinity selection with U2, U4 and SL RNA-specific 2'OMe RNA oligonucleotides (lanes 4–6) proteins were released, separated by a 15% SDS–polyacrylamide gel and visualized by silver staining. Proteins selected with an unrelated oligonucleotide served as a mock control (lane 3). In addition, proteins present in 10 μl of *T. brucei* S100 (lane 1) and in 10 μl of the DEAE fraction (lane 2) were analyzed. Protein markers are indicated in kDa on the right.

(1) Resuspend the beads in 100 μl 9 M urea solution and incubate the samples at room temperature for 30 min by gentle shaking, then centrifuge at 1500 g for 1 min. Take the supernatant into a fresh tube and save it.
(2) Repeat elution of the proteins by adding 100 μl 6 M urea solution to the beads, incubating for 30 min at room temperature (by shaking the sample) and a short spin at 1500 g. Collect the supernatant and mix it with the first urea eluate.
(3) Spin down the pooled fractions to remove all beads (high speed, 1 min). Transfer the supernatant into a new Eppendorf tube.

(4) Precipitate proteins by adding 5 volumes of cold acetone and place the samples at −20 °C. Efficient precipitation of proteins takes a minimum of 1 h.
(5) Warm up the samples to room temperature. Collect the pelleted proteins by centrifugation for 30 min at maximum speed in an Eppendorf microcentrifuge at room temperature (urea might precipitate at 4 °C).
(6) Wash the pellet with 80% ethanol at room temperature (add 1 ml ethanol to the pellet, invert the tubes a few times and centrifuge the samples for 5–10 min at maximum speed in a microcentrifuge at room temperature).
(7) Repeat Step 6 once more to remove all urea.
(8) Dissolve the protein pellet in small volume of Laemmli gel loading buffer. To be able to load all selected proteins into one well of a protein gel, use 20 μl sample buffer for dissolving the samples.
(9) Analyze the protein composition of the selected samples by SDS–PAGE. Proteins can be detected by Coomassie or silver staining, or sequential staining with Coomassie and silver (double staining). After silver staining a minor amount of released RNA might be also visible on the protein gel. If antibodies are available, the identity of proteins can be checked by Western blotting. Other proteins may be identified by mass-spectrometric analysis.

Second, *RNAs* are released from the affinity matrix by the following procedure:

(1) After releasing the proteins, wash the 100 μl packed beads in 1 ml HS-WB (rotating the suspension on a wheel for 5 min).
(2) Pellet the beads by a short centrifugation (1500 g, 1 min) and discard the supernatant.
(3) Mix the packed SAg beads with 100 μl 2 × proteinase K buffer. Incubate the slurry at 80 °C on a thermoshaker for 10 min, then centrifuge the beads at 1500 g for 1 min.
(4) Take off the supernatant from the beads and repeat the RNA elution with 100 μl of 1 × proteinase K buffer.
(5) Pool together the two eluted fractions and do a short centrifugation at maximal speed for 1 min to remove all beads.
(6) Then treat the eluted RNA samples with proteinase K. Add 1 mg/ml proteinase K to the solutions and incubate at 50 °C for 1 h.
(7) Phenolize the samples: add 200 μl phenol/chloroform/isoamylalcohol into the tubes and incubate by strong shaking for 10 min at room temperature.
(8) Centrifuge the samples at maximal speed for 10 min. Transfer the aqueous phase into fresh tubes and ethanol-precipitate the RNAs in the presence of 20 μg glycogen. Store the samples at −20 °C for 1 h.
(9) Collect the precipitated RNAs by a 30-min high-speed centrifugation at 4 °C, wash the pellets in cold 70% ethanol, dry and dissolve them in minimal volume (10 μl) RNA gel sample buffer.
(10) Analyze the RNAs on a denaturing polyacrylamide gel by silver staining (for low amounts of RNAs, use Northern blotting with specific DNA oligonucleotide probes).

Notes

(1) How much of the oligonucleotides and of SAg beads should be used for affinity selections? It is advisable to do first small-scale selections and to optimize the ratio of oligonucleotide to beads and extract. Also, vary the amount of oligonucleotide and determine the efficiency of selections (e.g. by primer extension). The minimal amount of SAg that can be handled is 10–20 µl of packed beads. The capacity for binding free biotin and biotinylated molecules is indicated on the product's data sheet. Do not use large amounts of SAg without need, since this might cause background problems.

(2) How can one prove that the selected proteins are specific for the targeted RNP? It is important to do mock selection with an unrelated oligonucleotide (or without oligonucleotide) to be able to distinguish between specific and contaminating proteins. When the background is too high, the stringency of washing the selected complex can be increased by raising the salt concentration or increasing the non-ionic detergent concentration in the washing buffer (HS-WB; NP-40 or Triton X-100 up to 1%). The oligonucleotide itself or the antisense oligonucleotide–RNA hybrid can bind proteins in the extract as well. Higher oligonucleotide amounts may lead to extensive contamination with various nucleic acid-binding proteins. To exclude this, the extract can be RNase treated before affinity selection to remove the target RNA (as a negative control). It is very useful to select the same particle with alternative antisense oligonucleotides complementary to different regions of the targeted RNA and to compare the resulting protein patterns.

(3) If the affinity purification has to be performed on a large scale, it is sometimes easier to do several small-scale purifications in parallel, instead of scaling up to a single large-scale purification. Furthermore the background of selection is strongly increased, when more than 100 µl of packed SAg beads are used in the same tube.

(4) It is possible to purify different RNPs from the same extract sequentially [37].

(5) We have used DEAE chromatography to enrich the total cell extract for spliceosomal snRNPs. However, DEAE chromatography imposes some stringency on the RNA–protein interactions and many RNPs are reduced to stable core complexes during the high-salt elution step (above 300 mM salt).

41.3.1.4 Affinity Selection of RNPs for Functional Studies by Displacement Strategy

Three different strategies have been developed to release affinity-selected RNP particles under native conditions for functional studies: (1) elution with free biotin from immobilized monomeric avidin [38, 39], (2) selection by a short biotinylated 2'OMe oligonucleotide in high-salt buffer at 4 °C and release at 37 °C in low-salt buffer [37], and (3) use of a chimeric 2'OMe RNA/DNA oligonucleotide partially complementary to the RNA target for selection and a displacement oligonucleotide for elution [24, 25, 28]. This displacement strategy, which is quite generally applicable, will be presented in detail below.

The affinity oligonucleotide is made of 2'OMe RNA (at least 10 nt in length), perfectly complementary to the target RNA, and extended with a DNA tail (any length), which is not complementary. The DNA tail can be positioned 3' or 5' relative to the 2'OMe RNA part of the oligonucleotide, depending on which orientation is less sterically hindered by the protein components of the targeted RNP.

Through a biotin residue, placed usually on the DNA part, this chimeric oligonucleotide can be immobilized to an affinity matrix. Cell extracts are incubated with the chimeric oligonucleotide as described earlier (see Section 41.3.1.3). Following binding and extensive washing steps, the selected RNPs are eluted from the affinity matrix by addition of a so-called displacement oligonucleotide, complementary to the affinity oligonucleotide over its entire length. Since the displacement oligonucleotide can form a thermodynamically more stable duplex with the affinity oligonucleotide than the target RNA, the entire RNP is released from the affinity matrix (for a schematic, see Fig. 41.2A). Efficient elution requires at least 2-fold molar excess [24] of the displacement oligonucleotide over the affinity oligonucleotide. In some cases displacement works without a non-specific DNA tail on the affinity oligonucleotide [26, 27].

Example: Affinity selection and release of functional *T. brucei* U1 snRNPs

In this case crude cell lysate has been used directly for affinity purification without prior enrichment of snRNPs. First, the biotinylated chimeric 2'OMe RNA/DNA oligonucleotide was bound to target RNPs in extract, followed by immobilizing the complexes formed on the affinity matrix and release with the displacement oligonucleotide. NAg beads were used instead of streptavidin because of their lower non-specific binding properties in crude cell lysates. In our specific example the affinity oligonucleotide was a chimeric 2'OMe RNA/DNA oligonucleotide (Tb-U1-5' chimeric) biotinylated at the 3' end [28]. The displacement oligonucleotide was Tb-U1 displace [28], complementary to the Tb-U1-5' chimeric oligonucleotide. As a mock control the displacement oligonucleotide was replaced by another one with an unrelated sequence (see Fig. 41.2B–D).

Procedure

(1) For preparative-scale U1 snRNP affinity selection 50 ml *T. brucei* total cell lysate (corresponding to approximately to 1×10^{11} cells equivalents) is briefly centrifuged (10 min at 10 000 g) to remove aggregates. Transfer the supernatant (10 ml aliquots) into five fresh tubes.
(2) Mix 15 µg biotinylated chimeric oligonucleotide in 500 µl HS-WB. Add a 100-µl aliquot of this solution in each aliquot of cell extract.
(3) Incubate and slowly rotate the tubes at 30 °C for 1 h.
(4) Transfer 5×100 µl Neutravidin Agarose suspension into five 1.5-ml Eppendorf tubes. Collect the beads by a short spin (1500 g, 1 min).
(5) Pre-block the NAg beads as described earlier (see Section 41.3.1.2.).
(6) Wash the beads 3 times in HS-WB by resuspending 50 µl packed beads in 0.5 ml wash buffer.
(7) Spin down the extract after the 1-h incubation to remove the aggregates

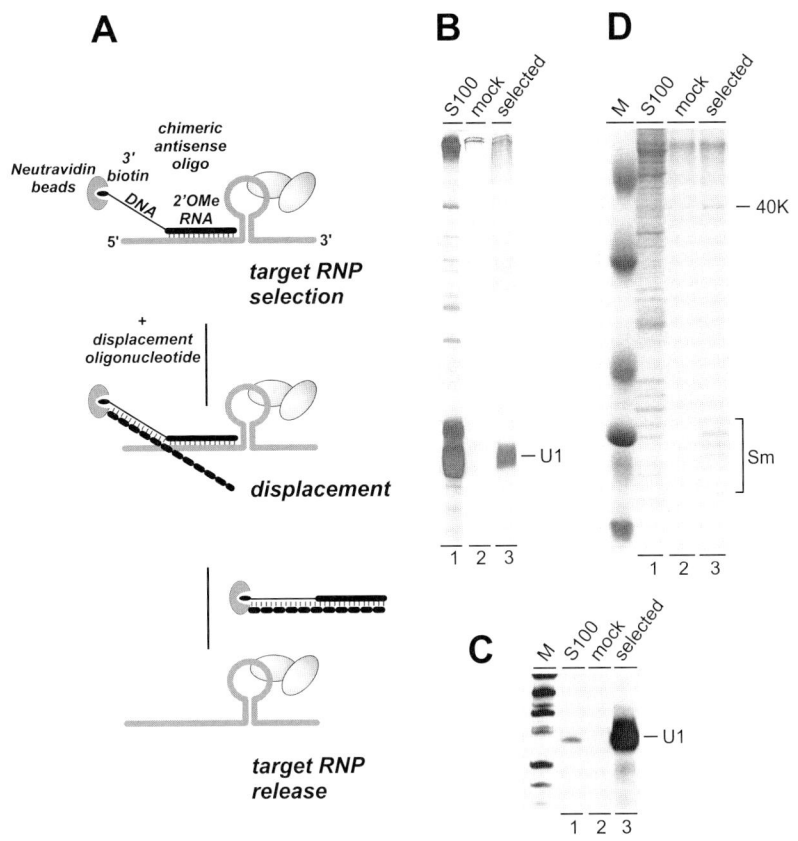

Fig. 41.2. Affinity selection of *T. brucei* U1 snRNP by the displacement strategy. (A) Schematic illustration of the displacement strategy (target RNP selection, displacement and target RNP release). For affinity selection a specific antisense 2'OMe RNA oligonucleotide is used, which is extended with an unpaired 3' terminal DNA tail. This oligonucleotide is immobilized through a biotin group to an affinity matrix (e.g. Neutravidin beads) and used for selection of the target RNP. Binding of the chimeric oligonucleotide to the target RNP is reversed by a displacement DNA oligonucleotide, which pairs with the chimeric oligonucleotide through its entire length. Thereby the affinity oligonucleotide is dissociated from its target, and the selected RNP is released from the affinity matrix. (B–D) Following this strategy, the U1 snRNP was purified from 50 ml of *T. brucei* S100 extract (lanes 1, a 20-µl sample is shown) by affinity selection with Neutravidin beads and a biotinylated chimeric 2'OMe RNA/DNA oligonucleotide directed against the 5'-terminal sequence of U1 snRNA. The affinity-selected material was released by an antisense displacement oligonucleotide (lanes 3) or, as a mock control, by an unrelated oligonucleotide (lanes 2). Following affinity purification, RNA and protein components were analyzed separately. (B) Detection of RNA by silver staining. The position of U1 snRNA is indicated on the right. Note that the *T. brucei* U1 snRNA (B, lane 3) runs in the tRNA region of total RNA on the gel. (C) The *T. brucei* U1 snRNA was identified by Northern blot analysis, using a U1-specific oligonucleotide probe. M, DIG-labeled DNA marker V (Roche 85598025; sizes of marker fragments: 57, 64, 80, 89, 104 and 123 nt). (D) The affinity-selected proteins were detected by Coomassie staining. The Sm proteins and a U1-specific 40-kDa protein are labeled on the right. M, Rainbow marker (Amersham Life *Science* RPN 755; sizes of markers: 6.5, 14.3, 21.5, 30 and 46 kDa).

(10 min at 10 000 g). Collect the supernatant into new tubes and place them on ice.

(8) Transfer each aliquot of the blocked, washed NAg beads into an aliquot of cell extract by using a small volume of the extract for the transfer. The mixture is slowly rotated at 4 °C for 2 h.

(9) Collect the NAg beads by a short centrifugation (1500 g, 1 min). Remove and save the supernatant (for checking the efficiency of U1 RNP depletion). Transfer the pelleted NAg beads into 1.5-ml Eppendorf tubes using 0.5 ml HS-WB.

(10) Incubate the beads in HS-WB for 5 min by slowly rotating at 4 °C.

(11) Spin down the beads (1500 g, 1 min) and remove the supernatant.

(12) Repeat washing the beads in 0.5 ml HS-WB 3 times (Steps 10 and 11).

(13) Resuspend the 50 µl packed NAg in each tube in 100 µl HS-WB and transfer the suspension into a single Eppendorf tube. Spin down the beads again. The packed volume is now 250 µl in total.

(14) Add 25 µg displacement oligonucleotide to 250 µl HS-WB and mix the solution with the pelleted NAg beads by slowly rotating on a turning wheel for 30 min at 30 °C.

(15) Collect the beads by a short centrifugation (1500 g, 1 min) and transfer the supernatant (this contains the eluted U1 snRNP) into a new Eppendorf tube.

(16) Repeat the elution step once. Resuspend the beads in 250 µl HS-WB containing 25 µg of the displacement oligonucleotide and incubate the suspension at 30 °C for 30 min by slowly rotating the tube on a wheel.

(17) Spin down the beads (1500 g, 1 min) and combine the supernatant (= second eluate) with the first eluate.

(18) Spin down the pooled fractions (maximal speed, 1 min) to get rid of all beads. The pooled fractions containing the eluted U1 snRNP particles can be used directly for functional or structural studies.

Notes

(1) The salt concentration of the cell lysate and the HS-WB used for RNP purification should be adjusted according to the stability of the targeted particle. At least core snRNP particles are usually stable under high-salt conditions.

(2) In some cases the presence of large amounts of displacement oligonucleotide might be disturbing for the functional assay (e.g. when the purified U1 snRNP is used for binding to RNAs containing a 5' splice site). In such cases the displacement oligonucleotide can be depleted from the eluted material through a biotinylated DNA oligonucleotide complementary to the displacement oligonucleotide ("anti-displacement oligo"). This step is best done on a small NAg column containing immobilized anti-displacement oligonucleotide in excess to the displacement oligonucleotide [28].

(3) Under the relatively mild conditions of the affinity purification by displacement the integrity of the RNP particles should be preserved, only some loosely associated protein components might be lost. To check for integrity of a particle after affinity purification, the eluted samples should be analyzed by native gel electrophoresis or glycerol gradient centrifugation.

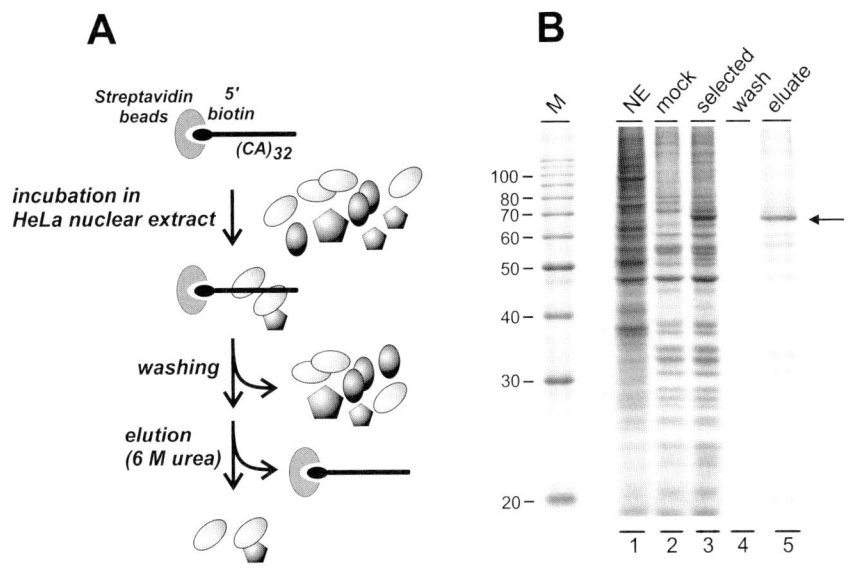

Fig. 41.3. Affinity purification of CA-repeat RNA binding proteins. (A) Schematic representation of the affinity purification strategy. 5′-biotinylated (CA)$_{32}$ RNA was pre-bound on SAg beads and incubated with HeLa cell nuclear extract. Unbound and loosely bound material was removed by stringent washing and tightly bound proteins were recovered under denaturing conditions (6 M urea). (B) The protein composition of total nuclear extract (lane 1), mock-selected material (lane 2), (CA)$_{32}$ RNA affinity-selected material (lane 3), a 2 M KCl wash fraction after affinity purification (lane 4) and the 6 M urea eluate (lane 5) was analyzed by 12.5% SDS–PAGE and Coomassie staining. The arrow points to the specifically selected 65-kDa CA-repeat RNA-binding hnRNP L protein [30]. Marker proteins and their sizes (in kDa) are shown on the left.

41.3.2
Affinity Purification of Specific RNA-binding Proteins by Biotinylated RNAs

In the following we present a fast and simple one-step affinity purification method for the isolation of specific RNA-binding proteins. A biotinylated RNA with a putative protein-binding sequence is immobilized on an affinity matrix and incubated with a crude cell lysate. Following extensive washing the specifically bound proteins are recovered from the affinity matrix by elution with urea (for a schematic, see Fig. 41.3A). For RNA–protein complex formation we recommend to use HPLC-purified, 5′- or 3′-biotinylated RNA oligonucleotides or internally biotinylated transcripts, which can be generated by *in vitro* transcription with T7 or SP6 RNA polymerase in the presence of biotinylated nucleotides (e.g. Biotin-16-UTP from Roche). If you biotinylate the RNA molecule during *in vitro* transcription, you have to find empirically the optimal incorporation level, since too many internal biotin residues may introduce steric hindrance, resulting in lower protein binding capacity.

Example: Isolation of CA-repeat RNA binding factor from HeLa cell nuclear extract using biotinylated RNA oligonucleotide

In this case we have used a 5′-biotinylated RNA oligonucleotide, 5′-biotin-(CA)$_{32}$-3′, which was first immobilized on SAg beads [30]. As a mock control the affinity selection was performed in the absence of bound RNA oligonucleotide. RNA–protein complexes were formed during a 1-h incubation at 30 °C. Then the non-specifically bound proteins were removed by washing in HS-WB with progressively increasing KCl concentration (up to 2 M). Affinity-selected proteins were eluted with 6 M urea and analyzed on a SDS–polyacrylamide gel (see Fig. 41.3B).

Procedure

(1) Prepare a crude nuclear extract from HeLa cells in the presence of protease inhibitors [35].
(2) Take 20 µl pre-blocked SAg beads (1:1 slurry in HS-WB) into a 1.5-ml Eppendorf tube.
(3) Add 200 µl HS-WB containing 6 µg of the 5′-biotinylated RNA oligonucleotide to the SAg suspension. Bind the oligonucleotide during a 5-h incubation at 4 °C with rotation.
(4) Collect the SAg beads by a short spin in Eppendorf centrifuge, discard the supernatant.
(5) Take up the pelleted beads in 1 ml HS-WB containing 400 mM KCl and incubate the sample for 5 min at 4 °C with rotation.
(6) Spin down the beads (1500 g, 1 min) and repeat washing in 1 ml HS-WB 3 times (Steps 4 and 5).
(7) Incubate the SAg beads with 150 µl HeLa cell nuclear extract (pre-cleared by a short high-speed centrifugation) at 30 °C for 1 h by slow rotation.
(8) Collect the SAg beads by a short spin (1500 g, 1 min). Remove the supernatant.
(9) Add 1 ml HS-WB containing 400 mM KCl to the pelleted beads and rotate the sample for 5 min at 4 °C.
(10) Centrifuge the SAg suspension at 1500 g for 1 min. Remove the supernatant.
(11) Repeat Steps 9–10 four times with HS-WB containing increasing amounts of KCl (0.6, 0.8, 1.2 and 2 M) to reduce the non-specific background.
(12) Elute the specifically bound proteins with 15 µl of 6 M urea, as described (see Section 41.3.1.3).
(13) Collect washing fractions as well as the eluate and use 5 µl of each for analysis by SDS–PAGE.

Notes

Binding affinities of RNA–protein interactions vary within a wide range and RNA–protein complexes may be disrupted during conventional purification procedures. A critical parameter is therefore the salt concentration of the binding/washing buffers. Make sure that the composition of these buffers is compatible with the application. For example, high salt concentrations can lead to disruption of protein binding; on the other hand, low-salt conditions may increase background binding of proteins.

41.4
Troubleshooting

The biotin-based affinity selection methods allow the rapid purification of RNA–protein complexes for protein identification or structural and functional assays. An intrinsic problem using biotinylated 2′OMe RNA oligonucleotides is that the target RNP might not be sufficiently exposed to the affinity oligonucleotide to allow efficient binding. Another obvious limitation to quantitative RNP affinity selection is that RNP particles might be heterogeneous in nature and that only a certain complex would be accessible to oligonucleotide hybridization.

To optimize the experimental setup for efficient affinity purification we refer to the Notes under Section 41.3.1.3, 41.3.1.4 and 41.3.2. In the following we discuss a few general checkpoints and give some additional recommendations.

41.4.1
Biotinylated 2′OMe RNA Oligonucleotides

- Purity. The synthesized biotinylated RNA oligonucleotide should be purified to avoid non-specific binding.
- Length. The oligonucleotide should be long enough for specific binding to target RNA.
- Sequence. The oligonucleotide sequence should be complementary to the target RNA in such regions, which are not protected by proteins.
- Biotinylation. The number and location of biotin residues can be variable, but the biotin groups should be accessible for binding to affinity matrices.

41.4.2
Extracts and Buffers

To avoid degradation of target RNA–protein complex the cell lysates should be prepared in the presence of RNase and protease inhibitors. The composition of all buffers should be adjusted according to the specific requirements for the formation of the RNA–protein complex of interest.

41.4.3
Optimization of the Experimental Conditions: When Yields are Low

- Ensure that the affinity oligonucleotide is biotinylated or that the internally biotinylated RNA used for selection of specific RNA binding proteins does not contain too many biotin residues that prevent protein binding.
- The affinity oligonucleotide might be inadequate. Try affinity purification with another biotinylated oligonucleotide complementary to a different region of the target RNA.
- It is useful to test binding/washing buffers with different stringencies to determine the optimal purification strategy for the target RNP.

- Addition of appropriate cofactors (such as Mg^{2+}) may support RNA–protein interactions.
- Check the cell lysate for the presence of the RNA–protein complex of interest (e.g. by primer extension for the RNA and by Western blotting for the proteins). Ensure that the target RNP is not degraded in the extract. Add specific inhibitors to all buffers (proteins: protease inhibitors like PMSF, leupeptin, aprotinin; RNA: RNase inhibitors like RNasin; phosphorylated proteins: phosphatase inhibitors like sodium fluoride).
- Ensure that the analysis method for RNAs and proteins is adequate. Use more sensitive methods like primer extension for RNAs and silver staining or Western blotting for proteins.

41.4.4
Optimization of the Experimental Conditions: When Non-specific Background is Too High

- Check for the specificity of the RNA oligonucleotide used for affinity purification.
- Increase the stringency of binding/washing buffers by increasing the salt and/or detergent concentration.
- Addition of blocking reagents (such as BSA or tRNA) to binding reactions may help to reduce background binding.
- Do not use unnecessarily large amounts of affinity beads and oligonucleotide for affinity selections.

References

1. O. PUIG, F. CASPARY, G. RIGAUT, B. RUTZ, E. BOUVERET, E. BRAGADO-NILSSON, M. WILM, B. SÉRAPHIN, Methods 2001, 24, 218–229.
2. J. NILSSON, S. STAHL, J. LUNDEBERG, M. UHLÉN, P.-A. NYGREN, Protein Expr. Purif. 1997, 11, 1–16.
3. C. SRISAWAT, D. R. ENGELKE, RNA 2001, 7, 632–641.
4. C. SRISAWAT, D. R. ENGELKE, Methods 2002, 26, 156–161.
5. K. HARTMUTH, H. URLAUB, H.-P. VORNLOCHER, C. L. WILL, M. GENTZEL, M. WILM, R. LÜHRMANN, Proc. Natl. Acad. Sci. USA 2002, 99, 16719–16724.
6. R. DAS, Z. ZHOU, R. REED, Mol. Cell 2000, 5, 779–787.
7. Z. ZHOU, J. SIM, J. GRIFFITH, R. REED, Proc. Natl. Acad. Sci. USA 2002, 99, 12203–12207.
8. Z. ZHOU, L. J. LICKLIDER, S. P. GYGI, R. REED, Nature 2002, 419, 182–185.
9. P. J. GRABOWSKI, P. A. SHARP, Science 1986, 233, 1294–1299.
10. BINDEREIF, M. R. GREEN, EMBO J. 1987, 6, 2415–2424.
11. J. RAPPSILBER, U. RYDER, A. I. LAMOND, M. MANN, Genome Res. 2002, 12, 1231–1245.
12. U. RYDER, B. S. SPROAT, A. I. LAMOND, Nucleic Acids Res. 1990, 18, 7373–7379.
13. S. M. BARABINO, B. S. SPROAT, U. RYDER, B. J. BLENCOWE, A. I. LAMOND, EMBO J. 1989, 8, 4171–4178.
14. B. J. BLENCOWE, B. S. SPROAT, U. RYDER, S. BARABINO, A. I. LAMOND, Cell 1989, 59, 531–539.
15. B. J. BLENCOWE, A. I. LAMOND, Methods Mol. Biol. 1999, 118, 275–287.

16 B. S. Sproat, A. I. Lamond, B. Beijer, P. Neuner, U. Ryder, *Nucleic Acids Res.* **1989**, *17*, 3373–3386.
17 A. M. Iribarren, B. S. Sproat, P. Neuner, I. Sulston, U. Ryder, A. I. Lamond, *Proc. Natl. Acad. Sci. USA* **1990**, *87*, 7747–7751.
18 M. Majlessi, N. C. Nelson, M. M. Becker, *Nucleic Acids Res.* **1998**, *26*, 2224–2229.
19 D. A. Wassarman, J. A. Steitz, *Mol. Cell. Biol.* **1991**, *11*, 3432–3445.
20 Z. Palfi, A. Günzl, M. Cross, A. Bindereif, *Proc. Natl. Acad. Sci. USA* **1991**, *88*, 9097–9101.
21 B. Lee, G. Matera, D. C. Ward, J. Craft, *Proc. Natl. Acad. Sci. USA* **1996**, *93*, 11471–11476.
22 C. L. Will, C. Schneider, R. Reed, R. Lührmann, *Science* **1996**, *284*, 2003–2005.
23 R. S. Pillai, C. L. Will, R. Lührmann, D. Schümperli, B. Müller, *EMBO J.* **2001**, *20*, 5470–5479.
24 J. Lingner, T. R. Cech, *Proc. Natl. Acad. Sci. USA* **1996**, *93*, 10712–10717.
25 G. Schnapp, H.-P. Rodi, W. J. Rettig, A. Schnapp, K. Damm, *Nucleic Acids Res.* **1998**, *26*, 3311–3313.
26 Y. W. Fong, Q. Zhou, *Nature* **2001**, *414*, 929–933.
27 Z. Yang, Q. Zhu, K. Luo, Q. Zhou, *Nature* **2001**, *414*, 317–322.
28 Z. Palfi, W. S. Lane, A. Bindereif, *Mol. Biochem. Parasitol.* **2002**, *121*, 233–243.
29 R. Reed, *Proc. Natl. Acad. Sci. USA* **1990**, *87*, 8031–8035.
30 J. Hui, K. Stangl, W. S. Lane, A. Bindereif, *Nat. Struct. Biol.* **2003**, *10*, 33–37.
31 A. Günzl, Z. Palfi, A. Bindereif, *Methods* **2002**, *26*, 162–169.
32 G. M. Lamm, B. J. Blencowe, B. S. Sproat, A. M. Iribarren, U. Ryder, A. I. Lamond, *Nucleic Acids Res.* **1991**, *19*, 3193–3198.
33 M. Cross, A. Günzl, Z. Palfi, A. Bindereif, *Mol. Cell. Biol.* **1991**, *11*, 5516–5526.
34 J. D. Dignam, R. M. Lebowitz, R. G. Roeder, *Nucleic Acids Res.* **1983**, *11*, 1475–1489.
35 K. A. Lee, A. Bindereif, M. R. Green, *Gene Anal. Tech.* **1988**, *5*, 22–31.
36 C. Wersig, A. Bindereif, *Mol. Cell. Biol.* **1992**, *12*, 1460–1468.
37 H. O. Smith, K. Tabiti, G. Schaffner, D. Soldati, U. Albrecht, M. L. Birnstiel, *Proc. Natl. Acad. Sci. USA* **1991**, *88*, 9784–9788.
38 B. P. Glover, C. S. McHenry, *Cell* **2001**, *105*, 925–934.
39 D. R. Kim, C. S. McHenry, *J. Biol. Chem.* **1996**, *271*, 20690–20698.

42
Immunoaffinity Purification of Spliceosomal and Small Nuclear Ribonucleoprotein Complexes

Cindy L. Will, Evgeny M. Makarov, Olga V. Makarova and Reinhard Lührmann

42.1
Introduction

The ability to isolate RNP complexes under native conditions is often a prerequisite for subsequent functional and ultrastructural studies. A number of affinity selection techniques have been employed to purify human small nuclear ribonucleoprotein (snRNP) and/or spliceosomal complexes from nuclear extract. These include immunoaffinity chromatography [1, 2], affinity selection using biotinylated antisense oligonucleotides [3], selection of pre-mRNAs with randomly incorporated biotinylated nucleotides [4, 5], or binding of aptamer-tagged pre-mRNA by either viral MS2 protein fused to the maltose binding protein [6, 7] or tobramycin [8]. Immunoaffinity chromatography with anti-peptide antibodies has proven to be highly effective for the isolation of both native snRNP and spliceosomal complexes. This procedure entails the initial production of antibodies against a peptide of one of the components of the RNP complex of interest. Antibodies are then affinity-purified and immobilized on a solid matrix [typically Protein A–Sepharose (PAS) beads]. RNP complexes are bound by the immobilized antibody via the targeted protein and the matrix is washed to remove contaminating, unbound material. Highly purified complexes are then eluted under native conditions by the addition of an excess of the cognate peptide.

42.2
Generation of Anti-peptide Antibodies: Peptide Selection Criteria

To choose a peptide for the production of anti-peptide antibodies, several criteria should be taken into consideration. First, the peptide should be 15–18 amino acids in length and should not contain an internal cysteine (see below). When choosing the sequence of the peptide, conserved motifs [e.g. RNA recognition motifs (RRMs), zinc fingers, etc.] and regions rich in a particular amino acid should be avoided (e.g. leucine-rich, proline-rich, etc.) as the generated antibodies will likely crossreact with many other proteins containing similar motifs. Likewise, regions known to interact with other proteins or RNA in your RNP complexes should be avoided, as they likely will not be accessible for interaction with an antibody. Optimal regions are those found on the surface of the protein. If the atomic structure of your protein is known, then surface amino acids can be chosen with some certainty. An alternative approach is to use a computer program (e.g. Protean from DNA Star) that provides information about amino acid hydrophobicity and thus the probability that a stretch of amino acids lies on the surface of your protein. These programs also provide information about the antigenicity of a particular region. Typically, regions containing prolines have a high antigenicity index and the presence of one or two internal prolines is thus desirable. Amino acids on the surface of a protein are typically charged and thus most candidate peptides contain multiple charged residues. However, one should avoid peptides that are extremely positively or negatively charged, as they may be difficult to synthesize and/or purify. The termini of a protein, in particular the C-terminus, often prove to be very effective for immunization purposes, providing they meet the criteria described above. After selecting the amino acid sequence, a cysteine residue should be added to the N or C-terminus. Prior to immunization, the peptide is normally coupled to a carrier protein such as keyhole limpet hemocyanin or ovalbumin (see Protocol 1) via the cysteine's sulfhydryl group using, for example, *m*-maleimidobenzoyl-*N*-hydroxysuccinimide ester (MBS). If the chosen peptide sequence is more C-terminal in your protein, the cysteine should be added to the N-terminus of your peptide and vice versa. As the success rate for the production of antibodies suitable for immunoaffinity purification is only moderate, it is advisable to carry out immunizations with more than one peptide sequence.

Protocol 1: Coupling of peptides to ovalbumin with MBS

- PBS (pH 7.0): 20 mM potassium phosphate (pH 7.0) and 130 mM NaCl.
- Protein sample buffer: 60 mM Tris–HCl (pH 6.8), 1 mM EDTA, 16% glycerol, 2.0% SDS (w/v), 0.1% bromophenol blue, 50 mM DTE. Add DTE directly prior to use.

(1) Dissolve 25 mg (sufficient to couple six peptides) of ovalbumin (Sigma) in 3.75 ml PBS (pH 7.0) and pipette 3.0 ml into a 12-ml plastic, capped tube.
(2) Directly prior to use, dissolve 50 mg of fresh MBS (Pierce) in 1.0 ml of DMSO. Combine 20 µl of the MBS solution with 280 µl DMSO and add immediately to the ovalbumin solution. Incubate for 30 min at room temperature.
(3) Apply the ovalbumin solution to a PD10 column (Amersham Biosciences) equilibrated with PBS (pH 7.0). Collect 10 1.0-ml fractions and perform a Bradford protein concentration determination (or a similar method) with 4 µl of each fraction to determine which contains ovalbumin. Typically, ovalbumin elutes in fractions 3–5. Combine the three most highly concentrated fractions.
(4) Pipette 0.5 ml of the MBS-treated ovalbumin solution into a microfuge tube

and add 150 µl of peptide (20 mg/ml dissolved in H$_2$O). Incubate for 30 min at room temperature.
(5) Add 350 µl of PBS (pH 7.0) to the peptide/ovalbumin solution and load onto a PD10 column equilibrated with PBS (pH 7.0). Collect six 1.0-ml fractions.
(6) Combine 4 µl of each fraction with 8 µl protein sample buffer and heat at 85 °C for 5 min. Analyze on a 13% polyacrylamide–SDS gel and visualize by staining with Coomassie. As a control, analyze 1 µl of the MBS-treated ovalbumin solution. If the peptide has been successfully coupled, multiple bands that run just above the MBS-treated ovalbumin will be observed. Peptide–ovalbumin conjugate typically peaks in fractions 4 and 5. Pool the two most highly concentrated fractions and use for immunization. The end concentration of the peptide-ovalbumin conjugate solution is around 0.5–0.75 mg/ml.

Protocol 2: Affinity purification of anti-peptide antibodies

- Peptide beads: 1.0 mg of peptide was covalently coupled to 1.0 ml of Sulfolink Coupling Gel exactly as described by the manufacturer (Pierce).
- PBS (pH 8.0): 20 mM potassium phosphate (pH 8.0) and 130 mM NaCl.

(1) Combine 8.0 ml of rabbit serum containing anti-peptide antibodies with 800 µl 10 × PBS (pH 8.0) and filter by passing through a 0.45-µm membrane. Wash 1.0 ml (bed volume) peptide beads twice with 10.0 ml PBS in a 15-ml capped tube, add the rabbit serum and incubate 2–3 h or overnight at 4 °C with head-over-tail rotation.
(2) Pellet beads by centrifuging for 1 min at 200 g (e.g. at 1000 r.p.m. in a Heraeus Megafuge 1.0R centrifuge). Wash the beads 5 times with 10.0 ml PBS. After the last wash transfer the beads to a plastic 10-ml column (e.g. a Poly-Prep Chromatography Column from Bio-Rad) and allow the buffer to drain from the column.
(3) To elute bound antibodies, apply 0.5 ml of 100 mM glycine (pH 2.5) to the column and collect the flow-through in a microfuge tube containing 30 µl of 1 M Tris (pH 9.5). Repeat 4 times, collecting each eluate in a separate tube. Mix well to ensure that the pH of the eluates has been neutralized. Check for the presence of antibody in each fraction by Bradford assay (or a similar method) and pool the two most highly concentrated fractions (typically eluates 2 and 3).
(4) Transfer the eluate to dialysis membrane (molecular weight cut-off 6000–8000) and dialyze at 4 °C against 1 l of PBS (pH 8.0) for 4 h, changing the buffer once after 2 h. Typically, 0.1–0.3 mg of affinity-purified antibody are recovered from 8 ml of serum, but this amount may vary considerably depending on the titer of the antibody. The activity of the eluted antibodies should subsequently be tested by Western blotting. Note that some antibodies loose significant activity when briefly exposed to extremely low pH. In those cases, an alternative method for the release of antibody from the peptide beads (such as incubating briefly with 3.5 M MgCl$_2$) should be used. To avoid loss of antibody activity, Steps 3 and 4 should be performed as quickly as possible.

(5) Peptide beads, which typically can be used multiple times, should be washed 3 times with PBS (pH 8.0) and stored at 4 °C in H$_2$O containing 0.02% sodium azide.

42.3
Immunoaffinity Selection of U4/U6.U5 Tri-snRNPs

To purify U4/U6.U5 tri-snRNPs from HeLa nuclear extract, antibodies were raised against a C-terminal peptide (amino acids 484–497; CAEFLKVKGEKSGLM) of the human U4/U6-specific 61K protein [9]. Antibodies were affinity-purified using a SulfoLink column (Pierce) containing the immobilized cognate peptide (Protocol 2) and then bound to PAS beads (CL-4B; Amersham Biosciences). BSA and tRNA were included to block non-specific binding sites on the PAS beads. HeLa nuclear extract or a mixture of spliceosomal snRNPs immunoaffinity-purified from HeLa nuclear extract with anti-m$_3$G antibodies [10] was then incubated with the anti-61K charged PAS beads. After extensive washing, bound snRNPs (which include 25S U4/U6.U5 tri-snRNPs and 13S U4/U6 snRNPs) were eluted with an excess of the 61K peptide used for immunization. To separate the U4/U6 and U4/U6.U5 complexes, the eluate was fractionated on a linear glycerol gradient (Fig. 42.1). Density gradient centrifugation not only allows for the separation of these complexes, but also enhances the purity of the isolated complexes, as contaminating proteins, as well as the peptide used for elution, normally migrate at the very top of the gradient. Stoichiometric amounts of U4, U5 and U6 snRNAs as well as a set of known tri-snRNP proteins: 220K, 200K, 116K, 110K, 102K, 90K, 65K, 61K, 60K, 40K, 27K, 20K, 15K, Sm and LSm proteins peak in fractions 17–19 (Fig. 42.1), demonstrating that an intact tri-snRNP complex has been isolated. In addition, equimolar amounts of U4 and U6 snRNAs and a set of U4/U6-specific proteins (90K, 61K, 60K, 20K, 15.5K and Sm-proteins) peak in fraction 8–10, consistent with the presence of 13S U4/U6 snRNPs (Fig. 42.1). As evident from Fig. 42.1, tri-snRNPs represent the majority (around 90–95%) of the purified complexes, whereas 13S U4/U6 snRNPs comprise maximally 10% of the eluted material. These results are consistent with our previous observation that the majority of the 61K protein is associated with the 25S U4/U6.U5 tri-snRNPs in HeLa nuclear extract [9]. On average, 75 µg (50 pmol) of 25S U4/U6.U5 tri-snRNPs can be recovered from 1.4 ml HeLa nuclear extract or from 1.5 mg of purified spliceosomal snRNPs.

Protocol 3: Immunoaffinity purification of U4/U6.U5 tri-snRNPs

- IP250 buffer: 20 mM HEPES (pH 7.9), 250 mM NaCl, 1.5 mM MgCl$_2$.
- IP150 buffer: 20 mM HEPES (pH 7.9), 150 mM NaCl, 1.5 mM MgCl$_2$, 0.5 mM DTT.
- Roeder C buffer: 20 mM HEPES (pH 7.9), 250 mM NaCl, 1.5 mM MgCl$_2$, 0.2 mM EDTA, 0.5 mM DTT, 0.5 mM PMSF and 5% glycerol (w/v).

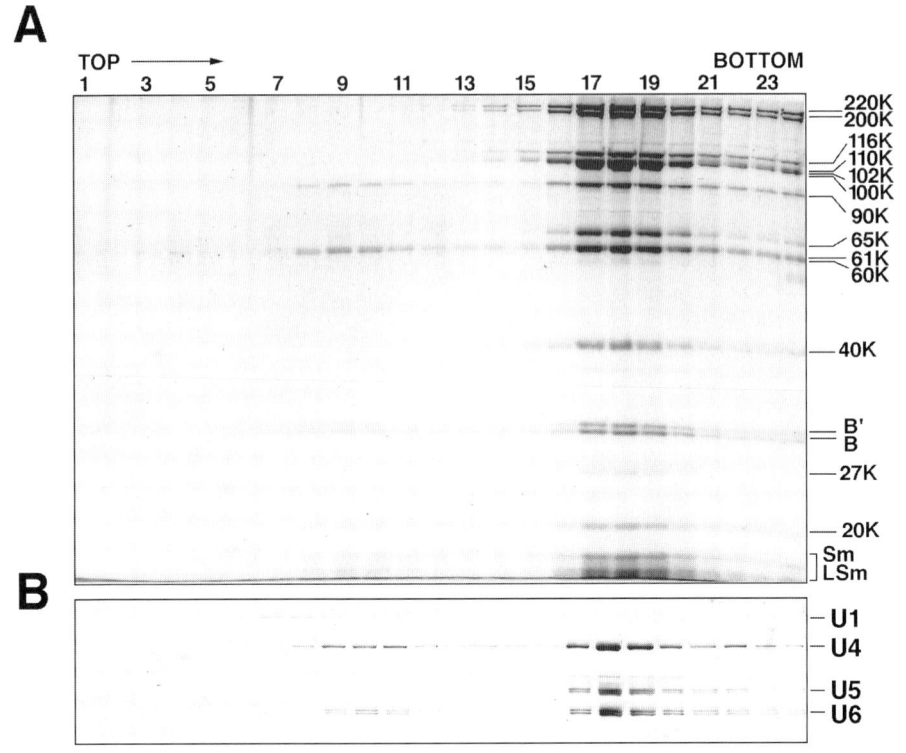

Fig. 42.1. Immunoaffinity-purified 25S U4/U6.U5 tri-snRNPs. Human 25S U4/U6.U5 tri-snRNPs were isolated from a mixture of anti-m$_3$G purified, HeLa spliceosomal snRNPs with anti-peptide antibodies against the U4/U6 61K protein and subjected to 10–30% glycerol gradient centrifugation. Distribution of (A) protein and (B) snRNA across the gradient. Proteins were separated by SDS–PAGE and visualized by staining with Coomassie. RNA was fractionated by denaturing PAGE and stained with silver. Gradient fraction numbers are indicated at the top and the identities of tri-snRNP proteins are shown on the right.

- Roeder D buffer: 20 mM HEPES (pH 7.9), 100 mM KCl, 1.5 mM MgCl$_2$, 0.5 mM DTT, 0.5 mM PMSF and 10% glycerol (w/v).
- 6 × SDS loading buffer: 0.25 M Tris (pH 6.8), 2.5% SDS (w/v), 0.25 M DTT, 20% glycerol, 0.2% bromophenol blue.
- RNA loading buffer: 95% formamide, 0.02% xylene cyanol, 0.02% bromophenol blue.
- HeLa nuclear extract: Extract was prepared according to [11].
- Anti-m$_3$G immunoaffinity-purified HeLa snRNPs:

Total snRNPs were immunoaffinity-selected from HeLa nuclear extract using the anti-m$_3$G monoclonal antibody H20 (Synaptic Systems) and eluted with 7-methylguanosine in Roeder C buffer as previously described [10].

(1) Combine 250 μl (bed volume) of PAS beads with 250 μg of affinity-purified anti-61K antibodies in 1.0 ml of PBS (pH 8.0) containing 0.5 mg acetylated BSA (Sigma; B2518) and 50 μg total yeast tRNA (Sigma; R8759). Incubate at

4 °C for 1.5 h with head-over-tail rotation. Note that before use the tRNA was extracted with phenol/chloroform and precipitated with ethanol.
(2) Wash the PAS beads twice with 1 ml of PBS (pH 8.0) and subsequently twice with 1 ml of IP250 buffer containing 0.05% NP-40.
(3) Dilute 1.5 mg of anti-m$_3$G immunoaffinity-purified HeLa snRNPs (in 0.8–2.0 ml Roeder C buffer) with an equal volume of IP250 buffer containing 0.05% NP-40 and incubate with the anti-61K charged PAS beads at 4 °C for 2 h with head-over-tail rotation. Note that if the concentration of the snRNPs is less than 0.75 mg/ml, dilution with IP250 buffer is no longer necessary. Alternatively, dilute 1.4 ml of HeLa nuclear extract (in Roeder D buffer adjusted to 150 mM KCl) with an equal volume of IP150 buffer containing 0.05% NP-40 and incubate with the anti-61K charged beads as described above.
(4) Wash the beads 6 times with 1 ml of IP250 buffer containing 0.05% NP-40. During the last wash, transfer the beads to a fresh tube.
(5) Elute for 40 min at 4 °C with 800 µl of IP250 buffer containing 5% glycerol and 0.5 mg/ml of the 61K peptide.
(6) Prepare two 4-ml, linear 10–30% glycerol gradients containing IP150 buffer in 11 × 60-mm polyallomer centrifuge tubes. Load 400 µl of the eluate onto each gradient and centrifuge in a Sorvall TH660 rotor (or the equivalent) for 13 h at 30 000 r.p.m. Fractionate the gradient from the top by hand into 24 aliquots of 175 µl.
(7) To determine the distribution of 25S U4/U6.U5 tri-snRNPs and 13S U4/U6 snRNPs across the gradient, analyze the RNA and/or protein content of each fraction of one of the gradients. For protein analysis, add 10 µl of 6 × SDS loading buffer to 60 µl of each fraction and heat for 10 min at 85 °C. Load onto a 10% polyacrylamide–SDS gel (1-mm thick) with a Hamilton syringe and run until the dye reaches the bottom of the gel. Visualize the proteins by staining with Coomassie brilliant blue (Fig. 42.1). For RNA analysis, extract 60 µl of each fraction from one gradient with an equal volume of phenol/chloroform and transfer the supernatant to a new tube. Precipitate with 2.5 volumes of 100% ethanol and wash once with 80% ethanol. Dry in a vacuum desiccator and dissolve the pellet in 7 µl RNA sample buffer, heating for 3 min at 95 °C prior to loading the gel. Analyze the RNA on a 14% polyacrylamide–7 M urea gel and visualize by staining with silver [10]. 25S U4/U6.U5 tri-snRNPs typically peak in fractions 17–19, whereas 13S U4/U6 snRNPs are found in fractions 8–10.

42.4
Immunoaffinity Purification of 17S U2 snRNPs

To purify 17S U2 snRNPs from HeLa nuclear extract, antibodies were raised against a region near the C-terminus (amino acids 444–458; CMLRPPLPSEGPG-NIP) of the 17S U2 SF3a66 protein [1]. Antibodies were affinity-purified (Protocol 2), bound to PAS beads and covalently attached to the beads using DMP (Protocol 4). 17S U2 snRNPs begin to dissociate at salt concentrations above 200 mM and thus HeLa nuclear extract was first dialyzed to reduce the concentration of KCl to 150 mM and then incubated with the anti-SF3a66 charged PAS beads. After extensive washing, bound snRNPs were eluted with an excess of the SF3a66 peptide used for immunization. To separate the 17S U2 snRNPs from contaminating U1 snRNPs and nuclear proteins, the eluate was fractionated on a linear glycerol gradient (Fig. 42.2). U2 snRNA, as well as known 17S U2 associated proteins such as subunits of SF3b and SF3a peaked in fractions 13–15, demonstrating the isolation of an intact U2 particle. On average, around 100 µg (100 pmol) of 17S U2 snRNPs can be recovered from 12.5 ml HeLa nuclear extract. The anti-SF3a66 charged PAS beads can be regenerated (i.e. freed of bound peptide) and used repeatedly.

Protocol 4: Immunoaffinity purification of 17S U2 snRNPs

- G150 Buffer: 20 mM HEPES (pH 7.9), 150 mM KCl, 1.5 mM $MgCl_2$, 5% glycerol, 0.5 mM DTE, 0.5 mM PMSF.
- Anti-SF3a66 antibodies coupled to PAS: antibodies were affinity-purified as described in Protocol 2 and covalently coupled to PAS beads with dimethylpimelimidate (DMP) as previously described [10].
- HeLa nuclear extract: extract was prepared according to [11].
- Protein sample buffer: 60 mM Tris–HCl (pH 6.8), 1 mM EDTA, 16% glycerol, 2.0% SDS (w/v), 0.1% bromophenol blue, 50 mM DTE. Add DTE directly prior to use.
- RNA loading buffer: 95% formamide, 0.02% Xylene cyanol, 0.02% bromophenol blue.

(1) Dialyze 12.5 ml freshly prepared HeLa nuclear extract against 1.0 liter G150 buffer for 4 h at 4 °C, changing the buffer once after 2 h. Dilute with 1 volume of G150 buffer and centrifuge at 4 °C in a 30-ml corex tube at 10 000 g (e.g. 9500 r.p.m. in a Sorvall SS34 rotor) for 10 min to pellet any precipitates.
(2) All of the following steps should be carried out at 4 °C. Pipette 500 µl PAS coupled with 250 µg of affinity-purified anti-SF3a66 antibodies into a small glass column and wash with 10 ml of G150 buffer. Using a peristaltic pump (e.g. Amersham P1 pump), pass the nuclear extract over the anti-SF3a66 column at a flow rate of 1–2 ml/h (i.e. overnight). Alternatively, combine the anti-SF3a66 charged PAS beads and nuclear extract in a 50-ml capped plastic tube and incubate overnight with head-over-tail rotation.
(3) Wash the PAS beads 3 times with 10 ml of G150 buffer and transfer to a 1.5-ml microfuge tube. Wash 2 times with 1.0 ml G150 buffer and remove as much liquid as possible.
(4) To elute bound snRNPs, incubate with 0.5 ml G150 buffer containing 0.4 mg/ml SF3a66 peptide for 30 min with head-over-tail rotation. Microfuge at 3 000 r.p.m. for 1.0 min and transfer the supernatant to a fresh tube. Repeat the elution twice with a fresh batch of peptide and pool the eluates. Microfuge 1 min at 13 000 r.p.m. and transfer to a new microfuge tube to ensure that the eluate is free of PAS beads.

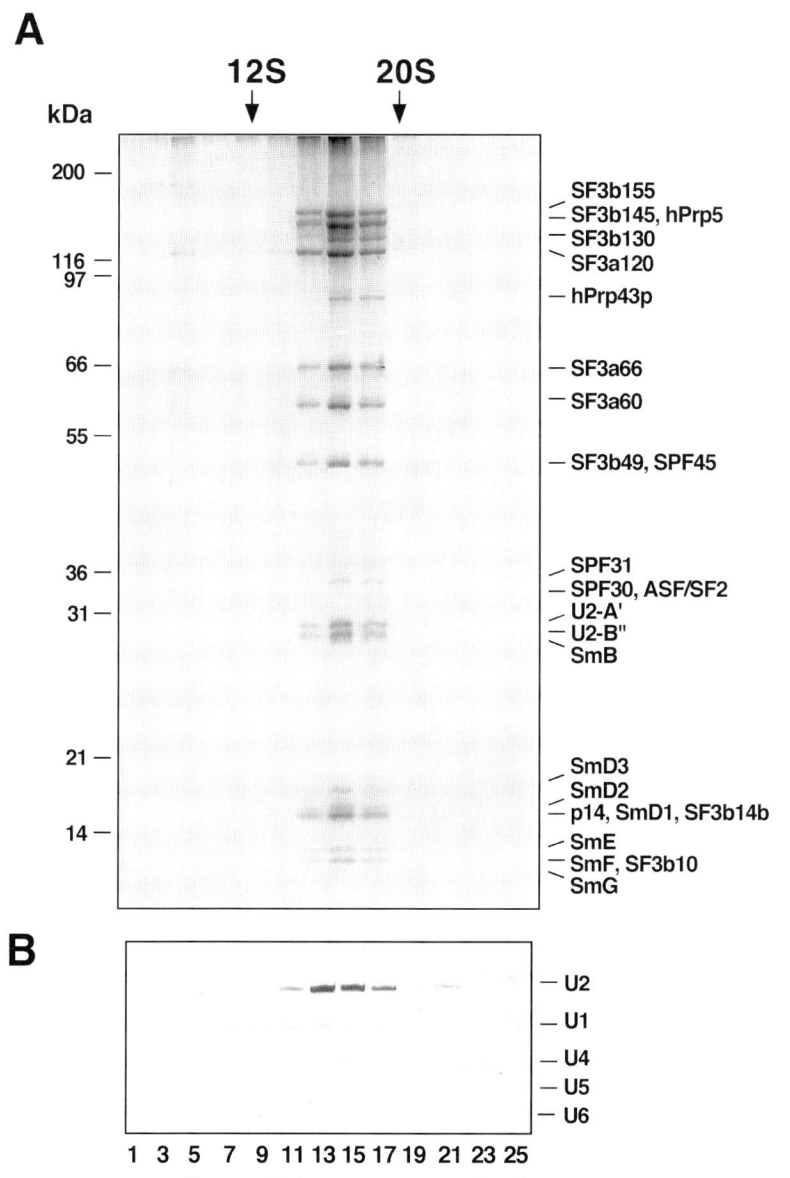

Fig. 42.2. Immunoaffinity-purified 17S U2 snRNPs. Human 17S U2 snRNPs were immunoaffinity-purified from HeLa nuclear extract with anti-peptide antibodies against the SF3a66 protein and subjected to 10–30% glycerol gradient centrifugation. Distribution of (A) protein and (B) snRNA across the gradient. Proteins were separated by SDS–PAGE and visualized by staining with Coomassie. RNA was fractionated by denaturing PAGE and stained with silver. Gradient fraction numbers are indicated at the bottom and the peak positions of 12S U1 and 20S U5 snRNPs, run in parallel, are indicated at the top. The identity of the major 17S U2 proteins is indicated on the right.

(5) Prepare six 4-ml, linear 10–30% glycerol gradients containing G150 buffer in 11 × 60-mm polyallomer centrifuge tubes. Load 200 μl of the eluate onto each gradient and centrifuge in a Sorvall TH660 rotor (or the equivalent) for 17 h at 27 000 r.p.m. (75 000 g). By hand, fractionate the gradient from the top into twenty-seven 150 μl aliquots.

(6) To determine the distribution of 17S U2 snRNPs across the gradient, analyze the RNA and/or protein content of each odd-numbered fraction of one of the gradients. Extract 150 μl of each fraction with an equal volume of phenol/chloroform and transfer the supernatant to a new tube. Add 5 volumes of acetone to the phenol phase (to precipitate protein) and 2.5 volumes of 100% ethanol plus 1/10 volume 3 M sodium acetate (pH 5.2) to the aqueous phase (to precipitate RNA). Mix well, incubate for 2 h at −20 °C or 30 min at −80 °C and microfuge for 10 min at 13 000 r.p.m. Wash the pellets once with 80% ethanol and dry in a vacuum desiccator. Dissolve each RNA pellet in 6 μl RNA sample buffer and heat for 3 min at 95 °C. Analyze RNA on a 10% polyacrylamide–7 M urea gel and visualize by staining with silver [10]. Dissolve each protein pellet in 10 μl protein sample buffer and heat 5 min at 85 °C. Analyze the proteins on a 10/13% polyacrylamide–SDS gel and visualize by staining with Coomassie. 17S U2 snRNPs typically peak in fractions 13–15 of the gradient.

(7) To regenerate the anti-SF3a66 antibodies coupled to PAS, transfer the beads to a glass column and wash with 20 ml of 10 mM $NaPO_4$ (pH 7.2). To avoid loss of activity, antibodies should be incubated only briefly with 3.5 M $MgCl_2$. Elute bound peptide by washing for 5 min with 5.0 ml of 3.5M $MgCl_2$ in 10 mM $NaPO_4$ (pH 7.2). Wash 3 times with 10 ml PBS containing 0.02% sodium azide and store at 4 °C.

42.5
Approaches for the Isolation of Native, Human Spliceosomal Complexes

Spliceosomes are comprised of the U1, U2, U4, U5 and U6 snRNPs, and a large number of non-snRNP splicing factors. The spliceosomal snRNPs and non-snRNP splicing factors associate with the pre-mRNA in an ordered manner [12]. First the U1 snRNP interacts followed by the stable association of U2 snRNP to form spliceosomal complex A. The pre-assembled 25S U4/U6.U5 tri-snRNP is then recruited to form complex B which is structurally rearranged to form the catalytically activated spliceosome (B*). The latter then catalyses the first transesterification reaction, generating complex C. After the second step of splicing, the mRNA is released, the post-spliceosomal complex disassembles and the snRNPs are recycled for new rounds of splicing. Thus, during their formation/catalytic activity, spliceosomes go through many intermediate assembly/functional stages (i.e., E, A, B, B* and C complex).

Two general methods have been employed to isolate native spliceosomal complexes. In the first, the pre-mRNA is targeted by adding an aptamer sequence to its 3′ end (e.g. the hairpin structures bound by tobramycin or the MS2 protein).

Either prior to or after allowing spliceosomal complexes to form under splicing conditions in HeLa nuclear extract, the aptamer is bound by its ligand (e.g. tobramycin or MS2 protein fused to the maltose binding protein), which binds or is covalently attached to a solid support. After washing, complexes can then be eluted under native conditions by the addition of an excess of the aptamer or ligand. This approach has been successfully used to isolate either a mixture of spliceosomal complexes [6], or spliceosomes at a specific assembly stage such as the A or C complex [7, 8]. A general problem of *in vitro* splicing is that it is not possible to synchronize spliceosome assembly. Therefore, at a given time point a heterogeneous population of spliceosomal complexes will be assembled on the pre-mRNA. Thus, if a specific spliceosomal complex is to be isolated, additional measures must be undertaken to isolate a more homogeneous population. One approach is to stall spliceosome assembly at a given point. For example, an accumulation of spliceosomal complex C can be achieved by mutating the 3′ splice site of the pre-mRNA [7, 13]; however, complexes formed on such a pre-mRNA substrate cannot be chased through the catalytic steps of splicing and thus the functional integrity of the purified complexes cannot be verified. Alternatively, splicing can be carried out for only very short periods of time, so that predominantly early spliceosomal complexes such as E and A will have time to form [8].

A second approach for the isolation of spliceosomal complexes is to target a component of the spliceosome other than the pre-mRNA. To isolate a specific spliceosomal complex, ideally such a component should transiently interact with the spliceosome at a specific stage of its assembly/function. Anti-peptide antibodies raised against such a protein would allow immunoprecipitation and subsequent peptide-induced elution of only those spliceosomal complexes that contain the targeted protein. This should result in the isolation of a more homogeneous population of native complexes and circumvent some of the problems (e.g. splicing complex heterogeneity or potential steric hindrance due to the presence of the aptamer) arising when the pre-mRNA is targeted for affinity selection. Indeed, using anti-peptide antibodies directed against spliceosomal proteins that are transiently, but stably associated with the spliceosome, we have been able to isolate activated (B*) spliceosomes [2], as well as spliceosomal complex B that lacks the U1 snRNP (designated BΔU1) (O. V. Makarova et al., submitted).

42.6
Isolation of Activated Spliceosomes by Immunoaffinity Selection with Anti-peptide Antibodies against the SKIP Protein

To isolate activated spliceosomes, anti-peptide antibodies were raised against a C-terminal peptide (amino acids 516–531; CRPSDSSRPKEHEHEGK) of the splicing factor SKIP (Ski oncogene interacting protein), which is stably integrated into the spliceosome first at the time of its activation (i.e. after B, but prior to C complex formation). The strategy for the isolation of activated spliceosomes with anti-SKIP antibodies is depicted in Fig. 42.3. First, we incubated pre-mRNA with HeLa nu-

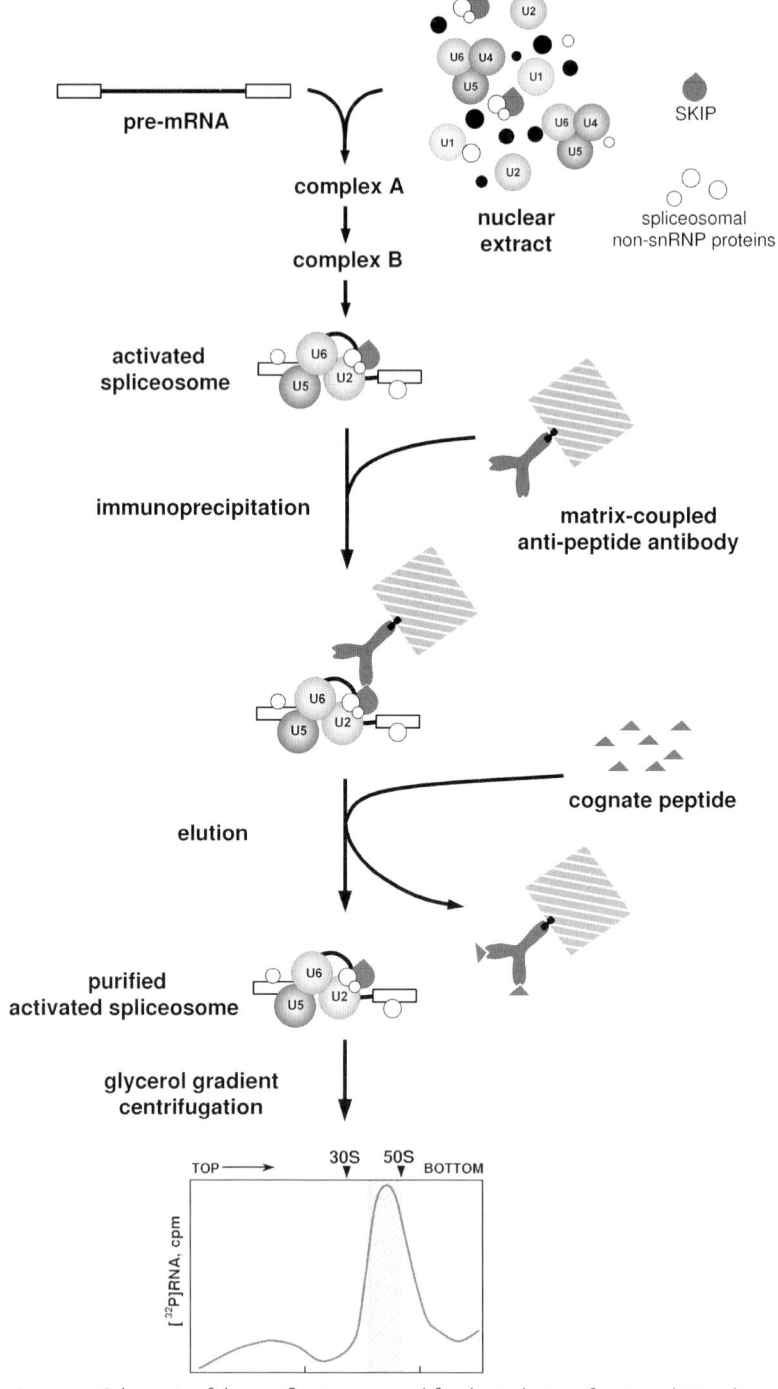

Fig. 42.3. Schematic of the purification protocol for the isolation of activated B* spliceosomes.

clear extract, allowing splicing complexes to form. The incubation time was limited to 10 minutes, as at this time point the first catalytic step of splicing had not yet occurred (i.e. splicing intermediates were not observed) and thus spliceosomal complex C, which also contains the SKIP protein, had not yet formed [2]. We intentionally carried out the purification under stringent conditions so that only stably bound proteins would be present. Therefore, after allowing for splicing complex formation, we added heparin to the reaction (general considerations for using heparin are discussed below) and incubated with PAS beads charged with affinity-purified anti-SKIP antibodies. Heparin not only immediately stops the splicing reaction, but, more importantly, significantly reduces aggregation of spliceosomal complexes. In addition, in the presence of heparin, less stably bound snRNPs will dissociate from the spliceosomal complexes. After extensive washing, spliceosomal complexes containing the SKIP protein were eluted from the beads under native conditions with an excess of the cognate peptide. Consistent with the presence of activated spliceosomes, the eluate contained equimolar amounts of pre-mRNA, and U2, U5 and U6 snRNAs, but lacked U1 and U4 which are known to dissociate at the time of activation (Fig. 42.4). The eluted complexes were further purified by density gradient centrifugation. Complexes containing equimolar amounts of pre-mRNA, U2, U5 and U6 peaked in the 45S region of the gradient (not shown; see [2]), indicating that intact activated spliceosomes had been isolated.

Although antibody-bound spliceosomal complexes are eluted under native conditions by competition with antigenic peptide, the addition of a peptide (even a non-cognate one) can trigger release of unspecifically bound material from the affinity matrix. During incubation of a standard *in vitro* splicing reaction at 30 °C, spliceosomal and other nuclear components aggregate, leading to an increase in the opacity of the solution. While these aggregates can be removed by brief centrifugation, up to 30% (depending on a the final salt concentration and time of incubation) of the pre-mRNA added to the reaction potentially will be lost (i.e. it will be present in the pelleted material). The addition of heparin to a final concentration of 0.5 mg/ml, followed by incubation at 30 °C for an additional 5–10 min prior to immunoaffinity selection of spliceosomal complexes, minimizes aggregation and thus significantly reduces nonspecific (background) precipitation/binding. Heparin is a heterogeneous mixture of variable sulfated polysaccharide chains composed of repeating units of D-glucosamine and either L-iduronic or D-glucuronic acids with molecular weight ranges from 6 to 30 kDa. There are several different preparations of heparin commercially available that differ in terms of their purity, average molecular weight and anticoagulant activity. Significantly, although heparin preparations with almost identical characteristics can be obtained from different manufacturers, their ability to dissociate RNP complexes and/or aggregates can differ considerably. Thus, to obtain reproducible results and also to meaningfully compare different complexes isolated in the presence of heparin, all experiments should be performed with the same batch (stock solution) of heparin. In addition to heparin treatment, dilution of the splicing reaction at least 8-fold prior to immunoprecipitation is also crucial for obtaining highly purified complexes. Pre-cleaning – incubation of the diluted splicing reaction with PAS beads pre-blocked

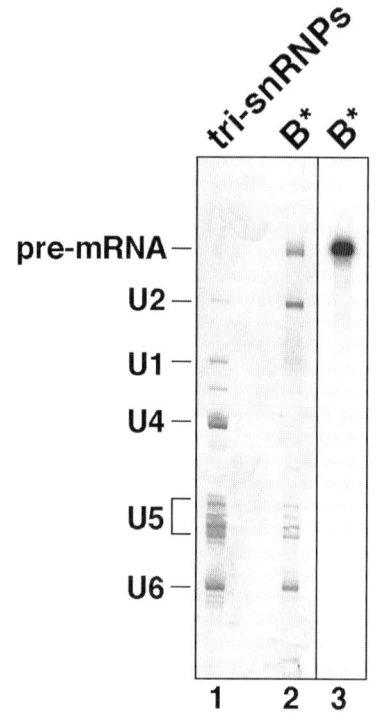

Fig. 42.4. Immunoaffinity purification of activated B* spliceosomes. Human activated spliceosomes were immunoaffinity-purified from HeLa splicing extract with anti-peptide antibodies against the SKIP protein. The RNA compositions of the eluate (lane 2) and that of 25S U4/U6.U5 tri-snRNPs (lane 1) were analyzed by denaturing PAGE and visualized by staining with silver. The radioactive pre-mRNA was also detected by autoradiography (lane 3). The identities of the RNAs are indicated on the left.

with BSA and tRNA (but lacking antibody) – also decreases background, as ribosomes (an often observed contaminant of purified spliceosomal complexes) bind non-specifically to the matrix during this procedure. However, pre-cleaning is less crucial than adding heparin or diluting the splicing reaction.

Successful immunoaffinity purification of spliceosomal complexes is also dependent on choosing the proper ratio of the amount of antibody, PAS matrix and nuclear extract. While it is desirable to use as little PAS as possible to minimize non-specific binding, using too little PAS can decrease the yield of spliceosomal complexes. After optimizing the amount of antibody (see below), the optimal amount of PAS can be empirically determined in titration experiments where the amount of antibody is kept constant; note that PAS maximally binds 20 mg antibody/ml. To isolate activated spliceosomes, we combine 1.0 ml of antibody-charged PAS with 4.0 ml of a splicing reaction containing 20 nM of MINX pre-mRNA. Under these splicing conditions, up to 75% of the MINX substrate will be converted into splicing complexes. We generally add 0.3 mg of affinity-purified antibody to 1.0 ml PAS. For most antibodies, the addition of more than 0.5 mg of affinity-purified antibodies per 1 ml of PAS does not increase the yield of spliceosomal complexes. However, since the activity and specificity of different antibodies can vary considerably, the relative amounts of antibody bound to PAS should be empirically determined for each antibody.

Protocol 5: Immunoaffinity purification of activated spliceosomes

- HeLa nuclear extract: extract was prepared according to [11].
- Heparin: sodium salt, 150 000 U/g, $M = 20\,000$ (Carl Roth, Germany). Stock solution of 5.0 mg/ml in H_2O.
- IP buffer: 20 mM HEPES (pH 7.9), 150 mM NaCl, 1.5 mM $MgCl_2$, 0.5 mM DTT.
- $1 \times$ SDS loading buffer: 40 mM Tris (pH 6.8), 0.4% SDS (w/v), 40 mM DTT, 3.3% glycerol, 0.03% bromophenol blue.

(1) Combine 1.0 ml (bed volume) of PAS beads with 300 µg of affinity-purified anti-SKIP antibodies in 3.0 ml of PBS (pH 8.0) containing 0.5 mg/ml acetylated BSA (Sigma) and 50 µg/ml total yeast tRNA (Sigma). Incubate at 4 °C for 1.5 h with head-over-tail rotation. Note that before use the tRNA was extracted with phenol/chloroform and precipitated with ethanol.

(2) To isolate 2–4 pmol spliceosomal complexes (i.e. an amount sufficient to visualize proteins separated by SDS–PAGE by Coomassie staining), prepare on ice a 4.0-ml splicing reaction containing 40% HeLa nuclear extract, 3.25 mM $MgCl_2$, 2 mM ATP, 20 mM creatine phosphate (Roche) and 20 nM MINX ^{32}P-labeled pre-mRNA (specific activity: 2000–4000 c.p.m./pmol) in a 15-ml plastic capped tube. Pipette 1.0 ml of the splicing reaction into four 1.5-ml microfuge tubes.

(3) Incubate at 30 °C for 10 min and add heparin to a final concentration of 0.5 mg/ml. To set up the optimal conditions for heparin treatment, several concentrations of heparin (e.g. 0.3–0.7 mg/ml) and varying incubation times (5–10 min) should be initially tested for each batch. Under optimal conditions, a high yield of immunoprecipitated complexes with a low background is observed. We estimate the background as the amount of ribosomal RNA relative to the amount of pre-mRNA and/or snRNAs detected in the eluate after denaturing PAGE and subsequent silver staining.

(4) Continue the incubation at 30 °C for an additional 5 min. All subsequent procedures should be carried out at 4 °C.

(5) To "pre-clean" the splicing reaction, pool the reactions and pipette 1.33 ml into three 15-ml, plastic capped tubes. Dilute 10-fold with IP buffer containing 0.05% NP-40 and incubate each tube for 1 h with head-over-tail rotation with 330 µl (bed volume) of PAS pre-blocked for 1 h with 0.5 mg/ml of BSA and 50 µl/ml of total yeast tRNA.

(6) Pellet the PAS beads by centrifuging briefly and transfer the supernatants to

43
Northwestern Techniques for the Identification of RNA-binding Proteins from cDNA Expression Libraries and the Analysis of RNA–Protein Interactions

Ángel Emilio Martínez de Alba, Michela Alessandra Denti and Martin Tabler

43.1
Introduction

In a living cell, different classes of biological macromolecules serve different functions. RNA plays a central role in gene expression. In most cases RNA needs to associate with specific proteins to form ribonucleoproteins (RNPs), such as ribosomes and spliceosomes, wherein the RNA often provides the actual catalytic component. RNA molecules can provide specific and non-specific binding sites for proteins. Thus, RNA-binding proteins participate in the processes of synthesizing, processing, eliciting, modifying, stabilizing, protecting or packaging RNA. They carry RNA molecules between cells and to their destinations within cells, mediate interactions of RNA with other macromolecules or act catalytically on RNA. In addition, all RNA viruses exploit RNA–protein interactions as a means of regulating their infectivity and replication. To gain mechanistic insight into how RNA-binding proteins can regulate those basic cellular processes, one must identify and characterize the proteins that interact with RNA sequences or structural motifs. Therefore, it is important to have experimental tools for the identification and characterization of RNA-binding proteins. Moreover, understanding how RNA-binding proteins specifically interact in a direct or indirect read-out with their target RNAs to form functional complexes is a key problem in molecular biology, from splicing to protein synthesis and from viral replication to embryo development and genetic diseases.

A number of genetic systems have been developed to identify RNA-binding proteins or RNA-binding peptides by screening cDNA libraries. In one of the first examples, phage display was used to identify RNP domain variants with altered specificities [1]. Subsequently, systems have been reported in which RNA-binding domains are used to activate transcription in a yeast three-hybrid assay [2, 3] or in a mammalian Tat-fusion system [4]. Alternatively, translation [5–8] or transcription termination was altered in bacteria [9]. In biological systems without well-developed genetics or transformation systems, RNA-binding proteins have to be

Handbook of RNA Biochemistry: Student Edition. Edited by R. K. Hartmann, A. Bindereif, A. Schön, E. Westhof
Copyright © 2009 WILEY-VCH Verlag GmbH & Co. KGaA, Weinheim
ISBN: 978-3-527-32534-4

identified by other approaches, which are still difficult and laborious. For example, UV crosslinking can be applied to screen for proteins directly interacting with RNAs *in vivo* [10]. UV light induces formation of covalent bonds between nucleic acids and proteins at their contact points, thereby "freezing" the interaction and facilitating the identification of RNA–protein interactions in RNP complexes. Further, classical biochemical purification schemes can be followed; however, they rarely yield protein quantities sufficient for direct N-terminal amino acid sequence determination using Edman degradation. A further limitation of the biochemical approach is the transient and reversible nature of many RNA–protein interactions, which limits the possibility for biochemical purification of RNA–protein complexes. The lack of a simple method has proved a major difficulty for the identification of RNA-binding proteins. In this chapter, we describe how the specific interactions of RNA and proteins can be used in an efficient and straightforward technique for the identification of cDNA clones expressing RNA-binding proteins. This methodology is conceptually related to the identification of DNA-binding proteins [11, 12], although it is quite different in the actual experimental procedure. The procedure makes use of immobilized recombinant proteins and a labeled RNA ligand. In view of the involvement of proteins that are detected by an RNA ligand, this procedure is usually referred to as the "Northwestern" technique. The Northwestern methodology is rapid, inexpensive and does not require the biochemical purification of the protein or the preparation of antibodies. Furthermore, a simple *in vitro*-synthesized RNA transcript can be employed without the need to fuse it to an unrelated RNA, which may influence the secondary structure and binding affinity. Moreover, once the RNA–protein interaction has been detected by the former approach, this methodology may be used to determine more stringent Northwestern library screening conditions. This method can be used not only for the identification of RNA-binding proteins, but also for the study and the characterization of RNA–protein interactions. Several different RNA–protein interactions have been identified and assayed successfully using this methodology without the need for labor-intensive biochemical purification and conventional cloning approaches [13–16]. This method is particularly useful for detection of RNA-binding proteins of low abundance in cDNA libraries, but it can be also used for directly studying the interaction of known RNA and proteins.

This chapter provides two protocols that differ primarily in the goal required. They also differ in the way the proteins are immobilized onto a nitrocellulose membrane. In the first protocol described below, the recombinant proteins are produced by bacteriophages and adsorbed to the membrane, whereas in the second protocol the proteins (recombinant or not) are electroblotted onto the membrane. Both protocols can be used to analyze RNA–protein interactions; however, Protocol 2 is more flexible and versatile to tackle this question. This chapter supplies a detailed description of the Northwestern procedure to ensure optimal conditions for a successful application of this technique.

43.2
Methods

43.2.1
Preparation of Probes and Buffers

43.2.1.1 Preparation of ^{32}P-labeled RNA Probes

Probe integrity is an important issue for a low level background (see Section 43.3.1). Thus, probes must remain generally intact throughout the entire assay. Therefore, high-quality, RNase-free reagents and careful handling are required from the beginning to the end. RNA probes of high specific activity should be made by *in vitro* transcription reactions using bacteriophage-encoded T7 or T3 or SP6 RNA polymerase. We routinely radiolabel with [α-^{32}P]UTP so that approximately 1:10 uridylate residues in the final RNA transcript is derived from the radioisotope source. We have not investigated the use of non-radioactive RNA probes in Northwestern assays as the presence of a chemical group in the RNA backbone could interfere with its interaction with the protein of interest. Moreover, it is recommended that the intactness of *in vitro* transcripts before and after the assay is verified by 5% PAGE containing 8 M urea (see Section 43.3.1).

(1) For transcription reactions mix 1 µl 10 mM of each ATP, CTP, GTP, 1 µl 100 µM UTP, 2 µl 10 × transcription buffer (0.4 mM Tris–HCl, pH 8.0, 100 mM DTT, 60 mM MgCl$_2$ and 20 mM spermidine), 2 µl [α-^{32}P]UTP (specific activity 800 Ci/mmol; Amersham Biosciences, Buckinghamshire, UK or Izotop, Budapest, Hungary), 1 µl ribonuclease inhibitor (20 U/µl), 1 µl SP6 or T7 or T3 RNA polymerase (50 U/µl) and 12 µl of RNase-free water per 20 µl reaction.
(2) Incubate the mixture at 37 °C for 1.5–2 h. After incubation, the template is digested with 1 µl RNase-free DNase (10 U/µl) for 10 min at 37 °C.
(3) The RNA synthesized is then separated from the unincorporated NTPs by chromatography through a 2-ml column of Biogel A 0.5 m (BioRad, Hercules, USA) in 10 mM Tris–HCl, 1 mM EDTA pH, 8.0 or a pre-packed spin column containing Sephadex G-50 (Roche, Basel, Switzerland) according to the manufacturer's recommendation.
(4) Quantification of the probe can be achieved by Cerenkov counting of 1 µl in a scintillation counter.

Typical RNA probe yields, following the given protocol, are 1.4×10^7 c.p.m.

43.2.1.2 Preparation of Blocking RNA

(1) To prepare the blocking RNA, 1 g of *Torula* yeast RNA (Sigma-Aldrich, Munich, Germany) is dissolved in 10 ml TE buffer, pH 7.5, and incubated with 5 µg of proteinase K for 2 h at 37 °C.
(2) RNA is then cleaned by phenol/chloroform extraction (1:1, v/v, adjusted to pH 8.0). The resulting aqueous phase is finally extracted with chloroform/isoamyl

alcohol (24:1, v/v), which removes the bulk of the residual phenol and RNA is recovered by ethanol precipitation with 3 volumes of 100% ethanol for 5 min at room temperature.
(3) After precipitation, the sample is centrifuged at 3500 g for 20 min and the pellet is washed in 70% ethanol.
(4) The pellet is resuspended in diethylpyrocarbonate (DEPC)-treated water to give a final concentration of 50 mg/ml and is dispensed in 1-ml aliquots that are stored at −20 °C.

43.2.1.3 Preparation of the Northwestern Buffer

A single buffer (SB buffer) containing 15 mM HEPES–KOH, pH 7.9, 50 mM KCl, 0.1% (v/v) Ficoll 400-DL/PVP-40, 0.01% (v/v) Nonidet P-40, 0.1% $MnCl_2$, 0.1% $ZnCl_2$, 0.1 mM EDTA and 0.5 mM DTT is used during the entire Northwestern protocols described below. This single-buffer system not only improves the reproducibility of the assay, but also results in an increase of the signal-to-background ratio.

(1) For the preparation of 1 l of SB buffer KCl, Ficoll 400-DL/PVP-40, Nonidet P-40, $MnCl_2$, $ZnCl_2$ and an appropriate volume of de-ionized water are mixed in a 1-l bottle. Then, DEPC to a final concentration of 0.1% (v/v) is added and the buffer is mixed at room temperature continuously by incorporating a stir-bar in the bottle and placing it on a magnetic stirrer and it is autoclaved the next day.
(2) The buffer is cooled down to 4 °C and HEPES–KOH, pH 7.9, EDTA, pH 8.0 and DTT are added. Finally, DEPC-treated water is added up to 1 l.

Please note that DTT cannot be autoclaved, and consequently the stock solution has to be prepared in DEPC-treated water, filter sterilized and stored at −20 °C. Similarly, HEPES and EDTA cannot be DEPC-treated, thus the stock solutions have to be prepared with DEPC-treated water, autoclaved at 120 °C for 20 min and stored at 4 °C. The rest of the stock solutions are prepared in the same way as described for HEPES and EDTA and stored at room temperature. High-quality, RNase-free reagents and careful handling are required during all steps.

43.2.2
Protocol 1: Northwestern Screening for Identification of RNA-binding Proteins from cDNA Expression Libraries

Northwestern, also called RNA-ligand screening, was initially described by Sägesser et al. [13]. It is a straightforward and relatively simple method and allows the identification of poorly expressed RNA-binding proteins from bacteriophage-based cDNA expression libraries without the requirement of laborious and expensive purifications of proteins from tissues. The experimental setup of the Northwestern screening is depicted in Fig. 43.1.

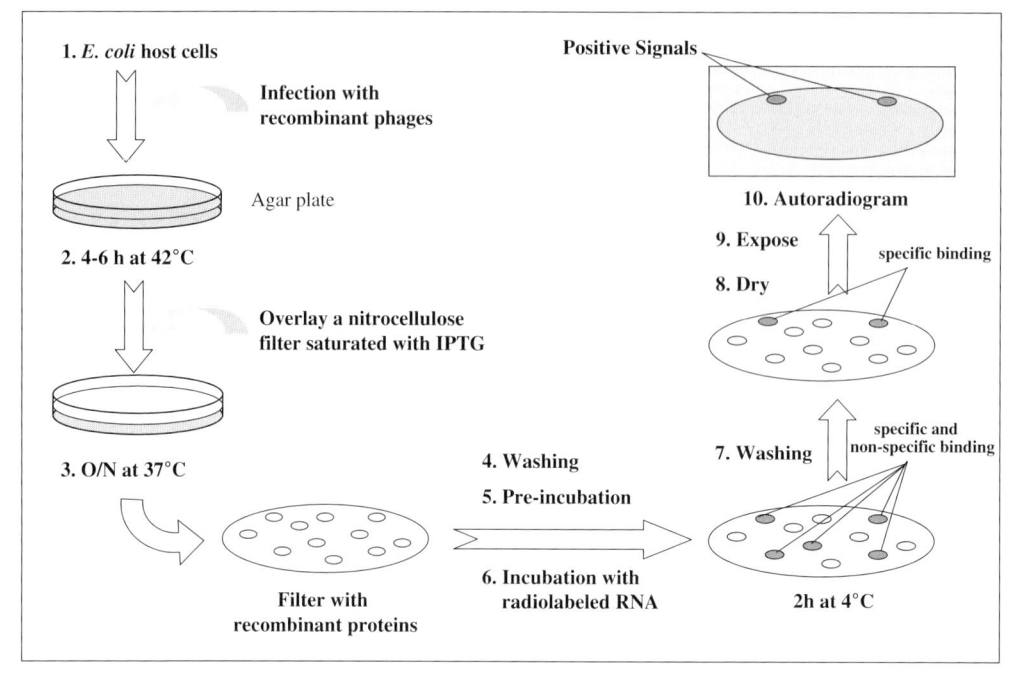

Fig. 43.1. Schematic outline of the Northwestern screening procedure for the identification of cDNA clones that express RNA-binding proteins.

43.2.2.1 Preparation of the Host Plating Culture

The appropriate *Escherichia coli* strain is usually supplied with the library and depends on the phage vector. For instance, "XL1-blue" is used as host strain for the lytic growth of λ-ZAPII and recombinant derivatives.

(1) A single colony of the appropriate host *E. coli* strain is inoculated into 50 ml LB medium supplemented with 0.2% maltose, 10 mM $MgSO_4$ and 10 μg/ml tetracycline, in a sterile 125-ml flask. $MgSO_4$ and maltose treatment allows for optimal adsorption of the λ phage to host bacteria.
(2) The culture is incubated at 30 °C overnight with vigorous shaking.
(3) The cells are collected by centrifugation in a sterile conical tube for 10 min at 4000 g and 4 °C.
(4) The supernatant is decanted and the cell pellet is carefully resuspended in sterile ice-cold 10 mM $MgSO_4$ solution to obtain an OD_{600} of 1.0. It is important not to vortex.
(5) Cells can be stored on ice till used or at 4 °C for 4 days without loss of viability.

43.2.2.2 Plating of the cDNA Phage Expression Library

(1) Bacterial cells are infected by mixing 100 μl of above $MgSO_4$-treated bacterial suspension with 10 μl bacteriophage [about 5×10^3 plaque-forming units

(p.f.u.)] in sterile glass tubes and incubated for 30 min at 37 °C to allow the bacteriophage to adsorb to the cells.

(2) For each plate, 2.5 ml of molten top agar (sterile NZYM medium [17]: 1% NZ amine [casein hydrolysate], 0.5% NaCl, 0.5% bacto-yeast extract and 0.2% $MgSO_4$, containing 0.7% electrophoresis grade agarose) is added to one tube at a time, shortly vortexed and immediately poured onto 90-mm bottom agar Petri dish (sterile NZYM medium [17]: 1% NZ amine, 0.5% NaCl, 0.5% bacto-yeast extract, and 0.2% $MgSO_4$ and 1.2% agar).

(3) Swirling of the plates quickly after pouring the top agar will help in the uniform distribution of the bacterial lawn. The melted top agar should be cooled to 50 °C before use. Higher temperatures will kill the host bacteria.

(4) The top agar is allowed to harden at room temperature for 10 min and then the plates are incubated at 42 °C. For the first 30 min the plates are incubated upside down with open lids to avoid humidity condensation, and then incubation is continued for 4–6 h with closed lids until pinpoint plaques appear. It is important not to pile the plates in the incubator in order to ensure a homogeneous temperature among the plates and hence a simultaneous appearance of the plaques in all plates.

43.2.2.3 Adsorbing Recombinant Proteins to Nitrocellulose Membranes

The type of membrane that is used during the assay may influence the signal-to-background ratio. In our hands, the best results concerning sharpness of the signal were obtained with nitrocellulose membrane filter [Immobilon-NC (HAHY); Millipore, Bedford, USA]. Similar results can be obtained with Hybond C membrane (Amersham, Buckinghamshire, UK), although the handling is more difficult as this membrane is more rigid. Another important issue is the size of the membrane; large membranes produce a non-homogeneous background and might prevent the detection of a positive signal. Small membranes (82 mm diameter) are therefore recommended. They generate a homogeneous background allowing the detection of positive signals even at the membrane periphery and, most importantly, the appearance of false positives is negligible. Moreover, the filters are also easier to handle.

(1) While the plates are incubating, the required number of circular nitrocellulose membranes is soaked in a filter-sterilized solution of 20 mM isopropyl-β-thiogalactopyranoside (IPTG) to impregnate them; 50 ml IPTG solution is used to submerge no more than 12 membranes.

(2) The membrane filters are transferred onto Whatman paper 3MM and labeled with a ballpoint pen or a lead pencil on the side of the membrane that will not touch the surface of the plates so that it will be possible to identify the appropriate plates after screening. It is important that the labeling is done quickly so that membranes are slightly damp (not dried or dripping wet) when applied to the surface of the plates.

(3) The IPTG-impregnated nitrocellulose filter circles are carefully overlaid onto the agar plates containing pinpoint plaques with the help of sterile forceps.

The filters should not be removed or adjusted when contacting lawn surface; similarly, air bubbles should be avoided.

(4) The plates with the filters are incubated upside down at 37 °C for 12–16 h. Fusion recombinant proteins will be induced and adsorbed to the nitrocellulose filter upon cell lysis.

43.2.2.4 Incubation with an RNA Ligand

Membranes have to be labeled prior to removal in order to denote their orientation relative to the Petri dish.

(1) At least three asymmetric locations close to the edges are marked by puncturing the membrane filters and the agarose under it with a needle previously imbibed in black ink.

(2) The membrane filters are removed from the plate very carefully avoiding picking up any top agar along with the nitrocellulose filters. If it is not possible to remove the filter cleanly, the plates should be refrigerated for 2 h at 4 °C to chill. This usually prevents top agar from sticking to the nitrocellulose filter.

(3) After removal, the filters are air-dried on Whatman 3MM paper for 5 min with the surface that has been in contact with the phage lawn up and transferred in the same orientation to a crystallizing dish (120 mm diameter, 60 mm high) with about 100 ml of SB cool buffer. The dish should be baked at 200 °C overnight before use to destroy any remaining ribonuclease activity. In case that more than one membrane will be incubated with the same probe, it is important to submerge one membrane at a time in the solution to ensure even contact of all the filters with the solution. All the subsequent steps are carried out at 4 °C; this low temperature will prevent RNA degradation due to trace amounts of ribonucleases. The Petri plates are stored at 4 °C for later use. Utilization of more than 14 membranes per dish is not recommended.

(4) The dish is placed on the platform of a gyratory shaker at 4 °C and the membrane filters are washed four times for 5 min immersed in 100 ml SB buffer in order to remove bacterial debris and material loosely bound to the membranes.

(5) After quantitatively decanting the SB buffer, the membranes are pre-incubated in 50 ml SB buffer supplemented with 100 µl *Torula* yeast blocking RNA (50 mg/ml) to mask the non-specific binding sites. Special precaution must be taken to avoid bubbles and to ensure uniform contact of all the membranes with the solution.

(6) After the addition of 1.4×10^7 c.p.m. of the ^{32}P-labeled RNA probe the incubation is continued for 120 min with gentle agitation on a platform shaker. By this plaque-lift procedure, not only recombinant proteins become transferred to the nitrocellulose membranes, but also proteins originating from lysed *E. coli* cells (see Sections 43.3.2 and 43.3.3). Most likely, these are proteins that bind RNA in a sequence-unspecific fashion, thus creating a general background. The inclusion of carrier RNA during pre-incubation and incubation steps avoids that protein interacting with RNA in a sequence-unspecific man-

ner associate with the radioactively labeled RNA ligand. Thus, the background mentioned is reduced.

43.2.2.5 Washing of Membranes

It is critical to monitor the integrity of the RNA probe during the binding step. This can be done by analysis of a 200-μl aliquot of the binding mixture in a vertical 5% polyacrylamide (30:1 acrylamide/bisacrylamide) slab gel, cast between two glass plates. Electrophoresis of the gel, which contains urea at a final concentration of 8 M, is carried out in TBE buffer (50 mM Tris base, 50 mM boric acid and 1 mM EDTA, pH 8.3) at a constant voltage of 15 V/cm. The gel is pre-run for warming up to 45 °C. The samples have to be denatured prior loading by heating at 95 °C for 5 min and immediately cooled on ice. The aliquot must be taken after the probe is incubated with the membranes for the prescribed time. Extensive degradation of the probe will reduce signal strength and result in failure of the assay (see Section 43.3.1).

(1) Non-specifically bound radioactivity is removed by washing the membranes four times for 5 min each in 100 ml SB buffer. The background levels of radioactivity can be monitored using a handheld Geiger counter.
(2) After quantitatively decanting the SB buffer, the membranes filters are transferred to Whatman 3MM paper and allowed to air-dry completely.
(3) Once the membranes have been thoroughly dried, they are wrapped in plastic foil and exposed to X-ray film (Kodak X-Omat AR; Amersham Biosciences, Buckinghamshire, UK) at −70 °C with the support of a Dupont Lightning-Plus intensifying screen. Asymmetric marks from the membrane filter may be transferred to the X-ray film to facilitate alignment of the film to the filters for positive clone localization. The exposure times will vary with the concentration of the RNA probe used and the strength of the specific signal. Generally, overnight exposures should be sufficient to detect positive signals.
(4) The film is developed and aligned with the membrane circles to locate signals representing the position of potentially positive plaques on the stored plates.

43.2.2.6 Identification of True Positives

To confirm the identity of a selected phage plaque as a true positive, clones have to be picked and plaque-purified to homogeneity by a secondary screening round (Figs 43.2 and 43.3). The uniformity of purified plaques is confirmed in a tertiary screening.

(1) The broad end of a sterilized Pasteur pipette is used to core the region around the selected plaque from the plate.
(2) The agar plug is transferred to a micro-centrifuge tube containing 1 ml of SM buffer to elute the phage overnight at 4 °C.
(3) Two drops of chloroform are added to inhibit bacterial growth.

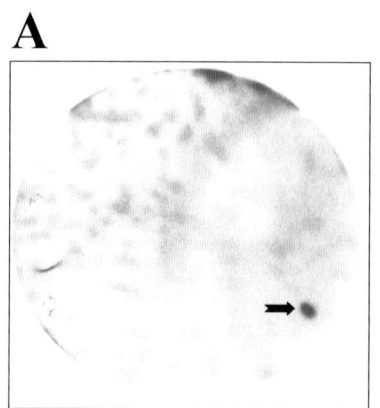

Fig. 43.2. Detection of PSTVd RNA-binding protein Virp1. A tomato cDNA expression library was plated at about 5000 p.f.u./plate and subjected to a primary screening, using a longer-than-unit-length PSTVd RNA transcript as a radioactively labeled probe [16]. (A) The black arrow shows the signal corresponding to λVirp1 clone after a primary screening. (B) The clone picked from the primary screening was plated at a density of 25 p.f.u./plate and PSTVd RNA-binding properties of λVirp1 clone were confirmed by secondary screening. Plaque lift and binding assays with ^{32}P-labeled RNA transcripts were performed as described in the text. Reprinted from [16], with permission.

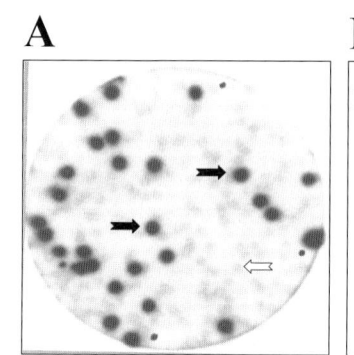

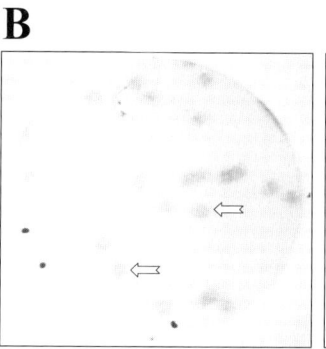

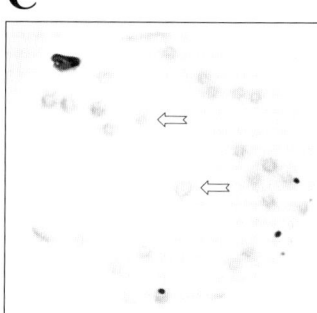

Fig. 43.3. Binding specificity of Virp1 and PSTVd RNA. A mixture (1:1) of λVirp1 and λZAPII phage was plated out and tested for binding with different RNAs. Only when (A) the PSTVd RNA form was used as RNA probe, could positive signals be discriminated from those due to background. When the same mixture was plated at a lower density of 25 p.f.u./plate and allowed to interact with either (B) potato U1 RNA or (C) human U1 RNA, only background signals were visible. In (B) and (C) the number of signals visible on the autoradiograph was the same as the number of plaques per plate. The black arrows show the signals corresponding to λVirp1 plaques and PSTVd interaction (positive signals) while empty arrows show the background signals. Reprinted from [16], with permission.

(4) After overnight elution of the phage particles from the agar plug, the new phage stock is plated to obtain up to 200 plaques per NZYM agar plate.
(5) The screening procedure is repeated following the protocols described above. True positive clones should contain an amplified number of positives on these secondary filters.
(6) A well-isolated positive plaque from the secondary screen is picked and if necessary, the screening procedure is repeated a third time with approximately 50 plaques per plate to obtain a well-isolated phage plaque.

43.2.3
Protocol 2: Northwestern Techniques to Detect and Analyze RNA–Protein Interactions

As outlined in Section 43.1, Northwestern analysis can be used for the identification of immobilized proteins using RNA probes and also as a tool for protein characterization, for example the determination of its approximate molecular weight.

43.2.3.1 Protein Sample Preparation

Biological Samples
Samples have to be immediately frozen in liquid nitrogen after collection, and turned to powder with a mortar and pestle.

(1) The total soluble protein is extracted in 3–4 ml of extraction buffer (100 mM Tris–HCl, pH 8.0, 200 mM NaCl, 1 mM EDTA, 3 mM $MgCl_2$, 1 mM DTT, 0.5 mM, 10 µM leupeptin, 10 µM aprotinin, 10 µM pepstatin and 10% glycerol) per gram of wet tissue.
(2) The mixture is vigorously shaken by vortexing for 1–2 min and the sample is placed on ice for 30 min.
(3) The mixture is centrifuged at full speed (approximately 10 000 g) for 30 min in a microcentrifuge at 4 °C.
(4) After centrifugation of the homogenate, the supernatant has to be carefully withdrawn and dispensed in 200-µl aliquots that are stored immediately at −80 °C.

Samples from E. coli
Northwestern technology can also be used for the delineation of an RNA–protein interaction once the protein has been cloned (Figs 43.4 and 43.5), in the same way that, for instance, the electrophoretic mobility shift assay is used, but without the need of protein purification, thus, overcoming a problem that arises in some cases, where the protein purification is a bottleneck. The specificity of the interaction (detected by this or other method) can be ascertained by using different unlabeled

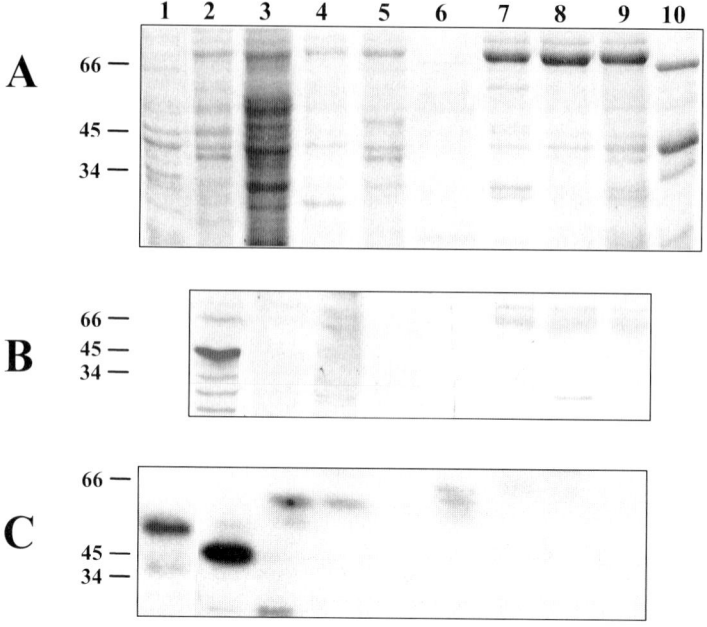

Fig. 43.4. Detection of RNA-binding protein by Northwestern blot assay in animal tissues. (A) Total extracts from various tissues (lane 2, ovary; lane 3, liver; lane 4, testes; lane 5, stomach; lane 6, intestine; lane 7, heart; lane 8, lung; lane 9, kidney) of the newt *Triturus cristatus carnifex* (corresponding to about 2 mg of wet tissue) were separated by electrophoresis on a 12% SDS–polyacrylamide gel, which was Coomassie stained. Total protein extract from *E. coli* expressing the recombinant NORA protein [14] was loaded as a positive control (lane 1). Lane 10: protein size marker. Molecular weights of markers in kDa are shown on the left. (B) A duplicate gel was run in parallel (lanes 1 and 10 omitted), transferred to nitrocellulose membrane and subjected to immunoblot analysis with a NORA-specific rabbit polyclonal antibody [14]. (C) A third identical gel was transferred to nitrocellulose membrane and probed with ^{32}P-labeled Sat2 transcript in SB buffer containing 0.1 mg/ml yeast RNA. Only in ovarian extracts could a strong signal be detected. The size of the signal was identical regardless whether the antibody or the RNA probe was used. Reprinted from [14], with permission.

competitor RNAs. Also mutant RNAs or proteins, as well as subfragments of both the protein and the RNA moiety, can be readily tested [18].

(1) 1 ml of recombinant *E. coli* cells harboring the desired protein is centrifuged at full speed in a microcentrifuge for 2 min.
(2) The supernatant is removed and discarded using a vacuum pump or an equivalent system to remove all traces of medium.
(3) The bacterial pellets are stored at −80 °C or resuspended in 100 µl extraction buffer (100 mM Tris–HCl, pH 8.0, 200 mM NaCl, 1 mM EDTA, 3 mM $MgCl_2$, 1 mM DTT, 0.5 mM, 10 µM leupeptin, 10 µM aprotinin, 10 µM pepstatin and 10% glycerol) by vortexing and stored at −80 °C.

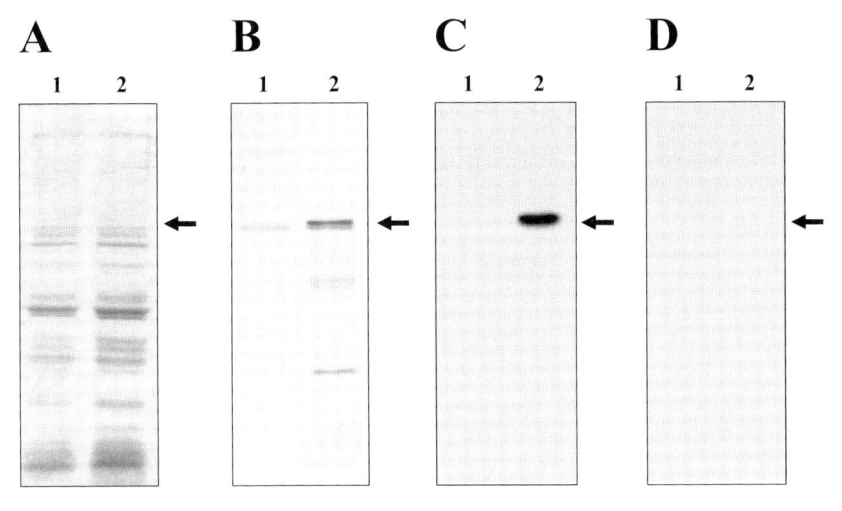

Fig. 43.5. Detection of RNA-binding protein by Northwestern blot assay in bacterial extracts. Crude cell extracts from *E. coli* (lane 1) or from *E. coli* expressing Virp1 protein [16] (lane 2) were separated on 10% SDS–PAGE gels, transferred to nitrocellulose membranes and stained (A) with Coomassie brilliant blue. (B) A duplicate gel was run in parallel, blotted to nitrocellulose membrane and detected by the Virp1-specific rabbit polyclonal antibody [16]. (C) A third identical gel was transferred to nitrocellulose membrane and probed with ^{32}P-labeled PSTVd (+) RNA in SB buffer containing 0.1 mg/ml yeast RNA. (D) A fourth gel, identical to the other three, was transferred to nitrocellulose membrane and probed with ^{32}P-labeled Human U1 (+) RNA in SB buffer containing 0.1 mg/ml yeast RNA. The black arrow indicates the Virp1 protein.

43.2.3.2 Protein Electrophoresis and Transfer

(1) 10 µl of cell extract is mixed with 10 µl loading buffer (100 mM Tris–HCl, pH 6.8, 200 mM DTT, 4% SDS, 0.1% bromophenol blue and 20% glycerol) and heated for 10 min at 95 °C prior to loading. Protein rainbow marker can be loaded along with the samples as a size reference.
(2) Protein samples are then subjected to SDS–PAGE following standard conditions. The acrylamide/bisacrylamide composition and percentage in the gel depends on the expected size of the proteins.
(3) Proteins are then electroblotted to nitrocellulose membrane (Protran Nitrocellulose; Schleicher & Schuell BioScience, Dassel, Germany). Care should be taken so that the membrane is labeled for blot orientation. A sandwich for the electrophoresis transfer cell is made as follows: face black, sponge, two layers of Whatman paper 3MM, gel, nitrocellulose membrane, two layers of Whatman paper 3MM, sponge, face red. Care should be taken so that the electrode of the correct polarity is applied, black (negative) to red (positive).
(4) The electroblotting is performed in a Bio-Rad mini trans-blot electrophoresis transfer cell (Bio-Rad, Hercules, USA) or equivalent system at either 30 V for

16 h or 100 V for 1 h. The transfer is done at 4 °C in Tris/glycine/methanol buffer (25 mM Tris, 200 mM glycine and 20% methanol). Note, that there is no SDS added to the transfer buffer; the SDS that is bound to the proteins during electrophoresis is sufficient for providing a negative charge during the electroblotting. Moreover, we have observed that further inclusion of SDS, notably increases the background in the Northwestern assay. The buffer is mixed continuously by incorporating a stir-bar in the transfer cell and placing the apparatus on a magnetic stirrer. The whole unit is placed in a refrigerator or in a cold room during the protein transfer.

(5) The efficiency of protein transfer to the membrane can be monitored after disassembling the sandwich by staining the gel with Coomassie blue R250. Alternatively, and more conveniently, staining of the nitrocellulose membrane is done with 1% Ponceau S solution for 5 min at room temperature. Washing the membrane for 2 min in de-ionized water will visualize the transferred proteins. Apart from monitoring the transfer efficiency, staining the membrane will help to locate and mark at this point the proteins of the size marker or any other transferred protein with a lead pencil. The stain can be completely removed by washing for longer times in deionizer water. Use of clean gloves and sterile forceps for membrane handling is recommended.

43.2.3.3 Incubation of the Membranes with an RNA Probe

Following transfer, ensure that no remainders of the acrylamide gel stick on the nitrocellulose blot. If that is the case, the acrylamide has to be carefully cleaned off to avoid later background problems (see Section 43.3.3). Then, place the membrane (protein face up) on a crystallizing dish (120 mm diameter, 60 mm high) containing about 100 ml of SB cool buffer. Care should be taken to submerge one membrane at a time in the solution to ensure regular and uniform contact of all the filters with the solution.

(1) The nitrocellulose membranes containing the transferred proteins are washed two times with an excess of SB buffer.
(2) After quantitatively decanting the SB buffer, the membranes are pre-incubated in 50 ml SB buffer supplemented with 100 µl *Torula* yeast blocking RNA (50 mg/ml), to mask the non-specific binding sites. As mentioned above (see Section 43.2.4.4), special precaution must be taken to avoid bubbles and to ensure uniform contact of all the membranes with the solution.
(3) After the addition of about 1.4×10^7 c.p.m. of the ^{32}P-labeled RNA probe, the incubation continues for 120 min under gentle agitation on a platform shaker.

43.2.3.4 Washing of Membranes and Autoradiography

(1) Once an aliquot has been collected for a qualitative analysis, the membranes are washed in order to remove unbound RNA probe. The washes are done in the same crystallizing dish for 15 min at 4 °C, with sufficient SB buffer to cover the membrane and with gentle agitation on a platform shaker. This step

is repeated a further 2 times till the background levels of radioactivity are minimal; this can be followed using a handheld Geiger counter.
(2) The membranes are then briefly air-dried on 3MM Whatman paper, wrapped in Saran Wrap (Down Chemical Company, Midland, USA) and exposed to X-ray film (Kodak X-Omat AR; Amersham Biosciences, Buckinghamshire, UK) at −70 °C with the support of a Dupont Lightning-Plus intensifying screen.
(3) The film is developed and aligned with the membrane to locate the proteins from the molecular weight marker so that it will be possible to estimate the size of the RNA-binding proteins that are detected by autoradiography due to their capacity to interact with the RNA probe (Figs 43.4 and 43.5).

43.3
Troubleshooting

The Northwestern technique described here is potentially applicable to any RNA–protein interaction. This methodology is simple to use and provides a versatile and reliable tool for the identification of λ bacteriophage cDNA clones that express RNA-binding proteins as well as for the analysis of any RNA–protein interaction of interest. The time required for performing a successful Northwestern screening should be less than 2 weeks in the hands of an experienced investigator. However, there may be some instances in which identification and analysis of RNA–protein interactions is not achieved. Some likely explanations for such a failure and possible solutions are as follows.

43.3.1
Probe Quality

As mentioned before, successful detection of RNA–protein interactions by Northwestern techniques depends to a large extent on the integrity of the RNA probe. Intactness of the input RNA probe as well as of the RNA probe after incubation with the membrane has to be checked on a denaturing polyacrylamide gel. Degradation of the RNA probe will reduce the specific signal strength and, what is even more important, it will increase the number of false positives or result in failure of the assay. The typical outcome of a Northwestern experiment in which the RNA probe is extensively degraded is a uniformly dark film with white plaques or bands (a negative-like appearance, Fig. 43.6A). Similarly, when the probe is partially degraded the film will be dark (Fig. 43.6B) and, despite the fact that the signals are still visible, true positives can be more difficult to be detected under such conditions. Therefore, great care has to be taken to avoid any degradation of the RNA probe. The use of high-quality reagents and RNA-free water for all steps will help to overcome to a certain extent RNA probe degradation by spurious nucleases. However, if degradation of the RNA probe occurs during the Northwestern assay, all the solutions should be tested for the presence of ribonucleases and replaced by fresh ones if contaminated.

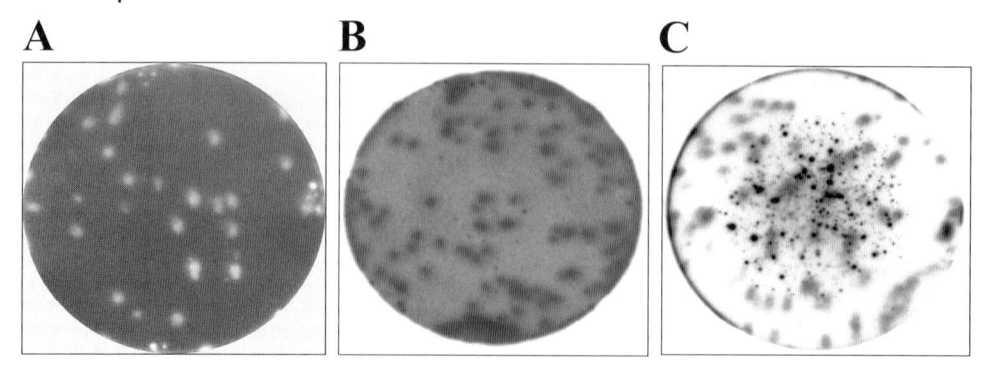

Fig. 43.6. Parameters influencing the background signal. A mixture (1:1) of λZ-NORA [14] and λZAPII phage was plated out. Incubation was done in SB buffer, but Sat2 RNA probes [14] from two independent transcript preparations were tested for binding. (A) Extensive degradation resulted in a very strong background signal and "white" plaques. (B) Slight degradation of the RNA probe gave an increased background. (C) Due to a large number of filters used for the Northwestern assay, non-specifically bound RNA was not properly washed off, leading to a typical spotty filter with low signal-to-background noise ratio that compromises the success of the assay.

43.3.2
Background Signals

During the Northwestern procedure, not only are recombinant proteins transferred to the nitrocellulose membranes, but also proteins originating from lysed *E. coli* cells (unless purified protein is used for the Northwestern analysis). Among these proteins there might be some that could bind RNA in a sequence-unspecific fashion, thus creating a general background and interfering with the objective to study RNA-binding proteins. The inclusion of blocking RNA during pre- and incubation steps avoids interaction of *E. coli* proteins with the radioactively labeled RNA ligand. If many false positives are observed in the screening or many transferred proteins bind nonspecifically to the probe in the Northwestern assay, it is necessary to check the integrity and the actual concentration of the blocking RNA by running it on an agarose gel.

43.3.3
Signal-to-Background Ratio

A good signal-to-background noise ratio is crucial for a successful application of this Northwestern protocol. Hence, a low ratio can significantly affect the strength of the signal and severely deteriorate the detection limit of the Northwestern assay. Various parameters could deteriorate the signal-to-background noise ratio apart from the RNA integrity previously mentioned (see Section 43.3.1). Care should be taken during incubation and washing steps. Nitrocellulose membranes have to be

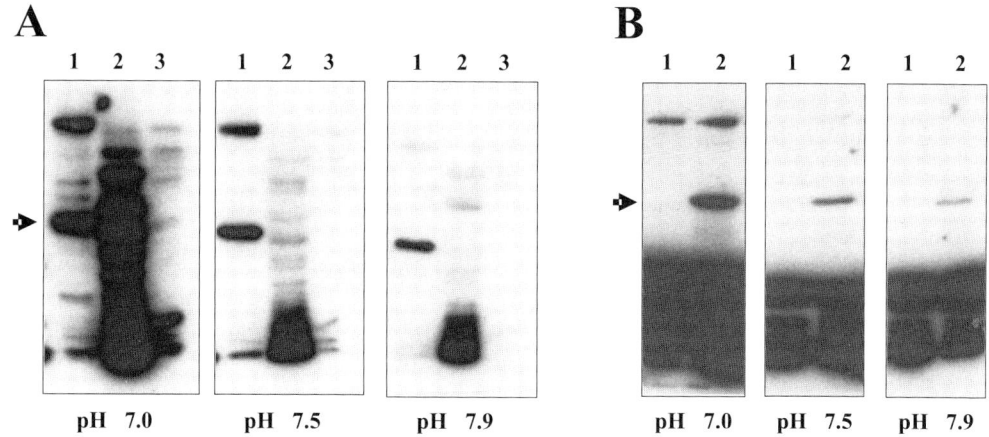

Fig. 43.7. Influence of the pH on the background signal. (A) Total protein extracts from various tissues (lane 1, ovary; lane 2, liver; lane 3, testes) of the newt *T. cristatus carnifex* (corresponding to about 2 mg of wet tissue) were separated by electrophoresis on three identical 10% SDS–polyacrylamide gels, transferred to nitrocellulose membranes and probed with ^{32}P-labeled Sat2 transcript [14] in SB buffer with three different pH values. (B) Crude cell extracts from *E. coli* (lane 1) or expressing NORA protein [14] (lane 2) were separated on three identical 10% SDS–PAGE gels, transferred to nitrocellulose membranes and probed with ^{32}P-labeled Sat2 transcript [14] in SB buffer with three different pH values. The black arrow indicates the NORA protein [14].

completely soaked in the solution and special precaution must be taken to ensure uniform contact of each and every membrane. In case that the membranes stick to each other during the incubation step the non-specifically bound RNA will not be washed off, leading to a low signal-to-background noise ratio that will drastically compromise the success of the Northwestern procedure (spotty film, Fig. 43.6C). The pH of the SB buffer is also a parameter that should be taken into consideration: pH adjustment is essential since it has been noticed that a pH of the SB buffer below 7.5 correlates with a considerable background increase, both for *E. coli* and tissue-extracted proteins (Fig. 43.7). The interaction between the protein of interest and its counterpart RNA will have an optimal pH, thus the SB buffer pH may affect it. However, we have observed that an increase in the pH, affects the specific RNA–protein interaction to a lower extent when compared to the non-specific interaction (Fig. 43.7). In our experience the use of 15 mM HEPES buffer, pH 7.9 gives the best signal-to-background ratio to allow detection of specific RNA–protein interactions. Similarly, the presence of transition metals, Mn^{2+} and Zn^{2+}, in the SB buffer greatly increases the signal-to-background ratio. The inclusion of transition metals apart from decreasing the non-specific interaction decreases also the specific interaction; thus, it might be convenient to optimize the concentration of transition metals when working with weak interactions.

43.3.4
Protein Conformation

The protein domains required for RNA interaction may undergo conformational alterations during the gel electrophoresis and/or when immobilized on nitrocellulose membranes, thus prohibiting RNA–protein interaction detection and analysis. In order to increase the signals obtained in the Northwestern analysis, the transferred proteins can be subjected to a denaturation/renaturation protocol [12]. Many proteins do not require the denaturation/renaturation procedure to be detected; however, in some instances, it can help to enhance the specific binding of the RNA probe to the membrane-bound proteins. The proteins have to be denatured for 10 min using 6 M guanidine hydrochloride in renaturation buffer (20 mM HEPES–KOH, pH 7.5, 25 mM NaCl and 1 mM DTT) and gradually renaturated by consecutive incubations for 10 min in large volumes of renaturation buffer containing 3, 1.5, 0.75, 0.375 and 0.187 M guanidine hydrochloride at room temperature with gentle shaking, and a final incubation in renaturation buffer. This procedure can also be performed using 8 M urea as denaturing agent [19].

43.3.5
Weak Binding Signals

The nature of the signal itself can be very weak. The identification of positive clones can be compromised when the strength of the signal is weak. One way to overcome this problem can be to use an RNA probe with a high specific activity that will help to rapidly detect the signal. Additionally, and more importantly, the plaque density per plate can be decreased, which will help to improve the strength of the signal.

43.3.6
False Positives

False positives can arise during the initial Northwestern screening. It is therefore essential to verify the identification of a cDNA clone that expresses an RNA-binding protein by a secondary and even a tertiary round of screening. However, a disproportionate number of false positives can lead to a laborious and discouraging effort. One reason for an unexpected high number of false positives could be the probe quality. In case of probe degradation, random sticking of the radiolabeled RNA to the nitrocellulose membrane will be the cause of such a problem. Once the probe has been checked and intactness confirmed, another possible explanation for a large false positive occurrence could be the presence of non-incorporated radionucleotides in the RNA probe. The RNA synthesized is separated from the unincorporated NTPs by gel-filtration through a pre-packed spin column containing Sephadex G-50 or a similar device. In some instances, e.g. when the reaction it is not carefully applied to the centre of the matrix, but to the sides of the column, non-incorporated NTPs can flow around, rather than through the matrix.

43.3.7
Quality of the cDNA Library

For Northwestern screening, the quality of the cDNA expression library is crucial for successful detection of RNA–protein interactions. Therefore, it is recommended to verify the titer of pre-made libraries before proceeding with the Northwestern screening. A titer of 10^8–10^9 p.f.u./ml is the recommended library concentration, which allows also the detection of rare cDNAs that are represented in one out of a million phages. Additionally, it is convenient to confirm the recombination frequency of the library; this can be achieved by applying the blue and white screening protocol provided by the supplier. The ratio of blue to colorless plaques is an indication of the proportion of the library that consists of non-recombinant bacteriophages. A good representative library should be in the range of 75–80% recombinant clones.

43.3.8
Fading Signals

Diffusion and fading of the signal has been observed when the filters are exposed to the X-ray film at room temperature, rather than at $-70\ °C$. This is probably due to the fact that the interaction between the labeled RNA and the protein is weak (non-covalent). Thus, it is recommended to always expose the filters at low temperatures (-40 to $-80\ °C$), even when an intensifying screen is not utilized.

43.3.9
Supplementary

Supplementary information concerning standard protocols, including details on working with λ phage, gel electrophoresis, protein transfer and autoradiography, which may be useful to complement the protocols provided in this chapter, can be found in [17].

References

1. I. A. LAIRD-OFFRINGA, J. G. BELASCO, Proc. Natl Acad. Sci. USA **1995**, 92, 11859–11863.
2. U. PUTZ, P. SKEHEL, D. KUHL, Nucleic Acids Res. **1996**, 24, 4838–4840.
3. D. J. SENGUPTA, B. ZHANG, B. KRAEMER, P. POCHART, S. FIELDS, M. WICKENS, Proc. Natl. Acad. Sci. USA **1996**, 93, 8496–8501.
4. R. TAN, A. D. FRANKEL, Proc. Natl. Acad. Sci. USA **1998**, 95, 4247–4252.
5. C. JAIN, J. G. BELASCO, Cell **1996**, 87, 115–125.
6. H. KOLLMUS, M. W. HENTZE, H. HAUSER, RNA **1996**, 2, 316–323.
7. S. WANG, H. L. TRUE, E. M. SEITZ, K. A. BENNETT, D. E. FOUTS, J. F. GARDNER, D. W. CELANDER, Nucleic Acids Res. **1997**, 25, 1649–1657.
8. E. PARASKEVA, A. ATZBERGER, M. W. HENTZE, Proc. Natl. Acad. Sci. USA **1998**, 95, 951–956.

9 K. Harada, S. S. Martin, A. D. Frankel, *Nature* **1996**, *380*, 175–179.
10 J. A. Daròs, R. Flores, *EMBO J.* **2002**, *21*, 749–759.
11 H. Singh, J. H. LeBowitz, A. S. Baldwin, P. A. Sharp, *Cell* **1988**, *52*, 415–423.
12 C. R. Vinson, K. L. LaMarco, P. F. Johnson, W. H. Landschulz, S. L. McKnight, *Genes Dev.* **1988**, *2*, 801–806.
13 R. Sägesser, E. Martínez, M. Tsagris, M. Tabler, *Nucleic Acids Res.* **1997**, *25*, 3816–3822.
14 M. A. Denti, A. E. Martínez de Alba, R. Sägesser, M. Tsagris, M. Tabler, *Nucleic Acids Res.* **2000**, *28*, 1045–1052.
15 M. Ta, S. Vrati, *J. Virol.* **2000**, *74*, 5108–5115.
16 A. E. Martínez de Alba, R. Sägesser, M. Tabler, M. Tsagris, *J. Virol.* **2003**, *77*, 9685–9694.
17 J. Sambrook, E. F. Fritsch, T. Maniatis, *Molecular Cloning: A Laboratory Manual*, Cold Spring Harbor Laboratory, Cold Spring Harbor Press, NY, **1989**.
18 M. Gozmanova, M. A. Denti, I. N. Minkov, M. Tsagris, M. Tabler, *Nucleic Acids Res.* **2003**, *31*, 5534–5543.
19 D. St. Johnson, N. H. Brown, J. G. Gall, M. Jantsch, *Proc. Natl. Acad. Sci. USA* **1992**, *89*, 10979–10983.

IV.2
RNA–Protein Interactions *in vivo*

44
Fluorescent Detection of Nascent Transcripts and RNA-binding Proteins in Cell Nuclei

Jennifer A. Geiger and Karla M. Neugebauer

44.1
Introduction

Fluorescent *in situ* hybridization (FISH) has become a powerful and widely used method for examining the distribution of specific RNAs in cells and tissues. Through hybridization of a fluorescently labeled DNA or RNA probe complementary to the desired RNA, one can broadly determine expression patterns in whole animals. At the subcellular level, one can examine the localization of particular RNAs to cellular compartments; such mRNA localization is a well-known mechanism for locally regulating gene expression. At the highest resolution, RNA FISH can pinpoint sites of gene transcription, because the highest concentration of specific RNA species in the cell is, generally speaking, at the site of synthesis. This has facilitated studies of nuclear organization and the regulation of nuclear activities, such as transcription itself, RNA processing and RNA transport. Thus, RNA FISH techniques are applicable to a number of biological questions, and each application carries with it special requirements of sensitivity and resolution.

While it is possible to determine the location of specific genes by hybridization of fluorescent probes to DNA targets, this technique does not address whether the gene is active or not. In addition, access to DNA sequences for hybridization requires denaturation of the sample and this often seriously compromises protein localization. *In situ* hybridization to RNA targets within the cell requires less harsh conditions. Moreover, the intensity of the RNA FISH signal is proportional to the RNA target synthesized at the site, offering the potential for quantitative studies of gene expression. For these reasons, visualization of nascent RNA transcripts within cell nuclei has become an important approach for addressing a variety of cell biological questions.

The ability to detect specific nascent RNAs at their sites of transcription by FISH has made possible experiments determining patterns of gene expression at the single-cell level. For example, entire sets of genes activated early in the G_1 phase of the cell cycle can be analyzed, using specific probes coupled to distinct fluorochromes [1]. Comparison of transcription levels of two alleles of the same gene has also been possible [2]. Beyond transcription, detection of pre-mRNAs and their spliced mRNA products at sites of transcription has helped to establish the co-transcriptional nature of pre-mRNA processing events, such as splicing [3–9]. Correlations between pre-mRNA processing events and subnuclear structures have also been the subject of intensive study. For example, transcriptionally active intron-containing genes frequently cluster around nuclear speckles, which contain high concentrations of pre-mRNA splicing factors [4, 6–8, 10–12], although counter-examples have also been reported [5]. Recently, detection of (pre)-mRNAs at sites of transcription and/or within nuclear speckles upon perturbation of RNA synthesis, RNA processing (e.g. splicing and polyadenylation) and nuclear export has provided evidence for novel pathways of mRNA surveillance [11, 13–15].

The utility of FISH as an approach to these questions depends (1) on the assay being sensitive enough to detect nascent RNAs of choice, (2) on the FISH conditions being compatible with immunodetection of proteins and (3) on the imaging resolution being sufficiently high to distinguish spatially between different subnuclear structures. Regarding sensitivity, it can be difficult to access RNA molecules within fixed cells, e.g. some protocols employ protease digestion to increase accessibility of the probes for RNA targets [3] and this treatment is clearly incompatible with protein localization studies. Other strategies include use of high temperatures and/or high concentrations of formamide, which can cause redistribution of relevant protein epitopes. Some protocols achieve higher signal-to-noise levels at the transcription site by pre-permeabilization of cells before fixation; in this scenario, RNAs not tethered to chromatin must partially diffuse away, decreasing background but also resulting in the potential loss of relatively soluble proteins [1]. In this chapter, we focus on methods that permit simultaneous detection of transcription sites by FISH and protein localization by immunocytochemistry (see Fig. 44.1).

44.2
Description of the Methods

44.2.1
Overview

Methods described in this chapter are designed to achieve two goals: (1) localization of sites of transcription within nuclei and (2) immunodetection of nuclear proteins, both under conditions that preserve protein localization in cells. Identification of conditions that satisfy these goals involved empirically determining the effects of time, temperature and formamide concentration on hybridization, as well as alternative methods in probe preparation. Two methods of probe preparation have been useful in our lab. In the first, nick-translation of plasmids containing relevant cDNA inserts followed by limited DNase I digestion produces small (around 100–400 bp) fragments into which nucleotides conjugated to biotin or digoxygenin have been incorporated. This is a standard method (presented here as a protocol that can be easily followed and trouble-shot), which offers the advantage

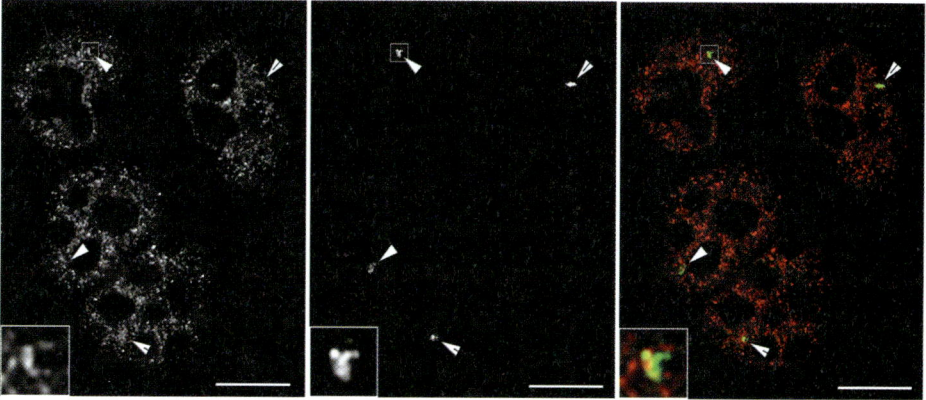

Fig. 44.1. Co-localization of the active rat *pem* gene locus with the splicing factor SF2 in Pem-HeLa cells [9]. (Left panel) Monoclonal antibody AK103 (anti-SF2) was diluted 1:12 in protein block and subsequently detected with TRITC-conjugated goat anti-mouse IgG (Jackson ImmunoChemicals) diluted 1:100 in Protein Block. (Center panel) Biotinylated DNA probe was generated by PCR of pem intron 3 (2.4-kb product) followed by DNase I digestion and incubated for 2 h in hybridization mix containing 20% formamide with the sample prepared as described. (Right panel) Merge of immunostaining (red) and FISH (green). Images were collected in z-stacks with the DeltaVision microscope system and deconvolved, and projections of nuclear reconstructions are shown. Under these conditions, the subnuclear distribution of SF2 in tiny dots throughout the nucleoplasm is indistinguishable from the staining pattern obtained without FISH. Alternative FISH protocols cause the SF2 localization pattern to be more "speckled" and less dot-like [9]. Arrowheads indicate detection of the integrated pem locus by FISH (one per cell). Scale bar: 10 μm; inset magnified ×4.

that overlapping hybridizing species, including fragments of the plasmid backbone itself, can help to amplify the signal. The second method, developed in our lab, is a derivative of the nick-translation protocol, in which large regions of the cloned DNA of interest are amplified by PCR. Nucleotide analogs are incorporated during the PCR amplification, and the PCR product is subjected to limited DNase I digestion to produce the appropriately sized probes. With this method, one can produce an intron-specific probe(s) from a cloned gene, by designing the appropriate primers. As pre-mRNA is most highly concentrated at the site of synthesis and introns are drastically lower in abundance throughout the nucleus and cytoplasm, use of these intron probes can significantly reduce signal away from the site of transcription without the disadvantages inherent in pre-permeabilization.

44.2.2
Preparation of Fluorescent DNA Probes for *In Situ* Hybridization

44.2.2.1 Method 1: Nick Translation of Plasmid DNA
In nick translation, single-stranded cuts are made in the double-stranded plasmid, providing entry points for the polymerase. The frequency of the cuts within the

plasmid will determine the average size of the probe – under-digestion will produce fragments that are too big (no signal or high background) to access the targets and over-digestion will destroy the probe (no signal). Therefore, it is critically important to titrate the DNase I used in the reaction. Note that the efficiency of DNase I, even among batches from the same vendor, can vary greatly.

Nick translation reaction mix
1 × nick translation buffer (NTB)[1]
dNTPs (0.05 mM of each dATP, dCTP, dGTP and 0.01 mM dTTP)
0.05 mM biotin-conjugated dUTP (Roche)
10 mM DTT (Invitrogen)
Plasmid DNA (1 µg)
15 U DNA polymerase I (Promega)
ddH$_2$O to 50 µl final volume
DNase I (Invitrogen, 92 U/µl)[2]
[1]10 × NTB is 0.5 M Tris–HCl, pH 7.8, 50 mM MgCl$_2$ and 0.5 mg/ml BSA (nuclease-free).
[2]Vary amounts between 0.1 and 10 U per reaction.

Combine reaction contents on ice, *adding the DNase I last*. Mix well by tapping and inverting the tube, briefly spin down and incubate at 15 °C for 2 h. Remove 5 µl for gel analysis and transfer remaining nick translation reactions directly to −20 °C.

Boil 5 µl of each reaction for 10 min at 95 °C, snap-cool and run on a 2% agarose gel to check the sizes. Ideally, you will have a smear ranging from 100 to 400 bp with the maximum intensity centered around 200 bp. Select the DNase I reaction(s) that best fit this size distribution and purify the probe on G-50 Sephadex (we use mini Quick Spin DNA Columns from Roche Applied Sciences with column buffer 1 × STE: 10 mM Tris–HCl, pH 8.0, 100 mM NaCl and 1 mM EDTA) to remove unincorporated nucleotides. If you pick more than one reaction, can you pool them and then purify (each column can hold a total of 75 µl). It is also possible to pool probes after purification. After the column, if you added 45 µl, you will have around 50 µl of probe solution. Store at −20 °C and keep on ice, as DNase may not have been completely inactivated by the column buffer. Note that boiling of the probe before the *in situ* procedure ensures inactivation of DNase I at this later step.

44.2.2.2 Method 2: PCR Amplification and DNase I Digestion

(1) PCR conditions should be verified for each different template and primer set. We typically use 3.5 min extension time (for a 2.5 kb product) and 30 cycles of amplification.

50 µl PCR reaction mixture
1 × Taq buffer
5 µg nuclease-free BSA

3 mM MgCl$_2$
0.2 mM (dATP, dGTP, dCTP)
0.15 mM dTTP
0.05 mM biotinylated dUTP (Roche)
0.06 µg plasmid template
150 ng of each primer
ddH$_2$O to 50 µl final volume
500 U *Taq* polymerase (Promega)

(2) DNase I titration. The purpose of DNase digestion of the PCR-generated probe is to produce fragments capable of penetrating the cell and gaining access to target sequences. This step must be done carefully (please see above). First, dilute the DNase I (fresh each time) 1:20 000 in ddH$_2$O for the following reaction:

50 µl DNase I reaction
10 µl PCR product (from Step 1)
5 µl 10 × NTB (see above)
ddH$_2$O to 50 µl final volume
0–10 µl diluted DNase I (Invitrogen, 92 U/µl)

Incubate the reaction for 2 h at 15 °C, freeze immediately at −20 °C following incubation. Follow the instructions from Section 44.2.2.1 above for analysis of DNA fragment sizes, purification and storage.

44.2.3
Performing Combined Immunocytochemistry and FISH

Start with tissue culture cells grown directly on glass coverslips in the appropriate medium to around 50% confluency.

(1) Fixation and immunostaining. Aspirate medium and rinse cells once in PBS. Fix at room temperature for 10 min with 4% paraformaldehyde (PFA) in 0.1 M PIPES (pH 6.9), 2 mM MgCl$_2$ and 1.25 mM EGTA. Remove PFA and rinse coverslips 3× in PBS. Permeabilize the cells for 10 min at room temperature (0.2% Triton X-100, MgPBS: PBS plus 10 mM MgCl$_2$), followed by 3× rinse in MgPBS. After permeabilization and washing, block the sample for 10 min in 5% normal goat serum (NGS) in MgPBS. For incubations, invert coverslips onto a 50-µl droplet of solution spotted onto Parafilm; always cover with a humid chamber to prevent coverslips from drying out. Wick off NGS (or any subsequent buffer) by tilting the coverslip onto Kimwipe tissue such that the edge of the coverslip (not the cells!) contacts the tissue. Add primary antibody diluted in Protein Block (3% BSA in PBS and 0.1 mM β-glycerophosphate if you are working with phospho-epitopes) and incubate 1 h at room temperature. Wash off the primary antibody by dipping coverslips 3× in PBS, followed by blocking 5 min in Protein Block. Next incubate in secondary antibody for

30 min. To wash, again dip 3× in MgPBS and block for 10 min in Protein Block. This blocking step is crucial to remove unspecific binding of the secondary antibody.

(2) Fix the immunostaining. Once the cells have been sufficiently washed and blocked after staining, fix the antibodies for 5 min in 4% PFA followed by 3× rinse in MgPBS. This post-fixing step is the most crucial step in preserving the antibody-staining pattern during the following *in situ* hybridization procedure.

(3) Prepare cells for hybridization. On a rocker, wash the coverslips for 10 min in MgPBS and 2 × 10 min in 2 × SSC. This can be done by placing the coverslips in a tissue culture dish (e.g. one coverslip per well in a six-well plate).

(4) *In situ* hybridization to RNA. Precipitate the labeled DNA probe in the following manner for use in the RNA *in situ* hybridization. In a microfuge tube, mix:

1–5 µl probe solution (see above)
2 µl tRNA (10 mg/ml)
ddH$_2$O to 50 µl final volume
5 µl 3 M NaAc pH 5.2
100 µl −20 °C EtOH

Vortex mixture for 10 s, place immediately in −80 °C freezer for at least 30 min. Centrifuge at 10 000 g or above, 4 °C for 15 min. Carefully remove supernatant without dislodging the DNA pellet. Wash the pellet in 500 µl 70% EtOH, centrifuge 2 min at 4 °C. Carefully remove supernatant. Dry the pellet in a Speed Vac for 15 min at room temperature. Add 10 µl of 40–100% deionized formamide (Ambion). Denature the probe at 85 °C for 10 min, snap-cool and keep on ice until ready to use. During this time, prepare the hybridization master mix: 4 × SSC, 2% BSA, 20% dextran sulfate (heat to 37 °C, vortexing a few times before use). Add 10 µl of the master mix to the microfuge tube containing the probe/formamide mixture. Pipette up and down to mix, avoiding bubbles. Spot the 20 µl mixture on a clean glass slide, invert the coverslip carefully onto the spot, avoiding bubbles. Place the glass slide in a sealed humid chamber and incubate 2–18 h at 42 °C.

(5) Post-hybridization washes and probe detection. Float the coverslip off of the slide with 2 × SSC and return it to the six-well plate. Rinse 1× with 2 × SSC, wash for 30 min in 2 × SSC, 20–50% formamide (the same final concentration used in the hybridization reaction) at 37 °C, followed by 30 min 2 × SSC and 30 min 1 × SSC washes, both with rocking at room temperature. Do not skimp on the washes – they are absolutely necessary to remove unspecific background! Incubate the cells for 5 min in FISH blocking solution (2 × SSC, 0.25% BSA). Add FITC-conjugated streptavidin (Jackson ImmunoLaboratories) diluted 1:500 in FISH blocking solution, incubate 30 min at room temperature. Dip coverslips 3× in 2 × SSC; incubate again in blocking solution for 5 min. Transfer coverslips back into the six-well plate for the final washes on a rocker; 3 × 10 min (4 × SSC, 0.1% Triton X-100), 3 × 10 min 4 × SSC. Dip coverslips 1× in ddH$_2$O, wick off moisture and mount. Seal with nail polish and view with fluorescent microscope. Figure 44.1 shows the results of an *in situ* hybrid-

ization experiment in which the subnuclear immunostaining pattern for a splicing factor is preserved.

44.2.4
Troubleshooting

We have found that several parameters affect the quality of both the immunostaining and *in situ* hybridization, and these can be varied. First, immunostaining is unpredictably (and sometimes drastically) affected by the formamide concentration. Use the minimum concentration necessary to get a reasonable hybridization efficiency and good signal-to-noise ratio. We use 20% formamide, but this will not work for all applications. Keeping hybridization times short is also a crucial factor in preserving protein staining but can reduce hybridization efficiency. Use the shortest hybridization time possible in combination with the lowest percentage of formamide possible. These parameters must be tested empirically for each cell type, antigen of interest and target RNA.

Problems with the *in situ* hybridization itself (even at high formamide concentrations and long hybridization times) can often be resolved by re-examining the size of the probe by agarose electrophoresis (see above). Some protocols recommend titrating the DNase by digesting the plasmid (not the nick translation reaction) in NTB. We have found that this leads to insufficient digestion of the probe, producing high background and no signal.

Acknowledgments

We thank Tom Misteli for helpful advice on *in situ* hybridization when we began these studies. This work was supported by National Science Foundation grant (MCB-9806046), by a Research Project Grant (RPG-00-110-01-MGO) from the American Cancer Society and by the Max-Planck-Gesellschaft.

References

1 J. M. LEVSKY, S. M. SHENOY, R. C. PEZO, R. H. SINGER, *Science* **2002**, *297*, 836–840.
2 M. WIJGERDE, F. GROSVELD, P. FRASER, *Nature* **1995**, *377*, 209–213.
3 R. W. DIRKS, F. M. VAN DE RIJKE, S. FUJISHITA, M. VAN DER PLOEG, A. K. RAAP, *J. Cell Sci.* **1993**, *104*, 1187–1197.
4 L. F. JIMENEZ-GARCIA, D. L. SPECTOR, *Cell* **1993**, *73*, 47–59.
5 G. ZHANG, K. L. TANEJA, R. H. SINGER, M. R. GREEN, *Nature* **1994**, *372*, 809–812.
6 S. HUANG, D. L. SPECTOR, *J. Cell Biol.* **1996**, *133*, 719–732.
7 Y. XING, C. V. JOHNSON, P. T. MOEN, JR, J. A. MCNEIL, J. LAWRENCE, *J. Cell Biol.* **1995**, *131*, 1635–1647.
8 Y. XING, C. V. JOHNSON, P. R. DOBNER, J. B. LAWRENCE, *Science* **1993**, *259*, 1326–1330.
9 T. MISTELI, D. L. SPECTOR, *Mol. Cell* **1999**, *3*, 697–705.
10 K. P. SMITH, P. T. MOEN, K. L. WYDNER, J. R. COLEMAN, J. B. LAWRENCE, *J. Cell Biol.* **1999**, *144*, 617–629.

11 L. S. SHOPLAND, C. V. JOHNSON, J. B. LAWRENCE, *J. Struct. Biol.* **2002**, *140*, 131–139.
12 L. S. SHOPLAND, C. V. JOHNSON, M. BYRON, J. MCNEIL, J. B. LAWRENCE, *J. Cell Biol.* **2003**, *162*, 981–990.
13 N. CUSTODIO, M. CARMO-FONSECA, F. GERAGHTY, H. S. PEREIRA, F. GROSVELD, M. ANTONIOU, *EMBO J.* **1999**, *18*, 2855–2866.
14 P. HILLEREN, T. MCCARTHY, M. ROSBASH, R. PARKER, T. H. JENSEN, *Nature* **2001**, *413*, 538–542.
15 T. H. JENSEN, K. PATRICIO, T. MCCARTHY, M. ROSBASH, *Mol. Cell* **2001**, *7*, 887–898.

45
Identification and Characterization of RNA-binding Proteins through Three-hybrid Analysis

Felicia Scott and David R. Engelke

45.1
Introduction

Macromolecular complexes between RNA and proteins play an integral role in many cellular processes. RNA–protein interactions have been studied using a variety of biochemical assays such as filter binding, electrophoretic mobility shift assays (EMSA) and RNA footprinting. A disadvantage of these *in vitro* techniques is that they do not allow direct identification of the cognate genes encoding the proteins or RNAs of interest; moreover, studies of interactions formed *in vitro* are limited by how well the *in vitro* incubation recapitulates the cellular processes that influence the interaction *in vivo*. The application of the *in vitro* biochemical techniques is often hindered by the relative low abundance of many RNA-binding proteins that make them difficult to detect, thereby making exogenous protein expression and purification mandatory for many of the *in vitro* biochemical analyses. In some instances, where an RNA or protein are known to interact *in vivo*, the protein and RNA complex cannot be detected by *in vitro* techniques due to the instability of the complex or folding issues *in vitro*. Many of these disadvantages have been overcome by the development of several methods using molecular genetics to study RNA–protein interactions.

We focus here on one such genetic method – the yeast three-hybrid system. The three-hybrid system is a modification of the widely employed genetic screen for detecting protein–protein interactions – the yeast two-hybrid system [1]. The three-hybrid system includes a chimeric RNA that fuses a "bait" RNA with an anchoring RNA. A specific RNA–protein interaction results in transcription of two independent reporter genes. The presence or absence of an interaction can be monitored by cell growth, colony color and enzymatic activity. One potential drawback of the three-hybrid technique is that it does not yield quantitative information (such as K_ds and stoichiometry) that can be gained from many biochemical approaches. One major advantage of this technique is that the RNA–protein interaction can be analyzed *in vivo* independent of its biological function. The three-hybrid system also has the added benefit in that a clone encoding the protein of interest is obtained directly when a library of cDNAs is used to screen for RNA-binding activities. Published examples of successful three-hybrid screens include: the discovery of proteins that bind a known RNA sequence, confirmation of a putative RNA–protein interaction, mutational analysis of interacting RNAs and proteins, and the discovery and analysis of multiprotein complexes (see the following review for a list of examples prior to 2001 [2]; [3–11]). Our lab has successfully used the two- and three-hybrid systems in the study of the biomolecular interactions among the 10 subunits (one RNA and nine protein subunits) of RNase P from *Saccharomyces cerevisiae* [6, 12].

For this discussion, we will provide a general overview of the three-hybrid system including the most commonly used plasmids, strains and protocols, as well as discuss the strengths and limitations of the system. Many of the strengths and limitations of the two-hybrid system are also common in the three-hybrid system. For a detailed discussion of these aspects relative to the two-hybrid system, see earlier reviews [13–17].

45.2
Basic Strategy of the Method

Two different three-hybrid systems were developed and published independently. Although the two systems are based on the same strategy, they vary in the details [18, 19]. The differences include the DNA-binding domain fusion protein, the bridge-forming RNA-binding protein and its RNA target site. The basic strategy of the three-hybrid method is illustrated in Fig. 45.1. The yeast three-hybrid system is based on the fact that many transacting transcriptional regulators are made up of physically separable, functionally independent domains. These regulators behave as bifunctional units, often containing a DNA-binding domain that binds to a specific promoter sequence and an activation domain that directs the RNA polymerase II complex to transcribe a downstream gene. Both domains are required to activate a gene. In principle, any activation domain can be associated with any DNA-binding domain since it the DNA-binding domain that provides specificity. In the three-hybrid system, the DNA binding sites are located upstream of reporter genes in the yeast chromosome. The first fusion protein consists of a DNA binding domain linked to a known RNA-binding domain. This RNA-binding domain interacts with its RNA target site in a hybrid RNA molecule. The other portion of the RNA molecule is the "bait" that interacts with a second fusion protein composed of another RNA-binding domain fused to a transcriptional activation domain. Thus, the hybrid RNA functions as a bridge between two fusion proteins. When this tripartite complex forms at the promoter of the reporter gene, transcription of the reporter gene is activated. Reporter gene expression can be monitored by growth phenotype or enzymatic activity.

The components of the three-hybrid system that are most commonly used were developed by Sengupta et al. [19] in the laboratory of Marvin Wickens at the University of Wisconsin. The DNA binding site consists of a 17-nt recognition site (operator) for the *Escherichia coli* LexA protein. Multiple LexA operators are located in the upstream promoters in the two reporter genes, *LacZ* (eight operators) and

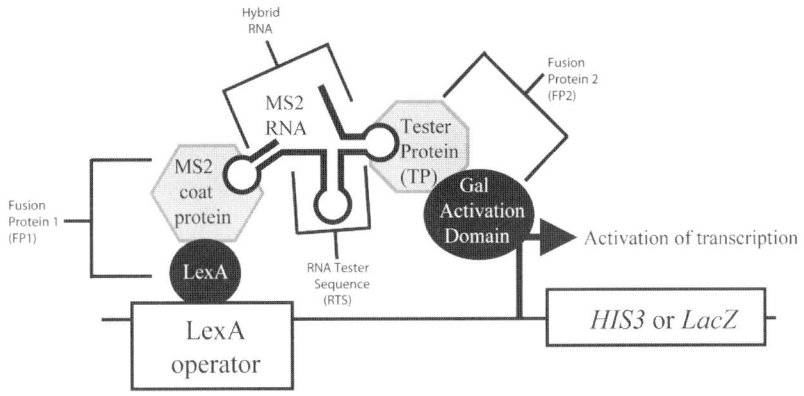

Fig. 45.1. The basic strategy of the three-hybrid system and the components used most often. The first fusion (FP1) contains the entire LexA protein fused to the MS2 coat binding protein. In the most commonly used strains, FP1 is integrated into the yeast genome. A hybrid RNA containing two MS2 RNA sites and the RNA tester sequence (RTS). The MS2 coat protein binds to a single MS2 RNA site as a dimer. The second fusion protein (FP2) contains the activation domain of GAL4 fused to the tester protein or cDNA library open reading frames. A productive interaction in the three-hybrid system leads to transcriptional activation of two reporter genes, *HIS3* and *LacZ*. Each reporter gene is under the control of multiple LexA operators. The LexA protein binds to each operator as a dimer. Adapted from Zhang et al. [34].

HIS3 (four operators). The entire prokaryotic LexA protein binds as a dimer to the LexA operator. Fusion protein 1 (FP1) consists of LexA protein fused to the bacteriophage MS2 coat protein. In the most commonly used yeast host, the gene for the LexA–MS2 fusion has been integrated into the chromosome. The MS2 protein binds as a dimer to a stem–loop structure. The hybrid RNA contains two MS2 coat protein binding sites linked to the RNA tester sequence (RTS). Fusion protein 2 (FP2) consists of the transcriptional activation domain of the yeast *GAL4* transcription factor fused to an RNA-binding tester protein (TP).

Alternate versions of the three-hybrid system are also available. These include modifications where the equivalent of the LexA–MS2 fusion protein and MS2 RNA-binding sites are replaced by (1) the Rev responsive element (RRE) sequence in the *env* gene and the RRE-RNA-binding protein RevM10 [18], (2) the NRE in *hunchback* mRNA's 3'-untranslated region (UTR) and its protein binding partner, Pumilio [20], or (3) hY5 RNA and its protein binding partner, Ro60 [21]. Similarly, the LexA DNA-binding domain can be replaced by the Gal4 DNA-binding domain [7, 18].

45.3
Detailed Components

The components of the three-hybrid system that are discussed in detail below can be obtained from the Wickens lab at the University of Wisconsin-Madison (www.biochem.wisc.edu/wickens/3H/). The RNA-Protein Hybrid™ Kit is commercially available from Invitrogen (www.invitrogen.com). The kit from Invitrogen contains modified vectors for producing the hybrid RNAs and FP2 with additional markers and cloning sites. To use the general procedure below, the researcher should be familiar with basic molecular biology techniques such as DNA ligations, *E. coli* transformations and restriction enzyme analysis, as well as basic yeast molecular biology and microbiological techniques. Suggested sources for additional information on these topics are *Current Protocols in Molecular Biology* [22], *Molecular Cloning: A Laboratory Manual* [23] and *Guide to Yeast Genetics and Molecular Biology* [24].

45.3.1
Yeast Reporter Strain

The genotype of the yeast reporter strain L40-coat is *MATa, ura3-52, leu2-3, 112, his3Δ200, trp1Δ1, ade2, LSY2::(lexA op)$_4$-HIS3, ura3::(LexA-op)$_8$-lacZ, LexA–MS2 coat (TRP1)*. The gene encoding the LexA–MS2 coat protein fusion has been integrated into the chromosome. The strain R40coat is identical to L40-coat except it is of the opposite mating type (MATα). A canavanine-resistant derivative of L40-coat, L40-coat-can, is also available. This strain carries a *can1* allele and becomes canavanine sensitive when transformed with pACTII/CAN (see below). The strains are auxotrophic for uracil, histidine, adenine and leucine. The strains are available from M. Wickens, University of Wisconsin. Two plasmids, one encoding the activation domain fusion protein (FP2) and the other encoding the hybrid RNA, need to be transformed into this strain for detection of the RNA–protein interaction.

45.3.2
Plasmids

Two interaction-specific constructs must be made by the user of the three-hybrid system, an RNA hybrid (containing the tester RNA sequence) and the activation domain FP2 (containing the protein of interest or cDNA library fused to GAL4 activation domain). The details of their construction along with technical considerations are discussed below. For maps and more detailed descriptions of the plasmids, see www.biochem.wisc.edu/wickens/3H/.

45.3.3
Hybrid RNA

The four most common plasmids for producing hybrid RNAs are pIII/MS2-1, pIII/MS2-2, pIIIA/MS2-1 and pIIIA/MS2-2. Each is a multicopy plasmid that can be propagated in either yeast or bacteria. pIII/MS2-1 and pIII/MS2-2 are derived from pIIIEx426RPR [25] and use the *S. cerevisiae* promoter for RNA polymerase III transcription of the RNase P RNA gene (*RPR1*). RPR1 RNA is normally retained in the nucleus [26, 27] and it is likely that this also holds true for most hybrid RNAs made from this promoter. The RNA tester sequence is inserted at

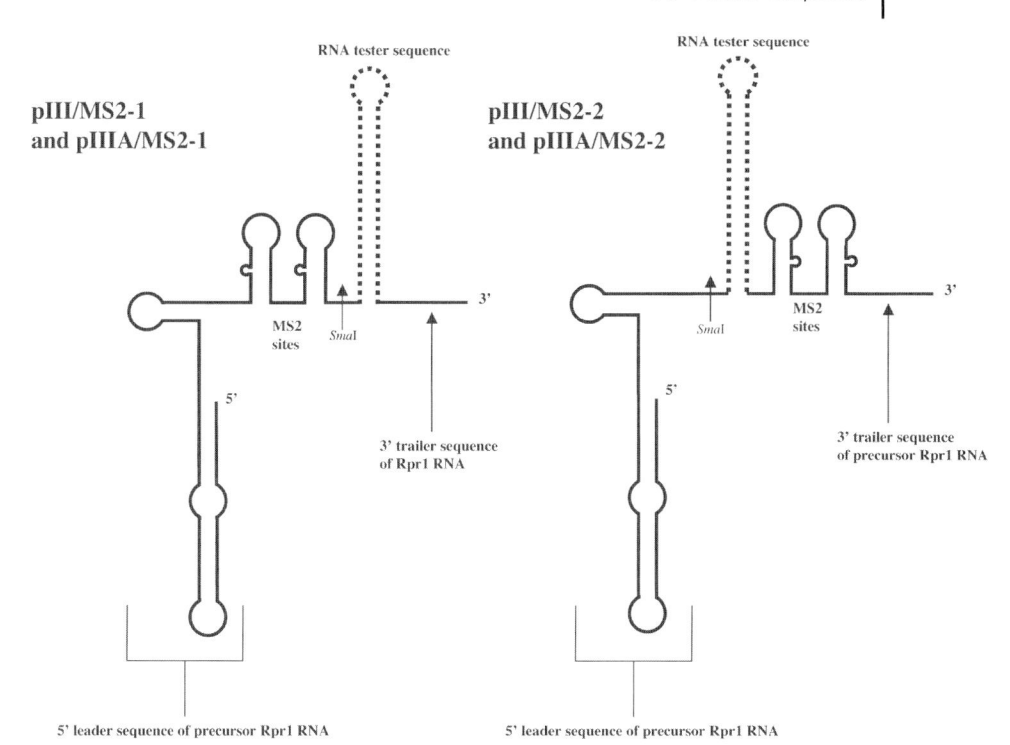

Fig. 45.2. The orientation of the RNA tester sequence relative to the two MS2 RNA sites is based on the plasmid used for generating the hybrid RNA. The RNA tester sequence is inserted at the *Sma*I or *Sph*I restriction sites in the pIIIA plasmids, or *Sma*I restriction site in the pIII plasmids. The *Sph*I restriction site is not unique in the pIII plasmids. The designation 2-1 or 2-2 indicates that the two plasmids differ only in the relative position of the restriction sites and the MS2 binding sites. Adapted from Zhang et al. [34].

the *Sma*I or *Sph*I restriction sites in the pIIIA plasmids or at the *Sma*I restriction site in the pIII plasmids. The *Sph*I restriction site is not unique in the pIII plasmids. The designation 2-1 or 2-2 indicates that the two plasmids differ only in the relative position of the restriction sites and the MS2 binding sites (Fig. 45.2). Both RNA plasmid series carry the *URA3* selectable marker. pIIIA/MS2-1 and pIIIA/MS2-2 are similar to pIII/MS2-1 and pIII/MS2-2, but they also carry the *ADE2* gene. Screening for the retention of *ADE2* gene can be used to help eliminate false positives that are RNA independent (see discussion below on false positives).

The hybrid RNA contains the following features: (1) the yeast 84-nt RNase P RNA leader sequence, (2) two tandem MS2 coat protein binding sites, (3) the RNA tester sequence, and (4) 41 nt of the 3′ terminus of RNase P RNA. Two MS2 coat protein binding sites are used since binding to adjacent sites by the MS2 coat protein is cooperative [28, 29]. The MS2 33-nt recognition site also contains a nucleotide change that enhances binding of the coat protein. MS2 coat protein binds this hairpin–loop RNA structure with a K_d of 10^{-9} to 10^{-10} M [30, 31]. The hybrid

RNAs are transcribed by the RNA polymerase III promoter and terminated at the TTTTT pol III terminator. The expression of small RNA molecules from the *RPR1* promoter can produce several thousand copies of an RNA molecule per cell [25]. In addition, transcripts from the RNase P promoter remain in the nucleus, which is where the three-hybrid interaction needs to take place.

45.3.3.1 Technical Considerations for the Hybrid RNA

The three-hybrid system is a sensitive method for detecting relatively weak and transient RNA–protein interactions. Such interactions may not be biochemically detectable, but may be critical for proper functioning of biological systems. The affinities of the RNA–protein interactions that have been detected using the three-hybrid system have K_ds in the 10^{-11} to 10^{-6} M range. The minimal affinity that is required to yield a detectable transcriptional activation has not been determined. However, affinity is not the only determinant of whether an RNA–protein interaction is detectable. The abundance, the conformation and the cellular location of the hybrid RNA and FP2 can also influence transcriptional activation in the system.

There are several technical constraints on the production of hybrid RNAs. The use of the *RPR1* promoter limits the sequence of the RNA tester sequence that can be analyzed. In general, the RNA tester sequence should not contain more than four uridines (preferably not more than three) in succession since transcription by RNA polymerase III can be terminated at these sites. To circumvent this sequence limitation, an alternative system using an RNA polymerase II promoter has been developed [18] and used successfully in the analysis of hnRNP-C1/C2 interactions with synthetic and naturally occurring uridine-rich sequences [7].

The size of the RNA tester sequence is also an important determinant of success when using the three-hybrid system. Typically, RNA inserts less than 200 nt yield higher signals than longer inserts. However, positive interactions have been detected for RNA tester sequences up to 1600 nt in length [32].

The MS2 and RNA tester sequence can be placed in either position relative to one another; however, the relative position of the RNA tester sequence and the MS2 sites can make a difference in terms of the transcriptional activation. For example, in the case of the three-hybrid test using the iron-responsive element (IRE) and the iron-regulatory protein 1 (IRP), placing the IRE downstream of the MS2 sites results in two- to 3-fold less transcription than does the opposite arrangement [19]. Although successful screens have been carried out with both arrangements of the MS2 sites, we have had the most success and experience with pIIIA/MS2-2 [6, 12]. Three main considerations influence the relative placement of these sites: (1) potential RNA polymerase III terminators, (2) RNA folding or the predicted secondary structure of your RNA and (3) position of the RNA tester sequence in its native context.

The MS2 coat protein binding site forms a stable stem–loop structure. The placement of two of these structures side-by-side should limit the formation of alternative structures by the RNA tester sequence. However, in some instances, alternative structures can form that interfere with the interaction of the hybrid RNA with the LexA–MS2 fusion protein (FP1) or the activation domain fusion protein

(FP2). Cassidy and Maher overcame this technical difficulty in their three-hybrid screen using NF-κB and an RNA aptamer by inserting a 13-bp GC-rich clamp to lock in the correct conformation of their RNA aptamer [3, 33].

To aid in the optimization of a particular cloning strategy, it might be helpful to use RNA folding programs such as mfold (www.bioinfo.rpi.edu/applications/mfold/) or mulfold (the Macintosh version of mfold; www.cgal.icnet.uk/macsoft.html) to predict the secondary structure of your RNA of interest fused to the MS2 recognition sites; however, these programs are often not indicative of the RNA structure *in vivo*. After obtaining clones with the desired sequence, Northern blot analysis can be used to confirm the integrity and expression level of the hybrid RNA.

45.3.4
Activation Domain FP2

There are two fusion proteins in the three-hybrid system. FP1, the LexA–MS2 "bait" for the hybrid RNA, remains constant. FP2 varies based on the interaction tested. The pACTII plasmid consists of the polypeptide of interest fused to the *GAL4* activation domain to create FP2. pACTII, a multicopy plasmid, encodes the *GAL4* activation domain followed by the hemagglutinin (HA) epitope tag {YPYDVPDYA} and a polylinker for cloning test polypeptide-coding sequences. The HA epitope tag permits immunodetection of the fusion protein by Western blotting using antibodies that are commercially available (Santa Cruz Biotechnology, Chemicon or Roche). Antibodies against the HA epitope tag can also be used for co-immunoprecipitation experiments intended to verify identified interactions. We have also used commercial antibodies against LexA and the Gal4 activation domain for immunodetection of the fusion proteins (Santa Cruz Biotechnology); however, many of these antibodies are not entirely robust and specific. The fusion protein is expressed from the constitutively expressed *ADH1* promoter and transcription is terminated by the *ADH1* terminator. The plasmid can be propagated in yeast (*LEU2* marker) and *E. coli* (ampicillin resistance).

The plasmid pACTII-CAN has also been constructed using the pACTII plasmid backbone [34]. The yeast *CAN1* gene has been inserted at the unique SalI restriction site on pACTII. The *CAN1* gene encodes an arginine permease that causes cells to die in media containing the arginine analog canavanine. Thus, canavanine selection leads to the loss of the activation domain plasmid similar to the loss of the hybrid RNA plasmid in the presence of 5-fluoro-orotic (5-FOA).

When designing a cloning strategy, remember that the gene of interest must be cloned in-frame with the sequence encoding the *GAL4* activation domain and the HA epitope. Clones should be sequenced to confirm that the gene is cloned in the correct orientation and in-frame with the *GAL4* activation domain.

45.3.4.1 Technical Considerations for the Activation Domain FP2
Sometimes RNA–protein interactions that occur normally *in vivo* are not detected by the three-hybrid system. This lack of detection might have several causes. First,

if high expression levels of the activation domain fusion protein are toxic to the reporter strain, the transformants may not grow. To circumvent this problem, you may want to (1) reclone the protein gene of interest on a low-copy-number plasmid, (2) put the gene of interest under an inducible promoter and/or (3) try using subfragments of the protein. There are a variety of plasmids (low-copy-number or inducible) available from the two-hybrid system that can easily be swapped into the three-hybrid system (see reviews [13–17]). It is important to make sure that the selectable marker is compatible with the three-hybrid system. As an alternative approach to alleviating the toxicity problem, it is possible to clone fragments of the gene interest instead of the full-length gene into the pACTII plasmid and test the fragments for RNA-binding activity.

Second, the expressed protein must be stable and the expression must be high enough to generate an interaction signal. Expression levels and stability of the proteins can be checked by Western blotting using commercially available antibodies against the HA epitope tag. Nevertheless, it is important not to place too much emphasis on these tests. In some cases, detection of the epitope has failed (the epitope may not be available because of the folding of the fusion protein or low expression may prevent detection), but the three-hybrid test has been successful. Also, successful detection of the expressed fusion protein does not imply that the interaction domain is accessible and properly folded; thus, the screen could fail in spite of the presence of the fusion protein. The best option is to utilize an antibody against your specific protein if it is available.

A number of groups have successfully used the three-hybrid system to screen cDNA libraries to identify new interactions for their RNA of interest [35–37]. We have had success with a yeast cDNA library created in pACTII (a gift from Steve Elledge). Other cDNA libraries are also available that are compatible with the three hybrid system (see, e.g. [17]). One factor that needs to be taken into account when screening with cDNA libraries is that you have to scale-up the transformation protocol (see below) in order to cover the complete complexity of the library, especially for rare messages. The degree of scale-up will depend on your library of choice.

The major technical challenge when using cDNA libraries is identifying those interactions that are biologically relevant. The strength of a positive signal is not always an indicator of relevance, since artificial interactions can be strong and a specific while biologically relevant RNA–protein interaction might be weak. It is essential to further verify true positives using an independent method such *in vitro* binding assays, co-immunoprecipitation, co-localization and genetic methods.

45.3.5
Positive Controls

The control plasmids for the three-hybrid system that are most commonly used are the pIIIA/IRE-MS2 and pAD-IRP plasmids. The plasmid pIIIA/IRE-MS2 expresses a hybrid RNA containing the rat ferritin light chain IRE [38, 39] fused to the MS2

RNA. The IRE RNA forms a hairpin loop structure that serves as the recognition site for the IRP [39]. Plasmid pAD-IRP expresses the rabbit IRP fused to the *GAL4* activation domain. When pIIIA/IRE-MS2 and pAD-IRP are co-transformed into the *L40-coat*, robust transcriptional activation of *HIS3* and *LacZ* occurs. If you test for growth on histidine-minus media to assess *HIS3* activation, 3-aminotriazole (3-AT) should be included to eliminate low levels of basal expression (see discussion below on 3-AT). For the pIIIA/IRE-MS2 and pAD-IRP, transformants are usually plated directly on media containing 5 mM 3-AT [19].

45.4
Protocols

An overview of the experimental protocol is outlined in Fig. 45.3. The protocol for testing a known or putative RNA–protein interaction is relatively straightforward. The RNA of interest and the gene encoding the putative RNA-binding protein are cloned into the appropriate plasmids. The plasmid containing the hybrid RNA and the plasmid containing the FP2 are transformed into the yeast strain. Once transformants are obtained, the cells are assayed for transcriptional activity of the reporter genes: *HIS3* and/or *LacZ*.

We used this approach to identify proteins that bound directly to the RNase P RNA subunit [6]. Similarly, an RNA–protein interaction between mouse telomerase RNA and a newly cloned telomerase protein (TP1) was confirmed by this approach [40].

In principle, this approach provides an easy way to delineate which portions of the RNA or proteins that are required for binding. In our case, we were able to use this approach to demonstrate that Pop1p, the largest protein subunit of RNase P, bound specifically and directly to the P3 subdomain of RNase P RNA, a 54-nt helix–internal loop–helix structure [12]. The best controls for specificity of RNA binding are the comparison of binding to the wild-type RNA with one containing point mutations that compromise *in vivo* function without necessarily affecting overall structure. Lee et al. [8] have used this approach along with site-directed mutagenesis to identify which amino acids in the Rous sarcoma virus nucleocapsid bind to the RNA packaging signal.

45.4.1
Transformation of Yeast

Plasmids can be introduced into yeast by transformation of spheroplasts, transformation of chemically treated cells or electroporation. We use a modification of the lithium acetate method [22] for the transformation of the L40-coat reporter strain. Yeast cells are grown in rich media, treated with lithium acetate and incubated in the presence of plasmid DNA. After transformation the cells are plated on minimal media that lacks the appropriate nutrients for plasmid selection.

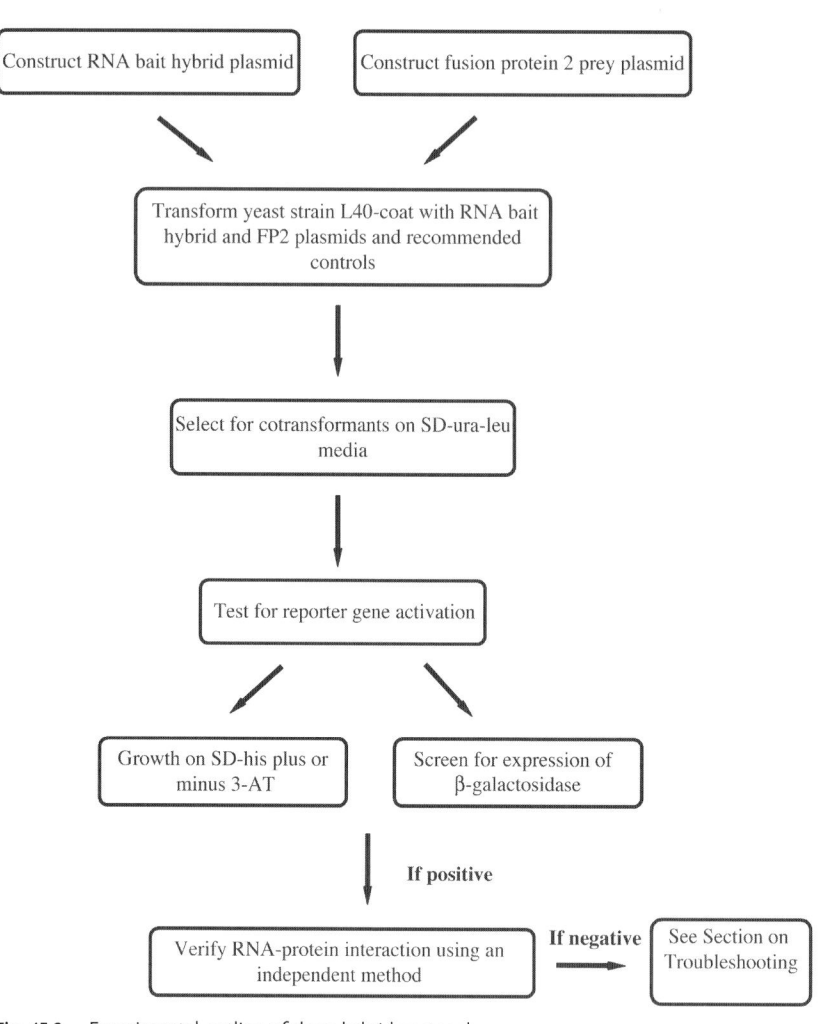

```
┌─────────────────────────────────┐         ┌─────────────────────────────────┐
│ Construct RNA bait hybrid plasmid│         │Construct fusion protein 2 prey plasmid│
└─────────────────────────────────┘         └─────────────────────────────────┘
                    ↘                       ↙
              ┌────────────────────────────────────────┐
              │ Transform yeast strain L40-coat with RNA bait │
              │ hybrid and FP2 plasmids and recommended │
              │                 controls               │
              └────────────────────────────────────────┘
                                 ↓
              ┌────────────────────────────────────────┐
              │ Select for cotransformants on SD-ura-leu│
              │                 media                  │
              └────────────────────────────────────────┘
                                 ↓
                    ┌─────────────────────────────┐
                    │ Test for reporter gene activation │
                    └─────────────────────────────┘
                         ↙                    ↘
        ┌───────────────────────┐    ┌───────────────────────┐
        │ Growth on SD-his plus or│    │ Screen for expression of│
        │      minus 3-AT       │    │     β-galactosidase   │
        └───────────────────────┘    └───────────────────────┘
                                 ↓ If positive
       ┌──────────────────────────────────┐  If negative  ┌──────────────────┐
       │ Verify RNA-protein interaction using an │──────────→│ See Section on   │
       │         independent method       │               │ Troubleshooting  │
       └──────────────────────────────────┘               └──────────────────┘
```

Fig. 45.3. Experimental outline of three-hybrid protocol.

(1) Inoculate 5 ml of YPD with a single yeast colony of the L40-coat strain. Grow overnight to saturation at 30 °C.

(2) The night before transformation, inoculate a 2-l sterile flask containing 300 ml YPAD medium with an appropriate amount of the saturated culture (25–100 μl) and grow overnight at 30 °C to an OD_{600} of 0.3–0.5 (1×10^7 cells/ml). For 2- to 3-fold higher transformation efficiency, dilute at this point to 2×10^6 cells/ml in fresh YPAD medium and grow for 2–4 h. It is important to perform this step when high-efficiency transformation is needed for cDNA and genomic libraries.

(3) Pellet the cells by centrifugation at 5000 r.p.m. (GSA rotor) for 5 min at room temperature.
(4) Resuspend the pellet in 10 ml of sterile water and transfer to 50-ml conical tube.
(5) Repeat Step 3.
(6) Pour off water and resuspend cells in 1.5 ml freshly prepared buffered lithium acetate solution (10 mM Tris–Cl, 1 mM EDTA, 100 mM lithium acetate made from stocks of 100 mM Tris–Cl (pH 7.5), 10 mM EDTA and 1 M lithium acetate (adjusted to pH 7.5 with dilute acetic acid). Incubate cells for 15 min at 30 °C without agitation before adding the transforming DNA.
(7) For each transformation, mix 50 µg sheared, heat-denatured salmon sperm DNA (carrier DNA) with 1–5 µg of each plasmid DNA in a sterile 1.5-ml microcentrifuge tube. Keep the total volume of DNA at 20 µl or below. Maximum transformation efficiency is achieved by boiling and chilling the carrier DNA (not the plasmid) immediately prior to use. Heat for 5–10 min followed by cooling on ice.
(8) Add 200 µl of yeast suspension from Step 6 to each microcentrifuge tube.
(9) Add 1.2 ml polyethylene glycol (PEG, MW 3350) solution (40% PEG, 10 mM Tris–Cl, 1 mM EDTA and 100 mM lithium acetate made fresh from stocks of 50% PEG 3340, 100 mM Tris–Cl (pH 7.5), 10 mM EDTA and 1 M lithium acetate (adjusted to pH 7.5 with dilute acetic acid) to each tube. Mix gently by inverting or vortexing.
(10) Incubate at 30 °C for 30 min without agitation.
(11) Add 150 µl dimethylsulfoxide (DMSO). Heat shock for 20 min at 42 °C.
(12) Pellet cells by centrifugation for 5 s at high speed. Remove supernatant and discard.
(13) Resuspend the cell pellet in 200–1000 µl of TE. Plate 100 µl of the cell suspension on SD-ura-leu plates, selecting for the hybrid RNA plasmid and the GAL4 activation domain FP2 plasmid.
(14) Incubate at 30 °C for 3–5 days until colonies of transformants are visible.
(15) Restreak colonies to obtain single colony isolates on SD-ura-leu plates.

Transformation of the reporter strain with a single plasmid yields 10^4–10^6 transformants/µg plasmid DNA. If both the hybrid RNA plasmid and the pACTII fusion protein plasmid are introduced at the same time, the transformation efficiency can decrease by a factor of 10. Instead of introducing both plasmids at the same time, you may want to transform the RNA hybrid plasmid first, especially when using cDNA libraries. This strain can be selected on SD-ura media. The creation of this strain first may be wise for several reasons. It allows you to transform the plasmid for the fusion protein of interest as well as control plasmids containing the GAL4 activation domain only (no insert) or a fusion protein that is unrelated to your RNA of interest. Second, if no transformants are obtained in the case of the double transformation, you do not know if it is due to the toxicity of your protein to the cell or a technical failure with the transformation procedure.

45.4.2
Assaying for *HIS3* Expression

Once transformants containing the RNA bait and protein prey (FP2) plasmids have been confirmed, the tests for reporter gene activation are performed. The first test we routinely performed is the ability to grow on media lacking histidine plus or minus 3-AT (Sigma). 3-AT is a competitive inhibitor of the *HIS3* gene product. Cells producing more His3p can survive higher concentrations of 3AT. We typically test transformants on SD-his-leu-ura plates containing increasing concentrations of 3-AT (0, 1, 5, 10 and 20 mM) to select for stronger interactions; thus, increasing the stringency to help eliminate weak activators. However, a less stringent selection is often preferable when the strength of the interaction is unknown. Higher concentrations of 3-AT have also been used successfully. For example, the stem loop binding protein binds histone mRNA at 25 mM 3-AT [37].

Typically, single colonies that have all the relevant plasmids (grown on SD-leu-ura plates) are streaked onto SD-his-leu-ura plates and grown for 2–3 days. Single colonies from the SD-his-leu-ura are streaked again to SD-his-leu-ura plates plus or minus increasing concentrations of 3-AT. Cells are allowed to grow for 3–5 days at 30 °C. The ability to grow in the presence of 3-AT is determined by growth of individual colonies throughout the streak, not by the smear of yeast from the initial streak. It is not unusual to see a few large colonies on 3-AT, which are artifacts that may be caused by mutations in the *HIS3* gene.

45.4.3
Assaying for β-Galactosidase Activity

β-Galactosidase activity can be assayed by measuring the conversion of a lactose analog to a chromogenic or luminescent product. The assay can be performed using colonies permeabilized on either a filter or cell lysate. The filter assay yields qualitative results, whereas the liquid assay is more quantitative. Our lab routinely uses the filter assay and the protocol is presented below. The level of β-galactosidase can be quantified in yeast using one of two different colorimetric assays. For a detailed protocol of these assays see Zhang et al. [34] or Bartel and Fields [13].

Filter assay

(1) Patch a single colony that has been growing on SD-leu-ura onto a fresh SD-leu-ura plate. Grow at 30 °C for 2–3 days.
(2) Lay a dry nitrocellulose filter paper (BA-S85; Schleicher & Schuell) on top of the cells growing on the SD-leu-ura plate in Step 1. Apply gentle pressure to the nitrocellulose filter to transfer cells to the filter.
(3) Lift the filter from the plate and place the filter colony side up on an aluminum foil boat. Immerse the filter and boat into liquid nitrogen for 10–20 s.
(4) Allow filter to thaw at room temperature, colony side up, for a few minutes.

Alternatively, instead of lysing the cells in liquid nitrogen, place the filter in a plastic Petri dish, cover and incubate for 10 min at $-80\,°C$ in a freezer.

(5) In the lid of a Petri dish, place 1.5 ml of buffer (60 mM Na_2HPO_4, 40 mM NaH_2PO_4, 10 mM KCl, 1 mM $MgSO_4$ and 50 mM 2-mercaptoethanol, pH 7.0) plus 30 µl of X-gal (50 mg/ml in N,N-dimethylformamide) onto a Whatmann number 1 filter circle (90 mm).
(6) Place nitrocellulose filter containing lysed cells colony side up onto Whatmann filter circle containing assay buffer. Place bottom of Petri dish over the nitrocellulose filter and Whatmann filter circle. Seal the chamber with parafilm and incubate at 30 °C.
(7) Monitor the color reaction for 30 min to overnight.
(8) Stop the reaction by floating nitrocellulose filter upright on 1–2 ml 100 mM EDTA on Saran Wrap.
(9) Dry nitrocellulose filter at room temperature in a new Petri dish.

We normally include the control strain that contains the plasmids pIIIA/IRE and pAD-IRP in every experiment. A strong interaction such as the pIIIA/IRE and pAD-IRP usually turns blue in 30 min to 1 h. With extended incubation times, weak interactions will also yield a blue color reaction.

45.5
Troubleshooting

Many of the problems that lead to an unsuccessful three-hybrid screen have been discussed under technical considerations for the hybrid RNA (Section 45.3.3.1) or activation domain fusion protein (Section 45.3.4.1). However, there are some problems that can be overcome with the proper controls and these will be discussed here.

In some instances, when a protein and RNA are known to interact by other biochemical or genetic methods, a directed test of the RNA–protein interaction in the three-hybrid system yields a negative result. The negative result might be due to improper folding of the protein or RNA and/or lack of expression of the protein or RNA.

There are several types of interactions that will yield a "false" positive result in the three-hybrid system. Most of these false positives are indicative of an RNA–protein interaction, although the interaction might not be biologically relevant. For example, high affinity, non-specific interactions can exist where some proteins may interact specifically with a limited number of RNAs within the cell, but bind many more RNAs in the three-hybrid system in a non-specific manner. In addition, bridging interactions can be registered as positive in the three-hybrid system. In this case the protein appears to interact with the RNA, but may in fact interact with a yeast cellular protein that binds directly to the RNA. This scenario becomes more of a problem when the activation domain FP2 is a yeast protein. For example, yeast She3p gave a positive result when tested with a portion of the *ASH1* mRNA's 3′-UTR in the three-hybrid system, but this positive result was due to its interaction with She2p which binds the hybrid RNA. If the Shep2p was removed from the cell genetically by gene deletion, the yeast She3p no longer gave a positive result in the three-hybrid assay [41].

Occasionally, RNA hybrid molecules will transactivate the reporter gene in the absence of a protein fusion protein. In cDNA library screening, the large majority of false positives can be RNA-independent. This is most likely due to direct protein–protein interactions with the LexA–MS2 coat protein. To circumvent this problem, a colony color assay was developed that helps eliminate this class of false positives [42]. The L40-coat strain is an *ade2* mutant so under growth conditions where adenine becomes limiting after a few days the yeast colonies accumulate a red pigment as a result of the metabolic block in the adenine biosynthetic pathway and the cells turn pink to red. To take advantage of this colony color assay, the tester RNA should be cloned into the pIIIA/MS2-1 or pIIIA/MS2-2 RNA hybrid vectors. Because the pIIIA plasmids carry the wild type *ADE2* gene, the yeast cells containing these plasmids remain white. The colony color difference enables one to distinguish between RNA-independent and RNA-dependent positives. After transforming the L40-coat strain with both the RNA plasmid and the cDNA library. Plate the transformants on media selecting for the cDNA plasmid (*LEU2*) and for *HIS3* reporter expression but not for the RNA plasmid (SD-his-leu plates). If activation of the reporter gene is independent of the RNA, a small percentage of the cells will lose the RNA plasmid since it has not been selected for, and the colony will become pink/red or sectored (white colonies with pink/red sectors). If the His$^+$ phenotype is RNA-dependent, all colonies will remain white.

The RNA-dependent or RNA-independent interactions can also be confirmed by subsequent selections against the RNA-producing plasmid using 5-FOA [42]. Due to the conversion of 5-FOA to the toxic 5-fluoro-deoxyuracil by the *URA3* gene product, yeast cells expressing the *URA3* gene are severely growth inhibited on media containing 5-FOA. In contrast, cells lacking the *URA3* gene product can grow in the presence of 5-FOA if uracil is provided in the growth media. Since the plasmids expressing the hybrid RNA carry the *URA3* marker, cells that have lost the plasmid can be selected by plating the primary HIS$^+$ transformants on media containing 0.1% 5-FOA. The 5-FOA screen can be used with either the pIII or pIIIA plasmids. In the case of an RNA-dependent interaction, cells lacking the RNA plasmid will no longer be able to activate the reporter.

Protein-independent false positives have also been observed in the three-hybrid system [34]. These include hybrid RNAs that activate transcription when bound to a promoter. The frequency of these activating RNAs in a genomic library can be quite high. The methods used to identify and eliminate these protein-independent false positives depend upon the pACTII plasmid used to generate the activation domain FP2. If the plasmid is derived from pACTII, the plasmid is cured by growing a transformant overnight in YPD (to allow for loss of the pACTII plasmid) followed by plating on SD-ura to select for the RNA plasmid. These URA$^+$ colonies are replica-plated onto SD-leu plates to determine which colonies lack the *LEU2* marker on the activation domain plasmid. If the activation domain plasmid is derived from pACTII/CAN, the plasmid can be cured by streaking colonies onto a SD-arg plus canavanine. Following curing of the plasmid by either method, colonies should be assayed for β-galactosidase activity and/or growth on histidine deficient media. Once an RNA–protein interaction has been confirmed the system can be modified in various ways to test for mutations in the RNA or protein that eliminate binding or deletional studies design to identify the minimal RNA sequence needed for an RNA–protein interaction.

45.6
Additional Applications

Several studies suggest that the three-hybrid method could potentially be used to identify target RNAs of specific RNA-binding proteins. Sengupta et al. [43] adapted the three-hybrid system to identify RNA ligands for the yeast Snp1, a homolog of the human U1-70K, protein using an RNA library constructed from short fragments of genomic DNA from yeast that were transcribed in yeast together with the RNA-binding sites for the MS2 coat protein. Venables et al. [44] used a similar approach to identify testes RNA sequences that bind DazIp, an RNA-binding protein encoded by a region on the Y chromosome implicated in infertility. In addition, the potential use of this application can be appreciated in light of genome-sequencing projects that have identified many putative RNA-binding proteins, based on the presence of known RNA-binding motifs. Moreover, biochemical and genetic techniques have identified other RNA-binding proteins, for which the RNA recognition site has not been identified. For example, the three-hybrid system was used to confirm that several brain mRNAs of unknown function that bound human fragile X mental retardation protein *in vitro* could bind *in vivo* [5].

The three-hybrid system may provide an *in vivo* method to identify and optimize artificial RNA ligands for proteins that might be useful as therapeutic targets. Cassidy et al. [3, 33] utilized the three-hybrid test to identify and optimize an RNA aptamer that functions as a decoy for the transcription factor NF-κB. In this case, a combination of *in vitro* selected RNA aptamers and *in vivo* genetic selections were crucial for obtaining RNA aptamers that could inhibit transcriptional activation by NF-κB.

The three-hybrid system may permit the rapid screening for therapeutic inhibitors of a known RNA–protein interaction, such as those involved in viral replication, transcription or assembly. Although no published reports of the use of the three-hybrid system to screen chemical inhibitors of a RNA–protein interaction is available, several parameters would have to be satisfied in order for the assay to work. First, the yeast cell must be permeable to the chemical inhibitor. Second, the chemical inhibitor cannot be toxic to the yeast cell. Third, the chemical inhibitor must be specific for the RNA–protein interaction tested. The use of RNA decoys or small peptides as inhibitors of specific RNA–protein interactions can be easily tested in the three-hybrid system by the introduction of an additional plasmid expressing the RNA decoy or peptide. The use of RNA decoys to disrupt known viral RNA–protein interactions such as Tat-TAR [45] and Rev-RRE [46] has been published.

Multicomponent complexes have also been studied using the three-hybrid system. A practical limitation in analyzing complexes containing many proteins is the number of plasmids and markers required. There are several types of multicomponent interactions that have been analyzed successfully in the three-hybrid system: (1) independent interactions in which each protein binds to its own site on the RNA, and is unaffected by the presence of the other proteins and (2) bridged interactions, in which one protein tethers another protein to the RNA, in this case both proteins are required for the interaction. The interaction of the bI4 group I intron both of its splicing partners, a tRNA synthetase and an intron-encoded maturase is an example of a multicomponent complex in which the RNA has two independent protein-binding partners [32]. In the case of RNase P, we have shown that Pop1p and Pop4p can also bind independently to the RNase P RNA [6]. An example of a bridge interaction that has been analyzed using the three-hybrid system is the interaction of She2p and She3p with the 3′-UTR of *ASH1* mRNA [41].

45.7
Summary

The three-hybrid assay provides the same versatility and usefulness to the study of RNA–protein interactions as the two-hybrid system provides for the study of protein–protein interactions. It will be interesting to see the further adaptations of the three-hybrid system beyond its current applications. The possibility of using the three-hybrid system to identify RNA targets for putative RNA-binding proteins predicted by genomic sequencing is also intriguing. In addition, the possibility of using the three-hybrid system as a rapid screen for therapeutic compounds that inhibit RNA–protein interactions involved in the viral life cycle may lead to the development of a whole new class of anti-viral agents.

Acknowledgments

This work was supported by National Institutes of Health grant GM34869 to D. R. E.

References

1 S. Fields, O. Song, *Nature* **1989**, *340*, 245–246.
2 D. S. Bernstein, N. Buter, C. Stumpf, M. Wickens, *Methods* **2002**, *26*, 123–141.
3 L. A. Cassiday, L. J. Maher, III, *Proc. Natl. Acad. Sci. USA* **2003**, *100*, 3930–3935.
4 A. Cohen, R. Reiner, N. Jarrous, *Nucleic Acids Res.* **2003**, *31*, 4836–4846.
5 N. Dolzhanskaya, Y. J. Sung, J.

Conti, J. R. Currie, R. B. Denman, *Biochem. Biophys. Res. Commun.* **2003**, *305*, 434–441.

6 F. Houser-Scott, S. Xiao, C. E. Millikin, J. M. Zengel, L. Lindahl, D. R. Engelke, *Proc. Natl. Acad. USA* **2002**, *99*, 2684–2689.

7 N. Koloteva-Levine, M. Amichay, O. Elroy-Stein, *FEBS Lett.* **2002**, *523*, 73–78.

8 E.-G. Lee, A. Alidina, C. May, M. L. Linial, *J. Virol.* **2003**, *77*, 2010–2020.

9 R. Pudi, S. Abhiman, N. Srinivasan, S. Das, *J. Biol. Chem.* **2003**, *278*, 12231–12240.

10 E. Maniataki, M. Tabler, M. Tsagris, *RNA* **2003**, *9*, 346–354.

11 B. Mazumder, P. Sampath, V. Seshadri, R. K. Maitra, P. E. DiCorleto, P. L. Fox, *Cell* **2003**, *115*, 187–198.

12 W. A. Ziehler, J. Morris, F. Houser-Scott, C. Millikin, D. R. Engelke, *RNA* **2001**, *7*, 565–575.

13 P. L. Bartel, S. Fields, *Methods Enzymol.* **1995**, *254*, 241–262.

14 P. L. Bartel, C.-T. Chien, R. Sternglanz, S. Fields, *Biotechniques* **1993**, *14*, 920–924.

15 P. L. Bartel, C.-T. Chien, R. Sternglanz, S. Fields, Using the two-hybrid system to detect protein–protein interactions, in: *Cellular Interactions in Development: A Practical Approach*, D. A. Hartley (ed.), Oxford University Press: Oxford, **1993**, pp. 153–179.

16 R. Brent, R. L. Finley Jr, *Annu. Rev. Genet.* **1997**, *31*, 663–704.

17 P. N. MacDonald, ed. *Methods in Molecular Biology: Two-hybrid Systems*, Humana Press, Totowa, NJ, **2001**.

18 U. Putz, P. Skehel, D. Kuhl, *Nucleic Acids Res.* **1996**, *24*, 4838–4840.

19 D. J. Sengupta, B. Zhang, B. Kraemer, P. Pochart, S. Fields, M. Wickens, *Proc. Natl. Acad. Sci. USA* **1996**, *93*, 8496–8501.

20 J. Sonoda, R. P. Wharton, *Genes Dev.* **1999**, *13*, 2704–2712.

21 P. Bouffard, E. Barbar, F. Brière, G. Boire, *RNA* **2000**, *6*, 66–78.

22 F. M. Ausubel, R. Brent, R. E. Kingston, D. D. Moore, J. G. Seidman, J. A. Smith, K. Struhl (eds), *Current Protocols in Molecular Biology*. Saccharomyces cerevisiae, Greene/Wiley-Interscience, New York, **1996**.

23 J. Sambrook, E. F. Fritsch, T. Maniatis, *Molecular Cloning: A Laboratory Manual*. 2nd edn. Cold Spring Harbor Laboratory Press, Cold Spring Harbor, NY, **1989**.

24 C. Guthrie, G. R. Fink, *Guide to Yeast Genetics and Molecular Biology (Methods in Enzymology)*, J. N. Abelson and M. I. Simon (eds), Academic Press, San Diego, CA, **1991**.

25 P. D. Good, D. R. Engelke, *Gene* **1994**, *151*, 209–214.

26 E. Bertrand, F. Houser-Scott, A. Kendall, R. H. Singer, D. R. Engelke, *Genes Dev.* **1998**, *12*, 2463–2468.

27 C. Srisawat, F. Houser-Scott, E. Bertrand, S. Xiao, R. H. Singer, D. R. Engelke, *RNA* **2002**, *8*, 1348–1360.

28 V. J. Bardwell, M. Wickens, *Nucleic Acids Res.* **1990**, *18*, 6587–6594.

29 G. W. Witherell, H.-N. Wu, O. C. Uhlenbeck, *Biochemistry* **1990**, *29*, 11051–11057.

30 O. C. Uhlenbeck, J. Carey, P. J. Romaniuk, P. T. Lowary, D. Beckett, *J. Biomol. Struct. Dyn.* **1983**, *1*, 539–552.

31 P. T. Lowary, O. C. Uhlenbeck, *Nucleic Acids Res.* **1987**, *15*, 10483–10493.

32 S. B. Rho, S. A. Martinis, *RNA* **2000**, *6*, 1882–1894.

33 L. A. Cassiday, L. J. Maher, III, *Biochemistry* **2001**, *40*, 2433–2438.

34 B. Zhang, B. Kraemer, D. Sengupta, S. Fields, M. Wickens, *Methods Enzymol.* **2000**, *318*, 399–419.

35 B. Zhang, M. Gallegos, A. Puoti, E. Durkin, S. Fields, J. Kimble, M. Wickens, *Nature* **1997**, *390*, 477–483.

36 Z.-F. Wang, M. L. Whitfield, T. C. Ingledue, Z. Dominiski, W. F. Marzluff, *Genes Dev.* **1996**, *10*, 3028–3040.

37 F. Martin, A. Schaller, S. Eglite, D. Schümperli, B. Müller, *EMBO J.* **1997**, *16*, 769–778.

38 N. Aziz, H. N. Munro, *Proc. Natl. Acad. Sci. USA* **1987**, *84*, 8478–8482.

39 R. D. Klausner, T. A. Rouault, J. B. Harford, Cell **1997**, *72*, 19–28.
40 L. Harrington, T. McPhail, V. Mar, W. Zhou, R. Oulton, Amgen EST Program, M. B. Bass, I. Arruda, M. O. Robinson, Science **1997**, *275*, 973–977.
41 R. M. Long, W. Gu, E. Lorimer, R. H. Singer, P. Chartrand, EMBO J. **2000**, *19*, 6592–6601.
42 Y. W. Park, S. L. Tan, M. G. Katze, Biotechniques **1999**, *26*, 1102–1106.
43 D. J. Sengupta, M. Wickens, S. Fields, RNA **1999**, *5*, 596–601.
44 J. P. Venables, M. Ruggiu, H. J. Cooke, Nucleic Acids Res. **2001**, *29*, 2479–2483.
45 P. R. Bohjanen, Y. Liu, M. A. Garcia-Blanco, Nucleic Acids Res. **1997**, *25*, 4481–4486.
46 K. Konopka, N. S. Lee, J. Rossi, N. Duzgunes, Gene **2000**, *255*, 235–244.

46
Analysis of Alternative Splicing *In Vivo* using Minigenes

Yesheng Tang, Tatyana Novoyatleva, Natalya Benderska,
Shivendra Kishore, Alphonse Thanaraj and Stefan Stamm

46.1
Introduction

After the initial experiment about 10 years ago [1], minigenes have been widely applied to study alternative splicing *in vivo*. Currently, original constructs from at least 78 different genes have been reported in the literature and were analyzed in living cells. From these original constructs several hundred mutated versions were constructed and analyzed. The analysis of alternative splicing *in vivo* using minigenes has been proven to be an extremely robust and reproducible method. It was applied to test unknown factors for their involvement in alternative splicing, as well as to study regulatory elements, signal transduction pathways and basic splicing patterns of genes of interest.

46.2
Overview of the Method

The experimental outline is shown in Fig. 46.1. First, a minigene has to be constructed by either cloning genomic fragments or PCR amplified genomic DNA under the control of a suitable promoter. Next, this construct is transfected into eukaryotic cells where the concentration of regulatory splicing factors are changed. Regulatory factors can either be elevated by co-transfecting an increasing amount of cDNA expressing the factor or decreased by RNA interference. In the next step, the splicing pattern of the minigene is analyzed. This is done either by analyzing the RNA by RT-PCR or by indirect methods, such as the splicing-dependent formation of a protein. The major advantage of the method is that almost every construct tested splices upon transfection. In contrast to *in vitro* splicing methods, the capacity of the cloning vector is the only limit for the intron length. Furthermore, different cell types can be tested. The major disadvantage is that the method is prone to indirect effects, since intact cells are studied.

Handbook of RNA Biochemistry: Student Edition. Edited by R. K. Hartmann, A. Bindereif, A. Schön, E. Westhof
Copyright © 2009 WILEY-VCH Verlag GmbH & Co. KGaA, Weinheim
ISBN: 978-3-527-32534-4

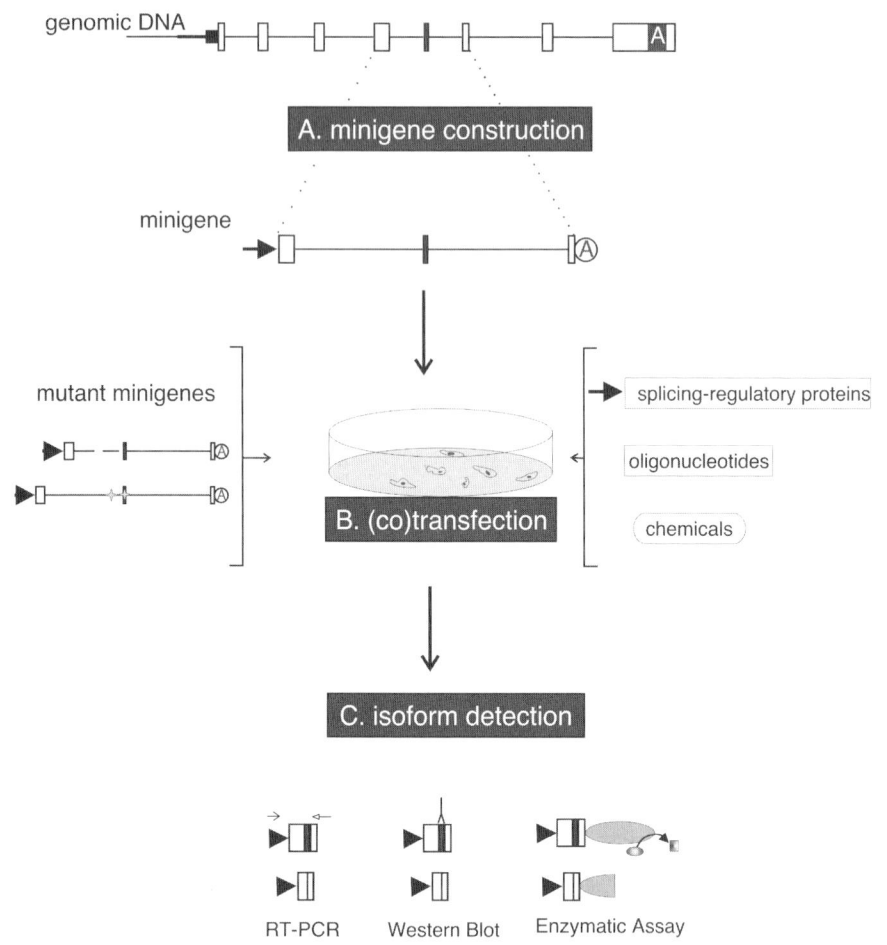

Fig. 46.1. Overview of the method. (A) Minigenes are constructed by amplification of a part of genomic DNA containing the alternative exons and its flanking constitutive exons. Cloning is achieved by long-range PCR using exon-specific primers. In almost all cases, the part amplified does not contain the promoter (black box, left) or the polyadenylation site (A, right). The minigene is then cloned (dotted lines) into an expression vector, containing a different promoter (arrow) and polyadenylation site Ⓐ. (B) The minigene is then analyzed by transfecting it into the cells of choice. Cis-acting mutations can be introduced by either deletion or point mutagenesis (left). In addition, putative trans-acting factors, such as splicing regulatory proteins, RNA and DNA oligonucleotides or low-molecular-weight substances (right) can be added into the assay. (C) The different isoforms can be determined by either RT-PCR using at least one primer specific for the minigene (open arrows), by Western blot, if appropriate antibodies are available, or with enzymatic assays, if the alternative exon generates a frameshift. For enzymatic assays, a reporter protein is cloned in frame with the alternative exon. If the exon is used, an activity such as luciferase can be measured. If the alternative exon is skipped, this activity is not present, due to the frameshift inserted by the alternative exon.

46.3
Methods

46.3.1
Construction of Minigenes

Most minigenes are amplified from genomic DNA, and contain the alternatively spliced exon and its flanking constitutive exons. In the majority of cases, these parts can be amplified by long-range PCR. If the introns are too long, several kilobases flanking the exons are amplified and ligated together (see Protocol 1).

An alternative to cloning large pieces of DNA is the usage of exon trap vectors, such as the exon trap cloning system (Mobitec, Goettingen, Germany). Here an alternative exon is cloned between two constitutive exons, usually derived from insulin. This chimeric gene is then analyzed similar to a genomic construct. This approach has two advantages: (1) often exon trap constructs are easier and faster to clone and (2) if these constructs behave like the endogenous gene, it is clear that the regulatory region is confined to the cloned exon. However, it has often been found that there is a discrepancy between alternative exons flanked by either their normal or heterologous contexts [2].

46.3.2
Transfection of Cells

The minigenes can be transfected into the cell line of choice by standard methods, including calcium phosphate, electroporation or liposome transfer. We routinely use HEK293 cells transfected with the calcium phosphate method to analyze alternative splicing *in vivo* (Protocol 2). When studying tissue-specific alternative splicing, usually several cell lines are screened to find a cell line that recapitulates the alternative splicing pattern observed *in vivo*. As a result, minigenes have been analyzed in numerous cell lines, including HeLa, HEK293, primary neurons, HepG2, CHO (see Table 46.1 and additional material under http://www.ebi.ac.uk/asd/minigene/index.html). Once such a system is established, two major questions can be addressed: which *cis*-sequences are necessary for the regulation and what *trans*-acting factors are involved (Fig. 46.1). *Cis*-acting sequences are usually determined by mutagenesis, either at specific sites or through deletion of larger parts. In a number of studies mutations resembling human mutations were analyzed by minigenes [3]. The role of *trans*-acting factors is usually studied by increasing their amount through co-transfection. Since splicing factors mostly work in a concentration-dependent manner [4], a correlation between alternative exon usage and amount of *trans*-acting factor is a good indication for regulation of this exon by that particular factor. However, since indirect effects will occur, controls and additional experiments have to be performed. Frequently observed indirect effects are sequestration, influence of mRNA stability or interference with the general splicing machinery.

Although in most cases regulatory factors are increased through co-transfections,

Tab. 46.1. Overview of existing minigenes.

A: minigenes containing one cassette exon

name	species	tissue specificity	minigene	cell line	ref
GABA$_A$γ2, 24 nt exon	R	neuronal / non-neuronal		HeLa	[26]
clathrin light chain B, exon EN	R	neuronal / non-neuronal		primary rat neuronal cell, rat primary glia cells	[27]
src, exon N1	M	neuronal / non-neuronal		neuronal LAN5, neuroblastoma, HeLa, HEK293	[28]
NCAM, exon 18	M	neuroblastoma		N2A, non-muscle fibroblast, myoblast	[29]
MHC-B, exon N30	H	neuronal		neuronal retinoblastoma Y79	[30]
fibronectin, EDI	H	HeLa cells		HeLa	[31]

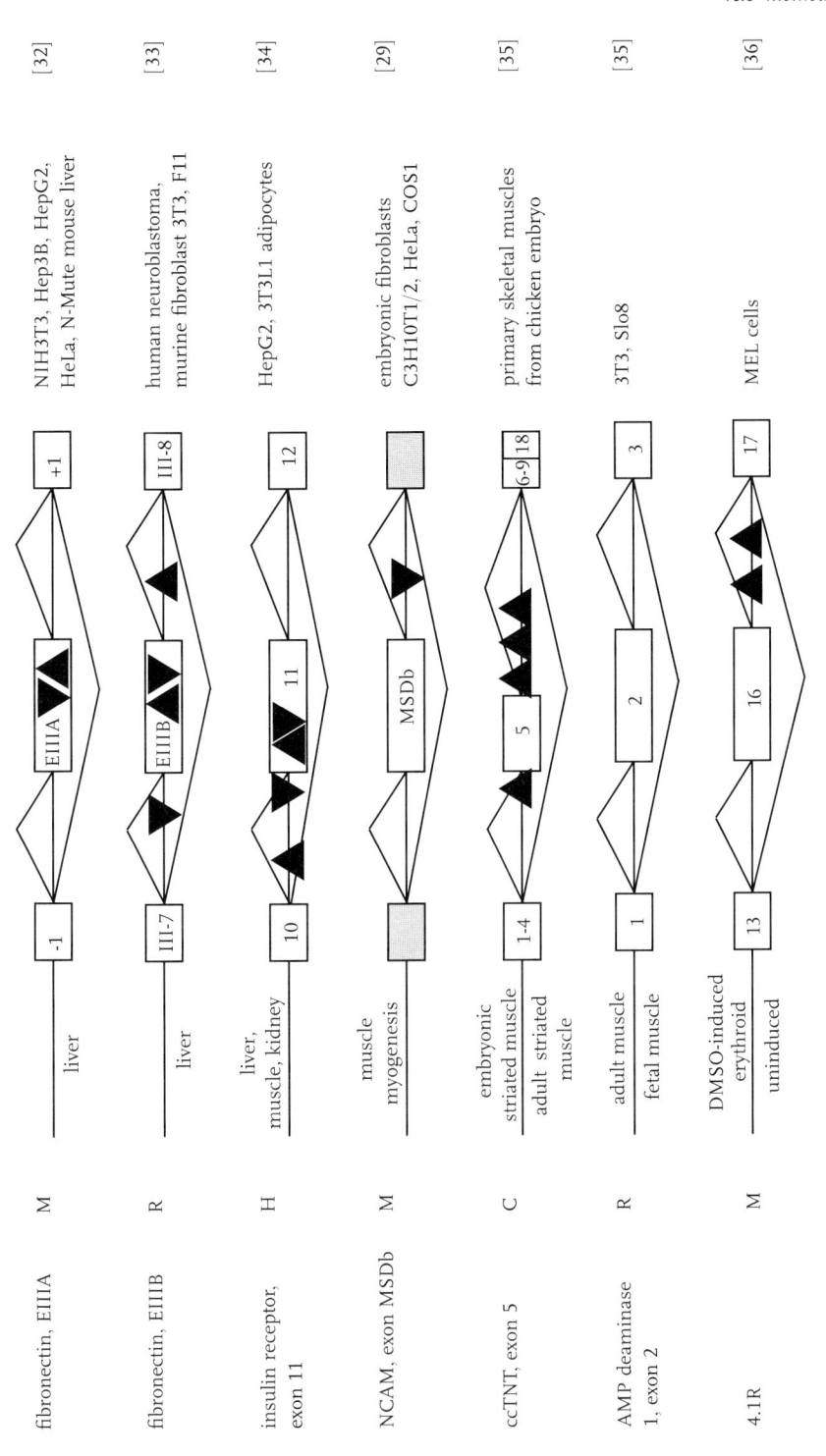

Gene	Species	Tissue/Condition	Splicing Pattern	Cell Lines	Ref
fibronectin, EIIIA	M	liver	-1 / EIIIA / +1	NIH3T3, Hep3B, HepG2, HeLa, N-Mute mouse liver	[32]
fibronectin, EIIIB	R	liver	III-7 / EIIIB / III-8	human neuroblastoma, murine fibroblast 3T3, F11	[33]
insulin receptor, exon 11	H	liver, muscle, kidney	10 / 11 / 12	HepG2, 3T3L1 adipocytes	[34]
NCAM, exon MSDb	M	muscle myogenesis	MSDb	embryonic fibroblasts C3H10T1/2, HeLa, COS1	[29]
ccTNT, exon 5	C	embryonic striated muscle / adult striated muscle	1-4 / 5 / 6-9 18	primary skeletal muscles from chicken embryo	[35]
AMP deaminase 1, exon 2	R	adult muscle / fetal muscle	1 / 2 / 3	3T3, Slo8	[35]
4.1R	M	DMSO-induced erythroid / uninduced	13 / 16 / 17	MEL cells	[36]

Tab. 46.1 (continued)

name	species	tissue specificity	minigene	cell line	ref
CASR	H	fetal kidney, HEK293 parathyroid, liver, thyroid, adult kidney	II — III — IV	HEK293, lymphoblastoid cell	[37]
caspase-2	H	skeletal muscles, brain ovaries, thymus, and spleen	8 — 9 — 10	HeLa, 293T	[38]
CFTR	H	FNMut promoter α-globin promoter	— 9 —	Hep3B	[39]
CYP3A5	H		5 — 6 — 7	CaCo-2	[40]
APP	H / M	neuron peripheral tissues, non-neuron	14 — 15 — 16	NTH3T3, P19, N2A, AtT20	[41]
hnRNPA1	H / M	HeLa cells	7 — 7B —	HeLa	[42]

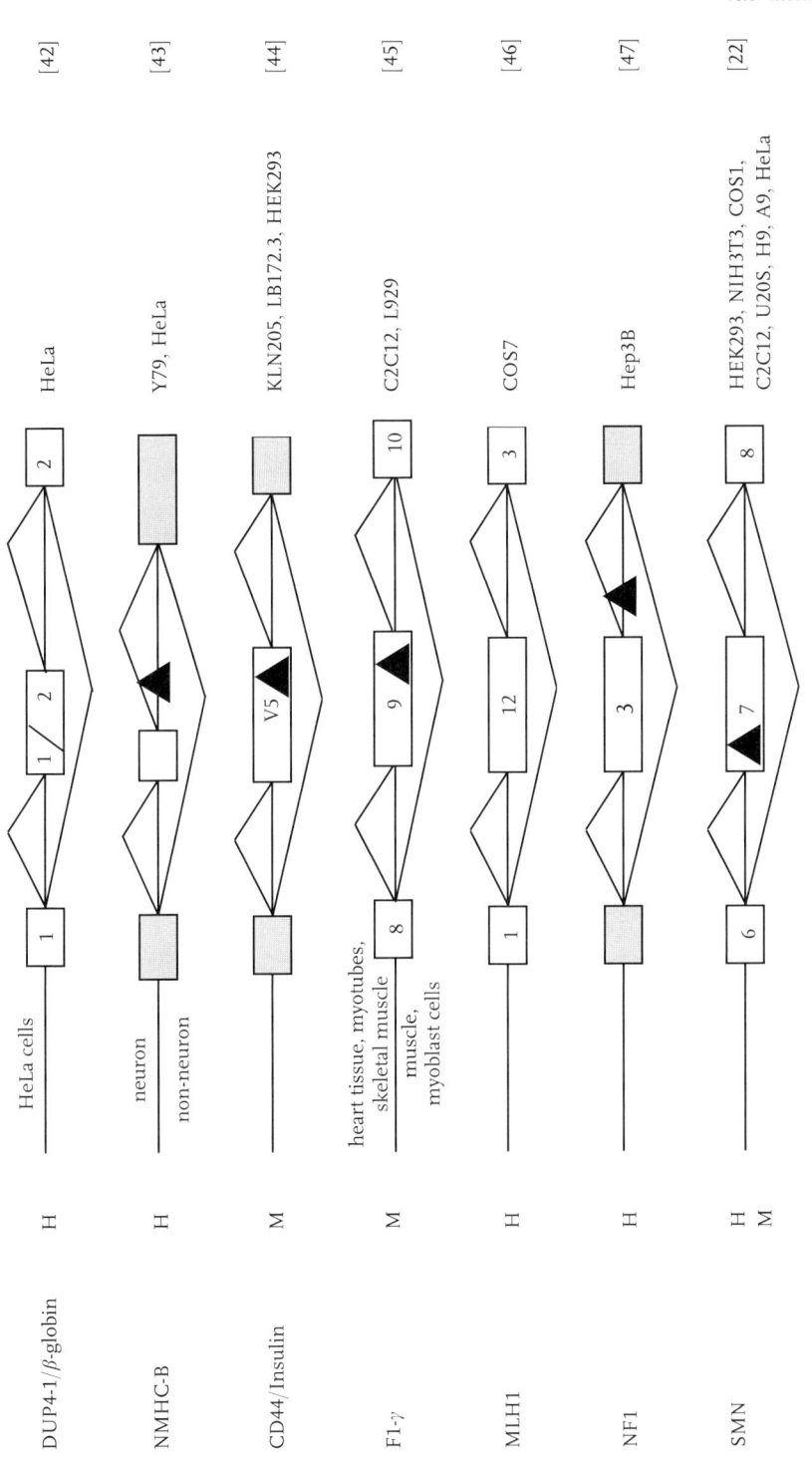

Gene	Species	Tissue/Cells	Construct (exons)	Cell lines	Ref
DUP4-1/β-globin	H	HeLa cells	1 — 1/2 — 2	HeLa	[42]
NMHC-B	H	neuron / non-neuron	□ — □▲ — □	Y79, HeLa	[43]
CD44/Insulin	M		□ — V5▲ — □	KLN205, LB172.3, HEK293	[44]
F1-γ	M	heart tissue, myotubes, skeletal muscle muscle, myoblast cells	8 — 9▲ — 10	C2C12, L929	[45]
MLH1	H		1 — 12 — 3	COS7	[46]
NF1	H		□ — 3▲ — □	Hep3B	[47]
SMN	H / M		6 — 7▲ — 8	HEK293, NIH3T3, COS1, C2C12, U2OS, H9, A9, HeLa	[22]

Tab. 46.1 (continued)

name	species	tissue specificity	minigene	cell line	ref
spastin exon 5	H	lymphocytes COS1 cells		COS1	[48]
spastin exon 9	H	lymphocytes COS1 cells		COS1	[48]
spastin exon 11	H	lymphocytes COS1 cells		COS1	[48]
tau exon 10	R / H	neuron COS1 cells, PC12 cells / brain		rat PC12, rat AR42J, monkey COS1, N2A, HeLa RB	[49]
tau exon 2	H	COS, NT2 and SKN cells		SKN, COS, N-Tera2 (NT2)	[50]
tau, exon 6	H	skeletal muscle, brain		Chinese hamster	[51]

46.3 Methods | 763

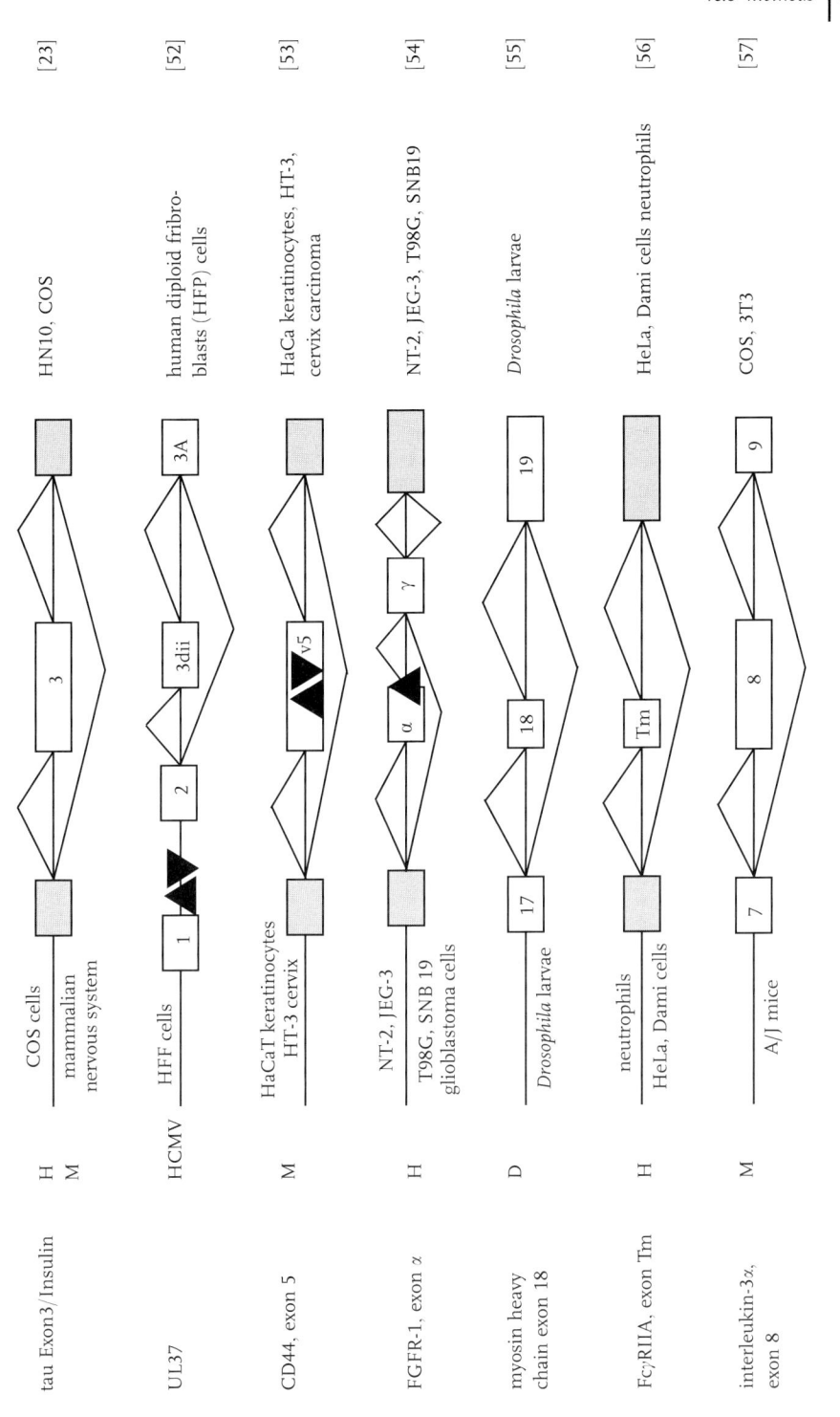

Gene	Species	Source cells	Diagram	Expression cells	Ref
tau Exon3/Insulin	H / M	COS cells / mammalian nervous system		HN10, COS	[23]
UL37	HCMV	HFF cells		human diploid fibroblasts (HFP) cells	[52]
CD44, exon 5	M	HaCaT keratinocytes HT-3 cervix		HaCa keratinocytes, HT-3, cervix carcinoma	[53]
FGFR-1, exon α	H	NT-2, JEG-3 T98G, SNB 19 glioblastoma cells		NT-2, JEG-3, T98G, SNB19	[54]
myosin heavy chain exon 18	D	*Drosophila* larvae		*Drosophila* larvae	[55]
FcγRIIA, exon Tm	H	neutrophils HeLa, Dami cells		HeLa, Dami cells neutrophils	[56]
interleukin-3α, exon 8	M	A/J mice		COS, 3T3	[57]

Tab. 46.1 (continued)

name	species	tissue specificity	minigene	cell line	ref
DHFR, exon 2A, reporter	HA	inactive DHFR / active DHFR		Chinese hamster	[58]
HIV-1, exon 6D	HIV-1			CEM CD4T-cell, HeLa-Tat	[59]
SRp20, exon 4	H			murine B, lymphoma K46	[60]
PPT, exon 4	R			NIH3T3, P19, N2A, AtT20	[61]

B: minigenes containing multiple cassette exons

name	species	tissue specificity	minigene	cell line	ref
fast skeletal TnT, exons 4–7	R		31 combinations of inclusion	COS, HeLa, nonmuscle cells	[62]
CD45, exons 4–6	H / M	B-cells / T-cells		B-cells, T-cells, thymoma cells (EL4, NIH3T3)	[63]

spastin	H		COS1	[48]

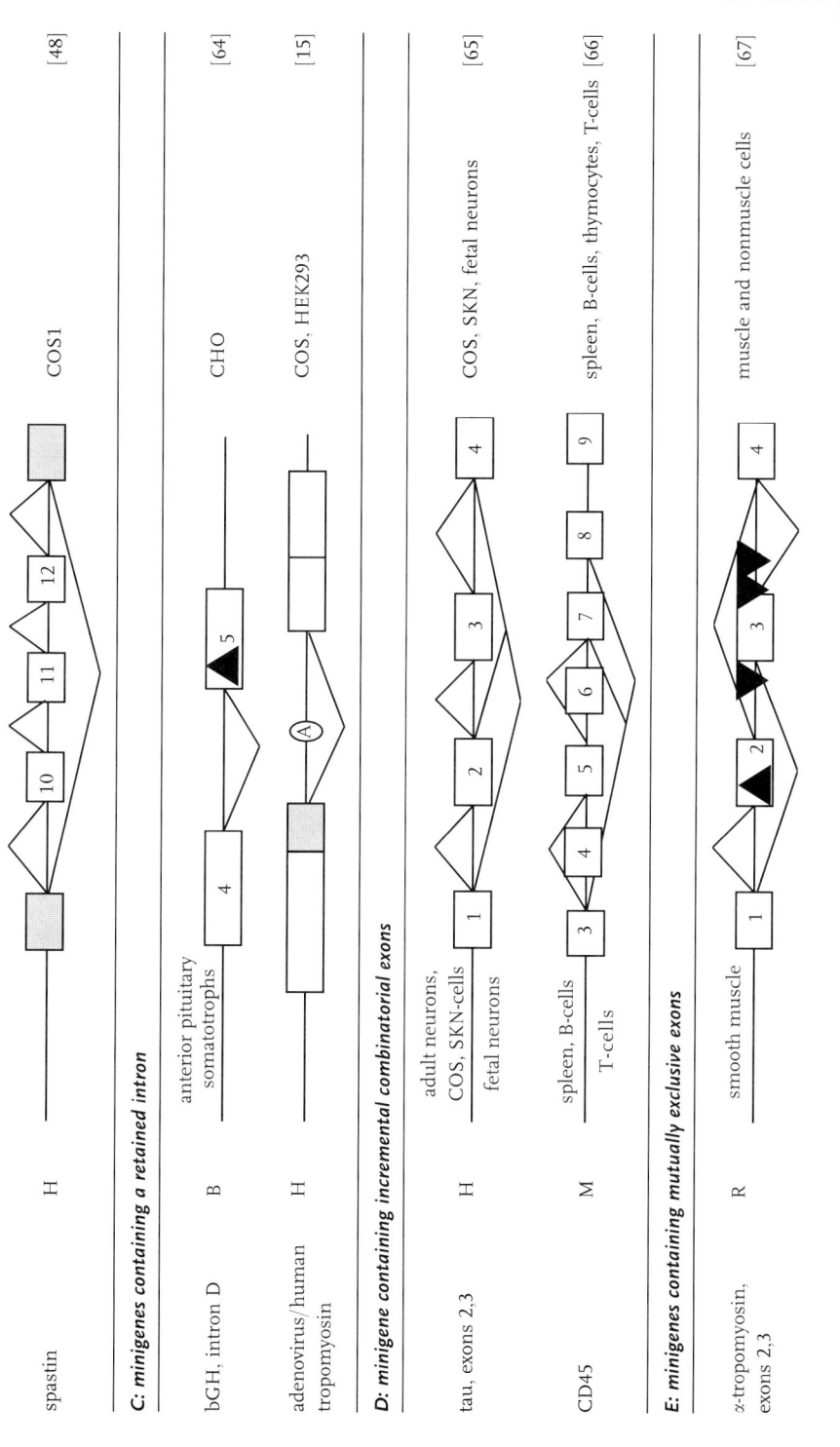

C: minigenes containing a retained intron

bGH, intron D	B		anterior pituitary somatotrophs	[64]
adenovirus/human tropomyosin	H		COS, HEK293	[15]

D: minigene containing incremental combinatorial exons

tau, exons 2,3	H		adult neurons, COS, SKN-cells fetal neurons	[65]
CD45	M		spleen, B-cells T-cells	spleen, B-cells, thymocytes, T-cells [66]

E: minigenes containing mutually exclusive exons

α-tropomyosin, exons 2,3	R		smooth muscle muscle and nonmuscle cells	[67]

Tab. 46.1 (continued)

name	species	tissue specificity	minigene	cell line	ref
α_1-tropomyosin, exons NM, SK	H	non-muscle / skeletal muscle	4 — NM / SK — 6	COS1, myoblasts	[68]
β-tropomyosin, exons 6A,B	C	smooth muscle+ non-muscle / skeletal muscle	5 — 6A / 6B — 7	mouse and quail muscle cells, HeLa	[69]
β-tropomyosin, exons 6,7	R	smooth muscle+ non-muscle / skeletal muscle	5 — 6 / 7 — 8	HeLa	[70]
pyruvate-kinase M, exons 9,10	H	skeletal muscle, heart and brain	8 — 9 / 10 — 11	dRLh-84cells and hepatocytes	[71]
albumin, exons G, H	R		F — G / H / I — J	nonhepatic cells, COS1, Hepatoma cells, HLE	[72]
MLC, exons 1/3	R	promoter upstream of exon 2 / promoter dependent splicing / promoter upstream of exon 1	2 — 3 — 4 — 5 / 1 — 3 — 4 — 5	HeLa	[73]

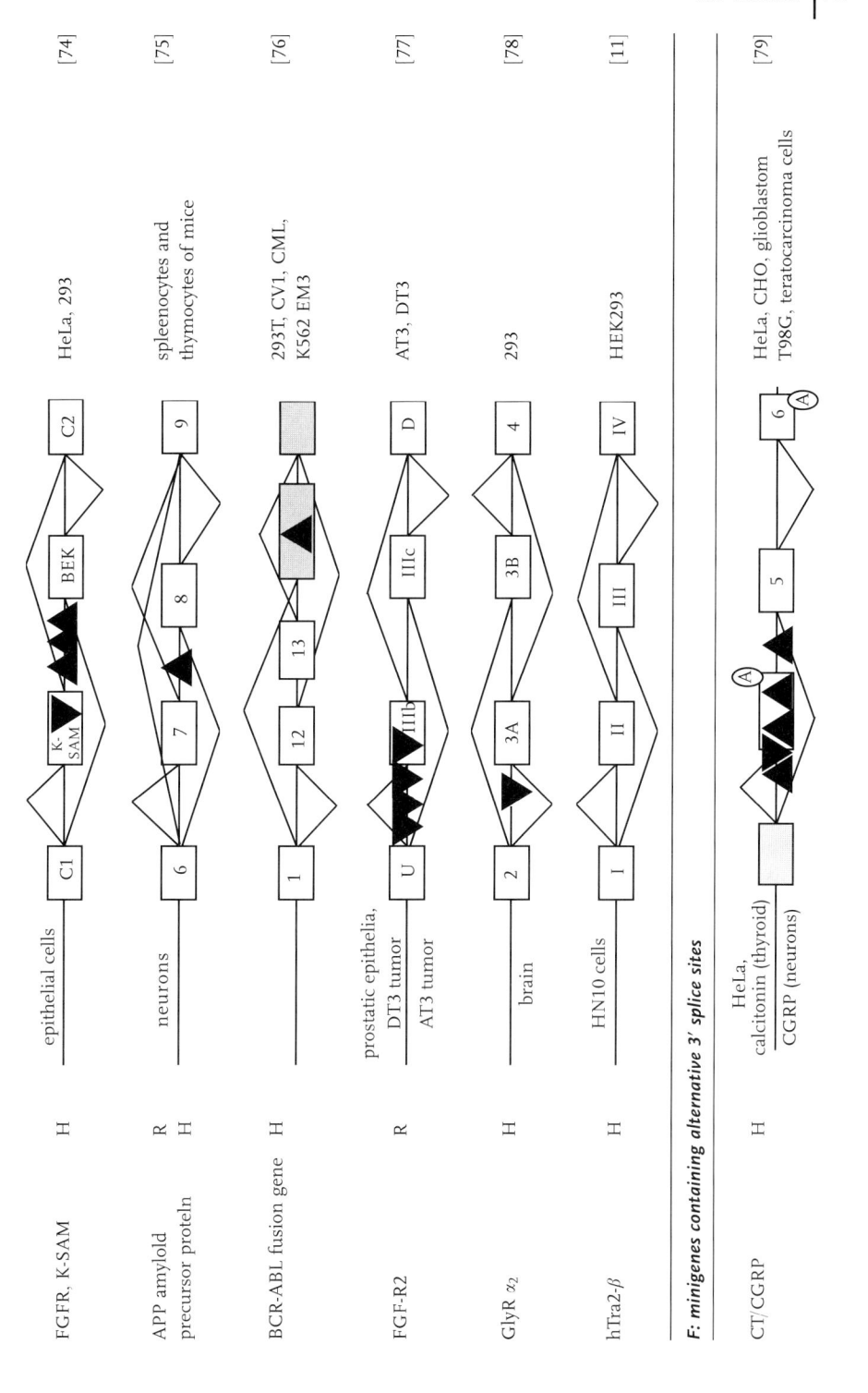

Gene	Species	Tissue/Context	Diagram	Cells	Ref
FGFR, K-SAM	H	epithelial cells	C1 — K-SAM — BEK — C2	HeLa, 293	[74]
APP amyloid precursor protein	R H	neurons	6 — 7 — 8 — 9	spleenocytes and thymocytes of mice	[75]
BCR-ABL fusion gene	H		1 — 12 — 13 — ☐	293T, CV1, CML, K562 EM3	[76]
FGF-R2	R	prostatic epithelia, DT3 tumor / AT3 tumor	U — IIIb — IIIc — D	AT3, DT3	[77]
GlyR α₂	H	brain	2 — 3A — 3B — 4	293	[78]
hTra2-β	H	HN10 cells	I — II — III — IV	HEK293	[11]

F: minigenes containing alternative 3′ splice sites

Gene	Species	Tissue/Context	Diagram	Cells	Ref
CT/CGRP	H	HeLa, calcitonin (thyroid) / CGRP (neurons)	☐ — 5 — 6 — Ⓐ	HeLa, CHO, glioblastom T98G, teratocarcinoma cells	[79]

Tab. 46.1 (continued)

name	species	tissue specificity	minigene	cell line	ref
dsx RO	D	female / male		Drosophila KC cells	[80]
M-tra	D	female / male		Drosophila melanogaster	[81]
BPV-1	BPV-1			HeLa	[82]
CFTR	H	nasal epithelial, lymphoblastoid cells		COS7, IB3	[83]
calcitonin/DHFR	HA			293	[84]
calcitonin with PNP	H	neuronal cells / thyroid cells		MTC, T98G, glioblastoma, HepG2	[85]

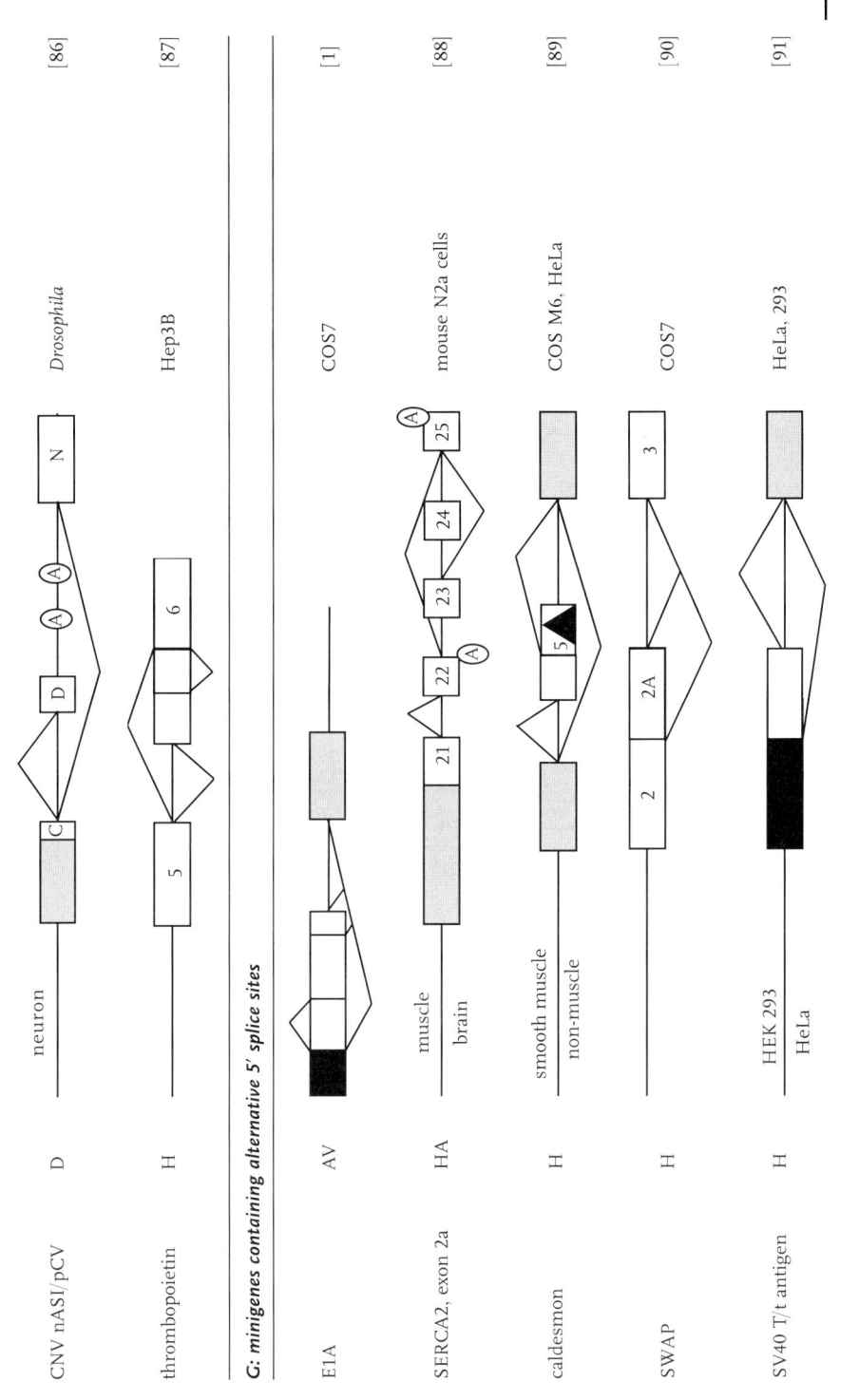

CNV nASI/pCV	D		neuron	*Drosophila*	[86]
thrombopoietin	H			Hep3B	[87]

G: minigenes containing alternative 5′ splice sites

E1A	AV			COS7	[1]
SERCA2, exon 2a	HA		muscle / brain	mouse N2a cells	[88]
caldesmon	H		smooth muscle / non-muscle	COS M6, HeLa	[89]
SWAP	H			COS7	[90]
SV40 T/t antigen	H		HEK 293 / HeLa	HeLa, 293	[91]

Tab. 46.1 (continued)

name	species	tissue specificity	minigene	cell line	ref
β-globin, β-thalassemic allele	H			HeLa, neuron	[92]
Adenovirus/human tropomyosin	H			COS, 293	[15]

H: minigene containing alternative 3'- and 5'-splice sites

fibronectin, exon IIICS	H	liver, adult: skipping of CS1 and CS5		COS7; HeLa	[93]

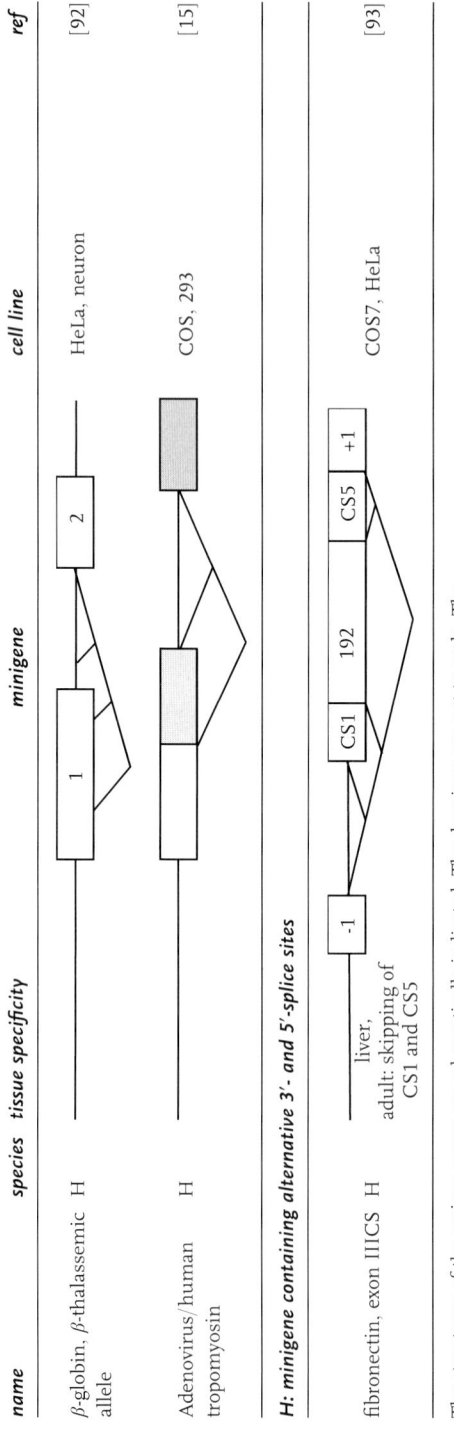

The structures of the various genes are schematically indicated. The drawings are not to scale. The various genes are sorted according to their splicing mechanism (panels A–H). Stimulatory effects of *trans*-acting factors binding to exonic splicing enhancers (ESE) or intronic enhancers (ISE) are indicated by an upward triangle. An inhibitory effect by *trans* acting factors binding to exonic (ESS) or intronic (ISS) splicing silencers is indicated with a downward triangle. (A) Alternative polyadenylation site. Cells used to analyze the minigenes are indicated. The collection of minigenes can be viewed under: http://www.ebi.ac.uk/asd/minigene/index.html – in this site, regulatory factor names are indicated if known and the minigenes are linked to the alternative exon database [21, 24]. Exons shaded in grey are heterologous exons, usually from exon trap-vectors. Sp: species, R: rat, M: mouse, H: human, C: chicken, D: *Drosophila*, HA: hamster, B: bovine, AV: adenovirus. For space reasons, only one reference per minigene is given.

an increasing number of experiments are reported that use oligonucleotides or RNA interference to decrease the amount of *trans*-factors [5]. Furthermore, the modification of *trans*-acting factors through phosphorylation has been studied by employing the appropriate kinases [2, 6, 7]. Finally, the effect of putative drugs on splice site selection has been determined using minigenes [8].

46.3.3
Analysis

46.3.3.1 RT-PCR

By far the most experiments are analyzed by RT-PCR (Protocol 3). To achieve reproducibility, it is important that mRNA is not damaged during isolation. In our experience this is best achieved with commercially available spin-column-based kits (e.g. RNeasy; Qiagen). At least one primer is chosen to be specific for the minigene to avoid amplification of the endogenous gene. RT-RCR should be performed with the lowest amount of cycles possible to ensure a linear relation between mRNA isoforms and amplified signals. A frequently occurring problem is the amplification of minigene-DNA if the construct is short enough to allow amplification. This can be avoided by shortening the extension time or by adding *Dpn*I (New England Biolabs) into the reverse transcription reaction. *Dpn*I cuts GATC sequences in double-stranded DNA when the adenosine is methylated but does not cut non-methylated single-stranded DNA or cDNA. We found that commercially available preparations of *Dpn*I (New England Biolabs) are essentially RNase-free and do not interfere with reverse transcription. The primers for amplification are usually chosen in flanking alternative exons. One primer has to be minigene-specific to avoid amplification of the endogenous gene and is usually complementary to the 5′ end of the transcript derived from the vector. As a result of this amplification, two products corresponding to exon skipping and inclusion are formed. PCR products are separated on agarose or acrylamide gels, the intensity of the bands is quantified and their ratio determined. The detection of the PCR products can be done by ethidium bromide staining (Fig. 46.2), or by labeling the primers with ^{32}P or a fluorescent oligonucleotide [3]. The isoforms are detected by UV light, autoradiography or by running them on a DNA sequencing machine, respectively. Numerous alternatively spliced mRNAs have been quantified from different tissues using real-time PCR with boundary spanning TaqMan probes [9] or molecular beacons and scorpion primers [10]. However, so far none of these methods have been applied to minigene analysis. In our hands, we found TaqMan probes and beacons unreliable in detecting splice variants. As an alternative, we employ primers that amplify only one isoform and quantify them in parallel using SYBR-green in the PCR reaction [11].

46.3.3.2 Other Analysis Methods

RNase protection assays [12] have been used to analyze splicing patterns in minigenes. The quantitative nature of this assay is the major advantage. However, the method is much more laborious than RT-PCR. An interesting recent development

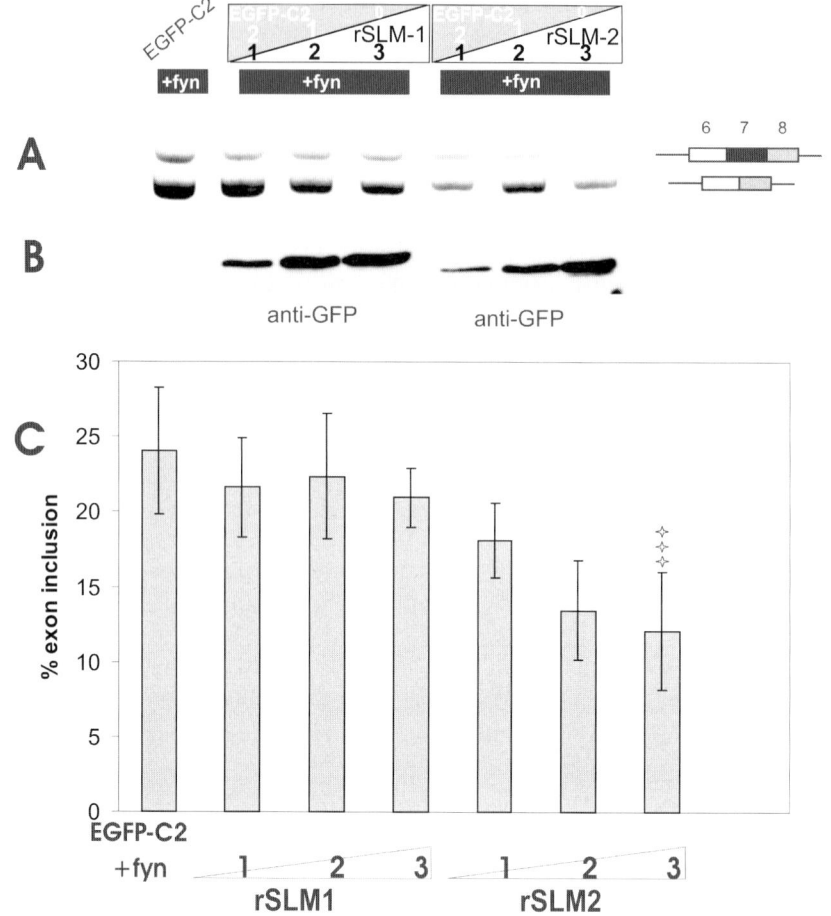

Fig. 46.2. Example of RT-PCR analysis. Alternative splicing of the SMN2 minigene [22] was analyzed with an increasing amount of trans-acting factor rSLM-1 and rSLM-2 [23] in the presence of the tyrosine kinase p59^{fyn}. (A) The transfection scheme is schematically indicated on the top. The amount of rSLM-1 or rSLM-2 was increased from 0 to 3 μg, as indicated by the black numbers. The total amount of transfected DNA was balanced with the parental; "empty" vector pEGFP-C2, as indicated by the white numbers. In all experiments, 0.5 μg of p59^{fyn} was present (black bars). The bands show a representative ethidium bromide-stained agarose gel. The SMN2 alternative splicing pattern is schematically indicated on the right. (B) Western blot analysis showing the expression of EGFP-tagged rSLM-1 (left) and rSLM-2 (right). Note that the protein levels increase proportionally to the amount of transfected DNA. An antibody against GFP was used. (C) Statistical evaluation of four independent experiments. In the presence of p59^{fyn}, the difference between in the splicing pattern of SMN2 is statistical significant with 3 μg of either rSLM-1 or rSLM-2 ($p < 0.05$, $t = -2.52$). As in (A), numbers under the columns indicate the micrograms of cDNA transfected.

is the analysis of alternative splicing by array formats [13]. Finally, assay systems have been developed that rely on the proteins generated by alternative splicing of minigenes. These different isoforms can be detected by Western blots [11] if specific antibodies are available (Fig. 46.1C). Related to this method are chimeric minigenes that express enhanced green fluorescent protein (EGFP) fusions depending on alternative exon usage. Because splicing events of these constructs can be detected by EGFP fluorescence, they are suitable for FACS analysis and genetic screening [14]. A direct measurement of alternative splicing is possible when luciferase reporters are employed, which has been used to determine signal transduction pathways [6, 7] regulating alternative splicing (Fig. 46.1C). To account for variations in transfection efficiency, double reporter assays have been developed [15].

46.3.4
Necessary Controls

The analysis of alternative splicing with minigenes requires several controls. First, it is important to determine the transfection efficiency in each experiment, which can be easily done by fluorescent microscopy when GFP-tagged constructs are used. When assaying the influence of *trans*-acting factors, Western blots need to be performed to determine whether an increase of cDNA expression constructs really cause an increase of protein generated. A first step to elucidate whether a *trans*-acting factor is acting directly on the pre-mRNA of interest is to determine whether it co-immunoprecipitates with the pre-mRNA (Protocol 4), which has even been performed with endogenous genes [16]. To account for unspecific effects, the amount of cDNA transfected and the amount of promoter-encoding DNA should be maintained constant. This is usually done by adding "empty" parental vector DNA. In most cases, minigene analyses are quite robust. However, the changes of alternative splicing *in vivo* are often relatively small, around 2- to 3-fold. It is therefore necessary to determine the statistical significance of the experiments by calculating the standard deviations and performing Student's test, which can be done on the World Wide Web (http://www.physics.csbsju.edu/stats/t-test.html).

46.3.5
Advantages and Disadvantages of the Method

The major disadvantage of the method is its susceptibility to indirect effects, such as sequestration of factors, as well as to changes in the cells used, e.g. their growth rate, proliferation status and density. These disadvantages are compensated by numerous advantages. In contrast to *in vitro* work, there are no length requirements for the minigene constructs that therefore resemble more their physiological counterparts. Essentially all transfectable cells can be used, which allows the analysis of tissue-specific splicing. In contrast, *in vitro* analysis requires nuclear extracts from fibroblasts, mostly HeLa cells. Finally, the use of minigenes in living cells enables the investigator to analyze dynamic processes, e.g. the influence of signal transduction pathways on alternative splicing.

46.3.6
Related Methods

The basic method has been modified in various ways. Minigenes have been used to determine binding sites for *trans*-acting factor by *in vivo* SELEX (Systematic Evolution of Ligands by EXponential enrichment) [17]. Here, a randomized sequence is cloned into an alternative exon and the complete mixture is transfected with a *trans*-acting factor. After RT-PCR, the mixture of alternative exons is isolated and recloned for a second round. This method was successfully used to identify AC-rich splicing enhancer [18]. Other modifications of the minigene approach include the use of kinases and DNA or RNA oligonucleotides to either phosphorylate or remove regulatory factors.

46.4
Troubleshooting

Transfection Efficiency

The most crucial parameter for the success of an *in vivo* splicing experiment is the transfection efficiency, especially when co-transfections with putative *trans*-acting factors are performed. We therefore usually employ EGFP-tagged cDNA expression constructs in co-transfection experiments that allow an easy monitoring of the transfection efficiency. Cell lines that allow high transfection efficiency should be used whenever possible. We routinely use HEK293 cells and reach 90% transfection efficiency. When using the calcium phosphate method, reasons for lower efficiencies are usually dense seeding of cells, a high passage number of cells or a deviation of the pH of the transfection solution. A wrong pH value during transfection can be caused by the wrong composition of the HBS buffer, that should have pH 7.05 or by the wrong CO_2 atmosphere that should be 3% during transfection.

Reproducibility

In vivo splicing assays are generally well reproducible when several parameters are kept constant. For transfection, cells should be always plated at the same density. It is also important to keep the time between seeding and transfection, as well as the actual transfection time constant.

Unreproducible Bands

Sometimes, we observed sporadic, unreproducible bands whose appearance depended on the transfection time. When using new minigenes, the transfection time should therefore be optimized.

Autoregulation

Several splicing factors seem to autoregulate their expression levels e.g. tra2-beta1, hnRNA1 and SF2/ASF. This can result in a substitution of the endogenous protein by the transfected cDNA, which means that the concentration of this splicing factor will not be dramatically changed. The autoregulation needs some time to occur and, if observed, the time between transfection and cell harvesting can be shortened. Therefore, it is best to perform the analysis in transient transfection systems at specific time points. Western blots are needed to monitor effects on *trans*-acting factors.

DNA Contamination
As with all PCR-based methods, DNA contaminations are a major problem. It is therefore advisable to make aliquoted stocks of all solutions. If a contamination occurs, the affected stock solutions are discarded. In all experiments, controls without any added cDNA or RNA have to be routinely included. If possible, the laboratory space where the reverse transcription and the setup of the PCR reaction are performed should be separated from the analysis of the PCR products.

Heterodimers
Often, the simultaneous generation of two PCR products that differ only in the presence or absence of short exonic sequences results in the formation of a heteroduplex that consists of two DNA strands differing by this exonic sequence. The heteroduplex usually migrates as a third PCR product. In our hands, heteroduplex formation increases when the reaction products are stored for longer time or if too many cycles in the PCR amplification are used. These parameters should therefore be minimized. If the problem persists, primers can be chosen that amplify each isoform individually.

46.5
Bioinformatic Resources

Prior to starting experiments employing minigenes, a bioinformatics analysis should be performed. If the *trans*-acting factors are unknown, putative splicing enhancer and their *trans*-acting factors can be determined by the ESE finder (http://exon.cshl.edu/ESE/) [19]. The RNA workbench (http://www.ebi.ac.uk/asd-srv/wb.cgi) compares sequences to know *trans*-acting factor binding sites and therefore generates similar ideas that can be experimentally tested. Alternative exons have been compiled in several databases (reviewed in [20]) that are useful to find related regulated genes. Finally, the compilation of minigenes in Table 46.1 is available at the website of the alternative splicing database consortium at (http://www.ebi.ac.uk/asd) and is linked to the alternative splicing database [21]. The datasets can be searched with query sequences.

46.6
Protocols

Protocol 1: Minigene construction
Artificial chromosome systems and the genomic subclones in λ phages are the best templates for the construction of minigenes. The genomic clones containing the alternatively spliced exon(s) of interest with the flanking constitutive exons are veri-

fied by Southern blot hybridization using standard procedures [25]. In the absence of any suitable genomic clones, genomic DNA prepared by standard protocols can serve as template for PCR amplification. However, the PCR amplification from the genomic DNA is often more difficult. If the genomic region of interest is surrounded by repeat elements, nested or touchdown PCR replaces the conventional PCR for better amplification. In case the intronic sequence flanking the alternative spliced exon is too large to be PCR amplified, part of the flanking introns will be amplified and ligated together. Often, sequence comparison between various species can narrow down the intronic regulatory sequences. In most cases, about 2 kb around each exon are sufficient for regulation.

Restriction sites absent from the PCR fragment or the genomic clones are introduced with the PCR primers (Fig. 46.1A). They should be placed in the most 5′ part of the primer. The part of the primers complementary to the genomic clone should have an annealing temperature between 62 and 65 °C to ensure specificity of the reaction. The suitable primers with appropriate annealing temperatures and desired GC content can be designed with freely available web interfaced primer3 software under http://www-genome.wi.mit.edu/genome_software/other/primer3.html.

The following reaction conditions are used as a starting point for amplification:

Long-range PCR amplification is performed according to the protocol supplied by the manufacturer (SAWADY Long PCR System, Peqlab Biotechnologie). For target sizes less than 30 kb, the following reaction setup is used:

36.5 µl	H$_2$O
5 µl	10 × long-range PCR buffer
2.5 µl	10 mM dNTPs
4.5 µl	25 mM MgCl$_2$
1 µl	template DNA (PAC, BAC or phage clones) (10 pg/µl)
0.5 µl	of a mixture of *Taq* and a high-fidelity thermostable polymerase with proofreading activity

The reactions are assembled on ice and amplification is performed using the following thermocycler settings: initial denaturation for 2 min at 93 °C; 10 cycles with 10 s denaturation at 93 °C, extension at 68 °C (allow 30–60 s extension per 1 kb); 15–20 cycles with 10 s denaturation at 93 °C, 30 s annealing at 65 °C, extension at 68 °C. Increase the extension time (30–60 s per 1 kb) for 20 s every cycle to compensate for enzyme inactivation; final extension for 7 min at 68 °C. The high annealing temperature of 68 °C is necessary for product specificity. Analyze 5–10 µl from the PCR reaction on a 0.8% agarose gel.

The gel purification and cloning of the PCR product into the pCR-XL-TOPO vector (Invitrogen) is performed according to the manufacturer's protocol with the following modification: mix the cloning reaction by adding 0.5 µl pCR-XL-TOPO vector to 2 µl of the gel-purified PCR product. After incubation for 5 min at room temperature, use the entire reaction for bacterial transformation.

Finally, the minigene is recloned from the pCR-XL-TOPO vector into an eukaryotic expression vector using the unique restriction sites introduced by the PCR primers. We found that SV40 promoters or CMV promoters work well for minigene analysis in many cell lines.

Protocol 2: Transient transfection using calcium phosphate

- Transient transfection of adherent HEK293 cells is performed using the calcium phosphate method on 35 mM plates (six-well tissue culture plate). The day before transfection, 3.0×10^5 cells per plate are seeded in 3 ml DMEM/10% FCS. This leads to approximately 40–60% confluency on the day of transfection. After splitting, the cells are incubated at 37 °C in 5% CO_2 for 17–24 h.
- Splicing assays are based on the titration of increasing amounts of plasmid DNA expressing a splicing factor to a constant concentration of minigene DNA. To avoid "squelching" effects, the "empty" parental expression plasmid containing the promotor is added to ensure a constant amount of transfected DNA (Fig. 46.2A).
- The standard assay employs five reactions, each containing 1–2 μg of minigene DNA and an increasing amount of plasmid DNA expressing a splicing factor. The amount of minigene used has to be optimized in several trial experiments. 0, 0.5, 1, 1.5 and 2 μg of splicing factor DNA is a good starting point for the titration of splicing factors. The appropriate amount of empty vector (2, 0.5, 1, 1.5 and 0 μg) is added to ensure that equal amounts of DNA are transfected. The DNA solutions are brought to a total volume of 75 μl with water and 25 μl 1 M $CaCl_2$ are added. While mixing the DNA/$CaCl_2$ solution with a vortex, 100 μl of $2 \times$ HBS is added dropwise.
- 1 M $CaCl_2$ solution: Dissolve 5.4 g $CaCl_2 \cdot 6H_2O$ in 20 ml H_2O, sterilize by filtration, store at -20 °C; $2 \times$ HBS: Dissolve 1.6 g NaCl (280 mM), 0.074 g KCl (10 mM), 0.027 g $Na_2HPO_4 \cdot 2H_2O$ (1.5 mM), 0.2 g dextrose (12 mM) and 1 g HEPES (50 mM) in 90 ml H_2O. Adjust pH to 7.05 with NaOH, then bring up to a total volume of 100 ml with H_2O. Sterilize by filtration, store at -20 °C.
- The mixture is incubated for 10–20 min at room temperature to allow the calcium phosphate–DNA precipitate to form.
- The precipitates are resuspended by pipetting and the complete mixture is added dropwise to the cultured cells.
- The dishes are incubated at 37 °C in 3% CO_2 overnight.
- After the incubation, a fine precipitate is visible on the cells. The transfection efficiency can be estimated by fluorescence microscopy if an EGFP-tagged construct is used and should be at least 50% with HEK293 cells. If the splicing factor itself is not EGFP tagged, the use of pEGFP-C2 (Clontech, Heidelberg, Germany) as an "empty" vector can help to monitor the transfection.

Protocol 3: RT-PCR analysis

- RNA is isolated 17–24 h after transfection using an RNeasy mini kit (Qiagen, Hilden, Germany) following the manufacturer's instructions. RNA is eluted in

40 µl RNase-free H$_2$O. The optimal transfection time has to be determined empirically.
- It is important to quantify the mRNA; 4 µl of the RNA is dissolved in 1 ml of water, the OD at 260 nm is measured. The value obtained is multiplied by 10 and this reading gives the concentration in µg/µl.
- Best results are achieved when reverse transcription and PCR are performed immediately after the RNA purification, thus avoiding freezing of the RNA or reverse-transcription reaction.
- For reverse transcription, 1 µg of isolated RNA is mixed with 5 pmol antisense minigene-specific primer in 0.5 µl H$_2$O, 2 µl 5 × RT buffer, 1 µl 100 mM DTT, 1 µl 10 mM dNTP, 3 µl H$_2$O, 0.25 µl RNase inhibitor and 0.25 µl H$^-$ reverse transcriptase (Superscript or Promega). In one sample the RNA is substituted with water as a control. After a brief centrifugation, the tubes are incubated for 45 min in a 42 °C water bath. Optionally 0.2 µl *Dpn*I (New England Biolabs) can be included in each sample to destroy vector DNA.
- During this incubation period, the PCR mixture is prepared. It consists of 50 pmol of sense and antisense primer each, 100 µl 10 × PCR buffer, 20 µl 10 mM dNTPs in a total of 1000 µl water. The optimal MgCl$_2$ concentration for amplification has to be determined empirically in trial experiments and is usually in a range of a final concentration of 1.5–3.0 mM.
- For six reactions, 1 µl *Taq* polymerase is added to 300 µl PCR mixture; 2 µl of the reverse transcriptase reaction are added to 50 µl of this mix and PCR is performed.
- The PCR program must be optimized for each minigene in trial experiments as we found that often identical programs show variations of amplification products when different thermocycler models are used. Using the Gene Amp PCR Systems 9799 from Perkin Elmer Applied Biosystems, we use the following profile for the SMN2 minigene (Fig. 46.2): initial denaturation for 4 min at 94 °C; 30 cycles: 20 s denaturation at 94 °C, annealing at 62 °C for 20 s, extension at 72 °C for 20 s, after 30 cycles final extension at 72 °C for 5 min and cooling to 4 °C. This program can be used as a starting point when optimizing a new reaction.
- The PCR reaction products are analyzed on a 0.3–0.4-cm thick 2% agarose TBE gel.

Protocol 4: Co-immunoprecipitation of pre-mRNA with *trans*-acting factors
In this method, a *trans*-acting factor is co-immunoprecipitated with the pre-mRNA generated by a minigene. This experiment is a first step to determine whether the factor regulates this minigene *in vivo*.
Cells are transfected as described in Protocol 2.

Preparation of cytoplasmic and nuclear extracts

- After transfection, place plates on ice.
- Wash 2× with cold PBS.

- Add 1 ml PBS–EDTA (PBS–1 mM EDTA) and scrape cells. Transfer to 1.5-ml tube.
- Pellet at 3000 r.p.m. for 5 min.
- Resuspend in 250–500 µl (at least 6× the packed cell volume) Harvest buffer (10 mM HEPES, pH 7.9, 50 mM NaCl, 0.5 M Sucrose, 0.1 mM EDTA, 0.5% Triton X-100, freshly add to harvest buffer just before use: 1 mM DTT, 10 mM tetrasodium pyrophosphate, 100 mM NaF, 17.5 mM β-glycerophosphate, 1 mM PMSF, 4 µg/ml aprotinin, 2 µg/ml pepstatin A).
- Incubate on ice for 5 min.
- Pellet at 1000 r.p.m. in a swinging bucket rotor for 10 min to pellet nuclei.
- Transfer the supernatant to a new tube. For best results, clear at 14 000 r.p.m. for 15 min and transfer supernatant to a new tube (= cytoplasmic/membrane extract).
- Wash/resuspend pellet in 500 µl buffer A (10 mM HEPES, pH 7.9, 10 mM KCl, 0.1 mM EDTA, 0.1 mM EGTA, freshly add to buffer A just before use: 1 mM DTT, 1 mM PMSF, 4 µg/ml aprotinin, 2 µg/ml pepstatin A).
- Pellet at 1000 r.p.m. in a swinging bucket rotor for 5 min.
- Remove and discard supernatant.
- Add 4 volumes of buffer C (10 mM HEPES, pH 7.9, 500 mM NaCl, 0.1 mM EDTA, 0.1 mM EGTA, 0.1% IGEPAL(NP-40), freshly add to buffer C just before use: 1 mM DTT, 1 mM PMSF, 4 µg/ml aprotinin, 2 µg/ml pepstatin A). For a more concentrated extract, use 2 volumes of 2 × buffer C.
- Vortex 15 min at 4 °C (in the cold room). Initially, the pellet is dispersed by the highest speed and then dissolved by medium speed of the vortex.
- Pellet at 14 000 r.p.m. for 10 min at 4 °C.
- Transfer the supernatant to new tube (= nuclear extract).

RNA precipitation

- Add the appropriate amount of an antibody against the *trans*-acting factor. If the factor is EGFP tagged, add 1.5 µl of anti-GFP antibody and 50 µl of Protein A–Sepharose, incubate overnight, at 4 °C.
- Wash 5× with RIPA buffer (150 mM NaCl, 50 mM Tris–HCl, pH 7.4, 0.1% SDS, 1% NP-40, 0.5% deoxycholate). Save a portion of this sample for Western blot to test for immunoprecipitation.
- Resuspend the beads in RNase-free DNase buffer (50 mM Tris–HCl, pH 7.4, 6 mM $MgCl_2$, 3 mM $CaCl_2$) with 20 U of DNase I and 20 U of RNase inhibitor at 37 °C for 1 h.
- Extract with phenol/chloroform at 55 °C: adjust the volume of the aqueous phase containing the RNA to 150 µl with TE, pH 8.0 then add 150 µl phenol/chloroform and mix thoroughly. Place at 55 °C for 10 min and then centrifuge for 10 min full speed. Take upper layer and transfer to new tubes.
- Extract with chloroform: add 100 µl of chloroform, invert 2–6 times and spin 2–3 min at room temperature. Take upper layer and transfer to new tubes.
- Ethanol precipitation: add 1/10 volume of 3 M NaAc, pH 5.2 and 2–2.5 volumes

of cold 100% EtOH. Incubate at −80 °C for 20 min (or overnight if required). Centrifuge at 14 000 g for 25 min at 4 °C. Carefully remove ethanol.
- Wash pellet with 300 µl of 70% EtOH, centrifuge at 14 000 g for 5 min at 4 °C. Carefully remove ethanol and dry pellet at room temperature.
- Dissolve in 30–50 µl RNase-free water (or TE pH 8.0).
- Perform RT-PCR as in Protocol 3.

References

1 J. F. Caceres, S. Stamm, D. M. Helfman, A. R. Krainer, Science **1994**, 265, 1706–1709.
2 A. M. Hartmann, D. Rujescu, T. Giannakouros, E. Nikolakaki, M. Goedert, E. M. Mandelkow, Q. S. Gao, A. Andreadis, S. Stamm, Mol. Cell. Neurosci. **2001**, 18, 80–90.
3 M. Nissim-Rafinia, O. Chiba-Falek, G. Sharon, A. Boss, B. Kerem, Hum. Mol. Genet. **2000**, 2000, 1771–1778.
4 C. W. Smith, J. Valcarcel, Trends Biochem. Sci. **2000**, 25, 381–388.
5 M. Vacek, P. Sazani, R. Kole, Cell. Mol. Life Sci. **2003**, 60, 825–833.
6 S. Weg-Remers, H. Ponta, P. Herrlich, H. König, EMBO J. **2001**, 20, 4194–4203.
7 N. Matter, P. Herrlich, H. König, Nature **2002**, 420, 691–695.
8 C. Andreassi, J. Jarecki, J. Zhou, D. D. Coovert, U. R. Monani, X. Chen, M. Whitney, B. Pollok, M. Zhang, E. Androphy, A. H. M. Burghes, Hum. Mol. Genet. **2001**, 10, 2841–2849.
9 I. I. Vandenbroucke, J. Vandesompele, A. D. Paepe, L. Messiaen, Nucleic Acids Res. **2001**, 29, E68–E68.
10 M. Taveau, D. Stockholm, M. Spencer, I. Richard, Anal. Biochem. **2002**, 305, 227–235.
11 P. Stoilov, R. Daoud, O. Nayler, S. Stamm, Hum. Mol. Genet. **2004**, 13, 509–524.
12 K. Rhodes, K. Hall, K. E. Lee, H. Razzaghi, M. Breindl, Gene Expr. **1996**, 6, 35–44.
13 J. M. Yeakley, J. B. Fan, D. Doucet, L. Luo, E. Wickham, Z. Ye, M. S. Chee, X. D. Fu, Nat. Biotechnol. **2002**, 20, 353–358.

14 P. Sheives, K. W. Lynch, RNA **2002**, 8, 1473–1481.
15 M. T. Nasim, H. M. Chowdhury, I. C. Eperon, Nucleic Acids Res. **2002**, 30, e109.
16 J. Ule, K. B. Jensen, M. Ruggiu, A. Mele, A. Ule, R. B. Darnell, Science **2003**, 302, 1212–1215.
17 T. A. Cooper, Methods Mol. Biol. **1999**, 118, 405–417.
18 L. R. Coulter, M. A. Landree, T. A. Cooper, Mol. Cell. Biol. **1997**, 17, 2143–2150.
19 L. Cartegni, J. Wang, Z. Zhu, M. Q. Zhang, A. R. Krainer, Nucleic Acids Res. **2003**, 31, 3568–3571.
20 T. A. Thanaraj, S. Stamm, Prog. Mol. Sub. Biol. **2003**, 1–31.
21 T. A. Thanaraj, S. Stamm, F. Clark, J. J. Riethoven, V. Le Texier, J. Muilu, Nucleic Acids Res. **2004**, 32 (Database Issue), D64–D69.
22 C. L. Lorson, E. Hahnen, E. J. Androphy, B. Wirth, Proc. Natl. Acad. Sci. USA **1999**, 96, 6307–6311.
23 O. Stoss, M. Olbrich, A. M. Hartmann, H. König, J. Memmott, A. Andreadis, S. Stamm, J. Biol. Chem. **2001**, 276, 8665–8673.
24 S. Stamm, J. Zhu, K. Nakai, P. Stoilov, O. Stoss, M. Q. Zhang, DNA Cell Biol. **2000**, 19, 739–756.
25 J. Sambrook, D. W. Russell, Molecular Cloning: A Laboraory Manual, Cold Spring Harbor Laboratory Press, Cold Spring Harbor, NY, **2001**.
26 M. Ashiya, P. J. Grabowski, RNA **1997**, 3, 996–1015.
27 S. Stamm, D. Casper, J. Dinsmore, C. A. Kaufmann, J. Brosius, D. M.

Helfman, *Nucleic Acids Res.* **1992**, *20*, 5097–5103.
28 R. C. Chan, D. L. Black, *Mol. Cell Biol.* **1997**, *17*, 2970.
29 H. Kawahigashi, Y. Harada, A. Asano, M. Nakamura, *Biochim. Biophys. Acta* **1998**, *1397*, 305–315.
30 S. Kawamoto, *J. Biol. Chem.* **1996**, *271*, 17613–17616.
31 M. Caputi, G. Casari, S. Guenzi, R. Tagliabue, A. Sidoli, C. A. Melo, F. E. Baralle, *Nucleic Acids Res.* **1994**, *22*, 1018–1022.
32 A. F. Muro, M. Caputi, R. Pariyarath, F. Pagani, E. Buratti, F. E. Baralle, *Mol. Cell. Biol.* **1999**, *19*, 2657–2671.
33 W. Chen, L. A. Culp, *Exp. Cell. Res.* **1996**, *223*, 9–19.
34 A. Kosaki, J. Nelson, N. J. Webster, *J. Biol. Chem.* **1998**, *273*, 10331–10337.
35 T. A. Cooper, *Mol. Cell. Biol.* **1998**, *18*, 4519–4525.
36 M. Deguillien, S. C. Huang, M. Moriniere, N. Dreumont, E. J. Benz, Jr, F. Baklouti, *Blood* **2001**, *98*, 3809–3816.
37 L. D'Souza-Li, L. Canaff, N. Janicic, D. E. Cole, G. N. Hendy, *Hum. Mutat.* **2001**, *18*, 411–421.
38 J. Cote, S. Dupuis, Z. Jiang, J. Y. Wu, *Proc. Natl. Acad. Sci. USA* **2001**, *98*, 938–943.
39 F. Pagani, E. Buratti, C. Stuani, M. Romano, E. Zuccato, M. Niksic, L. Giglio, D. Faraguna, F. E. Baralle, *J. Biol. Chem.* **2000**, *275*, 21041–7.
40 F. C. Chou, S. J. Tzeng, J. D. Huang, *Drug Metab. Dispos.* **2001**, *29*, 1205–9.
41 C. Bergsdorf, K. Paliga, S. Kreger, C. L. Masters, K. Beyreuther, *J. Biol. Chem.* **2000**, *275*, 2046–2056.
42 M. J. Simard, B. Chabot, *Mol. Cell Biol.* **2000**, *20*, 7353–7362.
43 N. Guo, S. Kawamoto, *J. Biol. Chem.* **2000**, *275*, 33641–33649.
44 H. König, H. Ponta, P. Herrlich, *EMBO J.* **1998**, *17*, 2904–2913.
45 M. Ichida, Y. Hakamata, M. Hayakawa, E. Ueno, U. Ikeda, K. Shimada, T. Hamamoto, Y. Kagawa, H. Endo, *J. Biol. Chem.* **2000**, *275*, 15992–6001.
46 A. Stella, A. Wagner, K. Shito, S. M. Lipkin, P. Watson, G. Guanti, H. T. Lynch, R. Fodde, B. Liu, *Cancer Res.* **2001**, *61*, 7020–7024.
47 M. Baralle, D. Baralle, L. De Conti, C. Mattocks, J. Whittaker, A. Knezevich, C. Ffrench-Constant, F. E. Baralle, *J. Med. Genet.* **2003**, *40*, 220–222.
48 I. K. Svenson, A. E. Ashley-Koch, P. C. Gaskell, T. J. Riney, W. J. Cumming, H. M. Kingston, E. L. Hogan, R. M. Boustany, J. M. Vance, M. A. Nance, et al., *Am. J. Hum. Genet.* **2001**, *68*, 1077–1085.
49 Q. S. Gao, J. Memmott, R. Lafyatis, S. Stamm, G. Screaton, A. Andreadis, *J. Neurochem.* **2000**, *74*, 490–500.
50 K. Li, M. C. Arikan, A. Andreadis, *Brain Res. Mol. Brain Res.* **2003**, *116*, 94–105.
51 M. L. Wei, A. Andreadis, *J. Neurochem.* **1998**, 1346–1356.
52 Y. Su, J. R. Testaverde, C. N. Davis, W. A. Hayajneh, R. Adair, A. M. Colberg-Poley, *J. Gen. Virol.* **2003**, *84*, 29–39.
53 K. W. Lynch, A. Weiss, *Mol. Cell. Biol.* **2000**, *20*, 70–80.
54 G. J. Cote, E. S. Huang, W. Jin, R. S. Morrison, *J. Biol. Chem.* **1997**, *272*, 1054–1060.
55 N. K. Hess, S. I. Bernstein, *Dev. Biol.* **1991**, *146*, 339–344.
56 M. A. Keller, S. E. McKenzie, D. L. Cassel, E. F. Rappaport, E. Schwartz, S. Surrey, *Gene Expr.* **1995**, *4*, 217–225.
57 M. Ichihara, T. Hara, M. Takagi, L. C. Cho, D. M. Gorman, A. Miyajima, *EMBO J.* **1995**, *14*, 939–950.
58 I.-T. Chen, L. A. Chasin, *Mol. Cell. Biol.* **1994**, *14*, 2140–2146.
59 M. P. Wentz, B. E. Moore, M. W. Cloyd, S. M. Berget, L. A. Donehower, *J. Virol.* **1997**, *71*, 8542–8551.
60 H. Jumaa, J.-L. Guenet, P. J. Nielsen, *Mol. Cell. Biol.* **1997**, *17*, 3116–3124.
61 H.-C. Kuo, F.-U. H. Nasim, P. J. Grabowski, *Science* **1991**, *251*, 1045–1050.
62 R. E. Breitbart, B. Nadal-Ginard, *Cell* **1987**, *49*, 793–803.

63 D. M. Rothstein, H. Saito, M. Streuli, S. F. Schlossman, C. Morimoto, *J. Biol. Chem.* **1992**, *267*, 7139–7147.
64 W. P. Dirksen, Q. Sun, F. M. Rottman, *J. Biol. Chem.* **1995**, *270*, 5346–5352.
65 A. Andreadis, J. A. Broderick, K. S. Kosik, *Nucleic Acids Res.* **1995**, *23*, 3585–3593.
66 E. L. Virts, O. Diago, W. C. Raschke, *Blood* **2003**, *101*, 849–855.
67 G. C. Roberts, C. Gooding, H. Y. Mak, N. Proudfoot, C. W. J. Smith, *Nucleic Acids Res.* **1998**, *1998*, 5568–5572.
68 I. R. Graham, M. Hamshere, I. C. Eperon, *Mol. Cell Biol.* **1992**, *12*, 3872–3882.
69 L. Balvay, A. M. Pret, D. Libri, D. M. Helfman, M. Y. Fiszman, *J. Biol. Chem.* **1994**, *269*, 19675–19678.
70 J. S. Grossman, M. I. Meyer, Y. C. Wang, G. J. Mulligan, R. Kobayashi, D. M. Helfman, *RNA* **1998**, *4*, 613–625.
71 M. Takenaka, K. Yamada, T. Lu, R. Kang, T. Tanaka, T. Noguchi, *Eur. J. Biochem.* **1996**, *235*, 366–371.
72 K. Sugiyama, Y. Hitomi, H. Adachi, H. Esumi, *Cancer Lett.* **1994**, *83*, 221–227.
73 M. E. Gallego, G. B. Nadal, *Mol. Cell. Biol.* **1990**, *10*, 2133–2144.
74 F. Del Gatto-Konczak, M. Olive, M. C. Gesnel, R. Breathnach, *Mol. Cell. Biol.* **1999**, *19*, 251–60.
75 A. Shibata, M. Hattori, H. Suda, Y. Sakaki, *Gene* **1996**, *175*, 203–208.
76 B. D. Lichty, S. Kamel-Reid, *Biochem. J.* **2000**, *348*, 63–69.
77 R. P. Carstens, E. J. Wagner, M. A. Garcia-Blanco, *Mol. Cell. Biol.* **2000**, *20*, 7388–7400.
78 A. D. Polydorides, H. J. Okano, Y. Y. L. Yang, G. Stefani, R. B. Darnell, *Proc. Natl. Acad Sci. USA* **2000**, *97*, 6350–6355.
79 G. J. Cote, *Biochem. Biophys. Res. Commun.* **1994**, *200*, 993–998.
80 K. Inoue, K. Hoshijima, I. Higuchi, H. Sakamoto, Y. Shimura, *Proc. Natl. Acad. Sci. USA* **1992**, *89*, 8092–8096.
81 J. Valcarcel, R. Singh, P. D. Zamore, M. R. Green, *Nature* **1993**, *362*, 171–175.
82 Z. M. Zheng, M. Huynen, C. C. Baker, *Proc. Natl. Acad. Sci. USA* **1998**, *95*, 14088–14093.
83 I. Aznarez, E. M. Chan, J. Zielenski, B. J. Blencowe, L. C. Tsui, *Hum. Mol. Genet.* **2003**, *12*, 2031–2040.
84 K. Harada, A. Yamada, D. Yang, K. Itoh, S. Shichijo, *Int. J. Cancer* **2001**, *93*, 623–628.
85 M. Messina, D. M. Yu, G. W. Both, P. L. Molloy, B. G. Robinson, *J. Clin. Endocrinol. Metab.* **2003**, *88*, 1310–1318.
86 M. J. Lisbin, J. Qiu, K. White, *Genes Dev.* **2001**, *15*, 2546–2561.
87 M. Romano, R. Marcucci, F. E. Baralle, *Nucleic Acids Res.* **2001**, *29*, 886–894.
88 L. Van den Bosch, J. Eggermont, H. De Smedt, L. Mertens, F. Wuytack, R. Casteels, *Biochem. J.* **1994**, *302*, 559–566.
89 M. B. Humphrey, J. Bryan, T. A. Cooper, S. M. Berget, *Mol. Cell. Biol.* **1995**, *15*, 3979–3988.
90 M. Sarkissian, A. Winne, R. Lafyatis, *J. Biol. Chem.* **1996**, *271*, 31106–31114.
91 H. Ge, J. L. Manley, *Cell* **1990**, *62*, 25–34.
92 A. R. Krainer, G. C. Conway, D. Kozak, *Cell* **1990**, *62*, 35–42.
93 H. J. Mardon, G. Sebastio, F. E. Baralle, *Nucleic Acids Res.* **1987**, *15*, 7725–7733.

IV.3
SELEX

47
Artificial Selection: Finding Function amongst Randomized Sequences

Ico de Zwart, Catherine Lozupone, Rob Knight,
Amanda Birmingham, Mali Illangasekare, Vasant Jadhav,
Michal Legiewicz, Irene Majerfeld, Jeremy Widmann
and Michael Yarus

47.1
The SELEX Method

In vitro selection–amplification, frequently called SELEX (Systematic Evolution of Ligands by EXponential enrichment), allows for the selection of rare nucleic acid sequences with specific catalytic or binding activity from a large pool of randomized, mostly inactive molecules. The crux of the technique is the cyclic use of selective enrichment (selection) followed by nucleic acid replication (amplification) for active molecules [1]. Here, we briefly describe aspects of the SELEX procedure, focusing on RNA selection. A generic selection cycle is shown (Fig. 47.1), along with appropriate advice for individual steps. Further information on SELEX protocols in particular can be found in an earlier review [2] or in specific papers on particular selections, cited below.

An initial pool of RNA molecules is generated by the transcription of template DNA molecules obtained from chemical synthesis. Template DNA molecules contain a randomized tract flanked at both ends by constant regions, required for PCR primer hybridization, and also a T7 RNA polymerase promoter sequence. The randomized region is typically composed of between 25 and 100 nt, although as many as 220 [3] and as few as 16 nt [4] have been employed for special purposes. Single-stranded DNA longer than 120 nt is difficult to obtain commercially, but can be assembled from shorter sequences [3]. Synthetic DNA molecules are first made double-stranded by either PCR amplification or Klenow extension (for general protocols, see [5, 6]).

Between 10^{13} and 10^{15} DNA sequences are typically PCR amplified for 4–10 cycles in a volume of 0.5–5 ml, which potentially generates up to 10–100 copies of each template sequence. The minimum number of cycles needed to convert the entire pool of synthetic single-stranded template DNA to a double-stranded product (visualized on an ethidium bromide stained gel) should be employed. Approximately 50–1000 pmol of PCR DNA is transcribed in the presence of an

Handbook of RNA Biochemistry: Student Edition. Edited by R. K. Hartmann, A. Bindereif, A. Schön, E. Westhof
Copyright © 2009 WILEY-VCH Verlag GmbH & Co. KGaA, Weinheim
ISBN: 978-3-527-32534-4

[α-^{32}P]nucleotide triphosphate and the product gel purified to give an initial pool of 1–10 nmol RNA. The radioactive label allows for the quantification of small amounts of RNA characteristic of the selected pools. The combined amplification due to PCR (around 4–10 times) and transcription (around 4–20 times) should give numerous copies of each sequence, although some sequences may be present in smaller numbers due to chance and amplification biases. The amounts quoted above are designed to give good sequence representation without requiring large amounts of reagents, which rapidly become expensive and difficult to handle in larger experiments. At concentrations greater than around 400 µg/ml (around 10 µM for a 120mer), randomized RNA solutions begin to precipitate in the presence of divalent metal ions, which sets a practical limit on the amount of RNA that can be used.

RNA is folded and subjected to the selection conditions (described below for affinity and chemical selection). RNA that meets the selection requirements is precipitated, reverse transcribed, PCR amplified and transcribed, and the selection cycle is repeated. If required, after one or two cycles a negative selection may be incorporated in order to remove RNAs that survive the selection but do not have the desired activity. Selection progress is monitored by determining the fraction of the RNA recovered from the selection step in successive cycles. Once sufficient activity is observed, the selected pool is usually cloned and sequenced to facilitate more resolved analysis of individual molecules.

47.2
Understanding a Selection

The success of a selection is frequently measured by the number of selection cycles required to purify an activity or by the variety of RNAs in the final pool. Critical factors that influence the outcome of a selection include the number of indepen-

---

Fig. 47.1. An overview of the various steps in a selection. *Initial PCR*. This PCR reaction is carried out using around 0.1–1 nmol template DNA and 5–10 nmol primers in 1–5 ml. This large-scale reaction may require higher total dNTP (0.8–1.6 mM) and magnesium (4–6 mM) concentrations than is typical for most PCR applications. It is worthwhile optimizing these concentrations on a smaller scale (20–50 µl) and then scaling up when optimal conditions are found. It is also worthwhile optimizing the annealing temperature and magnesium concentration for the PCR reactions later in the selection cycle that begin with around 0.05–1 pmol template DNA. *Transcription*. The initial transcription reaction should contain enough PCR DNA to maintain sequence diversity. For example, if the initial PCR resulted in around 10-fold amplification, then around 30% of the initial PCR DNA can be used. In subsequent cycles, a portion of the PCR reaction can be used for transcription, with the remainder being stored in case the cycle needs to be repeated or for future reference. *Reverse transcription*. Because of the small amounts of RNA involved, the precipitation of RNA after the selection step prior to reverse transcription should be done with glycogen or carrier RNA. In the first cycle, all the RT-DNA should be used as in the PCR reaction, while in later cycles some can be stored in case the PCR reaction fails. Selection PCR: The number of cycles required to generate enough PCR DNA should be determined using a small-scale test reaction. Mutagenic PCR [48] can be used in later cycles.

47.2 Understanding a Selection

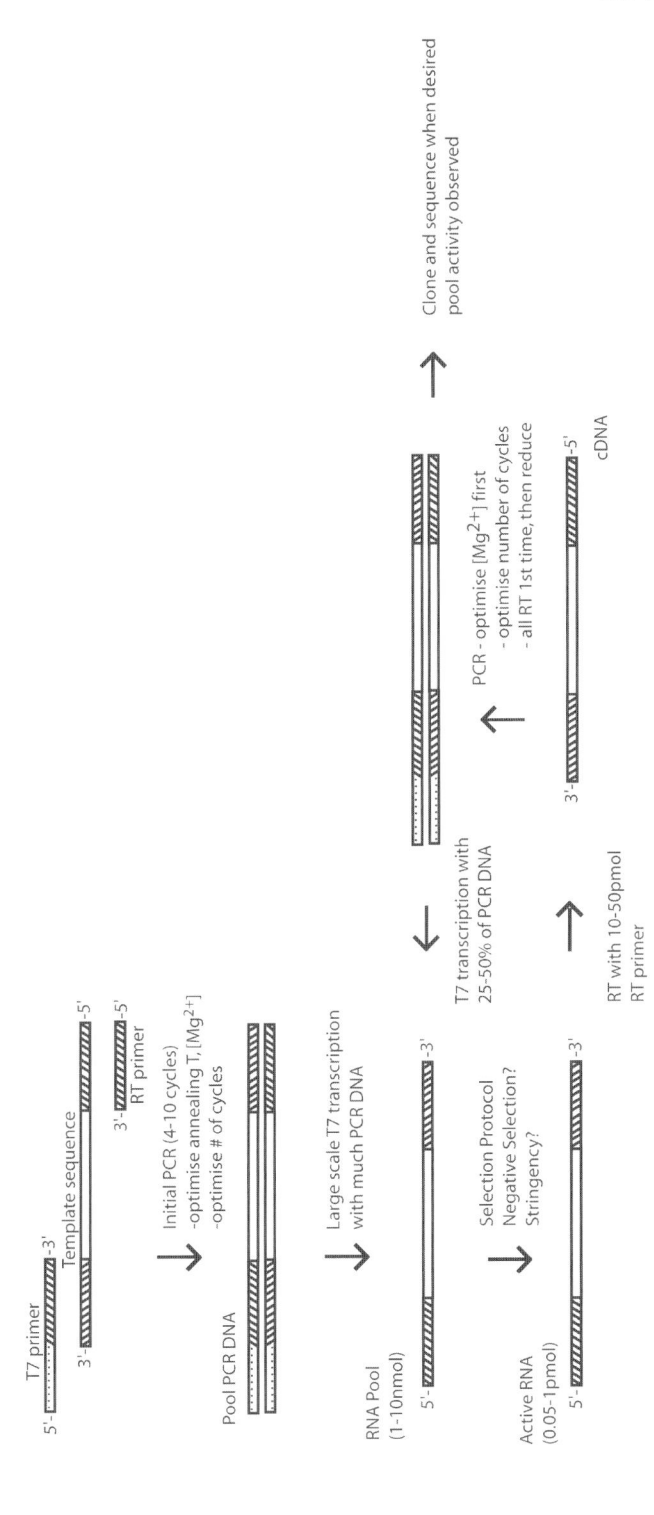

Fig. 47.1

T7 primer 5'- ----3'
Template sequence 3'- ▨▨▨-5'
RT primer 3'- ▨▨-5'

↓ Initial PCR (4–10 cycles)
  - optimise annealing T, [Mg^{2+}]
  - optimise # of cycles

Pool PCR DNA ▨▨===▨▨

↓ Large scale T7 transcription with much PCR DNA

RNA Pool (1–10 nmol) 5'- ———— -3'

↓ Selection Protocol
  Negative Selection?
  Stringency?

Active RNA (0.05–1 pmol) 5'- ———— -3'

↑ RT with 10–50 pmol RT primer

cDNA ▨▨=== -5' / 3'- ▨▨

↑ PCR - optimise [Mg^{2+}] first
  - optimise number of cycles
  - all RT 1st time, then reduce

↑ T7 transcription with 25–50% of PCR DNA

→ Clone and sequence when desired pool activity observed

dent RNA sequences in the initial pool that meet the selection criteria, the relative efficiency of recovery for these RNAs, the stringency of the selection step and the extent to which the amplification of selected RNAs is biased.

47.2.1
Sequence Motif Representation and Abundance

An important factor in a selection is the abundance of particular sequences in the randomized pool. Various studies have addressed the fraction of unique RNA sequences of a given length that can be studied on the laboratory scale, the size and structure of active sites that one can expect to retrieve reliably from a selection, and methods by which motif representation can be maximized by the design of the randomized pool employed in the selection.

The probability of finding a motif within a randomized sequence having N fixed nucleotides is $1/4^N$. As the length N of a motif increases, it ultimately becomes vanishingly rare in any pool. Complete representation of all contiguous sequences up to 23mers is possible in a typical selection where a maximum around 10^{15} different sequences (with a random region of 23 nt) is realistically achievable. Increasing the length of the random region enables slightly longer contiguous sequences to be searched (25mers for a 50-nt random region) because many trials to find the 25mers are included within the longer random region. However, increasing the number of RNA molecules 10-fold adds only 1.66 nt to the potential length of the represented $n$mer [2] and thus increasing pool size is a very demanding strategy for expediting selection.

Absolute requirement for every nucleotide in a long sequence is unlikely; instead important regions are often separated by variable spacers. When the spacers can have any length or sequence, there are many ways to find such a segmented motif [7–9]. This type of split motif, rather than a contiguous active site, is commonly found in selected aptamers, such as those that bind ATP [10] and FMN [11]. The "UAUU" motif repeatedly found in isoleucine aptamers described in Section 47.2 (and see Fig. 47.3C) is one example; it is composed of two modules, which form on either side of an internal loop. Calculations for the abundance of modular motifs in a random pool show that motifs that are broken into several evenly divided modules will be the most abundant. In a pool with 100 randomized positions and around 10^{15} molecules, motifs up to a size of 33 nt divided into four similar sized modules should be exhaustively represented. A maximum of 32 required nucleotides are represented for three similar sized modules, 30 nt for two modules and 28 nt for one module in a pool with the same characteristics [7, 9].

Calculations also suggest that sequence motif representation will increase significantly with the size of the randomized tract [8, 9]. The addition of arbitrary sequences, however, can reduce the probability that the active motif, if present, will fold into the proper structure. This can be appreciated by conceptualizing the pool as rare examples of the motif desired (almost always present as a single copy in a few molecules) embedded in randomized regions that do not have the motif. As the surrounding randomized tracts grow, the probability that an alternative structure will disturb the fold of the active motif increases. An experiment where excess sequence was added to a selected ligase ribozyme suggests that this adverse factor is smaller than the statistical advantages of larger randomized regions [8]. However, other experiments ([12] and unpublished observation) have indicated that longer randomized tracts may have unexpectedly negative effects on a selection. The general outcome of the conflict between the more efficient search for a motif in longer randomized regions and interference by surrounding sequences is not yet understood, since calculating the probability that a motif will fold properly is complex. Additionally, new evidence suggests that even in well-studied ribozymes such as the hammerhead, the identity of apparently "variable" sequences can actually be constrained, so that the full site can elude enumeration [13]. Consequently, calculations of motif abundance should be treated with caution. One consequence is that because SELEX is only searching a limited sequence space exhaustively, failed selections do not necessary indicate that RNA is not capable of the task. In fact, the wide range of RNA activity exhibited by RNA containing a few tens of essential nucleotides is an impressive testimony to the extent of RNA's catalytic and binding capabilities. Selection, conversely, has occasionally resulted in the isolation of large motifs of a size calculated to be very rare [3]. This could either be the occasional observation of a rare event, or more interestingly, it could be that though molecules of high complexity are rare, a large fraction of such RNA molecules are functional [3].

At present then, the jury is still out on the optimal length of the random region, if any, required for a selection. Indeed, it is likely to be different for different selections [12]. In light of this, an initial pool of around 10^{14} sequences with 50–80 random nucleotides is a versatile starting pool for many selections. This length is both readily accessible through chemical synthesis and searches enough sequence space to allow for the isolation of many functional RNAs, as evidenced by the obvious success of a variety of selections that have used starting conditions in this range.

47.2.2
The Recovery Efficiency of Different RNAs

Because only a relative few sequences will ever be characterized, a selection is always a contest between RNA molecules for representation in the final pool. The contest can have different outcomes at each cycle. We illustrate with a model that features an abundant simple motif that is less efficiently recovered from the selection technique, and a rarer complicated motif that is more effective and efficiently recovered. As shown (Fig. 47.2), under simplified assumptions, a more abundant yet indifferently performing motif is predicted to purify faster (to 98% of the pool after three rounds of selection) and can dominate the pool for several selection cycles. Eventually, the model predicts that the better performing motif will dominate the selection; in this example, composing 88% of the pool after cycle 9 of selection and reducing the prevalence of the less active motif to unobservable levels after cycle 10.

The model, even though it utilizes simplified assumptions that could be changed, explains why non-functional or subfunctional RNAs are a frequent component of active selected pools, since even RNA with very limited activity can be

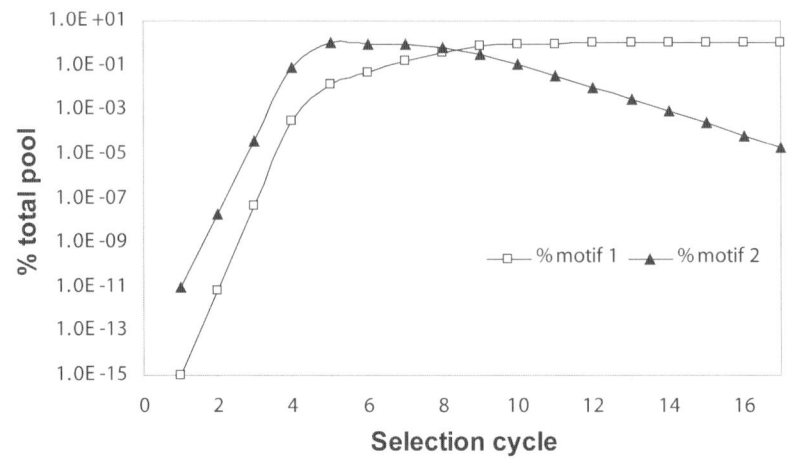

Fig. 47.2. A graph illustrating a simple model of purification of two functional motifs in the presence of a competing background. The model was derived as follows. Motif 1 (M1) is a complicated, better performing motif and Motif 2 (M2) a more abundant, but poorly performing motif. Since M1 requires a longer sequence, it is present only once in the starting pool and thus the fraction of M1 in 10^{15} molecules (P_{M1}) is 1×10^{-15}. E_{M1} is the fraction of the M1 molecules recovered in the selection step. For calculations, this value was set to 0.7. The fraction of M2 (P_{M2}) in the starting pool is higher to represent the larger number of molecules present for a simpler motif. In this model, $P_{M2} = 1 \times 10^{-11}$, representing 10 000 sequences. The efficiency for M2 (E_{M2}) was set to 0.2, as might be the case for a less effective RNA molecule. The background (B) represents molecules that are isolated by the experiment for spurious reasons and not because of any association with the ligand. The background was estimated to be 0.01% of the initial pool, a common figure for the first round of real selections. The value of P_{Mx} can be calculated for each selection cycle (C): $P_{Mx(c)} = (P_{Mx(c-1)})(E_{Mx})/\{[(P_{Mx(c-1)})(E_{Mx})] + (P_{My(c-1)})(E_{My}) + B\}$. The simple less active motif is purified to observable levels (7.4%) after four rounds of selection and is dominant until the sixth round of selection, when it is still predicted to compose 95% of the pool. Subsequently, the more complicated, better performing motif dominates the selection, excluding the simpler motif from observation in a practical experiment around cycle 10.

purified. Even biases in amplification alone may be sufficient for relative success, particularly in the early rounds of selection.

One lesson from the model is that, counter to intuition, the molecule that is purified first in a selection may not be the best performing RNA in the experiment. Less functional molecules will likely appear first, because they may be smaller and will thus be more abundant in randomized sequences. This suggests that if the most effective RNA is desired, the experiment should not be terminated when activity is first observed. Selection should continue for a few cycles after the observation of initial activity, perhaps in combination with increasing stringency, to allow a potentially more complex functional molecule to rise to observable levels within the final pool. However, if the simplest functional motif is desired, e.g. for incorporation into a larger structure, then selection should stop with the appearance of activity, since this motif could virtually disappear in later cycles.

47.2.3
Stringency

Selection stringency refers to the extent of discrimination between functional and non-functional sequences in the selection step. The effects of stringency on a selection have been previously described [1] in the context of selection using a filter-binding assay. In general, selections conducted under low stringency take many cycles before success because the purification of functional molecules is delayed by failure to exclude non-functional background molecules. Conversely, if conditions are too stringent, all functional molecules can be lost in the initial stages and the selection may fail outright. A commonly used protocol is thus to increase the stringency over the course of an experiment. The methodology for controlling stringency varies with the selection step employed. Affinity chromatography and ribozyme selections are discussed in Sections 47.3.1 and 47.4.5, respectively.

47.2.4
Amplification and Transcription Biases

The extent to which biases in the efficiency of reverse transcription, PCR amplification and transcription may influence the results of a selection are not well understood. That these techniques do have inherent biases, however, has been well demonstrated. For instance, the formation of RNA secondary structures can inhibit reverse transcription either by resulting in strong reverse transcription stops within the sequence or by interfering with primer annealing. Substantial bias in PCR amplification efficiency has been demonstrated for sequences with the same primer binding sites and as little as a single point mutation [14]. Another PCR effect, where the negative strand of DNA competes with primers for hybridization, specifically inhibits the amplification of particular molecules based on their level of enrichment in later cycles of selection [15].

Moreover, there is abundant internal evidence of amplification bias in the data of many selections. We have observed differences of 100-fold in the abundance of apparently equally functional motifs after selection. The range observed would probably be greater save for practical limits on the total number of clones sequenced. These differential abundances are most easily accounted for by differential amplification. In addition, even when thousands of sequences that are functional should be present in the initial pool (as for a small motif), one repeatedly isolates the descendants of only one or a few parents. Thus equally functional sequences survive selection in a biased manner. Because of such biases, the optimization of reverse transcription and amplification protocols prior to selection is a worthwhile endeavor so that the minimum number of PCR cycles is employed to obtain sufficient PCR product for transcription of RNA for the next cycle.

Further details of the SELEX technique will vary according to the type of activity sought. In the following discussion, we describe in more detail selections for RNAs that bind small molecules and for ribozymes. We also discuss various methods for the analysis of the isolated RNAs.

47.3
Isolation of RNAs that Bind Small Molecules

In the following text, we demonstrate methodology for finding and characterizing small-molecule aptamers using affinity chromatography. We use the selection for isoleucine binding RNAs [4, 16] to illustrate concepts.

In affinity chromatography selection, functional molecules are purified by applying the RNA molecules to a column matrix with the ligand of interest covalently attached. Bound RNA is eluted from the column with free ligand after unbound RNA molecules have been removed. Ligand-eluted molecules are amplified and the fractionation is iterated until a significant fraction of the pool is eluted with free ligand. This affinity elution ensures that the selected aptamer binds the free ligand, rather than the (necessarily modified) form of it immobilized on the column.

Our column matrix usually contains between 0.2 and 0.4 ml of Sepharose with the ligand attached at a concentration of 0.5–10 mM. The ligand should be attached so that the essential functional groups are available for RNA binding. In the isoleucine selections, for instance, we attached L-isoleucine to EAH Sepharose 4B (Pharmacia) with its carboxylic acid group so that the positively charged amino group would be available to aid in RNA binding. This emulates the chemistry of an amino acid activated for peptide synthesis.

The selection buffer composition should provide an optimal environment for binding or an environment designed to test for binding under certain conditions. In our initial isoleucine selection, the buffer was 10 mM HEPES, pH 7.0, 100 mM NaCl, 7.5 mM $MgCl_2$, and 0.1 mM each of $ZnCl_2$, $CaCl_2$ and $MnCl_2$. The divalent metal ions are potential cofactors for the selected RNAs [17].

After washing with selection buffer, bound RNA molecules are removed with around 5 column volumes of selection buffer with the same concentration of ligand as in the Sepharose matrix. The amount of selection buffer added before specific elution is generally between 5 and 20 column volumes, but can be varied based on the desired stringency and predicted dissociation constants (K_D) of the resulting aptamers (see Section 47.3.1).

The progress of the selection is monitored by collecting fractions of around 0.5 column volumes and counting the radioactivity (Fig. 47.3A). Selection is terminated when a significant fraction of the RNA pool is in the eluted fraction, producing an elution peak. The columns can usually be recycled, washing between selection cycles with more than 15 column volumes of high salt buffer to remove

Fig. 47.3. Isoleucine selection. (A) Profiles from an isoleucine selection [4] with 26 randomized positions showing the elution profiles at cycle 1, 3 and 6. By cycle 3, the RNA displays delayed elution from the column. The pool was cloned at cycle 6 when an elution peak was observed. (B) Representative profile for a counterselection: RNAs from the first three fractions (totaling 250 µl or 1.25 column volumes) are precipitated for addition to the ligand–matrix. (C) Summary structure of the isoleucine aptamer [4]. Conserved nucleotides (greater than 70%) are indicated and non-conserved nucleotides are indicated with an "N".

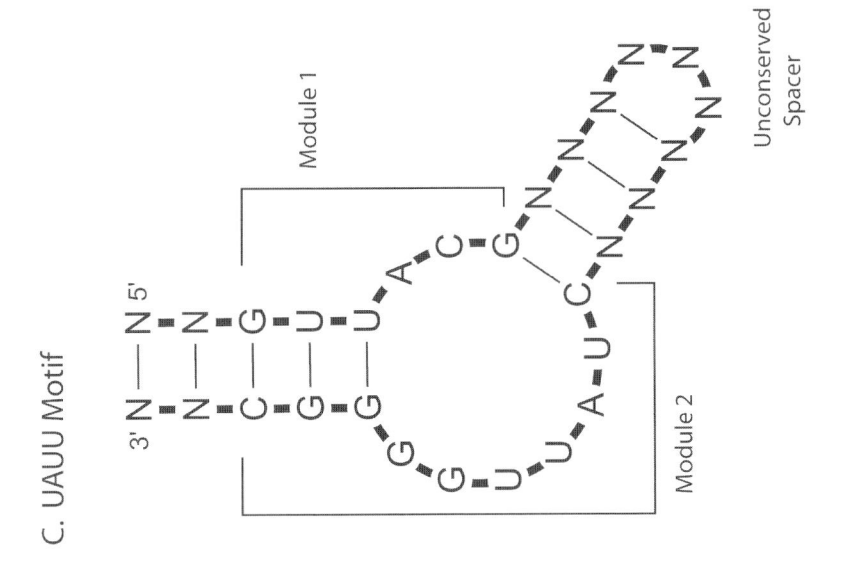

47.3 Isolation of RNAs that Bind Small Molecules

C. UAUU Motif

A. Ligand-selection profile

B. Counterselection profile

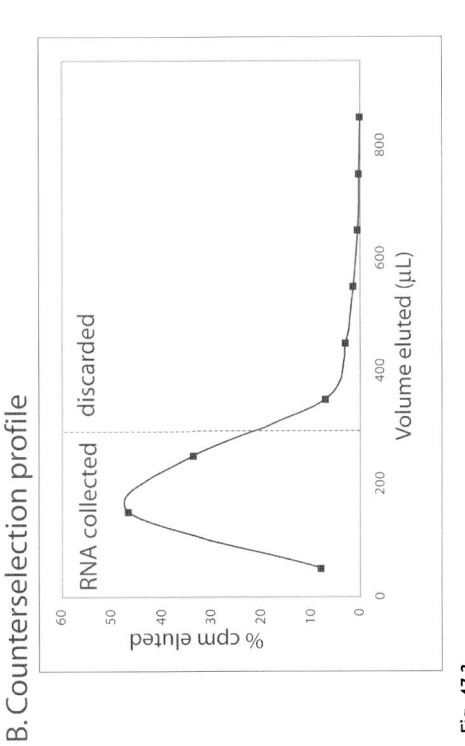

Fig. 47.3

molecules still bound to the column and then equilibrated with more than 10 column volumes of selection buffer before the next cycle of selection.

In any sufficiently long selection, sequences with affinity for the chromatographic material will probably be selected. This includes the column walls and the supporting disc at column exit. A negative selection should be performed between selection cycles in order to remove RNA with affinity for the column. The negative selection is typically done by passing the RNA through a column containing the matrix and ligand support, but without the attached ligand. In the isoleucine selection we used acetylated EAH Sepharose 4B (Pharmacia). Selection buffer is applied in 0.5 column volume fractions and the first couple of fractions (typically around 66% of the total RNA) are collected, precipitated, and refolded before the selection step (Fig. 47.3B).

Using this fractionation technique, we repeatedly recovered aptamers from unrelated parents containing an RNA motif, called the "UAUU" motif (Fig. 47.3C), that binds free isoleucine with an estimated K_D of 0.2–0.5 mM.

47.3.1
Stringency and K_D

Stringency plays a major role in the outcome of a selection; a recommended protocol is to conduct selections with low stringency in early cycles, so that rare functional molecules are not spuriously lost, and with higher stringency later, so that non- and subfunctional molecules can be more quickly removed. In affinity chromatography, stringency is regulated by the volume of selection buffer applied to the column before functional RNAs are eluted, the concentration of immobilized ligand in the column, and the concentration of free ligand in the elution buffer. This concept is illustrated by the equation used to calculate dissociation constants from affinity chromatography data [18, 19]. The dissociation constant (K_C) for an RNA and a ligand–column matrix is:

$$K_C = L_C V_n / (V_e - V_n) \tag{1}$$

where L_C is the concentration of the ligand within the column bed, V_n is the median elution volume when the RNA does not interact with the column (operationally defined as the volume at which 50% of the randomized pool elutes) and V_e is the median elution volume of the RNA specifically interacting with the column.

The volume required to elute 50% of the RNA from the column (V_e) is thus a linear function of both the concentration of the ligand in the column matrix (L_C) and dissociation constant of RNA for column sites (K_C). This equation can be used to estimate the amount of selection buffer to apply to a column before elution with free ligand. For example, the measured K_C of the "UAUU" motif for the isoleucine column is around 500 µM. With 10 mM isoleucine on the column and assuming a V_n of 0.2 ml (the column volume), 50% of the RNA with the "UAUU" motif would remain on the column after washing with 4.2 ml ($21*V_n$) of column buffer. Thus, washing with 4 ml of column buffer would be suitable after a number of rounds of selection, but potentially too stringent in the first round, when there may be very few copies of each RNA, and the chance of washing a particular RNA off the column is far higher. Typically, we elute bound RNA after washing with around 5 column volumes in the early rounds of a selection and this is increased to 10–20 column volumes in later cycles.

V_e represents the point at which 50% of the RNA can potentially still be recovered from the column after elution with free ligand during the selection step. The amount that is actually recovered from a selection column, however, depends primarily on the concentration of free ligand, L, in the elution buffer and the K_D of the RNA for the free ligand versus the ligand–matrix. In a selection there is a trade-off between maximizing the recovery of molecules that bind to free ligand (achieved with high L), and minimizing the isolation of molecules with affinity for the ligand-matrix only (low L). As a compromise, we typically use the same concentration of ligand in the elution buffer as in the matrix. In later cycles, we typically elute RNA with a smaller volume of buffer containing ligand as an additional means of maximizing the isolation of RNAs that prefer the free ligand to the column bound ligand.

The expected K_D for aptamers with unknown activity can be estimated based on the observed activity of previously described aptamers. Selected small molecule binding RNAs have estimated dissociation constants (K_D) ranging from 0.0008 µM for the antibiotic Tobramycin to 12 000 µM for the amino acid valine [20]. In general, RNA binds strongly to planar aromatic molecules, molecules with hydrogen-bond donors or acceptors, and positively charged groups, and binds weakly to non-planar molecules with a neutral to negative overall charge or largely hydrophobic character [20].

47.3.2
Selection for Multiple Targets in One Column

Although the discussion thus far has applied to a selection for affinity to a single target, computer simulations predict that SELEX is capable of generating different aptamers for different ligand targets in a heterogeneous mixture, regardless of large variations in aptamer–ligand activities [21]. For example, we have searched for aptamers that bind four different amino acids, histidine, tryptophan, valine and glutamine, by binding and eluting the RNA with a 4-amino-acid matrix and eluant. Using this technique we simultaneously isolated aptamers for histidine and tryptophan but did not isolate a previously found aptamer for valine that had a very high K_D (Majerfeld and Yarus, unpublished). This suggests that this technique can isolate aptamers for multiple targets but there are some limitations to the range of ligand–target activities that can be explored.

47.3.3
Characterizing Motif Activity

Members of recovered sequence families can be screened for activity by evaluating RNA from a representative clone under experimental conditions. This will indicate whether each transcript is present because it elutes with the ligand, because it has affinity for the matrix only, or for spurious reasons such as superior amplification.

The K_D for an individual RNA is frequently desired information. We often use isocratic affinity chromatography to measure the K_D [18, 19], as follows:

$$K_D = L(V_{eL} - V_n)/(V_e - V_{eL}) \qquad (2)$$

where L is the free ligand concentration and V_{eL} is the median elution volume of the RNA in the presence of L. Although some of the immobilized ligand in the column may not be available for binding, this does not affect the K_D for free ligand, whose measurement requires only that equilibrium is maintained. This technique is convenient since columns are already available from the selection. Another technique to determine K_D is equilibrium dialysis, which is optimal for strongly binding RNAs but requires prohibitively high RNA concentrations for many small-molecule aptamers, which tend to have relatively weak interactions with the ligand.

Median elution volumes are determined by applying radiolabeled RNA to the column, applying selection buffer with ligand (for V_{eL}) or without ligand (for V_e) in around 0.1 ml fractions and determining the volume required to wash 50% of the total counts from the column. Using more than one concentration of ligand to measure K_D yields a more accurate estimate. K_D measurements derived in this manner have been shown to agree with equilibrium dissociation constants determined by other methods [22, 23].

In the case of the isoleucine aptamer, the K_D was also measured for similar amino acids in order to get an index of specificity. In this case, the binding of the RNA to isoleucine–Sepharose competed with free amino acids in the buffer such as glycine, leucine, and valine, exhibiting K_D's in the 1–4 mM range for these ligands [16].

Requirements for motif activity can also be evaluated by column chromatography using different buffer environments. For instance, in the case of the isoleucine aptamer, we were interested in the divalent metal requirements for activity. Sequentially removing each divalent metal ion from the buffer indicated that Zn^{2+} was required in addition to Mg^{2+} for isoleucine binding.

Functional aptamers can be further characterized in order to determine their mechanism for activity. As these techniques are not specific to affinity chromatography, they are discussed below in Section 47.5.

47.4
Techniques for Selecting Ribozymes

One of the most demanding applications of SELEX is expansion of the catalytic repertoire of RNA by the isolation of new ribozymes. We have focused on isolating RNAs that are capable of carrying out particular types of biochemical reactions, such as the four reactions of translation, and the synthesis and utilization of coenzymes. However, the techniques described here are amenable to the selection of any ribozyme (see other chapters in this handbook). We will illustrate selection of ribozymes by describing the selection of an acyl-CoA synthase ribozyme [24], while digressing to illustrate other novel and interesting selection strategies.

Two methods are available for the purification of catalytic RNAs: selection for self-modification or selection for binding to a transition state analog. By far the most successful strategy is selection for self-modification and we will emphasize this. The selection for transition state analog affinity has yielded a nucleic acid catalyst that catalyses the metallation of porphyrin [25] as well as one that catalyses the isomerization of a bridged biphenyl compound [26]. This technique has been very successful in isolation of catalytic antibodies [27].

As the name suggests, the key to self-modification is that active RNAs must chemically alter themselves when they perform the selected reaction. They thereby become chemically or physically distinguishable from inactive RNAs, so they can be purified. Thus, the RNA molecule itself must be one of the substrates for the reaction. The means by which the RNA is modified, if necessary, so that it can act as a substrate, and the method of selecting the active RNAs are the key to a successful selection. Some common themes link many previous experiments.

47.4.1
Making the RNA a Substrate for the Reaction

In some cases, the transcribed RNA can be used without any further modifications by exploiting the inherent reactivity of the RNA molecule. In other cases the RNA must first be modified since RNA lacks the requisite functional groups required. For instance, in the selection for an acyl-CoA synthase ribozyme, the reaction of interest is thioester formations between the thiol group of coenzyme A and an acyl adenylate (Fig. 47.4). This requires that coenzyme A be attached to the RNA pool prior to the selection reaction. To do this, a previously selected ribozyme [28], capable of attaching a variety of phosphorylated molecules to its $5'$ end via a diphosphate linkage, was modified to cap a randomized RNA pool with coenzyme A (Fig. 47.4). This strategy is unusual since it uses a ribozyme to modify the RNA pool; RNA-catalyzed cofactor attachment was specifically chosen to show that the chemistry was appropriate for a RNA world. However, there are a number of other ways in which RNA can be modified if required. For instance, an alternate method for making a CoA-RNA pool used dephospho-CoA as an AMP analog to initiate transcription [29] under the control of an A-initiating T7 RNA polymerase promoter.

47.4.2
The Inherent Reactivity of RNA

Many selections do not require modified RNA, since they utilize RNA's intrinsic reactivity. The $5'$-triphosphate can be used as an electrophile, while the $2'$-hydroxyls within the RNA and the $2',3'$-diol at the $3'$ end can act as nucleophiles. Since a number of metabolic reactions involve the reaction of a triphosphate with a nucleophile, reaction at the $5'$ end of RNA has been used frequently.

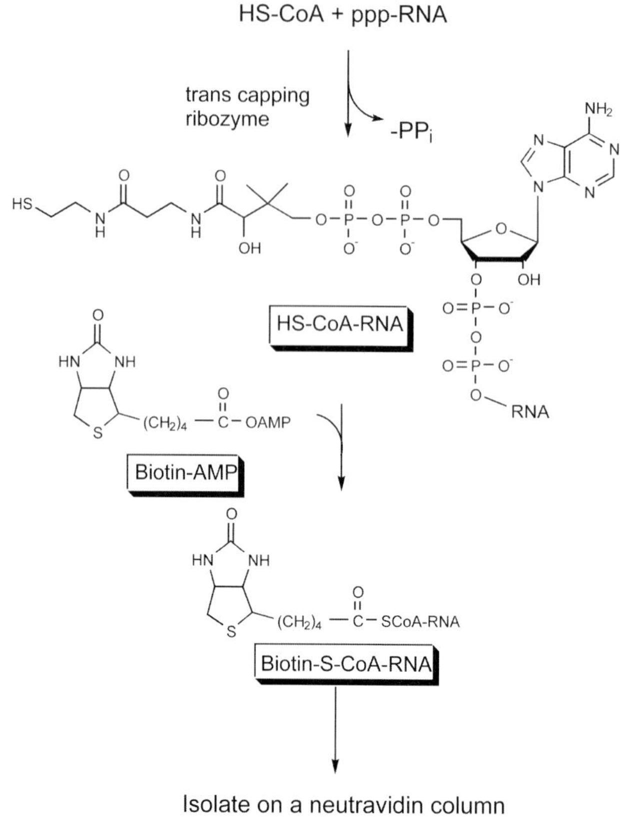

Fig. 47.4. Selection strategy for the selection of an acyl-CoA synthase ribozyme.

At the 5′ end, modifications can be introduced during transcription with T7 polymerase by adding suitable guanosine derivatives as transcription initiators. T7 polymerase usually initiates transcription with GTP, but will accept analogs as well. If the analogs lack a triphosphate they cannot be incorporated elsewhere in the RNA transcript. This method has been used extensively to introduce guanosine monophosphorothioate (GMPS) at the 5′ end of the RNA, allowing for subsequent modifications directed at the sulfur [30, 31]. More elaborate molecules have been incorporated into RNA in this manner [32–34]. Modifications can also be introduced at either end by ligation using T4 ligase, a strategy used in the selection for nucleotide synthase ribozymes [35]. The commercial availability of a wide variety of modified oligonucleotides makes this a relatively simple method for attaching functional groups to RNA. Modified nucleotides can be incorporated within the RNA transcript by adding modified nucleotide triphosphates to the transcription reaction [36–38]. However, the disadvantage of this strategy is that the site of incorporation cannot be controlled. Finally, the 3′ end of RNA is chemically unique and can be selectively oxidized with periodate to give the dialdehyde. This can be fur-

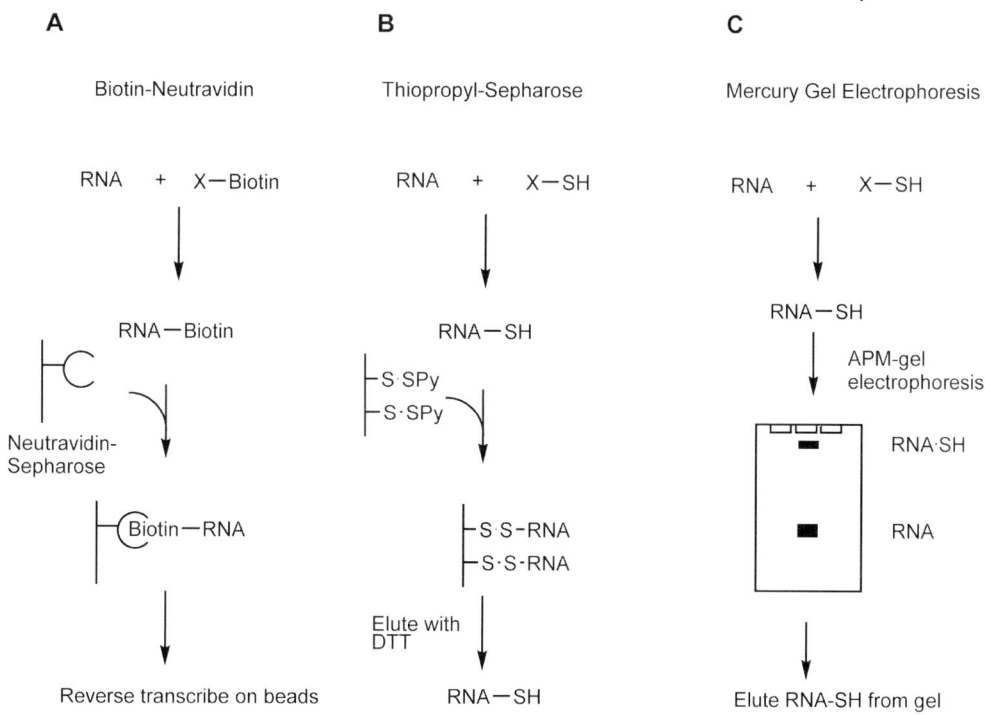

Fig. 47.5. Three common selection strategies: biotin–neutravidin (A), thiopropyl Sepharose (B) and mercury gel electrophoresis (C).

ther reacted with amines, allowing for the potential addition of a variety of functionalities to the 3′ end.

47.4.3
Selecting Active RNAs

The selection reaction is designed so that the active RNAs can be distinguished from inactive RNAs. In the acyl-CoA synthase selection, this depended on biotinyl-AMP as the substrate, which is a fatty acyl adenylate analog and also enables selection using the affinity of biotin for neutravidin. Biotinylated substrates have been widely used in selection reactions because of the ease of isolation using neutravidin or streptavidin conjugates, such as neutravidin–Sepharose and streptavidin-coated magnetic beads ([39, 40] and Fig. 47.5A). This method has been used to isolate RNAs that carry out several other acyl transfer reactions [31, 41, 42].

Another common method for isolating RNAs is to incorporate a thiol or thione into the substrate. RNAs that incorporate the substrate can then be selected using thiopropyl-activated Sepharose (Fig. 47.5B) or mercury gel electrophoresis

[43] (Fig. 47.5C), which exploits the affinity of mercury for thiols and thiones. Ribozymes for amino acid activation [44], nucleotide synthesis [35] and RNA polymerization [45], a 5′ kinase [46], and the capping ribozyme used in the acyl-CoA synthase selection have been selected using this method.

These are the most common methods for identifying RNAs that attach small substrates to themselves, since the materials required are readily available or can easily be synthesized. However, a variety of other methods have also been used, and we list some here for possible inspiration when planning new selections.

Illangasekare [47] used HPLC in the selection for a ribozyme that aminoacylates its 2′,3′ terminus, mimicking the reaction catalyzed by the aminoacyl-tRNA synthetases. RNAs capable of incorporating phenylalanine at their 2′,3′ terminus can be distinguished from inactive RNAs by the presence of a primary amine. The RNA pool was reacted with napthoxyacetyl-NHS, which quantitatively reacts at the amino group and appends a large hydrophobic napthoxy acetyl residue to the phenylalanine-RNA. Active RNAs were fractionated from inactive RNAs using reverse-phase HPLC, which could resolve RNAs carrying a napthoxy residue from those without the napthoxy group.

An early and diverse class of ribozymes were ligases, which catalyze the formation of a phosphodiester bond between two RNAs. The strategy used to select these RNAs was for active RNAs to attach a primer to themselves, enabling only active RNAs to be competent for the subsequent RT-PCR amplification steps. Bartel and Szostak used this strategy for their ligase selection [3]. RNAs capable of attaching an oligonucleotide to their 5′ end were isolated on an affinity column using an oligonucleotide sequence complementary to the ligated oligonucleotide. Active RNAs were then further enriched by PCR using a forward primer complementary to that ligated. PCR using the original primers provided the PCR pool for transcription for the next round of selection.

A last important class of ribozymes are the nucleases, which cleave phosphodiester bonds. Gel electrophoresis has been used to isolate RNAs that cleave themselves into smaller fragments [48, 49], as well as circular RNAs that linearize themselves [50]. Another method of isolating nuclease ribozymes is to attach the RNA to a solid support, with the active RNAs being isolated by virtue of their ability to release themselves by self-cleavage [51].

47.4.4
Negative Selections

Incorporation of a negative selection is often useful during the selection process to eliminate RNAs that survive the selection process in a spurious or unexpected way. For example, in the selection for acyl-CoA synthase ribozymes, the 2′-hydroxyls of RNA can act as nucleophiles that compete with the thiol. Given that there may be up to 200 2′-hydroxyls in an RNA, such a side-reaction can quickly become prevalent. One strategy is to carry out the negative selection exactly as for a positive selection, but to do so prior to modifying the RNA pool. In the acyl-CoA synthase selection, the negative selection was carried out by reaction with biotinyl-AMP prior to capping the RNA pool with coenzyme A. RNA from the negative selection was fractionated on a neutravidin column, and those that were not biotinylated were eluted, capped and used for the positive selection.

47.4.5
Stringency

Selection stringency can be increased by decreasing either the reaction time or the concentration of added substrate. In combination with mutagenic PCR [52], increasing the stringency allows for the isolation of more active catalysts from partially enriched pools. Mutagenic PCR introduces a small number of mutations to each sequence (around 0.066% per nucleotide per PCR cycle) and enables the exploration of sequence space surrounding active sequences in an attempt to find the local peak of activity. These techniques are often introduced after 3–5 cycles, when the danger of losing rare sequences is greatly diminished. Given the strong effect of stringency, the effort put into its optimization before beginning a selection will usually be repaid during the selection and characterization of products.

47.4.6
Analysis of the Product

One of the most difficult parts of a selection for active ribozymes is positively identifying the product of the reaction. Typically, reacted RNAs are digested with a nuclease (e.g. P1, RNase A, U2) and the reacted products are characterized using HPLC and mass spectrometry and compared with authentic standards. However, because the products are usually still attached to a nucleotide, obtaining the correct standards for comparison can be difficult.

Finding proper standards was a major problem in the selection for an acyl-CoA synthase ribozyme, since the product of the reaction was biotin–CoA–ppRNA. Digestion with P1 nuclease results in dephospho biotinyl-CoA, for which no standard could be obtained. To solve this problem, active RNAs were incubated with acetyl-AMP or butyl-AMP. P1 digestion of the product results in dephospho acetyl-CoA and dephospho butyl-CoA, respectively. These could be compared with standards obtained from the P1 digestion of authentic acetyl-CoA and butyl-CoA. Considering the strategy for analyzing the selection products when designing a selection strategy is therefore critical.

47.4.7
Determining the Scope of the Reaction

In many cases the reaction being selected for is not quite the reaction of interest, since the substrate must contain a tag to distinguish active RNAs from inactive ones. It is necessary then to determine how specific the RNA is for the substrate with which it was selected. The acyl-CoA synthase ribozyme incorporates acetyl-, butryl- and biotinyl-moieties, suggesting that is broadly specific for AMP-activated carboxylates. The capping ribozyme used to attach coenzyme A to the RNA pool was selected using thioUMP, yet will accept many phosphate-bearing molecules [24]. Another example is a ribozyme that activates carboxylate groups by forming acyl adenylates, initially selected using 3-mercaptopropionic acid with the free thiol group as a tag [44]. The isolate studied, KK13, will activate almost any amino acid, which was the original activity desired. 3-Mercaptopropionic acid was used for the selection instead of cysteine, however, because the α-amino group of cysteine made the product of the reaction with cysteine too unstable to survive the selection process. The selection was also carried out at pH 4.5 to stabilize the adenylate of 3-mercaptopropionic acid, since it is also too unstable at higher pH.

The above examples would suggest that most newly selected ribozymes are not particularly specific. However, this is not always the case. A self-aminoacylating ribozyme, RNA 77, selected using AMP-phenylalanine as the substrate is more selective by greater than 10^4-fold for AMP-phenylalanine than other non-cognate amino acids, such as in AMP-alanine, -isoleucine, -serine and -glutamine [53].

47.5
Sequence Analysis

Having performed a successful selection, the goal is often to find out how the RNA molecules perform the task selected. This is done using a combination of techniques including computational analysis, structural probing with chemicals and enzymes, reselection from partially randomized or "doped" pools, and kinetics. The first step is to clone individual RNA sequences to find out what the surviving sequences have in common. Between 30 and 100 sequences are usually cloned from the final pool of functional RNA molecules. Typically these sequences will consist of families of related sequences, as well as a number of unrelated unique sequences. Sequence families are composed of sequences with high similarity derived from a single common ancestor. As there might be many ways to catalyze a particular reaction or bind a particular target, several unrelated sequence families can be present.

47.5.1
Identifying Related Sequences

Sequences that are nearly identical are identified and grouped into families using computed multiple sequence alignment. Primer sequences should be removed before doing any alignments, since they will otherwise become the salient points of alignment. ClustalW (http://www.ebi.ac.uk/clustalw/ [54]) and PileUp [55] are probably most familiar, with ClustalW giving generally better results. Other algorithms, such as TCoffee [56], are increasingly being used. Typically, the alignment algorithm will insert large runs of gaps in the regions that share no sequence similarity. Therefore this is an iterative process; use the initial alignment as a guide to choose families, and then align each family by itself, by running the program again on just that family.

After global sequence alignment has been completed, divergent sequences are examined to determine if they share common sequence or structure motifs. Conserved sequences define independent derivations of the same active site, which can give an estimate of the abundance of particular kinds of "solutions" to a given functional task. When using all motif finders, both primer sequences and duplicate or near-similar sequences should be removed. This is typically done by looking at the pairwise distance between sequences and removing any that are less than 75% different. Most common alignment programs perform poorly at this task since they search for global rather than local alignments, and so find spurious matches and miss real ones. Local alignment methods such as BLAST [57] can sometimes find long motifs, but are insufficiently sensitive to find short, inexact ones. Dedicated motif finders that use a probabilistic approach, such as MEME (http://meme.sdsc.edu/meme/website/intro.html) and the Gibbs Motif Sampler (http://bayesweb.wadsworth.org/cgi-bin/gibbs.8.pl) perform better. MEME requires that you specify the number of 'best' motifs to search for, while the Gibbs Motif Sampler requires you to specify the length of the motifs you expect to find, making neither of them automatic. However, these programs were initially designed for protein sequences rather than RNA.

We have developed a motif finder specifically for RNA based on the probability that similar sequences in different RNAs would have appeared by chance. The probability of finding a specified number of copies of one particular motif in a set of sequences can be approximated by the Poisson distribution, which depends on how often a match would be found in a single attempt and on the number of places where the match could be found in a longer sequence. To determine whether the presence of a particular motif (containing N nt) is significant, we calculate the probability that any of the 4^N possible motifs would be found as frequently as that particular motif, since the sequence of the motif is usually not known a priori. To illustrate the results, Fig. 47.6 shows the probability of finding motifs a certain number of times in a typical data set of 20 sequences each with a randomized region of 50 nt. Finding an exact four-base motif in every sequence would be highly significant, but finding an exact three-base motif or a four-base motif with one mismatch would be expected by chance. The motif finder assesses the significance of each motif and is available at http://bayes.colorado.edu/theme. It allows individual motifs to be highlighted on the sequences and provides a convenient method of displaying one or more motifs on the sequences.

Motifs that are virtually statistically indistinguishable from random motifs are called subtle motifs [58]. Because functional RNA motifs typically consist of many short modules [4, 7] it is often impossible to find some of the pieces of the motif using sequence information only. One approach for finding subtle motifs is to use information from a sequence alignment: a detailed analysis of this problem has been published [59] as well as an algorithm for performing the search [58]. However, this approach is extremely slow and requires that the sequences be sufficiently similar to align, which is often not the case for sequences from SELEX. A

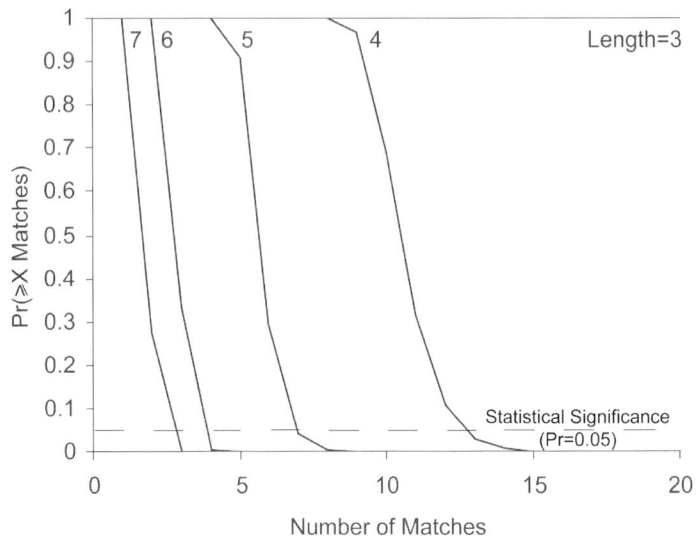

Fig. 47.6. The probability of finding a specified number of occurrences of the same motif as a function of motif length. Here, we show the probability of finding at least X copies (X ranging from 1 to 20) of a motif of length 3–7, assuming a typical data set of 20 sequences with 50 randomized positions each. The dashed line shows the region where none of the possible motifs would be expected to appear as often as observed at a significance level of 0.05. For example, it is over 99% probable that at least one triplet appears in every sequence, but for quartets the corresponding probability drops to below one in a million. For degenerate motifs, count only the specified positions (e.g. a motif of seven with three unspecified bases follows the same curve as an exact four-base motif).

better approach would be to use folding, restricting the search for subtle motifs (such as the tetramer in the isoleucine site, Fig. 47.3C) to regions in the secondary structure that are close to an obvious motif (such as the isoleucine octamer). This approach might be able to find all the parts of the motif even in sequences that align poorly.

47.5.2
Predicting Structure

Additional information can be obtained by analyzing the secondary structure of the selected molecules. The secondary structures of single sequences can be estimated with thermodynamic algorithms such as Mfold [60] and the Vienna package [61], which try to find the minimum free energy (most stable) structure. These packages can be run from the command line, but also have web interfaces, available at http://www.bioinfo.rpi.edu/applications/mfold/old/rna/form1.cgi and http://rna.tbi.univie.ac.at/cgi-bin/RNAfold.cgi, respectively. These programs predict about 70–80% of base pairs correctly in the 'best' structure based on analysis of tRNA and rRNA [62]. For multiple sequence alignments, such as sequence families often found after selections, programs like FoldAlign [63] and MWM [64] are reported to give more accurate results; however, they are computationally expensive and can be difficult to compile.

We have developed an algorithm called BayesFold [65] that objectively calculates structure predictions by integrating several different criteria, including folding energy, mismatches, mutual information, and chemical mapping. Because it is based on Bayes' Theorem, BayesFold does not require the user to pick arbitrary weights for the criteria and can deliver probability scores for each predicted structure. It is highly accurate on even moderately sized alignments, missing only one base pair in the "best" structure produced from alignments of between 10 and 85 bacterial tRNAs.

A web interface to BayesFold is available at http://bayes.colorado.edu/fold/. This interface has been specifically designed for ease-of-use, allowing the display of any sequence in an alignment threaded through any predicted secondary structure, as well as interactive exploration of which structures are supported by which criteria. Undesired criteria can be excluded from the calculations. The program also supports the display of data on the structure diagram, including mismatches and motifs. In particular, the ability to display motifs across a set of structures makes it easy to see whether important parts of the molecule are consistently placed. Publication-quality graphics can be produced directly from the web browser.

47.5.3
Chemical and Enzymatic Mapping

The most accessible way to refine a computed secondary structure is to use chemical and enzymatic mapping to find out which bases are paired and which are unpaired. Common agents and their specificities are Pb^{2+} and S1 nuclease (any unpaired base), V1 nuclease (any paired base), DMS (unpaired A and C), CMCT (unpaired U and G) and kethoxal (unpaired G). Levels of chemical or enzyme and short reaction times are chosen so that fewer than 10% of molecules are modified. Treatment of end-labeled RNA with Pb^{2+}, S1 or V1 results in RNA cleavage; subsequently, the cleavage sites can be mapped by running the reactions on a sequencing length gel alongside NaOH and T1 ladders, with the cleavage sites appearing as dark bands. DMS, CMCT and kethoxal modify RNA in a way that prevents subsequent primer extension by a reverse transcriptase; sites where the polymerase stops appear as dark bands (one nucleotide shorter than the actual position of modification) when the primer is end-labeled and analyzed on a sequencing length gel.

Modification experiments can also be used to find active sites, by performing the mapping with and without substrate, either looking for sites that are protected by the substrate (i.e. the substrate blocks access, directly or indirectly) or where modification prevents binding. For example, RNA is modified, passed through an activity assay such as an affinity column, and partitioned into functional and non-functional fractions. These can then be separately analyzed for the position of modification. Detailed protocols for these experiments are available [66, 67]. Protections and interferences often reveal the specific nucleotides that are implicated in activity.

47.5.4
Finding Minimal Requirements

Finding minimal requirements for a structure helps to distinguish active parts of the motif from internal spacers. Terminal random truncation experiments using alkaline hydrolysis [2] can be used to determine the sequence from the 5′ or 3′ end that is not essential for motif activity. Minimization by non-homologous random recombination [68] can determine whether spacer regions are needed by shuffling essential segments of a modular motif.

A frequently useful method is to do a doped reselection [69]. A functional parent sequence is chosen, and a pool is constructed that has a high percentage of the original nucleotide at each position (e.g. 70–90%) and a small percentage of each of the other 3 nt. The pool is then reselected; any positions that do not vary from the parent can be assumed to be important. This technique can also refine secondary structure predictions because covariations often reveal the paired regions in great detail. The optimal level of doping depends on whether the goal is to make every possible variant of the original site or to search a wider volume of sequence space for better solutions [70]. A web form is available to assist with doping calculations at: http://bayes.colorado.edu/doped_pools/poolsInput.html.

The information obtained from the above experiments can be used to construct smaller sequences incorporating the required motifs in order to provide minimal structures that still have the requisite activity. For example, a self-aminoacylating RNA could be cut down to only 29 nt from an original 95 nt [71]. However, some caution is warranted: Kvorova et al. [13] found that minimal hammerhead ribozymes were active *in vitro* but not *in vivo*, and that additional non-conserved loops were required for activity at physiological Mg^{2+} concentrations.

47.5.5
Three-dimensional Structural Modeling

Three-dimensional structural modeling from first principles is still fairly rudimentary. Pseudoknots, base triples and tertiary interactions are rather difficult to find and require large sequence alignments (dozens of closely related sequences at minimum). Doped reselections are often the only realistic way to provide enough covariation to detect tertiary interactions. The structures of several small aptamers and ribozymes have been solved by NMR and/or X-ray crystallography: these can be found in NDB, the nucleic acid database (http://ndbserver.rutgers.edu). Some information about active sites can often be gleaned by making targeted mutations that disrupt base pairing or higher-order interactions, while more sophisticated techniques, such as NAIM (nucleotide analog interference mapping), can define the role of individual atoms by elucidating the effects of related nucleotide analogs at the active site [72].

Typically a combination of techniques is used, depending on the level of analysis desired. However, every successful selection yields many more unique sequences than can be studied, so that there are surely many untold treasures hidden in laboratory freezers all over the world.

Acknowledgments

We thank Jason Carnes, Teresa Janas and James Watt for useful comments on the manuscript. Support from NASA and the NIH made this work possible.

References

1. D. Irvine, C. Tuerk, L. Gold, *J. Mol. Biol.* **1991**, *222*, 739–761.
2. J. Ciesiolka, M. Illangasekare, I. Majerfeld, T. Nickles, M. Welch, M. Yarus, S. Zinnen, *Methods Enzymol.* **1996**, *267*, 315–335.
3. D. P. Bartel, J. W. Szostak, *Science* **1993**, *261*, 1411–1418.
4. C. Lozupone, S. Changayil, M. Yarus, *RNA* **2003**, *9*, 1315–1322.
5. F. Ausubel, R. Brent, R. Kingston, D. Moore, J. Seidman, J. Smith, K. Struhl, *Current Protocols in Molecular Biology*, Wiley, New York, **2001**.
6. J. Sambrook, E. F. Fritsch, T. Maniatis, *Molecular Cloning: A Laboratory Manual*, 2nd edn, Cold Spring Harbour Laboratory Press, Cold Spring Harbour, NY, **1989**.
7. M. Yarus, R. D. Knight, in: *The Genetic Code and the Origin of Life*, L. R. d. Pouplana (ed.), Landes Bioscience, Georgetown, TX, **2003**.
8. P. C. Sabeti, P. J. Unrau, D. P. Bartel, *Chem. Biol.* **1997**, *4*, 767–774.
9. R. D. Knight, M. Yarus, *RNA* **2003**, *9*, 218–230.
10. M. Sassanfar, J. W. Szostak, *Nature* **1993**, *364*, 550–553.
11. C. T. Lauhon, J. W. Szostak, *J. Am. Chem. Soc.* **1995**, *117*, 1246–1257.
12. T. M. Coleman, F. Huang, *Chem. Biol.* **2002**, *9*, 1227–1236.
13. A. Khvorova, A. Lescoute, E. Westhof, S. D. Jayasena, *Nat. Struct. Biol.* **2003**, 708–712.
14. R. Barnard, V. Futo, N. Pecheniuk, M. Slattery, T. Walsh, *Biotechniques* **1998**, *25*, 684–691.
15. F. Mathieu-Daude, J. Welsh, T. Vogt, M. McClelland, *Nucleic Acids Res.* **1996**, *24*, 2080–2086.
16. I. Majerfeld, M. Yarus, *RNA* **1998**, *4*, 471–478.
17. M. Yarus, *FASEB J.* **1993**, *7*, 31–39.
18. B. M. Dunn, I. M. Chaiken, *Proc. Natl. Acad. Sci. USA* **1974**, *71*, 2382–2385.
19. G. J. Connell, M. Yarus, *Science* **1994**, *264*, 1137–1141.
20. D. S. Wilson, J. W. Szostak, *Annu. Rev. Biochem.* **1999**, *68*, 611–647.
21. B. Vant-Hull, A. Payano-Baez, R. H. Davis, L. Gold, *J. Mol. Biol.* **1998**, *278*, 579–597.
22. M. Famulok, J. W. Szostak, *Angew. Chem. Int. Ed. Engl.* **1992**, *31*, 979–988.
23. B. M. Dunn, I. M. Chaiken, *Biochemistry* **1975**, *14*, 2343–2349.
24. V. R. Jadhav, M. Yarus, *Biochemistry* **2002**, *41*, 723–729.
25. N. Kawazoe, N. Teramoto, H. Ichinari, Y. Imanishi, Y. Ito, *Biomacromolecules* **2001**, *2*, 681–686.
26. J. R. Prudent, T. Uno, P. G. Schultz, *Science* **1994**, *264*, 1924–1927.
27. P. Wentworth, K. D. Janda, *Curr. Opin. Chem. Biol.* **1998**, *2*, 138–144.
28. F. Huang, M. Yarus, *Proc. Natl. Acad. Sci. USA* **1997**, *94*, 8965–8969.
29. F. Huang, *Nucleic Acids Res.* **2003**, *31*, e8.
30. A. Jenne, M. Famulok, *Chem. Biol.* **1998**, *5*, 23–34.

31 B. Zhang, T. R. Cech, *Nature* **1997**, *390*, 96–100.
32 B. Seelig, A. Jaschke, *Bioconjug. Chem.* **1999**, *10*, 371–378.
33 G. Sengle, A. Jenne, P. S. Arora, B. Seelig, J. S. Nowick, A. Jaschke, M. Famulok, *Bioorg. Med. Chem.* **2000**, *8*, 1317–1329.
34 L. Zhang, L. Sun, Z. Cui, R. L. Gottlieb, B. Zhang, *Bioconjug. Chem.* **2001**, *12*, 939–948.
35 P. J. Unrau, D. P. Bartel, *Nature* **1998**, *395*, 260–263.
36 T. Ohtsuki, M. Kimoto, M. Ishikawa, T. Mitsui, I. Hirao, S. Yokoyama, *Proc. Natl. Acad. Sci. USA* **2001**, *98*, 4922–4925.
37 J. A. Piccirilli, T. Krauch, S. E. Moroney, S. A. Benner, *Nature* **1990**, *343*, 33–37.
38 N. K. Vaish, A. W. Fraley, J. W. Szostak, L. W. McLaughlin, *Nucleic Acids Res.* **2000**, *28*, 3316–3322.
39 N. M. Green, *Adv. Protein Chem.* **1975**, *29*, 85–133.
40 C. Wilson, J. W. Szostak, *Nature* **1995**, *374*, 777–782.
41 P. A. Lohse, J. W. Szostak, *Nature* **1996**, *381*, 442–444.
42 T. W. Wiegand, R. C. Janssen, B. E. Eaton, *Chem. Biol.* **1997**, *4*, 675–683.
43 G. L. Igloi, *Biochemistry* **1988**, *27*, 3842–3849.
44 R. K. Kumar, M. Yarus, *Biochemistry* **2001**, *40*, 6998–7004.
45 W. K. Johnston, P. J. Unrau, M. S. Lawrence, M. E. Glasner, D. P. Bartel, *Science* **2001**, *292*, 1319–1325.
46 J. R. Lorsch, J. W. Szostak, *Nature* **1994**, *371*, 31–36.
47 M. Illangasekare, G. Sanchez, T. Nickles, M. Yarus, *Science* **1995**, *267*, 643–647.
48 V. K. Jayasena, L. Gold, *Proc. Natl. Acad. Sci. USA* **1997**, *94*, 10612–10617.
49 N. K. Vaish, P. A. Heaton, F. Eckstein, *Biochemistry* **1997**, *36*, 6495–6501.
50 T. Pan, O. C. Uhlenbeck, *Biochemistry* **1992**, *31*, 3887–3895.
51 J. Wrzesinski, M. Legiewicz, B. Smolska, J. Ciesiolka, *Nucleic Acids Res.* **2001**, *29*, 4482–4492.
52 R. C. Cadwell, G. F. Joyce, *PCR Methods Appl.* **1994**, *3*, S136–140.
53 M. Illangasekare, M. Yarus, *Proc. Natl. Acad. Sci. USA* **1999**, *96*, 5470–5475.
54 J. D. Thompson, D. G. Higgins, T. J. Gibson, *Nucleic Acids Res.* **1994**, *22*, 4673–4680.
55 Genetics Computer Group, *Program Manual for the GCG Package*, 7th edn, Genetics Computer Group, Madison, WI, **1991**.
56 C. Notredame, D. G. Higgins, J. Heringa, *J. Mol. Biol.* **2000**, *302*, 205–217.
57 S. F. Altschul, W. Gish, W. Miller, E. W. Myers, D. J. Lipman, *J. Mol. Biol.* **1990**, *215*, 403–410.
58 U. Keich, P. A. Pevzner, *Bioinformatics* **2002**, *18*, 1374–1381.
59 U. Keich, P. A. Pevzner, *Bioinformatics* **2002**, *18*, 1382–1390.
60 M. Zuker, *Nucleic Acids Res.* **2003**, *31*, 3406–3415.
61 I. Hofacker, W. Fontana, P. Stadler, S. Bonhoeffer, M. Tacker, P. Schuster, *Monatsh. Chem.* **1994**, *125*, 167–188.
62 M. Zuker, *Curr. Opin. Struct. Biol.* **2000**, *10*, 303–310.
63 J. Gorodkin, S. L. Stricklin, G. D. Stormo, *Nucleic Acids Res.* **2001**, *29*, 2135–2144.
64 J. E. Tabaska, R. B. Cary, H. N. Gabow, G. D. Stormo, *Bioinformatics* **1998**, *14*, 691–699.
65 R. D. Knight, M. Yarus, *RNA* **2004**, in press.
66 C. Ehresmann, F. Baudin, M. Mougel, P. Romby, J. P. Ebel, B. Ehresmann, *Nucleic Acids Res.* **1987**, *15*, 9109–9128.
67 A. Krol, P. Carbon, *Methods Enzymol.* **1989**, *180*, 212–227.
68 J. A. Bittker, B. V. Le, D. R. Liu, *Nat. Biotechnol.* **2002**, *20*, 1024–1029.
69 E. H. Ekland, D. P. Bartel, *Nucleic Acids Res.* **1995**, *23*, 3231–3238.
70 R. D. Knight, M. Yarus, *Nucleic Acids Res.* **2003**, *31*, e30.
71 M. Illangasekare, M. Yarus, *RNA* **1999**, *5*, 1482–1489.
72 S. P. Ryder, L. Ortoleva-Donnelly, A. B. Kosek, S. A. Strobel, *Methods Enzymol.* **2000**, *317*, 92–109.

48
Aptamer Selection against Biological Macromolecules: Proteins and Carbohydrates

C. Stefan Vörtler and Maria Milovnikova

48.1
Introduction

Over the last decade, combinatorial approaches have brought new solutions to chemical and biochemical problems due to the analysis of large sets of molecules in parallel (a library) rather than testing individual members one at a time. Aptamers are one product of such strategies, representing *de novo* generated macromolecules with the ability to bind a chosen target, as implied by the Latin word root *aptus* = to fit. Generation of binding ability requires foremost the formation of a binding pocket or surface with stabilizing ionic, hydrogen-bonding or hydrophobic interactions, features that can be provided by macromolecules. The two salient requirements of *in vitro* selection approaches are methods to separate active from inactive library members (the partitioning or selection step) and to subsequently regenerate the library, enriched in binders, for a next round of selection. The latter is based on replicability and the direct connection of phenotype with genotype, criteria that are fulfilled by nucleic acid polymers [1], making RNA and DNA the first and still most important molecules in the construction of complex libraries as a source for the isolation of aptamers with tailored properties [2–4]. Related is the selection for catalytic RNA/DNA molecules, described for RNA in Chapter 47. In comparison, isolation of peptide aptamers was developed much later [5], and requires additional techniques like phage, ribosome and mRNA display to overcome the missing direct linkage between phenotype and genotype, either by confining the DNA to the phage body presenting the phenotype (compartmentalization), formation of a stalling ribosome or covalent attachment of peptide product to its mRNA template [6–10].

Targets for successful aptamer selections have included inorganic ions [11–13], small organic molecules like malachite green [14] or theophylline [15], antibiotics [13] and metabolites like ATP [16, 17] or AdoMet [18], biological macromolecules like peptides [19], proteins [3] or carbohydrates [20], supramolecular structures such as viral particles [21] or ribosomes [22], and whole cells, e.g. Trypanosomes [23] or red blood cell ghosts [24]. Aptamers often undergo a substantial conformational rearrangement upon binding to their target, a principle also providing the functional basis of natural riboswitches [25, 26] and switch ribozymes [27] involved in the regulation of gene expression. Several reviews document the diversity of the field [1, 28–33], while other articles focus on technical aspects [11, 34, 35] or introduce the reader to relevant online databases (Aptamer Database at http://aptamer.icmb.utexas.edu [36] and SELEX_DB at http://wwwmgs.bionet.nsc.ru/mgs/systems/selex/ [37]). Since each selection is unique in its requirements, this chapter can just provide an overview of the technique, its potentials and critical steps, based on two target types presented. Careful design by searching the literature and aptamer databases for targets related to one's own selection aims remains essential.

What to expect and not to expect from aptamers? First, aptamers bind their targets with high specificity. This is exemplified by differentiation among protein kinase C isoforms [38] or a single methyl group addition in caffeine compared to theophylline resulting in a 4 orders of magnitude weaker binding to an anti-theophylline aptamer [15]. Second, they have high affinities for their targets, with K_D values in the range of low micromolar to low nanomolar, in exceptional cases even picomolar for protein targets, while aptamers against small organic molecules display usually a higher K_D in the micromolar range. Third, aptamers allow for chemical modifications in order to stabilize against nucleases, e.g. by incorporation of 2'-fluoro- or 2'-amino-nucleotide analogs during transcription [39, 40] or chemical synthesis [41]. Also, selection efficiency may be improved by equipping the library molecules with photoactivatable bases that permit crosslinking of target–aptamer complexes [42]. Fourth, aptamers can be engineered to reduce their size and affinity tags, fluorophores [43] or modules for allosteric regulation may be added. They work as chiral HPLC phases [44], sensors [45, 46] or signal transducers [47–49], and even in array-based technologies [50] and high-throughput assays [51]. Fifth, optimization or adaptation of binding can be performed by re-selection using a partly randomized library based on the initial aptamer sequence [subtractive SELEX (Systematic Evolution of Ligands by EXponential enrichment)] [52]. Sixth, selection processes can be automated and further expedited by *in vitro* transcription/translation of target proteins with a biotin tag for immobilization [53]. Finally, the selection process is often rapid compared with the generation of antibodies, and particularly aptamer production after selection stands out in terms of reproducibility and robustness of the process. Moreover, since selections are performed *in vitro*, any selection scheme and target can be used.

48.2
General Strategy

Combinatorial macromolecular libraries can be subjected to screening or selection/evolution processes. The first requires multi-tube (or multi-well) high-throughput assays, while the latter is a one-tube reaction, explaining why this approach is suitable for any small laboratory with basic equipment. The library is passed through a sequence of steps (the selection cycle) aimed at separating active from inactive molecules (Fig. 48.1). The cycle is repeated (counting in selection rounds) until the de-

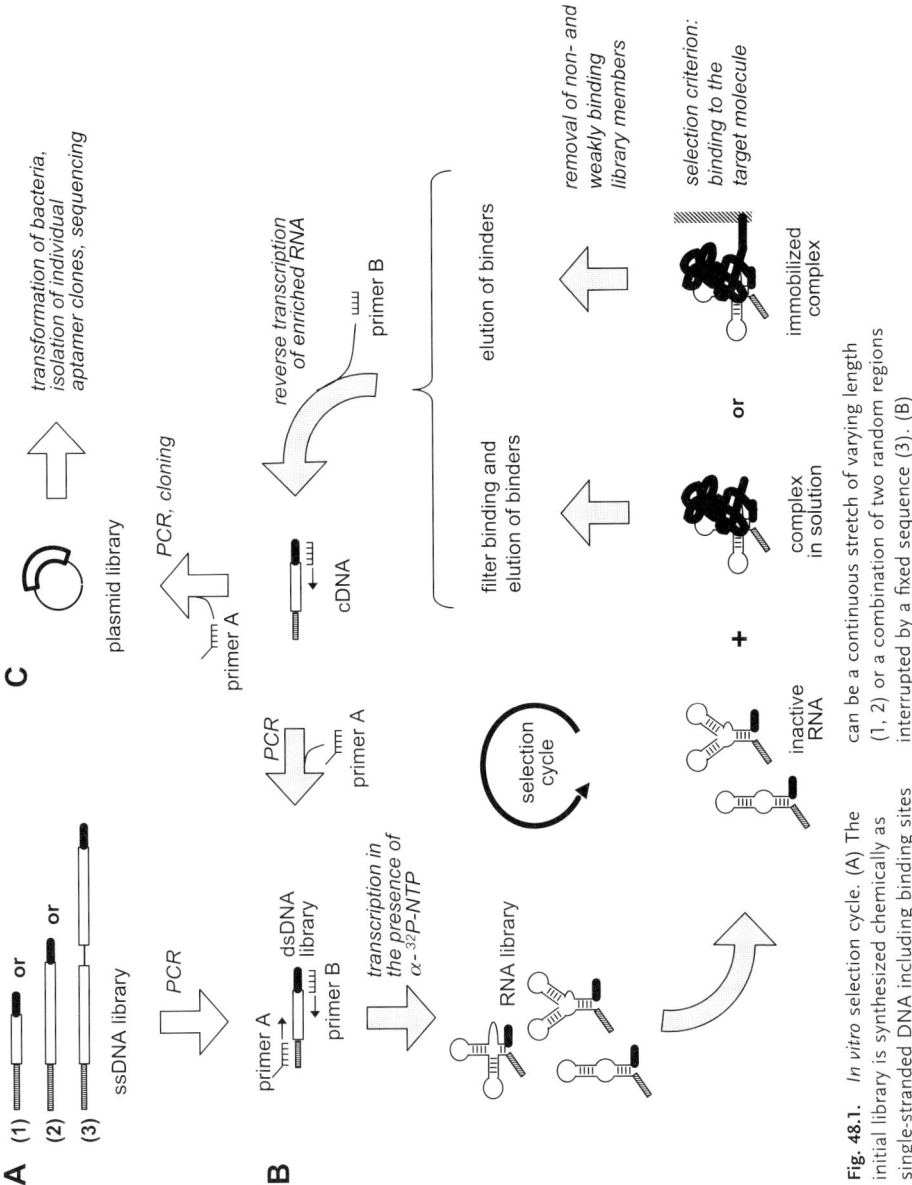

Fig. 48.1. *In vitro* selection cycle. (A) The initial library is synthesized chemically as single-stranded DNA including binding sites for primer A and B; primer A includes a T7 RNA polymerase promoter to initiate RNA transcription. The variable region (white boxes) can be a continuous stretch of varying length (1, 2) or a combination of two random regions interrupted by a fixed sequence (3). (B) Selection cycle. (C) Cloning of library members to retrieve individual aptamer sequences.

sired enrichment of functional variants is observed. With increasing cycle number, more and more functional molecules in addition to non-functional molecules are removed by performing the partitioning process under increasingly stringent conditions. SELEX is one technical term describing this process, based on the Darwinian principle of the survival of the fittest. *"In vitro* evolution" implies more than simple "selection" since the principles of mutation and adaptation are concurrently introduced by randomization techniques like mutagenic PCR or DNA shuffling. In contrast to *in vitro* selection, functional molecules finally isolated by *in vitro* evolution techniques are unlikely to have been present in the initial library since they were generated (evolved) in the course of the experiment. However, for most applications with the rather simple goal of isolating aptamers that bind a ligand with reasonable affinity and specificity, smaller variant libraries and an appropriate *in vitro* selection protocol will suffice.

48.2.1
Choosing a Suitable Target

Generally, targets with positively charged surface areas are expected to produce enrichment faster than those carrying negative or no charges. Among biological macromolecules, proteins are excellent targets based on their surface and charge distribution properties. Carbohydrates seem to be more problematic and much fewer cases of successful selections against carbohydrates have been reported. This is possibly a result of their uncharged nature and less rigid conformation resulting in the absence of distinct binding pockets and surface features. Nucleic acids as aptamer targets suffer from the likely selection of antisense binders since the strongest interaction is driven by Watson–Crick complementarity, as illustrated by the selection of aptamers that recognize their RNA targets predominantly via kissing loop interactions [31, 54, 55]. Lipids are again a challenging target due to their hydrophobicity, although first examples of membrane-associating aptamers have been described [56, 57]. In the following, we will focus on proteins and carbohydrates as targets, and will highlight specific problems and important aspects associated with these two target types.

48.2.1.1 **Protein Targets**
Nucleic acid-binding proteins were among the first aptamer targets, with the goal to study the binding requirements of their highly adapted binding sites [3]. Meanwhile, selections against many more proteins, representing a variety of shapes and functions, have been successful, with over 100 protein and peptide targets listed in the aptamer database. A very important aspect is the state or form of the protein used as target as well as the context in which it is presented during selection. For example, when the target is a cell surface receptor, intact cells [23], a membrane fraction containing the protein [24], isolated protein or recombinant variants, proteolytic fragments or just peptide stretches could be used. However, strategies have to be adapted accordingly. Complex assemblies like cells present many targets, giving rise to different aptamers with a limited chance to find those specific for the

protein of interest. Such approaches require 15–25 rounds of selection to remove less specific background binding, as well as specialized techniques like deconvolution SELEX. In the latter method, aptamers are photo-crosslinked to their protein targets, the complexes formed are resolved by SDS–PAGE and electroblotted onto nitrocellulose to remove non-covalently associated RNAs. The protein–RNA complex band of interest is then excised from the membrane, followed by protein digestion and PCR amplification of the retained RNA [24]. While often requiring a more elaborate selection procedure, aptamers successfully selected against targets presented in a complex context can be expected to work properly in their natural environment. Selection with a purified protein can never take this complexity into account, but has the advantage that a biochemically more defined target is used with a higher probability of successful selection in fewer cycles from an environment better to control. Most selections use this strategy, which requires rigorous purification of the target protein. In case of protein degradation, expression or solubility problems, isolated proteolytic fragments or derived peptides offer an alternative (see Chapter 50). Thus, aptamers may be directed against a specific domain or even a single loop, provided that such a structural element is part of an accessible epitope in the intact protein.

A cloned protein, overexpressed in soluble form with an affinity tag, is usually an optimal starting point for *in vitro* selection. Since tags may interfere with folding and function in a position-dependent manner, C- and N-terminally tagged protein variants should be prepared in parallel. In addition, different types of tags may be tested as well (see Section 48.2.2). Affinity tags permit efficient purification in a short time and the tag can be utilized to immobilize the protein during selection. Otherwise, one is restricted in the selection design mainly to nitrocellulose filter binding, the method employed for the majority of proteins isolated from natural sources. The protein preparation has to be as pure as possible, since contaminants will affect the selection outcome. A combination of affinity, ion-exchange and gel-permeation chromatography is strongly advised, followed by analytical PAGE with overloading the lanes to identify impurities. Around 10 mg of pure protein suffices for selection and characterization. To reduce the possibility of artifacts, a single batch of the protein should be used throughout the selection procedure.

48.2.1.2 Carbohydrate Targets

Carbohydrates are the most ubiquitous and prominently exposed molecules on the surface of living cells. Cell–cell recognition and cell activation throughout development and maturation of a living organism depend on this class of macromolecules, and carbohydrate patterns frequently vary according to specific stages of cellular differentiation and development; likewise, alterations in carbohydration are often related to diseases (e.g. cancer or retrovirus infection). Aptamers offer a versatile tool to analyze and interfere with carbohydrate-mediated recognition processes. However, carbohydrate recognition is generally characterized by exceptionally weak binding with K_D values in the millimolar range. Specialized natural proteins, the lectins, overcome this by clustering of carbohydrate recognition units, which results in micromolar K_D values. Binding to highly charged nucleic acids requires structural properties not common to carbohydrates. Moreover, as conventional biochemical methodology is adapted to charged macromolecules, characterization of aptamer–carbohydrate interaction requires adapted techniques. Incorporation of positive charges reduces this problem, giving aminoglycosides a unique position among oligosaccharides by making them potent natural RNA binders that, for example, block RNA function on the ribosome and thus act as antibiotics. The spatial arrangement of the functional hydroxy and amino groups in aminoglycosides largely defines if specific rather than simply counter-charge driven RNA binding can be achieved. The specific binding mode of natural aminoglycosides is illustrated by their blockage of prokaryotic ribosomes while leaving eukaryotic ribosomes largely unaffected [58–60]. Ionic interactions and even pseudo-base pairs between RNA bases and sugar rings apparently dominate aminoglycoside–RNA interactions, whereas hydrophobic intercalations seem to play only a marginal role, explaining why plain carbohydrates have a low RNA binding capacity.

Details of target preparation are beyond the scope of this article. A basic problem is the high structural variability of oligosaccharides compared to proteins. Anomer, epimer, enantiomer, diastereomer, furanose/pyranose, α/β linkage and branched forms exist. Isolation from natural sources, *de novo* chemical as well as combined chemical/enzymatic syntheses are feasible, but difficult, and labor- and cost-intensive [61, 62]. Thus, aptamer selections published so far have concentrated on small fragments or commercially available polymers like dextran or Sephadex [20, 63, 64].

48.2.2
Immobilization of the Target

For recombinant proteins, immobilization is easiest achieved by incorporation of a small terminal tag, such as $(His)_6$-, Strep- or Nano-tag [65], which permits coupling to tag-specific substituted agarose or silica bead materials. Larger constructs, like GST-fusion proteins, carry the risk to select aptamers against the tag rather than the target. One should further bear in mind that any immobilization matrix represents a potential aptamer target itself (see Section 48.3). It is good practice to assay activity of the immobilized protein before selection to avoid that aptamers against inactive or denatured targets may be isolated. Untagged proteins, like those purified from natural sources, are selected by isolating the complex formed in solution, e.g. by nitrocellulose filter binding.

Carbohydrates can be chemically coupled to succinimidyl-activated dextran or silica beads [66] after introduction of a reactive amine during synthesis or post-synthetically. Often, *cis*-diol oxidation of sugar rings to aldehydes followed by coupling of amino functions is employed.

48.2.3
Selection Assays

Aptamers exploit surface features of the given target to achieve specific and tight binding. The required biochemistry is straightforward compared to selection for catalytic function. However, proper experimental design must ensure that only target-specific library members are enriched. One strategy is self-immobilization of binders to a solid phase-anchored target, followed by washing steps to remove bound molecules of lower affinity and elution of entire target–aptamer complexes or affinity elution of aptamers alone by addition of free target (Protocol 4). More target material is needed for the latter elution variant, but it ensures higher specificity due to involving two successive binding events. Alternatively, target–aptamer complexes formed in solution are separated from unbound library members by adsorption to nitrocellulose (Protocol 5) or using an electrophoretic mobility shift assay (EMSA, Protocol 9). It is important to keep in mind that all steps of the selection procedure represent selection criteria contributing to its outcome, including target preparation and presentation, as well as reverse transcriptase (RT), PCR and transcription reactions (Section 48.3).

48.2.4
Design and Preparation of the Library

Polynucleotides are ideal library molecules due to their replicability, inherent flexibility to adapt and bind to many surface topologies as well as the ease of ^{32}P-labeling for tracing purposes. Non-radioactive labeling (e.g. with fluorescent tags) may provide an alternative, but carries the danger of affecting the binding event and biasing the selection outcome. We will focus here on RNA libraries, since they include all points relevant to dealing with DNA as well. An initial single-stranded DNA library is chemically synthesized, enzymatically amplified and transcribed into the starting RNA pool (Protocol 1). Constant regions at the 5′ and 3′ ends are included for transcription and amplification, but may be omitted if methods for their post-selectional addition are developed [67]. Follow PCR rules to optimize these primer-binding sites. In the simplest design, a randomized region of up to 80 nt is introduced between flanking primer binding sites. *A priori*, it is hard to estimate the size of the sought-after binding motif, although some guidelines have been summarized [11]. The diversity (or complexity) of the randomized region defines the library and depends on the total number of nucleotides randomized (n), and the probability p with which the four bases may occur at a given random position. Usually each nucleotide should have the same probability, allowing to calculate the expected number of sequence variants within the library as $N = 4^n$ (the sequence space). Diversity can be further increased by including chemically modified nucleotide analogs during transcription [16, 17, 68]. Alternatively, a less complex library can be constructed by favoring incorporation of the natural nucleotide at a given position during chemical synthesis, with limited incorporation of the other three nucleotides (Protocol 1). Such biased or doped pools are useful to search for sequence variations around a natural RNA motif or to optimize aptamers in binding affinity by re-selection (see Protocol 7, Comment 3). Libraries of low complexity can further be obtained by mutagenic PCR of a non-randomized template [69, 70].

An experimentally manageable library contains millimolar amounts of RNA, since at higher concentrations RNAs tend to precipitate as Mg^{2+} salts. This corresponds to 10^{12}–10^{15} different sequences in an Eppendorf tube, thus exceeding the number of variants obtained by other combinatorial methods like phage display by 3–6 orders of magnitude, but still representing only a subset of the theoretically possible sequence space. Nevertheless, diversity usually suffices as a given RNA sequence can adopt a multitude of conformations, referred to as the "conformational hell" that causes general problems in RNA biochemistry [71]. In turn, however, this implies that a favorable binding conformation can be formed by many sequences, with a few of them likely to be present in the explored fraction of the theoretical sequence space.

48.3
Running the *In Vitro* Selection Cycle

Addition of the prepared RNA library to the target starts the first round of selection (Protocol 4). Its outcome is determined by the type of selection assay (each associated with different matrix-binding effects) and incubation conditions, subsumed under the term stringency, which include absolute concentrations as well as ratio of target to library, buffer composition, and incubation temperature and time.

The concentration of an individual sequence depends on the library complexity. One nmol of RNA with one random position corresponds to 0.25 nmol (151×10^{12} copies) per individual sequence. Having 10 random positions, each of the $4^{10} = 1.05 \times 10^6$ library members is present roughly with 1 fmol or 600×10^6 copies. For a library of medium complexity (40 randomizations), at least 10 nmol of library as a millimolar RNA solution should be used to cover a substantial fraction of the sequence space. Target concentration should always be higher than library concentration to provide sufficient binding sites and to reduce matrix binding. A buffer should be chosen which is low in divalent metal ion concentration, usually Mg^{2+}, to approximate physiological conditions and to minimize metal ion-induced RNA hydrolysis. Nonetheless, divalent metal ions, such as Mg^{2+}, Ca^{2+} or Mn^{2+}, are required for RNA structure formation. Also, buffer pH affects the net charge of a target protein. More positive surface charges lead to stronger, but rather less specific binding of the negatively charged library molecules. Similar argumentation holds for the ionic strength, with a requirement for monovalent ion concentrations of at least 100 mM to reduce unspecific binding. Temperature affects the binding kinetics as well as protein and library stability. Incubations to achieve complex formation are best performed at a temperature to be used for aptamer application. Unwanted binding, for example to the immobilization matrix, can never be excluded as the matrix concentration is usually much higher than that of the immobilized target [12, 20]. This problem can be counteracted by pre-incubation (negative selection) with matrix material either unsubstituted (pre-selection) or derivatized with a protein or carbohydrate different from the target (counter-selection). A yet more powerful strategy is alternating between different selection assays, for example performing even rounds on an agarose matrix and odd rounds on nitrocellulose filters.

Selections are generally started under conditions of lower stringency (mainly defined by a high target to library concentration ratio) which is then increased in later rounds. Thereby, a broad spectrum of binding molecules is first enriched to several copies per sequence, counteracting early loss of variety, e.g. due to adsorption to tube walls as a consequence of low target availability, before high stringency rounds later permit to isolate the best binders. However, this approach in turn increases the chance to accumulate weak target binders in early low stringency rounds, illustrating the dilemma of selection experiments.

Since association rates are likely to be fast for the majority of competing RNA variants, tight binding (low K_D values) will primarily depend on a low off-rate (k_{off}) and binding specificity is therefore determined by differences in k_{off}. Thus, high stringency is most efficiently reached by extensive washing rather than shortened incubation times. The latter is rather a measure of precaution to exclude RNAs which have to undergo slow rearrangements to reach a productive binding state. It is important to keep in mind that any change in stringency will affect non-target binding as well. For example, prolonged incubation times may increase the background of matrix binders or the fraction of lost sequences due to constant unspecific adsorption, for example to tube walls. In addition, all enzymatic steps involved in the cycle potentially affect the outcome, as illustrated by RT-PCR-induced predominance of certain sequences [72].

In initial rounds, just a small fraction of the library, 2–3% or less, can be expected to bind to the target. Progress is monitored by measuring target-bound radioactivity, with increasing levels of retained material expected from round 2–3 onwards, but taking into account that every change in stringency will change the selection course. Also, prolongation of the selection process beyond a certain point will lead to progressive elimination of binding sequences. After recovery of the target-binding fraction in each round, the enriched library is regenerated by RT-PCR and T7 RNA polymerase transcription, and the next cycle is started. A fraction of the amplified DNA of each round must be kept for record, either for later analysis or to be able to repeat a cycle under identical or altered selection conditions. After the final cycle, the library is ligated into a cloning vector, followed by transformation into bacteria to isolate individual members.

48.4
Analysis of the Selection Outcome

Three levels of analysis are recommended. Since many of the suggested experiments will be covered in the chapters devoted to RNA structure probing (Chapters 10, 13 and 15), not all protocols will be detailed here. Starting point is a sequence comparison of the isolated clones. General strategies are described in the literature [73] and sequence analysis tools for fast alignments can be found on the Internet (http://www.expasy.org) or in commercial software packages (like DNAstar, Vector NTI). Depending on the number of selection cycles performed, either a very diverse set of sequences or a few sequence families, each comprising several related sequences, can be expected. Most important are stretches of conserved positions found across several aptamers (a consensus motif), indicating likely target interaction sites and allowing arrangement into families. Sequence analysis is then followed by prediction of secondary structure using Zuker's algorithms (www.bioinfo.rpi.edu/applications/mfold [74]) or other folding tools (www.imb-jena.de/RNA.html). As different programs might result in different structure predictions, a comparison is recommended. The structure model needs to be validated experimentally by RNA secondary structure probing, footprinting of the target–aptamer complex, boundary experiments to identify minimal binding sequences (Protocol 13) and possibly 2–4 cycles of re-selection after partial randomization of a defined aptamer sequence (Protocol 7, Comment 3) [75]. The latter is a powerful tool to prove the importance of a binding motif, but allows for optimization of the original motif as well. Frequently, more avidly binding aptamers are identified by re-selection [76]. Finally, the third line of characterization concerns quantification of target–aptamer interaction by determining an apparent equilibrium binding constant, K_D, using double filter binding, EMSA or surface plasmon resonance (SPR) measurements (Protocols 9–12). It is recommended to validate results by using two different techniques.

48.5
Troubleshooting

It is in the nature of an *in vitro* selection experiment that its outcome can never be foreseen, being actually the hallmark of this approach, as new unconsidered solutions are found. Still, selections can fail completely. This is usually already recognized during the selection process, when no enrichment occurs. However, enrichment can be very modest (below 10%) if the background of other sequences is high and merely a subfraction of binders reproducibly folds into their binding-competent conformation. For protein targets, the inability to retrieve an aptamer is most likely a technical problem rather than a general property of the target. The same holds for positively charged carbohydrates, while uncharged carbohydrates are at the front line of the technology. Relevant questions to be addressed during the selection process include: Is DNA produced in the RT-PCR step? Does it have the correct size and represent one clear band? Is the DNA transcribed into RNA, again of expected size without fragmentation? Does the RNA have a high affinity toward the matrix or filters used? Was the stringency too high in early rounds? Is the target correctly immobilized, active and accessible?

Problems often encountered in individual steps are:

(1) No or multiple bands obtained by RT-PCR. Running a control RT-PCR without adding RT will exclude DNA contamination problems. Varying the cycle number is the first starting point to reduce PCR artifacts, followed by changes in the PCR extension time, template input and primer concentration. Standard PCR optimization rules should be followed, particularly with respect to the design of primers, which should be at least 20 nt long and void of internal secondary structures and complementarities to minimize dimerization.

(2) No or additional smaller RNA fragments obtained in the transcription reaction. Often T7 RNA polymerase encounters sequence-dependent pausing sites where it drops off the template, giving rise to smaller RNA products which run below the main product. Changes in the transcript length pattern with increasing rounds of selection may indicate enrichment of truncated RNA species better adapted to the selection conditions. RNase contamination, often feared, is actually rather rare; backbone hydrolysis at fragile sequence stretches due to the presence of Mg^{2+} in the incubation mixture is far more likely.

For all steps, "good RNA working practice" should be followed. RNases are water soluble and contamination can be avoided even without DEPC treatment (see Chapter 4): by wearing gloves, rinsing glassware extensively in warm followed by Millipore or double-distilled water before use, avoiding speaking while pipetting solutions, using freshly opened bags of plastic-ware which is usually RNase-free, although not guaranteed, avoiding dirt and dust deposits on storage bags and working spaces, usage of pipettes exclusively assigned to RNA work, possibly equipped with special filter-tips to avoid contamination.

48.6
Protocols

Protocol 1: Preparing a nucleic acid library

Design of the template
Efficient and specific primer binding sites need to be coupled with one or more random sequence stretches and an RNA polymerase promoter, usually that for T7 RNA polymerase (5′-TAATACGACTCACTATA-3′; Fig. 48.1). Start the coding sequence with at least two G residues for efficient transcription initiation. Such a pool of PCR products is then used as template for run-off transcription. Terminal restriction sites introduced by the PCR primers may later be utilized for cloning purposes, providing an alternative to the TA cloning after selection (Protocol 8). Reduce the cross-contamination risk within your laboratory by altering primer sequences for each selection and include a unique restriction site next to the ones used for cloning to identify the origin of the selected molecules.

Synthesis of the initial DNA library
The initial library needs to be synthesized chemically. Order from a commercial vendor or use an in-house DNA synthesis facility. A fifth port is needed for synthesis of a complete random stretch in order to connect a vial with a pre-made amidite mixture in the molar ratio of dA:dC:dG:dT = 3:3:2:2, correcting for differences in coupling efficiencies [77]. If a partially biased library derived from one particular sequence is preferred, for example to explore the sequence space around a given RNA motif for optimization of binding, two strategies can be followed. (1) Dope the dA port vial with small amounts of one, two or all three of the other nucleotide amidites, proceed accordingly with the other ports, and run synthesis of the paren-

tal sequence [78]. This requires either an eight-port synthesizer or two changes of vials during synthesis, from the constant region to partial random, and from the partial random to the second constant region. Purify the DNA product by denaturing PAGE (dPAGE) after complete deprotection. (2) Alternatively, use a mutagenic PCR (Protocol 7).

Large-scale PCR to prepare the initial pool
For transcription by T7 RNA polymerase, at least the promoter region of the single-stranded DNA obtained by chemical synthesis needs to be converted to a double-stranded form [79]. A large-scale PCR will produce fully double-stranded DNA, eliminate sequences non-amplifiable due to synthesis artifacts or incomplete protection group removal and increase copy number per sequence. For aptamer selections with a modest complexity of the library (around 40-nt random region), complete coverage of sequence space is impossible, but not essential for the isolation of functional molecules, thus a 1-ml scale suffices. However, if specialized strategies aim at including all possible variants, which is feasible in the case of short randomized stretches, calculate the required amount of DNA for full coverage of sequence variants by dividing the number of sequences within the library (sequence space $N = 4^n$ with n equal to the number of randomized positions) by Avogadro's number. Between 3 and 5 times this molar amount should then be produced in the initial PCR to ensure that no variant gets lost. In some cases, PCR reactions were even up-scaled in such a way that manual cycling in water baths was required when reaction volumes exceeded thermocycler capacities [34, 35].

(1) For a standard library, follow Protocol 7 to set up 10 100-µl PCR reactions in parallel, limited to 15 cycles; analyze the outcome on agarose gels. If insufficient amounts of DNA are produced, continue for a few more cycles.
(2) Combine reactions, mix and aliquot out in 20-µl fractions for storage at −20 °C.
(3) Use one aliquot to start the initial round by RNA transcription (Protocol 3); keep the remaining aliquots for later analyses or selections.

Characterization of the initial pool
The complexity of the initial pool is verified by primer extension with a radiolabeled primer, following the RT protocol (Protocol 7) and including 0.5 mM of one ddNTP at a time in four parallel reactions, or by RNA sequencing with RNase T1 (Chapter 9). An equal distribution of all four nucleotides in the random sequence tract is expected and serves as reference to monitor enrichment during selection rounds. Additionally, cloning members from the initial library (analyze around 20 clones) should provide a set of completely unrelated sequences.

Protocol 2: Preparing the target

Preparation of a protein target
The reader is referred here to excellent monographs addressing the problems and pitfalls of protein purification in much detail [80]. In short:

(1) Clone the target protein into an expression vector providing a C- or N-terminal tag, and transform into an expression strain such as *Escherichia coli* BL21.
(2) Monitor growth of two 50-ml LB cultures until an $OD_{600\,nm}$ of around 0.6 is reached and induce target protein expression in one flask by adding IPTG to 1 mM. Continue growth monitoring over 4 h by taking 5-ml aliquots of induced and non-induced cultures at different time points for protein analysis. Harvest cells by centrifugation at 3000 g for 2 min, remove liquid and store pellets at $-20\,°C$ until use.
(3) Analyze expression by SDS–PAGE after boiling cells directly in SDS sample buffer or by using a small-scale cell disruption protocol (see Protocol 2, Comments), which allows one to determine protein localization in the soluble fraction and, after SDS extraction, also in the pelleted material.
(4) Test immobilization with the small-scale protein preparation. For a His_6-tagged target, wash 50 µl Ni-NTA-Agarose (Qiagen) slurry 3 times with 500 µl protein buffer, add 50 µl protein solution, incubate 60 min on ice with slow agitation and recover the supernatant. Then wash 3 times with 500 µl protein buffer containing 20 mM imidazol and elute twice with 50 µl protein buffer containing 250 mM imidazol. Use 20 µl for SDS–PAGE.
(5) Apply the identified optimal growth conditions to prepare target protein in large scale from a 2–10-l culture. To optimize protein purity, we strongly recommend an anion-exchange chromatography step prior to affinity purification, followed by a final chromatographic step such as gel filtration.
(6) Determine concentration and, if possible, activity of the protein, which should be unaltered in the presence of the tag. Purity should be as high as possible and can be monitored by overloading an SDS–PAA gel or by using sensitive staining methods such as silver staining [81]. Dialyze the purified protein into a suitable protein storage buffer, ideally the selection buffer, and supplement with glycerol to 50% (v/v) for storage at $-20\,°C$.

Preparation of a carbohydrate target

(1) Test the purity of targets, either obtained from a commercial supplier, isolated from natural sources or synthesized chemically, using for example HPLC and/or ESI- or MALDI-MS. Further purify if needed, as each contaminant increases the risk to drive selection in an unwanted direction.
(2) Prepare the carbohydrate for immobilization by tethering it to a spacer molecule containing an activated thiol (e.g. the SPDP reagent from PIERCE, USA), which in turn is used for disulfide coupling to thiopropyl-Sepharose (Protocol 6). Spacer attachment is rather straightforward, particularly if primary amino functions are available as in the case of aminoglycosides. Detailed chemistry is beyond the scope of this article, but is well documented in the literature [66].

Solutions
SDS sample buffer: 65 mM Tris–HCl, pH 6.8, 10% glycerol, 2.3% (w/v) SDS, 5% (v/v) β-mercaptoethanol, 0.23% (w/v) bromophenol blue.

Protein buffer: 50 mM Tris–HCl, pH 8.1, 5 mM EDTA, 5% glycerol, 1 mM β-mercaptoethanol, 20 µM proteinase inhibitor phenylmethylsulfonyl fluoride (PMSF); the latter might modify amino acid side chains and is required only in early steps to irreversibly block serine-type proteases.

Comments

(1) Generally, the target protein, when overexpressed in *E. coli* strains, should be visible as a strong or even predominant band in crude lysates. If not, try protein enrichment by affinity chromatography via its tag, although binding to the matrix might be impaired due to the complex composition and high total protein concentration of crude lysates. Alternatively, modify the expression protocol by reduction of growth temperature, variation of inducer concentration, by changing the tag position or even by employing a different expression construct. Note that recombinant target proteins can be detected with tag-specific antibodies, such as anti-His$_6$-antibodies (Sigma, Qiagen).

(2) Cell disruption is a critical step since the target protein may be damaged during this process. Work on ice or in a cold room and as fast as possible. For small-scale protein isolation: resuspend the cell pellet derived from a 12-ml culture in 50 µl Protein buffer, transfer to an Eppendorf tube, and add step-by-step 10 µl 100 mg/ml lysozyme, 20 µl 50 mg/ml sodium deoxycholate, 10 µl 100 mg/ml DNase I and 20 µl 1 M MgCl$_2$, followed by 10 min of incubation at 20 °C (room temperature), during which viscosity decreases, and final centrifugation for 15 min at 4000 g to remove cell debris. The supernatant is used directly for functional tests or analysis by SDS–PAGE. For larger-scale preparations, cell disruption by nitrogen decompression (manufactured by Parr bomb; www.parrinst.com) is recommended, since the technique is simple, fast and avoids oxidation as well as warming of the sample. Alternatives include the French press, microfluidizer or ultrasound treatment, each followed by centrifugation steps at 30 000 and 100 000 g.

Protocol 3: *In vitro* transcription using T7 RNA polymerase

Setup

		Final concentration
8 µl	80 mM HEPES–KOH, pH 7.5	80 mM
2.2 µl	1 M MgCl$_2$	22 mM
4 µl	100 mM ATP	4 mM
4 µl	100 mM CTP	4 mM
4 µl	100 mM GTP	4 mM
4 µl	100 mM UTP	4 mM
1 µl	100 mM spermidine	1 mM
2 µl	0.5 M DTT	10 mM
2 µl	0.5% Triton X-100	0.01%

0.5 µl	40 U/µl RNasin (Promega)	0.2 U/µl
10 µl	20 U/µl T7 RNA polymerase	2 U/µl
x µl	DNA template	0.125 µM
to 100 µl	with double-distilled H$_2$O (ddH$_2$O)	

For radioactive labeling, supplement the above reaction with 1 µCi [α-^{32}P]ATP, but maintain the cold ATP concentration at 4 mM for the sake of high transcription yields.

Procedure

(1) Incubate 1.5–3 h at 37 °C; prolonged incubations may result in RNA degradation (see Chapter 1). Withdrawal of aliquots at different time points permits to analyze the course of RNA production and to adapt reaction conditions accordingly.
(2) Add 1.5 µl DNase I (molecular biology quality, RNase-free; Roche) and continue incubation for 1 h at 37 °C to avoid amplification of contaminating DNA template rather than functional RNA in subsequent RT-PCR reactions.
(3) Add 16 µl 0.5 M EDTA (pH 7.7) to chelate Mg^{2+} in order to dissolve precipitated Mg^{2+}-pyrophosphate complexes.
(4) At this point, samples may be either shock-frozen in liquid nitrogen or a mixture of dry ice/isopropanol for storage at −20 °C, or purified directly by 8% dPAGE (Protocol 10). For large-scale transcription reactions and before dPAGE purification, extract once with 100 µl phenol to remove protein and 3 times with 100 µl ether, followed by ethanol precipitation to concentrate the RNA (see Protocol 4, Step 6); alternatively, ultrafiltration may be employed for RNA concentration, e.g. using a microcon device (Millipore). Ether is preferred over chloroform due to its higher volatility and since it partitions above the aqueous phase, allowing for easy removal without changing tubes.

Comments

(1) Check nucleotide solutions for pH, which should be close to neutral; determine their concentration by UV spectroscopy, especially if self-prepared from salts. A low pH will block transcription. DTT should be prepared at a 0.5 M stock concentration, aliquoted and kept frozen at −20 °C to minimize oxidation.
(2) Optimization of conditions pays off, particularly for preparative transcription reactions. The Mg^{2+} concentration should be 2–4 mM above the total NTP concentration to provide enough Mg^{2+} ions for RNA structure formation, NTP binding and polymerase catalysis, while remaining below inhibitory levels, requirements usually fulfilled for Mg^{2+} concentrations of 14–22 mM.
(3) Minor 3′ end heterogeneities generated by T7 RNA polymerase (see Chapters 1 and 2) are usually not critical for aptamer selections, as termini lie within the constant primer binding sites that are restored at the end of each selection round. For more details on T7 transcription, see Chapter 1.

Protocol 4: Selection against immobilized protein targets

Procedure

(1) Prepare 200 µl Ni-NTA-Agarose slurry (Qiagen) in a reaction tube by washing twice with 1000 µl and resuspension in 200 µl selection buffer.
(2) Transfer one-third of the slurry to a fresh tube, add sufficient protein to saturate binding sites and incubate 1 h on ice. Keep the supernatant as well as the solutions of three washes with 500 µl selection buffer to determine the fraction of bound protein. If the supernatant contains no protein, repeat the procedure to ensure saturation of the affinity material. Keep the matrix with bound protein on ice until use.
(3) Denature 1.5 nmol of the ^{32}P-labeled RNA library (100 000–500 000 c.p.m.) in 80 µl 10 mM NaCl by incubation for 3 min at 80 °C, place the tube in a styrofoam rack at room temperature for controlled refolding over a period of 10 min and add 20 µl 5 × selection buffer; for pre-selection to remove matrix binders, combine this solution with the other two-thirds of washed agarose slurry and incubate for 10 min. Rotate or shake the tubes to avoid sedimentation of the agarose.
(4) Transfer the supernatant obtained from the pre-selection step to the one-third of agarose slurry with bound target protein for affinity binding, incubate for 10 min, remove the supernatant and wash 3 × with 100 µl selection buffer. Keep all fractions for analysis (see below).
(5) Elute bound RNA either by incubation with 100 µl 250 mM imidazol, pH 7 in selection buffer, displacing the entire RNA–protein complex; alternatively, perform affinity elution with a more than 10-fold excess of free over immobilized protein, which increases the stringency since only molecules binding to immobilized as well as free target will be eluted.
(6) Prepare the eluted RNA for RT-PCR reactions and storage as follows: extract once with phenol and 3-times with ether, add 1/10 volume 3 M NaAc, pH 5.6 and 2.5 volumes 100% EtOH to the aqueous phase, and place the tube at -20 °C for at least 2 h or overnight. For high salt and low volume samples, add 1–2 volumes of double-distilled H_2O (ddH_2O) before extraction to avoid phase separation problems. Remove the supernatant after centrifugation for 30 min at 4 °C and 10 000–15 000 g, briefly dry the pellet for 5 min in a Speed Vac or vacuum desiccator and redissolve in 40 µl ddH_2O by leaving the sample for 10 min at room temperature. Avoid more extensive RNA drying, since dehydrated RNA cannot be redissolved without disintegration. Shock-freeze the RNA solution in liquid nitrogen or in a dry ice/isopropanol bath and store at -20 °C.

Buffers
1 × Selection buffer: 50 mM Tris–HCl, pH 7.5, 100 mM NaCl, 7 mM $MgCl_2$.

Monitoring enrichment
Quantify the RNA contained in eluted and wash fractions, as well as that retained on pre- and selection matrices; this is achieved by Cerenkov counting of the entire reaction tube in a liquid scintillation counter or by spotting an aliquot onto a 3MM

Whatman filter, followed by liquid scintillation counting. Also save and count aliquots of input material before the pre-selection and selection steps to be able to calculate the percentages of bound radioactivity. In the beginning, the fraction of target-bound RNA may even fall below 1%, but a steady increase should be visible after rounds 3–5. Adjust the selection strategy accordingly, for example by including a second pre-selection step if matrix rather than target binding is increasing. Additionally, monitor the decrease in library complexity during enrichment as mentioned in Protocol 1.

Comments

(1) Handle agarose matrices with care. Use yellow tips for pipetting, cut 5 mm above the tip with a sterile scalpel to widen the opening for reduction of shearing forces. Centrifuge with low speed to avoid damage of the matrix resulting in leakage of bound material. Use of siliconized tubes reduces surface adsorption. Spin columns work equally well and are commercially available for self-packing. Alternatively, a simple RNase-free two-part system that prevents any leakage of radioactive RNA is rapidly assembled and works equally well: with a thin razor blade, make a small vertical cut into the bottom of a small polypropylene reaction tube (e.g. 400-μl tube 7518.1; Roth, Germany); fill in the agarose slurry and insert the tube into a conventional 1.5-ml Eppendorf tube whose closed lid has been briefly touched with the heated end of a Pasteur pipette to create an opening just fitting the small tube in diameter. By use of a needle, jab a second hole in the 1.5-ml tube's lid to allow air circulation and collect all liquid at the bottom of the large tube after modest centrifugation at 1000 g.

(2) Selection conditions must be adapted to each target. The buffer should provide enough ionic strength to reduce non-specific adsorption, and contain Mg^{2+} ions for RNA structure formation. Vary salt and Mg^{2+} concentration to adjust selection stringency. Lowering the Mg^{2+} concentration may disrupt more labile three-dimensional RNA structures. Inclusion of unspecific competitors (any other natural RNA such as tRNA) should drive the selection to more specific aptamers. Likewise, specific competitors (natural ligands or binding partners of the target) are expected to introduce a bias toward high-affinity binders or may have the effect of restricting the number of aptamer-accessible epitopes on the protein. High pool to target concentration ratios carry the danger of eliminating many binders in early rounds and increase the risk to select matrix binders. Better start with a low ratio, and increase stringency by gradually reducing the number of available target molecules from one selection round to the next.

Protocol 5: Selection by filter binding of protein–aptamer complexes

Procedure

(1) Follow Protocol 4, Step 3, for denaturation and refolding of the library. Incubate library and target to form target–protein complexes.

(2) Filter the target–library mixture over nitrocellulose filters as described in Protocol 11.
(3) Elute retained material by submerging the filter in 500–1000 µl elution buffer in a reaction tube, place 5 min at 95 °C and remove the supernatant; add another 500–1000 µl elution buffer and repeat the heating step. Combine both elution fractions, add glycogen to 60 ng/µl, extract with phenol/ether and recover the RNA by ethanol precipitation (Protocol 4). Continue with RT-PCR (Protocol 7).

Buffers
$1 \times$ Selection buffer: 50 mM Tris–HCl, pH 7.5, 100 mM NaCl, 7 mM $MgCl_2$.
$1 \times$ Elution buffer: 7 M urea, 50 mM HEPES–NaOH, pH 7.5, 10 mM EDTA.

Comments

(1) Ensure that the filter has the capacity to bind all complexes by filtering the same amount of target protein as used in the selection step and quantifying the amounts of protein retained on the filter and in the flow-through. This control experiment will also reveal if the target protein is only inefficiently retained on nitrocellulose filters despite sufficient binding capacity of the membrane. In such cases, nitrocellulose filter binding would be inappropriate for the selection assay.
(2) Conceivable alternative elution and recovery methods are (i) incubation of the filter in 4 M guanidinium thiocyanate for 15 min at 75 °C, (ii) proteolytic degradation of the target protein or (iii) dissolving the entire filter in a suitable organic solvent such as acetone, followed by addition of an aqueous phase and phenol/ether extraction.

Protocol 6: Selecting against carbohydrate targets
Target immobilization is achieved by coupling an activated carbohydrate to thiopropyl-Sepharose via a disulfide bond. This coupling is position-specific and allows reversible elution of RNA–target complexes. Both features are of advantage compared with coupling to epoxy-activated Sepharose, e.g. as employed in selections against tobramycin or neomycin [82].

Procedure

(1) Swell 2 g of activated thiopropyl-Sepharose 6B matrix (Amersham Biosciences) by incubation for 1 h in degassed $1 \times$ Activation buffer supplemented with 20 mM DTT. Monitor the release of 2-thiopyridone via absorption at 342 nm as the reaction progresses, which correlates with the formation of free SH groups. Wash the matrix extensively over a sintered glass filter until absorption between 190 and 500 nm is no longer detectable. Split the matrix in three aliquots.

(2) Couple the thiopyridyl-activated carbohydrate target to one-third of the matrix by overnight incubation in 1 × Activation buffer. Monitor again the coupling process and thus substitution level via release of 2-thiopyridone (change in $OD_{342\,nm}$). Wash as described above. Use the remaining two-thirds of the matrix to prepare the pre-selection material, either by coupling the spacer employed to activate the carbohydrate target (washing procedure as above) or by straight deactivation. Deactivation is achieved by reacting the free SH groups with 500 mM iodoacetate in 100% EtOH at room temperature for 1 h. Wash extensively with 100% EtOH. Store reduced and modified Sepharose material under 5% (v/v) EtOH in H_2O or buffer at 4 °C. No loss of activity is observable over 10 weeks or even longer storage periods.

(3) Use the pre-selection and carbohydrate-coupled matrices as described for the protein selection according to Protocol 4. Bound RNA is recovered either by affinity elution with free carbohydrate, by reduction and release of the carbohydrate-RNA with 20 mM DTT in selection buffer or by adding the entire column material directly to the RT reaction (Protocol 7).

Buffer

1 × Activation buffer: 100 mM Tris–HCl, pH 7.5 at 25 °C, 250 mM NaCl, 1 mM EDTA.

Comments

(1) Embedding a carbohydrate moiety within a network of modified agarose requires stringent pre- and counter-selection techniques to enrich aptamers against the target. Alternate use of this and other immobilization surfaces or an entire switch to other matrices, such as glass beads or microtiter plates, may mitigate the problem of matrix binder enrichment. However, available data are too limited to recommend such alternative surfaces.

Protocol 7: Regeneration of the enriched library

RT of the selected RNA

Setup for a 20 µl RT reaction:

		Final concentration
5 µl	10 µM primer A	2.5 µM
10 µl	selected RNA library in ddH_2O	

Denaturing and reannealing step

2 µl	20 U/µl AMV reverse transcriptase	2 U/µl
1 µl	dNTP mix, 10 mM each	0.5 mM each
2 µl	10 × RT buffer	1×
	add ddH_2O to a final volume of 20 µl	

(1) Add primer A (Fig. 48.1) to the selected RNA library, denature for 3 min at 80 °C and anneal the primers by placing the reaction tubes for 15 min in a styrofoam rack or a thermal incubator at 37 °C.
(2) Supplement the additional components and incubate for 1.5 h at 42 °C. As in contrast to PCR only one round of elongation occurs, optimal primer design and annealing conditions are of particular importance.

PCR to amplify the DNA library
Setup for a 100 µl PCR reaction:

		Final concentration
5 µl	10 µM primer A	0.5 µM[1]
5 µl	10 µM primer B	0.5 µM
x µl	DNA library as template	≈ 1 fM[2]
2 µl	dNTP mix, each dNTP 10 mM, pH 7.0	0.2 mM
10 µl	10 × PCR buffer	1×
0.4 µl	5 U/µl *Taq* DNA polymerase	20 U/ml
	with ddH$_2$O to 100 µl	

[1] In late selection rounds only; see Step (1) below
[2] For round 1, use initial chemically synthesized DNA library; from round 2, use RT reaction (see above).

(1) Setup the reaction, always adding template last to avoid cross-contamination. For RT-PCR of early selection rounds, use the entire RT reaction mix as template and omit primer A as it is provided in sufficient amounts with the RT reaction mix. The RT reaction mix may make up one-fifth of the PCR reaction volume.
(2) Start cycling as exemplified for a library with 40 randomized positions, which we have used for protein and carbohydrate selections. For optimal PCR performance, the cycling program must be adjusted in each selection round due to changes in template nature and yields.
step 1: 95 °C 1 min (denaturation, use 5–10 min in case of plasmid DNA)
step 2: 95 °C 30 s (denaturation)
step 3: 55 °C 30 s (annealing)
step 4: 72 °C 20 s (polymerization) – go back to step 2, run 20–25 cycles
step 5: 72 °C 5 min
step 6: 4 °C storage until further processing
(3) Analyze PCR progress on 2% agarose gels by taking a 5 µl aliquot every fifth PCR cycle to avoid formation of artifacts indicated by multiple products. Prior to starting a new selection round, a 10-µl aliquot is stored at −20 °C as reference.
(4) Recover amplified DNA by 1 × phenol/3 × ether extraction and ethanol precipitation as described in Protocol 4, Step 6.

Buffer system
Usually buffers are supplied with purchased enzymes. Basic compositions which work well are:

1 × RT buffer: 50 mM Tris–HCl, pH 8.8 (25 °C), 50 mM KCl, 6 mM MgCl$_2$.
1 × PCR buffer: 10 mM Tris–HCl, pH 8.8 (25 °C), 50 mM KCl, 1.5 mM MgCl$_2$.

Comments

(1) To avoid contaminations, use filter-containing tips for pipetting and work at a freshly cleaned place or, if available, under a laminar flow cabinet or sterile work bench. Include a negative control (omission of template), as well as a positive control using a different template resulting in a product of similar size as that expected for amplification of the library. Native 8% acrylamide gels provide better resolution then agarose in case of blurred and diffuse PCR bands.
(2) For the selection process, we prefer *Taq* DNA polymerase over high-fidelity DNA polymerases.
(3) Mutagenic PCR offers a fast possibility to screen sequence space around a given sequence (theoretical background discussed in [75]). Either adjust the PCR reaction to special salt conditions that reduce the fidelity of the polymerase [83, 84] or dope the dNTP-mixture with mutagenic nucleotides that have an extended base-pairing potential [70]. Run 3–6 rounds of re-selection with this mutagenized library using the initial selection setup.

Protocol 8: Isolation of individual aptamers from enriched pools
This is the only step in the whole process that requires *in vivo* techniques.

Cloning the library into a vector

(1) Load RT-PCR amplified pool DNA (Protocol 7) onto a 2% agarose gel, excise the corresponding PCR product after staining and extract it by passive diffusion or by use of a gel extraction kit such as QIAEX II (Qiagen).
(2) The purified PCR products are easiest cloned into a linearized vector with single 3′ thymidine overhangs (e.g. the pGEM-T vector, Promega), since *Taq* DNA polymerase adds a single deoxyadenosine, in a template-independent fashion, to the 3′ ends of amplified fragments. Incubate around 2 ng of the eluted DNA, 16 ng vector and 400 U T4 DNA ligase in a 10-µl reaction for 16 h at 14 °C. Alternatively, use the restriction sites included at the end of the primers for conventional cloning (see Protocol 1).
(3) Add the 10-µl reaction to 300 µl competent cells in a reaction tube, incubate 30 min on ice, then for 1 min at 42 °C, place immediately on ice, add 1 ml LB medium without antibiotic and incubate 60 min at 37 °C under shaking. Harvest bacteria by gentle centrifugation (1000 g), resuspend in 100 µl LB and plate on a LB agar plate with suitable antibiotic for overnight growth at 37 °C.

Isolation of individual aptamer sequences

(1) Ideally, apply blue–white screening to identify colonies with inserts, using an appropriate cloning vector such as pGEM-T. Additionally, colony PCR allows

us to determine insert size as outlined below. Pick 40 white colonies from the transformation plate; for this purpose, dip with a sterile yellow tip into a colony, then touch with the tip a new LB-antibiotic plate placed on a numbered square grid for replica plate growth and finally inoculate 3 ml LB-antibiotic broth by ejecting the tip into the medium. Incubate agar plates and shake cultures overnight at 37 °C.

(2) The replica plate is used the next morning to run colony PCRs; to collect rather constant amounts of cell material, stab a pipette tip vertically into the colony, wash the bacteria directly into a PCR mixture by pipetting up and down, and run the PCR as described in Protocol 7.

(3) If clones contain an insert of expected size, 300 µl of the corresponding overnight culture (see Step 1) is used to prepare glycerol stocks by adding the same volume of 75% (v/v) sterile glycerol for storage at −80 °C, while the remaining 2.7 ml are used for plasmid preparation and subsequent sequencing. Keep 100 ng of plasmid DNA as back-up for later transformations.

Comments

(1) Always perform control transformations with intact pGEM-T or other high copy vectors (more than 2000 colonies expected) to check if cells are competent.

(2) Insert length may vary, as internal deletions will produce shorter sequences. These clones likely contain active aptamers and will thus provide important information about binding motifs and their length.

(3) Since cloned inserts include the T7 promoter, prepared plasmids can be used directly for *in vitro* run-off transcription after linearization at a 3′-terminal restriction site, either within the insert as part of the primer sequence or provided by the vector; alternatively, the insert DNA may be reamplified by PCR with ca. 1 ng of plasmid as template following Protocol 7.

Protocol 9: Electrophoretic mobility shift assay (EMSA)

Setup

Native PAGE is best performed in a small (such as the Hoefer or Bio-Rad mini systems) or medium-size (20 × 20 cm) gel system using 6–10% acrylamide gels (Protocol 10) for good separation, depending on size and charge of the complex. It is further recommended to use a buffer system without monovalent ions to reduce current and heat production, and to employ a setup with temperature control to avoid irreversible complex dissociation due to overheating. The concentration of ^{32}P-labeled RNA is kept constant (low nanomolar range; 200–1000 Cerenkov c.p.m. suffice in most cases), while the target concentration is varied in a first experiment to cover 3 orders of magnitude up to micromolar concentrations (e.g. 10 nM, 100 nM, 1 µM and 10 µM). A sharp transition should be visible, indicating the concentration range which needs to be assayed in more detail.

Sample preparation and assay

(1) Denature 10 000 c.p.m. of ^{32}P-labeled aptamer RNA in 10 mM NaCl for 1 min at 90 °C and transfer the tube to a styrofoam rack kept at room temperature; after 3 min add incubation buffer to initiate RNA folding and incubate for another 7 min.

(2) While renaturation is in progress, prepare for each target concentration a reaction tube on ice with buffer, protein and additional components. Include control reactions, such as RNA without target, as well as target plus RNA denatured prior to gel loading by adding 20 mM EDTA and heating for 1 min at 90 °C. Add RNA last, mix gently and incubate for 15–60 min on ice or at any other temperature of choice.

(3) Pre-run the gel for at least 30 min. For electrophoresis at 4 °C, use a system that permits buffer cooling or clamp an aluminum cooling plate to the front of the gel rather than simply performing the experiment in a cold room. Add no more than one-fifth reaction volume of Native loading buffer to the samples and rinse the gel pockets with a syringe immediately before gel loading.

(4) Electrophoresis conditions depend on acrylamide concentration and size of the target–RNA complex. In case of a 100-nt aptamer and a 40-kDa protein, electrophoresis for 2 h at 100 V in an 8% native polyacrylamide gel works well (Fig. 48.2). After electrophoresis, expose a phosphoimager screen to the gel, provided that enough radioactivity is present to allow for a short-term exposure (2–10 h, depending on complex nature) in order to limit diffusion and to avoid image blurring; otherwise, dry the gel in a gel dryer for 45 min at 80 °C before phosphoimaging.

Buffers

Gel shift buffer: 25 mM Tris-acetate, pH 7.4, 5 mM magnesium acetate; well working system which is used for complex formation and electrophoresis; the pH can be raised to 8.4 to increase gel mobility, although this enhances the risk of RNA degradation.

Native loading buffer: Gel shift buffer plus 30% glycerol (v/v), 0.25% (w/v) each bromophenol blue and xylene cyanol (see Comments below).

Comments

(1) EMSA experiments exploit the so-called cage effect that prevents irreversible dissociation during electrophoresis [85]. Organic dyes like bromophenol blue have been suspected to affect complex association/dissociation equilibria. Omission of dyes from the loading buffer may improve complex stability, while the progress of electrophoresis can be monitored by loading an adjacent lane with a dye-containing buffer.

(2) EMSA and nitrocellulose filter binding assays are best performed in parallel,

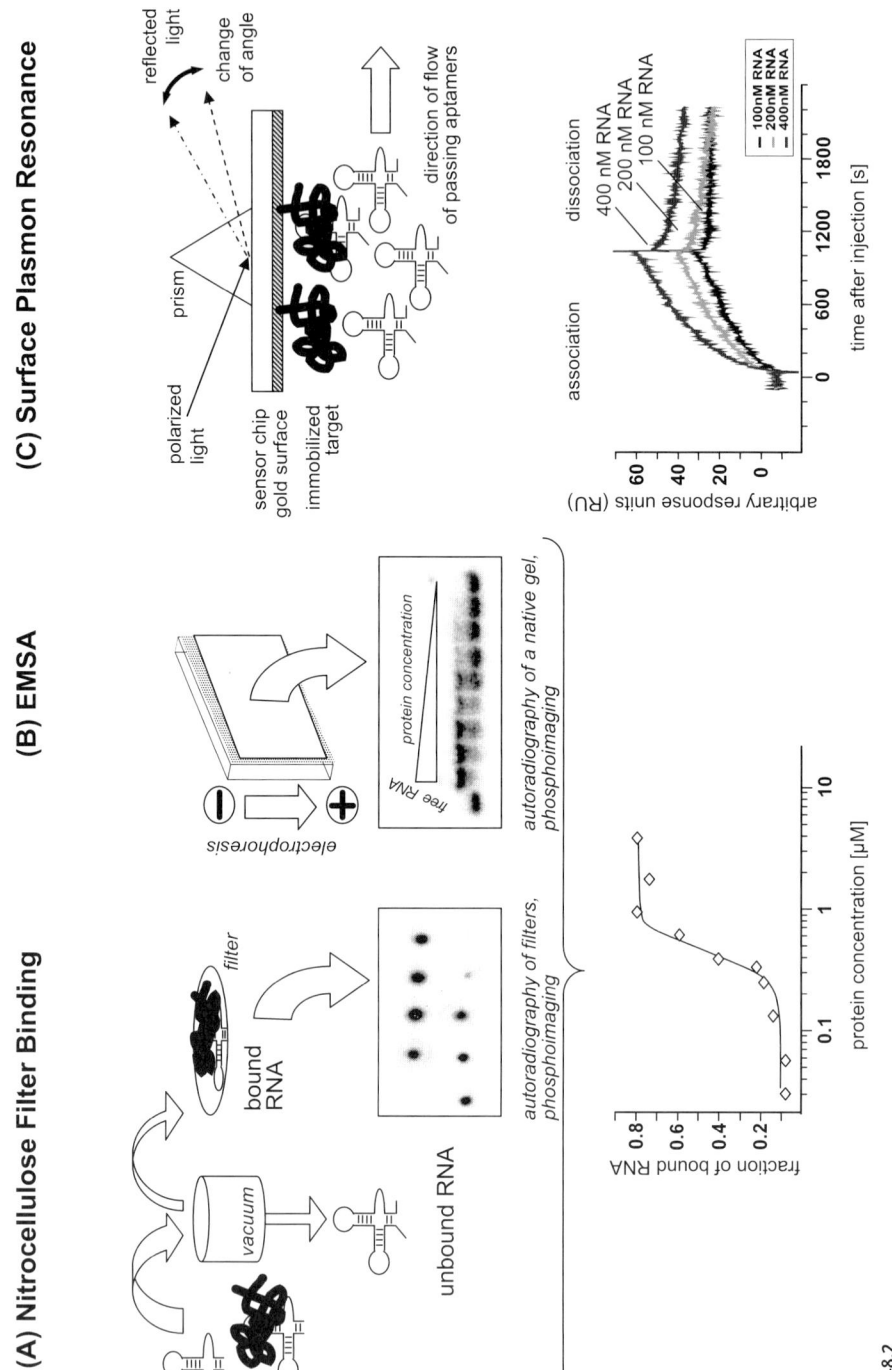

Fig. 48.2

using 6 µl of a 20-µl reaction setup for the native PAGE experiment and the remainder for filter binding (Protocol 11).
(3) The EMSA technique may also be used to isolate aptamer-target complexes during selections. However, the approximate position of the shifted complex needs to be known to excise and elute the appropriate region of the gel. Therefore, EMSA is best suited as an alternate selection assay in later selection rounds when shifted bands become more prominent and discrete due to enrichment of affine binders.

Protocol 10: PAGE

Procedure

Depending on the experimental question, different setups are recommended:

	Size (cm)	Acrylamide (%)	Electrophoresis (V/h)
Analytic dPAGE	20 × 20 × 0.02	8–20	1000/1–1.5
Preparative dPAGE	20 × 20 × 0.1	8	500/2–4
Native PAGE	20 × 20 × 0.02	8	250/2–4[1]

[1] Native PAGE may require gel cooling during electrophoresis (see Protocol 9)

(1) Prepare solutions of the desired acrylamide concentration by mixing acrylamide with urea stock solutions.
(2) Silanize the gel-facing side of both glass plates with 2 ml 2% (v/v) dichlorodimethylsilane (Sigma) in chloroform by dispensing the liquid with a Kimwipe or other paper tissue in a fume cupboard, wearing gloves; repeat this twice and finally wipe the plate once with 100% EtOH. This treatment is critical for smooth gel pouring, reduced trapping of air bubbles and proper disassembly after electrophoresis as the gel will not stick to the plates.
(3) For analytic gels, assemble the sandwich of glass plates, clamp together with ear-plate up, tilt slightly and allow the gel solution to enter slowly across the entire width between the glass plates by capillary forces. Insert comb, place horizontally until polymerized. Thicker preparative gels require a bottom spacer and are prone to leak at the corners, which can be prevented by polymerizing a few milliliters of the acrylamide solution barely visible at the bottom and adding the rest once the bottom layer has solidified.

Fig. 48.2. Characterization of aptamer binding. (A) In nitrocellulose filter binding assays, radioactively labeled aptamer RNA is retained on the filter via binding to its target protein. Trace amounts of RNA are incubated with increasing amounts of protein. Bound RNA can be quantified by phosphoimaging or liquid scintillation counting. (B) EMSA exploiting slower migration of aptamer–target protein complexes compared with free aptamer during native PAGE; decreasing amounts of protein (from left to right) are incubated with trace amounts of radiolabeled aptamer; first lane on the left: RNA without protein. Quantitative analysis of data derived from (A) and (B) should produce binding curves of the type shown at the bottom. (C) SPR utilizes angular changes of reflected light on the chip surface due to mass changes, resulting in an ascending phase of the response curve during binding and a descending curve phase during washing.

(4) Run a pre-electrophoresis to remove charged molecules left from the polymerization reaction, such as ammonium peroxodisulfate. For dPAGE, 30 min at 30 W is recommended to also warm the gel up.

(5) Rinse the pockets well, especially for dPAGE, add loading buffer to samples and run electrophoresis until the dye markers indicate sufficient resolution in the separation range of interest. For dPAGE, the gel temperature should not exceed 60 °C, since glass plates might break at higher temperatures.

(6) Preparative amounts of RNA can be visualized directly by UV shadowing (254 nm) as described in Chapter 3. Briefly, cut the edges of the band under UV light with a sterile razor blade or scalpel to mark their position, but fully excise the gel piece in the absence of UV light. Store as such at -20 °C or start elution by crushing the gel piece with a yellow tip, add 0.3 M NaOAc, pH 5.6, until the gel material is covered, incubate for 30 min on ice, shortly centrifuge in a desktop centrifuge, recover the supernatant and add another fresh aliquot of 0.3 M NaOAc as above. After three repetitions, usually 80% of the RNA has been eluted. Precipitate by adding 2.5 volumes EtOH.

Solutions

20% Acrylamide stock solution: 19% (w/v) acrylamide, 1% (w/v) bisacrylamide, 8 M urea in $1 \times$ TBE.
Urea stock solution: 8 M urea in $1 \times$ TBE.
$10 \times$ Electrophoresis buffer ($10 \times$ TBE): 900 mM Tris, 900 mM boric acid, 20 mM EDTA (pH is 8.3).
Denaturing loading buffer: 8 M urea, 50 mM EDTA, 0.25% (w/v) bromophenol blue, 0.25% (w/v) xylene cyanol.
Native loading buffer: any incubation buffer plus 50% (v/v) glycerol.

Comments

(1) Use pre-made acrylamide solutions, such as 38% (w/v) acrylamide, 2% (w/v) bisacrylamide (Roth, Germany), to avoid handling of acrylamide powder. Solutions should be used up within 3–4 months, since acrylamide can deaminate to acrylic acid. Check conductivity if unsure.

(2) Use conditions for native PAGE as described in Protocol 9.

Protocol 11: Nitrocellulose filter binding

Procedure

Various assay formats exist, from 96-well microtiter plates to simple vacuum manifolds with round filter holders, usually 2.5 cm in diameter. The latter simple setup is used in the protocol described below:

(1) Wet 0.45-μm nitrocellulose filters (Schleicher & Schuell) in incubation buffer for at least 5 min.

(2) Follow EMSA Protocol 9 to prepare e.g. 10 μl samples. Ideally, perform EMSA in parallel, which allows for comparison of results.

(3) Place a filter on the filter holder, ensuring proper and constant suction (around 300 Torr using a water jet pump from Brand), pre-wash the filter with 100 µl buffer, then immediately pipette 4 µl of the reaction sample onto the center of the filter and wash again with 100 µl buffer.
(4) Remove filter carefully with flat-tip tweezers, repeat filter binding with a second 4-µl aliquot of the same sample to obtain two data points, allow filters to dry 20 min under red-light and count by liquid scintillation or exposition of a phosphoimager plate.
(5) After each double filtration step, put incubation tubes on ice until the last filter has dried, spot 1 µl of the remainder of each sample on a fresh nitrocellulose filter, dry without filtration and count directly. This value is used to calculate total radioactivity in the essay.

Buffers
Use buffers as described for EMSA protocol.

Comments

(1) Nitrocellulose is brittle, should not be touched with hands and marked solely at the edges to avoid artifacts due to changes in binding capacity. The material will also age, particularly if exposed to sunlight, resulting in reduced binding affinity. Although in use since the 1960s and commonly applied to characterize RNA–protein interactions [86], the exact nature of the molecular principles underlying binding remains unclear. However, binding is dependent on hydrophobic interactions with the protein. RNA alone should usually bind to a much lesser extent than proteins, although nucleic acids can form hydrophobic interactions, which explains why nitrocellulose binders are often enriched during selections. It should further be noted that some proteins bind inefficiently to nitrocellulose, which may be a possible explanation in cases of failure to retain a protein–RNA complex on nitrocellulose filters.
(2) For exact quantitation, a double-filter experiment is suggested [87], which places a second DEAE filter disk (e.g. Whatman DE81 or Amersham Hybond-N^+) beneath the nitrocellulose filter. By this means, all RNA material that is not retained on the top filter is trapped on the second filter, making it possible to directly determine the ratio of bound to unbound RNA.
(3) Use high-precision pipettes for small assay volumes or double the assay volume to achieve more accurate pipetting.

Protocol 12: Surface Plasmon Resonance (SPR) for K_D determination

Procedure
SPR is an elegant technique to directly follow association and dissociation kinetics during binding events over a wide range ($k_{on} = 10^3$ to 10^7 M^{-1} s^{-1}, $k_{off} = 10^{-6}$ to 10^{-1} s^{-1}, $K_D = 1$ mM to 10 pM [88]), particularly if approaches such as nitrocellulose filter binding (no protein involved, no filter retention) or EMSA (lack of charge, small size of the complex) fail. Instruments like the basic Biacore X and

its more advanced successors [89] have increasingly come into use and detailed descriptions exist [90, 91]. More recent application surveys are an excellent starting point to develop own experimental setups and problem solutions [92, 93].

(1) Immobilize the target on a commercially available sensor chip by, if possible, use of a coupling chemistry identical to that employed in the selection, such as a cyclodextrane-derivatized gold surface presenting nitrilotriacetic acid functionalities (Biacore). Use slow flow rates (5 µl/min) for long contact times. After mounting the chip in the instrument and before sample injection, prewash the chip for 20 min with 350 mM EDTA in SPR buffer to remove interfering metals, then 20 min with SPR buffer, and finally 20 min with 500 µM $NiCl_2$ in SPR buffer.
(2) Bind His-tagged protein by flushing the measurement cell only with 200 nM protein in SPR buffer until a reasonable level of arbitrary response units is reached (around 1500 RU), then wash with SPR buffer until a stable baseline is observed. As unspecifically adsorbed protein will be removed, about one-third of the initial response units will remain on the chip.
(3) Prior to RNA loading, increase flow rate to 20–30 µl/min to reduce surface adsorption and sample dilution effects, as well as mass transfer representing repeated association and dissociation events. Wait for equilibrium, as inferred from a stable baseline.
(4) Prepare RNA analyte with a concentration of 0.1 and 1 µM in SPR buffer as a starting point for aptamer analysis. The smallest recommended sample size is 45 µl including dead volumes in valves and tubing. If possible, use larger sample volumes and prolong association phases.
(5) Load sample loop using the two bubble technique (see Comments below) against the direction of flow and inject by maintaining the high flow rate; follow association and dissociation kinetics until a stable signal is observed.
(6) Analyze the data by fitting association and dissociation curves with a program such as BIAevaluation (Biacore, Sweden) or Clamp [94], paying special attention to the binding model chosen as starting condition of the fit [95].
(7) For quantitative kinetic analysis, repeat injection with multiple samples covering a broad concentration range (roughly between 0.1 and up to 100 times the expected K_D).

Buffer
1 × SPR buffer: 10 mM HEPES–NaOH, pH 7.4, 100 mM NaCl, 7 mM $MgCl_2$.

Comments

(1) The technique is based on interaction of the binder with the target, affecting the refractive index of the surface. Binding events are measured at a surface, which can result in effects not observed free in solution. Chips can be stored overnight at 4 °C in SPR buffer, but it is best to complete the measurement in one sweep as immobilized ligand could be lost over time.

(2) SPR is very sensitive and thus prone to artifacts. Running tests for each individual binding reaction to be assayed is essential. Particular attention should be given to flow rate, temperature, concentrations of target and binder, measurement time, general buffer composition, addition of detergents to avoid precipitation, and system clogging [95]. For RNA work, the concentration of Mg^{2+} as well as the pH are particularly critical. Include controls, such as a non-specifically binding protein, nucleic acid or carbohydrate.

(3) Clean chip and tubings regularly with the program provided by the manufacturer to prevent microbial growth and artifacts due to denatured deposits. However, avoid stripping or inactivation of immobilized target.

(4) Two bubble technique. Set pipette to 45 µl, take up sample, remove pipette with tip from the liquid, adjust pipette volume to 48 µl, thereby soaking up a bit of air producing a bubble, dip pipette tip back into the liquid and adjust pipette volume to 55 µl, thus taking up more sample, again remove tip from the liquid and adjust pipette volume to 58 µl, soaking up a second air bubble. In the yellow tip should appear, from top to bottom: major sample amount – air bubble – small sample amount – air bubble. The tandem air bubbles will effectively prevent dilution of the major sample aliquot with buffer while passing through the instrument; in addition, the two bubbles will cause a twin signal indicating chip loading.

(5) Alternatively to sample loading with a loop, particularly if no plateau and thus equilibrium is reached in the response curve due to limited sample volume, add the analyte to the flow buffer [96].

Protocol 13: Structural characterization of aptamers and their binding motifs

Aptamer characterization requires RNA end-labeling with high specific activity, which is achieved by labeling transcribed and gel-purified aptamers 5′ or 3′ with [γ-^{32}P]ATP and T4 polynucleotide kinase (PNK) or [^{32}P]pCp and T4 RNA ligase, respectively (see Chapter 3 and [97]). As a start, determine the size of the minimal binding motif and gather information on the tertiary structure of the free and target-bound aptamer. Reliable, fast and direct probing techniques, such as cleavage by RNases V1, S1 and T1 as well as lead-induced hydrolysis, are recommended for the initial characterization. For structure probing with nucleases, lead ions and chemical reagents, the reader is referred to Chapters 10, 11, 13 and 15.

Boundary experiment to identify minimal binding motifs

(1) Produce an alkaline hydrolysis ladder by addition of 4 µl (50 000–100 000 Cerenkov c.p.m.) of 5′-end-labeled RNA to 36 µl of 50 mM NaHCO$_3$ and incubation at 90 °C. The exact amount of time needs to be determined for each RNA. Recover by ethanol precipitation with glycogen as co-precipitant.

(2) Use the selection assay setup, e.g. target protein immobilization on a Ni-NTA matrix, to analyze complex formation between RNA and target; incubate the partially hydrolyzed RNA pool with the matrix-bound target protein, e.g. as de-

scribed for the selection in Protocol 4. Unbound material is removed by 3–6 washing steps, and bound fragments are specifically eluted with imidazol or free target protein.

(3) After phenol/ether extraction (Protocol 3), RNA is precipitated with ethanol, re-dissolved in 8 µl H_2O and analyzed by 12% dPAGE. Boundaries defining the essential binding regions are identified by a sharp drop or, at best, absence of ladder intensity below a certain fragment size in the elution fraction. Quality of the picture depends on the stringency of washing steps.

(4) Repeat analysis with 3′-end-labeled material to have results from both ends, making the interpretation more meaningful.

Comments

(1) As low concentrations of RNA are used in boundary and probing experiments, include molecular biology grade glycogen (Merck) as co-precipitant and, if possible, unlabeled, unspecific RNA to reduce surface adsorptions which can also be reduced by use of siliconized reaction tubes.

(2) Incubation conditions (buffer, temperature) should be closely adapted to those used during selection and should further take into account the later application context of the aptamer as well as requirements for RNA structure formation (e.g. sufficient amounts of Mg^{2+}).

(3) Of particular interest are the contact sites between target and aptamer, which requires to compare aptamer structure in the presence and absence of target.

Acknowledgments

We like to thank Mathias Sprinzl for corrections, comments and ongoing support, and Dmitry Agafonov, Alexandra Wolfrum as well as Irina Häcker for critical reading of the manuscript. Part of the research described was supported by the Deutsche Forschungsgemeinschaft grant "Molekulare Mechanismen der Translationstermination" to M. S. and S. V.

References

1 T. Hermann, D. J. Patel, Science 2000, 287, 820–825.
2 A. D. Ellington, J. W. Szostak, Nature 1990, 346, 818–822.
3 C. Tuerk, L. Gold, Science 1990, 249, 505–510.
4 D. L. Robertson, G. F. Joyce, Nature 1990, 344, 467–468.
5 J. H. Blum, S. L. Dove, A. Hochschild, J. J. Mekalanos, Proc. Natl. Acad. Sci. USA 2000, 97, 2241–2246.
6 H. R. Hoogenboom, Methods Mol. Biol. 2002, 178, 1–37.
7 R. H. Hoess, Chem. Rev. 2001, 101, 3205–3218.
8 R. Liu, J. E. Barrick, J. W. Szostak, R. W. Roberts, Methods Enzymol. 2000, 318, 268–293.
9 T. T. Takahashi, R. J. Austin, R. W.

Roberts, *Trends Biochem. Sci.* **2003**, *28*, 159–165.

10 C. Schaffitzel, A. Pluckthun, *Trends Biochem. Sci.* **2001**, *26*, 577–579.

11 J. Ciesiolka, M. Illangasekare, I. Majerfeld, T. Nickles, M. Welch, M. Yarus, S. Zinnen, *Methods Enzymol.* **1996**, *267*, 315–335.

12 H. P. Hofmann, S. Limmer, V. Hornung, M. Sprinzl, *RNA* **1997**, *3*, 1289–1300.

13 Y. Wang, J. Killian, K. Hamasaki, R. R. Rando, *Biochemistry* **1996**, *35*, 12338–12346.

14 C. Baugh, D. Grate, C. Wilson, *J. Mol. Biol* **2000**, *301*, 117–128.

15 R. D. Jenison, S. C. Gill, A. Pardi, B. Polisky, *Science* **1994**, *263*, 1425–1429.

16 Z. Huang, J. W. Szostak, *RNA* **2003**, *9*, 1456–1463.

17 N. K. Vaish, R. Larralde, A. W. Fraley, J. W. Szostak, L. W. McLaughlin, *Biochemistry* **2003**, *42*, 8842–8851.

18 D. H. Burke, L. Gold, *Nucleic Acids Res.* **1997**, *25*, 2020–2024.

19 D. Faulhammer, B. Eschgfaller, S. Stark, P. Burgstaller, W. Englberger, J. Erfurth, F. Kleinjung, J. Rupp, V. S. Dan, W. Schroder, S. Vonhoff, H. Nawrath, C. Gillen, S. Klussmann, *RNA* **2004**, *10*, 516–527.

20 C. Srisawat, I. J. Goldstein, D. R. Engelke, *Nucleic Acids Res.* **2001**, *29*, E4.

21 W. Pan, R. C. Craven, Q. Qiu, C. B. Wilson, J. W. Wills, S. Golovine, J. F. Wang, *Proc. Natl. Acad. Sci. USA* **1995**, *92*, 11509–11513.

22 S. Ringquist, T. Jones, E. E. Snyder, T. Gibson, I. Boni, L. Gold, *Biochemistry* **1995**, *34*, 3640–3648.

23 M. Homann, H. U. Goringer, *Nucleic Acids Res.* **1999**, *27*, 2006–2014.

24 K. N. Morris, K. B. Jensen, C. M. Julin, M. Weil, L. Gold, *Proc. Natl. Acad. Sci. USA* **1998**, *95*, 2902–2907.

25 W. C. Winkler, R. R. Breaker, *ChemBioChem* **2003**, *4*, 1024–1032.

26 M. Mandal, B. Boese, J. E. Barrick, W. C. Winkler, R. R. Breaker, *Cell* **2003**, *113*, 577–586.

27 W. C. Winkler, A. Nahvi, A. Roth, J. A. Collins, R. R. Breaker, *Nature* **2004**, *428*, 281–286.

28 S. E. Osborne, I. Matsumura, A. D. Ellington, *Curr. Opin. Chem. Biol.* **1997**, *1*, 5–9.

29 E. N. Brody, L. Gold, *J. Biotechnol.* **2000**, *74*, 5–13.

30 P. Burgstaller, A. Jenne, M. Blind, *Curr. Opin. Drug Discov. Dev.* **2002**, *5*, 690–700.

31 J. J. Toulme, F. Darfeuille, G. Kolb, S. Chabas, C. Staedel, *Biol. Cell* **2003**, *95*, 229–238.

32 B. E. Eaton, *Curr. Opin. Chem. Biol.* **1997**, *1*, 10–16.

33 M. Famulok, S. Verma, *Trends Biotechnol.* **2002**, *20*, 462–466.

34 R. C. Conrad, L. Giver, Y. Tian, A. D. Ellington, *Methods Enzymol.* **1996**, *267*, 336–367.

35 K. A. Marshall, A. D. Ellington, *Methods Enzymol.* **2000**, *318*, 193–214.

36 J. F. Lee, J. R. Hesselberth, L. A. Meyers, A. D. Ellington, *Nucleic Acids Res.* **2004**, *32*, D95–100.

37 J. V. Ponomarenko, G. V. Orlova, A. S. Frolov, M. S. Gelfand, M. P. Ponomarenko, *Nucleic Acids Res.* **2002**, *30*, 195–199.

38 R. Conrad, L. M. Keranen, A. D. Ellington, A. C. Newton, *J. Biol. Chem.* **1994**, *269*, 32051–32054.

39 H. Aurup, D. M. Williams, F. Eckstein, *Biochemistry* **1992**, *31*, 9636–9641.

40 T. Fitzwater, B. Polisky, *Methods Enzymol.* **1996**, *267*, 275–301.

41 W. Kusser, *J. Biotechnol.* **2000**, *74*, 27–38.

42 D. Smith, B. D. Collins, J. Heil, T. H. Koch, *Mol. Cell Proteomics* **2003**, *2*, 11–18.

43 J. J. Li, X. Fang, W. Tan, *Biochem. Biophys. Res. Commun.* **2002**, *292*, 31–40.

44 M. Michaud, E. Jourdan, A. Villet, A. Ravel, C. Grosset, E. Peyrin, *J. Am. Chem. Soc.* **2003**, *125*, 8672–8679.

45 R. A. Potyrailo, R. C. Conrad, A. D. Ellington, G. M. Hieftje, *Anal. Chem.* **1998**, *70*, 3419–3425.

46 R. Nutiu, Y. Li, *J. Am. Chem. Soc.* **2003**, *125*, 4771–4778.
47 N. Hamaguchi, A. Ellington, M. Stanton, *Anal. Biochem.* **2001**, *294*, 126–131.
48 S. Jhaveri, M. Rajendran, A. D. Ellington, *Nat. Biotechnol.* **2000**, *18*, 1293–1297.
49 M. Rajendran, A. D. Ellington, *Nucleic Acids Res.* **2003**, *31*, 5700–5713.
50 E. N. Brody, M. C. Willis, J. D. Smith, S. Jayasena, D. Zichi, L. Gold, *Mol. Diagn.* **1999**, *4*, 381–388.
51 X. Fang, A. Sen, M. Vicens, W. Tan, *ChemBioChem* **2003**, *4*, 829–834.
52 C. Wang, M. Zhang, G. Yang, D. Zhang, H. Ding, H. Wang, M. Fan, B. Shen, N. Shao, *J. Biotechnol.* **2003**, *102*, 15–22.
53 J. C. Cox, A. Hayhurst, J. Hesselberth, T. S. Bayer, G. Georgiou, A. D. Ellington, *Nucleic Acids Res.* **2002**, *30*, E108.
54 J. B. Tok, J. Cho, R. R. Rando, *Nucleic Acids Res.* **2000**, *28*, 2902–2910.
55 S. da Rocha Gomes, E. Dausse, J. J. Toulme, *Biochem. Biophys. Res. Commun.* **2004**, *322*, 820–826.
56 A. Khvorova, Y. G. Kwak, M. Tamkun, I. Majerfeld, M. Yarus, *Proc. Natl. Acad. Sci. USA* **1999**, *96*, 10649–10654.
57 T. Janas, M. Yarus, *RNA* **2003**, *9*, 1353–1361.
58 P. Pfister, S. Hobbie, Q. Vicens, E. C. Bottger, E. Westhof, *ChemBioChem* **2003**, *4*, 1078–1088.
59 Q. Vicens, E. Westhof, *Biopolymers* **2003**, *70*, 42–57.
60 Q. Vicens, E. Westhof, *ChemBioChem* **2003**, *4*, 1018–1023.
61 P. H. Seeberger, W. C. Haase, *Chem. Rev.* **2000**, *100*, 4349–4394.
62 K. M. Koeller, C. H. Wong, *Chem. Rev.* **2000**, *100*, 4465–4494.
63 S. Jeong, T. Eom, S. Kim, S. Lee, J. Yu, *Biochem. Biophys. Res. Commun.* **2001**, *281*, 237–243.
64 M. M. Masud, M. Kuwahara, H. Ozaki, H. Sawai, *Bioorg. Med. Chem.* **2004**, *12*, 1111–1120.
65 K. Terpe, *Appl. Microbiol. Biotechnol.* **2003**, *60*, 523–533.
66 T. Feizi, F. Fazio, W. Chai, C. H. Wong, *Curr. Opin. Struct. Biol.* **2003**, *13*, 637–645.
67 A. Vater, F. Jarosch, K. Buchner, S. Klussmann, *Nucleic Acids Res.* **2003**, *31*, e130.
68 T. Fitzwater, B. Polisky, *Methods Enzymol.* **1996**, *267*, 275–301.
69 A. R. Kore, N. K. Vaish, J. A. Morris, F. Eckstein, *J. Mol. Biol.* **2000**, *301*, 1113–1121.
70 M. Zaccolo, D. M. Williams, D. M. Brown, E. Gherardi, *J. Mol. Biol.* **1996**, *255*, 589–603.
71 O. C. Uhlenbeck, *RNA* **1995**, *1*, 4–6.
72 N. K. Vaish, P. A. Heaton, F. Eckstein, *Biochemistry* **1997**, *36*, 6495–6501.
73 J. P. Davis, N. Janjic, B. E. Javornik, D. A. Zichi, *Methods Enzymol.* **1996**, *267*, 302–314.
74 M. Zuker, *Nucleic Acids Res.* **2003**, *31*, 3406–3415.
75 R. Knight, M. Yarus, *Nucleic Acids Res.* **2003**, *31*, e30.
76 I. Hirao, Y. Harada, T. Nojima, Y. Osawa, H. Masaki, S. Yokoyama, *Biochemistry* **2004**, *43*, 3214–3221.
77 D. P. Bartel, J. W. Szostak, *Science* **1993**, *261*, 1411–1418.
78 E. T. Peterson, J. Blank, M. Sprinzl, O. C. Uhlenbeck, *EMBO J.* **1993**, *12*, 2959–2967.
79 J. F. Milligan, O. C. Uhlenbeck, *Methods Enzymol.* **1989**, *180*, 51–62.
80 J. C. Janson, L. Ryden (eds), *Protein Purification: Principles, High Resolution Methods and Applications*, 2nd ed., Wiley-VCH, Weinheim, **1998**.
81 J. X. Yan, R. Wait, T. Berkelman, R. A. Harry, J. A. Westbrook, C. H. Wheeler, M. J. Dunn, *Electrophoresis* **2000**, *21*, 3666–3672.
82 S. T. Wallace, R. Schroeder, *Methods Enzymol.* **2000**, *318*, 214–229.
83 R. C. Cadwell, G. F. Joyce, *PCR Methods Appl.* **1994**, *3*, S136–S140.
84 R. C. Cadwell, G. F. Joyce, *PCR Methods Appl.* **1992**, *2*, 28–33.
85 J. Carey, *Methods Enzymol.* **1991**, *208*, 103–117.
86 J. Carey, P. T. Lowary, O. C. Uhlenbeck, *Biochemistry* **1983**, *22*, 4723–4730.

87 I. Wong, T. M. Lohman, *Proc. Natl. Acad. Sci. USA* **1993**, *90*, 5428–5432.
88 B. Johnsson, S. Lofas, G. Lindquist, *Anal. Biochem.* **1991**, *198*, 268–277.
89 C. L. Baird, D. G. Myszka, *J. Mol. Recognit.* **2001**, *14*, 261–268.
90 T. A. Morton, D. G. Myszka, *Methods Enzymol.* **1998**, *295*, 268–294.
91 D. G. Myszka, *Methods Enzymol.* **2000**, *323*, 325–340.
92 R. L. Rich, D. G. Myszka, *J. Mol. Recognit.* **2003**, *16*, 351–382.
93 R. L. Rich, D. G. Myszka, *J. Mol. Recognit.* **2001**, *14*, 273–294.
94 D. G. Myszka, T. A. Morton, *Trends Biochem. Sci.* **1998**, *23*, 149–150.
95 D. G. Myszka, *J. Mol. Recognit.* **1999**, *12*, 279–284.
96 D. G. Myszka, M. D. Jonsen, B. J. Graves, *Anal. Biochem.* **1998**, *265*, 326–330.
97 C. Brunel and P. Romby, *Meth. Enzymol.* **2000**, *318*, 3–21.

49
In Vivo SELEX Strategies

Thomas A. Cooper

49.1
Introduction

SELEX (Systematic Evolution of Ligands by EXponential enrichment) uses iterative selection strategies to enrich for specific RNA molecules from randomized pools based on binding or an RNA-mediated activity [1, 2]. The procedure is typically performed using recombinant protein in a cell-free system to identify optimal sequences bound by RNA-binding proteins that have been associated with RNA-processing functions. Other uses include identification of RNA aptamers that bind small molecules or proteins of interest to human disease [3]. The use of SELEX has expanded such that strategies that select for catalytic function have identified catalytic RNAs [4]. It is of interest to expand the application of a SELEX approach to include identification of RNA sequences that promote RNA-processing events *in vivo*.

The focus of this chapter is the use of a functional SELEX strategy to identify exonic elements that enhance pre-mRNA splicing *in vivo*. Exonic elements that enhance splicing are called exonic splicing enhancers (ESEs) and elements that repress splicing are exonic splicing silencers (ESSs). The selection strategy described in this chapter is particularly well suited to identify exonic sequences that enhance splicing since they are "captured" in an mRNA by splicing. Initially two major classes of ESEs had been defined, i.e. purine- and AC-rich [5], and it was thought that ESEs were in a small minority of exons. In contrast, recent results from bioinformatic analyses has identified a diverse array of ESEs indicative of a greater variety and prevalence than previously realized [6]. It has also become clear that mutations that disrupt the function of ESEs and that lead to aberrant splicing are a frequent cause of human disease [7–9].

Several laboratories have used functional SELEX strategies to identify sequences that promote splicing in cell-free (*in vitro*) splicing assays that utilize HeLa nuclear extracts [10–13]. *In vitro* transcribed RNAs containing two exons separated by an intron are readily spliced *in vitro*. In these approaches, a poorly spliced downstream exon contains a randomized region of up to 20 nt. An RT-PCR-based strategy is devised to selectively amplify the exonic sequences of the spliced mRNAs, and then amplify and clone the selected sequences back into the pre-mRNA context and continue additional rounds of splicing and RT-PCR amplification. Two groups used this approach to identify both purine-rich and non-purine-rich sequences that enhanced splicing in HeLa nuclear extracts [10, 11]. Among the sequences identified were binding sites for members of the SR family of splicing factors. Several SR protein family members have been shown to bind to ESEs and enhance splicing [14].

Cytoplasmic extracts prepared from HeLa cells are not competent for *in vitro* splicing and addition of individual SR proteins as recombinant proteins restores splicing activity [15]. This complementation assay was combined with a functional SELEX approach to identify the preferred ESE targets for individual SR proteins. Recombinant proteins for four SR proteins were used to complement cytoplasmic extracts for splicing of transcripts containing randomized nucleotides in the position of a required ESE of the downstream exon [12, 13]. Consensus ESE sequences required for enhanced splicing by each of the four SR proteins were derived and have been used to develop an ESE prediction program (http://exon.cshl.org/ESE). This program has been used to identify ESEs that when mutated result in human disease [7].

49.2
Procedure Overview

This chapter describes an extension of the approaches described above to identify ESEs that function *in vivo*. A minigene plasmid expressing a pre-mRNA containing a poorly spliced internal alternative exon with a randomized sequence is transiently transfected and selective RT-PCR is used to amplify only those mRNAs that include the exon. Multiple cycles of transfection and selective RT-PCR enrich for exon sequences that enhance inclusion of the alternative exon. This approach has been described previously [16, 17] and has recently been used to identify exonic nucleotides within *SMN* exon 7 critical for exon inclusion [18]. The selection scheme is outlined in Fig. 49.1. A synthetic single-stranded DNA template containing the randomized region (13 nt in this case) flanked by restriction sites is made double-stranded and amplified by up to 5 cycles of PCR. The PCR-amplified cassette is digested with the flanking restriction enzymes and directionally ligated into the alternative exon of the plasmid minigene. The exon is constructed to be poorly recognized in the absence of an ESE. This can be accomplished by either modifying exon size (smaller is weaker) or by modifying the splice site sequences away from consensus [19]. The ligation reaction is then transiently transfected directly into cultured cells and total RNA is extracted after 40–48 h. RT-PCR is used to selectively amplify spliced mRNA that includes the randomized exon using oligos that prime across the exon junctions (Fig. 49.2). The PCR product is digested by the restriction enzymes whose sites are incorporated into the single-stranded template to excise the selected randomized cassette. This digestion product contains a population of sequences that has been selected for enhanced exon inclusion

```
                GGACGTAGGGTCGAC
       5'GGACGTAGGGTCGACGTT-(N)n-GAATGGATCCGTCGTGACTGGGAAAAC 3'
                                    3' CAGCACTGACCCTTTTG 5'
```

↓ Step 1: amplify randomized cassette

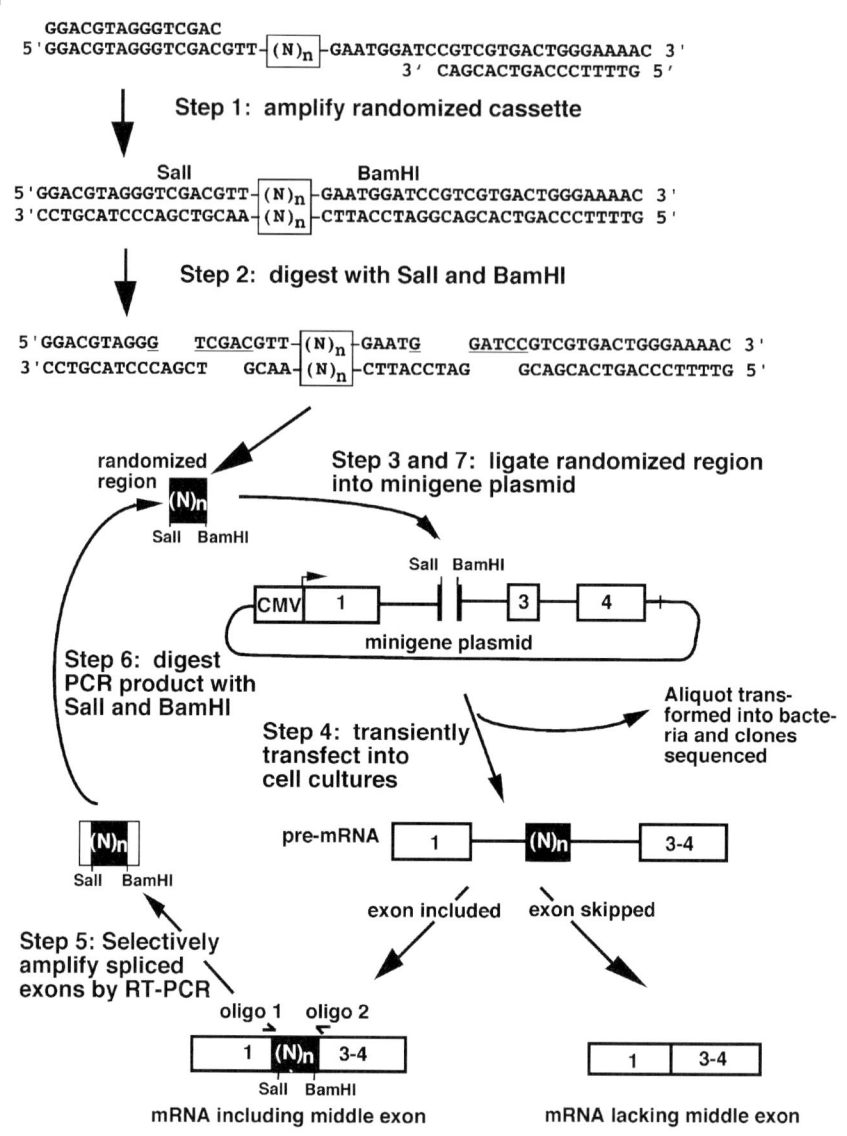

```
                  SalI                BamHI
       5'GGACGTAGGGTCGACGTT-(N)n-GAATGGATCCGTCGTGACTGGGAAAAC 3'
       3'CCTGCATCCCAGCTGCAA-(N)n-CTTACCTAGGCAGCACTGACCCTTTTG 5'
```

↓ Step 2: digest with SalI and BamHI

```
       5'GGACGTAGGG    TCGACGTT-(N)n-GAATG    GATCCGTCGTGACTGGGAAAAC 3'
       3'CCTGCATCCCAGCT    GCAA-(N)n-CTTACCTAG    GCAGCACTGACCCTTTTG 5'
```

Fig. 49.1. Iterative procedure to enrich for exon sequences that enhance splicing. The randomized region is represented as $(N)_n$.

as well as non-enhancing sequences spliced due to low levels of background splicing inherent in most minigenes. The excised randomized cassette is then ligated back into the minigene exon for additional cycles of selection.

The cassette exons obtained during the first 3–4 rounds of selection are evaluated with regard to: (1) sequence and (2) enhancer activity. To identify sequence

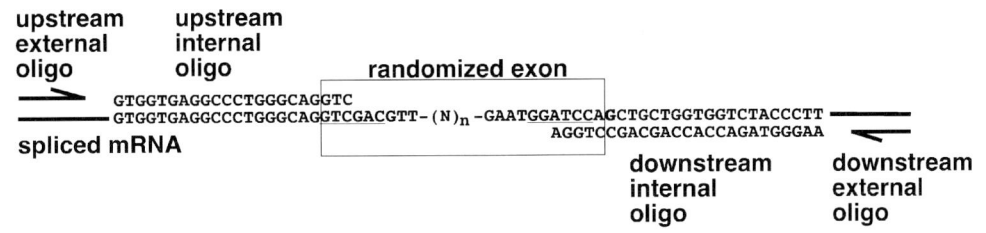

```
upstream       upstream
external       internal
oligo          oligo           randomized exon
        GTGGTGAGGCCCTGGGCAGGTC
         ──────▶
               GTGGTGAGGCCCTGGGCAGGTCGACGTT-(N)ₙ-GAATGGATCCAGCTGCTGGTGGTCTACCCTT ────────
spliced mRNA                                   AGGTCCGACGACCACCAGATGGGAA
                                                                                      ◀──────
                                                    downstream        downstream
                                                    internal          external
                                                    oligo             oligo
```

Fig. 49.2. Amplification of the randomized exon included in the mRNA using double-nested RT-PCR. The mRNA sequence is represented by the sequence and thick lines on either side. The randomized exon is boxed. Following reverse transcription, PCR 1 is performed using the external primers. An aliquot of PCR 1 is then amplified using the internal primers. Note that the internal primers are designed such that the last 2–4 nt anneal within the randomized exon. These oligonucleotides will prime DNA synthesis only on correctly spliced mRNAs and not on plasmid DNA, unspliced pre-mRNA or mRNAs with a skipped randomized exon. Restriction sites are underlined.

motifs that are enriched during selection, an aliquot of the ligation reaction from each round is transformed into bacteria; plasmid DNA is prepared from 20–30 individual colonies for sequencing. Plasmid DNA from individual clones is transiently transfected and the level of exon inclusion is assayed by RT-PCR to determine the splicing enhancer activity of individual cassettes. The level of inclusion of selected exons is compared to that of non-selected exons to determine whether the procedure enriched for splicing enhancers. The ability of selected cassettes to enhance splicing of a different alternative exon should be tested to determine whether the selected sequence has intrinsic enhancer activity, independent of the minigene used for its selection.

We have used this procedure in a fibroblast cell line to identify an AC-rich motif that enhances splicing [16]. Now that the feasibility of the approach has been established, it would be of interest to perform *in vivo* selection in several differentiated and undifferentiated cell types to reveal the presence of cell-specific ESEs or in cells that overexpress an exogenous protein known to mediate enhancer activity to identify enhancer sequences that are preferred by individual proteins *in vivo*. It is also possible to use the procedure in cells deficient in particular proteins (either by RNAi or using cultures made null by homologous recombination). The latter could be useful to identify the roles of individual proteins that act cooperatively within exons to regulate exon inclusion.

49.2.1
Design of the Randomized Exon Cassette

The double-stranded cassette containing the randomized region is generated from low-cycle PCR using oligonucleotides complementary to the constant regions (Fig. 49.1). The oligo containing the randomized region must contain several features (illustrated in Fig. 49.1). First, priming sites for oligonucleotides located upstream and downstream of both restriction sites to allow PCR amplification of a double-stranded cassette (Step 1, Fig. 49.1). Second, it contains two different restriction endonuclease sites on either side of the randomized region for directional cloning (Fig. 49.1, *Sal*I and *Bam*HI sites). These sites must be unique to the minigene plasmid (see below), have incompatible overhangs (for directional cloning and to prevent recircularization of the minigene plasmid), and should cut and ligate efficiently. It is most convenient if both enzymes are optimally active in the same buffer. Position the restriction sites within the oligonucleotide such that the fragment containing the randomized region will be a different size than the other two fragments generated from digestion of the double-stranded cassette with both restriction enzymes. The restriction endonuclease digestion product containing the randomized region is gel-isolated and quantitated. Third, make certain that the constant regions within the selected exon do not contain sequences that can affect splicing such as known splicing enhancers, potential cryptic splice sites or in-frame translation stop codons.

When designing the randomized region, keep in mind that as the number of randomized nucleotides increases, the number of molecules containing a particular sequence decreases in a constant amount of DNA. In our initial studies, we used a relatively low number of random positions due to the concern that a large variable region would result in insufficient copies of any one sequence to allow detection by RT-PCR. Now that feasibility of the approach has been demonstrated and transfection efficiencies have greatly improved, it would be worthwhile testing larger randomized sequences. A second consideration is the total size of the exon. As the size of a cassette exon increases, the efficiency of exon recognition increases resulting in higher levels of exon inclusion [20–22]. Therefore, larger exons are included at a relatively high level independent of the exon sequence. Efficient exon inclusion can be compensated by decreasing the match of the 3' and/or 5' splice sites to consensus. Another solution is to include the randomized region near the 3' splice site of a terminal exon that is inefficiently spliced (see Fig. 49.3D) similar to the approaches used *in vitro* [10–13].

49.2.2
Design of the Minigene

This section is geared to an internal cassette alternative exon as shown in Figs 49.1 and 49.3(A). An advantage of this exon is that it selects for sequences that mediate exon inclusion via either the 3' or 5' splice sites or via enhancing interactions across the exon which define an exon [23]. Alternative exon architectures can be utilized that will theoretically select for sequences that enhance either 3' splice sites (Fig. 49.3C and D) or 5' splice sites (Fig. 49.3B and E).

The exon that is to receive the randomized region should have the appropriate restriction sites for directional cloning of the randomized cassette into an exon. These restriction sites must be unique to plasmid. It is useful to clone a large stuffer fragment into these sites so that plasmid cut with both enzymes can be distinguished from single cut and uncut plasmid. Low levels of contaminating plas-

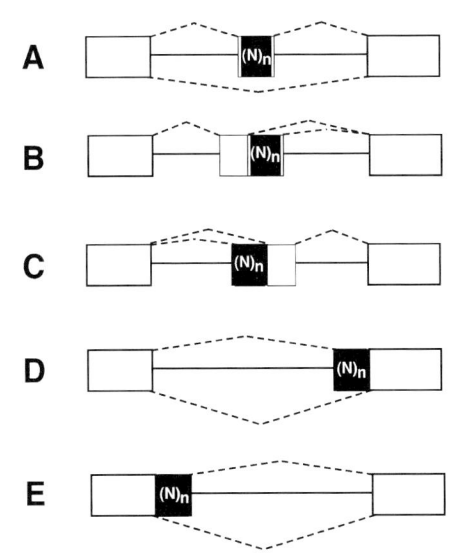

Fig. 49.3. Different pre-mRNA architectures used for selection. The randomized region is indicated by $(N)_n$. Splicing patterns are indicated by dashed lines.

mid containing a potentially "spliceable" sequence will generate a major contaminant of the selected sequences.

The stringency of selection can be adjusted by adjusting the basal level of exon inclusion in the absence of enhancer sequences. An exon that is completely skipped will theoretically select for stronger exonic enhancers than one that has a low basal level of inclusion. Note, however, that it is difficult to induce exon inclusion of particularly weak exons no matter how strong the ESE. We prefer to have a low background level of exon inclusion to ensure that the exon is able to be spliced and to allow for selection of weaker ESE motifs. Establishing the desired balance of exon inclusion and exclusion may require modifying features such as exon size and strength of the 5' and 3' splice sites [19].

To determine the background level of exon inclusion in the absence of selected sequences, it is best to use pooled non-selected exons. This is accomplished by transforming an aliquot of the initial ligation (Fig. 49.1) into high-efficiency competent bacteria according to standard procedures to the point of adding SOC and incubating at 37 °C. Instead of plating out the transformed cells on agar plates, use these bacteria to inoculate 100 ml of media including the appropriate antibiotic to prepare plasmid DNA for transfection. It is advisable to plate the transformed bacteria and pick individual clones containing non-selected exons; confirm that the majority contain the complete randomized cassette by sequencing. Another approach to determine background level of exon inclusion is to randomly pick colonies, prepare plasmids for transfection, transfect them individually and average the level of exon inclusion. It is not unusual to obtain one or two out of 20 randomly picked non-selected plasmids that exhibit relatively high levels of exon inclusion demonstrating the strength of the effects of random sequence on the level of inclusion of an exon that is balanced between skipping and inclusion.

The presence of premature stop codons has dramatic effects on mRNA half-life and on splicing of the resident exon [24]. To limit analysis to the effects of ESEs, it is best not to have a natural open reading frame in the minigene mRNA.

Ubiquitously active transcription enhancers (such as RSV or CMV) are most useful as they allow use of the minigene in almost any cell type. Cell-specific promoters are useful to ensure expression in the desired cell of a mixed cell population such as primary cultures or a cell line induced to differentiate.

49.2.3
RT-PCR Amplification

The RT-PCR procedure performs two functions in this protocol: (1) detect low amounts of RNA with very little background and (2) selectively amplify only those mRNAs that contain the alternative exon. These goals are accomplished by nested PCR (Fig. 49.2). First, cDNA is synthesized using random hexamers (an alternative is to use an mRNA-specific primer that primes cDNA synthesis from a site located downstream from the external PCR primer pair; see Fig. 49.2). Then mRNAs that include and exclude the alternative exon are amplified using oligonucleotides that anneal to the upstream and downstream exons (upstream and downstream external oligos, respectively; Fig. 49.2). Finally, a double-nested PCR reaction is performed using the upstream and downstream internal oligos. Both internal oligos will anneal only to correctly spliced mRNAs and prime DNA synthesis from within the alternative exon (see Fig. 49.2), thereby selectively amplifying exons that are spliced into the mRNA. PCR products from mRNAs that skip the randomized exon or from unspliced pre-mRNA are not amplified. In addition, the specificity of nested primers means that a large number of cycles can be used which is useful to amplify DNA from small amounts of RNA. Because the restriction sites are maintained, the PCR product is then digested and cycled through multiple rounds of ligation, transfection and amplification.

49.2.4
Monitoring for Enrichment of Exon Sequences that Function as Splicing Enhancers

There are several approaches to monitor the success of the selection. The first is to determine whether identifiable sequence motifs are enriched after several rounds of selection. For example, we found clear enrichment of two different motifs after two and three rounds [16]. To obtain individual clones from each round of selection, 5 µl of the 100-µl ligation reaction containing randomized exon and minigene vector is transformed into competent bacteria (Fig. 49.1). Miniprep DNA is prepared from twenty to thirty individual colonies and sequenced.

To determine whether the procedure is enriching for *bona fide* splicing enhancers, individual clones are tested by transient transfection and the level of exon inclusion is determined by a quantitative RT-PCR assay [16], primer extension or RNase protection.

Alternatively, the levels of exon inclusion can be determined directly on the pools of RNA from each round using a quantitative RT-PCR assay. However, since cells express a high background level of mRNAs that lack the exon even in the absence of insert, this assay does not give reliable results. The reason for this is that transfected linear plasmid molecules (that fail to take up insert) become blunt-ended and circularized in cultured cells. This is demonstrated by transfection of only vector DNA without ligase which results in expression a high level of mRNAs that lack the middle exon. While pools from sequential rounds of selection will show a relative enhanced exon inclusion, the background from recircularized plasmid makes it difficult to reliably quantitate the level of exon inclusion.

49.2.5
Troubleshooting

One major hurdle is to obtain high expression levels of spliced mRNA in transfected cultures. This requires high ligation and transfection efficiencies. The ligation efficiency of the minigene vector and insert preparations should be tested at a small scale prior to use. Ligation efficiencies can be optimized using a bacterial transformation assay that contains 50 ng of vector and a 1- to 5-fold molar excess of insert in a 10-μl reaction. Ligation efficiency is defined as the number of colonies obtained per nanogram vector DNA. Important are both low background in the absence of insert and efficient ligation when insert is included. A particularly labile component in the ligation reaction is ATP. Ligase buffer supplied by companies often includes ATP and it is best to store aliquots at $-20\,°C$. Each aliquot is used for 1 month.

If indicated, more vector DNA can be used for the ligation to increase the RNA expressed and the number of sequences available for selection. The amount of DNA used is limited primarily by the ability to isolate large amounts of cut minigene vector DNA.

Several alternative transfection reagents are commercially available and can give strikingly different efficiencies in different cell lines. It is often useful to talk to the suppliers to determine optimal conditions for individual cell lines.

The number of PCR cycles necessary to detect spliced mRNA will be high. For example, our nested RT-PCR procedure required a total of 80 cycles [16]. The extremely high sensitivity of PCR makes it necessary to establish the rules listed below.

(1) Use dedicated equipment for PCR including a pipetman with aerosol resistant tips, tube racks, microfuges and vortexers. Do not use these for plasmid DNA.
(2) Include "no RNA", "no DNA" and "minus reverse transcriptase" controls for each experiment to detect contamination.
(3) Water is a common source of plasmid contamination. Purchase bottled water if necessary.

(4) Aliquot reagents, mark one tube as "in use" and use only that tube until it is finished.
(5) If oligos are suspected to be contaminated with plasmid DNA, they can be reordered or purified by gel isolation on 10% non-denaturing acrylamide gel. Visualize the oligos by UV shadowing, cut out the piece of acrylamide containing the band, place it in a 1.5-ml Eppendorf tube and grind it using a blue tip. Add 1 ml of water and incubate with shaking overnight at 37 °C. Filter the solution through a 0.2-µm Millex filter (Millipore) using a 3-ml syringe.

49.3
Protocols

Protocol 1: Preparation of the randomized cassette

(1) 100 ng of the single-stranded oligonucleotide containing the randomized region is amplified and converted to double-stranded DNA in a standard PCR reaction using oligos that flank the randomized region (Fig. 49.1, Step 1). Note that the amount of starting material is flexible and determines the number of random sequences represented. Prepare a 100-µl reaction containing 500 ng of each flanking oligo, $1 \times$ Vent DNA polymerase buffer [New England Biolabs: $10\times = 10$ mM KCl, 10 mM $(NH_4)_2SO_4$, 20 mM Tris–HCl (pH 8.8 at 25 °C), 2 mM $MgSO_4$, 0.1% Triton X-100], 0.2 mM dGATC and 1 U Vent DNA polymerase (New England Biolabs). The PCR conditions depend on the T_m of the oligonucleotides and should be adjusted based on the quality of the final product as assayed on a non-denaturing acrylamide gel. Typically we use no more than 5 cycles. The last step of PCR is 5 min at 72 °C to complete synthesis of the second strand.

(2) Following the PCR reaction, add EDTA to 2 mM and NaCl to 0.2 M and extract the PCR product once with an equal volume of phenol/chloroform/isoamyl alcohol (25:25:1) and then once with an equal volume of chloroform/isoamyl alcohol (25:1). Precipitate with 2.5 volumes of ethanol and wash the pellet once in 70% ethanol. The pellet is vacuum-dried and dissolved in 20 µl of TEE (10 mM Tris, pH 7.5, 0.1 mM EDTA).

(3) Digest the PCR product with the appropriate restriction enzymes. Plan on using approximately 0.5 pmol of the randomized cassette DNA (7.4 ng of a 24-bp fragment) for every 1 µg of vector (of 6.5 kb) in the ligation, and 1 µg of vector per round of selection. The restriction digest is loaded directly onto a 7% non-denaturing polyacrylamide gel (20:1 acryl:bis). Visualize the digestion products by staining in ethidium bromide and then gel-isolated as follows. Use a razor blade to cut a piece of acrylamide containing the band. Transfer to a 1.5-ml Eppendorf tube and cut into pieces about 1 mm². Add 400 µl of acrylamide elution buffer (0.5 M NH_4OAc, 10 mM 0.5 M EDTA) and incubate at 37 °C with shaking overnight. Spin 12 000 g for 5 min to pellet gel fragments and transfer the supernatant (completely free of gel fragments) to a clean tube and add 16 µl

of 5 M NaCl and 1 ml 100% ethanol, mix well and centrifuge (12 000 g) for 15 min. Residual soluble acrylamide from the gel precipitates in ethanol and acts as a carrier so there is no need to add glycogen. Dissolve the pellet in 10–20 µl TEE and run an aliquot on an acrylamide gel alongside known amounts of size marker to estimate the DNA concentration.

Protocol 2: Vector preparation

(1) Cut 20 µg of plasmid DNA with the appropriate enzymes. Load the restriction digest directly onto a 0.9% agarose gel. Use a single well comb or make a large well from a standard comb by placing tape over several teeth. Do not overload the gel since this can lead to smearing of the bands and contamination of the desired double-cut DNA with uncut plasmid DNA.
(2) Isolate the DNA using GeneClean (Bio 101) or QIAEX II Gel Extraction Kit (Qiagen).
(3) Check the recovery and approximate concentration of the isolated DNA by running an aliquot on a minigel beside known amounts of marker DNA.

Protocol 3: Ligation of randomized exon cassette into the minigene vector and transfection into cultured cells

(1) Ligate 1 µg of gel-isolated minigene plasmid vector and a 2-fold molar excess of insert at 15 °C overnight in 100-µl ligation reaction using 1 × ligation buffer and 400 U of T4 DNA ligase (New England Biolabs); 10 × ligation buffer (supplied by New England Biolabs): 500 mM Tris–HCl, pH 7.8, 100 mM $MgCl_2$, 100 mM DTT, 10 mM ATP, 500 µg/ml BSA. Store 200-µl aliquots at −20 °C.
(2) Transfect the ligation reaction directly into two 60-mm plates of QT35 cells in which cells were plated 4–18 h earlier at a density of 10^6 cells per plate. Conditions for transfection vary depending on the cell line. It is important to express as many pre-mRNA molecules as possible so that a large pool of sequences is available for selection. The number of expressed RNAs is determined primarily by the ligation efficiency (the fraction of isolated vector that recircularizes with insert) and the transfection efficiency (the fraction of ligated molecules that make it to the nucleus and are transcribed). The conditions for ligation and transfection should be optimized as outlined in Section 49.2.5 ("Troubleshooting").
(3) The transfection protocol will depend on the cell type used. In our original protocol, transfection was by the calcium phosphate protocol [16, 17]. We have since switched to FuGENE 6 (Roche).
(4) Harvest cells 40–48 h after transfecting the DNA ligation reaction.

Protocol 4: RNA extraction and DNase treatment

(1) Harvest RNA 40–48 h following transfection. Wash the plates once with 2 ml of cold 1 × PBS (10 × PBS: 136.9 mM NaCl, 2.7 mM KCl, 10.0 mM Na_2HPO_4,

1.4 mM KH$_2$PO$_4$; bring pH to 7.4 with HCl; dilute 1:10 with ddH$_2$O and filter sterilize to make 1 × working stock). Let the plates drain at a 45° angle for 1 min and aspirate the remaining liquid at the bottom edge of the plate.

(2) Add 650 µl of solution A (see below) to each of the two plates then scrape off cells with a policeman and pool both plates into one 1.5-ml Eppendorf tube. Add 210 µl of chloroform/isoamyl alcohol (25:1). Vortex for at least 20 s with great agitation, making sure that the phases mix. [RNA extraction solution A (52.5 ml) = 25 ml of phenol (H$_2$O-saturated), 25 ml of solution B, 2.5 ml of 2 M NaOAc (pH 4.0) and 180 µl β-mercaptoethanol. Can be stored at 4 °C for more than 4 months. RNA extraction solution B (50 ml) = 4 M guanidinium thiocyanate, 25 mM sodium citrate then add ddH$_2$O to 45 ml. Adjust pH to 7.0 with 1N HCl. Add ddH$_2$O to 50 ml. Can be stored at 4 °C for more than 4 months.]

(3) Place the tubes on ice for 20 min then spin (12 000 g) for 20 min at 4 °C.

(4) Transfer the upper (aqueous) phase to a new 1.5-ml Eppendorf tube containing 870 µl of isopropanol. Vortex to mix well and store at −20 °C for at least 1 h.

(5) Spin tubes (12 000 g) for 20 min at 4 °C. Pour off the supernatant and wash the pellet by vortexing with 1 ml of cold 75% ethanol (stored at −20 °C). Spin for 10 min. Pour off the supernatant immediately after the centrifuge stops; otherwise the pellet may dislodge from the tube and pour out. Repeat this washing procedure a second time, then vacuum-dry the pellet.

(6) Redissolve the pellet in 50 µl of the DNase cocktail by gently vortexing for about 30 s, being careful not to introduce bubbles. Incubate at 37 °C for 30 min. For each sample, the DNase cocktail contains: 10 µl of 5 × in vitro transcription buffer (supplied by Promega: 5× = 200 mM Tris–HCl, pH 7.5 at 37 °C, 30 mM MgCl$_2$, 10 mM spermidine, 50 mM NaCl), 2.5 µl of 100 mM DTT, 0.2 µl of RNasin (40 U/µl, Promega), 0.4 µl of DNase [Worthington DPFF DNase at 1.0 mg/ml (2 U/µl) in 10 mM Tris–HCl, pH 7.5 in 10-µl aliquots stored at −80 °C] and 36.8 µl of ddH$_2$O.

(7) Make up cocktail of stop solution to add following the incubation. For each reaction add 2 µl of 0.5 M EDTA, 4 µl of 5 M NaCl and 44 µl of ddH$_2$O. Extract once with phenol/chloroform/isoamyl alcohol and once with chloroform/isoamyl alcohol. Ethanol-precipitate the aqueous layer using 260 µl of ethanol. Redissolve the pellet in 384 µl ddH$_2$O then add 16 µl of 5 M NaCl and 1 ml of 100% ethanol, and vortex to mix. Store this precipitated RNA at −20 °C as a suspension in 70% ethanol; do not pellet the RNA.

Protocol 5: RT-PCR

(1) Vortex each RNA sample to resuspend the RNA precipitate and remove 40 µl. This works out to be about 1/5 of a 60-mm plate or 10–15 µg of total RNA (from QT35 cell cultures prepared as described above). Spin down the RNA and vacuum-dry the pellet.

(2) Make up the reverse transcriptase cocktail and use 20-µl aliquots to dissolve each RNA pellet. Then, for each reaction combine 2.0 µl of 10 × magnesium-free Taq PCR buffer [supplied by Promega: 10× = 50 mM KCl, 10 mM Tris–

HCl (pH 9.0 at 25 °C), 0.1% Triton X-100], 2.4 µl of 25 mM MgCl$_2$, 2.0 µl of 1 µg/µl BSA, 2.0 µl of 10 mM dGATC, 1.0 µl of 100 pmol/µl hexamers, 1.0 µl of 100 mM DTT, 5 U of RNasin (Promega), 2 U of AMV reverse transcriptase (Life Sciences) and H$_2$O to 20 µl.

(3) Once the RNA is dissolved, incubate at room temperature for 10 min then transfer the tubes to 42 °C for 1 h. Following the reverse transcriptase reaction, heat the tubes at 95 °C for 5 min in a heating block then immediately plunge into an ice/water slurry.

PCR 1

(1) Make up a cocktail for 80-µl PCR reactions (per reaction): 8.0 µl of 10 × magnesium-free *Taq* buffer, 6.4 µl of 25 mM MgCl$_2$, 1.6 µl of BSA (1 µg/µl), 0.5 µl of Taq polymerase (Promega), 200 ng of upstream external oligo, 200 ng of downstream external oligo and H$_2$O to 80 µl. Note that additional deoxyribonucleotides (dGATC) is not necessary since the final concentration is 0.2 mM from the reverse transcriptase reaction.

(2) Mix and add 80 µl of the cocktail to each of the 20-µl reverse transcriptase reactions using a fresh pipette tip for each tube; transfer the reactions to 500-µl tubes. Add 2 drops of light mineral oil and run a PCR program using appropriate conditions. RT-PCR conditions must be optimized empirically for sensitivity with low background by varying the number of cycles, the annealing and reaction temperatures, and the MgCl$_2$ concentration. PCR 1 is typically for 20 cycles.

PCR 2 (nested PCR reaction)

(1) Make up cocktail for 80-µl PCR reactions (per reaction): 8.0 µl of 10 × *Taq* PCR buffer, 6.4 µl of 25 mM MgCl$_2$, 0.64 µl of 25 mM dGATC, 1.6 µl of 1 µg/µl BSA stock, 200 ng of upstream internal oligo, 200 ng of downstream internal oligo, 2.5 U of Taq polymerase and H$_2$O to 79 µl.

(2) Mix and add 79 µl of the cocktail to 500 µl tubes. Add 1 µl of PCR 1, 2 drops of light mineral oil, and run a PCR program using appropriate temperatures and number of cycles. Note that for our nested PCR [16], it was necessary to use 79 °C for annealing and polymerization, since at 76 °C the reaction generated significant background and low amounts of the correct product after 60 cycles. At 81 °C, no PCR product was formed, probably due to an absence of annealing. At 79 °C, the reaction produced large amounts of only the correct product.

(3) Add EDTA to 2 mM and NaCl to 0.2 M, and extract the PCR product once with an equal volume of phenol:chloroform:isoamyl alcohol (25:25:1) and once with an equal volume of chloroform/isoamyl alcohol (25:1). Precipitate with 2.5 volumes of ethanol and wash the pellet once in 70% ethanol.

(4) Digest the PCR product with the appropriate restriction enzymes and isolate the fragment from a 7% non-denaturing polyacrylamide gel as in Step 3 of Protocol 1.

(5) Repeat for the desired number of rounds of selection.

Acknowledgments

This work was initiated by Mark Landree and completed by Lydia Coulter, and was supported by the National Institutes of Health.

References

1 C. Tuerk, *Methods Mol. Biol.* **1997**, *67*, 219–230.
2 C. Tuerk, L. Gold, *Science* **1990**, *249*, 505–510.
3 E. N. Brody, L. Gold, *J. Biotechnol.* **2000**, *74*, 5–13.
4 F. J. Schmidt, *Mol. Cells* **1999**, *9*, 459–463.
5 T. A. Cooper, W. Mattox, *Am. J. Hum. Genet.* **1997**, *61*, 259–266.
6 W. G. Fairbrother, R. F. Yeh, P. A. Sharp, C. B. Burge, *Science* **2002**, *297*, 1007–1013.
7 L. Cartegni, S. L. Chew, A. R. Krainer, *Nat. Rev. Genet.* **2002**, *3*, 285–298.
8 J. F. Caceres, A. R. Kornblihtt, *Trends Genet.* **2002**, *18*, 186–193.
9 N. A. Faustino, T. A. Cooper, *Genes Dev.* **2003**, *17*, 419–437.
10 T. D. Schaal, T. Maniatis, *Mol. Cell. Biol.* **1999**, *19*, 1705–1719.
11 H. C. Tian, R. Kole, *Mol. Cell. Biol.* **1995**, *15*, 6291–6298.
12 H. X. Liu, M. Zhang, A. R. Krainer, *Genes Dev.* **1998**, *12*, 1998–2012.
13 H. X. Liu, S. L. Chew, L. Cartegni, M. Q. Zhang, A. R. Krainer, *Mol. Cell. Biol* **2000**, *20*, 1063–1071.
14 B. R. Graveley, *RNA* **2000**, *6*, 1197–1211.
15 X. D. Fu, *RNA* **1995**, *1*, 663–680.
16 L. R. Coulter, M. A. Landree, T. A. Cooper, *Mol. Cell. Biol.* **1997**, *17*, 2143–2150.
17 T. A. Cooper, *Methods Mol. Biol.* **1999**, *118*, 405–417.
18 N. N. Singh, E. J. Androphy, R. N. Singh, *Biochem. Biophys. Res. Commun.* **2004**, *315*, 381–388.
19 T. A. Cooper, *Methods Mol. Biol.* **1999**, *118*, 391–403.
20 R. Xu, J. Teng, T. A. Cooper, *Mol. Cell. Biol.* **1993**, *13*, 3660–3674.
21 D. L. Black, *Genes Dev.* **1991**, *5*, 389–402.
22 Z. Dominski, R. Kole, *Mol. Cell. Biol.* **1991**, *11*, 6075–6083.
23 S. M. Berget, *J. Biol. Chem.* **1995**, *270*, 2411–2414.
24 L. E. Maquat, *Curr. Biol.* **2002**, *12*, R196–197.

50
In Vitro Selection against Small Targets

*Dirk Eulberg, Christian Maasch, Werner G. Purschke
and Sven Klussmann*

50.1
Introduction

Aptamers have been raised against a plethora of small targets, e.g. organic dyes, by so-called SELEX (Systematic Evolution of Ligands by EXponential enrichment) experiments [1, 2]. In the early days of *in vitro* selection, one important driver to select against a target was to study the ability of nucleic acids to form stable, binding pocket- or crevice-like three-dimensional (3-D) structures in a system that was as simple as possible. The theories about the origins of life fuelled many selection experiments; every new, functional nucleic acid structure corroborated the versatility of nucleic acids – which consist of only four different building blocks! The fact alone that aptamers could even be readily isolated against targets that do not naturally bind nucleic acids or polyanions, e.g. hydrophobic molecules such as the amino acids isoleucine or valine, triggered many SELEX experiments [3, 4].

The molecular basis by which small molecules inhibit RNA function or interact with RNA was addressed by performing *in vitro* selections against antibiotics with the objective to understand RNA–small molecule recognition, information which is of extreme importance for attempts to develop novel drugs targeted against pathogen-specific RNA [5, 6].

The comparable properties of aptamers and antibodies regarding affinity and specificity suggested the use of aptamers as monitoring or typing tools. Two examples are (1) the isolation of an aptamer against 7,8-dihydro-8-hydroxy-2′-deoxyguanosine (oxodG), one of the most common mutagenic lesions produced in DNA by oxidative damage, which can be applied to identify and quantify this modification in DNA [7], or (2) a cAMP-binding aptamer which can be used to quantify this second messenger in solution [8].

Intracellular processes may be elucidated using aptamers which bind fluorophores like sulforhodamine B or fluorescein. Such intracellularly expressed aptamers have been successfully employed to detect transcription localization by their ability to sequester free fluorophore in solution [9].

Handbook of RNA Biochemistry: Student Edition. Edited by R. K. Hartmann, A. Bindereif, A. Schön, E. Westhof
Copyright © 2009 WILEY-VCH Verlag GmbH & Co. KGaA, Weinheim
ISBN: 978-3-527-32534-4

Another fascinating field of nucleic acid biochemistry is the area of catalytically active nucleic acids, the ribozymes or deoxyribozymes. Binding of the substrate molecule is the first step in any macromolecule-catalyzed reaction; when aiming to design specific biochemical ribozyme activities, the experimental strategy consequently is to initially generate an aptamer for the desired substrate. Small targets like arginine, ATP or S-adenosyl methionine can be used to produce aptamers as highly affine substrate-binding building blocks, which then serve as the starting material to evolve catalytic nucleic acids [10–12]. In the same way, allosteric domains can be generated by using aptamers that were selected against an allosteric effector of interest. Such a nucleic acid effector module may readily be conjugated to a catalytic nucleic acid domain [13] to a give an allosteric ribozyme, also known as aptazyme.

Finally, nucleic acid ligands with therapeutic and/or diagnostic potential, which bind peptide targets, have been isolated. It could be shown that such a receptor-mimetic bioactive antagonist, for the pharmacologically relevant nonapeptide gonadotropin-releasing hormone I (GnRH), does effectively bind and neutralize GnRH *in vivo* [14].

Compared to selection experiments against proteins, a major difference when selecting oligonucleotides aimed at small targets is that the resulting aptamers have to bind and clamp the small target in a receptor-like fashion; the required size of the isolated minimal binding motifs may therefore be larger in comparison to protein binders. Moreover, the potential contact surface to an aptamer is small. It can be hypothesized that the binding of peptides, for example, may require a rigid 3-D scaffold, since small peptides are, in general, intrinsically flexible structures and usually exist in solution as an equilibrium of multiple conformers [15]. For all these reasons, resulting binding affinities may be negatively influenced by the conformational entropy lost upon binding of a small flexible molecule.

Perhaps the most important experimental details affecting the success of *in vitro* selections against small targets are the modes of target immobilization. The smaller the target is, the greater the danger of isolating aptamers which require not only the target itself for binding, but also structures involved in immobilization, i.e. the linker and/or the matrix.

The key step of every *in vitro* selection experiment is the process of partitioning (Fig. 50.1) – the phase of the selection where the properties of the binders to be selected are defined. The input RNA pool, which may contain 10^{15} or even more different species, is brought into contact with the target of interest, often immobilized on an affinity chromatography column. The rare RNA species which interact with the immobilized target will remain bound to the column; the largest part of the RNA population, on the other hand, will simply pass through. As aptamers with the highest affinities possible are desired, weak binders are removed from the column by repeated washing. In theory, only specific binders should survive the process, to then be eluted in the final stage – either by applying denaturing conditions to the column, or by applying a buffer containing free target in excess concentration, thus competitively stripping the bound species off the immobilized target. In practice, however, the situation is often more complex and unpredictable.

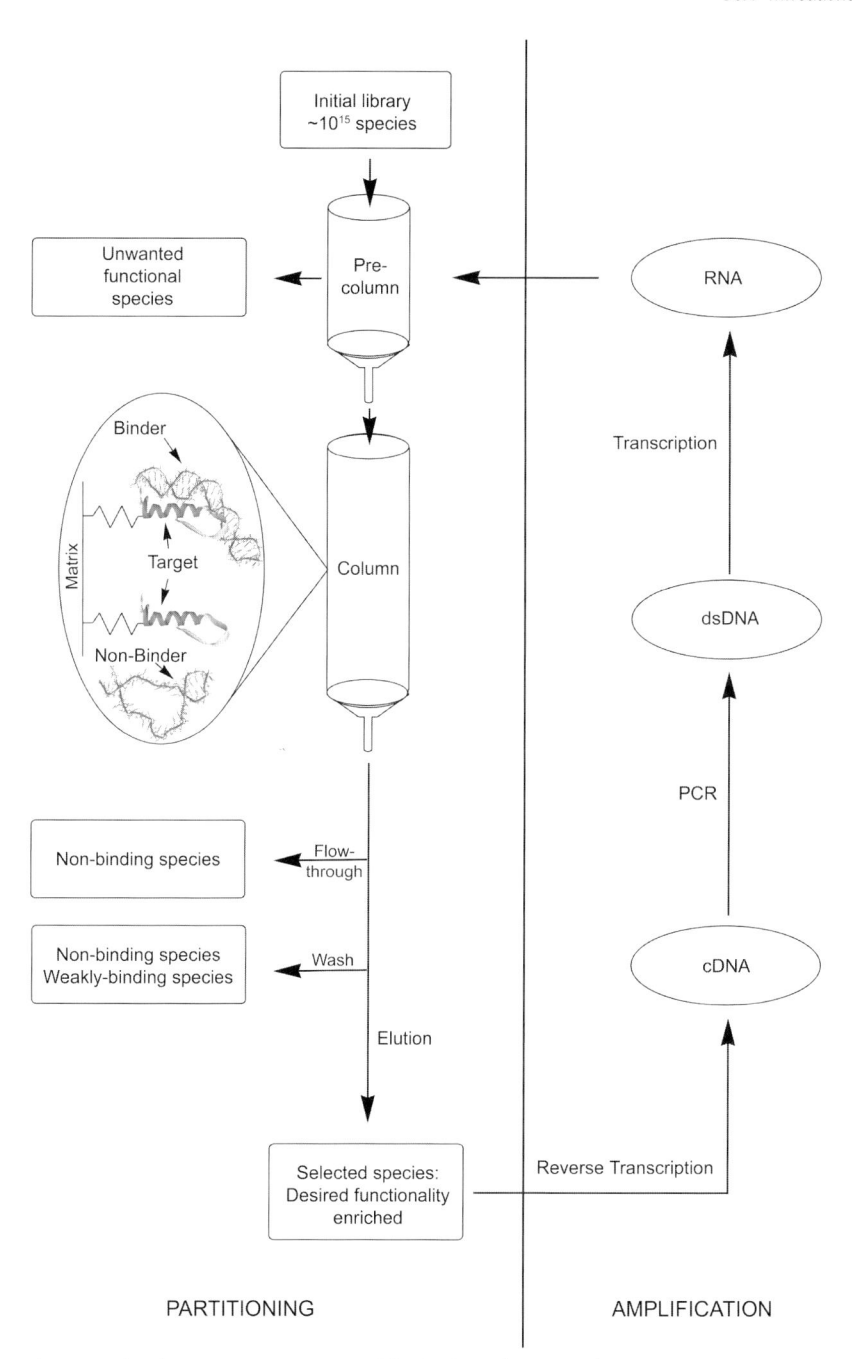

Fig. 50.1. A schematic representation of the *in vitro* selection cycle.

Target molecules must be immobilized for partitioning, and the introduction of chromatography material brings surfaces into the system which will bind RNA molecules, either specifically or non-specifically. Specific matrix binders can be avoided by pre-selecting the RNA pool with non-derivatized column material and by discarding the column-bound material, using only the flow-through. Non-specifically retained RNA molecules represent the background binding which is always present, and which should remain on a constant low level in a smoothly running *in vitro* selection until specific binders get enriched and predominate after several rounds of selection. In the following sections, all steps of an *in vitro* selection experiment will be discussed in detail, with particular emphasis on selection against small targets. Target immobilization, partitioning and assessment of binding constants will be the focus of this chapter.

50.2
Target Immobilization

The aptamer species finally selected are those which derive the maximum benefit from the experimental setup. This does not necessarily mean that the selected aptamers are high-affinity binders to the intended target: unwanted or unforeseen selection pressure may lead to unwanted or unforeseen results. Since the selection process requires immobilization of the target to a surface, the danger of selecting for species which bind to the immobilization matrix or recognize target and matrix cooperatively is always present. Therefore, the most important parameter for the design of a successful *in vitro* selection experiment, especially when selecting against small targets, is the optimal presentation of the target during the binding reaction.

To avoid potential problems, the choice of the optimal immobilization matrix as well as the coupling chemistry are key parameters. Two partitioning methods that have been successfully employed for selecting protein-binding aptamers are unfortunately not suited for small molecules:

- Passive, hydrophobic adsorption on plastic surfaces: most small targets do not exhibit hydrophobic regions to the extent required, and even if they were to display hydrophobic regions, small targets would probably not be accessible for an aptamer.
- Filtration through nitrocellulose filter membranes: complexes between aptamers and small targets may not be retained on the filters.

Nevertheless, a cornucopia of useful media for the immobilization of small ligands exists. Usually, these media are commercially available as non-magnetic or paramagnetic particles composed of many different polymers. Such particles are offered with a wide variety of activated surfaces, which may influence binding between target and aptamer.

Length and nature of the spacer connecting the target and the solid support are critical parameters, especially when working with very small targets. The binding site of aptamers may be located deeply within the molecule, and matrices, prepared by coupling small ligands directly to the support, may suffer from problems of steric hindrance. On the one hand, it is desirable to attach the target via a long spacer to ensure optimal accessibility; on the other hand, the longer the spacer is, the more it will change the overall characteristics of the immobilized target with respect to charge or polarity, and may thus reduce the specificity of the separation. Unfortunately, no general rule exists as to which matrix and coupling technique eventually will lead to a successful *in vitro* selection experiment.

The key step of each *in vitro* selection experiment is the partitioning of binding from non-binding species. The more efficient this separation works, the better the prospects for isolation of the desired binders. Low background binding of the nucleic acid starting pool to the matrix is a prerequisite for success. These parameters have to be checked experimentally before starting the selection experiments. Another important aspect is the binding capacity of the matrix, which can be determined during the immobilization of the target to the activated matrix. The immobilized target concentration per matrix volume can be assessed by measuring the difference between the target concentration in the coupling solution before and after the immobilization process via HPLC, photometry, or other suited methods.

Partitioning of ligand-binding from non-binding species during *in vitro* selection is comparable to a standard affinity chromatography. The preparation of some useful matrices will be described in detail in this section. This includes covalent immobilization procedures on three different, widely used matrix types and the non-covalent immobilization via biotin–avidin interaction.

50.2.1
Covalent Immobilization

50.2.1.1 Epoxy-activated Matrices

Epoxy-activated matrices, such as Amersham's Epoxy-activated Sepharose 6B, ImmunoPure Epoxy-activated Agarose (Pierce) or paramagnetic Dynabeads M-270 Epoxy (Dynal) can be used for covalent immobilization of ligands via amino groups at pH 9–11, thiol groups at pH 7.5–8.5 or hydroxyl groups at pH 11–13 (Fig. 50.2A). For Epoxy-activated Sepharose 6B, the functional epoxy group is linked to the polysaccharide support via a 12-atom hydrophilic spacer which makes it suitable for immobilization of small molecules; the active group density is in the range of 30 μmol/ml of drained coupling material.

Protocol: Epoxy-activated Sepharose 6B (Amersham)

Gel preparation (around 3.5 ml matrix)

(1) Suspend 1 g of lyophilized matrix in 10 ml RNase-free water, incubate for 10 min at room temperature. Do not stir the suspension with a magnetic stirring

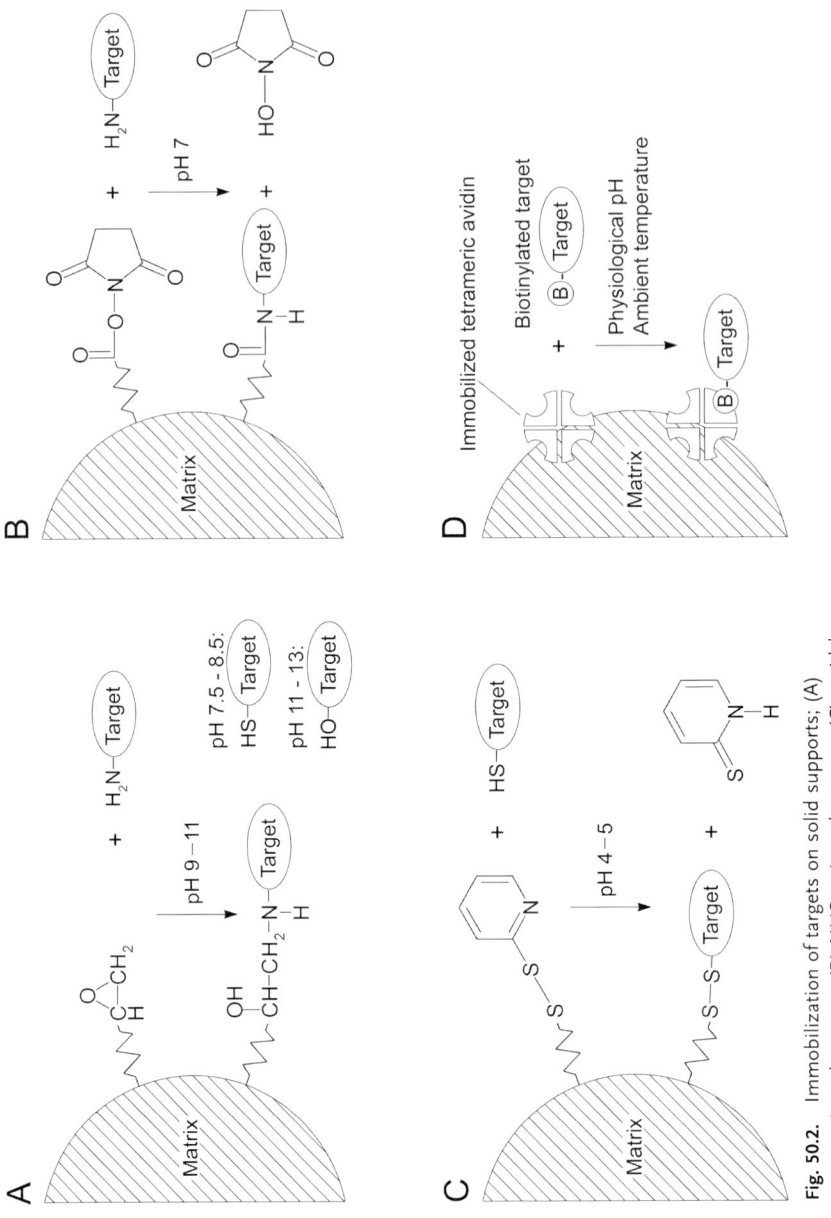

Fig. 50.2. Immobilization of targets on solid supports; (A) epoxy-activated support, (B) NHS-activated support, (C) pyridyl disulfide-activated support and (D) support with immobilized tetrameric avidin.

bar, as this may grind the agarose to fine particles. The powder will swell to a volume of approximately 3.5 ml.
(2) Wash the swollen gel in a sintered glass filter by adding at least 200 ml of RNase-free water over 1 h.

Ligand coupling

(1) Prepare 1 ml 200 mM ligand solution in 100 mM coupling buffer (carbonate, phosphate or borate buffered). Amino group-containing buffers like Tris or other nucleophiles like thiols, hydroxyls or phenols must not be present. The reaction proceeds most efficiently at pH 9–13. In case of coupling hydroxyl groups and if the ligand is alkali-stable, couple at pH 13 in NaOH solution. If low concentrations of ligand on the matrix are desired, ligand concentrations of 2 mM or less may be used.
(2) Add the ligand solution to 1 ml of swollen, drained gel in a closed reaction tube and mix for 16 h at 37 °C or room temperature. Mix gently, for example on a rotating wheel. Do not use a magnetic stirrer.
(3) Wash the matrix twice with at least 5 volumes of coupling buffer to remove excess ligand. Save an aliquot of the ligand solution before and after the coupling reaction to determine the coupling efficiency.
(4) Suspend the matrix in one volume 1 M ethanolamine, pH 8 to block remaining active groups.
(5) Incubate at room temperature for 4–16 h.
(6) Wash the matrix with 5 volumes 0.1 M acetate buffer/0.5 M NaCl, pH 4, and then with 5 volumes of 0.1 M Tris–HCl buffer/0.5 M NaCl, pH 8.
(7) Repeat Step 6 at least twice.

The matrix material for pre-selection can be prepared by using 200 mM glycine solution in coupling buffer instead of ligand solution, or by simply starting at Step 4 of the coupling protocol with pre-swollen gel.

50.2.1.2 NHS-activated Matrices

NHS-activated matrices allow for rapid, gentle and stable coupling and are highly selective for primary amino groups under physiological conditions. Such media – like Bio-Rad's Affi-Gel 10 or Amersham's NHS-activated Sepharose – are N-hydroxysuccinimide esters of a derivatized polymer support (Fig. 50.2B). Both media contain a 10-atom spacer arm. The active ester content is 15 μmol/ml gel for Affi-Gel and around 10 μmol/ml for NHS-activated Sepharose. The media are provided fully swollen and solvent-stabilized in isopropyl alcohol.

Protocol: Affi-Gel 10 (Bio-Rad)

Ligand coupling

(1) Shake the vial and transfer the desired amount of slurry to a glass-fritted funnel. Drain the supernatant solvent.

(2) Wash the gel with 3 volumes ice-cold, RNase-free water. Do not allow the gel bed to go dry, especially when working under vacuum. The time between solvent removal and addition of ligand solution should not exceed 20 min.

(3) Transfer the gel to a tube and add 0.5–1 gel volumes ice-cold ligand in coupling buffer. Ligand concentration can be 20 mM or less, depending on the desired ligand concentration on the matrix. As coupling buffer, an acetate, carbonate, MOPS, or HEPES buffered solution, pH 6.5–8.5 is recommended. The coupling buffer should be 10–100 mM to ensure pH stability. Amino group-containing buffers like Tris must not be used.

(4) Incubate the suspension under gentle agitation on a rotating wheel or shaker for 1 h at room temperature or 4 h at 4 °C. Save an aliquot of the ligand solution before and after the coupling reaction to determine the coupling efficiency.

(5) Block unreacted active esters by addition of 0.1 volumes of 1 M ethanolamine, pH 8.

(6) Incubate for 1 h at room temperature.

(7) Wash the gel 5 times with 5 volumes of coupling buffer or water.

(8) Wash the gel 3 times with 3 volumes of the buffer that will be used during *in vitro* selection.

Matrix material for pre-selection can be prepared by using 100 mM ethanolamine solution (pH 8) instead of ligand solution.

50.2.1.3 Pyridyl Disulfide-activated Matrices

Pyridyl disulfide-activated matrices such as Thiopropyl Sepharose 6B (Amersham) or Sigma's Thiopropyl-activated Agarose react with ligands that contain thiol groups under mild conditions to form mixed disulfides (Fig. 50.2C). The active group content for Thiopropyl Sepharose 6B is 18–31 µmol/ml gel; the medium contains a four-atom spacer arm. Such matrices are especially suited for the immobilization of peptides, which can be easily synthesized with an additional cysteine residue at the N- or C-terminus as coupling site.

In contrast to other immobilization methods, coupling of ligands via disulfide bridges is reversible. These matrices therefore facilitate the elution of any binding RNA molecule together with the ligand by reduction of disulfide bonds with dithiothreitol (DTT) or β-mercaptoethanol.

Protocol: Thiopropyl Sepharose 6B (Amersham)

Gel preparation (around 3 ml matrix)

(1) Suspend 1 g of lyophilized matrix in 10 ml RNase-free water, incubate for 10 min at room temperature. Do not stir the suspension with a magnetic stirring bar as this may grind the agarose to fine particles.

(2) Wash the swollen gel on a sintered glass filter by adding approximately 200 ml of RNase-free water over 15 min.

Ligand coupling

(1) Prepare 100 mM Tris, acetate or phosphate coupling buffer containing 100–500 mM NaCl, pH 7.5. EDTA may be added to a final concentration of 1 mM to remove trace amounts of heavy metal ions which may catalyze oxidation of thiols.
(2) Degas the coupling buffer to avoid oxidation of thiols.
(3) Add 1 volume of 20 mM or lower concentrated ligand solution in coupling buffer to the swollen, drained gel in a closed reaction tube and mix gently for 1 h at room temperature. Mix gently, e.g. on a rotating wheel or shaker. Do not use a magnetic stirrer.
(4) Transfer the material to a chromatography column and drain the ligand solution. Save an aliquot of the ligand solution before and after the coupling reaction to determine the coupling efficiency. If monitoring non-bound substances photometrically, the contribution of released thiopyridone to the absorbance value has to be taken into account.
(5) Wash with 30 volumes of 100 mM sodium acetate, pH 6.
(6) Block remaining reactive groups by suspending the matrix in 7 ml of 5 mM β-mercaptoethanol and mix gently at room temperature for 45 min.
(7) Drain the gel and wash with 30 volumes of the buffer that will be used during *in vitro* selection.

Matrix material for pre-selection can be prepared as described above, but the ligand must be omitted.

50.2.2
Non-covalent Immobilization

Covalent immobilization of ligands to solid supports is easily achieved using commercially available activated chromatography media as described above. Using such ligand-derivatized materials for partitioning is a straightforward process. However, a selection parameter that can only be controlled with great difficulty is the concentration of ligand immobilized on the matrix. An alternative to covalent coupling is the very effective, non-covalent avidin–biotin system, which allows for a reaction between the biotinylated selection target and RNA pool in solution. Capturing biotinylated molecules by immobilized streptavidin, a tetrameric 60-kDa protein, proceeds rapidly and efficiently under a wide range of pH, temperature and solvent conditions as well as in the presence of denaturants (Fig. 50.2D). "Fishing" of target-binding RNA species can therefore be accomplished in the desired selection buffer without interfering with the RNA–target interaction. Thus, the researcher is completely free in the choice of reaction partner concentrations. The only necessity is the attachment of biotin to the target of interest. This can easily be achieved with any target molecule bearing a primary amino group: by reaction with commercially available, NHS-activated and linker-equipped biotin such as Sulfo-NHS-LC-Biotin from Pierce or Biotin-NHS from Calbiochem.

If working with immobilized streptavidin, the binding capacity towards the biotinylated selection target must be determined initially by incubating a constant concentration of biotinylated target with different amounts of avidin matrix in the buffer used during the selection. In physiological buffer, binding will be completed at room temperature or 37 °C after 15 min. The measurement of unbound target in the supernatant enables the determination of the binding capacity. Among the suppliers of immobilized streptavidin are Roche (Streptavidin Magnetic Particles), Pierce (ImmunoPure Immobilized Streptavidin) and Sigma (Streptavidin-Agarose).

50.3
Nucleic Acid Libraries

In general, combinatorial chemistry allows the generation of a vast number of different molecules by the repeated execution of a limited number of synthetic steps. SELEX is the *in vitro* selection of oligonucleotides from combinatorial nucleic acid libraries. The complexity of these libraries is defined by one or more randomized regions within the oligonucleotides. Randomization can easily be accomplished by offering a mixture of all four building blocks during the coupling steps of the solid phase synthesis. Usually, the random regions are flanked by constant sequences, rendering the molecules accessible to enzymatic amplification procedures.

50.3.1
Library Design

Nucleic acid libraries for the generation of RNA pools have three distinct functional features (Fig. 50.3):

- A 5′-terminal promoter region to initiate *in vitro* transcription. The commonly used RNA polymerase for *in vitro* selection is that of phage T7, requiring the

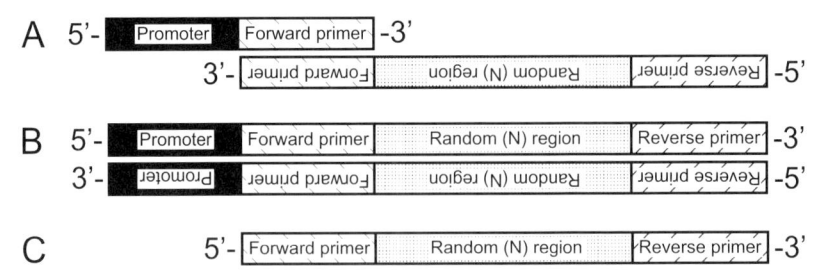

Fig. 50.3. Structure of RNA library intermediates. (A) and (B) annealed promoter-bearing forward primer and chemically synthesized DNA library [(−)strand] before (A) and after (B) fill-in or PCR reaction; (C) RNA which results from *in vitro* transcription of template B.

double-stranded 17-bp T7 promoter core region; other commercially available RNA polymerases are SP6 or T3 RNA polymerase. The promoter part is generally not synthesized as part of the library, but is added during the fill-in reaction or PCR amplification of the chemically synthesized template DNA (Fig. 50.3; see Section 50.3.2).

- Forward and reverse primer regions for PCR amplification. These regions usually comprise 15–20 nt and are designed to ensure optimal PCR by minimizing primer-dimer formation and mispriming between the fixed regions. General rules for PCR primer design should be applied; an optimal annealing temperature above 50 °C is recommended.
- Central randomized region. This region is the source of the vast complexity which is necessary to start any *in vitro* selection. When designing the length of this region, the following aspects should be considered: The practical limits of molecular biology (around 10^{15} molecules are the maximum that can be readily handled in a reaction tube) have already been reached when the four bases are randomly incorporated at 25 positions and each variant is represented once ($4^{25} \approx 10^{15}$ molecules, corresponding to roughly 1 nmol). However, solutions for the binding problem to be solved (that is structures capable of specific interaction with the target) may not be contained in an RNA pool with only 25 randomized positions. Only larger RNAs may be capable of shaping the required stable tertiary structure. Commonly used random regions therefore contain between 30 or more than 100 nt, even if only a minute fraction of the theoretically possible complexity will be represented in practice. As the selection against small targets results in receptor mimics that often depend on a complex 3-D structure, longer randomized stretches are preferable if the aptamer length is not of importance.

The successfully used pool AL.60 [16] may serve as an example for library design; the T7 promoter core sequence is underlined:

Pool AL.60 (−)strand	5′-TCAGCTGGACGTCTTCGAAT-N_{60}-TGTCAGGAGCTCGAATTCCC-3′
RNA AL.60 (+)strand	5′-GGGAATTCGAGCTCCTGACA-N_{60}-ATTCGAAGACGTCCAGCTGA-3′
AL.T7F (forward primer)	5′-TTC<u>TAATACGACTCACTATAG</u>GG-AATTCGAGCTCCTGACA-3′
AL.R (reverse primer)	5′-TCAGCTGGACGTCTTCGAAT-3′

50.3.2
Starting Pool Preparation

Preparation of input material for *in vitro* selection is performed by generating dsDNA from the chemically synthesized (−) ssDNA, followed by *in vitro* transcription into RNA. Transcription-ready dsDNA may either be produced by performing 1–3 PCR cycles (Protocol A) or by a one-step fill-in reaction of the ssDNA, using

DNA polymerases like *Taq* or the Large (Klenow) Fragment of *Escherichia coli* DNA polymerase I (Protocol B).

Protocol A: Large-scale PCR

ssDNA template	1 nmol
10 × PCR buffer	100 µl [Invitrogen; 200 mM Tris–HCl (pH 8.3), 500 mM KCl]
MgCl$_2$	2.5 mM
dNTPs	200 µM each
Forward primer	5 µM
Reverse primer	5 µM
Betain	0.75 M
Taq polymerase	50 U (Invitrogen; 5 U/µl)
Final volume	1 ml

(1) Divide the reactions into 100-µl aliquots.
(2) Perform 1–3 PCR cycles. Incubation for 1 min at appropriate denaturing, annealing and extension temperature is generally sufficient, with a 5-min final polymerization step.
(3) Ethanol-precipitate, wash with 70% ethanol, and dry the generated dsDNA.
(4) Use the dsDNA pellet as template in a 1 ml *in vitro* transcription reaction (see Section 50.4.3). If radiolabeled RNA is desired for the *in vitro* selection procedure, an appropriate amount of an α-^{32}P-labeled NTP may be included in the transcription mix.

Protocol B: Fill-in reaction with Klenow polymerase

ssDNA template	1 nmol
10 × Klenow buffer	100 µl [100 mM Tris–HCl (pH 7.5), 50 mM MgCl$_2$]
dNTPs	200 µM each
Forward primer	3 µM
Betain	0.75 M
Final volume	1 ml

(1) Denature the template for 5 min at 95 °C.
(2) Place on ice for 5 min.
(3) Add DTT to a final concentration of 2.5 mM.
(4) Add 100 U DNA polymerase I, Large Fragment (Klenow; New England Biolabs).
(5) Incubate at 37 °C for 1 h.
(6) Ethanol precipitate, wash with 70% ethanol and dry the generated dsDNA.
(7) Use the dsDNA pellet as template in a 1 ml *in vitro* transcription reaction (see Section 50.4.3). If radiolabeled RNA is desired for the *in vitro* selection procedure, an appropriate amount of an α-^{32}P-labeled NTP may be included in the reaction.

50.4
Enzymatics

In this section, the enzymatic part of *in vitro* selection experiments will be discussed, i.e. the generation of input material for the following round of selection. For RNA selections, amplification is achieved by a series of enzymatic reactions including reverse transcription, PCR, and *in vitro* transcription (Fig. 50.1). These reactions convert the selected RNA species into cDNA, into dsDNA and back again into RNA.

While the exclusive goal of the amplification is to generate sufficient material for the next round of selection, one should remember that in practice every enzymatic step introduces an unwanted selection pressure beside the wanted selection parameters. Care has to be taken to keep this pressure at a minimum. Depending on their nucleotide sequence, different molecules are amplified more or less efficiently by a given enzymatic system. It can be assumed that molecules which form complex and rigid 3-D structures may serve as poorer templates during the enzymatic procedures. These restrictions may limit or even reduce the library complexity and can hardly be eliminated. One possibility to counteract, however, is to include nucleic acid melting-point-lowering agents in the reactions, e.g. dimethylsulfoxide, betaine or commercially available solutions like Epicentre Enhancer and Qiagen Q-Solution. Such reagents help to dissolve secondary structures, thus reducing the amount of break-off products. For the same reason, the highest possible temperature should be applied during annealing in the PCR and RT reactions.

Exemplary protocols, which usually give satisfactory results for reverse transcription, PCR, and *in vitro* transcription, are presented in this section. Various aspects of these protocols have undergone optimization and should therefore serve as a good starting point, although they should by no means be taken as a final protocol. Different pools require different conditions for optimal amplification. Furthermore, different binders within a given pool may again vary in their amplification, which makes it necessary to find the conditions that ensure amplification across the broadest range possible for every pool. In essence, the reaction conditions for any step in the amplification should be adjusted to support the maximum level of diversity in the pool – which sometimes may be difficult in practice.

50.4.1
Reverse Transcription

Reverse transcription is the first enzymatic step that follows the partitioning of bound from unbound RNA. The most commonly used forms of reverse transcriptase are avian myeloblastosis virus (AMV) and Moloney murine leukemia virus (M-MLV) RTase, both of which lack $3' \rightarrow 5'$ exonuclease activity. The exemplary protocol given below uses Superscript II RTase (Invitrogen), an engineered version of M-MLV RTase with reduced RNase H activity and increased thermal stability.

Protocol: Reverse transcription
The eluted and purified RNA is annealed to the reverse primer as follows:

RNA	0.1–5 pmol
Reverse primer	100 pmol
Betain	1.5 M
Final volume	10 μl

(1) Denature at 95 °C for 5 min.
(2) Place on ice for 5 min.
(3) Add the following:

5 × First Strand buffer	4 μl [Invitrogen; 250 mM Tris–HCl (pH 8.3), 375 mM KCl, 15 mM $MgCl_2$]
dNTPs	500 μM each
DTT	10 mM
Superscript II RTase	200 U (Invitrogen; 200 U/μl)
Final volume	20 μl

(4) Mix and incubate at 42 °C for 30 min.
(5) Inactivate the RTase by incubating 10 min at 65 °C.
(6) Ethanol-precipitate; wash with 70% ethanol and dry the pellet.
(7) Dissolve the nucleic acid pellet in 10 μl H_2O.

50.4.2
PCR

The main purpose of the PCR step is the production of dsDNA as transcription template for the next selection round. Additionally, the T7 promoter, which is necessary for *in vitro* transcription, is attached to the selected sequences. PCR protocols that are widely used for *in vitro* selections do not differ significantly from standard protocols. However, since selections take many cycles of amplification from the first to the last selection round, it is imperative to keep the number of PCR cycles per round as low as possible, thus avoiding the generation of amplification artifacts. As for every PCR experiment, the primers have to be carefully chosen to avoid the formation of potential primer-dimers. Melting temperatures of reverse (3′) primer and the hybridizing part of the forward (5′) primer should match and allow for an annealing temperature of at least 50 °C.

Protocol: PCR

cDNA	0.05–1 pmol
10 × *Taq* buffer	10 μl [Invitrogen; 200 mM Tris–HCl (pH 8.4), 500 mM KCl]
$MgCl_2$	2.5 mM
5′ primer	1–3 μM
3′ primer	1–3 μM
dNTPs	200 μM each

Betain	0.75 M
Taq polymerase	5 U (Invitrogen; 5 U/μl)
Final volume	100 μl

As input cDNA, 20–100% of the reverse transcribed RNA should be used (see Section 50.4.1). Optimally, not more than 1 pmol cDNA should be amplified per 100 μl reaction volume. As already mentioned, the cycling parameters will depend on the pool and should be optimized with the initial pool. The following profile may serve as a starting point for many pools:

1	94 °C	2 min	
2	94 °C	1 min	denaturing
3	50–72 °C	1 min	annealing
4	72 °C	1 min	polymerization
5	72 °C	5 min	

Repeat Steps 2–4 for 6–20 cycles

The number of cycles needed to generate enough product for the following *in vitro* transcription will mainly depend on the amount of template. To avoid performing more cycles than necessary, and thus risking the introduction of artifacts, it is better to start with a small number of cycles and only perform additional cycles if required (to be checked by PAGE). A more sophisticated method involves multiple analytical PCR reactions with a small cDNA aliquot to determine the minimal cycle number required to obtain a certain amount of product, followed by the preparative PCR under those conditions.

50.4.3
In Vitro Transcription

In vitro transcription of the dsDNA templates produced in the PCR step results in RNA, thus closing the amplification part of the selection round. The predominantly employed T7 RNA polymerase does not only transcribe DNA to the RNA level, but can produce up to one hundred RNA copies of each dsDNA template. Therefore, the transcription step contributes significantly to the amplification efficiency. A 100 μl-reaction containing 100 pmol dsDNA template typically yields 2–4 nmol RNA, with even higher amplification factors for lower template concentrations. The reaction is relatively time-consuming and is therefore most conveniently incubated overnight; however, yields obtained after 3–6 h will be sufficient in most cases. The following protocol works reliably and can be transferred to most pools.

Protocol: *In vitro* transcription

10 × Reaction buffer	10 μl [800 mM HEPES–KOH (pH 7.5); 220 mM MgCl$_2$, 10 mM spermidine]
dsDNA template	10–100 pmol
DTT	10 mM

NTPs	4 mM each
BSA	120 µg/ml
Betain	0.75 M
[α-^{32}P]ATP	as required
T7 RNA polymerase	100 U (Promega; 50 U/µl)
RNase Out	40 U (Invitrogen; 40 U/µl)
Final volume	100 µl

(1) Incubate at 37 °C for 3–16 h.
(2) Add 20 U RNase-free DNase I (Sigma) to digest the dsDNA template.
(3) Incubate further 15 min at 37 °C.
(4) Add EDTA to 20 mM and incubate 10 min at 65 °C to terminate the reaction.
(5) Purify the RNA by denaturing 8% PAGE.

50.5
Partitioning

The process of separating desired from non-desired molecules is the central part of an *in vitro* selection experiment. As depicted in Fig. 50.1, the input population is split up in different subpopulations:

- Unwanted functional species which are removed by pre- or counter-selection.
- Non-binding and weakly binding species that do not bind to the affinity column or can be removed by washing.
- Selected species that bind tightly to the target and can be amplified for the next selection round after elution.

Every selection is started by denaturation and refolding of the RNA pool under buffer and temperature conditions to be used during the selection. This step is aimed at favoring the generation of equilibrium structures over randomly coiled structures that may result from RNA preparation.

The RNA population is then applied to a column that binds unwanted functional species. In the case of a simple pre-selection, the RNA is passed through a column containing non-derivatized matrix material which detains matrix-binding species. Counter-selection can be applied to ensure that the desired binders discriminate between their target – which should be bound tightly – and a structurally related molecule. The column used for counter-selection will be derivatized with the structurally related molecule instead of the target, and will only let non-binding species pass, which then are selected for target binding.

The next step is to bring the RNA pool and the target into contact. Retrieving the desired RNA–target complexes from non-binding RNA is difficult in selection experiments against small targets: the physicochemical differences between the free RNA molecules (typical molecular weight around 30 kDa) and a complex of RNA and target are minuscule when the targets are at least one order of magnitude smaller than the RNA. Therefore, partitioning is often carried out "on-column" using pre-immobilized target molecules.

Another approach takes advantage of the avidin–biotin interaction (Section 50.2.2), using biotinylated target molecules and capturing them from the mixture with immobilized avidin after the binding reaction. A benefit of this strategy is that the concentration of the reaction partners can be freely chosen, thus enabling the researcher to establish more stringent reaction conditions which, in theory, favor binders with high affinity. A further positive characteristic is the possibility to extend the interaction time between pool and target, while keeping the contact time to the matrix short. An extended incubation time is necessary to reach binding equilibrium if the concentration of target binders is low, which is normally the case during the first selection rounds. Furthermore, extended incubation may also be advantageous in later rounds, if very low target concentrations are applied to select only binders with the highest affinity.

Short contact time of pool and column material lessens the chance to enrich unwanted matrix binders. Target-binding species are retained on the affinity chromatography material, whereas non-binding RNAs are found in the unbound fraction. By washing under selection conditions, non-binding and weakly binding species are removed; slightly more stringent washing may help to reduce background binding.

Finally, the bound RNAs must be eluted from the immobilized matrix. Basically, two principles can be exploited to achieve this: (1) application of denaturing conditions or (b) competitive affinity elution by offering non-immobilized target molecules. For the latter, the matrix is incubated with an excess of free target, which competitively elutes the bound nucleic acids, leaving the 3-D structure of the RNA intact. Since the binding of an aptamer to its target depends on the aptamer 3-D structure and on non-covalent interactions between both binding partners, denaturing agents like urea or guanidinium thiocyanate can be alternatively used to elute even molecules that bind to the target with extremely high affinity. Such molecules might get lost when applying competitive affinity elution, because the off-rate for binders of said quality can be so slow that equilibrium may not be reached under practical experimental conditions.

Stringency is a critical selection parameter. In order to obtain highly affine aptamers, it is necessary to continuously adjust the conditions during the consecutive selection rounds by starting with relatively high target concentrations and decreasing the target level in the course of selection. This supports the enrichment of highly affine binders, whereas the chance for weak binders to survive the selection process is minimized. Modifying the washing parameters (volume, time, ionic strength) is an additional possibility to increase stringency. All these parameters must be adjusted according to the proportion of RNA specifically bound to the target. The bound fraction can be measured by using radioactively labeled pools. The general rule is: as soon as the amount of the bound molecules increases relative to the previous round, the stringency in the next round is increased by lowering the target concentration and/or intensifying the washing procedure.

Special attention must also be given to the conditions under which the binding

Tab. 50.1. Reagents used for partitioning.

Reagents	
SB (selection buffer)	20 mM Tris, pH 7.4; 150 mM NaCl; 5 mM KCl; 1 mM MgCl$_2$; 1 mM CaCl$_2$; 0.005% Triton X-100
10 × SB without Mg^{2+}/Ca^{2+}	200 mM Tris, pH 7.4; 1.5 M NaCl; 50 mM KCl
10 × Mg^{2+}/Ca^{2+}	10 mM MgCl$_2$; 10 mM CaCl$_2$
AES (affinity elution solution)	0.1–1 mM target in SB
DES (denaturing elution solution)	4 M guanidinium thiocyanate; 25 mM sodium citrate, pH 7; 0.5% sarcosyl; 100 mM β-mercaptoethanol

reaction is performed: since the structure of aptamers is temperature- and/or ion-dependent, it is essential to perform the selection process at the relevant temperature and in the buffer in which the aptamers should be eventually applied. If functionality at or near physiological conditions is required, phosphate, Tris or HEPES buffers around 20 mM and pH 7.5 with 100–250 mM NaCl and 5 mM KCl can be used. The divalent cations Mg^{2+} and Ca^{2+}, which usually are included at about 1 mM in a physiological selection buffer, are of special relevance for nucleic acid structure formation. Finally, a detergent may be included at concentrations between 0.005 and 0.5%. The presence of detergents in the selection process is beneficial to ensure a reduction of unwanted hydrophobic interactions between the RNA and the selection matrix.

In the following, three exemplary protocols for the selection process are given. In protocol A, the concept of pre- or counter-selection is exemplified; protocol B describes the selection against pre-immobilized target and competitive affinity elution; finally, protocol C lists the steps of a selection against a target in solution, followed by denaturing elution. Reagent compositions of solutions used are specified in Table 50.1.

Protocol A: Pre- or counter-selection

(1) Fill pure chromatography matrix (see Section 50.2.1) without target into an expendable column. Use the same matrix amount that is used in the following selection step. A simple pre-selection is carried out with non-derivatized column material to remove unwanted matrix-binding RNA species; for a counter-selection, the material is loaded with the molecule which should not be recognized by the desired aptamers. The column thus removes such unwanted binders from the pool.
(2) Rinse the column with ten column volumes of selection buffer (SB).
(3) Denature the nucleic acid pool in SB without Mg^{2+}/Ca^{2+} and without Triton X-100 for 1–5 min at 70–95 °C and place on ice.
(4) Bring the pool to SB conditions with 10 × Mg^{2+}/Ca^{2+} and 0.5% Triton X-100 and re-fold for at least 5 min at the selection temperature.

Tab. 50.2. Exemplary selection: pre-immobilized target.

Round	c_{Target} on matrix (µM)	Matrix volume (µl)	n_{Target} (nmol)	n_{RNA} (nmol)	Ratio target:RNA	RNA bound (%)
1	100	800	80	2	40	1
2	100	400	40	2	20	1.3
3	100	200	20	2	10	4
4	100	50	5	1	5	8
5	10	100	1	1	1	4
6	10	50	0.5	1	0.5	12

(5) Apply the pool to the column and collect flow-through and two column volumes of SB before adding it to the affinity matrix carrying the target. A distinct loss of material is tolerated to minimize the chances for selection of matrix binders.

Protocol B: Covalently immobilized target/affinity elution

(1) Fill an appropriate amount of the chromatographic affinity matrix (see Section 50.2.1), loaded with the target at the desired concentration, into a small expendable column. See Table 50.2 for an example of a typical selection course.
(2) Equilibrate the affinity matrix with 10 column volumes of SB.
(3) Denature and re-fold the pool as described in Protocol A, Steps 3 and 4, and apply the material to the column. See Table 50.2 for appropriate amounts of target and pool. Beginning with round 3, the pool is pre-selected against the pure matrix without target before every selection step (Protocol A); as soon as the pool is pre- or counter-selected (after round 3), the denaturing/refolding procedure will only be carried out prior to the pre- or counter-selection step.
(4) Wash the affinity column with SB until only around 1% of the pool remains on the matrix. Yields of 1% or less can be expected in the initial rounds; more RNA will bind as soon as target binding sequences become enriched during the course of the selection (see Table 50.2). Washing conditions may be changed toward increased stringency by using high-salt conditions (e.g. 3 M NaCl in SB) or buffered 1 M urea solution.
(5) Pass three column volumes of AES (see Table 50.1) through the column and collect the eluate. If too much pool remains bound to the matrix, the column can be incubated with AES for a period of time, instead of simply passing it, to increase the elution efficiency. Alternatively, AES with at least 10× higher concentrated target can be used.
(6) Add 10 µg yeast RNA as carrier and extract the eluate with phenol/chloroform.
(7) Add 0.1 volumes 3 M sodium acetate, pH 5.2, and one volume ice-cold isopropanol. Incubate at −20 °C for 30 min and collect the RNA by centrifugation. Wash the pellet with 70% ethanol before starting the enzymatic amplification steps.

Tab. 50.3. Exemplary selection: target in solution.

Round	n_{Target}; c_{Target} (nmol); (μM)	Reaction volume (μl)	n_{RNA}; c_{RNA} (nmol); (μM)	Ratio target:RNA	RNA bound with target (%)	RNA bound without target (%)
1	8; 50	160	4; 25	2	0.9	1.3
2	4; 25	160	2; 12.5	2	1.1	1.1
3	4; 10	400	2; 5	2	1.4	0.8
4	2; 10	200	1; 5	2	4	0.5
5	1; 1	1000	1; 1	1	2	0.5
6	0.5; 0.25	2000	1; 0.5	0.5	3	0.7
7	0.25; 0.25	1000	0.5; 0.5	0.5	4	0.4
8	0.1; 0.1	1000	0.5; 0.5	0.2	2	0.6
9	0.1; 0.1	1000	0.5; 0.5	0.2	9	0.6

Protocol C: Biotinylated target/denaturing elution

(1) Denature and re-fold the RNA pool as described in Protocol A, Steps 3 to 4. Beginning with round 3, the pool is pre-selected against the pure matrix without target before every selection step (Protocol A); as soon as the pool is pre- or counter-selected, the denaturing/re-folding procedure will only be carried out prior to the pre- or counter-selection step. Furthermore, a binding reaction without target as background control is introduced to the selection scheme from round 3 onwards. Determination of the ratio of RNA binding in the target reaction versus the background control (i.e. the signal/control ratio) is helpful for monitoring the progress during the selection and adjustment of the stringency of the process.

(2) Combine the biotinylated target and the RNA pool in a defined volume of SB to reach appropriate concentrations for pool and target. See Table 50.3 for an example of a typical selection course.

(3) Incubate the binding reaction for 2–16 h at the desired temperature.

(4) Wash and equilibrate streptavidin agarose beads with ten matrix volumes of SB.

(5) Add the pre-equilibrated streptavidin agarose beads to the binding reaction. For efficient immobilization of RNA–target complexes use 3 times the amount of matrix that is necessary to completely immobilize the amount of target present in the binding reaction.

(6) Incubate for 30 min with agitation to keep the beads in suspension for efficient capture of the RNA–target complexes.

(7) Briefly centrifuge the reaction tube to pellet the beads.

(8) Withdraw supernatant and wash the beads by resuspension in appropriate volumes of SB until around 1% of RNA remains bound to the matrix. Yields of 1% or less can be expected in the initial rounds; more RNA will bind as soon as target binding sequences become enriched during the course of the selection (see Table 50.3).

(9) Elute the bound RNA by shaking the beads in five volumes of DES (see Table 50.1) for 10 min at 37 °C in the presence of 10 µg yeast total RNA.
(10) Collect the RNA and repeat the elution at 55 °C.
(11) Extract the combined eluates with phenol/chloroform.
(12) Add 0.1 volumes 3 M sodium acetate, pH 5.2, and 1 volume ice-cold isopropanol. Incubate at −20 °C for 30 min and collect the RNA by centrifugation. Wash the pellet with 70% ethanol before starting the enzymatic amplification steps. The use of isopropanol is imperative here, as guanidinium thiocyanate will co-precipitate when employing ethanol, which can inhibit the following enzymatic reactions.

50.6
Binding Assays

Methods for the determination of the binding characteristics are important for tracking the selection progress, as well as a detailed analysis of individual binders obtained by cloning the enriched pool and subsequent sequencing. Most conveniently, cloning and sequencing services are provided commercially. For detailed information on cloning procedures, the researcher is referred to Sambrook et al. [17].

In the following, three procedures for the dissociation constant determination of RNA–small molecule complexes are described in detail.

50.6.1
Equilibrium Dialysis

Equilibrium dialysis is a specific application of dialysis, used to study small molecule–macromolecule interactions [18, 19]. The availability of a labeled ligand that is small enough to be dialyzed through a membrane, while the RNA aptamer is excluded, is the prerequisite for the determination of a dissociation constant by equilibrium dialysis. The amount of ligand bound to the macromolecule of interest is then determined by dialyzing the free ligand through the membrane, while the macromolecules as well as macromolecule–ligand complexes are retained on their side of the membrane. It is essential that the concentration of the labeled binding partner is much lower than the RNA concentration, to ensure that the free RNA concentration is not significantly affected by ligand binding: [RNA] ≫ [labeled ligand]. An equilibrium dialysis apparatus consists of a pair of chambers with volumes of 50 µl or more, separated by a dialysis membrane with the required molecular weight cut-off. When working with aptamers with a length of more than 50 nt and ligands with less than 2 kDa, a 6- to 8-kDa cut-off membrane is a good choice.

Protocol

(1) Denature the RNA in the binding buffer without Mg^{2+}/Ca^{2+} and without Triton X-100 for 1–5 min at 70–95 °C and place on ice. The binding buffer used

for the experiments should contain at least 100 mM salts to compensate for the Gibbs-Donnan effect which can perturb the equilibrium.

(2) Bring the pool to binding buffer conditions with 10 × Mg^{2+}/Ca^{2+} and 0.5% Triton X-100 and re-fold for at least 5 min at the selection temperature.

(3) Prepare a dilution series of the RNA in binding buffer. The range of the dilution series depends on the expected dissociation constant of the examined RNA; for low micromolar binding, 0.05–50 µM represent a suitable range.

(4) Prepare a solution of 1–50 nM of radioactively labeled ligand in binding buffer. The necessary ligand concentration is dependent on the specific radioactivity; choose the ligand concentration as low as possible.

(5) Set up dialysis units, load RNA dilution series in one chamber and the labeled ligand in the other chamber. Include a non-RNA control to correct for background (= small, non-diffusing fraction of labeled ligand).

(6) Incubate the dialysis units for 24 h at the desired temperature. Rotating the chambers at around 10 r.p.m. shortens the time required to reach equilibrium

(7) Measure radioactivity in the ligand (r_{LC}) and RNA (r_{RC}) chamber of each dialysis unit by scintillation counting.

The fraction of bound ligand is calculated using $F = (r_{RC} - r_{LC})/(r_{RC} + r_{LC})$. Assuming that (1) the RNA–ligand dissociation constant K_D is identical for a labeled and an unlabeled ligand, and (2) complex stoichiometry is 1:1, the generated data can be fitted by a non-linear least squares regression analysis to a standard binding equation to obtain the K_D, with software such as GraFit (Erithacus Software Ltd) [20].

50.6.2
Equilibrium Filtration Analysis

Equilibrium filtration analysis is a rapid alternative to equilibrium dialysis and takes advantage of the same physical phenomenon [21, 22]. It is even more convenient, as disposable, commercially available ultrafiltration devices can be used. As for equilibrium dialysis, the concentration of the labeled binding partner must be much lower than the RNA concentration. Appropriate filtration devices are Microcon YM-10 ultrafiltration units (Millipore) with a nominal molecular weight cut-off of 10 kDa.

Protocol

(1) Denature the RNA in the binding buffer without Mg^{2+}/Ca^{2+} and without Triton X-100 for 1–5 min at 70–95 °C and place on ice.

(2) Bring the pool to binding buffer conditions with 10 × Mg^{2+}/Ca^{2+} and 0.5% Triton X-100 and re-fold for at least 5 min at the selection temperature.

(3) Make up a dilution series of the RNA in binding buffer. The range of the dilution series depends on the expected dissociation constant of the examined RNA; for low micromolar binding, 0.05–50 µM represent a suitable range.

(4) Mix 225 µl of the RNA dilutions, including a non-RNA control, with 25 µl of radiolabeled ligand in binding buffer. Final concentration of ligand is approximately 1–50 nM.
(5) Equilibrate for 2 h at the desired temperature.
(6) Load 200 µl into a Microcon YM-10 spin filter.
(7) Centrifuge at 13 000 g for 30 s to saturate the membranes.
(8) Fit filter unit with fresh collection tube, discard old tube.
(9) Centrifuge at 13 000 g for 90 s. In order to skew the aptamer and ligand–aptamer complex concentration in the retentate not more than necessary, the filter units must not be centrifuged longer than required to obtain enough filtrate for determination of radioactivity (around 25–50 µl).
(10) Determine the radioactivity in 25-µl aliquots from the top (retentate; r_T) and bottom (filtrate; r_B).

Only unbound ligand molecules pass through the membrane, while ligand–aptamer complexes, as well as free aptamer molecules, are retained. Hence, bottom counts correspond to the free ligand concentration, while the top counts represent total (bound plus free) ligand concentration. The fraction of bound ligand is calculated using $F = (r_T - r_B)/(r_T + r_B)$. A K_D can be calculated from the generated data under the same assumptions and with the same software as described in Section 50.6.1.

50.6.3
Isocratic Competitive Affinity Chromatography

The method of isocratic affinity chromatography allows for assessment of dissociation constants based on observation of the interaction of an aptamer with an affinity column in the presence or absence of free ligand. The aptamer which is to be characterized is loaded onto the affinity column in the absence and presence of free ligand, followed by elution of the aptamer with pure binding buffer or with binding buffer containing free ligand, respectively (Fig. 50.4). From the volumes of buffer with and without free ligand necessary for elution of the aptamer, the dissociation constant of the aptamer–free ligand complex can be calculated [10, 23, 24].

Protocol

(1) Denature and re-fold two 1–10 pmol aliquots ^{32}P-labeled aptamer and a similar amount of unselected, ^{32}P-labeled pool RNA in the binding buffer without Mg^{2+}/Ca^{2+} and without Triton X-100 for 1–5 min at 70–95 °C and place on ice.
(2) Bring the three solutions to binding buffer conditions with $10 \times Mg^{2+}/Ca^{2+}$ and 0.5% Triton X-100 and re-fold for at least 5 min at the selection temperature.
(3) Add ligand for elution in appropriate concentration to the first aptamer aliquot

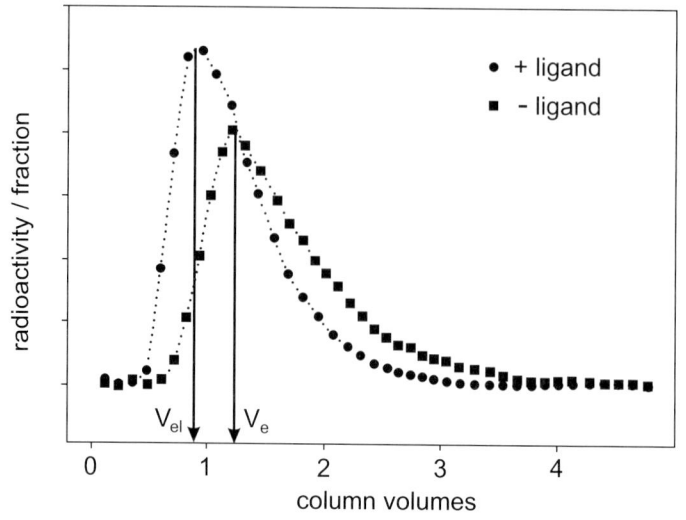

Fig. 50.4. Isocratic competitive affinity chromatography. Schematic exemplary elution profile of labeled aptamer from an affinity column. V_{el}, median elution volume in the presence of free ligand; V_e, median elution profile in the absence of free ligand.

and an identical volume of binding buffer to the second aptamer aliquot and the pool RNA. The ligand concentration should be within plus/minus one order of magnitude around the expected K_D.
(4) Incubate at room temperature for 2 h to obtain binding equilibrium before loading onto an affinity column equilibrated with binding buffer.
(5) Apply the sample containing ligand and aptamer to the column and elute the RNA isocratically with several column volumes of binding buffer containing the ligand in the chosen concentration.
(6) Collect the flow-through in fractions and measure radioactivity in the fractions to determine the elution volume (V_{el}).
(7) Equilibrate the column and repeat the procedure with the second aptamer aliquot (without ligand). Elute isocratically with pure binding buffer (V_e).
(8) Equilibrate the column and repeat the procedure with the unselected pool RNA. Elute isocratically with pure binding buffer (V_n).

The dissociation constant K_D can be calculated as follows:

Affinity for immobilized ligand $K_D = L_c(V_n/V_e - V_n)$
Affinity for ligand in solution $K_D = L[(V_{el} - V_n)/(V_e - V_{el})]$

With:
L_c = the concentration of immobilized affinity ligand within the column bed
L = the concentration of free affinity ligand used to elute the RNA

V_{el} = the median elution volume of RNA eluted in the continuous presence of free ligand (Fig. 50.4)

V_e = the median elution volume measured in the absence of free ligand in the buffer (Fig. 50.4)

V_n = the volume at which a random RNA population of similar molecular size, having no column interaction, would elute (void volume)

References

1 A. D. Ellington, J. W. Szostak, *Nature* **1990**, *346*, 818–822.
2 C. Tuerk, L. Gold, *Science* **1990**, *249*, 505–510.
3 I. Majerfeld, M. Yarus, *Struct. Biol.* **1994**, *1*, 287–292.
4 I. Majerfeld, M. Yarus, *RNA* **1998**, *4*, 471–478.
5 S. T. Wallace, R. Schroeder, *RNA* **1998**, *4*, 112–123.
6 D. H. Burke, D. C. Hoffman, A. Brown, M. Hansen, A. Pardi, L. Gold, *Chem. Biol.* **1997**, *4*, 833–843.
7 S. M. Rink, J.-C. Shen, L. A. Loeb, *Proc. Natl. Acad. Sci. USA* **1998**, *95*, 11619–11624.
8 M. Koizumi, R. R. Breaker, *Biochemistry* **2000**, *39*, 8983–8992.
9 L. A. Holeman, S. L. Robinson, J. W. Szostak, C. Wilson, *Fold. Des.* **1998**, *3*, 423–431.
10 G. J. Connell, M. Illangesekare, M. Yarus, *Biochemistry* **1993**, *32*, 5497–5502.
11 M. Sassanfar, J. W. Szostak, *Nature* **1993**, *364*, 550–553.
12 D. H. Burke, L. Gold, *Nucleic Acids Res.* **1997**, *25*, 2020–2024.
13 J. Tang, R. R. Breaker, *Chem. Biol.* **1997**, *4*, 453–459.
14 B. Wlotzka, S. Leva, B. Eschgfäller, J. Burmeister, F. Kleinjung, C. Kaduk, P. Muhn, H. Hess-Stumpp, S. Klussmann, *Proc. Natl. Acad. Sci. USA* **2002**, *99*, 8898–8902.
15 J. Rizo, L. M. Gierasch, *Annu. Rev. Biochem.* **1992**, *61*, 387–418.
16 S. Leva, A. Lichte, J. Burmeister, P. Muhn, B. Jahnke, D. Fesser, J. Erfurth, P. Burgstaller, S. Klussmann, *Chem. Biol.* **2002**, *9*, 351–359.
17 J. Sambrook, E. F. Fritsch, T. Maniatis, *Molecular Cloning: A Laboratory Manual*, Cold Spring Harbor Laboratory Press, Cold Spring Harbor, NY, **1989**.
18 K. P. Williams, X.-H. Liu, T. N. M. Schumacher, H. Y. Lin, D. A. Ausiello, P. S. Kim, D. P. Bartel, *Proc. Natl. Acad. Sci. USA* **1997**, *94*, 11285–11290.
19 D. Nieuwlandt, M. Wecker, L. Gold, *Biochemistry* **1995**, *34*, 5651–5659.
20 K. A. Connors, *Binding Constants*, Wiley, New York, **1987**.
21 R. D. Jenison, S. C. Gill, A. Pardi, B. Polisky, *Science* **1994**, *263*, 1425–1429.
22 J. H. Davis, J. W. Szostak, *Proc. Natl. Acad. Sci. USA* **2002**, *99*, 11616–11621.
23 D. Scarabino, A. Crisari, S. Lorenzini, K. Williams, G. P. Tocchini-Valentini, *EMBO J.* **1999**, *18*, 4571–4578.
24 J. Ciesiolka, J. Gorski, M. Yarus, *RNA* **1995**, *1*, 538–550.

51
SELEX Strategies to Identify Antisense and Protein Target Sites in RNA or Heterogeneous Nuclear Ribonucleoprotein Complexes

Martin Lützelberger, Martin R. Jakobsen and Jørgen Kjems

51.1
Introduction

The recognition of RNA elements in the complex environment of the cell is a key issue, both for studying basic molecular biology of the cell and for the development of therapeutic strategies that will interfere with gene expression. A cornerstone in this line of research is the SELEX (Systematic Evolution of Ligands by EXponential enrichment) procedure where ribonucleic acids that bind tightly to a ligand of interest are identified through successive rounds of binding, fractionation and amplification of an RNA library. In SELEX, as originally developed, the library consists of 10^{14}–10^{15} random sequences [1]. An extension of this method, named genomic SELEX, where natural DNAs are used as input sequences, was originally invented by Singer et al. [2]. Both methods involve a PCR amplification step which requires that the amplified sequences are flanked by fixed sequence primer annealing sites. A T7 RNA polymerase promoter is included in one of the flanking sequences in order to produce RNA by *in vitro* T7 transcription. The protocol described here resembles the genomic SELEX protocol described by Singer et al. [2], but with the notable difference that the libraries produced here contain RNAs with a fixed size of 20 nt.

The RNA library may contain sequences derived from whole genomic DNA, cDNA or smaller genetic entities such as plasmids or viruses. In addition to the library construction, we include protocols for two lines of its application that we envision are the most important:

(1) The characterization of accessible target sites for antisense oligonucleotide annealing in highly structured RNA or heterogeneous nuclear ribonucleoprotein (hnRNP) complexes.
(2) The characterization of binding sites for a specific protein within RNA derived from a single gene or complete genomes.

However, the RNA library generated by this approach may directly or in a modified form have many other potential applications.

51.1.1
Applications for Antisense

The application of various types of antisense technologies, such as antisense oligonucleotides, ribozymes, DNAzymes and RNAi may be severely inhibited by RNA structure and RNA-binding proteins in the cell. In the absence of any detailed structural information on the target mRNA, antisense design has traditionally been based on the principle of trial and error. Finding the best target can therefore be a cumbersome and expensive exercise. Several theoretical and practical approaches to determine accessible regions in RNA have been described [reviewed in 3, 4]. The most popular approach involves the use of chemicals and enzymes reactive to RNA to elucidate the exposed regions (described in Chapter 10), but this method has proven inadequate for rational antisense design. Approaches to measure the susceptibility to oligonucleotide annealing include the mapping of hybrids formed between the mRNA and random DNA oligonucleotides using RNase H [5–7] or primer extension [8]. Alternatively, a method has been described where labeled mRNA is annealed to an immobilized array of a complete set of antisense oligonucleotides, but this is costly and not practical if multiple mRNAs are going to be investigated [9]. In the protocol below we describe the selection of 20-nt long sense or antisense ribonucleotides from a library derived from plasmids or whole genomes that are able to bind efficiently to a particular target RNA.

The advantages of this approach are that the length of the RNAs in the library is similar to that of therapeutically relevant antisense oligonucleotides and that the selection can be performed under any given condition. Moreover, the use of natural sequences as input DNA lowers the complexity of libraries significantly, enabling the selection to occur in a few rounds and in relatively small volumes.

51.1.2
Selecting Protein-binding Sites

The realization that the expression of many mammalian genes is highly regulated at the level of RNA processing by a large number of proteins and hnRNP complexes has called for improved high throughput methods for characterization of the RNA sequences involved. The genomic SELEX method previously described [2] is useful for this purpose [10], but the larger and variable size of the RNA fragments (100–300 nt) in these libraries makes it harder to interpret the result. The 20mer sequences in our library are sufficient to accommodate the binding site for most proteins that recognize a primary RNA sequence motif.

51.2
Construction of the Library

The key steps in generation of the library, outlined in Fig. 51.1, are as follows:

(1) Generation of 200- to 1000-bp fragments by random degradation of the input DNA, which may be either plasmid or genomic DNA.

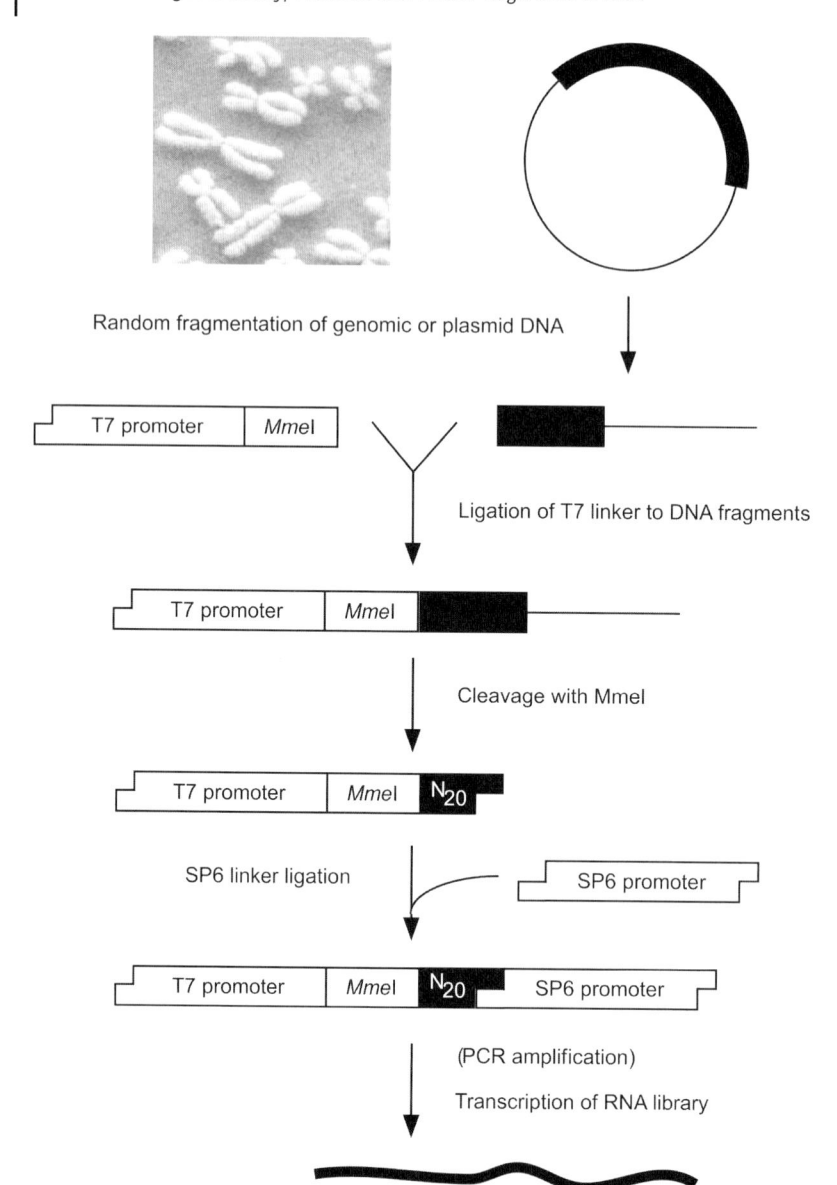

Fig. 51.1. Flowchart diagram showing the different steps in the construction of the library. Fragments from genomic or plasmid DNA are generated using ultrasound or DNase I treatment. The fragments are ligated to a T7 linker, consisting of the T7 promoter sequence and the recognition site for the restriction enzyme *Mme*I. After digestion of the ligated products with *Mme*I, which cuts 20/18 bp downstream of its recognition site, fragments are ligated to an SP6 linker containing the SP6 promoter sequence. The library can be amplified by PCR or directly used for T7 transcription, producing transcripts of 66 nt.

(2) Ligation of an upstream linker to the DNA fragments, containing a T7 promoter sequence and an *Mme*I site.
(3) Cleavage by *Mme*I restriction enzyme, leaving 20 nt of genomic DNA attached to the T7 linker.
(4) Ligation of a downstream linker, containing an SP6 promoter sequence.
(5) *In vitro* transcription, using the ligated product as template.

51.2.1
Generation of Random DNA Fragments from Genomic or Plasmid DNA

To break nucleotide bonds, the DNA can be treated with ultrasound or DNase I. The advantage of ultrasound is that it breaks the nucleotide bonds at non-specific positions, whereas DNase I, which binds to the minor groove, has a slight preference to cleave within AT-rich sequences [11], which might influence the random distribution of the fragments. Thus, we recommend sonication as the method of choice for the generation of DNA fragments. However, DNase I treatment may sometimes be preferred for small plasmid DNA, which is usually more difficult to break by ultrasound treatment. If a DNase I reaction buffer containing Mn^{2+} is used, blunt ends are produced and subsequent Klenow polymerase treatment is unnecessary.

If the library is constructed from a DNA fragment inserted into a plasmid, the amount of starting material is usually not critical to generate enough molecules representing all positions of the insert. However, taking into account that some material will be lost during enzymatic reactions, phenol extraction and precipitation, about 10–20 µg of plasmid DNA is a good starting point.

If the library is made from cDNA or genomic DNA, the complete coverage of the genome must be ensured. Hence, the pool of molecules should contain a sufficient set of overlapping subfragments, so that each nucleotide in the genome is represented by more than one molecule. Thus, for a library made from human genomic DNA, at least 20 µg should be used as starting material: Given that the human genome has about 3×10^9 bp, no less than 3×10^9 fragments should be generated to have each nucleotide of the genome represented at least once at a 5′-terminal position. If the fragments produced by sonication have an average size of 600 bp, 10 µg of them would equal 26 pmol or 1.6×10^{13} molecules, which is a 10 000-fold excess of ends over the total number of base pairs in the human genome. We suggest using 50 µg of genomic DNA for the construction of a human SELEX library.

51.2.2
Preparing RNA Libraries from Plasmid, cDNA or Genomic DNA

When a sufficient amount of DNA fragments has been produced, they are ligated to an upstream linker that contains a T7 promoter sequence and a recognition site for the restriction enzyme *Mme*I (Fig. 51.1). Since the ligation efficiency of blunt end fragments largely depends on their relative concentrations, we suggest to adjust the ratio between linker DNA and genomic fragments in such a way that for each linker molecule at least two genomic DNA fragment are present. This will "drive" the reaction and decrease the formation of linker dimers.

MmeI belongs to the class-II restriction endonucleases and has the unique feature to cleave DNA 20/18 nt downstream of its non-palindromic recognition sequence (TCCRAC), thereby generating a 2 nt 3′ NN overhang [12, 13]. The fragments cleaved with MmeI are gel-purified and subsequently ligated to a downstream linker containing an SP6 promoter sequence. The SP6 promoter is included to be able to produce RNA from the antisense strand. However, for the applications described here, it is not used as a promoter.

Both the T7 and SP6 linkers have a 1-nt 3′ overhang (see Protocols) to avoid the formation of linker concatemers, which enhances the ligation efficiency and prevents the inclusion of linker sequences into the library.

Depending on its concentration, the purified ligation product may be directly used for T7 transcription. If the library is amplified by PCR, conditions must be adjusted carefully, that is the amplification should be done with as few cycles as possible to avoid creating a bias. The inclusion of enhancing agents such as dimethyl sulfoxide (DMSO), which facilitates strand separation, might be advantageous for the amplification of GC-rich sequences. In Fig. 51.2 (lane 5), an example of a PCR-amplified library is given. It is usually not necessary to purify the PCR products prior to T7 transcription, provided that products of discrete size have been obtained.

With Protocol 3 about 20–40 pmol RNA can be produced from a single 50-µl PCR reaction. If a higher yield is required, the reaction may be scaled up accordingly. It is convenient to include [α-^{32}P]UTP in the transcription reaction to quantitate the yield and to trace the RNA in the subsequent selections.

51.3
Identification of Optimal Antisense Annealing Sites in RNAs

To characterize sites in RNA or RNP complexes that are susceptible to RNA annealing, an N_{20} RNA library is prepared starting from a plasmid containing the sequence encoding the target RNA as input material. The library is subsequently mixed with the radioactively labeled target RNA or RNP complex under physiological conditions to maintain the native structure in the RNA or RNP target. After a short annealing step, the target RNA or RNP complex is separated on a native gel from the unbound fraction of the library and eluted from the gel. The bound oligoribonucleotides are selectively amplified by RT-PCR using primers complementary to the flanking T7 and SP6 linker sequences.

Since a T7 RNA polymerase promoter sequence is included in the upstream linker, the PCR products can be used directly as templates for the next round of transcription. After 3–5 successive rounds of selection, the RT-PCR products are cloned into a plasmid and 50–100 inserts are sequenced. Using a plasmid to generate the input RNA, in our hands 3 rounds of selection usually are sufficient.

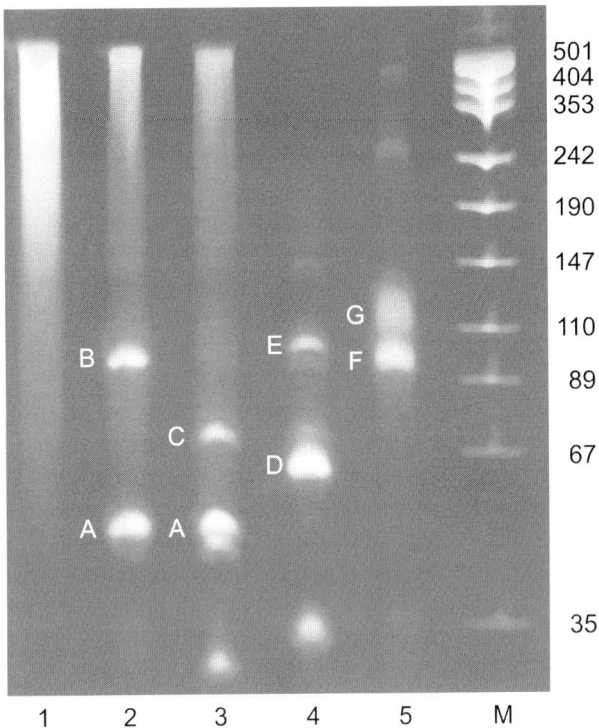

Fig. 51.2. An example of a gel showing products of the individual steps in the construction of a human genome-wide library of 20mer RNAs. Small aliquots of each reaction were taken and separated on a 6% acrylamide/bisacrylamide gel (19:1 in 1 × TBE). Lane 1, sonicated human genomic DNA; lane 2, products of T7 linker ligation; lane 3, *Mme*I digestion after T7 linker ligation; lane 4, products of SP6 linker ligation; lane 5, PCR-amplified library; lane M, pUC18 vector digested with *Hpa*II. The bands marked with capital letters correspond to: A, T7 linker; B, T7 linker dimer; C, *Mme*I-digested fragments; D, SP6 linker dimer; E, final ligation product (library); F, PCR-amplified library; G, product formed by denaturing and re-annealing of single-stranded library molecules that differ in their variable parts and therefore form a bulge. The T7 and SP6 linker sequences of these molecules are double-stranded, which is sufficient for *in vitro* transcription.

Using highly structured RNA, for instance the HIV-1 5′-untranslated region, we find that only 1–2% of the potential antisense oligonucleotides are selected when using the conditions described below [14]. We have also noticed that different sequences in the final pool generally are represented in stoichiometric amounts, probably reflecting that they bind the target RNA in a non-competitive fashion. To select for oligonucleotides with fast binding kinetics, we generally shorten the library-annealing step from 30 min in the initial selection to only 5 min in the final round.

By varying the setup it is possible to select for oligonucleotides that bind to RNP complexes, or inhibit protein binding or dimerization of RNA molecules. In

Protocol 4, the selection of oligonucleotides that bind structured RNAs of 300–1000 nt is described.

51.4
Identification of Natural RNA Substrates for Proteins and Other Ligands

The selection of optimal RNA binding sites for a particular ligand (e.g. protein) can be performed by mixing the library, prepared from plasmid DNA, a cDNA library or genomic DNA, with purified protein and separating the complexes from the non-bound RNA by filter binding, immunoprecipitation, protein tag-purification or native gel electrophoresis. Protocol 5 uses the latter approach, which allows us to distinguish between single protein and multimeric complexes. Native gel electrophoresis also has the advantage that an enrichment of RNA molecules with affinity to the purification matrix can be avoided.

In contrast to SELEX procedures using random libraries, only 3–5 rounds of selection are sufficient to obtain high-affinity RNA-binding sites. For natural RNA binding proteins, the binding buffer and stringency of the selection conditions must be optimized. Addition of competitor RNA during the initial round of selection is usually not required, since the majority of library molecules do not bind to the ligand. However, addition of 10-fold excess of competitor RNA in the last selection rounds might improve the stringency of the selection. Reamplification of the RNA pool follows the same scheme as for antisense RNA selection.

It is important to realize the limitations of using an N_{20} library for the selection of RNA-binding sites for a ligand. It is only applicable to ligands that recognize less than 20-nt primary sequences. For ligands that recognize more complex RNA secondary and tertiary structural elements it is better to use the genomic SELEX protocol described by Singer et al. [2].

51.5
Cloning, Sequencing and Validating the Selected Inserts

To verify the library, we recommend sequencing of at least 10 randomly picked clones from the unselected pool. It should be verified whether the inserts have the correct length and whether they originate from different locations of the starting material. Furthermore, the nucleotide frequencies of the 20mers should be evaluated to rule out that a bias has been created by PCR amplification of the library.

Cloning of the RNA inserts can be performed using the *Pst*I and *Eco*RI sites in the flanking sequences (see Materials). Alternatively, one can use the TA cloning procedure (e.g. with the TOPO TA cloning Kit from Invitrogen) to clone the PCR fragments directly. If a large number of 20mers has to be sequenced, it is recommended to blunt the ends of RT-PCR products using Klenow polymerase, ligate them into concatemers and ligate these into a plasmid. This approach will cut down the number of sequencing reactions up to 10-fold.

To ease the sequence analysis we recommend using the EMBOSS suite [15], which is freely available for UNIX platforms from http://emboss.sourceforge.net. It contains tools, such as the vectorstrip program, to extract the 20mers from raw sequence data, which is particularly useful for the analysis of a large number of sequences. For the conversion of the sequence information into aligned sequences of the selected inserts we recommend the ClustalX toolkit, available from ftp://ftp.ebi.ac.uk/pub/software/unix/clustalx.

51.6 Troubleshooting

51.6.1 Sonication of Plasmid DNA does not Yield Shorter Fragments

The sonication of plasmid DNA usually requires more power than chromosomal DNA of high molecular weight, since plasmids are small and covalently closed circular molecules. If you do not obtain the desired fragment length with ultrasound, use DNase I treatment instead.

51.6.2 Inefficient Ligation

For efficient ligation of the SP6 linker to the *Mme*I fragments, it is crucial that it has a truly random NN overhang. Most custom-made oligonucleotides are synthesized by solid-phase chemistry in the 3′–5′ direction, starting with a column containing the 3′ nucleotide temporarily immobilized on a solid support. Based on our experience, some companies have difficulties to produce oligonucleotides with a true "N" at the 3′ end. To avoid problems, make sure that your distributor is able to synthesize such oligonucleotides, otherwise order a set of four primers, each ending with a different nucleotide, and mix them.

The NEB 10 × T4 ligation buffer already contains 10 mM ATP. Addition of ATP is usually not necessary, but recommended if the 10 × reaction buffer went through a couple of freeze–thaw cycles. To avoid such problems, store the buffer in small aliquots.

The addition of crowding agents, such as polyethylene glycol, to increase the effective concentration of the reactants might also improve the ligation efficiency, especially if the amount of DNA fragments is limited.

51.6.3 Inefficient *Mme*I Digestion

We find that *Mme*I cleavage is inefficient if the enzyme is used in excessive amounts. Do not use more units than described in Protocol 2. The enzymatic activity of *Mme*I is completely inhibited in the presence of 100 mM KCl [12]. Therefore it is important to precipitate the DNA and to wash the pellet with 70% ethanol to get rid of the salt after the ligation step.

51.6.4
The Amplification of the Unselected Library is Inefficient

It is recommended to use only the gel-purified ligated product as a template. We observed that the carry-over of non-ligated SP6 linker inhibits the amplification of the unselected library, since the antisense oligonucleotide of the SP6 linker is able to anneal to any library molecule containing an SP6 promoter sequence. However, due to its "NN" end most of the annealed molecules will have an unpaired 3′ end which cannot be extended by *Taq* polymerase.

51.6.5
The Library Appears to be Non-random in the Unselected Pool

It is important not to cleave DNA with any restriction enzymes before sonication. Otherwise, sequences flanking the restriction sites will become over-represented. If linker sequences appear in the variable part of the RNA, the ratio of linker to DNA fragments should be lowered.

51.6.6
The Selected RNAs do not Bind Native Protein

There is always the risk in SELEX that RNAs are selected which bind to the matrix or purification tag. Although this seems to be a minor problem when using RNA libraries derived from genomic sequences, it is preferable to alternate between different tags or matrices. For instance, when using a protein to select RNA we recommend using two different tags (e.g. His- and GST-tagged proteins) at alternating selection steps.

51.7
Protocols

Oligonucleotides

- T7 linker, containing a *Pst*I site, T7 promoter sequence (underlined) and an *Mme*I site (small letters): sense: 5′-AGCCTGCAG<u>TAATACGACTCACTATAG</u>G-GATCGCTTAGtccgac-3′; antisense: 3′-GTCGGACGTC<u>ATTATGCTGAGTGA-TATC</u>CCTAGCGAATCaggctg-5′.
- SP6 linker, containing the SP6 promoter sequence (underlined) and an *Eco*RI site (small letters): sense: 5′-CTTC<u>TATAGTGTCACCTAAAT</u>gaattcG-3′; antisense: 3′-NNGAAG<u>ATATCACAGTGGATTTA</u>cttaag-5′.
- SP6 primer: 5′-Cgaattc<u>ATTTAGGTGACACTATAG</u>-3′.

Reagents

- DNA polymerase I (Klenow fragment), 5 U/µl, New England Biolabs (NEB).
- DNase I, RNase-free, 2 U/µl, Ambion.
- T4 polynucleotide kinase (T4 PNK), 10 U/µl, NEB.
- T4 DNA ligase, 6 U/µl, NEB.
- *Mme*I restriction endonuclease, 2 U/µl, NEB.
- *Taq* DNA polymerase, 5 U/µl, Invitrogen.
- 3 mM *S*-adenosylmethionine hydrochloride, NEB.
- T7 RNA polymerase, 50 U/µl, NEB.
- AMV reverse transcriptase, 10 U/µl, Promega.
- RNasin, RNase inhibitor, 30 U/µl, Promega.
- 2 × RNA binding buffer: 50 mM sodium cacodylate, pH 7.5, 250 mM KCl, 5 mM $MgCl_2$.
- 2 × Native loading buffer: 10% glycerol, 0.025% xylene cyanol, 0.025% bromophenol blue, 10 mM $MgCl_2$.
- Denaturing RNA loading buffer: 50 mM Tris–HCl, pH 7.9, 10 mM EDTA, 0.025% bromophenol blue, 0.025% xylene cyanol, 80% formamide.
- 4% acrylamide/bisacrylamide (19:1), 8 M urea in 1 × TBE.
- 6% acrylamide/bisacrylamide (19:1) in 1 × TBE.
- 6% acrylamide/bisacrylamide (19:1) in 1 × TBM.
- 10 × TBM buffer: 1 M Tris–HCl, 0.9 M boric acid, 50 mM $MgCl_2$.
- Gel extraction buffer (store in the dark at room temperature): 0.75 mM ammonium acetate, 0.1% SDS, 10 mM magnesium acetate, 0.1 mM EDTA, pH 7.0.
- 2 × hnRNP A1 binding buffer: 20 mM HEPES–KOH, pH 7.6, 200 mM KCl, 4 mM $MgCl_2$, 1 mM EDTA, 2 mM DTT, 20% glycerol.
- hnRNP A1 buffer: 20 mM HEPES–KOH, pH 7.9, 150 mM KCl, 1 mM DTT, 0.2 mM EDTA, 20% glycerol.

Protocol 1: Generation of random DNA fragments from genomic or plasmid DNA

Sonication

(1) Adjust the volume of the DNA solution to at least 500 µl. For example, dilute 50 µg DNA in 500 µl H_2O.
(2) Sonicate the genomic DNA for 30 min. Chill the DNA on ice during the whole process to prevent heating of the sample. With a Branson Ultrasonic Sonifier II, Model W-250, equipped with a microtip, we are setting the output to 10 W and a duty cycle of 10%, so that the total time of ultrasound treatment is about 3 min. As a rule of thumb, it is better to sonicate for a long time with a low output than for a short time with high intensity. For sonication of plasmid DNA, increase the power to 15 W and sonicate for 10 min with a 50% duty cycle. An example of sonicated human genomic DNA is shown in Fig. 51.2, lane 1.

(3) Take 5 µl aliquots in intervals of 5 min to check the efficiency of sonication.
(4) Load the aliquots on a 6% acrylamide/bisacrylamide gel (19:1, 1 × TBE). Run the gel until the xylene cyanol (XC) migrated 2/3 of the gel length. Stain the gel with ethidium bromide.
(5) If the average fragment length is about 600 bp, ethanol-precipitate the DNA and proceed to the next step. The range of fragment sizes should be between 1 kbp and 200 bp.
(6) The sonicated DNA fragments are treated with Klenow polymerase to ensure that they have blunt ends.

Klenow reaction, 40 µl:

5 µl sonicated DNA, 50 µg
4 µl 10 mM dNTP, 2.5 mM each
4 µl 10 × reaction buffer, NEB: 100 mM Tris–HCl, pH 7.5, 50 mM $MgCl_2$, 75 mM DTT
2 µl DNA polymerase I (Klenow fragment), 10 U
25 µl H_2O

(7) Incubate at 30 °C for 20 min.
(8) Fill up with water to 100 µl and extract with 100 µl phenol/chloroform (1:1). Centrifuge briefly and transfer the supernatant into a fresh tube leaving the interphase behind.
(9) Repeat the extraction with 100 µl chloroform.
(10) Precipitate the DNA with 2.5 volumes 96% ethanol and 1/10 volume 3 M sodium acetate, pH 6.5.
(11) Centrifuge for 20 min with 10 000 g at 4 °C. Wash the DNA pellet with 70% ethanol, and dry in a vacuum concentrator. Dissolve the DNA in 20 µl H_2O.

DNase I treatment

It is strongly recommended to optimize the incubation time for the DNase I treatment. This can be achieved by taking out small aliquots at different time points and terminating the reaction by addition of 50 mM EDTA. DNase I-digested fragments in a range between 200 bp and 1 kbp nucleotides are suitable for library construction.

(1) *DNase I reaction, 100 µl:*

10 µg plasmid DNA
50 µl 2 × DNase I buffer: 250 mM Tris–HCl, pH 7.6, 2.5 mM DTT, 125 mM NaCl, 5 mM $CaCl_2$, 50 mM $MnCl_2$ (add $MnCl_2$ just before use!)
39 µl H_2O
1 µl DNase I, 2 U

(2) Incubate at 37 °C until the desired range of fragments is produced (1–30 min time course). The reaction is terminated by addition of 10 µl 50 mM EDTA, followed by phenol extraction, ethanol precipitation and gel purification as described below. Dissolve the DNA in 20 µl H_2O.

Protocol 2: Construction of the RNA library

Phosphorylating the 5′ ends of DNA fragments

(1) All DNA fragments must have a 5′ phosphate end in order to be ligated to the T7 linker.

T4 PNK reaction, 30 µl:

20 µl DNA fragments from Protocol 1
3 µl 10 × PNK buffer, NEB: 700 mM Tris–HCl, pH 7.6, 100 mM $MgCl_2$, 50 mM DTT
3 µl 10 mM ATP
2 µl H_2O
2 µl T4 PNK, 20 U

(2) Incubate at 37 °C for 1 h.
(3) Fill up with water to 100 µl and extract with 100 µl phenol/chloroform (1:1).
(4) Precipitate the DNA as described above and dissolve the DNA in 20 µl H_2O.

Preparing the double-stranded T7 linker

(1) Before the two primers are annealed, the antisense primer must be phosphorylated at its 5′ end.

T4 PNK reaction, 10 µl:

1 µl 10 × PNK buffer, NEB: 700 mM Tris–HCl, pH 7.6, 100 mM $MgCl_2$, 50 mM DTT
1 µl 10 mM ATP
2 µl (100 pmol/µl) antisense primer
5 µl H_2O
1 µl T4 PNK, 10 U

(2) Incubate for 1 h at 37 °C.
(3) Fill up with water to 100 µl and extract with 100 µl phenol/chloroform (1:1). Precipitate as described above and dissolve the DNA in 7 µl H_2O.
(4) To anneal the two T7 linker oligonucleotides, add to the phosphorylated antisense oligonucleotide:

2 µl sense oligonucleotide, 100 pmol/µl
1 µl 1 M KCl

(5) Heat the mixture to 90 °C in a heating block for 2 min. Let it cool down to room temperature for 10 min, but do not place on ice.

Ligation of the T7 linker to the DNA fragments

(1) *T4 DNA ligation, 50 µl:*

20 µl 5′-phosphorylated blunt-ended DNA fragments
10 µl double-stranded T7 linker

5 μl 10 × T4 DNA ligase buffer, NEB: 500 mM Tris–HCl, pH 7.5, 100 mM MgCl$_2$, 100 mM DTT, 10 mM ATP, 250 μg/ml BSA
12 μl H$_2$O
3 μl T4 DNA ligase, 18 U

(2) Incubate the reaction overnight at 14–16 °C.
(3) Fill up with water to 100 μl and extract with 100 μl phenol/chloroform (1:1).
(4) Precipitate as described above and dissolve the DNA in 20 μl H$_2$O.

MmeI cleavage

(1) Before starting the *Mme*I cleavage reaction, save 2 μl of the non-digested material to load it next to the preparative *Mme*I reaction products on the gel. This will help to identify the cleavage products and to verify the ligation and digestion efficiency. An example of the cleavage products generated by *Mme*I is shown in Fig. 51.2 (compare lanes 2 and 3).

*Mme*I restriction reaction, 30 μl:

0.5 μl 3 mM S-adenosylmethionine hydrochloride
3.0 μl 10 × *Mme*I reaction buffer (NEB buffer 4): 500 mM potassium acetate, 200 mM Tris acetate, pH 7.9, 100 mM magnesium acetate, 10 mM DTT
18.0 μl ligated products
7.0 μl H$_2$O
1.5 μl *Mme*I, 3 U

(2) Incubate for 1 h at 37 °C.
(3) Separate the reaction products on a 6% acrylamide/bisacrylamide gel (19:1, 1 × TBE).
(4) Cut out the band of interest and elute the fragments from the gel piece overnight in 300 μl TE buffer. For a more efficient extraction of the fragments, we strongly recommend electroelution.
(5) Extract with 300 μl phenol/chloroform (1:1) and precipitate as described above.
(6) Dissolve the DNA in 10–20 μl H$_2$O.

Preparation and ligation of the SP6 linker

(1) 5′-phosphorylate the sense primer of the SP6 linker and anneal it to the SP6 antisense primer as described for the T7 linker.
(2) Ligate the eluted *Mme*I fragments to the SP6 linker using the same ligation conditions as described above. Depending on the amount of *Mme*I fragments obtained, the reaction volume may be reduced to 25 μl or less to increase the ligation efficiency.
(3) Gel-purify the ligated products on a 6% acrylamide/bisacrylamide gel (19:1, 1 × TBE) and precipitate the DNA as described above. An example of an SP6 linker ligation is shown in Fig. 51.2, lane 4.

PCR amplification of the library

(1) *PCR reaction, 50 µl:*

   2.0 µl T7 linker, sense primer (25 pmol/µl)
   2.0 µl SP6 primer (25 pmol/µl)
   4.0 µl 10 mM dNTP mix (2.5 mM each)
   2.0 µl 50 mM $MgCl_2$
   5.0 µl 10 × *Taq* DNA polymerase reaction buffer, Invitrogen: 200 mM Tris–HCl, pH 8.4, 500 mM KCl
   5.0 µl DMSO
   5.0 µl eluted library DNA
   24.5 µl H_2O
   0.5 µl *Taq* DNA polymerase, 2.5 U

(2) Amplify the DNA with: 92 °C for 3 min; 25–30 cycles of 92 °C for 30 s, 49 °C for 30 s, 72 °C for 30 s; 72 °C for 7 min.
(3) Ethanol-precipitate the PCR product. Dissolve the DNA in 10 µl H_2O.

Protocol 3: T7 transcription of the library

(1) Mix all components of the reaction at room temperature to avoid precipitation.

   T7 transcription reaction, 25 µl

   9.0 µl H_2O
   2.5 µl 10 × T7 transcription buffer, NEB: 400 mM Tris–HCl, pH 7.9, 60 mM $MgCl_2$, 20 mM spermidine, 100 mM DTT
   1.0 µl 300 mM DTT
   0.5 µl RNasin RNase inhibitor, 15 U
   2.5 µl NTP mix (2 mM UTP, 5 mM each ATP, CTP, GTP)
   5.0 µl DNA (PCR product from Protocol 2)
   2.5 µl [α-^{32}P]UTP, 3000 Ci/mmol, 20 mCi/ml
   2.0 µl T7 RNA polymerase, 100 U

(2) Incubate for 2 h at 37 °C.
(3) Add 2 µl RNase-free DNase I (4 U) and continue the incubation for 30 min.
(4) Save 1 µl to determine the total Cerenkov counts per minute (c.p.m.) of UTP used (*cpm1*).
(5) Add 10 µl denaturing RNA loading buffer and heat the mixture for 3 min to 90 °C. Chill on ice and load the sample on a denaturing 4% acrylamide/bisacrylamide gel (19:1) with 8 M urea in 1 × TBE.
(6) When the bromophenol blue has migrated approximately two-thirds of the gel length, disassemble the gel and wrap it into plastic foil to prevent dehydration. Place tracking tape in the corners so the film can be aligned to the gel after development and expose an X-ray film.

(7) Cut out the desired band and place the gel piece into 400 µl (approximately 5 volumes) gel extraction buffer. Elute the RNA for at least 4 h under vigorous shaking at room temperature.

(8) Add 400 µl isopropanol to the supernatant and precipitate the RNA overnight at −20 °C. Spin for 30 min with 10 000 g at 4 °C. Wash the pellet with 70% ethanol and dry it in a vacuum concentrator.

(9) Dissolve the RNA in 20–30 µl water, measure the c.p.m. of 1 µl solution and calculate the total c.p.m. ($cpm2$). With the conditions above, the amount of RNA (pmol) can be calculated as follows: ($cpm2$ × pmol UTP in 25 µl transcription mixture)/($cpm1$ × number of U residues per transcript) = ($cpm2$ × 5000)/($cpm1$ × 19). Transcripts (66 nt, if PCR products are used directly; see above) include four U residues encoded by the T7 linker (including *Mme*I recognition site), five U residues on average in the variant region and 10 U residues encoded in the SP6 linker sequence, which is a total number of 19 U residues.

(10) Dilute the RNA library to the desired concentration.

Protocol 4: Identification of accessible sites in a structured RNA for optimal antisense annealing

SELEX assay

(1) Prepare 1 pmol of radioactively labeled target RNA (internally or end-labeled) and dissolve it in 10 µl of 1 × RNA binding buffer.

(2) Denature the RNA for 2 min at 85 °C and cool the reaction slowly to room temperature over 20 min. Chill the sample on ice for 5 min. Save 2 µl for native gel analysis.

(3) *RNA-binding reaction, 20 µl:*

8 µl ^{32}P-labeled target RNA
5 pmol RNA library
Adjust the volume to 20 µl using 1 × RNA binding buffer.

(4) Incubate the reaction at 37 °C for 30 min (the incubation time may be lowered to 5 min in the final round). Add 20 µl 2 × native loading buffer and separate the RNA on a native 6% acrylamide/bisacrylamide gel containing 5 mM $MgCl_2$.

(5) Disassemble the gel as described in Protocol 3 and expose X-ray film at 4 °C. Cut out the band corresponding to target RNA–antisense RNA complexes. Depending on the size of the target RNA, complexes may migrate with slightly slower mobility than the naked RNA.

(6) Elute and precipitate the RNA as described in Protocol 3 and dissolve it in 10 µl H_2O. Proceed to the RT-PCR amplification step outlined below.

Protocol 5: SELEX selection of protein-binding sites

The selection of binding sites on RNA may be performed with any RNA binding ligand. In the example below, binding sites for hnRNP A1 protein are selected.

Binding reaction

(1) *Binding reaction, 50 μl:*

   0.5 μl RNasin, RNase inhibitor, 15 U
   5 pmol RNA library
   5 pmol hnRNP A1

(2) Adjust the volume to 25.0 μl with H_2O and 2 × hnRNP A1 binding buffer to a final concentration of 1 × hnRNP A1 binding buffer. Dilute the protein in hnRNP A1 buffer, if necessary.
(3) Mix and incubate for 15 min at room temperature.
(4) Load the mixture on a native 6% acrylamide/bisacrylamide 1 × TBE gel using 0.5 × TBE as running buffer. Do not mix the samples with loading buffer, but put an aliquot of the latter into an empty well next to the samples. Run the gel at 4 °C with 300 V for 90 min until the bromophenol blue has migrated approximately two-thirds of the gel length.
(5) Disassemble the gel as described in Protocol 3 and expose an X-ray film at 4 °C. Cut out the desired band and elute the RNA under vigorous shaking in 400 μl (approximately 5 volumes) extraction buffer for at least 4 h at room temperature.
(6) Extract the supernatant with 400 μl phenol/chloroform (1:1).
(7) Precipitate the RNA as described above. Dissolve the RNA in 10 μl H_2O and proceed to the RT-PCR amplification step outlined below.

RT-PCR

(1) Add 2 pmol of SP6 primer to 2–5 μl eluted RNA. Fill up with water to 8 μl and heat the mixture in a heating block to 70 °C for 5 min. Chill on ice immediately for 5 min.
(2) Add to the annealed RNA/primer mixture:

   5.0 μl 5 × AMV reverse transcriptase reaction buffer: 250 mM Tris–HCl, pH 8.3, 250 mM KCl, 50 mM $MgCl_2$, 2.5 mM spermidine, 50 mM DTT
   2.5 μl dNTP mix (2.5 mM each)
   1.0 μl RNasin, RNase inhibitor, 30 U, Promega
   2.5 μl sodium pyrophosphate, 40 mM (pre-warmed to 42 °C)
   1.0 μl AMV reverse transcriptase, 10 U, Promega
   5.0 μl H_2O

(3) Incubate at 42 °C for 60 min.
(4) Use 2–5 μl as a template for PCR using the same conditions as for amplification of the library.
(5) The PCR fragments may either be used for a new round of transcription or for cloning and sequencing.

References

1 C. Tuerk, L. Gold, *Science* **1990**, *249*, 505–510.
2 B. S. Singer, T. Shtatland, D. Brown, L. Gold, *Nucleic Acids Res.* **1997**, *25*, 781–786.
3 S. P. Ho, D. H. Britton, Y. Bao, M. S. Scully, *Methods Enzymol.* **2000**, *314*, 168–183.
4 M. Sohail, E. M. Southern, *Adv. Drug Deliv. Rev.* **2000**, *44*, 23–34.
5 W. F. Lima, V. Brown-Driver, M. Fox, R. Hanecak, T. W. Bruice, *J. Biol. Chem.* **1997**, *272*, 626–638.
6 K. R. Birikh, Y. A. Berlin, H. Soreq, F. Eckstein, *RNA* **1997**, *3*, 429–437.
7 S. P. Ho, Y. Bao, T. Lesher, R. Malhotra, L. Y. Ma, S. J. Fluharty, R. R. Sakai, *Nat. Biotechnol.* **1998**, *16*, 59–63.
8 H. T. Allawi, F. Dong, H. S. Ip, B. P. Neri, V. I. Lyamichev, *RNA* **2001**, *7*, 314–327.
9 N. Milner, K. U. Mir, E. M. Southern, *Nat. Biotechnol.* **1997**, *15*, 537–541.
10 S. Lindtner, B. K. Felber, J. Kjems, *RNA* **2002**, *8*, 345–356.
11 J. E. Herrera, J. B. Chaires, *J. Mol. Biol.* **1994**, *236*, 405–411.
12 J. Tucholski, P. M. Skowron, A. J. Podhajska, *Gene* **1995**, *157*, 87–92.
13 J. Tucholski, J. W. Zmijewski, A. J. Podhajska, *Gene* **1998**, *223*, 293–302.
14 M. R. Jakobsen, C. K. Damgaard, E. S. Andersen, A. J. Podhajska, J. Kjems, *Nucleic Acids Res.* **2004**, *32*, e67.
15 P. Rice, I. Longden, A. Bleasby, *Trends Genet.* **2000**, *16*, 276–277.

Part V
RNAi

52
Gene Silencing Methods for Mammalian Cells: Application of Synthetic Short Interfering RNAs

Matthias John, Anke Geick, Philipp Hadwiger,
Hans-Peter Vornlocher and Olaf Heidenreich

52.1
Introduction

The recent discovery of gene-specific silencing in mammalian cells mediated by around 20–25-nt double-stranded RNA, better known as short interfering RNA (siRNA), has become the basis for the development of new tools for studying gene and protein function *in vitro* and *in vivo* [1]. Once introduced into cells, an endogenous and highly conserved multienzyme machinery recognizes both the siRNA and its cognate target sequence on either a mRNA or viral RNA. Subsequently, those targets are inactivated by a cleavage event with subsequent nucleolytic degradation of the corresponding RNA, leading to a reduced expression of the target gene. This process is called RNA interference (RNAi) [2, 3].

Frequently, siRNA-based protocols are applied for specific inhibition of gene expression in cell culture studies. Two aspects of siRNA-based methods are critical for obtaining high silencing efficiencies. First, selecting the right target sequence within the gene of interest is important, but definitive rules for identifying the best sequence have yet to be determined. Second, the efficiency of a cell transfection procedure for a particular cell line has dramatic effects on the observed degree of target gene depletion. Several commercial suppliers have already tested large numbers of siRNA sequences against many mammalian target genes. Those validated and ready-to-use siRNAs allow scientists to skip the process of target site optimization. RNAi works in a wide variety of cell types. For most cell lines, protocols and transfection kits for introducing siRNA are also available.

The ease of use and the high success rate of siRNA in gene silencing experiments make this technology very attractive. Our goal in this chapter is to provide some practical advice and additional information for those that are new in the field of mammalian RNAi.

52.2
Background Information

Applying siRNA of 20–25 nt in length leads to a strong reduction of either cellular or viral gene expression in mammalian cells [1, 4, 5]. The combined studies in plants, *Drosophila melanogaster*, *Caenorhabditis elegans* and mammalian systems have resulted in the following model of RNAi in mammalian systems [2, 3, 6] (Fig. 52.1).

Extended stretches of dsRNA are recognized by dicer, an RNase III-like nuclease. Dicer cleaves the dsRNA into smaller pieces of an average length of around 22 nt. These short dsRNAs are then bound by a multiprotein complex, RISC (RNA-Induced Silencing Complex). RISC is able to find homologous sequences among different RNA species, most importantly mRNA, and possesses helicase and nuclease activity. RISC uses the siRNA as a guide for the identification of complementarity between siRNA and target RNA. The identified target RNA is a substrate for endonucleolytic cleavage leading to its further degradation by exo- and endonucleases. This pathway may represent a cellular defense mechanism against viral infection and transposon replication.

The discovery of micro-RNAs (miRNA) suggests that RNAi is also involved in the complex regulation of developmental genes. In mammals, miRNA genes are conserved and expressed in a tissue-specific manner [7–9]. The miRNA precursors are processed by dicer, generating ssRNA or dsRNA of 20–24 nt. This short miRNA is able to interfere with gene expression in either of two ways.

First, miRNA bind to 3′-non-translated sequences of mRNAs. Due to several mismatches, the target sequence is not perfectly complementary to the miRNA. Those mismatches are likely the trigger of a process that results in translational inhibition of the mRNA without degradation of the RNA.

Second, miRNA that is perfectly complementary to its target sequence on the mRNA act like a siRNA and induces a sequence-specific endonucleolytic cleavage. Work with recombinant reporter genes has shown that changing the target sequence causes a given miRNA to act as a translational inhibitor (imperfect complementarity with target site) or as a siRNA-like inducer of nuclease activity (perfect complementarity with target site) [10, 11].

Mimicking this feature of miRNAs, it was possible to create vectors that express short hairpin RNAs (shRNA) intracellularly that induce a strong gene silencing effect (see below, Table 52.1).

Long dsRNAs cause a general arrest of translation and an unspecific degradation of RNA ("interferon response") by activating protein kinase R and RNase L, respectively. Therefore, they are not suitable for gene-specific silencing in mammalian cells. siRNAs induce neither of these activities due to their small size [4]. Nonetheless, the induction of interferon response to some extent was also observed in some cases for endogenously expressed shRNA [12]. These contradictory results necessitate further research in order to clarify the overall potency and limitations of siRNAs in gene expression studies.

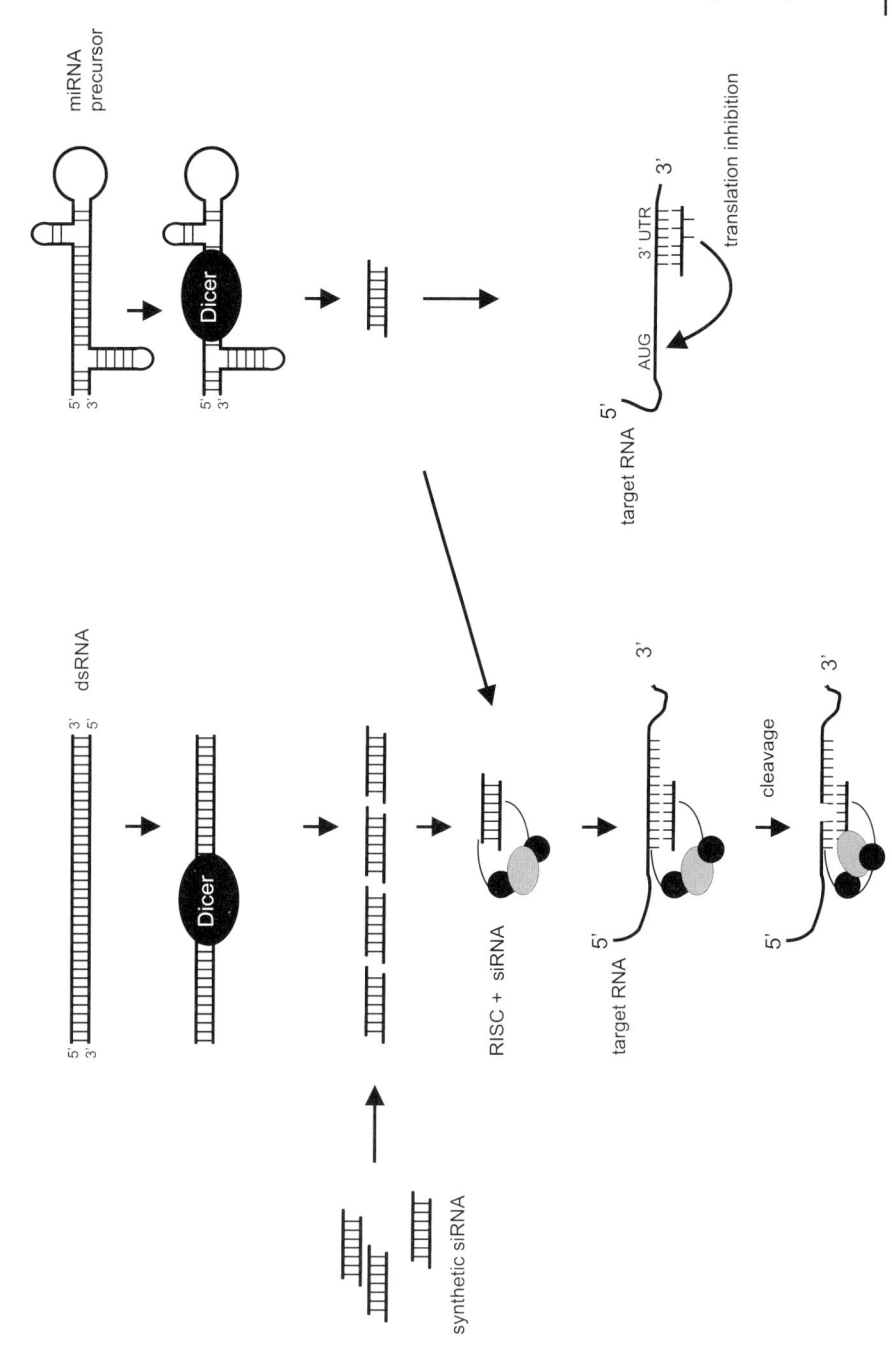

Fig. 52.1. A model for RNAi in mammalian cells. The siRNAs lead to strand-specific cleavage of their cognate target site on RNA transcripts (left), whereas miRNAs are not perfectly complementary to their target site and trigger a translational block on mRNA (right). In the case where a perfect target site is available, the miRNA behaves like siRNA and induces RNA cleavage. See text for details.

Tab. 52.1. Examples of vector systems.

RNA expressed	Promotor	Inducible	Reference
PCR products			
s + as	U6	–	29
shRNA	U6	–	29
Plasmids			
shRNA	tRNA^{Val}	–	22
shRNA	tRNA^{Met}	–	23
shRNA	CMV IE–U6	–	24
s + as	U6	–	26
shRNA	H1	–	27
shRNA	U6	–	26, 30, 31
shRNA	U6-tetO	Tet	32
shRNA	U6-tetO; 7SK-tetO	Tet	33
shRNA	H1-tetO	Tet	34
Lentiviral vectors			
shRNA	U6	–	35, 36
shRNA	H1	–	37
Adenoviral vectors			
shRNA	CMV	–	25

The intracellular expression of siRNA is achieved by transcription of separate sense and antisense strand (s + as) or shRNA. Vector systems are either PCR-generated DNA, plasmids or recombinant viral genomes. Some systems exploit tetracycline regulation for inducible transcription of the siRNA (tetO).

52.3
Ways to Induce RNAi in Mammalian Cells

The induction of siRNA-mediated RNAi in mammalian cells can be achieved by:

- Microinjection of siRNA
- Transfection of synthetic siRNA with cationic lipids
- Electroporation of synthetic siRNA
- Transfection of PCR-generated DNA modules containing siRNA or shRNA expression cassettes
- Transfection of plasmids containing siRNA or shRNA expression cassettes
- Infection of cells by recombinant viruses containing siRNA or shRNA expression cassettes

One straightforward approach combines synthetic siRNA with transfection technologies for transferring the molecules into the cells. This results in transient effects lasting 3–5 days. Alternatively, the integration of siRNA expression cassettes in the host genome mediated by either plasmids or recombinant viral vectors (e.g. lentiviral vectors) allows sustained suppression of target genes.

This chapter contains a basic protocol for transfection of chemically synthesized siRNA and an alternate protocol for electroporating siRNA. The possibilities to express siRNA molecules inside cells are summarized. However, we refer to other sources for detailed protocols as this is beyond the scope of this chapter.

52.3.1
Important Parameters

52.3.1.1 siRNA Design

Most researchers exploiting RNAi mechanisms in mammals use chemically synthesized siRNAs that follow the basic design suggested by Tuschl et al. [1]. The siRNAs are made of two antiparallel, complementary sequences of 19 nt. The sense strand is identical with the target mRNA. The antisense strand is complementary to both the sense strand and the mRNA. Additionally, two deoxythymidine nucleotides are incorporated into the siRNA at both 3' ends. This results in 2-nt overhangs. The introduction of short DNA moieties reduces synthesis costs and may contribute to increased stability in culture medium [1] (Fig. 52.2).

Although this siRNA design is currently used in most RNAi studies, other variants are equally or even more effective in comparative studies. The length of the double strand may vary between 19 and 22 nt, and the 2-nt overhang may include ribouridine or any other ribonucleotide instead of deoxythymidine [4, 13].

Coupling of fluorophores allows the detection of siRNA by fluorescence microscopy or fluorescence-activated cell sorting (FACS). Customized siRNA with fluorophores are commercially available.

52.3.1.2 Target Site Selection

Target sites on mRNA, non-coding RNA (e.g. rRNA) or viral RNA are subject to RNAi. The cognate sequence of a siRNA is either located in the untranslated 5' region of an mRNA, the open reading frame or the 3' end of the message. Sequences within introns never serve as siRNA target sites as the gene silencing machinery works only in the cytoplasm and not in the nucleus of mammalian cells [14].

The efficiency of individual target sites on a given RNA may vary considerably without obvious reasons, but it is assumed that stable secondary structures within the RNA may function less efficiently as siRNA target sites [15, 16]. Although the basic mechanisms of RNAi are not yet completely understood, commercial suppliers of chemically synthesized siRNA are developing algorithms for best target site selection. Simple versions of those programs are freely available and accessible via the World Wide Web. Use of more accurate programs is currently restricted to

synthetic siRNA   5'   CUGGACUUCCAGAAGAACATT   3'
                  3'   TTGACCUGAAGGUCUUCUUGU   5'

Fig. 52.2. Example of an siRNA targeting the human lamin a/c gene with 19-nt double-stranded region and a 2-nt deoxythymidine overhang at the 3' end on both strands [1].

purchasers of siRNA from commercial sources (e.g. www.dharmacon.com, www.qiagen.com or www.ambion.com).

A BLAST search of the siRNA sequence against a mouse or human gene for both mRNA and genomic sequences bank is highly recommended in order to avoid off-target effects. We recommend to identify siRNA sequences with at least three or four mismatches to any non-target mRNA in order to avoid unwanted siRNA-mediated cleavage [17].

52.3.1.3 Preparation of siRNA Samples

All commercial manufacturers offer synthesized siRNA in a highly purified form with annealed strands and at a defined concentration. Some researchers might prefer ordering individual strands or synthesizing the RNA in-house. Gel purification under denaturing conditions is similar for short RNA and DNA. A purification protocol for short oligonucleotides is given in reference [18]. Dharmacon (Lafayette, CO, USA) synthesizes RNA with a proprietary technology that results in RNA of sufficiently high purity that precludes the necessity for additional purification steps by the customer.

After purification, the RNA concentration is accurately determined by UV spectroscopy at 260 nm. The molar extinction coefficient of each RNA sequence has to be calculated individually, due to the limited lengths of 21–23 nt. An easy-to-use program for this purpose is the biopolymer calculator http://paris.chem.yale.edu/extinct.html (website for computing molecular properties of RNA strands).

For the annealing of equimolar amounts of sense and antisense strand, the concentration of double-stranded RNA is adjusted to 20 µM in 100–500 µl final volume of annealing buffer (100 mM NaCl, 20 mM Na_2HPO_4/NaH_2PO_4, pH 6.8). RNA samples are incubated for 3 min at 90 °C in a stirred water bath. The samples are allowed to cool down to room temperature over 3–4 h. The quality of the annealing procedure can be monitored by vertical gel electrophoresis. About 1.5 µg of both single strands and the double stranded siRNA are loaded on a nondenaturing 10% polyacrylamide gel with Tris–borate–EDTA as running buffer. The RNA is visualized by ethidium bromide staining. Samples of properly annealed siRNA run slower compared to sense and antisense strand and show no signs of remaining single stranded RNA.

siRNA stock solutions should be stored at −80 °C. Smaller samples of siRNA can be stored at −20 °C for at least 6 months and repeated freezing–thawing cycles do not decrease their quality. We do not keep solutions with highly diluted siRNA.

52.3.2
Transfection of Mammalian Cells with siRNA

The transfection of mammalian cells with siRNA is similar to DNA transfection protocols. However, methods based on calcium phosphate complexes are rarely used for siRNA, although adapted protocols have been reported [19].

Special transfection reagents were developed in order to achieve optimal transfection efficiencies with siRNA:

- Mirus TKO (Mirus, Madison, WI, USA)
- Oligofectamine (Invitrogen, Carlsbad, CA, USA)
- RNAiFect (Qiagen, Hilden, Germany)
- Saint-Mix (Synvolux, Groningen, Netherlands)
- SiPort (Ambion, Austin, TX, USA)

We recommend testing several transfection kits in order to achieve optimal results with siRNA. Each cell line may behave differently. A summary of successful transfection protocols is listed in [20].

For some cell lines, we observe that DNA transfection reagents such as Lipofectamine Plus of Lipofectamine 2000 (Invitrogen) may be more suitable. In addition to a high transfection efficiency, a low toxicity of the transfection process should be obtained in order to get reproducible results.

Materials

Mammalian cells to be transfected (e.g. HeLa-S3)
Complete medium
Trypsin solution (0.05% Trypsin/0.02% EDTA in PBS, pH 7.2)
Culture dishes (12-well dishes)
1.5-ml polypropylene tubes (Eppendorf)
siRNA at a stock concentration of 20 µM
Liposome transfection reagent Lipofectamine Plus (Invitrogen)

Protocol and notes

(1) Cultivate cells to late log phase stage in complete medium.
(2) Trypsinize, wash and resuspend cells in complete medium.
(3) Seed 1 ml of 1×10^4 to 1×10^5 cells (depending of cell type) in each of 2-cm diameter well. Incubate for about 24 h at 37 °C and 5% CO_2.
(4) Replace medium by 500 µl fresh medium 4 h before adding the transfection reagents.
(5) For each well, mix 1.25 µl 20 µM siRNA with 43.75 µl serum-free medium in a 1.5-ml reaction tube. Add 5 µl PLUS reagent. *In addition to specific siRNA, all experiments have to include at least one siRNA unrelated to the target mRNA in order to evaluate non-specific or toxic effects.*
(6) Take a second 1.5-ml reaction tube and dilute 2.5 µl Lipofectamine reagent in 47.5 µl serum-free medium.
(7) Incubate the two tubes for 10 min at room temperature.
(8) Add the Lipofectamine solution to the siRNA solution and mix gently. Incubate for 5 min at room temperature.

(9) Remove the complete medium from each well and replace it with 400 µl of serum-free medium.
(10) Add the 100 µl siRNA transfection mix. The final concentration of siRNA is 50 nM. *The optimal siRNA concentration varies depending on cell line or target gene, but is probably between 1 and 100 nM.*
(11) After 2 h incubation at 37 °C and 5% CO_2, remove the transfection medium and replace it with 1 ml complete medium.
(12) Analyze RNA or protein expression after 16–48 h incubation. *Potent siRNAs are able to decrease target RNA or protein levels between 4- and 20-fold. Depending on the gene of interest, cell line used or the gene product, the incubation time may have to be extended to 72 h. We recommend a combined analysis of target RNA and protein abundance. As the half-life of proteins are frequently longer than that of their cognate mRNA, an efficient inhibition of mRNA expression is not necessarily immediately reflected in a reduction of protein level.*

52.3.3
Electroporation of Mammalian Cells with siRNA

Many mammalian cell types can be easily and efficiently electroporated. Tests with fluorescently labeled siRNA and subsequent analysis in fluorescence activated cell sorting indicate that up to 100% of cells can be transfected. Electroporation is a method particularly important for cell lines grown in suspension. Such cells are often difficult to transfect with cationic lipids as the transfection aid. In addition, electroporation is a simple technique enabling transfection of many adherent cell types as well. Some cell lines (e.g. NIH 3T3 fibroblasts), however, are inefficiently transfected using electroporation protocols and, thus, a lipofection protocol has to be tested as an alternative.

Materials

Mammalian cells to be transfected
Complete medium
Trypsin solution (0.05% Trypsin/0.02% EDTA in PBS, pH 7.2)
Culture dishes
Square wave electroporator (e.g. EPI 2500, Fischer, Heidelberg, Germany)
4 mm electroporation cuvettes (e.g. Equibio ECU104)
siRNA at a stock concentration of 20 µM

Protocols and notes

(1) Cultivate cells to late log phase in complete medium.
(2) Spin the cell suspension for 5 min at 300 g at room temperature. *Adherent cells are trypsinized and washed once with culture medium before centrifugation.*
(3) Resuspend cells in complete medium at a final concentration of $1–2 \times 10^7$ cells/ml.

(4) Transfer 100–800 μl of cell suspension into an electroporation cuvette at room temperature.
(5) Add siRNA to cell suspension at a final concentration of 100–200 nM. The attachment of a fluorophore to the siRNA allows the determination of electroporation efficiency. Fluorescein-labeled siRNA should be avoided, since they cannot be efficiently detected when located in the cytoplasm [21].
(6) Mix well by flicking the cuvette.
(7) Place cuvette into the holder of the electroporator and electroporate at the desired voltage and time setting. *For hematopoietic cell lines such as BAF3, HL60, K562 or U937 varying voltages from 300 to 400 V and a constant time of 10 ms are recommended. To establish the electroporation conditions, we routinely electroporate 10^6 cells in 100 μl medium containing 100 nM to 1 μM of the corresponding siRNA.*
(8) Remove the cuvette from the holder and keep it for 15 min at room temperature.
(9) Return the electroporated cells to a culture vessel and dilute the cell suspension to the desired cell density.
(10) Analyze RNA, protein expression or phenotype changes. *Decreased RNA levels are visible within 16 h and may last up to 72 h after electroporation. The reduction of protein levels depends on the half-life of the targeted protein. Minimal protein levels are observed between 24 and 96 h after electroporation.*

52.3.4
Induction of RNAi by Intracellular siRNA Expression

A different approach for inducing a gene silencing effect is the intracellular synthesis of siRNA. The expression cassette is either part of a PCR product, a plasmid or a viral vector system (Fig. 52.3). The core of each expression system starts with a short human polymerase III promoter sequence, usually H1 or U6. Those promoters are particularly effective in accurately transcribing very short RNA sequences *in vivo*. Alternatively, promotors for tRNA precursors (tRNA^{Val}, tRNA^{Met}) have also been used successfully [22, 23]. Very strong viral promotors, e.g. the cytomegalovirus immediate early promotor (CMV IE), lead to high expression levels of shRNA and improved gene silencing effects in certain cell types [24, 25]. Modifications of the promoter systems by inserting bacterial tetracycline operators (tetO) allow the induction of RNA expression by adding tetracycline or doxycycline to the medium. The minigene itself codes for an siRNA sequence in either sense or antisense orientation and two expression units are required for obtaining a functional siRNA. The sense and antisense strands need to hybridize to generate the siRNA [26]. Alternatively, the gene codes for a short hairpin RNA sequence. After transcription, the shRNA folds into a hairpin structure that is a substrate of dicer. The final processing product resembles regular siRNAs and induces gene silencing [19, 27, 28]. A termination site of multiple thymidines defines the end of each expression cassette (Fig. 52.4). Due to the constitutive expression of the siRNA, a sustained inhibition of target gene expression is achievable. Plasmid or viral vectors

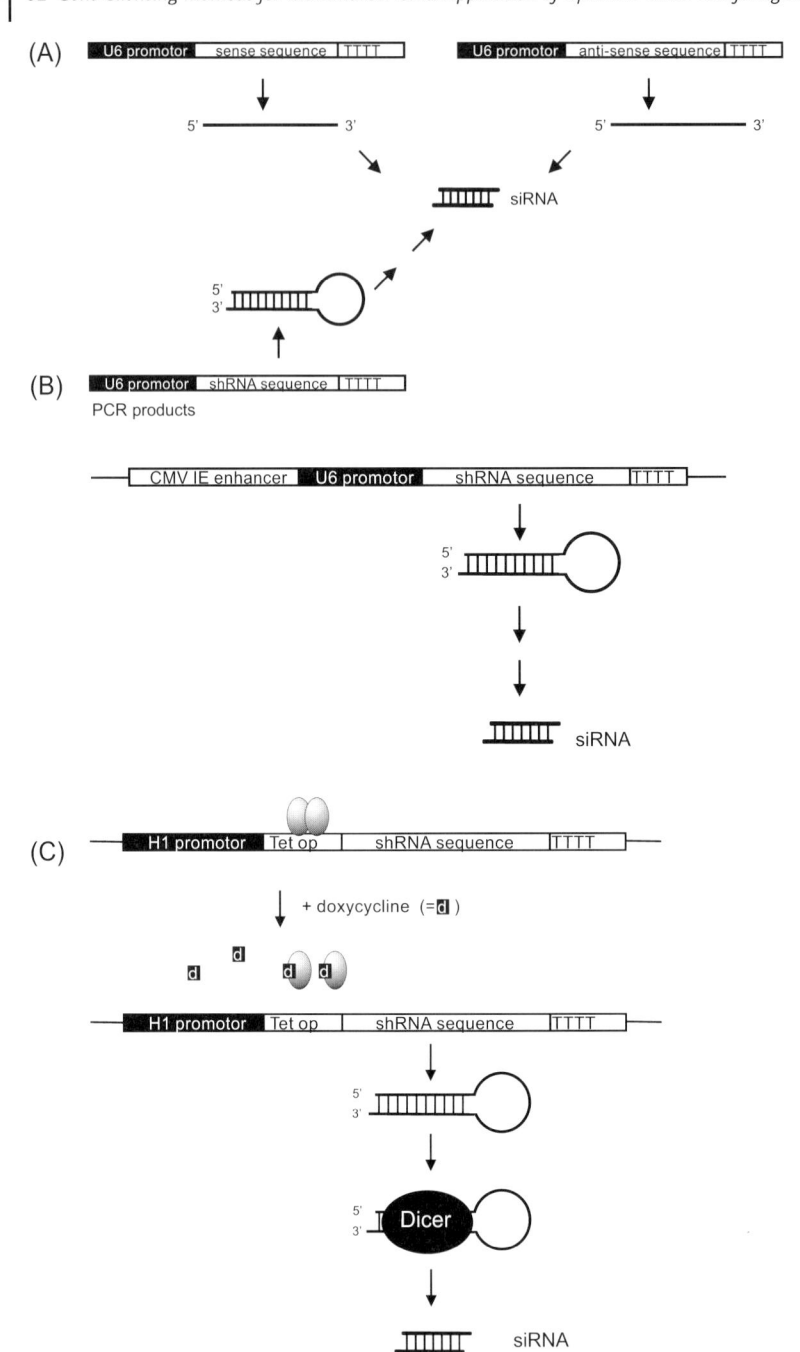

Fig. 52.3. Examples of intracellular siRNA expression systems. (A) PCR products [29]. (B) Enhanced expression by fusing a CMV immediate early enhancer to a U6 promotor [24]. (C) Inducible expression system by placing the H1 promotor under control of a Tet operator [34].

```
                sense strand              loop          antisense strand   RNA Pol III
                                                                           terminator
...TGAGAAGUCTCCCAGTCAGTTCAAGAGACTGACTGGGAGACTTCTCATTTTT...
...ACTCTTCAGAGGGTCAGTCAAGTTCTCTGACTGACCCTCTGAAGAGTAAAAA...
```

                           ↓

```
              5' UGAGAAGUCUCCCAGUCAG U U C
                                         A
                  U ACUCUUCAGAGGGUCAGUC   A            shRNA
              3' UUUUU U                G A G
```

                           ↓

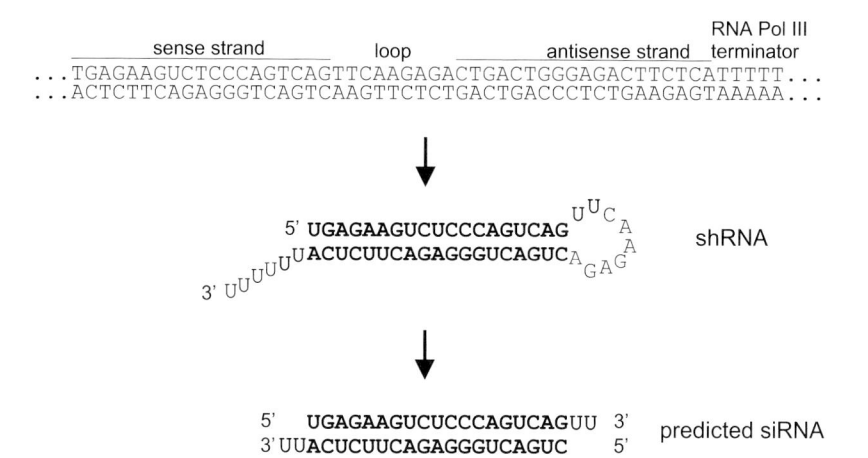

```
              5'  UGAGAAGUCUCCCAGUCAGUU  3'
              3' UUACUCUUCAGAGGGUCAGUC   5'           predicted siRNA
```

Fig. 52.4. Generation of a siRNA targeting the human CDH1 gene by expressing a short hairpin RNA precursor and processing of the molecule by dicer [27].

Tab. 52.2. Troubleshooting.

Problem	Possible cause	Solution
No gene silencing	siRNA was not properly annealed	repeat annealing and check on non-denaturing gel
	siRNA was degraded	check oligoribonucleotide integrity by 5'-labeling and sequencing gel analysis
		discard and use a new siRNA batch
	low transfection efficiency	try other transfection methods and use fluorescent siRNA for verifying transfection efficiency
		vary siRNA concentration in your experiment
	siRNA was targeting an intron sequence	choose a new sequence outside the intron region
	target site was inaccessible	try other target sequence
	target site and siRNA sequence differed	check whether your sequence information about the target site is correct for your particular cell line
	wrong time window	analyze target gene at different time intervals
Gene silencing observed with negative control siRNA	siRNA concentrations were too high	try lower siRNA concentrations
	siRNA affected more than one gene	repeat BLAST search
	siRNA were toxic for the cells	try other transfection reagent or reduce siRNA concentration

can also be used for the generation of stable cell lines. Gene silencing plasmids are commercially available (Ambion, Imgenex) and are delivered with detailed protocols for the proper setup of a hairpin-like RNA expression system.

52.4
Troubleshooting

Possible problems, causes and solutions are given in Table 52.2.

References

1 S. M. Elbashir, J. Harborth, W. Lendeckel, A. Yalcin, K. Weber, T. Tuschl, *Nature* **2001**, *411*, 494–498.
2 G. J. Hannon, *Nature* **2002**, *418*, 244–251.
3 G. J. Hannon, *RNAi. A Guide to Gene Silencing*, Cold Spring Harbor Laboratory Press, Cold Spring Harbor, NY, **2003**.
4 N. J. Caplen, S. Parrish, F. Imani, A. Fire, R. A. Morgan, *Proc. Natl. Acad. Sci. USA* **2001**, *98*, 9742–9747.
5 C. D. Novina, M. F. Murray, D. M. Dykxhoorn, P. J. Beresford, J. Riess, S.-K. Lee, R. G. Collman, J. Lieberman, P. Shankar, P. A. Sharp, *Nat. Med.* **2001**, *8*, 681–686.
6 P. A. Sharp, *Genes Dev.* **2001**, *15*, 485–490.
7 M. Lagos-Quintana, R. Rauhut, W. Lendeckel, T. Tuschl, *Science* **2001**, *294*, 853–858.
8 M. Lagos-Quintana, R. Rauhut, A. Yalcin, J. Meyer, W. Lendeckel, T. Tuschl, *Curr. Biol.* **2002**, *12*, 735–739.
9 D. S. Schwarz, P. D. Zamore, *Genes Dev.* **2003**, *16*, 1025–1031.
10 Y. Zeng, E. J. Wagner, B. R. Cullen, *Mol. Cell* **2002**, *9*, 1327–1333.
11 J. G. Doench, C. P. Petersen, P. A. Sharp, *Genes Dev.* **2003**, *17*, 438–442.
12 C. A. Sledz, M. Holko, M. J. de Veer, R. H. Silverman, B. R. G. Williams, *Nat. Cell Biol.* **2003**, *5*, 834–839.
13 H. Hohjoh, *FEBS Lett.* **2002**, *521*, 195–199.
14 Y. Zeng, B. R. Cullen, *RNA* **2002**, *8*, 855–860.
15 T. Holen, M. Amarzguioui, M. T. Wiiger, E. Babaie, H. Prydz, *Nucleic Acids Res.* **2002**, *30*, 1757–1766.
16 R. Kretschmer-Kazemi Far, G. Sczakiel, *Nucleic Acids Res.* **2003**, *31*, 4417–4424.
17 M. Amarzguioui, T. Holen, E. Babaie, H. Prydz, *Nucleic Acids Res.* **2003**, *31*, 589–595.
18 A. Ellington, J. D. Pollard, in: *Current Protocols in Molecular Biology*, F. M. Ausubel, R. Brent, R. E. Kingston, D. D. Moore, J. G. Seidman, J. A. Smith, K. Struhl (eds), Wiley, New York, **1998**.
19 P. J. Paddison, A. A. Caudy, E. Bernstein, G. J. Hannon, D. S. Conklin, *Genes Dev.* **2002**, *16*, 948–958.
20 M. T. McManus, P. A. Sharp, *Nat. Genet.* **2002**, *3*, 737–747.
21 J. Dunne, B. Drescher, H. Riehle, B. D. Young, J. Krauter, O. Heidenreich, *Oligonucleotides* **2003**, *13*, 376–380.
22 H. Kawasaki, K. Taira, *Nucleic Acids Res.* **2003**, *31*, 700–707.
23 D. Boden, O. Pusch, F. Lee, L. Tucker, P. R. Shank, B. Ramratnam, *Nucleic Acids Res.* **2003**, *31*, 5033–5038.
24 X. G. Xia, H. Zhou, H. Ding, E. B. Affar, Y. Shi, Z. Xu, *Nucleic Acids Res.* **2003**, *31*, e100.
25 H. Xia, Q. Mao, H. M. Paulson, B. L. Davidson, *Nat. Biotechnol.* **2002**, *20*, 1006–1010.
26 N. S. Lee, T. Dohjima, G. Bauer, H. Li, M.-J. Li, A. Ehsani, P.

Salvaterra, J. Rossi, *Nat. Biotechnol.* **2002**, *20*, 500–505.

27 T. R. Brummelkamp, R. Bernards, R. Agami, *Science* **2002**, *296*, 550–553.

28 G. Sui, C. Soohoo, E. B. Affar, F. Gay, Y. Shi, W. C. Forrester, Y. Shi, *Proc. Natl. Acad Sci. USA* **2002**, *99*, 5515–5520.

29 D. Castanotto, H. Li, J. Rossi, *RNA* **2002**, *8*, 1454–1460.

30 C. P. Paul, P. D. Good, I. Winer, D. R. Engelke, *Nat. Biotechnol.* **2002**, *20*, 505–509.

31 C. P. Paul, P. D. Good, S. X. L. Li, A. Kleihauer, J. Rossi, D. R. Engelke, *Mol. Ther.* **2003**, *7*, 237–247.

32 S. Matsukura, P. A. Jones, D. Takai, *Nucleic Acids Res.* **2003**, *31*, e77.

33 F. Czauderna, A. Santel, M. Hinz, M. Fechtner, B. Durieux, G. Fisch, F. Leenders, W. Arnold, K. Giese, A. Klippel, J. Kaufmann, *Nucleic Acids Res.* **2003**, *31*, e127.

34 M. Wetering van de, I. Oving, V. Muncan, M. T. P. Fong, II. Brantjes, D. Leenen van, F. C. P. Holstege, T. R. Brummelkamp, R. Agami, H. Clevers, *EMBO Rep.* **2003**, *4*, 609–615.

35 X.-F. Qin, D. S. An, I. S. Y. Chen, D. Baltimore, *Proc. Natl. Acad. Sci. USA* **2003**, *100*, 183–188.

36 S. A. Stewart, D. M. Dykxhoorn, D. Palliser, H. Mizuno, E. Y. Yu, D. S. An, D. M. Sabatini, I. S. Y. Chen, W. C. Hahn, P. A. Sharp, R. A. Weinberg, C. D. Novina, *RNA* **2003**, *9*, 493–501.

37 G. Tiscornia, O. Singer, M. Ikawa, I. M. Verma, *Proc. Natl. Acad. Sci. USA* **2003**, *100*, 1844–1848.

Appendix: UV Spectroscopy for the Quantitation of RNA

UV spectroscopy provides the almost universal basis for nucleic acid quantitation. Reliable determination of RNA/DNA concentration is critical for many applications, such as ribozyme kinetics or measurements of equilibrium constants in RNA-ligand interactions. A recent publication [1] has readdressed the "old" molar extinction (absorbance) coefficients determined by UV spectroscopy for nucleoside-5′-monophosphates ($\varepsilon_{260\,nm}$ values are essentially identical for 5′- and 3′-NMPs). This study, utilizing NMR for accurate determination of concentration and detection of impurities, revealed that $\varepsilon_{260\,nm}$ values, particularly those for pC and pdC, had previously been overestimated. Thus, concentrations of single-stranded RNA (ssRNA) and single-stranded DNA (ssDNA) based on complete hydrolysis and the "old" extinction coefficients were underestimated – for mixed non-repetitive sequences by about 4% and 2–3%, respectively. The authors thus recommend revised rules of thumb for such sequences: 1 A_{260} unit = 38 μg/ml for ssDNA and 1 A_{260} unit = 37 μg/ml for ssRNA (free acid) due to the higher ε_{260} of pU versus pT. This correlation changes with the type of counter ion present (e.g. multiply 37 μg/ml with a factor of 1.07 for the sodium salt of the ssRNA). Nonetheless, the value of 37 μg/ml roughly matches information given in [2]. 1 A_{260} unit = 38 μg/ml for ssRNA) and on many company homepages (1 A_{260} unit = 40 μg/ml for ssRNA; e.g. Roche, Ambion, Amersham Biosciences). For two classes of nucleic acids, however, this rule of thumb may generate errors far exceeding 10%: (1) oligonucleotides with strong sequence bias (e.g. purine clusters) and (2) structured RNAs, as discussed in the following.

A central issue is hypochromicity due to base stacking interactions. Hypochromicity effects increase with the number of neighboring purines. For ε_{260} determination of oligonucleotides, a strategy often applied therefore is to subtract the ε values of individual mononucleotides from those of dinucleotide neighbors along the sequence (nearest neighbor estimates). A concise guideline of this method is available at the homepage of Integrated DNA Technologies (Coralville, IA, USA), http://www.idtdna.com/Support/Technical/CalculatingMolarExtinctionCoefficient/Page1.aspx. The *Biopolymer Calculator* at http://paris.chem.yale.edu/extinct.frames.html utilizes this algorithm and is thus recommended for online ε_{260} prediction of oligonucleotides. Subsequent correction by factors derived from the revised ε_{260} values of mononucleotides [1] should then provide a more accurate basis for

UV-based quantitation of oligonucleotides. For principles of oligonucleotide quantitation, the reader may also consult the homepage of IBA (Göttingen, Germany; technical info) at http://www.iba-go.com/tec_info.html.

In the case of structured RNAs, the classical approach is alkaline or ribonuclease hydrolysis of the RNA to mononucleotides and subsequent calculation of RNA concentration on the basis of ε_{260} values for the individual mononucleotides. To obtain ε_{260} of the intact RNA, the sum of mononucleotide ε_{260} values is then corrected by a factor corresponding to the ratio of A_{260} values of the RNA sample before and after hydrolysis. Alternative RNA quantitation approaches are phosphate analysis [3] or a fluorometric assay [4]. For example, the hydrolysis approach resulted in an ε_{260} value of 3.46×10^6 M^{-1} cm^{-1} for a 409-nt transcript of *Bacillus subtilis* RNase P RNA [5], which corresponds to an average ε_{260} value of about 8460 M^{-1} cm^{-1} per nucleotide and equals 38.5 μg/ml per A_{260} unit for the free acid and about 41 μg/ml for the ammonium salt. Likewise, ε_{260} was determined as 6.43×10^5 M^{-1} cm^{-1} (equal to 8350 M^{-1} cm^{-1} per nucleotide; about 39 μg/ml per A_{260} unit for the free acid) for a 77-nt long tRNA^{Asp} transcript [6]. Similarly, an experimental ε_{260} of 3.2×10^6 M^{-1} cm^{-1} was obtained for the 388-nt L-21 *ScaI* transcript of the *Tetrahymena* ribozyme (about 39.5 μg/ml per A_{260} unit for the free acid; [7]). Thus, for these types of RNAs, despite their assumed extensive secondary structure, the experimentally determined ε_{260} values are in the same range as those given for ssRNA. Why? In some cases, the explanation may simply be substantial RNA unfolding due to absorbance measurement in water. A more sophisticated explanation could be related to two counteracting phenomena: (1) The ε_{260} values for mononucleotides at neutral pH decrease in the order pA: 15.02 > pG: 12.08 > pU: 9.66 > pC: 7.07 (in mM^{-1} cm^{-1}; [1]). Thus, overrepresentation of A (and G) residues should lead to overestimation of RNA concentration when applying the rule of thumb of 1 A_{260} = 40 μg/ml. (2) This effect is at least partially compensated by enhanced hypochromicity due to increased stacking of purines. Purines, and A residues in particular, are overrepresented in *B. subtilis* RNase P RNA (28.4% A, 28.4% G, 23% U, 20.3% C) and the L-21 *ScaI* ribozyme (29.4% A, 27.1% G, 25.3% U, 18.3% C). Accordingly, a concentration overestimation effect is evident for these two RNAs, explaining why they give values slightly below 40 μg/ml per A_{260} unit for the free acid. All in all, we think it will be satisfactory for many applications to calculate the concentration of structured undenatured RNA on the basis of 1 A_{260} unit = 41 μg/ml for the free acid, already taking into account the 4% correction of ε_{260} suggested in [1].

There are, of course, exceptions to the rule. For example, a correlation of about 51 μg/ml per A_{260} unit (free acid) was obtained for a 23meric RNA-stem loop structure based on ε_{260} determined by phosphorus quantitation [8]. For precise concentration measurement of structured RNA molecules, we therefore propose to routinely use one of two straightforward experimental protocols to determine ε_{260}. A prerequisite is to know the RNA sequence. The molecular weight of the RNA (free acid, sodium, ammonium or potassium salt, 5′-monophosphorylated or 5′-dephosphorylated) can be calculated with the help of the *Biopolymer Calculator* (see above) in order to obtain the correlation of μg/ml per A_{260} unit.

First approach:

(1) Measure the A_{260} of your RNA sample at room temperature in a defined buffer, for example 50 mM ammonium acetate (pH 7.0). Trace amounts of components that stimulate activity of hydrolysis enzymes (step 2), such as $ZnSO_4$, may be added.
(2) Add nuclease P1 (or a mixture of RNases A, T1 and T2); incubate until the absorbance at 260 nm no longer increases over time (at least 1 h); measure A_{260} at the endpoint of hydrolysis [5, 7].
(3) Calculate the RNA concentration based on the A_{260} value at the endpoint of hydrolysis using the extinction coefficients of the individual mononucleotides according to [1].
(4) The A_{260} difference of the intact (step 1) and hydrolyzed (step 3) RNA solution defines the difference in ε_{260} between your RNA in this particular buffer and the sum of ε_{260} values of individual mononucleotides.

The second approach [9] essentially differs from the first in step 2. Instead of digestion to mononucleotides, the RNA is completely denatured at high temperatures (>80 °C, in the absence of divalent metal ions and generally at low ionic strength, e.g. at 20 mM Tris-HCl, pH 7.0).

We advise against measuring RNA absorbance in water, because varying traces of salts may be brought in from preceding preparation steps, making the folding state of the RNA uncertain (strategies and protocols for desalting of RNA and salt exchange are described in Chapters 4, 7 and 27). We also like to note that theoretical predictions of ε_{260} values by the *Biopolymer Calculator* (see above) are not reliable for structured RNA.

Finally, although seemingly trivial, it should be mentioned that a major source of erroneous RNA quantitation will be insufficient separation of nucleotides from RNA samples after transcription. Also, one should be aware that UV absorption of nucleic acids is pH-sensitive and that reliable measurement requires an A_{260} of at least 0.1 or better 0.15, making cuvettes with small volumes rather attractive when amounts of RNA are limited. Ionic strength, nature of ions, presence of EDTA or denaturants such as urea and formamide, and temperature influence UV absorption by affecting the folding state of the RNA. In addition, the quality of the UV spectrophotometer will play a role, such as the accuracy of cuvette positioning.

References

1 M. J. CAVALUZZI, P. N. BORER (2004) Revised UV extinction coefficients for nucleoside-5'-monophosphates and unpaired DNA and RNA. *Nucleic Acids Res.* 32, e13.

2 J. SAMBROOK, and D. W. RUSSELL (2001) Molecular Cloning. A Laboratory Manual, Edn 3. Cold Spring Harbor Laboratory Press, Cold Spring Harbor, NY.

3 J. H. MURPHY, T. L. TRAPANE (1996) Concentration and extinction coefficient determination for oligonucleotides and analogs using a general phosphate analysis. *Anal. Biochem.* 240, 273–282.

4 L. J. Jones, S. T. Yue, C. Y. Cheung, V. L. Singer (1998) RNA quantitation by fluorescence-based solution assay: RiboGreen reagent characterization. *Anal Biochem.* **265**, 368–374.

5 X. W. Fang, X. J. Yang, K. Littrell, S. Niranjanakumari, P. Thiyagarajan, C. A. Fierke, T. R. Sosnick, T. Pan (2001). The *Bacillus subtilis* RNase P holoenzyme contains two RNase P RNA and two RNase P protein subunits. *RNA* **7**, 233–241.

6 J. A. Beebe, C. A. Fierke (1994) A kinetic mechanism for cleavage of precursor tRNA(Asp) catalyzed by the RNA component of *Bacillus subtilis* ribonuclease P. *Biochemistry* **33**, 10294–10304.

7 A. J. Zaug, C. A. Grosshans, T. R. Cech (1988) Sequence-specific endoribonuclease activity of the *Tetrahymena* ribozyme: enhanced cleavage of certain oligonucleotide substrates that form mismatched ribozyme-substrate complexes. *Biochemistry* **27**, 8924–8931.

8 J. A. Cowan, T. Ohyama, D. Wang, K. Natarajan (2000) Recognition of a cognate RNA aptamer by neomycin B: quantitative evaluation of hydrogen bonding and electrostatic interactions. *Nucleic Acids Res.* **28**, 2935–2942.

9 T. Pan, T. R. Sosnick (1997) Intermediates and kinetic traps in the folding of a large ribozyme revealed by circular dichroism and UV absorbance spectroscopies and catalytic activity. *Nat. Struct. Biol.* **4**, 931–938.

Index

a
absorbance   70, 104
ac^{4}C   146, 149
acetonitrile   96 f
5' adaptor   632
adenosine deaminase   478
 – binding to RNA   478
S-adenosyl methionine   807
affinity   87
affinity chromatography   854
affinity column   667 ff, 674
 – CNBr-activated sepharose 4B   668
 – DNA   674
 – NHS-tRNA   668
 – tRNA   667
affinity matrix   678 ff
 – monomeric avidin   678 ff
 – neutravidin agarose   678 ff
 – streptavidin-agarose   678 ff
 – streptavidin-coated magnetic bead   678 ff
affinity purification   676 ff, 680 ff, 796
 – neutravidin   796
 – displacement strategy   685
 – of specific RNA-binding protein   689
affinity selection, *see* affinity purification
affinity tag   609 f
 – FLAG   610
 – His$_6$ 610
AFMM, *see* automated frequency matching methodology
Äkta OligoPilot 10   99
alkaline hydrolysis ladder   160, 835
alternative splicing   658, 755
 – annotation   658
AMBER   565 f
aminoglycoside   349, 447 f, 812
 – condition testing   448
A-minor motif   267
ammonium bicarbonate   102
ammonium hydroxide   101

c7-AMPαS   295
AMV   186, 648
 – reverse *trans*criptase   233
aniline   144 f, 232
 – cleavage   145, 232
antibiotic resistance gene   616
 – terminator   616
antibody   175, 182 ff, 184, 364, 733 ff
 – in Western blotting   184
 – in immunocytochemistry   733 ff
anti-peptide antibody   694 ff
 – affinity purification   696 ff
 – peptide selection criteria   694
 – U2 snRNP   699 f
 – U4/U6 snRNP   697 ff
 – U4/U6.U5 tri-snRNP   697 ff
 – U5 snRNP   701
antisense mechanism   624
antisense RNA   597, 599 f, 608, 610
 – *cis*-encoded   600, 610
 – screening   610
 – *trans*-encoded   600
antisense technology   95, 879
2APαS   266
aptamer   95, 608, 790, 807 ff, 835, 853 f
 – affinity   808
 – carbohydrate   807
 – comparison with antibody   808
 – contact surface   854
 – crosslinking   808
 – database   808
 – 7,8-dihydro-8-hydroxy-2'-deoxyguanosine (oxodG)   853
 – *in vitro* selection   807
 – isolation of   790
 – membrane-associating   810
 – module   626
 – protein   807
 – specificity   808
 – structural analysis   835

aptazyme 854
apyrase 147
Arabidopsis thaliana 672
arylazide 76, 79, 83
 – 3'-addition 83
ascorbic acid 346
[γ-^{32}P]ATP 26, 138, 147, 156, 177, 186, 232, 308, 649
ATPase 66
automated frequency matching methodology 566
autoradiography 347
avian myeloblastosis virus, *see* AMV
avidin 86, 857 f, 861 f, 869
Avogadro's number 818
azidophenacyl 78, 82
 – crosslinking radius 82
 – photo-crosslinking agent 82
azidophenacyl bromide 82
 – attachment 82

b

bacterial culture 636
 – growth 636
 – harvest 636
bacteriophage 610
 – infection by T4 phage 54
base modification 143
base pair breathing 221
base triple 299
betaine 865
binding force, determination by SFS 481
BioEdit 500
biotin 86, 88, 808, 861
 – attachment, 3'-terminal 88
 – NHS-activated 861
biotin hydrazide 88
biotinamidocaproyl hydrazide 87 ff
biotin-XX-hydrazide 88
biotinylated oligodeoxyribonucleotide 625
biotinylation 91
 – efficiency 91
BLAST search 602, 624
BLASTN search 608 f
blue/white screening 638, 827
BPB 142
1-(*p*-bromoacetamidobenzyl)-EDTA 239
bromophenol blue 135, 156, 160, 163, 177, 206, 233, 241, 382
Brönsted base 214
BS190 599
BS203 599
bulge 220

c

calf intestinal alkaline phosphatase (CIAP) 43, 126, 137, 178, 273
cap analog 47
cap structure 175
capillary blotting 631
capillary gel electrophoresis 107
cation exchange chromatography 19
cDNA 598, 633, 638
 – cloning 598
 – PCR amplification 638
 – synthesis 596, 608
cDNA library 607, 647, 713 ff
 – from sized RNA fraction 647
 – by directional cloning 607
cell disruption 820
cellular extract 178
Cerenkov radiation 222
cesium salts, in density gradient centrifugation 429 ff
chaotrope 103, 440
CHARMM 565 f
chemical genetics 53
chemical modification 78
chemical modification interference 259, 287
 – chemical synthesis 287
chemical phylogeny 266
chemical probing 160, 164, 230, 628
chloroform 121, 224
chloroform extraction 15, 48 ff, 69, 83, 89, 165, 224, 251, 282, 306
chromatography 18, 80, 87, 90, 107, 110, 135, 156, 175, 282, 297 f, 369, 811, 819, 875
 – affinity matrix 135
 – affinity 667 ff, 676 ff, 854 ff, 819
 – anion exchange 80, 134
 – cation exchange 18, 110
 – gel exclusion 90
 – gel filtration 107, 156, 282, 811
 – immunoaffinity 175, 694 ff
 – ion-pairing reversed-phase 156
 – isocratic competitive affinity 875
 – Merck EMD fractogel 18
 – monoQ 156
 – Ni-NTA 297 f, 819
 – Sephadex G-50 90, 282
 – spin column 90, 369
 – streptavidin 135
 – TLC, *see* thin-layer chromatography
CIAP, see *calf intestinal alkaline phosphatase*
2',3'-*cis*-diol 78
citrus exocortis viroid (CEV) 59

cloning   596 ff, 606, 614 ff, 633, 644 ff, 827, 884
– TOPO TA cloning kit   633, 884
Clustal   504, 582
ClustalW   581
ClustalX   885
CMCT   153 ff, 161, 164, 168, 185 ff, 628
comparative analysis   491 ff, 540
– alignment   540
– compensatory mutation   493
conformational change
– detection by Pb^{2+} cleavage   221
contig   608
controlled pore glass (CPG)   96
cordycepin see 3'-deoxyadenosine   138
cosmid library   599
COVE   579
crosslinking   173, 259 f, 354 ff, 358, 368 ff, 374 ff, 649, 808
– assignment of protein-RNA crosslinking site   358
– by UV   354
– chemical   374
– conditions   356
– cysteine   355
– histidine   355
– intra-RNA crosslink   358
– leucine   355
– methionine   355
– naked (protein-free) RNA   368
– phenylalanine   355
– primer extension   378, 381
– ribozyme-substrate complex   375
– RNA and protein   173 ff, 354 ff
– RNP particle   358
– tyrosin   355
crystal growth   442
crystal screen   442, 445 ff
– initial screen   445
– factorial plan   446
– crystallization robot   447
– simplex   446
crystallization   438 ff
CsrA   601, 609
CsrB   600 ff, 609
CsrC   600 f
2',3'-cyclic phosphate   24 ff, 54 ff, 89
1-cyclohexyl-3-(2-morpholinoethyl)carbodiimide p-toluenesulfonate, see CMCT
3'-cytidine monophosphate   139
cyanine   466, 470
cytoplasmic extract   175 f

d
[α-^{32}P]dATP   139
DEAE filter disk   833
7-deazaAαS   266
degP   600
density gradient centrifugation   428 ff, 432, 435
– buoyant density   435
– fraction of snRNP   428 ff
– glycerol   429
– sedimentation marker   432
3'-deoxyadenosine   138
deoxyoligonucleotide   135
– biotinylated   135
deoxyribozyme   879
DEPC, see diethyl pyrocarbonate
dephosphorylation   9, 33, 38, 43, 49, 89, 273, 306
– protocol   33, 49, 89
desiccator   822
dextran   812
dialdehyde   78
dialysis   19 f, 107
DICER   614, 898 ff
dichlorodimethylsilane   831
diethyl pyrocarbonate (DEPC)   69 ff, 144, 155, 817
dimethyl formamide (DMF)   121
dimethyl pimelinidate dihydrochloride   371
dimethyl sulfate (DMS)   144, 164, 185 ff, 229, 287
dimethyl sulfoxide (DMSO)   58, 101, 153, 160, 865, 882
dinucleotide   9, 148 f
– 2'-O-methylated   149
directional cloning   607 f
diphosgene   118, 121 f
disulfide crosslink   112, 117 ff
DMF, see dimethyl formamide
DMS, see dimethyl sulfate
DMSO, see dimethyl sulfoxide
DNA   5, 125, 480, 881
– instability   125
– visualization by SFM   480
10-23 DNA enzyme   29
DNA ligase, see T4 DNA ligase
DNA polymerase   138 f, 648
– T7   139
DNA polymerase I   345, 647, 862
– active site   345
DNA purification   638
DNA splint   37 ff, 53 ff
DNAMAN   653

DNase I   15, 40, 46, 66, 176, 178, 224, 231, 732 ff, 820 f, 880 f, 881, 885, 887 f, 891
DNAzyme, see 10-23 DNA enzyme
double-stranded RNA, see RNA, double-stranded
DROSHA   614
DsrA   600, 609, 624
dye, fluorescent   86 ff, 454, 457 ff
dynamics   453, 548, 560
– molecular mechanics (MM)   548, 560

e
EDTA   324
– binding of thiophilic metal ion   324
effector RNA (eRNA), see small untranslated RNA
electric field   402
– effect on RNA conformation   402
electroblotting   631
electrophoresis   134 f, 137, 139, 440
– loading buffer   135
– polymer concentration   137
– two-dimensional   134
electrophoretic mobility shift assay (EMSA)   91, 126, 173, 178 ff, 813, 816, 828 ff
– protocol   828
electrostatic interaction   567
EMBOSS   885
end group analysis   143 ff, 148
5'-end-labeling   26, 308, 835
3'-end-labeling   26, 308
endonuclease   345
energy dot plot   524
enzyme
– kinetic analysis   328
– radioactive labeling   351
eosin-5-thiosemicarbazide   88
EPR spin probe   118, 120
equilibrium dialysis   250 f, 254, 256, 873 f
– Gibbs-Donnan effect   874
– of RNA and Mg^{2+} 251
equilibrium filtration analysis   874
ERPIN   578 ff
ethanol precipitation   16, 26, 29, 47, 66 ff, 79, 83 f, 89, 122, 135, 157 ff, 224, 272 ff, 368 ff, 647, 821 ff
ethanolamine   371, 859 ff
ether extraction   821 ff
ethidium bromide   14, 68, 599
ethyl acetate   71
ethyl-dichlorophosphite   80
ethylene diamine   78, 83
3'-exonuclease   345
exponential phase   604, 607, 635 f

expression cassette   5
expression pattern   596
extract
– cytoplasmic   778 f
– nuclear   773, 778 f

f
Fe^{2+} 239, 241
– tethered to RNA and protein   239
– chelated to tetracycline   240 ff
Fe^{2+}-mediated cleavage   240 ff, 345 ff
– by hydroxyl radical   238 ff, 345 ff
– inhibiton by aminoglycoside   349
– Mg^{2+} competition   241 ff, 347
– *Tetrahymena* ribozyme   239 ff
$FeCl_2$ 240
Fenton cleavage   238 ff, 346
– of *Tetrahymena* ribozyme   240
fhlA   624
FISH, see fluorescence
fluorescein-5-thiosemicarbazide   87 ff
fluorescence   86 ff, 453 ff
– fluorescence correlation spectroscopy (FCS)   455, 489
– fluorescence *in situ* hybridization (FISH)   729 ff
– fluorescence *in vivo* hybridization (FIVH)   461
– fluorescence recovery after photobleaching (FRAP)   454
– fluorescence resonance energy transfer (FRET)   86, 456 ff
– fluorophore   88, 453 ff
– labeling   86, 89, 91 ff
– multiparameter fluorescence detection (MFD)   465
– photobleaching   89, 454, 470
footprinting   174, 178, 185, 259 f
force field   565 ff
forced dialysis   250 ff
– of RNA and Mg^{2+} 251
formamide   123, 382
FPLC   17, 439
free energy   415, 423, 494, 513 ff, 536
free Gibbs enthalpy   422
French press, in cell disruption   820
FRET, see fluorescence
functional genomics   618
Fur   605, 623

g
β-galactosidase   600, 621, 626
– fusion protein   600 f
galK   624

gel electrophoresis 620, 633
- two-dimensional 620
- see also polyacrylamide gel electrophoresis
gel elution 6, 16, 26, 40, 47, 68, 70 f, 305, 307, 653, 827,
- Biotrap 16, 71
- buffer 134
- diffusion method 16, 71, 307, 653, 827
- electroelution 16, 71, 166
- Elutrap 16
- from gel 157
- passive diffusion 40
- yield 47
gel retardation assay 301, 313
gel-shift assay 628
gene silencing 897 ff
glass milk 638
glgC 601
glycerol gradient centrifugation 429 ff
glycogen 16, 137, 157, 176, 178, 186, 232, 240, 243, 381 ff
C7-GMPαS 295
GMPS, see guanosine 5'-monophosphorothioate 79
gonadotropin-releasing hormone I 854
green fluorescent protein (GFP) 621
GROMOS 565
group I intron 260, 265 f, 267 f, 269, 299
group II intron 257, 260, 265, 269 ff, 297
- ai5γ 270, 275 f, 287, 290
- secondary structure 270, 274
- tertiary interaction 269, 274, 277, 287, 290
GST-tag 886
guanidine-HCl 255
guanosine 6, 9, 11 ff, 79 ff, 265 f, 268 f, 294, 296, 306 ff
guanosine 5'-monophosphorothioate 79 f
- purification 80 f
- mass chromatography 81 f
- synthesis 80 f
guanosine 5'-p-nitrophenylphosphate 80

h

H_2O_2 238, 240, 243 f, 346 f, 351
hairpin ribozyme 113
hammerhead ribozyme 23 ff, 59 f, 64 f, 89, 113, 119, 215, 298
- cis-cleaving 23, 29, 64 f
- cleavage reaction 215
- trans-acting 26
HDV 23

HDV ribozyme 23 ff, 208 ff
- cis-cleaving 23, 28 f
- footprinting 208 ff
- trans-cleavage 26
HeLa cell 176, 179 f, 182 ff, 354, 429, 689 f
helical stacking 125
helicase 259, 269
helicase, RNA 259, 269, 272 f, 277 ff
- NPH-II 269, 277 ff
heparin 179, 182
hepatitis delta virus, see HDV
heterochromatin 614
- formation 614
hexamine cobalt chloride 58, 61
Hfq 599, 606, 609, 611, 618, 625
high peformance liquid chromatography, see HPLC
Hill equation 331
- metal ion cooperativity 331 f, 335
His_6-tag 812, 819 f, 886
HIV RNA 180 ff
- EMSA supershift 180, 182 f
H-NS, histone-like protein 600, 609
homo-dinucleotide 296
homonucleotide band compression 298
homo-oligonucleotide 296, 539
Hoogsteen edge 266, 539
HPLC 97, 103 ff, 110, 123 f, 819
- anion exchange 97, 103 f, 124, 440
- cation exchange 110
- desalting column 103
- equilibration time 123
- flow rate 103
- isocratic 123
- linear gradient 104, 123
- reversed phase 97, 105, 123 f
- salt gradient 104
HQS, see 8-hydroxyquinoline-5-sulfonic acid
HU, histone-like protein 609
hybrid selection 135 f
hybridization 631 f, 649, 729 ff
- dot-blot 649
- fluorescent in situ 729 ff
hydrazine 144 f
hydrophobic intercalation 812
hydroxyl radical, see Fe^{2+} mediated cleavage
N-hydroxysuccinimidyl 4-azidobenzoate 83 f
N-hydroxysuccinimidyl ester 78
8-hydroxyquinoline-5-sulfonic acid 251, 255, 257
- emission spectrum 255
- excitation spectrum 255
- in determination of Mg^{2+} concentration 251

i

IGR, see intergenic region
immunoaffinity purification 694 ff
  – of small nuclear ribonucleoprotein complexes 694 ff
  – of spliceosomal complexes 694 ff
immunocytochemistry 730, 733 ff
immunoprecipitation 355, 362 ff, 599, 609 f, 703, 705, 884
  – protocol 370 ff
in vitro selection 783 ff, 807 ff, 853 ff, 858, 869, 871, 878 ff,
  – automation 808
  – carbohydrate 811
  – counter-selection 814
  – cycle 809, 814, 855
  – immobilization matrix 858
  – library derived from natural nucleic acid sequences 878
  – nucleic acid library 808, 813
  – of deoxyribozyme 807
  – of ribozyme 807
  – pre-selection 814, 856
  – protein 810
  – protocol 871
  – selection assay 810, 854, 869
  – stringency 815, 822 f, 869
  – target 810 f
  – target immobilization 812, 854, 856, 859, 861, 871
  – see also SELEX
in vivo modification 193, 230
inosine 265 ff, 294 f
intergenic region 596 f, 602, 608
iodoethane 261
  – iodine cleavage 261
ionic strength 402
IPTG induction 19, 617
isoamyl alcohol 47, 224, 368
4-isocyanato TEMPO, as EPR probe 118, 120
isopycnic ultracentrifugation 429, 434
IstR 604, 618, 624, 628
istR locus 618
ITPαS 283, 296 f, 305 f

k

kethoxal 161, 167 ff, 185, 232, 628
Klenow fragment, see DNA polymerase I

l

lacI (repressor) gene 617
lacZ, see β-galactosidase
ladder
  – acid 142
  – alkaline 156, 160
lag phase 607
lead(II)-induced cleavage 157, 165 ff, 214 ff, 628
lectin 811
library 601, 607
  – cDNA 608 f
  – cosmid 599
  – genomic 600 f
  – phage 599
ligand coupling, see in vitro selection, target immobilization
ligation, see RNA
liquid chromatography 439
  – see also HPLC, FPLC
liquid scintillation counting 50
log phase 607
logarithmic growth 609
luciferase 621
lyophilization 105, 110
lysozyme 163, 231, 820

m

M1 RNA, see also RNase P 598, 601 f
M13 DNA 282
m^5C 149
$m^6A\alpha S$ 266, 271 ff
m^7G 146
MacConkey lactose plate 600
machine learning approach 604
magnesium-8-quinolinol complex 251
manip 538, 540
mass spectroscopy 107, 110, 126, 626, 819
  – electrospray ionization 107
  – ESI 819
  – identification of protein 620
  – matrix assisted laser desorption ionization (MALDI) 107, 819
  – sequence identification 620
matrix 857 ff
  – epoxy-activated 857 f
  – NHS-activated 859 f
  – pyridyl disulfide-activated 860 f
melting point-lowering agents 865
melting temperature (T_m) 135, 140, 399 f, 409
β-mercaptoethanol 195, 231
messenger RNA, see mRNA
metabolome 615, 620
metal ion 214 f, 220, 238, 267, 295, 319, 323, 346
  – affinity 329, 334
  – and RNA catalysis 319
  – Ca^{2+} 215, 220
  – Cd^{2+} 267, 295
  – cooperativity 336

- divalent   346
- Eu^{3+} 215, 220
- Fe^{2+} 238, 346
- hard Lewis acid   295
- inner-sphere coordination   295
- Mg^{2+} 215, 220
- Mn^{2+} 215, 220, 267, 295
- Pb^{2+} 215
- RNA cleavage   214
- soft Lewis acid   295
- specificity switch   320
- stock solution   323
- Tb^{3+} 220
- thiophilic   267, 295, 319 ff
- Zn^{2+} 215

metal ion binding   319, 333
- anti-cooperative   333
- cooperative   333
- independent   333

metal ion binding site   205, 220, 239, 319, 345
- determination by Fe^{2+} 239
- determination by lead(II)   155
- determination by terbium(III)   205
- determination by thiophilic metal ion rescue   319
- mapping in proteins   345

metalloenzyme   346, 349
mfold   35, 72, 278, 501, 518, 520, 628, 653
Mg^{2+} 250 ff, 255
- binding affinity   252 f
- determination by fluorescence   251
- determination by HQS   251, 255
- stoichiometry of RNA binding   250 ff

Mg^{2+} binding sites   256 f
- calculation of   256
- in RNase P   257

MicF   599 f, 628
microarray   596, 603, 605, 619, 625, 655 ff
- high-density oligonucleotide probe
- index array   657
- RASL (RNA-mediated annealing, selection and ligation)   655 ff
- target design   658

Microcon column   282
Microcon filter   252
Microcon YM-10   874 f
microfluidizer, in cell disruption   820
microRNA   607, 614, 898 f
microscopy   454, 546
- confocal   454
- cryo-electron microscopy   546

mimic
- of transition state   113

minigene   755 ff, 841

- construction of   757, 775 f
- transfection of   757 ff

minor groove edge   266 f, 296
minor groove modification   272, 296
miRNA, see microRNA
modeling   536
modification   133, 146 ff
- of natural nucleotides   133, 146 ff

modified nucleotide   7, 14 f, 75, 77 ff, 146 ff
- co-transcriptional incorporation   7, 14 f, 77 f
- identification   146

molecular sieve   101
5'-monophosphorothioate nucleotide   80
- chemical synthesis   80

MP170 RNA   599
MP200 RNA   599
mRNA   605, 608, 655
- expression analysis   605
- leader   603, 608
- splice variant   655
- trailer   603, 608

MS2 coat protein   742
Mud-lac phage   621 f
mutagenesis   346, 601 f, 616
- chemically induced   616
- homologous recombination   616
- λ phage recombination   616
- random genome insertion   601
- replacement of chromosomal gene by marker   616
- site-directed   346
- transposon   602, 616

n

NaBH$_4$ 144 f
NAIM   259 ff, 265, 267 ff, 273, 279 f, 283 f, 286, 289, 294 ff, 311, 314, 629
- analysis of potassium ion binding   268
- analysis of RNA structural motif   262
- calculation example   286, 314
- chemogenetic approach   294
- description of method   262, 294 f
- interference pattern   266 f, 269, 289, 295, 301, 303
- interference suppression, NAIS   263, 268 f, 276, 287, 289 f, 300 f, 303 f
- interpretation of results   284, 298
- iodine cleavage   260 f, 262, 283, 297
- library of modified RNA molecules   260
- nucleoside analog thiotriphosphate   260
- nucleotide analog   262, 265, 295
- phosphorothioate tag   260, 296
- quantification   280, 286, 311, 314

– resolution   260
– RNA-protein binding assay   297
– selection of functional RNA molecules   259, 262, 297
– selection step   273, 297
– transcription termination assay   279
NAIS, nucleotide analog interference suppression, *see* NAIM
NAMD   565
NAP column   16, 66, 107, 110
nascent RNA   729
native PAGE technique   438
ncRNA, *see* non-coding RNA
neomycin   349, 824
Neurospora Varkud satellite ribozyme   26
nick translation   731 ff
nickel complex   155
nitrilo tri-acetic acid   137 ff
– inhibition of phosphatase   138
nitrocellulose   811, 813
nitrocellulose filter binding assay   829 ff, 884
nitrocellulose membrane   715 f
nitrogen decompression, in cell disruption   820
4-nitrophenyl phosphodichloridate   80
NMR spectroscopy   107, 125, 259 f
non-coding RNA (ncRNA)   491, 577, 579, 595, 643 ff
– C/D box snoRNA   579
– chromosome maintenance   643
– DNA replication   643
– regulation of transcription   643
– RNA processing   643
– 6S RNA   491 ff
– signal recognition particle RNA   577
– *see also* small untranslated RNA (sRNA)
Nonidet P-40   198, 364, 371
non-messenger RNA, *see* small untranslated RNA
Northern analysis   627, 649
Northern blot   596, 602, 604 f, 606 f, 629 ff, 649 ff
– protocol   629 ff, 649 ff
Northwestern technique   710 ff
NPH-II   269, 277 f, 279
nuclear extract   176, 179 f, 191, 428 ff, 679, 689 f, 773, 778 f
nuclear magnetic resonance, *see* NMR spectroscopy
nuclease contamination   235
nuclease P1   141, 147 f
– cleavage of phosphorothioate isomer   326
nucleic acid library   813 f, 862

– complexity   862
– diversity   813
– randomized   862 f
nucleic acid polymer   807
nucleophilic attack   26, 215 f, 261 f
nucleoside, chemical phosphorylation   80
nucleoside monophosphate   147
– preparation of   147
nucleotide analog, *see* NAIM
nucleotide analog interference mapping, *see* NAIM
nucleotide analog interference supression, *see* NAIM
nucleotide modification   75, 112 ff, 117
– commercially available   113 ff
– co-transcriptional   75
– disulfide crosslink agent   117
– EPR active probe   117
– fluorescent probe   117
– post-transcriptional   75

o

OLIGO version 4.0   72
OmpC   599
OmpF   599
orthophosphate labeling   596, 598, 606
oxyR   624
OxyS, *oxyS*   598, 607, 609, 620, 624 f, 628

p

P1 phage transduction   616
PAA, *see* polyacrylamide
PAGE, *see* polyacrylamide gel electrophoresis
PALINGOL   579
PATSCAN   579
PATSEARCH   579
Pb^{2+}, *see* lead (II)
pCp   56 f, 138 ff
– RNA 3' labeling   58, 139, 177, 598
– synthesis   139
[^{32}P]pCp   58, 139, 177 f, 598
PCR   29, 813, 882
– mutagenic   810, 813, 827
– overlap extension   29 f, 31 f
PEG, *see* polyethylene glycol
peptide
– mRNA display   807
– phage display   807, 814
– ribosome display   807
periodate   87
– oxidation of RNA   78, 86 ff
Pfu DNA polymerase   31
– proofreading activity   31
Pfold   581 f

phage
- Lambda (λ) 601
- MudII 600
- phagemid 600
phage display 807, 814
phage polymerase, see RNA polymerase
phase diagram 443
phase lock gel (PLG) tube 635 f
phenol extraction 15, 49, 93, 159, 164, 168, 224, 231, 273 ff, 306, 821 ff
phenol/chloroform extraction 46 ff, 69, 83, 89, 159, 165, 231, 368, 633, 647 ff, 888 ff
phenotypic array 618
phenylmethylsulfonyl fluoride 18, 177
phosphatase 66, 156
   - see also calf intestinal alkaline phosphatase
phosphodiester 345
phosphodiester bond 214 f
   - formation 37 f
phosphoramidite 70, 95 f, 288
phosphoramidite strategy 112
phosphorothioate 76, 79, 82, 260 f
   - linkage 260 f
   - modification 295
   - modified RNA 82
phosphorothioate stereoisomer 326, 339
   - effect of impurity on reaction rate 339
   - identification by nuclease cleavage 326
phosphorothioate nucleotide analog 7, 266, 321
   - coordination of Mg^{2+} 321
   - incorporation into RNA 325
   - in thiophilic metal ion rescue 321
5'-phosphorylation 9, 38, 43, 49, 64, 67, 369, 889
   - non-radioactive 43, 49, 889
   - of DNA 889
   - protocol 43, 49, 67
   - radioactive 49
   - site-specific labeling 43
phosphoImager 198, 308, 632
photoaffinity crosslinking 374
photo-crosslinking 75, 79
photoreactive group 75
phylogenetic analysis 498, 502, 596
plaque hybridization 599
plasmid 599, 601, 616 f, 645, 827
   - antibiotic marker 617
   - ColE1/pUC 617
   - F 600, 616
   - high copy 601, 617
   - host range 617
   - low copy 601

- maintenance function 617
- multicopy 599 f
- origin of replication 617
- P1 616
- p15A 616
- pACYC184 600
- pBR322 600
- pGEM-T 827
- pKS 645
- promoter/operator combination 617
- pSC101* 616
- pSPORT 1 645
- single-copy 616
PMSF, see phenylmethylsulfonyl fluoride
poly(A) polymerase 138 f, 349, 607, 644 ff
poly(A) tail 59, 345
poly(A)-specific ribonuclease 345, 347
   - enzymatic activity 347
   - metal ion binding site 345 ff
polyacrylamide gel 402, 426
   - polymer concentration 402
   - pore size 402
polyacrylamide gel electrophoresis (PAGE) 6, 16, 22, 26, 46, 66, 90, 123, 142, 156, 206, 307, 382, 441, 647, 867
polyadenylation site 59
polyanion 250
polyethylene glycol (PEG) 44, 58, 885
polynucleotide kinase, T4, see T4 polynucleotide kinase
polyvinyl alcohol 44
polyvinyl pyrrolidine 44
post-synthetic RNA labeling 112 f, 115 ff, 119 f, 123, 125 f
   - 2'-amino group 116 f, 123
   - 2'-O-(2-aminoethyl) modification 119
   - 5'- and 3'-amino modifier 117
   - 5-alkylamino modified pyrimidine 117
   - aliphatic isocyanate 120, 123
   - bimane 120
   - cholesterol 120
   - fluorescein 119
   - fluorescent pyrene label 119
   - glutathione 120
   - nitrophenol 120
   - nitroxide spin-label 120, 126
   - Pd-catalyzed coupling reaction 116
   - phosphoramidate 116
   - phosphorothioate 116
   - photocrosslinking reagent 119
   - pyrene 120
   - rhodamine 119
   - 2'-thioureido modification 119
   - 4-thiouridine 125

precursor tRNA   65, 67 f, 297, 299 ff, 377 ff
– crosslinking   377 ff
primer extension   188, 223, 232, 355, 357, 368, 879
probe, biotin-labeled   87
promoter   3 f, 22, 29 f, 39, 597, 601 f, 604, 617
– λ P$_L$ 601, 617
– σ^{70} promoter consensus   602
– σ^{70} promoter   604
– control   617
– induction   617
– leakiness   617
– pBAD   617
– phage-specific   39
– *plac*   617
– *ptac*   617
– regulatable   601
– repression   617
– T7 class II   4
– T7 class III   4, 9, 22, 30
– temperature sensitivity   617
prop-2-yl trimethylsilyl ether   101 f
protease inhibitor   18 f
– leupeptin   18 f
– phenylmethylsulfonyl fluoride (PMSF)   18 f, 177
protecting group   96, 122
– N-phenoxyacetyl (pac)   96
– *N-tert*-butylphenoxyacetyl (tac)   96
– photo-cleavable   117
– solid-phase synthesis   122
– 2'-trifluoroacetyl   122
protein   485
– coupling to SFS surface   485
protein A-sepharose (PAS)   195 ff, 198, 364, 650
protein G-sepharose   195 ff, 198
protein kinase   351
– bovine heart   351
protein kinase R   898
protein labeling   620
– fluorescent dye   620
– *in vivo* labeling by [^{35}S]methionine   620
proteinase K   43, 356 f, 364
proteome   615
proteomics   620
proton sponge   121 f
PrrB   601
pseudo-base pair   812
pseudouridine   146
pseudouridylation   614, 643
PSTVd RNA   401 ff
– analysis by TGGE   401 ff
PSTVd transcript   406

– metastable conformation   406
– sequential folding   405 ff
PurαS   266
pyrophosphatase   8, 15, 66, 79, 137
– tobacco   137
– yeast   79
pyrophosphate   8, 15, 17, 46, 66
– precipitate   15, 17, 66

q
QIAEX II gel extraction kit   638
QRNA   604

r
RACE, *see rapid amplification of cDNA ends*
radical scavenger   245
radius of gyration   385
rapid amplification of cDNA ends (RACE)   59, 599, 617, 627, 632 ff
– of small untranslated RNA   627
RcsA   600
recG   600
regulatory RNA, *see* small untranslated RNA
regulatory sequence
– prediction   623
reporter gene   621 f, 626 f
reporter gene fusion   621 ff
– Mud-lac fusion library   621 f
restriction endonuclease   5, 31, 601, 271, 644, 646, 648, 881, 883 f
– 5'-overhanging end   271
– *Bam*HI   646
– *Bgl* II   31
– blunt end   271
– *Eco*R I   884
– *Fok* I   5
– *Hpa* II   883
– *Mme* I   881 ff
– *Not* I   644, 646
– *Pst* I   884
– *Sal* I   646
reverse transcriptase   59, 156, 160, 186, 225, 233, 255, 357 ff, 382, 606, 608, 633, 637, 645, 648, 651, 813, 825, 865, 887
– avian myeloblastosis virus (AMV)   156, 186, 865
– C-tailing   645
– moloney murine leukemia virus (M-MLV)   865
– protocol   637, 825
– RT-PCR   651
– thermostable   608, 633
rhodamine   466 f, 470
ribonuclease   69

– contamination   69
ribonucleoprotein (RNP)   172, 187 ff, 354 ff, 365, 428, 667, 710 ff
– assembly *in vitro*   365
– chemical probing   187 ff
– enzymatic probing   187 ff
– in pre-mRNA processing   354
– sedimentation analysis   428 ff
riboregulator, *see* small untranslated RNA
ribose methylation   143
ribosomal RNA   595
riboswitch   608, 808
– natural   808
ribozyme   23, 26, 28, 34, 40, 42, 64, 92, 95, 260, 265, 267, 273, 295, 299, 328, 808, 854, 879
– allosteric   854
– 3' cassette   23
– 5' cassette   23
– *cis*-cleaving   23, 64
– folding interference   28
– group I intron   260, 265, 323
– group II intron   260, 265, 273, 323
– hairpin ribozyme   265
– hammerhead, *see* hammerhead ribozyme
– RNase P   26 ff, 216 f, 265, 294 ff, 323
– switch ribozyme   808
– temperature cycling procedure   34
– *Tetrahymena*   299
– *trans*-cleaving   26
ring nitrogen   335
– in metal ion rescue   335
RISC   898
RNA
– abundance in the cell   598
– acceptor   38, 54
– activating agent   95
– adaptor ligation   637, 651
– affinity matrix   667 ff
– cap   38, 40, 46
– catalytic   319, 595, 787
– chemical modification   112, 144, 152, 187
– chemical synthesis   70 f, 95 ff, 113, 287 f
– chimeric   36 f
– circular   38, 59
– class   643
– donor   38, 54
– double-stranded   480
– electrophoretic mobility, in TGGE   401
– end group   26
– Hoogsteen edge   266, 539
– hybridization   136, 188, 403, 632, 649 f
– hydrogen bonding   266, 295 f, 298 f

– immobilization   87, 667
– immunostaining   403
– intramolecular circularization   59
– isolation from cells   224, 231, 636, 647
– ligand interaction   221
– methylation   143, 614, 643
– module   537 f
– nucleotide analysis   146
– packing motif   267
– periodate-oxidized   86
– phosphate exchange   138
– protected nucleobase   98
– protecting group   95 f, 122
– protecting group, photo-cleavable   117
– pseudouridylation   614, 643
– purification   134
– quantification   70, 630, 910
– regulatory   595
– single-stranded   220
– stability   119 f
– tobramycin affinity tag   626
– TOM   95
– transition curve in TGGE   399
– turnover   598
4S RNA   598
4.5S RNA   598 f, 616
6S RNA   491 ff, 597 f, 608
6S RNA gene   597
6Sa RNA   599
RNA cleavage   214 ff, 217, 222
– enzymatic   126 f, 140 ff, 157, 189, 326
– mechanism   214
– metal ion-induced   217
– Pb^{2+}-induced   214, 222
– protocol   222
– single-stranded region   214
RNA degradation   59, 69 f, 72, 618, 624, 626 f, 630
– on agarose gel   630
RNA denaturation   144, 255, 389, 401
– by guanidine-HCl   255
– in TGGE   401
– SAXS study   389
RNA extraction   134, 195, 231, 244, 607
RNA footprinting   191, 205 ff, 208 ff, 238 ff
– by hydroxyl radical   238
– by terbium(III)   205
– of U5 snRNA   191
RNA fragmentation   214
– Pb^{2+}-induced   214
RNA interference   614, 879, 897
RNA labeling   75 ff, 86 ff, 112 ff, 598
– biotin labeling   86 ff
– crosslinking reagents   78, 82, 118

- EPR spin label   118, 120
- fluorescence labeling   86 ff
- 3'-^{32}P-end-labeling   138 f, 156, 222
- 5'-^{32}P-end-labeling   137 ff, 156, 178, 222
- see also T4 polynucleotide kinase
- see also T4 RNA ligase

RNA library   264, 271, 273, 294, 301, 305, 783, 813 f, 862 f, 889
- 5'-^{32}P-end-labeled   301
- reference pool   305

RNA ligase, see T4 RNA ligase

RNA ligation   36 ff, 44 f, 50, 53 ff, 63, 288, 635, 637, 645, 651

RNA oligonucleotide   485, 677 ff
- biotinylated   677
- coupling to SFS tip   485
- 2'-O-methyl   677
- chimeric 2'-O-methyl/DNA   677 ff

RNA polymerase   3, 7 f, 15, 17, 20, 22, 38, 46, 53, 65, 77 ff, 156, 176 f, 272, 279 f, 296, 306, 817, 820
- and ribose modification   7
- multi-subunit   280
- mutant T7   7, 77, 272, 296
- preparation   17
- SP6   4, 176 f
- specific activity   20
- T7   3, 7 f, 15, 17, 20, 22, 38, 46, 53, 77, 79, 156, 177, 272
- Y639F mutant   7, 15, 272, 296

RNA processing   614, 627

RNA renaturation   157, 166, 177 f, 305, 328, 378, 382

RNA sequencing   140 ff, 146
- CL3 mix   141
- counting ladder   143, 146
- enzymatic cleavage   143
- enzymatic hydrolysis   141
- ladder, acid   141 ff
- of tRNA   143
- RNA denaturation   144
- S7 mix   141
- T1 mix   141
- terminal nucleotide   141

RNA staining
- silver staining   403
- toluidine blue staining   134 ff

RNA structure   172 ff, 214 ff, 405, 513, 539 f, 551
- branched structure   401
- conformational change   233, 399
- crystal structure   259
- detection by hybridization   405 ff
- secondary structure   144, 160, 167 f, 205 ff, 482, 513, 539, 604
- structural motif   262
- tertiary structure   205 ff, 233, 239, 259, 537, 540, 551, 564 ff

RNA structure probing   151 ff, 159 ff, 172, 185 ff, 193, 205, 229, 232, 235, 238 ff, 374, 628
- by CMCT   153 ff
- by DEPC   154 f, 162
- by dimethyl sulfate   152 ff, 229
- by hydroxyl radical   238
- by kethoxal   153 ff
- by lead (II)   155, 214 ff
- by terbium(III)   205
- enzymatic   151 ff
- in vitro   151, 185, 222 ff
- in vivo   151, 163, 193, 223 ff, 229, 235
- of RNase P RNA   218 f
- of RNA III   163 ff
- of thrS mRNA   159
- primer extension   161, 188, 225, 235
- reverse transcription   153, 161 f, 225, 232

RNA synthesis   3, 95, 99
- automated   70, 100, 287
- chemical synthesis   36, 53, 65, 70 f, 95 ff, 113, 287 f
- solid-phase synthesis   70, 95, 97 f, 101, 117, 122

RNAalifold   581 f

RNA-aminoglycoside interaction   221

RNA-antibiotic interaction   217

RNA-binding protein   172 ff, 354, 480, 667 ff, 710 ff, 737
- identification by Northwestern technique   710
- three-hybrid analysis   737 ff
- visualization by SFM   477 f, 480, 483 f

RNAfold   518, 520, 527 ff
- dot plot   521 ff

RNAi, see RNA interference

RNAML   524

RNAMLview   524

RNAMOTIF   578 f, 581, 583, 585 ff, 592

RNAP, see RNA polymerase

RNA-protein complex   180 ff, 221, 259, 269, 354 ff, 385, 428 ff, 667, 676 ff, 694 ff, 710 ff, 810 f, 878
- K_D determination   180
- supershift method   180

RNA-protein interaction   172 ff, 217, 259, 269, 354 ff, 667 ff, 710 ff, 737 ff, 878

RNA-RNA interaction   217

RNase   133, 667

- contamination 72, 109, 133, 226, 313, 365, 817
- endogenous RNase H 45
- inhibitor 46, 68, 235, 633
- protection 771
- RNase III 610, 614, 618, 627
- RNase A 92, 146, 195, 235
- RNase CL3 141 f
- RNase E 627
- RNase H 38, 40ff, 45, 48, 647, 879
- RNase L 898
- RNase P, *see there*
- RNase T1 92, 141, 153 ff, 156, 168, 185 ff, 189, 195 ff, 326, 818
- RNase T2 141, 146, 153 f, 156, 185 ff, 189
- RNase U2 141
- RNase V1 153 f, 156 ff, 167, 185 ff, 189
- RNase Z 667 ff
- site-specific cleavage 38, 41 f, 48

RNase P 26 ff, 34, 65, 218 f, 223, 257, 295, 377 ff, 389 ff, 393, 598, 601, 616
- aberrant cleavage 28
- cleavage fidelity 28
- crosslinking 377 ff
- *E. coli* 28, 34, 219
- Mg^{2+} binding 257
- oligomerization state 389 f
- Pb^{2+}-induced cleavage 218 f
- primer extension 378
- SAXS study 389
- scattering profile 390 f
- sequence requirement 27 f
- structure 393
- substrate complex 223
- *T. thermophilus* 28, 34
- thermostable 28

RNasin 821
RNAStar 628
RNomics 596, 606 f, 610, 614, 643, 646 f, 649
- cDNA library 646
- RNA (sub)class 643, 647
- sequence analysis 649

RNP, *see* ribonucleoprotein
Rp, Sp (stereo)isomer 267, 295, 326
RP-HPLC, *see* HPLC, reversed phase
RpoS 600, 624
RprA 600, 609
RsmA 609
RsmZ 609
C7-RTPαS 283
RT-PCR 633, 645, 651, 771, 777 f, 841
- protocol 651, 777 f
run-off transcription 3, 5, 280 ff
RyeA 604

RyeB, *ryeB* 603 f, 608
RyhB 605, 609, 618, 623 f, 625

S

S1 nuclease 185, 199
SAP, *see* shrimp alkaline phosphatase
SAXS, *see* small angle X-ray scattering
scanning force microscopy (SFM) 475 ff, 483 f
- buffer condition 484
- analysis of protein binding 478
- experimental setup 476
- of dsRNA 477
- of mRNA 477
- resolution of imaging 479
- RNA-protein interaction 477 f

scanning force spectroscopy (SFS) 475, 481 ff, 485
- coupling of protein to surface 485
- coupling of RNA to tip 485
- double-stranded DNA 482
- force-distance curve 482
- functionalized surface 485
- functionalized tip 485
- protein-nucleic acid interaction 485

Scatchard plot 256
SCFG 579 f
sdh mRNA 618
SDS-PAGE 18 f, 347 ff, 366, 432, 672 f, 683 f, 689 f, 696 ff, 720 ff, 811, 819 f
- *see also* polyacrylamide gel electrophoresis

SELEX (systematic evolution of ligand by exponential enrichment) 783 ff, 789, 792, 794, 798, 800, 808, 811, 840 ff, 853, 878
- genomic 878
- *in vivo* 840
- negative selection 784, 792, 798 ff
- randomized exon cassette 843
- selection of ribozyme 794 ff
- sequence analysis 800 ff
- stringency 789, 792, 815, 822 f, 869
- *see also in vitro* selection

self-splicing intron 233 ff, 241, 273 f, 295, 297
Sephadex 16, 106, 254, 812
Sep-Pak cartridge 71
sequence alignment 493, 500 ff, 582 ff, 800 ff
sequence analysis 815 f, 885
sequence conservation 504, 507 f, 597, 602
sequence motif 506 f, 508, 577, 786
sequencing 133 ff, 598 f
- nuclease fingerprinting 598

SFM, *see* scanning force microscopy
SFS, *see* scanning force spectroscopy
shearing 601
short hairpin RNA (shRNA) 898 ff

shotgun cloning   596 ff, 614
shrimp alkaline phosphatase   43
shRNA, see short hairpin RNA
Sigma brilliant blue   135
sigma factor   600
signal recognition particle (SRP)   268, 577 ff
silver staining   347 f, 351, 366 ff, 432, 436, 599
simulation   547 ff, 560 ff, 683 ff, 698 ff
– molecular dynamics   548, 560 ff
– Monte Carlo   548, 554, 556
single-stranded RNA   220, 480
– Pb^{2+}-induced cleavage   220
– visualization by SFM   480
siRNA, see small interfering RNA
size exclusion chromatography   254
– of RNA and Mg^{2+} 254
Sm proteins   362, 365 f
small angle X-ray scattering   385 ff, 394 f
– data analysis   386, 389
– experimental setup   388
– radiation damage   395
– RNase P holoenzyme-substrate complex   392 ff
– sample requirement   388
– scattering background   387, 389 f, 395
small interfering RNA (siRNA)   95, 607, 614, 897 ff, 901 ff
– design   901
– electroporation   904 f
– intracelluar expression   905
– lipofectamine   903
– preparation of   902
– target site selction   901
– transfection of mammalian cell   902
small non-messenger RNA   597
small nuclear RNA (snRNA)   175 ff, 188, 354 ff, 428 ff, 610, 694 ff
small nuclear ribonucleoprotein (snRNP)   175 ff, 188, 354 ff, 428 ff, 610, 694 ff
– crosslinking   354 ff
– probing   187 ff
small nucleolar RNA (snoRNA)   172, 175, 180 f, 193 ff, 614, 643, 647, 650 ff
– probing   194
small RNA   595, 597
small untranslated RNA   595 ff, 614 ff, 643 ff
– abundance   623
– affinity purification   626
– associated protein   625
– base pairing with target RNA   624
– co-immunoprecipitation   609 f
– complementation plasmid   615
– co-purification with protein   609

– co-purification with target RNA   609
– deletion of gene   615 f, 618 f, 621
– detection of gene   615
– effect on neighboring gene   616, 618,624
– effector RNA (eRNA)   597
– expression profile   606
– function   597 ff, 614 ff
– growth rate-specific expression   607
– homolog   603, 605 f, 609
– identification   595
– interaction with protein   609
– intergenic location of gene   596 f, 603, 606, 608, 616, 634 f
– interacting site   624
– mapping of protein-binding domain   628 f
– mapping of structure   628 f
– motif search   653
– null mutant strain   616
– phylogenetic conservation   605
– phylogenetic conservation, of neighboring gene   624
– prediction of regulatory sequence   623
– processing   596 f, 599, 603, 606
– promoter   597, 602 f, 605, 610
– proteomics   620
– radiolabeling   596 f, 598 f
– regulatory function   614
– screening   596 f, 599, 600 f, 602, 604, 609
– secondary structure   598, 604, 608 f, 611, 617, 628 f,
– shotgun cloning   606
– stability   598,609, 625
– stabilization by protein   609, 625
– stress-induced   607
– synthesis rate   598
– transcript stability   623
– transcription start site   617, 623, 627, 632, 634 f, 639
– transcriptional regulation   619, 623
– trans-encoded   599 f, 624 f
– turnover   598
– up-regulation   598
snake venom phosphodiesterase   126, 326
– cleavage of phosphorothioate isomer   326
snmRNA, see small non-messenger RNA
snoRNA, see small nucleolar RNA
snRNA, see small nuclear RNA
snRNP, see small nuclear ribonucleoprotein
sodium ascorbate   238, 240, 243, 247
sodium azide   103
sonication   19 f, 820, 881, 885 f, 887 f
– of DNA   887 f
– in cell disruption   820
SOS response   618, 627

Speed Vac system   71, 102 ff, 105 f, 282, 288
spin-labeling reagent   121
   – 4-isocyanato TEMPO   121
spliceosome   175, 190 f, 367, 428 f, 434, 694, 697 ff
splicing   192 f, 273 f, 428 f, 614, 655 ff, 677, 702 f, 705 f, 755 ff
   – alternative splicing   655 ff, 755 ff, 841, 843 f, 846
   – exonic splicing enhance (ESE)   840
   – exonic splicing silencer (ESS)   840
   – microarray   655 ff
splint labeling   138, 140
Spot 42   598, 609, 624 f
SmpB   626
sra   602
SraC   604, 632, 635
SraI, see RyhB
SroA   608
SroG   608
SsrA, see tmRNA   626
Staphylococcus aureus nuclease   141, 142 f
stationary phase   597, 600, 604, 607 f, 634 f, 636,
stochastic context-free grammars (SCFG)   579
StpA   609
streptavidin   86 f, 91 f, 93, 861 f
   – agarose bead   297, 872
   – coated matrix   625
stress response   601, 618, 624, 627
sulfur, metal binding geometry   320 ff, 322
surface plasmon resonance (SPR)   87, 816, 830 f, 833
   – protocol   833

t
T4 DNA ligase   37 f, 44 f, 50, 648, 887 ff
   – acceptor   37 ff, 40, 45
   – "disrupter" oligonucleotide   45
   – DNA splint   37 ff, 44, 50
   – donor   37 ff
   – K_m for polynucleotides   38
   – ligation   36 ff
   – ligation protocol   50, 889
   – sequence specificity   38
   – three-way ligation   37
   – two-way ligation   37
   – Weiss unit   44
T4 DNA polymerase   648
T4 PNK, see T4 polynucleotide kinase
T4 polynucleotide kinase   26, 43, 54, 65, 67, 89, 138 f, 146 ff, 176, 178, 232, 273, 282, 308, 369, 598 f, 644 ff, 649, 835, 887
   – 5' end-labeling   26, 138, 308, 369, 598, 835, 887
   – phosphatase activity   26, 89
T4 polynucleotide ligase, see T4 DNA ligase
T4 RNA ligase   53 f, 56 ff, 63 ff, 139, 156, 176, 178, 308, 316, 598, 633 f, 644 ff, 652, 835
   – acceptor substrate   54, 56, 61, 63
   – accessibility of ligation site   63
   – 3'-amino ATP   59
   – 2-aminopurine riboside triphosphate   59
   – circularization   54, 56, 59
   – DNA splint   61, 63
   – donor adenylation   56
   – donor substrate   54, 56, 61, 63
   – donor tandem   58
   – 3'-end labeling   58, 139, 308 f, 316, 598, 835
   – 5'-end-labeling   59
   – intermolecular ligation   56, 59 ff
   – polyhomoribonucleotides   56
   – primer-RNA concatemer   653
   – *quasi*-intramolecular reaction   59, 61
   – reaction conditions   57
   – reaction mechanism   54 ff
   – RNA-DNA duplex   59, 63
   – substrate specificity   56 ff
T7 promoter   4 ff, 29 ff, 39
T7 transcription   3 ff, 22 ff, 29, 37 ff, 42, 46, 53, 65, 77 ff, 89, 177, 306, 820, 821, 867 f
   – aberrant   14
   – 5'-adenosine   9
   – average rate   3
   – 5'-biotinylated ApG   8
   – cap   38, 46 f
   – *cis*-cleaving ribozyme   6 f
   – *cis*-hammerhead   89
   – 3'-dephospho-coenzyme A   8
   – 3' end heterogeneity   22, 37, 42, 53, 89
   – 5' end heterogeneity   6, 22, 42, 53
   – 3' end homogeneity   6, 22 ff, 40
   – 5' end homogeneity   6, 22 ff, 37
   – error frequency   3
   – 2'-fluoro-modified   14 f
   – initiator dinucleotide   9, 11, 40
   – initiator (oligo)nucleotide   7, 9, 47, 79
   – internal labeling   9
   – 2'-*O*-methyl approach   29, 40, 42
   – modified substrates   7
   – non-templated nucleotide addition   6, 22, 40, 42, 45,
   – large scale preparative   9 ff, 38 f
   – RNase H-based strategy   29, 40 ff
   – small scale   8 ff
   – template   3, 5 ff, 22, 28 ff, 39 ff, 46, 66

– vectors 5
TA-cloning kit 31, 633, 638
TAP, see tobacco acid pyrophosphatase
taq DNA polymerase 31, 633, 638 f, 644 ff, 826 ff, 864, 887
– non-templated A residue 31
Tb^{3+}, see terbium(III)
TBE buffer 16, 17, 47, 69, 177
td group I intron 233
– RNA structure probing 233
temperature-gradient gel electrophoresis (TGGE) 398 ff, 403 ff, 409, 411
– analysis of conformational transition 399
– electrophoretic mobility of RNA 401
– instrument 400
– mutant analysis 400, 409
– of 5S rRNA 403 ff
– of 7S RNA 403 ff
– of PSTVd 403 ff
– of total RNA 403 ff
– protein-RNA complexes 409 ff
– psbA mRNA protein complex 411
– transition temperature 401 f
template strand 3
terbium(III) 205 ff
terbium(III) footprinting, see terbium(III) mediated cleavage
terbium(III) mediated cleavage 205 f
– buffer 206
– quantification 208
– reagent 206
terminal transferase 61
termination signal 3, 602, 604
– Rho-independent 602, 604
tetrabutylammonium fluoride (TBAF) 71
tetracycline 241, 244, 246
– binding site 241
– binding to rRNA 244
tetrahydrofurane (THF) 71
tetraloop 220, 267
TGGE, see temperature-gradient gel electrophoresis
thallium ion rescue 268
– analysis of potassium ion binding 268
thermoscript RT-PCR system 633
thin-layer chromatography (TLC) 81, 140, 146 ff
– marker mix 147 ff
– marker nucleotide 147
– plates 147
– polyethyleneimine plate 81, 149
– solvent 147
– two-dimensional 146 f

6-thioguanosine (S^{6}G) 76, 80, 268, 374 f, 377 f
thionucleotide, identification by TLC 149
thiophilic metal ion rescue 319 ff, 337
– buffer condition 324
– in ribozymes 321
– k_{rel} 337
– reaction rate 329
thiophosgene 121
thiophosphoryl chloride 80
thio-urea 240, 243
4-thiouridine (S^{4}U) 76, 115, 118, 120, 125, 195, 374 ff, 377 f
three-hybrid analysis 737 ff
tisAB 624
TLC, see thin layer chromatography
tmRNA 598 f, 605, 616, 626
tobacco acid pyrophosphatase (TAP) 609, 627, 632 f, 639
– protocol 637
tobramycin 626, 824
toluidine blue 20, 134, 137
transcription 3 ff, 22 ff, 36 f, 40, 46, 53, 65, 79, 177, 271, 597, 610, 710, 712, 817, 820 f, 862 ff, 893, see also T7 transcription
– 5'-end modification 6 ff, 46 ff, 79
– antisense to protein-encoding gene 610
– initiation site 632, 639
– in vivo 70, 853
– premature termination 22
– primary transcript 22 ff, 639
– priming 77, 79, see also T7 transcription
– protocol 8 ff, 46, 65, 867 f, 712
– run-off 3, 5, 53, 817, 828
– SP6 177, 712
– T7, see T7 transcription
– termination 22, 597, 710
– yield 4 f, 10, 17, 29, 40
transcription elongation complex 279
transcription initiation signal, see promoter
transcriptional profiling, see microarray
transcriptome 615
transesterification 26, 56
transition state 113, 212, 215, 320, 334, 340
transposon 610
trap bag 101
trichloroacetic acid (TCA) 99
triethylamine 101
5-trifluoromethyl-2'-deoxyuridine 125
trinucleotide repeat 221
tri-snRNP 191, 355, 367
trityl-off RNA 288
TRIzol 629, 633, 636, 647
tRNA 26 ff, 28, 34, 54 ff, 61, 65 ff, 143, 303, 595, 604, 667 ff

- 7-bp acceptor stem 28
  affinity column 667 ff
- anticodon loop 54, 61
- class I 28
- D loop 61
- discriminator base 28
- misfolding 34
- processing 29, 667, 672
- production of 25 f, 65 ff
- sequence analysis 143
- T arm 303
- T loop 61
- variable arm 28

tRNA cassette 28
tRNA precursor transcript 27

u

U1 70K protein 355, 362 f
U1 snRNA 355 f, 360 f, 365
- Sm site 360 f, 365

U1 snRNP 175, 355, 358, 361 f, 364 f, 428
U1A protein 356
U3 sno RNP 180
- K_D determination 180

U4 snRNA 355 f
- human 356

U4/U6-specific protein 61K 355
U5 snRNA 189, 191, 365
- footprinting 191

U5 snRNP 175, 365
ultrafiltration 107, 282, 821, 874
ultrasound, see sonication
UptR, uptR 600 f
[α-^{32}P]UTP 13, 177
UV crosslinking 195, 200, 202, 354 ff, 631
- buffer 196
- HIV RNA 196 ff
- RNA and protein 195 ff

UV irradiation 354 f, 358, 367, 380
UV lamp 380
UV melting 415 ff
- strand break in RNA 358, 364

UV shadowing 6, 40, 48, 68, 71, 124, 272, 307, 368, 832
- UV-induced damage of RNA 307

UV spectrophotometer 417
UV spectroscopy 16, 48, 68, 70, 106, 125, 307, 415, 821, 910 ff
UV thermal denaturation 125 f
- hypochromicity 125

v

van t'Hoff expression 422
VS ribozyme, see neurospora varkud satellite ribozyme

w

Watson-Crick base pair 296
Western blot 184 f, 202
Wincott procedure 288

x

X-Gal 623, 626, 638
X-ray crystallography 107, 126, 268, 345
X-ray scattering 385 ff
- see also small angle X-ray scattering

xylene cyanol 16, 47, 69, 135, 142, 156, 160, 163, 177, 206, 233, 241, 382

y

yammp 547 ff, 554, 558
yeast RNA 141, 144, 148, 176, 179 ff, 186, 200
yeast tRNA^{Phe} 214, 216
ygfA 597